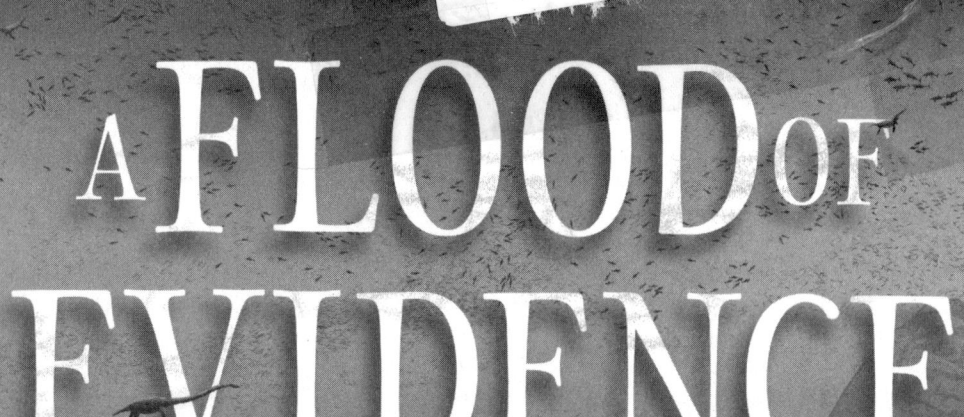

A FLOOD OF EVIDENCE

40 Reasons Noah and the Ark Still Matter

Ken Ham and Bodie Hodge

First printing: September 2016
Seventh printing: August 2021

Master Books®, P.O. Box 726, Green Forest, AR 72638
Master Books® is a division of the New Leaf Publishing Group, Inc.

ISBN: 978-0-89051-978-3
ISBN: 978-1-61458-561-9 (digital)
Library of Congress Number: 2016913219

Cover by John Taylor

Photo and image credits: Thanks to Dan Lietha, Maria Murphy, Bodie Hodge, Ken Ham, Answers in Genesis, Wikipedia Commons, Renton Maclachlan (Te Wairoa items), Dr. Tommy Mitchell, Dan Stelzer, and Larry Smith.

Printed in the United States of America

Please visit our website for other great titles:
www.masterbooks.com

For information regarding author interviews,
please contact the publicity department at (870) 438-5288.

Master Books®
A Division of New Leaf Publishing Group
www.masterbooks.com

contents

NOAH'S ARK

CONCLUDING REMARKS

Preface

Bodie Hodge and Ken Ham

Why Another Book on the Flood and the Ark?

There are hosts of books, videos, film, and Internet information about Noah's ark. Some of it is quite good. Much of it is miserable! Some are blatantly inaccurate and unbiblical (consider the *Noah* film with Russell Crowe and directed by Darren Aronofsky, which was primarily an attack on the Bible). Some ark and Flood information is too technical for many readers. Some ark and Flood resources are so basic that readers lose interest.

What this book was designed to do was filter through the massive amounts of information and give you *what you need to know*. We tried to do it in a way that keeps you from getting caught up in too much technical data and debates, and yet gives you some "meat to chew on" without being too basic. If you wanted to move on to a deeper level of technical debate after this book (which builds on some of the chapters), then at least you would have a foundation to dive into the subjects further.

Also, we wanted this to be a book that stands on the *authority of the Bible* as our absolute and final standard on issues. We didn't want a book to rest on the *opinions of men* like so many resources do, but to treat the Bible for what it is — the truth. We are unashamed of this and up-front about this as you can see.

In doing so, we hope this book also points you to the Lord Jesus Christ of which the Flood (the first part of the book) and Noah's ark (the second part of the book) is a picture. The Flood was a judgment on sin and the ark was a means of salvation from that catastrophe. All you had to do was enter the door of the ark to be saved.

In the same way, Jesus Christ ultimately judges sin for all eternity. Yet Jesus Christ describes Himself as *the door* by which all must enter to be saved from their sin. The shed blood of Jesus Christ, who took the punishment for sin upon Himself, is the *only* means of salvation.

> Nor is there salvation in any other, for there is no other name under heaven given among men by which we must be saved (Acts 4:12).

Introduction

1 The Flood and Noah's Ark: An Introduction – Getting Down to the Basics

Ken Ham and Bodie Hodge

Introduction

> That there was a flood 4,000 years ago is not provable. In fact, the evidence, for me at least, as a reasonable man, is overwhelming that it couldn't possibly have happened; there's no evidence for it. — Bill Nye[1]

Evolutionist Bill Nye repeatedly attacked the global Flood recorded in the Bible at the historic debate with Ken Ham over creation-evolution at the Creation Museum on February 4, 2014. Did you ever wonder why he did this? The reason was simple. He *had* to. In fact, anyone who believes in an old earth, Christian or secular, cannot believe in a global Flood.

In any scenario that involved millions and billions of years of earth history, a catastrophe like the Flood of Noah's day would be devastating! Thinking clearly, the secular idea of millions of years is used in an attempt to explain rock layers (that contain fossils). The fossil layers were supposedly laid down slowly over long periods of time *without any major catastrophes.*

Yet the global Flood of Noah's day would account for the majority of those rock layers quickly.[2] So millions of years would vanish in light of a catastrophe like that! This is why Bill Nye (and anyone else) arguing for an old earth must fight against the famous Flood of biblical proportions!

---

1. Ken Ham and Bodie Hodge, *Inside the Nye-Ham Debate* (Green Forest, AR: Master Books, 2014), p. 344.
2. Of course we have had some rock layers form since that time from volcanoes, local floods, and so on, but the bulk of the rock layers that contain fossils came from Noah's Flood.

Importance of the Flood

Is the issue of Noah's Flood important? Absolutely. It may be one of the key battles in the authority of Scripture in our day and age. Sadly, even certain Christians join with non-Christians to attack the truth of a global Flood. For example, popular old-earth Christian Dr. Hugh Ross attacks the ark and the global Flood by saying:

> Not even an ark of steel armor plate could survive the rigors of a Flood gone global in Forty Days nor of the devastating effects of tens of thousands of feet of erosion in forty days and similar uplift within a year's time.[3]

> No viable scientific evidence has ever been found for a recent, global Flood.[4]

Surprising to many, the attacks on the ark and Noah's worldwide Flood are coming from both *outside* the Church and *inside* the Church. So the Flood is an important issue. Join with us on this adventure to investigate issues and questions surrounding the Flood and ark of Noah to help lay to rest these erroneous claims.

The Church and a Global Flood

It is not enough to merely teach a global Flood — we need to teach the next generation how to defend it (this dives into the field of apologetics[5] — defending the faith). The Flood is documented and discussed in the Bible in Genesis 6–8 as well as other portions of Scripture (that refer back to the Flood event) quoting Jesus Christ or Paul the Apostle or a Psalmist. The universal Flood really occurred, and Scripture clearly teaches this truth.

The secular culture has been taught to deny the global Flood of Noah, however, many people in our churches (and culture) now struggle with the fact of the world-covering Flood! To put it bluntly, the churchgoer has been influenced by the secular belief system — evolutionism — that opposes the reality of the biblical Flood. Many in the Church succumb to this secular peer pressure and deny the global Flood.

3. Hugh Ross, *The Genesis Question* (Colorado Springs, CO: NavPress, 1998), p. 156.
4. Ibid., p. 157.
5. Apologetics is derived from the Greek word *apologia,* which means "to give a defense or give an answer."

Therefore, *in the churches* across the Western world, there is a significant problem. Consider the original study of the general public by America's Research Group led by Britt Beemer in 2015 to help estimate and prepare for the number of people who would visit the Ark Encounter in Kentucky.

For instance, when asked "Do you believe Noah's ark was actually built or only a legend?" the results for those who said, "Yes , it was actually built" were as follows (by age group):

60s — 86%
50s — 74%
40s — 81%
30s — 81%
20s — 52%[6]

Notice how the generation currently in their 20s overwhelmingly dropped! This is the impact of secular schooling, most media and secular museums, and their insistence on millions of years and a denial of the Flood.

What about church attendees? As far back as the year 2000, George Barna, who works with church statistics, revealed that the number of people who do not believe in absolute truth in the pews was staggering. One writer concluded:

> Can you believe that 27% of all the people who attend an evangelical church do not believe in absolute truth?[7]

Consider this statistic. God is the absolute truth and so is the Bible (his written revelation) . . . including the account of Noah's ark and the Flood! And yet, over 1 in 4 in our evangelical churches do not really believe it. More recently, Britt Beemer, of America's Research Group, found that:

> One in six said their pastor said something to make them believe that the Book of Genesis contained myths and legends that we know are untrue.[8]

> Over 20% (one in five) said their pastors taught that there is no problem, biblically speaking, if Christians believe in an earth that is millions or billions of years old.[9]

6. Ken Ham with Jeff Kinley, research by Britt Beemer, *Ready to Return* (Green Forest, AR: Master Books, 2015), p. 103–104.
7. Carl Kerby, "WDJS, not just WWJD," May 11, 2000, based on research by George Barna, https://answersingenesis.org/jesus-christ/wdjs-not-just-wwjd/.
8. Ham with Kinley, *Ready to Return*, p. 30.
9. Ibid., p. 30.

Keep in mind that the idea that the earth is millions of years old comes from rejecting the idea of a *global* Flood of Noah's day. Then all the Flood sediment is seen as evidence of slow gradual accumulations over millions of years! Regarding kids who attended church, Beemer found that:

> Eighty-three percent said their science teachers taught them that the earth was millions or billions of years old.[10]

What this means is that the secular, atheistic world is being more effective at training the church kids than the Church! It is time to reverse this trend! Approximately 90 percent of kids in church homes attend state schools where they are indoctrinated with secular thought (think naturalism and atheism) for about 40 hours a week — including secular morality (moral relativism)! Churches are doing well if they are allowed to teach kids within their doors for 3 hours a week!

Essentially, many Christians are sending their children to be taught a different religion in state school systems and then we wonder why our kids walk away from Church! Is there any doubt as to why the next generation is walking after the religion of secular humanism (think atheism, agnosticism, secularism, naturalism, etc.) and its evil tenants, and leaving the Church?

You need to understand that the foundation of Christianity is under attack in our day and age. These attacks are focused on Genesis 1–11, including the Flood and Noah's ark — in state schools, secular media, and secular museums . . . and even in many churches and Christian schools! We need to be prepared for the secular claims so that we can be ready to give a defense for the hope that is in us for our Christian faith surrounding the Flood and Noah's ark (e.g., 2 Corinthians 10:4–5[11]; 1 Peter 3:15[12]). Now let us embark on this journey of *what you need to know.*

---

10. Ibid., p. 28.
11. For the weapons of our warfare are not carnal but mighty in God for pulling down strongholds, casting down arguments and every high thing that exalts itself against the knowledge of God, bringing every thought into captivity to the obedience of Christ.
12. But sanctify the Lord God in your hearts, and always be ready to give a defense to everyone who asks you a reason for the hope that is in you, with meekness and fear.

2 The Genesis Account of the Flood

Bodie Hodge

Genesis 6:1–9:20 with a few clarifying comments for the reader.

Genesis 6

1 Now it came to pass, when men began to multiply on the face of the earth, and daughters were born to them,

2 that the sons of God saw the daughters of men, that they were beautiful; and they took wives for themselves of all whom they chose.

3 And the Lord said, "My Spirit shall not strive with man forever, for he is indeed flesh; yet his days shall be one hundred and twenty years." **[This is not a longevity limitation, as many people lived well beyond this for a thousand years, but is instead the countdown to the Flood.]**

4 There were giants **[Hebrew: *nephilim*, which is related to the verb "to fall"; not necessarily giants here as context simply doesn't tell us; all we know is that the later *nephilim* in Numbers 13:33[1] were giant in stature by their immediate context and their placement with other giant peoples in the land of Canaan]** on the earth in those days, and also afterward, when the sons of God came in to the daughters of men and they bore children to them. Those were the mighty men who were of old, men of renown.

1. "There we saw the giants (the descendants of Anak came from the giants); and we were like grasshoppers in our own sight, and so we were in their sight."

5 Then the LORD saw that the wickedness of man was great in the earth, and that every intent of the thoughts of his heart was only evil continually.

6 And the LORD was sorry that He had made man on the earth, and He was grieved in His heart [**not that God was sorry in His unchanging nature or *will*, but of his state of *work* with man. Matthew Poole rightly asserts about this figure of speech that it is "a common figure called *anthropopathia*, whereby also eyes, ears, hands, nose, &c. are ascribed to God; and it signifies an alienation of God's heart and affections from men for their wickedness"**].

7 So the LORD said, "I will destroy man whom I have created from the face of the earth, both man and beast, creeping thing and birds of the air, for I am sorry that I have made them."

8 But Noah found grace in the eyes of the LORD.

9 This is the genealogy of Noah. Noah was a just man, perfect in his generations. Noah walked with God.

10 And Noah begot three sons: Shem, Ham, and Japheth.

11 The earth also was corrupt before God, and the earth was filled with violence.

12 So God looked upon the earth, and indeed it was corrupt; for all flesh had corrupted their way on the earth [**both man and land-dependent animals**].

13 And God said to Noah, "The end of all flesh has come before Me, for the earth is filled with violence through them; and behold, I will destroy them with the earth.

14 "Make yourself an ark of gopherwood [**either a type of wood or a way the wood was worked (think of plywood or pressed wood or a planking style)**]; make rooms in the ark, and cover it inside and outside with pitch [**likely sap/plant-based as opposed to petroleum based, which is largely a product from the Flood**].

15 "And this is how you shall make it: The length of the ark shall be three hundred cubits, its width fifty cubits, and its height thirty cubits [**ancient cubits tended to vary from about 18 inches (short cubits) to 21 inches (long cubits)**].

16 "You shall make a window for the ark [**Hebrew *tsohar*, even though it is translated as window, it actually means "noon"**

or "midday." It was a feature that runs the length of the top and middle of the ark to allow lighting and ventilation like a "ridge vent" on a house today], and you shall finish it to a cubit from above; and set the door of the ark in its side. You shall make it with lower, second, and third decks.

17 "And behold, I Myself am bringing floodwaters on the earth, to destroy from under heaven all flesh in which is the breath of life; everything that is on the earth shall die.

18 "But I will establish My covenant with you; and you shall go into the ark — you, your sons, your wife, and your sons' wives with you. [**This is a guarantee to Noah that his family would survive the persecution he would receive for following the Lord to be faithful at building the ark and being a preacher of righteousness.**]

19 "And of every living thing of all flesh you shall bring two of every sort into the ark, to keep them alive with you; they shall be male and female.

20 "Of the birds after their kind, of animals after their kind, and of every creeping thing of the earth after its kind, two of every kind will come to you to keep them alive. [**Kinds are not necessarily to be equated with our modern concept of species, but in most instances are *family* level (or *genus* or *species* in some instances) or in rare cases, an *order* level.**]

21 "And you shall take for yourself of all food that is eaten, and you shall gather it to yourself; and it shall be food for you and for them."

22 Thus Noah did; according to all that God commanded him, so he did.

Genesis 7

1 Then the LORD said to Noah, "Come into the ark, you and all your household, because I have seen that you are righteous before Me in this generation.

2 "You shall take with you seven each of every clean animal, a male and his female; two each of animals that are unclean, a male and his female;

3 "also seven each of birds of the air, male and female, to keep the species alive on the face of all the earth.

4 "For after seven more days I will cause it to rain on the earth forty days and forty nights, and I will destroy from the face of the earth all living things that I have made."

5 And Noah did according to all that the LORD commanded him.

6 Noah was six hundred years old when the floodwaters were on the earth.

7 So Noah, with his sons, his wife, and his sons' wives, went into the ark because of the waters of the flood.

8 Of clean animals, of animals that are unclean, of birds, and of everything that creeps on the earth,

9 two by two they went into the ark to Noah, male and female, as God had commanded Noah.

10 And it came to pass after seven days that the waters of the flood were on the earth.

11 In the six hundredth year of Noah's life, in the second month, the seventeenth day of the month, on that day all the fountains of the great deep were broken up, and the windows of heaven were opened.

12 And the rain was on the earth forty days and forty nights. **[Any rain, after this and until the 150th day when the rain was restrained per Genesis 8:2, was not *on the earth*, which was now submerged for a time but the rain was falling on the waters that were currently covering the earth.]**

13 On the very same day Noah and Noah's sons, Shem, Ham, and Japheth, and Noah's wife and the three wives of his sons with them, entered the ark —

14 they and every beast after its kind, all cattle after their kind, every creeping thing that creeps on the earth after its kind, and every bird after its kind, every bird of every sort.

15 And they went into the ark to Noah, two by two, of all flesh in which is the breath of life.

16 So those that entered, male and female of all flesh, went in as God had commanded him; and the LORD shut him in.

17 Now the flood was on the earth forty days. The waters increased and lifted up the ark, and it rose high above the earth.

18 The waters prevailed and greatly increased on the earth, and the ark moved about on the surface of the waters.

19 And the waters prevailed exceedingly on the earth, and all the high hills under the whole heaven were covered.

20 The waters prevailed fifteen cubits upward, and the mountains were covered. [**Clearly, this is a global Flood.**]

21 And all flesh died that moved on the earth: birds and cattle and beasts and every creeping thing that creeps on the earth, and every man.

22 All in whose nostrils was the breath of the spirit of life, all that was on the dry land, died [**that is, all land-dwelling, air-breathing animals that were not on board the ark had died**].

23 So He destroyed all living things which were on the face of the ground: both man and cattle, creeping thing and bird of the air. They were destroyed from the earth. Only Noah and those who were with him in the ark remained alive.

24 And the waters prevailed on the earth one hundred and fifty days.

Genesis 8

1 Then God remembered Noah, and every living thing, and all the animals that were with him in the ark. [**Not that an all-knowing God forgot, but this is *standing against* those that died in the Flood in the context immediately above — Genesis 7:21–23.**] And God made a wind to pass over the earth, and the waters subsided.

2 The fountains of the deep and the windows of heaven were also stopped, and the rain from heaven was restrained.

3 And the waters receded continually from the earth. At the end of the hundred and fifty days the waters decreased.

4 Then the ark rested in the seventh month, the seventeenth day of the month, on the mountains of Ararat. [**Take note that this is not necessarily Mt. Ararat, the active volcano that we know today.**]

5 And the waters decreased continually until the tenth month. In the tenth month, on the first day of the month, the tops of the

mountains were seen [***visible***; **by either water level reducing and/or vapor/fog clearing or perhaps a combination of both**].

6 So it came to pass, at the end of forty days, that Noah opened the window of the ark which he had made.

7 Then he sent out a raven, which kept going to and fro until the waters had dried up from the earth [**a hardy bird that can survive the rough conditions that were still present outside the ark**].

8 He also sent out from himself a dove, to see if the waters had receded from the face of the ground.

9 But the dove found no resting place for the sole of her foot, and she returned into the ark to him, for the waters were on the face of the whole earth. So he put out his hand and took her, and drew her into the ark to himself.

10 And he waited yet another seven days, and again he sent the dove out from the ark.

11 Then the dove came to him in the evening, and behold, a freshly plucked olive leaf was in her mouth; and Noah knew that the waters had receded from the earth.

12 So he waited yet another seven days and sent out the dove, which did not return again to him anymore.

13 And it came to pass in the six hundred and first year, in the first month, the first day of the month, that the waters were dried up from the earth; and Noah removed the covering of the ark and looked, and indeed the surface of the ground was dry [**note this is just the surface of the ground**].

14 And in the second month, on the twenty-seventh day of the month, the earth was dried [**note this is the whole earth**].

15 Then God spoke to Noah, saying,

16 "Go out of the ark, you and your wife, and your sons and your sons' wives with you.

17 "Bring out with you every living thing of all flesh that is with you: birds and cattle and every creeping thing that creeps on the earth, so that they may abound on the earth, and be fruitful and multiply on the earth."

18 So Noah went out, and his sons and his wife and his sons' wives with him.

19 Every animal, every creeping thing, every bird, and whatever creeps on the earth, according to their families, went out of the ark. [**They did not come out two-by-two, but now by their *families*, so there probably was some breeding on the ark — this may be where the term "breed like rabbits" originates!**]

20 Then Noah built an altar to the Lord, and took of every clean animal and of every clean bird, and offered burnt offerings on the altar [**interestingly, this leaves a breeding stock of male and female clean animals for each of Noah's three sons**].

21 And the Lord smelled a soothing aroma. Then the Lord said in His heart, "I will never again curse the ground for man's sake, although the imagination of man's heart is evil from his youth; [**note the differences between this and Genesis 6:5**] nor will I again destroy every living thing as I have done.

22 "While the earth remains, seedtime and harvest, cold and heat, winter and summer, and day and night shall not cease."

Genesis 9:1–20

1 So God blessed Noah and his sons, and said to them: "Be fruitful and multiply, and fill the earth. [**Knowing that God blessed Noah's sons here, which included Ham, could be one of the reasons Noah didn't curse Ham, but instead Ham's son Canaan later in Genesis 9:25.**[2]]

2 "And the fear of you and the dread of you shall be on every beast of the earth, on every bird of the air, on all that move on the earth, and on all the fish of the sea. They are given into your hand.

3 "Every moving thing that lives shall be food for you. I have given you all things, even as the green herbs. [**The original diet was changed. Now God first permits meat to be eaten.**]

4 "But you shall not eat flesh with its life, that is, its blood.

5 "Surely for your lifeblood I will demand a reckoning; from the hand of every beast I will require it, and from the hand of man. From the hand of every man's brother I will require the life of man.

6 "Whoever sheds man's blood, by man his blood shall be shed; for in the image of God He made man.

2. Then he said: "Cursed be Canaan; a servant of servants he shall be to his brethren."

7 And as for you, be fruitful and multiply; bring forth abundantly in the earth and multiply in it."

8 Then God spoke to Noah and to his sons with him, saying:

9 "And as for Me, behold, I establish My covenant with you and with your descendants after you,

10 "and with every living creature that is with you: the birds, the cattle, and every beast of the earth with you, of all that go out of the ark, every beast of the earth.

11 "Thus I establish My covenant with you: Never again shall all flesh be cut off by the waters of the flood; never again shall there be a flood to destroy the earth." [**Again, this shows it was a unique event — a global flood, as we have local floods all the time.**]

12 And God said: "This is the sign of the covenant which I make between Me and you, and every living creature that is with you, for perpetual generations:

13 "I set My rainbow in the cloud, and it shall be for the sign of the covenant between Me and the earth.

14 "It shall be, when I bring a cloud over the earth, that the rainbow shall be seen in the cloud; [**not that rainbows were impossible before this, but like bread used in communion, it now has significance and meaning**]

15 "and I will remember My covenant which is between Me and you and every living creature of all flesh; the waters shall never again become a flood to destroy all flesh.

16 "The rainbow shall be in the cloud, and I will look on it to remember the everlasting covenant between God and every living creature of all flesh that is on the earth." [**The rainbow now became a symbol and reminder, like bread and wine for communion.**]

17 And God said to Noah, "This is the sign of the covenant which I have established between Me and all flesh that is on the earth."

18 Now the sons of Noah who went out of the ark were Shem, Ham, and Japheth. And Ham was the father of Canaan.

19 These three were the sons of Noah, and from these the whole earth was populated.

20 And Noah began to be a farmer, and he planted a vineyard.

The Flood and the History of "Millions of Years"

3 Why Was It Necessary to Kill All Those *Innocent* People in the Flood?

Ken Ham and Bodie Hodge

Have you heard these questions or even asked them yourself?

- Why would God send a Flood to kill all those people — they weren't that bad?
- How could God kill the innocent children in the Flood — they didn't do anything wrong?
- Why would God kill the animals because of man's actions?

We are surprised how often we have heard these questions. Even within the Church, people struggle with the fact that God sent a Flood to kill everybody (except Noah's family). There are some views floating around that the pre-Flood people really weren't that bad and that God overreacted to judge the world with a global Flood. Let's look at these serious questions in more detail.

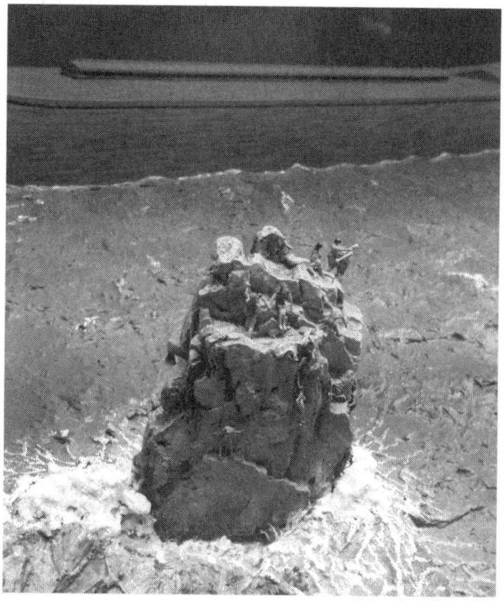

Diorama at the Creation Museum showing people stranded as the ark is floating.

Were People Innocent at the Time of the Flood?

There is this idea that people are "basically good" and "basically innocent." If you have been in this camp, get ready for a shocking statement — no one is basically good or innocent! Only God is good (Luke 18:19[1]). But why? Couldn't a good God have created a good world?

God *did* make a good world. It was perfect (Deuteronomy 32:4[2]) and very good (Genesis 1:31[3]). But we no longer live in *that* world, but instead, we live in a world that has been cursed due to sin and we have to deal with death and suffering — which is the punishment for sin! This occurred in Genesis 3 (see also Romans 8:19–22[4]). This is why death, suffering, natural evils (disasters like tornadoes, earthquakes, and hurricanes), and atrocities happens. It is a result of judgment on sin. Essentially, we have been given *a taste* of what life is like without God.

When Adam (and Eve) sinned, the punishment was ultimately death (Genesis 2:17,[5] 3:19[6]). Because of this, we all sin too as we were in Adam when he sinned and so we have sin nature (yes, we all came from Adam). This is why we are prone to sin too. We have sinful flesh (Romans 8:3[7]) and sinful minds from the moment we are conceived/fertilized (Genesis 8:21,[8] Psalm 51:5,[9] Jeremiah 17:9,[10] Colossians 1:21,[11] etc.).

1. So Jesus said to him, "Why do you call Me good? No one is good but One, that is, God."
2. He is the Rock, His work is perfect; for all His ways are justice, a God of truth and without injustice; righteous and upright is He.
3. Then God saw everything that He had made, and indeed it was very good. So the evening and the morning were the sixth day.
4. For the earnest expectation of the creation eagerly waits for the revealing of the sons of God. For the creation was subjected to futility, not willingly, but because of Him who subjected it in hope; because the creation itself also will be delivered from the bondage of corruption into the glorious liberty of the children of God. For we know that the whole creation groans and labors with birth pangs together until now.
5. "But of the tree of the knowledge of good and evil you shall not eat, for in the day that you eat of it you shall surely die."
6. "In the sweat of your face you shall eat bread till you return to the ground, for out of it you were taken; for dust you are, and to dust you shall return."
7. For what the law could not do in that it was weak through the flesh, God did by sending His own Son in the likeness of sinful flesh, on account of sin: He condemned sin in the flesh.
8. And the Lord smelled a soothing aroma. Then the Lord said in His heart, "I will never again curse the ground for man's sake, although the imagination of man's heart is evil from his youth; nor will I again destroy every living thing as I have done."
9. Behold, I was brought forth in iniquity, and in sin my mother conceived me.
10. The heart is deceitful above all things, and desperately wicked; who can know it?
11. And you, who once were alienated and enemies in your mind by wicked works, yet now He has reconciled.

And this is why everyone eventually dies. Really, when people ask why God judged the people of Noah's day with death, it's the wrong question — the point is everyone is under the judgment of death. The right question to ask is why it was the time for these people of Noah's day to die.

Ever since sin with Adam and Eve in the Garden of Eden, people have been prone to evil and sin. So is anyone innocent because we are all sinners? No. This is confirmed in Romans 3:23[12] as well, since we have all sinned and fallen short of the glory of God.

With this foundation in mind, what were the people like before the Flood? God tells us:

> Then the LORD saw that the wickedness of man was great in the earth, and that every intent of the thoughts of his heart was only evil continually (Genesis 6:5).

> The earth also was corrupt before God, and the earth was filled with violence. So God looked upon the earth, and indeed it was corrupt; for all flesh had corrupted their way on the earth (Genesis 6:11–12).

No one was innocent in the sight of God anyway. Now imagine the cannibalism, rape, murder, child sacrifice, and so on. Yet God was patient. He gave a 120-year countdown to the Flood (Genesis 6:3[13]) and even sent a preacher of righteousness (2 Peter 2:5[14]). There was no excuse not to repent. The judgment was righteous.

But What About the Children — Aren't They Innocent?

A Christian lady once pleaded with me that she, "just couldn't see how God could send the Flood knowing that innocent children would die in the Flood." As we have seen before, none are innocent (Romans 3:23[15]). The punishment for sin is death, and even at conception a baby can die as we sadly see quite often with miscarriages and rampant abortion in today's secular culture. We have to remember that we are all already judged with death.

12. For all have sinned and fall short of the glory of God.
13. And the LORD said, "My Spirit shall not strive with man forever, for he is indeed flesh; yet his days shall be one hundred and twenty years."
14. And did not spare the ancient world, but saved Noah, one of eight people, a preacher of righteousness, bringing in the flood on the world of the ungodly.
15. For all have sinned and fall short of the glory of God.

It would be impossible for someone who is sinless to die — even our sinless Christ could not have died had He not taken our sin upon Him. Christ became sin for us (2 Corinthians 5:21[16]) and bore the punishment that we deserve for sin on the Cross — through His shed blood (1 Peter 2:24[17]).

There is a misconception that children are innocent of sin, but this is not true. They can be wicked, violent, corrupt, and have evil thoughts all the time. Parents often concur that this is true! This is why parents are so important when training their children. But we know from the Bible that all were corrupt, evil, violent, and so on. This includes children. And children can further develop these habits from their parents if not trained in righteousness. If the Word of God was not impressed upon children, then false humanistic views were impressed upon the kids (Deuteronomy 11:19[18])!

But let's just consider for a moment a theoretical situation. *If* there were righteous children — why would they not be on the ark? Noah was the only one found righteous at the onset of the Flood (Genesis 7:1[19]). But further, who was it who kept these children from entering the door of the ark? It was their unrighteous parents or guardians. So one should not misplace the responsibility of the parents/guardians and try to say God was the one who didn't send a means of salvation for the children! Keep in mind that the children were not innocent by the time the Flood came. And their parents, being unrighteous themselves, were surely encouraging them in their sin!

God did send a means of salvation with the ark for the righteous. But it was the parents of those children who were so unrighteous, that they refused to even give their children an opportunity but impressed upon them sin as well — that is what we expect from those whose every thought is evil all the time.

Consider another issue. How many children were alive at the time of the Flood? When a civilization becomes evil, who usually pays the price first? It

16. For He made Him who knew no sin to be sin for us, that we might become the righteousness of God in Him.
17. Who Himself bore our sins in His own body on the tree, that we, having died to sins, might live for righteousness — by whose stripes you were healed.
18. You shall teach them to your children, speaking of them when you sit in your house, when you walk by the way, when you lie down, and when you rise up.
19. Then the LORD said to Noah, "Come into the ark, you and all your household, because I have seen that you are righteous before Me in this generation."

is the children and babies. How many ancient cultures killed or sacrificed children? Far too many!

In the Bible, we find evil people like Pharaoh of Egypt trying to kill baby boys through the midwives when they were born. Then Pharaoh threw them into the Nile — little did Pharaoh know that his own son and his army of Egyptians would later be drowned by God in the Red Sea! Herod murdered children. Canaanites (Leviticus 18) and later the Israelites (Jeremiah 32:35[20]) sacrificed their children to Molech, a false god. A few others include Carthage and the Incan and Aztec cultures.

In our own evil day, the German Nazis, led by Adolf Hitler, killed adults and children in terrible manners to build their "thousand-year Reich." And even the USA, China, England, France, and many other countries regularly practice abortion, the sacrifice of babies for their own selfish purposes in hopes to make their own lives better, which is exactly what ancient sacrifices were supposed to do too! Many of those who accuse the Christian God of being a "genocidal god" because of the Flood, support the genocidal killing of millions of children in their mother's wombs, which is a double standard fallacy!

With a broken heart, we see sacrifice of children in our day all the time, and yet the culture before the Flood was so bad that God judged them with the Flood. Were there even any children left? People were trying to make a name for *themselves*, not their children (Genesis 6:4,[21] "men of renown")! If the whole world were murderers, the population of the world cuts in half in one day, and who were prime targets? The children most likely were the first to receive the brunt of this attack! Evil people rarely want the responsibility of raising children; the kids often get in their way.

Regardless, our Creator God, who is a just God, had every right to enact His judgment as He determined.

Why Did Animals Get Judged?

We often believe that animals are entirely innocent creatures to their actions. But are they? The serpent in the Garden of Eden, an instrument of Satan,

20. And they built the high places of Baal which are in the Valley of the Son of Hinnom, to cause their sons and their daughters to pass through the fire to Molech, which I did not command them, nor did it come into My mind that they should do this abomination, to cause Judah to sin.

21. There were giants on the earth in those days, and also afterward, when the sons of God came in to the daughters of men and they bore children to them. Those were the mighty men who were of old, men of renown.

was still cursed for its involvement (Genesis 3:14[22]). In the Law of Moses, if a man or woman had sexual relations with an animal, both they and the animal were to be put to death (e.g., Leviticus 20:15–16[23]). If an ox kills someone in Mosaic Law, the animal must be put to death (Exodus 21:28–29[24]). Let us put it this way: animals get their due as well.

The Bible says:

> So God looked upon the earth, and indeed it was corrupt; for all flesh had corrupted their way on the earth (Genesis 6:12).

This means that all animals (i.e., all flesh) *had* corrupted themselves on the earth, not just humanity. Animals were originally to be vegetarian (Genesis 1:30[25]), and many were likely corrupting themselves by being meat eaters. Many animals may have been corrupted to be homosexual in their actions (defying their ordinance to be fruitful and multiply in Genesis 1) or attacking people (some of these corruptions could be by the influence of demons, consider Mark 5:12[26]).

Let us also not forget that man had dominion over the animals (Genesis 1:26–28[27]). So if the dominion of man falls, so do the things under that dominion. When Hitler made a decision to begin WWII, that decision affected his whole dominion. Did that mean that the animals within Hitler's dominion were not affected by his decision? By no means. When those who

22. So the LORD God said to the serpent: "Because you have done this, you are cursed more than all cattle, and more than every beast of the field; on your belly you shall go, and you shall eat dust all the days of your life."
23. If a man mates with an animal, he shall surely be put to death, and you shall kill the animal. If a woman approaches any animal and mates with it, you shall kill the woman and the animal. They shall surely be put to death. Their blood is upon them.
24. If an ox gores a man or a woman to death, then the ox shall surely be stoned, and its flesh shall not be eaten; but the owner of the ox shall be acquitted. But if the ox tended to thrust with its horn in times past, and it has been made known to his owner, and he has not kept it confined, so that it has killed a man or a woman, the ox shall be stoned and its owner also shall be put to death.
25. "Also, to every beast of the earth, to every bird of the air, and to everything that creeps on the earth, in which there is life, I have given every green herb for food"; and it was so.
26. So all the demons begged Him, saying, "Send us to the swine, that we may enter them."
27. Then God said, "Let Us make man in Our image, according to Our likeness; let them have dominion over the fish of the sea, over the birds of the air, and over the cattle, over all the earth and over every creeping thing that creeps on the earth." So God created man in His own image; in the image of God He created him; male and female He created them. Then God blessed them, and God said to them, "Be fruitful and multiply; fill the earth and subdue it; have dominion over the fish of the sea, over the birds of the air, and over every living thing that moves on the earth."

had dominion (man) were to be judged with the global Flood, their whole dominion fell (including animals). But consider this oft-overlooked point: the animal kinds aboard the ark were saved from judgment too.

Conclusion

Keep in mind that God is a righteous judge. He knew the hearts of everyone who was judged in the Flood and it was a *righteous* judgment. Who are we to proclaim, in any measure, that the judgment on wicked and violent people prior to the Flood was unjust? And by what standard would we be judging? It would be our own arbitrary, fallible, sinful standard.

> I see the treacherous, and am disgusted, because they do not keep Your word. Consider how I love Your precepts; revive me, O Lord, according to Your lovingkindness. The entirety of Your word is truth, and every one of Your righteous judgments endures forever (Psalm 119:158–160).

So was the Flood necessary? Yes! And it was a *righteous* judgment on sinners who refused to repent and wallowed in their sin. They loved sin more than they loved God, so there was no longer any reason for them to continue to reap the good benefits from a good God. A very violent group of anti-God people were met with a very violent Flood . . . a fitting means of cleansing vindication for those they killed and harmed for so long.

4 Flood Legends

Bodie Hodge

When we start with the Bible, the Flood of Noah's day was an actual historical event about 4,350 years ago. Outside the Bible, do we expect that cultures around the world retained this history to one degree or another? Yes we do. Let me start with some context.

Babel

After the Flood, there was another significant event when God confused the language of the whole world, because people were trying to defy God. They were refusing to listen to God's command to fill the earth (Genesis 9:1[1]). Instead they came together and built a city and tower at Babel to try to keep them from filling the earth (Genesis 11:4[2]).

So God came down and confused their language, and different families came out of Babel speaking differently. These language families continued to change into the various languages we have today.[3]

But what is significant about the splitting at Babel with regard to the Flood? It simply means that people went in various directions, taking their history of the world (Genesis 1–11) to different parts of the globe.

1. So God blessed Noah and his sons, and said to them: "Be fruitful and multiply, and fill the earth."
2. And they said, "Come, let us build ourselves a city, and a tower whose top is in the heavens; let us make a name for ourselves, lest we be scattered abroad over the face of the whole earth."
3. For more on this subject see Bodie Hodge, *Tower of Babel* (Green Forest, AR: Master Books, 2013).

We expect details of this history to be lost in some instances, embellished in some instances, major events lost all together, names changed (it was language change, after all), local animals to be inserted (e.g., duck instead of a dove as in Genesis 8:11[4]), local mountains to replace the mountains of Ararat (Genesis 8:4[5]), etc. to replace places in the accounts and so on.

Thus, we expect that these cultures retain some of the true history of the Flood one way or another. But we seriously doubt they will have kept the entire account error free. Of course, the Bible records the true account because the Holy Spirit, who is God, inspired the account through Moses.[6]

How Many Flood Legends?

So what do have? There are somewhere between 300 and 500 Flood legends worldwide. I prefer to say "over 300," and you'll see why in a moment. But why the range in the first place? Simply put, we keep finding more.

Allow me to explain. You can research and easily find about 200 Flood legends without extensive research. Dr. John Morris of the Institute for Creation Research has done this.[7] So has Japanese researcher Nozomi Osanai for her Master's thesis at Wesley Biblical Seminary.[8] In fact, *I've* done this too! Though there were more, I read over 200 flood accounts when researching a family book that I co-authored with Laura Welch on flood legends (*The Flood of Noah, Legends and Lore of Survival,* Master Books, 2014).

In years past, various researchers would tally a few of these together in books on Native American history, Aboriginal Australian culture, or various other regional cultures. In other cases, you would find one legend here or there. You need to understand that we are just now living in an era when many of these are being tallied together. Few books and resources were

4. Then the dove came to him in the evening, and behold, a freshly plucked olive leaf was in her mouth; and Noah knew that the waters had receded from the earth.
5. Then the ark rested in the seventh month, the seventeenth day of the month, on the mountains of Ararat.
6. Terry Mortenson and Bodie Hodge, "Did Moses Write Genesis?" in *How Do We Know the Bible Is True? Volume 1*, Ken Ham and Bodie Hodge, gen. eds. (Green Forest, AR: Master Books, 2011), p. 85–102.
7. John Morris, "Why Does Nearly Every Culture Have a Tradition of a Global Flood?" *Acts & Facts* 30 (9), 2001, http://www.icr.org/article/why-does-nearly-every-culture-have-tradition-globa/.
8. Nozomi Osanai, "A Comparative Study of the Flood Accounts in the Gilgamesh Epic and Genesis," Wesley Biblical Seminary, reprinted on Answers in Genesis, August, 3, 2005, https://answersingenesis.org/the-flood/flood-legends/flood-gilgamesh-epic/.

all-encompassing when it comes to Flood legends. And this makes sense as more come to light each year.

For example, I met two missionaries who visited the Creation Museum while on furlough from their missions work. They are missionaries who are the first contact with certain tribes in the Amazon. At the time, they were the first in contact with two different tribes. I asked if those tribes had a Flood or creation legend. They said the tribes did and they even had a legend where the languages split apart too! The missionaries told me that they were unaware if any of their co-missionaries with these tribes had documented this yet. Of course, I encouraged them to make sure to do so in the future so there is a record of it. But this outlines that new Flood legends (as well as Babel and creation legends) are still being ascertained even now.

Even though people did tally some Flood accounts in the past (1918 and 1931),[9] let's just look at more modern numbers. In 1961, researchers Drs. Henry Morris and John Whitcomb looked into this subject saying there were "scores and hundreds"[10] of these legends. Later, others used round numbers like 200. Even later, as research afforded, people recorded what they could find as actual numbers. For example, Dr. Duane Gish and Gloria Clanin, researching for children's books, were confident enough to state 270 Flood stories in 1992[11] and 1996.[12]

Since this time, numbers have gone up again to a point that Laura Welch (Senior Editor for Christian publisher Master Books) and I were able to confidently state from over 300 to upward of 500 legends in the family book on Flood legends. Even evolutionists have concurred that the number of Flood stories could be as high as 500.[13]

Our preference is to be safer with the count and say "over 300" as opposed to say "nearly 500." Why, you might ask? The reason is that most of the legends I found clearly spoke of a *global* Flood. But there were a handful that we could not discern if they were speaking of a global Flood or a local Flood. So it wouldn't surprise me if a small portion of Flood legends are discussing a

9. For example, Bryan Nelson, *The Deluge Story in Stone* (Augsburg, MN, 1931) or Sir James Frazer, *Folk-Lore in the Old Testament* (London, England: Macmillan & Co., Ltd., 1918).
10. John Whitcomb and Henry Morris, *The Genesis Flood* (Phillipsburg, NJ: P&R Publishing, 1961), p. 48.
11. Duane Gish, *Dinosaurs by Design* (Green Forest, AR: Master Books, 1992), p. 74.
12. Gloria Clanin, *In the days of Noah* (Green Forest, AR: Master Books, 1996), p. 60.
13. Robert Schoch, *Voyages of the Pyramid Builders* (New York, NY: Jeremy P. Tarcher/Penguin, 2003), p. 249.

flood in their history, but not necessarily the deviated accounts of the Flood of Noah. So I prefer to err on the side of a conservative number, not the maximum number. I would encourage others to be diligent in this as well.

What Are Some of These Flood Legends?

Whole volumes of books could be written if we put every account into one volume. Many accounts are nearly chapter-length in duration! So instead, here are excerpts and recounts from a few shorter examples from various parts of the world to "whet your appetite."

Aztec Legend of the Flood

> A man named Tapi lived a long time ago. Tapi was a very pious man. The creator told Tapi to build a boat that he would live in. He was told that he should take his wife, a pair of every animal that was alive into this boat. Naturally everyone thought he was crazy. Then the rain started and the flood came. The men and animals tried to climb the mountains but the mountains became flooded as well. Finally the rain ended. Tapi decided that the water had dried up when he let a dove loose that did not return.[14]

Hawaiian Legend of the Flood

> Hawaiians have a flood story that tells of a time when, long after the death of the first man, the world became a wicked, terrible place. Only one good man was left, and his name was Nu-u. He made a great canoe with a house on it and filled it with animals. In this story, the waters came up over all the earth and killed all the people; only Nu-u and his family were saved.[15]

Chinese Legend of the Flood

> Another flood story is from China. It records that Fuhi, his wife, three sons, and three daughters escaped a great flood and were the only people alive on earth. After the great flood, they repopulated the world.[16]

---

14. Flood Legends from around the world, http://www.nwcreation.net/noahlegends.html.
15. Monty White, "Flood Legends," March 29, 2007, https://answersingenesis.org/the-flood/flood-legends/flood-legends/.
16. Ibid.

Miao Legend of the Flood

So it poured forty days in sheets and in torrents.
Then fifty-five days of misting and drizzle.
The waters surmounted the mountains and ranges.
The deluge ascending leapt valley and hollow.
An earth with no earth upon which to take refuge!
A world with no foothold where one might subsist!
The people were baffled, impotent and ruined,
Despairing, horror stricken, diminished and finished.
But the Patriarch Nuah was righteous.
The Matriarch Gaw Bo-lu-en upright.
Built a boat very wide. Made a ship very vast.
Their household entire got aboard and were floated,
The family complete rode the deluge in safety.
The animals with him were female and male.
The birds went along and were mated in pairs.
When the time was fulfilled, God commanded the waters.
The day had arrived, the flood waters receded.
Then Nuah liberated a dove from their refuge,
Sent a bird to go forth and bring again tidings.
The flood had gone down into lake and to ocean;
The mud was confined to the pools and the hollows."
There was land once again where a man might reside;
There was a place in the earth now to rear habitations.
Buffalo then were brought, an oblation to God,
Fatter cattle became sacrifice to the Mighty.
The Divine One then gave them His blessing;
Their God then bestowed His good graces.[17]

Tanzania Legend of the Flood

Once upon a time the rivers began to flood. The god told two people to get into a ship. He told them to take lots of seed and to take lots of animals. The water of the flood eventually covered the mountains. Finally the flood stopped. Then one of the men, wanting to know if the water had dried up let a dove loose. The dove

17. Ernest Truax, "Genesis According to the Miao People," *Acts & Facts* 20 (4) 1991, http://www.icr.org/article/genesis-according-miao-people/.

returned. Later he let loose a hawk which did not return. Then the men left the boat and took the animals and the seeds with them.[18]

Names of Noah . . . and His Wife?

Another fascinating aspect of Flood legends is to see the various names of Noah and his wife. Recall that after Babel many patriarchs at Babel often had new names applied to them from the various languages that came out of Babel. Here are 20 (of the many) of variants of Noah's name.

Names of Noah in Various Ancient Cultures

	Name	Culture	Authority/Reference
1	Noah*	Israel	Genesis 6–9.
2	Noes/Noe	Germany and Scandinavia	Wright, ed., *Reliiquae Antiquae*, 1841–1845, copy at London's Guildhall Library, Aldermanbury, p. 173.**
3	Noeh	Ireland	Annals of Clonmacnois
4	Nuah	Maio (China)	E.A. Truax, "Genesis According to the Miao People," *Acts & Facts*. 20 (4) 1991.
5	Deucalion	Greece	e.g., Apollodorus, 1.7.2.
6	Titan	Celtic	B. Sproul, *Primal Myths* (New York: HarperOne Publisher, 1979).
7	Ziusudra	Sumeria	D. Hammerly-Dupuy, *Some Observations on the Assyro-Babylonian and Sumerian Flood Stories* (Lima, Peru: Colegio Union, 1968).
8	Atrahasis	Babylonia	S. Dalley, *Myths from Mesopotamia* (Oxford: Oxford University Press, 1989).
9	Xisuthrus	Chaldea	G. Smith, "The Chaldean Account of the Deluge," *Transactions of the Society of Biblical Archaeology*, 2:213–34. 1873.

18. "Flood Legends from Around the World," http://www.nwcreation.net/noahlegends.html.

10	Tumbainot (and wife Naipande)	East Africa (Masai people)	J.G. Frazier, *The Golden Bough* (Hertfordshire: Wordsworth Editions Ltd., 1993).
11	Nama	Central Asia	U. Holmburg, "Finno-Ugric, Siberian," in C.J.A. MacCulloch, ed., *The Mythology of All Races v. IV* (Boston, MA: Marshall Jones, 1927).
12	Manu	India	T.H. Gaster, *Myth, Legend, and Custom in the Old Testament* (New York, NY: Harper & Row, 1969).
13	Nol	Pacific Island (Loyalty Islands)	Gaster, p. 107.
14	Nu'u	Hawaii	D.B. Barrere, *The Kumuhonua Legends: A Study of Late 19th Century Hawaiian Stories of Creation and Origins,* Pacific Anthropological Records, No. 3, Bishop Museum, Honolulu, HI, 1969, p. 19–21.
15	Kunyan	Alaska	Gaster, p. 117–118.
16	Wissaketchak	Cree (Native Americans)	Frazier, p. 309–310.
17	Nanaboujou	Ottawa (Native Americans)	Frazier, p. 308.
18	Montezuma	Papago (Native Americans)	Gaster, p. 114–115.
19	Tezpi	Mexico (Michoacan)	Gaster, p. 122.
20	Marerewana	Guyana (South America)	Gaster, p. 126.

* Several cultures actually still retained the name "Noah."

** See also: MS. Cotton, Otho. B. XI., cit. Magoun, p. 249; Assersius, *De Rebus Gestis Alfredi* (Ed. Stevenson, Oxford, 1904), Cap I; Vetustissima Regum Septentrionis Series Langfethgatal dicta, *Scriptores Rerum Danicarum Medii AEvi*, Ed., Jacobus Langebek, Vol. I, Hafniae, 1772, p. 1–6.

We even find names of Noah's wife in some instances. There are well over 100 accounts that give a name for Noah's wife.[19] Keep in mind that the Bible doesn't specifically reveal Noah's wife's name. Listed below are a few examples of Noah's wife's name given in Flood legends. In some cases, you can recognize similarities (e.g., Emzara, Amzurah, and Noyemzar; or Haykêl and Haikal).

Names of Noah's Wife in Various Ancient Cultures

	Name	**Culture**	**Reference/ Authority**
1	*Emzara*, daughter of Rake'el, son of Methuselah	Judea	*Book of Jubilees*
2	*Haykêl*, the daughter of Namûs (or Namousa), the daughter of Enoch, the brother of Methuselah	Arabia	*Kitab al-Magall* (the Book of Rolls)
3	*Noyemzar* (variants: Nemzar, or Noyanzar)	Armenia	The Bible Cyclopedia*
4	Set	Latium (Italy)	*Inventiones Nominum*
5	*Naamah*, the daughter of Lamech and sister of Tubal-Cain (from Genesis 4:22**)	Judea	*Genesis Rabba* midrash
6	*Haykêl*, the daughter of Namûs (or Namousa), the daughter of Enoch, the brother of Methuselah	Syria	*Book of the Cave of Treasures*
7	*Dalila* (Variant *Dalida*)	Angles and Saxons (Germany and England)	Dialogue of *Solomon and Saturn*
8	*Haikal*, the daughter of Abaraz, of the daughters of the sons of Enos	Ethiopia/Cush	*Conflict of Adam and Eve with Satan*

19. E.g., Francis Lee Utley, "The One Hundred and Three Names of Noah's Wife," The University of Chicago Press, *Speculum*, Vol. 16, No. 4 (Oct., 1941), p. 426–452.

9	*Gaw Bo-lu-en*	China (Maio)	Edgar Truax on Maio recitations***
10	*Haykêl*, the daughter of Namûs (or Namousa), the daughter of Enoch, the brother of Methu-selah	Upper Egypt	*Patriarch Eutychius of Alexandria*
11	*Phiapphara*	Angles and Sax-ons (Germany and England)	Ælfric of Eynsham's translation of the first 7 books of the Bible (Heptateuch)
12	*Amzurah*, the daughter of Barakil, another son of Mehujael	Persia (Iran)	Persian historian Muhammad ibn Jarir al-Tabari
13	*Percoba*	Ireland	*Codex Junius*

* *The Bible Cyclopedia*, Volume 2 (London: Harrison and Co.; John W. Parker, 1843), p. 735.

** "And as for Zillah, she also bore Tubal-Cain, an instructor of every craftsman in bronze and iron. And the sister of Tubal-Cain was Naamah."

*** E.A. Truax, "Genesis According to the Miao People," *Acts & Facts*, 20 (4) 1991.

Conclusion

Flood legends are an excellent confirmation of what we expected to find in a biblical worldview. Consider the converse. In an evolutionary story with millions of years where there was supposedly no global flood, there shouldn't be any global flood stories. So why would *anyone* have a massive global flood account in their history? Is it because they live near a river and see little floods? This gets even more troubling when you realize that many of these legends are by cultures in super dry deserts and mountainous regions where Floods simply do not occur!

Flood legends are also useful in witnessing the true history of the world as a foundation to the Gospel. You can use these Flood, creation, or Tower of Babel legends[20] from cultures' own histories as a steppingstone to the truth in Genesis. Commend these people's ancestors for trying to hold on to that foundational history, as it is important to set the foundation of the Gospel.

20. For more Tower of Babel legends see Bodie Hodge, *Tower of Babel* (Green Forest, AR: Master Books, 2013), p. 221–226.

It is a powerful way to witness to people as a foundation to point to the work of Christ on the Cross. Genesis can be used as a steppingstone to reveal that we are all connected to our Creator who came to save us from sin and death that was founded in that same history from our common grandparents, Adam and Eve.

5 A Changing View of the Flood Evidence

Ken Ham and Bodie Hodge

Analogy to Get Started

Let's say you eyewitnessed a catastrophe such as a volcano blowing its top — like Mount St. Helens did in 1980 (and erupting again in 1982). Then you, being a trustworthy and qualified person, documented it properly (wrote down what happened in detail). You documented that the eruption laid down rock layers quickly in a matter of hours for example or that it carved out canyons in a subsequent eruption in 1982 in about one day.

Now let's say that 100 years later, geologists decide that Mount St. Helens never did erupt and there was no catastrophe that affected things around it. Then these same scientists decide to report that the rock layers (from the eruption) were *not* laid down by an eruption but instead were laid down slowly over long periods of time . . . perhaps millions of years.

Furthermore, they decide the canyons were actually evidence of slow gradual erosion, but definitely not from volcanic action! After all, they don't see Mount St. Helens erupting, so they assume it never had.

Then they criticize you for "making up fairy tales about Mount St. Helens" when you "*falsely*" claim that it actually erupted! Let's say they go so far as to get laws passed to refuse to allow anyone to consider that Mount St. Helens actually did erupt in 1980 in public arenas like state schools or museums that are state funded.

Sounds ludicrous, doesn't it? Do you realize this follows exactly what happened with regard to God's eyewitness account of the Flood of Noah and its evidence in our modern time? Let's take a look.

HOW DID PEOPLE VIEW THE FLOOD PRIOR TO THE 1800S?

God's account is the eyewitness account of the Flood. This Flood finds its confirmation all over the world in historical accounts. Having seen immense Flood legends in the historical accounts of cultures all over the world, it is rather safe to say that cultures all over the world held to a global Flood in their past. This is expected since the Bible is true.[1] As Noah's descendants scattered across the globe (Genesis 11:1–9), these people took their history of the Flood with them and it varied and deviated due to the effects of sin.

Ancient historians like Josephus the Jew, Berosus the Chaldean, Hieronymus the Egyptian, Mnaseas, and Nicolaus of Damascus (Josephus even mentions these last four) discussed a powerful flood that occurred in their past.

Ancient Greek historians like Xenophanes, Herodotus, Eratosthenes, and Strabo all commented on fossils being from a significant *water* event in the past (not always to the extent of biblical proportions but they understood the point). Naturally, Christians and Jews (and even Muslims) were still following what Genesis says occurred in the past regarding a global Flood, and likewise held to the Flood as the mechanism for the bulk of the fossil formation.

The connections between fossils and the Flood are obvious. From a biblical perspective, we expect to find trillions of things buried by a water-based sediment in the rock layers. For thousands of years, Flood layers were readily viewed as just that . . . Flood layers. Rock layers that contain fossils are primarily from the Flood of Noah and this was not really in much dispute in ages past.

Now, we want to add a caveat. Did you notice that we said "rock layers that contain fossils are *primarily* from the Flood"? Why did we say that? The reason is that we have had some fossil layers *since* the time of the Flood. There have been local catastrophes that have added layers of fossils in some local regions. This could be from local floods, tsunamis, volcanoes, earthquakes, and so on. Of course, the bulk came from the judgment of the Flood in Noah's day.

Due to respect and authority for Genesis, it was readily seen throughout the Western World, Middle East, and many other places as the true

1. Keep in mind that if an evolutionary history was true, there should be *no* global flood legends!

source of history including the Flood. But even though the majority readily recognized that the Flood easily explains rock layers that contained fossils, something major happened! All of this was about to change.

The Great Turning Point

In the late 1700s and early 1800s, there was a movement to leave the Bible out of the subject of rock layers. In other words, they wanted to leave God out of it. God is the ultimate eyewitness. Not only did He create all things, knows all things, upholds all things, and has always been there, but He cannot lie either! What greater person to document things of the past than God? Through Moses, God the Holy Spirit recorded the events of the past regarding the Flood.

Certain people attacked God and said not to trust what God said. Instead, they rejected that a global Flood occurred by appealing to fallible and imperfect ideas of man making guesses about the past.

When the debate had simmered down, the predominant thought in the 1800s was that all the fossil rock layers were laid down slowly and gradually (without any catastrophes) over millions of years. They said the fossil layers were made by slow accumulations throughout supposed long ages. This is called "uniformitarianism." This is because things supposedly formed in a *uniform* fashion, without catastrophes (i.e., slowly over a long time).

In other words, these people said to reject God's eyewitness account (making man out to be greater than God, by the way) and say that Noah's Flood is a farce (not true). Having seen many geological catastrophes in my own lifetime, it is a *hard pill to swallow* to believe that there were no catastrophes in the past. But nonetheless, people actually fell for this strange idea!

Believe it or not, the secularists have even succeeded in passing laws that forbid the freedom of teaching about the global Flood in the state classroom or by state-sponsored places for fear of getting fired and sued! Sounds ludicrous, but this has occurred! The secularists did to the Flood exactly what occurred in our example at the opening of this chapter about Mount St. Helens!

So, to summarize, fossils layers from the Flood of Noah's day came under attack and people arbitrarily decided there were no major catastrophes (and certainly not a global flood) in the past. They decided that these rock layers that have fossils (fossiliferous) were laid down slowly and gradually over millions of years.

The secularists openly claimed God got it wrong and then pushed for legislation that made it illegal for the state or state schools or state-sponsored places to teach the global Flood from the Bible. Does this mean that the rock layers really were laid down slowly over millions of years and there was no global Flood? By no means! It just means some people are forced to teach only the secular humanistic religion (the common religion of the day that teaches that man supersedes God).

Furthermore, it means that many kids are not allowed to hear the truth, but only a false religious view that is being imposed on generations of unsuspecting kids. The rest of us (i.e., Christians) are not so limited, but have an obligation and freedom to teach the truth. It is time to get back to the truth and stop worrying about the false secular religion that has permeated our society.

6 Biblically, How Old Is the Earth?

Bodie Hodge

The question of the age of the earth has produced heated discussions on Internet debate boards, classrooms, TV, radio, and in many churches, Christian colleges, and seminaries. The primary sides are:

- Young-earth proponents (biblical age of the earth and universe of about 6,000 years)[1]
- Old-earth proponents (secular age of the earth of about 4.5 billion years and a universe about 14 billion years old)[2]

The difference is immense! Let's give a little history of where the *biblical calculation* came from. We will discuss the old earth development in the subsequent chapter.

Where Did a Young-Earth Worldview Come From?

Simply put, it came from the Bible. Of course, the Bible doesn't say explicitly anywhere, "The earth is 6,000 years old." Good thing it doesn't; otherwise it would be out of date the following year. But we wouldn't expect an all-knowing God to make that kind of a mistake.

God gave us something better. In essence, He gave us a "birth certificate." For example, using a personal birth certificate, a person can calculate

1. Not all young-earth creationists agree on this age. Some have slight variation to this number.
2. Some of these old-earth proponents accept molecules-to-man biological evolution and so are called theistic evolutionists. Others reject neo-Darwinian evolution but accept the evolutionary timescale for astronomical and geological evolution, and hence agree with the evolutionary order of events in history.

how old he is at any point. It is similar with the earth. Genesis 1 says that the earth was created on the first day of creation (Genesis 1:1–5). From there, we can begin to calculate the age of the earth.

Let's do a rough calculation to show how this works. The age of the earth can be estimated by taking the first five days of creation (from earth's creation to Adam), then following the genealogies from Adam to Abraham in Genesis 5 and 11, then adding in the time from Abraham to today.

Adam was created on day 6, so there were five days before him. If we add up the dates from Adam to Abraham, we get about 2,000 years, using the Masoretic Hebrew text of Genesis 5 and 11.[3] Whether Christian or secular, most scholars would agree that Abraham lived about 2,000 B.C. (4,000 years ago). So a simple calculation is:

5 days
+ ~2000 years
+ ~4000 years
―――――――――
~6000 years

At this point, the first five days are negligible. Quite a few people have done this calculation using the Masoretic text (which is what most English translations are based on) and, with careful attention to the biblical details, have arrived at the same time frame of about 6,000 years, or about 4000 B.C. Two of the most popular, and perhaps best, are a recent work by Floyd Jones[4] and a much earlier book by James Ussher[5] (1581–1656). See table 1.

Table 1: Jones and Ussher

Name	Age Calculated	Reference and Date
James Ussher	4004 B.C.	*The Annals of the World,* A.D. 1658
Floyd Nolan Jones	4004 B.C.	*The Chronology of the Old Testament*, A.D. 1993

The misconception exists that Ussher and Jones were the only ones to arrive at a date of 4000 B.C.; however, this is not the case at all. Jones[6] lists several

3. Bodie Hodge, "Ancient Patriarchs in Genesis," Answers in Genesis, https://answersingenesis.org/bible-characters/ancient-patriarchs-in-genesis/.
4. Floyd Nolan Jones, *Chronology of the Old Testament* (Green Forest, AR: Master Books, 2005).
5. James Ussher, *The Annals of the World* (Green Forest, AR: Master Books, 2003), translated by Larry and Marion Pierce.
6. Jones, *Chronology of the Old Testament,* p. 26.

chronologists who have undertaken the task of calculating the age of the earth based on the Bible, and their calculations range from 5501 to 3836 B.C. A few are listed in table 2 with a couple of newer ones and their references.

Table 2: Chronologists' Calculations

Chronologist	When Calculated?	Date B.C.
Julius Africanus	**c. 240**	**5501**
George Syncellus	**c. 810**	**5492**
John Jackson	**1752**	**5426**
Dr William Hales	**c. 1830**	**5411**
Eusebius	c. 330	5199
Benjamin Shaw	2004	* 4954
Marianus Scotus	c. 1070	4192
L. Condomanus	n/a	4141
Jim Liles	2013	** 4115
Thomas Lydiat	c. 1600	4103
M. Michael Maestlinus	c. 1600	4079
J. Ricciolus	n/a	4062
Jacob Salianus	c. 1600	4053
H. Spondanus	c. 1600	4051
Martin Anstey	1913	4042
W. Lange	n/a	4041
E. Reinholt	n/a	4021
J. Cappellus	c. 1600	4005
E. Greswell	1830	4004
E. Faulstich	1986	4001
D. Petavius	c. 1627	3983
Frank Klassen	1975	3975
Becke	n/a	3974
Krentzeim	n/a	3971
W. Dolen	2003	3971
E. Reusnerus	n/a	3970

J. Claverius	n/a	3968
C. Longomontanus	c. 1600	3966
P. Melanchthon	c. 1550	3964
J. Haynlinus	n/a	3963
A. Salmeron	d. 1585	3958
J. Scaliger	d. 1609	3949
M. Beroaldus	c. 1575	3927
A. Helwigius	c. 1630	3836

* Benjamin Shaw, "The Genealogies of Genesis 5 and 11 and their Significance for Chronology," BJU, December, 2004. Dr. Shaw states the date as "about 5000 B.C." in Appendix I, but the specific date is derived from adding 1,656 years (the time from creation to the Flood) to his date of the Flood, which is stated as 3298 B.C. on p. 222.

** Jim Liles, Earth's Sacred Calendar (Tarzana, CA: Bible Timeline, 2013).

As you will likely note from table 2, the dates are not all 4004 B.C. There are several reasons chronologists have different dates,[7] but two primary reasons:

1. Some used the Septuagint or another early translation, instead of the Hebrew Masoretic text. The Septuagint is a Greek translation of the Hebrew Old Testament, done about 250 B.C. by about 70 Jewish scholars (hence it is often cited as the LXX). It is good in most places, but appears to have a number of inaccuracies. For example, one relates to the Genesis chronologies where the LXX indicates that Methuselah would have lived past the Flood, without being on the ark!
2. Several points in the biblical timeline are not straightforward to calculate. They require very careful study of more than one passage. These include exactly how much time the Israelites were in Egypt and what Terah's age was when Abraham was born. (See Jones' and Ussher's books for a detailed discussion of these difficulties.)

The first four in table 2 (bolded) are calculated from the Septuagint (others give certain favoritism to the LXX too), which gives ages for the patriarchs' firstborn much higher than the Masoretic text or the Samarian Pentateuch (a version of the Old Testament from the Jews in Samaria just before Christ). Because of this, the Septuagint adds in extra time. Though

7. Others would include gaps in the chronology based on the presences of an extra Cainan in Luke 3:36. But there are good reasons this should be left out.

the Samarian and Masoretic texts are much closer, they still have a few differences. See table 3.

Table 3: Septuagint, Masoretic, and Samarian Early Patriarchal Ages at the Birth of the Following Son

Name	Masoretic	Samarian Pentateuch	Septuagint
Adam	130	130	230
Seth	105	105	205
Enosh	90	90	190
Cainan	70	70	170
Mahalaleel	65	65	165
Jared	162	62	162
Enoch	65	65	165
Methuselah	187	67	167
Lamech	182	53	188
Noah	500	500	500

Using data from table 2 (excluding the Septuagint calculations and including Jones and Ussher), the average date of the creation of the earth is 4045 B.C. This still yields an average of about 6,000 years for the age of the earth.

Extra-biblical Calculations for the Age of the Earth

Cultures throughout the world have kept track of history as well. From a biblical perspective, we would expect the dates given for creation of the earth to align much closer to the biblical date than billions of years.

This is expected since everyone was descended from Noah and scattered from the Tower of Babel. Another expectation is that there should be some discrepancies about the age of the earth among people as they scattered throughout the world, taking their uninspired records or oral history to different parts of the globe.

Under the entry "creation," *Young's Analytical Concordance of the Bible*[8] lists William Hales' accumulation of dates of creation from many cultures, and in most cases Hales says which authority gave the date. See table 4.

8. Robert Young, *Young's Analytical Concordance to the Bible* (Peadoby, MA: Hendrickson, 1996), referring to William Hales, *A New Analysis of Chronology and Geography, History and Prophecy*, Vol. 1 (1830), p. 210.

Table 4: Selected Dates for the Age of the Earth by Various Cultures

Culture	Age, B.C.	Authority listed by Hales
Spain by Alfonso X	6984	Muller
Spain by Alfonso X	6484	Strauchius
India	6204	Gentil
India	6174	Arab Records
Babylon	6158	Bailly
Chinese	6157	Bailly
Greece by Diogenes Laertius	6138	Playfair
Egypt	6081	Bailly
Persia	5507	Bailly
Israel/Judea by Josephus	5555	Playfair
Israel/Judea by Josephus	5481	Jackson
Israel/Judea by Josephus	5402	Hales
Israel/Judea by Josephus	4698	University History
India	5369	Megasthenes
Babylon (Talmud)	5344	Petrus Alliacens
Vatican (Catholic using the Septuagint)	5270	N/A
Samaria	4427	Scaliger
German, Holy Roman Empire by Johannes Kepler*	3993	Playfair
German, reformer by Martin Luther	3961	N/A
Israel/Judea by computation	3760	Strauchius
Israel/Judea by Rabbi Lipman	3616	University History

* Luther, Kepler, Lipman, and the Jewish computation likely used biblical texts to determine the date.

Historian Bill Cooper's research in *After the Flood* provides dates from several ancient cultures.[9] The first is that of the Anglo-Saxons, whose history has 5,200 years from creation to Christ, according to the Laud and Parker Chronicles. Cooper's research also indicated that Nennius' record of the

9. Bill Cooper, *After the Flood* (UK: New Wine Press, 1995), p. 122–129.

ancient British history has 5,228 years from creation to Christ. The Irish chronology has a date of about 4000 B.C. for creation, which is surprisingly close to Ussher and Jones! Even the Mayans had a date for the Flood of 3113 B.C.

This meticulous work of many historians should not be ignored. Their dates of only thousands of years are good support for the biblical date somewhere in the neighborhood of about 6,000 years, but not the supposed billions of years claimed by many today.

7 Origin of the Old-Earth Worldview

Ken Ham and Bodie Hodge

> Knowing this first: that scoffers will come in the last days, walking according to their own lusts, and saying, "Where is the promise of His coming? For since the fathers fell asleep, all things continue as they were from the beginning of creation." For this they willfully forget: that by the word of God the heavens were of old, and the earth standing out of water and in the water, by which the world that then existed perished, being flooded with water (2 Peter 3:3–6).

Attacking the Flood

Prior to the late 1700s, precious few believed in an old earth. The approximate 6,000-year age for the earth was challenged only rather recently, beginning in the late 18th century. These opponents of the biblical chronology essentially left God out of the picture. Three of the old-earth advocates included Comte de Buffon, who thought the earth was at least 75,000 years old. Pièrre LaPlace imagined an indefinite but very long history, and Jean Lamarck also proposed long ages.[1]

However, the idea of millions of years really took hold in geology when men like Abraham Werner, James Hutton, William Smith, Georges Cuvier, and Charles Lyell used their interpretations of geology as the standard,

1. Terry Mortenson, "The Origin of Old-earth Geology and its Ramifications for Life in the 21st Century," *TJ* 18, no. 1 (2004): 22–26, online at www.answersingenesis.org/tj/v18/i1/oldearth.asp.

rather than the Bible. Werner estimated the age of the earth at about one million years. Smith and Cuvier believed untold ages were needed for the formation of rock layers. Hutton said he could see no geological evidence of a beginning of the earth; and building on Hutton's thinking, Lyell advocated "millions of years."

From these men and others came the secular consensus view that the geologic layers were laid down slowly over long periods of time based on the rates at which we see them accumulating today. Hutton said:

> The past history of our globe must be explained by what can be seen to be happening now. . . . No powers are to be employed that are not natural to the globe, no action to be admitted except those of which we know the principle.[2]

This viewpoint is called *naturalistic uniformitarianism*, and it excludes any major catastrophes such as Noah's Flood. Though some, such as Cuvier and Smith, believed in multiple catastrophes separated by long periods of time, the uniformitarian concept became the ruling dogma in geology.

Thinking biblically, we can see that the global Flood in Genesis 6–8 would wipe away the concept of millions of years, for this Flood would explain massive amounts of fossil layers. Most Christians fail to realize that a global Flood could rip up many of the previous rock layers and redeposit them elsewhere, destroying the previous fragile contents. This would destroy any evidence of alleged millions of years anyway. So the rock layers can theoretically represent the evidence of either millions of years or a global Flood, but not both. Sadly, by about 1840 even most of the Church elite had accepted the dogmatic claims of the secular geologists and rejected the global Flood and the biblical age of the earth.

After Lyell, in 1899 Lord Kelvin (William Thomson) calculated the age of the earth, based on the cooling rate of a molten sphere instead of water (Genesis 1:2[3]). He calculated it to be a maximum of about 20–40 million years (this was revised from his earlier calculation of 100 million years in 1862).[4]

2. James Hutton, *Theory of the Earth* (trans. of Roy. Soc. of Edinburgh, 1785); quoted in A. Holmes, *Principles of Physical Geology* (UK: Thomas Nelson & Sons Ltd., 1965), p. 43–44.
3. The earth was without form, and void; and darkness was on the face of the deep. And the Spirit of God was hovering over the face of the waters.
4. Mark McCartney, "William Thompson: King of Victorian Physics," *Physics World*, December 2002, physicsworld.com/cws/article/print/16484.

With the development of radiometric dating in the early 20th century, the age of the earth expanded radically. In 1913, Arthur Holmes' book, *The Age of the Earth*, gave an age of 1.6 billion years.[5] Since then, the supposed age of the earth has expanded to its present estimate of about 4.5 billion years (and about 14 billion years for the universe). But there is growing scientific evidence that radiometric dating methods are completely unreliable (more on this in subsequent chapters).[6]

Table 1 Summary of the Old-Earth Proponents for Long Ages

Who?	Age of the Earth	When Was This?
Comte de Buffon	78 thousand years old	1779
Abraham Werner	1 million years	1786
James Hutton	Perhaps eternal, long ages	1795
Pièrre LaPlace	Long ages	1796
Jean Lamarck	Long ages	1809
William Smith	Long ages	1835
Georges Cuvier	Long ages	1812
Charles Lyell	Millions of years	1830–1833
Lord Kelvin	20–100 million years	1862–1899
Arthur Holmes	1.6 billion years	1913
Clair Patterson	4.5 billion years	1956

Christians who have felt compelled to accept the millions of years as fact and try to fit them into the Bible need to become aware of this evidence. Today, secular geologists will allow some catastrophic events into their thinking as an explanation for what they see in the rocks. But uniformitarian thinking is still widespread and secular geologists will seemingly never entertain the idea of the global catastrophic Flood of Noah's day.

5. Terry Mortenson, "The History of the Development of the Geological Column," in *The Geologic Column*, eds. Michael Oard and John Reed (Chino Valley, AZ: Creation Research Society, 2006).
6. For articles at the layman's level, see www.answersingenesis.org/home/area/faq/dating.asp. For a technical discussion, see Larry Vardiman, Andrew Snelling, and Eugene Chaffin, eds., *Radioisotopes and the Age of the Earth*, Vol. 1 and 2 (El Cajon, CA: Institute for Creation Research; Chino Valley, Arizona: Creation Research Society, 2000 and 2005). See also "Half-Life Heresy," *New Scientist*, October 21 2006, p. 36–39, abstract online at www.newscientist.com/channel/fundamentals/mg19225741.100-halflife-heresy-accelerating-radioactive-decay.html.

The age of the earth debate ultimately comes down to this foundational question: Are we trusting man's imperfect and changing ideas and assumptions about the past? Or are we trusting God's perfectly accurate eyewitness account of the past, including the creation of the world, Noah's global Flood, and the age of the earth?

Conclusion

When we start our thinking with God's Word, we can calculate that the world is about 6,000 years old.

The age of the earth ultimately comes down to a matter of trust — it's a worldview issue. Will you trust what an all-knowing God says on the subject or will you trust imperfect man's assumptions and imaginations about the past that regularly are changing?

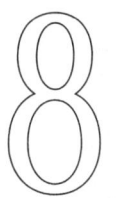

8 Age of the Earth Battle: Millions of Years or a Global Flood

Ken Ham and Bodie Hodge

Introduction

Who would have thought that rock layers would be the source of such an immense debate? After all, they are just rock layers, right? But this is really what the age of the earth debate boils down to — a global Flood or slow, gradual millions of years. We look at the same evidence — rock layers — but we look at them from two different viewpoints.

When it comes to the age of the earth debate, many think that it has something to do with radiometric dating. But dating methods are a new idea that people have tried to use as a secondary point to back up this idea (belief) of rock layers being evidence of millions of years — more on this in the next chapter.

But there are two opposing timescales based on these two opposing views of the rock layers. Let's dive into this.

Timescale

Geological timescales are everywhere! You can hardly miss them. They are in children's books, textbooks, laymen books, technical journals, and so on. They typically look like the one on the following page.

Naturally, there are some that can get technical. But take note that they assume certain layers took millions of years to form. Early geological scales, in the 1800s for example, did not have a time stamped on them.

Even timescales in the past look different from the ones today. Once they started saying these rock layers were "eras of time" they started putting age

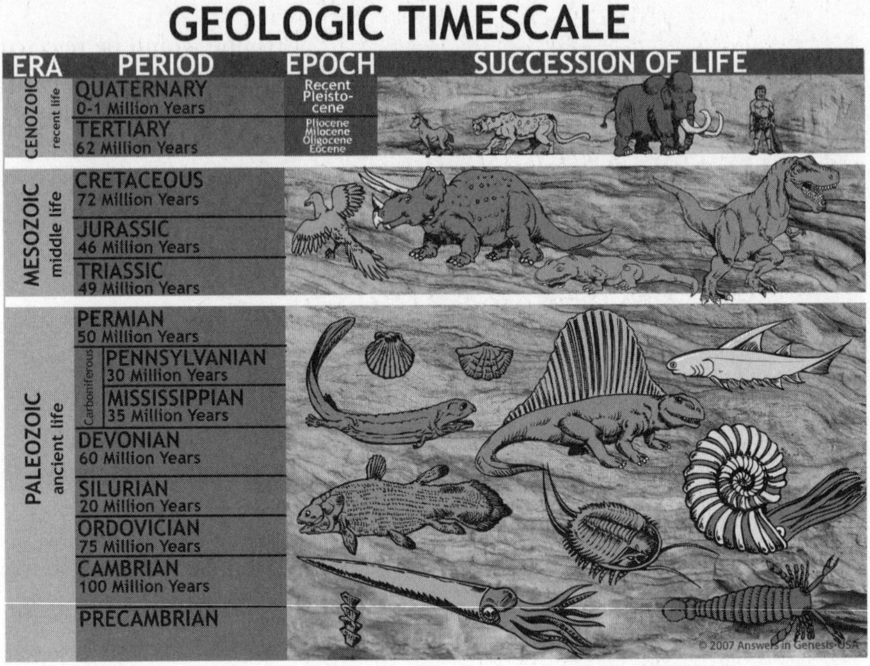
GEOLOGIC TIMESCALE
ERA
PERIOD
EPOCH
SUCCESSION OF LIFE
CENOZOIC
recent life
QUATERNARY
0-1 Million Years
TERTIARY
62 Million Years
Recent
Pleisto-cene
Pliocene
Miocene
Oligocene
Eocene
MESOZOIC
middle life
CRETACEOUS
72 Million Years
JURASSIC
46 Million Years
TRIASSIC
49 Million Years
PALEOZOIC
ancient life
PERMIAN
50 Million Years
Carboniferous
PENNSYLVANIAN
30 Million Years
MISSISSIPPIAN
35 Million Years
DEVONIAN
60 Million Years
SILURIAN
20 Million Years
ORDOVICIAN
75 Million Years
CAMBRIAN
100 Million Years
PRECAMBRIAN
© 2007 Answers in Genesis-USA

dates on them. Then they kept adjusting the dates to get older and older ones. They kept pushing earth's age back further and further to get more time in hopes that they would have enough time for biological evolution. Even with the time they have today, it still isn't even enough time (as molecules-to-man evolution is impossible anyway), but they have slowed down changing it in our day and age. But this doesn't mean it won't change tomorrow!

Timescale Discussion

Keep in mind that secular timescales are based on the idea that there were no significant catastrophes that occurred in the past. They further assume that these rock layers are formed because of slow deposition over millions and billions of years. We don't accept this assumption that there were no major catastrophes in the past — especially in light of the global Flood in Genesis 6–8. This assumption of no catastrophes (particularly no global Flood) is just a bad arbitrary guess — and really a deliberate ploy in an attempt to discredit the Bible.

For the moment, consider what would happen if a massive catastrophe did occur like a global Flood or an asteroid strike. The entire timescale would be called into question, as previous rock layers would be eradicated and new rock layers would occur quickly. So it would not match this idea of slow gradual changes. This is why any major catastrophe would be rejected in the secular world, because it would destroy the idea of millions and billions of years. They have to have this long timescale to even attempt to propose their biological evolutionary ideas.

You need to understand that when people reject major catastrophes like the Flood, they assume that the rock layers formed slowly based on uniform rates we see *today*. This is called "uniformitarianism" or "gradualism." For example, they say rock layers form when dust, sand, or dirt slowly accumulates over the course of a year or sediment from a river or the edge of a sea accumulates into a layer over the course of a year. Then these layers add up over the course of long ages and capture living creatures to form rock layers with fossils.

In the same way, they assume that a river carves out a canyon at a slow rate like what we see on a typical day, and over long ages it finally carves out these canyons to a significant degree. Any catastrophes would upset this, and even though we see catastrophes today (tsunamis, local floods, earthquakes, volcanoes, etc.), these are largely ignored to develop the long timescales.

Do Creationists Agree with the Rock Layers on the Timescale?

Believe it or not, creationists and evolutionists actually agree on something. We agree on the actual rock layers that exist. That is not in dispute. We all agree on the rock layers and that they can be divided into types in much the same way. Our disagreement rises with the *timing* of the formation of these rock layers!

Where the evolutionists say the fossiliferous rock layers "were laid down slowly over millions of years without any major catastrophes," the creationists say that "most of these rock layers were deposited by the Flood of Noah's day." So it is the same evidence based on the same observations, but we simply have two different interpretations of that evidence. Creationists start with God's Word, which informs us of the global Flood (e.g., Genesis 7:19–20[1]) that was a major catastrophe about 4,300 years ago. A creationist's geologic timescale would be much closer to the one on the following page.

1. And the waters prevailed exceedingly on the earth, and all the high hills under the whole heaven were covered. The waters prevailed fifteen cubits upward, and the mountains were covered.

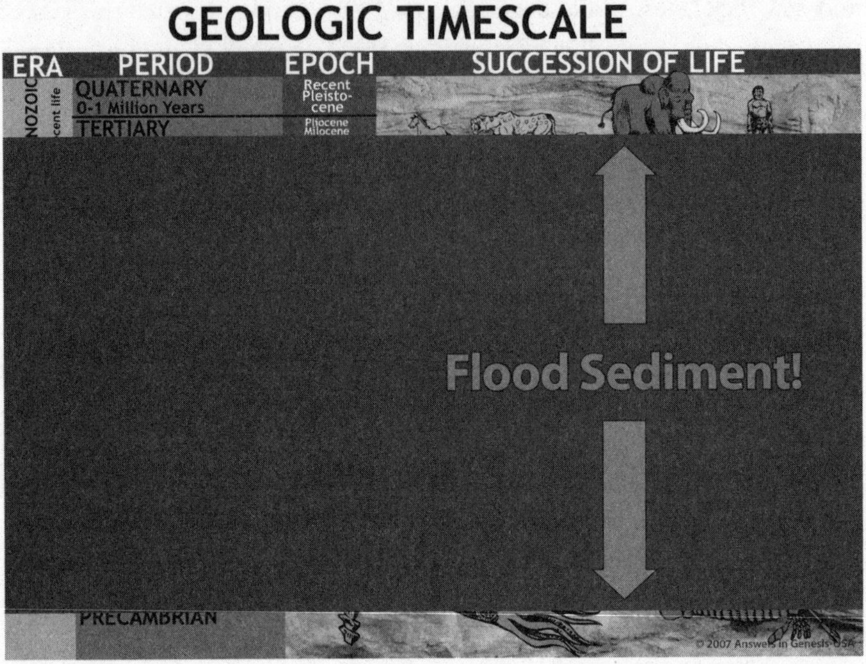

Now there would be rock layers formed *since* the Flood (not over millions of years) from tsunamis, local floods, volcanoes, etc., but most of the rock layers that have fossils come from the global Flood of Noah's day. So we agree on the rock layers, but the *timing* is different.

Catastrophism

In the 1800s, there were people in the secular world who held to what is called "catastrophism." They disagreed with the uniformitarians' ideas that there were no catastrophes, and yet they disagreed with the Bible that discussed the biblical global Flood. They held that there *were* major catastrophes from time to time, like a multitude of global or massive floods. But they had these separated out by eons of time. This view all but died a tragic death and uniformitarian ideas dominated until recently with a new revival of this view.

Just now — in our day and age — we have secular geologists who entertain the idea that there may have been several major catastrophes in the past. These ideas are again becoming popular. For example, many secular geologists (scientists who study rock layers) and paleontologists (scientists who study fossils) have been asserting this idea of a *mass extinction event* about "65 million years ago" that finished off the dinosaurs. Furthermore, they will often discuss several major extinction events — many including water.

The good thing is that these geologists recognize that a catastrophe best explains the fossiliferous rock layers. The bad thing is that they refuse to acknowledge that the Genesis Flood was that event. Furthermore, few notice the implication of how even a single catastrophe in the past *destroys the geologic timescale.* The rock layers are still the rock layers, but the *timing* then must be in error if you have a major catastrophe to lay down a particular rock layer instead of millions of years!

In other words, catastrophists who believe in multiple catastrophes over the course of millions of years *cannot have millions of years* in the rock layers themselves, by which they try to appeal for long ages! A catastrophe destroys the idea of rock layers being slowly and gradually formed over millions of years. And yet, many catastrophists are more than willing to continue to use the geologic timescale that their own view refutes! This is a major inconsistency!

Grand Canyon

When talking about rock layers and timing, it is difficult to avoid the Grand Canyon and the Grand Staircase (which extends farther out and above the immediate Grand Canyon layers). This is an ideal example of the debate over timing. Was the canyon formed:

1. slowly and gradually over millions of years and then carved out by the Colorado River slowly over millions of years or
2. by a catastrophe that laid down the rock layers and a catastrophe that carved out the rock layers?

The uniformitarian would argue it was a lot of time and a little bit of water, whereas the creationist would argue it was a lot of water and little bit of time! We have the same rock layers, same fossils, same Grand Canyon, but two different interpretations of that evidence. The Flood makes sense of depositing and carving those rock layers quickly.

Slow, gradual accumulations make little sense because many of the rock layers at the Grand Canyon have no evidence of erosion between the rock layers, which in some cases supposedly have millions of years separating

these rock layers. For example, the flat boundary between Hermit Shale and the Coconino Sandstone are supposed to have millions of years of erosion between them, but they don't.

Now as a note, creationists disagree over when the Grand Canyon was officially carved out. Both positions agree that the canyon was carved via a catastrophe; but where many have this at the end stages of the Flood, others have it as the result of a breached dam either at the end of the Flood[2] or during the early post-Flood era. Either way, it was formed as a consequence of the global Flood.

2. For more on the Grand Canyon formation please see Andrew Snelling and Tom Vail, "When and How Did the Grand Canyon Form?" *The New Answers Book 3*, Ken Ham, gen. ed. (Green Forest, AR: Master Books, 2010), p. 173–186.

9 A Lesson in Radiometric Dating – Semi-Technical

Bodie Hodge

Introduction

We've seen how the secularists have used rock layers that have fossils to build the geologic timescale and how that relates to the age of the earth. Radiometric dating was the next installment that took the age of the earth from the supposed *millions of years* to the *billions of years*! But is radiometric dating the knockout punch for the age of the earth? Not at all, though many seem to falsely think so. Let's break this idea down.

Uniformitarian Methods for Dating the Age of the Earth

Radiometric dating methods are one particular form of *uniformitarian* dating method. Don't let the word uniformitarian scare you! Uniformitarian dating methods are simply assuming something (i.e., a rate) has been uniform in the past; that is, unchanging. For example, if we wanted to estimate how old the earth was by one of the many uniformitarian methods, we might select sodium influx into the oceans. Here is how it works:

A. We assume there was no sodium (e.g., salt) in the oceans to begin with.
B. We measure how much sodium is eroding into the oceans today by rivers, volcanoes, and so on.
C. We measure how much sodium is leaving the oceans (like sea spray from ocean storms that come inland and leave some sodium inland).

D. We assume that there has never been any *significant* catastrophes in the past to make major changes to sodium influx into the oceans (in other words we assume these rates have always been the same).

E. Then we calculate how long it would take for the oceans to arrive at their sodium level today (how salty the ocean has become).

In this example, we really made some wild assumptions didn't we? We assume there were no catastrophes in the past. And next, we assumed that the ocean had no salt to begin with. Of course, these assumptions are in error. When God gathered the seas on day 3 and let dry land appear (Genesis 1:9–10[1]), He could well have made sure they had a certain degree of sodium or saltiness. This would immediately upset this dating method.

Furthermore, the global Flood in Genesis 6–8 was a significant catastrophe that surely increased the sodium influx tremendously. Let's not forget about famines and droughts that plagued the Old Testament world that surely played a role as well, such as reducing influx of sodium or reducing the sea spray. The point is that catastrophes have been an important part of our past that throws a "monkey wrench" into *any* uniformitarian dating method.

Radiometric Dating

Radiometric dating, perhaps the most popular form of uniformitarian dating, was the culminating factor that led to the belief in *billions* of years for earth history for the secular humanists. A radiometric dating method requires a radioactive (an element that wants to break down into another element) material A (the parent) into material B (the daughter). For example, a radioactive form of potassium (the parent) wants to break down into argon (the daughter).

Any radiometric dating model or other uniformitarian dating method can and does have problems as referenced before dealing with their assumptions. All uniformitarian dating methods require assumptions for extrapolating present-day processes back into the past. The assumptions related to radiometric dating can specifically be seen in these questions:

- Initial amounts?
- Was any parent amount added?

1. Then God said, "Let the waters under the heavens be gathered together into one place, and let the dry land appear"; and it was so. And God called the dry land Earth, and the gathering together of the waters He called Seas. And God saw that it was good.

- Was any daughter amount added?
- Was any parent amount removed?
- Was any daughter amount removed?
- Has the rate of decay changed?

If the assumptions are accurate, then uniformitarian dates should agree with radiometric dating across the board for the same event. However, radiometric dates often disagree with one another or do not match other uniformitarian dating methods for the age of the earth, such as the influx of salts into the ocean, the rate of decay of the earth's magnetic field, and the growth rate of human population.[2]

The late Dr. Henry Morris compiled a list of 68 uniformitarian estimates for the age of the earth by Christian and secular sources.[3] The current accepted secular age of the earth is about 4.54 billion years, based on radiometric dating of a group of meteorites,[4] so keep this in mind when viewing Table 1.

Table 1: Uniformitarian Estimates Other than Radiometric Dating Estimates for Earth's Age Compiled by Morris

	0 – 10,000 years	>10,000 – 100,000 years	>100,000 – 1 million years	>1 million – 500 million years	>500 million – 4 billion years	>4 billion – 5 billion years
Number of uniformitarian methods*	23	10	11	23	0	0

* When a range of ages is given, the maximum age was used to be generous to the evolutionists. In one case, the date was uncertain, so it was not used in this tally, so the total estimates used were 67. A few on the list had reference to Saturn, the sun, etc., but since biblically the earth is older than these, dates related to them were used.

As you can see from Table 1, uniformitarian maximum ages for the earth obtained from other methods are nowhere near the 4.5 billion years esti-

2. Russell Humphrey, "Evidence for a Young World," *Impact*, June 2005, online at http://www.answersingenesis.org/docs/4005.asp.
3. Henry M. Morris, *The New Defender's Study Bible* (Nashville, TN: World Publishing, 2006), p. 2076–2079.
4. C.C. Patterson, "Age of Meteorites and the age of the Earth," *Geochemica et Cosmochemica Acta*, 10(1956):230–237.

mated by radiometric dating; of the other methods, only two calculated dates were as much as 500 million years. So they do not match up and most methods disagree with the age of the earth being 4.5 billion years!

So why do people reject most radiometric and uniformitarian dates, and only trust the dates that give extremely long ages (like potassium argon, uranium-lead, or rubidium-strontium)? It is because of their presupposition to billions of years before they even look at the subject.

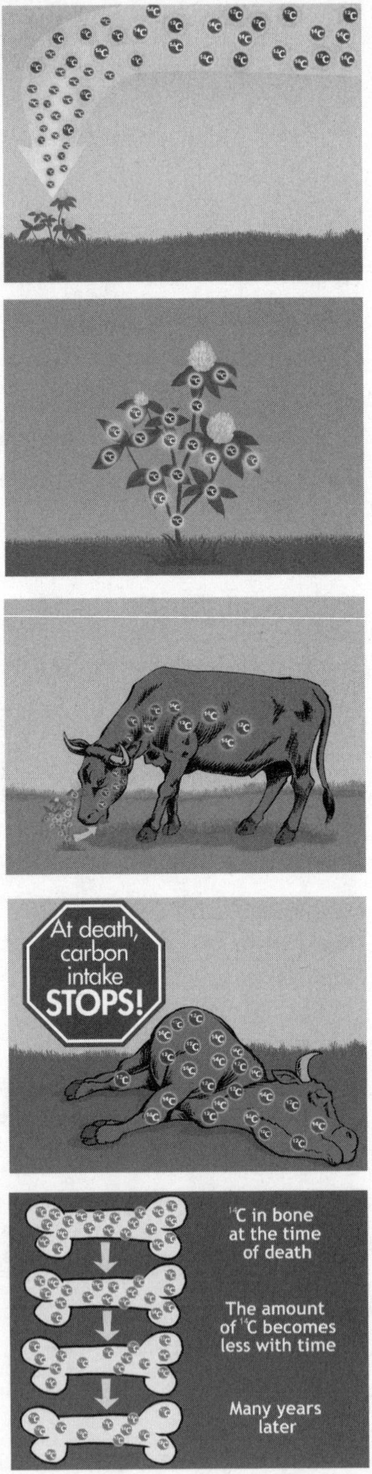
At death, carbon intake STOPS!
^{14}C in bone at the time of death
The amount of ^{14}C becomes less with time
Many years later

Carbon Dating

Right up front, carbon-14 (^{14}C) cannot give dates of millions or billions of years as many think. This is a common misconception in the general public. It can only give calculations of thousands of years (50,000–100,000 years as a theoretical maximum). The results from some radiometric dating methods completely undermine those from the other radiometric methods; ^{14}C dating is one such example.

How does ^{14}C work? As long as an organism is alive it takes in ^{14}C and ^{12}C (normal carbon) from the atmosphere (where it is assumed to be constant) through breathing and diet; however, when it dies, the carbon intake stops. Since ^{14}C is radioactive (decays into ^{14}N), the amount of ^{14}C in a dead organism gets less and less over time.

Carbon-14 dates are determined from the measured ratio of radioactive carbon-14 to normal carbon-12 ($^{14}C/^{12}C$). Used on samples that came from once living or growing creatures, such as wood or bone, the measured $^{14}C/^{12}C$ ratio is compared with the ratio in things today. Carbon-14 has a derived half-life of 5,730 years (again an estimate based on lab experiments), so the ^{14}C in any organic material supposedly 100,000 years old should have decayed into nitrogen.[5]

Some things, such as wood trapped and encased in lava flows, said to be millions of years old by other radiometric dating methods, still have ^{14}C in them per geologist Dr. Andrew Snelling's research.[6] If the items were millions of years old, then they shouldn't have any traces of ^{14}C.

Coal and diamonds, which are found in or sandwiched between rock layers allegedly millions and billions of years old, have been shown to have ^{14}C ages of only tens of thousands of years.[7] So which date, if any, is correct? The older date should immediately be thrown out. The diamonds or coal can't be millions of years old if they have any traces of ^{14}C still in them. This shows that these dating methods are completely unreliable and indicates that the presumed assumptions in the methods are erroneous.

Potassium-Argon Dating

Similar kinds of problems are seen in the case of potassium-argon dating, which has been considered one of the most reliable methods. When a lava flow solidifies, it is supposed to start the radiometric-dating clock. We know when many lava flows occurred — the exact year in fact. Dr. Andrew Snelling points out several of these problems with potassium-argon, as seen in Table 2 with volcanoes when we knew their exact eruption year.[8]

5. This does not mean that a ^{14}C date of 50,000 or 100,000 would be entirely trustworthy. I am only using this to highlight the mistaken assumptions behind uniformitarian dating methods.
6. Andrew Snelling, "Conflicting 'Ages' of Tertiary Basalt and Contained Fossilized Wood, Crinum, Central Queensland Australia," *Technical Journal* 14, no. 2 (2005): 99–122.
7. John Baumgardner, "^{14}C Evidence for a Recent Global Flood and a Young Earth," in *Radioisotopes and the Age of the Earth: Results of a Young-Earth Creationist Research Initiative*, ed. Vardiman et al. (Santee, CA: Institute for Creation Research; Chino Valley, AZ: Creation Research Society, 2005), p.587–630.
8. Andrew Snelling, "Excess Argon: The 'Achilles' Heel' of Potassium-Argon and Argon-Argon Dating of Volcanic Rocks," *Impact*, January 1999, online at www.icr.org/index.php?module=articles&action=view&ID=436.

Table 2: Potassium-argon (K-Ar) Dates in Error

Volcanic Eruption	When the Rock Formed	Date by (K-Ar) Radiometric Dating
Mt. Etna basalt, Sicily	122 B.C.	170,000–330,000 years old
Mt. Etna basalt, Sicily	A.D. 1972	210,000–490,000 years old
Mount St. Helens, Washington	A.D.1986	Up to 2.8 million years old
Hualalai basalt, Hawaii	A.D.1800–1801	1.32–1.76 million years old
Mt Ngauruhoe, New Zealand	A.D.1954	Up to 3.5 million years old
Kilauea Iki basalt, Hawaii	A.D.1959	1.7–15.3 million years old

These dates simply don't even come close! If we cannot trust the dates on known volcanoes, why trust the method elsewhere?

Rubidium Strontium Dating

One popular method used for *really old* age dating is rubidium-strontium, where a radioactive element (rubidium; Rb) changes into another element (strontium; Sr).[9] More specifically, it is the specific isotopes of ^{87}Rb and ^{87}Sr or ^{86}Sr. For the lay reader, don't let this terminology scare you — it is simply a form of rubidium (the parent element) that changes into strontium (the daughter element).

It is claimed that the half-life of this change from Rb to Sr is about 48–49 billion years. In other words, it is claimed that for half of the Rb to change into Sr, it would take about 49 billion years. Yes, you read this correctly; it would take about 3 times the estimated age of the universe to change half of the Rb in the universe to Sr. So it is curious where all the strontium came from in the universe since only minute amounts could have been generated from Rb if the universe is only 13–15 billions years. Strontium is, after all, the 15th most common element on earth, yet rubidium is the 23rd most common element.[10] But this problem is not an issue for a God who created the universe element rich.

9. Strontium is named for a village in Scotland, where it was first discovered.
10. K.K. Turekian and K.H. Wedepohl, "Distribution of the Elements in Some Major Units of the Earth's Crust," *Geological Society of America Bulletin* 72 (2): 175–192, 1961; W.C. Butterman and R.G. Reese, "Mineral Commodity Profiles: Rubidium," accessed June 23, 2014, http://pubs.usgs.gov/of/2003/of03-045/of03-045.pdf.

Obviously, an experiment cannot be observed or repeated to verify this half-life in its fullness, so where did this *guess* come from? After all, there are other isotopes of Rb like ^{83}Rb or ^{86}Rb and in total there are about 32 isotopes, though only two are naturally occurring. But these other isotopes decay in days — that is, a few have observable half-lives of less than 90 days, some even less than 35 days, and in many cases less than one day. So why is it assumed that the half-life of ^{87}Rb is about 48–49 billion years?

Simple — they observe in a lab how much ^{87}Rb is there one day and then measure how much ^{87}Rb is there another day and then calculate the half-life by extrapolating. For the technical person that is [$t_{1/2} = t * \ln(2)/\ln(N_0/N_t)$]; where *t* is time between observations, N_0 is how much was there originally and N_t is the amount of the substance after this amount of time.

But the problem here is simple: how do we know that the half-life really is this long (claimed to be 48–49 billion years)? We really should not detect many decay changes (alpha and beta particle discharges, for example) over days, weeks, and a few years by observation. And yet this is extrapolated from tens to the power of 8! This would not be acceptable in other disciplines.

Allow me to use a layman example. If I polled 5 people on a question and then extrapolated that to say all 6–7 billion people on earth agree . . . would that be acceptable? Not at all! Even if we had a few years of observations, this amount of time is still far too short to give us a reasonable sampling of N_t, let alone to calculate a proper rate. But there are further problems.

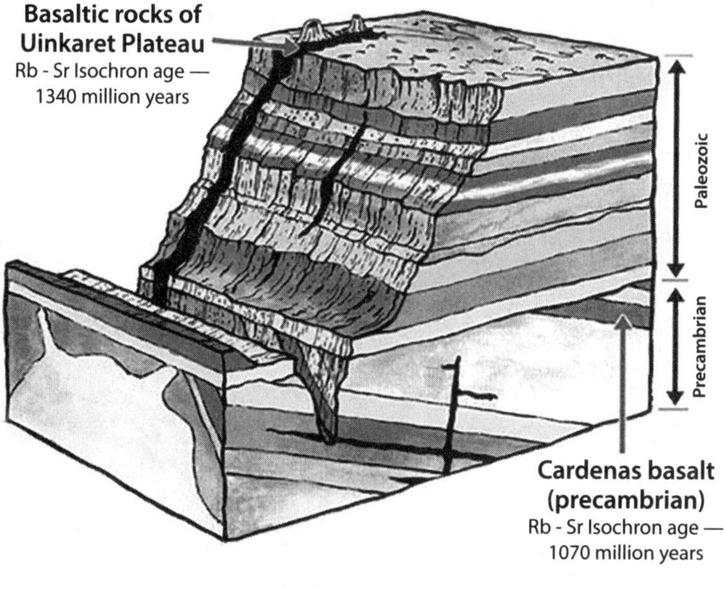

Rubidium-Strontium gives dates we know are wrong. A great example is at the Grand Canyon. At the bottom of the Grand Canyon is a rock layer that is Precambrian called the Cardenas Basalt. It obviously existed before the fossiliferous rock layers of the Grand Canyon were deposited well above it. At the top of the Grand Canyon there is a lava flow that ran down the canyon into the Colorado River. This lava flow is clearly a later, recent formation.

These two volcanic layers have been dated using Rb-Sr. This lava flow on top of the Grand Canyon (calculated at 1.34 billion years) came out to be over 250 million years older than the basalt at the bottom (calculated at 1.07 billion years)![11] This is blatantly impossible. Not that these ages are remotely close, as the volcano is sitting on top of rock layers from the Flood of Noah and is less than 4,500 years old! So the Rb-Sr date is off by more than a billion years! And yet to make things worse, Native Americans may have witnessed this eruption according to some accounts and yet they are not billions of years old!

Conclusion

There are other radiometric methods like uranium-lead, thorium-lead, or samarium-neodymium that secularists try to use to profess certain dates, but they all run into the same problems derived from their assumptions. This brief look at these methods here raises a critical question. If radiometric dating fails to get an accurate date on something of which we *do* know the true age, then how can it be trusted to give us the correct age for rocks that had no human observers to record when they formed?

If the methods don't work on rocks of known age, it is most unreasonable to trust that they work on rocks of unknown age. It is better to trust the Word of the God who created the world, knows its history perfectly, and has revealed sufficient information in the Bible for us to understand that history and the age of the creation.

11. Steve Austin, ed., *Grand Canyon Monument to Catastrophe* (Santee, CA: Institute for Creation Research, 1994), p. 111–131.

10 Can't Christians Just Take the Idea of "Long Ages" and Insert Them into the Bible Somewhere?

Ken Ham and Bodie Hodge

Introduction

Many Christians mix their religion based on God's Word with the secular humanistic religion (man's word), particularly regarding the supposed millions of years for earth's history. They are really taking aspects of the false religion and mixing it with their Christianity.

In most cases, this is done by taking the secular beliefs about origins like big bang, millions of years, and biological evolution, and reinterpreting God's Word in an attempt to "fit them in." Regrettably, many Christians accept man's views of evolution (in geology, biology, astronomy, chemistry, etc.) and try to mix it with Scripture.

The Four "Evolutions"

Let me clarify something. Most people, when they hear the word "evolution," really think of the *"general theory of evolution (GTE)"* or, in laymen's terms, "molecules-to-man," "electron-to-engineer," or "goo-to-you." But there are really four types of evolution that make up that word:

> ***Cosmological/Astronomical evolution:*** including big-bang models (essentially everything evolved from nothing)[1]
>
> ***Geological evolution:*** millions of years of slow, gradual accumulations of rock layers (instead of mostly being rock layers from the Flood)

1. Some may hold to other variant evolutionary forms like an infinitely regressive universe or steady-state model but most adhere to one of the big-bang models.

Chemical evolution: life came from matter (non-life), otherwise called "abiogenesis"

Biological evolution: a single, simple life form gave rise to all other life forms down through the ages

Within the Church, there are those who try to fit astronomical evolution, geological evolution, chemical evolution, or biological evolution into the Bible. Some Christians accept one, some, or all of these in their attempts to accommodate the secular beliefs with God's Word. Usually, it is geological evolution. This acceptance undermines the authority of the Word of God as it takes man's ideas about the past to supersede God's Word in Genesis. This is ultimately undermining the Gospel too, as it undercuts the WORD from which the Gospel comes.

Where Do People Try to Put Millions of Years?

Virtually all Christians who have bought into an old earth (that is millions and billions of years of long ages) place the millions of years *prior* to Adam.

We have genealogical lists that connect Adam to Christ (e.g., Luke 3). For the *old-earth Christians*, it would be blatantly absurd to try to insert millions and billions of years into these genealogies and say that Adam and Eve were made at the beginning of creation.[2]

Instead, old-earth creationists (as they are often denoted[3]) take these long ages and insert them somewhere prior to Adam; hence creation week has been a divisive point in Christianity ever since the idea of long ages such as millions of years became popular in the 1800s. Here are some of the differing positions within the Church — but all have one common factor — endeavoring to somehow fit millions of years into the Bible.

Gap Theories (incorporating geological and astronomical evolution)

1. **Pre-time gap.** This view adds long ages prior to God creating in Genesis 1:1.[4] The pre-time gap falls short for a number of

2. In Mark 10:6, Jesus says: "But from the beginning of the creation, God 'made them male and female.' "
3. In many other cases, those Christians who adhere to long ages are called "compromised Christians" since they are compromising by mixing these two religions' origins accounts (humanism and Christianity). Properly, this is syncretism, where they synchronize the religions of Christianity and humanism into one blended religion.
4. In the beginning God created the heavens and the earth.

reasons, such as having death before sin (discussed in detail in the next chapter), allowance of man's ideas about millions of years to supersede God's Word, having *time* before time existed, and the like. As another example, how can one have millions of years of time prior to the creation of time? It is quite illogical.

2. **Ruin-reconstruction gap.** This is the most popular gap idea, which adds long ages between Genesis 1:1[5] and Genesis 1:2.[6] Scottish pastor Thomas Chalmers popularized it in the early 1800s as a response to long ages, which was becoming popular. This idea is promoted in the Scofield and Dake Study Bibles and is often associated with a Luciferian fall and flood — but that would make Lucifer (Satan) in his sinful state very good and perfect. After God created Adam, God said everything He made was "very good" (Deuteronomy 32:4[7]; Genesis 1:31[8]).[9]
3. **Modified gap/precreation chaos gap.** This view adds long ages between Genesis 1:2[10] and 1:3,[11] and it is primarily addressed in the International Conference on Creation.[12] It has many of the same problems already mentioned in the first two gaps already discussed.
4. **Soft gap.** This also includes a gap between Genesis 1:2[13] and 1:3,[14] but unlike previous views, it has no catastrophic events or destruction of a previous state. Furthermore, it merely proposes that God

5. Ibid.
6. The earth was without form, and void; and darkness was on the face of the deep. And the Spirit of God was hovering over the face of the waters.
7. He is the Rock, His work is perfect; for all His ways are justice, a God of truth and without injustice; righteous and upright is He.
8. Then God saw everything that He had made, and indeed it was very good.
9. K. Ham, "What About the Gap & Ruin-Reconstruction Theories?" in *The New Answers Book*, K. Ham, gen. ed. (Green Forest, AR: Master Books, 2006); for a technical response see also, W. Fields, *Unformed and Unfilled* (Collinsville, IL: Burgener Enterprises, 1997).
10. The earth was without form, and void; and darkness was on the face of the deep. And the Spirit of God was hovering over the face of the waters.
11. Then God said, "Let there be light"; and there was light.
12. One refutation of this view is in Andrew Snelling, ed., *Proceedings of the Sixth International Conference on Creationism* (Dallas, TX: Institute for Creation Research, 2008), "A Critique of the Precreation Chaos Gap Theory," by John Zoschke.
13. The earth was without form, and void; and darkness was on the face of the deep. And the Spirit of God was hovering over the face of the waters.
14. Then God said, "Let there be light"; and there was light.

created the world this way and left it for long periods of time in an effort to get starlight here. In essence, this view has a young earth and an old universe. The problem is that stars were created after the proposed gap (day 4), and it is unnecessary to make accommodations for long ages to solve the so-called starlight problem. Getting distant starlight to earth is not a problem for an all-powerful God. It is only a problem in a strict naturalistic view.

5. **Late gap.** This view has a gap between chapters 2 and 3 of Genesis. In other words, some believe that Adam and Eve lived in the Garden for long ages before sin. This view has problems too. For example, Adam and Eve were told by God to be "fruitful and multiply" in Genesis 1:28,[15] and waiting long ages to do so would have been disobeying God's Word. This doesn't make sense. In addition, there is the problem of Adam only living 930 years as recorded in Genesis (Genesis 5:5[16]).[17]

When someone tries to put a large gap of time in the Scriptures when it is not warranted by the text, this should throw up a red flag to any Christian. In many gap theory models, Satan allegedly rebels between Genesis 1:1–2[18] (or otherwise in the first three verses of Scripture). Consider the theological problem of Satan, in his sinful state being called "very good" in Genesis 1:31.[19] This would make an evil Satan very good. In fact, this would make *sin* very good too. Satan could not have fallen into sin until after this declaration in Genesis 1:31.[20]

Day Age Models (each model adheres to geological and astronomical evolution)

1. **Day-age.** This idea was popularized by Hugh Miller in the early 1800s after walking away from Thomas Chalmers' idea of

15. Then God blessed them, and God said to them, "Be fruitful and multiply; fill the earth and subdue it; have dominion over the fish of the sea, over the birds of the air, and over every living thing that moves on the earth."
16. So all the days that Adam lived were nine hundred and thirty years; and he died.
17. Bodie Hodge, *The Fall of Satan* (Green Forest, AR: Master Books, 2011), p. 23–26, https://answersingenesis.org/bible-characters/adam-and-eve/when-did-adam-and-eve-rebel/.
18. In the beginning God created the heavens and the earth. The earth was without form, and void; and darkness was on the face of the deep. And the Spirit of God was hovering over the face of the waters.
19. Then God saw everything that He had made, and indeed it was very good. So the evening and the morning were the sixth day.
20. Ibid.

the gap theory, and prior to his suicide. This model basically stretched the days of creation out to be millions of years long. Of course, lengthening the days in Genesis to accommodate the secular evolutionist view of history simply doesn't match up with what is stated in Genesis 1.[21]

2. **Progressive creation.** This is a modified form of the day-age idea (really in many ways it's similar to theistic evolution) led by Dr. Hugh Ross, head of an organization called Reasons to Believe. He appeals to nature (actually the secular interpretations of nature) as the supposed 67th book of the Bible, and then uses these interpretations to supersede what the Bible says. Recall that nature is cursed according to Genesis 3 and Romans 8. Dr. John Ankerberg is also a leading supporter of this viewpoint.[22] This view proposes that living creatures go extinct repeatedly over millions of years, but God, from time to time, makes new kinds and new species all fitting with a (geologically and cosmological/astronomically) evolutionary view of history.[23] Things are out of order in creation week in the progressive creation view, and death before sin is devastating to this position.

Theistic Evolutionary Models (each variant basically adheres to geological, astronomical, and biological evolution)

1. **Theistic evolution (evolutionary creation).** Basically, the idea of Genesis 1–11 is thrown out or heavily reinterpreted to allow for evolutionary ideas to supersede the Scriptures. Harvard botany professor Asa Gray was a contemporary of Darwin and promoted this idea, but Darwin opposed Gray's mixing of Christianity with evolution since they were two opposing views. They wrote several letters to one another. Charles Hodge and Benjamin B. Warfield of Princeton Theological Seminary in the mid-to-late

---

21. T. Mortenson, "Evolution vs. Creation: The Order of Events Matters!" *Answers in Genesis*, April 4, 2006, https://answersingenesis.org/why-does-creation-matter/evolution-vs-creation-the-order-of-events-matters/.
22. J. Seegert, "Responding to the Compromise Views of John Ankerberg," *Answers in Genesis*, March 2, 2005, https://answersingenesis.org/reviews/tv/responding-to-the-compromise-views-of-john-ankerberg/.
23. K. Ham and T. Mortenson, "What's Wrong with Progressive Creation?" in K. Ham, gen. ed., *The New Answers Book 2* (Green Forest, AR: Master Book, 2008), p. 123–134.

1800s also advocated the mixing of Christianity with evolution. Today, this view is heavily promoted by a group called BioLogos. Basically, they accept the prevailing evolutionist (false) history including the big bang and then add a demoted form of God to it. BioLogos writers have different ways of wildly reinterpreting Genesis to accommodate evolution into Scripture.

2. **Framework hypothesis.** Dr. Meredith Kline (1922–2007), who accepted many evolutionary ideas, popularized this view in America.[24] It is very common in many seminaries today. Those who hold to framework treat Genesis 1 as a literary device (think poetic or semi-poetic), with the first three days paralleling and equating to the last three days of creation. These days are not seen as 24-hour days but are taken as metaphorical or allegorical to allow for ideas like evolution/millions of years to be entertained. Hence, Genesis 1 is treated as merely being a literary device to teach that God created everything (essentially in 3 days[25]).[26] However, Genesis 1 is not written as poetry but as literal history.[27]
3. **Cosmic Temple.** Dr. John Walton agrees the language of Genesis 1 means ordinary days, but since he believes in evolution had to do something about it. Walton proposes that Genesis 1 has nothing to do with material origins but instead is referring to

24. It was originally developed in 1924 by Professor Arnie Noordtzij in Europe, which was a couple of decades before Dr. Kline jumped on board with framework hypothesis.
25. "For in six days the LORD made the heavens and the earth, the sea, and all that is in them, and rested the seventh day. Therefore the LORD blessed the Sabbath day and hallowed it" (Exodus 20:11). "It is a sign between Me and the children of Israel forever; for in six days the LORD made the heavens and the earth, and on the seventh day He rested and was refreshed" (Exodus 31:17).
26. T. Chaffey and B. McCabe, "What is Wrong with the Framework Hypothesis?" *Answers in Genesis*, June 11, 2011, https://answersingenesis.org/creationism/old-earth/whats-wrong-with-the-framework-hypothesis/.
27. Hebrew expert Dr. Steven Boyd writes: "For Genesis 1:1–2:3, this probability is between 0.999942 and 0.999987 at a 99.5% confidence level. Thus, we conclude with statistical certainty that this text is narrative, not poetry. It is therefore statistically indefensible to argue that it is poetry. The hermeneutical implication of this finding is that this text should be read as other historical narratives." Dr. Steven Boyd, Associate Professor of Bible, The Master's College, *Radioisotopes and the Age of the Earth*, Volume II, Editors Larry Vardiman, Andrew Snelling, and Eugene Chaffin (Dallas, TX: Institute for Creation Research, 2005), p. 632. We would go one step further than Dr. Boyd, who left open the slim possibility of Genesis not being historical narrative, and say it *is* historical narrative and all doctrines of theology, directly or indirectly, are founded in the early pages of Genesis — though we appreciated Dr. Boyd's research.

what he calls "God's Cosmic Temple." By relegating Genesis 1 to be disconnected from material origins of earth, then he is free to believe in evolution and millions of years.

Conclusion

Note that each compromise position has one common factor — attempting to fit millions of years into the Bible. That's why there are so many such compromise positions, with very creative ways of attempting to add in the supposed long ages. The only position that works is the one in which Genesis is taken as literal history!

Each old-earth Christian worldview has no choice but to demote a global Flood to a local flood in order to accommodate the alleged millions of years (geological evolution) of rock layers (a global Flood would have destroyed these layers and laid down new ones).[28]

Also, the compromise views that accept the big-bang idea have accepted a view that contradicts Scripture. They have adopted a model to explain the universe without God (which is what the big bang is — a model that requires no God). So if God is added to the big-bang idea, then really . . . God didn't do anything because the big bang dictates that the universe really created itself.[29]

Each old-earth view also has an insurmountable problem in regard to the issue of death before sin (discussed in the next chapter) that undermines both the authority of God's Word and the gospel.[30] The idea of millions of years came out of naturalism — the belief that the fossil-bearing rock layers were laid down slowly and gradually over millions of years before man.

This idea of long ages was meant to do away with the belief that Noah's Flood was responsible for most of the fossil-bearing sedimentary layers. There is no reason for Christians to adopt the atheistic, naturalism religious view (secular humanism) and mix it with their Christianity.

---

28. J. Lisle and T. Chaffey, "Defense — A Local Flood?" in *Old Earth Creation on Trial* (Green Forest, AR: Master Books, 2008), p. 93–106, https://answersingenesis.org/the-flood/global/defensea-local-flood/.
29. J. Lisle, "Does the Big Bang Fit with the Bible?" in K. Ham, gen. ed., *The New Answers Book 2* (Green Forest, AR: Master Books, 2008), p. 103–110, https://answersingenesis.org/big-bang/does-the-big-bang-fit-with-the-bible/.
30. B. Hodge, *The Fall of Satan* (Green Forest, AR: Master Books, 2011), p. 68–76.

11 Biblically, Could Death Have Existed Before Sin?

Ken Ham and Bodie Hodge

Death and sin — these are two things today's society seems to want to avoid in a conversation! In today's secular culture, kids have been taught for generations that death goes back for millions of years. But there is a huge contrast when you open the pages of Scripture beginning in Genesis.

There is no greater authority than God on any issue (consider Hebrews 6:13[1]). Since God is the authority regarding the past (as well as the authority on all matters), then it is logical that the Bible should be the authority on the issue of death and its relationship with sin. Let's get a big picture of sin and death and how they are related in the Bible.

Everything Was Perfect Originally

> Then God saw everything that He had made, and indeed it was very good. So the evening and the morning were the sixth day (Genesis 1:31).

> He is the Rock, His work is perfect; for all His ways are justice, a God of truth and without injustice; righteous and upright is He (Deuteronomy 32:4).

When God finished creating at the end of day 6, God declared everything "very good." It was very good — it was perfect. God's work of creation is perfect and this is verified in Deuteronomy 32:4. We would expect nothing less of a perfect God.

1. For when God made a promise to Abraham, because He could swear by no one greater, He swore by Himself.

What was this "perfect" or "very good" creation like? Were animals dying? Was man dying? Let's look more closely at what the Bible teaches.

A very good creation

Everything Was Vegetarian Originally

> And God said, "See, I have given you every herb that yields seed which is on the face of all the earth, and every tree whose fruit yields seed; to you it shall be for food. Also, to every beast of the earth, to every bird of the air, and to everything that creeps on the earth, in which there is life, I have given every green herb for food"; and it was so (Genesis 1:29–30).

From Genesis 1:29–30 we know that living things like animals and man were not eating meat originally. So meat-eaters today were all vegetarian *originally*, which indicates that death (bloodshed) was *not* part of the original creation. One would not expect a God of life to be a *god of death*. When we look at heaven in Revelation 21–22, there will be no death, pain, or suffering.

One might object and say, "But plants could have died!" But the person asking fails to realize that plants were not "alive" in the biblical sense of *nephesh chayyah* — only animals and man. Since plants are not alive by God's standard, then they can't die.

Plants were made for the purpose of food. They are like solar-powered, self-replicating, biological machines for the purpose of food, clothes, medicine, etc. So plants (microbes and cells too) are ruled out as an option of death before the Fall.[2] The Bible only describes humans and animals as living; and of course, man is distinguished from animals, as human life is made in the image of God.

If a Christian wants to side with the humanistic (e.g., atheistic) view of the world where death existed for millions of years and try to use the majority of the fossil layers as their evidence of slow gradual accumulation instead of a global Flood, they have major problems.[3]

2. Michael Todhunter, "Do Leaves Die?" September 6, 2006, http://www.answersingenesis.org/articles/am/v1/n2/do-leaves-die.
3. Bodie Hodge, "How Old Is the Earth?" May 30, 2007, AiG website, http://www.answersingenesis.org/articles/2007/05/30/how-old-is-earth.

The fossil layers contain examples of many animals that had eaten other animals and the remains are still found in stomach contents.[4] So this rules out many of the rock layers as being evidence of million of years, because the Lord declared that animals were originally vegetarian. The Flood of Noah's day is a much better explanation of the rock layers with examples of animals having eaten other animals, as such would have happened *after* sin.

Death Is a Punishment

> And the LORD God commanded the man, saying, "Of every tree of the garden you may freely eat; but of the tree of the knowledge of good and evil you shall not eat, for in the day that you eat of it you shall surely die" (Genesis 2:16–17).

God gave the command in Genesis 2:16–17 that sin would be punishable by death. This is significant when we look at the big picture of death. If death in *any* form was around prior to God's declaration in Genesis 1:31 that everything was "very good," then death would be very good too — hence it would not be a punishment at all! So death (and bloodshed, suffering, and disease) shouldn't be in the original perfect creation but only as a *punishment* for sin.

Some have pointed out that this passage is not referring to animal death. In one sense, we agree with them; this verse was not directed toward animals (more on this in a moment). But by the same logic, it was not directed toward Eve or to the rest of us (Adam's descendants), but Eve died and so do we! This passage is in reference to Adam's death, but as will be seen in a moment, this sin also affected Eve with death. Even sinful angels and Satan's sin is counted against them as eternal death in hell (e.g., Matthew 25:41,[5] 2 Peter 2:4,[6] Revelation 20:10,[7] 20:13–14[8]). This shows the all-encompassing effect of the sin-death relationship.

---

4. As an example see Ryan McClay, "Dino Dinner Hard to Swallow," Answers in Genesis website, January 21, 2005, http://www.answersingenesis.org/docs2005/0121dino_dinner.asp.
5. Then He will also say to those on the left hand, "Depart from Me, you cursed, into the everlasting fire prepared for the devil and his angels."
6. For if God did not spare the angels who sinned, but cast them down to hell and delivered them into chains of darkness, to be reserved for judgment.
7. The devil, who deceived them, was cast into the lake of fire and brimstone where the beast and the false prophet are. And they will be tormented day and night forever and ever.
8. The sea gave up the dead who were in it, and Death and Hades delivered up the dead who were in them. And they were judged, each one according to his works. Then Death and Hades were cast into the lake of fire. This is the second death.

Adam Knew What "Die" Meant

Some people have brought up the objection that if there was no death existing in the world, then how did Adam know what God meant in Genesis 2:17. God, the author of language, programmed Adam with language when He created him, as we know they conversed right from the start on day 6 (see Genesis 2). Since God makes things perfectly, Adam knew what death meant — even if he did not have *experiential* knowledge of it. Since God makes things perfectly, Adam knew what the word *death* meant in the language God had given him, and probably understood it better than any of us!

Sin Brought Animal Death

The first recorded death and passages referring to death as a reality came with sin in Genesis 3 when the serpent, Eve, and Adam all were disobedient to God. Please note that what happened is the first *hint* that things will die:

> So the LORD God said to the serpent: "Because you have done this, you are cursed more than all cattle, and more than every beast of the field; on your belly you shall go, and you shall eat dust all the days of your life (Genesis 3:14).

Genesis 3:14 indicates that animals, which were cursed along with the serpent, would no longer live forever but have a limited life (*all the days of your life*). This is the first hint of animal death, though this is not a "knock-down, drag-out" case but merely a hint. Since animals were cursed in this verse, it also affects them and they too will die. Though this particular verse doesn't rule out animal death prior to sin, its placement with sin and the Curse in Genesis 3 may very well be significant.

The first recorded death of animals was in Genesis 3:21, when God covered Adam and Eve with coats of *skins* to replace their fig leaf coverings they quickly assume would hide their nakedness.

> Also for Adam and his wife the LORD God made tunics of skin, and clothed them (Genesis 3:21).

Abel apparently mimicked something like this when he sacrificed from his flocks (fat portions) in Genesis 4:4 and Noah after the Flood in Genesis 8:20, and Abraham and the Israelites did this as well, giving sin offerings of lambs, doves, etc.

The Lord's sacrifice to make coats of skins for Adam and Eve

Noah offering sacrifices

The punishment for sin was death, so something had to die. Rightly, Adam and Eve deserved to die, but we serve a God of grace, mercy, and love. And out of His love and His mercy, He basically gave us a "grace" period to repent.

The Lord sacrificed animals to cover this sin. It was not enough to *take away* sin, but merely offered a temporary covering. This shows how much more valuable mankind is than animals (see also Matthew 6:26[9], 12:12[10]).

The punishment from an infinite God is an infinite punishment and animals are not infinite. They simply cannot take that punishment. We needed

9. Look at the birds of the air, for they neither sow nor reap nor gather into barns; yet your heavenly Father feeds them. Are you not of more value than they?
10. Of how much more value then is a man than a sheep? Therefore it is lawful to do good on the Sabbath.

a perfect and infinite sacrifice that could take the infinite punishment from an infinite God. Jesus Christ, the Son of God, who is infinite, could take that punishment. These animal sacrifices were foreshadowing Jesus Christ who was the ultimate, perfect, infinite sacrifice for our sins on the Cross. Hebrews reveals:

> And according to the law almost all things are purified with blood, and without shedding of blood there is no remission (Hebrews 9:22).

This is why Jesus had to die and this is why animals were sacrificed to cover sin. These passages make it clear that animal death has a relationship with *man's* sin. Consider the fact that animal death first came about *after* man's sin. It was a direct result of human sin. Also, it is the very basis and foundation of the gospel.

If there was millions of years of death and bloodshed of animals before sin, then what would the shedding of blood have to do with the remission of sin? It just doesn't work to try to fit millions of years of death and bloodshed before Adam sinned.

Sin Brought Human Death

This same type of proclamation that animals will ultimately die (*all the days of your life*) is mimicked in Genesis 3:17 where man would also die (*all the days of your life*). Like the animals, man would die fulfilling what was said in Genesis 2:17 (for in the day that you eat of it you shall *surely die*).

> Then to Adam He said, "Because you have heeded the voice of your wife, and have eaten from the tree of which I commanded you, saying, 'You shall not eat of it': Cursed is the ground for your sake; in toil you shall eat of it all the days of your life" (Genesis 3:17).

Some have stated that they believe this was only a *spiritual* death, but God made it clear in Genesis 3:19 by further describing it as "returning to dust" from which they came, which makes it clear it was not excluding a physical death.

> In the sweat of your face you shall eat bread till you return to the ground, for out of it you were taken; for dust you are, and to dust you shall return (Genesis 3:19).

Even Paul, when speaking of human death specifically says:

> Therefore, just as through one man sin entered the world, and death through sin, and thus death spread to all men, because all sinned (Romans 5:12).
>
> The last enemy that will be destroyed is death (1 Corinthians 15:26).
>
> Nevertheless death reigned from Adam to Moses, even over those who had not sinned according to the likeness of the transgression of Adam, who is a type of Him who was to come (Romans 5:14).
>
> For if by the one man's offense death reigned through the one, much more those who receive abundance of grace and of the gift of righteousness will reign in life through the One, Jesus Christ (Romans 5:17).

If the death God mentions is only spiritual, then why did Jesus have to die physically — or rise physically? If the Curse meant only spiritual death, then the gospel is undermined.

It is true that Adam and Eve didn't die the *exact* day they ate, as some seem to think Genesis 2:17 implies. The Hebrew is *die-die* (*muwth-muwth*), which is often translated as "surely die" or literally as "dying you shall die," which indicates the beginning of dying (i.e., an ingressive sense).

At that point, Adam and Eve began to die and would return to dust. If they were meant to have died right then, the text should have used *muwth* only once, as is used in the Hebrew meaning "dead, died, or die" and not "*beginning to* die" or "*surely* die."

Does the Bible Teach Death Before Sin?

The Bible tells us very clearly from many passages that there was no death before sin. In fact, there are *no Bible verses* indicating there was death prior to sin.

The only reason some people try to insert death before sin is to try to fit man's ideas of "millions of years" of death that is found in the fossil record into the Bible. But this makes a mockery of God's statement that everything was very good in Genesis 1:31.[11]

11. Then God saw everything that He had made, and indeed it was very good. So the evening and the morning were the sixth day.

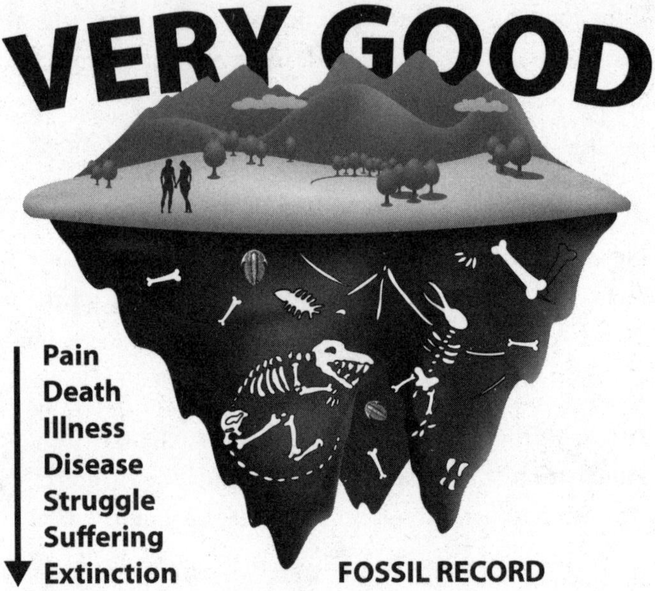

Death before sin is a problem for a perfect creation at the end of day 6.

Death, animals eating other animals, thorns, cancer, tumors, and so on are not very good, and yet these are found in those fossil layers. But if one argues that these are part of a perfect world, then do they expect these things in heaven — which is a restoration that is to be a perfect new creation once again?

This leads to compromising what God plainly says to accommodate fallible man's ideas. Besides, the Scriptures reveal a global Flood in Genesis 6–8, *after sin*, which explains the vast majority of fossil layers. It is better to trust what God says:

> It is better to trust in the LORD than to put confidence in man (Psalm 118:8).

Keep in mind that having death before also undermines the very gospel where Jesus Christ stepped into history to conquer sin and death. In doing so, He had graciously offered the free gift of salvation to all who receive him.

Two Views of Death

There are primarily two views of history (man's opinions/secular and God's view in the Bible) with two different authorities (man's fallible reason *apart* from God, and a perfect God) arguing over the past.

Biblical view of death

According to the Bible, a perfect God created a perfect creation and because of man's sin, death and suffering came into the world — death is an enemy. But through Christ we look forward to a time when there will be no more pain or death or suffering (Revelation 21:4). According to man's ideas about the past, death is actually the glorified hero in the story.

In a secular worldview, there has always been death and always will be. So when *Christians* try to incorporate secular history of millions of years into their theology, was there really change when Adam and Eve sinned? And what will heaven really be like then?

A secular view of death

Summary

In the fossil remains in the rock layers, there is evidence of death, suffering, thorns, carnivory, cancer, and other diseases like arthritis. So *all* old-earth worldviews have to then accept death, suffering, bloodshed, thorns, carnivory, and diseases like cancer before Adam's sin. Now after God created Adam, He said everything He made was "very good" (Genesis 1:31). This is confirmed as a *perfect* creation by the God of life in Deuteronomy 32:4 since every work of God is perfect.

But if one has accepted the millions-of-years idea to explain the fossil record, then millions of years of death, bloodshed, disease, thorns, suffering, and carnivory existed before man. But as the Bible makes clear, it was Adam's sin that caused death (Genesis 2:16–17, 3:19), suffering (e.g., Genesis 3:16–17), thorns, (Genesis 3:18) and the whole reason why we need a new heavens and a new earth (e.g., Isaiah 66:22;[12] 2 Peter 3:13;[13] Revelation 21:1[14]) — because what we have now are cursed and broken (Romans 8:22[15]).

Also, originally, the Bible makes it clear in Genesis 1:29–30[16] that man and animals were vegetarian — however, the fossil record has many evidences of animals eating animals. Genesis 1:30 is verified as a strictly vegetarian diet since man was not permitted to eat meat until after the Flood in Genesis 9:3,[17] which was directly contrasted to the command in Genesis 1:29.

To accept millions of years also means God called diseases like cancer (of which there is evidence in the fossil record) "very good." And because ". . . without shedding of blood there is no remission" (Hebrews 9:22[18]), then allowing the shedding of blood millions of years before sin would *undermine* the atonement. Really, believing in millions of years blames God for death and disease instead of blaming our sin from which Christ came to rescue us.

---

12. "For as the new heavens and the new earth which I will make shall remain before Me," declares the LORD, "so your descendants and your name remain.
13. Nevertheless, we, according to His promise, look for new heavens and a new earth in which righteousness dwells.
14. Now I saw a new heaven and a new earth, for the first heaven and the first earth had passed away. Also there was no more sea.
15. For we know that the whole creation groans and labors with birth pangs together until now.
16. And God said, "See, I have given you every herb that yields seed which is on the face of all the earth, and every tree whose fruit yields seed; to you it shall be for food. Also, to every beast of the earth, to every bird of the air, and to everything that creeps on the earth, in which there is life, I have given every green herb for food"; and it was so.
17. Every moving thing that lives shall be food for you. I have given you all things, even as the green herbs.
18. And according to the law almost all things are purified with blood, and without shedding of blood there is no remission.

One cannot deny biblically that there is a relationship between human sin and animal death. Just briefly look at the sacrifices of animals required for human sin throughout the Old Testament. This sacrifice began in the Garden of Eden (the first blood sacrifice as a covering for their sin, a picture of what was to come in the lamb of God who takes away the sin of the world), that points to Jesus Christ, the ultimate and final sacrifice:

> . . . for this He did once for all when He offered up Himself (Hebrews 7:27).

Our hope is that these Christians (who have bought into an old earth) will return to the plain teachings in the Bible and stop mixing God's Word with secular beliefs that clearly contradict God's revelation and undermine the gospel by blaming God for death instead of our sin.

12 | Global or Local Flood?

Ken Ham and Bodie Hodge

Introduction

> The flood was indeed a river flood. . . . The language of Genesis allows for a regional flood. . . . The parts of modern Iraq which were occupied by the ancient Sumerians are extremely flat. The floodplain, surrounding the Tigris and the Euphrates rivers, covers over 50,000 square miles which slope toward the gulf at less than one foot per mile. . . . Drainage is extremely poor and flooding is quite common, even without large rainstorms during the summer river-level peak (when Noah's flood happened).[1]

A Christian who believes that Noah's Flood was local and did not cover the entire globe penned these words. In fact, the idea of a small regional flood in Noah's day is often promoted by Christians who mix their religion with "millions of years" of supposed naturalistic history!

As a reminder, you need to understand that the idea of millions of years comes from the idea that rock layers all over the world were laid down slowly over long ages without any major catastrophes. In other words, the idea of millions of years is predicated on the idea that there could NOT have been a global Flood. Otherwise, a global Flood would disrupt rock layers that exist and rearrange the sediment and lay down new rock layers!

1. Don Stoner, "The Historical Context for the Book of Genesis," Revision 2011-06-06, Part 3: Identifying Noah and the Great Flood, http://www.dstoner.net/Genesis_Context/Context.html#part3.

What Does the Bible Say?

Did Noah experience a local flood, which left only a few sediment layers, as floods do today? God's record is clear: the water covered the entire globe and killed all the animals on earth. Such unique conditions are the only consistent way to explain worldwide fossil-bearing layers thousands of feet deep.

Scripture is clear about the historic reality of a global Flood in Noah's day. Genesis 7:17–23 specifically says:

> Now the flood was on the earth forty days. The waters increased and lifted up the ark, and it rose high above the earth. The waters prevailed and greatly increased on the earth, and the ark moved about on the surface of the waters. And the waters prevailed exceedingly on the earth, and all the high hills under the whole heaven were covered. The waters prevailed fifteen cubits upward, and the mountains were covered. And all flesh died that moved on the earth: birds and cattle and beasts and every creeping thing that creeps on the earth, and every man. All in whose nostrils was the breath of the spirit of life, all that was on the dry land, died. So He destroyed all living things which were on the face of the ground: both man and cattle, creeping thing and bird of the air. They were destroyed from the earth. Only Noah and those who were with him in the ark remained alive (Genesis 7:17–23).

The Scripture is clear that "all the high hills under the whole heaven were covered" as "the waters prevailed fifteen cubits [that is about ~26 feet[2], or ~8 m] upward." All air-breathing land animals and people that were outside the ark that lived on the earth also died (Genesis 7:22–23).

Today, many people, including Christians, unfortunately do not accept the biblical account of a worldwide flood because they have been taught that most rocks and fossils were deposited over millions of years (and therefore not by a global Flood). Until the 1800s, most people from the Middle East to the Western World believed what the Bible records about creation and the global Flood. The secular idea of millions of years did not gain extensive

2. Using the long cubit of about 20.4 inches.

popularity until the 1830s, under the influence of a man named Charles Lyell — who opposed a global Flood!

Based on how slowly some rock layers seem to form today (assuming no catastrophes), Lyell rejected the Bible's claims and declared that the earth's many rock layers must have been laid down slowly over millions of years. But he never witnessed the actual formation of the earlier rocks to see whether a unique, one-time global Flood unlike anything we observe today could lay the majority of the rock layers with fossils.

Lyell's claim was based on his own preconceptions and belief in the religion of naturalism, not his observations. Lyell's idea took hold in Western universities and spread throughout the Western World.

As a response, many Christians simply tried to add this idea of long ages to the Bible. What these Christians should have done was stand on the authority of the Bible and defend the global Flood, which can easily account for the bulk of fossil-bearing rock layers we find all over the world. Naturally, we have had some rock layers since the time of the Flood with local catastrophes such as volcanoes or local floods. But the bulk of the rock layers with fossils came from the Flood of Noah.

Some Christians have tried to put millions of years of rock formation before the global Flood to explain the bulk of the rock layers that contain fossils. But the problem is that the Flood waters would have ripped up a number of these old rock layers and laid down new ones! So this compromise not only fails to explain the rock layers but also dishonors the clear claims of Scripture. The global Flood makes perfect sense, and it is foolish to stray from God's Word just because some men disagree.

Although there is tremendous physical evidence (fossil laden sedimentary strata over the earth) of a global flood, ultimately it is a matter of trust in a perfect God who created everything (Genesis 1:1[3]), knows everything (Colossians 2:3[4]), has always been there (Revelation 22:13[5]), and cannot lie (Titus 1:2[6]). The only alternative is to trust imperfect, fallible human beings who can only speculate on the past (see Romans 3:4[7]).

3. In the beginning God created the heavens and the earth.
4. In whom [Christ] are hidden all the treasures of wisdom and knowledge.
5. I am the Alpha and the Omega, the Beginning and the End, the First and the Last.
6. In hope of eternal life which God, who cannot lie, promised before time began.
7. Certainly not! Indeed, let God be true but every man a liar. As it is written: "That You may be justified in Your words, and may overcome when You are judged."

Local Flood Problems

Additionally, there are many problems with the claim that Noah's Flood was local. For instance:

- Why did God tell Noah to build an ark? If the Flood had been only local, Noah and his family could have just moved to higher ground or over a local mountain range or hills to avoid the floodwaters.
- The wicked people that the Flood was intended to destroy could have escaped God's judgment in the same manner. They could have used small boats or floating debris to swim to the edge of the flood and survive.
- Why would Noah have to put birds on the ark when they could have flown over the hills to safe ground?
- Why would animals be required to be on the ark to keep their kinds alive on the earth (Genesis 7:2–3[8]), if representatives of their kinds existed all over the earth outside of the alleged local Flood area?
- Did God fail at His stated task where He said that He would destroy all land animals on the earth since the Flood was *local* (Genesis 6:17[9])?
- Why would a flood take place *over the course* of about a year if it were local?
- Why did Noah remain on the ark for about seven months after coming to rest from a little river flood? Does a local flood really have about five months of rising and five months of falling in a river valley? Such a flood would merely carve out a deep valley and wash Noah downstream to the ocean!
- How could the ark have landed in the mountains of Ararat far upstream (and up in the mountains above) of the alleged river valley when all flow is going to take the ark in the opposite direction?
- The Flood occurred about 1,656 years after creation. If all people outside the ark were judged and drowned in this little

8. You shall take with you seven each of every clean animal, a male and his female; two each of animals that are unclean, a male and his female; also seven each of birds of the air, male and female, to keep the species alive on the face of all the earth.
9. And behold, I Myself am bringing floodwaters on the earth, to destroy from under heaven all flesh in which is the breath of life; everything that is on the earth shall die.

> local river flood (e.g., Genesis 7:23,[10] Matthew 24:39[11]) then they were all still living in this one little region on earth. Why didn't the Lord previously confuse their languages and scatter them for disobeying his command in Genesis 1 to be fruitful and multiply (Genesis 1:28[12])? It only took about 100 years or so after the Flood for God to judge mankind for not scattering at Babel.[13]

The proposal of a local Flood for Genesis 6–8 simply doesn't make sense of the context.

Rainbow Promise

Another problem presents itself. If the Flood were local, then God would be a liar, for God promised in Genesis 9:11[14] never to send a Flood like the one He just did to destroy the earth again. Yet the world has seen many local floods. Why the rainbow promise? The Bible says:

> Thus I establish My covenant with you: Never again shall all flesh be cut off by the waters of the flood; never again shall there be a flood to destroy the earth." And God said: "This is the sign of the covenant which I make between Me and you, and every living creature that is with you, for perpetual generations: I set My rainbow in the cloud, and it shall be for the sign of the covenant between Me and the earth. It shall be, when I bring a cloud over the earth, that the rainbow shall be seen in the cloud; and I will remember My covenant which is between Me and you and every living creature of all flesh; the waters shall never again become a flood to destroy all flesh. The rainbow shall be in the cloud, and I will look on it to remember the everlasting covenant between God and every living creature of all flesh that is on the earth." And God said to Noah,

10. So He destroyed all living things which were on the face of the ground: both man and cattle, creeping thing and bird of the air. They were destroyed from the earth. Only Noah and those who were with him in the ark remained alive.
11. And did not know until the flood came and took them all away, so also will the coming of the Son of Man be.
12. Then God blessed them, and God said to them, "Be fruitful and multiply; fill the earth and subdue it; have dominion over the fish of the sea, over the birds of the air, and over every living thing that moves on the earth."
13. Bodie Hodge, *Tower of Babel* (Green Forest, AR: Master Books, 2013), p. 37–42.
14. Thus I establish My covenant with you: Never again shall all flesh be cut off by the waters of the flood; never again shall there be a flood to destroy the earth.

"This is the sign of the covenant which I have established between Me and all flesh that is on the earth" (Genesis 9:11–17).

This rules out the idea of a local flood. Some have commented that they think rainbows didn't exist until this point in Genesis 9. However, the Bible doesn't say this. Like bread and wine used in communion, so a rainbow now takes on the meaning as designated by God. The main reason some have suggested rainbows didn't exist was their assumption there was no rain before the Flood. Let us discuss this idea, and where it came from.

Rain Before the Flood?

In Genesis 2, which is largely a breakdown of what occurred on day 6 of creation, we read:

> This is the history of the heavens and the earth when they were created, in the day that the LORD God made the earth and the heavens, before any plant of the field was in the earth and before any herb of the field had grown. For the LORD God had not caused it to rain on the earth, and there was no man to till the ground; but a mist went up from the earth and watered the whole face of the ground (Genesis 2:4–6).

Where it states that "God had not caused it to rain on the earth" is projected into the future until the time of the Flood. The Bible never mentions rain from this point until the Flood. Is this warranted? Not necessarily.

The context of this passage is looking back over creation week, not projecting forward. The *field* crops had not grown yet (herbs and crops of the field). Man did not exist yet either, so this is early on day 6. Though apparently God had planted the garden already, it just hadn't grown yet (Genesis 2:8[15]). That did not occur until after God made Adam and placed him in the Garden (Genesis 2:9[16]). A mist had come up from the ground and watered the face of the ground.

What we know is that this was the case when man was created. What we don't know is whether or not the situation remained this way until the

15. The LORD God planted a garden eastward in Eden, and there He put the man whom He had formed.
16. And out of the ground the LORD God made every tree grow that is pleasant to the sight and good for food. The tree of life was also in the midst of the garden, and the tree of the knowledge of good and evil.

Flood. It would seem to defy the laws of physics that God Himself sustains. Dr. Tommy Mitchell sums this up nicely when he writes on the "no rain" subject as an argument we should avoid:

> The passage describes the environment before Adam was created. This mist may have been one of the primary methods that God used to hydrate the dry land He created on Day Three. Furthermore, while this mist was likely the watering source for that vegetation throughout the remainder of Creation Week, the text does not require it to be the only water source after Adam's creation.
>
> Some argue that this mist eliminated the need for rain until the time of the Flood. However, presence of the mist prior to Adam's creation does not preclude the existence of or the need for rain after he was created.
>
> Genesis 2:5–6 reveals that before the Sixth Day of Creation Week, God had watered the plants He made with a mist, but had not yet caused rain or created a man to till the ground. To demand that rain didn't happen until after the Flood from this passage has no more logical support than to claim, from the passage, that no one farmed until after the Flood."[17]

Enough said.

17. Tommy Mitchell, "There Was No Rain Before the Flood," October 19, 2010, https://answersingenesis.org/creationism/arguments-to-avoid/there-was-no-rain-before-the-flood/.

13 The Religious Attack of Humanism's "Millions of Years" . . . in My Era

Bodie Hodge

Growing up in the 80's and 90's, my state schooling constantly imposed the idea of millions of years on my impressionable mind. Little did I know that I was subtly being taught the religion of humanism.

Humanistic teachers, who taught evolution, millions of years, and big bang, had told me that there was a separation of religion from schools. They said that schools were neutral. But that was merely a lie, which was usually directed toward kids going to church, to make us think that what we were learning at school was perfectly compatible with any religion, including Christianity.

What most of these humanistic teachers don't tell you is that they are really attacking the Bible and want to influence you to start dismissing parts of the Bible (e.g., creation, belief in the biblical God, etc.). In other words, the humanistic teachers were fine with saying that you can believe [the humanistic elements] they were teaching and still be a "Christian," as long as you don't really believe what the Bible says.

Not all humanistic teachers were so subtle either. Some of the teachers I had openly attacked God and the Bible while promoting the religion of humanism. And let's not forget the handful of good Christian teachers who were stuck in the difficult position of being in an education system that is now set up to attack the Bible. They need our prayers by the way — they are like missionaries next to the humanists who now dominate our education system. But in retrospect, I was being deceived at my state schools not just regarding origins, but I also had humanistic morality imposed on me, humanistic views of sexuality, etc.

As I looked back, what these secular humanists really meant by separation of religion and state (so called 'separation of church and state'), was that they were trying to keep *Christianity* out of the classroom, while forcing their religion onto unsuspecting kids. It was a subtle attack on Christians, and many parents didn't know the wiser that they were sending their kids to be taught to be radical humanists!

The humanists are clever though. They understood that if you control the minds of the next generation, then you are discipling them to be the next generation of humanists. So they have crept in under the radar, if you will, to make it sound as though their religion is not a religion and then use the state (and your tax money) to fund their preaching projects in state schools. While cap-stoning what humanists were already doing in school, leading humanist John Dunphy stated in 1983:

> I am convinced that the battle for humankind's future must be waged and won in the public school classroom by teachers who correctly perceive their role as the proselytizers of a new faith: a religion of humanity that recognizes and respects the spark of what theologians call divinity in every human being. These teachers must embody the same selfless dedication as the most rabid fundamentalist preachers, for they will be ministers of another sort, utilizing a classroom instead of a pulpit to convey humanist values in whatever subject they teach, regardless of the educational level — preschool day care or large state university. The classroom must and will become an arena of conflict between the old and the new — the rotting corpse of Christianity, together with all its adjacent evils and misery, and the new faith of humanism.[1]

This quote came out when I was in grade school. I see reflections of this too. For example, I was in grade school when we were told we were no longer permitted to say the *Pledge of Allegiance* because it mentions God (thus, this promotes the anti-God religion of humanism). In our extracurricular activities like basketball and football, we used to pray to the Lord for safety for our team and the opposing team. I recall when the superintendent came out and suppressed this and said we are no longer allowed to

1. J. Dunphy, "A Religion for a New Age," *The Humanist*, Jan.–Feb. 1983, p. 23, 26; as cited by Wendell R. Bird, *Origin of the Species — Revisited*, Vol. II (New York, NY, Philosophical Library, c. 1989), p. 257.

pray — thus, our extracurricular activates were going to operate in accordance with humanism.

Many reading this may recall when prayer, the Bible, or creation was taken out of schools. More recently the *Ten Commandments* have been ripped out of public places and nativity scenes are being attacked and removed. But religion was not removed — merely replaced with the religion of humanism. We see the religion of humanism rampant like a disease in textbooks, state schools, museums, movies, law, and media, much of which is funded by taxes.

There is a concerted attack on Christian kids to try to subtly train them to be humanists! It is a government-funded program (i.e., state-funded schools, state-funded museums, etc.) to force this religion onto the youth. They tried to do it to me. This is nothing new — Hitler, prior to WWII did this with the kids in Germany, and raised himself an army of Nazis in one generation! Nazism is just one variant of humanism, just like atheism, evolutionism, agnosticism, and secularism. The Nazis and Hitler held firmly to humanism's tenants of naturalism, evolution, and millions of years.

Interestingly, education is *meaningless* in a humanistic worldview where everything is just rearranged chemicals doing what chemicals do! Consider that snails do not develop education systems to train in philosophy, science, history, literature, etc. Education is required by God's command to train (e.g., Exodus 18:20[2]; Proverbs 22:6[3]) and predicated on the fact that man is made in the image of a logical, knowledgeable God (Genesis 1:26–27[4]). In other words, education is a *Christian* institution that humanism infiltrates and tries to destroy. The humanists have been doing an effective job for a number of years at destroying education. We are seeing the fruits of it in our culture today.

My guess is that most Christian parents and grandparents in today's era do not realize that schools are openly teaching this religion, thinking that schools are *neutral*. But they are not. They teach philosophy from a secular humanistic perspective, they teach biology, earth science, history,

2. "And you shall teach them the statutes and the laws, and show them the way in which they must walk and the work they must do.
3. Train up a child in the way he should go, and when he is old he will not depart from it.
4. Then God said, "Let Us make man in Our image, according to Our likeness; let them have dominion over the fish of the sea, over the birds of the air, and over the cattle, over all the earth and over every creeping thing that creeps on the earth." So God created man in His own image; in the image of God He created him; male and female He created them.

sex education, and so on from a humanistic perspective. It was in my textbooks as a kid and it is even more pronounced in the textbooks today!

If you have ever been taught big bang, steady states, abiogenesis (life accidentally came from matter/non-life), millions of years (no-global Flood), evolution, and the like, then you too were taught the religion of humanism!

Humanism teaches that man is the ultimate authority on all subjects (hence, God would *not* be seen as the ultimate authority). Humanism holds to a position of naturalism — that nature is all that exists. Hence, once again, God and the spiritual are not seen to exist. Humanism teaches things like evolution — that God had nothing to do with the creation of the kinds of life we find. Humanistic teachers and textbooks teach that life came from inanimate matter by accident — which is just another attack claiming God had nothing to do with life!

But I suggest one of the biggest attacks by humanists is the idea of "millions of years." Even many Christians in our generation have succumbed to this false tenet and stand with non-Christians to oppose Genesis! This may be one of the biggest problems within the Church today.

This religious idea of long ages was imposed on me from an early age in school from many different angles. I recall teacher after teacher discussing this alleged "truth" from grade school, then high school, to state university. As an example, one of my grade textbooks says:

> "that in the past 2 million years, many ice glaciers. . . ."[5]
> "About 200 million years ago. . . ."[6]
> "Over millions of years. . . ."[7]
> "Many millions of years later. . . ."[8]
> "For millions of years, most of earth. . . ."[9]
> "over millions of years, the parts of simple animals began to. . . ."[10]
> "about 500 million years ago."[11]
> "for a hundred million years."[12]

5. Albert Piltz and Roger Van Bever, *Discovering Science 5* (Columbus, OH: Charles E. Merrill Publishing Co., 1970), p. 205.
6. Ibid., p. 313.
7. Ibid., p. 313.
8. Ibid., p. 313.
9. Ibid., p. 314.
10. Ibid., p. 316.
11. Ibid., p. 317.
12. Ibid., p. 318.

> "more than 500 million years."[13]
> "had become extinct 70 million years ago."[14]
> "lived on the earth fifty million years ago."[15]

This textbook also had subtle hints at long ages as well that simply permeate the mind of the undiscerning, which is what I was in those early formative years.

> "Antarctic ice is more than 100,000 years old."[16]
> "the last Ice Age began to melt away about 10,000 years ago."[17]
> "After a long time, birds developed."[18]
> "Some animals became extinct long before man lived on earth."[19]

Little did I know, or my parents know, that I was being educated in the religion of humanism, and specifically hit hard with *millions of years*, for 40 hours a week for most of my life by the time I was 18. Then I was grilled even harder with this concept at the university by the time I had finished with my master's degree at the age of 24!

Throughout my years I was indoctrinated to believe that it took millions of years to form oil (petroleum), coal, diamonds, precious gemstones, fossils, rock, petrified wood, canyons, etc. Knowing that this humanistic story of millions of years is just a false religion that had been imposed on me . . . *does* it take millions of years to form these things? Let's investigate and put this religious viewpoint to the test.

---

13. Ibid., p. 318.
14. Ibid., p. 319.
15. Ibid., p. 323.
16. Ibid., p. 206.
17. Ibid., p. 205.
18. Ibid., p. 322.
19. Ibid., p. 324.

14 | Doesn't It Take Millions of Years to Form Rock and Rock Layers?

Bodie Hodge

I once had a student jump up and say, "But it takes millions of years to form rock." My response was, "What experiments have been run over the course of millions of years that proves it takes millions of years to form rock?"

There have never been any observations or experiments run over the course of millions of years. It is mere fantasy to think otherwise. Sadly, this student merely assumed what he had been indoctrinated with. He presumed that it took millions of years to form rock because that is what he had been inundated with and that we must have been "off of our rocker" to think otherwise. Naturally, I hadn't bought into his false religious belief on this issue.

It was he who arbitrarily had the false belief that something took millions of years to form. My response was to reveal that his belief was *mere opinion* (blind belief) that did not have any merit. His arbitrary opinion was just that . . . arbitrary and therefore fallacious.

Scientifically, we can prove that it doesn't take millions of years to form rock. The easiest example is concrete. Concrete can form in hours to days — we merely reproduce the chemical process to make certain rock (concretions). So it doesn't take millions of years to form rock, just the right conditions.

Following are a lot of examples of rapid rock. In 1975, a clock was found embedded in rock near the South Jetty at Westport, Washington.[1] This clock was not millions of years old! Another example is a recent marine spark plug encased in rock.[2]

1. The Clock in a Rock, *Creation*, 19, no 3 (June 1997), https://answersingenesis.org/geology/catastrophism/the-clock-in-the-rock/.
2. "Sparking Interest in Rapid Rocks," Creation, 21, no 4 (September 1999), https://answersingenesis.org/geology/geologic-time-scale/sparking-interest-in-rapid-rocks/.

Petrifying Well at Knaresborough

A ship's bell was encased in a natural marine concretion.[3] The bell came from a wooden ship named *Isabella Watson.* The ship sank in 1852 near Victoria, Australia. It is currently in the possession of the Maritime Archaeological Unit of Heritage Victoria.

In Spray Mine, which had been closed for 50 years, a hat that had been left had turned to stone.[4] This is the original *hard hat* I guess!

In the region of Yorkshire, England, there is a famous place that has attracted tourists since the 1600s. It is the famous Petrifying Well at Knaresborough and it is famous for converting teddy bears to stone in about three to five months![5]

When volcanoes go off, they can form rock quickly too. Even many items that were buried under the hot ash can petrify and turn to stone quickly. Mt. Tarawera blew its top in New Zealand in June of 1886 and killed over 150 people. After some excavation of one buried village called Te Wairoa, they found many items petrified, including a bowler hat, a ham, and a bag of flour.[6]

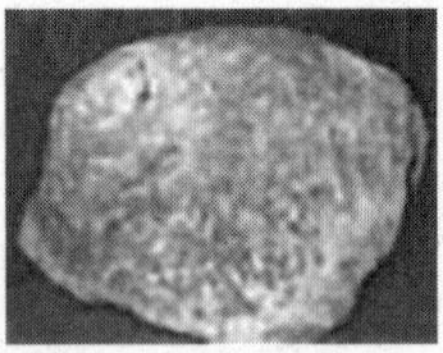

3. "Bell-ieve It: Rapid Rock Formation Rings True," *Creation*, vol. 20, no. 2, March 1998: 6, https://answersingenesis.org/geology/catastrophism/bell-ieve-it-rapid-rock-formation-rings-true/.
4. John Mackay, "Fossil Bolts and Fossil Hats," *Creation Ex Nihilo*, vol. 8, Nov., p. 10.
5. M. White, "The Amazing Stone Bears of Yorkshire," Answers in Genesis, June 1, 2002, http://www.answersingenesis.org/articles/cm/v24/n3/stone-bears.
6. Renton Maclachlan, "Tarawera's Night of Terror," *Creation*, vol. 18, no. 1, December 1995: 16–19, https://answersingenesis.org/geology/catastrophism/taraweras-night-of-terror/.

These few examples show proof that it doesn't take millions of years to form rock — just the right conditions.

Rock Layers - Do They Take Millions of Years to Accumulate?

Many of us have been taught that it takes millions of years of gradual accumulations to form the sedimentary rock layers. For example, we are told it took about 72–79 million years to lay down the Cretaceous rock layer. The secularists claim it was laid down from about 145 million years ago (give or take many millions of years) until about 66 million years ago. The Live Science website has more specific claims:

> The Cretaceous Period was the last and longest segment of the Mesozoic Era. It lasted approximately 79 million years, from the minor extinction event that closed the Jurassic Period about 145.5 million years ago to the Cretaceous-Paleogene (K-Pg) extinction event dated at 65.5 million years ago.[7]

Most of the rock layers that contain fossils are given huge amounts of time to accumulate. See the secular chart on the following page for the typical durations. To reiterate, this concept is called "uniformitarianism." In other words, the secularists assume the layers were formed *uniformly* over long, millions of years, ages. Because of this, they must reject that catastrophes occurred in the past, i.e., global Flood of Noah's day. This is where the geologic timescale (the chart above) came from — assuming that each layer required millions of years to form and that catastrophes didn't occur.

Mt. St. Helens Explosive Throwdown

The Mount St. Helens eruption (1980) laid down rock layers quickly — in a matter of hours and days. Some of these accumulated rock layers extended over 600 feet (~183 meters) in depth since the volcano exploded.[8] Differing types of sedimentary layers occurred sitting one on top of the other — some due to various types of flows (e.g., mudflows, hot or pyroclastic flows), others due to air fall, and so on. The point is that catastrophes can form rock layers quickly.

---

7. Mary Bagley, "Cretaceous Period: Animals, Plants & Extinction Event," Livescience.com, January 7, 2016, http://www.livescience.com/29231-cretaceous-period.html.
8. Dr. Steve Austin, "Why Is Mount St. Helens Important to the Origins Controversy?" *New Answers Book 3*, gen. ed., Ken Ham (Green Forest, AR: Master Books, 2010), https://answersingenesis.org/geology/mount-st-helens/why-is-mount-st-helens-important-to-the-origins-controversy/.

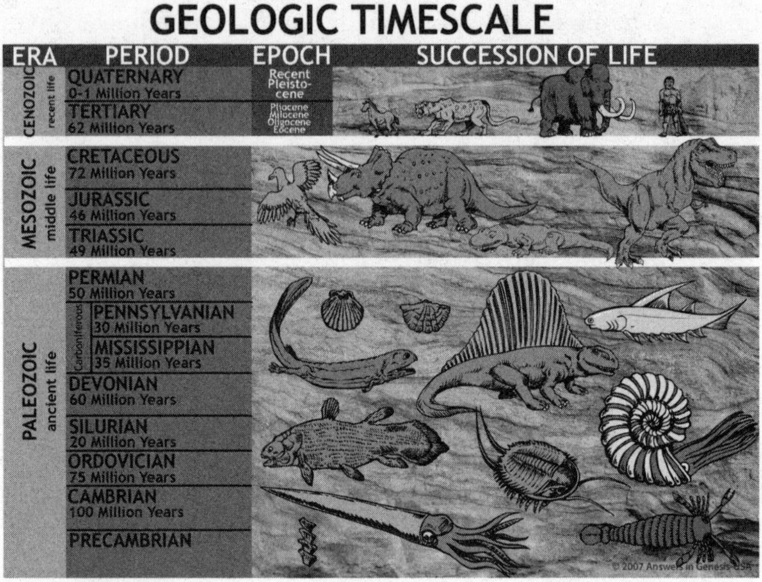

If this small catastrophic volcano did this in a short time, imagine what the catastrophic global Flood of Noah's day did all over the world.

Laminae Sediments (Varves) Quickly

Laminae (singular *lamina*) are basically fine layers of sedimentary rock — usually less than 1 cm (~4/10 inch). Researchers using laboratory tests have shown that many layers of lamina form quickly and at the same time in both air and water.[9] The phenomenon occurs in moments, not long periods of time.

Furthermore, these laminae are very similar to the tiny layers of sediment we see in nature. These tiny layers in nature, like lake deposits, are called "varves," and supposedly each one takes one year to form. I've always been stumped by the secular idea that every year at the same time, sedimentary deposits suddenly begin a *new* layer! One researcher, Mike Oard reports:

> A rhythmite is a repeating sequence of two or more lamina, which are thin layers. . . . Multiple *turbidites*, the deposits of fast-moving bottom flows of sediment, can form thick sequences of varve-like rhythmites quickly.[10]

9. Guy Berthault, "Experiments on Laminations of Sediments," April 1, 1988, https://answersingenesis.org/geology/sedimentation/experiments-on-lamination-of-sediments/.
10. Mike Oard, "Are There Half a Million Years in the Sediments of Lake Van?" *Answers in Depth*, Answers in Genesis, May 9, 2007, https://answersingenesis.org/geology/sedimentation/are-there-half-a-million-years-in-the-sediments-of-lake-van/.

In other words, currents moving downhill underwater carrying sediment make varves or laminae quickly. These types of examples can be observed with:

> . . . flowing sediment with different particle sizes, shapes, and densities; flooding into lakes; underwater slides; turbidity currents; snow-melt events; underflows from a muddy bottom layer; overflows from a muddy layer floating at the top of a lake; interflows from a muddy layer at intermediate depths; and multiple rapid blooms of microorganisms within one year.[11]

Varves do not require long periods of time. Multiple varve layers can occur quickly and have been observed.

***Bent* Rock Layers?**

Carbon Canyon

Doesn't rock break when it bends? Yes. Rock is a ceramic material and is brittle. So if you bend rocks it shatters (think of a glass or a ceramic plate being bent in half). And yet we find rock layers that have bent or folded layers.

These are often seen at the Grand Canyon. Perhaps the most popular is a sequence of sedimentary layers of the Tapeats Sandstone, which is a layer at the bottom of the Grand Canyon. Geologists have traced this 90-degree fold and the rock formation doesn't break or fracture! There are no signs of heat deformation either — so it is not metamorphic rock. The only signs of cracking are from drying, which is to be expected.[12]

The layers were still soft and pliable sediment when they were kinked into that angle during folding. *Then* they were solidified into solid rock. This means that each of those folded layers was laid down at the same time.

11. Ibid.
12. Andrew Snelling, "Rock Layers Folded, Not Fractured," *Answers Magazine*, April–June 2009, p. 80–83, https://answersingenesis.org/geology/rock-layers/rock-layers-folded-not-fractured/.

We also see this at another place in the Grand Canyon (many in fact). Another worth mentioning is Monument Fold. Rafters traveling down the Canyon come in close contact with this bent rock. It makes a "Z" shape.

Monument Fold

Other bent rock formations can be seen in various parts of the world. Here is one along a Pennsylvania highway in the Appalachian Mountains in the eastern United States.

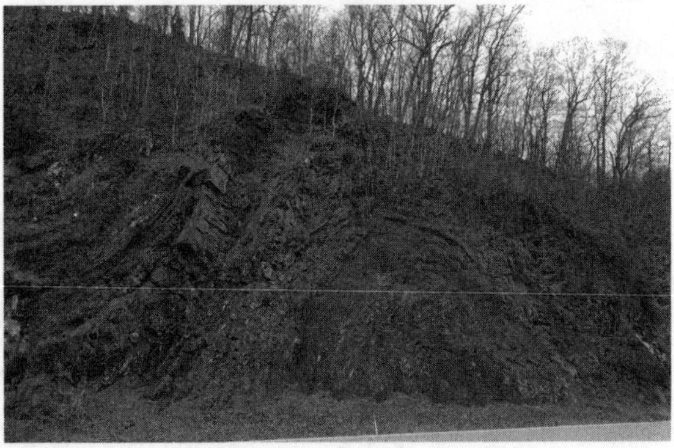

Appalachian Mountains

You can see how this thrust came from below and the sediment simply bent around it and then solidified into rock.

Polystrate Fossils

Another classic example of reducing the alleged long ages of rock layers comes from studying polystrate fossils. *Poly* means "many" and *strate* means "strata" or rock layers. So when we find fossils that extend through multiple rock layers, these are evidences that the rock layers are not as old as we have been led to believe.

Trees make great examples, even though there are others. Tree fossils often extend through several layers of rock. Are we to believe that the tree just sat there waiting for thousands and perhaps, millions of years to get covered up with sediment to transform into a fossil? Not at all. The tree would rot, decay, or otherwise disintegrate.

As we begin to conclude, it is obvious that rock layers are *not* evidence of millions of years. They can be made quickly in labs, we see it with catastrophes like a volcano, we observe it with flowing water that has sediment, and so forth. Bent rock layers and polystrate fossils also confirm rock layer formation to be rapid events.

A New Trend

So how does the world respond? An intriguing thing is happening in our culture right now. There is a shift going on.

Formerly, those who believed the rock layers were evidence of millions of years (uniformitarians) said the "millions of years" was to be counted *within* the rock layer. This idea was solidified in the minds of many scientists as a result of a lawyer, Charles Lyell, in the 1830s.

Lyell argued that rock layers were laid down slowly and gradually over millions of years. As a result, the secular idea is that the Cretaceous rock layer was supposedly laid down slowly and gradually over 72–79 million years — as well as all the dates on the popular geologic timescale.

This view has dominated since the mid-1800s *until recently* among the secularists. But all that is changing in our modern times. A secular scientist, Dr. Warren D. Allmon, says:

> Indeed geology appears at last to have outgrown Lyell. In an intellectual shift that may well rival that which accompanied the widespread acceptance of plate tectonics, the last 30 years have witnessed an increasing acceptance of rapid, rare, episodic, and "catastrophic" events.[13]

One cannot argue that catastrophes didn't occur in the past, because we see them all the time — like Mt. St. Helens, Hurricane Katrina, and so on. Therefore, many secularists have been shifting away from the idea that fossil layers were laid down slowly and gradually. They have been agreeing that catastrophes did occur in the past. They are now openly disagreeing with Lyell but still agree with the geological timescale and millions of years. So where do they put the "millions of years" if they don't put them where Lyell put them?

An Elephant in the Room

Let's just call out the *elephant in the room*. If the layers were not laid down slowly over millions of years (because the secularists now agree that

13. W. Allmon, "Post-Gradualism," review of *The New Catastrophism*, by Derek V. Ager (New York: Cambridge University Press, 1993); Science, vol. 262, October 1, 1993, p. 122.

catastrophes laid these fossil deposits down quickly), why do they still believe in the geological timescale and millions of years? Here is why.

Instead of the rock layers being where the millions of years were supposedly contained, they now say the millions of years is *in between* rapidly deposited layers. Thus, they claim that the space between the Jurassic rock layer and the Cretaceous is about 125 million years ago. Then there are a few "snapshots" of the Cretaceous over 72–79 million years due to a few catastrophes that laid down rock layers. Then, secularists claim the space between the Cretaceous and the Tertiary is 65 million years ago.

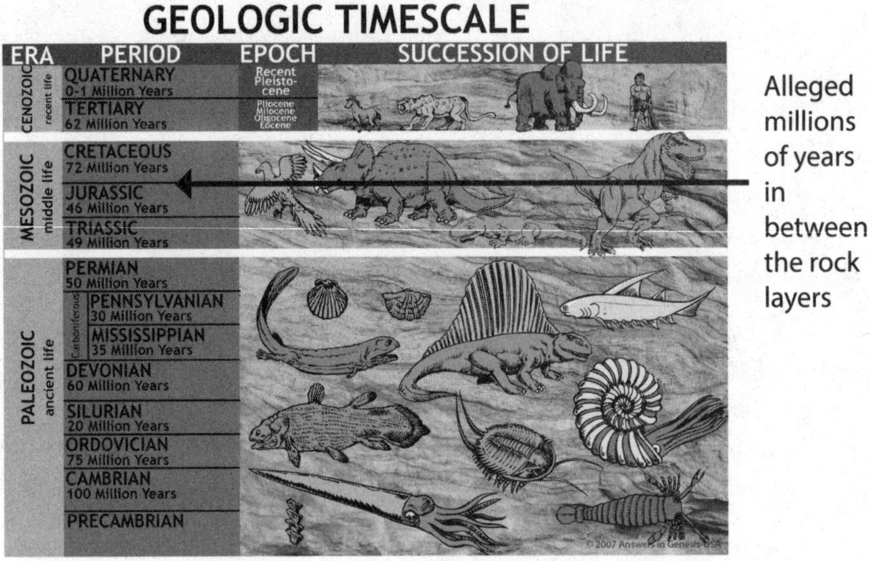

You might be thinking, "Hmm, that settles the issue of the elephant in the room." But it doesn't. It actually introduces a second elephant in the room. Here is why.

There is now *no evidence* for millions of years whatsoever. The evidence used to be the rock layers. However, the secularists are shifting to agree the rock layers are laid down quickly by catastrophes. So the secular view is that the imaginary space between the rock layers *is* the evidence for millions of years. Thus, they have no evidence for millions of years. Leaving open the idea of catastrophes that lays down sediment quickly destroys "millions of years."

15 Doesn't It Take Millions of Years to Form Canyons?

Bodie Hodge

Introduction

I still have my sixth grade textbook. It says:

> During millions of years, the Colorado River has worn away thousands of feet of rock and soil to form the Grand Canyon in Arizona.[1]

Other canyons supposedly formed over millions of years. Consider Copper Canyon in Mexico where the *USA Today* travel advisor to the canyon states:

> Copper Canyon's deep ravine was cut out of the mountainous region over the course of millions of years by the force of rushing water from six separate rivers.[2]

Or perhaps the beautiful Verdon Gorge (*Les Gorges du Verdon*) in France? This literally translates into *Grand Canyon* in English by the way, but is not to be confused with the Grand Canyon in the USA. The rafting site for Verdon Gorge casually states:

> From that point, we take some time to marvel at the work of millions of years of erosion on the impressive limestone cliffs surrounding us between the successive class II and III rapids.[3]

1. Albert Piltz and Roger Van Bever, *Discovering Science 5* (Columbus OH: Charles E. Merrill Publishing Co., 1970), p. 196.
2. Richard Kalinowski, "Mexico's Copper Canyon," accessed September 28, 2015, http://traveltips.usatoday.com/mexicos-copper-canyon-1388.html.
3. "Discovery of the Verdon," accessed September 28, 2015, http://www.rafting-verdon.com/gb/details-sheets-packages/Verdon-6-days-spring.html.

Hosts of references could be used of canyons all over the world that state that the canyons in question take millions of years of slow gradual erosion to carve them. Secularists have no scientific observations of its formation. They can't repeat it either. So they merely look at the rivers running through these canyons today (in the present) and *assume there are no significant catastrophes* in the past for millions of years and then draw the conclusion that these layers were slowly carved out over that time.

Did you catch that? You need to understand that for them to make such a statement, they are assuming that significant catastrophes, like major floods, didn't occur in the past! Sounds like a crazy assumption doesn't it? Well . . . it is. When a catastrophe occurs, things change quickly! We see major catastrophes all the time all over the world (tsunamis, earthquakes, floods, hurricanes, etc.). Why assume there weren't any in the past . . . for millions of years?

Do Canyons Really Take Millions of Years to Form?

Mt. St. Helens, a volcano in Washington State in the USA erupted in 1980 and it went off again in 1982. For once, scientists had the opportunity to see what a catastrophe could accomplish on a descent-sized scale! What was observed?

Drs. Steven A. Austin and John Morris pointed out:

> How long would it take to erode a 100-foot (30 meter) deep canyon, in hard basaltic rock? On Mount St. Helens, such a canyon was formed rapidly as rock avalanched from the crater followed by other episodic, catastrophic processes.[4]

The Little Grand Canyon

Drs. Austin and Morris also pointed out that the new drainage of the North Fork of the Toutle River occurred quickly too! Based on observations of the 1982 eruption, they write:

> In a single day, the new drainage channel of the North Fork of the Toutle River was established westward through the debris dam by a catastrophic mudflow! The Canyon produced by the mud has been called "The Little Grand Canyon" because it

4. John Morris and Steven A. Austin, *Footprints in Ash* (Green Forest, AR: Master Books, 2003), p. 70.

> appears to be one-fortieth scale model of the Grand Canyon of Arizona.[5]

Interestingly, the topography of the Little Grand Canyon is very similar to the features we see at the Grand Canyon as well. It is more reasonable to believe that larger canyons like the Grand Canyon formed by similar (but bigger) processes as this catastrophic Little Grand Canyon. I find it fascinating that if someone had no idea that a catastrophe formed this large canyon, they might mistakenly assume it took long ages, perhaps millions of years, to form . . . especially if they assumed that catastrophes of this magnitude never occurred in the past.

Canyon Lake Gorge

In 2002, the Guadalupe River in Texas had extensive flooding. Water began pouring over Canyon Lake spillway (a reservoir) due to massive rain. In July of that year, it was estimated that about 70,000 cubic feet per second was exiting over the spillway, which was much more than the normal 350 cubic feet per second!

The Guadalupe River flood carved out a gorge about 1 mile (1.6 kilometers) in length. It was over 50 feet (15 meters) deep in places. Keep in mind this was carving through solid limestone! This rapid canyon exposed rock layers from the Flood of Noah's day while at the same time helping disprove this idea that it takes millions of years to form canyons! Researchers state:

> A narrow gorge is sometimes inferred to represent slow persistent erosion, whereas Canyon Lake Gorge was formed in a matter of days.[6]

The online encyclopedia, Wikipedia, has entries that tend to change rather often and is biased toward the religion of secularism. Even so, they write:

> Typically a steep-walled, narrow gorge is inferred to represent slow persistent erosion. But because many of the geological formations of Canyon Lake Gorge are virtually indistinguishable from other formations which have been attributed to long term (slower) processes, the data collected from Canyon Lake Gorge

5. Ibid., p. 75.
6. Michael P. Lamb and Mark A. Fonstad, "Rapid Formation of a Modern Bedrock Canyon by a Single Flood Event," *Nature Geoscience*, June 20, 2010, p. 4, DOI: 10.1038/NGEO894.

> lends further credence to the hypothesis that some of the most spectacular canyons on Earth may have been carved rapidly during ancient megaflood events.[7]

Notice that the religion of secular humanism still reigns supreme in this quote. The encyclopedia refuses to give the possibility of a global Flood (Noah's Flood) being the triggering factor (as well as subsequent factors resulting from the Flood) for many of the great canyon's formations. Instead they appeal to "megafloods." But regardless, major floods and other catastrophes destroy the idea of millions of years and long ages.

These are not the only rapidly forming canyons. Providence Canyon in Georgia was a rapidly formed canyon now dubbed Georgia's "Little Grand Canyon."[8] Another in Walla Walla, Washington, formed in six days (Burlingame Canyon)![9] This list could continue! What we know from observation is that it doesn't take millions of years to form canyons, just the right catastrophic conditions!

---

7. Wikipedia, Entry: Canyon Lake Gorge, accessed September 28, 2015, https://en.wikipedia.org/wiki/Canyon_Lake_Gorge.
8. Rebecca Gibson, "Canyon Creation," September, 1, 2000, https://answersingenesis.org/geology/natural-features/canyon-creation/.
9. John Morris, "A Canyon in Six Days," September 1, 2002, https://answersingenesis.org/geology/natural-features/a-canyon-in-six-days/.

16 | Aren't Coal and Oil Millions of Years Old?

Bodie Hodge

Coal

I grew up in Western Illinois. We used to have a lot of coal mines in our area. We even had a few lumps of coal that we would burn in our wood/coal stove in the farmhouse in which I grew up.

From time to time at school, we had videos shown in our classroom that taught that coal was a fossil fuel and took millions of years to form. This idea still persists today. As one example, the California Energy Commission writes:

> There are three major forms of fossil fuels: coal, oil and natural gas. All three were formed many hundreds of millions of years ago before the time of the dinosaurs — hence the name fossil fuels.[1]

The site continues:

> More and more rock piled on top of more rock, and it weighed more and more. It began to press down on the peat. The peat was squeezed and squeezed until the water came out of it and it eventually, over millions of years, it turned into coal, oil or petroleum, and natural gas.[2]

1. Energy Story, chapter 8: "Fossil Fuels — Coal, Oil and Natural Gas," California Energy Commission, 1994–2012, http://www.energyquest.ca.gov/story/chapter08.html.
2. Ibid.

Has the *California Energy Commission* actually repeated an experiment over the course of millions of years to prove that it takes millions of years to form coal (or these other things like oil)? No. Have they had any observations over millions of years to prove it takes millions of years to form coal? No. With this in mind, I want you to understand that this statement of theirs is not scientific, as science is based on observations and repeatability. This is their religious conviction that it takes millions of years to make things like coal and oil.

Does it really take millions of years to form coal? Based on research by geologist Dr. Steve Austin at Mt. St. Helens,[3] Dr. Gary Parker and his wife Mary write:

> In just minutes and months, Mount St. Helens and Spirit Lake produced a coal-like sediment pattern once thought to take millions of years to form.[4]

Furthermore, Argonne National Laboratory has proven that it doesn't take millions of years to form coal . . . but instead months.[5] They used wood, specifically lignin that makes up much of the wood, water, and clay that was acidic. Then they heated it to about 300°F (150°C) for about 1–9 months and it formed black coal. Since then, others have been able to use processes for the coalification of peat in short periods of time.[6]

So it doesn't take millions of years to form coal. This can be observed and repeated. So scientifically, we know that it doesn't take millions of years to form coal.

C14 in Coal (Semi-Technical)

But there is more. An interesting thing about coal is that is it is made of *carbon*. If coal layers are supposed to be millions of years old, then there

3. For more, see John Morris and Steven A. Austin, *Footprints in the Ash* (Green Forest, AR: Master Books, 2003), p. 78–89.
4. Gary and Mary Parker, *The Fossil Book* (Green Forest, AR: Master Books), p. 15.
5. R. Hayatsu, R.L. McBeth, R.G. Scott, R.E. Botto, R.E. Winans, *Organic Geochemistry*, vol. 6 (1984), p. 463–471.
6. W.H. Orem, S.G. Neuzil, H.E. Lerch, and C.B. Cecil, "Experimental Early-stage Coalification of a Peat Sample and a Peatified Wood Sample from Indonesia," *Organic Geochemistry* 24(2):111–125, 1996; A.D. Cohen and A.M. Bailey, "Petrographic Changes Induced by Artificial Coalification of Peat: Comparison of Two Planar Facies (Rhizophora and Cladium) from the Everglades-Mangrove Complex of Florida and a Domed Facies (Cyrilla) from the Okefenokee Swamp of Georgia," *International Journal of Coal Geology* 34:163–194, 1997; S. Yao, C. Xue, W. Hu, J. Cao, C. Zhang, "A Comparative Study of Experimental Maturation of Peat, Brown Coal and Subbituminous Coal: Implications for Coalification," *International Journal of Coal Geology* 66:108–118, 2006.

should be no carbon-14 (^{14}C) left in it — ^{14}C can only give dates of thousands of years, not millions of years, otherwise it should all be decayed away. Now as a caveat, ^{14}C dates have hosts of problems such as living creatures that date to outrageously old dates. For example, we find incorrect dates for:

- Living mollusks supposedly 23,000 years old[7]
- Living snails were supposedly 27,000 years old[8]
- A freshly killed seal was supposedly 1,300 years old[9]
- Dinosaur bones with ^{14}C[10]

People may offer excuses for these, but the fact is that they came out with inaccurate dates. If we can't trust ^{14}C dating on dates we know, how can we trust it on dates we don't know? That would be illogical. The measured half-life of ^{14}C is 5,730 years in a laboratory under certain conditions, though this has never been fully observed and there may be things that affect the rate of decay even more than we already know. But for the sake of argument, let's assume this half-life is accurate for now.

Based on the half-life of ^{14}C, that means we should have no ^{14}C in any sample that is about 50,000–100,000 years (theoretically). Geophysicist Dr. John Baumgardner has a listing of 90 samples from secular literature of things *supposedly older than 100,000 years old in the secular reasoning*, and yet they have measurable ^{14}C in them![11] This means these things cannot be that old!

In fact, anything that is supposed to be a million years or more should have no ^{14}C. In other words, coal, which is supposed to be many millions of years old, should have no ^{14}C in it whatsoever. And yet, we find ^{14}C all over in it![12] We find C14 in things supposedly millions and billions of years old such as:

7. *Science*, vol. 141 (1963), p. 634–637.
8. *Science*, vol. 224, (1984), p. 58–61.
9. *Antarctic Journal*, vol. 6 (Sept.–Oct. 1971), p. 211.
10. Brian Thomas, "Carbon-14 Found in Dinosaur Fossils," ICR, July 6, 2015, http://www.icr.org/article/carbon-14-found-dinosaur-fossils/.
11. L. Vardiman, A.A. Snelling, and E.F. Chaffin (eds.), *Radioisotopes and the Age of the Earth, Vol. 2: Results of a Young-earth Creationist Research Initiative* (El Cajon, CA: Institute for Creation Research and Chino Valley, AZ: Creation Research Society, 2005), p. 596–597.
12. Ibid., p. 587–630.

- Diamonds[13]
- Meteorites[14]
- Coal[15]

Yes, even coal is replete with ^{14}C, so it *cannot* be millions of years old.

Oil/Petroleum

What about oil or petroleum? The California Energy Commission also included petroleum/oil items as millions of years old too. Even the Live Science website says of oil:

> Nature has been transmuting dead life into black gold for millions of years using little more than heat, pressure and time, scientists tell us.[16]

Sinclair Oil Corporation even utilizes an Apatosaurus on their emblem, relating the idea that the oil was supposedly from millions of years ago.

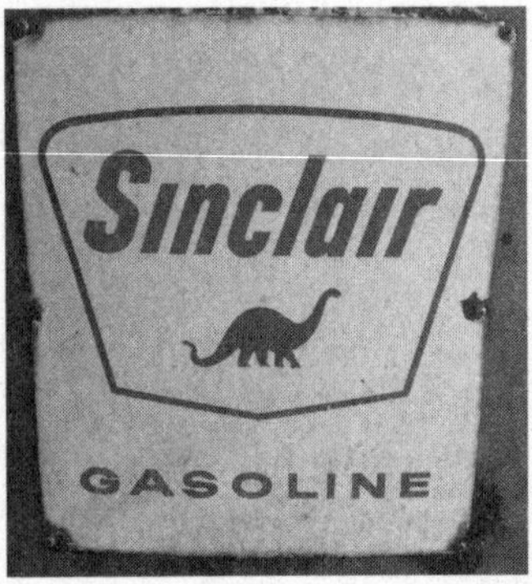

> Sinclair Oil's advertising writers struggled with this question in 1930. They needed a vehicle for their message. At that time, Sinclair's oils and lubricants came from Pennsylvania crude oil more than 270 million years old. Those oils perked and mellowed in the ground when dinosaurs wandered the

13. Ibid., p. 609–614.
14. Many meteorites contain Cosmogenic Carbon 14, which means they really can't be millions or billions of years old (some via internal sampling). The secularists assume that the C-14 became part of them after entering the atmosphere and remaining on the ground and try to use this for terrestrial dates. How does a meteorite take in C-14; they are not plants and animals that utilize this as part of their diet. For example see: A. Snelling, "Radioisotope Dating of Meteorites: I. The Allende CV3 Carbonaceous Chrondrite," *Answers Research Journal*, vol. 7, 2014, p. 103–145, https://cdn-assets.answersingenesis.org/doc/articles/pdf-versions/radioisotope_dating_Allende.pdf. See also Pillinger et al, "The Meteorite from Lake House," 74th Annual Meteoritical Society Meeting, 2011, http://www.lpi.usra.edu/meetings/metsoc2011/pdf/5326.pdf.
15. L. Vardiman, A.A. Snelling, and E.F. Chaffin (eds.), *Radioisotopes and the Age of the Earth, Vol. 2: Results of a Young-earth Creationist Research Initiative* (El Cajon, CA: Institute for Creation Research and Chino Valley, AZ: Creation Research Society, 2005), p. 587–630.
16. K. Than, "The Mysterious Origin and Supply of Oil," Live Science, October 10, 2005, http://www.livescience.com/9404-mysterious-origin-supply-oil.html.

> earth. So Sinclair's ad writers developed a campaign using about a dozen different dinosaurs: the tyrannosaurus rex, triceratops and the apatosaurus (brontosaurus) were but a few.[17]

The sauropod (Apatosaurus) won out and so millions of people have seen this emblem all over the USA since the 1930s.

But does oil/petroleum take millions of years to form? Not at all. Animal wastes are also being used to make oil. Geologist Dr. Andrew Snelling writes:

> Turkey and pig slaughterhouse wastes are daily trucked into the world's first biorefinery, a thermal conversion processing plant in Carthage, Missouri. On peak production days, 500 barrels of high-quality fuel oil better than crude oil are made from 270 tons of turkey guts and 20 tons of pig fat.[18]

Oil can be made from brown coal in a matter of days too. Dr. Snelling continues:

> Thus, for example, it has been demonstrated in the laboratory that moderate heating of the brown coals of the Gippsland Basin of Victoria, Australia, to simulate their rapid deeper burial,

17. Christopher Bennett, "Why Do Sinclair Gas Stations Use a Dinosaur in Their Logo?" *Winona Daily News*, March 15, 2007, http://www.winonadailynews.com/news/why-do-sinclair-gas-stations-use-a-dinosaur-in-their/article_566e7613-d6f3-5f68-bdc5-9f90ade1b29a.html.
18. Dr. Andrew Snelling, "The Origin of Oil," Answers 2, no. 1, p. 74–77, http://www.answersingenesis.org/articles/am/v2/n1/origin-of-oil.

> will generate crude oil and natural gas similar to that found in reservoir rocks offshore in only 2–5 days.[19]

Algae can also be made into oil . . . but in minutes. Researcher Brian Thomas reports:

> Researchers at the Pacific Northwest National Laboratory (PNNL) in Washington State have pioneered a new technology that makes diesel fuel from algae — and their cutting-edge machine produces the fuel in just minutes. . . . Simply heat pea-green algal soup to 662°F (350°C) at 3,000 psi for almost 60 minutes.[20]

But now it is quicker than that! The same lab has been able to make crude oil from algae in 30 minutes. The news report said:

> Scientists at the Pacific Northwest National Laboratory are claiming success in perfecting a method that can transform a pea-soupy solution of algae into crude oil by pressure cooking it for about 30 minutes. The process, called hydrothermal liquefaction, also works on other streams of organic matter, such as municipal sewage.[21]

Even sewage sludge is being used to make oil (diesel grade fuel) in Australia.[22] Perhaps it is time to cast this false religious idea that it takes millions of years to form *oil and flush it down the toilet!* Okay, so that was "tongue in cheek" humor. But seriously, this idea of oil taking millions of years to form is false — and industry is not wasting time in capitalizing on the truth.

19. Ibid.
20. B. Thomas, "One-Hour Oil Production," ICR, January 13, 2014, https://www.icr.org/article/7874/.
21. Christopher Helman, Green Oil: Scientists Turn Algae Into Petroleum In 30 Minutes, Forbes.com, 12/23/2013, http://www.forbes.com/sites/christopherhelman/2013/12/23/green-oil-scientists-turn-algae-into-petroleum-in-30-minutes/.
22. Australian Stock Exchange Release, Environmental Solutions International Ltd, Osborne Park, Western Australia, Oct. 25, 1996. Media Statement, Minister for Water Resources, Western Australia, October 25, 1996.

17 What about Stalagmites and Stalactites – Do They Take Millions of Years?

Bodie Hodge

A Warm Introduction to the Subject

Being that I was originally from Illinois, I often ventured across the border to Missouri and went caving. One of my favorite trips was exploring random caves while on a float trip down the Meramec River. There were hosts of caves to explore. I even visited the famous Meramec Caverns, which is beautiful.

Of course, for anyone who likes caves, remember one simple thing. Do not enter a warm cave. Let me repeat this . . . do not enter a warm cave. While on this float trip in the 1990s, a group of us (toward the end of our trip) floated our canoe into what looked like a watery entrance of a cave. And of course . . . our flashlights had finally stopped working. So we went in based on the vague light entering the cave's entrance. The canoe finally struck ground and we meandered on strange damp ground in that *warm* cave. In fact — the ground *crunched.*

We realized we needed light since things could not be discerned at all at this depth of the cave. So one of my buddies pulled out a handy lighter that had miraculously not gotten wet. We ignited it. We found ourselves surrounded by more spiders, more strange insects, and snakes than I had ever seen in one condensed spot in my entire life. We shrieked, the lighter went out, and we were all darting for the water in what seemed like dense darkness. And we didn't care if we made it in the canoe, we just wanted out.

I think this was a natural reaction to the creatures that were now crawling on us. And I'm pretty sure I stepped on at least two snakes trying to exit that cave before I jumped in the water and began dragging the canoe toward the cave's entrance with the two gents in tow!

Needless to say, we were all a little hesitant to EVER enter another cave. That was until we inquired and a cave guide at Meramec Caverns told us to be careful entering a *warm* cave. Ahhh, *cold* . . . that was the key to a good cave adventure — it keeps out the creepy crawlies! Since that time, I have been in many caves in several states, but I always take note of the temperature the moment that I take that first step into one.

Now to the *Hardened* Question

So how many times have I entered a cave and a guide has told me that things like stalactites and stalagmites were millions of years old? It has been too many to count. For the reader, stalactites are the formations on top (think of it like this: they hold on to the roof of the cave *tight* being stalag-*tites*). The formations that form on the bottom of the cave and work their way up are called stalagmites. Now there are other formations, like a column (when a stalagtite and stalagmite grow together) or straws or flowstone, and so on. All cave formations are basically summed up in the technical term *speleothem*.

But does it take millions of years to form things like stalactites and stalagmites? Knowing that all cave formations have occurred since the time of the Flood (in some cases as resultant actions of the end stages of the Flood) then they cannot be millions of years old.

Let's look at this a little closer. A cave specialist reported an instance in which it only took days for a stalactite to grow several inches. The article reads:

> Jerry Trout (cave specialist with the Forest Service) says that through photo monitoring, he has watched a stalactite grow several inches in a matter of days.[1]

In Jenolan Caves in Australia, a bottle was left in the cave in early 1950s. The bottle is coated in a layer of calcite and continues to get thicker. It is already a bottle stalagmite![2]

1. Marilyn Taylor, "Descent," *Arizona Highways*, January 1993, p. 11.
2. Editors, "Bottle Stalagmite," Answers in Genesis, March 1, 1995, originally published in *Creation*, vol. 17, no. 2, March 1995:6, https://answersingenesis.org/geology/caves/bottle-stalagmite/.

Cave geologist, Dr. Emil Silvestru comments:

> In the Cripple Creek Gold Mine in Colorado, stalagmites and stalactites over a meter long have grown in less than 100 years![3]

I have personally had the opportunity to see a stalagmite that had formed rapidly. I was touring Cosmic Caverns in northern Arkansas with my family, and draping over our footpath was a huge stalagmite. The guide told us to step around it and as I did, the guide made a comment about how that large stalagmite formed the previous year due to excessive rainstorms (in 2011). So I stopped and took a picture.

In a general sense, a cave can be active (called a *live* cave) or inactive (called a *relict* cave). Active means they have water action that transports minerals, and we see some growth of certain features in the cave (some have a little action, some have more). Inactive (relict) is a dry cave that has little to no growth as the water table is likely below the cave or is diverted from passing through the cave. It may be in an arid region with little to no rain to seep down through it. All *inactive* caves that have cave features were at one time *active* caves and have since "dried" out.

This particular cave in Arkansas obviously saw an increase in water to move mineral due to heavy rains that year that affected the cave and made it more active to grow that stalagmite. But with all this, it doesn't take millions of years to form stalactites and stalagmites.

3. Emil Silvestru, *The Cave Book* (Green Forest, AR: Master Books, 2008), p. 46.

18 | Doesn't It Take Millions of Years to Form Precious Gems?

Bodie Hodge

Diamonds

Gemstones are often touted as being *millions* of years old, and in some cases *billions* of years old — diamonds for example. But do they really take this long to form? *Physics.org* seems to think so. They write:

> In addition, the scientists show that diamonds take millions of years to grow. Moreover, diamonds are often half as old as the Earth.[1]

Of course, no scientist has ever *shown* this to be the case over the course of millions and billions of years! The article continues:

> The investigated diamonds are from Yakutia in Russia and show that in this region they formed in two important periods in the past: 1 billion years ago and 2 billion years ago. Many individual diamonds record growth in both periods proving for the first time that diamonds take millions of years to form.[2]

Do diamonds really take millions of years to form? Two companies, Apollo Diamond Incorporated in Massachusetts and Gemesis Corporation in Florida, are manufacturing diamonds![3] They use two different methods, no less.

1. "Diamonds Grow Like Trees, but Over Millions of Years," Physics.org, September 16, 2013, http://phys.org/news/2013-09-diamonds-trees-millions-years.html.
2. Ibid.
3. Greg Hunter and Andrew Paparella, "Lab-Made Diamonds Just Like Natural Ones," September 9, 2015, ABC News Internet Ventures, produced for *Good Morning America*, http://abcnews.go.com/GMA/story?id=124787.

The article reports:

> The new manmade diamonds from Apollo Diamond are created in a few days by a machine in an industrial park near Boston through a process called chemical vapor deposition.[4]
>
> At Gemesis, which is based in Sarasota, Fla., synthetic diamonds are created through a high-pressure, high-temperature technique that mimics the geologic conditions under which natural diamonds are formed. In a capsule placed under high temperature and pressure, graphite — a form of carbon — breaks down into atoms and travels through a metal solvent to bond to a tiny diamond seed, crystallizing layer by layer. Three or four days later, the stone that is formed is then removed from the chamber and cut and polished into a synthetic diamond.[5]

In both cases, the methods make diamonds *in days*. The article says of the manufactured diamonds:

> "They are real," said Linares. "They meet every measure of being diamond. From the aesthetic point of view, from the scientific point of view, they are diamonds and so therefore they have all the properties of diamond.[6]

But there is more . . . you can now take the remains of your cremated loved one and have their remaining carbon pressed into a diamond! Companies like *LifeGem* or *Algordanza* are currently doing this. The whole process only takes months. Writing about *Algordanza's* work, Rae Ellen Bichell comments:

> Swiss company Algordanza takes cremated human remains and — under high heat and pressure that mimic conditions deep within the Earth — compress them into diamonds. . . . Each year, the remains of between 800 and 900 people enter the facility. About three months later, they exit as diamonds, to be kept in a box or turned into jewelry.[7]

---

4. Ibid.
5. Ibid.
6. Ibid.
7. Rae Ellen Bichell, "From Ashes To Ashes To Diamonds: A Way To Treasure The Dead," NPR, January 19, 2014, http://www.npr.org/2014/01/19/263128098/swiss-company-compresses-cremation-ashes-into-diamonds.

There are also companies (like *DNA-2Diamonds* or *Pet-Gems*) that specialize in turning your beloved pet's ashes or hair into diamonds. The point is, it doesn't take millions or billions of years to form diamonds, but rather days and weeks.

Other Gemstones

Opals were once thought to take millions of years to form. In discussions with people over the years, I've heard them drop the date to as low as 30,000 years, but I've heard nothing lower than that. But the norm is to teach millions of years. One gemologist writes in an article called, "The Truth About Opals":

> Like most precious gemstones, opals take millions of years to form. . . . Over millions of years, these silica layers accumulate, forming beautiful polychromatic opal deposits for lucky miners to discover.[8]

Again, this is not the case. Researcher Len Cram has figured out how to make opal! He does it in a matter of weeks. Dr. Andrew Snelling writes:

> A committed Christian, Len [Cram] has discovered the secret that has enabled him to actually "grow" opals in glass jars stored in his wooden shed laboratory, and the process takes only a matter of weeks![9]

When it comes to gemstones (like rubies, topaz, garnet, cubic zirconia, emeralds, etc.), most can easily be made in a laboratory. It is such a thriving business now that much jewelry actually utilizes these newly made gems. They are called "synthetic" gemstones and can be found at jewelry shops across many parts of the world.

---

8. K. Jetter, GIA graduate gemologist and accredited jewelry designer, "The Truth About Opals," originally published in *Elite Traveler*, Santa Fe, New Mexico, http://www.katherinejetter.com/history-of-opals.
9. Dr. Andrew Snelling, "Creating Opals," *Creation ex nihilo* 17, no. 1, Dec. 1994: 14–17.

In fact, this process is nothing new. *Flame fusion* is a process used in the late 1800s and early 1900s to grow rubies. These types of rubies (called "Verneuil Rubies") are named for one of the scientists who improved the method.[10] Germany was growing emeralds in the 1930s by a *flux-grown* method.[11]

Gemstones are often produced naturally and quickly too when a volcano erupts. This is due to the rapid heat and pressure. At Mount St. Helens, which blew its top in 1980 and went off again in 1982, a magnificent array of gemstones was produced! The Mount St. Helens Gift shop website openly states:

> Volcanoes are an incubator for many of the World's treasures. Other gems commonly associated with volcanic origins include Emerald, Diamond, Garnet, Peridot, and Topaz.[12]

The *GemSelect* website also confirms the origin of most gemstones in *igneous* rock layers (think volcanic). Their site says:

> The long list of gemstones formed from igneous rock include the chrysoberyl group, all quartz (including amethyst, citrine and ametrine), beryl (emerald, morganite and aquamarine), garnet, moonstone, apatite, diamond, spinel, tanzanite, tourmaline, topaz and zircon.[13]

Dr. Andrew Snelling researched the idea of rapid diamond production in the earth in which diamonds were then transported to the surface with magmas to form the famous Argyle diamond deposit in Australia. Interestingly, the Aborigines witnessed this volcanic event in years past (showing it is recent) and this account passed down through the tribe. Dr. Snelling concludes:

> The diamond crystals themselves are thus carried rapidly from their place of formation deep in the earth into these pipes,

10. Editors, "Synthetic and Artificial Gemstone Growth Methods: In-Depth Treatment Information," Jewelry Television online, July 2012, http://www.jtv.com/library/synthetic-artificial-gemstone-methods.html.
11. Ibid.
12. Editors, Mount St. Helens Gift Shop Website, http://www.mt-st-helens.com/obsidianite.html, downloaded April 7, 2014.
13. Editors, "How Gemstones Are Formed," GemSelect, accessed August 18, 2015, http://www.gemselect.com/other-info/gemstone-formation.php.

> where we find them today along with the shattered remains of the magmas that brought them up to the earth's surface.
>
> This evidence for the rapid formation of diamond deposits confirms that there are extremely rapid and catastrophic geological processes which evolutionary geologists have been forced to concede do occur. Furthermore, the eyewitness testimony from the Australian Aborigines, distorted by verbal transmission and the 'mists of time', undoubtedly points to their having seen the explosive eruption that produced the Argyle diamond deposit, which places its formation therefore in the very recent post-Flood period.[14]

So it doesn't take millions of years to form gemstones including diamonds. Even natural diamonds can be formed quickly by the heat and pressures in the earth and quickly formed and deposited by volcanic eruptions.

14. Andrew Snelling, "Diamonds — Evidence of Explosive Geological Processes," *Creation*, vol. 16, no. 1, December 1993: 42–45, https://answersingenesis.org/geology/rocks-and-minerals/diamonds-evidence-of-explosive-geological-processes/.

19 Isn't Petrified Wood Millions of Years Old?

Bodie Hodge

Introduction

The Live Science website proclaims:

> Petrified wood forms when fallen trees get washed down a river and buried under layers of mud, ash from volcanoes and other materials. Sealed beneath this muck deprives the rotting wood from oxygen — the necessary ingredient for decay. As the wood's organic tissues slowly break down, the resulting voids in the tree are filled with minerals such as silica — the stuff of rocks. Over millions of years, these minerals crystallize within the wood's cellular structure forming the stone-like material known as petrified wood.[1]

Does it really take millions of years to form petrified wood? Once again, keep in mind this has never been observed or repeated over the course of millions of years, so it is really just a religious belief about the past. What do we really observe?

Rapid Petrified Wood

Researchers have proven it doesn't take millions of years to form petrified wood. In 1986, Hamilton Hicks received his patent for developing a process

1. Michelle Bryner, "How Long Does It Take to Make Petrified Wood?" Live Science website, November 20, 2012, http://www.livescience.com/32316-how-long-does-it-take-to-make-petrified-wood.html.

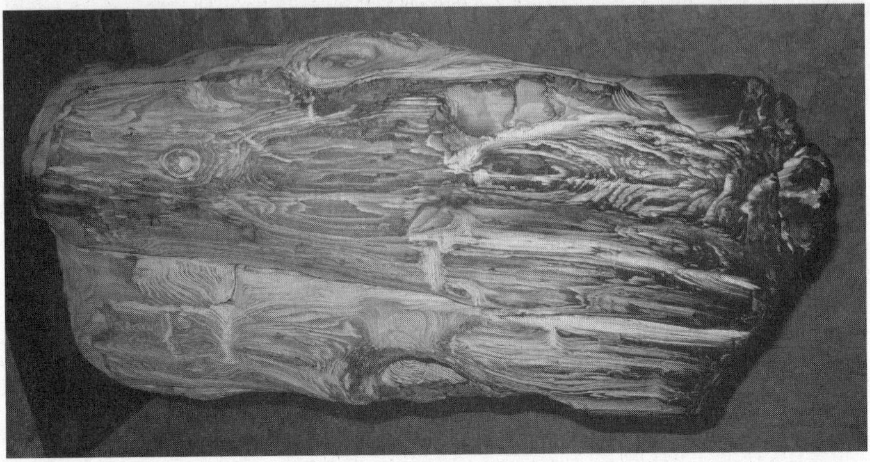

to petrify wood quickly with a mineral and acid-rich solution.[2] The solution is applied to wood and the patent states:

> When applied to wood or wood cellulose products, the observable action is hardening, density increase and apparent petrification similar to that occurring in naturally petrified wood.[3]

This solution is similar to what is found naturally and Hicks' patent points out:

> It is possible to use natural or volcanic mineral water into which the commercial sodium silicate solution (water glass) is dissolved, or to artificially mineralize water, by mixing it with mineral clay or gypsum, for example.[4]

In other words, this type of solution could be produced naturally, such as utilizing volcanic mineral water. It is interesting that most natural petrified wood is found associated with volcanic rock layers (i.e., which was occurring during and after the Flood). These are ideal conditions to produce petrified wood quickly.

This process is now used to manufacture petrified wood, and even you can be a recipient of it. You can go to your local home store or flooring store and order some petrified wood flooring too if you want to renovate your floors!

2. Hamilton Hicks, Sodium silicate composition, United States Patent Number 4,612,050, September 16,1986, http://www.google.com/patents/US4612050.
3. Ibid.
4. Ibid.

In the article "Petrified Wood in Days," we read about petrified wood experiments by Yongsoon Shin, at Pacific Northwest National Laboratory (PNNL). The researchers state:

> Back at PNNL, they gave a 1 centimeter cube of wood a two-day acid bath, soaked it in a silica solution for two more (for best results, repeat this step up to three times), air-dried it, popped it into an argon-filled furnace gradually cranked up to 1,400 degrees centigrade to cook for two hours, then let cool in argon to room temperature. Presto. Instant petrified wood, the silica taking up permanent residence with the carbon left in the cellulose to form a new silicon carbide, or SiC, ceramic.[5]

Yes this is hot — 1,400 degrees! I've had people point out that temperatures that hot are not likely in the natural world. So they assumed that making petrified wood via this route would be impossible. But this is not necessarily the case. Recall, most petrified wood is found associated with volcanic rock layers where it does get around these temperatures and even hotter! Certain magma has been calculated to be around 1,560 degrees centigrade (e.g., komatiite melts) at eruption![6]

There are also examples of petrified wood occurring as a mere product of nature — even without the heat! As an example, Dr. Andrew Snelling recounts:

> From the other side of the world comes a report of the chapel of Santa Maria of Health (Santa Maria de Salute), built in 1630 in Venice, Italy, to celebrate the end of The Plague. Because Venice is built on water saturated clay and sand, the chapel was constructed on 180,000 wooden pilings to reinforce the foundations. Even though the chapel is a massive stone block structure, it has remained firm since its construction. How have the wooden pilings lasted over 360 years? They have petrified! The chapel now rests on "stone" pilings![7]

5. Editors, "Instant Petrified Wood," Physics.org, January 25, 2005, http://phys.org/news/2005-01-petrified-wood-days.html.
6. For example, see E.G. Nisbet, M.J. Cheadle, N.T. Arndt, and M.J. Bickle, "Constraining the Potential Temperature of the Archaean Mantle: A Review of the Evidence from Komatiites," *Elsevier* journal, vol. 30, Issues 3–4, September 1993, p. 291–307, http://www.sciencedirect.com/science/article/pii/002449379390042B.
7. Andrew Snelling, " 'Instant' Petrified Wood," *Creation*, vol. 17, no. 4, September 1995, p. 38-40, online September 1, 1995, https://answersingenesis.org/fossils/how-are-fossils-formed/instant-petrified-wood/.

These few examples discussed[8] prove it doesn't take the *alleged* millions of years to form petrified wood. Volcanic activity from the Flood and its aftermath (especially the mountain-building phase of the Flood) as well as naturally occurring mineral waters in certain places can cause the petrified wood we find without the appeal to *millions of years.*

8. For more examples, please see the previous reference by Dr. Andrew Snelling and his excellent research into this subject.

20 Doesn't It Take Millions of Years to Form Fossils?

Ken Ham and Bodie Hodge

Much of the world has been inundated to believe that fossils are preserved remains of creatures that often lived *millions or billions of years* ago. While fossils *are* preserved remains or impressions, they *are not* millions of years old. Millions of years are not required for fossil formation.

Nevertheless, we are often led to believe that long ages are a prerequisite for fossil formation. The *Oxford University Museum of Natural History* (OUMNH) writes:

> An animal dies and its body sinks to the sea floor. . . . The skeleton continues to be buried as sediment is added to the surface of the sea floor. As the sea floor sinks, pressure increases in the lower layers of sediment and it turns it into hard rock.[1]

Just so we are clear, we don't observe fossils forming at the bottom of lakes and ocean floors. But as the evolutionary story unfolds, this bone is finally buried and the bone is left in a mold (British spelling is "mould") within the new rock layers. OUMNH continues:

> Water rich in minerals enters the mould, and fills the cavity. The minerals deposited in the mould form a cast of the mould. . . . Millions of years later, the rock surrounding the skeleton rises to the Earth's surface (this happens during mountain building, earthquakes and other earth processes). The rock is worn away

1. Oxford University Museum of Natural History, The Learning Zone, "How Do Fossils Form?" 2006, http://www.oum.ox.ac.uk/thezone/fossils/intro/form.htm.

> by wind and rain, and the fossil is now exposed, waiting to be found![2]

You see how "millions of years" are thrust into the discussion of fossil formation. We see this elsewhere too. For example, in *A Guide to Dinosaurs* (which is geared toward the youth), they use a dinosaur instead of a fish as their example of the formation of fossils. They write:

> Below the surface of a lake a dead dinosaur's flesh rots away or is eaten by aquatic creatures. Layers of silt build up over the dinosaur's bones and prevent them from being washed away. Weighted down by sediment, the dinosaur bones are slowly replaced by minerals. Millions of years later, seismic disturbances bring the fossilized bones to the surface.[3]

They just had to insert "millions of years." Once again, though, we do not see fossils forming slowly at the bottom of lakebeds.

HOW TO MAKE A FOSSIL

Fossils do not require long ages to form. In fact, they *must* form quickly, otherwise the organism's softer tissues and even bones suffer decay (shells or teeth enamel naturally take longer to disintegrate). The photo is of a bone left to the elements in just a short time (note the deterioration).

Fossils can be made in labs and it is done quickly. Wood fossils can be made *in days* in the lab.[4] Turning hard material (e.g., bones) into fossils is easy in a lab setting, but in 1993, scientists were even able to make fossils from *soft* animal tissues! *New York Times'* Science Watch reports:

2. Ibid.
3. Christopher Brochu et al, *A Guide to Dinosaurs* (San Francisco, CA: Fog City Press, 1997–2004), p. 19.
4. Geoff Brumflel, "Furnace Creates Instant Fossils," Nature.com, January 28, 2005, http://www.nature.com/news/2005/050128/full/news050124-14.html.

> Scientists have for the first time produced fossils of soft animal tissues in a laboratory. In the process they discovered that most of the phosphate required for the fossilization of small animal carcasses comes from within the animal itself.[5]

Notice that the creature already provided the necessary chemicals to fossilize itself. This makes sense since many of the fossils we find were buried quickly and left to themselves. As expected, the lab fossils are actually better specimen than natural ones and only takes weeks to months to form.

Fossil formation is actually a rapid event and many today, even secularists, are finally conceding this. It requires rapid burial to seal out the oxygen (so it doesn't entirely decay away). Then mineral-rich water takes out the organic material and replaces it with minerals like limestone, etc.

It is a rapid event as witnessed by the fact that we even find fossils that preserved soft tissue — like the case of massive numbers of jellyfish being fossilized. Jellyfish decay quickly, and yet their fragile tissue was rapidly fossilized before disintegration. Dr. Gary Parker writes:

> Jellyfish often wash ashore, but in a matter of hours they have turned into nondescript "blobs" (although watch out — the stinging cells continue to work for quite a while!). To preserve the markings and detail of the Ediacara jellyfish, the organisms seem to have landed on a wet sand that acted as a natural cement. The sand turned to sandstone before the jellyfish had time to rot, preserving the jellyfish's markings, somewhat as you can preserve your handprint if you push it into cement during that brief time when it's neither too wet nor too dry. Indeed, the evolutionist who discovered the Ediacara jellyfish said the fossils must have formed in less than 24 hours. He didn't mean one jellyfish in 24 hours; he meant millions of jellyfish and other forms had fossilized throughout the entire Ediacara formation, which stretches about 300 miles or 500 km from South Australia into the Northern Territory, in less than 24 hours! In short, floods form fossils fast![6]

---

5. "Man-Made Fossils," *New York Times*, Science Watch, March 9, 1993, http://www.nytimes.com/1993/03/09/science/science-watch-man-made-fossils.html.
6. Gary Parker, *Creation Facts of Life*, Chapter 3: "How Fast," January 1, 1994, https://answersingenesis.org/fossils/how-are-fossils-formed/how-fast/.

Another reason we know that fossils can be made quickly is . . . the rapid number of fakes hitting the market! Let's not forget the famous faked fossil Archaeorapter that was found to be a fraud and published in the *National Geographic* in the year 2000. The problem of faked fossils is all too common and now researchers are turning to CT Scans, x-rays, and other techniques to spot faked and manipulated fossils. The *Scientific American* states:

> Another much more serious problem, however, is posed by forged, faked and manipulated specimens — such as *National Geographic*'s Archaeoraptor — which are becoming increasingly common.[7]
>
> The problem of faked fossils in China is serious and growing.[8]
>
> An investigative report published in *Science* in 2010 revealed that as many as 80 percent of marine reptile fossils on display in Chinese museums had been altered or manipulated.[9]

People are getting better at making fake fossils, which is why technical papers are now being written on how to spot the fakes! But being frank, if a poor person who has little concern for God and His Law can fake or manipulate a fossil to sell and make incredible money — they are going to try it.

Conditions During Flood for Fossilization

The conditions during the Flood were ideal for fossil formation. Even though many things surely rotted and decayed, many other specimens were rapidly buried and fossilized. Genesis 6:13 points out:

> And God said to Noah, "The end of all flesh has come before Me, for the earth is filled with violence through them; and behold, I will destroy them with the earth.

The Flood was not merely a mass of water; but a collection of mud/sediment (earth) that was utilized to destroy the pre-Flood world for their sin. So we

7. John Pickert, How Fake Fossils Pervert Paleontology [excerpt], *Scientific American*, November 15, 2014, http://www.scientificamerican.com/article/how-fake-fossils-pervert-paleontology-excerpt/.
8. Ibid.
9. Ibid.

expect fossils and we even expect a general trend of order. Some of these factors include elevation, sorting power of water, and buoyancy.

Obviously, things living at a lower level have a better chance of being buried and fossilized, hence why about 95 percent of fossil layers consist of marine organisms.[10] There is also the natural sorting power of water that separates creatures' burial in the Flood.

Another factor is that reptiles and amphibians tend to sink, so they are more likely to be fossilized than other creatures like mammals. It also makes sense why mammals and birds are less likely to be fossilized since they are lighter (more buoyant, less dense) and many tend to float. Thus, they would be better candidates for rotting and decaying. Of course, there are exceptions to this but we would expect a general trend in the fossil layers.[11]

Where Did the Organic Material Go That Was Originally Part of the Creatures That Were Fossilized?

Most oil deposits that we have are a result of the Flood. There may be other factors (e.g., oil production from bacteria), but most of it came from the Flood and the conditions thereof. Think about the fossils of marine organisms, plants and trees, algae, land creatures, etc. When they fossilize, their organic material is removed by water and replaced by minerals (e.g., limestone) to turn it into rock. Where does all that organic material go?

It seeps down with the water into pockets in the earth. Then it separates from the water into pools or deposits. What remains is primarily a mixture of hydrocarbons, gases, and water. We call this " crude oil."

Rapid Fossils

There are hosts of examples of rapid fossils. Let's entertain a few shall we?

There is a fossil at the Creation Museum of a horseshoe crab that was walking (footprints fossilized too) and then it was stopped and fossilized dead in its tracks.

---

10. Andrew Snelling, "Where Are All the Human Fossils?" *Creation* 14(1):28–33, December 1991; John Morris, *The Young Earth* (Green Forest, AR: Master Books, 2002), p. 71.
11. For more on the order of fossils see Andrew Snelling, "Doesn't the Order of Fossils in the Rock Record Favor Long Ages?" in Ken Ham, gen. ed., *The New Answers Book 2* (Green Forest, AR: Master Books, 2008), p. 341–354, https://answersingenesis.org/fossils/fossil-record/doesnt-order-of-fossils-in-rock-favor-long-ages/.

Horseshoe crab fossil

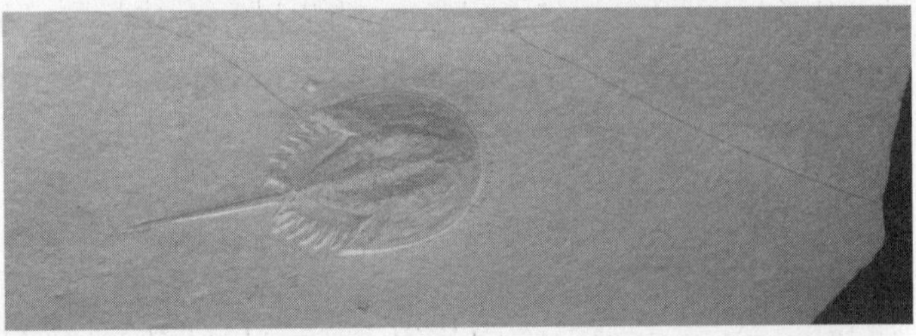

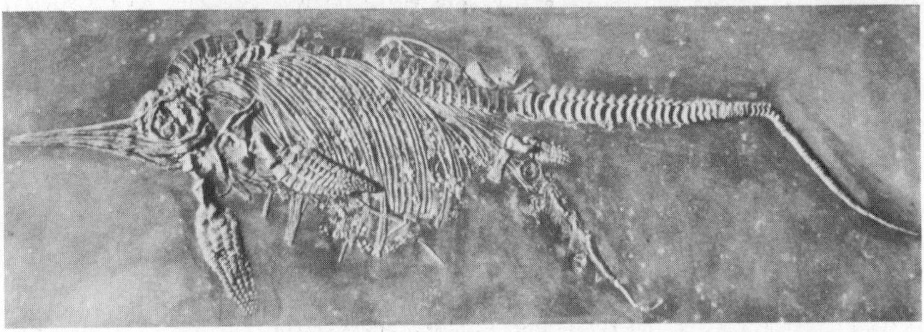

A marine reptile called an ichthyosaur was buried and fossilized so fast that it didn't finish the birthing process.

This fish didn't get an opportunity to finish its dinner (on display at the Creation Museum).

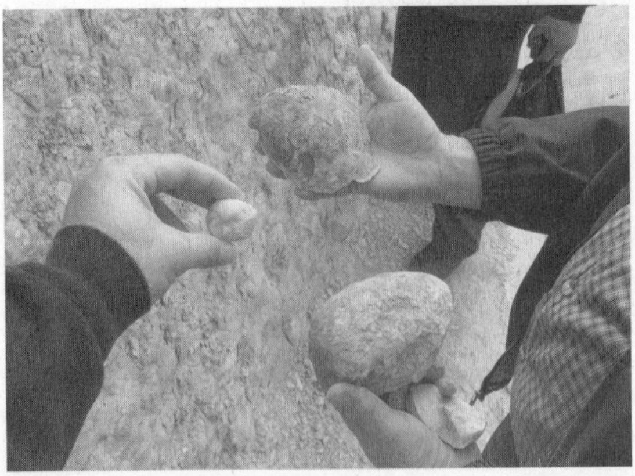

Dozens of fragile closed shells that were buried rapidly and fossilized before they could open

The sample of dozens of fragile closed shells that were buried rapidly and fossilized before they could open, which is what shellfish typically do, are in the Andes Mountains of Peru above Cusco, which is over two miles high. A local guide took us to the site. We left the fossils in the care of a local church in Cusco.

A group of turtles were buried and fossilized so fast that all nine pairs were still stuck in the process of mating.[12] Being critical of the report's slow gradual explanation of fossil formation, these turtle pairs are better explained by catastrophe and rapid events. This happens often, showing the speed at which burial and fossilization took place.

Picture of mating turtles, taken at Genesis Expo and Fossil Shop in England

These few examples should whet your appetite for the immense numbers of fossils that display rapid burial and fossilization.

Thorn Fossils . . . and Their Significance

One type of fossil needs to be mentioned that has incredible theological significance. It is thorn fossils. We find a number of these in the fossil record like *Sawdonia* — which has nasty, vicious thorns. Others like *Psilophyton crenulatum* are found in rock layers that the secular world considers to be 350–400 million years old (Devonian rock layer).[13]

Reading the Bible, we find that thorns did not exist until the curse after Adam and Eve sinned, which was a matter of a few thousand of years ago (Genesis 3:18[14]). So how can these thorns be millions of years old (which many Christians sadly teach)? They cannot! As an example, Devonian rock is not millions of years old, but rock formed *during* the Flood of Noah's day a few thousand years ago. Having thorns buried in Flood sediment, which occurred after Adam sinned, makes perfect sense.

12. Brian Switek, "Sex Locked in Stone," *Nature*, 20, June 2012, http://www.nature.com/news/sex-locked-in-stone-1.10850.
13. Wilson N. Stewart and Gar W. Rothwell, *Paleobotany and the Evolution of Plants* (Cambridge, UK: Cambridge University Press, 1993), p. 172–176.
14. Both thorns and thistles it shall bring forth for you, and you shall eat the herb of the field.

Let's evaluate this from a big picture theological angle. First, Genesis 1:29–30[15] teaches that man and animals were originally vegetarian (before Adam's sin). How do we know this for sure? Humans weren't told they could eat meat until after the Flood in Genesis 9:3.[16] This later verse makes it clear that mankind was originally vegetarian, but this changed after the Flood. Verse 30 of Genesis 1 (about animals' diet) is worded in the same basic way as verse 29 (man's diet), so it confirms that originally the animals were vegetarian too.

Second, at the end of the creation week, God described everything He had made as "very good" (Genesis 1:31[17]). Third, Genesis 3 makes it clear that the animals (v. 14[18]) and the ground (v. 17[19]) were cursed. And verse 18 makes it clear that thorns came into existence after sin and the Curse: "Both thorns and thistles it [the ground] shall bring forth for you."

Recall that the idea that things have been around for millions of years came from the belief that the fossil record was laid down slowly over millions of years, long before man's existence. So when Christians accept millions of years, they must also accept that the fossil layers were laid down before Adam — before the first human sin.

Yet the fossil record contains fossil thorns — claimed by evolutionists to be hundreds of millions of years old. How could that be if thorns came *after* Adam's sin?

Coupled with this, the fossil record also contains lots of examples of animals that ate other animals — bones in their stomachs, teeth marks on bones, and so on. But according to the Bible, animals were vegetarian before sin. Furthermore, the fossil record contains well-documented examples of diseases, such as brain tumors, cancer, and arthritis. But if these existed before man, then God called such diseases "very good."

15. And God said, "See, I have given you every herb that yields seed which is on the face of all the earth, and every tree whose fruit yields seed; to you it shall be for food. Also, to every beast of the earth, to every bird of the air, and to everything that creeps on the earth, in which there is life, I have given every green herb for food"; and it was so.
16. Every moving thing that lives shall be food for you. I have given you all things, even as the green herbs.
17. Then God saw everything that He had made, and indeed it was very good. So the evening and the morning were the sixth day.
18. So the LORD God said to the serpent: "Because you have done this, you are cursed more than all cattle, and more than every beast of the field; on your belly you shall go, and you shall eat dust all the days of your life."
19. Then to Adam He said, "Because you have heeded the voice of your wife, and have eaten from the tree of which I commanded you, saying, 'You shall not eat of it': Cursed is the ground for your sake; in toil you shall eat of it all the days of your life."

Taking all this into consideration, it seems obvious that bloodshed, death of animals and man, disease, suffering, and thorns came *after* sin. So the fossil record had to be laid down after sin, too. Noah's Flood would easily account for the majority of the fossils.

But what does this have to do with a gospel issue? The Bible calls death an "enemy" (1 Corinthians 15:26[20]). When God clothed Adam and Eve with coats of skins (Genesis 3:21[21]), this was the first death witnessed in Scripture — the death and bloodshed of an animal (a direct result of human sin!).

Elsewhere in Scripture we learn that without the shedding of blood there is no remission of sins (Hebrews 9:22[22]), and the life of the flesh is in the blood (Leviticus 17:11[23]). Because Adam sinned, a payment for sin was needed. Because sin's penalty was death, then death and bloodshed were needed to atone for sin. So Genesis 3:21 would describe the first blood sacrifice as a penalty for sin — looking forward to the One who would die "once for all" (Hebrews 10:10–14[24]).

An Indirect Salvation Issue

Many Christians believe in millions of years and are truly born again. Their belief in millions of years doesn't affect their salvation. But what does it do? It affects how *other* people, such as their children or others they teach, view Scripture.

Their example can be a stumbling block to others. For instance, telling young people they can reinterpret Genesis to fit in millions of years sets a deadly example: they can start outside Scripture and add ideas into Scripture.

We suggest that such people can, over time, get the idea that the Bible is not God's infallible Word. This creates doubt in God's Word — and doubt often leads to unbelief. Eventually they can reject Scripture altogether. Since

20. The last enemy that will be destroyed is death.
21. Also for Adam and his wife the Lord God made tunics of skin, and clothed them.
22. And according to the law almost all things are purified with blood, and without shedding of blood there is no remission.
23. For the life of the flesh is in the blood, and I have given it to you upon the altar to make atonement for your souls; for it is the blood that makes atonement for the soul.
24. By that will we have been sanctified through the offering of the body of Jesus Christ once for all. And every priest stands ministering daily and offering repeatedly the same sacrifices, which can never take away sins. But this Man, after He had offered one sacrifice for sins forever, sat down at the right hand of God, from that time waiting till His enemies are made His footstool. For by one offering He has perfected forever those who are being sanctified.

the gospel comes from a book they don't trust or believe is true, they can easily reject the gospel itself.

So the age of the earth and universe is not a salvation issue *per se* — somebody can be saved even without believing what the Bible says on this issue. But it is a salvation issue indirectly. Christians who compromise on millions of years can encourage *others* toward unbelief concerning God's Word and the gospel.

The Israelites sacrificed animals over and over again, as a ceremonial covering for sin. But Hebrews 10:4[25] tells us that the blood of bulls and goats can't take away our sin — we are not physically related to animals. We needed a perfect human sacrifice. So all this animal sacrifice was looking forward to the One called the Messiah (Jesus Christ).

Now if there was death and bloodshed of animals before sin, then this undermines the atonement. Also, if there were death, disease, bloodshed, and suffering before sin, then such would be God's fault — not our fault! Why would God require death as a sacrifice for sin if He were the one responsible for death and bloodshed, having created the world with these bad things in place?

One of today's most-asked questions is how Christians can believe in a loving God with so much death and suffering in the world. The correct answer is that God's just Curse because of Adam's sin resulted in this death and suffering. *We* are to blame. God is not an unloving or incompetent Creator of a "very bad" world. He had a loving plan from eternity to rescue people from sin and its consequence of eternal separation from God in hell.

So to believe in millions of years is a gospel issue. This belief ultimately impugns the character of the Creator and Savior and undermines the foundation of the soul-saving gospel.

25. For it is not possible that the blood of bulls and goats could take away sins.

21 Does It Take Long Ages to Grow Coral, Lay Down Ice Layers (Cores), and Grow Tree Rings?

Bodie Hodge

Coral

The National Oceanic and Atmospheric Administration states:

> How old are today's reefs? The geological record indicates that ancestors of modern coral reef ecosystems were formed at least 240 million years ago. The coral reefs existing today began growing as early as 50 million years ago. Most established coral reefs are between 5,000 and 10,000 years old.[1]

Bill Nye, in the famous debate with Ken Ham in February of 2013, also concurred when discussing zooxanthellae coral. Nye says:

> And when you look at it closely you can see that they live their entire lives, they lived typically 20 years, sometimes more than that if the water conditions are correct. And so we are standing on millions of layers of ancient life.[2]

To the untrained eye, one might think Bill Nye just presented a devastating argument. But this is a fallacious means of arguing. Nye is assuming naturalism to try to prove naturalism. He has assumed there were no catastrophes in the past and that rates and processes have always been identical (this is called "uniformitarianism" and is based on naturalism) to make the claims

1. "From Polyp to Colony: Coral Reefs," NOAA Coral Reef Conservation Program, Nation Ocean Service, September 3, 2005, http://coralreef.noaa.gov/aboutcorals/coral101/polypcolony/.
2. http://www.youngearth.org/index.php/archives/rmcf-articles/item/21-transcript-of-ken-ham-vs-bill-nye-debate.

at hand. Then he proceeds to use this to definitively state the naturalistic position of millions of years. This fallacy is called *affirming the consequent* and is arguably the most common fallacy that evolutionists commit. Nye has proven nothing, but has merely assumed what he is trying to prove and thus is being arbitrary and self-refuting.

But does it take long ages for coral growth and accumulation? Dr. John Whitmore, a geologist and professor at Cedarville University, writes (with his references included):

> Corals which build coral reefs have been reported to grow as much as 99 to 432 mm per year.[3] Large coral accumulations have been found on sunken World War II ships only after several decades.[4] *Acropora* colonies have reached 60–80 cm in diameter in just 4.5 years in some experimental rehabilitation studies.[5] At the highest known growth rates, the Eniwetok Atoll (the thickest known reef at 1400 m) would have taken about 3,240 years to rise from the ocean floor.[6]

Dr. Whitmore goes on to point out that it doesn't require long ages to form coral reefs but thousands of years given the right conditions and starting point. This makes sense in light of a global Flood that "reset" things in the oceans as a new starting point to grow coral reefs on all sorts of underwater features such as volcanic platforms, chalk beds, and so forth.

Ice cores

Speaking of Bill Nye, he also used ice cores in an effort to discredit the idea of a global Flood. Ice cores are cores of ice that have been drilled out of ice sheets such as those on Greenland or Antarctica. These compacted ice layers came from previous storms of ice and snow to build them up.

Nye points out there are some cores with upwards of 680,000 layers that are to be assumed in them — and he purports that it took long ages to form them. Now we agree there are certainly a lot of these ice layers. Of course,

3. A.A. Roth, *Origins* (Hagerstown, MD: Review and Herald Publishing Association, 1998), p. 237.
4. S.A. Earle, "Life springs from death in Truk Lagoon," *National Geographic* 149(5):578–603, 1976.
5. H.E. Fox, "Rapid Coral Growth on Reef Rehabilitation Treatments in Komodo National Park, Indonesia," *Coral Reefs* 24:263, 2005.
6. John Whitmore, "Aren't Millions of Years Required for Geological Processes?" in *The New Answers Book 2*, Ken Ham, gen. ed. (Green Forest, AR: Master Books, 2008), p. 240–241.

the layers toward the bottom — due to compression — have molecular diffusion (where multiple layers fuse together). Thus, they have to be *estimated* as to how many layers might have been there (it is true that there is long age assumption here too, of which one needs to be careful). But generally, we agree that there are a lot of ice layers. But remember — the ice layers don't come with labels on them telling us their age!

Nye makes an interpretation of these observable *and* estimated ice layers. He says:

> And we find certain of these cylinders to have 680,000 layers. 680,000 snow winter-summer cycles.[7]

Nye assumed these layers were divided by winter-summer cycles. Apparently, there miraculously becomes an observable division between ice layers, where all the ice and snow from storms in one year get compacted into one single layer and all the ice and snow from storms the next year get compacted into the next layer, and so on!

I have personally seen multiple ice layers in *one winter* in Kentucky! It was simply several ice storms and snow storms that piled one on top of the other. Sometimes it was merely due to various phases within the same storm (such as ice-snow-ice in the same storm). The lower layers sometimes became more compact with the layers on top of it — and yet this was merely *one winter* in Kentucky.

Many people around the world could concur that multiple storms will produce multiple layers of snow and ice and can and have observed this every winter. In places like Kentucky, these ice layers melt off each year, but in places where ice sheets are growing (some places in Antarctica and Greenland, for example), they do not melt off but continue to accumulate.

The point is that multiple storms can be observed to produce multiple ice layers. So why assume each of these layers are winter-summer cycles? That would go against scientific observation to say the least. Such abundant ice layers are not dependent upon a winter-summer cycle. Bill Nye has made a fallible assumption — one that does not fit with what we observe.

But let's discuss this further. Dr. Larry Vardiman (PhD in atmospheric science from Colorado State University), while working with the Institute for Creation Research, and Mike Oard, an expert meteorologist (M.Sc.

---

7. http://www.youngearth.org/index.php/archives/rmcf-articles/item/21-transcript-of-ken-ham-vs-bill-nye-debate.

in Atmospheric science from the University of Washington) and board member of the Creation Research Society and former National Weather Service meteorologist, are both weather scientists who have dealt with this issue extensively. Neither would yield to the ice cores as being evidence for long ages or winter-summer cycles.

Dr. Vardiman has pointed out that WWII planes that were buried just 50 years before in Greenland, were already under 250 feet of ice when found![8] This is not 250 ice layers, but 250 *feet* of minuscule ice layers! Dr. Vardiman points out that if uniformitarian rates were to be assumed from this data alone, it would take less than 1,000 years to form the entire Greenland ice sheet! Dr. Vardiman did further calculations and found that a few thousands years were all that would be required for the entire sheet.[9]

Mr. Oard has also addressed the ice cores in great detail from a technical perspective, addressing oxygen content which is to be expected since climates and atmospheric changes do occur and have occurred since the time of the Flood.[10] Such things are actually perfectly expected in a biblical creationist's framework.

Tree Rings: Bristlecone Pines

Bill Nye also brought up bristlecone pines in the debate and arbitrarily asserts that some of them are over 6,000 years old in an effort to discredit the age of the earth based on God's Word. And to back this up, Nye made another arbitrary assertion that a tree (Old Tjikko) is specifically 9,550 years old. That tree is a Norway spruce, not a bristlecone pine, and its age is based on carbon dating, not tree rings.

Please note that when skeptics, like Nye or others, make a claim like this without proving it (i.e., "just take my word for it"), it is arbitrary. An arbitrary claim is simply that, *arbitrary*, and carries no weight in an argument or debate.

Let me explain where the so-called dates for these trees come from. They come from tree rings. In simple form, people add up tree rings (this involves

8. Editors, "Deep Layers," Answers in Genesis, October 26, 2002, https://answersingenesis.org/evidence-against-evolution/deep-layers/.
9. Larry Vardiman, Ph.D., "Ice Cores and the Age of the Earth," *Acts & Facts* 21 (4) 1992, http://www.icr.org/article/ice-cores-age-earth/.
10. Michael Oard, "Do Greenland Ice Cores Show over One Hundred Thousand Years of Annual Layers?" December 1, 2001, http://www.answersingenesis.org/articles/tj/v15/n3/greenland.

assumptions too), as there can be thousands of them. Then one has to try to cross match them with others to determine a supposed unbroken chronology. Of course, there is a degree of guesswork here.

So what is the big deal? Simple: there is the assumption that each tree ring is a yearly cycle and that it would be impossible to have more than one per year. It would be like every March 1st, all these trees decide to make a new ring.

But the answer is rather easy: the tree rings are actually from *growth* cycles, not necessarily from *yearly* cycles. Many types of trees are observed to have multiple growth cycles even in one year with favorable conditions. So, depending on previous wet-dry cycles (or otherwise good or bad growth periods within a year) there could be multiple growth cycles, thus multiple rings.[11]

Bristlecone pines are no exception. The ones that tend to have more rings and live longer are those in arid, higher altitudes where little rain occurs each year. Let me explain; in the dry arid areas, growth stops until the next rain or watering, then the tree can again begin to grow. But when it becomes dry and arid again, it ceases growing once more.

So growth is dependent upon getting water, not a calendar date (yearly). Having great numbers of rings simply means greater numbers of growth cycles in the past. To say that each ring is a yearly ring would require skeptics to prove that each year in the past, there was only one rainfall per year — a truly bold and unprovable assumption.

The point to take here is that when someone says the Bristlecone Pines are of a certain age, it is *not* due to direct observation, but by interpretation based on un-provable assumptions. Mark Matthews, writing about Bristlecone Pines, says:

> Perhaps the best evidence that some BCPs can grow multiple rings per year is the fact that it has already been demonstrated. Lammerts, a creationist, induced multiple ring growth in sapling BCPs by simply simulating a two-week drought.[12] Some dismiss this evidence, saying that while multiplicity has been demonstrated in young BCPs, it hasn't been demonstrated in mature BCPs and

11. Mark Matthews, "Evidence for Multiple Ring Growth Per Year in Bristlecone Pines," *Journal of Creation* 20(3):95–103, December 2006.
12. W.E. Lammerts, "Are the Bristle-cone Pine Trees Really So Old?" *Creation Research Society Quarterly* 20(2):108–115, 1983.

> therefore may not occur in mature BCPs.[13] While this hypothesis could be true, surely the burden of proof should be on those who propose that what happens in immature trees doesn't happen in mature trees.
>
> An expert in the genus *Pinus* didn't seem to have any problem believing that White Mountain BCPs grew multiple rings per year. In his book, *The Genus Pinus*, Nicholas Mirov states, "Apparently a semblance of annual rings is formed after every rather infrequent cloudburst."[14] If an expert like Mirov readily accepted multiplicity in these BCPs, then perhaps the doubters of this notion should at least give the evidence a serious examination.[15]

The Bristlecone Pines, with the dry climate as it is today, doesn't afford multiple tree rings readily, but to assume the climate has always been identical to today is without warrant, even by the long ager's standard.

In short, trees with high numbers of tree rings are not a problem with an age of the earth about 6,000 years and Flood about 4,350 years ago. It is merely arbitrary to say they took long ages.

---

13. V.C. LaMarche Jr. and T.P. Harlan, "Accuracy of Tree Ring Dating of Bristlecone Pine for Calibration of the Radiocarbon Time Scale," *Journal of Geophysical Research* 78(36):8849–8858, 1973.
14. Ibid.
15. Matthews, "Evidence for Multiple Ring Growth Per Year in Bristlecone Pines."

22 Was There an Ice Age That Followed the Flood?

Ken Ham and Bodie Hodge

Introduction

Creationists and evolutionists essentially agree there was an ice age. Creationists argue there was one major Ice Age that followed the Flood of Noah. In the secular world, they believe in a multitude of ice ages going back for what seems like an eternity. As a point of clarification, when creationists typically discuss the post-Flood Ice Age, it is denoted in caps, whereas the supposed secular ice ages are not capped to distinguish which is being discussed.

Creationists hold to an Ice Age that was triggered by the Flood. The Flood occurred about 2348 B.C., which is about 4,300 years ago.[1] The secularists' most recent ice age was supposedly about 10,000 years ago (by their dating system).

How Does an Ice Age Occur?

An ice age does not occur by simply making the earth cold. If the earth became cold, you would have a cold earth, not an ice age.

Instead, an ice age occurs when you have warm oceans to get extra evaporation and thus, extra-accumulated snowfall in winter, *and* cool summers so that the accumulated snow and ice does not get a chance to melt off. Then the following winter, additional accumulation piles on and it builds up into an ice age. Warm oceans and cool summers are the primary reason for an ice age, even though other factors are involved.

1. According to Ussher's date.

How Did the Flood Trigger the Ice Age?

Warm Oceans

The Flood would generate immense amounts of heat as evidenced from its onset with the springs of the great deep bursting forth. Continental movements would generate heat; volcanic activity occurring while mountain building was occurring generates heat; and so forth. The point is that it heats the ocean water significantly. Naturally, the ocean would have more evaporation, subsequently causing immense fog, excessive clouds, and storms with more rain, ice, and snowfall than what we get currently.

Cool Summers

What about cooler summers? The Flood explains this as well. But first, a little volcano knowledge is required.

For 200 years, we have known how volcanoes affect our climate. When a volcano erupts, it sends ash particles and dioxides (such as sulfur dioxide) into the atmosphere. If the eruption is powerful enough, it sends these things to the upper atmosphere (stratosphere).

When these sub-microscopic items get to that height it is difficult for them to wash out. It takes a long time. So they linger and cause all sorts of problem for the climate, simply because they reflect sunlight back to space causing the temperature of the globe to cool.

As an example, Mount St. Helens, a relatively small volcano, caused a drop of 0.1 degree Celsius in global temperature.[2] Remember that Mount St. Helens was a small volcano acting alone, so we didn't expect much of a change; but notice that the global temperature went down for a short time.

Larger volcanoes of the past have had much more damaging effects. Some have dropped the global temperature by 1 degree Celsius (e.g., El Chichon), which is quite significant![3] Mt. Tambora blasted in 1815 and caused summer to cease in the Northern Hemisphere in 1816. It is called "the year without a summer," and it was estimated to drop the global temperature by 3 degrees Celsius![4]

2. Jack Williams, "The Epic Volcano Eruption That Led to the 'Year Without a Summer,' " *The Washington Post*, April 24, 2015, https://www.washingtonpost.com/news/capital-weather-gang/wp/2015/04/24/the-epic-volcano-eruption-that-led-to-the-year-without-a-summer/.
3. Ibid. Keep in mind that those arguing for a global warming and climate change see only tenths of a degree change, which is quite common in fluctuations that usually match the suns output — but consider a tenth of degree versus an entire degree with this volcano!
4. Ibid.

As you can see, volcanoes that send particles and dioxides into the upper atmosphere can cause severe weather problems — specifically causing summers to be cooler. Most volcanoes we have in modern times are acting alone.

But consider the mountain-building period of the Flood of Noah's day (e.g., Genesis 8:4,[5] Psalm 104:8–9,[6] etc.) involving immense volcanic activity *acting in conjunction* for more than half of the year and surely some volcanic activity that was post-Flood too — which would extend the effects. The point is that immense amounts of fine ash and dioxides were put in the upper atmosphere to linger for hundreds and hundreds of years.

The result was a lot of reflected sunlight and cooler summers back to back for extended amounts of time. Initially, you get accumulation at the poles, and then it extends downward from the North Pole and upward from the South Pole. Then you get more that pile on top of each other and compacts lower layers into ice layers (some layers even combine with each other when the ice gets deep enough and this is called *molecular diffusion*). Some of these ice layers glaciate. Some glaciers move horizontally or downhill as a result of the weight of the ice above them.

Warm oceans and cool summers are the key to the Ice Age.

When Did the Ice Age Peak and Retreat?

Creationists believe the Flood triggered the Ice Age. But that doesn't mean the Ice Age was in full effect immediately. It took time to accumulate up to a maximum (called maximum glaciation). Then, it took time to wane.

Surely, there were some minor fluctuations during the Ice Age where increases and decreases in ice occurred. During the Ice Age, there were times when it advanced and retreated even though the general trend was a growing ice extent. Conversely, there were times when ice sheets were growing when the general trend was reducing.

Even in later times, these fluctuations are felt. For example, there is the Little Ice Age where growth of glaciers was occurring in medieval times. This brings us to two important questions: When did the Ice Age peak? And when did it end (finish its retreat)?

---

5. Then the ark rested in the seventh month, the seventeenth day of the month, on the mountains of Ararat.
6. The mountains rose; the valleys sank down to the place which You established for them. You set a boundary that they may not pass over, so that they will not return to cover the earth (NASB).

Frankly, the Bible doesn't tell us. Thus, creation scientists construct various scientific models to try to answer the question. Naturally, not all models agree with each other.

When did the Ice Age end? Some might argue that it never really ended, since we still have glaciers and ice sheets today (even the Greenland and Antarctic ice sheets are still growing — others are waning!).

This answer doesn't really help us much, so let's refine the question. When did the retreat of the Ice Age finally get to a point of approximate equilibrium? In other words, when did the ice and snow melt off to a point that it remains relatively stable (not growing and not reducing much). This depends on when the peak of the Ice Age was, and so it brings us back to the first question.

Some weather experts (Dr. Jake Hebert,[7] Dr. Larry Vardiman,[8] and retired meteorologist Mike Oard[9]) working with weather data have independently suggested a build up and peak of about 500 years after the Flood, with about 200 or so years for the ice to melt off and retreat to equilibrium.

A competing model by geologist Dr. Andrew Snelling and writer/editor Mike Matthews, based on radiometric dating, have suggested a peak about 250 years after the Flood and about 100 years after that to equalize.[10] Either way, it is a matter of hundreds of years after the Flood. Keep in mind that models are not absolute and are subject to change.

One thing we would like to see an expert research in more detail is based on observations we see today. Some ice sheets are growing while others are retreating. Is it possible that the ice was growing and retreating in different areas, causing some areas to be affected by the Ice Age at one time and other areas affected by it later? After all, what we see in the rock record is an overall ice extent, but did this peak occur *all at once* in the past? Perhaps future research would be helpful.

---

7. Jake Hebert, "Ice Cores, Seafloor Sediments, and the Age of the Earth," Part 2, *Acts & Facts* 43 (7), 2014, http://www.icr.org/article/8181.
8. See Larry Vardiman, "An Analytical Young-Earth Flow Model of the Ice Sheet Formation During the 'Ice-Age,' " in *Proceedings of the Third International Conference on Creationism*, Robert Walsh, ed. (Pittsburg, PA: Creation Science Fellowship, Inc., 1994), p. 561–568; Larry Vardiman, "Ice Cores and the Age of the Earth," *Acts & Facts* 21 (4), 1992, http://www.icr.org/article/ice-cores-age-earth/.
9. Mike Oard, *An Ice Age Caused by the Genesis Flood* (El Cajon, CA: Institute for Creation Research, 1990), p. 23–38.
10. Andrew Snelling and Mike Matthews, "When was the Ice Age in Biblical History?" *Answers* magazine, vol. 8 no. 2, April–June, 2013, p. 46–52.

23 Do We Find Human Fossils with Dinosaur Fossils?

Ken Ham and Bodie Hodge

Humans and Dinosaur Fossils Together: The Misconception

Land animals (which includes dinosaurs) and man were made on day 6 of the creation week. So we lived at the same time.

Often, people believe that if human bones aren't found with dinosaur bones, then they *didn't* live together. This is a false assumption. If human bones aren't found buried directly with dinosaur bones, it simply means they weren't buried together during the Flood.

As the floodwaters advanced during the global Flood, humans would have fled to higher ground, swam, or held on to floating debris for as long as possible. Also, human corpses tend to bloat and therefore float on the water's surface. Hence, it makes sense that very few, if any, humans would be buried by sediment. Instead, they would have rotted and decayed without fossilization.

It is expected that marine creatures and plants were the first things buried and fossilized, since they are at a lower elevation and couldn't escape the sediment and water. When we look at the fossil record, statistically we find:

95% of all fossils were marine organisms.
95% of the remaining 5% were algae, plants/trees.
95% of the remaining 0.25% were invertebrates, including insects.
The remaining 0.0125% were vertebrates, mostly fish.[1]

1. John Morris, *The Young Earth* (Green Forest, AR: Master Books, 2002), p. 70; Andrew Snelling, "Where Are All the Human Fossils?" *Creation* 14(1):28–33, December 1991.

So we shouldn't expect to find many human fossils at all. There is still the possibility of finding human fossils in the lower levels of Flood sediments, but the creation/Flood model doesn't require it.

Remember, we don't find human bones buried with coelacanths either, but we live together today. (Coelacanths are a type of fish, which scientists claimed to have gone extinct millions of years ago but have recently been found alive.) Some were even enjoying them for dinner!

Pre-Flood Population

Estimates for the pre-Flood population are based on very little information since Genesis 1 doesn't give extensive family size and growth information. We know that Noah was in the tenth generation of his line and it was about 1,650 years after creation. Genesis also indicates that in Noah's lineage children were being born when their fathers were between the ages of 65 (Enoch to Methuselah) to well over 500 (Noah to his three sons).

How many generations were there in other lineages? We don't know. We know the line from Adam to Noah was living upward of 900 years, but we can't be certain everyone lived this long. How often and how many children were born? We don't know. What were the death rates? We don't know.

Despite this lack of information, some estimates have been done. Tom Pickett gives a range of about 5 to 17 billion people.[2] This is based on various population growth rates and generations of 16–22 prior to the Flood. Dr. John Gill, the famous Baptist expositor, in the 1700s was also open to 11 billion people or higher.[3]

Recall that Noah was only in the tenth generation, so this may be well beyond the higher end of the population maximum. The late Henry Morris had conservative estimates as low as 235 million people. He also calculated rates based on modern population growth, giving about 3 billion people.[4]

John Morris reports estimates that there were about 350 million people pre-Flood.[5] Based on these estimates, pre-Flood populations may have ranged from the low hundreds of millions to 17 billion people. Considering that the

2. Tom Pickett, "Population of the PreFlood World," http://www.ldolphin.org/pickett.html, accessed 8/21/2006.
3. John Gill, *Exposition of Genesis*, notes on Genesis 7:21, 1748–1763, http://www.studylight.org/commentary/genesis/7-21.html#geb.
4. Henry Morris, *Biblical Cosmology and Modern Science* (Grand Rapids, MI: Baker Book House, 1970), p. 77–78.
5. John Morris, *The Young Earth* (Green Forest, AR: Master Books, 2002), 11th printing, p. 71.

world was violent (Genesis 6:11–13[6]) and wicked (Genesis 6:5[7]) to such a horrible extreme before the Flood for 120 years, how many people were left at the onset of the Flood? In other words, if the entire world consisted of murderers, then the world's population could be cut in half in one day! It is possible that the population was much lower than we might think. (See chapter 39 for a more detailed discussion of the pre-Flood population.)

Were All Humans Fossilized?

During the 2004 tsunami in Southeast Asia, the Associated Press reported that although many humans were killed during the catastrophe, surprisingly very few livestock animals were killed.[8] Based on this evidence, it is possible that land animals may have had a better chance of survival as the Flood began to devastate and overtake the coastlines than humans did. Initially, people swept out to sea would not be candidates for fossilization. Additionally, inland people could try to flee to higher ground, float, or latch onto debris, reducing their probability of fossilization too.

As sad as it was, the tsunami of 2004 was a good example of the destructiveness of water — even though it was a relatively small flood. According to the United Nation's Office of the Special Envoy for Tsunami Recovery, nearly 43,000 of the approximate 230,000 people that died, were never found — and we know exactly where they were lost — but definitely *not* fossilized.[9]

Were All Humans Evenly Distributed in the Flood Sediment?

We know humans have a tendency to live in groups like towns, villages, and cities. People were probably not evenly distributed before the Flood. Before the Flood, a city was recorded in Genesis 4:17.[10] In accordance with this,

6. The earth also was corrupt before God, and the earth was filled with violence. So God looked upon the earth, and indeed it was corrupt; for all flesh had corrupted their way on the earth. And God said to Noah, "The end of all flesh has come before Me, for the earth is filled with violence through them; and behold, I will destroy them with the earth. (Scripture in this chapter is from the New King James Version of the Bible unless otherwise noted.)
7. Then the Lord saw that the wickedness of man was great in the earth, and that every intent of the thoughts of his heart was only evil continually.
8. Gemunu Amarasinghe, "Tsunami Kills Few Animals in Sri Lanka," Associated Press, December 30, 2004, http://www.livescience.com/animalworld/tsunami_wildlife_041230.html, accessed 8/25/2006.
9. "The Human Toll," http://www.tsunamispecialenvoy.org/country/humantoll.asp, accessed June 6, 2006.
10. And Cain knew his wife, and she conceived and bore Enoch. And he built a city, and called the name of the city after the name of his son — Enoch.

most of the population today lives within 100 miles of the coastline. One report says:

> Already nearly two-thirds of humanity — some 3.6 billion people — crowd along a coastline, or live within 150 kilometers of one.[11]

This is further confirmation that the pre-Flood civilizations probably were not evenly distributed either. If man wasn't evenly distributed, then the likelihood of man being evenly distributed in Flood sediment becomes extremely remote.

How Much Flood Sediment Is There?

John Woodmorappe's studies indicate that there are about 700 million cubic kilometers, which translates to about 168 million cubic miles of Flood sediment.[12] Dr. John Morris states that there is about 350 million cubic miles of Flood sediment.[13] However, this number may be high since the total volume of water on the earth is estimated at about 332.5 million cubic miles according to the U.S. Geological Survey.[14]

So, a small human population and massive amounts of sediment are two prominent factors why we haven't found human fossils in Flood sediments. It also may simply be that we haven't found the sediment where humans were living and were buried.

Let's Think about the Question

Again, people mistakenly believe that if human bones aren't found with dinosaur bones, then they didn't live together. Let's think about it this way instead: if human bones aren't found buried with dinosaur bones, it simply means they weren't *buried* together.

A great example is that of the coelacanth. Coelacanth fossils are found in layers below dinosaurs.[15] It was thought the Coelacanth became

11. "Coastal Policy," http://coastalpolicy.blogspot.com/2005_02_01_archive.html.
12. John Woodmorappe, *Studies in Flood Geology* (Dallas, TX: Institute for Creation Research, 1999), p. 59. This number actually comes from *International Geology Review* 24(11) 1982, A.B. Ronov, "The Earth's Sedimentary Shell," p. 1321–1339.
13. Morris, *The Young Earth*, p. 71.
14. "The World's Water," U.S. Geological Survey, http://ga.water.usgs.gov/edu/earthwherewater.html.
15. Lynn Dicks, "The Creatures Time Forgot," *New Scientist*, October 23, 1999; 164: (2209) p. 36–39.

extinct about 70 million years ago (by secular reckoning) because their fossils are not found after this time. However, in 1938 living populations were found in the Indian Ocean![16]

Humans are not buried with crocodiles, but we live together. Humans are not buried with ginkgo trees but exist at the same time. The list can go on! This shows that the fossil record is not complete, not necessarily that things did or did not co-exist.

If human and dinosaur bones are found in the same geologic layers in the future, it would be consistent with the biblical view. In fact, it would be more of a problem for those who accept the geologic layers as evidence for millions of years. If the fossil layers really represented millions of years, then finding a human and dinosaur fossilized in the same layers would cause problems because in the old-earth view, man wasn't supposed to be that old, or dinosaurs that young.

As biblical creationists, we don't *require* that human and dinosaur fossils have to be found buried together. Whether they are found together or not does not affect the biblical view.

16. Rebecca Driver, "Sea Monsters . . . More Than a Legend?" *Creation* magazine 19(4):38–42, September 1997, http://www.answersingenesis.org/creation/v19/i4/seamonsters.asp.

24 | Was There Originally One Continent?

Bodie Hodge

Introduction

> Then God said, "Let the waters under the heavens be gathered together into one place, and let the dry land appear"; and it was so. And God called the dry land Earth, and the gathering together of the waters He called Seas. And God saw that it was good (Genesis 1:9–10).

Most creationists believe there was one continent originally, in light of these verses. However, we need to be careful because the text doesn't specifically say this. It says the *waters* (under the heaven) were gathered into one place.

At this point of gathering with the land appearing, one could rightfully assume the earth has taken its familiar shape of a sphere, leaving open the option that this watery mass may *not* have been a sphere beforehand (perhaps close though). From children's books to scientific models, we assume the earth was a nearly perfect sphere of water initially, but that is an assumption.

Keep in mind these initial created waters had no *specific* form according to Genesis 1:2,[1] merely a surface. The waters were since separated (Genesis 1:6-7[2]). The portion of waters below the heavens were then gathered together into our seas.

1. The earth was formless and void, and darkness was over the surface of the deep, and the Spirit of God was moving over the surface of the waters (NASB).
2. Then God said, "Let there be an expanse in the midst of the waters, and let it separate the waters from the waters." God made the expanse, and separated the waters which were below the expanse from the waters which were above the expanse; and it was so (NASB).

Consider God's poetic description of the gathering of the waters into a "heap" or a "storehouse" as in Psalm 33:6–9:

> By the word of the LORD the heavens were made, and all the host of them by the breath of His mouth. He gathers the waters of the sea together as a heap; He lays up the deep in storehouses. Let all the earth fear the LORD; let all the inhabitants of the world stand in awe of Him. For He spoke, and it was done; He commanded, and it stood fast.

So the gathering stage may have been more significant than we have often been led to think. However, we would not be adamant about this interpretation, but leave it open.

Pangaea and Rodinia?

Regarding the continent though, it is possible to have more than one continent with a situation where waters are still in one place. Even so, some have proposed an initial supercontinent that looked like Pangaea going back to a creationist, Antonio Snider, in the 1800s.[3] In this model, Pangaea breaks apart into the continents we have today during a catastrophic breakup during the Flood.

Others, including some creationists, have models with a supercontinent that looks like Rodinia[4] (one of the alleged supercontinent reconstructions that secularists have *prior* to Pangaea based on radiometric dating[5]). The creationist form of the model has Rodinia breaking up during the Flood, coming back together as Pangaea, and then breaking up again late in the Flood.[6]

Still others have left open the option that we simply do not know the size or shape of an original continent(s) especially in light of a global Flood that destroyed the surface of the earth (Genesis 6–8; 2 Peter 3:5–6[7]). These potential models play off the fact that so much happened in the Flood that it may be too difficult to reconstruct what the original earth looked like (e.g.,

3. A. Snider, *Le Création et ses Mystères Devoilés* (Paris, France: Franck and Dentu, 1859).
4. *Rodinia* is the Russian word for "The Motherland."
5. More specifically, it is based on radiometric dates of A-type granites and radiometric dates of fold mountains called orogenic belts.
6. Andrew Snelling, "Noah's Lost World," https://answersingenesis.org/geology/plate-tectonics/noahs-lost-world/.
7. For this they willfully forget: that by the word of God the heavens were of old, and the earth standing out of water and in the water, by which the world that then existed perished, being flooded with water.

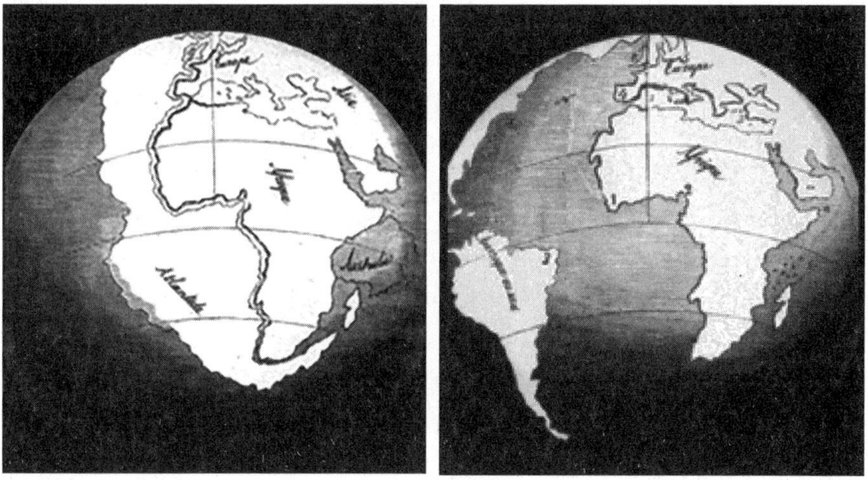

Maps made in 1858 by geographer Antonio Snider, showing his version of how the American and African continents may have once fit together, then later separated.

could it have originally been something half way between Pangaea and what we have today for example).

One continent is a possibility and makes sense, but keep in mind this is not exactly what the text says, so it would be wise not to attack those who are open to more than one, though most agree that it was the Flood of Noah's day that broke apart what was originally made to what we have today.

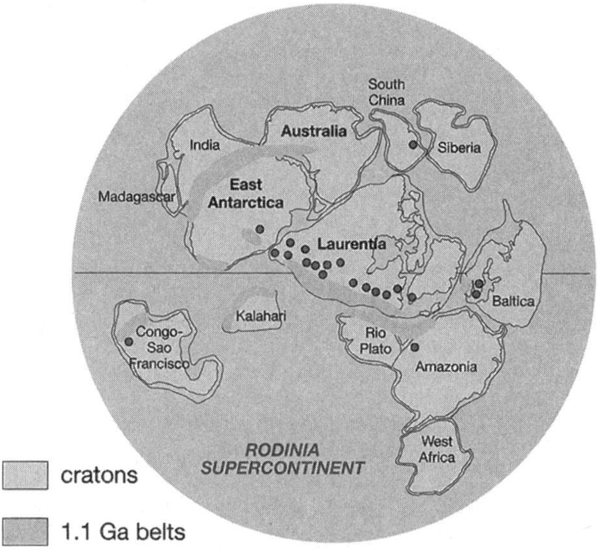

A proposed reconstruction of the supercontinent Rodinia

About the breakup to what we have today, the text of Scripture gives us some clues. By the 150th day of the Flood, the mountains of Ararat existed

(Genesis 7:24–8:4[8]). These mountains (as well as the others in the Alpide stretch of mountain ranges that go from Europe to Asia) appear to have been built by the continental collisions of the Arabian, African, Indian, and Eurasian plates. Thus, continental movement for these mountains and plates may well have been largely stopped by the 150th day.[9]

This makes sense as the primary mechanisms for the Flood (*springs of the great deep and windows of heaven*) were stopped on the 150th day as well. Thus, it triggered the waters to now be in a recessional stage as the valleys go down (e.g., ocean basins etc.). This is subsequent to the mountains rising, which had already been occurring up to the 150th day (e.g., mountain ranges and continent extending above the waters) at this stage of the Flood (Psalm 104:6–9[10]).

At creation, was the land *under* the surface of the waters or was it made uniquely on day 3?

There are two models/positions on this.

1. The first is that the land was under the surface of the water the whole time (from Genesis 1:1) and the water was moved out of the way as the land raised through the waters and then the land dried out to become dry.
2. The earth was pure water on day 1. Then the waters were separated and those below were gathered into one place (on day 2) and then the dry land was made separately and uniquely (being that it is creation week) or that some of the water transformed into dry land directly.

---

8. And the waters prevailed on the earth one hundred and fifty days. Then God remembered Noah, and every living thing, and all the animals that were with him in the ark. And God made a wind to pass over the earth, and the waters subsided. The fountains of the deep and the windows of heaven were also stopped, and the rain from heaven was restrained. And the waters receded continually from the earth. At the end of the hundred and fifty days the waters decreased. Then the ark rested in the seventh month, the seventeenth day of the month, on the mountains of Ararat.
9. Naturally this causes a problem for the *Rodinia-to-Pangaea-to-today* scenario. If Rodinia breaks into Pangaea and late in the Flood Pangaea is still under water and needs to break apart into what we have today, then the mountains of Ararat should not have existed so early in the Flood on the 150th day of the Flood.
10. You covered it with the deep as with a garment; the waters were standing above the mountains. At Your rebuke they fled, at the sound of Your thunder they hurried away. The mountains rose; the valleys sank down to the place which You established for them. You set a boundary that they may not pass over, so that they will not return to cover the earth (NASB).

If the waters were truly void as Genesis 1:2[11] says, then the idea of a land mass in the midst from the very beginning, may not be the best solution — especially if the midst of the water was where the point of water separation was to occur on day 2! It makes more sense that the land was made later *on* day 3 out of water. This comes from 2 Peter 3:5–6[12] that says the earth was formed "out of water and by water" and these waters were used later in the Flood!

But one point can be made: the land was dry, *not wet,* when it appeared. This was so important that God stated it multiple times, in Genesis 1:9[13] and 1:10[14] as well as Psalm 95:5.[15] This is a good argument for a supernatural fiat of the appearance of the land, not a wet sedimentary flow on day 3 that was naturally pushed up through the waters and later became dry, unless God supernaturally dried it. But this is another assumption that would be required that is not in the text. Again, that would be wetland that *became* dry.

When the land appeared in Genesis 1 on day 3, the land that was being separated from the water was *dry,* not wet. The text in Genesis says that the waters were gathered into one place (i.e., in heaps and storehouses) *and then* the dry land appeared. It says nothing of water running off of the land as it rises; otherwise, "wet" land would have appeared and then *become dry*. The response from God, "and it was so" seems to refute the idea that it was wet, and then became dry.

But really, all we can be certain about is that the *dry land appeared on day 3.* And keep in mind that models are not Scripture and subject to change.

---

11. The earth was without form, and void; and darkness was on the face of the deep. And the Spirit of God was hovering over the face of the waters.
12. For this they willfully forget: that by the word of God the heavens were of old, and the earth standing out of water and in the water, by which the world that then existed perished, being flooded with water.
13. Then God said, "Let the waters under the heavens be gathered together into one place, and let the dry land appear"; and it was so.
14. And God called the dry land Earth, and the gathering together of the waters He called Seas. And God saw that it was good.
15. The sea is His, for He made it; and His hands formed the dry land.

25 How Long Did It Take for Noah to Build the Ark?

Bodie Hodge

Let's dispel something up front. Some confuse God's statement in Genesis 6:3 as describing the time it took Noah to build the ark. It states:

> And the LORD said, "My Spirit shall not strive with man forever, for he is indeed flesh; yet his days shall be one hundred and twenty years."

However, these 120 years are a countdown to the Flood.[1] In other words, mankind's violence had reached its peak and God declared that 120 years was the "drop dead" date for mankind who is a mortal being (Genesis 6:3–7[2]). From a quick look, these 120 years would seem to be the absolute maximum for the time given to build the ark, but the Scriptures reveal much more, allowing us to be more accurate.

1. Some have also described this as the longevity of mankind. For a number of generations after the Flood, people lived to be much older than this (e.g., Isaac lived to 180 years), so it is not referring to longevity.
2. And the LORD said, "My Spirit shall not strive with man forever, for he is indeed flesh; yet his days shall be one hundred and twenty years." There were giants on the earth in those days, and also afterward, when the sons of God came in to the daughters of men and they bore children to them. Those were the mighty men who were of old, men of renown. Then the LORD saw that the wickedness of man was great in the earth, and that every intent of the thoughts of his heart was only evil continually. And the LORD was sorry that He had made man on the earth, and He was grieved in His heart. So the LORD said, "I will destroy man whom I have created from the face of the earth, both man and beast, creeping thing and birds of the air, for I am sorry that I have made them."

For example, Noah was 500 years old when Japheth, the first of his sons, was born (Genesis 5:32[3]). And yet Noah's second son, Shem, had his first son two years after the Flood, when he was 100 (Genesis 11:10[4]).[5] This means that Shem was 98 years old when the Flood came and it also means that Shem was born when Noah was 502 years old. So for Noah to begin having children at 500 means that Japheth was indeed the older brother, as per Genesis 10:21,[6] being born when Noah was 500. Ham is mentioned as the youngest of Noah (Genesis 9:24[7]).

When God finally gave Noah instructions to build the ark, it was not at the beginning of the 120-year countdown. God told Noah that he, his wife, and his three sons and their wives (Genesis 6:15–18[8])[9] would go aboard the ark at this same time.

Deducing that Shem was born 98 years before the Flood, it could be no more than this. But even more so, Ham hadn't been born yet! If we were to assume the same time between Ham and Shem as between Japheth and Shem, then Ham could have been born around 96 years before the Flood.

Although the Bible is silent on the exact timing, it is reasonable to assume that some time elapsed for the three sons to grow up and find wives. I would be most comfortable giving a tentative range of anywhere from 20 to 40 years, making Ham no less than 16 at his marriage.

So if we think about this logically and tabulate it, we would end up with a tentative range of about 55 to 75 years for a reasonable *maximum* time to build the ark (see table). Of course, it could be less than this depending on the ages of Noah's sons when they took wives.

---

3. And Noah was five hundred years old, and Noah begot Shem, Ham, and Japheth.
4. This is the genealogy of Shem: Shem was one hundred years old, and begot Arphaxad two years after the flood.
5. Shem is often listed first (e.g., Genesis 6:10, 7:13) due to *importance*, much as Abraham is listed first; yet, Shem was not the oldest. From the lineage of Shem and Abraham came Christ.
6. And children were born also to Shem, the father of all the children of Eber, the brother of Japheth the elder.
7. So Noah awoke from his wine, and knew what his younger son had done to him.
8. And this is how you shall make it: The length of the ark shall be three hundred cubits, its width fifty cubits, and its height thirty cubits. You shall make a window for the ark, and you shall finish it to a cubit from above; and set the door of the ark in its side. You shall make it with lower, second, and third decks. And behold, I Myself am bringing floodwaters on the earth, to destroy from under heaven all flesh in which is the breath of life; everything that is on the earth shall die. But I will establish My covenant with you; and you shall go into the ark — you, your sons, your wife, and your sons' wives with you.
9. *Youngest* is used in ESV, NAS, and other translations for the word qatan, which means small, young, or insignificant.

Years until the Flood	Event	Bible reference
120	Countdown to the Flood begins	Genesis 6:3
100	Noah had Japheth, the first of his sons, when he was 500 years old	Genesis 5:32[1], 10:21[2]
98	Noah had Shem who was 100 two years after the Flood	Genesis 11:10[3]
? Perhaps 95 or 96, the same time between Japheth and Shem	Ham was the youngest one born to Noah and was aboard the ark, so he was born prior to the Flood	Genesis 9:24[4]; Genesis 7:13[5]
? Perhaps 20-40 years for all of the sons to be raised and find a wife	Each son was old enough to be married before construction on the ark began	Genesis 6:18[6]
~ 55–75 years (estimate)	Noah was told to build the ark, for he, his wife, his sons, and his sons' wives would be aboard the ark	Genesis 6:18[7]
Ark Completed		
?	Gather food and put it aboard the ark	Genesis 6:21[8]
7 days	Loading the ark	Genesis 7:2-3[9]
0	Noah was 600 when the floodwaters came on the earth.	Genesis 7:6[10]

Footnotes for Table

1. And Noah was five hundred years old, and Noah begot Shem, Ham, and Japheth.
2. And children were born also to Shem, the father of all the children of Eber, the brother of Japheth the elder.
3. This is the genealogy of Shem: Shem was one hundred years old, and begot Arphaxad two years after the flood.
4. So Noah awoke from his wine, and knew what his younger son had done to him.
5. On the very same day Noah and Noah's sons, Shem, Ham, and Japheth, and Noah's wife and the three wives of his sons with them, entered the ark.
6. But I will establish My covenant with you; and you shall go into the ark — you, your sons, your wife, and your sons' wives with you.
7. Ibid.
8. And you shall take for yourself of all food that is eaten, and you shall gather it to yourself; and it shall be food for you and for them.
9. You shall take with you seven each of every clean animal, a male and his female; two each of animals that are unclean, a male and his female; also seven each of birds of the air, male and female, to keep the species alive on the face of all the earth.
10. Noah was six hundred years old when the floodwaters were on the earth.

Keep in mind that Noah may have researched the subject for years or worked to get funds/supplies to build the ark. Hosts of things could have occurred prior to Noah and his family "breaking ground" on the ark.

We know that the ark was completed prior to loading the animals that the Lord brought to Noah (Genesis 6:22–7:4[10]) and that they had to take time to gather food and store it aboard the ark (Genesis 6:21[11]). So carefully considering the text, we can conclude that the construction of the ark did not involve the 120 years mentioned in Genesis 6:3 but 75 years at the most.

10. Thus Noah did; according to all that God commanded him, so he did. Then the LORD said to Noah, "Come into the ark, you and all your household, because I have seen that you are righteous before Me in this generation. You shall take with you seven each of every clean animal, a male and his female; two each of animals that are unclean, a male and his female; also seven each of birds of the air, male and female, to keep the species alive on the face of all the earth. For after seven more days I will cause it to rain on the earth forty days and forty nights, and I will destroy from the face of the earth all living things that I have made.
11. And you shall take for yourself of all food that is eaten, and you shall gather it to yourself; and it shall be food for you and for them.

26 Timeline of the Flood

Bodie Hodge

What was the duration of the Flood you might ask? The Bible gives us the answer. It was from the 2nd month, the 17th day of the month, of Noah's 600th year until the next year (Noah's 601st year) on the 2nd month, the 27th day of the month. So it was one year and 10 days, by Noah's calendar year. So how many total days was that? That will depend on what calendar Noah was using!

What Calendar?

Obviously, Noah was not using our modern Gregorian calendar that came into effect in its present form in A.D. 1582. No one used the 365-day year until the Egyptians, according to ancient historian Herodotus [in book 2, line 4]. Prior to this, the Egyptians used a 360-day calendar with an intercalary month thrown in from time to time to bring it back to where it should be.

The Egyptians were descendants of Noah's grandson Mizraim, so this was well after the Flood.[1] Most early ancient calendars used a 360-day year with an intercalary month every few years — from Egypt to the Mayans.[2] This may have been a carryover from Noah through Babel.

It is possible that the Flood account used the calendar that Moses was accustomed to since he was the one who gave us the inspired text of Genesis (likely from pre-existing texts and by the power of the Holy

1. Bodie Hodge, *The Tower of Babel* (Green Forest, AR: Master Books, 2013), p. 122–124.
2. James Ussher, *The Annals of the World, The Epistle to the Reader, 1656*, translated by Larry and Marion Pierce (Green Forest, AR: Master Books, 2003), p. 9.

Spirit).[3] If so, that would mean that the dates had to be translated from Noah's system of timekeeping to the one Moses and the Israelites used before the captivity — which was a Canaanite calendar. It makes sense that they used a Canaanite calendar, as it was surely a carryover from the sojourn of his ancestors Abraham, Isaac, and Israel in Canaan, though Moses was surely familiar with the ancient Egyptian calendar that had 360 days too, since he was educated in Egypt's elite royal house.

Much later, the Israelites adopted the Babylonian calendar upon their captivity beginning with Nebuchadnezzar (e.g., the month of *Tammuz* is in the Jewish calendar is named for the Babylonian "god" Tammuz). The Canaanite month names used in Scripture prior to the captivity (with their roughly corresponding Babylonian names) are:

- *Abib* (Exodus 13:4,[4] 23:1,5[5] 34:18,[6] and Deuteronomy 16:1[7]) was later called *Nisan*, 30 days
- *Ziv* (1 Kings 6:1,[8] 6:37,[9] 29 days) was later called *Iyyar*
- *Ethanim* (1 Kings 8:2,[10] 30 days) which was later called *Tishri*
- *Bul* (1 Kings 6:38,[11] usually 29 days) which was later called *Marcheshvan* or simply *Cheshvan* or *Heshvan*

If we look at the text of Scripture, the Flood account in Genesis 6–8 never uses month names that Moses and other Israelites used later (Canaanite or Babylonian names). Instead, Moses chose *not* to insert names from the Canaanite

3. Terry Mortenson and Bodie Hodge, "Did Moses Write Genesis?" in *How Do We Know the Bible Is True?* Volume 1, Ken Ham and Bodie Hodge, gen. eds. (Green Forest, AR: Master Books, 2011), p. 85–102.
4. On this day you are going out, in the month Abib.
5. You shall keep the Feast of Unleavened Bread (you shall eat unleavened bread seven days, as I commanded you, at the time appointed in the month of Abib, for in it you came out of Egypt; none shall appear before Me empty).
6. The Feast of Unleavened Bread you shall keep. Seven days you shall eat unleavened bread, as I commanded you, in the appointed time of the month of Abib; for in the month of Abib you came out from Egypt.
7. Observe the month of Abib, and keep the Passover to the LORD your God, for in the month of Abib the LORD your God brought you out of Egypt by night.
8. And it came to pass in the four hundred and eightieth year after the children of Israel had come out of the land of Egypt, in the fourth year of Solomon's reign over Israel, in the month of Ziv, which is the second month, that he began to build the house of the LORD.
9. In the fourth year the foundation of the house of the LORD was laid, in the month of Ziv.
10. Therefore all the men of Israel assembled with King Solomon at the feast in the month of Ethanim, which is the seventh month.
11. And in the eleventh year, in the month of Bul, which is the eighth month, the house was finished in all its details and according to all its plans. So he was seven years in building it.

calendar but left them numbered (first month, second month, etc.). Since the dates are referenced to Noah's age, it makes the most sense that Moses kept the dating system that was utilized on the ark while penning Genesis 7–8.

Keep in mind that the later Israelite calendar (i.e., the Babylonian calendar) was lunar.[12] The months alternated with 29 or 30 days. In the Bible, we find that 150 days was equivalent to 5 months based on the context in the Flood account (Genesis 7:24–8:4[13]). This would yield month-lengths of 30 days each, not 29 and 30 days alternating for 5 months (as in a lunar calendar).

So with this in mind, it makes the most sense to stick with the common 360-day calendar that many ancients used that had 30-day months. James Ussher states in *The Epistle to the Reader* of his treatise *The Annals of the World*:

> Moreover, we find that the years of our forefathers, the years of the ancient Egyptians and Hebrews, were the same length as the Julian year. It consisted of twelve months containing thirty days each. (It cannot be proven that the Hebrews used lunar months before the Babylonian captivity.) Five days were added after the twelfth month each year. Every four years, six days were added after the twelfth month. {*Diod. Sic., l. 1. c. 50. s. 2. 1:177} {*Strabo, l. 17. c. 1. s. 46. 8:125} {*Strabo, l. 17. c. 1. s. 29. 8:85} {*Herodotus, l. 2. c. 4. 1:279} {#Ge 7:11,24 8:3-5,13,14}.[14]

Furthermore, applying any lunar calendar to the duration of the onset of the Flood until the date that the ark struck the mountains of Ararat (the 2nd month, 17th day of the month to the 7th month, 17th day of the month) would have the ark landing in the mountains of Ararat two to three days *before* the waters receded, subsided, and the rain and great deep

12. My friend and astronomer Dr. Danny Faulkner prefers the lunar calendar to be applied to the Flood account, though I respectfully disagree. But nonetheless, I encourage him in his research. See: Danny Faulkner, How Long Did the Flood Last?, *Answers Research Journal,* 8 (2015):253–259, May 13, 2015, https://answersingenesis.org/the-flood/how-long-did-the-flood-last/.
13. And the waters prevailed on the earth one hundred and fifty days. Then God remembered Noah, and every living thing, and all the animals that were with him in the ark. And God made a wind to pass over the earth, and the waters subsided. The fountains of the deep and the windows of heaven were also stopped, and the rain from heaven was restrained. And the waters receded continually from the earth. At the end of the hundred and fifty days the waters decreased. Then the ark rested in the seventh month, the seventeenth day of the month, on the mountains of Ararat.
14. James Ussher, *The Annals of the World*, revised and updated by Larry and Marion Pierce (Green Forest, AR: Master Books, 2003), p. 9.

were restrained (the Flood mechanisms). Based on the context and theological issues, I humbly suggest that the best calendar to use is the 360-day calendar, which was often used in the Bible.[15]

TIMELINE OF THE FLOOD

The following tentative table utilizes a 360-day calendar as most ancient calendars had in the Middle East (and elsewhere). This understanding of the Flood is assumed to *exclude* an intercalary month.

An examination of the Flood account in Genesis 6–8 gives some time-related milestones that form the overall structure in the progression of the yearlong global Flood. Table 1 briefly summarizes these milestones that can help us understand some of the geologic details of the Flood.

Table 1: Timeline of Flood Duration

Timeline (days)	Duration	Month/Day	Description	Bible reference*
0	Initial reference point	600th year of Noah's life: 2nd month, 17th day of the month	The fountains of the great deep broke apart and the windows of heaven were opened; it began to rain. This happened on the 17th day of the 2nd month. Noah actually entered the ark seven days prior to this.	Genesis 7:11
40	40 days and nights	3rd month, 27th day of the month	Rain fell for 40 days then it covered the earth's highest places (at that time) by over ~20 feet (15 cubits) and began the stage of Flooding until the next milestone.** At this time, the ark was lifted up.	Genesis 7:11–12 Genesis 7:17–20
150	150 days (including the initial 40 days)	7th month, 17th day of the month	The water rose to its highest level (covering the whole earth) sometime between the 40th and 150th day, and the end of these 150 days was the 17th day of the 7th month. The ark rested on the mountains of Ararat. On the 150th day, the springs of the great deep were shut off, and the rain from above ceased, and the water began continually receding.	Genesis 7:24–8:5

15. For examples, the Persians still used a 360-day calendar as witnessed in Esther 1:4 where a 180 days equated with a half year feast; John used 3 and half years as 42 months with 1260 days, which uses a 360-day calendar (Revelation 11:2-3, 12: 6, 13:5-7), etc.

150 + 74 = 224	74 days	10th month, 1st day of the month	The tops of the mountains became visible on the tenth month, first day.	Genesis 8:5
224 + 40 = 264	40 days	11th month, 11th day	After 40 more days, Noah sent out a raven.	Genesis 8:6
264 + 7 = 271	7 days	11th month, 18th day of the month	The dove was sent out seven days after the raven. It had no resting place and returned to Noah.	Genesis 8:6–12
271 + 7 = 278	7 days	11th month, 25th day of the month	After seven more days, Noah sent out the dove again. It returned again, but this time with an olive leaf in its beak.	Genesis 8:10–11
278 + 7 = 285	7 days	12th month, 2nd day of the month	After seven more days, Noah sent out the dove again, and it did not return.	Genesis 8:12
314	29 days	601st year of Noah's life: 1st month, 1st day of the month	Noah removed the cover of the ark on the first day of the first month. The *surface* of the earth was dried up and Noah could verify this to the extent of what he could see.	Genesis 8:13
370 (371 if counting the first day and last day as full days)	56 days	2nd month, 27th day of the month	The *earth* was dry and God commanded Noah's family and the animals to come out of the ark. From the first day of the year during the daylight portion there were 29.5 more days left in the month plus 26.5 more days left in the second month until the exit.	Genesis 8:14–17 Genesis 7:11

* References not listed in the footnotes for the table — see chapter 2 for the text of the Flood account.

** Some argue from the Hebrew that the ark officially rose off the surface on the 40th day, e.g., William D. Barrick and Roger Sigler, "Hebrew and Geologic Analysis of the Chronology and Parallelism of the Flood: Implications for the Interpretations of the Geologic Record," in *Proceedings of the Fifth International Conference on Creationism*, ed. Robert L. Ivey Jr., (Pittsburg, PA: Creation Science Fellowship, 2003), p. 397–408.

Because the biblical account is a reliable record of earth history, it is to be expected that these milestones would be significant in correlating the prominent geological features preserved in the rock record. For example, we are told that the onset of the Flood was triggered by the breaking up of the fountains of "the great deep."

This would imply a violent beginning to the Flood, as springs or fountains of water burst forth to spew vast quantities of water and perhaps other material onto the surface from deeper inside the earth. Furthermore, because this subterranean water and other materials bursting forth is mentioned *first* in Genesis 7:11[16] and 8:2,[17] this may suggest that the majority of the water for the Flood came from that source and perhaps helped to supply the waters that are referred to as falling through "the windows of heaven."[18]

The springs of the great deep were likely the trigger that ultimately resulted in continental scale breaking up of the earth's crust. The bursting forth of subterranean waters would probably produce tsunamis (granting the ocean depth was sufficient) and would therefore seem to also imply that the Flood began with catastrophic means. Thus, this description of the onset of the Flood provides clues as to where we should look in the geologic record for the pre-Flood/Flood boundary.

Of course the issue of pre-Flood sedimentation needs to be discussed. Rivers, such as Hiddekel, Gihon, Tigris, and Euphrates, would have been carrying some sediment for about 1,650 years. It is also possible for other smaller catastrophes to have occurred during this time — e.g., volcanoes. So the question really becomes, were these sediments disturbed and/or redistributed during the Flood or were they buried *in situ*?

Another milestone with geological implications is day 150. At this stage of the Flood we are told that the ark came to rest in the mountains of Ararat. This implies that modern mountain building, at least in what we now call the Middle East, had begun (see also Psalm 104:8–9).[19]

16. In the six hundredth year of Noah's life, in the second month, the seventeenth day of the month, on that day all the fountains of the great deep were broken up, and the windows of heaven were opened.
17. The fountains of the deep and the windows of heaven were also stopped, and the rain from heaven was restrained.
18. A.A. Snelling, "A Catastrophic Breakup: A Scientific Look at Catastrophic Plate Tectonics," *Answers*, 2:2 (2007), p. 44–48.
19. Psalm 104 begins with a reference to the events of creation week and goes on to mention ships (vs. 26) and Lebanon (vs. 16), which are near the time of the Psalmist. So logically, other events in history since creation, such as the Flood, should be expected within the Psalm as it continues. It should be obvious that vs. 6–9 are referring to the Flood since verse 9 specifically says the water will not *return* to cover the earth, which refers to God's post-Flood declarations in Genesis 9:11, 15 and Isaiah 54:9. Had verses 6–9 been referring to creation week, then God would have erred since water did return to cover the earth during the Flood. For these and other reasons, Psalms 104:8a should be rendered from Hebrew into English as: "The mountains rose and valleys sank down . . .", which several translations

Furthermore, if our current understanding of mountain building is correct, for the mountains of Ararat to have been formed requires the Eurasian Plate, African Plate, and Arabian Plate to be colliding with one another (perhaps with some contribution from movement of the Indian Plate).

The biblical account also indicates that on day 150 the springs of the great deep were stopped and the windows of heaven were closed, so from then on the waters began to steadily recede. We might therefore expect to see in the geologic record evidence of a transition perhaps from larger scale sediment layers to smaller scale geologic effects as well as higher concentration of basin, abyssal plain, and continental shelf sedimentation.

Yet another milestone is day 314 (see Table 1). By this time during the Flood event the biblical account indicates that the water had receded from off the continental land surfaces sufficiently for the surface of the landscape to essentially be dry, at least in the areas as far as Noah could observe.

Then finally, by day 370 the earth's continental land surfaces were dry from the Flood waters. Thus, it can be noted that the recessional stage of the Flood (when the waters were receding) lasted about five and half months, while the Flood's inundatory stage (when the waters were rising) lasted exactly five months. The recessional stage lasted almost the same length of time as it took for the water to overtake the earth globally. The Flood event finished with another two months needed to complete the drying process.

After the Flood ended on day 370 (with the proclamation for Noah to exit the ark), it would seem that the hydrological cycle had already been re-established with renewed regularity, as indicated by the rain clouds through which Noah saw the rainbow and the set times for seed time and harvest in accordance with the seasonal cycle of rain (Genesis 8:22,[20] 9:12–17). Other milestones throughout the Flood account could of course be highlighted, but these are ones that are most related to geological and weather processes and should suffice for this brief tentative overview.

---

have concluded (e.g., Latin Vulgate A.D. 405, Geneva Bible A.D. 1599, Tyndale/Coverdale A.D. 1535, New American Standard A.D. 1971 and 1995, English Standard Version A.D. 2001, Holman Christian Standard Bible A.D. 2004, etc.).

20. While the earth remains, seedtime and harvest, cold and heat, winter and summer, and day and night shall not cease.

27 Where Is the Flood Rock?

Ken Ham and Bodie Hodge

The answer to this is simple; and yet I'm surprised so many people miss it. The flood rock is all over the world, sometimes miles deep. It is the majority of the fossil layers in the "backyard" of most people reading this. These fossil layers can be found in most parts of the world (e.g., some have been scraped off by glaciers since the Flood, etc.). Many basins and mountains consist of them (e.g., the Appalachian mountains). Naturally, we have had some layering and erosion since the time of the Flood and that brings us to a debate.

One of the most hotly debated subjects in creationist literature is the placement of what is Flood rock and what is post-Flood rock. Of course, the debate breaks into a boil in some technical conferences, especially when interpreting what is found in various rock layers toward the top of the fossil sediments. You need to understand that creationists generally agree upon the majority of the fossil layers (e.g., the pre-Flood boundary, the Ice Age rock being Pleistocene, etc.), but it is those pesky layers on which creationists don't agree surrounding the post-Flood boundary that causes the controversy.

Furthermore, there are implications depending on where the post-Flood boundary is.[1] In some creationists' eyes, particular fossil deposits or rock layers could be seen as Flood rock, while others have it as post-Flood rock. The difference here is immense . . . while some say the animal remains found in those fossil layers are from the Flood, others would have to find migration routes from the ark to these deposits very soon after the Flood.

1. Layers discussed here will be layers that encompass the Flood and post-Flood, not pre-Flood layering unless noted.

As this simple example reveals, the technical differences can be large and creationists can be very intense about this subject. As a preface before we get into the post-Flood boundary discussion, let's first review the relevant biblical passages, geological layers, and the pre-Flood boundary so we can properly understand the debate at hand.

A Few Milestones

Many of the milestones in the previous chapter are significant to geological features. For example, the onset of the Flood was triggered by the springs of the great deep bursting forth. In light of this, we would expect a violent beginning to the Flood (bursting forth).

We would also expect that most of the water coming from springs as "water bursting forth" would *help supply* the waters that are classified as the windows of heaven. The bursting forth would likely begin a series of rapid catastrophic events that lead to massive sedimentation as well as continental scale break up. This particular milestone will play a roll when looking at the pre-Flood boundary.

Another milestone in geology is day 150. This is when the ark strikes the mountains of Ararat. This shows that mountain building, at least involving plates that collide in the Middle East, had begun. For the mountains of Ararat to have been formed requires that the Eurasian Plate, Arabian Plate, and African Plate (perhaps with some influence of a few other continental plates) have come together. On day 150, the Bible also reveals that the waters subsided, then the rain stopped, the waters then begin to steadily recede, the springs of the great deep are stopped, the windows of heaven stopped, and therefore, should transition to smaller scale effects, as opposed to continental scale layers.

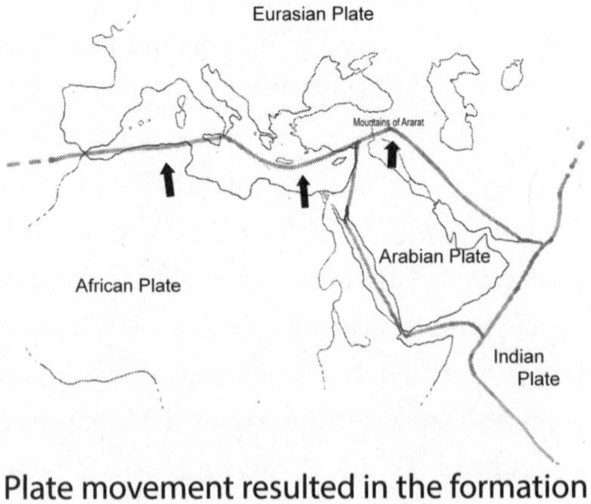

Plate movement resulted in the formation of the mountains of Ararat.
(Drawing: Bodie Hodge)

Another milestone is day 314. The water from the earth had receded in order for the surface of the earth to be dry (see also Psalm 104:8–10[2]) in Noah's view. And finally by day 370, the earth was completely dry, likely including the muddy areas that would have still persisted under the surface on day 314. But if we take note, the recessional stage of the Flood took about 5½ months, and the Flood's inundatory stage was exactly 5 months. So the recession stage took nearly the same time as it took for the water to prevail on the earth. Then there was another couple of months to complete the drying process.

After day 370, it seems the hydrological cycle was then more uniform with regularity (Genesis 8:22,[3] Genesis 9:12–14[4]) with the coming of clouds (Noah saw the rainbow which require droplets of moisture) and set times for seedtime and harvest (which require rain). Of course, other milestones throughout the Flood account can be found but this should suffice for a brief overview.

Geological Layers

The geological layers are generally stacked in a vertical column in books and articles that looks similar to the chart.

Even though most of these layers are exposed at various places around the earth, they are often stacked, because these layers have a general trend to be found on top of another layer when they are found together, though this is not always the case. For example, when one finds Cambrian layers, they are normally above Pre-Cambrian layers; Ordovician layers are generally above Cambrian layers; and so on. Though there are some exceptions, this is the *general* trend.

The names of the geological layers[5] were usually given based on a name of the area or something similar where they were found. For example, the Devonian layer was named for Devonshire (Devon), England. Cambrian

2. The mountains rose; the valleys sank down To the place which You established for them. You set a boundary that they may not pass over, So that they will not return to cover the earth. He sends forth springs in the valleys; They flow between the mountains. (NASB).
3. While the earth remains, seedtime and harvest, cold and heat, winter and summer, and day and night shall not cease.
4. And God said: "This is the sign of the covenant which I make between Me and you, and every living creature that is with you, for perpetual generations: I set My rainbow in the cloud, and it shall be for the sign of the covenant between Me and the earth. It shall be, when I bring a cloud over the earth, that the rainbow shall be seen in the cloud.
5. The classification of geological layers was often done by determining what fossils it contains. Logically, this can be very confusing and lead to some vicious circular arguments, but for the sake of what has been classified, we will continue to use the names for those layers.

was named for Cambria, the Latin name for Wales. Ordovician and Silurian were named for Ordices and Silures, early peoples in Britain and Wales, respectively. Permian was named for the Perm district in Russian and so on.[6]

Recent
Pleistocene
Pliocene
Miocene
Oligocene
Eocene
Paleocene
Cretaceous
Jurassic
Triassic
Permian
Pennsylvanian
Mississippian
Devonian
Silurian
Ordovician
Cambrian
Precambrian

So the names of the actual layers have very little to do with long ages; however, other names have been given to groups of these layers that have long-age implications. These are:

Eons

1. Era
2. Period
3. Epoch

Some of the names under these subcategories have been changing and continue to change, such as "Tertiary" or "Quaternary," which are viewed as out-of-date technical names, though they are still commonly used. The layers themselves would be better understood as depths, but more precisely as rock layers that contain particular sediment type and/or particular fossils, not necessarily representations of millions of years.

For the most part, both creationists and evolutionists generally agree that layers below another layer are older and the layers above are younger. However, we disagree by orders of magnitude. Typically for creationists, the deeper the Flood layers the earlier it was laid down during the Flood year, not millions of years of separation.

The Pre-Flood Boundary

With very few exceptions, creationists agree that the Flood in Genesis 6–8 explains the vast majority of rock layers that contain fossils on earth. Naturally, smaller rock layer units have formed since the time of the Flood, such as volcanoes and other local catastrophes (e.g., the Mount St. Helens sediment). Even so, most creationists have a general agreement about the placement of the pre-Flood boundary. This boundary is between the Pre-Cambrian and Cambrian layers. Of course, there are again exceptions,

6. "Table of Geological Periods," Information Please® Database, Pearson Education, Inc., 2007, http://www.infoplease.com/ipa/A0001822.html, accessed 12/18/2008.

depending on some specific places being studied.

When looking at the Tapeats Sandstone (the lowest sedimentary layer from the Flood in the Grand Canyon), it sits immediately above the Great Unconformity, which is the boundary where the Flood layers begin. Below the Tapeats are inverse rock layers that contain virtually no fossils. It makes sense that these lower non-fossiliferous layers were inverted at the onset of the Flood and then flood layers began to accumulate above it.

In fact, the vast rock layers of the Flood that encompass large areas, even entire continents, began to form at this time. This is what would be expected from a global scale Flood.

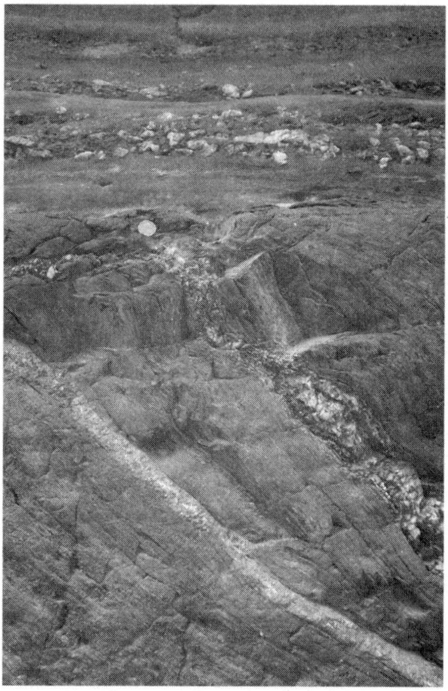

The Great Unconformity in the Grand Canyon reveals the beginning of the Flood layers with a Cambrian layer (Tapeats Sandstone) sitting immediately above Precambrian rock (Vishnu Schist) — quarter for scale

Post-Flood Boundary Controversy

If only the post-Flood boundary could have been determined so easily! With the Flood's steady decline from day 150 until day 314, and then further drying out until day 370, it can be a disaster trying to figure out what aspects of geology occurred during and after the Flood, particularly the end stages of the event. But even then we need a definitive point in the Flood account in Genesis to determine what the end of the Flood is. Most would recognize that when the Lord called Noah off the ark (day 370), then that is the official end of the Flood.

Based on mapping and boreholes of the geological layers, some appear to stretch across continents, or at least a vast majority of a continent. Most creationists recognize these as Flood layers. These dominate layers from the Cambrian to the end of the Cretaceous where it strikes the Tertiary sediment (Paleocene, Eocene, Oligocene, Miocene, and Pliocene). This is also known as the K/T boundary because the German for Cretaceous is Kreide, hence Kreide-Tertiary or K/T. After that, we begin to see smaller regional scale deposits (though still quite voluminous). From here, the situation gets difficult.

For example, during the slow recessional stage there would be sedimentary runoff that settled in some areas, breached dams that could burst, settling in stagnant areas, volcanoes that would go off during mountain building, mountain building that *caused* further sedimentation, and so on. Regional effects would surely begin during the recessional stage.

But also, *after* the recessional stage of the Flood, there could be sedimentary runoff that settled in some areas due to breached dams that burst, volcanoes that went off, and so on. A lot was happening from the moment the ark struck the mountains of Ararat and when Noah was called off the ark. This dangerous time is likely the reason that Noah wasn't called off the ark sooner!

Further regional affects could surely take place after the Flood as well (in fact, we see regional residual affects today). In the post-Flood era, there could be settling sediments, plate movement, drying effects, warm oceans causing rough weather, volcanic activity that is still proceeding, breached dams of water being released. But for the most part, the world was safe enough for Noah and his family to disembark and the Lord to give the declaration for them to be fruitful and fill the earth.

But here is the question . . . how can we be certain if one of these breached dams, volcanoes, or other regional effects happened prior to day 370 or soon after? This is where the debate heats up and why creationists tend to disagree with each other so much!

Technical articles (too many to cite) propose different positions of the post-Flood boundary in various places around the world and based on various aspects. Depending on which deposit, some say the boundary is as far down as the end of the Cretaceous (some even place it a little farther down *in* the Cretaceous layers, below the K/T boundary).[7] Others place it as high as the middle of the Pleistocene in some areas. Still others have it in other places (mostly in between). The various placements of the post-Flood

7. There are some holding to a Anglo-European model who try to place the post-Flood boundary at Pre-Cambrian and Cambrian boundary and claim all the sedimentation above that point has accumulated since then. However, most creationists rightly reject this, but that is not for discussion in this chapter.

boundary are *scientific* models, which are built on theological models, which are built on biblical models, which are taken from the text of Scripture.

So what does this really mean with regard to biblical authority? It means scientific models are not absolute and debate is *encouraged* on the subject in an *iron sharpening iron* (Proverbs 27:17) fashion. For creationists, scientific models are about debate and questioning, and in the future can change or be discarded. If one or both of the most popular placements of the Flood boundary are one day discarded, it in no way affects the truth of Scripture, which is absolute. This short article on the subject is not meant to be a treatise to battle for one or the other, but to educate on the views and the implications of the views.

Where Is the Post-Flood Boundary?

In the *flood* of creationist literature thereof, we find that for a hundred years most look in their backyard to try to determine the post-Flood boundary. Let me put it like this: Australian geologists tend to look at Australian geology, Americans tend to look at American geology, and Europeans tend to look at European geology. Naturally, there are exceptions to this.

But researchers tend to look at the size of the rock layers and the extent, and try to determine if they are large enough layers to identify as being from the Flood, or if they are more regional then they may be post-Flood. They look at the fossils in the layers and try to determine if they would be from the Flood or post-Flood (i.e., if kangaroos are found in the tertiary rock layers in Australia, then it seems likely these are post-Flood rock layers, as kangaroos migrated to Australia after the Flood to where they live today).[8]

In short, researchers look at the geology on various continents, and based on man-made criteria, they arbitrarily try to determine if it is from the Flood. For example, one such paper was published that did this very thing, but using suites of criteria.[9] But is this the best method?

8. Others, however, object and point out that there is a 16.7% change of any animal kind repopulating and thriving in the same continental area where they lived pre-Flood. This comes from the odds of them returning to that particular continent out of six continents after the Flood. So this is expected for some creatures anyway.
9. John Whitmore and Paul Garner, "Using Suites of Criteria to Recognize Pre-Flood, Flood and Post-Flood Strata in the Rock Record with Application to Wyoming (USA)," Andrew Snelling, editor, *Proceedings of the Sixth International Conference on Creationism* (Pittsburg, PA: Creation Science Fellowship, Inc. and Dallas TX: Institute for Creation Research, 2008), p. 425–448.

Let's start with the Bible. We know that the mountains of Ararat were formed in the Flood, by the 150th day. The water receded and never returned to overtake them. So of what are the mountains of Ararat made? This should give us a clue as to what was made *well before* the end of the Flood; in fact, we would know what was formed by 150th day of the Flood (less than halfway through the Flood).

Two different geologists (who simply do not have a hand in this debate since they are secular), who were searching for oil deposits, have mapped the regions that include the mountains of Ararat and beyond extensively.[10] What we find are layers intrinsic to the formation of the mountains of Ararat (Armenia and Anatolia regions) that include:

1. Permian
2. Lower-Middle Triassic
3. Middle Triassic-Middle Cretaceous
4. Paleocene-Lower Eocene
5. Lower Eocene
6. Middle Eocene
7. Middle-Upper Miocene

For much of this, the Eocene and Miocene rock layers are *inverted* and pushing up the Cretaceous and Triassic rock layers. In other words, without the Eocene and Miocene rock layers, the mountains of Ararat cannot exist!

What can we glean from this? It means that Miocene and Eocene rock layers existed by day 150 in the mountains of Ararat. These layers are tertiary sediments much higher than the K/T boundary. What we can know is that these Eocene and Miocene rock layers were formed prior to the post-Flood period. Does this mean that other Miocene and Eocene and other tertiary sediments (outside the mountains of Ararat) are also Flood deposits? It is possible. But I will leave the researchers to debate this further.

10. Y. Yilmaz, "Alochthonous Terranes in the Tethyan Middle East: Anatolia and the Surrounding Regions," *Philosophical Transactions of the Royal Society*, London, A 331, 611–624 (1990); G.C. Schmidt, "A Review of Permian and Mesozoic Formations Exposed Near the Turkey/Iraq Border at Harbol," Mobil Exploration Mediterranean, Inc. Ankara, MTA *Bulletin of the Mineral Research and Exploration Institute*, no. 62, 1964, p. 103–119.

Conclusion

So where should new creationists sit on this issue?

	Recent	Regional Layers
	Pleistocene (Ice Age)	
End of the Flood?	Pliocene	
	Miocene (Mts. of Ararat)	
	Oligocene	
	Eocene (Mts. of Ararat)	
	Paleocene	
End of the Flood?	Cretaceous (Mts. of Ararat)	Worldwide Layers
	Jurassic	
	Triassic (Mts. of Ararat)	
	Permian (Mts. of Ararat)	
	Pennsylvanian	
	Mississippian	
	Devonian	
	Silurian	
	Ordovician	
Beginning of the Flood	Cambrian	
	Precambrian	Pre-Flood Layers

We should keep in mind that scientific models that deal with the placement of the post-Flood boundary are good but not perfect, as the Scriptures are. Also, we need to keep in mind that there is a friendly debate in this area and there are implications. Answers in Genesis has tended to hold to a higher placement of the post-Flood boundary from the K/T boundary (end of the Cretaceous) or above but also recognizes there is a debate going on.

As for the details, we want to encourage both sides to have a kind debate as they hash out various regional formations to determine if they are Flood (day 150–day 370) or post-Flood deposits, but we think the biblical clue gives us a brilliant foundation by which future research is to be developed.

Noah's Ark

28 What about the Size of Noah's Ark?

Bodie Hodge

> If you are wise, you are wise for yourself, and if you scoff, you will bear it alone (Proverbs 9:12).

People often scoff at Noah's ark, saying it was not big enough to hold the animals. Others scoff at the ark because they say it was too big and thus couldn't be seaworthy! Did you realize that scoffers scoff for the sake of scoffing sometimes? Let's look at the size of Noah's ark, and leave the scoffers to bear it alone.

Ark Size

God tells us the dimensions of Noah's ark:

> And this is how you shall make it: The length of the ark shall be three hundred cubits, its width fifty cubits, and its height thirty cubits (Genesis 6:15).

Unlike the measure of cubits, we either use a metric (e.g., *decimal* based like centimeters, meters, kilometers, liters) or English/Imperial/US Customary system of units (e.g., often *fraction* based like inches, feet, miles, gallons). So while the Bible tells us that the length of Noah's ark was 300 cubits, its width 50 cubits, and its height 30 cubits, we must first ask, "How long is a cubit?"

The answer, however, is not so precise, because ancient people groups assigned different lengths to the term "cubit" (Hebrew word אמה [*ammah*]), the primary unit of measure in the Old Testament. This unit of measure was also used for the ark of the covenant (Exodus 25:10[1]), the Altar (Exodus 38:1[2]), Goliath (1 Samuel 17:4[3]), and Solomon's Temple (1 Kings 6:2[4])?

The length of a cubit was based on the distance from the elbow to the fingertips, so it varied between different ancient groups of people. Here are some samples from Egypt, Babylon, and ancient Israel:

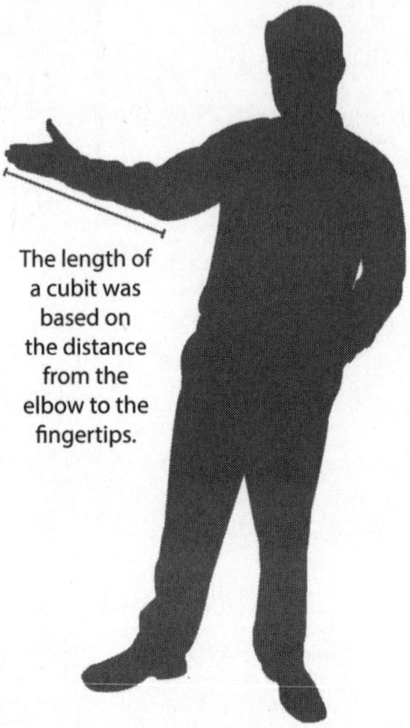
The length of a cubit was based on the distance from the elbow to the fingertips.

Culture	Inches (Centimeters)
Hebrew (short)	17.5 (44.5)
Egyptian	17.6 (44.7)
Common (short)	18 (45.7)
Babylonian (long)	19.8 (50.3)
Hebrew or royal (long)	20.4 (51.8)
Egyptian (long)	20.6 (52.3)

When Noah came off the ark, naturally it was his cubit measurement that existed — the one he had used to construct the ark. Unfortunately, the exact length of this cubit is unknown. After the nations were divided years later at the Tower of Babel, different cultures (people groups) adopted different sized cubits. So it requires some logical guesswork (based on historical research) to reconstruct the most likely length of the original cubit.

Since the Babel dispersion was so soon after the Flood, it is reasonable to assume that builders of that time were still using the cubit that Noah used. Moreover, we would expect that the people who settled near Babel would have retained or remained close to the original cubit. Yet cubits from that region (the ancient Near East) are generally either a common (short) or a long cubit. Which one is most likely to have come from Noah?

1. And they shall make an ark of acacia wood; two and a half cubits shall be its length, a cubit and a half its width, and a cubit and a half its height.
2. He made the altar of burnt offering of acacia wood; five cubits was its length and five cubits its width — it was square — and its height was three cubits.
3. And a champion went out from the camp of the Philistines, named Goliath, from Gath, whose height was six cubits and a span.
4. Now the house which King Solomon built for the LORD, its length was sixty cubits, its width twenty, and its height thirty cubits.

In large-scale construction projects, ancient civilizations typically used the long cubit (about 19.8–20.6 inches [52 cm]). The Bible offers some input in 2 Chronicles 3:3,[5] which reveals that Solomon used an older (long) cubit in construction of the Temple.

Most archaeological finds in Israel are not as ancient as Solomon. More modern finds consistently reveal the use of a short cubit, such as confirmed by archaeologists while measuring Hezekiah's tunnel. However, in Ezekiel's vision, an angel used "a cubit plus a handbreadth," an unmistakable definition for the long cubit (Ezekiel 43:13[6]). The long cubit appears to be God's preferred standard of measurement. Perhaps this matter did not escape Solomon's notice either.

The original cubit length is uncertain. It was most likely one of the long cubits (about 19.8–20.6 inches). If so, the ark was actually bigger than the size described in most books today (prior to the opening of the Ark Encounter), which usually use the short cubit of 18 inches.

Using the short cubit (18 inches), Noah's ark would have been about:

- 450 feet (137 meters) by 75 feet (22.9 meters) by 45 feet (13.7 meters)

Short cubit ark put to a box shape

Whereas using the long cubit of about 20.4 inches, Noah's ark would have been about:

5. This is the foundation which Solomon laid for building the house of God: The length was sixty cubits (by cubits according to the former measure) and the width twenty cubits.
6. These are the measurements of the altar in cubits (the cubit is one cubit and a handbreadth): the base one cubit high and one cubit wide, with a rim all around its edge of one span. This is the height of the altar.

- 510 feet (155 meters) long by 85 feet (25.9 meters) by 51 feet (15.5 meters)

Long cubit ark put to ship shape

Again, the long (older) cubit is to be preferred, so Noah's vessel was likely about this length, give or take a little. Also take note that these dimensions for length and width are to be measured from the water level, which is called the "draft level."[7]

Could a Wooden Ship This Big Be Possible?

Many scoffers have attacked the fact that a wooden ship with the dimensions of Noah's ark (which was 300 cubits long; that is about 450 feet or 510 feet, depending on the short or long cubit) could survive the Flood.[8] For example, Bill Nye, a skeptic, compared the ark to a wooden ship called the *Wyoming*, which was one of the biggest wooden sailing ships built with 1800s technology.

The *Wyoming* was a sailing ship, not a floater like Noah's ark. Interestingly, it was about the same length of Noah's ark if you use the short cubit (18-inch cubit as opposed to the 20.4-inch cubit). It was 450 feet long if you count the jib-boom. Bill Nye pointed out that the *Wyoming* was not a

7. Draft is simply where the water level is on the outside of a ship, based on how heavy it is.
8. T. Chaffey, "Bill Nye the Straw Man Guy and Noah's Ark," Answers in Genesis, http://www.answersingenesis.org/articles/2014/02/28/bill-nye-straw-man-guy-noahs-ark.

great seaworthy vessel and it twisted and sank, killing all 14 on board. Ergo, Noah's ark was basically "an impossibility."

We beg to differ. The *Wyoming* was a great example to show that the ark was indeed possible. Besides, using one example of a wooden ship (when there is much historic information about massive wooden ships built in the past) to claim Noah's ark couldn't have been built to survive the Flood is a very poor argument indeed.

The *Wyoming* was used and sailed for *14 years* before she sank. Noah's ark only needed a maximum of about 110 days of floating time.[9] Second, the *Wyoming* used 1800s technology, and by then mankind had lost a great deal of technology about boat building. Ark researcher and mechanical engineer Tim Lovett writes:

> Ancient shipbuilders usually began with a shell of planks (strakes) and then built internal framing (ribs) to fit inside. This is the complete reverse of the familiar European method where planking was added to the frame. In shell-first construction, the planks must be attached to each other somehow. Some used overlapping (clinker) planks that were dowelled or nailed, while others used rope to sew the planks together. The ancient Greeks used a sophisticated system where the planks were interlocked with thousands of precise mortise and tenon joints. The resulting hull was strong enough to ram another ship, yet light enough to be hauled onto a beach by the crew.[10]

It makes sense that this ancient technology was passed down through the Flood to the coastline/maritime peoples (Genesis 10:5[11]) and had been lost by the Age of Exploration. Lovett writes:

> At first, historians dismissed ancient Greek claims that the *Tessarakonteres* was 425 feet (130 m) long. But as more information was learned, the reputation of these early shipbuilders grew markedly. One of the greatest challenges to the construction of large

9. B. Hodge, "Biblical Overview of the Flood Timeline," Answers in Genesis, http://www.answersingenesis.org/articles/2010/08/23/overview-flood-timeline; Also, see chapter 26 in this book.
10. T. Lovett with B. Hodge, "What Did Noah's Ark Look Like?" in K. Ham, gen. ed., *The New Answers Book 3* (Green Forest, AR: Master Books, 2010), p. 17–28.
11. From these the coastland peoples of the Gentiles were separated into their lands, everyone according to his language, according to their families, into their nations.

> wooden ships is finding a way to lay planks around the outside in a way that will ensure little or no leaking, which is caused when there is too much movement between the planks. Apparently, the Greeks had access to an extraordinary method of planking that was lost for centuries, and only recently brought to light by marine archaeology.[12]

Larry Pierce (translator of James Ussher's *Annals of the World*) discusses several ancient wooden ships of antiquity that have been recorded in ancient history.[13] One of these ships was the *Leontifera*, and based on its description of rowers, the ship was about 400–500 feet long (120–150 meters). Those mentioned by Pierce were:

- *Leontifera* (warship): ~400–500 feet (~120–150 meters)
- Ships of Demetrius: ~400 feet (~120 meters)
- Ptolemy Philopator's warship: 420 feet (130)

Unlike Noah's ark, which was a floater, these ships required extra space just to man oars, sails, and the like. These sizes show that ancients possessed the brilliant capability to build large wooden ships that were known to ram other ships in war and stay afloat. That technology had all been lost throughout the ages.

Was Noah Unskilled at Ship Building?

There is the claim that Noah and his family were unskilled at shipbuilding and so the feat of building the ark of such a size would be impossible. The Bible reveals that Noah had 500 years under his belt before he was given the instruction to build the ark. Unlike shipbuilders in the ancient world (post-Flood) or even up through the Age of Exploration (~A.D. 1900), few would have a resume such as Noah had!

Noah also had 50–75 years (estimate) to research and build the ark.[14] I trust that few people in the past 3,000 years would ask someone to build them a ship and give them upward of 75 years to do it! The point is, if Noah

12. Tim Lovett, "Thinking Outside the Box," Answers in Genesis, https://answersingenesis.org/noahs-ark/thinking-outside-the-box/.
13. Larry Pierce, "The Large Ships of Antiquity," Answers in Genesis, https://answersingenesis.org/noahs-ark/the-large-ships-of-antiquity/.
14. B. Hodge, "How Long Did It Take for Noah to Build the Ark?" http://www.answersingenesis.org/articles/2010/06/01/long-to-build-the-ark; also see chapter 25 in this book.

wasn't an expert already, then he could easily have had the time to become one. It is the same with his family.

Second, how does anyone know that God didn't give Noah specific instruction on how to build it? The Bible simply doesn't say, but this is a possibility. The Bible tells us that God gave him the dimensions and what wood to use (as well as a few other specifics). It would be outrageous to believe that an all-knowing God wouldn't know how to design an ark to survive a Flood.

Next, it is possible that Noah had help. He could easily have contracted with people that may have been experts on certain things. Furthermore, other family and friends could have been helping until they died, such as Methuselah, Lamech, or others (Genesis 5:30,[15] 6:11[16]). There could have been righteous people helping until they were killed or died.

What we know was that Noah (with his family) was all that was left to be saved by God at the time of the Flood. But Noah knew this. He was comforted in knowing that his family would not be killed and would be aboard the ark as the Lord revealed this to him well in advance (Genesis 6:18[17]).

---

15. After he begot Noah, Lamech lived five hundred and ninety-five years, and had sons and daughters.
16. The earth also was corrupt before God, and the earth was filled with violence.
17. But I will establish My covenant with you; and you shall go into the ark — you, your sons, your wife, and your sons' wives with you.

29 What Was the Shape of Noah's Ark?

Bodie Hodge

Introduction

Many readers have surely seen a multitude of images of Noah's ark. In the last chapter, we just saw two. Sadly, many are quite fanciful, unrealistic, and even laughable, in their proportions. We often call these "bathtub arks" because they look like an overloaded bathtub with today's animals (and usually with giraffes sticking out through the chimney).

Figure 1: False arks *do* play a role.

We suggest these types of "mythical arks" are actually quite dangerous. Many kids are raised with them. Although they look cute, it sends a subtle and false message that Noah's ark was not a real ship and certainly couldn't fit all the animals needed on board.

Even atheists proclaimed that the idea of the ark being a mythological account was the first step in their move to the religion of atheism: "I remember the moment in 1995 that I realized that the story of Noah's ark was just an ancient hand-me-down ridiculous mythological tale." Interestingly, the comment was accompanied by an unrealistic Noah's ark image (see figure 1).[1]

1. Linda LaScola, "Leaving Religion: Like Crumbling Jenga Blocks," http://www.patheos.com/blogs/rationaldoubt/2015/09/leaving-religion-in-stages-like-crumbling-jenga-blocks/.

Ancient Arks

At Answer in Genesis, we received a letter from Harlan and Stacy Hutchins with a printed image of an engraving done in London in 1760 by a man named P. Simms. Mr. Hutchins came across this engraving while working as an antique map and print dealer.

At first glance, you would recognize the antiquity of such an image. But it is what was *on* the image that got me excited! It was a treasure indeed — an engraving of Noah's ark.

What Makes This Ark Unique?

This ark was not like the common inaccurate bathtub arks that "float" around in today's culture. Instead, Simms took the time to think about Noah's ark, its dimensions, and its construction. A copy of the image can be seen on the following page, along with detailed portions.

The first thing you will notice is that Simms' ark is much closer to the biblical proportions that were given: 300 by 50 by 30 cubits (Genesis 6:15). Mr. Simms simply squares them off.

I'm surprised many illustrators and researchers today have failed to attain this basic information, considering it is given in the Scriptures. Instead, they proceed to depict bathtub or otherwise inaccurate arks.[2]

Another feature you will notice is that Simms' ark clearly has three decks, as revealed in Genesis 6:16. Simms also placed rooms in the ark (Genesis 6:14) and placed some animals inside for scale. As a teaching point, he also made some of the walls "see through" so observers can get an idea of the scale. Such things are mysteriously absent from many of the bathtub arks and other inaccurate depictions of the ark today.

I can tell Simms also thought about the "window." The window is difficult for us to understand, being that it is translated as "window," but the Hebrew word is typically translated as "noon" or "midday" elsewhere in Scripture. The "window" was something meant to allow ventilation and lighting in the ark. Though Simms placed several windows in the ark "a cubit from above" (Genesis 6:16) to serve this purpose, they are not at the noon or midday position. Today, we prefer a centrally located "midday" window finished a cubit from *above* the roof — a closer match to the biblical

2. Of course, there is debate over which cubit to use, be it the shorter or the longer cubit, but for more information on that subject see, "How Long Was the Original Cubit?" http://www.answersingenesis.org/articles/am/v2/n2/original-cubit.

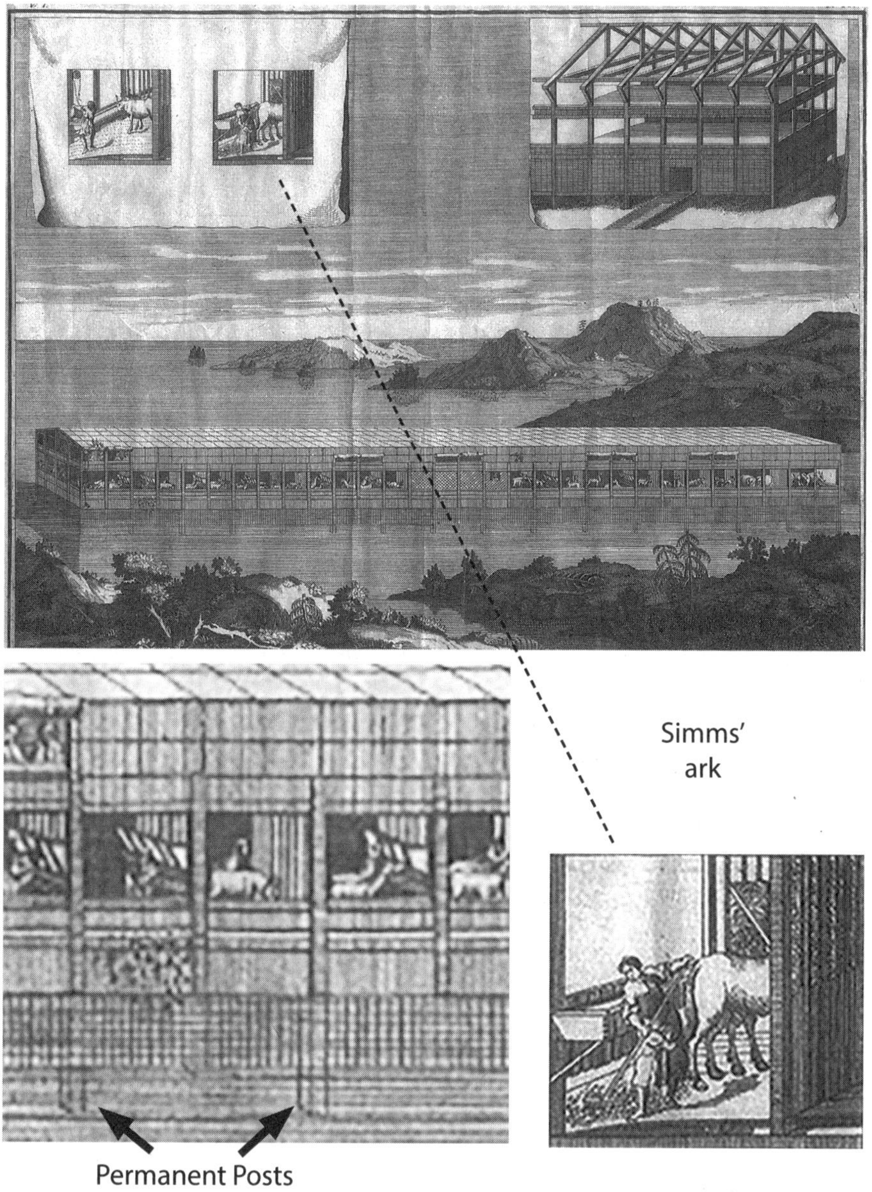

description and a more water-resistant design resembling a deck hatch. But in 1760, Simms was clearly trying to let Scripture guide his design, unlike modern bathtub mythological arks!

He also thought about the roof construction. In the upper right, he postulated how the ark may have been constructed with an angled roof and overhang to prevent water leakage from water sitting on the roof. He even considered the footings of how the ark could have been elevated on certain pillars to allow water underneath until the time of liftoff.

However, Simms' considerations for the water were hardly necessary. Water would have had no trouble getting underneath the ark. The pillars (which happened to be about as high as a human) would have allowed humans access underneath the ark. At the world famous *Ark Encounter* something similar has been done. This is much more important for a wooden vessel where planks need to be accessible, unlike the modern system of construction where modules are welded together in a single weld line across the ship. Many ships today are actually built above ground *and then* placed in the water. Noah probably did not have that luxury, and so Simms considered the pillars a viable option.

The ark may have been too large for picking up and placing into the water like a small boat (not to mention that Noah was shut in the ark by the Lord), so the only options comparable to modern shipbuilding are either launching on a slipway or flooding in a dry-dock. In Noah's case, rising floodwater is very similar to flooding a dry-dock, except not nearly as controllable.

One interesting detail in Simms' ark is the permanent posts. They are integral in the structure, so it looks like Mr. Simms was expecting the pillars to act the same way when the ark was beached at the end of the voyage.

It would be difficult to ensure that so many extended posts remained waterproof, as they would be very vulnerable to any bumps and scrapes. A glancing blow against something hard, and they would loosen up somewhat and let the water in. But at least he was thinking about it and coming up with ideas.

He even considers feeding, watering (even a bucket and pulley system), loading (with the ramp), perhaps human living quarters (middle thatched area on the image), and cleaning options too. Though obviously not perfect, he thought about it in his image.

Mr. Simms' ark is unique, because many ark pictures from Simms' time equated the ark with the ocean vessels *of his day* (think of the sailing ships carrying people to the Americas during the age of exploration).

Other Thoughtful Ark Depictions

Simms was not alone, nor the first to think through the ark's features. Back in 1554, Johannes Buteo published the first serious study of Noah's ark, investigating its construction, capacity, and the number of animals and food and provisions required.[3] His illustration was still rather basic. But even so, his ark was standing against some incredibly false shapes of Noah's ark permeating in his day. His ark was the basic rectangular box-shaped ark we often see.

More than a century later, Athanasius Kircher published his *Arca Noë* in 1675, with lavish illustrations depicting a rectangular ark of biblical proportions.

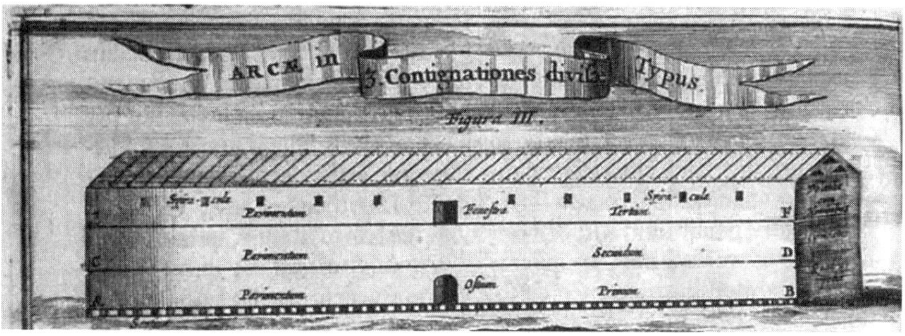

Kircher ark depiction

In 1707, a German Bible had an image of Noah's ark that also had been carefully considered. Famed Baptist commentator Dr. John Gill included it in his commentary on Genesis in 1748–63[4] (see following page).

Of course, there have been structural studies, animal studies, and so on for Noah's ark since the 1960s, mostly inspired by the recent return to biblical authority that was triggered by the book *The Genesis Flood* in the early 1960s by Drs. Henry Morris and John Whitcomb.

Since Simms and these others, more research has been done on ancient ships and technology of wooden vessels. Structural studies and designs have been discussed in detail in modern times. Some details about Noah's ark are believed to be different from these earlier images.

Was Noah's Ark Box-Shaped?

The dimensions for Noah's ark are given in length, width, and height. Does that mean it was shaped like a box? The dimensions given in Genesis 6:15

3. Tim Griffith and Natali Monnette, translators, "Johannes Buteo's On the Shape and Capacity of Noah's Ark," *Issues in Creation*, no. 2, February 28, 2008.
4. John Gill, *An Exposition of the First Book of Moses, Called Genesis* (Springfield, MO: Particular Baptist Press, 2010) (originally Aaron Ward Publishing, London, England, 1763–1766), p. 121.

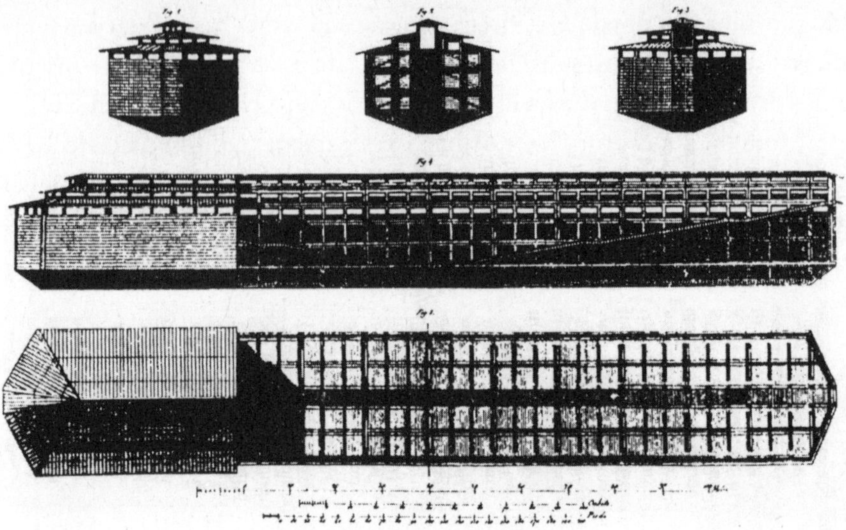

The John Gill ark depiction

are 300 x 50 x 30 cubits, which translate on the *long* or *royal* cubit as roughly 510 x 85 x 51 feet (see the previous chapter for a discussion on cubits). Some argue for a box shape while others argue for something that looked more like a ship. The basis for this comes from the scant information given in Genesis:

> Make yourself an ark of gopherwood; make rooms in the ark, and cover it inside and outside with pitch. And this is how you shall make it: The length of the ark shall be three hundred cubits, its width fifty cubits, and its height thirty cubits. You shall make a window for the ark, and you shall finish it to a cubit from above; and set the door of the ark in its side. You shall make it with lower, second, and third decks (Genesis 6:14–16).

This ratio of dimensions is extremely well placed. According to researchers, this ratio sits nearly central between strength, comfort, and stability.[5] Even some ocean-going vessels were designed based on this ratio because it is ideal.

But keep in mind that the Bible *doesn't* say the ark was a box shape. A box-shape *is assumed* by taking the length, width, and height of the ark and merely making it into a box. However, this was likely not the case, as ship dimensions are typically given as length, width, and height. This is common

5. Gon et al, "Safety Investigation of Noah's Ark in a Seaway," *Technical Journal*, vol. 8, no. 1, April 1994, p. 26–36, https://answersingenesis.org/noahs-ark/safety-investigation-of-noahs-ark-in-a-seaway/.

for other things as well. For example, a Corvette ZR1's dimensions were once given as the following:

Overall length (in / mm): 176.2 / 4476
Overall width (in / mm): 75.9 / 1928
Overall height (in / mm): 49 / 1244

Are we to assume the Corvette is a box shape given only its length, width, and height? Not at all. Noah's ark was a ship; therefore, it likely had features that ships would commonly have. These are not at all unreasonable assumptions.

Noah was 500–600 years old and knew better than to make a simple box that would have had significant issues in a global Flood (e.g., forces on the sharp corners would be too destructive, it could capsize if it was not facing into the wind and waves, and so on). The common way that a floating boat would keep from being turned perpendicular with the wind and waves is by having features on the front and back that naturally utilize the wind and waves to point it into the wind and waves. This would eliminate the need for sharp corners as well.

The word for ark in Hebrew is *tebah*, which is also the word for the bulrush basket in which Moses was laid as a child and then placed in the Nile (Exodus 2:3[6]).[7] Furthermore, *tebah* is not the same word used for the ark of the testimony or ark of the covenant — that Hebrew word is *'arown*.

The word *tebah* in some modern lexicons yields the definition of *box*, but is more properly *chest* and related to the Egyptian word for *coffin*. I'm sure that being in the ark for nearly a year must have felt like being in a coffin! Recognized Hebrew lexicons such as *Brown, Driver, and Briggs* yield the meaning of *tebah* regarding Noah's ark and the basket of Moses as a *vessel*, not a box.[8]

Why has the box shape persisted for so long? Well, it is *easy* — and little thought was required to do a box/rectangle. Besides, for the better part of the past, the question was about fitting animals on board the ark, so the structure wasn't seen as that big of a deal.

---

6. But when she could no longer hide him, she took an ark of bulrushes for him, daubed it with asphalt and pitch, put the child in it, and laid it in the reeds by the river's bank.
7. "Moses baskets" have become a novelty in certain places today and they are not a box shape either, but rather rounded. Their dimensions are commonly 30 in. x 12 in. x 9 in. Of course, since the Bible does not provide the dimensions of the ark in which Moses was placed we cannot state with any certainty what the size may have been.
8. *The Brown Driver and Briggs Hebrew and English Lexicon*, 9th printing (Peabody MA: Hendrickson Publishers, 2005), p. 1061.

Bodie Hodge on a "bathtub ark" ride

The response was usually, "Let's make a box and then see if we can fit the required animals on the ark." After years of research, that question has been answered sufficiently (see the next chapter for highlights). The skeptical questions are now shifting. Instead of asking how the animals fit on the ark (that question does still persist among those not learned on the subject), they ask how the box-shaped ark could survive the Flood. They point out that sharp corners would not be able to handle the stresses and would break, and the ark would flood and sink, for example.

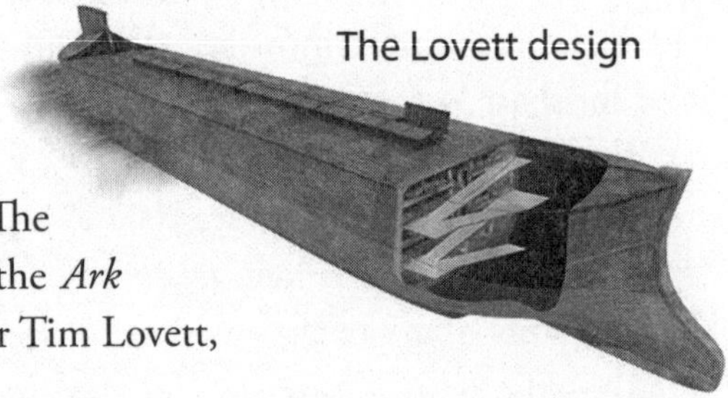

The Lovett design

So in recent times, this research has been focused on the shape and structure of Noah's ark. This has led to what is called the Lovett design. The basic design that was used for the *Ark Encounter* is from ship researcher Tim Lovett, a mechanical engineer.

Is this the best design? Great question. But it is one of the most thought-through structural designs throughout the ages. And based on experimental data, it is stable in wind and waves and even rights itself in rough waters/windy air so that it doesn't capsize. The *Ark Encounter* design is a variation of the Lovett design.

The Ark Encounter (Photo: Ken Ham)

30 Did the Animals Fit on Noah's Ark?

Bodie Hodge

> Of the birds after their kind, of animals after their kind, and of every creeping thing *of the earth* after its kind, two of every kind will come to you to keep them alive (Genesis 6:20, emphasis mine).

The first order of business is to point out that it was representative kinds of land-based creatures on the ark. No aquatic creatures like fish or trilobites needed to be on the ark. Much sea life died in the Flood — it was a violent Flood after all. But the kinds of *water*-based creatures could easily survive the *watery* Flood.

Kinds vs. Species

To understand how many animals went on the ark, some preliminary information is needed. This brings us to the question of "What is a kind?"

Kinds are not necessarily species, genus, family, or order by our modern classification system. In a general sense, it is probably *closer* to the family level in most instances, but a few can still be at a species (like humans), genus, or even order level depending on what we are looking at.

Kinds are like the dog sort (including dingoes, wolves, coyotes, domestic dogs, etc.), cat sort (including lions, tigers, cougars, bobcats, domestic cats, etc.), horse sort (ponies, Clydesdales, donkeys, zebras, etc.), and so on. There is variation within these kinds especially since the Flood, but not evolution where one kind changes into a totally different kind over long periods of time — which is not observed anyway (e.g., amoebas turning into dogs).

Species is an arbitrary dividing line and becomes a paradox, hence the famous "*Species Problem*" when trying to define it. Some claim that if creatures can interbreed, then they are the same "species"; but clearly that is not always the case, since various species of dogs can and do interbreed — yet they are still considered different species.

The same is true of cats. For instance, we have a blinx (bobcat father with a linx mother hybrid), ligers (lion father-tiger mother mix), tigons (offspring of a male tiger and female lion), etc.

As stated, most kinds are closer to a family level, but in some cases they could be genus, species, or in the event of the elephant kind, is likely at the level of order, including the extinct families of elephants such as mammoths and mastodons, for example.

The point of all of this is that one cannot try to tally *species* on board Noah's ark or they will have a gross exaggeration. All the land-dwelling, air-breathing species we have today are the descendants of the representative kinds that were aboard the ark. Ocean-based creatures are not required on the ark (e.g., trilobites). But let's look at the other information in Scripture that further limits what went onboard Noah's ark.

Flesh, Insects, Land-Dwelling, and Air-Breathing . . . What?

> And all flesh died that moved on the earth: birds and cattle and beasts and every creeping thing that creeps on the earth, and every man. All in whose nostrils was the breath of the spirit of life, all that was on the dry land, died. So He destroyed all living things which were on the face of the ground: both man and cattle, creeping thing and bird of the air. They were destroyed from the earth. Only Noah and those who were with him in the ark remained alive (Genesis 7:21–23).

What we learn in this passage is that the animals aboard the ark were further limited by being *air-breathing through the nostrils* as well as being *land-dwelling*. Naturally, creatures that are air-breathing and yet lived in the ocean need not be required (e.g., whales and dolphins). They can survive outside the ark.

Insects use tubes in their exoskeleton to take in oxygen, not nostrils. Therefore, many researchers do not have insects onboard Noah's ark.[1] They

1. Even if insects were on the ark, due to their extremely small size, they could have easily fit. If a million individuals were taken, it would still be a small footprint in the ark.

could easily survive on driftwood; or at the very least their eggs could survive on rafted wood and debris.

We also learn that all *flesh* that moved on the earth had died (Genesis 7:21, see also Genesis 6:13–19). This Hebrew word for flesh is *basar*. Insects were never considered flesh or *basar* in the Bible, which is a further confirmation that their residence wasn't required on the ark during the Flood. Furthermore, the Hebrew terms used for *behemah* and *remes*[2] in Genesis 6:19–20[3] were never used of insects in the Bible.

So we are left with land-dwelling, air-breathing (through their nostrils) kinds that were considered flesh or *basar*. What are the numbers?

Minimum Figures

What are the minimum figures for animals on the ark? Researcher Arthur Jones, writing in the *Creation Research Society Quarterly* in 1973, simply put these qualifications at a family level and did the numbers.[4] He arrived at about 1,000 families (and equated this with kinds). This would be about 2,000 individuals taken on the ark.

This includes the number of clean animals that came in by sevens. Even so, this is likely a minimum figure as some kinds are at a genus and species level (in a few instances). Consider one simple reason. We have found more animals since 1970. Many are within current families, but some are not. So naturally the numbers will have gone up a little. Thus, Jones' numbers would be a minimum figure.

Maximum Figures

Researcher John Woodmorappe decided to go a different route in 1996. He wanted to calculate maximum figures for animals and also included food and even water on board Noah's ark.[5]

To do this, Woodmorappe decided to use the smaller-sized ark (based on the short cubit of 18 inches). This would be an ark about 450 feet long.

2. *The Theological Wordbook of the Old Testament* says of *remes*: "The root encompasses all smaller animals but seems to exclude the large grazing animals, whales, birds, and insects," William White (Chicago, IL: Moody Press, 1999), p. 850.
3. And of every living thing of all flesh you shall bring two of every sort into the ark, to keep them alive with you; they shall be male and female. Of the birds after their kind, of animals [*behemah*] after their kind, and of every creeping thing [*remes*] of the earth after its kind, two of every kind will come to you to keep them alive.
4. Arthur J. Jones, "How Many Animals in the Ark?" *Creation Research Society Quarterly*, vol. 10, no. 2, September 1973, p. 102–108.
5. John Woodmorappe, *Noah's Ark: A Feasibility Study* (Dallas, TX: ICR Publications, 2009).

Then, instead of a *family* level for the kind (although he recognized the kind was closer to the family level), he used a *genus* level for all the kinds! So instead of one dog kind, there would be more than ten dog kinds represented in his numbers since there are more than 10 levels of genus within the dogs!

For the sake of maximum figures, John Woodmorappe still did 14 *of each genus* of the clean animals, which again was still not that many!

What he found was under 8,000 kinds and about 16,000 (15,745) individuals maximum, based on this genus level and calculations. With a smaller ark and this maximum number of animals and their required floor space/cages/rooms, this came out to be about 46.8% of the ark used to hold the animals! Remember, this is maximum figure. But Woodmorappe didn't stop there.

Then, Woodmorappe calculated various types of foodstuffs on the ark for this maximum number of animals. He calculated various numbers, but it is doubtful that Noah took *loose hay* (which takes a lot of space) as a feeding stuff for all animals to eat! The reasonable numbers are:

	Foodstuff	Ark space required
1	Lightly compressed dried hay pellet	16.3 %
2	Doubly compressed hay pellet	12.5%
3	Pelleted horse food	7.0%
4	Dried fruits/vegetables	6.8%

Let's use the highest number here (16.3%). Even though Noah and his family could have harvested water for part of their duration on the ark, Woodmorappe decided to calculate fresh water and put it in storage on board the ark. He found that it would take about 9.4% of the ark.

Putting all this together with the smaller ark and maximum number of animals, maximum amount of food, and maximum amount of water, this would be about 72.5% of the ark. Of course, this number could reduce depending on the types of foodstuff. Keep in mind that Noah likely used the long (older) cubit and took far fewer animals.

During the voyage, some animals reproduced (surely rabbits, right?). When God called the animals off the ark, it was now by their *families*, not two by two (Genesis 8:19[6]).

6. Every animal, every creeping thing, every bird, and whatever creeps on the earth, according to their families, went out of the ark.

Ark Encounter Figures

The Ark Encounter is basically using the Lovett design, which utilizes the older long cubit (20.4 inches), and this makes a longer and bigger ark (510 feet long). So the ark has much more space. This needs to be kept in mind with Ark Encounter figures.

When it comes to the animals on board, the Ark Encounter uses "reasonable maximums" — in engineering this is basically equated with a *safety factor or over engineered!* The project is intentionally trying to err on the high end but not to the degree that Woodmorappe did with a genus level. The purpose is to avoid being accused of trying to have intentionally lower numbers to fit on the ark.

So the Ark Encounter is using 14 of the clean animals and 14 of *all* birds. They also *split* animals when determining kinds as opposed to *lumping.* In other words, the current secular "family" lists are all being equated with a separate kind, unless otherwise warranted. More on this in a moment.

A few discussion points need to be understood prior to giving the tentatively final Ark Encounter numbers.

2 or 4 and 7 or 14?

Let's discuss 7 and 14 in more detail. Was it 2 of each unclean kind or 4 of each sort that went on the ark? Was it 7 of each clean animal and bird of the air kind or 14 of each that went on the ark?

> You shall take with you seven each of every clean animal, a male and his female; two each of animals that are unclean, a male and his female; also seven each of birds of the air, male and female, to keep the species alive on the face of all the earth (Genesis 7:2–3).

Furthermore, the animals went on by pairs, a male and its mate, and in some cases (i.e., the clean kinds) 7 went on board. The list of clean birds and animals in the Old Testament is found in Leviticus 11 and Deuteronomy 14.[7]

The Ark Encounter is erring on the side of caution in having 14 of each clean animal and *all* birds/winged creatures to maximize the number

---

7. Clean animals in the New Testament now differ in that Christ declared all foods clean by His power (e.g., Mark 7:19). This was demonstrated by both Peter (Acts 10:9–16) and Paul (e.g., 1 Corinthians 6:12–13; 1 Timothy 4:1–4; and Colossians 2:16), thus returning essentially to the diet given in Genesis 9:3 not by merely going backward to eat all foods clean *and unclean,* but by eating *only clean* foods, as all food is now made clean.

of animals on the ark, though they are only taking 2 of the unclean animals. However, this may not be the *actual* case.

The reason for the confusion is the way it is worded. Genesis 7:2–3, 7:9,[8] and 7:15[9] use wording that is uncommon in English. When the word is translated as *two* or as *seven*, the Hebrew behind it is [two two] or [seven seven in Genesis 7:2-3]. This type of construct is not normal to English, though, it is common in Hebrew.

Let me explain. The Hebrew is essentially "two two" and "seven seven," which is translated as "two" and "seven," denoting that they are paired when possible — remember Noah sacrificed of the clean animals after the Flood (Genesis 8:20[10]).[11] These are called supernumary animals — those meant for sacrifice, not requiring a female mate for the purpose of reproduction after the Flood.

Think of it as if we were to say "*pair of two.*" "Pair of two" still indicates only 2 (not two plus two). We see this pairing in Exodus 36:30[12] (i.e., "two sockets under each of the boards" is "two two" implying that it was two total, but that it was paired with the socket above).

It is indicative of pairing. In other words, [two two] would be 2 in the form of a pair (i.e., a male and his female). Likewise, it would be 7 in the form of pairs when possible with a supernumary sacrificial animal. So it isn't doubled or multiplied. Why do I suggest this?

First, the immediate context called out 2 of each kind twice in Genesis 6:19-20[13] and Genesis 7:2 in reference to the same critters that are also denoted later as "two two." Thus, we know that two two is merely 2 in a paired form. Both "two" and "two two" were in reference to the same pair, male and female of unclean creatures.

---

8. Two by two they went into the ark to Noah, male and female, as God had commanded Noah.
9. And they went into the ark to Noah, two by two, of all flesh in which is the breath of life.
10. Then Noah built an altar to the LORD, and took of every clean animal and of every clean bird, and offered burnt offerings on the altar.
11. John Calvin writes: "Moreover, the expression, by sevens, is to be understood not of seven pairs of each kind, but of three pairs, to which one animal is added for the sake of sacrifice." John Calvin, *Commentaries on the First Book of Moses called Genesis*, translated by John King, reprint (Grand Rapids, MI: Baker Books, 2003), commentary notes on Genesis 7:2, p. 267.
12. So there were eight boards and their sockets — sixteen sockets of silver — two sockets under each of the boards.
13. Of the birds after their kind, of animals after their kind, and of every creeping thing of the earth after its kind, two of every kind will come to you to keep them alive.

As discussed elsewhere in Scripture, it uses the construct of [two two] as 2 (not doubled) as well. Exodus 36:29–30[14] also discusses pairs or couples of sockets under a board. So if you are looking for maximum figures, you can go with 14 and 4, but it makes more sense to remain with 2 and 7. Famed commentator H.C. Leupold concurs when he writes:

> The Hebrew expression "take seven seven" means "seven each" (Koenig's *Syntax* 85; 316b; Gesenius' *Grammatik* rev. by Kautzsch 134q). Hebrew parallels support this explanation. In any case, it would be a most clumsy method of trying to say "fourteen." Three pairs and one supernumery makes the "seven." As has been often suggested, the supernumary beast was the one Noah could conveniently offer for sacrifice after the termination of the Flood.[15]

It would have been very easy to say 14 (Hebrew *arba*) if that were meant. Seven animals has made sense to most commentators throughout the years. Why 7 clean? After the Flood, Noah sacrificed of each of the clean animals and perhaps this was to leave a breeding stock for Noah's three sons. This is a possibility.

7 of all Birds or 7 of the Clean Birds?

Another debate that presents itself at the Ark Encounter is the question of how many birds? The Bible says:

> You shall take with you seven each of every clean animal, a male and his female; two each of animals that are unclean, a male and his female; also seven each of birds of the air, male and female, to keep the species alive on the face of all the earth (Genesis 7:2–3).

Commentators recognize that when it says in Genesis 7:3, "seven each of birds of the air" that it is tied to the backdrop of *clean* creatures in Genesis 7:2. Contextually, this doesn't mean all birds came by 7 but instead that it is limited to 7 of the *clean* birds. Leupold writes:

---

14. And they were coupled at the bottom and coupled together at the top by one ring. Thus he made both of them for the two corners. So there were eight boards and their sockets — sixteen sockets of silver — two sockets under each of the boards.
15. H.C. Leupold, *Exposition of Genesis*, commentary notes on Genesis 7:2, as quoted by Drs. Henry Morris and John Whitcomb, *The Genesis Flood* (Phillipsburg, NJ: Presbyterian and Reformed Publishing Company, 1961), 44th printing, p. 65.

> In v. 3 the idea of "the birds of the heavens" must, of course, be supplemented by the adjective "clean," according to the principle laid down in v. 2.[16]

Dr. John Gill agrees when discussing the birds. He says:

> That is, of such as were clean; seven couple of these were to be brought into the ark, for the like use as of the clean beasts, and those under the law."[17]

Commentator Matthew Poole agrees:

> Of clean fowls, which he leaves to be understood out of the foregoing verse.[18]

Even so, the Ark Encounter is inserting 14 of each bird regardless of whether it was clean or unclean. So this increases the number dramatically to help maximize the numbers on the ark.

Splitting and Lumping

Another factor that the Ark Encounter has used when researching kinds is *splitting* rather than *lumping*. What does that mean? When it comes to a group of animals that have many similarities, should they be lumped together as one kind or should they be left separated into separate kinds?

Although other factors may be involved, the key biblical factor to determine if two creatures are the same kind is if they can interbreed (though some may have lost the ability). But if they can have offspring together, then they are the same kind.

For example, a coyote, wolf, dingo, and a Great Dane can interbred. Thus, they are the same kind, so they can be lumped together. Camels and llamas can interbreed, so they can be lumped together. Finches and sparrows can interbreed, so they can be lumped together.

But some animals have never been brought into contact with each other to know if they can have offspring — this is clearly the case with creatures that have died out (e.g., dinosaur kinds!). Consider that there are several

16. Ibid.
17. John Gill, *An Exposition of the First Book of Moses, Called Genesis* (Springfield, Missouri: Particular Baptist Press, 2010 edition (original publication: London: Aaron Ward publisher, 1763–1766), p. 128.
18. Matthew Poole, "Commentary on Genesis 7:3," *Matthew Poole's English Annotations on the Holy Bible*, 1685, //www.studylight.org/commentaries/mpc/genesis-7.html.

sauropod dinosaur families because some have longer tails/necks or are buried in Flood sediment that is higher up in the rock record.[19] Realistically, they could be one kind, but we cannot get hybridization data for them since they are no longer around, so we are stuck with the secular family divisions. Thus, they are split into several families.

Therefore, even if they have many similarities and might be part of the same kind, they were split and each family was left as its own kind, unless data is found to lump them together. For example, all bats may be of the same kind and at minimum could only require two on board the ark. However, since studies of interbreeding among bats has not been documented as of the opening of the Ark Encounter, then they potentially have over 300 of them. This also causes numbers to be significantly (and purposely) higher at the Ark Encounter.

Ark Encounter Final Numbers

Using the maximums discussed, the Ark Encounter estimation is about 1,400 kinds. This is translated to about 6,700 individuals on the ark. This number maximizes the animals but should likely go down with more research into new hybrids as they come to light.

Realistic Figures Are Much Less

Realistic numbers are clearly much lower. For example, having only 7 individuals of the *clean* birds rapidly reduced the number of birds/winged creatures on the ark. Bringing certain animals together that may be in the same kind, like sauropods, would reduce the number too. Also having *7 instead of 14* of the clean birds and beasts, and so on, would reduce the number of kinds and significantly reduce the number of individuals on the ark. The actual is likely less than 3,200 creatures.

Jones' minimum number of about 1,000 kinds and the Ark Encounter's maximum number of about 1,400 kinds gives us a decent range. From here, researchers can build and hone in on more realistic numbers.

---

19. In the secular mindset, these layers would be separated by millions of years and thus is also used so separate sauropods into different families. Creationists don't necessarily agree to this standard but for the sake of a starting point, the Ark Encounter is using this suite of criteria.

31 How Could Noah Get and Care for All the Animals?

Bodie Hodge

Gathering the Animals – The Easy Part!

Some have pondered how Noah was capable of finding all the animals that needed to be on the ark. The task would be unthinkable (especially when you are in the middle of building the ark and gathering food for the ark's voyage). However, this was an easy task for Noah because *God* brought the animals to Noah.

> And of every living thing of all flesh you shall bring two of every sort into the ark, to keep them alive with you; they shall be male and female. Of the birds[1] after their kind, of animals after their kind, and of every creeping thing of the earth after its kind, two of every kind will come to you to keep them alive (Genesis 6:19–20).

God brought them in pairs, a male and its mate (female) of each kind (think dog kind, cat kind, elephant kind, etc.). Many are aware that seven of the clean animals and birds went aboard the ark. The Bible says:

> You shall take with you seven each of every clean animal, a male and his female; two each of animals that are unclean, a male and his female; also seven each of birds of the air, male and female, to keep the species alive on the face of all the earth (Genesis 7:2–3).

1. Bird here is *owph* in Hebrew, which means winged creature, so it would also include bats, pterosaurs, etc.

Interestingly, the Bible never says God brought seven of each kind to Noah, so it is possible they already had the 5 extra clean animals and birds (which is a small list per Leviticus 11 and Deuteronomy 14) in their possession. The clean animals could be used for sacrifice, leather, milk, and so forth. Recall, that it was not until after the Flood that meat eating was permissible for man (Genesis 9:3[2]).

Noah likely had been given the numbers by the Lord so that he could prepare the cages/rooms for them in the ark. It also allowed him to know the approximate amount of food to gather.

Loading the Animals

The animals went on the ark in pairs — a male and its mate. The loading could have taken upward of a week. Scripture says:

> Then the LORD said to Noah, "Come into the ark, you and all your household, because I have seen that you are righteous before Me in this generation. You shall take with you seven each of every clean animal, a male and his female; two each of animals that are unclean, a male and his female; also seven each of birds of the air, male and female, to keep the species alive on the face of all the earth. For after seven more days I will cause it to rain on the earth forty days and forty nights, and I will destroy from the face of the earth all living things that I have made (Genesis 7:1–4).

So Noah had listened to the Lord and built the ark just as God commanded. It was ready. Noah had it stocked with food (Genesis 6:21[3]) and likely initial water too.

Then the Lord gave permission to come aboard with the animals. Essentially, this was like a permit. When builders prepare a building, it has to be up to code. Noah's work on the ark had to be up to code as well. God agreed that Noah was in compliance (Genesis 6:22: "Thus Noah did; according to all that God commanded him, so he did"). Therefore, Noah received permission to inhabit the ark.

This is when Noah began the loading process (as well as taking their personal belongings) on the ark. He took the animals by twos, a male and its

2. "Every moving thing that lives shall be food for you. I have given you all things, even as the green herbs.
3. And you shall take for yourself of all food that is eaten, and you shall gather it to yourself; and it shall be food for you and for them.

mate (Genesis 7:8–9[4]). Keep in mind that of the clean animals' sacrifice was required after the Flood. So this would be the extra male[5] without a mate since it would not be required to mate and repopulate the earth afterward.

Those that entered the ark *in pairs* were for the purpose of repopulating their kind after the Flood. This is why the female is paired with it. But the sacrificial clean animals were obviously not required to go in pairs as they had no mates and their function was not to keep their kinds alive after the Flood (Genesis 7:3[6]).

Sacrifice costs the *sacrificer* something. If not, it really isn't a sacrifice. King David later recognized this (2 Samuel 24:24[7]). It makes sense that the sacrificial animals were those owned by the family and were taken aboard the ark. In other words, it was not to be counted from the pairs that God brought to Noah. The Bible simply doesn't give us details of *when* the sacrificial animals boarded, just as the Bible doesn't tell exactly when the food was loaded.

It is possible the sacrificial animals were among the first-born offspring of the clean animals taken aboard the ark. When the animals came off the ark, they came by their families (Genesis 8:19[8]), implying that many had likely borne offspring.

Caring for the Animals on the Ark

Caring for the animals aboard the ark may have been easier than initially thought. If operational designs were planned correctly for food and water dispensing, waste management, etc., this task would be quite easy.

Keep in mind that the ark is *not* a zoo, nor is it a permanent vessel meant for 20 years of service. Automatic watering systems and feeding systems could make the job almost too easy. Some animals could have hibernated.

---

4. Of clean animals, of animals that are unclean, of birds, and of everything that creeps on the earth, two by two they went into the ark to Noah, male and female, as God had commanded Noah
5. In most cases in the Old Testament, the sacrificial animals were male. There are a few cases where it could be either male or female (e.g., Genesis 15:9; Leviticus 3:1–6), but they are the exception, not the norm.
6. Also seven each of birds of the air, male and female, to keep the species alive on the face of all the earth.
7. Then the king said to Araunah, "No, but I will surely buy it from you for a price; nor will I offer burnt offerings to the Lord my God with that which costs me nothing." So David bought the threshing floor and the oxen for fifty shekels of silver.
8. Every animal, every creeping thing, every bird, and whatever creeps on the earth, according to their families, went out of the ark.

As a farmer in my (Bodie) youth, my experiences lend a little light to this subject. I never scooped poop from the hogs, cattle, chickens, etc., everyday. In fact, cleaning out the barn or chicken coup was rarely done. The point is that there are ways of collecting waste from animals or at least neutralizing it so that you don't have to scoop waste all the time. When properly done, once a year was sufficient. Who knows what sophisticated ideas people had developed before the Flood.

With Noah's voyage being about a year, they may not have had to scoop the waste overboard at all. It would surprise me if Noah and his family had to clean out the waste even once. They could have had drop floors for waste to pass through. They could have thick straw covering the floors to collect waste and cancel out effects of ammonia, and so forth. Noah's ark was only needed for about a year and then it was time to exit.

Didn't the Dinosaurs Eat the Other Animals on the Ark?

Carnivorous animals, as we know them today, weren't always like that. Sharks eat kelp, spiders eat pollen, bears eat more plants than animals, and some cats have refused to eat meat. Originally, all animals were vegetarian (Genesis 1:30[9]), so having a vegetarian diet on the ark is not a problem. In fact, the animals that God brought to the ark may have been animals that were still vegetarian — even though flesh of all sorts — including animals — was corrupted on the earth (Genesis 6:12[10]).

Even if some ark inhabitants had begun to be carnivores by eating meat by the time of the ark's voyage, they could still eat vegetarian foods when they became hungry enough. Keep in mind that animals were separated into their rooms/cages, so trying to eat other animals would be a moot point anyway.

Conclusion

Getting the animals was the easy part, since God did that. Loading the animals and caring for them during the voyage was likely much easier than we are commonly led to believe. These issues really aren't a big deal when we think through the subject.

---

9. "Also, to every beast of the earth, to every bird of the air, and to everything that creeps on the earth, in which there is life, I have given every green herb for food"; and it was so.
10. So God looked upon the earth, and indeed it was corrupt; for all flesh had corrupted their way on the earth.

32 Could Noah Really Build the Ark by Himself?

Ken Ham and Bodie Hodge

"How could Noah build that ark *by himself*?" is a question we hear far too often! Research into the subject reveals a lot of information. First, Noah wasn't by himself. He had three sons — Shem, Ham and Japheth. The Bible says:

> And Noah was five hundred years old, and Noah begot Shem, Ham, and Japheth (Genesis 5:32).

Noah's Three Sons . . . Triplets?

No, they were not triplets. Genesis 5:32 indicates that Noah was 500 when his first son was born. Listings such as these are rarely an indication of ages but shows that Noah *began* having children when He was 500. We see the same thing when Terah gave birth to his sons, Abram, Nahor, and Haran (Genesis 11:26[1]).

The listing of children often started with the most important one (through Shem we receive the blessing of Christ). Those with a greater godly legacy are listed first, being the most important. Consider that Jesus has a name above all other names (Philippians 2:9–10[2]).

By doing further research in the Bible, we know the relative ages of each of Noah's sons.

---

1. Now Terah lived seventy years, and begot Abram, Nahor, and Haran.
2. Therefore God also has highly exalted Him and given Him the name which is above every name, that at the name of Jesus every knee should bow, of those in heaven, and of those on earth, and of those under the earth.

> Unto Shem also, the father of all the children of Eber, the brother of Japheth the elder, even to him were children born (Genesis 10:21; KJV).
>
> When Noah awoke from his wine and knew what his youngest son [Ham] had done to him (Genesis 9:24; ESV).

Genesis 10:21 indicates that Japheth was the oldest and thus was born when Noah was 500 years old. We are informed that Ham is the youngest, as indicated in Genesis 9:24, after Ham's inappropriate actions to his father. Therefore, Shem had to be born in between Japheth and Ham. Shem wasn't born as a triplet or twin of Japheth when Noah was 500, as shown by Genesis 7:6 and Genesis 11:10.

> Noah was six hundred years old when the floodwaters were on the earth (Genesis 7:6).
>
> These are the generations of Shem. When Shem was 100 years old, he fathered Arphaxad two years after the flood (Genesis 11:10; ESV).

Noah was 600 when the floodwaters came on the earth, and two years later Shem was 100. Therefore, Shem had to be born to Noah when he was 502. We are not sure of Ham's exact age in Scripture, but he had to be born after Shem. Thus, Genesis 5:32 introduces us to Noah's sons all together, and other passages give more detail about their birth order and age. Now back to the point at hand — these three sons were there with Noah to help as well.

Lamech and Methuselah

Also consider that Lamech and Methuselah — Noah's father and grandfather were around to help too — that is, for a certain amount of time, though both died prior to the Flood. Lamech died about five years before the Flood and Methuselah died the year of the Flood.

Methuselah, being raised by a godly parent, Enoch (who walked with God and was translated to heaven without death), was surely righteous as well (Hebrews 11:5[3]). His death surely preceded the Flood unlike what is portrayed in the unbiblical movie *Noah* with Russell Crowe and Darren Aronofsky.

3. By faith Enoch was taken away so that he did not see death, "and was not found, because God had taken him"; for before he was taken he had this testimony, that he pleased God.

Some have suggested that Methuselah died immediately before the Flood and hence a seven-day mourning period was in order. This possibly makes sense of why Noah was given seven days until the Flood came (e.g., Genesis 7:4[4]). Whether this is true or not, we cannot be certain.

God's instruction for Noah and His family to board the ark seven days in advance was for several reasons. Obviously, one reason was to complete the final phase of loading the animals (Genesis 7:2–9), and a second was a final test of faith for Noah and his family, the final boarding being on the seventh day (Genesis 7:11–16).

Keep in mind that it was common for prominent people to be honored with designated times of mourning after they passed (e.g., Genesis 27:41,[5] 50:4–5[6]; Deuteronomy 34:8[7]; 2 Samuel 11:27[8]). However, there were surely many who had mourning periods that are simply not mentioned in the Bible. Was this a mourning period for Methuselah? Whether this is true or not, we cannot be certain, but it would make a good "cap" to the life of Methuselah.

Regardless, Methuselah and Lamech would have been available for help for some time during the ark's construction. Their ability to do certain things may have been limited by their age however.

Righteous People Who Were Killed?

Let us also not forget the potential number of people who were righteous due to Noah's preaching (2 Peter 2:5[9]) who may have been killed or murdered leading up to the Flood. These are a possibility.

---

4. "For after seven more days I will cause it to rain on the earth forty days and forty nights, and I will destroy from the face of the earth all living things that I have made."
5. So Esau hated Jacob because of the blessing with which his father blessed him, and Esau said in his heart, "The days of mourning for my father are at hand; then I will kill my brother Jacob."
6. And when the days of his mourning were past, Joseph spoke to the household of Pharaoh, saying, "If now I have found favor in your eyes, please speak in the hearing of Pharaoh, saying. 'My father made me swear, saying, "Behold, I am dying; in my grave which I dug for myself in the land of Canaan, there you shall bury me." Now therefore, please let me go up and bury my father, and I will come back.' "
7. And the children of Israel wept for Moses in the plains of Moab thirty days. So the days of weeping and mourning for Moses ended.
8. And when her mourning was over, David sent and brought her to his house, and she becam`e his wife and bore him a son. But the thing that David had done displeased the Lord.
9. And did not spare the ancient world, but saved Noah, one of eight people, a preacher of righteousness, bringing in the flood on the world of the ungodly.

With the passing of Methuselah, and the recent passing of Lamech, we pause to realize that there were not many righteous people left on earth. After all, fewer than ten people were saved on the ark. Consider Abraham's discussion with the Lord over the destruction of Sodom (Genesis 18:26–32). Abraham did not proceed to a number fewer than ten righteous people when pleading for Sodom. He may have believed that judgment would come if there were fewer than ten — perhaps a reflection of his knowledge of the Flood.

Methuselah and Lamech had recently died, and this left eight. So, judgment was coming, but the Lord also prepared a means of salvation for Noah and his family on the ark, just as He did by sending the angels to rescue Lot and his family from Sodom.

Hired Hands/Contractors

Also, there is nothing wrong with Noah having hired hands or contractors to do specific work, gather wood, process wood, make specialized tools, food delivery, and so forth.

Noah could easily contract with unrighteous farmers or carpenters of his day to produce a certain amount of food or joists (etc.) since he was too busy with other aspects of the ark. I've often thought that if the Lord had blessed Noah as he did people like Abraham, Isaac, and Jacob, or David or Solomon, then Noah may have had the wealth to hire the manpower to get things done correctly and in a timely fashion.

Noah's Special Blessing

In a world full of wickedness and violence, did it ever worry Noah about the danger his family would be in during the trials of building the ark? Noah was righteous and standing against that wicked world. One would think he was a "target" or perhaps his family was a "target" for evil motives and possibly murder. But consider this wonderful verse of assurance given to Noah:

> But I will establish My covenant with you; and you shall go into the ark — you, your sons, your wife, and your sons' wives with you (Genesis 6:18).

This was told to Noah when he was first informed to build the ark. Noah was told that his wife, his sons, and his sons' wives would be on board the ark. Noah knew this information in advance. His family would not be murdered or killed during this ordeal. This prophecy given to Noah was a blessing of reassurance that his family would be safe. Thus, Noah had no need of concern or worry over his family. He knew in advance that they would be on the ark.

Some have wrestled with this verse, holding that it means that Noah knew that his preaching would be in vain and that the ark was not to be built as a sufficient rescue vessel for mankind, just designed for his family alone, though we don't see it that way. We see it as a guarantee that Noah could know that *at least* his family would be on it.

Think of it this way. In the same way, Christ's sacrifice on the Cross is sufficient means of salvation for all to receive the blood of Christ to be saved (1 John 2:2[10]; Hebrews 7:27[11]; Romans 6:10[12]). That doesn't mean all will receive (e.g., Matthew 7:13[13]) in the same way it doesn't mean that all would enter the door of the ark, but Noah was comforted in knowing that at the very least, his immediate family was going to be aboard.

10. And He Himself is the propitiation for our sins, and not for ours only but also for the whole world.
11. Who does not need daily, as those high priests, to offer up sacrifices, first for His own sins and then for the people's, for this He did once for all when He offered up Himself.
12. For the death that He died, He died to sin once for all; but the life that He lives, He lives to God.
13. "Enter by the narrow gate; for wide is the gate and broad is the way that leads to destruction, and there are many who go in by it.

33 Was There a Canopy Around the Earth until the Flood?

Bodie Hodge

Preliminary Comments

If there is one thing you need to know about biblical creationists . . . they can be divided on a subject. This isn't necessarily a bad thing. Though we all have the same heart to follow Christ and do the best we can for the sake of biblical authority and the cause of Christ, we can have differences when it comes to details of models used to explain various aspects of God's creation. Models though are subject to change, but what Scripture teaches is not subject to change!

When divisions occur over scientific models, this helps us dive into an issue in more detail and discover if that model is good, bad, needs revision, and so on. But note over *what* we are divided; it is not the Word of God nor is it even theology — it is a division over a *scientific model.*

This is where Christians can rightly be divided on a subject and still do so with Christian love, which I hope is how each Christian would conduct themselves. These "iron-sharpening-iron" dealings on a model can occur while still promoting a heart for the gospel (Proverbs 27:17[1]).

The debate over a canopy model is no different — we are all brothers and sisters in Christ trying to understand *what the Bible says and what it doesn't say* on this subject (2 Timothy 2:15[2]). It is the Bible that reigns

1. As iron sharpens iron, so a man sharpens the countenance of his friend.
2. Be diligent to present yourself approved to God, a worker who does not need to be ashamed, rightly dividing the word of truth.

supreme on the issue, and our scientific analysis on the subject will always be subservient to the Bible's text.

Introduction

For those familiar with creation, you may also be familiar with a particular scientific model that has dominated creation circles for about 50 years. This model has been denoted as the canopy theory or canopy model(s). In this science model, there was supposed to be a canopy of water (solid, liquid, or gas) that may have surrounded the earth from creation until its alleged dissipation at the Flood.

The reason for this model was to try to explain a better atmosphere prior to the Flood and possible health benefits for man, animals, and plants. After all, man before the Flood was living to great ages. So this model was proposed based on the "waters above" in Genesis 1.

For those familiar with creation, you may also know that many people no longer hold to this particular model. And you may be asking the question, what is going on with the canopy model? Let's evaluate the situation. Furthermore, since this is still a sensitive subject to many, a little more time will be spent in this chapter.

What Is the Canopy Model(s)?

There are several canopy models, but they all have one thing in common.[3] They all interpret the "waters above" the expanse (firmament) in Genesis 1:7 as some form of water-based canopy surrounding the earth that endured from creation until the Flood.

> Then God said, "Let there be a firmament [expanse] in the midst of the waters, and let it divide the waters from the waters." Thus God made the firmament [expanse], and divided the waters which were under the firmament [expanse] from the waters which were above the firmament [expanse]; and it was so (Genesis 1:6–7).

3. This is not to be confused with canopy ideas that have the edge of water at or near the end of the universe (e.g., white hole cosmology), but instead the models that have a water canopy in the atmosphere, e.g., like those mentioned in John C. Whitcomb and Henry M. Morris, *The Genesis Flood*, originally published c. 1960. Also see J.C. Dillow, *The Waters Above: Earth's Pre-Flood Vapor Canopy* (see footnote 4 in this chapter) or John C. Whitcomb, *The World that Perished.*

Essentially, the waters above are believed to have formed either a vapor, water (liquid), or ice canopy around the earth. It is the vapor canopy that seemed to dominate all of the proposed models.[4] It is suggested that this canopy was responsible for several things such as keeping harmful radiation from penetrating the earth, increasing the surface atmospheric pressure of oxygen, keeping the globe at a consistent temperature for a more uniform climate around the globe, and providing one of the sources of water for the Flood.

Some of these factors, like keeping radiation out and increasing the surface atmospheric pressures of oxygen, were thought to allow for human longevity to be increased from its present state (upward of 900 years or so as described in Genesis 5). So this scientific model was an effort to explain several things, including the long human lifespan prior to the Flood.

Other potential issues solved by the models were to destroy the possibility of large-scale storms with reduced airflow patterns for less extreme weather possibilities, have a climate without rain (such as Dillow's model, see below) but instead merely dew every night, and reduce any forms of barrenness like deserts and ice caps. It would have higher atmospheric pressure to possibly help certain creatures fly that otherwise may not be able to.

A Brief History of Canopy Models

Modern canopy models can be traced back to Dr. Henry Morris and Dr. John Whitcomb in their groundbreaking book *The Genesis Flood* in 1961. This book triggered a return to biblical authority in our age, which is highly commendable, and much is owed to their efforts. In this volume, Whitcomb and Morris introduce the possibility of a vapor canopy as the waters above.

The canopy models gained popularity thanks to the work of Dr. Joseph Dillow,[5] and many creationists have since researched various aspects of these scientific models such as Dr. Larry Vardiman with the Institute for Creation Research.

Researchers have studied the possibility of solid canopies, water canopies, vapor canopies, thick canopies, thin canopies, and so on. Each model

4. This is in large part due to the influence of Joseph Dillow, whose scientific treatise left only the vapor models with any potential: "We showed that only a vapor canopy model can satisfactorily meet the requirements of the necessary support mechanism." J.C. Dillow, *The Waters Above: Earth's Pre-Flood Vapor Canopy*, revised edition (Chicago, IL: Moody Press, 1981), p. 422.
5. J.C. Dillow, *The Waters Above: Earth's Pre-Flood Vapor Canopy*, revised edition (Chicago, IL: Moody Press, 1981).

has the canopy collapsing into history at the time of the Flood. Researchers thought it could have provided at least some of the water for the Flood and was associated with the 40 days of rain coming from the "windows of heaven" mentioned along with the fountains of the great deep at the onset of the Flood (Genesis 7:11).

However, the current state of the canopy models have faded to such an extent that most researchers and apologists have abandoned the various models. Let's take a look at the biblical and scientific reasons behind the abandonment.

Biblical Issues

Though both will be discussed, any biblical difficulties that bear on the discussion of the canopy must *trump* scientific considerations, as it is the authority of the Bible that is supreme in all that it teaches.

Interpretations of Scripture Are Not Scripture

The necessity for a water-based canopy about the earth is not directly stated in the text. It is an *interpretation* of the text. Keep in mind that it is the *text* that is inspired, not our interpretations of it.

Others have interpreted the water's above as something entirely different from a water-based canopy about the earth. Most commentators appeal to the waters above as simply being the clouds, which are water droplets (not vapor) in the atmosphere, for they are simply "waters" that are above.

Most do not limit this interpretation as simply being the clouds, but perhaps something that reaches deep into space and extends as far as the *Third Heaven* or *Heaven of Heavens*. For example, expositor Dr. John Gill in the 1700s said:

> The lower part of it, the atmosphere above, which are the clouds full of water, from whence rain descends upon the earth; and which divided between them and those that were left on the earth, and so under it, not yet gathered into one place; as it now does between the clouds of heaven and the waters of the sea. Though Mr. Gregory is of the opinion, that an abyss of waters above the most supreme orb is here meant; or a great deep between the heavens and the heaven of heavens.[6]

6. John Gill, *Exposition of the Bible*, Genesis 1:7.

Gill agrees that clouds were inclusive of these waters above but that the waters also extend to the heaven of heavens, at the outer edge of the universe. Matthew Poole noted this possibility as well in his commentary in the 1600s:

> . . . the expansion, or extension, because it is extended far and wide, even from the earth to the third heaven; called also the firmament, because it is fixed in its proper place, from whence it cannot be moved, unless by force.[7]

Matthew Henry also concurs that this expanse extends to the heaven of heavens (third heaven):

> The command of God concerning it: Let there be a firmament, an expansion, so the Hebrew word signifies, like a sheet spread, or a curtain drawn out. This includes all that is visible above the earth, between it and the third heavens: the air, its higher, middle, and lower, regions — the celestial globe, and all the spheres and orbs of light above: it reaches as high as the place where the stars are fixed, for that is called here the firmament of heaven Ge 1:14,15, and as low as the place where the birds fly, for that also is called the firmament of heaven, Ge 1:20.[8]

The point is that a canopy model about the earth is simply that . . . an interpretation. It should be evaluated as such, not taken as Scripture itself. Many respected Bible interpreters do not share in the interpretation of the "waters above" being a water canopy in the upper atmosphere of earth.

Stars for Seasons and Light and Other Implications

Another biblical issue crops up when we read in Genesis 1:14–15

> Then God said, "Let there be lights in the firmament [expanse] of the heavens to divide the day from the night; and let them be for signs and seasons, and for days and years; and let them be for lights in the firmament [expanse] of the heavens to give light on the earth"; and it was so.[9]

The stars are intended by God to be used to map seasons. And they were also to "give light on the earth." Though this is not much light, it does help

7. Matthew Poole, *A Commentary on the Holy Bible*, Genesis 1:7.
8. Matthew Henry, *A Commentary on the Whole Bible*, Genesis 1:7.
9. See also Genesis 1:17.

significantly during new moon conditions — that is, if you live in an area not affected by light pollution.

Water

If the canopy were liquid water, then in its various forms like mist or haze it would inhibit seeing these stars. How could one see the stars to map the seasons? It would be like a perpetually cloudy day. The light would be absorbed or reflected back to space much the way fog does the headlights of a car. What little light is transmitted through would not be sufficiently discernable to make out stars and star patterns to map seasons. Unlike a vapor canopy, clouds are moving and in motion; one can still see the stars to map seasons when they move through. Furthermore, if it was water why didn't it fall?[10]

Ice

If it were ice, then it *is* possible to see the stars but they would not appear in the positions one normally sees them, but still they would be sufficient to map seasons. But ice, when kept cool (to remain ice), tends to coat at the surface where other water molecules freezes to it (think of the coating you see on an ice cube left in the freezer). This could inhibit visibility, as evaporated water from the ocean surface would surely make contact — especially in a sin-cursed and broken world.

Vapor

If an invisible vapor canopy existed in our upper atmosphere, then it makes the most sense, but there could still turn out to be a problem. As cooler vapor nears space, water condenses and begins to haze, though as long as the vapor in the upper atmosphere is kept warm and above the dew point, it could remain invisible. But there are a lot of "if's." In short, the stars may not serve their purpose to give light on the earth with some possibilities within these models.

Consider that if there were a water *vapor* canopy, what would stop it from interacting with the rest of the atmosphere *that is vapor*? Gasses mix to equilibrium and that is the way God upholds the universe.[11] If it was a vapor, then why is it distinguished from the atmosphere, which is vapor?

10. Would one appeal to the supernatural? If so, it defeats the purpose of this scientific model that seeks to explain things in a naturalistic fashion.
11. Again, would one appeal to the supernatural? If so, it defeats the purpose of this scientific model that seeks to explain things in a naturalistic fashion.

The Bible uses the terms *waters* above, which implies that the temperature is between 32°F and 212°F (0°C and 100°C). If it was meant to be vapor, then why say "waters" above? Why not say vapor (*hebel*), which was used in the Old Testament?

Where Were the Stars Made?

If the canopy really was part of earth's atmosphere, then all the stars, sun, and moon would have been created within the earth's atmosphere.

Why is this? A closer look at Genesis 1:14[12] reveals that the "waters above" may very well be much farther out — if they still exist today.

The entirety of the stars, including our own sun (the greater light) and moon (lesser light) were made "*in* the expanse." Further, they are obviously not in our atmosphere. Recall that the waters of verse 7 are above the expanse. If the canopy were just outside the atmosphere of the young earth, then the sun, moon, and stars would have to be in the atmosphere, according to verse 14.

Further, the winged creatures were flying *in the face of* the expanse (Genesis 1:20[13]; the NKJV accurately translates the Hebrew), and this helps reveal the extent of the expanse. It would likely include aspects of the atmosphere as well as space. The Bible calls the firmament "heaven" in Genesis 1:8,[14] which would include both. Perhaps our understanding of "sky" is similar or perhaps the best translation of this as well.

Regardless, this understanding of the text allows for the stars to be in the expanse, and this means that any waters above, which is beyond the stars, is not limited to being in the atmosphere. Also, 2 Corinthians 12:2[15] discusses three heavens, which are likely the atmosphere (airy heavens), space (starry heavens), and the heaven of heavens (Nehemiah 9:6[16]).

Some have argued that the prepositions in, under, above, etc. are not in the Hebrew text but are determined from the context, so the meaning

12. Then God said, "Let there be lights in the firmament of the heavens to divide the day from the night; and let them be for signs and seasons, and for days and years."
13. Then God said, "Let the waters abound with an abundance of living creatures, and let birds fly above the earth across the face of the firmament of the heavens."
14. And God called the firmament Heaven. So the evening and the morning were the second day.
15. I know a man in Christ who fourteen years ago — whether in the body I do not know, or whether out of the body I do not know, God knows — such a one was caught up to the third heaven.
16. You alone are the Lord; You have made heaven, the heaven of heavens, with all their host, the earth and everything on it, the seas and all that is in them, and You preserve them all. The host of heaven worships You.

in verses 14 and 17 is vague. It is true that the prepositions are determined by the context, so we must rely on a proper translation of Genesis 1:14.[17] Virtually all translations have the sun, moon, and stars being created *in* the expanse, not *above,* as any canopy model would require.

In Genesis 1, some have attempted to make a distinction between the expanse in which the birds fly (Genesis 1:20[18]) and the expanse in which the sun, moon, and stars were placed (Genesis 1:7[19]); this was in an effort to have the sun, moon, and stars made in the second expanse. This is not a distinction that is necessary from the text, and is only necessary if a canopy is assumed.

From the Hebrew, the birds are said to fly "across the face of the firmament of the heavens." Looking up at a bird flying across the sky, it would be seen against the face of both the atmosphere and the space beyond the atmosphere — the "heavens." The proponents of the canopy model must make a distinction between these two expanses to support the position, but this is an arbitrary assertion that is only necessary to support the view and is not described elsewhere in Scripture.

Expanse (Firmament) Still Existed Post-Flood

Another issue that is raised from the Bible is that the waters above the heavens were mentioned *after* the Flood, when it was supposedly gone.

> Praise Him, you heavens of heavens, and you waters above the heavens! (Psalm 148:4).
>
> So an officer on whose hand the king leaned answered the man of God and said, "Look, if the LORD would make windows in heaven, could this thing be?" And he said, "In fact, you shall see it with your eyes, but you shall not eat of it" (2 Kings 7:2; see also 2 Kings 7:19[20]).
>
> "Bring all the tithes into the storehouse, that there may be food in My house, and try Me now in this," says the LORD of

17. Then God said, "Let there be lights in the firmament of the heavens to divide the day from the night; and let them be for signs and seasons, and for days and years."
18. Then God said, "Let the waters abound with an abundance of living creatures, and let birds fly above the earth across the face of the firmament of the heavens."
19. Thus God made the firmament, and divided the waters which were under the firmament from the waters which were above the firmament; and it was so.
20. Then that officer had answered the man of God, and said, "Now look, if the LORD would make windows in heaven, could such a thing be?" And he had said, "In fact, you shall see it with your eyes, but you shall not eat of it."

> hosts, "If I will not open for you the windows of heaven and pour out for you such blessing that there will not be room enough to receive it" (Malachi 3:10).

The biblical authors wrote these in a post-Flood world in the context of other post-Flood aspects. So it appears that the "waters above" and "windows of heaven" are in reference to something that still existed after the Flood. The "waters above" can't be referring to a long-gone canopy that dissipated at the Flood and still be present after the Flood. This is complemented by:

> The fountains of the deep and the windows of heaven were also stopped, and the rain from heaven was restrained (Genesis 8:2)

Genesis 8:2 merely points out that the two sources were stopped and restrained, not necessarily *done away* with. The verses above suggest that the windows of heaven remained after the Flood. Even the "springs of the great deep" were stopped but did not entirely disappear, but there may have been residual waters trapped that have slowly oozed out since that time, clearly not in any gushing spring-like fashion.[21]

Is a Canopy Necessary Biblically?

Finally, is a canopy necessary from the text? At this stage, perhaps not. It was promoted as a scientific model based on a possible interpretation of Genesis 1 to deal with several aspects of the overall biblical creation model developed in the mid-1900s. I don't say this lightly for my brothers and sisters in the Lord who may still find it appealing. Last century, I was introduced to the canopy model and found it fascinating. For years, I had espoused it, but after further study, I began leaning against it, as did many other creationists.

Old biblical commentators were not distraught at the windows of heaven or the waters not being a canopy encircling the earth. Such an interpretation was not deemed necessary in their sight. In fact, this idea is a recent addition to scriptural interpretation that is less than 100 years old. The canopy model was a scientific interpretation developed in an effort to help explain certain aspects of the text to those who were skeptical of the Bible's accounts of earth history, but when it comes down to it, it is not necessary and even has some serious biblical issues associated with it.

---

21. I would leave open the option that this affected the ocean sea level to a small degree, but the main reasons for changing sea level were via the Ice Age.

Scientific Issues (Semi-Technical)

Clearly, there are some biblical issues that are difficult to overcome. Researchers have often pointed out the scientific issues of the canopy model, as well. A couple will be noted below.

This is no discredit to the *researchers* by any means. The research was valuable and necessary to see how the model may or may not work with variations and types. The development and testing of models is an important part of scientific inquiry and we should continue to do so with many models to help us understand the world God has given us. So I appreciate and applaud all the work that has been done, and I further wish to encourage researchers to study other aspects to see if anything was missed.

Temperatures

To answer the question about how the earth regulates its temperature without a canopy, consider that it may not have been that much different than the way it regulates it today — by the atmosphere and oceans. Although there may have been much water underground prior to the Flood, there was obviously enough at or near the surface to sustain immense amounts of sea life. We know this because of the well-known figure that nearly 95 percent of the fossil record consists of shallow-water marine organisms. Was the earth's surface around 70 percent water before the Flood? That is a question creationist researchers still debate.

An infinitely knowledgeable God would have no problem designing the earth in a perfect world to have an ideal climate (even with variations like the cool of the day — Genesis 3:8[22]) where people could have filled the earth without wearing clothes (Genesis 2:25,[23] 1:28[24]). But with a different continental scheme, that contained remnants of a perfect world (merely cursed, not rearranged by the Flood yet), it would surely have been better equipped to deal with regulated temperatures and climate.

A vapor canopy, on the other hand, would cause major problems for the regulation of earth's temperature. A vapor canopy would absorb both

22. And they heard the sound of the LORD God walking in the garden in the cool of the day, and Adam and his wife hid themselves from the presence of the LORD God among the trees of the garden.
23. And they were both naked, the man and his wife, and were not ashamed.
24. Then God blessed them, and God said to them, "Be fruitful and multiply; fill the earth and subdue it; have dominion over the fish of the sea, over the birds of the air, and over every living thing that moves on the earth."

solar and infrared radiation and become hot, which would heat the surface by conduction downward. The various canopy models have therefore been plagued with heat problems from the greenhouse effect. For example, solar radiation would have to decrease by around 25 percent to make the most plausible model work.[25] The heat problem actually makes this model very problematic and adds an additional problem rather than helping to explain the environment before the Flood.[26]

The Source of Water

The primary source of water for the Flood was the springs of the great deep bursting forth (Genesis 7:11[27]). This water in turn likely provided some of the water in the "windows of heaven" in an indirect fashion. There is no need for an ocean of vapor above the atmosphere to provide for extreme amounts of water for the rain that fell during the Flood.

For example, if Dillow's vapor canopy existed (40 feet of precipitable water) and collapsed at the time of the Flood to supply, in large part, the rainfall, the latent heat of condensation would have boiled the atmosphere! And a viable canopy would not have had enough water vapor in it to sustain 40 days and nights of torrential global rain as in Vardiman's model (2–6 feet of precipitable water). Thus, the vapor canopy doesn't adequately explain the rain at the Flood.

Longevity

Some have appealed to a canopy to increase surface atmospheric pressures prior to the Flood. The reasoning is to allow for better healing as well as living longer and bigger as a result. However, increased oxygen (and likewise oxidation that produces dangerous free radicals), though beneficial in a few respects, is mostly a detriment to biological systems. Hence, antioxidants (including things like catalase and vitamins E, A, and C) are very important to reduce these free radicals within organisms.

---

25. For more on this see "Temperature Profiles for an Optimized Water Vapor Canopy" by Dr. Larry Vardiman, a researcher on this subject for over 25 years at the time of writing that paper, http://static.icr.org/i/pdf/technical/Temperature-Profiles-for-an-Optimized-Water-Vapor-Canopy.pdf.
26. Another issue is the amount of water vapor in the canopy. Dillow's 40 feet of precipitable water, the amount collected after all the water condenses, has major heat problems. But Vardiman's view has modeled canopies with 2 to 6 feet of precipitable water with better temperature results and we look forward to seeing his latest results.
27. In the six hundredth year of Noah's life, in the second month, the seventeenth day of the month, on that day all the fountains of the great deep were broken up, and the windows of heaven were opened.

Longevity (and the large size of many creatures) before and after the Flood is better explained by genetics through the bottlenecks of the Flood and the Tower of Babel as opposed to pre-Flood oxygen levels due to a canopy. Not to belabor these points, this idea has already been discussed elsewhere.[28]

Pre-Flood Climate

Regardless of canopy models, creationists generally agree that climate before the Fall was perfect. This doesn't mean the air was stagnant and 70°F every day, but instead had variations within the days and nights (Genesis 3:8[29]). These variations were not extreme but very reasonable.

Consider that Adam and Eve were told to be fruitful and multiply and fill the earth (Genesis 1:27–28[30]). In a perfect world where there was no need for clothes to cover sin (this came after the Fall), we can deduce that man should have been able to fill the earth without wearing clothes, hence the extremes were not as they are today or the couple would have been miserable as the temperatures fluctuated.

Even after the Fall, it makes sense that these weather variations were minimally different, because the general positions of continents and oceans were still the same. But with the global Flood that destroyed the earth and rearranged continents and so on, the extremes become pronounced — we now have ice caps and extremely high mountains that were pushed up from the Flood (Psalm 104:8[31]). We now have deserts that have extreme heat and cold and little water.

Biblical Models and Encouragement

I continue to encourage research and the development of scientific and theological models. However, a good grasp of all biblical passages that are relevant

28. Ken Ham,ed., *New Answers Book 2* (Green Forest, AR: Master Books, 2008), p. 159–168; Bodie Hodge, Tower of Babel (Green Forest, AR: Master Books, 2013), p. 205–212.
29. And they heard the sound of the LORD God walking in the garden in the cool of the day, and Adam and his wife hid themselves from the presence of the LORD God among the trees of the garden.
30. So God created man in His own image; in the image of God He created him; male and female He created them. Then God blessed them, and God said to them, "Be fruitful and multiply; fill the earth and subdue it; have dominion over the fish of the sea, over the birds of the air, and over every living thing that moves on the earth."
31. The mountains rose; the valleys sank down to the place which You established for them (NASB).

to the topic must precede the scientific research and models, and the Bible must be the ultimate judge over all of our conclusions.

The canopy model may have a glimmer of hope still remaining, and that will be left to the proponents to more carefully explain, but both the biblical and scientific difficulties need to be addressed thoroughly and convincingly for the model to be embraced. So we do look forward to future research.

In all of this, we must remember that scientific models are not Scripture, and it is the Scripture that we should defend as the authority. While we must surely affirm that the waters above were divided from the waters below, something the Bible clearly states, whether or not there was a canopy must be held loosely lest we do damage to the text of Scripture or the limits of scientific understanding.

34 Did Noah Need Oxygen on the Ark?

Bodie Hodge

Introduction

If the waters of the Flood covered the highest mountains, does that mean the ark's inhabitants needed oxygen supplies?

Why would someone ask this question? Let's back up and look at this from a big picture. Consider what the Bible says about the voyage of the ark:

> The water prevailed more and more upon the earth, so that all the high mountains everywhere under the heavens were covered. The water prevailed fifteen cubits higher, and the mountains were covered. (Genesis 7:19–20)[1]

People then look at the earth *today* and note that the highest mountain is Mt. Everest, which stands just over 29,000 feet above sea level. Then they put two and two together and say that Noah's ark floated at least 15 cubits above Mt. Everest — and at such high altitude, people need oxygen!

It sounds like a straightforward argument, doesn't it? But did you notice that I emphasized the word *today*? In light of this, the solution is quite simple: the Flood did not happen on today's earth, but rather on the earth of nearly 4,300 years ago. The world today is not the same as it was before the Flood, or even *during* the Flood. For instance, if the mountains, continents, and ocean basins of today's earth were more leveled out, the planet's surface

1. Scripture in this chapter is from the New American Standard Bible (NASB).

water alone would cover the earth to a calculated 1.66 miles deep — about 8,000 feet. Yet when I visited Cusco, Peru, which is around 11,000 feet above sea level, I didn't need an oxygen tank.

Furthermore, atmospheric air pressure is relative to sea level. So as rising sea levels pushed the air column higher, the air pressure at sea level would stay the same. So again, oxygen tanks are unnecessary.

Psalm 104:6–9: Creation or the Flood?

Beginning on day 150 of the Flood, mountains began overtaking the water again as the mountain-building phase had begun (Genesis 8:2–4[2]). Poetic Psalm 104 gives further hints of this possible mountain building as the valley basins sank down:

> You covered it with the deep as with a garment; the waters were standing above the mountains. At Your rebuke they fled, at the sound of Your thunder they hurried away. The mountains rose; the valleys sank down to the place which You established for them. You set a boundary that they may not pass over, so that they will not return to cover the earth (Psalm 104:6–9).

This section of the Psalm is obviously speaking of the Flood, as water would no longer *return* to cover the earth — if this passage were speaking of creation week (as some commentators have stated), then God would have erred when the waters covered the whole earth during the Flood.

Consider this overview of the entire Psalm as it continues down through history:

Psalm 104:1–5	Creation Week
Psalm 104:6–9	Flood
Psalm 104:10–35	Post-Flood

It makes sense that, because the Psalm is referring to the earth and what is in it, it begins with earth history (creation week). But mentions of donkeys (verse 11) and goats (verse 18) show variation within the created kind, which shows this would have taken place *after* the Flood. Also, a post-Flood

2. The fountains of the deep and the floodgates of the sky were closed, and the rain from the sky was restrained; and the waters receded steadily from the earth, and at the end of one hundred and fifty days the waters decreased. In the seventh month, on the seventeenth day of the month, the ark rested upon the mountains of Ararat.

geographic location is named (Lebanon, verse 16) as well as ships (verse 26) that indicate this Psalm was not looking strictly at creation week.

Lost in Translation?

While everyone agrees that Psalm 104:1–5[3] is referring to creation week, what of the argument — made by many commentators from the 1600s onward — that attributes Psalm 104:6–9 to creation week also? One could suggest that much of this is due to the translation being viewed. Two basic variants of the translation of the Hebrew in Psalm 104:8 read:

1. They went up over the mountains and went down into the valleys.
2. Mountains rose and the valleys sank down.

In fact, a variety of translations yield some variant of one of these two possibilities.

Table 1. Translations of Psalm 104:8a[4]

Translation	Agrees with: "They went up over the moun-tains and went down into the valleys"	Agrees with: "Mountains rose and the valleys sank down"
New American Standard		X
New International Version	X	
King James Version	X	
New King James Version	X	
English Standard Version		X
Holman Christian Standard		X
English translation of the Septuagint	X	

3. Bless the Lord, O my soul! O Lord my God, You are very great: You are clothed with honor and majesty, covering Yourself with light as with a cloak, stretching out heaven like a tent curtain. He lays the beams of His upper chambers in the waters; He makes the clouds His chariot; He walks upon the wings of the wind; He makes the winds His messengers, flaming fire His ministers. He established the earth upon its foundations, so that it will not totter forever and ever.
4. Data was taken from two sources: Charles Taylor, "Did Mountains Really Rise According to Psalm 104:8?" *TJ* 12(3), 1998, 312–313, and Online Bible, Larry Pierce, February 2009, or looked up separately.

Revised Version (UK)	X	
Amplified Bible		X
Good News Bible	X	
New English Bible	X	
Revised Berkley		X
J.N. Darby's		X
Living Bible		X
New Living Translation		X
Jerusalem Bible	X	
R.G. Moulton	X	
Knox Version		X
The Holy Scriptures according to the Masoretic Text (a new translation by the Jewish Publication Society)		X
Revised Standard Version		X
Young's Literal Translation	X	
King James 21st Century Version	X	
Geneva Bible		X
New Revised Standard Version	X	
Webster's Bible	X	
New International Children's Version		X
Interlinear Bible		X

Obviously, there is no consensus on translation among these English versions. Looking at other languages, we see how the Hebrew was translated.

Table 2. Some Foreign Translations of Psalm 104:8a[5]

Foreign translation	Agrees with: "They went up over the mountains and went down into the valleys"	Agrees with: "Mountains rose and the valleys sank down"

5. Ibid.

Luther's German		X
Menge's German		X
French Protestant Bible (Version Synondale)		X
Italian Edizione Paoline		X
Swedish Protestant		X
Spanish Reins Valera		X
Latin Vulgate (by Jerome)		X
La Bible Louis Segond 1910 (French)		X
Septuagint (Koine Greek)		X

Notice that there doesn't seem to be a discrepancy. Of course, there are many translations, so one cannot be dogmatic, but the point is that foreign translations agree with "mountains rising and valleys sinking down."

Hebrew

In Hebrew, which reads right to left, the phrase in 104:8a is basically *biq'ah yarad har 'alah*. Translated into English, the phrase in question is:

biq'ah	yarad	har	'alah
valleys	down go/sink	mountains	up go/rise

Take note that there are no prepositions like "over" or "into." It is literally "up go mountains, down go valleys." It makes sense why many translations, including non-English translations, use the phrase "mountains rose and the valleys sank down" — this is what it should be.

Why Would Commentators Miss This?

Commentaries could easily misinterpret this passage if they were based on translations that agree with "they went up over the mountains and went down into the valleys." For example, the most popular English translation for several hundred years, the King James Version, reads this way.

Furthermore, from a logical perspective, water doesn't flow uphill over mountains, but rather the opposite. Given language like this, commentators likely attributed this to a miraculous event during creation week, when many miracles were taking place anyway; also, creation week was referenced earlier in the chapter. Of course, the problems came when reading the rest of the context. One excellent commentator, Dr. John Gill, to whom I have appealed many times, regarding verse 9 and the waters not returning to cover the earth, stated:

> That they turn not again to cover the earth; as they did when it was first made, Psalm 104:6 that is, not without the divine leave and power; for they did turn again and cover the earth, at the time of the flood; but never shall more.[6]

Gill was forced to conclude that the waters did return to cover the earth, and he justified their return on "divine leave and power"! Yet this would mean that God breaks promises. Because we know that God does not break promises, this must be referring to the end of the Flood; so on this point, I disagree with Gill.

That said, we should understand the difficulty in commenting on the passage: it is a psalm of praise to God, and thus it is not as straightforward as literal history. It is difficult to determine where the shift from creation to the Flood occurs and where the shift from Flood to post-Flood occurs. However, there are a few more hints in the text.

A Few More Comments

We should use clear passages in Scripture to help interpret unclear passages. Consider that God's "rebuke" would not exist in a perfect world, where nothing would need rebuking or correcting. Remember, a perfect God created a perfect world — Genesis 1:31,[7] Deuteronomy 32:4.[8] One should expect nothing less of such a God. It was due to man's sin that the world is now imperfect and fallen.

Therefore, during creation week when everything was good, there would be no need for any rebuking. If Psalm 104:6–9 were referring to creation week (specifically day 3), then why the rebuke in Psalm 104:7? This implies an imperfect, *not* very good creation. But if Psalm 104:6–9 is referring to

6. http://biblehub.com/commentaries/psalms/104-9.htm.
7. God saw everything that He had made, and behold, it was very good. And there was evening and there was morning, the sixth day.
8. "The Rock! His work is perfect, for all His ways are just, a God of faithfulness and without injustice, righteous and upright is He.

the Flood, then of course a rebuke would exist in a fallen world where the judgment of water had overtaken the earth.

Additionally, note that Psalm 104:9 is clearly referencing Genesis 9:8–16 in saying that the waters would not return to cover the earth.

Some have asked how mountains and valleys could move up and down when the foundations are identified as immovable in Psalm 104:5. Keep in mind that mountains and valleys are not the foundation, but like the seas that sit well above the foundation. In fact, continents shifting and mountains rising do no damage to the foundation that is immovable.

Last, note that when the land appeared in Genesis 1 on day 3, the land that was being separated from the water was *dry*, not wet. The text in Genesis says that the waters were gathered into one place *and then* the dry land appeared. It says nothing of water flowing over the land to make it wet; otherwise, wet land would have appeared and then *became dry*, but during the Flood, the land was indeed overtaken by water that eventually stood above the land.

Conclusion

The Hebrew phrase in Psalm 104:8a is the basis for the correct translation of mountains rising and valleys sinking. This shows that mountains and valleys during the Flood were not the same height as they are today. Even today mountains and valleys are changing their height; volcanic mountains, for instance, can grow very quickly, such as Surtsey (a new island) or Paricutin (a volcanic mountain in Mexico that formed in 1943).

Therefore, with mountains and continents leveled out and ocean basins nowhere near the depth they are today, it makes perfect sense that Noah was not at the height of modern-day Mt. Everest. Noah and those aboard the ark would not have required oxygen.

35 Where Did the Water for the Flood Come From, and Where Did It Go?

Ken Ham AND Bodie Hodge

Introduction

Just like the last chapter, one of the main reasons people ask this question is because of Mt. Everest. Yes, you read this correctly. It is from Mt. Everest, which is the highest mountain peak in the world (over 29,029 feet, or 8,848 meters).

The question is often framed like this, "Where did all the water come from to cover Mt. Everest and where did all that water go?" You need to understand that when people ask this question, they are making the assumption that the Flood didn't affect the earth or the terrain such as Mt. Everest today. They merely assume that Mt. Everest was basically the same height before and after the Flood.

Misconception about the Flood

Once again, this is a rather naïve understanding of the Flood *and* Mt. Everest. It is *because of* the Flood that Mt. Everest now exists in the first place! In other words, Mt. Everest didn't exist before the Flood but is the result of mountainous uplift associated with the Flood that formed Mt. Everest, as well as other mountain ranges and peaks we have today![1]

When people falsely presume that Mt. Everest existed before the Flood and was the same basic height as today, they struggle with understanding where all the water was to cover that peak! But this is based on a false understanding of the Flood and mountains that formed as a result of the Flood.

1. We have seen a handful of mountains grow or form since the Flood like the volcano Paricutin in Mexico. Though these few are minute in comparison to the majority that formed in the Flood.

Is the Water (and Mountains) a Problem?

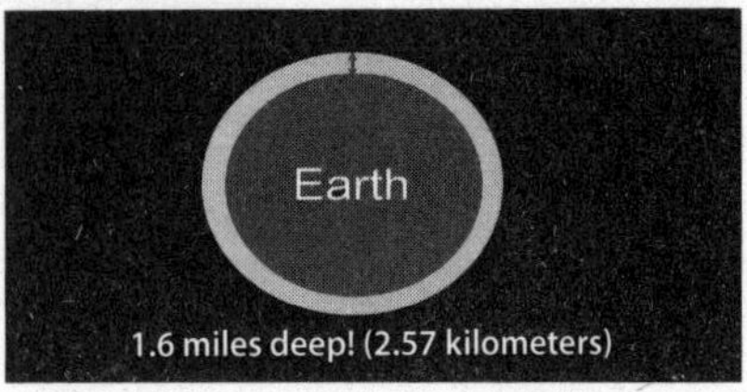

If we take ocean basins and bring them up and take mountain ranges and continents and bring them down to a level position, there is enough water to cover the earth 1.6 miles deep (2.57 km deep), so there is plenty of water on the earth for a global Flood.

Yet there was only the need for the highest underwater peak *during the Flood* to be covered by 15 cubits (22.5 feet or ~6.8 meters based on the short cubit to 25.5 feet or ~7.8 meters based on the long cubit) per Genesis 7:20.[2]

By the 150th day of the Flood (Genesis 8:3–4[3]), the mountains were already in the process of being built, likely by tectonic plate collisions on the earth's outer shell. For example, when two plates collide (called "convergent plate boundaries"), one plate goes up and the other plate goes under (called a "subduction zone").[4]

We know the mountains of Ararat existed by the 150th day of the Flood; thus, we also know that the plate collisions that formed this range were essentially completed at this time. Of course, residual effects likely occurred for some time — we still feel some earthquakes today!

By the 150th day, much of the rapid plate movement would largely begin to cease since the mechanism to move such plates had now ceased as well — springs of the great deep, rain, and windows of heaven were restrained (Genesis 8:2) and thus geological mechanisms associated with these would begin to cease as well.

The mountains of Ararat are part of the larger mountain chain called the *Alpide Belt* or *Alpine-Himalayan Belt.* This range extends from Spain and North Africa, through the Alps and Middle Eastern ranges (like the mountains of Ararat), and through the Himalayas down the Malay Peninsula and Indonesia, almost reaching Australia. It makes sense that these Alpine mountain ranges were all formed about the same time during the Flood's mountain-building,

2. The waters prevailed fifteen cubits upward, and the mountains were covered.
3. And the waters receded continually from the earth. At the end of the hundred and fifty days the waters decreased. Then the ark rested in the seventh month, the seventeenth day of the month, on the mountains of Ararat.
4. There are other plate interactions like "transform plate movements" (when plates go beside each other), or "divergent plate boundaries" (when they are going away from each other) — think mid-Atlantic Ridge.

which coincides with the valley sinking phase (ocean basins going down).

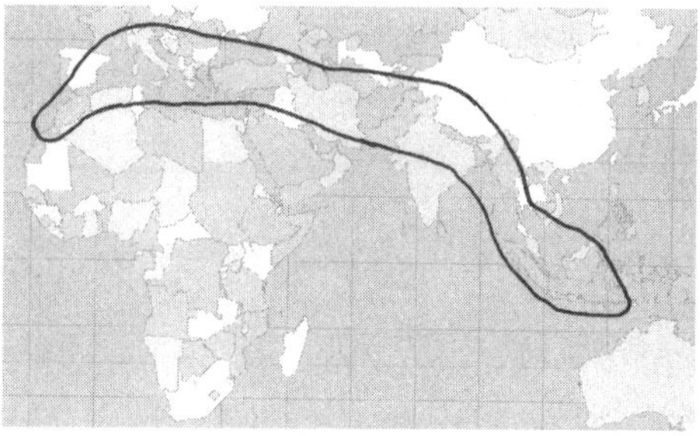
The Alpide Belt

Psalm 104:6–9 says:

> You covered it with the deep as with a garment; the waters were standing above the mountains. At Your rebuke they fled, at the sound of Your thunder they hurried away. The mountains rose; the valleys sank down to the place which You established for them. You set a boundary that they may not pass over, so that they will not return to cover the earth (NASB).

Mt. Everest's initial formation, being in the Alpide Belt, was about the same time as the mountains of Ararat or very shortly thereafter — since they involve much the same continental collisions. Naturally, Mt. Everest would be pushed up near its current height, though a little less. Mt. Everest still rises a little each year, merely due to residual effects that extend back to the Flood.

So Where Did All the Water Come From?

The majority of the water was subterranean (came from under the earth). The Bible says:

> In the six hundredth year of Noah's life, in the second month, the seventeenth day of the month, the same day were all the fountains of the great deep broken up, and the windows of heaven were opened (Genesis 7:11; KJV).

> The fountains also of the deep and the windows of heaven were stopped, and the rain from heaven was restrained (Genesis 8:2; KJV).

The fountains or springs of the great deep were the source of the majority of the waters for the Flood. The *windows of heaven* were also a source of water, though fully saturated clouds would not be sufficient for the initial 40 days rain until the ark raised up (Genesis 7:17,[5]) nor the continued rain until the

5. Now the flood was on the earth forty days. The waters increased and lifted up the ark, and it rose high above the earth.

150th day (Genesis 8:2).[6] So its contribution was likely smaller initially but continually renewed from the springs of the great deep that supplied water to the sky.

Also, there may be more to the meaning of "windows of heaven." This phrase is used in a metaphorical fashion in Malachi. The text says:

> Bring the whole tithe into the storehouse, so that there may be food in My house, and test Me now in this," says the LORD of hosts, "if I will not open for you the windows of heaven and pour out for you a blessing until it overflows (Malachi 3:10; NASB).

It basically means that something was poured out from God in abundance — whether blessing, as in the case of Malachi, or water, as in the case of the Flood! Think of it like this: The springs of the great deep burst forth and it was "raining cats and dogs!" In our English language, this phrase "raining cats and dogs" doesn't mean it was raining literal cats and dogs, but instead raining immensely. This could easily be the meaning of "windows of heaven" in Genesis 7:11 and 8:2, though, the windows of heaven may be in reference to the physical clouds and vapor in the sky too. Either option yields to the text.

Where Did All the Water Go?

The obvious answer is that the water is in the oceans, lakes, rivers, and so on that we have today! Some water may have seeped and again become subterranean (we do observe *springs* today), though most of that initial subterranean water from the Flood has now been exhausted. The continental shifting during the Flood did away with the original water "storehouses" that supplied the fountains of the great deep.

Any initial oceans before the Flood were likely much more shallow with a few deep areas. Keep in mind that about 95 percent of all fossils are from shallow marine organism — so this makes sense. Our current post-Flood oceanography has some areas that are shallow, but most is quite deep.

Consider that oceans cover about 70 percent of the earth surface today. At one point the whole earth was covered with the Floodwater. It was very kind of the Lord to give us 30 percent of land surface back.

---

6. See the chapter in this volume as to why an alleged canopy of water or vapor is not what is in view here.

36 How Did Animals Get to Places All over the World like Australia?

Bodie Hodge

Obviously, when animals came off the ark in the Middle East, they could progress throughout Europe, Asia, and Africa rather easily, since those areas are connected by land. Of course, they would have had hurdles such as mountains and rivers to go around or cross, among other issues with terrain.[1]

Naturally, these obstacles would not be such a hindrance for birds and other flying creatures. It only took starlings (a type of bird) about 100 years to cover the entire North American continent when about 60 were released in New York City in 1890. With this in mind, it probably did not take long for many places to be populated with flying creatures after the Flood. Many birds can transverse great distances over lakes, seas, and oceans.

Some birds and other flying creatures may have *lost* the ability to fly due to mutations or breeding (particularly inbreeding) since the Flood. This could have occurred *after* migrating long distances. On the farm where I grew up, we had some giant white turkeys. They were so large, awkward, and had such poor feathers that with one look at them you would be able to discern they could never fly. In fact, I've thought the same thing when I first saw an emu, kiwi, ostrich, and some others — surely, they could never fly! But is their current look the result of mutations and breeding? It is possible.

---

1. Keep in mind that the weather patterns and climates for places may not have been established yet and were still under constant flux at this time, so many rivers may have been smaller in places and larger in others (like areas that later became deserts). For example, many areas that are deserts today would not have been so right after the Flood. Instead, they were likely the results of later weather patterns (e.g., Genesis 12:10, 26:1).

We also had wild turkeys where I lived. I often frightened them so that they took flight, and they could fly very well. I once spooked a few turkeys into flight to such a degree that one of them flew immensely high; I could hardly see it! It was much higher than the high-flying vultures that dominated our area. And it flew so far that it was miles before it descended.

The giant white turkeys that could never fly resulted from recent breeding of wild turkeys to be large supermarket specials. Considering this, it is easy to see how other birds may have *lost* their ability to fly. I would also leave this option open with other flightless birds.

Land Animals

In general, placental animals would move slower than marsupials, which can collect their young (e.g., in pouches) and continue migrating. Many placental animals need to stop and settle for a time to raise their young but, theoretically, great varieties of land animals could have gone to any region of Europe, Asia, and Africa. Still, this doesn't mean they did, nor does it mean they thrived there (they may have died out). Is it possible that kangaroos made it to Europe and died out? It is possible, and we would leave open such an option. What we do know is that kangaroos have thrived in Australia, where they currently live. In other words, marsupials can travel farther and faster than many placentals. This may help explain why marsupials dominated Australia.

Some may have migrated to certain areas but not to others. In other instances, some of these animals may have made it to a particular area and become extinct for various reasons — ultimately due to sin, of course! One objection to this is that we should find fossils of them if they lived in an area, but this is fallacious.[2] Paul Taylor states the following regarding this subject on fossils:

> But the expectation of such fossils is a presuppositional error. Such an expectation is predicated on the assumption that fossils form gradually and inevitably from animal populations. In fact, fossilization is by no means inevitable. It usually requires sudden, rapid burial. Otherwise the bones would decompose before permineralization. One ought likewise to ask why it is that, despite the fact that

2. Many fossils can be determined if they are in situ or were transported. But many are still difficult to ascertain. If something is found fossilized someplace, the one thing we can be certain about is its final burial place.

> millions of bison used to roam the prairies of North America, hardly any bison fossils are found there. Similarly, lion fossils are not found in Israel even though we know that lions once lived there.[3]

Even recently, researchers at IUCN (International Union for Conservation of Nature) have tentatively declared that the West African Black Rhinoceros has gone extinct.[4] However, other rhinoceroses have continued in different areas of Africa and Asia. This is a good example of how a species can die out while another member of the same kind remains in other parts around the world. Could this have happened with other animals?[5] Surely it has.

Ice Age and Land Bridges to the Americas and Australia?

But how did they get to Australia? How did animals get to the Americas or remote islands?

Most creationists believe there was an Ice Age (see chapter 22).[6] If the globe simply cools down, that does not cause an ice age; instead, it causes a cool globe. An ice age requires warmer oceans and cool summers. Here is why: with warmer oceans there is more evaporation, which provides greater accumulation of snow during winter months.

With cool summers, it does not heat up enough to melt off the previous winter's snow accumulation. So during the next winter even more snow layers accumulate, their weight causing the previous winter's snow layers to compact into ice layers, thus eventually causing an ice age.

Most believe the Flood of Noah triggered the Ice Age. The rising magmas, lavas, and hot waters associated with continental plate movements would have caused ocean temperatures to rise. Also, fine ash from volcanic eruptions probably lingered in the upper atmosphere in post-Flood years, which, unlike a greenhouse effect, would reduce the sunlight for cooler summers. So the mechanism for such a rare event was in place due to Genesis 6–8.

---

3. *The New Answers Book 1*, Ken Ham, Gen. Ed., chapter by Paul Taylor: "How Did Animals Spread All Over the World from Where the Ark Landed?", Master Books, Green Forest, AK, 2006, page 144.
4. Matthew Knight, Western Black Rhino Declared Extinct, CNN, November 6, 2013, http://www.cnn.com/2011/11/10/world/africa/rhino-extinct-species-report/.
5. There are slim possibilities that they could be living elsewhere in remote areas, but the point is that many animals are going extinct even today, despite conservation efforts.
6. Ice Age is capitalized so as not to confuse it with the alleged numerous ice ages that supposedly occurred several times hundreds of millions of years ago, for which there isn't unequivocal evidence. The only Ice Age was triggered by the global Flood in Noah's day.

But what happens in an ice age? A lot of water is taken out of the ocean and deposited on land, so the ocean level drops.[7] This exposes land bridges. One well-known land bridge was the one that crossed what we call today "the Bering Strait" from Alaska to Russia, so it is easily feasible for animals to have walked from Asia to North and South America.

Other land bridges could also have connected the British Isles to the mainland. Much of the North Sea was formerly known as *Doggerland* or *Dogger Hills,* including later shallow danger areas as it appears on old maps such as the one included here.

(Map by Guillaume De L'Isle in 1730 entitled *Les Isles Britannique ou font le Royaumes D'Angleterre*)

Other land bridges could have connected Japan to Korea, and potentially Japan to the mainland as well. It is possible that Australia could have been connected to Southeast Asia, although today this route is much deeper and may not have been open as long.

That area is known for tectonic activity, and we still see consequences of plate movements from the many earthquakes in Southeast Asia (e.g., consider the earthquake and resultant tsunami in 2004). So the depth today may well be a result of activity *since* the Flood and Ice Age. But let's look at this in more detail.

Other Natural Effects

We also need to keep in mind that tectonic activity has been occurring since the time of the Flood, causing earthquakes and other issues, and we have seen examples even today when faults shift — some even cause tsunamis. Two large earthquakes and resultant tsunamis have recently occurred in Eastern Japan in 2011 and Southeast Asia in 2004 that were due to ocean floor shifting.

7. Some estimates as low as 350 feet (~107 meters) lower than the current ocean level.

It is possible that some land bridges sank or were destroyed by these movements of the earth and ocean floor. Could this be the case with the connection between Australia and Southeast Asia since the time of the Ice Age? It seems to be a bit deeper today than what it likely was many years ago. Also, Scripture often records earthquakes (e.g., Amos 1:1[8]; Matthew 28:2[9]), and we need to keep in mind that many are not felt in other parts of the world — even very large ones! If tectonic activity reduced this bridge so that it wasn't open as long, this helps explain the following concept.

If this Southeast Asia–Australia land bridge was not in existence for as long as others, like the Bering Strait Bridge, that could explain why marsupials dominated the continent. Recall that marsupials travel farther faster compared to some placental animals, which lag behind. Marsupials could have made it across the land bridge during the migratory period, prior to the arrival of most placentals.

This makes much more sense than the common evolutionary model where marsupials evolved in Australia, which can't explain why marsupials like opossums came to North and South America. The common explanation that Australia and South America were linked is much harder to believe than a short-lived land bridge to Southeast Asia. Furthermore, if South America and Australia were linked (barring any global Flood, as the secularists teach), then why doesn't South America abound with marsupials?

The Ice Age may also have contributed something else to animal migrations. Generally speaking, reptiles are found in larger numbers and greater varieties in warmer climates, potentially like most dinosaurs, and would not thrive as well in the cold. It makes sense that they strayed from colder areas, died out, or their numbers were at least reduced. It also makes sense that mammals would thrive in colder climates.

Many believe that the post-Flood era has more extreme weather patterns (e.g., colder winters, higher elevations in area). In line with this, Adam and Eve did not wear clothes originally in the Garden of Eden, and God's creation was declared to be "very good." So, hypothetically speaking, it makes sense that people should have been able to fill the earth without much need for clothes, if any. After sin and the Curse, things changed, but the continent(s)

8. The words of Amos, who was among the sheepbreeders of Tekoa, which he saw concerning Israel in the days of Uzziah king of Judah, and in the days of Jeroboam the son of Joash, king of Israel, two years before the earthquake.

9. And behold, there was a great earthquake; for an angel of the Lord descended from heaven, and came and rolled back the stone from the door and sat on it (KJV).

and arrangements were not affected at the Curse (that we know of). So at least the pre-Flood topography was a little closer to a perfect world. The world after the Flood is a demolished remnant of the pre-Flood created continent(s).

Our current arrangements of continents and topography makes some areas colder and some hotter, due to different elevation, latitude, etc. Then there is Antarctica sitting at the South Pole! Many ideal habitats could have been completely eradicated during the Flood, never to be replaced. Insects that grew large in the past now die out by winter in many parts of the world before reaching maturity. Therefore, it is possible that the ones that matured more quickly, although they were smaller, laid their eggs prior to winter, and thus had a better chance of surviving.

Even desertification may have been triggered by changes in the weather. The new conditions could have wiped out populations in those areas or permitted a select few to survive. With variations of creatures after the Flood, they had to find a new niche or die out. The Ice Age and new weather patterns surely helped solidify where they lived and flourished from that time until now.

Could animals have migrated to a part of the world they were previously familiar with (latitude and longitude)? We've always wondered this. If a continent ended up at a particular place on the globe, and migratory animals thrived in those former areas before continental movement, is it possible that some attempted to migrate back to that original latitude and longitude? We would leave that option open.

In some cases, animals could have ridden on floating debris to make it to islands or other far-reaching places. Consider tsunamis, hurricanes, or other storms that force animals near coasts to grab onto things for their survival. They may be whisked out to sea only to arrive at another place to make their home.

What We See Today

Let's not forget another major factor to animal distribution — humans! Humans have been involved since the Flood. In fact, due to the ark, land animals and birds exist today.

Although rats had already traveled to many parts of the world, by the age of exploration (A.D. 1400–1800), these stowaways were easily distributed around the world in all the European exploits and trade. They were commonplace on most ships and ended up all over the world because men accidentally transported them. Think how many insects were surely taken to various places in the same manner.

Throughout history, people have brought plants and animals to new locations, and those organisms have become permanent populations, interacting with the original creatures. For example, it is claimed that the Romans brought pheasants (members of the chicken kind) to England, and they have since been regular inhabitants of various habitats. In fact, the Romans redistributed organisms from one side of the Roman Empire to the other.[10] When we were in Australia, and went out to Green Island, we found out that the coconuts that grew there were planted to provide food for shipwrecked people. Horses, wild boars, fallow deer, and wild goats are well-known examples of animals introduced to North America.

The point is that many animals and plants have been redistributed to places all over the world by mankind. Many were pets and went wild (such as dingoes); many were introduced as potential food sources (e.g., pigs), and so on. Imagine how much of this redistribution was done prior to the years when we actually started keeping track!

Conclusion

We know that a host of factors were involved with getting animals to various places. In fact, there are likely options that were not explored in this chapter.

The Bible gives us a framework in which to interpret this topic even though little is given by way of specifics. When it comes to answering questions like this, it is always best to uphold the Bible as our authority and reject ideas that are inconsistent with God's Word.

10. BBC Nature Wildlife website, "Pheasant," 2012, http://www.bbc.co.uk/nature/life/Common_Pheasant.

37 | Has Noah's Ark Been Found?

Bodie Hodge

Introduction

As with many questions, there are always debates, and the questions surrounding the search for Noah's ark are no different. However, one debate most people are probably *somewhat* familiar with, or have at the very least considered, is "Has Noah's Ark been found?" There is actually much more to this than meets the eye.

Entire volumes could be written on this subject of the ark, and some have been written already. However, the aim here is to provide some concise answers to the best of our ability to the many questions about the ark in an overview format.

Biblical Data

The Bible gives some information about the ark[1]:

- Its overall dimensions were 300 by 50 by 30 cubits (Genesis 6:15). Using the short or common cubit (~18 inches), it would have been about 450 feet long; or using a longer royal cubit (~20.4 inches), it would have been around 510 feet long[2]

1. Footnotes of the passages in the bulleted tabulation will not be given — most can be looked up easily in chapter 2.
2. Bodie Hodge, "How Long Was the Original Cubit?" *Answers* magazine, April–June 2007, p. 82, http://www.answersingenesis.org/articles/am/v2/n2/original-cubit.

- It was made of wood (gopher)[3] — Genesis 6:14
- It was covered with pitch inside and out — Genesis 6:14
- The ark had rooms — Genesis 6:14
- It had three decks — Genesis 6:16
- The ark had a covering — Genesis 8:13
- It had a window (Hebrew: *tsohar*, which means "noon"), which was finished to a cubit from above (think of something like a "ridge vent" on houses today for ventilation and lighting) and could be opened and shut (though Noah did not open it until 40 days after they landed on the mountains of Ararat — Genesis 6:16, 8:6
- The ark was made/fabricated, and done so with godly fear — Genesis 6:14–15, 6:22; Hebrews 11:7
- One of its purposes was to house land-dwelling, air-breathing animals during the Flood with a male-female pair from each of the representative kinds[4] of the unclean animals and seven individuals (or pairs — the meaning is debated) of the clean animals (likely three breeding pairs of these clean animals, as well as sacrificial individuals for after the Flood) — Genesis 6:20, 7:2–3, 21–23, 8:20
- Eight people survived on the ark: Noah, Shem, Ham, Japheth, and their respective wives — Genesis 7:7, 13; 2 Peter 2:5; 1 Peter 3:20
- It had a door, which was likely in the center deck as implied by the wording "lower, second, and third decks"; that is, one deck was lower than the door — Genesis 6:16
- The Lord shut the door to the ark from the outside (and it is probable that it too was sealed with pitch like the rest of the ark; otherwise, the rest of the pitch was pointless with these untreated seals) — Genesis 7:14, 16

---

3. Scholars have debated whether this was a particular type of wood or a means of processing wood (similar to the process of making plywood or pressed wood). Many ancient ships of antiquity had intricate wood that had been processed to make it stronger and more durable. Was that technology passed down through Noah and his sons? It is possible.
4. It is important to note that a kind is not necessarily what we know today as a "species." For more information, see Georgia Purdom and Bodie Hodge, "What are 'Kinds' in Genesis?" in *The New Answers Book 3*, Ken Ham, ed. (Green Forest, AR: Master Books, 2010), p. 39–48.

- The unrighteous sinners who did not go on the ark did not realize their doom, even up to the day that Noah boarded the ark — Matthew 24:38; Luke 17:27
- The ark was lifted off the ground by or on the fortieth day of the Flood and then floated high above land surface on the waters — Genesis 7:17
- It landed in the mountains of Ararat on the 150th day of the Flood (confirmed by calculating from Genesis 7:11 with a 360-day year) — Genesis 8:3–4
- The ark survived the Flood and Noah's family and the animals came out of the ark — Genesis 8:18–19
- They had remained on the ark for 370 days (or 371, depending on whether half days are rounded as full days or not) — Genesis 7:11, 8:14–16
- Noah's family left the ark and settled where there was fertile soil for Noah, who became a farmer — Genesis 8:19, 9:1, 20. This first settlement would have been in an east/west direction from Babel, the later place of rebellion — Genesis 11:2[5]

Notice that very little information is given about the ark's resting place (simply "mountains of Ararat"). However, there are some deductions and inferences that can be made from the Scriptures, which leads to the debate over the ark's landing site.

Where Are the Mountains of Ararat?

If someone had asked me years ago which mountain Noah's ark landed on, my response would have been a naïve, "Mt. Ararat, of course, because that is what the Bible says." However, a reading of Genesis 8:4 reveals no such thing. Instead, the text says the "mountains of Ararat," which refers to a range of mountains, not a specific mountain.

And this raises an important point. Christians always need to check information with the Scriptures. Let God be the authority, rather than man, on any subject. Believers know Noah's ark existed, and they can be certain of that because of God's Word, regardless of whether or not any remains of the

5. There could be more information from the biblical text, but this should be sufficient to give us the relevant highlights of ark information from the Bible.

ark are found. The all-knowing God says in His Word that the ark existed. There is no greater authority on this subject to whom one can appeal.

So where are the mountains of Ararat? The mountains of Ararat form a mountain range named after the Urartu people who settled in that region after the dispersion event at the Tower of Babel. In Hebrew, Ararat and Urartu are even spelled the same way. Hebrew does not have written vowels, so both are essentially spelled *rrt.*

Josephus, a Jewish historian living about 2,000 years ago, said that Armenia was made up of the descendants of Hul through Aram and Shem.[6] Armenia is the *later name* of the region of Urartu/Ararat, which is a specific part of the Armenian highlands. So it is understandable why Josephus used the later name, whereas Moses used the earlier name.

When Moses wrote Genesis around 1491–1451 B.C.,[7] he had been educated in Egypt as royalty (and he had been inspired by the Holy Spirit), so it is to be expected that he understood the geography of the peoples in the Middle East. In fact, other Bible writers like Isaiah and Jeremiah, who lived well after Moses but well before Josephus, were also familiar with the Ararat land and people:

> Now it came to pass, as he was worshiping in the house of Nisroch his god, that his sons Adrammelech and Sharezer struck him down with the sword; and they escaped into the land of Ararat. Then Esarhaddon his son reigned in his place (Isaiah 37:38).

> Set up a banner in the land, blow the trumpet among the nations! Prepare the nations against her, call the kingdoms together against her: Ararat, Minni, and Ashkenaz. Appoint a general against her; cause the horses to come up like the bristling locusts (Jeremiah 51:27).

This ancient region is basically in the eastern part of modern-day Turkey, Armenia, and western Iran.

The Debate over Which Mountain

One of the most heated debates on this subject, though, is over which specific *mountain* the ark landed on within the mountain range. Of course, the

6. Bodie Hodge, "Josephus and Genesis Chapter Ten," Answers in Genesis, http://www.answersingenesis.org/articles/aid/v4/n1/josephus-and-genesis-chapter-ten#fnList_1_1.
7. James Ussher, *The Annals of the World*, Larry and Marion Pierce, eds. (Green Forest, AR: Master Books, 2003), p.39–47.

Bible does not say the ark landed on a specific mountain, but this is inferred. It is possible it landed in a lower area within the mountains of Ararat. However, the two most popular sites are:

Mt. Ararat (Agri Dagh)
Mt. Cudi (or Cudi Dagh; Cudi sounds like "Judi")

Many ark landing sites have been proposed over the years. One that has been rejected as a geological formation by most scholars in recent years is the Durupinar or Akyayla site in Turkey, near the Iran and Turkey border. That site consists of something akin to a boat-shaped feature that is readily recognizable (think of a football field-sized "footprint" in the shape of a boat).[8] The area contains several of these geological features and that is really all that it is.

Other sites that have attained some popularity but have been largely rejected by archaeologists, geologists, and researchers are Mt. Salvalon and Mt. Suleiman in Iran. It is unreasonable for these mountains to be included in the region of Ararat. There are other problems associated with them too.[9]

Ararat

The discussion following will focus on the debate over these two primary alleged resting places, Cudi and Ararat. Key verses in the Scriptures need to be consulted before proceeding:

> Then the ark rested in the seventh month, the seventeenth day of the month, on the mountains of Ararat. And the waters decreased continually until the tenth month. In the tenth month, on the first day of the month, the tops of the mountains were seen (Genesis 8:4–5).

The tops of the surrounding mountains were seen 74 days after the ark landed in the mountains of Ararat. This gives the impression that the mountain the ark landed on was much higher than the others. So the obvious choice is Mt. Ararat, which today towers excessively over all the other mountains in the region.[10]

8. Dr. Andrew Snelling, "Special Report: Amazing 'Ark' Exposé," *Creation ex nihilo*, Sept. 1, 1992, p. 26–38, http://www.answersingenesis.org/articles/cm/v14/n4/special-report-amazing-ark-expose.
9. Gordon Franz, "Did the BASE Institute Discover Noah's Ark in Iran?" Associates For Biblical Research, February 16, 2007, http://www.biblearchaeology.org/post/2007/02/Did-the-BASE-Institute-Discover-Noahs-Ark-in-Iran.aspx#Article.
10. This does not take into account the fact that some mountains of the region may have been raising and lowering during this transitional period of the Flood (Psalms 104:8–9).

Mt. Ararat is a large volcano that extends to a height of 16,854 feet! This is higher than any mountain in the 48 contiguous United States (Alaska does have a few mountains that are taller). Lesser Ararat (also known as Little Ararat) is another volcano that stands adjacent to Mt. Ararat and is 12,782 feet high, which is similar in height to a number of impressive peaks in the Rocky Mountains in the United States.

Many say that if the ark landed on Mt. Ararat, then it would have taken another two and one-half months for the water to reveal other surrounding mountain peaks. This seems logical. In fact, this is one reason some scholars argue that Mt. Ararat is the resting place for the ark.

Nevertheless, this is not the main reason why the search for the ark has focused on Ararat. The primary reason is because of the eyewitness accounts of ark sightings in recent times. B.J. Corbin wrote a book on the search for Noah's ark, which is helpful to anyone wanting to find out the details of various expeditions on Ararat. The book also discusses Mt. Cudi, the other proposed site. In the preface of the second edition, Corbin states:

> The only major reason to consider Mount Ararat is because of the few documented eyewitnesses. . . . There is a number of intriguing statements from individuals who indicate that there may be a barge-like or boat-like structure high on modern day Mount Ararat. These statements are really the primary basis for the search on Mount Ararat.[11]

Corbin, who has also been involved in the ark search on Ararat, confirms that the primary reason for the search on Ararat is because of the eyewitness accounts. There have been quite a few accounts including many reputable people in the 20th century, and Corbin in the preface to his book documents these as well. Furthermore, Ararat is covered with ice and glaciers all year, so this is an ideal hiding place (i.e., more difficult to locate) for an ark.

Even in some older literature, such as in the writings of Byzantine historian Philostorgius in the fifth century, Ararat was suggested as the ark's landing site. After the 13th century A.D., more sources affirm this mountain as the landing site.[12]

---

11. B.J. Corbin, *The Explorers of Ararat* (Long Beach, CA: Great Commission Illustrated Books, 1999), p. 8.
12. Richard Lanser, "The Case for Ararat," *Bible and Spade*, Fall 2006, p. 114–118.

Considering the scriptural basis of the highest mountain, the eyewitness accounts, and the historical sources, why would anyone look elsewhere for the landing site?

The Debate Gets Heated

On the other side of the debate, there are some objections to consider. First, even with all the eyewitness accounts of purportedly seeing something like the ark on Ararat, there has never been anything of substance ever found or documented to prove the ark landed on Ararat.

Also, the Bible does not explicitly say that it was *only* due to the water's recession (which all sides agree is indeed a factor) as to why mountaintops were seen. The text says "the tops of the mountains were seen" (Genesis 8:5). This involves two things: water level (1) *and* visibility (2).

This second factor that is often overlooked is the conditions that may affect visibility. The warmer ocean water (which is expected from the Flood with continental shifting, rising basalts from the mantle, and possibly some nuclear decay would surely generate heat and volcanism) gives off vapors and mists that form low-lying fog and clouds. Hence, visibility would likely be rather low. Genesis 8:5 may well be discussing the state of visibility and atmospheric condition regarding clouds and fog from the heated ocean just as much at it discusses water level.

One way or another, this passage (Genesis 8:5[13]) cannot be so easily used to affirm a landing spot on the highest peak. It *may* still be the highest peak, but one cannot be dogmatic. Another factor needs to be considered here too — if it were the highest peak, what was the highest peak *at this time*?

One common objection is that if the ark landed at such a high altitude, how did the animals get off the ark and make their way down from this deadly mountain? And how did man and the animals at that high altitude survive all that time without sufficient oxygen after striking ground (day 150) until being called off the ark (day 370)? Oxygen tanks would not be necessary when floating on the surface of the water, because oxygen percentages are based on sea level (about 21 percent at sea level). If the ark were at 16,000 feet above sea level, then when the water receded, oxygen would be

13. And the waters decreased continually until the tenth month. In the tenth month, on the first day of the month, the tops of the mountains were seen.

a requirement because serious problems can occur due to lack of oxygen at altitudes over 12,000 feet.[14]

Another oft-used argument is that pillow lavas should be found on Mt. Ararat if it formed underwater. For those unfamiliar with pillow lavas, they are formed when a volcanic eruption occurs underwater. The lavas that come in contact with water cause it to harden quickly in masses that look "like a pillow."[15]

Some believe there may possibly be some pillow lavas on Ararat, as reported by Corbin[16] and through observation attributed specifically to Clifford Burdick. However, if this volcano was formed in the Flood before day 150 when the ark ran aground, then such pillow lavas should have extensively covered it. But this is not the case. Rather, there is a severe lack of evidence that this mountain was ever covered by water. There are some pillow lavas on Ararat at very high altitudes (e.g., 14,000 feet)[17], but the same characteristic features of pillow lavas also form when lavas meet ice and snow, which may be a better explanation of these specific pillow lavas at high altitudes on Ararat where it is capped in snow and ice.[18]

Another argument must also be considered: Mt. Ararat and Lesser Ararat are volcanoes. They have been identified as having been formed after the Flood because they sit on top of fossil-bearing sediment from the Flood.[19] Classed as Pleistocene rock, Ararat is regarded by most creation researchers as post-Flood continuous with the Ice Age that followed the Flood.[20]

---

14. It is possible that this volcano was much smaller originally and later post-Flood eruptions are what caused it to become so large and so high. But if this were the case, eruptions should have burned the wooden vessel to oblivion, so no remains of the ark should ever be found on Ararat. It is possible that petrification of the wood could take place at such temperatures; however, being coated in pitch, which is typically rather flammable, and being made of seasoned dry wood, it makes more sense that the ark would be burned in the presence of volcanic heat, not petrified.
15. There are also other underwater geological evidences that should be present such as interbedded water-deposited volcaniclastics and pyroclastics, but these do not cover the mountain either.
16. Corbin, *The Explorers of Ararat*, p. 326.
17. Ibid., p. 326.
18. "Ararat," NoahsArkSearch.Com, http://www.noahsarksearch.com/ararat.htm.
19. Y. Yilmaz, Y. Güner, and F. Şaroğlu, "Geology of the Quaternary Volcanic Centres of the East Anatolia," *Journal of Volcanology and Geothermal Research* 85 (1998): 173–210.
20. For more on the post-Flood Ice Age see Michael Oard, "Where Does the Ice Age Fit?" in *The New Answers Book 1*, ed. Ken Ham (Green Forest, AR: Master Books, 2006).

By this argument, these volcanoes *did not exist* at the time the ark landed. When viewing these volcanoes from above, one can readily see the lava and volcanic flow from the volcanoes *overlaying* the foothills and plains that make up part of the region of the mountains of Ararat. From the account of Scripture, the mountains of Ararat were made by day 150 of the Flood (Genesis 8:4) and the ark landed on day 150 of the Flood (Genesis 8:4), so these volcanoes had to come *after* both the mountain formation and ark landing to have their volcanic flows sitting aloft on the foothills of the mountains of Ararat today.[21]

Furthermore, fossils are readily found within the mountains of Ararat, but they are rare or absent entirely on Mt. Ararat. Some claim to have found some, but there is no documentation for *in situ* (in their original place) fossils on Ararat. The layers on Ararat are volcanic, not sedimentary.

Habermehl has reviewed the search for Noah's ark.[22] Though we do not agree with all of Habermehl's assertions,[23] she does provide a thorough review of evidences and arguments regarding Ararat and Cudi.

Cudi

The other potential mountain that has long been proposed is Mt. Cudi. Crouse and Franz point out that this mountain has gone by various names

---

21. It is possible these volcanos were smaller at the time of the Flood and further eruptions have covered or destroyed any remains of the ark at the previous height of the mountains, but if this were the case, the ark did not come to rest on Ararat as we know it, nor would we know if it were taller than any other mountain in the range at that time.
22. Anne Habermehl, "A Review of the Search for Noah's Ark," in *Proceedings of the Sixth International Conference on Creationism*, ed. Andrew A. Snelling (Pittsburg, PA: Creation Science Fellowship; Dallas, TX: Institute for Creation Research, 2008), p. 485–502.
23. As one example, she holds the position that Noah and his family settled rather close to the ark and hence uses Genesis 11:2 as a basis to relocate Babel to an east-west direction of the ark landing site. Many scholars have pointed out the fallacy in this east-west direction, as this is in reference to Noah's *first* settlement after the Flood (see footnote 7 or Adam Clarke's Commentary on Genesis 11:2, http://www.sacred-texts.com/bib/cmt/clarke/gen011.htm). Noah's initial settlement is unknown, but it was a place that was fertile enough to farm. Noah and his family were also able to live in tents. One cannot assume this was essentially still at the ark landing site, as Noah and his immediate family were told to come off the ark (Genesis 8:16) and fill the earth (Genesis 9:1.) It was not until Noah had (in some cases) great, great, great, great grandsons that the rebellion occurred at Babel. Also, why live in tents when there is a huge wooden mansion to live in (i.e., the ark) or, at the very least, wood enough to build a proper shelter? Furthermore, Noah had his pick of the new world, so why remain at or near the rough mountainous area of the ark landing site and not find a place to start a new beginning, especially somewhere suitable for farming?

such as Judi, Cardu, Quardu, Kardu, Ararat, Nipur, Gardyene, and others.[24] Cudi, being in the mountains of the Ararat region, also sits in a "specified" range of mountains known as the Gordian, Kurdish, Gordyene, and others.

This is important to know, as many ancient sources say the ark landed on this specific portion of the mountains. Both Ararat and Cudi are in the basic region of where the Urartu lived, but whereas Ararat is referred to in some early literature (5th century at the earliest) as the ark's landing site, Mt. Cudi is referred to as the landing site in many more and far earlier sources.

In *Bible and Spade*, there were cases presented for Ararat (Lanser) and for Cudi (Crouse and Franz), along with other pertinent articles on the subject.[25] Crouse and Franz did an extensive historical review referring to numerous ancient and modern sources that point to Cudi. These include direct and indirect allusions to Cudi from Jewish (e.g., Josephus, Targums, Book of Jubilees, and Benjamin of Tudela), Christian (e.g., Theophilus of Antioch of Syria, Julius Africanus, Eusebius, and several others), pagan (e.g., Berossus and The Epic of Gilgamesh), and Muslim sources (e.g., Koran [Qur'an], Al-Mas'udi, Zakariya ibn Muhammad al Qazvini).

Cudi is much lower in elevation, being about 6,800 feet high, so it would not have been so difficult to herd animals down the mountain. There would have been no problems with low oxygen levels, and this mountain is not a volcano that is resting upon the top of the mountains of Ararat (like volcanic Ararat is). But it was easily in a place where pieces could be looted or taken as relics. According to Crouse and Franz, the Muslims claimed to have taken the last of the major beams for use in a mosque.[26]

The legends and lore associated with this mountain still persist in the area as well. Christians, Jews, Muslims, and others still came together for a yearly celebration in honor of the sacrifices made by Noah after the Flood as

24. Bill Crouse and Gordon Franz, "Mount Cudi — True Mountain of Noah's Ark," *Bible and Spade*, Fall 2006, p. 99–111, http://www.biblearchaeology.org/publications/bas19_4.pdf.
25. Ibid.
26. One also has to consider the amount of deterioration the wooden vessel underwent over 4,350 years. If kept frozen or in a dry, arid climate, a wooden ark could last quite a long time. However, in mid-temperate areas with alternating wet-dry conditions, it should not last long at all (think of a barn in the Midwest; one must work hard to keep such a thing for even 200 years). Being coated with pitch helps, but even that is not a perfect preservative. A perfectly engineered ark would have the pitch's usefulness end at the end of the Flood (~370 days).

recorded by a historian nearly 100 years ago (W.A. Wigram).[27] There is even a place on Cudi that is the traditional landing spot of the ark on a particular ridge. So is this the absolute landing site? We simply do not know.

Conclusion

Has Noah's ark been found? The obvious answer is that people would not be asking this question if Noah's ark really had been found! It would likely be the find of a lifetime.

Both Ararat and Cudi have had their share of popularity over the years. And both have strong supporters on their side. When viewing the evidence through the lenses of Scripture, the more logical choice is that of Cudi, not modern-day, volcanic Mt. Ararat that sits on top of fossil-bearing sediment from the Flood.

But would we be dogmatic that Cudi was the landing spot? Not at all. The Bible simply does not say, and though many ancient sources point to Cudi, these sources are not absolute, while Scripture is. The fact is that there has been no indisputable evidence of Noah's ark having been found anywhere (outside of Scripture, which itself is sufficient proof that the ark existed, as there is no greater authority on any subject than God). But is such external evidence needed? Not at all.

To summarize, there was so much more that could have been discussed, but with such limitations, a brief overview of the debate is the best that can be hoped for in a single chapter of a book. My hope is that this brief introduction will encourage you to learn more about the subject, and that you will give glory to God when doing so. Much more research on the topic of the ark's landing site needs to be done, be it on Ararat, Cudi, or other places.

Would undisputed evidence of the ark be of value? Absolutely. But is it necessary for one's faith? Not in the least. So do not forget this point: the Bible is true, and Christ is who He says He is, regardless of whether anyone finds the remains of the ark or not.

Further Reading:

1. Bible and Spade Debate: http://www.biblearchaeology.org/publications/bas19_4.pdf.
2. *The Explorers of Ararat*, B.J. Corbin (Long Beach, CA: Great Commission Illustrated Books, 1999).

27. Habermehl, "A Review of the Search for Noah's Ark."

3. Noah's Ark Search website: http://www.noahsarksearch.com/.
4. Rick Lanser of the Associate for Biblical Research has published a four-part series on the group's website entitled "The Landing-Place of Noah's Ark: Testimonial, Geological and Historical Considerations," parts 1–4, available at http://www.biblearchaeology.org/category/flood.aspx.

I would like to extend a special thanks to Dr. Andrew Snelling for his guidance on this chapter.

38 | What about the Window, Pitch, Door, "Gopher" Wood, and Other Ark Features?

Ken Ham and Bodie Hodge

The size and shape of Noah's ark are not the only things unbelievers tend to attack. I've seen hosts of attacks on a variety of features and aspects of the ark. Some attacks are completely without warrant and not worth comment. However, some need to be addressed. So let's "dive in!"

Window of the Ark

The Bible says there is a window on the ark. The Bible says:

> You shall make a window for the ark, and you shall finish it to a cubit from above; and set the door of the ark in its side. You shall make it with lower, second, and third decks (Genesis 6:16).

> So it came to pass, at the end of forty days, that Noah opened the window of the ark which he had made (Genesis 8:6).

Many criticisms of the ark designs were summed up in this question: "Where is the little window at?" As you look at the design, you might concur, and say, "I don't see the window either, so what is going on."

This is because most people don't realize what the "window" is. The Hebrew word for window is *tsohar*. It means "noon" or "midday." We simply translate it as "window." This *window* is something that runs along the *overhead* position (top, middle) of the ark that allows lighting and ventilation into the ark, while keeping bad weather out!

We do something similar when we build many houses today. We put a ridge vent on the top to do the very same thing.

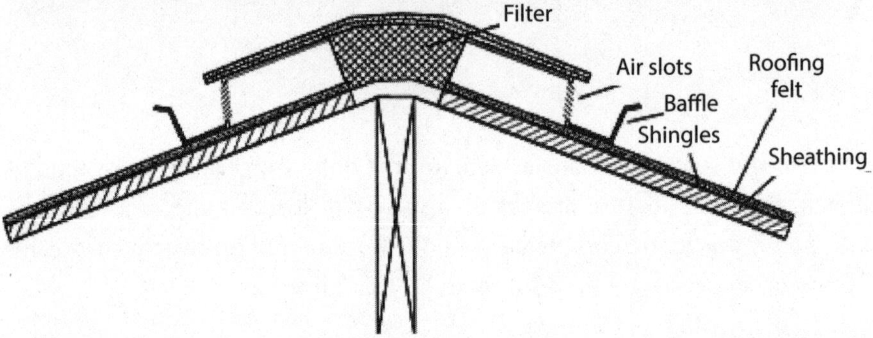

So the window is what you see along many ark images that run the course of the top middle and that is finished to "a cubit from above." Noah can easily open this area up for a better view too when he is analyzing the post-Flood world.

Door of the Ark

Few unbelievers question that there was a door in the ark, so that is not the criticism. But we do receive other questions like, *In what level of the three decks of the ark was the door? Was it under the water level? How did Noah seal up the door of the ark with pitch, when he was inside of it?* Let's briefly *swim* through these questions!

What Level Was the Door Placed?

The Bible says:

> You shall make a window for the ark, and you shall finish it to a cubit from above; and set the door of the ark in its side. You shall make it with lower, second, and third decks (Genesis 6:16).

Obviously, the door was in the *side* of the ship, not the front nor back. Also by the terminology of "lower," second, and third decks, it sounds as though the door was on the second level. So as you come into the ark at the door, there is one level that is "lower."

Was the Door under the Water Level?

It makes more sense to have the door above the *draft* level (sometimes spelled *draught*). *Draft* is how much of the ship is below the water level. A ship can't float if the water depth is less than the *draft*; otherwise the ship runs aground.

With this in mind, would the door of the ark on the second level be above the draft? The dimensions of the ark, based on the long cubit, are 510 x 85 x 51 feet. This is 51 feet of total vertical height of the ark from the keel to the window.

Just doing a remedial calculation for three decks, this is 51 divided by 3, which equals 17 feet per deck. If the lower deck is 17 feet, then the door must be higher than 17 feet (this is a minimum). Is this feasible?

Let's compare this to a big, heavy cruise ship. The *Emerald Princess* cruise ship is an 113,000-ton ship. Its height is 195 feet but it only has a draft of 26 feet![1] Naturally, it can vary depending on weight. But only about 13 percent of this ship is under the water line!

If this same percentage (13 percent) is applied to the ark, then that would only be about 6–7 feet of draft! Even with variation in the weight, this is still no problem, being well above 17 feet. Though there are many factors to calculate a proper draft, I expect it to be a bit more than this. But the second level of the ark for the door is perfectly reasonable.

Some have suggested that draft was 15 cubits (half of the ark) because of Genesis 7:20,[2] where the water covered the mountains by at least 15 cubits, though we're not convinced that we can directly relate this to the draft. Even so, let's say the draft level had the door under the water line. It was *the Lord* who shut them in. So it is still *not* a problem, as such a remedial job by the Creator God would be done perfectly (Deuteronomy 32:4[3]).

---

1. Linda Garrison, "Emerald Princess Cruise Ship," About Travel website, accessed, October 9, 2015, http://cruises.about.com/od/princesscruises/ig/Emerald-Princess.-1wQ/index.htm.
2. The waters prevailed fifteen cubits upward, and the mountains were covered.
3. He is the Rock, His work is perfect; for all His ways are justice, a God of truth and without injustice; righteous and upright is He.

How can Noah seal up the door of the ark with pitch (Genesis 6:14, to waterproof it) . . . when he was inside of the ark?

This question is rather easy to answer. Consider what we just read (it was the Lord's job to shut them in!) and what the Bible says:

> So those that entered, male and female of all flesh, went in as God had commanded him; and the LORD shut him in (Genesis 7:16).

The Lord shut Noah and his family in the ark. So *Who* sealed up the door with pitch to shut Noah in? The Lord. Is the answer that simple? Yes. The Bible doesn't say, but Noah likely did the final pitching for the *inside* of the door, since the ark was coated with pitch inside and out (Genesis 6:14[4]).

Pitch

Where did the pitch come from to seal up the ark in the first place? We make pitch today mainly from petroleum products, but it can also be made from plants.

Regarding petroleum pitch, is this a problem? Many unbelievers tend to think it is. They argue that Christians point out that petroleum was largely formed during the Flood. But the Flood hadn't occurred yet, so where did this pitch come from?

Petroleum Pitch

It is possible that a certain amount of petroleum was in existence from creation or developed due to pre-Flood processes. And Noah could have utilized this, although we tend to think Noah didn't. More on this in a moment.

The bulk of the petroleum we have today was from the Flood. Think of it like this — organic material from the ocean and land was buried by sediment-rich water (think vegetation, microorganism, algae and animals, shells, fish, etc.).[5] As this water seeped down into the earth, it transported the organic material elsewhere and replaced the organic material (in things like trees, shells, and so on) to produce fossils. These fossils are now the

4. Make yourself an ark of gopherwood; make rooms in the ark, and cover it inside and outside with pitch.
5. Coal wasn't fossilized into things like limestone and thus it retained much of the organic material.

remains of once organic things that are now made of material like limestone. The organic material from micro-organisms, marine organisms, trees, algae, etc. accumulates in trapped pools under the surface and is essentially what we call crude oil/petroleum. They are simply hydrocarbons.

The Flood buried the world's supply of most organic material at the time of Noah so that is where most petroleum that we tap into comes from today. Though pitch can be made from petroleum (and coal tar), Noah likely used a different kind of pitch.

Plant Pitch

Pitch can also be made from plants. It is also called *resin* and in some cases *rosin*. A number of ancient wooden ships were covered with pitch, and even canoes were often sealed with plant pitch like that made from balsam fir. This is quite logical with wooden ships to help seal them up.

The Bible points out that when Noah was building the ark, he was to cover it with pitch, inside and outside. Why inside? During longer construction periods, having things pitched on the inside helped preserve the integrity of the wood from weathering. This could be one reason for the command as well as the fact that internal pitching adds another layer of sealing from the water outside the ark.

There were plenty of plants pre-Flood as witnessed from the Flood sediment, so plant-based pitch was likely what Noah used as opposed to petroleum-based pitch.

Gopher Wood

What is gopher wood? Great question. In the past, many thought this may have been a type of tree (think of oak, pine, hickory, etc.). However, many now lean against that interpretation.

The prevailing thought now is that it was a way the wood was *processed.* Processed wood is nothing new. Though many of you may be familiar with some processed wood and may not realize it! Did you know that *plywood* or *pressed wood/particleboard* or *wood veneer* is merely processed wood. Most houses built today utilize processed wood — even processed particleboard beams such as floor joists!

Working with wood like this was common in the past. Certain ancient ships also utilized incredible types of wood processing technology. There have been ancient ships found with some almost unbelievable planking styles that helped make the ship structurally sound, lighter, and even proved to be superior in waterproofing! Ark researcher Tim Lovett comments:

> Ancient shipbuilders usually began with a shell of planks (strakes) and then built internal framing (ribs) to fit inside. This is the complete reverse of the familiar European method where planking was added to the frame. In shell-first construction, the planks must be attached to each other somehow. Some used overlapping (clinker) planks that were dowelled or nailed, others used rope to sew the planks together. The ancient Greeks used a sophisticated system where the planks were interlocked with thousands of precise mortise and tenon joints. The resulting hull was strong enough to ram another ship, yet light enough to be hauled onto a beach by the crew.[6]

The Age of Exploration by Europeans lost much of this technology when ships were made via carvel planking techniques, which were simple and quick but prone to problems. We suggest that many of these incredible planking styles may have been carry-overs from the technology Noah and his three sons had. It could easily have been passed along to subsequent generations. Keep in mind that Noah lived 350 years after the Flood (Genesis 9:28[7]) and Shem lived 500 years after the Flood (Genesis 11:10–11[8]). I am not sure how long Japheth and Ham lived after the Flood; it could have been as long or even much longer than Shem!

"Free for All" in the Ark?

Certain portrayals of the ark have animals in a "free for all" going all over the ark — bathtub arks especially! However, the Bible indicates the ark had rooms. The Bible says:

> Make yourself an ark of gopherwood; make rooms in the ark, and cover it inside and outside with pitch (Genesis 6:14).

6. Tim Lovett, "Thinking Outside the Box," Answers in Genesis, https://answersingenesis.org/noahs-ark/thinking-outside-the-box/.
7. And Noah lived after the flood three hundred and fifty years.
8. This is the genealogy of Shem: Shem was one hundred years old, and begot Arphaxad two years after the flood. After he begot Arphaxad, Shem lived five hundred years, and begot sons and daughters.

So the animals were divided in the ark. Many could have shared rooms with others. Originally, all animals were vegetarian (Genesis 1:30[9]), and even though all flesh had corrupted itself on the earth (Genesis 6:12[10]), the Lord brought specific animals to Noah for survival. It makes more sense that God sent animals to Noah that were still in accordance with Genesis 1:30,[11] thus conflicts of carnivory may not have been what many think they were on board the ark. Nonetheless, the animals were separated into rooms/cages.

Some animals can destroy wood cages easily — like rabbits and woodpeckers! I grew up on a farm, and rabbits would chew through wood cages quickly without the wire mesh affixed in the cages. But having cages reinforced with iron is not a problem. Iron working had preceded the ark for quite some time (Genesis 4:22[12]). Thus, cages with some iron reinforcements are not a problem. Keep in mind that many animals could have hibernated as well during the voyage for safekeeping.

Iron Tools

If the Iron Age were after the Flood with Noah's descendants, then where would Noah get the tools necessary to build the ark? This is a misconception based on the world's story of origins. In the world's story, people have to become smart enough to develop the technology to do iron working. This is based on the evolutionary story where man supposedly evolved from apes and had to get smarter.

However, in the Christian account, iron and bronze working was well before the Flood (Genesis 4:22[13]). This technology passed through the Flood and what occurred next was why some peoples lost this technology and had to regain it — that was the scattering at Babel.

Not every family at Babel knew iron or bronze working; just like today, not every family alive knows iron working or bronze working! So at the scattering at Babel, those who had a family member who knew the art could

9. "Also, to every beast of the earth, to every bird of the air, and to everything that creeps on the earth, in which there is life, I have given every green herb for food"; and it was so.
10. So God looked upon the earth, and indeed it was corrupt; for all flesh had corrupted their way on the earth.
11. See footnote 9.
12. And as for Zillah, she also bore Tubal-Cain, an instructor of every craftsman in bronze and iron. And the sister of Tubal-Cain was Naamah.
13. Ibid.

easily retain it for his descendants wherever they initially moved from Babel. Once they were on their feet and settled in, they could pick it back up.

Others, who did not know the art of iron or bronze working, would lose that technology for their families until later contact with those who did have active iron or bronze working that would permit them to regain this technology. But with all this in mind, Noah likely used iron or bronze-based tools.

Plants/Seeds on the Ark?

Did Noah take plants and seeds on the ark? The Bible does not say directly. Though the foodstuffs for Noah's family and many animals were surely grains, those are seeds (Genesis 6:21[14]), and many of those could have been utilized after the Flood.

Noah surely planned ahead. He knew they would come off the ark and have to begin again. If you were thrust into that situation, would you take seeds and plants (bulbs, saplings, some potted planted, domestic plants, etc.) to be ready to start again? I suggest Noah was prepared by taking seeds and plants for the new world (particularly food plants).

Keep in mind that Noah became a farmer once he settled. The Bible says:

> And Noah began to be a farmer, and he planted a vineyard (Genesis 9:20).

Seeds would have been extremely valuable at this stage of human history. Surely Noah was prepared.

14. "And you shall take for yourself of all food that is eaten, and you shall gather it to yourself; and it shall be food for you and for them."

39 How Many People Died in the Flood?

Ken Ham and Bodie Hodge

Introduction

For years, we've heard this question. We think it is more of a *curious* question rather than a question that attacks the authority of God's Word. Whatever answer we give is speculative because the Bible simply doesn't tell us — though an educated guess may be in order here.

Biblical Information

The Bible reveals information that is useful to the discussion. For example, the Bible indicates there are ten generations from Adam to Noah. We also know that the timeline from Adam to the Flood was 1,656 years.[1] We also know that the lineage from Adam through Cain to Naamah (in Genesis 4) was only eight generations.

We also know that generation times seemed far slower than today — consider that Noah did not have children until he was 500 years old! Other patriarchs often waited to have children as well.

Popular Estimates

As you can see from the following table, estimates for the pre-Flood population are based on very little information, since Genesis 1–6 doesn't give extensive family size and growth information. Genesis also indicates that in

1. By adding up the ages of the sons from creation week to Noah's oldest son Japheth (born when Noah was 500 years old), we get 1,556. The Flood began when Noah was 600 years old, so add 100 years to 1,556 and you get 1,656.

Patriarchal Tables*

Patriarch	Age	Age of Son**	Bible Reference
Adam	930	130	Genesis 5:3–4
Seth	912	105	Genesis 5:6–8
Enosh	905	90	Genesis 5:9–11
Cainan	910	70	Genesis 5:12–14
Mahalalel	895	65	Genesis 5:15-17
Jared	962	162	Genesis 5:18–20
Enoch	365 (translated)	65	Genesis 5:21–23
Methuselah	969	187	Genesis 5:25–27
Lamech	777	182	Genesis 5:28–31
Noah	950	500	Genesis 5:32, 9:29

* Bible verse footnotes are not given for the table.

** In the lineage of Adam to Japheth; Japheth was Noah's oldest born when Noah was 500 years old. Shem was born when Noah was 502 being that he was 100, two years after the Flood.

Noah's lineage children were being born when their fathers were between the ages of 65 (Enoch to Methuselah) to 500 (Noah to the first of his three sons).

How many generations were there in other lineages outside of Cain's recorded lineage to Naamah? We don't know. We know the line from Adam to Noah was living upward of 900 years, but we can't be certain everyone lived this long. How often and how many children were born? We don't know? What were the death/mortality rates? We don't know.

Despite this lack of information, some estimates have been done. Tom Pickett gives a range of about 5 to 17 billion people.[2] This is based on various population growth rates and generations of 16–22 prior to the Flood. Recall that Noah was in the 10th generation and Naamah was the 8th generation, so this may be well beyond the higher end of the population maximum.

Consider that we have had about 100 generations or less *since* Noah and our world population is only about 7 billion in A.D. 2015,[3] so these numbers seem considerably high.

2. Tom Pickett, "Population of the Pre-Flood World," www.ldolphin.org/pickett.html.
3. I, Bodie, have a continuous lineage (long and short) from Adam to Noah, through Japheth down through Woden to Alfred the Great down to Edward the Longshanks and his son Thomas the Earl of Norfolk down to my mother. The long yields Noah as my 90th great grandfather; the short yields Noah as my 74th great grandfather, both through Japheth.

More reasonable numbers come from Dr. Henry Morris who had conservative estimates as low as 235 million people. He also calculated rates based on modern population growth, giving about 3 billion people.[4] John Morris reports estimates that there were about 350 million people pre-Flood.[5] Based on these estimates, pre-Flood populations could have peaked from the low hundred millions to 3 billion people.

These would be reasonable peak or maximum numbers pre-Flood, though we still tend to think these are rather high for the population at the time of the Flood. As we have seen, there were only eight generations from Adam through Cain to Naamah. So if this stat were used instead of ten generations, the numbers would go down much further.

Did you realize that one of the few commands given to mankind through Adam was "to be fruitful and multiply" (Genesis 1:28[6])? If man had intense disobedience to God (e.g., Genesis 6:5,[7] 6:12[8]) 120 years prior to the Flood (Genesis 6:3[9]), are we to believe they were still being obedient to God's command here? Likely not!

Consider Joshua's Generation

We know of Joshua's genealogy, and he was contemporaneous with Moses, yet younger. His lineage is revealed in 1 Chronicles 7:22–27:

> Then Ephraim [1st generation] their father mourned many days, and his brethren came to comfort him. And when he went in to his wife, she conceived and bore a son; and he called his name Beriah [2nd gen.], because tragedy had come upon his house. Now his daughter was Sheerah, who built Lower and Upper Beth Horon and Uzzen Sheerah; and Rephah [3rd gen.] was his son, as well as Resheph, and Telah [4th gen.] his son, Tahan [5th gen.]

4. H. Morris, *Biblical Cosmology and Modern Science* (Grand Rapids, MI: Baker Book House, 1970), p. 77–78.
5. J. Morris, *The Young Earth* (Green Forest, AR: Master Books, 2002), p. 71.
6. Then God blessed them, and God said to them, "Be fruitful and multiply; fill the earth and subdue it; have dominion over the fish of the sea, over the birds of the air, and over every living thing that moves on the earth."
7. Then the Lord saw that the wickedness of man was great in the earth, and that every intent of the thoughts of his heart was only evil continually.
8. So God looked upon the earth, and indeed it was corrupt; for all flesh had corrupted their way on the earth.
9. And the Lord said, "My Spirit shall not strive with man forever, for he is indeed flesh; yet his days shall be one hundred and twenty years."

> his son, Laadan [6th gen.] his son, Ammihud [7th gen.] his son, Elishama [8th gen.] his son, Nun [9th gen.] his son, and Joshua [10th gen.] his son.

Joshua was the tenth generation from Joseph (having his son Ephraim as the first generation). So Joshua, who was contemporaneous with Moses and yet was the tenth generation, led the fourth generation and their descendants into conquest of the Promised Land. But *note* that Joshua was the tenth generation — and Noah was in the tenth generation too.

The Bible reveals how many males of fighting age existed among the Israelites at the Exodus. There were 603,550 males over 20 years of age in Numbers 1:1–3,[10] 2:32,[11] though this is an exceptional growth rate.

Consider the Lord's prophetic promise to Abraham and then Isaac was that *God Himself* would increase them (Genesis 13:16,[12] 22:17,[13] 26:4[14]; Exodus 1:7[15]; Deuteronomy 1:10[16]) — and this came true.

God is the one responsible for multiplying Abraham's descendants, and this exceeding increase came to Israel. The Egyptians recognized this and wanted to do something about this population explosion occurring with the Israelites — hence enslaving them and trying to kill their baby boys in an effort to control them!

So this was an exceptional growth rate discussed in the Bible, but this would yield a population (if ~equal male to female) just over 1.2 million

---

10. Now the Lord spoke to Moses in the Wilderness of Sinai, in the tabernacle of meeting, on the first day of the second month, in the second year after they had come out of the land of Egypt, saying: "Take a census of all the congregation of the children of Israel, by their families, by their fathers' houses, according to the number of names, every male individually, from twenty years old and above — all who are able to go to war in Israel. You and Aaron shall number them by their armies."
11. These are the ones who were numbered of the children of Israel by their fathers' houses. All who were numbered according to their armies of the forces were six hundred and three thousand five hundred and fifty.
12. And I will make your descendants as the dust of the earth; so that if a man could number the dust of the earth, then your descendants also could be numbered.
13. Blessing I will bless you, and multiplying I will multiply your descendants as the stars of the heaven and as the sand which is on the seashore; and your descendants shall possess the gate of their enemies.
14. And I will make your descendants multiply as the stars of heaven; I will give to your descendants all these lands; and in your seed all the nations of the earth shall be blessed.
15. But the children of Israel were fruitful and increased abundantly, multiplied and grew exceedingly mighty; and the land was filled with them.
16. The Lord your God has multiplied you, and here you are today, as the stars of heaven in multitude.

people and their children in these ten generations. This almost sets an extreme upper limit, as the Lord was not increasing the people before the Flood, as He did with the Israelites. Thus, we tentatively suggest the pre-Flood population was far less than this at its peak — perhaps just a few hundred thousand. Allow us to elaborate.

Obedient to "Be Fruitful and Multiply"?

Childbirths were likely reduced — people were making a name for themselves being *men of renown* (Genesis 6:4[17]), which means they would hardly care or have time for any children! With people living lives that are about themselves, they rarely care for children — even in our culture today!

With pre-Flood wickedness (Genesis 6:5[18]) abounding (which often includes sodomy/homosexuality — Genesis 13:13[19]; see also Judges 19–20), this would naturally reduce the possibility of children and population growth. This would be especially true if this population reduction were consistent in the century leading up to the Flood.

Let us also not forget who gets the brunt of an evil culture. It is almost always the children! This is how Pharaoh dealt with the Israelites — killing baby boys (Exodus 1:16,[20] 1:22[21]). The Canaanites were sacrificing their children (Leviticus 18:21,[22] 20:2–5[23]). This is what we saw with Herod —

17. The Nephilim were on the earth in those days, and also afterward, when the sons of God came in to the daughters of men, and they bore children to them. Those were the mighty men who were of old, men of renown (NASB).
18. Then the Lord saw that the wickedness of man was great in the earth, and that every intent of the thoughts of his heart was only evil continually.
19. But the men of Sodom were exceedingly wicked and sinful against the Lord.
20. And he said, "When you do the duties of a midwife for the Hebrew women, and see them on the birthstools, if it is a son, then you shall kill him; but if it is a daughter, then she shall live."
21. So Pharaoh commanded all his people, saying, "Every son who is born you shall cast into the river, and every daughter you shall save alive."
22. And you shall not let any of your descendants pass through the fire to Molech, nor shall you profane the name of your God: I am the Lord.
23. Again, you shall say to the children of Israel: "Whoever of the children of Israel, or of the strangers who dwell in Israel, who gives any of his descendants to Molech, he shall surely be put to death. The people of the land shall stone him with stones. I will set My face against that man, and will cut him off from his people, because he has given some of his descendants to Molech, to defile My sanctuary and profane My holy name. And if the people of the land should in any way hide their eyes from the man, when he gives some of his descendants to Molech, and they do not kill him, then I will set My face against that man and against his family; and I will cut him off from his people, and all who prostitute themselves with him to commit harlotry with Molech.

A sacrificed young girl in Peru

killing baby boys (Matthew 2:16[24]). Pagan cultures often sacrificed their children.

This is what we saw in modern times as well with Hitler when he killed upward of 13 million Jews, Poles, Slavs, and Gypsies, and their children were not exempt! This is what we see in our own modern secular culture where millions of babies are aborted (murdered through child sacrifice) every year through state-funded organizations like Planned Parenthood! The population of the USA has lost over 55 million people to abortion since the Supreme Court permitted the murder of babies in 1973 (and counting).

Population Increasing or Decreasing at the Flood?

With these types of things occurring in the pre-Flood world for at least 120 years, it makes more sense that the population was declining not rising during this period. But what of other factors?

24. Then Herod, when he saw that he was deceived by the wise men, was exceedingly angry; and he sent forth and put to death all the male children who were in Bethlehem and in all its districts, from two years old and under, according to the time which he had determined from the wise men.

Did you realize the world was violent — very violent (Genesis 6:12–13[25])? Every thought was evil continually and the wickedness and violence (Genesis 6:5–6[26]) was unrestricted. Just imagine if half of the world were murderers . . . the world's population would cut in half in one day! Our humble suggestion is that the pre-Flood world's population was quite low — far less than suggested estimates listed above.

Was the Ark a Sufficient Means of Salvation?

We have had a number of people ask this question over the years. How does this relate to the pre-Flood population? It goes something like this, "If the ark was a type of salvation from the Flood, then how can we say it was a *sufficient* means of salvation if the population was too big to fit on the ark?"

In other words, the claim is that the ark wouldn't have been able to hold all the pre-Flood population; therefore, they claim that God's means of salvation was not adequate. "Couldn't people have rightly stated that the ark wouldn't have been good enough to save them?"

Besides the fact that this excuse is merely trying to blame God (for enacting justice upon their sin no less!), there are a couple of problems with this. Only eight people survived the Flood on the ark, so if someone pre-Flood really wanted to complain with that excuse, then they need to realize there was plenty of space for that one complainer — had they repented and become righteous!

But let's evaluate the other aspect. Was the ark sufficient to hold the pre-Flood world's population? Again, all we have are estimates. Is it possible that an unknown-sized population could have fit inside the ark, giving them no excuse? From the inductive argument given, the probability that the population size was rather small (perhaps only a matter of thousands), it is highly possible they could fit on the ark.[27]

But let's not forget one thing — God knew how many people would survive on the ark. So for the pre-Flood unrepentant sinners there is no excuse anyway.

25. So God looked upon the earth, and indeed it was corrupt; for all flesh had corrupted their way on the earth. And God said to Noah, "The end of all flesh has come before Me, for the earth is filled with violence through them; and behold, I will destroy them with the earth."
26. Then the Lord saw that the wickedness of man was great in the earth, and that every intent of the thoughts of his heart was only evil continually. And the Lord was sorry that He had made man on the earth, and He was grieved in His heart.
27. But consider another factor — if the world had repented of their sin, would the Flood have been necessary? Although certain "if" question are simply that . . . "if."

So Was the Ark Designed Simply for Those Eight People?

The Bible doesn't tell us. What we know is that when Noah was told to build the ark, the text says:

> But I will establish My covenant with you; and you shall go into the ark — you, your sons, your wife, and your sons' wives with you (Genesis 6:18).

Does this imply that *only* these eight people were given the privileged possibility of survival? The text did not say the ark was for them *alone*. It was for those eight people *at least*. Consider the same type of phrasing ("shall go into") in Amos:

> Their king shall go into captivity, He and his princes together," says the LORD (Amos 1:15).

Does this mean that others within the king's dominion had no possibility of going into captivity *with* the king and his family? By no means. We know they did!

What we learn subsequently is that no others came into the ark (confirmed by 2 Peter 2:5[28]), but we cannot say they didn't have the opportunity or possibility. The onus for their absence on the ark is entirely on their own sinful heads.

But some object and ask, what about the "innocent children" — if there were any left — at the time of the Flood? First they weren't innocent (Romans 3:23[29]). But again, the onus would be on the parents and guardians who refused to allow their evil children the possibility of survival on the ark!

Glance more closely at Genesis 6:18.[30] What the Bible says is that Noah *knew* that his wife, his sons, and his sons wives would indeed be on the ark. This passage is not only instructional but also comforting.

With this information told to Noah in advance, he had a reassurance that his family would not succumb to the murderous actions of the evil people on earth at the time. That was a guarantee for Noah's comfort like a "hedge of protection" around his family during this time. And this short bit

28. And did not spare the ancient world, but saved Noah, one of eight people, a preacher of righteousness, bringing in the flood on the world of the ungodly.
29. For all have sinned and fall short of the glory of God.
30. But I will establish My covenant with you; and you shall go into the ark — you, your sons, your wife, and your sons' wives with you.

of information was revealed to Noah in advance, even though God always knew how many would truly board the ark.

Noah, being a preacher of righteousness, was not preaching in vain (2 Peter 2:5[31]). Keep in mind *the heart* of Noah when doing this preaching . . . Noah surely lost brothers and sisters in the Flood. Consider:

> After he begot Noah, Lamech lived five hundred and ninety-five years, and had sons and daughters (Genesis 5:30).

There may have been some pre-Flood people who became righteous and were murdered well before the Flood began. Noah's preaching was surely a sign that he knew that there was a possibility of others becoming righteous. As it was, only the righteous warranted that final invitation from the Lord to board the ark (Genesis 7:1[32]).

31. See footnote 28.
32. Then the Lord said to Noah, "Come into the ark, you and all your household, because I have seen that you are righteous before Me in this generation.

Concluding Remarks

40 | Our Real Motive for Building Ark Encounter

Ken Ham

Throughout this book, we have discussed the Flood, the ark, and the alleged millions of years that permeate our culture. This is coupled with the building of a full-sized replica of Noah's ark — the Ark Encounter. I believe its opening is a historic moment in Christendom. It's the opening of one of the greatest Christian outreaches of our era: the life-size Noah's ark in Northern Kentucky.

Ark Encounter

As I read many of the secularist attacks on the Ark (and occasional criticisms from self-described Christians), I saw one theme coming up over and over again: our motive! Most secularists, who are in rebellion against God, just can't get their head around why we would build a replica of this massive wooden ship as described in the Bible. Many claim we must be doing it for the money!

Well, those of you who know Answers in Genesis understand that, while money is certainly needed to build and then maintain such a massive project and to construct future phases, money is not our motive in the slightest degree.

Some critics who say they are Christians declare that we're building an idol, supposedly because we are worshiping the Bible and not God! (We were even accused of that bizarre claim when building the Creation Museum.)

Of course, anyone who has visited the museum (and the same will be true for the life-size Ark) understands that this Bible-upholding center is not an idol in any way. The Creation Museum and Ark Encounter direct people to the Word of God and the gospel of Jesus Christ.

Others have accused us of building the Ark out of pride, claiming we just want to build something for the sake of getting our name in the news! Amazing!

The Judas Proposition: The Money Should Be Spent on the Poor!

Others (some claiming to be Christians) say we shouldn't have built the Ark but should have spent the money on feeding the poor. There seem to be those habitual complainers who insist the money not be "wasted" this way. Such people either don't understand or don't seem to care about the millions who will be reached with the most important food in the universe — the spiritual food of the saving gospel — the very message that their eternal life depends on.

Before I address that, it is interesting to note that there are many projects underway in this nation that cost as much or enormous amounts more than the Ark Encounter. For instance, this project in Louisiana is costing about the same as the Ark project:

> Students at Louisiana State University will soon be able to soak up the sun in a manmade "lazy river," part of an $85 million leisure project under way despite the school's desperate financial situation.[1]

So where are all the naysayers complaining that the money for this "lazy river" should be spent on the poor? Or consider:

> One piece of art (Triptych) sold for $142 million.[2]

Where were all the naysayers saying the money would be best spent on the poor for that deal? In fact, there are thousands of multi-million dollar projects going on across the United States — and throughout the world. But it seems the Ark project is singled out — why?

I think it's as simple as this — it's the message! The Ark project (like the Creation Museum) is a professional, powerful, and gracious way to present the truth of God's Word and the gospel.

Yes, we do need to help the poor. And most Christians like me do that personally and through various ministries. Recently I encouraged people to give to the relief efforts in Nepal after the recent massive earthquake, and

1. Aalia Shaheed, "LSU's $85M 'Lazy River' Leisure Project Rolls on, Despite School's Budget Woes," Fox News, May 17, 2015, http://www.foxnews.com/us/2015/05/17/lsu-85m-lazy-river-leisure-project-rolls-on-despite-school-budget-woes.html.
2. Carol Vogel, "At $142.4 Million, Triptych Is the Most Expensive Artwork Ever Sold at an Auction," *The New York Times*, November 12, 2013, http://www.nytimes.com/2013/11/13/arts/design/bacons-study-of-freud-sells-for-more-than-142-million.html?_r=2.

Ken Ham and his daughter, Renee Hodge, in front of the Ark Encounter.

we provided a link to the donation page of Gospel for Asia, which has relief work going on there.

More directly, last year, through AiG's Vacation Bible School program, hundreds of thousands of meals were provided to needy children around the world. AiG worked with the Children's Hunger Fund — and we set a record for providing such meals.

At Answers in Genesis, our mission is to "proclaim the absolute truth and authority of the Bible with boldness, relate the relevance of a literal Genesis to the church and world today, and obey God's call to deliver the message of the gospel." So while it is important to help the poor and needy meet their physical needs (which we do), it is even more important to help meet their spiritual need — the need to come to know Jesus Christ, the Savior of the world — because lives — and eternity — hang in the balance.

The Ark Encounter will help us do that in a powerful, non-threatening way by simply sharing the truth of God's Word with visitors at the Ark concerning the historicity of Noah's ark, the Genesis Flood, and other authentic accounts of history revealed in the Scriptures, including the account of redemption weaved throughout the Bible.

Our motivation for the Ark project is to reach as many people as we can worldwide with the saving gospel message:

> And they sang a new song, saying, "Worthy are you to take the scroll and to open its seals, for you were slain, and by your blood you ransomed people for God from every tribe and language and people and nation" (Revelation 5:9; ESV).

Actually, the more I read such comments from the anti-Ark complainers, who are obviously inconsistent, it reminds me of how important the Ark Encounter is and how much the enemy doesn't want it happening!

For those people who say the money for building the Ark should be given to the poor, would these same critics give the money to the poor they have saved for their retirement? I bring this up because 65% of the funds to build the Ark are from a bond offering where people who support God's Word decided to use some of their funds to invest in interest-paying bonds for the sake of the Kingdom! Actually, for some of those who complain about money in regard to the poor, I wonder if they care about the poor very much. I'm reminded of the Judas Proposition found in Matthew 26:9–10, Mark 14:3–6, and John 12:3–8. John 12:3–8 says:

> Mary therefore took a pound of expensive ointment made from pure nard, and anointed the feet of Jesus and wiped his feet with her hair. The house was filled with the fragrance of the perfume. But Judas Iscariot, one of his disciples (he who was about to betray him), said, "Why was this ointment not sold for three hundred denarii and given to the poor?" He said this, not because he cared about the poor, but because he was a thief, and having charge of the moneybag he used to help himself to what was put into it. Jesus said, "Leave her alone, so that she may keep it for the day of my burial. For the poor you always have with you, but you do not always have me" (John 12:1–8; ESV).

Usually when people mimic the Judas Proposition, it shows they really don't care about the poor but, like Judas, have other motives.

What Is Our Real Motive for Building Themed Attractions Like the Creation Museum and the Ark Encounter?

A number of years ago in Australia, my wife Mally and I attended the commencement ceremony held at a secular university as one of our family members was graduating. A local judge gave the commencement address. Her speech went something like this: "Students, you are graduating from university. You're thinking of your future. Eventually you will die. So what do you do until you're dead?"

At this point, I turned to Mally and said, "Wow, this is going to be a message of hope and encouragement!"

Well, the judge went on to say, "In my life, there were books that greatly influenced my life." She named *Zen and the Art of Motorcycle Maintenance* and *The Hitchhiker's Guide to the Galaxy* (where a computer came up with the meaning of life as the number 42). The judge explained how these books influenced her life. She then encouraged students to find what would influence their lives to be impactful until they are dead!

She then sat down to a standing ovation by the faculty! I turned again to Mally and said, "If I were a student, I would feel compelled to jump off a cliff right now and get life over and done with." What a message of meaninglessness, hopelessness, and purposelessness she offered.

As a Christian, doesn't a speech like that make you want to stand up and declare to the audience that there is a message of real hope — not only for this life, but for eternity?"

What the judge presented was the ultimate message of the world. And sadly, it is being given daily to millions of students in public schools, universities, and through most of the media and the entertainment industry! Doesn't your heart ache when you think about this hopelessness? No wonder the suicide rate is rising in America and the Western World. No wonder younger generations turn to sexual immorality, drunkenness, and drugs.

I remember saying to Mally after the commencement speech, "I wish we had a way to get the Bible's teaching of the hope of the gospel to these students. How can we get out the message of truth concerning God's Word and our hope in Christ to this lost world?"

Really, what I shared that day with Mally sums up our motive! You see, every human being is one of our family — we're all related going back to Noah (and then back to Adam). We're all sinners in need of salvation. We're all under the judgment of death. But God reminds us:

> The Lord is . . . not willing that any should perish but that all should come to repentance (2 Peter 3:9).

> If you confess with your mouth that Jesus is Lord and believe in your heart that God raised Him from the dead, you will be saved (Romans 10:9; ESV).

Our real motive for building the Creation Museum, and the Ark, can be summed up in these verses:

> Go into all the world and preach the gospel to every creature (Mark 16:15).

> Go therefore and make disciples of all the nations (Matthew 28:19).

> But sanctify the Lord God in your hearts, and always be ready to give a defense to everyone who asks you a reason for the hope that is in you, with meekness and fear (1 Peter 3:15).
>
> Contend earnestly for the faith (Jude 3).
>
> Do business till I come (Luke 19:13).

Yes, our motive is to do the King's business until He comes. And that means preaching the gospel and defending the faith, so that we can reach as many souls as we can with the greatest message of purpose, hope, and meaning — that even though we rebelled against our Creator, He provided a way as a free gift so we can spend eternity with Him — through His shed blood on the Cross.

Oh, how we want to see as many as possible repent and receive this free gift of salvation! As a corollary to more people getting saved, there will be more who are motivated by Christ to help the poor.

I can't even describe how I feel right now contemplating that millions of souls will hear the most important message of all — not one of hopelessness from a human judge, but a message of hope from the holy, righteous Judge who, despite our sin, wants us to spend eternity with Him! Wow! Now that's a motive to build an Ark.

We need your prayers and support to keep the park's doors open to millions of guests who will learn the truth of God's Word and its life-changing gospel message.

The Ark Encounter at night

Photo Credits:

John Goodge, via Wikimedia Commons, page 165, lower
Bodie Hodge, page 110, upper and lower; 119; 120; 121, right and left; 125; 129; 134; 138; 142, 1st, 3rd, and 4th; 143; 185; 288
Tommy Mitchell, page 109
Maria Murphy, page 300

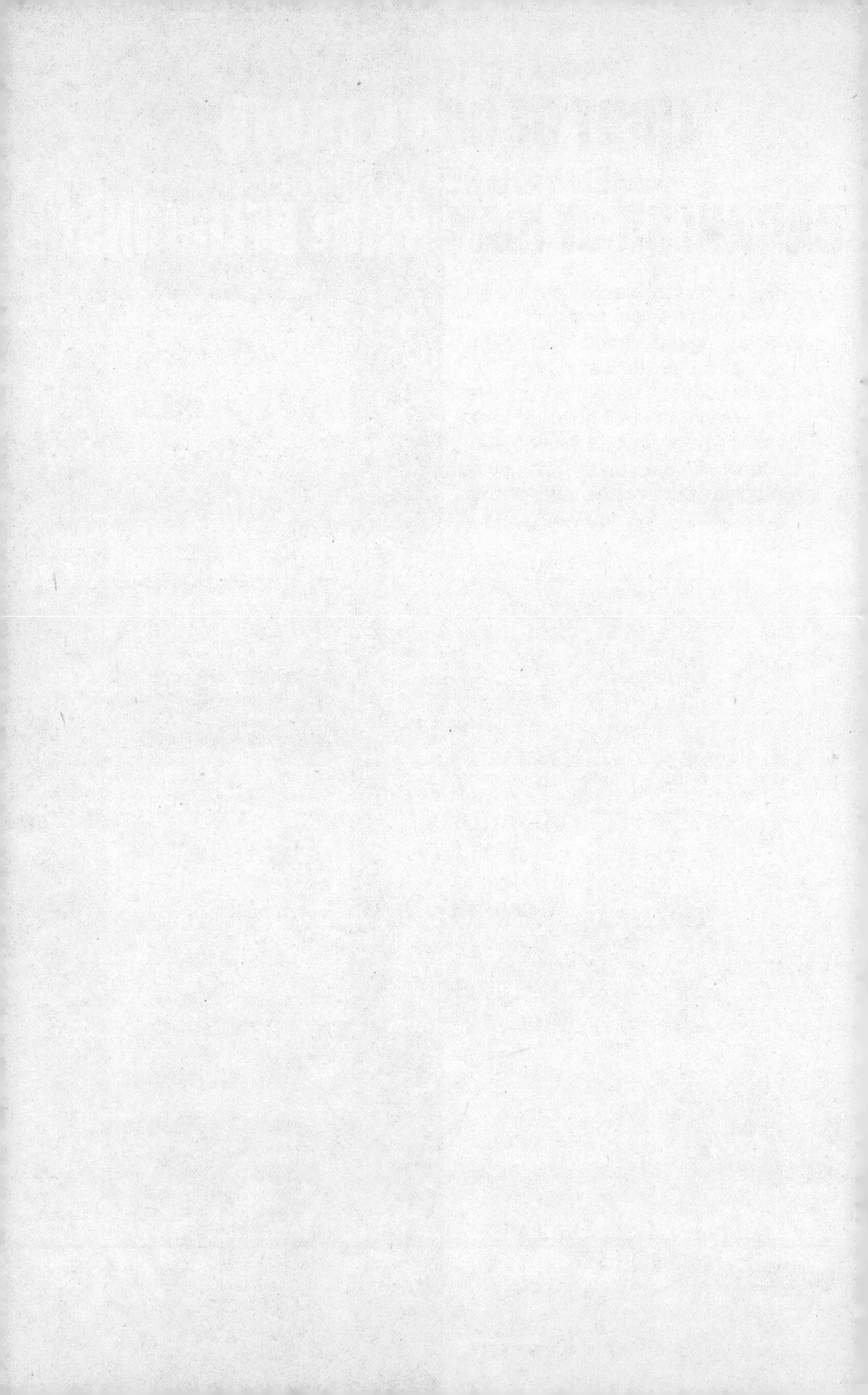

The New Answers Book

Over 25 Questions on Creation/Evolution and the Bible

First printing: November 2006
Twenty-ninth printing: June 2021

Master Books®, P.O. Box 726, Green Forest, AR 72638.

Master Books® is a division of the New Leaf Publishing Group, Inc.

ISBN: 978-0-89051-509-9
ISBN: 978-1-61458-016-4 (digital)
Library of Congress Number: 2006937546

Cover design by Left Coast Design, Portland, Oregon
Interior design by Diane King
Compiled by Bodie Hodge and Gary Vaterlaus

Please consider requesting that a copy of this volume be purchased by your local library system.

Printed in the United States of America

Please visit our website for other great titles:
www.masterbooks.com

For information regarding author interviews, please contact the publicity department at (870) 438-5288.

Master Books®
A Division of New Leaf Publishing Group
www.masterbooks.com

Acknowledgments and special thanks

To Dr. John Baumgardner, Dr. John Whitmore, Dr. Don DeYoung, Dr. Larry Vardiman, Dr. Danny Faulkner, Dr. Bob Compton, Dr. Gary Parker, Dr. Jason Lisle, Dr. Georgia Purdom, Dr. Terry Mortenson, Ken Ham, Bodie Hodge, Mike Matthews, and Stacia McKeever for reviewing chapters of this book.

To Dan Lietha for many of the illustrations used in this book. To Dr. John Baumgardner for the illustrations in the chapter on plate tectonics. To Mike Oard for the illustrations in the chapter on the Ice Age. (All other illustrations are noted on the illustration, figure, or photograph.)

To Roger Patterson for developing the glossary of terms.

I would recommend that every person seeking the truth of God's Word read this book. *The New Answers Book* provides well-documented answers to tough questions asked by many unbelievers as well as Christians. It is a wealth of information that belongs in every library.

— John D. Morris
President, Institute for Creation Research

Even a young teen can read and understand the 27 different topics in this wonderful, eye-opening book. It is excellent for educational purposes or as a ministry resource.

— Ray Comfort
Living Waters Publications

Ken Ham is a gifted thinker and a gift to the Christian community. He is not only a biblical thinker, but a powerhouse communicator in the debate on creation v. evolution. Read this book, then train your children to have a biblical world view like Ken does.

— Dennis Rainey
President, FamilyLife

Contents

1

Is There Really a God?

KEN HAM & JASON LISLE

God—an Eternal, Uncreated Being?

In our everyday experience, just about everything seems to have a beginning. In fact, the laws of science show that even things which look the same through our lifetime, such as the sun and other stars, are, in reality, running down. The sun is using up its fuel at millions of tons each second—since the sun cannot last forever, it had to have a beginning. The same can be shown to be true for the entire universe.

So when Christians claim that the God of the Bible created all the basic entities of life and the universe, some will ask what seems to be a logical question: "Who created God?"

The very first verse in the Bible declares: "In the beginning God … ." There is no attempt in these words to prove the existence of God or imply in any way that God had a beginning. In fact, the Bible makes it clear in many places that God is outside time. He is eternal, with no beginning or end. He also knows all things, being infinitely intelligent.[1]

Is it logical, though, to accept the existence of such an eternal being? Can modern science, which has produced our technology of computers, space shuttles, and medical advances, even allow for such a notion?

[1] Psalm 90:2; 106:48; 147:5. Notice that only things which have a beginning have to have a cause.

What Would We Look For?

What evidence would we expect to find if there really is an infinite God who created all things as the Bible claims? How would we even recognize the hand of such an omnipotent (all-powerful) Creator?

The Bible claims that God knows all things—He is omniscient! Therefore, He is infinitely intelligent. To recognize His handiwork, one would have to know how to recognize the evidence of the works of His intelligence.

How Do We Recognize the Evidence of Intelligence?

Why do scientists become so excited when they discover stone tools together with bones in a cave? The stone tools show signs of intelligence. The scientists recognize that these tools could not have designed themselves but that they are a product of intelligent input. Thus, the researchers rightly conclude that an intelligent creature was responsible for making these tools.

In a similar way, one would never look at the Great Wall of China, the U.S. Capitol building in Washington, D.C., or the Sydney Opera House in Australia and conclude that such structures were formed after explosions in a brick factory.

Neither would anyone believe that the presidents' heads on Mt. Rushmore were the products of millions of years of erosion. We *can* recognize design, the evidence of the outworkings of intelligence. We see man-made objects all around us—cars, airplanes, computers, stereos, houses, appliances, and so on. And yet, at no time would anyone ever suggest that such objects were just the products of time and chance. Design is everywhere. It would never enter our minds that metal, left to itself, would eventually form into engines, transmissions, wheels, and all the other intricate parts needed to produce an automobile.

This "design argument" is often associated with the name of William Paley, an Anglican clergyman who wrote on this topic in the late eighteenth century. He is particularly remembered for his example of the watch and the watchmaker. In discussing a comparison between a stone and a watch, he concluded that "the watch must have had a maker; that there must have existed, at some time and at some place or other, an artificer or artificers, who formed it for the purpose which we find it actually to answer; who comprehended its construction, and designed its use."[2]

Paley thus believed that, just as the watch implied a watchmaker, so too does design in living things imply a Designer. Although he believed in a God who created all things, his God was a Master Designer who is now remote from His Creation, not the personal God of the Bible.[3]

Today, however, a large proportion of the population, including many leading scientists, believe that all plants and creatures, including the intelligent engineers who make watches, cars, etc., were the product of an evolutionary process—not a Creator God.[4] But this is not a defensible position, as we will see.

Living Things Show Evidence of Design!

The late Isaac Asimov, an ardent anti-creationist, declared, "In man is a three-pound brain which, as far as we know, is the most complex and orderly arrangement of matter in the universe."[5] It is much more complex than the most complicated computer ever built. Wouldn't it be logical to assume that if man's highly intelligent brain designed the computer, then the human brain was also the product of design?

Scientists who reject the concept of a Creator God agree that all living

---

[2] W. Paley, *Natural Theology: or Evidences of the Existence and Attributes of the Deity, Collected from the Appearances of Nature*, reprinted in 1972 by St. Thomas Press, Houston, Texas, 3.

[3] I. Taylor, *In the Minds of Men*, TFE Publishing, Toronto, Canada, 1991, 121.

[4] This is the process by which life is supposed to have arisen spontaneously from nonlife. Over long periods of time, different kinds of animals and plants have then supposedly developed as a result of small changes, resulting in an increase in genetic information. For instance, evolutionists propose that fish developed into amphibians, amphibians into reptiles, reptiles evolved into birds and mammals. Man eventually evolved from an ancestor shared with apes.

[5] I. Asimov, In the game of energy and thermodynamics you can't even break even, *Smithsonian*, June 1970, 10.

things exhibit evidence of design. In essence, they accept the design argument of Paley, but not Paley's Designer. For example, Dr. Michael Denton, a non-Christian medical doctor and scientist with a doctorate in molecular biology, concludes:

> It is the sheer universality of perfection, the fact that everywhere we look, to whatever depth we look, we find an elegance and ingenuity of an absolutely transcending quality, which so mitigates against the idea of chance.
>
> Alongside the level of ingenuity and complexity exhibited by the molecular machinery of life, even our most advanced artifacts appear clumsy. We feel humbled, as neolithic man would in the presence of twentieth-century technology.
>
> It would be an illusion to think that what we are aware of at present is any more than a fraction of the full extent of biological design. In practically every field of fundamental biological research ever-increasing levels of design and complexity are being revealed at an ever-accelerating rate.[6]

Dr. Richard Dawkins, holder of the Charles Simonyi Chair of Public Understanding of Science at Oxford University, has become one of the world's leading evolutionist spokespersons. His fame has come as the result of the publication of books, including *The Blind Watchmaker*, which defend modern evolutionary theory and claim to refute once and for all the notion of a Creator God. He said, "We have seen that living things are too improbable and too beautifully 'designed' to have come into existence by chance."[7]

There is no doubt that even the most ardent atheist concedes that design is evident in the animals and plants that inhabit our planet. If Dawkins rejects "chance" in design, what does he put in place of "chance" if he does not accept a Creator God?

Who—or What—Is the Designer Then?

Design obviously implies a designer. To a Christian, the design we see all around us is totally consistent with the Bible's explanation: "In the beginning God created the heavens and the earth" (Genesis 1:1), and "For by him

[6] M. Denton, *Evolution: A Theory in Crisis*, Adler & Adler Publishers, Bethesda, Maryland, 1986, 32.

[7] R. Dawkins, *The Blind Watchmaker*, W.W. Norton & Co., New York, 1987, 43.

[Jesus Christ] all things were created that are in heaven and that are in earth, visible and invisible, whether thrones or dominions or principalities or powers. All things were created through him and for him" (Colossians 1:16).

However, evolutionists like Richard Dawkins, who admit the design in living things, reject the idea of any kind of a Designer/God. In reference to Paley, Dawkins states:

> Paley's argument is made with passionate sincerity and is informed by the best biological scholarship of his day, but it is wrong, gloriously and utterly wrong. The analogy between telescope and eye, between watch and living organism, is false.[8]

Why? It is because Dawkins attributes the design to what he calls "blind forces of physics" and the processes of natural selection. Dawkins writes:

> *All appearance to the contrary*, the only watchmaker in nature is the blind forces of physics, albeit deployed in a very special way. A true watchmaker has foresight: he designs his cogs and springs, and plans their interconnections, with future purpose in his mind's eye. Natural selection, the blind, unconscious, automatic process which Darwin discovered, and which we now know is the explanation for the existence and apparently purposeful form of all life, has no purpose in mind. It has no mind and no mind's eye. It does not plan for the future. It has no vision, no foresight, no sight at all. If it can be said to play the role of watchmaker in nature, it is the blind watchmaker [emphasis added].[9]

Dawkins does, however, concede that "the more statistically improbable a thing is, the less can we believe that it just happened by blind chance. Superficially the obvious alternative to chance is an Intelligent Designer."[10]

Nonetheless, he rejects the idea of an "Intelligent Designer" and instead offers this "answer":

[8] Ibid., 5.

[9] Ibid., 5.

[10] R. Dawkins, The necessity of Darwinism, *New Scientist* **94**:130, 1982.

> The answer, Darwin's answer, is by gradual, step-by-step transformations from simple beginnings, from primordial entities sufficiently simple to have come into existence by chance. Each successive change in the gradual evolutionary process was simple enough, relative to its predecessor, to have arisen by chance.
>
> But the whole sequence of cumulative steps constitutes anything but a chance process, when you consider the complexity of the final end product relative to the original starting point. The cumulative process is directed by nonrandom survival. The purpose of this chapter is to demonstrate the power of this cumulative selection as a fundamentally nonrandom process.[11]

Basically, then, Dawkins is doing nothing more than insisting that natural selection[12] and mutations [13] together provide the mechanism for the evolutionary process. He believes these processes are nonrandom and directed. In reality, this is just a sophisticated way of saying that evolution is itself the designer.

---

[11] Dawkins, *The Blind Watchmaker*, 43.

[12] Dr. Gary Parker, a creationist, argues that natural selection does occur, but operates as a "preservative" and has nothing to do with one organism changing into another. "Natural selection is just one of the processes that operates in our present corrupted world to insure that the created kinds can indeed spread throughout the Earth in all its ecologic and geographic variety (often, nowadays, in spite of human pollution)." G. Parker, *Creation: Facts of Life*, Master Books, Green Forest, Arkansas, 1994, 75.

"[Richard] Lewontin is an evolutionist and outspoken anticreationist, but he honestly recognizes the same limitations of natural selection that creation scientists do: '… natural selection operates essentially to enable the organisms to maintain their state of adaptation rather than to improve it.' Natural selection does not lead to continual improvement (evolution); it only helps to maintain features that organisms already have (creation). Lewontin also notes that extinct species seem to have been just as fit to survive as modern ones, so he adds: '… natural selection over the long run does not seem to improve a species' chances of survival, but simply enables it to "track," or keep up with, the constantly changing environment.'"

"It seems to me that natural selection works only because each kind was created with sufficient variety to multiply and fill the earth in all its ecologic and geographic variety." G. Parker, *Creation: Facts of Life*, 84–86.

[13] "After all, mutations are only changes in genes that already exist," G. Parker, *Creation: Facts of Life*, 103.

"In an article paradoxically titled 'The Mechanisms of Evolution,' Francisco Ayala defines a mutation as 'an error' in DNA." G. Parker, *Creation: Facts of Life*, 99.

Does Natural Selection Produce Design?

Life is built on information. A great amount of this information is contained in that molecule of heredity, DNA, which makes up the genes of an organism. Therefore, to argue that natural selection and mutations are the basic mechanisms of the evolutionary process, one must show that these processes produce the information responsible for the design that is evident in living things.

Anyone who understands basic biology recognizes, of course, as Darwin did, that natural selection is a logical process that one can observe. However, natural selection only operates on the information that is already contained in the genes—it does not produce new information.[14] Actually, this is consistent with the Bible's account of origins, in that God created distinct kinds of animals and plants, each to reproduce after its own kind.

It is true that one can observe great variation in a kind and see the results of natural selection. For instance, wolves, coyotes, and dingoes have developed over time as a result of natural selection operating on the information found in the genes of the wolf/dog kind. But the point is that no new information was produced—these varieties of dogs have resulted from a rearrangement, sorting out, and separation of the information in the original dog kind. One kind has never been observed to change into a totally different kind with information that previously

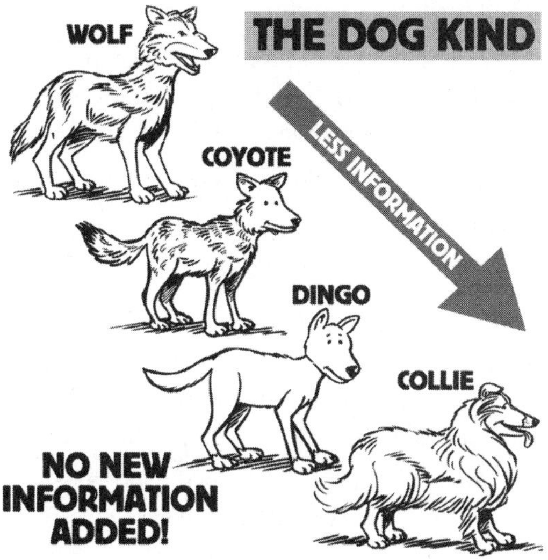

[14] L.P. Lester and R.G.Bohlin, *The Natural Limits to Biological Change*, Probe Books, Dallas, 1989, 175–176.
E. Noble et al., *Parasitology: The Biology of Animal Parasites*, Lea & Febiger, Philadelphia, 1989. Chapter 6: "Evolution of Parasitism?" 516, states, "Natural selection can act only on those biologic properties that already exist; it cannot create properties in order to meet adaptational needs."

did not exist.[15] Without intelligent input to increase information, natural selection will not work as a mechanism for evolution.

Denton confirms this when he states:

> It cannot be stressed enough that evolution by natural selection is analogous to problem solving without any intelligent guidance, without any intelligent input whatsoever. No activity which involves an intelligent input can possibly be analogous to evolution by natural selection.[16]

Without a way to increase information, natural selection will not work as a mechanism for evolution. Evolutionists would agree with this, but they believe that mutations somehow provide the new information for natural selection to act upon.

Can Mutations Produce New Information?

Actually, scientists now know that the answer is "no!" Dr. Lee Spetner, a highly qualified scientist who taught information and communication theory at Johns Hopkins University, makes this abundantly clear in his scholarly and thoroughly researched book, *Not by Chance*:

> In this chapter I'll bring several examples of evolution, particularly mutations, and show that information is not increased. ... But in all the reading I've done in the life-sciences literature, I've never found a mutation that *added* information.[17]

> All point mutations that have been studied on the molecular level turn out to *reduce* the genetic information and not to increase it.[18]

> The NDT [neo-Darwinian theory] is supposed to explain how information of life has been built up by evolution. The essential biological difference between a human and a bacterium is in the information they contain. All other biological differences follow from that. The human genome has much more information than does the bacterial genome. *Information cannot be built up by mutations that lose it.*

[15] For instance, despite many unproved claims to the contrary by evolutionists, nobody has observed or documented a reptile changing into a bird. The classic example paraded by some evolutionists as an "in-between" creature, *Archaeopteryx*, has now been rejected by many evolutionists.

[16] M. Denton, *Evolution: A Theory in Crisis*, 317.

[17] L. Spetner, *Not By Chance*, The Judaica Press, Brooklyn, New York, 1997, 131–132.

[18] Ibid., 138.

> *A business can't make money by losing it a little at a time* [emphasis added].[19]

Evolutionary scientists have no way around this conclusion that many scientists, including Dr. Spetner, have now come to. Mutations do not work as a mechanism for the evolutionary process. Spetner sums it all up as follows:

> The neo-Darwinians would like us to believe that large evolutionary changes can result from a series of small events if there are enough of them. But if these events all *lose* information they can't be the steps in the kind of evolution the NDT is supposed to explain, no matter how many mutations there are. Whoever thinks macroevolution can be made by mutations that lose information is like the merchant who lost a little money on every sale but thought he could make it up in volume … . Not even one mutation has been observed that adds a little information to the genome. That surely shows that there are not the millions upon millions of potential mutations the theory demands. There may well not be any. The failure to observe even one mutation that adds information is more than just a failure to find support for the theory. It is evidence *against* the theory. We have here a serious challenge to neo-Darwinian theory [emphasis added].[20]

This is also confirmed by Dr. Werner Gitt, a director and professor at the German Federal Institute of Physics and Technology. In answering the question, "Can new information originate through mutations?" he said:

> This idea is central in representations of evolution, but mutations can only cause changes in *existing* information. There can be no increase in information, and in general the results are injurious. New blueprints for new functions or new organs cannot arise; mutations cannot be the source of new (creative) information [emphasis added].[21]

So if natural selection and mutations are eliminated as mechanisms to produce the information and design of living systems, then another source must be found.

But there are even more basic problems for those who reject the Creator God as the source of information.

---

[19] Ibid., 143.

[20] Ibid., 159–160.

[21] W. Gitt, *In the Beginning Was Information*, Master Books, Green Forest, Arkansas, 2006, 127.

More Problems!

Imagine yourself sitting in the seat of a 747 airplane, reading about the construction of this great plane. You are fascinated by the fact that this flying machine is made up of six million parts—but then you realize that not one part by itself flies. This realization can be rather disconcerting if you are flying along at 500 mph (805 km/h) at 35,000 feet (10,668 m).

You can be comforted, however, by the fact that even though not one part of an airplane flies on its own, when it is assembled as a completed machine, it does fly.

We can use the construction of an airplane as an analogy to understand the basic mechanisms of the biochemistry of cells that enable organisms to function.

Scientists have found that within the cell there are thousands of what can be called "biochemical machines." For example, one could cite the cell's ability to sense light and turn it into electrical impulses. But what scientists once thought was a simple process within a cell, such as being able to sense light and turn it into electrical impulses, is in fact a highly complicated event. For just this one example alone to work, numerous compounds must all be in the right place, at the right time, in the right

COMPLEXITY OF THE ANIMAL CELL

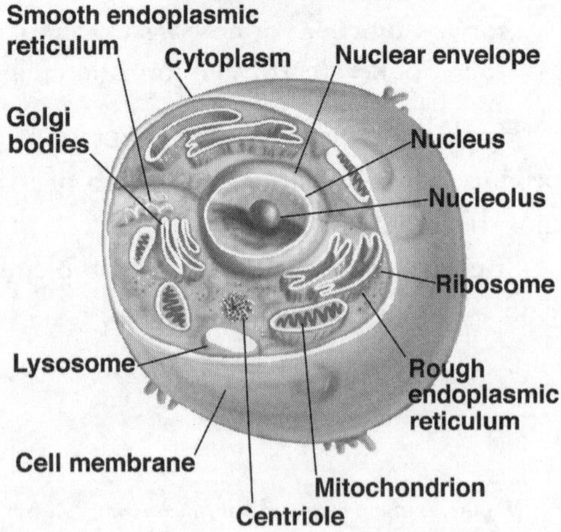

concentration—or it just won't happen. In other words, just as all the parts of a 747 need to be assembled before it can fly, so all the parts of these "biochemical machines" in cells need to be in place, or they can't function. And there are literally thousands of such "machines" in a single cell that are vital for it to operate.

What does this mean? Quite simply, evolution from chemicals to a living system is impossible.

Scientists now know that life is built on these "machines." Dr. Michael Behe, Associate Professor of Biochemistry at Lehigh University in Pennsylvania, describes these "biochemical machines" as examples of "irreducible complexity":

> Now it's the turn of the fundamental science of life, modern biochemistry, to disturb. *The simplicity that was once expected to be the foundation of life has proven to be a phantom;* instead, systems of horrendous, irreducible complexity inhabit the cell. The resulting realization that life was designed by an intelligence is a shock to us in the twentieth century who have gotten used to thinking of life as the result of simple natural laws. But other centuries have had their shocks, and there is no reason to suppose that we should escape them [emphasis added].[22]

To illustrate this further, consider swatting a mosquito.

Then think about this question: Why did the mosquito die? You see, the squashed mosquito has all the chemicals for life that an evolutionist could ever hope for in some primordial soup. Yet we know that nothing is going to evolve from this mosquito "soup." So why did the mosquito die? Because by squashing it, you *disorganized* it.

Once the "machinery" of the mosquito has been destroyed, the organism can no longer exist. At a cellular level, literally thousands of "machines" need to exist before life

[22] M.J. Behe, *Darwin's Black Box*, The Free Press, New York, 1996, 252–253.

ever becomes possible. This means that evolution from chemicals is *impossible.* Evolutionist Dawkins recognizes this problem of needing "machinery" to start with when he states:

> A Xerox machine is capable of copying its own blueprints, but it is not capable of springing spontaneously into existence. Biomorphs readily replicate in the environment provided by a suitably written computer program, but they can't write their own program or build a computer to run it. The theory of the blind watchmaker is extremely powerful given that we are allowed to assume replication and hence cumulative selection. But if replication needs complex machinery, since the only way we know for complex machinery ultimately to come into existence is cumulative selection, we have a problem.[23]

A problem indeed! The more we look into the workings of life, the more complicated it becomes, and the more we see that life could *not* arise by itself. Not only does life require a source of information, but the complex "machines" of the chemistry of life must be in existence *right from the start.*

A Greater Problem Still!

Some scientists and educators have tried to get around the above problems by speculating that as long as all the chemicals that make up the molecule of heredity (and the information it contains) came together at some time in the past, then life could have begun.

Life is built upon information. In fact, in just one of the trillions of cells that make up the human body, the amount of information in its genes would fill at least 1,000 books of 500 pages of typewritten information. Scientists now think this is hugely underestimated.

Where did all this information come from? Some try to explain it this way: imagine a professor taking all the letters of the alphabet, A–Z, and placing them in a hat. He then passes the hat around to students of his class and asks each to randomly select a letter.

[23] Dawkins, *The Blind Watchmaker*, 139–140.

It is easy for us to see the possibility (no matter how remote it seems) of three students in a row selecting B then A and finally T. Put these three letters together and they spell a word—BAT. Thus, the professor concludes, given enough time, no matter how improbable it seems, there is always the possibility one could form a series of words that make a sentence, and eventually compile an encyclopedia. The students are then led to believe that no intelligence is necessary in the evolution of life from chemicals. As long as the molecules came together in the right order for such compounds as DNA, then life could have begun.

On the surface, this sounds like a logical argument. However, there is a basic, fatal flaw in this analogy. The sequence of letters, BAT, is a word to whom? Someone who speaks English, Dutch, French, German, or Chinese? It is a word only to someone who knows the language. In other words, the order of letters is meaningless unless there is a language system and a translation system already in place to make the order meaningful.

In the DNA of a cell, the order of its molecules is also meaningless, except that in the biochemistry of a cell, there is a language system (other molecules) that makes the order meaningful. DNA without the language system is meaningless, and the language system without the DNA wouldn't work either. The other complication is that the language system that reads the order of the molecules in the DNA is itself specified by the DNA. This is another one of those "machines" that must already be in existence and fully formed, or life won't work!

Can Information Arise from Noninformation?

We have already shown that information cannot come from mutations, a so-called mechanism of evolution, but is there any other possible way information could arise from matter?

Dr. Werner Gitt makes it clear that one of the things we know for sure from science is that information *cannot* arise from disorder by chance. It *always* takes (greater) information to produce information, and ultimately information is the result of intelligence:

> A code system is always the result of a mental process (it requires an intelligent origin or inventor) . . . . It should be emphasized that matter as such is unable to generate any code. All experiences indicate that a thinking being voluntarily exercising his own free will, cognition, and creativity, is required.[24]

> There is no known natural law through which matter can give rise to information, neither is any physical process or material phenomenon known that can do this.[25]

> "There is no known law of nature, no known process and no known sequence of events which can cause information to originate by itself in matter.[26]

What Then Is the Source of the Information?

We can therefore conclude that the huge amount of information in living things must originally have come from an intelligence, which had to have been far superior to ours. But then, some will say that such a source would have to be caused by something with even greater information/intelligence.

However, if they reason this way, one could ask where even this greater information/intelligence came from. And then where did that one come from? One could extrapolate to infinity, unless there was a source of infinite intelligence, beyond our finite understanding. But isn't this what the Bible indicates when we read, "In the beginning God…"? The God of the Bible is not bound by limitations of time, space, or anything else.

Even Richard Dawkins recognizes this:

> Once we are allowed simply to postulate organized complexity, if only the organized complexity of the DNA/protein replicating engine, it is relatively easy to invoke it as a generator of yet more organized complexity. That, indeed, is what most of this book is about. But of course any God capable of intelligently designing something as complex as the DNA/protein replicating machine must have been at least as complex and organized as that machine itself.

> Far more so if we suppose him additionally capable of such advanced functions as listening to prayers and forgiving sins. To explain the origin of the DNA/protein machine by invoking a supernatural Designer is

24 Gitt, *In the Beginning Was Information*, 64–67.
25 Ibid., 79.
26 Ibid., 107.

> to explain precisely nothing, for it leaves unexplained the origin of the Designer. You have to say something like, "God was always there," and if you allow yourself that kind of lazy way out, you might as well just say "DNA was always there," or "Life was always there," and be done with it.[27]

So what is the logically defensible position? Is it that matter has eternally existed (or came into existence by itself for no reason) and then that, by itself, matter was arranged into information systems against everything observed in real science? Or did an eternal Being, the God of the Bible, the source of infinite intelligence,[28] create information systems for life to exist, which *agrees* with real science?

What we see in God's world agrees with what we read in God's Word.

If real science supports the Bible's claims about an eternal Creator God, then why isn't this readily accepted? Michael Behe answers with this:

> The fourth and most powerful reason for science's reluctance to embrace a theory of intelligent design is also based on philosophical considerations. Many people, including many important and well-respected scientists, just don't want there to be anything beyond nature. They don't want a supernatural being to affect nature, no matter how brief or constructive the interaction may have been. In other words … they bring an *a priori* philosophical commitment to their science that restricts what kinds of explanations they will accept about the physical world. Sometimes this leads to rather odd behavior.[29]

The crux of the matter is this: if one accepts there is a God who created us, then that God also owns us. If this God is the God of the Bible, He owns us and thus has a right to set the rules by which we must live. More important, He also tells us in the Bible that we are in rebellion against Him,

---

[27] Dawkins, *The Blind Watchmaker*, 141.

[28] Thus, it is capable of generating infinite information, and certainly the enormous, though finite, information of life.

[29] Behe, *Darwin's Black Box*, 243.

our Creator. Because of this rebellion (called sin), our physical bodies are sentenced to death; but we will live on forever, either with God or without Him in a place of judgment. But the good news is that our Creator provided a means of deliverance for our sin of rebellion, so that those who come to Him in faith and repentance for their sin can receive the forgiveness of a holy God and spend eternity with Him.

God Is the Foundation for Science and Reason

As stated before, the Bible takes God's existence as a given. It never attempts to prove the existence of God, and this for a very good reason. When we logically prove a particular thing, we show that it must be true because it follows logically from something *authoritative*. But there is nothing more authoritative than God and His Word. God knows absolutely everything. So it makes sense to base our worldview on what God has written in His Word.

Some people claim that it is unscientific to start from God's Word. But in reality, nothing could be further from the truth. A belief in God is actually foundational to logical thought and scientific inquiry. Think about it: why is logical reasoning possible? There are laws of logic that we use when we reason. For example, there is the law of noncontradiction, which states that you can't have "A" and "not-A" at the same time and in the same relationship. We all "know" that this is true. But *why* is it true, and *how* do we know it?

The Bible makes sense of this: God is self-consistent. He is noncontradictory, and so this law follows from God's nature. And God has made us in His image; so we instinctively know this law. It has been hard-wired into us. Logical reasoning is possible because God is logical and has made us in His image. (Of course, because of the Curse we sometimes make mistakes in logic.)

But if the universe were merely a chance accident, then why should logical reasoning be possible? If my brain is merely the product of mutations (guided only by natural selection), then why should I think that it can determine what is *true*? The secular, evolutionary worldview cannot account for the existence of logical reasoning.

Likewise, only a biblical worldview can really account for the existence of science—the study of the natural world. Science depends on the fact that the universe obeys orderly laws which do not arbitrarily change. But why should that be so? If the universe were merely an accident, why should it obey logical, orderly laws—or any laws at all for that matter? And why should these laws not be constantly changing, since so many other things change?

The Bible explains this. There are orderly laws because a logical Law-Giver upholds the universe in a logical and consistent way. God does not change; so He sustains the universe in a consistent way. Only a biblical worldview can account for the existence of science and technology.

Now, does this mean that a non-Christian is incapable of reasoning logically or doing science? Not at all. But he is being inconsistent. The non-Christian must "borrow" the above biblical principles in order to do science, or to think rationally. But this is inconsistent. The unbeliever must use *biblical ideas* in order to use science and reason, while he simultaneously denies that the Bible is true.

So Who Created God?

By very definition, an eternal Being has always existed—nobody created Him. God is the Self-Existent One—the great "I Am" of the Bible.[30] He is outside time; in fact, He created time. Think about it this way: everything that has a *beginning* requires a *cause*. The universe has a beginning and therefore requires a cause. But God has no beginning since He is beyond time. So God does not need a cause. There is nothing illogical about an eternal Being who has always existed even though it might be difficult to fully understand.

You might argue, "But that means I have to accept this by faith because I can't totally understand it."

We read in the book of Hebrews: "But without faith it is impossible to please Him, for he who comes to God must believe that He is, and that He is a rewarder of those who diligently seek Him" (11:6).

What kind of faith is Christianity then? It is not blind faith as some may think. In fact, it is the evolutionists who deny the Creator who have the blind "faith."[31] They have to believe in something (i.e., that information can arise from disorder by chance) which goes against real science.

But Christ, through the Holy Spirit, actually opens the eyes of

[30] See Exodus 3:14; Job 38:4; John 8:58; Revelation 1:18; Isaiah 44:6; Deuteronomy 4:39.

[31] See Matthew 13:15; John 12:40; Romans 11:8–10.

Christians so that they can see that their faith is real.[32] The Christian faith is a logically defensible faith. This is why the Bible makes it very clear that anyone who does not believe in God is without excuse: "For since the creation of the world His invisible *attributes* are clearly seen, being understood by the things that are made, *even* His eternal power and Godhead, so that they are without excuse" (Romans 1:20).

How Do We Know the Creator Is the God of the Bible?

You can believe fallible man's ideas that there is no God, or trust the perfect Word of God, the 66 books of the Bible, that says there is. The issue is simple; it is a matter of faith—God exists or God doesn't exist. The exciting thing about being a Christian is knowing that the Bible is not just another religious book, but it is the Word of the Creator God, as it claims.[33]

Only the Bible explains why there is beauty and ugliness; why there is life and death; why there is health and disease; why there is love and hate. Only the Bible gives the true and reliable account of the origin of all basic entities of life and the entire universe.

And over and over again, the Bible's historical account has been confirmed by archaeology, biology, geology, and astronomy. No contradiction or erroneous information has ever been found in its pages, even though it was written over hundreds of years by many different authors, each inspired by the Holy Spirit.

Scientists from many different fields have produced hundreds of books and tapes defending the Bible's accuracy and its claim that it is a revelation to us from our Creator. It not only tells us who we are and where we came from, but it also shares the good news of how we can spend eternity with our Lord and Savior. Take that first step and place your faith in God and His Word.

---

32 See Matthew 13:16; Acts 26:18; Ephesians 1:18; 1 John 1:1.

33 See Matthew 5:18; 2 Timothy 3:16; 2 Peter 1:21; Psalms 12:6; 1 Thessalonians 2:13.

2

Why Shouldn't Christians Accept Millions of Years?

TERRY MORTENSON

There is an intensifying controversy in the church all over the world regarding the age of the earth. For the first 18 centuries of church history the almost universal belief of Christians was that God created the world in six literal days roughly 4,000 years before Christ and destroyed the world with a global Flood at the time of Noah.

But about 200 years ago some scientists developed new theories of earth history, which proposed that the earth and universe are millions of years old. Over the past 200 years Christian leaders have made various attempts to fit the millions of years into the Bible. These include the day-age view, gap theory, local flood view, framework hypothesis, theistic evolution, and progressive creation.

A growing number of Christians (now called young-earth creationists), including many scientists, hold to the traditional view, believing it to be the only view that is truly faithful to Scripture and that fits the scientific evidence far better than the reigning old-earth evolutionary theory.

Many Christians say that the age of the earth is an unimportant and divisive side issue that hinders the proclamation of the gospel. But is that really the case? Answers in Genesis and many other creationist organizations think not.

In this chapter, I want to introduce you to some of the reasons we think that Christians cannot accept the millions of years without doing great damage to the church and her witness in the world. Other chapters in this book will go into much more detail on these issues.

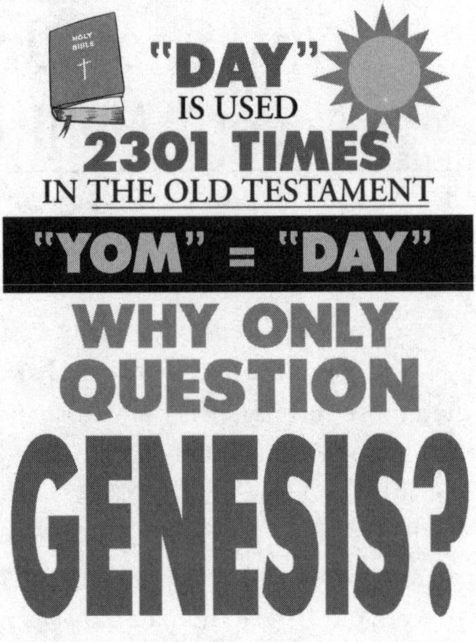

1. **The Bible clearly teaches that God created in six literal, 24-hour days a few thousand years ago.** The Hebrew word for day in Genesis 1 is *yom*. In the vast majority of its uses in the Old Testament it means a literal day; and where it doesn't, the context makes this clear.

2. **The context of Genesis 1 clearly shows that the days of creation were literal days.** First, *yom* is defined the first time it is used in the Bible (Genesis 1:4–5) in its two literal senses: the light portion of the light/dark cycle and the whole light/dark cycle. Second, *yom* is used with "evening" and "morning." Everywhere these two words are used in the Old Testament, either together or separately and with or without *yom* in the context, they always mean a literal evening or morning of a literal day. Third, *yom* is modified with a number: one day, second day, third day, etc., which everywhere else in the Old Testament indicates literal days. Fourth, *yom* is defined literally in Genesis 1:14 in relation to the heavenly bodies.

3. **The genealogies of Genesis 5 and 11 make it clear that the creation days happened only about 6,000 years ago.** It is transparent from the genealogies of Genesis 5 and 11 (which give very detailed chronological information, unlike the clearly abbreviated genealogy in Matthew 1)

and other chronological information in the Bible that the Creation Week took place only about 6,000 years ago.

4. **Exodus 20:9–11 blocks all attempts to fit millions of years into Genesis 1.** "Six days you shall labor and do all your work, but the seventh day is a sabbath of the LORD your God; in it you shall not do any work, you or your son or your daughter, your male or your female servant or your cattle or your sojourner who stays with you. For in six days the LORD made the heavens and the earth, the sea and all that is in them, and rested on the seventh day; therefore the LORD blessed the sabbath day and made it holy" (Exodus 20:9-11).

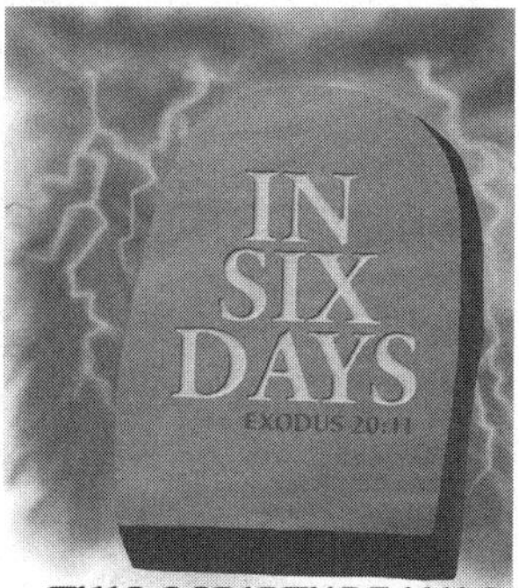

THIS SCRIPTURE WAS INSCRIBED BY GOD!
EXODUS 20:11

This passage gives the reason for God's command to Israel to work six days and then take a sabbath rest. *Yom* is used in both parts of the commandment. If God meant that the Jews were to work six days because He created over six long periods of time, He could have said that using one of three indefinite Hebrew time words. He chose the only word that means a literal day, and the Jews understood it literally (until the idea of millions of years developed in the early nineteenth century). For this reason, the day-age view or framework hypothesis must be rejected. The gap theory or any other attempt to put millions of years before the six days are also false because God says that in six days He made the heaven and the earth and the sea and *all* that is in them. So He made everything in those six literal days and nothing before the first day.

5. **Noah's Flood washes away millions of years**. The evidence in Genesis 6–9 for a global catastrophic flood is overwhelming. For example, the Flood was intended to destroy not only all sinful people but also all land animals and birds and the surface of the earth, which only a global flood could accomplish. The Ark's purpose was to save two of every kind of land animal and bird (and seven of some) to repopulate the earth after the Flood. The Ark was totally unnecessary if the Flood was only local. People, animals, and birds could have migrated out of the flood zone before it occurred, or the zone could have been populated from creatures outside the area after the Flood. The catastrophic nature of the Flood is seen in the nonstop rain for at least 40 days, which would have produced

massive erosion, mud slides, hurricanes, etc. The Hebrew words translated "the fountains of the great deep burst open" (Genesis 7:11) clearly point to tectonic rupturing of the earth's surface in many places for 150 days, resulting in volcanoes, earthquakes, and tsunamis. Noah's Flood would produce exactly the kind of complex geological record we see worldwide today: thousands of feet of sediments clearly deposited by water and later hardened into rock and containing billions of fossils. If the year-long Flood is responsible for most of the rock layers and fossils, then those rocks and fossils cannot represent the history of the earth over millions of years, as evolutionists claim.

6. **Jesus was a young-earth creationist.** Jesus consistently treated the miracle accounts of the Old Testament as straightforward, truthful, historical accounts (e.g., creation of Adam, Noah and the Flood, Lot and his wife in Sodom, Moses and the manna, and Jonah in the fish). He continually affirmed the authority of Scripture over men's ideas and traditions (Matthew 15:1–9). In Mark 10:6 we have the clearest (but not the only) statement showing that Jesus was a young-earth creationist. He teaches that Adam and Eve were made at the "*beginning* of creation," not billions of years after the beginning, as would be the case if the universe were really billions of years old. So, if Jesus was a young-earth creationist, then how can His faithful followers have any other view?

7. **Belief in millions of years undermines the Bible's teaching on death and on the character of God.** Genesis 1 says six times that God called the creation "good," and when He finished creation on Day 6, He called

everything "very good." Man and animals and birds were originally vegetarian (Gen. 1:29–30, plants are not "living creatures," as people and animals are, according to Scripture). But Adam and Eve sinned, resulting in the judgment of God on the whole creation. Instantly Adam and Eve died spiritually, and after God's curse they began to die physically. The serpent and Eve were changed physically and the ground itself was cursed (Genesis 3:14–19). The whole creation now groans in bondage to corruption, waiting for the final redemption of Christians (Romans 8:19–25) when we will see the restoration of all things (Acts 3:21; Colossians 1:20) to a state similar to the pre-Fall world, when there will be no more carnivorous behavior (Isaiah11:6–9) and no disease, suffering, or death (Revelation 21:3–5) because there will be no more Curse (Revelation 22:3). To accept millions of years of animal death before the creation and Fall of man contradicts and destroys the Bible's teaching on death and the full redemptive work of Christ. It also makes God into a bumbling, cruel creator who uses (or can't prevent) disease, natural disasters, and extinctions to mar His creative work, without any moral cause, but still calls it all "very good."

8. **The idea of millions of years did not come from the scientific facts.** This idea of long ages was developed by deistic and atheistic geologists in the late eighteenth and early nineteenth centuries. These men used anti-biblical philosophical and religious assumptions to interpret the geological observations in a way that plainly contradicted the biblical account of creation, the Flood, and the age of the earth. Most church leaders and

scholars quickly compromised using the gap theory, day-age view, local flood view, etc. to try to fit "deep time" into the Bible. But they did not understand the geological arguments and they did not defend their views by careful Bible study. The "deep time" idea flows out of naturalistic assumptions, not scientific observations.

9. **Radiometric dating methods do not prove millions of years.** Radiometric dating was not developed until the early twentieth century, by which time virtually the whole world had already accepted the millions of years. For many years creation scientists have cited numerous examples in the published scientific literature of these dating methods clearly giving erroneous dates (e.g., a date of millions of years for lava flows that occurred in the past few hundred years or even decades). In recent years creationists in the RATE project have done experimental, theoretical, and field research to uncover more such evidence (e.g., diamonds and coal, which the evolutionists say are millions of years old, were dated by carbon-14 to be only thousands of years old) and to show that decay rates were orders of magnitude faster in the past, which shrinks the millions of years to thousands of years, confirming the Bible.[1]

Conclusion

These are just some of the reasons why we believe that the Bible is giving us the true history of the world. God's Word must be the final authority on all matters about which it speaks—not just the moral and spiritual matters, but also its teachings that bear on history, archaeology, and science.

What is at stake here is the authority of Scripture, the character of God, the doctrine of death, and the very foundation of the gospel. If the early chapters of Genesis are not true literal history, then faith in the rest of the Bible is undermined, including its teaching about salvation and morality. I urge you to carefully read the other chapters in this book. The health of the church, the effectiveness of her mission to a lost world, and the glory of God are at stake.

[1] For the results of the RATE project, see Larry Vardiman, Andrew Snelling, and Eugene Chaffin, eds., *Radioisotopes and the Age of the Earth*, Vol. 2, Master Books, Green Forest, Arkansas, 2005; and Don DeYoung, *Thousands ... Not Billions*, Master Books, Green Forest, Arkansas, 2005.

3

Couldn't God Have Used Evolution?

KEN HAM

During the Scopes Trial in 1925, ACLU attorney Clarence Darrow placed William Jennings Bryan (seen as the man representing Christianity) on the stand and questioned him about his faith. In his questioning, Darrow pitted Bryan's faith in the Bible against his belief in modern scientific thinking. Darrow questioned Bryan about the meaning of the word "day" in Genesis. Bryan's answer rejected the clear teaching of Scripture, which indicates that the days of Genesis 1 are six actual days of approximately 24 hours. Bryan accepted modern evolutionary thinking instead when he said, "I think it would be just as easy for the kind of God we believe in to make the earth in six days as in six years or in six million years or in 600 million years. I do not think it important whether we believe one or the other."[1] This is not the first time a Christian has rejected the intended meaning of God's Word, and it certainly will not be the last.

Many Christians today claim that millions of years of earth history fit with the Bible and that God could have used evolutionary processes to create. This idea is not a recent invention. For over 200 years, many theologians have attempted such harmonizations in response to the work of people like Charles Darwin and Scottish geologist Charles Lyell, who helped popularize the idea of millions of years of earth history and slow geological processes.

[1] *The World's Most Famous Court Trial*, Second Reprint Edition, Bryan College, Dayton, Tennessee,1990, 296, 302–303.

When we consider the possibility that God used evolutionary processes to create over millions of years, we are faced with serious consequences: the Word of God is no longer authoritative, and the character of our loving God is questioned.

SCRIPTURAL IMPLICATIONS

Already in Darwin's day, one of the leading evolutionists saw the compromise involved in claiming that God used evolution, and his insightful comments are worth reading again. Once you accept evolution and its implications about history, then man becomes free to pick and choose which parts of the Bible he wants to accept.

From an Evolutionist's Perspective

The leading humanist of Darwin's day, Thomas Huxley (1825–1895), eloquently pointed out the inconsistencies of reinterpreting Scripture to fit with popular scientific thinking. Huxley, an ardent evolutionary humanist, was known as "Darwin's bulldog," as he did more to popularize Darwin's ideas than Darwin himself. Huxley understood Christianity much more clearly than did compromising theologians who tried to add evolution and millions of years to the Bible. He used their compromise against them to help his cause in undermining Christianity.

In his essay "Lights of the Church and Science," Huxley stated,

> I am fairly at a loss to comprehend how anyone, for a moment, can doubt that Christian theology must stand or fall with the historical trustworthiness of the Jewish Scriptures. The very conception of the Messiah, or Christ, is inextricably interwoven with Jewish history; the identification of Jesus of Nazareth with that Messiah rests upon the interpretation of the passages of the Hebrew Scriptures which have no evidential value unless they possess the historical character assigned to them. If the covenant with Abraham was not made; if circumcision and sacrifices were not ordained by Jahveh; if the 'ten words' were not written by God's hand on the stone tables; if Abraham is more or less a mythical hero, such as Theseus; the Story of the Deluge a fiction; that of the Fall a legend; and that of the Creation the dream of a seer; if all these definite and detailed narratives of apparently real events have no more value as history than have the stories of the regal period of Rome—what is to be said about the Messianic doctrine, which is so much less clearly enunciated: And

> what about the authority of the writers of the books of the New Testament, who, on this theory, have not merely accepted flimsy fictions for solid truths, but have built the very foundations of Christian dogma upon legendary quicksands?[2]

Huxley made the point that if we are to believe the New Testament doctrines, we must believe the historical account of Genesis as historical truth.

Huxley was definitely out to destroy the truth of the biblical record. When people rejected the Bible, he was happy. But when they tried to harmonize evolutionary ideas with the Bible and reinterpret it, he vigorously attacked this position.

> I confess I soon lose my way when I try to follow those who walk delicately among "types" and allegories. A certain passion for clearness forces me to ask, bluntly, whether the writer means to say that Jesus did not believe the stories in question or that he did? When Jesus spoke, as a matter of fact, that "the Flood came and destroyed them all," did he believe that the Deluge really took place, or not? It seems to me that, as the narrative mentions Noah's wife, and his sons' wives, there is good scriptural warranty for the statement that the antediluvians married and were given in marriage: and I should have thought that their eating and drinking might be assumed by the firmest believer in the literal truth of the story. Moreover, I venture to ask what sort of value, as an illustration of God's methods of dealing with sin, has an account of an event that never happened? If no Flood swept the careless people away, how is the warning of more worth than the cry of 'Wolf' when there is no wolf?[3]

Huxley then gave a lesson on New Testament theology. He quoted Matthew 19:4–5: "And He answered and said to them, 'Have you not read that He who made *them* at the beginning "made them male and female," and said, "For this reason a man shall leave his father and mother and be joined to his wife, and the two shall become one flesh"?'" Huxley commented, "If divine authority is not here claimed for the twenty-fourth verse of the second chapter of Genesis, what is the value of language? And again, I ask, if one may play fast and loose with the story of the Fall as a 'type' or 'allegory,' what becomes of the foundation of Pauline theology?"[4]

---

[2] T. Huxley, *Science and Hebrew Tradition*, D. Appleton and Company, New York, 1897, 207.

[3] Ibid., 232.

[4] Ibid., 235–236.

And to substantiate this, Huxley quoted 1 Corinthians 15:21–22: "For since by man *came* death, by Man also *came* the resurrection of the dead. For as in Adam all die, even so in Christ all shall be made alive."

Huxley continued, "If Adam may be held to be no more real a personage than Prometheus, and if the story of the Fall is merely an instructive 'type,' comparable to the profound Promethean mythos, what value has Paul's dialectic?"[5]

Thus, concerning those who accepted the New Testament doctrines that Paul and Christ teach but rejected Genesis as literal history, Huxley claimed "the melancholy fact remains, that the position they have taken up is hopelessly untenable."[6]

He was adamant that science (by which he meant evolutionary, long-age ideas about the past) had proven that one cannot intelligently accept the Genesis account of creation and the Flood as historical truth. He further pointed out that various doctrines in the New Testament are dependent on the truth of these events, such as Paul's teaching on the doctrine of sin, Christ's teaching on the doctrine of marriage, and the warning of future judgment. Huxley mocked those who try to harmonize evolution and millions of years with the Bible, because it requires them to give up a historical Genesis while still trying to hold to the doctrines of the New Testament.

What was Huxley's point? He insisted that the theologians had to accept evolution and millions of years, but he pointed out that, to be consistent, they had to give up the Bible totally. Compromise is impossible.

[5] Ibid., 236.
[6] Ibid., 236.

From the Teaching of Christian Leaders

B. B. Warfield and Charles Hodge, great leaders of the Christian faith during the 1800s, adopted the billions-of-years belief concerning the age of the earth and reinterpreted Genesis 1 accordingly. In regard to a discussion on Genesis 1 and the days of creation, Hodge said, "The Church has been forced more than once to alter her interpretation of the Bible to accommodate the discoveries of science. But this has been done without doing any violence to the Scriptures or in any degree impairing their authority."[7]

Even though much of Warfield's and Hodge's teachings were biblically sound, these two men helped unlock the door of compromise, which helped to begin to undermine biblical authority. Once Christians concede to the world that we don't have to take the words in Genesis as written but can use outside beliefs to reinterpret Scripture (e.g., concerning the age of earth), then the door has been unlocked to do this throughout the whole of Scripture. Once this door is unlocked, subsequent generations push it open even farther.

In a number of instances throughout the Bible, one sees compromise in one generation, and in the next, the compromise is usually much greater. It isn't long before the godly foundation is eroded (e.g., the kings of Israel; and idolatry in 2 Kings 14–16, especially in light of Exodus 20:4–6).

Warfield and Hodge taught that Scripture could and should be altered to agree with the newest "scientific" discoveries (which were really men's interpretations about the past) while they claimed that the authority of the other teachings in God's Word remained. But this thinking is faulty. How can one portion of God's Word be open to interpretation while the other portion is untouchable? It can't.

Adding evolution to God's creation has serious scriptural implications because it undermines and attacks the authority of the Word of God.

CHARACTER IMPLICATIONS

Another result of believing that God used evolution or that millions of years of earth history can fit into the Bible is that God's character comes into question.

---

[7] C. Hodge, *Systematic Theology*, Vol. 1, Wm. B. Eerdmans, Grand Rapids, Michigan, 1997, 573. Hodge was probably referring to the usual humanist spin-doctoring of the Galileo affair, but for a more accurate portrayal, see R. Grigg, The Galileo "twist," *Creation* **19**(4):30–32, 1997, and T. Schirrmacher, The Galileo affair: history or heroic hagiography? *TJ* **14**(1):91–100, 2000.

The book of Genesis teaches that death is the result of Adam's sin (Genesis 3:19; Romans 5:12, 8:18–22) and that all of God's creation was "very good" upon its completion (Genesis 1:31). All animals and humans were originally vegetarian (Genesis 1:29–30). But if we compromise on the history of Genesis by adding millions of years, we must believe that death and disease were part of the world before Adam sinned. You see, the (alleged) millions of years of earth history in the fossil record shows evidence of animals eating each other,[8] diseases like cancer in their bones,[9] violence, plants with thorns,[10] and so on. All of this supposedly takes place *before* man appears on the scene, and thus before sin (and its curse of death, disease, thorns, carnivory, etc.) entered the world.

Christians who believe in an old earth (billions of years) need to come to grips with the real nature of the god of an old earth—it is *not* the loving God of the Bible. Even many conservative, evangelical Christian leaders accept and actively promote a belief in millions and billions of years for the age of rocks. How could a God of love allow such horrible processes as disease, suffering, and death for millions of years as part of His "very good" creation?

Interestingly, the liberal camp points out the inconsistencies in holding to an old earth while trying to cling to evangelical Christianity. For instance, Bishop John Shelby Spong, the retired bishop of the Episcopal Diocese of Newark, states:

> The Bible began with the assumption that God had created a finished and perfect world from which human beings had fallen away in an act of cosmic rebellion. Original sin was the reality in which all life was presumed to live. Darwin postulated instead an unfinished and thus imperfect creation… . Human beings did not fall from perfection into sin as the Church had taught for centuries… . Thus the basic myth of Christianity that interpreted Jesus as a divine emissary who came to rescue the victims of the fall from the results of their original sin became inoperative.[11]

---

[8] E.g., ground up dinosaur bones were found in the fossil dung of another dinosaur. See *Nature* **393**(6686):680–682, 1998.

[9] D.H. Tanke and B.M. Rothschild, Paleopathology, P.J. Currie and K. Padian, eds., *Encyclopedia of Dinosaurs*, Academic Press, San Diego, 1997, 525–530.

[10] H.P. Banks, *Evolution and Plants of the Past*, Wadsworth Publishing Company, Belmont, California, 1970, 9–10.

[11] J.S. Spong, A call for a new Reformation, www.dioceseofnewark.org/jsspong/reform.html.

This is an obvious reference to the millions of years associated with the fossil record. The god of an old earth is one who uses death as part of creating. Death therefore can't be the penalty for sin and can't be described as the last enemy (1 Corinthians 15:26).

The god of an old earth cannot therefore be the God of the Bible who is able to save us from sin and death. Thus, when Christians compromise with the millions of years attributed by many scientists to the fossil record, they are, in that sense, seemingly worshipping a different god—the cruel god of an old earth.

People must remember that God created a perfect world; so when they look at this present world, they are not looking at the nature of God but at the results of our sin.

The God of the Bible, the God of mercy, grace, and love, sent His one and only Son to become a man (but God nonetheless), to become our sin-bearer so that we could be saved from sin and eternal separation from God. As 2 Corinthians 5:21 says, "For He has made Him who knew no sin, to be sin for us, that we might become the righteousness of God in Him."

There's no doubt—the god of an old earth destroys the gospel.

DOOR OF COMPROMISE

Now it is true that rejection of six literal days doesn't ultimately affect one's salvation, if one is truly born again. However, we need to stand back and look at the big picture.

In many nations, the Word of God was once widely respected and taken seriously. But once the door of compromise is unlocked, once Christian leaders concede that we shouldn't interpret the Bible as written in Genesis, why

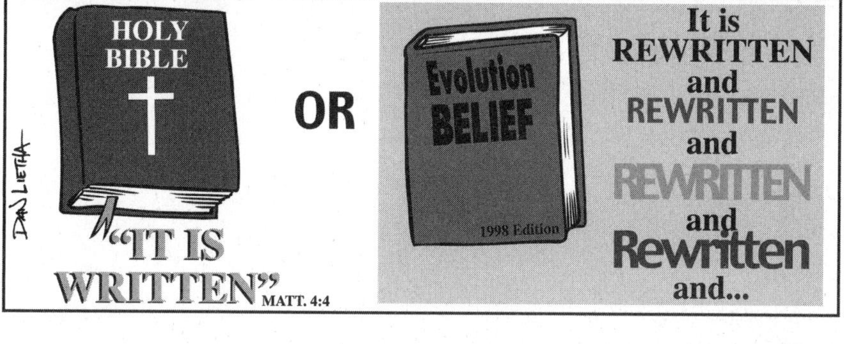
CREATIONWISE
WHICH ONE ARE YOU GOING TO TRUST?
HOLY BIBLE
"IT IS WRITTEN" MATT. 4:4
OR
Evolution BELIEF
1998 Edition
It is REWRITTEN and REWRITTEN and REWRITTEN and Rewritten and...
DAN LIETHA

should the world take heed of God's Word in any area? Because the church has told the world that one can use man's interpretation of the world, such as billions of years, to reinterpret the Bible, this Book is seen as an outdated, scientifically incorrect holy book not intended to be believed as written.

As each subsequent generation has pushed this door of compromise open farther and farther, they are increasingly not accepting the morality or salvation of the Bible either. After all, if the history in Genesis is not correct, how can one be sure the rest is correct? Jesus said, "If I have told you earthly things, and you do not believe, how will you believe if I tell you of heavenly things?" (John 3:12).

The battle is not one of young earth vs. old earth, or billions of years vs. six days, or creation vs. evolution—the real battle is the authority of the Word of God vs. man's fallible opinions.

Why do Christians believe in the bodily Resurrection of Jesus Christ? Because of the words of Scripture ("according to the Scriptures").

And why should Christians believe in the six literal days of creation? Because of the words of Scripture ("In six days the Lord made …").

The real issue is one of authority—is God's Word the authority, or is man's word the authority? So, couldn't God have used evolution to create? The answer is No. A belief in millions of years of evolution not only contradicts the clear teaching of Genesis and the rest of Scripture but also impugns the character of God. He told us in the book of Genesis that He created the whole universe and everything in it in six days by His word: "Then God said… ." His Word is the evidence of how and when God created, and His Word is incredibly clear.

4

Don't Creationists Deny the Laws of Nature?

JASON LISLE

The Word of God

Everything in the universe, every plant and animal, every rock, every particle of matter or light wave, is bound by laws, which it has no choice but to obey. The Bible tells us that there are laws of nature—"ordinances of heaven and earth" (Jeremiah 33:25). These laws describe the way God normally accomplishes His will in the universe.

God's logic is built into the universe, and so the universe is not haphazard or arbitrary. It obeys laws of chemistry which are logically derived from the laws of physics, many of which can be logically derived from other laws of physics and laws of mathematics. The most fundamental laws of nature exist only because God wills them to; they are the logical, orderly way that the Lord upholds and sustains the universe He has created. The atheist is unable to account for the logical orderly state of the universe. Why should the universe obey laws if there is no law-giver? But laws of nature are perfectly consistent with biblical creation. In fact, the Bible is the foundation for natural laws. So, of course, creationists do not deny these laws; laws of nature are exactly what a creationist would expect.

The Law of Life (Biogenesis)

There is one well-known law of life: the law of biogenesis. This law states simply that life always comes from life. This is what observational science tells us; organisms reproduce other organisms after their own kind. Historically, Louis Pasteur disproved one form of spontaneous generation; he showed that life comes from previous life. Since then, we have seen that this law is universal—with no known exceptions. This is, of course, exactly what we would expect from the Bible. According to Genesis 1, God supernaturally created the first diverse kinds of life on earth and made them to reproduce after their kind. Notice that molecules-to-man evolution violates the law of biogenesis. Evolutionists believe that life (at least once) spontaneously formed from nonliving chemicals. But this is inconsistent with the law of biogenesis. Real science confirms the Bible.

The Laws of Chemistry

Life requires a specific chemistry. Our bodies are powered by chemical reactions and depend on the laws of chemistry operating in a uniform fashion. Every living being has information stored on a long molecule called DNA. Life as we know it would not be possible if the laws of chemistry were different. God created the laws of chemistry in just the right way so that life would be possible.

The laws of chemistry give different properties to the various elements (each made of one type of atom) and compounds (made up of two or more types of atoms that are bonded together) in the universe. For example, when given sufficient activation energy, the lightest element (hydrogen) will react

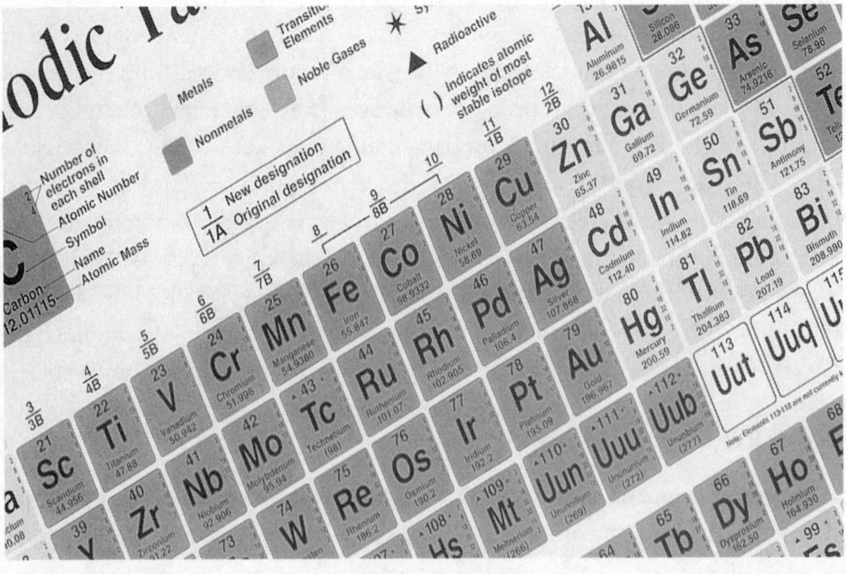

with oxygen to form water. Water itself has some interesting properties such as the ability to hold an unusually large amount of heat energy. When frozen, water forms crystals with six-sided symmetry (which is why snowflakes are generally six-sided). Contrast this with salt (sodium chloride) crystals which tend to form cubes. It is the six-fold symmetry of water-ice that causes "holes" in its crystal, making it less dense than its own liquid. That's why ice floats in water (whereas essentially all other frozen compounds sink in their own liquid.)

The properties of elements and compounds are not arbitrary. In fact, the elements can be logically organized into a periodic table based on their physical properties. Substances in the same column on the table tend to have similar properties. This follows because elements in a vertical column have the same outer electron structure. It is these outermost electrons which determine the physical characteristics of the atom. This periodic table did not happen by chance. Atoms and molecules have their various properties because their electrons are bound by the laws of quantum physics. In other words, chemistry is based on physics. If the laws of quantum physics were just a bit different, atoms might not even be possible. God designed the laws of physics *just right* so that the laws of chemistry would come out the way He wanted them to.

The Laws of Planetary Motion

The creation scientist Johannes Kepler discovered that the planets in our solar system obey three laws of nature. He found that planets orbit in ellipses (not perfect circles as had been previously thought) with the sun at one focus of the ellipse; thus, a given planet is sometimes closer to the sun than at other times. Kepler found that planets sweep out equal areas in equal times—in other words, planets speed up as they get closer to the sun within their orbit. And third, Kepler found the exact mathematical relationship between a planet's distance from the sun (a) as measured in AUs, and its orbital period (p) as measured in years; planets that are farther from the sun take much longer to orbit than planets that are closer (expressed as $p^2 = a^3$). Kepler's laws also apply to the orbits of moons around a given planet.[1]

As with the laws of chemistry, these laws of planetary motion are not fundamental. Rather, they are the logical derivation of other laws of nature. In fact, it was another creation scientist (Sir Isaac Newton) who discovered

[1] However, the constant of proportionality is different for the third law. This is due to the fact that the sun has a different mass than the planets.

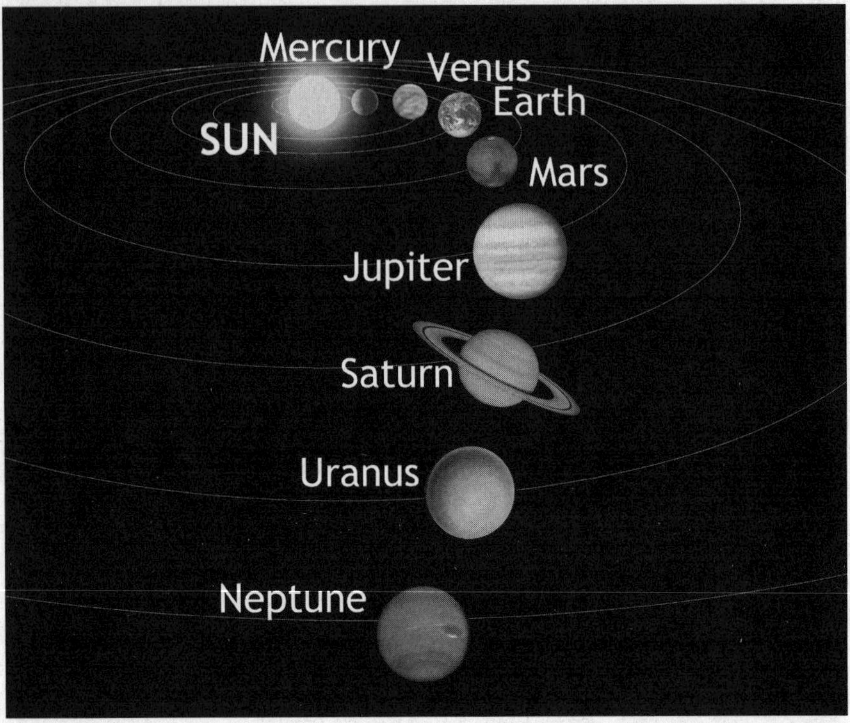

that Kepler's laws could be derived mathematically from certain laws of physics—specifically, the laws of gravity and motion (which Newton himself formulated).

The Laws of Physics

The field of physics describes the behavior of the universe at its most fundamental level. There are many different laws of physics. They describe the way in which the universe operates today. There are laws of physics that describe how light propagates, how energy is transported, how gravity operates, how mass moves through space, and many other phenomena. The laws of physics are usually mathematical in nature; some laws of physics can be described with a concise formula such as $E=mc^2$. The simple formula $F=ma$ shows how an object with mass (m) will accelerate (a) when a net force (F) is applied to it. It is amazing that every object in the universe consistently obeys these rules.

There is a hierarchy in physics: some laws of physics can be derived from other laws of physics. For example, Einstein's famous formula $E=mc^2$ can be derived from the principles and equations of special relativity. Conversely, there are many laws of physics that cannot be derived from other laws of physics; many of these are suspected to be derivative principles, but scientists have not yet deduced their derivation.

And some laws of physics may be truly fundamental (not based on other laws); they exist only because God wills them to. In fact, this *must* be the case for at least *one* law of physics (and perhaps several)—the most fundamental. (Logically, this is because if the most fundamental law were based on some other law, it would not be the most fundamental law.)

Universal Constants

Additionally, there are many physical constants of nature. These are parameters within the laws of physics which set the strengths of the fundamental forces (such as gravity), and the masses of fundamental particles (such as electrons). As with the laws of physics, some constants depend on others, whereas some constants are likely fundamental—God alone has set their value. These constants are essential for life. In many cases, if the fundamental constants had a slightly different value, life would not be possible. For example, if the strength of the electromagnetic coupling constant were slightly altered, molecules could not exist.

The Anthropic Principle

The laws of physics (along with their associated constants) are fine-tuned in just the right way so that life, particularly human life, is possible. This fact is called the "anthropic principle."[2] God created the fundamental laws of physics in just the right way, and gave the constants just the right values so that the other constants and derivative laws of physics would come out in just the right way, so that chemistry would work in the right way, so that the elements and compounds would have the right properties, so that life would be possible![3] It's an amazingly complex challenge—one that no mere human being has the intellectual capacity to solve.[4] In fact, there are many, many

---

[2] Anthropic comes from the Greek word for man: *anthropos*.

[3] Of course, there may be more than one possible solution. That is, it might be possible for God to create life that uses an entirely different chemistry, based on entirely different physics. God may have had considerable freedom in how He chose to create the universe. But it seems likely that there are many more possible (hypothetical) universes in which life is not possible than universes in which life is possible.

[4] A number of resources are available on the anthropic principle. See the secular book *The Anthropic Cosmological Principle* by J. Barrow, F. Tipler, and J. Wheeler, Oxford Univ. Press, New York, 1988.

aspects of this present universe that we still do not completely understand. The laws of nature which we have discovered and expressed mathematically are only imperfect models of reality. Our current understanding of the creation is imperfect. One is reminded of 1 Corinthians 13:12, which tells us that we now only "see through a glass darkly."

The Laws of Mathematics

Notice that the laws of physics are highly mathematical in nature. They would not work if there were not also laws of mathematics. Mathematical laws and principles include the rules of addition, the transitive property, the commutative properties of addition and multiplication, the binomial theorem, and many others. Like the laws of physics, some laws and properties of mathematics can be derived from other mathematical principles. But unlike the laws of physics, the laws of mathematics are abstract; they are not "attached" to any specific part of the universe. It is possible to imagine a universe where the laws of physics are different; but it is difficult to imagine a (consistent) universe where the laws of mathematics are different.[5]

The laws of mathematics are an example of a "transcendent truth." They *must* be true regardless of what kind of universe God created. This may be because God's nature is logical and mathematical. Thus, any universe He chose to create would necessarily be mathematical in nature. The secular naturalist cannot account for the laws of mathematics. Certainly, he would believe in mathematics and would use mathematics; but he is unable to account for the existence of mathematics within a naturalistic framework since mathematics is not a part of the physical universe. However, the Christian understands that there is a God beyond the universe and that mathematics reflects the thoughts of the Lord. Understanding math is in a sense "thinking God's thoughts after Him"[6] (though in a limited, finite way, of course).

Some have supposed that mathematics is a human invention; it is said that if human history had been different, an entirely different form of math would have been constructed—with alternate laws, theorems, axioms, etc. But such thinking is not consistent. Are we to believe that the universe did not obey mathematical laws before people discovered them? Did the planets

[5] Granted, there are different systems of starting definitions and axioms that allow for some variation in mathematical systems of thought (alternate geometries, etc.), but most of the basic principles remain unchanged.

[6] This phrase is attributed to the creation astronomer Johannes Kepler.

orbit differently before Kepler discovered that $p^2 = a^3$? Clearly, mathematical laws are something that human beings have *discovered*—not *invented*. The only thing that might have been different (had human history taken a different course) is the notation—the way in which we choose to express mathematical truths through symbols. But these truths exist regardless of how they are expressed. Mathematics is the "language of creation."

The Laws of Logic

All the laws of nature, from physics and chemistry to the law of biogenesis, depend on the laws of logic. Like mathematics, the laws of logic are transcendent truths. One cannot imagine that the laws of logic could be anything different than what they are. Take the law of noncontradiction as an example. This law states that you cannot have both "A" and "not A" at the same time and in the same relationship. Without the laws of logic, reasoning would be impossible. But where do the laws of logic come from?

The atheist cannot account for the laws of logic, even though he or she must accept that they exist in order to do any rational thinking. But according to the Bible, God is logical. Indeed, the law of noncontradiction is reflective of God's nature; God cannot lie (Numbers 23:19) or be tempted with evil (James 1:13) since these things contradict His perfect nature. Since we have been made in God's image, we instinctively know the laws of logic. We are able to reason logically (though because of finite minds and sin we don't always think entirely logically).

The Uniformity of Nature

The laws of nature are uniform. They do not change arbitrarily, and they apply throughout the whole cosmos. The laws of nature apply in the future just as they have applied in the past—this is one of the most basic assumptions in all of science. Without this assumption, science would be impossible. If the laws of nature suddenly and arbitrarily changed tomorrow, then past experimental results would tell us nothing about the future. Why is it that we can depend on the laws of nature to apply consistently throughout time? The secular scientist cannot justify this important assumption. But the Christian can; the Bible gives us the answer. God is Lord over all creation and sustains the universe in a consistent and logical way. God does not change, and so He upholds the universe in a consistent, uniform way throughout time (Jeremiah 33:25).

Conclusions

We have seen that the laws of nature depend on other laws of nature, which ultimately depend on God's will. Thus, God created the laws of physics in just the right way so that the laws of chemistry would be correct, so that life can exist. It is doubtful that any human would have been able to solve such a complex puzzle. Yet, God has done so. The atheist cannot account for these laws of nature, even though he agrees that they must exist, for such laws are inconsistent with naturalism. Yet they are perfectly consistent with the Bible. We expect the universe to be organized in a logical, orderly fashion and to obey uniform laws because the universe was created by the power of God.

5

What About the Gap & Ruin-Reconstruction Theories?

KEN HAM

Because of the accepted teachings of evolution, many Christians have tried to place a gap of indeterminate time between the first two verses of Genesis 1. Genesis 1:1–2 states: "In the beginning God created the heavens and the earth. The earth was without form, and void; and darkness was on the face of the deep. And the Spirit of God was hovering over the face of the waters."

There are many different versions as to what supposedly happened during this gap of time, but most versions of the gap theory place millions of years of geologic time (including billions of animal fossils) between the Bible's first two verses. This version of the gap theory is sometimes called the ruin-reconstruction theory.

Most ruin-reconstruction theorists have allowed the fallible theories of secular scientists to determine the meaning of Scripture and have, therefore, accepted the millions-of-years dates for the fossil record.

Some theorists also put the fall of Satan in this supposed period. But any rebellion of Satan during this gap of time contradicts God's description of His completed creation on Day 6 as all being "very good" (Genesis 1:31).

All versions of the gap theory impose outside ideas on Scripture and thus open the door for further compromise.

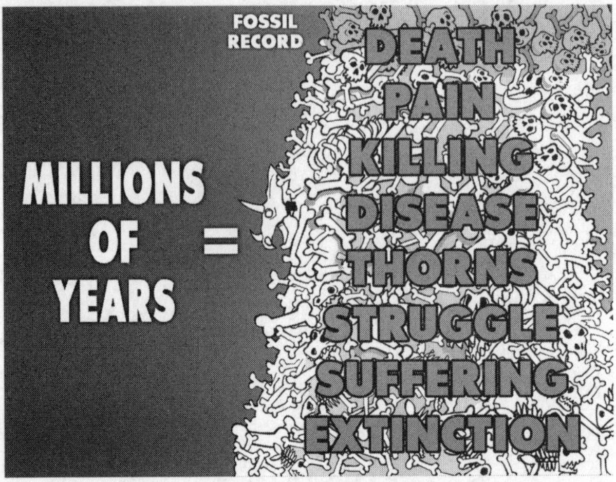
FOSSIL RECORD
MILLIONS OF YEARS =
DEATH
PAIN
KILLING
DISEASE
THORNS
STRUGGLE
SUFFERING
EXTINCTION

Where Did the Gap Theory Come From?

Christians have made many attempts over the years to harmonize the Genesis account of creation with accepted geology and its teaching of billions of years for the age of the earth. Examples of such attempts include the views of theistic evolution, progressive creation, and the gap theory.

This idea of the gap theory can be traced back to the rather obscure writings of the Dutchman Episcopius (1583–1643), but it was first recorded from one of the lectures of Thomas Chalmers.[1] Chalmers (1780–1847) was a notable Scottish theologian and the first modera-

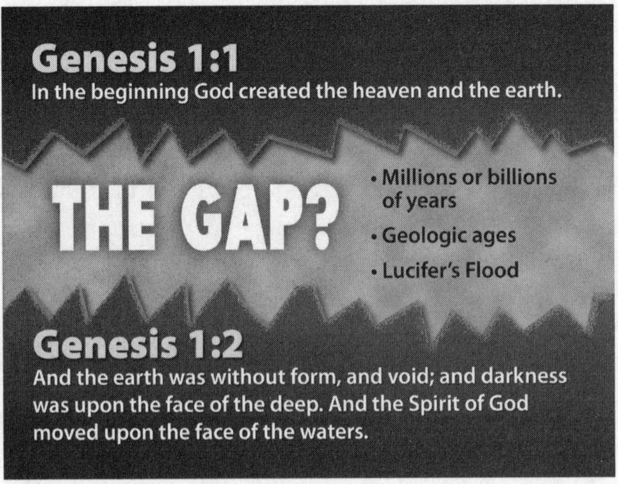
Genesis 1:1
In the beginning God created the heaven and the earth.
THE GAP?
• Millions or billions of years
• Geologic ages
• Lucifer's Flood
Genesis 1:2
And the earth was without form, and void; and darkness was upon the face of the deep. And the Spirit of God moved upon the face of the waters.

[1] I. Taylor, *In the Minds of Men: Darwin and the New World Order*, TFE Publishing, Toronto, Canada, 1984, 363.

tor of the Free Church of Scotland, and he was perhaps the man most responsible for the gap theory.[2] Rev. William Buckland, a geologist, also did much to popularize the idea.

Although Chalmers' writings give very little information about the gap theory,[3] many of the details are obtained from other writers, such as the nineteenth century geologist Hugh Miller, who quoted from Chalmers' lectures on the subject.[4]

The most notably influential nineteenth century writer to popularize this view was G. H. Pember, in his book *Earth's Earliest Ages*,[5] first published in 1884. Numerous editions of this work were published, the 15th edition appearing in 1942.[6]

The 20th-century writer who published the most academic defense of the gap theory was Arthur C. Custance in his work *Without Form and Void*.[7]

Bible study aids such as the Scofield Reference Bible, Dake's Annotated Reference Bible, and The Newberry Reference Bible also include the gap theory and have influenced many to accept this teaching. The basic reason for developing and promoting this view can be seen from the following very-telling quotes:

> Scofield Study Bible: "Relegate fossils to the primitive creation, and no conflict of science with the Genesis cosmogony remains."[8]

> Dake's Annotated Reference Bible: "When men finally agree on the age of the earth, then place the many years (over the historical 6,000) between Genesis 1:1 and 1:2, there will be no conflict between the Book of Genesis and science."[9]

These quotes are typical of the many compromise positions—accepting

[2] W.W. Fields, *Unformed and Unfilled*, Burgeners Enterprises, Collinsville, Illinois, 1976, 40.

[3] W. Hanna, ed., *Natural Theology*, Selected works of Thomas Chalmers, Vol. 5, Thomas Constable, Edinburgh, 1857, 146. The only thing Chalmers basically states concerning the gap theory in these writings is "The detailed history of creation in the first chapter of Genesis begins at the middle of the second verse."

[4] H. Miller, *The Testimony of the Rocks*, Boston, Gould and Lincoln, New York, 1867, 143.

[5] G.H. Pember, *Earth's Earliest Ages*, H. Revell Company, New York, 1900.

[6] Taylor, *In the Minds of Men*, 363.

[7] A.C. Custance, *Without Form and Void*, Brookville, Canada, 1970.

[8] C.I. Scofield, Ed., The Scofield Study Bible, Oxford University Press, New York, 1945. (Originally published as The Scofield Reference Bible; this edition is unaltered from the original of 1909.)

[9] F.H. Dake, Dake's Annotated Reference Bible, Dake Bible Sales, Lawrenceville, Georgia, 1961, 51.

so-called "science"[10] and its long ages for the earth, and incorporating them into Scripture.

A Testimony of Struggle

G. H. Pember's struggle with long geologic ages, recounted in *Earth's Earliest Ages*, has been the struggle of many Christians ever since the idea of millions of years for the fossil record became popular in the early nineteenth century. Many respected Christian leaders of today wrestle with this same issue.

Reading Pember's struggle helps us understand the implications of the gap theory. Pember, like today's conservative Christians, defended the authority of Scripture. He was adamant that one had to start *from* Scripture alone and *not* bring preconceived ideas *to* Scripture. He boldly chastened people who came to the Bible "filled with myths, philosophies, and prejudices, which they could not altogether throw off, but retained, in part at least, and mingled—quite unwillingly, perhaps—with the truth of God" (p. 5). He describes how the church is weakened when man's philosophies are used to interpret God's Word: "For, by skillfully blending their own systems with the truths of Scripture, they so bewildered the minds of the multitude that but few retained the power of distinguishing the revelation of God from the craftily interwoven teachings of men" (p. 7). He also said, "And the result is that inconsistent and unsound interpretations have been handed down from generation to generation, and received as if they were integral parts of the Scriptures themselves; while any texts which seemed violently opposed were allegorized, spiritualized, or explained away, till they ceased to be troublesome, or perchance, were even made subservient" (p. 8).

He then warns Christians, "For, if we be observant and honest, we must often ourselves feel the difficulty of approaching the sacred writings without bias, seeing that we bring with us a number of stereotyped ideas, which we have received as absolutely certain, and never think of testing, but only seek to confirm" (p. 8).

What happened to Pember should warn us that no matter how great a theologian we may be or how respected and knowledgeable a Christian leader, we, as finite, sinful human beings, cannot easily empty ourselves of preconceived

[10] Many people now equate the teaching of millions of years and evolution with science. However, these teachings are *not* science in the empirical (repeatable, testable) sense. Scientists have only the present to work with. To connect the present to the past involves interpretations based on unprovable assumptions.

ideas. Pember did exactly what he preached against, without realizing it. Such is the ingrained nature of the long-ages issue. He did not want to question Scripture (he accepted the six literal days of creation), but he did not question the long ages, either. So Pember struggled with what to do. Many of today's respected Christian leaders show the same struggle in their commentaries as they then capitulate to progressive creation or even theistic evolution.[11]

Pember said, "For, as the fossil remains clearly show not only were disease and death—inseparable companions of sin—then prevalent among the living creatures of the earth, but even ferocity and slaughter." He, therefore, recognized that a fossil record of death, decay, and disease before sin was totally inconsistent with the Bible's teaching. And he understood that there could be no carnivores before sin: "On the Sixth Day God pronounced every thing which He had made to be very good, a declaration which would seem altogether inconsistent with the present condition of the animal as well as the vegetable kingdom. Again: He gave the green herb alone for food 'to every beast of the field, and to every fowl of the air, and to every thing that creepeth upon the earth.' There were, therefore, no carnivora in the sinless world" (p. 35).

Pember taught from Isaiah that the earth will be restored to what it was like at first—no more death, disease, or carnivorous activity. However, because he had accepted the long ages for the fossil record, what was he to do with all this death, disease, and destruction in the record? He responded, "Since, then, the fossil remains are those of creatures anterior to Adam, and yet show evident tokens of disease, death, and mutual destruction, they must have belonged to another world, and have a sin-stained history of their own" (p. 35).

Thus, in trying to reconcile the long ages with Scripture, Pember justified the gap theory by saying, "There is room for any length of time between the first and second verses of the Bible. And again; since we have no inspired account of geological formations, we are at liberty to believe that they were developed just in the order which we find them. The whole process took place in pre-Adamite times, in connection, perhaps, with another race of beings, and, consequently, does not at present concern us" (p. 28).

With this background, let us consider this gap theory in detail. Basically, this theory incorporates three strands of thought:

1. A literal view of Genesis.
2. Belief in an extremely long but unidentified age for the earth.

[11] K. Ham, Millions of years and the "doctrine of Balaam," *Creation* **19**(3):15–17, 1997.

3. An obligation to fit the origin of most of the geologic strata and other geologic evidence between Genesis 1:1 and 1:2. (Gap theorists oppose evolution but believe in an ancient origin of the universe.)

There are many variations of the gap theory. According to the author Weston Fields, the theory can be summarized as follows, "In the far distant dateless past, God created a perfect heaven and perfect earth. Satan was ruler of the earth which was peopled by a race of 'men' without any souls. Eventually, Satan, who dwelled in a garden of Eden composed of minerals (Ezekiel 28), rebelled by desiring to become like God (Isaiah 14). Because of Satan's fall, sin entered the universe and brought on the earth God's judgment in the form of a flood (indicated by the water of 1:2), and then a global ice age when the light and heat from the sun were somehow removed. All the plant, animal, and human fossils upon the earth today date from this 'Lucifer's flood' and do not bear any genetic relationship with the plants, animals, and fossils living upon the earth today."[12]

Some versions of the gap theory state that the fossil record (geologic column) formed over millions of years, and then God destroyed the earth with a catastrophe (i.e., Lucifer's flood) that left it "without form and void."

Western Bible commentaries written before the eithteenth century (before the belief in a long age for the earth became popular) knew nothing of any gap between Genesis 1:1 and 1:2. Certainly some commentaries proposed intervals of various lengths of time for reasons relating to Satan's fall,[13] but none proposed a ruin-reconstruction situation or a pre-Adamite world. In the nineteenth century, it became popular to believe that the geological changes occurred slowly and roughly at the present rate (uniformitarianism[14]). With increased acceptance of uniformitarianism, many theologians urged

[12] Fields, *Unformed and Unfilled*, 7.

[13] Those who try to put the fall of Satan (not connected with millions of years) into this gap, need to consider that if all the angels were a part of the original creation, as Exodus 20:11 indicates and Colossians 1 seems to confirm, then *everything* God had created by the end of the sixth day was "very good." There could not have been *any* rebellion before this time. So Satan fell some time after Day 7.

[14] The term "uniformitarian" commonly refers to the idea that geological processes, such as erosion and sedimentation, have remained essentially the same throughout time, and so the present is the key to the past. But after the mid-nineteenth century, the application of the concept has been extended. Huxley said, "Consistent uniformitarianism postulates evolution as much in the organic as in the inorganic world." It is now assumed that a closed system exists, to which neither God nor any other nonhuman or nonnatural force has access (from J. Rendle-Short, *Man: Ape or Image*, Master Books, Green Forest, Arkansas, 1984, 20, note 4).

reinterpretation of Genesis (with ideas such as day-age, progressive creation, theistic evolution, and days-of-revelation).

Problems with the Gap Theory

Believing in the gap theory presents a number of problems and inconsistencies, especially for a Christian.

1. It is inconsistent with God creating *everything* in six days, as Scripture states.

Exodus 20:11 says, "For in six days the LORD made the heavens and earth, the sea, and all that is in them, and rested the seventh day. Therefore the LORD blessed the Sabbath day, and hallowed it." Thus the creation of the heavens and the earth (Genesis 1:1) and the sea and *all that is in them* (the rest of the creation) was completed in six days.[15] Is there any time for a gap?

THIS SCRIPTURE WAS INSCRIBED BY GOD!
EXODUS 20:11

2. It puts death, disease, and suffering before the Fall, contrary to Scripture.

Romans 5:12 says, "Therefore, just as through one man [Adam] sin entered the world, and death through sin, and thus death spread to all men, because all sinned." From this we understand that there could not have been human sin or death before Adam. The Bible teaches in 1 Corinthians 15 that Adam was the first man, and as a result of his rebellion (sin), death and corruption (disease, bloodshed, and suffering) entered the universe. Before Adam sinned, there could not have been any animal *(nephesh*[16]*)* or human death. Note also that there could not have been a race of men before Adam that died in Lucifer's flood because 1 Corinthians 15:45 tells us that Adam was the first man.

Genesis 1:29–30 teaches us that animals and man were originally created to eat plants, which is consistent with God's description of His creation as "very good." But how could a fossil record, which gives evidence of disease, violence, death, and decay (fossils have been found of animals

---

[15] See Chapter 8 for more details.

[16] The Bible speaks of animals and humans having or being *nephesh* (Hebrew), or soul-life, in various contexts suggesting conscious life. The death of a jellyfish, for example, may not be death of a *nephesh* animal.

apparently fighting and certainly eating each other), be described as "very good"? For this to be true, the death of billions of animals (and many humans) as seen in the fossil record must have occurred *after* Adam's sin. The historical event of the global Flood, recorded in Genesis, explains the presence of huge numbers of dead animals buried in rock layers, laid down by water all over the earth.

Romans 8:22 teaches that "the whole creation groans and travails in pain together until now." Clearly the whole of creation was, and is, subject to decay and corruption because of sin. When gap theorists believe that disease, decay, and death existed before Adam sinned, they ignore that this contradicts the teaching of Scripture.[17]

The version of the gap theory that puts Satan's fall at the end of the geological ages, just before the supposed Lucifer's flood that destroyed all pre-Adamic life, has a further problem—the death and suffering recorded in the fossils must have been God's fault. Since it happened before Satan's fall, Satan and sin cannot be blamed for it.[18]

3. The gap theory is logically inconsistent because it explains away what it is supposed to accommodate—supposed evidence for an old earth.

Gap theorists accept that the earth is very old—a belief based on geologic evidence interpreted with the assumption that the present is the key to the past. This assumption implies that in the past sediments containing fossils formed at basically the same rate as they do today. This process is also used by most geologists and biologists to justify belief that the geologic column represents billions of years of earth history. This geologic column has become the showcase of evolution because the fossils are claimed to show ascent from simple to complex life forms.

This places gap theorists in a dilemma. Committed to literal creation because of their acceptance of a literal view of Genesis, they cannot accept the conclusions of evolution based on the geologic column. Nor can they accept that the days in the Genesis record correspond to geologic periods. So they propose that God reshaped the earth and re-created all life in six literal days after Lucifer's flood (which produced the fossils); hence the name "ruin-reconstruction." Satan's sin supposedly caused this flood, and the resulting

---

17 See chapter 26; also, K. Ham, *The Lie: Evolution,* Master Books, Green Forest, Arkansas, 1987, 71–82.

18 H. Morris, Why the gap theory won't work, *Back to Genesis* No. 107, Institute for Creation Research, San Diego, California, 1997.

judgment upon that sin reduced the previous world to a state of being "without form and void."

While the gap theorist may think Lucifer's flood solves the problem of life before God's creation recorded in Genesis 1:2 and following, this actually removes the reason for the theory in the first place. If all, or most, of the sediments and fossils were produced quickly in one massive worldwide Lucifer's flood, then the main evidence that the earth is extremely old no longer exists, because the age of the earth is based on the assumed slow formation of earth's sediments.

Also, if the world was reduced to a shapeless, chaotic mess, as gap theorists propose, how could a reasonably ordered assemblage of fossils and sediments remain as evidence? Surely with such chaos the fossil record would have been severely disrupted, if not entirely destroyed. This argument also applies to those who say the fossil record formed over hundreds of millions of years before this so-called Lucifer's flood, which would have severely rearranged things.

4. The gap theory does away with the evidence for the historical event of the global Flood.

If the fossil record was formed by Lucifer's flood, then what did the global Flood of Noah's day do? On this point the gap theorist is forced to conclude that the global Flood must have left virtually no trace. To be consistent, the gap theorist would also have to defend that the global Flood was a local event. Custance, one of the major proponents of the gap theory, did just that, and he even published a paper defending a local flood.[19]

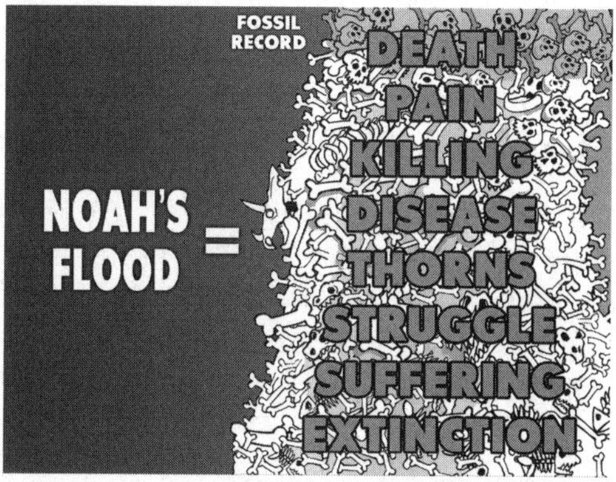

Genesis, however, depicts the global Flood as a judgment for man's sin (Genesis 6). Water flooded the earth for over a year (Genesis 6:17; 7:19–24) and only eight people, along with two of every kind (and seven of some) of air-breathing, land-dwelling animal survived (Genesis 7:23). It is more

[19] A.C. Custance, The Flood: local or global? *The Doorway Papers,* Zondervan, Grand Rapids, Michigan, Vol. 9, 1970.

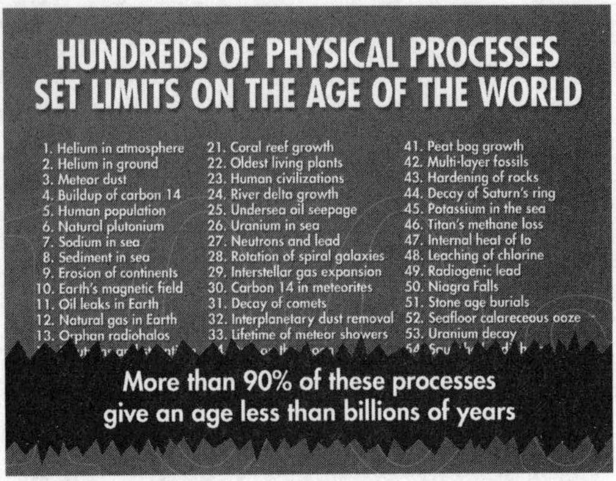

consistent with the whole framework of Scripture to attribute most fossils to the global Flood of Noah's day rather than to resort to a strained interpretation of the fall of Satan[20] and a totally speculative catastrophe that contributes nothing to biblical understanding or to science.

Sadly, in relegating the fossil record to the supposed gap, gappists have removed the evidence of God's judgment in the Flood, which is the basis for God's warning of judgment to come (2 Peter 3:2–14).

5. The gap theorist ignores the evidence for a young earth.

The true gap theorist also ignores evidence consistent with an earth fewer than 10,000 years of age. There is much evidence for this—the decay and rapid reversals of the earth's magnetic field, the amount of salt in the oceans, the wind-up of spiral galaxies, and much more.[21]

6. The gap theory fails to accommodate standard uniformitarian geology with its long ages.

Today's uniformitarian geologists allow for no worldwide flood of any kind—the imaginary Lucifer's flood or the historical Flood of Noah's day. They also recognize no break between the supposed former created world and the current recreated world.

7. Most importantly, the gap theory undermines the gospel at its foundations.

By accepting an ancient age for the earth (based on the standard uniformitarian interpretation of the geologic column), gap theorists leave the evolutionary system intact (which by their own assumptions they oppose).

Even worse, they must also theorize that Romans 5:12 and Genesis 3:3 refer only to spiritual death. But this contradicts other scriptures, such as

---

[20] This also impinges upon the perspicuity of Scripture—that is, that the Bible is clear and understandable to ordinary Christians in all that's important.

[21] D.R. Humphreys, Evidence for a young world, *Creation* **13**(3):46–50, 1991; also available as a pamphlet. See also www.answersingenesis.org/go/young.

1 Corinthians 15 and Genesis 3:22–23. These passages tell us that Adam's sin led to *physical* death, as well as spiritual death. In 1 Corinthians 15 the death of the Last Adam (the Lord Jesus Christ) is compared with the death of the first Adam. Jesus suffered physical death for man's sin, because Adam, the first man, died physically because of sin.

In cursing man with physical death, God also provided a way to redeem man through the person of His Son Jesus Christ, who suffered the curse of death on the Cross for us. He tasted "death for everyone" according to Hebrews 2:9. He took the penalty that should rightly have been ours at the hands of the Righteous Judge, and bore it in His own body on the Cross. Jesus Christ tasted death for all mankind, and He defeated death when He rose from the grave three days later. Men can be free from eternal death in hell if they believe in Jesus Christ as Lord and Savior. They then are received back to God to spend eternity with Him. That is the message of Christianity.

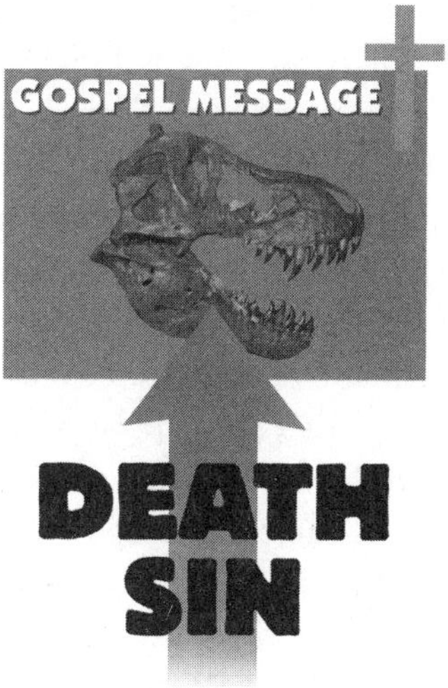

To believe there was death before Adam's sin destroys the basis of the Christian message. The Bible states that man's rebellious actions led to death and the corruption of the universe, but the gap theory undermines the reason that man needs a Savior.

A Closer Look at Genesis 1:1–2

The earliest available manuscript of Genesis 1:1–2 is found in the Greek translation of the Old Testament, called the Septuagint (LXX), which was prepared about 250–200 B.C. The LXX does not permit the reading of any ruin-reconstruction scenario into these verses, as even Custance admitted. A closer look at these verses reveals that the gap theory imposes an interpretation upon Genesis 1:1–2 that is unnatural and grammatically unsound. Like many attempts to harmonize the Bible with uniformitarian geology, the gap theory involves a well-meant but misguided twisting of Scripture.

Below are the five major challenges to the gap theory in interpreting Scripture. For a much fuller analysis, we recommend the book *Unformed and Unfilled* by Weston Fields, published by Burgener Enterprises, 1997.

Creating and Making (Hebrew: *Bara* and *Asah*)

It is generally acknowledged that the Hebrew word *bara*, used with "God" as its subject, means "to create"—in the sense of the production of something which did not exist before.

However, according to Exodus 20:11, God "made" *(asah)* the heavens and the earth and everything in them in six days. If God made everything in six days, then there is clearly no room for a gap. To avoid this clear scriptural testimony against any gap, gap theorists have alleged that *asah* does not mean "to create," but "to form" or even "re-form." They claim that Exodus 20:11 refers not to six days of creation but to six days of re-forming a ruined world.

Is there such a difference between *bara* and *asah* in biblical usage? A number of verses show that, while *asah* may mean "to do" or "to make," it can also mean "to create," which is the same as *bara*. For example, Nehemiah 9:6 states that God made (*asah*) "heaven, the heaven of heavens, with all their host, the earth and everything on it, the seas and all that is in them." This reference is obviously to the original *ex nihilo* (out of nothing) creation, but the word *asah* is used. (We may safely assume that no gappist will want to say that Nehemiah 9:6 refers to the supposed reconstruction, because if the passage did, the gappist would have to include the geological strata in the reconstruction, thereby depriving the whole theory of any power to explain away the fossil record.)

The fact is that the words *bara* and *asah* are often used interchangeably in the Old Testament; indeed, in some places they are used in synonymous parallelism (e.g., Genesis 1:26–27, 2:4; Exodus 34:10; Isaiah 41:20, 43:7).

Applying this conclusion to Exodus 20:11, 31:17, and Nehemiah 9:6, we see that Scripture teaches that God created the universe (everything) in six days, as outlined in Genesis 1.

The Grammar of Genesis 1:1–2

Many adherents of the gap theory claim that the grammar of Genesis 1:1–2 allows, and even requires, a time-gap between the events in verse 1 and the events in verse 2. Into this gap—believed by many to be billions of years—they want to place all the major geological phenomena that have shaped the world.

This is an unnatural interpretation, not suggested by the plain meaning of the text. The most straightforward reading of the verses sees verse 1 as a subject-and-verb clause, with verse 2 containing three circumstantial clauses (i.e., three statements that further describe the circumstances introduced by the principal clause in verse 1).

This conclusion is reinforced by the grammarian Gesenius. He says that the Hebrew conjunction *waw,* meaning "and" at the beginning of verse 2, is a "*waw* copulative," which compares with the old English expression "to wit." This grammatical connection between verses 1 and 2 thus rules out the gap theory. Verse 2 is in fact a description of the state of the originally created earth: "*And* the earth was without form and void" (Genesis 1:2a).[22]

"Was" or "Became"?

Gappists translate "the earth *was* without form and void" to be "the earth *became* (or, *had become*) without form and void." At stake is the translation of the Hebrew word *hayetah* (a form of the Hebrew verb, *hayah*, meaning "to be").

Custance, a supporter of the gap theory, claims that out of 1,320 occurrences of the verb *hayah* in the Old Testament, only 24 can certainly be said to bear the meaning "to be." He concludes that in Genesis 1:2 *hayetah* must mean "became" and not simply "was."

However, we must note that the meaning of a word is controlled by its context, and that verse 2 is circumstantial to verse 1. Thus "was" is the most natural and appropriate translation for *hayetah*. It is rendered this way in most English versions (as well as in the LXX). Furthermore, in Genesis 1:2 *hayetah* is not followed by the preposition *le*, which would have removed any ambiguity in the Hebrew and required the translation "became."

Genesis 1:2
And the earth was without form, and void; and darkness [was] upon the face of the deep. And the Spirit of God moved upon the face of the waters.

the earth WAS formless and void OR **the earth BECAME formless and void**

SEQUENTIAL CLAUSE, NOT CASUAL

Therefore, the CORRECT reading is:
And the earth WAS formless and void …

[22] The word "and" is included in the KJV translation but is translated "now" in the NIV and is not translated at all in the NKJV or the NASB.

Tohu and *Bohu*

The words *tohu* and *bohu*, usually translated "formless and void," are used in Genesis 1:2. They imply that the original universe was created unformed and unfilled and was, during six days, formed and filled by God's creative actions.

Gappists claim that these words imply a process of judgmental destruction and that they indicate a sinful, and therefore not an original, state of the earth. However, this brings interpretations from other parts of the Old Testament with very different contexts (namely, Isaiah 34:11 and Jeremiah 4:23) and imports them into Genesis 1.

Tohu and *bohu* appear together only in the three above-mentioned places in the Old Testament. However, *tohu* appears alone in a number of other places and in all cases simply means "formless." The word itself does not tell us about the cause of formlessness; this has to be gleaned from the context. Isaiah 45:18 (often quoted by gappists) is rendered in the KJV "he created it not in vain [*tohu*], he formed it to be inhabited." In the context, Isaiah is speaking about Israel, God's people, and His grace in restoring them. He did not choose His people in order to destroy them, but to be their God and for them to be His people. Isaiah draws an analogy with God's purpose in creation: He did not create the world for it to be empty. No, He created it to be formed and filled, a suitable abode for His creation. Gappists miss the point altogether when they argue that because Isaiah says God did not create the world *tohu,* it must have *become tohu* at some later time. Isaiah 45:18 is about God's *purpose* in creating, not about the original state of the creation.

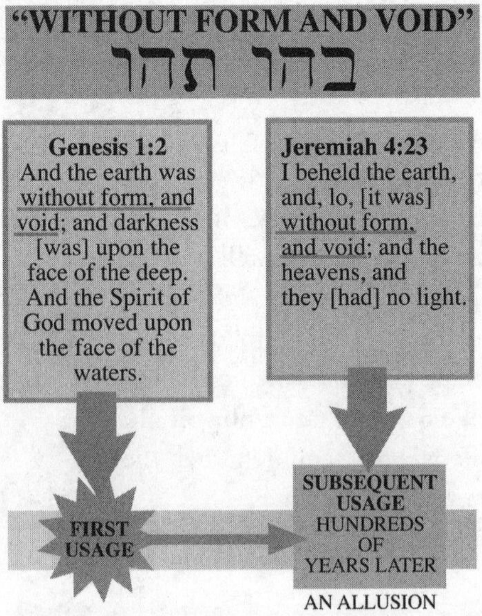
"WITHOUT FORM AND VOID"
בהו תהו
Genesis 1:2
And the earth was without form, and void; and darkness [was] upon the face of the deep. And the Spirit of God moved upon the face of the waters.
Jeremiah 4:23
I beheld the earth, and, lo, [it was] without form, and void; and the heavens, and they [had] no light.
FIRST USAGE
SUBSEQUENT USAGE HUNDREDS OF YEARS LATER
AN ALLUSION

Though the expression "*tohu* and *bohu*" in Isaiah 34:11 and Jeremiah 4:23 speaks of a formlessness and emptiness resulting from divine judgment for sin, this meaning is not implicit in the expression itself but is gained from the particular contexts in which it occurs. It is not valid therefore to infer that same meaning from Genesis 1:2, where the context does not suggest any judgment. As an analogy, we might think of a word like "blank" in reference to a computer screen. It can be blank because nothing has been typed on the keyboard, or it can be blank because the screen has been erased. The word "blank" does not suggest, in itself, the reason why the screen is blank. Likewise with "formless and void"—the earth began that way simply because it was not yet formed and filled, or it was that way because of judgment.

Theologians call the form of use of *tohu* and/or *bohu* in Isaiah 34:11 and Jeremiah 4:23 a "verbal allusion." These passages on judgment allude to the formless and empty earth at the beginning of creation to suggest the extent of God's judgment to come. God's judgment will be so complete that the result will be like the earth before it was formed and filled—formless and empty. This does not imply that the state of the creation in Genesis 1:2 was arrived at by some sort of judgment or destruction as imagined by gappists. As theologian Robert Chisholm, Jr. wrote, "By the way, allusion only works one way. It is unwarranted to assume that Jeremiah's use of the phrase in a context of judgment implies some sort of judgment in the context of Genesis 1:2. Jeremiah is not interpreting the meaning of Genesis 1:2."[23]

"Replenish"

Many gappists have used the word "replenish" in the KJV translation of Genesis 1:28 to justify the gap theory on the basis that this word means "refill." Thus, they claim that God told Adam and Eve to refill the earth, implying it was once before filled with people (the pre-Adamites). However,

[23] R.B. Chisholm, Jr., *From Exegesis to Exposition: A Practical Guide to Using Biblical Hebrew,* Baker Books, Grand Rapids, Michigan, 1998, 41.

this is wrong. The Hebrew word translated "replenish," *male*,[24] simply means "fill" (or "fulfill" or "be filled").

The English word "replenish" meant "fill" from the thirteenth to the seventeenth centuries; then it changed to mean "refill." When the KJV was published in 1611, the translators used the English word "replenish," which at that time meant only "fill," not "refill."[25]

The Straightforward Meaning of Genesis 1:1–2

The gap (or ruin-reconstruction) theory is based on a very tenuous interpretation of Scripture.

The simple, straightforward meaning of Genesis 1:1–2 is that, when God created the earth at the beginning, it was initially formless, empty, and dark, and God's Spirit was there above the waters. It was through His creative energy that the world was then progressively formed and filled during the six days of creation.

Consider the analogy of a potter making a vase. The first thing he does is gather a ball of clay. What he has is good, but it is unformed. Next, he shapes it into a vase, using his potter's wheel. Now the ball of clay is no longer formless. He then dries it, applies glaze, and fires it. Now it is ready to be filled—with flowers and water. At no time could one of the stages be considered evil or bad. It was just unfinished—unformed and unfilled. When the vase was finally formed and filled, it could be described as "very good."

Warning

Many sincere Christians have invented reinterpretations of Scripture to avoid intellectual conflicts with popular scientific ideas. The gap theory was one such reinterpretation designed to fit in with scientific concepts that arose in the early 1800s and are still popular today.

[24] *Strong's Concordance*, Hebrew word No. 4390.

[25] See C. Taylor, What does "replenish the earth" mean? *Creation* **18**(2):44–45, 1996, for more details on the history of the meaning of "replenish."

In reality, though, the gap theory was an effective anesthetic that put the church to sleep for over 100 years. When the children who learned this compromise position went on to higher education, they were shocked to discover that this theory explained nothing. Many of them then accepted the only remaining "respectable" theory—evolution—which went hand-in-hand with millions of years. The results were usually disastrous for their faith.

Today, other compromise positions, such as progressive creation or theistic evolution, have mostly replaced the gap theory.[26] The gappists, by attempting to maintain a literal Genesis but adhering to the long ages (millions of years), opened the door for greater compromise in the next generation—the reinterpretation of the days, God using evolution, etc.

But whether it is the gap theory, day-age/progressive creation, or theistic evolution, the results are the same. These positions may be acceptable in some churches, but the learned in the secular world will, with some justification, mock those who hold them because they see the inconsistencies.

In Martin Luther's day the church compromised what the Bible clearly taught, and he nailed his *Ninety-Five Theses* to the door of the church to call them back to the authority of God's Word. In the same way, the church today has, by and large, neglected what the Bible clearly says in Genesis 1–11. It's time to call the church back to the authority of God's Word beginning with Genesis.

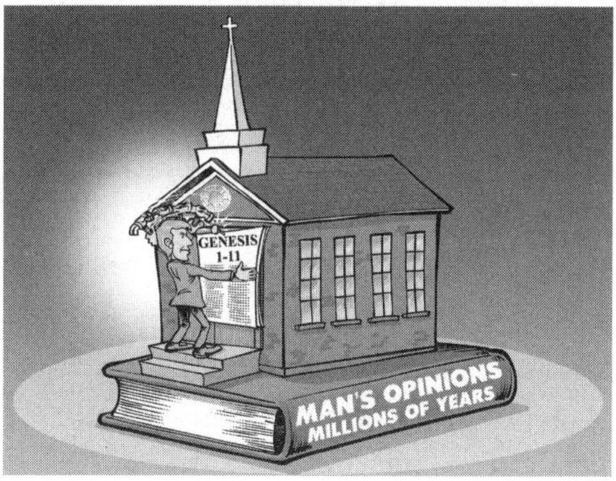

[26] A strange modern gap theory is found in *Genesis Unbound*, by J. Sailhamer, Multnomah Books, Sisters, Oregon, 1996. The author fits the supposed millions of years of geologic history into Genesis 1:1 and then claims the six days of creation relate to the Promised Land. He states his motivation for this novel approach on p. 29: "If billions of years really are covered by the simple statement, 'In the beginning God created the heavens and the earth,' then many of the processes described by modern scientists fall into the period covered by the Hebrew term 'beginning.' Within that 'beginning' would fit the countless geologic ages, ice ages, and the many global climatic changes on our planet. The many biological eras would also fit within 'the beginning' of Genesis 1:1, including the long ages during which the dinosaurs roamed the earth. By the time human beings were created on the sixth day of the week, the dinosaurs already could have flourished and become extinct—all during the 'beginning' recorded in Genesis 1:1." Many of the problems with the classical gap theory also apply to this attempt to fit millions of years into the Bible.

6

Cain's Wife—Who Was She?

KEN HAM

Is She the Most-Talked-About Wife in History?

We don't even know her name, yet she was discussed at the Scopes Trial, mentioned in the movies *Inherit the Wind*[1] and *Contact*,[2] and talked about in countries all over the world for hundreds of years.

Skeptics of the Bible have used Cain's wife time and again to try to discredit the book of Genesis as a true historical record. Sadly, most Christians have not given an adequate answer to this question. As a result, the world sees them as not being able to defend the authority of Scripture and thus the Christian faith.

For instance, at the historic Scopes Trial in Tennessee in 1925, William Jennings Bryan, the prosecutor who stood for the Christian faith, failed to answer the question about Cain's wife posed by the ACLU lawyer Clarence Darrow. Consider the following excerpt from the trial record as Darrow interrogates Bryan:

Q—Did you ever discover where Cain got his wife?

A—No, sir; I leave the agnostics to hunt for her.

---

1 This is a Hollywood version of the famous Scopes Trial. K. Ham, The wrong way round! *Creation* **8**(3):38-41, 1996; D. Menton, Inherit the Wind: an historical analysis, *Creation* **19**(1):35-38, 1997.

2 *Contact*, Warner Bros., released July 11, 1997. Based on Carl Sagan's *Contact*, Pocket Books, New York, 1985.

Q—You have never found out?

A—I have never tried to find.

Q—You have never tried to find?

A—No.

Q—The Bible says he got one doesn't it? Were there other people on the earth at that time?

A—I cannot say.

Q—You cannot say. Did that ever enter your consideration?

A—Never bothered me.

Q—There were no others recorded, but Cain got a wife.

A—That is what the Bible says.

Q—Where she came from you do not know.[3]

The world's press was focused on this trial, and what they heard has affected Christianity to this day—Christians can't defend the biblical record!

In recent times, this same example was taken up by Carl Sagan in his book *Contact* [2] (which was on the *New York Times* best-seller list) and used in the movie of the same name based upon this work.

In the book, we read the fictional character Ellie's account of how she could not get answers from a minister's wife, who was the leader of a church discussion group:

> Ellie had never seriously read the Bible before ... . So over the weekend preceding her first class, she read through what seemed to be the important parts of the Old Testament, trying to keep an open mind. She at once recognized that there were two different and mutually contradictory stories of Creation ... and had trouble figuring out exactly who it was that Cain had married.[4]

Sagan cleverly listed a number of common questions (including Cain's wife) that are often directed at Christians in an attempt to supposedly

[3] *The World's Most Famous Court Trial, Tennessee Evolution Case* (a word-for-word report), Bryan College (reprinted from the original edition), p. 302, 1990.

[4] C. Sagan, *Contact*, Pocket Books, New York, 1985, 19–20.

prove the Bible is full of contradictions and can't be defended. The truth is—most Christians probably couldn't answer these questions. And yet there *are* answers. But since churches lack in the teaching of apologetics,[5] particularly in regard to the book of Genesis, most believers in the church are not able to be "always be ready to give a defense to everyone who asks you a reason for the hope that is in you, with meekness and fear" (1 Peter 3:15).

Why Is It Important?

Many skeptics have claimed that for Cain to find a wife, there must have been other "races" of people on the earth who were not descendants of Adam and Eve. To many people, this question is a stumbling block to accepting the creation account of Genesis and its record of only one man and woman at the beginning of history. Defenders of the gospel must be able to show that all human beings are descendants of one man and one woman (Adam and Eve) because only descendants of Adam and Eve can be saved. Thus, believers need to be able to account for Cain's wife and show clearly she was a descendant of Adam and Eve.

[5] Apologetics—from the Greek word, ἀπολογία *(apologia),* meaning "to give a defense." The field of Christian apologetics covers the ability of Christians to give a defense of their faith in Jesus Christ and their hope in Him for salvation, as expressed in 1 Peter 3:15. This ability requires a thorough knowledge of Scripture, including the doctrines of the creation, Original Sin, Curse, Flood, Virgin Birth, life and ministry of Jesus of Nazareth, the Cross, Crucifixion, Resurrection, Ascension, promise of the Second Coming, and a new heaven and new earth. Then one needs to be able to explain logically and clearly these various doctrines in a way that justifies one's faith and hope in Jesus Christ.

In order to answer this question of where Cain got his wife, we first need to cover some background information concerning the meaning of the gospel.

The First Man

"Wherefore, as by one man sin entered into the world, and death by sin; and so death passed upon all men, for that all have sinned" (Romans 5:12).

We read in 1 Corinthians 15:45 that Adam was "the first man." God did not start by making a race of men.

The Bible makes it clear that *only* the descendants of Adam can be saved. Romans 5 teaches that we sin because Adam sinned. The death penalty, which Adam received as judgment for his sin of rebellion, has also been passed on to all his descendants.

Since Adam was the head of the human race, when he fell we who were in the loins of Adam fell also. Thus, we are all separated from God. The final consequence of sin would be separation from God in our sinful state forever. However, the good news is that there is a way for us to return to God.

Because a man brought sin and death into the world, the human race, who are all descendants of Adam, needed a sinless Man to pay the penalty for sin and the resulting judgment of death. However, the Bible teaches that "all have sinned" (Romans 3:23). What was the solution?

Jesus Christ
"The Last Adam"
1 Cor. 15:45
For as in Adam all die,
even so in Christ shall all be made alive.
1 Corinthians 15:22
"The First Adam"
1 Cor. 15:45
Which Adam is "non-essential" to the Gospel?

The Last Adam

God provided the solution—a way to deliver man from his wretched state. Paul explains in 1 Corinthians 15 that God provided another Adam. The Son of God became a man—a *perfect* Man—yet still our relation. He is called "the last Adam" (1 Corinthians 15:45) because he took the place of the first Adam. He became the new head and, because He was sinless, was able to pay the penalty for sin:

"For since by [a] man came death, by [a] Man also came the resurrection of the dead. For as in Adam all die, even so in Christ all shall be made alive" (1 Corinthians 15:21–22).

Christ suffered death (the penalty for sin) on the Cross, shedding His blood ("and without shedding of blood there is no remission," Hebrews 9:22) so that those who put their trust in His work on the Cross can come in repentance of their sin of rebellion (in Adam) and be reconciled to God.

Thus, only descendants of the first man Adam can be saved.

All Related

Since the Bible describes *all* human beings as sinners, and we are *all* related ("And He has made from one blood every nation of men to dwell on all the face of the earth," Acts 17:26), the gospel makes sense only on the basis that all humans alive and all that have ever lived (except for the first woman[6]) are descendants of the first man Adam. If this were not so, then the gospel could not be explained or defended.

Thus, there was only *one* man at the beginning—made from the dust of the earth (Genesis 2:7).

This also means that Cain's wife was a descendant of Adam. She couldn't have come from another race of people and must be accounted for from Adam's descendants.

The First Woman

In Genesis 3:20 we read, "And Adam called his wife's name Eve, because she was the mother of all living." In other words, all people other than Adam

[6] Eve, in a sense, was a descendant of Adam in that she was made from his flesh and thus had a direct biological connection to him (Genesis 2:21–23).

AFTER EDEN
by Dan Lietha
6
www.AnswersInGenesis.org
EVE, YOU'RE THE ONLY ONE FOR ME!
© 2000 AIG
In his first attempt to be romantic, Adam merely states the obvious.

are descendants of Eve—she was the first woman.

Eve was made from Adam's side (Genesis 2:21–24)—this was a unique event. In the New Testament, Jesus (Matthew 19:4-6) and Paul (Ephesians 5:31) use this historical and one-time event as the foundation for the marriage of one man and one woman.

Also, in Genesis 2:20, we are told that when Adam looked at the animals, he couldn't find a mate—there was no one of his kind.

All this makes it obvious that there was only *one* woman, Adam's wife, from the beginning. There could not have been a "race" of women.

Thus, if Christians cannot defend that all humans, including Cain's wife, can trace their ancestry ultimately to Adam and Eve, then how can they understand and explain the gospel? How can they justify sending missionaries to every tribe and nation? Therefore, one needs to be able to explain Cain's wife, to illustrate that Christians can defend the gospel and all that it teaches.

Who Was Cain?

Cain was the first child of Adam and Eve recorded in Scripture (Genesis 4:1). He and his brothers, Abel (Genesis 4:2) and Seth (Genesis 4:25), were part of the first generation of children ever born on this earth. Even though these three males are specifically mentioned, Adam and Eve had other children.

Cain's Brothers and Sisters

In Genesis 5:4 we read a statement that sums up the life of Adam and Eve: "After he begot Seth, the days of Adam were eight hundred years; and he had sons and daughters."

During their lives, Adam and Eve had a number of male and female children. In fact, the Jewish historian Josephus wrote, "The number of

Adam's children, as says the old tradition, was thirty-three sons and twenty-three daughters."[7]

Scripture doesn't tell us how many children were born to Adam and Eve, but considering their long life spans (Adam lived for 930 years—Genesis 5:5), it would seem logical to suggest there were many. Remember, they were commanded to "be fruitful, and multiply" (Genesis 1:28).

The Wife

If we now work totally from Scripture, without any personal prejudices or other extrabiblical ideas, then back at the beginning, when there was only the first generation, brothers would have had to marry sisters or there wouldn't have been any more generations!

We're not told when Cain married or many of the details of other marriages and children, but we can say for certain that Cain's wife was either his sister or a close relative.

A closer look at the Hebrew word for "wife" in Genesis reveals something readers may miss in translation. It was more obvious to those speaking Hebrew that Cain's wife was likely his sister. (There is a slim possibility that she was his niece, but either way, a brother and sister would have married in the beginning.) The Hebrew word for "wife" used in Genesis 4:17 (the first mention of Cain's wife) is *ishshah*, and it means "woman/wife/female."

> And Cain knew his wife [*ishshah*], and she conceived and bore Enoch. And he built a city, and called the name of the city after the name of his son—Enoch (Genesis 4:17).

The word *ishshah* is the word for "woman," and it means "from man." It is a derivation of the Hebrew word *iysh* and *enowsh,* which both mean "man." This can be seen in Genesis 2:23 where the name "woman" (*ishshah*) is given to one who came from Adam.

---

[7] F. Josephus, *The Complete Works of Josephus*, translated by W. Whiston, Kregel Publications, Grand Rapids, Michigan, 1981, 27.

> And Adam said: "This is now bone of my bones and flesh of my flesh; She shall be called Woman [*ishshah*], because she was taken out of Man [*iysh*]" (Genesis 2:23).

Thus, Cain's wife is a descendant of Adam/man. Therefore, she had to be his sister (or possibly niece). Hebrew readers should be able to make this connection easier; however, much is lost when translated.

Objections

GOD'S LAWS

Many people immediately reject the conclusion that Adam and Eve's sons and daughters married each other by appealing to the law against brother-sister marriage. Some say that you can't marry your relation. Actually, if you don't marry your relation, you don't marry a human! A wife is related to her husband before they are married because all people are descendants of Adam and Eve—all are of *one blood.* This law forbidding *close* relatives marrying was not given until the time of Moses (Leviticus 18–20). Provided marriage was one man for one woman for life (based on Genesis 1–2), there was no disobedience to God's law originally (before the time of Moses) when close relatives (even brothers and sisters) married each other.

Remember that Abraham was married to his half-sister (Genesis 20:12).[8] God's law forbade such marriages,[9] but that was some four hundred years later at the time of Moses.

BIOLOGICAL DEFORMITIES

Today, brothers and sisters (and half-brothers and half-sisters, etc.) are not currently permitted by law to marry and have children.

Now it is true that children produced in a union between brother and sister have a greater chance to be deformed. As a matter of fact, the closer the couple are in relationship, the *more* likely it is that any offspring will be deformed. It is very easy to understand this without going into all the technical details.

Each person inherits a set of genes from his or her mother and father. Unfortunately, genes today contain many mistakes (because of sin

---

[8] Another example would be Isaac's wife, Rebekah—she was Isaac's second cousin (Genesis 24:15).

[9] Leviticus 18–20.

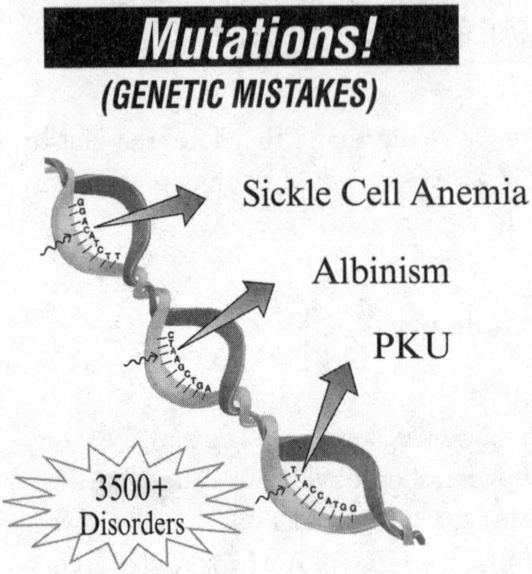

and the Curse), and these mistakes show up in a variety of ways. For instance, people let their hair grow over their ears to hide the fact that one ear is lower than the other. Or perhaps someone's nose is not quite in the middle of his or her face, or someone's jaw is a little out of shape. Let's face it, the main reason we call each other normal is because of our common agreement to do so!

The more closely related two people are, the more likely it is that they will have similar mistakes in their genes, inherited from the same parents. Therefore, brother and sister are likely to have similar mistakes in their genetic material. If there were to be a union between these two that produces offspring, children would inherit one set of genes from each of their parents. Because the genes probably have similar mistakes, the mistakes pair together and result in deformities in the children.

Conversely, the further away the parents are in relationship to each other, the more likely it is that they will have different mistakes in their genes. Children, inheriting one set of genes from each parent, are likely to end up with some of the pairs of genes containing only one bad gene in each pair. The good gene tends to override the bad so that a deformity (a serious one, anyway) does not occur. Instead of having totally deformed ears, for instance, a person may have only crooked ones. (Overall, though, the human race is slowly degenerating as mistakes accumulate generation after generation.)

However, this fact of present-day life did not apply to Adam and Eve. When the first two people were created, they were perfect. Everything God made was "very good" (Genesis 1:31). That means their genes were perfect—no mistakes. But when sin entered the world because of Adam (Genesis 3:6), God cursed the world so that the perfect creation then began to degenerate, that is, suffer death and decay (Romans 8:22). Over a long period of time, this degeneration would have resulted in all sorts of mistakes occurring in the genetic material of living things.

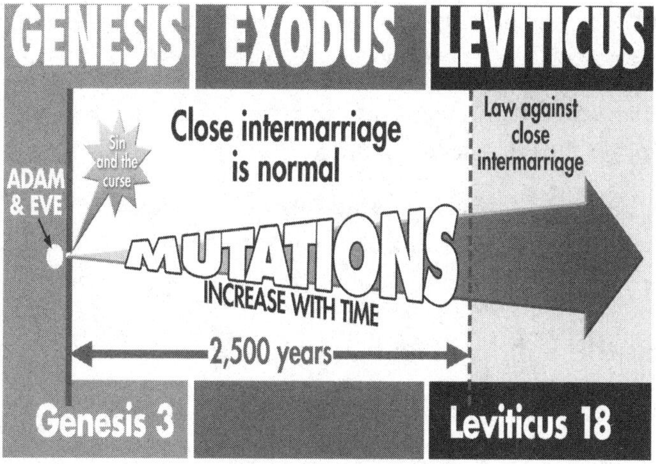

But Cain was in the first generation of children ever born. He, as well as his brothers and sisters, would have received virtually no imperfect genes from Adam or Eve, since the effects of sin and the Curse would have been minimal to start with. In that situation, brother and sister could have married (provided it was one man for one woman, which is what marriage is all about, Matthew 19:4–6) without any potential to produce deformed offspring.

By the time of Moses (about 2,500 years later), degenerative mistakes would have accumulated to such an extent in the human race that it would have been necessary for God to bring in the laws forbidding brother-sister (and close relative) marriage (Leviticus 18–20).[10]

(Also, there were plenty of people on the earth by now, and there was no reason for close relations to marry.)

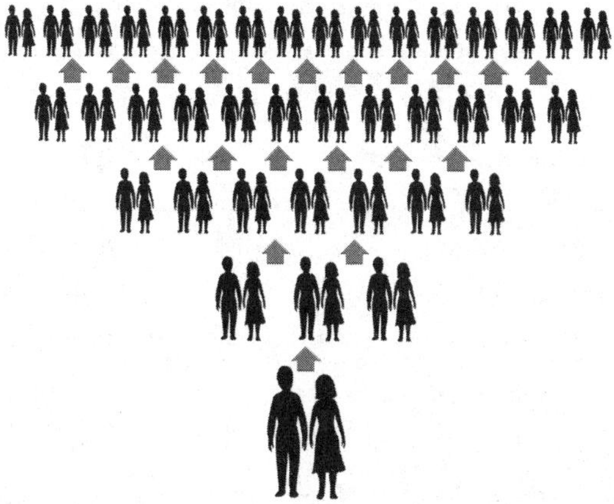

In all, there appear to be three interrelated reasons for the introduction of laws forbidding close intermarriage:

[10] Some have claimed this means God changed His mind by changing the laws. But God didn't change His mind—because of the changes that sin brought and because God never changes, He introduced new laws for our sake.

1. As we have already discussed, there was the need to protect against the increasing potential to produce deformed offspring.
2. God's laws were instrumental in keeping the Jewish nation strong, healthy, and within the purposes of God.
3. These laws were a means of protecting the individual, the family structure, and society at large. The psychological damage caused by incestuous relationships should not be minimized.

Cain and the Land of Nod

Some claim that the passage in Genesis 4:16–17 means that Cain went to the land of Nod and found a wife. Thus, they conclude there must have been another race of people on the earth who were not descendants of Adam, who produced Cain's wife.

> Then Cain went out from the presence of the LORD and dwelt in the land of Nod on the east of Eden. And Cain knew his wife, and she conceived and bore Enoch. And he built a city, and called the name of the city after the name of his son—Enoch.

From what has been stated above, it is clear that *all* humans, Cain's wife included, are descendants of Adam. However, this passage does not say that Cain went to the land of Nod and found a wife. John Calvin in commenting on these verses states:

> From the context we may gather that Cain, before he slew his brother, had married a wife; otherwise Moses would now have related something respecting his marriage.[11]

Cain was married *before* he went to the land of Nod. He didn't find a wife there but "knew" (had sexual relations with) his wife.[12]

This makes sense in light of what Nod is, too. Nod means "wandering" in Hebrew. So when Cain went to the land of Nod, he was literally going to the land of wandering, not a place full of people.

[11] J. Calvin, *Commentaries on The First Book of Moses Called Genesis*, Vol. 1, reprinted, Baker House, Grand Rapids, Michigan, 1979, 215.

[12] Even if Calvin's suggestion concerning this matter is not correct, there was still plenty of time for numerous descendants of Adam and Eve to move out and settle areas such as the land of Nod.

Who was Cain Fearful of (Genesis 4:14)?

Some claim that there had to be lots of people on the earth other than Adam and Eve's descendants; otherwise Cain wouldn't have been fearful of people wanting to slay him because he killed Abel.

First of all, one reason that someone would want to harm Cain for killing Abel is if that person was a close relation of Abel!

Secondly, Cain and Abel were born quite some time before the event of Abel's death. Genesis 4:3 states:

> And in the process of time it came to pass that Cain brought an offering of the fruit of the ground to the LORD.

Note the phrase "in the process of time." We know Seth was born when Adam was 130 years old (Genesis 5:3), and Eve saw him as a replacement for Abel (Genesis 4:25). Therefore, the time period from Cain's birth to Abel's death may have been 100 years or more—allowing plenty of time for other children of Adam and Eve to marry and have children. By the time Abel was killed, there may have been a considerable number of descendants of Adam and Eve involving several generations.

Where Did the Technology Come From?

Some claim that for Cain to go to the land of Nod and build a city, he would have required a lot of technology that must have already been in that land, presumably developed by other races.

Adam and Eve's descendants were very intelligent people. We are told that Jubal made musical instruments, such as the harp and organ (Genesis 4:21), and Tubal-cain worked with brass and iron (Genesis 4:22).

Because of intense evolutionary indoctrination, many people today have the idea that their generation is the most advanced that has ever been on this planet. Just because we have jet airplanes and computers doesn't mean we are the most intelligent or advanced. This modern technology is really a result of the accumulation of knowledge.

We must remember that our brains have suffered from 6,000 years of the Curse. We have greatly degenerated compared to people many generations ago. We may be nowhere near as intelligent or inventive as Adam and Eve's children. Scripture gives us a glimpse of what appears to be advanced technology almost from the beginning.

Cain had the knowledge and talent to know how to build a city!

Conclusion

One of the reasons many Christians cannot answer the question about Cain's wife is that they tend to look at today's world and the problems that would be associated with close relations marrying, and they do not look at the clear historical record God has given to us.

They try to interpret Genesis from our present situation rather than understand the true biblical history of the world and the changes that have occurred because of sin. Because they are not building their worldview on Scripture but taking a secular way of thinking to the Bible, they are blinded to the simple answers.

Genesis is the record of the God who was there as history happened. It is the Word of One who knows everything and who is a reliable Witness from the past. Thus, when we use Genesis as a basis for understanding history, we can make sense of evidence which would otherwise be a real mystery. You see, if evolution is true, science has an even bigger problem than Cain's wife to explain—namely, how could man ever evolve by mutations (mistakes) in the first place, since that process would have made everyone's children deformed? The mere fact that people can produce offspring that are *not* largely deformed is a testimony to creation, not evolution.

7

Doesn't Carbon-14 Dating Disprove the Bible?

MIKE RIDDLE

Scientists use a technique called radiometric dating to estimate the ages of rocks, fossils, and the earth. Many people have been led to believe that radiometric dating methods have proved the earth to be billions of years old. This has caused many in the church to reevaluate the biblical creation account, specifically the meaning of the word "day" in Genesis 1. With our focus on one particular form of radiometric dating—carbon dating—we will see that carbon dating strongly supports a young earth. Note that, contrary to a popular misconception, carbon dating is not used to date rocks at millions of years old.

Basics

Before we get into the details of how radiometric dating methods are used, we need to review some preliminary concepts from chemistry. Recall that atoms are the basic building blocks of matter. Atoms are made up of much smaller particles called protons, neutrons, and electrons. Protons and neutrons make up the center (nucleus) of the atom, and electrons form shells around the nucleus.

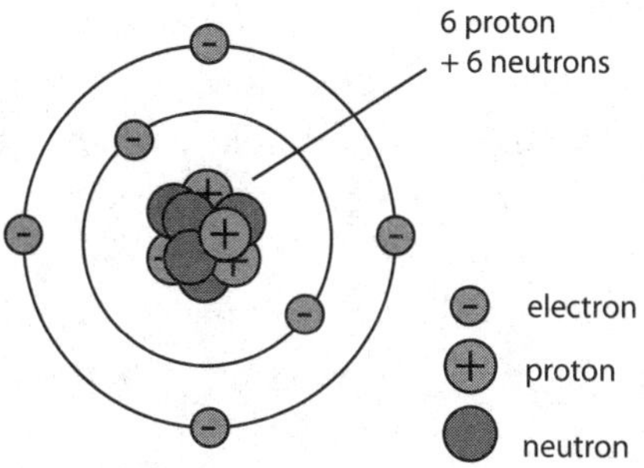

Carbon atom

The number of protons in the nucleus of an atom determines the element. For example, all carbon atoms have 6 protons, all atoms of nitrogen have 7 protons, and all oxygen atoms have 8 protons. The number of neutrons in the nucleus can vary in any given type of atom. So, a carbon atom might have six neutrons, or seven, or possibly eight—but it would always have six protons. An "isotope" is any of several different forms of an element, each having different numbers of neutrons. The illustration below shows the three isotopes of carbon.

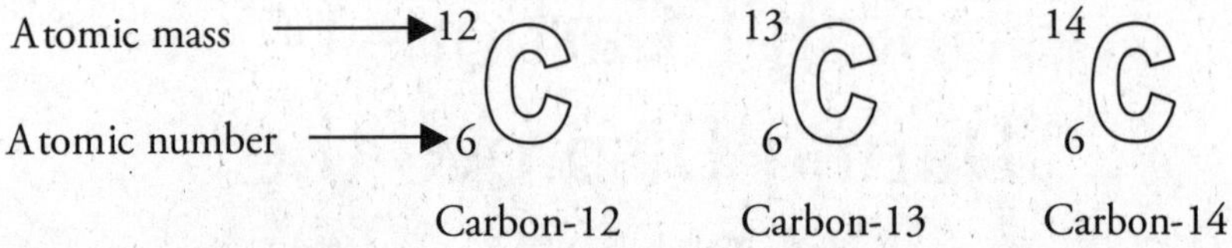

The atomic number corresponds to the number of protons in an atom. Atomic mass is a combination of the number of protons and neutrons in the nucleus. (The electrons are so much lighter that they do not contribute significantly to the mass of an atom.)

Some isotopes of certain elements are unstable; they can spontaneously change into another kind of atom in a process called "radioactive decay." Since this process presently happens at a known measured rate, scientists attempt to use it like a "clock" to tell how long ago a rock or fossil formed. There are two main applications for radiometric dating. One is for potentially dating fossils (once-living things) using carbon-14 dating, and the other is for dating rocks and the age of the earth using uranium, potassium and other radioactive atoms.

Carbon-14 Dating

Carbon-14 (^{14}C), also referred to as radiocarbon, is claimed to be a reliable dating method for determining the ages of fossils up to 50,000 to 60,000 years. If this claim is true, the biblical account of a young earth (about 6,000 years) is in question, since ^{14}C dates of tens of thousands of years are common.[1]

When a scientist's interpretation of data does not match the clear meaning of the text in the Bible, we should never reinterpret the Bible. God knows just what He meant to say, and His understanding of science is infallible, whereas ours is fallible. So we should never think it necessary to modify His Word. Genesis 1 defines the days of creation to be literal days (a number with the word "day" always means a normal day in the Old Testament, and the

[1] *Earth Science* (Teachers Edition), Prentice Hall, 2002, p. 301.

phrase "evening and morning" further defines the days as literal days). Since the Bible is the inspired Word of God, we should examine the validity of the standard interpretation of ^{14}C dating by asking several questions:

1. Is the explanation of the data derived from empirical, observational science, or an interpretation of past events (historical science)?
2. Are there any assumptions involved in the dating method?
3. Are the dates provided by ^{14}C dating consistent with what we observe?
4. Do all scientists accept the ^{14}C dating method as reliable and accurate?

All radiometric dating methods use scientific procedures in the present to interpret what has happened in the past. The procedures used are not necessarily in question. The interpretation of past events is in question. The secular (evolutionary) worldview interprets the universe and world to be billions of years old. The Bible teaches a young universe and earth. Which worldview does science support? Can carbon-14 dating help solve the mystery of which worldview is more accurate?

The use of carbon-14 dating is often misunderstood. Carbon-14 is mostly used to date once-living things (organic material). It cannot be used directly to date rocks; however, it can potentially be used to put time constraints on some inorganic material such as diamonds (diamonds could contain carbon-14). Because of the rapid rate of decay of ^{14}C, it can only give dates in the thousands-of-year range and not millions.

There are three different naturally occurring varieties (isotopes) of carbon: ^{12}C, ^{13}C, and ^{14}C. Carbon-14 is used for dating because it is unstable (radioactive), whereas ^{12}C and ^{13}C are stable. Radioactive means that ^{14}C will decay (emit radiation) over time and become a different element. During this process (called "beta decay") a neutron in the ^{14}C atom will be converted into a proton and an electron. By losing one neutron and gaining one proton, ^{14}C is changed into nitrogen-14 (^{14}N = 7 protons and 7 neutrons).

If ^{14}C is constantly decaying, will the earth eventually run out of ^{14}C? The answer is no. Carbon-14 is constantly being added to the atmosphere. Cosmic rays from outer space, which contain high levels of energy, bombard the earth's upper atmosphere.

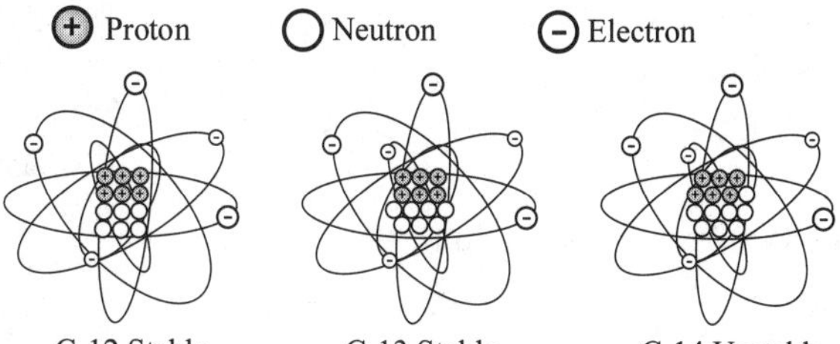

C-12 Stable C-13 Stable C-14 Unstable

These cosmic rays collide with atoms in the atmosphere and can cause them to come apart. Neutrons that come from these fragmented atoms collide with ^{14}N atoms (the atmosphere is made mostly of nitrogen and oxygen) and convert them into ^{14}C atoms (a proton changes into a neutron).

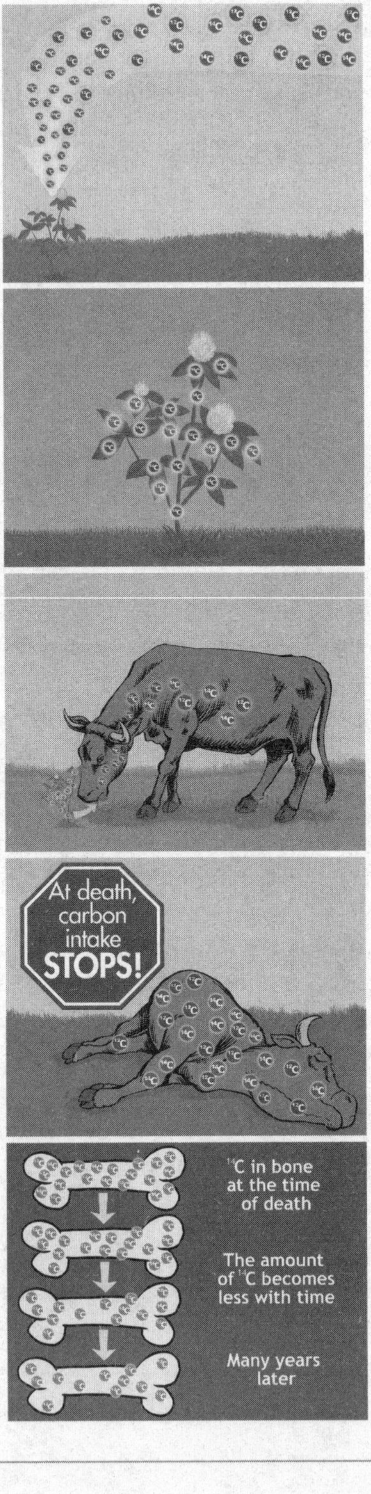
At death, carbon intake STOPS!
^{14}C in bone at the time of death
The amount of ^{14}C becomes less with time
Many years later

Once ^{14}C is produced, it combines with oxygen in the atmosphere (^{12}C behaves like ^{14}C and also combines with oxygen) to form carbon dioxide (CO_2). Because CO_2 gets incorporated into plants (which means the food we eat contains ^{14}C and ^{12}C), all living things should have the same ratio of ^{14}C and ^{12}C in them as in the air we breathe.

HOW THE CARBON-14 DATING PROCESS WORKS

Once a living thing dies, the dating process begins. As long as an organism is alive it will continue to take in ^{14}C; however, when it dies, it will stop. Since ^{14}C is radioactive (decays into ^{14}N), the amount of ^{14}C in a dead organism gets less and less over time. Therefore, part of the dating process involves measuring the amount of ^{14}C that remains after some has been lost (decayed). Scientists now use a device called an "Accelerator Mass Spectrometer" (AMS) to determine the ratio of ^{14}C to ^{12}C, which increases the assumed accuracy to about 80,000 years. In order to actually do the dating, other things need to be known. Two such things include the following questions:

1. How fast does ^{14}C decay?
2. What was the starting amount of ^{14}C in the creature when it died?

The decay rate of radioactive elements is described in terms of half-life. The half-life of an atom is the amount of time it takes for half of the atoms in a sample to decay. The half-life of ^{14}C is 5,730 years. For example, a jar starting full of ^{14}C atoms at time zero will contain half ^{14}C atoms and half ^{14}N atoms at the end of 5,730 years (one half-life). At the end of 11,460 years (two half-lives) the jar will contain one-quarter ^{14}C atoms and three-quarter ^{14}N atoms.

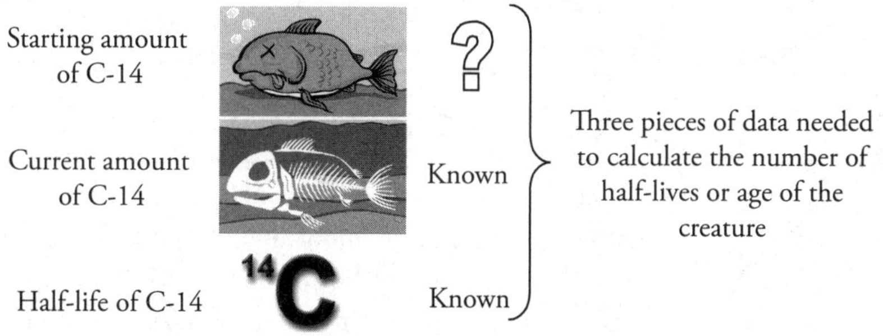

Since the half-life of ^{14}C is known (how fast it decays), the only part left to determine is the starting amount of ^{14}C in a fossil. If scientists know the original amount of ^{14}C in a creature when it died, they can measure the current amount and then calculate how many half-lives have passed. Since no one was there to measure the amount of ^{14}C when a creature died, scientists need to find a method to determine how much ^{14}C has decayed.

To do this, scientists use the main isotope of carbon, called carbon-12 (^{12}C). Because ^{12}C is a stable isotope of carbon, it will remain constant; however, the amount of ^{14}C will decrease after a creature dies. All living things take in carbon (^{14}C and ^{12}C) from eating and breathing. Therefore, the ratio of ^{14}C to ^{12}C in living creatures will be the same as in the atmosphere. This ratio turns out to be about one ^{14}C atom for every 1 trillion ^{12}C atoms. Scientists can use this ratio to help determine the starting amount of ^{14}C.

When an organism dies, this ratio (1 to 1 trillion) will begin to change. The amount of ^{12}C will remain constant, but the amount of ^{14}C will become less and less. The smaller the ratio, the longer the organism has been dead. The following illustration demonstrates how the age is estimated using this ratio.

Percent ^{14}C Remaining	Percent ^{12}C Remaining	Ratio	Number of Half-Lives	Years Dead (Age of Fossil)
100	100	1 to 1T	0	0
50	100	1 to 2T	1	5,730
25	100	1 to 4T	2	11,460
12.5	100	1 to 8T	3	17,190
6.25	100	1 to 16T	4	22,920
3.125	100	1 to 32T	5	28,650

T = Trillion

A Critical Assumption

A critical assumption used in carbon-14 dating has to do with this ratio. It is assumed that the ratio of ^{14}C to ^{12}C in the atmosphere has always been the same as it is today (1 to 1 trillion). If this assumption is true, then the AMS ^{14}C dating method is valid up to about 80,000 years. Beyond this number, the instruments scientists use would not be able to detect enough remaining ^{14}C to be useful in age estimates. This is a critical assumption in the dating process. If this assumption is not true, then the method will give incorrect dates. What could cause this ratio to change? If the production rate of ^{14}C in the atmosphere is not equal to the removal rate (mostly through decay), this ratio will change. In other words, the amount of ^{14}C being produced in the atmosphere must equal the amount being removed to be in a steady state (also called "equilibrium"). If this is not true, the ratio of ^{14}C to ^{12}C is not a constant, which would make knowing the starting amount of ^{14}C in a specimen difficult or impossible to accurately determine.

Dr. Willard Libby, the founder of the carbon-14 dating method, assumed this ratio to be constant. His reasoning was based on a belief in evolution, which assumes the earth must be billions of years old. Assumptions in the scientific community are extremely important. If the starting assumption is false, all the calculations based on that assumption might be correct but still give a wrong conclusion. In Dr. Libby's original work, he noted that the atmosphere did not appear to be in equilibrium. This was a troubling idea for Dr. Libby since he believed the world was billions of years old and enough time had passed to achieve equilibrium. Dr. Libby's calculations showed that if the earth started with no ^{14}C in the atmosphere, it would take up to 30,000 years to build up to a steady state (equilibrium).

> If the cosmic radiation has remained at its present intensity for 20,000 or 30,000 years, and if the carbon reservoir has not changed appreciably in this time, then there exists at the present time a complete balance between the rate of disintegration of radiocarbon atoms and the rate of assimilation of new radiocarbon atoms for all material in the life-cycle.[2]

Dr. Libby chose to ignore this discrepancy (nonequilibrium state), and he attributed it to experimental error. However, the discrepancy has turned out to be very real. The ratio of ^{14}C /^{12}C is not constant.

> The Specific Production Rate (SPR) of C-14 is known to be 18.8 atoms per gram of total carbon per minute. The Specific Decay Rate (SDR) is known to be only 16.1 disintegrations per gram per minute.[3]

What does this mean? If it takes about 30,000 years to reach equilibrium and ^{14}C is still out of equilibrium, then maybe the earth is not very old.

Magnetic Field of the Earth

Other factors can affect the production rate of ^{14}C in the atmosphere. The earth has a magnetic field around it which helps protect us from harmful radiation from outer space. This magnetic field is decaying (getting weaker). The stronger the field is around the earth, the fewer the number of cosmic rays that are able to reach the atmosphere. This would result in a smaller production of ^{14}C in the atmosphere in earth's past.

> The cause for the long term variation of the C-14 level is not known. The variation is certainly partially the result of a change in the cosmic ray production rate of radiocarbon. The cosmic-ray flux, and hence the production rate of C-14, is a function not only of the solar activity but also of the magnetic dipole moment of the Earth.[4]

> Though complex, this history of the earth's magnetic field agrees with Barnes' basic hypothesis, that the field has always freely decayed .... The field has always been losing energy despite its variations, so it cannot be more than 10,000 years old.[5]

---

[2] W. Libby, *Radiocarbon Dating*, Univ. of Chicago Press, Chicago, Illinois, 1952, 8.

[3] C. Sewell, "Carbon-14 and the Age of the Earth," 1999, www.rae.org/pdf/bits23.pdf.

[4] M. Stuiver and H. Suess, On the relationship between radiocarbon dates and true sample ages, *Radiocarbon* vol. 8, 1966, 535.

[5] D.R. Humphreys, The mystery of earth's magnetic field, ICR *Impact #292*, Feb 1, 1989. www.icr.org/article/292.

> Earth's magnetic field is fading. Today it is about 10 percent weaker than it was when German mathematician Carl Friedrich Gauss started keeping tabs on it in 1845, scientists say.[6]

If the production rate of ^{14}C in the atmosphere was less in the past, dates given using the carbon-14 method would incorrectly assume that more ^{14}C had decayed out of a specimen than what has actually occurred. This would result in giving older dates than the true age.

Genesis Flood

What role might the Genesis Flood have played in the amount of carbon? The Flood would have buried large amounts of carbon from living organisms (plant and animal) to form today's fossil fuels (coal, oil, etc.). The amount of fossil fuels indicates there must have been a vastly larger quantity of vegetation in existence prior to the Flood than exists today. This means that the biosphere just prior to the Flood might have had 500 times more carbon in living organisms than today. This would further dilute the amount of ^{14}C and cause the $^{14}C / ^{12}C$ ratio to be much smaller than today.

> If that were the case, and this C-14 were distributed uniformly throughout the biosphere, and the total amount of biosphere C were, for example, 500 times that of today's world, the resulting C-14/C-12 ratio would be 1/500 of today's level … .[7]

When the Flood is taken into account, along with the decay of the magnetic field, it is reasonable to believe that the assumption of equilibrium is a false assumption. Because of this false assumption, any age estimates using ^{14}C on organic material that dates from prior to the Flood will give much older dates than the true ages. Pre-Flood organic materials would be dated at perhaps ten times the true age.

The RATE Group Findings

In 1997 an eight-year research project was started to investigate the age of the earth. The group was called the RATE group (Radioisotopes and the Age of The Earth). The team of scientists included:

---

[6] J. Roach, *National Geographic News*, September 9, 2004.

[7] J.R. Baumgarder, C-14 evidence for a recent global Flood and a young earth, in L. Vardiman, A.A. Snelling, and E.F. Chaffin (Eds.), *Radioisotopes and the Age of the Earth: Results of a Young-Earth Creationist Research Initiative*, Institute for Creation Research, Santee, California, and Creation Research Society, Chino Valley, Arizona, 2005, 618.

Larry Vardiman, PhD Atmospheric Science
Russell Humphreys, PhD Physics
Eugene Chaffin, PhD Physics
Donald DeYoung, PhD Physics
John Baumgardner, PhD Geophysics
Steven Austin, PhD Geology
Andrew Snelling, PhD Geology
Steven Boyd, PhD Hebraic and Cognate Studies

The objective was to gather data commonly ignored or censored by evolutionary standards of dating. The scientists reviewed the assumptions and procedures used in estimating the ages of rocks and fossils. The results of the carbon-14 dating demonstrated serious problems for long geologic ages. For example, a series of fossilized wood samples that conventionally have been dated according to their host strata to be from Tertiary to Permian (40-250 million years old) all yielded significant, detectable levels of carbon-14 that would conventionally equate to only 30,000-45,000 years "ages" for the original trees.[8] Similarly, a survey of the conventional radiocarbon journals resulted in more than forty examples of supposedly ancient organic materials, including limestones, that contained carbon-14, as reported by leading laboratories.[9]

Samples were then taken from ten different coal layers that, according to evolutionists, represent different time periods in the geologic column (Cenozoic, Mesozoic, and Paleozoic). The RATE group obtained these ten coal samples from the U.S. Department of Energy Coal Sample Bank, from samples collected from major coalfields across the United States. The chosen coal samples, which dated millions to hundreds of millions of years old based on standard evolution time estimates, all contained measurable amounts of ^{14}C. In all cases, careful precautions were taken to eliminate any possibility of contamination from other sources. Samples, in all three "time periods," displayed

[8] A.A. Snelling, Radioactive "dating" in conflict! Fossil wood in ancient lava flow yields radiocarbon, *Creation Ex Nihilo* **20**(1):24–27, 1997; A.A. Snelling, Stumping old-age dogma: Radiocarbon in an "ancient" fossil tree stump casts doubt on traditional rock/fossil dating, *Creation Ex Nihilo* **20**(4):48–51, 1998; A.A. Snelling, Dating dilemma: Fossil wood in ancient sandstone: *Creation Ex Nihilo* **21**(3):39–41, 1992; A.A. Snelling, Geological conflict: Young radiocarbon date for ancient fossil wood challenges fossil dating, *Creation Ex Nihilo* **22**(2):44–47, 2000; A.A. Snelling, Conflicting "ages" of Tertiary basalt and contained fossilized wood, Crinum, central Queensland, Australia, *Creation Ex Nihilo Technical Journal* **14**(2):99–122, 2000.

[9] P, Giem, Carbon-14 content of fossil carbon, *Origins* **51**:6–30, 2001.

significant amounts of ^{14}C. This is a significant discovery. Since the half-life of ^{14}C is relatively short (5,730 years), there should be no detectable ^{14}C left after about 100,000 years. The average ^{14}C estimated age for all the layers from these three time periods was approximately 50,000 years. However, using a more realistic pre-Flood ^{14}C /^{12}C ratio reduces that age to about 5,000 years.

These results indicate that the entire fossil-bearing geologic column is much less than 100,000 years old—and even much younger. This confirms the Bible and challenges the evolutionary idea of long geologic ages.

> Because the lifetime of C-14 is so brief, these AMS [Accelerator Mass Spectrometer] measurements pose an obvious challenge to the standard geological timescale that assigns millions to hundreds of millions of years to this part of the rock layer.[10]

Another noteworthy observation from the RATE group was the amount of ^{14}C found in diamonds. Secular scientists have estimated the ages of diamonds to be millions to billions of years old using other radiometric dating methods. These methods are also based on questionable assumptions and are discussed elsewhere.[11] Because of their hardness, diamonds (the hardest known substance) are extremely resistant to contamination through chemical exchange. Since diamonds are considered to be so old by evolutionary

[10] J.R. Baumgardner, ibid., 587.

[11] M. Riddle, Does radiometric dating prove the earth is old?, in K.A. Ham (Ed.), *The New Answers Book*, Master Books, Green Forest, Arkansas, pp. 113–124, 2006.

standards, finding any ^{14}C in them would be strong support for a recent creation.

The RATE group analyzed twelve diamond samples for possible carbon-14 content. Similar to the coal results, all twelve diamond samples contained detectable, but lower levels of ^{14}C. These findings are powerful evidence that coal and diamonds cannot be the millions or billions of years old that evolutionists claim. Indeed, these RATE findings of detectable ^{14}C in diamonds have been confirmed independently.[12] Carbon-14 found in fossils at all layers of the geologic column, in coal and in diamonds, is evidence which confirms the biblical timescale of thousands of years and not billions.

> Because of C-14's short half-life, such a finding would argue that carbon and probably the entire physical earth as well must have a recent origin.[13]

Conclusion

All radiometric dating methods are based on assumptions about events that happened in the past. If the assumptions are accepted as true (as is typically done in the evolutionary dating processes), results can be biased toward a desired age. In the reported ages given in textbooks and other journals, these evolutionary assumptions have not been questioned, while results inconsistent with long ages have been censored. When the assumptions are evaluated and shown to be faulty, the results support the biblical account of a global Flood and young earth. Thus Christians should not be afraid of radiometric dating methods. Carbon-14 dating is really the friend of Christians, because it supports a young earth.

> The RATE scientists are convinced that the popular idea attributed to geologist Charles Lyell from nearly two centuries ago, "The present is the key to the past," is simply not valid for an earth history of millions or billions of years. An alternative interpretation of the carbon-14 data is that the earth experienced a global flood catastrophe which laid down most of the rock strata and fossils … . Whatever the source of the carbon-14, its presence in nearly every sample tested worldwide is a strong challenge to an ancient age. Carbon-14 data is now firmly on the side of the young-earth view of history.[14]

[12] R.E. Taylor, and J. Southon, Use of natural diamonds to monitor ^{14}C AMS instrument backgrounds, *Nuclear Instruments and Methods in Physics Research B* **259**:282–287, 2007.
[13] J.R. Baumgardner, ibid., 609.
[14] D. DeYoung, *Thousands…Not Billions*, Master Books, Green Forest, Arkansas, 2005, 61.

8

Could God Really Have Created Everything in Six Days?

KEN HAM

Why Is It Important?

If the days of creation are really geologic ages of millions of years, then the gospel message is undermined at its foundation because it puts death, disease, thorns, and suffering *before* the Fall. The effort to define "days" as "geologic ages" results from an erroneous approach to Scripture—reinterpreting the Word of God on the basis of the fallible theories of sinful people.

It is a good exercise to read Genesis 1 and try to put aside outside influences that may cause you to have a predetermined idea of what the word "day" may mean. Just let the words of the passage speak to you.

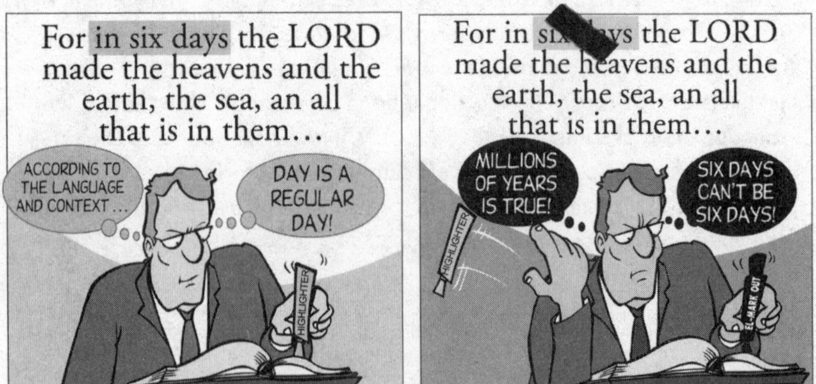

Taking Genesis 1 in this way, at face value, without doubt it says that God created the universe, the earth, the sun, moon and stars, plants and animals, and the first two people within six ordinary (approximately 24-hour) days. Being really honest, you would have to admit that you could never get the idea of millions of years from reading this passage.

The majority of Christians (including many Christian leaders) in the Western world, however, do not insist that these days of creation were ordinary-length days, and many of them accept and teach, based on outside influences, that they must have been long periods of time—even millions or billions of years.

How Does God Communicate to Us?

God communicates through language. When He made the first man, Adam, He had already "programmed" him with a language, so there could be communication. Human language consists of words used in a specific context that relates to the entire reality around us.

Thus, God can reveal things to man, and man can communicate with God, because words have meaning and convey an understandable message. If this were not so, how could any of us communicate with each other or with God?

Why "Long Days"?

Romans 3:4 declares: "Let God be true, and every man a liar."

In *every* instance where someone has not accepted the "days" of creation to be ordinary days, they have *not* allowed the words of Scripture to speak to them in context, as the language requires for communication. They have been influenced by ideas from *outside* of Scripture. Thus, they have set a precedent that could allow any word to be reinterpreted by the preconceived ideas of the person reading the words. Ultimately, this will lead to a communication breakdown, as the same words in the same context could mean different things to different people.

The Church Fathers

Most church fathers accepted the days of creation as ordinary days.[1] It is true that some of the early church fathers did not teach the days of creation as

[1] M. Van Bebber and P. Taylor, *Creation and Time: A Report on the Progressive Creationist Book by Hugh Ross,* Films for Christ, Mesa, Arizona, 1994.

ordinary days—but many of them had been influenced by Greek philosophy, which caused them to interpret the days as allegorical. They reasoned that the creation days were related to God's activities, and God being timeless meant that the days could not be related to human time.[2] In contrast to today's allegorizers, they could not accept that God took *as long as* six days.

Thus, the non-literal days resulted from extrabiblical influences (i.e., influences from *outside* the Bible), not from the words of the Bible.

This approach has affected the way people interpret Scripture to this day. As the man who started the Reformation said,

> The days of creation were ordinary days in length. We must understand that these days were actual days *(veros dies),* contrary to the opinion of the Holy Fathers. Whenever we observe that the opinions of the Fathers disagree with Scripture, we reverently bear with them and acknowledge them to be our elders. Nevertheless, we do not depart from the authority of Scripture for their sake.[3]

Today's Church Leaders

Many church leaders today do *not* accept the creation days as ordinary earth-rotation days. However, when their reasons are investigated, we find that influences from *outside* of Scripture (particularly belief in a billions-of-years-old universe) are the ultimate cause.

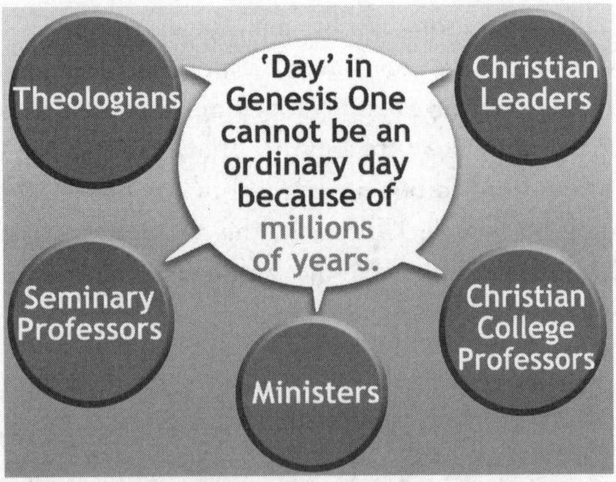

[2] G. Hasel, The "days" of creation in Genesis 1: literal "days" or figurative "periods/epochs" of time? *Origins* **21**(1):5–38, 1994.

[3] Martin Luther as cited in E. Plass, *What Martin Luther Says: A Practical In-Home Anthology for the Active Christian,* Concordia Publishing House, St. Louis, Missouri, 1991, 1523.

Again and again, such leaders admit that Genesis 1, taken in a straightforward way, seems to teach six ordinary days. But they then say that this cannot be because of the age of the universe or some other extrabiblical reason.

Consider the following representative quotes from Bible scholars who are considered to be conservative yet who do not accept the days of creation as ordinary-length days:

> From a superficial reading of Genesis 1, the impression would seem to be that the entire creative process took place in six twenty-four-hour days. ... This seems to run counter to modern scientific research, which indicates that the planet Earth was created several billion years ago.[4]

> We have shown the possibility of God's having formed the Earth and its life in a series of creative days representing long periods. In view of the apparent age of the Earth, this is not only possible—it is probable.[5]

It is as if these theologians view "nature" as a "67th book of the Bible," albeit with more authority than the 66 written books. Rather, we should consider the words of Charles Haddon Spurgeon, the renowned "prince of preachers," in 1877:

> We are invited, brethren, most earnestly to go away from the old-fashioned belief of our forefathers because of the supposed discoveries of science. What is science? The method by which man tries to conceal his ignorance. It should not be so, but so it is. You are not to be dogmatical in theology, my brethren, it is wicked; but for scientific men it is the correct thing. You are never to assert anything very strongly; but scientists may boldly assert what they cannot prove, and may demand a faith far more credulous than any we possess. Forsooth, you and I are to take our Bibles and shape and mould our belief according to the ever-shifting teachings of so-called scientific men. What folly is this! Why, the march of science, falsely so called, through the world may be traced by exploded fallacies and abandoned theories. Former explorers once adored are now ridiculed; the continual wreckings of false hypotheses is a matter of universal notoriety. You may tell where the learned have encamped by the debris left behind of suppositions and theories as plentiful as broken bottles.[6]

---

[4] G. Archer, *A Survey of Old Testament Introduction,* Moody Press, Chicago, 1994, 196–197.

[5] J. Boice, *Genesis: An Expositional Commentary,* Vol. 1, Genesis 1:1–11, Zondervan Publishing House, Grand Rapids, 1982, 68.

[6] C.H. Spurgeon, *The Sword and the Trowel,* 1877, 197.

Those who would use historical science (as propounded by people who, by and large, ignore God's written revelation) to interpret the Bible, to teach us things about God, have matters front to back. Because we are fallen, fallible creatures, we need God's written Word, illuminated by the Holy Spirit, to properly understand natural history. The respected systematic theologian Berkhof said:

> Since the entrance of sin into the world, man can gather true knowledge about God from His general revelation only if he studies it in the light of Scripture, in which the elements of God's original self-revelation, which were obscured and perverted by the blight of sin, are republished, corrected, and interpreted. ... Some are inclined to speak of God's general revelation as a second source; but this is hardly correct in view of the fact that nature can come into consideration here only as interpreted in the light of Scripture.[7]

In other words, Christians should build their thinking on the Bible, not on science.

The "Days" of Genesis 1

What does the Bible tell us about the meaning of "day" in Genesis 1? A word can have more than one meaning, depending on the context. For instance, the English word "day" can have perhaps 14 different meanings. For example, consider the following sentence: "Back in my grandfather's day, it took 12 days to drive across the country during the day."

Here the first occurrence of "day" means "time" in a general sense. The second "day," where a number is used, refers to an ordinary day, and the third refers to the daylight portion of the 24-hour period. The point is that words can have more than one meaning, depending on the context.

Back in my grandfather's day, it took 12 days to drive across the country during the day.
ARE WE THERE YET?

To understand the meaning of "day" in Genesis 1, we

---

[7] L. Berkhof, Introductory volume to *Systematic Theology*, Wm. B. Eerdmans, Grand Rapids, Michigan, 1946, 60, 96.

need to determine how the Hebrew word for "day," *yom,* is used in the context of Scripture. Consider the following:

- A typical concordance will illustrate that *yom* can have a range of meanings: a period of light as contrasted to night, a 24-hour period, time, a specific point of time, or a year.
- A classic, well-respected Hebrew-English lexicon[8] (a dictionary) has seven headings and many subheadings for the meaning of *yom*—but it defines the creation days of Genesis 1 as ordinary days under the heading "day as defined by evening and morning."
- A number and the phrase "evening and morning" are used with each of the six days of creation (Gen. 1:5, 8, 13, 19, 23, 31).
- Outside Genesis 1, *yom* is used with a number 359 times, and each time it means an ordinary day.[9] Why would Genesis 1 be the exception?[10]
- Outside Genesis 1, *yom* is used with the word "evening" or "morning"[11] 23 times. "Evening" and "morning" appear in association, but without *yom*, 38 times. All 61 times the text refers to an ordinary day. Why would Genesis 1 be the exception?[12]
- In Genesis 1:5, *yom* occurs in context with the word "night." Outside of Genesis 1, "night" is used with *yom* 53 times, and each time it means an ordinary day. Why would Genesis 1 be the exception? Even the usage of the word "light" with *yom* in this passage determines the meaning as ordinary day.[13]
- The plural of *yom*, which does not appear in Genesis 1, *can* be used to communicate a longer time period, such as "in those days."[14] Adding a

---

[8] F. Brown, S. Driver, and C. Briggs, *A Hebrew and English Lexicon of the Old Testament,* Clarendon Press, Oxford, 1951, 398.

[9] Some say that Hosea 6:2 is an exception to this because of the figurative language. However, the Hebrew idiomatic expression used, "After two days ... in the third day," meaning "in a short time," makes sense only if "day" is understood in its normal sense.

[10] J. Stambaugh, The days of creation: a semantic approach, *TJ* **5**(1):70–78, April 1991. Available online at www.answersingenesis.org/go/days.

[11] The Jews start their day in the evening (sundown followed by night), obviously based on the fact that Genesis begins the day with the "evening."

[12] Stambaugh, The days of creation: a semantic approach, 75.

[13] Ibid., 72.

[14] Ibid., 72–73.

USES OF "DAY" OUTSIDE OF GENESIS 1

"DAY" WITH NUMBER
359 times
(in plural or singular)

"EVENING" AND "MORNING" TOGETHER WITHOUT "DAY"
38 times

"EVENING" OR "MORNING" TOGETHER WITH "DAY"
23 times each

"NIGHT" WITH "DAY"
53 times

number here would be nonsensical. Clearly, in Exodus 20:11, where a number is used with "days," it unambiguously refers to six earth-rotation days.

- There are words in biblical Hebrew (such as *olam* or *qedem*) that are very suitable for communicating long periods of time, or indefinite time, but *none* of these words are used in Genesis 1.[15] Alternatively, the days or years could have been compared with grains of sand if long periods were meant.

Dr. James Barr (Regius Professor of Hebrew at Oxford University), who himself does not believe Genesis is true history, nonetheless admitted as far as the language of Genesis 1 is concerned that

> So far as I know, there is no professor of Hebrew or Old Testament at any world-class university who does not believe that the writer(s) of Gen. 1–11 intended to convey to their readers the ideas that (a) creation took place in a series of six days which were the same as the days of 24 hours we now experience (b) the figures contained in the Genesis genealogies provided by simple addition a chronology from the beginning of the world up to later stages in the biblical story (c) Noah's Flood was understood to be worldwide and extinguish all human and animal life except for those in the ark.[16]

15 Stambaugh, The days of creation: a semantic approach, 73–74.
16 J. Barr, personal letter to David Watson, April 23, 1984.

In like manner, nineteenth century liberal Professor Marcus Dods, New College, Edinburgh, said,

> If, for example, the word "day" in these chapters does not mean a period of twenty-four hours, the interpretation of Scripture is hopeless.[17]

Conclusion About "Day" in Genesis 1

If we are prepared to let the words of the language speak to us in accord with the context and normal definitions, without being influenced by outside ideas, then the word for "day" found in Genesis 1—which is qualified by a number, the phrase "evening and morning" and for Day 1 the words "light and darkness"—*obviously* means an ordinary day (about 24 hours).

In Martin Luther's day, some of the church fathers were saying that God created everything in only one day or in an instant. Martin Luther wrote,

> When Moses writes that God created Heaven and Earth and whatever is in them in six days, then let this period continue to have been six days, and do not venture to devise any comment according to which six days were one day. But, if you cannot understand how this could have been done in six days, then grant the Holy Spirit the honor of being more learned than you are. For you are to deal with Scripture in such a way that you bear in mind that God Himself says what is written. But since God is speaking, it is not fitting for you wantonly to turn His Word in the direction you wish to go.[18]

Similarly, John Calvin stated, "Albeit the duration of the world, now declining to its ultimate end, has not yet attained six thousand years. … God's work was completed not in a moment but in six days."[19]

Luther and Calvin were the backbone of the Protestant Reformation that called the church back to Scripture—*Sola Scriptura* (Scripture alone). Both of these men were adamant that Genesis 1 taught six ordinary days of creation—only thousands of years ago.

---

[17] M. Dods, *Expositor's Bible,* T & T Clark, Edinburgh, 1888, 4, as cited by D. Kelly, *Creation and Change,* Christian Focus Publications, Fearn, Scotland, 1997, 112.

[18] Plass, *What Martin Luther Says: A Practical In-Home Anthology for the Active Christian,* 1523.

[19] J. McNeil, Ed., *Calvin: Institutes of the Christian Religion 1,* Westminster Press, Louisville, Kentucky, 1960, 160–161, 182.

Why Six Days?

Exodus 31:12 says that God commanded Moses to say to the children of Israel:

> Six days may work be done, but on the seventh is the sabbath of rest, holy to the Lord. Whoever does any work in the Sabbath day, he shall surely be put to death. Therefore the sons of Israel shall keep the Sabbath, to observe the Sabbath throughout their generations, for an everlasting covenant. It is a sign between me and the sons of Israel forever. For in six days the Lord made the heavens and the earth, and on the seventh day He rested, and was refreshed (Exodus 31:15–17).

Then God gave Moses two tablets of stone upon which were written the commandments of God, written by the finger of God (Exodus 31:18).

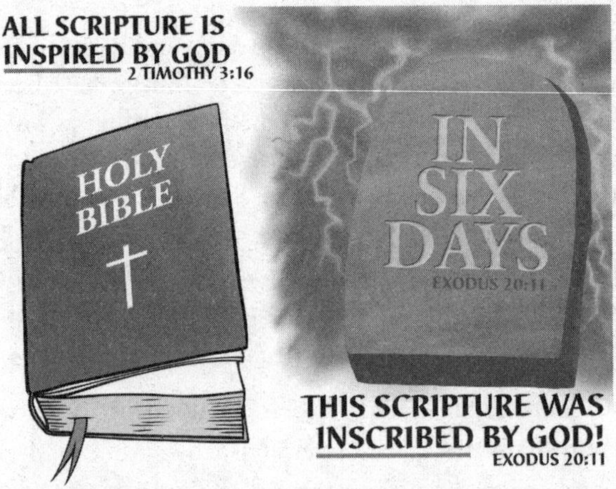

Because God is infinite in power and wisdom, there's no doubt He could have created the universe and its contents in no time at all, or six seconds, or six minutes, or six hours—after all, with God nothing shall be impossible (Luke 1:37).

However, the question to ask is, "Why did God take so long? Why as long as six days?" The answer is also given in Exodus 20:11, and that answer is the basis of the Fourth Commandment:

> For *in* six days the LORD made the heavens and the earth, the sea, and all that *is* in them, and rested the seventh day. Therefore the LORD blessed the Sabbath day and hallowed it.

The seven-day week has no basis outside of Scripture. In this Old Testament passage, God commands His people, Israel, to work for six days and

rest for one—thus giving us a reason why He deliberately took as long as six days to create everything. He set the example for man. Our week is patterned after this principle. Now if He created everything in six thousand (or six million) years, followed by a rest of one thousand or one million years, then we would have a very interesting week indeed.

Some say that Exodus 20:11 is only an analogy in the sense that man is to work and rest—not that it was to mean six literal ordinary days followed by one literal ordinary day. However, Bible scholars have shown that this commandment "does not use analogy or archetypal thinking but that its emphasis is 'stated in terms of the imitation of God or a divine precedent that is to be followed.'"[20] In other words, it was to be six literal days of work, followed by one literal day of rest, just as God worked for six literal days and rested for one.

Some have argued that "the heavens and the earth" is just earth and perhaps the solar system, not the whole universe. However, this verse clearly says that God made *everything* in six days—six consecutive ordinary days, just like the commandment in the previous verse to work for six consecutive ordinary days.

The phrase "heaven(s) and earth" in Scripture is an example of a figure of speech called a *merism*, where two opposites are combined into an all-encompassing single concept, in this case the totality of creation. A linguistic analysis of the words "heaven(s) and earth" in Scripture shows that they refer to the totality of all creation (the Hebrews did not have a word for "universe"). For example, in Genesis 14:19 God is called "Creator of heaven and earth." In Jeremiah 23:24 God speaks of Himself as filling "heaven and earth." See also Genesis 14:22; 2 Kings 19:15; 2 Chronicles 2:12; Psalms 115:15, 121:2, 124:8, 134:3, 146:6; and Isaiah 37:16.

[20] G. Hasel, The "days" of creation in Genesis 1: literal "days" or figurative "periods/epochs" of time? *Origins* **21**(1):29, 1994.

Thus, there is no scriptural warrant for restricting Exodus 20:11 to earth and its atmosphere or the solar system alone. So Exodus 20:11 does show that the whole universe was created in six ordinary days.

Implication

As the days of creation are ordinary days in length, then by adding up the years in Scripture (assuming no gaps in the genealogies[21]), the age of the universe is only about six thousand years.[22]

Refuting Common Objections to Six Literal Days

OBJECTION 1

"Science" has shown the earth and universe are billions of years old; therefore the "days" of creation must be long periods (or indefinite periods) of time.

ANSWER

a. The age of the earth, as determined by man's fallible methods, is based on unproven assumptions, so it is not proven that the earth is billions of years old.[23]

b. This unproven age is being used to force an interpretation on the language of the Bible. Thus, man's fallible theories are allowed to interpret the Bible. This ultimately undermines the use of language to communicate.

---

[21] J. Whitcomb and H. Morris, *The Genesis Flood,* Presbyterian and Reformed Publ., Phillipsburg, New Jersey, 1961, 481–483, Appendix II. They allow for the possibility of gaps in the genealogies because the word "begat" can skip generations. However, they point out that even allowing for gaps would give a maximum age of around 10,000 years.

[22] L. Pierce, The forgotten archbishop, *Creation* **20**(2):42–43, 1998. Ussher carried out a very scholarly work in adding up all the years in Scripture to obtain a date of creation of 4004 BC. Ussher has been mocked for stating that creation occurred on October 23—he obtained this date by working backward using the Jewish civil year and accounting for how the year and month were derived over the years. Thus, he didn't just pull this date out of the air but gave a scholarly mathematical basis for it. This is not to say this is the correct date, as there are assumptions involved, but the point is, his work is not to be scoffed at. Ussher did *not* specify the hour of the day for creation, as some skeptics assert. Young's *Analytical Concordance,* under "creation," lists many other authorities, including extrabiblical ones, who all give a date for creation of less than 10,000 years ago.

[23] See chapters 7 and 9 on these dating methods to see the assumptions involved. See also H. Morris and J. Morris, *Science, Scripture, and the Young Earth,* Institute for Creation Research, El Cajon, California, 1989, 39–44; J. Morris, *The Young Earth,* Master Books, Green Forest, Arkansas, 1996, 51–67; S. Austin, *Grand Canyon: Monument to Catastrophe,* Institute for Creation Research, El Cajon, California, pp. 1994, 111–131; L. Vardiman, ed., *Radio Isotopes and the Age of the Earth*, Vol. 2, Master Books, Green Forest, Arkansas, 2005.

c. Evolutionary scientists claim the fossil layers over the earth's surface date back hundreds of millions of years. As soon as one allows millions of years for the fossil layers, then one has accepted death, bloodshed, disease, thorns, and suffering before Adam's sin.

The Bible makes it clear[24] that death, bloodshed, disease, thorns, and suffering are a *consequence* of sin.[25] In Genesis 1:29–30, God gave Adam and Eve and the animals plants to eat (this is reading Genesis at face value, as literal history, as Jesus did in Matthew 19:3–6). In fact, there is a theological distinction made between animals and plants. Human beings and higher animals are described in Genesis 1 as having a *nephesh,* or life principle. (This is true of at least the vertebrate land animals as well as the birds and fish: Genesis 1:20, 24.) Plants do not have this *nephesh*—they are not "alive" in the same sense animals are. They were given for food.

Man was permitted to eat meat only after the Flood (Genesis 9:3). This makes it obvious that the statements in Genesis 1:29–30 were meant to inform us that man and the animals were vegetarian to start with. Also, in Genesis 9:2, we are told of a change God apparently made in the way animals react to man.

God warned Adam in Genesis 2:17 that if he ate of the "tree of the knowledge of good and evil" he would "die." The Hebrew grammar actually means, "dying, you will die." In other words, it would be the commencement of a process of physical dying (see Genesis 3:19). It also clearly involved spiritual death (separation from God).

After Adam disobeyed God, the Lord clothed Adam and Eve with "coats of skins" (Genesis 3:21).[26] To do this He must have killed and shed the blood of at least one animal. The reason for this can be summed up by Hebrews 9:22:

---

[24] K. Ham, *The Lie: Evolution,* Master Books, Green Forest, Arkansas, Introduction, 1987, xiii–xiv; K. Ham, The necessity for believing in six literal days, *Creation* **18**(1):38–41, 1996; K. Ham, The wrong way round! *Creation* **18**(3):38–41, 1996; K. Ham, Fathers, promises and vegemite, *Creation* **19**(1):14–17, 1997; K. Ham, The narrow road, *Creation* **19**(2):47–49, 1997; K. Ham, Millions of years and the "doctrine of Balaam," *Creation* **19**(3):15–17, 1997.

[25] J. Gill, *A Body of Doctrinal and Practical Divinity*, 1760. Republished by Primitive Baptist Library, Carthage, Illinois, 1980, 191. This is not just a new idea from modern scholars. In 1760 John Gill, in his commentaries, insisted there was no death, bloodshed, disease, or suffering before sin.

[26] All Eve's progeny, except the God-man Jesus Christ, were born with original sin (Romans 5:12, 18–19), so Eve could not have conceived when she was sinless. So the Fall must have occurred fairly quickly, before Eve had conceived any children (they were told to "be fruitful and multiply").

> And according to the law almost all things are purified with blood, and without shedding of blood there is no remission.

God requires the shedding of blood for the remission of sins. What happened in the garden was a picture of what was to come in Jesus Christ, who shed His blood on the Cross as the Lamb of God who took away the sin of the world (John 1:29).

Now if the Garden of Eden were sitting on a fossil record of dead things millions of years old, then blood was shed *before* sin. This would destroy the foundation of the Atonement. The Bible is clear: the sin of Adam brought death and suffering into the world. As Romans 8:19–22 tells us, the whole of creation "groans" because of the effects of the fall of Adam, and the creation will be liberated "from the bondage of corruption into the glorious liberty of the children of God" (Rom. 8:21). Also, bear in mind that thorns came into existence after the Curse. Because there are thorns in the fossil record, it had to be formed after Adam and Eve sinned.

The pronouncement of the death penalty on Adam was both a curse and a blessing. A curse because death is horrible and continually reminds us of the ugliness of sin; a blessing because it meant the consequences of sin—separation from fellowship with God—need not be eternal. Death stopped Adam and his descendants from living in a state of sin, with all its consequences, forever. And because death was the just penalty for sin, Jesus Christ suffered physical death, shedding His blood, to release Adam's descendants from the consequences of sin. The Apostle Paul discusses this in depth in Romans 5 and 1 Corinthians 15.

Revelation 21–22 makes it clear that there will be a "new heavens and a new earth" one day, where there will be "no more death" and "no more

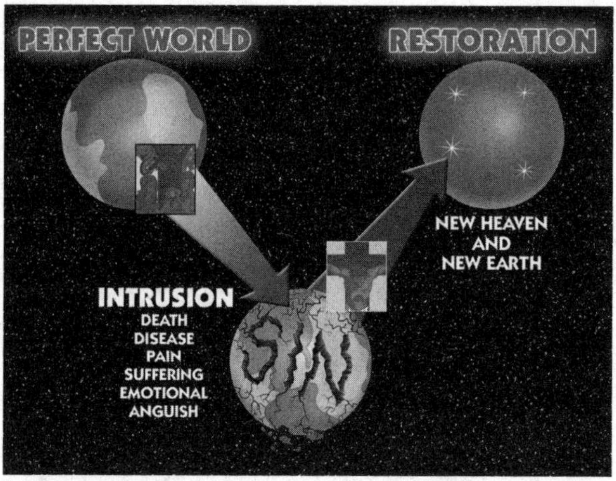

curse"—just like it was before sin changed everything. If there are to be animals as part of the new earth, obviously they will not be dying or eating each other, nor eating the redeemed people!

Thus, adding the supposed millions of years to Scripture destroys the foundations of the message of the Cross.

OBJECTION 2

According to Genesis 1, the sun was not created until Day 4. How could there be day and night (ordinary days) without the sun for the first three days?

ANSWER

a. Again, it is important for us to let the language of God's Word speak to us. If we come to Genesis 1 without any outside influences, as has been shown, each of the six days of creation appears with the Hebrew word *yom* qualified by a number and the phrase "evening and morning." The first three days are written the *same* way as the next three. So if we let the language speak to us, all six days were ordinary earth days.

b. The sun is not needed for day and night. What is needed is light and a rotating earth. On the first day of creation, God made light (Genesis 1:3). The phrase "evening and morning" certainly implies a rotating earth. Thus, if we have light from one direction, and a spinning earth, there can be day and night.

Genesis 1

verse

5 And God called the light Day, and the darkness he called Night. And the evening and the morning were the first day. 1

8b And the evening and the morning were the second day. 2

13 And the evening and the morning were the third day. 3

19 And the evening and the morning were the fourth day. 4

23 And the evening and the morning were the fifth day. 5

31b And the evening and the morning were the sixth day. 6

Where did the light come from? We are not told,[27] but Genesis 1:3 certainly indicates it was a created light to provide day and night until God made the sun on Day 4 to rule the day. Revelation 21:23 tells us that one day the sun will not be needed because the glory of God will light the heavenly city.

Perhaps one reason God did it this way was to illustrate that the sun did not have the priority in the creation that people have tended to give it. The sun did not give birth to the earth as evolutionary theories postulate; the sun was God's created tool to rule the day that God had made (Genesis 1:16).

Down through the ages, people such as the Egyptians have worshiped the sun. God warned the Israelites, in Deuteronomy 4:19, not to worship the sun as the pagan cultures around them did. They were commanded to worship the God who made the sun—not the sun that was *made* by God.

Evolutionary theories (the "big bang" hypothesis for instance) state that the sun came before the earth and that the sun's energy on the earth eventually gave rise to life. Just as in pagan beliefs, the sun is, in a sense, given credit for the wonder of creation.

It is interesting to contrast the speculations of modern cosmology with the writings of the early church father Theophilus:

> On the fourth day the luminaries came into existence. Since God has foreknowledge, he understood the nonsense of the foolish philosophers

[27] Some people ask why God did not tell us the source of this light. However, if God told us everything, we would have so many books we would not have time to read them. God has given us all the information we need to come to the right conclusions about the things that really matter.

who were going to say that the things produced on Earth came from the stars, so that they might set God aside. In order therefore that the truth might be demonstrated, plants and seeds came into existence before stars. For what comes into existence later cannot cause what is prior to it.[28]

OBJECTION 3

2 Peter 3:8 states that "one day is with the Lord as a thousand years," therefore the days of creation could be long periods of time.

ANSWER

a. This passage has *no* creation context—it is *not* referring to Genesis or the six days of creation.

b. This verse has what is called a "comparative article"—"as" or "like"—which is not found in Genesis 1. In other words, it is *not* saying a day *is* a thousand years; it is comparing a real, literal day to a real, literal thousand years. The context of this passage is the Second Coming of Christ. It is saying that, to God, a day is *like* a thousand years, because God is outside of time. God is not limited by natural processes and time as humans are. What may seem like a long time to us (e.g., waiting for the Second Coming), or a short time, is nothing to God, either way.

c. The second part of the verse reads "and a thousand years as one day," which, in essence, cancels out the first part of the verse for those who

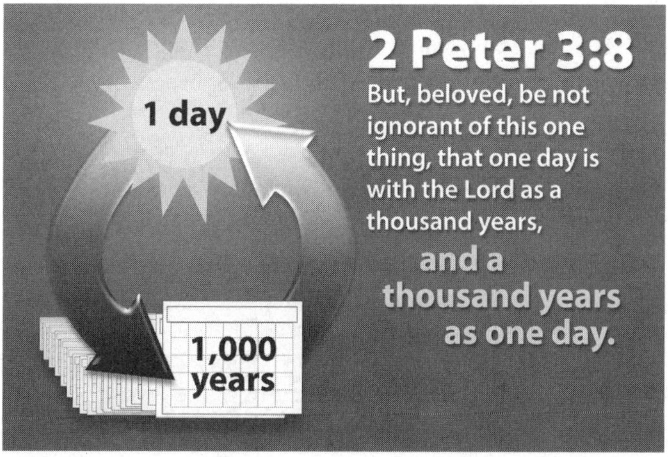

[28] L. Lavallee, The early church defended creation science, *Impact,* No. 160, p. ii, 1986. Quotation from *Theophilus "To Autolycus,"* 2.8, Oxford Early Christian Texts.

want to equate a day with a thousand years. Thus, it cannot be saying a day is a thousand years or vice versa.

d. Psalm 90:4 states, "For a thousand years in your sight are as yesterday when it is past, and as a watch in the night." Here a thousand years is being compared with a "watch in the night" (four hours[29]). Because the phrase "watch in the night" is joined in a particular way to "yesterday," it is saying that a thousand years is being compared with a short period of time—not simply to a day.

e. If one used this passage to claim that "day" in the Bible means a thousand years, then, to be consistent, one would have to say that Jonah was in the belly of the fish three thousand years, or that Jesus has not yet risen from the dead after two thousand years in the grave.

OBJECTION 4

Insisting on six solar days for creation limits God, whereas allowing God billions of years does not limit Him.

ANSWER

Actually, insisting on six ordinary earth-rotation days of creation is not limiting *God,* but limiting *us* to believing that God actually did what He tells us in His Word. Also, if God created everything in six days, as the Bible says, then surely this reveals the power and wisdom of God in a profound way—Almighty God did not *need* eons of time. However, the billions-of-years scenarios diminish God by suggesting that mere chance could create things or that God needed huge amounts of time to create things—this would be limiting God's power by reducing it to naturalistic explanations.

OBJECTION 5

Adam could not have accomplished all that the Bible states in one day (Day 6). He could not have named all the animals, for instance; there was not enough time.

ANSWER

Adam did not have to name *all* the animals—only those God brought to him. For instance, Adam was commanded to name "every beast of the field" (Genesis 2:20), not "beast of the earth" (Genesis 1:25). The phrase "beast of

[29] The Jews had three watches during the night (sunset to 10 pm; 10 pm to 2 am; 2 am to sunrise), but the Romans had four watches, beginning at 6 pm.

the field" is most likely a subset of the larger group "beast of the earth." He did not have to name "everything that creeps upon the earth" (Genesis 1:25) or any of the sea creatures. Also, the number of "kinds" would be much less than the number of species in today's classification.

When critics say that Adam could not name the animals in less than one day, what they really mean is they do not understand how *they* could do it, so Adam could not. However, our brain has suffered from 6,000 years of the Curse—it has been greatly affected by the Fall. Before sin, Adam's brain was perfect.

When God made Adam, He must have programmed him with a perfect language. Today we program computers to "speak" and "remember." How much more could our Creator God have created Adam as a mature human (he was not born as a baby needing to learn to speak), having in his memory a perfect language with a perfect understanding of each word. (That is why Adam understood what God meant when he said he would "die" if he disobeyed, even though he had not seen any death.) Adam may also have had a "perfect" memory (something like a photographic memory, perhaps).

It would have been no problem for this first perfect man to make up words and name the animals God brought to him and remember the names—in far less than one day.[30]

OBJECTION 6

Genesis 2 is a different account of creation, with a different order, so how can the first chapter be accepted as teaching six literal days?

---

30 Andrew Kulikovsky, How Could Adam Have Named All the Animals in a Single Day? Answers in Genesis, www.answersingenesis.org/docs2002/1112animals.asp.

ANSWER

Actually, Genesis 2 is not a *different* account of creation. It is a *more detailed* account of Day 6 of creation. Chapter 1 is an overview of the whole of creation; chapter 2 gives details surrounding the creation of the garden, the first man, and his activities on Day 6.[31]

Between the creation of Adam and the creation of Eve, the King James Version says, "Out of the ground the Lord God formed every beast of the field and every fowl of the air" (Genesis 2:19). This seems to say that the land beasts and birds were created between the creation of Adam and Eve. However, Jewish scholars did not recognize any such conflict with the account in chapter 1, where Adam and Eve were both created after the beasts and birds (Genesis 1:23–25). There is no contradiction, because in Hebrew the precise tense of a verb is determined by the context. It is clear from chapter 1 that the beasts and birds were created before Adam, so Jewish scholars would have understood the verb "formed" to mean "had formed" or "having formed" in Genesis 2:19 If we translate verse 19, "Now the Lord God had formed out of the ground all the beasts of the field," the apparent disagreement with Genesis 1 disappears completely.

Regarding the plants and herbs in Genesis 2:5 and the trees in Genesis 2:9 (compare with Genesis 1:12), the plants and herbs are described as "of the field" and they needed a man to tend them. These are clearly cultivated plants, not just plants in general (Genesis 1). Also, the trees (Genesis 2:9) are only the trees planted in the garden, not trees in general.

In Matthew 19:3–6 Jesus Christ quotes from both Genesis 1:27 and Genesis 2:24 when referring to the *same man and woman* in teaching the doctrine of marriage. Clearly, Jesus saw them as *complementary* accounts, *not* contradictory ones.

OBJECTION 7

There is no "evening and morning" for the seventh day of the Creation Week (Genesis 2:2). Thus, we must still be in the "seventh day," so none of the days can be ordinary days.

---

[31] Paul Taylor, Isn't the Bible Full of Contradictions? chapter 27 in Ken Ham, ed., *The New Answers Book 2* (Green Forest, AR: Master Books, 2008), 288–291; M. Kruger, An understanding of Genesis 2:5, *CEN Technical Journal* **11**(1):106–110, 1997.

ANSWER

Look again at the section entitled "Why Six Days?" on page 97. Exodus 20:11 is clearly referring to seven literal days—six for work and one for rest.

Also, God stated that He *"rested"* from His work of creation (not that He *is resting!*). The fact that He rested from His work of creation does not preclude Him from continuing to rest from this activity. God's work now is different—it is a work of sustaining His creation and of reconciliation and redemption because of man's sin.

The word *yom* is qualified by a number (Genesis 2:2–3), so the context still determines that it is an ordinary solar day. Also, God blessed this seventh day and made it holy. In Genesis 3:17–19 we read of the Curse on the earth because of sin. Paul refers to this in Romans 8:22. It does not make sense that God would call this day holy and blessed if He cursed the ground on this "day." We live in a sin-cursed earth—we are not in the seventh blessed holy day!

Note that in arguing that the seventh day is not an ordinary day because it is not associated with "evening and morning," proponents are tacitly agreeing that the other six days are ordinary days because they are defined by an evening and a morning.

Some have argued that Hebrews 4:3–4 implies that the seventh day is continuing today:

> For we who have believed do enter that rest, as He has said: "So I swore in My wrath, 'They shall not enter My rest,'" although the works were finished from the foundation of the world. For He has spoken in a certain place of the seventh day in this way: "And God rested on the seventh day from all His works... ."

However, verse 4 reiterates that God rested (past tense) on the seventh day. If someone says on Monday that he rested on Friday and is still resting, this would not suggest that Friday continued through to Monday! Also, only those who have believed in Christ will enter that rest, showing that it is a spiritual rest, which is compared with God's rest since the Creation Week. It is not some sort of continuation of the seventh day (otherwise *everyone* would be "in" this rest).[32]

Hebrews does *not* say that the seventh day of Creation Week is continuing today, merely that the rest He instituted is continuing.

[32] Tim Chaffey and Jason Lisle, *Old Earth Creationism on Trial* (Green Forest, AR: Master Books, 2008), 51–52.

OBJECTION 8

Genesis 2:4 states, "In the day that the Lord God made the earth and the heavens." As this refers to all six days of creation, it shows that the word "day" does not mean an ordinary day.

ANSWER

The Hebrew word *yom* as used here is *not* qualified by a number, the phrase "evening and morning," or light or darkness. In this context, the verse really means "in the time God created" (referring to the Creation Week) or "when God created."

Other Problems with Long Days and Similar Interpretations

- If the plants made on Day 3 were separated by millions of years from the birds and nectar bats (created Day 5) and insects (created Day 6) necessary for their pollination, then such plants could not have survived. This problem would be especially acute for species with complex symbiotic relationships (each depending on the other; e.g., the yucca plant and the associated moth[33]).
- Adam was created on Day 6, lived through Day 7, and then died when he was 930 years old (Genesis 5:5). If each day were a thousand years or millions of years, this would make no sense of Adam's age at death.
- Some have claimed that the word for "made" (*asah*) in Exodus 20:11 actually means "show." They propose that God showed or revealed the information about creation to Moses during a six-day period. This allows for the creation itself to have occurred over millions of years. However, "showed" is not a valid translation for *asah*. Its meaning covers "to make, manufacture, produce, do," etc., but not "to show" in the sense of reveal.[34] Where *asah* is translated as "show"—for example, "show kindness" (Genesis 24:12)—it is in the sense of "to do" or "make" kindness.

---

33 F. Meldau, *Why We Believe in Creation Not in Evolution,* Christian Victory Publ., Denver, Colorado, 1972, 114–116.

34 Nothing in Gesenius's *Lexicon* supports the interpretation of *asah* as "show"; See Charles Taylor's "Days of Revelation or creation?" (1997) found at www.answersingenesis.org/docs/188.asp.

- Some have claimed that because the word *asah* is used for the creation of the sun, moon, and stars on Day 4, and not the word *bara,* which is used in Genesis 1:1 for "create," this means God only revealed the sun, moon, and stars at this stage. They insist the word *asah* has the meaning of "revealed." In other words, the luminaries were supposedly already in existence and were only revealed at this stage. However, *bara* and *asah* are used in Scripture to describe the same event. For example, *asah* is used in Exodus 20:11 to refer to the creation of the heavens and the earth, but *bara* is used to refer to the creation of the heavens and the earth in Genesis 1:1. The word *asah* is used concerning the creation of the first people in Genesis 1:26—they did not previously exist. And then they are said to have been created *(bara)* in Genesis 1:27. There are many other similar examples. *Asah* has a broad range of meanings involving "to do" or "to make," which includes *bara* creation.

- Some accept that the days of creation are ordinary days as far as the language of Genesis is concerned but not as literal days of history as far as man is concerned. This is basically the view called the "framework hypothesis."[35] This is a very complex and contrived view which has been thoroughly refuted by scholars.[36]

The real purpose of the framework hypothesis can be seen in the following quote from an article by one of its proponents:

> To rebut the literalist interpretation of the Genesis creation "week" propounded by the young-earth theorists is a central concern of this article.[37]

- Some people want the days of creation to be long periods in an attempt to harmonize evolution or billions of years with the Bible's account of origins. However, the order of events according to long-age beliefs does not agree with that of Genesis. Consider the following table:

---

35 M. Kline, Because it had not rained, *Westminster Theological Journal* **20**:146–157, 1957–1958.

36 Kruger, An understanding of Genesis 2:5, 106–110; J. Pipa, From chaos to cosmos: a critique of the framework hypothesis, presented at the Far-Western Regional Annual Meeting of the Evangelical Theological Society, USA, April 26, 1996; Wayne Grudem's *Systematic Theology,* InterVarsity Press, Downers Grove, Illinois, 1994, 302–305, summarizes the framework hypothesis and its problems and inconsistencies.

37 M. Kline, Space and time in the Genesis cosmology, *Perspectives on Science & Christian Faith* **48**(1), 1996.

CONTRADICTIONS BETWEEN THE ORDER OF CREATION IN THE BIBLE AND EVOLUTION/LONG-AGES

Biblical account of creation	Evolutionary/long-age speculation
Earth before the sun and stars	Stars and sun before earth
Earth covered in water initially	Earth a molten blob initially
Oceans first, then dry land	Dry land, then the oceans
Life first created on the land	Life started in the oceans
Plants created before the sun	Plants came long after the sun
Land animals created after birds	Land animals existed before birds
Whales before land animals	Land animals before whales

Clearly, those who do not accept the six literal days are the ones reading their own preconceived ideas into the passage.

Long-Age Compromises

Other than the "gap theory" (the belief that there is a gap of indeterminate time between the first two verses of Genesis 1), the major compromise positions that try to harmonize long ages and/or evolution with Genesis fall into two categories:

1. "theistic evolution" wherein God supposedly directed the evolutionary process of millions of years, or even just set it up and let it run, and
2. "progressive creation" where God supposedly intervened in the processes of death and struggle to create millions of species at various times over millions of years.

All long-age compromises reject Noah's Flood as global—it could only be a local event because the fossil layers are accepted as evidence for millions of years. A global Flood would have destroyed this record and produced another. Therefore, these positions cannot allow a catastrophic global Flood that would form layers of fossil-bearing rocks over the earth. This, of course, goes against Scripture, which obviously teaches a global Flood (Genesis 6–9).[38] Sadly, most theologians years ago simply tried to add this belief to the Bible instead of realizing that these layers were laid down by Noah's Flood.

---

38 M. Van Bebber and P. Taylor, *Creation and Time: A Report on the Progressive Creationist Book by Hugh Ross*, 55–59; Whitcomb and Morris, *The Genesis Flood*, 212–330.

Does It Really Matter?

Yes, it does matter what a Christian believes concerning the days of creation in Genesis 1. Most importantly, all schemes which insert eons of time into, or before, creation undermine the gospel by putting death, bloodshed, disease, thorns, and suffering before sin and the Fall, as explained above (see answer to Objection 1). Here are two more reasons:

1. It is really a matter of how one approaches the Bible, in principle. If we do not allow the language to speak to us in context, but try to make the text fit ideas outside of Scripture, then ultimately the meaning of any word in any part of the Bible depends on man's interpretation, which can change according to whatever outside ideas are in vogue.

2. If one allows science (which has wrongly become synonymous with evolution and materialism) to determine our understanding of Scripture, then this can lead to a slippery slope of unbelief through the rest of Scripture. For instance, science would proclaim that a person cannot be raised from the dead. Does this mean we should interpret the Resurrection of Christ to reflect this? Sadly, some do just this, saying that the Resurrection simply means that Jesus' teachings live on in His followers.

When people accept at face value what Genesis is teaching and accept the days as ordinary days, they will have no problem accepting and making sense of the rest of the Bible.

Martin Luther once said:

> I have often said that whoever would study Holy Scripture should be sure to see to it that he stays with the simple words as long as he can and by no means departs from them unless an article of faith compels him to understand them differently. For of this we must be certain: no clearer speech has been heard on Earth than what God has spoken.[39]

Pure Words

God's people need to realize that the Word of God is something very special. It is not just the words of men. As Paul said in 1 Thessalonians 2:13, "You received it not as the word of men, but as it is, truly the word of God."

Proverbs 30:5–6 states that "every word of God is pure … . Do not add to His words, lest He reprove you and you be found a liar." The Bible cannot

[39] Plass, *What Martin Luther Says: A Practical In-Home Anthology for the Active Christian,* 93.

be treated as just some great literary work. We need to "tremble at his word" (Isaiah 66:2) and not forget:

> All Scripture is given by inspiration of God, and is profitable for doctrine, for reproof, for correction, for instruction in righteousness, that the man of God may be complete, thoroughly equipped for every good work (2 Timothy 3:16–17).

In the original autographs, every word and letter in the Bible is there because God put it there. Let us listen to God speaking to us through His Word and not arrogantly think we can tell God what He really means!

9

Does Radiometric Dating Prove the Earth Is Old?

MIKE RIDDLE

The presupposition of long ages is an icon and foundational to the evolutionary model. Nearly every textbook and media journal teaches that the earth is billions of years old.

> Using radioactive dating, scientists have determined that the Earth is about 4.5 billion years old, ancient enough for all species to have been formed through evolution.[1]

> The earth is now regarded as between 4.5 and 4.6 billion years old.[2]

The primary dating method scientists use for determining the age of the earth is radioisotope dating. Proponents of evolution publicize radioisotope dating as a reliable and consistent method for obtaining absolute ages of rocks and the age of the earth. This apparent consistency in textbooks and the media has convinced many Christians to accept an old earth (supposedly 4.6 billion years old).

What Is Radioisotope Dating?

Radioisotope dating (also referred to as radiometric dating) is the process of estimating the ages of rocks from the decay of radioactive elements in

[1] *Biology: Visualizing Life*, Holt, Rinehart, and Winston, Austin, Texas, 1998, 117.
[2] C. Plummer, D. Carlson, and D. McGeary, *Physical Geology*, McGraw Hill, New York, 2006, 216.

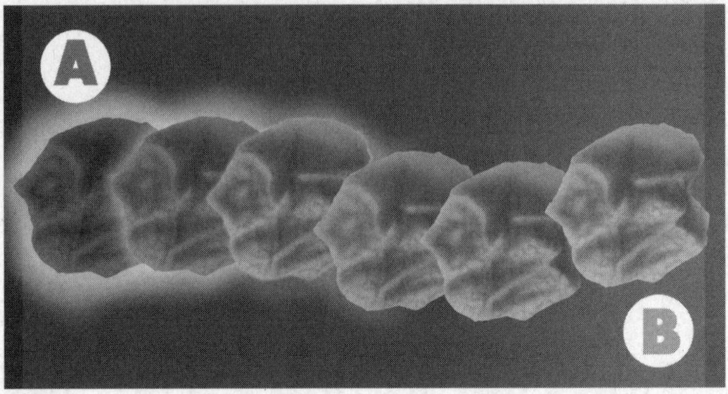

them. There are certain kinds of atoms in nature that are unstable and spontaneously change (decay) into other kinds of atoms. For example, uranium will radioactively decay through a series of steps until it becomes the stable element lead. Likewise, potassium decays into the element argon. The original element is referred to as the parent element (in these cases uranium and potassium), and the end result is called the daughter element (lead and argon).

The Importance of Radioisotope Dating

The straightforward reading of Scripture reveals that the days of creation (Genesis 1) were literal days and that the earth is just thousands of years old, and not billions. There appears to be a fundamental conflict between the Bible and the reported ages given by radioisotope dating. Since God is the Creator of all things (including the human ability to do science), and His Word is true ("Sanctify them by Your truth. Your word is truth," John 17:17), the true age of the earth must agree with His Word. However, rather than accept the biblical account of creation, many Christians have accepted the radioisotope dates of billions of years and attempted to fit long ages into the Bible. The implications of doing this are profound and affect many parts of the Bible.

How Radioisotope Dating Works

Radioisotope dating is commonly used to date igneous rocks. These are rocks which form when hot, molten material cools and solidifies. Types of igneous rocks include granite and basalt (lava). Sedimentary rocks, which contain most of the world's fossils, are not commonly used in radioisotope dating. These types of rocks are comprised of particles from many preexisting rocks which were transported (mostly by water) and redeposited somewhere else. Types of sedimentary rocks include sandstone, shale, and limestone.

The radioisotope dating clock starts when a rock cools. During the molten state it is assumed that the intense heat will force any gaseous daughter elements like argon to escape. Once the rock cools it is assumed that no more atoms can escape and any daughter element found in a rock will be the result of radioactive decay. The dating process then requires measuring how much daughter element is in a rock sample and knowing the decay rate (i.e., how long it takes the parent element to decay into the daughter element—uranium into lead or potassium

Uranium-238 (^{238}U) is an isotope of uranium. Isotopes are varieties of an element that have the same number of protons but a different number of neutrons within the nucleus. For example, carbon-14 (^{14}C) is a particular isotope. All carbon atoms have 6 protons but can vary in the number of neutrons. ^{12}C has 6 protons and 6 neutrons in its nucleus. ^{13}C has 6 protons and 7 neutrons. ^{14}C has 6 protons and 8 neutrons. Extra neutrons often lead to instability, or radioactivity. Likewise, all isotopes (varieties) of uranium have 92 protons. ^{238}U has 92 protons and 146 neutrons. It is unstable and will radioactively decay first into ^{234}Th (thorium-234) and finally into ^{206}Pb (lead-206). Sometimes a radioactive decay will cause an atom to lose 2 protons and 2 neutrons (called alpha decay). For example, the decay of ^{238}U into ^{234}Th is an alpha decay process. In this case the atomic mass changes (238 to 234). Atomic mass is the heaviness of an atom when compared to hydrogen, which is assigned the value of one. Another type of decay is called beta decay. In beta decay, either an electron is lost and a neutron is converted into a proton (beta minus decay) or an electron is added and a proton is converted into a neutron (beta plus decay). In beta decay the total atomic mass does not change significantly. The decay of ^{234}Th into ^{234}Pa (protactinium-234) is an example of beta decay.

Uranium-238
Thorium-234
Protactinium-234
Uranium-234
Thorium-230
Radium-226
Radon-222
Polonium-218
Lead-214
Bismuth-214
Polonium-214
Lead-210
Bismuth-210
Polonium-210
Lead-206 (stable)

Uranium to lead decay sequence

into argon). The decay rate is measured in terms of half-life. Half-life is defined as the length of time it takes half of the remaining atoms of a radioactive parent element to decay. For example, the remaining radioactive parent material will decrease by 1/2 during the passage of each half-life (1‡1/2‡1/4‡1/8‡1/16, etc.). Half-lives as measured today are very accurate, even the extremely slow half-lives. That is, billion-year half-lives can be measured statistically in just hours of time. The following table is a sample of different element half-lives.

Parent	Daughter	Half-life
Polonium-218	Lead-214	3 minutes
Thorium-234	Protactinium-234	24 days
Carbon-14	Nitrogen-14	5,730 years
Potassium-40	Argon-40	1.25 billion years
Uranium-238	Lead-206	4.47 billion years
Rubidium-87	Strontium-87	48.8 billion years

Science and Assumptions

Scientists use observational science to measure the amount of a daughter element within a rock sample and to determine the present observable decay rate of the parent element. Dating methods must also rely on another kind of science called historical science. Historical science cannot be observed. Determining the conditions present when a rock first formed can only be studied through historical science. Determining how the environment might have affected a rock also falls under historical science. Neither condition is directly observable. Since radioisotope dating uses both types of science, we can't directly measure the age of something. We can use scientific techniques in the present, combined with assumptions about historical events, to estimate the age. Therefore, there are several assumptions that must be made in radioisotope dating. Three critical assumptions can affect the results during radioisotope dating:

1. The initial conditions of the rock sample are accurately known.
2. The amount of parent or daughter elements in a sample has not been altered by processes other than radioactive decay.
3. The decay rate (or half-life) of the parent isotope has remained constant since the rock was formed.

The Hourglass Illustration

Radioisotope dating can be better understood using an illustration with an hourglass. If we walk into a room and observe an hourglass with sand at the top and sand at the bottom, we could calculate how long the hourglass has been running. By estimating how fast the sand is falling and measuring the amount of sand at the bottom, we could calculate how much time has elapsed since the hourglass was turned over. All our calculations could be correct (observational science), but the result could be wrong. This is because we failed to take into account some critical assumptions.

1. Was there any sand at the bottom when the hourglass was first turned over (initial conditions)?
2. Has any sand been added or taken out of the hourglass? (Unlike the open-system nature of a rock, this is not possible for a sealed hourglass.)
3. Has the sand always been falling at a constant rate?

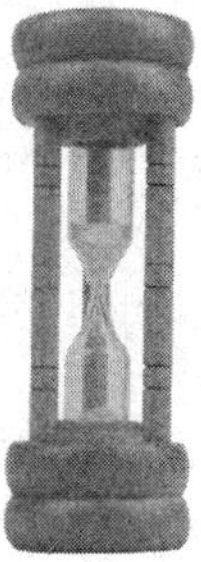

Since we did not observe the initial conditions when the hourglass time started, we must make assumptions. All three of these assumptions can affect our time calculations. If scientists fail to consider each of these three critical assumptions, then radioisotope dating can give incorrect ages.

The Facts

We know that radioisotope dating does not always work because we can test it on rocks of known age. In 1997, a team of eight research scientists known as the RATE group (Radioisotopes and the Age of The Earth) set out to investigate the assumptions commonly made in standard radioisotope dating practices (both single-sample and multiple-samples radioisotope dating). Their findings were significant and directly impact the evolutionary dates of millions of years.[3]

[3] L. Vardiman, A.A. Snelling and E.F. Chaffin (eds.), *Radioisotopes and the Age of the Earth: A Young-Earth Creationist Research Initiative*, Institute for Creation Research, Santee, California, and Creation Research Society, St. Joseph, Missouri, 2000; L. Vardiman, A.A. Snelling and E.F. Chaffin (eds.), *Radioisotopes and the Age of the Earth: Results of a Young-Earth Creationist Research Initiative*, Institute for Creation Research, Santee, California, and Creation Research Society, Chino Valley, Arizona, 2005; D. DeYoung, *Thousands … Not Billions*, Master Books, Green Forest, Arkansas, 2005.

A rock sample from the newly formed 1986 lava dome from Mount St. Helens was dated using Potassium-Argon dating. The newly formed rock gave ages for the different minerals in it of between 0.5 and 2.8 million years.[4] These dates show that significant argon (daughter element) was present when the rock solidified (assumption 1 is false).

Mount Ngauruhoe is located on the North Island of New Zealand and is one of the country's most active volcanoes. Eleven samples were taken from solidified lava and dated. These rocks are known to have formed from eruptions in 1949, 1954, and 1975. The rock samples were sent to a respected commercial laboratory (Geochron Laboratories in Cambridge, Massachusetts). The "ages" of the rocks ranged from 0.27 to 3.5 million years old.[5] Because these rocks are known to be less than 70 years old, it is apparent that assumption #1 is again false. When radioisotope dating fails to give accurate dates on rocks of known age, why should we trust it for rocks of unknown age? In each case, the ages of the rocks were greatly inflated.

Isochron Dating

There is another form of dating called isochron dating, which involves analyzing four or more samples from the same rock unit. This form of dating attempts to eliminate one of the assumptions in single-sample radioisotope dating by using ratios and graphs rather than counting atoms present. It does not depend on the initial concentration of the daughter element being zero. The isochron dating technique is thought to be infallible, because it supposedly eliminates the assumption about starting conditions. However, this

[4] S.A. Austin, Excess argon within mineral concentrates from the new dacite lava dome at Mount St Helens volcano, *Creation Ex Nihilo Technical Journal* **10**(3): 335–343, 1996.

[5] A.A. Snelling, The cause of anomalous potassium-argon "ages" for recent andesite flows at Mt Ngauruhoe, New Zealand, and the implications for potassium-argon "dating," in R.E. Walsh (ed.), *Proceedings of the Fourth International Conference on Creationism*, Creation Science Fellowship, Pittsburgh, Pennsylvania, pp. 503–525, 1998.

method has a different assumption about starting conditions and can also give incorrect dates.

If single-sample and isochron dating methods are objective and reliable they should agree. However, they frequently do not. When a rock is dated by more than one method it may yield very different ages. For example, the RATE group obtained radioisotope dates from ten different locations. To omit any potential bias, the rock samples were analyzed by several commercial laboratories. In each case, the isochron dates differed substantially from the single-sample radioisotope dates. In some cases the range was more than 500 million years.[6] Two conclusions drawn by the RATE group include:

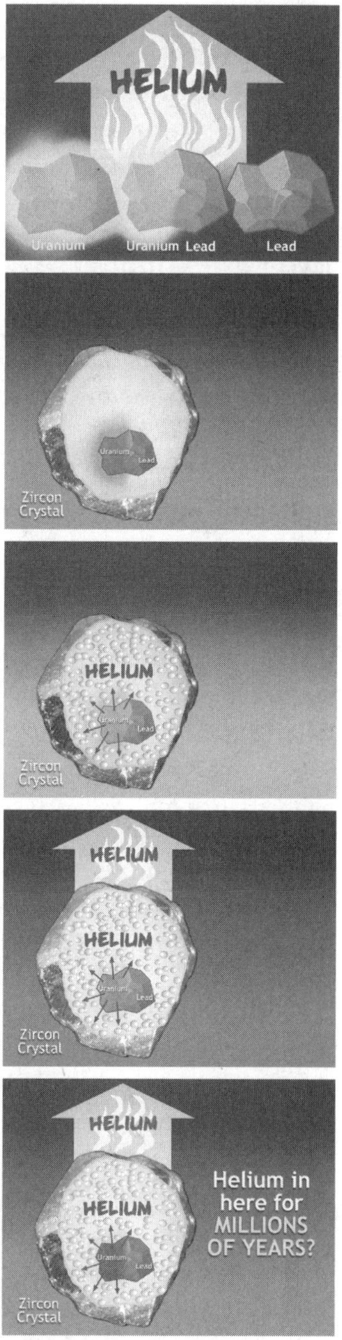

1. The single-sample potassium-argon dates showed a wide variation.
2. A marked variation in ages was found in the isochron method using different parent-daughter analyses.

If different methods yield different ages and there are variations with the same method, how can scientists know for sure the age of any rock or the age of the earth?

In one specific case, samples were taken from the Cardenas Basalt, which is among the oldest strata in the eastern Grand Canyon. Next, samples from the western canyon basalt lava flows, which are among the youngest formations in the canyon, were analyzed. Using the rubidium-strontium isochron dating

[6] A.A. Snelling, Isochron discordances and the role of inheritance and mixing of radioisotopes in the mantle and crust, in Vardiman et al., *Radioisotopes and the Age of the Earth*, pp. 393–524, 2005; D. DeYoung, *Thousands … Not Billions*, pp. 123–139, 2005.

method, an age of 1.11 billion years was assigned to the oldest rocks and a date of 1.14 billion years to the youngest lava flows. The youngest rocks gave a billion year age the same as the oldest rocks! Are the dates given in textbooks and journals accurate and objective? When assumptions are taken into consideration and discordant (disagreeing or unacceptable) dates are not omitted, radioisotope dating often gives inconsistent and inflated ages.

Two Case Studies

The RATE team selected two locations to collect rock samples to conduct analyses using multiple radioisotope dating methods. Both sites are understood by geologists to date from the Precambrian (supposedly 543–4,600 million years ago). The two sites chosen were the Beartooth Mountains of northwest Wyoming near Yellowstone National Park, and the Bass Rapids sill in the central portion of Arizona's Grand Canyon. All rock samples (whole rock and separate minerals within the rock) were analyzed using four radioisotope methods. These included the isotopes potassium-argon (K-Ar), rubidium-strontium (Rb-Sr), samarium-neodymium (Sm-Nd), and lead-lead (Pb-Pb). In order to avoid any bias, the dating procedures were contracted out to commercial laboratories located in Colorado, Massachusetts, and Ontario, Canada.

In order to have a level of confidence in dating, different radioisotope methods used to date a rock sample should closely coincide in age. When this occurs, the sample ages are said to be concordant. In contrast, if multiple results for a rock disagree with each other in age they are said to be discordant.

Beartooth Mountains Sample Results

Geologists believe the Beartooth Mountains rock unit to contain some of the oldest rocks in the United States, with an estimated age of 2,790 million years. The following table summarizes the RATE results.[7]

The results show a significant scatter in the ages for the various minerals and also between the isotope methods. In some cases, the whole rock age is greater than the age of the minerals, and for others, the reverse occurs. The potassium-argon mineral results vary between 1,520 and 2,620 million years (a difference of 1,100 million years).

[7] S.A. Austin, Do radioisotope clocks need repair? Testing the assumptions of isochron dating using K-Ar, Rb-Sr, Sm-Nd, and Pb-Pb isotopes, in Vardiman et al., *Radioisotopes and the Age of the Earth,* pp. 325–392, 2005; D. DeYoung, *Thousands … Not Billions*, pp. 109-121, 2005.

Dating Isotopes	Millions of Years	Type of Data (whole rock or separate mineral within the rock)
Potassium-Argon (single-sample)	1,520	Quartz-plagioclase mineral
	2,011	Whole rock
	2,403	Biotite mineral
	2,620	Hornblende mineral
Rubidium-Strontium (isochron)	2,515	5 minerals
	2,790	Previously published result based on 30 whole rock samples (1982)
Samarium-Neodymium (isochron)	2,886	4 minerals
Lead-Lead (isochron)	2,689	5 minerals

Bass Rapids Sill Sample Results

The 11 Grand Canyon rock samples were also dated commercially using the most advanced radioisotope technology. The generally accepted age for this formation is 1,070 million years. The RATE results are summarized in the table on the following page.[8]

The RATE results differ considerably from the generally accepted age of 1,070 million years. Especially noteworthy is the multiple whole rocks potassium-argon isochron age of 841.5 million years while the samarium-neodymium isochron gives 1,379 million years (a difference of 537.5 million years).

Possible Explanations for the Discordance

There are three possible explanations for the discordant isotope dates.

1. There may have been a mixing of isotopes between the molten rock and the rocks into which it intruded. There are ways to determine if this has occurred and can be eliminated as a possible explanation.

2. Some of the minerals have crystallized at different times as the rock formed and cooled. However, there is no evidence that molten rock

[8] A.A. Snelling, S.A. Austin, and W.A. Hoesch, Radioisotopes in the diabase sill (Upper Precambrian) at Bass Rapids, Grand Canyon, Arizona: an application and test of the isochron dating methods, in R.L. Ivey, Jr. (ed.), *Proceedings of the Fifth International Conference on Creationism*, Creation Science Fellowship, Pittsburgh, Pennsylvania, pp. 269–284, 203; S.A. Austin, in Vardiman et al., 2005, 325–392; D. DeYoung, 2005, 109–121.

Dating Isotopes	Millions of Years	Type of data (whole rock or separate mineral within the rock)
Potassium-Argon	841.5	11 whole rock samples (isochron)
	665 to 1,053	Model ages from single-sample whole rocks
Rubidium-Strontium (isochron)	1,007	Magnetite mineral grains from 7 rock samples
	1,055	11 whole rocks
	1,060	7 minerals
	1,070	Previously published age based on 5 whole rock samples (1982)
	1,075	12 minerals
Lead-Lead (isochron)	1,250	11 whole rocks
	1,327	6 minerals
Samarium-Neodymium (isochron)	1,330	8 minerals
	1,336	Magnetite mineral grains from 7 rock samples
	1,379	6 minerals

crystallizes and cools in the same place at such an incredibly slow pace. Rather, molten rocks crystallize and cool relatively rapidly, so therefore this explanation can be eliminated.

3. The radioactive decay rates have been different in the past than they are today. The following section will show that this provides the best explanation for the discordant ages.

New Studies

New studies by the RATE group have provided evidence that radioactive decay supports a young earth. One of their studies involved the amount of helium found in granite rocks. Granite contains tiny zircon crystals, which contain radioactive uranium (^{238}U), which decays into lead (^{206}Pb). During this process, for each atom of ^{238}U decaying into ^{206}Pb, eight helium atoms are formed and migrate out of the zircons and granite rapidly.

> Within the zircon[9] crystals, any helium atoms generated by nuclear decay in the distant past should have long ago migrated outward and escaped from these crystals. One would expect the helium gas to eventually diffuse upward out of the ground and then disappear into the atmosphere. To everyone's surprise, however, large amounts of helium have been found trapped inside zircons.[10]

The decay of ^{238}U into lead (^{206}Pb) is a slow process (half-life of 4.47 billion years). Since helium migrates out of rocks rapidly, there should be very little to no helium remaining in the zircon crystals.

Why is so much helium still in the zircons? One likely explanation is that sometime in the past the radioactive decay rate was greatly accelerated. The decay rate was accelerated so much that helium was being produced faster than it could have escaped, causing an abundant amount of helium to remain in the zircons in the granite. The RATE group has gathered evidence that at some time in history nuclear decay was greatly accelerated.

> The experiments the RATE project commissioned have clearly confirmed the numerical predictions of our Creation model.... The data and our analysis show that over a billion years worth of nuclear decay has occurred very recently, between 4000 and 8000 years ago.[11]

Confirmation of this accelerated nuclear decay having occurred is provided by adjacent uranium and polonium radiohalos that formed at the same time in the same biotite flakes in granites.[12] Radiohalos result from the physical damage caused by radioactive decay of uranium and intermediate daughter atoms of polonium, so they are observable evidence that a lot of radioactive decay has occurred during the earth's history. However, because the

---

[9] L. Vardiman, A. Snelling, and E. Chaffin, eds., *Radioisotopes and the Age of the Earth*, vol. 2, El Cajon, California, Institute of Creation Research and Chino Valley, Arizona: Creation Research Society, 2005, 74.

[10] DeYoung, *Thousands ... Not Billions*, 2005, 68.

[11] R. Humphreys, Young helium diffusion age of zircons supports accelerated nuclear decay, in Vardiman et al., *Radioisotopes and the Age of the Earth*, 2005, 74.

[12] A.A. Snelling, Radiohalos in granites: evidence of accelerated nuclear decay, in Vardiman et al., *Radioisotopes and the Age of the Earth*, 2005, 101–207; D. DeYoung, *Thousands ... Not Billions*, 2005, 81–97.

daughter polonium atoms are only short-lived (for example, polonium-218 decays within 3 minutes, compared to 4.47 billion years for uranium-238), the polonium radiohalos had to form within hours to a few days. But in order to supply the needed polonium atoms to produce these polonium radiohalos within that timeframe, the nearby uranium atoms had to decay at an accelerated rate. Thus hundreds of millions of years worth of uranium decay (compared to today's slow decay rate) had to have occurred within hours to a few days to produce these adjacent uranium and polonium radiohalos in granites.

The RATE group has suggested that this accelerated decay took place early during the Creation Week and then again during the Flood. Accelerated decay of this magnitude would result in immense amounts of heat being generated in rocks. Determining how this heat was dissipated presents a new and exciting opportunity for creation research.

Conclusion

The best way to learn about history and the age of the earth is to consult the history book of the universe—the Bible. Many scientists and theologians accept a straightforward reading of Scripture and agree that the earth is about 6,000 years old. It is better to use the infallible Word of God for our scientific assumptions than to change His Word in order to compromise with "science" that is based upon man's fallible assumptions. True science will always support God's Word.

> Based on the measured helium retention, a statistical analysis gives an estimated age for the zircons of 6,000 ± 2,000 years. This age agrees with literal biblical history and is about 250,000 times shorter than the conventional age of 1.5 billion years for zircons. The conclusion is that helium diffusion data strongly supports the young-earth view of history.[13]

It must also be concluded, therefore, that because nuclear decay has been shown to have occurred at grossly accelerated rates when molten rocks were forming, crystallizing and cooling, the radiometric methods cannot possibly date these rocks accurately based on the false assumption of constant decay through earth history at today's slow rates. Thus the radiometric dating methods are highly unreliable and don't prove the earth is old.

[13] DeYoung, *Thousands … Not Billions*, 2005, 76.

10

Was There Really a Noah's Ark & Flood?

KEN HAM & TIM LOVETT

The account of Noah and the Ark is one of the most widely known events in the history of mankind. Unfortunately, like other Bible accounts, it is often taken as a mere fairy tale.

The Bible, though, is the true history book of the universe, and in that light, the most-asked questions about the Ark and Flood of Noah can be answered with authority and confidence.

How Large Was Noah's Ark?

> The length of the ark shall be three hundred cubits, its width fifty cubits, and its height thirty cubits (Genesis 6:15).

Unlike many whimsical drawings that depict the Ark as some kind of overgrown houseboat (with giraffes sticking out the top), the Ark described in the Bible was a huge vessel. Not until the late 1800s was a ship built that exceeded the capacity of Noah's Ark.

The dimensions of the Ark are convincing for two reasons: the proportions are like that of a modern cargo ship, and it is about as large as a wooden ship can be built. The cubit gives us a good indication of size[1] Using the most likely cubit length, an ancient royal cubit of at least 20.4 inches (0.518 m),

[1] The cubit was defined as the length of the forearm from elbow to fingertip. Ancient cubits vary anywhere from 17.5 inches (45 cm) to 22 inches (56 cm), with the longer sizes dominating the major ancient constructions. Despite this, even a conservative 18-inch (46 cm) cubit describes a sizeable vessel.

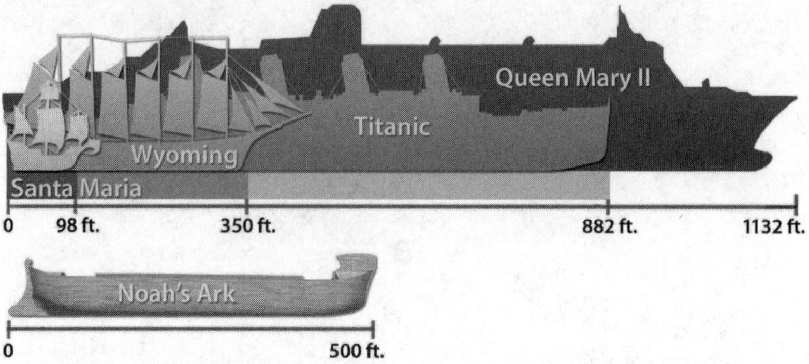

we know that the Ark must have been no less than 510 feet (155 m) long, 85 feet (26 m) wide, and 51feet (15.5 m) high. In the Western world, wooden sailing ships never got much longer than about 330 feet (100 m), yet the ancient Greeks built vessels at least this size 2,000 years earlier. China built huge wooden ships in the 1400s that may have been as large as the Ark. The biblical Ark is one of the largest wooden ships of all time—a mid-sized cargo ship by today's standards.

How Could Noah Build the Ark?

The Bible does not tell us that Noah and his sons built the Ark by themselves. Noah could have hired skilled laborers or had relatives, such as Methuselah and Lamech, help build the vessel. However, nothing indicates that they could not—or that they did not—build the Ark themselves in the time allotted. The physical strength and mental processes of men in Noah's day was at least as great (quite likely even superior) to our own.[2] They certainly would have had efficient means for harvesting and cutting timber, as well as for shaping, transporting, and erecting the massive beams and boards required.

If one or two men today can erect a large house in just 12 weeks, how much more could three or four men do in a few years? Adam's descendants were making complex musical instruments, forging metal, and building cities—their tools, machines, and techniques were not primitive.

History has shown that technology can be lost. In Egypt, China, and the Americas, the earlier dynasties built more impressive buildings or had finer art or better science. Many so-called modern inventions turn out to be re-inventions, like concrete, which was used by the Romans.

---

[2] For the evidence, see Dr. Donald Chittick, *The Puzzle of Ancient Man* (Newberg, OR: Creation Compass, 1998). This book details evidence of man's intelligence in early post-Flood civilizations.

Even accounting for the possible loss of technology due to the Flood, early post-Flood civilizations possessed all the engineering know-how necessary for a project like Noah's Ark. People were sawing and drilling wood in Noah's day, only a few centuries before the Egyptians were sawing and drilling granite; it is very reasonable! The idea that more primitive civilizations are further back in time is an evolutionary concept.

In reality, when God created Adam, he was perfect. Today, the individual human intellect has suffered from 6,000 years of sin and decay. The sudden rise in technology in the last few centuries has nothing to do with increasing intelligence; it is a combination of publishing and sharing ideas, and the spread of key inventions that became tools for investigation and manufacturing. One of the most recent tools is the computer, which compensates a great deal for our natural decline in mental performance and discipline, since it permits us to gather and store information as perhaps never before.

How Could Noah Round Up So Many Animals?

> Of the birds after their kind, of animals after their kind, and of every creeping thing of the earth after its kind, two of every kind will come to you, to keep them alive (Genesis 6:20).

This verse tells us that Noah didn't have to search or travel to faraway places to bring the animals on board. The world map was completely different before the Flood, and on the basis of Genesis 1, there may have been only one continent. The animals simply arrived at the Ark as if called by a "homing instinct" (a behavior implanted in the animals by their Creator) and marched up the ramp, all by themselves.

Though this was probably a supernatural event (one that cannot be explained by our understanding of nature), compare it to the impressive migratory behavior we see in some animals today. We are still far from understanding all of the marvelous animal behaviors exhibited in God's creation: the migration of Canada geese and other birds, the amazing flights of monarch butterflies, the annual travels of whales and fish, hibernation instincts, earthquake sensitivity, and countless other fascinating capabilities of God's animal kingdom.

Were Dinosaurs on Noah's Ark?

The history of God's creation (told in Genesis 1 and 2) tells us that all the land-dwelling creatures were made on Day 6 of Creation Week—the same day God made Adam and Eve. Therefore, it is clear that dinosaurs (being land animals) were made with man.

Also, two of every kind (seven of some) of land animal boarded the Ark. Nothing indicates that any of the land animal kinds were already extinct before the Flood. Besides, the description of "behemoth" in chapter 40 of the book of Job (Job lived after the Flood) only fits with something like a sauropod dinosaur. The ancestor of "behemoth" must have been on board the Ark.[3]

We also find many dinosaurs that were trapped and fossilized in Flood sediment. Widespread legends of encounters with dragons give another indication that at least some dinosaurs survived the Flood. The only way this could happen is if they were on the Ark.

Juveniles of even the largest land animals do not present a size problem, and, being young, they have their full breeding life ahead of them. Yet most dinosaurs were not very large at all—some were the size of a chicken (although absolutely no relation to birds, as many evolutionists are now saying). Most scientists agree that the average size of a dinosaur is actually the size of a large sheep or bison.

For example, God most likely brought Noah two young adult sauropods (e.g., apatosaurs), rather than two full-grown sauropods. The same goes for the elephant kind, the giraffe kind, and other animals that grow to be very large. However, there was adequate room on the Ark for most fully grown adult animals anyway.

As far as the number of different types of dinosaurs, it should be recognized that, although there are hundreds of names for different varieties (species) of dinosaurs that have been discovered, there are probably only 50 to 90 actual different kinds.

How Could Noah Fit All the Animals on the Ark?

> And of every living thing of all flesh you shall bring two of every sort into the ark, to keep them alive with you; they shall be male and female (Genesis 6:19).

In the book *Noah's Ark: A Feasibility Study*,[4] creationist researcher John Woodmorappe suggests that, at most, 16,000 animals were all that were needed to preserve the created kinds that God brought into the Ark. Woodmorappe used a "worst case" scenario of the biblical "kind" being equated to the genus level of classification.

---

3 For some remarkable evidence that dinosaurs have lived until relatively recent times, see chapter 12, "What Really Happened to the Dinosaurs?" Also read *The Great Dinosaur Mystery Solved* (Green Forest, AR: New Leaf Press, 2000). Also visit www. answersingenesis. org/go/dinosaurs.

4 John Woodmorappe, *Noah's Ark: A Feasibility Study* (Santee, CA Institute for Creation Research, 2003).

The Ark did not need to carry every kind of animal—nor did God command it. It carried only air-breathing, land-dwelling animals, creeping things, and winged animals such as birds. Aquatic life (fish, whales, etc.) and many amphibious creatures could have survived in sufficient numbers outside the Ark. This cuts down significantly the total number that needed to be on board.

Using a short cubit of 18 inches (46 cm) for the Ark to be conservative, Woodmorappe's conclusion is that "less than half of the cumulative area of the Ark's three decks need to have been occupied by the animals and their enclosures.[5] This meant there was plenty of room for fresh food, water, and even many other people. Noah's cubit was probably longer, like the royal cubits used in the pyramids of Egypt and elsewhere. For the Ark Encounter, we have used a royal cubit of 20.4 inches (51.8 cm) as a typical benchmark for ancient construction.

Another factor which greatly reduces the space requirements is the fact that the tremendous variety in species we see today did not exist in the days of Noah. Only the parent "kinds" of these species were required to be on board in order to repopulate the earth"[6] For example, only two dogs were needed to give rise to all the dog species that exist today.

Creationist estimates for the maximum number of animals that would have been necessary to come on board the Ark have ranged from a few thousand to 35,000, but they may be as few as 7,000 if the biblical kind is approximately the same as the modern family classification. Researchers for the Ark Encounter have determined that there would have been approximately 1,400 kinds of animals on the Ark. Taking into account that God brought pairs of the unclean animals—seven pairs of clean animals, and seven pairs of the flying creatures (including birds, bats, and pterosaurs)—there would have been about 7,000 individual animals on the Ark.

As stated before, Noah wouldn't have taken the largest animals onto the Ark; it is more likely he took juveniles aboard the Ark to repopulate the earth after the Flood was over. These younger animals also require less space, less food, and have less waste. Additionally, it was God's desire to have the animals

---

[5] Ibid.

[6] Here's one example: more than 200 different breeds of dogs exist today, from the miniature poodle to the St. Bernard—all of which have descended from one original dog "kind" (as have the wolf, dingo, etc.). Many other types of animals— cat kind, horse kind, cow kind, etc.— have similarly been naturally and selectively bred to achieve the wonderful variation in species that we have today. God "programmed" this variety into the genetic code of all animal kinds— even humankind! God also made it impossible for the basic "kinds" of animals to breed and reproduce with a different kind. For example, cats and dogs cannot breed to make a new type of creature. This is by God's design, and it is one fact that makes evolution impossible.

multiply after the Flood (Genesis 8:17), so taking the largest animals (often the oldest) would have been antithetical to this purpose of quickly replenishing the animal population.

How Did Noah Care for All the Animals?

Just as God brought the animals to Noah by some form of supernatural means, He surely also prepared them for this amazing event. Creation scientists suggest that God gave the animals the ability to hibernate, as we see in many species today. Most animals react to natural disasters in ways that were designed to help them survive. It's very possible many animals did hibernate, perhaps even supernaturally intensified by God.

Whether it was supernatural or simply a normal response to the darkness and confinement of a rocking ship, the fact that God told Noah to build rooms ("*qen*"—literally in Hebrew "nests") in Genesis 6:14 implies that the animals were subdued or nesting. God also told Noah to take food for them (Genesis 6:21), which tells us that they were not in a year-long coma either.

Were we able to walk through the Ark as it was being built, we would undoubtedly be amazed at the ingenious systems on board for water and food storage and distribution. As Woodmorappe explains in *Noah's Ark: A Feasibility Study*, a small group of farmers today can raise thousands of cattle and other animals in a very small space. One can easily imagine all kinds of devices on the Ark that would have enabled a small number of people to feed and care for the animals, from watering to waste removal.

As Woodmorappe points out, no special devices were needed for eight people to care for 16,000 animals. But if they existed, how would these devices be powered? There are all sorts of possibilities. How about a plumbing system for gravity-fed drinking water, a ventilation system driven by wind or wave motion, or hoppers that dispense grain as the animals eat it? None of these require higher technology than what we know existed in ancient cultures, and yet these cultures were likely well short of the skill and capability of Noah and the pre-Flood world.

How Could a Flood Destroy Every Living Thing?

> And all flesh died that moved on the earth: birds and cattle and beasts and every creeping thing that creeps on the earth, and every man. All in whose nostrils was the breath of the spirit of life, all that was on the dry land, died (Genesis 7:21–22).

Noah's Flood was much more destructive than any 40-day rainstorm ever could be. Scripture says that the "fountains of the great deep" broke open. In other words, earthquakes, volcanoes, and geysers of molten lava and scalding water were squeezed out of the earth's crust in a violent, explosive upheaval. These fountains were not stopped until 150 days into the Flood—so the earth was literally churning underneath the waters for about five months! The duration of the Flood was extensive, and Noah and his family were aboard the Ark for around a year.

Relatively recent local floods, volcanoes, and earthquakes—though clearly devastating to life and land—are tiny in comparison to the worldwide catastrophe that destroyed "the world that then existed" (2 Peter 3:6). All land animals and people not on board the Ark were destroyed in the floodwaters—billions of animals were preserved in the great fossil record we see today.

How Could the Ark Survive the Flood?

The description of the Ark is very brief—Genesis 6:14–16. Those three verses contain critical information including overall dimensions, but Noah was possibly given more detail than this. Other divinely specified constructions in the Bible are meticulously detailed, like the descriptions of Moses' Tabernacle or the temple in Ezekiel's vision.

The Bible does not say the Ark was a rectangular box. In fact, Scripture gives no clue about the shape of Noah's Ark other than the proportions—length, width, and depth. Ships have long been described like this without ever implying a block-shaped hull.

Moses used the obscure term *tebah*, a word that is only used again for the basket that carried baby Moses (Exodus 2:3). One was a huge wooden ship and the other a tiny wicker basket. Both float, both rescue life, and both are covered. But the similarity ends there. We can be quite sure that the baby basket did not have the same proportions as the Ark, and Egyptian baskets of the time were typically rounded. Perhaps *tebah* means "lifeboat."

For many years, biblical creationists have simply depicted the Ark as a rectangular box. This shape helped illustrate its size while avoiding the distractions of hull curvature. It also made it easy to compare volume. By using a short cubit and the maximum number of animal "kinds," creationists, as we've seen, have demonstrated how easily the Ark could fit the payload.[7] At the time, space was the main issue; other factors were secondary.

[7] To read a thorough study on this research, see *Noah's Ark: A Feasibility Study* by John Woodmorappe.

However, the next phase of research investigated sea-keeping (behavior and comfort at sea), hull strength, and stability. This began with a Korean study performed at the world-class ship research center (KRISO) in 1992.[8] The team of nine KRISO researchers was led by Dr. Hong, who is now director-general of the research center.

The study confirmed that the Ark could handle waves as high as 98 feet (30 m), and that the proportions of the biblical Ark are near optimal—an interesting admission from Dr. Hong, who believes evolutionary ideas, openly claiming "life came from the sea."[9] The study combined analysis, model wave testing, and ship standards, yet the concept was simple: compare the biblical Ark with 12 other vessels of the same volume but modified in length, width, or depth. Three qualities were measured—stability, hull strength, and comfort.

Ship Qualities Measured in the 1992 Korean Study

While Noah's Ark was an average performer in each quality, it was among the best designs overall. In other words, the proportions show a careful design balance that is easily lost when proportions are modified the wrong way. It is no surprise that modern ships have similar proportions—those proportions work.

Interesting to note is the fact that this study makes nonsense of the claim that Genesis was written only a few centuries before Christ and was based on flood legends such as the Epic of Gilgamesh. The Babylonian Ark is a cube shape, something so far from reality that even the shortest hull in the Korean study was not even close. But we would expect mistakes from other flood accounts, like that of Gilgamesh, as the account of Noah would have been distorted as it was passed down through different cultures.

Yet one mystery remained. The Korean study did not hide the fact that some shorter hulls slightly outperformed the biblical Noah's Ark. Further work by Tim Lovett, one author of this chapter, and two naval architects, Jim King and Dr. Allen Magnuson, focused attention on the issue of broaching— being turned sideways by the waves.

How do we know what the waves were like? If there were no waves at all, stability, comfort, or strength would be unimportant, and the proportions would not matter. A shorter hull would then be a more efficient

[8] Hong, et al., Safety Investigation of Noah's Ark in a seaway, *TJ* 8(1):26–36, April 1994. www.answersingenesis.org/tj/v8/i1/noah.asp.

[9] Seok Won Hong, Warm greetings from the Director-General of MOERI (former KRISO), Director-General of MOERI/KORDI, www.moeri.re.kr/eng/about/about.htm.

volume, taking less wood and less work. However, we can take clues from the proportions of the Ark itself. The Korean study had assumed waves came from every direction, giving shorter hulls an advantage. But real ocean waves usually have a dominant direction due to the wind, favoring a short, wide hull even more.

Another type of wave may also have affected the Ark during the Flood—tsunamis. Earthquakes can create tsunamis that devastate coastlines. However, when a tsunami travels in deep water it is imperceptible to a ship. During the Flood, the water would have been very deep—there is enough water in today's oceans to cover the earth to a depth of about 1.7 miles (2.7 km). The Bible states that the Ark rose "high above the earth" (Genesis 7:17). Launched from high ground by the rising floodwaters, the Ark would have avoided the initial devastation of coastlines and low-lying areas, and remained safe from tsunamis throughout the voyage.

After several months at sea, God sent a wind (Genesis 8:1), which could have produced very large waves since these waves can be produced by a strong, steady wind. Open-water testing confirms that any drifting vessel will naturally turn side-on to the waves (broach). With waves approaching the side of the vessel (beam sea), a long vessel like the Ark would be trapped in an uncomfortable situation; in heavy weather it could become dangerous. This could be overcome, however, by the vessel catching the wind (Genesis 8:1) at the bow and catching the water at the stern—aligning itself like a wind vane. These features appear to have inspired a number of ancient ship designs. Once the Ark points into the waves, the long, ship-like proportions create a more comfortable and controlled voyage. Traveling slowly with the wind, it had no need for speed, but the Bible does say the Ark moved about on the surface of the waters (Genesis 7:18).

However, not all waves are aligned with the local wind, and the Ark may have also encountered distant swells from any direction. The first line of defense came from the excellent proportions of the Ark, confirmed in a study by a world-class ship research center in Korea.[10] In addition, the outer keels that provide protection on land also improve roll damping in the waves, much like

[10] A study undertaken at KRISO, a Korean ship research center and published in 1994, which dealt with the effect of changing the proportions of the biblical Ark in a random sea. They concluded that Noah's Ark was near optimal and that it could handle waves up to 47.5 m at the roll limit. This value is given as significant wave height, which is the mean wave height (trough to crest) of the highest third of the waves ($H_{1/3}$). So some waves could be even higher and the Ark would still be safe. https://answersingenesis.org/noahs-ark/safety-investigation-of-noahs-ark-in-a-seaway/.

the bilge keels of a modern ship. So it is prudent for Noah to have used lots of ancient ingenuity in the project of his life.

Compared to a ship-like bow and stern, blunt ends are not as strong, have edges that are vulnerable to damage during launch and beaching, and give a rougher ride. Since the Bible gives proportions like that of a true ship, it makes sense that it should look and act ship-like. The below design is an attempt to flesh out the biblical outline using real-life experiments and archeological evidence of ancient ships.

While Scripture does not point out a wind-catching feature at the bow, the abbreviated account we are given in Genesis makes no mention of drinking water, the number of animals, or the way they got out of the Ark either.

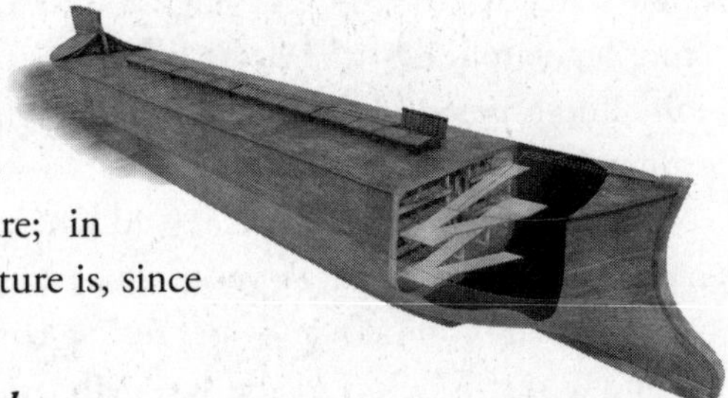

Nothing in this newly depicted Ark contradicts Scripture; in fact, it shows how accurate Scripture is, since the proportions are so realistic!

1. Something to catch the wind

Wind-driven waves would cause a drifting vessel to turn dangerously side-on to the weather. However, such waves could be safely navigated by making the Ark steer itself with a wind-catching obstruction on the bow. To be effective, this obstruction must be large enough to overcome the turning effect of the waves. While many designs could work, the possibility shown here reflects the high stems that were a hallmark of ancient ships.

2. A cubit upward and above

Any opening on the deck of a ship needs a wall (coaming) to prevent water from flowing in, especially when the ship rolls. In this illustration, the window "ends a cubit upward and above," as described in Genesis 6:16. The central position of the skylight is chosen to reflect the idea of a "noon light." This also means that the window does not need to be exactly one cubit. Perhaps the skylight had a transparent roof (even

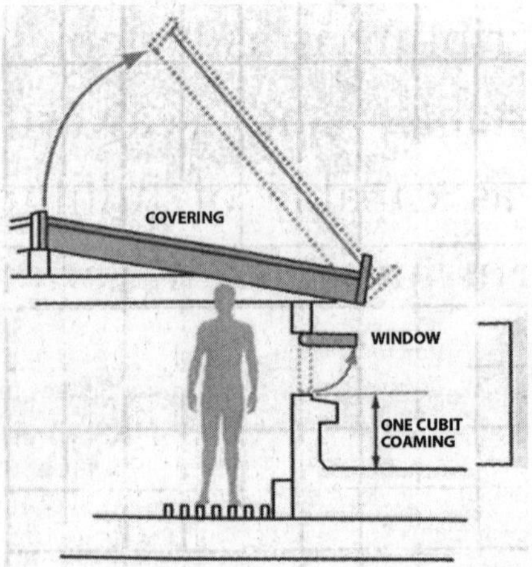

more a "noon light"), or the skylight roof could be opened (which might correspond to when "Noah removed the covering of the Ark"). While variations are possible, a window without coaming is not the most logical solution.

3. Mortise and tenon planking

Ancient shipbuilders usually began with a shell of planks (strakes) and then built internal framing (ribs) to fit inside. This is the complete reverse of the familiar European method where planking was added to the frame. In shell-first construction, the planks must be attached to each other somehow. Some used overlapping (clinker) planks that were dowelled or nailed; others used rope to sew the planks together. The ancient Greeks used a sophisticated system where the planks were interlocked with thousands of precise mortise and tenon joints. The resulting hull was strong enough to ram another ship, yet light enough to be hauled onto a beach by the crew. If this is what the Greeks could do centuries before Christ, what could Noah do centuries after Tubal-Cain invented forged metal tools?

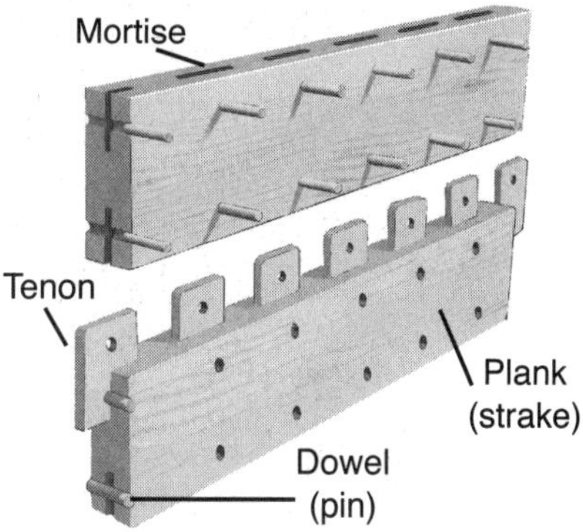

4. Ramps

Ramps help to get animals and heavy loads between decks. Running them across the hull avoids cutting through important deck beams, and this location is away from the middle of the hull where bending stresses are highest. (This placement also better utilizes the irregular space at bow and stern.)

5. Something to catch the water

To assist in turning the Ark to point with the wind, the stern should resist being pushed sideways. This is the same as a fixed rudder or skeg that provides directional control. There are many ways this could be done, but here we are reflecting the "mysterious" stern extensions seen on the earliest large ships of the Mediterranean.

Where Did All the Water Come From?

> In the six hundredth year of Noah's life, in the second month, the seventeenth day of the month, on that day all the fountains of the great deep were broken up, and the windows of heaven were opened. And the rain was on the earth forty days and forty nights (Genesis 7:11–12).

The Bible tells us that water came from two sources: below the earth and above the earth. Evidently, the source for water below the ground was in great subterranean pools, or "fountains" of fresh water, which were broken open by volcanic and seismic (earthquake) activity.[11]

Where Did All the Water Go?

> And the waters receded continually from the earth. At the end of the hundred and fifty days the waters decreased (Genesis 8:3).

Simply put, the water from the Flood is in the oceans and seas we see today. Three-quarters of the earth's surface is covered with water.

As even secular geologists observe, it does appear that the continents were at one time "together" and not separated by the vast oceans of today. The forces involved in the Flood were certainly sufficient to change all of this.

Scripture indicates that God formed the ocean basins, raising the land out of the water so that the floodwaters returned to a safe place. (Some theologians believe Psalm 104:7–9 may refer to this event.) Some creation scientists believe this breakup of the continent was part of the mechanism that ultimately caused the Flood.[12]

Some have speculated, because of Genesis 10:25, that the continental break occurred during the time of Peleg. However, this division is mentioned in the context of the Tower of Babel's language division of the whole earth (Genesis 10–11), so the context points to a dividing of the languages and people groups, not the land breaking apart.

If there were a massive movement of continents during the time of Peleg, there would have been another worldwide flood. The Bible indicates that the mountains of Ararat existed for the Ark to land in them (Genesis 8:4); so the Indian-Australian Plate and Eurasian Plate had to have already collided, indicating that the continents had already shifted prior to Peleg.

Was Noah's Flood Global?

> And the waters prevailed exceedingly on the earth, and all the high hills under the whole heaven were covered. The waters prevailed fifteen cubits upward, and the mountains were covered (Genesis 7:19–20).

Many Christians today claim that the Flood of Noah's time was only a local flood. These people generally believe in a local flood because they

---

[11] For deeper study on this, please see Nozomi Osanai, "A Comparative Study of the Flood Accounts in the Gilgamesh Epic and Genesis," www.answersingenesis.org/go/gilgamesh.

[12] See chapter 14 by Dr. Andrew Snelling for more details on this subject.

have accepted the widely believed evolutionary history of the earth, which interprets fossil layers as the history of the sequential appearance of life over millions of years.[13]

Scientists once understood the fossils, which are buried in water-carried sediments of mud and sand, to be mostly the result of the great Flood. Those who now accept millions of years of gradual accumulation of fossils have, in their way of thinking, explained away the evidence for the global Flood. Hence, many compromising Christians insist on a local flood.

Secularists deny the possibility of a worldwide Flood at all. If they would think from a biblical perspective, however, they would see the abundant evidence for the global Flood. As someone once quipped, "I wouldn't have seen it if I hadn't believed it."

Those who accept the evolutionary time frame, with its fossil accumulation, also rob the Fall of Adam of its serious consequences. They put the fossils, which testify of disease, suffering, and death, before Adam and Eve sinned and brought death and suffering into the world. In doing this, they also undermine the meaning of the death and Resurrection of Christ. Such a scenario also robs all meaning from God's description of His finished creation as "very good."

If the Flood only affected the area of Mesopotamia, as some claim, why did Noah have to build an Ark? He could have walked to the other side of the mountains and escaped. Most importantly, if the Flood were local, people not living in the vicinity of the Flood would not have been affected by it. They would have escaped God's judgment on sin.

In addition, Jesus believed that the Flood killed every person not on the Ark. What else could Christ mean when He likened the coming world judgment to the judgment of "all" men in the days of Noah (Matthew 24:37–39)?

In 2 Peter 3, the coming judgment by fire is likened to the former judgment by water in Noah's Flood. A partial judgment in Noah's day, therefore, would mean a partial judgment to come.

If the Flood were only local, how could the waters rise to 20 feet (6 m) above the mountains (Genesis 7:20)? Water seeks its own level; it could not rise to cover the local mountains while leaving the rest of the world untouched.

---

[13] For compelling evidence that the earth is not billions of years old, read *The Young Earth* by Dr. John Morris (Green Forest, AR: Master Books, 1994) and *Thousands . . . not Billions* by Dr. Don DeYoung (Green Forest, AR: Master Book, 2005). Also see www.answersingenesis.org/go/young.

A local Flood?

Even what is now Mt. Everest was once covered with water and uplifted afterward.[14] If we even out the ocean basins and flatten out the mountains, there is enough water to cover the entire earth by about 1.7 miles (2.7 km).[15] Also important to note is that, with the leveling out of the oceans and mountains, the Ark would not have been riding at the height of the current Mt. Everest; thus no need for such things as oxygen masks either.

There's more. If the Flood were a local flood, God would have repeatedly broken His promise never to send such a flood again. God put a rainbow in the sky as a covenant between God and man and the animals that He would never repeat such an event. There have been huge local floods in recent times (e.g., in Bangladesh, Indonesia, and Japan); but never has there been another global Flood that killed all life on the land.

Where Is the Evidence in the Earth for Noah's Flood?

> For this they willingly forget: that by the word of God the heavens were of old, and the earth standing out of water and in the water, by which the world that then existed perished, being flooded with water (2 Peter 3:5–6).

Evidence of Noah's Flood can be seen all over the earth, from seabeds to mountaintops. Whether you travel by car, train, or plane, the physical features of the earth's terrain clearly indicate a catastrophic past, from canyons

[14] Mount Everest is more than 5 miles (8 km) high. How, then, could the Flood have covered "all the mountains under the whole heaven?" Before the Flood, the mountains were not so high. The mountains today were formed only toward the end of, and after, the Flood by collision of the tectonic plates and the associated up-thrusting. In support of this, the layers that form the uppermost parts of Mt. Everest are themselves composed of fossil-bearing, water-deposited layers. For more on this, see chapter 14 on catastrophic plate tectonics.

[15] A.R. Wallace, *Man's Place in the Universe* (New York: McClure, Phillips & Co, 1903), p. 225–226; www.wku.edu/~smithch/wallace/S728-3.htm.

and craters to coal beds and caverns. Some layers of strata extend across continents, revealing the effects of a huge catastrophe.

The earth's crust has massive amounts of layered sedimentary rock, sometimes miles (kilometers) deep! These layers of sand, soil, and material—mostly laid down by water—were once soft like mud, but they are now hard stone. Encased in these sedimentary layers are billions of dead things (fossils of plants and animals) buried very quickly. The evidence all over the earth is staring everyone in the face.

Where Is Noah's Ark Today?

> Then the ark rested in the seventh month, the seventeenth day of the month, on the mountains of Ararat (Genesis 8:4).

The Ark landed in mountains. The ancient name for these mountains could refer to several areas in the Middle East, such as Mt. Ararat in Turkey or other mountain ranges in neighboring countries.

Mt. Ararat has attracted the most attention because it has permanent ice, and some people report to have seen the Ark. Many expeditions have searched for the Ark there. There is no conclusive evidence of the Ark's location or survival; after all, it landed on the mountains about 4,500 years ago. Also, it could easily have deteriorated, been destroyed, or been used as lumber by Noah and his descendants.

Some scientists and Bible scholars, though, believe the Ark could indeed be preserved—perhaps to be providentially revealed at a future time as a reminder of the past judgment and the judgment to come, although the same could be said for things like the Ark of the Covenant or other biblical icons. Jesus said, "If they do not hear Moses and the prophets, neither will they be persuaded though one rise from the dead" (Luke 16:31).

The Ark is unlikely to have survived without supernatural intervention, but this is neither promised nor expected from Scripture. However, it is a good idea to check if it still exists.

Why Did God Destroy the Earth That He Had Made?

> Then the Lord saw that the wickedness of man was great in the earth, and that every intent of the thoughts of his heart was only evil continually. But Noah found grace in the eyes of the Lord (Genesis 6:5, 8).

These verses speak for themselves. Every human being on the face of the earth had turned after the wickedness in their own hearts, but Noah, because of his righteousness before God, was spared from God's judgment, along with his wife, their sons, and their wives. As a result of man's wickedness, God sent judgment on all mankind. As harsh as the destruction was, no living person was without excuse.

God also used the Flood to separate and to purify those who believed in Him from those who didn't. Throughout history and throughout the Bible, this cycle has taken place time after time: separation, purification, judgment, and redemption.

Without God and without a true knowledge and understanding of Scripture, which provides the true history of the world, man is doomed to repeat the same mistakes over and over again.

How Is Christ Like the Ark?

> For the Son of Man has come to save that which was lost (Matthew 18:11).

As God's Son, the Lord Jesus Christ is like Noah's Ark. Jesus came to seek and to save the lost. Just as Noah and his family were saved by the Ark, rescued by God from the floodwaters, so anyone who believes in Jesus as Lord and Savior will be spared from the coming final judgment of mankind, rescued by God from the fire that will destroy the earth after the last days (2 Peter 3:7).

Noah and his family had to go through a doorway into the Ark to be saved, and the Lord shut the door behind them (Genesis 7:16). So we too have to go through a "doorway" to be saved so that we won't be eternally separated from God. The Son of God, Jesus, stepped into history to pay the penalty for our sin of rebellion. Jesus said, "I am the door. If anyone enters by Me, he will be saved, and will go in and out and find pasture" (John 10:9).

11

How Did Animals Spread All Over the World from Where the Ark Landed?

PAUL F. TAYLOR

An issue often used in an attempt to beat biblical creationists over the head is the worldwide distribution of animals. Such a distribution, say critics, proves that there could never have been a global Flood or an Ark. If the Ark landed somewhere in the Middle East, then all the animals would have disembarked at that point, including animals that we do not find in the Middle East today, or in the fossil record in that area. How did kangaroos get to Australia, or kiwis to New Zealand? How did polar bears get to North America and penguins to Antarctica?

Not a Science Textbook

Skeptics often claim, "The Bible is not a science textbook." This, of course, is true—because science textbooks change every year, whereas the Bible is the unchanging Word of God—the God who cannot lie. Nevertheless, the Bible can be relied upon when it touches on every scientific issue, including ecology. It is the Bible that gives us the big picture. Within this big picture, we can build scientific models that help us explain how past events may have come about. Such models should be held to lightly, but the Scripture to which they refer is inerrant. That is to say future research may cast doubt on an actual model, without casting doubt on Scripture.

With this in mind, the question needs to be asked, "Is there a Bible-based model that we can use to help explain how animals might have migrated from where the Ark landed to where they live today?" The answer is yes.

The Hard Facts

A biblical model of animal migration obviously must start with the Bible. From Genesis we can glean the following pertinent facts:

1. "And of every living thing of all flesh you shall bring two of every sort into the ark, to keep them alive with you; they shall be male and female. Of the birds after their kind, of animals after their kind, and of every creep- ing thing of the earth after its kind, two of every kind will come to you to keep them alive" (Genesis 6:19–20). The Bible is clear that representatives of all the *kinds* of air-breathing land animals and birds were present on the Ark. A technical term used by some creation scientists for these *kinds* is *baramin*—derived from the Hebrew words for *created kind*. Within these baramins is all the information necessary to produce all current species. For example, it is unlikely that the Ark contained two lions and two tigers. It is more likely that it contained two feline animals, from which lions, tigers, and other cat-like creatures have developed.

2. Another lesson from Genesis 6:20 is that the animals came to Noah. He did not have to go and catch them. Therefore, this preservation of the world's fauna was divinely controlled. It was God's intention that the fauna be preserved. The animals' recolonization of the land masses was therefore determined by God, and not left to chance.

3. "Then the ark rested in the seventh month, the seventeenth day of the month, on the mountains of Ararat" (Genesis 8:4). The Bible is clear that the Ark landed in the region of Ararat, but much debate has ensued over whether this is the same region as the locality of the present-day mountain known as Ararat. This issue is of importance, as we shall see. The Bible uses the plural "mountains." It is unlikely that the Ark rested on a point on the top of a mountain, in the manner often illustrated in children's picture books. Rather, the landing would have been among the mountainous areas of western Turkey, where present-day Mount Ararat is located, and eastern Iran, where the range extends.

4. It was God's will that the earth be recolonized. "Then God spoke to Noah, saying, 'Go out of the ark, you and your wife, and your sons

and your sons' wives with you. Bring out with you every living thing of all flesh that is with you: birds and cattle and every creeping thing that creeps on the earth, so that they may abound on the earth, and be fruitful and multiply on the earth.' So Noah went out, and his sons and his wife and his sons' wives with him. Every animal, every creeping thing, every bird, and whatever creeps on the earth, according to their families, went out of the ark" (Genesis 8:15–19). The abundance and multiplication of the animals was also God's will.

The biblical principles that we can establish then are that, after the Flood, God desired the ecological reconstruction of the world, including its vulnerable animal kinds, and the animals must have spread out from a mountainous region known as Ararat.

The construction of any biblical model of recolonization must include these principles. The model suggested on the following pages is constructed in good faith, to explain the observed facts through the "eyeglasses" of the Bible. The Bible is inspired, but our scientific models are not. If we subsequently find the model to be untenable, this would not shake our commitment to the absolute authority of Scripture.

The model uses the multiplication of dogs as an example of how animals could have quickly repopulated the earth. Two dogs came off Noah's Ark and began breeding more dogs. Within a relatively short time period, there would be an incredible number of dogs of all sorts of different shapes and sizes. These dogs then began to spread out from the Ararat region to all parts of the globe.

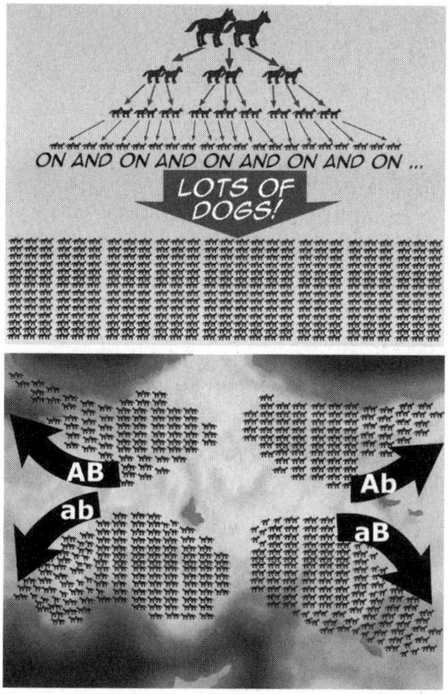

As these dogs spread around the world, variations within the dog kind led to many of the varieties we find today. But it is important to note that they are still dogs. This

multiplication of variations within a kind is the same with the many other kinds of animals.

One final comment must be made in this section. As I have used the word recolonization several times, I must emphasize that I am not re- ferring to the so-called *Recolonization Theory*. This theory will be discussed later.

Modern Recolonizations

One accusation thrown at biblical creationists is that kangaroos could not have hopped to Australia, because there are no fossils of kangaroos on the way. But the expectation of such fossils is a presuppositional error. Such an expectation is predicated on the assumption that fossils form gradually and inevitably from animal populations. In fact, fossilization is by no means inevitable. It usually requires sudden, rapid burial. Otherwise the bones would decompose before permineralization. One ought likewise to ask why it is that, despite the fact that millions of bison used to roam the prairies of North America, hardly any bison fossils are found there. Similarly, lion fossils are not found in Israel even though we know that lions once lived there.

Comparisons can be made with more modern recolonizations. For example, the *Encyclopædia Britannica* has the following to say about Surtsey Island and Krakatoa and the multiplication of species.

> Six months after the eruption of a volcano on the island of Surtsey off the coast of Iceland in 1963, the island had been colonized by a few bacteria, molds, insects, and birds. Within about a year of the eruption of a volcano on the island of Krakatoa in the tropical Pacific in 1883, a few grass species, insects, and vertebrates had taken hold. On both Surtsey and Krakatoa, only a few decades had elapsed before hundreds of species reached the islands. Not all species are able to take hold and become permanently established, but eventually the island communities stabilize into a dynamic equilibrium.[1]

There is little secret, therefore, how nonflying animals may have traveled to the outer parts of the world after the Flood. Many of them could have floated on vast floating logs, leftovers from the massive pre-Flood forests that were ripped up during the Flood and likely remained afloat for many decades on the world's oceans, transported by world currents. Others could later have been taken by people. Savolainen et al., have suggested, for example, that

1 https://www.britannica.com/science/community-ecology/Community-equilibrium-and-species-diversity.

all Australian dingoes are descended from a single female domesticated dog from Southeast Asia.[2] A third explanation of possible later migration is that animals could have crossed land bridges. This is, after all, how it is supposed by evolutionists that many animals and people migrated from Asia to the Americas—over a land bridge at the Bering Straits. For such land bridges to have existed, we may need to assume that sea levels were lower in the post-Flood period—an assumption based on a biblical model of the Ice Age.

Ice Age

As Michael Oard, a retired meteorologist and Ice Age researcher, has suggested,[3] an Ice Age may have followed closely after the Flood. In his detailed analysis, Oard proposed a mechanism of how the rare conditions required to form an Ice Age may have been triggered by the Flood, and shows how this explains the field evidence for an Ice Age.[4]

Severe climatic changes could have been the catalyst that encouraged certain species to migrate in certain directions. These severe changes could also have accounted for some of the many extinctions that occurred. Additionally, Oard's studies provide a model for how land bridges could have developed.

Oard points out that certain observed features from the Ice Age cause problems for the evolutionist, not the creationist. Thus, a creationist explanation of the Ice Age better explains the facts. An example of such an issue is that of disharmonious associations of fossils—fossils of creatures normally associated with different conditions (such as creatures with a preference for hot and cold climates) being found in close proximity.

> One of the more puzzling problems for uniformitarian theories of the ice age is disharmonious associations of fossils, in which species from different climatic regimes are juxtaposed. For example, a hippopotamus fossil found together with a reindeer fossil.[5]

Oard suggests that even with present topography, a number of significant land bridges would have existed to facilitate migrations if the sea level were

---

[2] Savolainen et al., A detailed picture of the origin of the Australian dingo, obtained from the study of mitochondrial DNA, *PNAS* (Proceedings of the National Academy of Sciences of the National of the United States of America) 101:12387–12390, August 2004.

[3] See chapter 16 of this book.

[4] Oard has published many articles in journals and on the AiG and ICR websites on these issues. For a detailed account of his findings, see his book: M. Oard, *An Ice Age Caused by the Genesis Flood* (El Cajon, CA: Institute for Creation Research, 2002).

[5] Ibid, p. 80.

only 180 ft. (55 m) below current levels. However, there is even evidence that the land in some places where land bridges would be necessary could have been higher still. Thus, land bridges facilitated by the Ice Age constitute a serious model to explain how some migrations could have been possible.

Some still remain skeptical about the idea of land bridges all the way to Australia. Nevertheless, by a combination of methods that we see today, including land bridges, there are rational explanations as to how animals may have reached the far corners of the world. Of course, we were not there at the time to witness how this migration may have happened, but those adhering to a biblical worldview can be certain that animals obviously did get to far places, and that there are rational ways in which it could have happened.

We should therefore have no problem accepting the Bible as true. Creationist scientific models of animal migration are equally as valid as evolutionary models, if not more so. The reason such models are rejected is that they do not fit in with the orthodox, secular evolutionary worldview.

It is not a problem for us to rationalize why certain animals do not appear in certain parts of the world. Why, for example, does Australia have such an unusual fauna, including so many marsupials? Marsupials are, of course, known elsewhere in the world. For example, opossums are found in North and South America, and fossilized marsupials have been found elsewhere. But in many places, climatic changes and other factors could lead to their extinction.

The lack of great marsupials in other continents need be no more of a problem than the lack of dinosaurs. As with many species today, they just died out—a reminder of a sin-cursed world. One proposed theory is that marsupials—because they bore their young in pouches—were able to travel farther and faster than mammals that had to stop to care for their young.

They were able to establish themselves in far-flung Australia before competitors reached the continent.

Similar statements could be made about the many unusual bird species in New Zealand, on islands from which mammals were absent until the arrival of European settlers.

Recolonization Theory

The most logical interpretation of the biblical record of the Flood and its aftermath would seem to suggest that the animals disembarked and then recolonized the planet. Comparisons with modern migrations and incidents such as Surtsey have suggested that this recolonization need not have taken long. A plain reading of Scripture suggests that the Ark landed in the mountains of Ararat, most likely in the region of modern Turkey and Central Asia. It is also our contention that the significant quantity of death represented by the fossil record is best understood by reference to the Genesis Flood (i.e., the majority of fossils formed as a result of the Flood).

More recently, a theory has developed among certain creationists in the UK and Europe that suggests that the fossil record is actually a record not of catastrophe but of processes occurring during recolonization. This theory is called the Recolonization Theory.[6]

Proponents of this theory suggest that the Flood completely obliterated the earth's previous crust so that none of the present fossils were caused by it. To accommodate fossilization processes, Recolonization Theory suggests that the age of the earth be stretched by a few thousand years. Some advocates of this view suggest an age of about 8,000 years for the earth, while others suggest figures as high as 20,000 years.

A detailed criticism of Recolonization Theory has previously been published by McIntosh, Edmondson, and Taylor,[7] and another by Holt.[8]

The principal error of this view is that it starts from supposed scientific anomalies, such as the fossil record, rather than from Scripture. This has led to the proposals among some Recolonizers, but not all, that there must be gaps in the genealogies recorded in Genesis 5 and 11, even though there is no need for such gaps. Indeed, the suggestion of gaps in these genealogies causes further doctrinal problems.[9]

Even the views of those Recolonizers who do not expand the genealogies contain possible seeds of compromise. Because the Recolonizers accept the geologic column, and because the Middle East has a great deal of what is called Cretaceous rock, it follows that the Middle East would need to be submerged after the Flood, at the very time of the Tower of Babel events in Genesis 11. This has led some of the Recolonizers to speculate that the Ark

6 Spelled "Recolonisation" in the UK, which is where the theory began.

7 A.C. McIntosh, T. Edmondson, and S. Taylor, "Flood Models: The need for an integrated approach," *TJ* 14(1):52–59, April 2000; A.C. McIntosh, T. Edmondson, and S. Taylor, Genesis and Catastrophe, *TJ* 14(1):101–109, April 2000. Recolonizers' disagreements with these article were answered in A.C. McIntosh, T. Edmondson, and S. Taylor, McIntosh, Taylor, and Edmondson reply to Flood Models, *TJ* 14(3):80–82, 2000, available online at https://answersingenesis.org/the-flood/mcintosh-taylor-and-edmondson-reply-to-flood-models/.

8 R. Holt, *Evidence for a Late Cainozoic Flood/post-Flood Boundary*, *TJ* 10(1):128–168, April 1996.

9 For more on this see Ken Ham, editor, *The New Answers Book 2* (Green Forest, AR: Master Books 2016), chapter 5, "Are There Gaps in the Genesis Genealogies?" https://answersingenesis.org/bible-timeline/genealogy/gaps-in-the-genesis-genealogies/.

actually landed in Africa, and therefore, that continent was the host to the events of Genesis 11 and 12. This would seem to be a very weak position exegetically and historically. Such exegetical weaknesses led Professor Andy McIntosh and his colleagues to comment, "Their science is driving their interpretation of Scripture, and not the other way round."[10]

Conclusions

We must not be downhearted by critics and their frequent accusations against the Bible. We must not be surprised that so many people will believe all sorts of strange things, whatever the logic.

Starting from our presupposition that the Bible's account is true, we have seen that scientific models can be developed to explain the post-Flood migration of animals. These models correspond to observed data and are consistent with the Bible's account. It is notable that opponents of biblical creationism use similar models in their evolutionary explanations of animal migrations. While a model may eventually be superseded, it is important to note that such biblically consistent models exist. In any event, we have confidence in the scriptural account, finding it to be accurate and authoritative.[11] The fact of animal migration around the world is illustrative of the goodness and graciousness of God, who provided above and beyond our needs.

10 A.C. McIntosh, T. Edmondson, and S. Taylor, McIntosh, Taylor, and Edmondson reply to Flood Models, *TJ* 14(3):80–82, 2000.

11 See *Inside Noah's Ark*, Laura Welch, editor (Green Forest, AR: Master Books, 2016), https://answersingenesis.org/store/product/inside-noahs-ark/?sku=10-2-480.

12

What Really Happened to the Dinosaurs?

KEN HAM

Dinosaurs are used more than almost anything else to indoctrinate children and adults in the idea of millions of years of earth history. However, the Bible gives us a framework for explaining dinosaurs in terms of thousands of years of history, including the mystery of when they lived and what happened to them. Two key texts are Genesis 1:24–25 and Job 40:15–24.

Are Dinosaurs a Mystery?

Many think that the existence of dinosaurs and their demise is shrouded in such mystery that we may never know the truth about where they came from, when they lived, and what happened to them. However, dinosaurs are only a mystery *if* you accept the evolutionary story of their history.

According to evolutionists: Dinosaurs first evolved around 235 million years ago, long before man evolved.[1] No human being ever lived with dinosaurs. Their history is recorded in the fossil layers on earth, which were deposited over millions of years. They were so successful as a group of animals that they eventually ruled the earth. However, around 65 million years ago,

[1] J. Horner and D. Lessem, *The Complete T. Rex,* Simon & Schuster, New York, 1993, 18; M. Norell, E. Gaffney and L. Dingus, *Discovering Dinosaurs in the American Museum of Natural History,* Nevraumont Publ., New York, 1995, 17, says that the oldest dinosaur fossil is dated at 228 million years.

something happened to change all of this—the dinosaurs disappeared. Most evolutionists believe some sort of cataclysmic event, such as an asteroid impact, killed them. But many evolutionists claim that some dinosaurs evolved into birds, and thus they are not extinct but are flying around us even today.[2]

There is no mystery surrounding dinosaurs if you accept the Bible's totally different account of dinosaur history.

According to the Bible: Dinosaurs first existed around 6,000 years ago.[3] God made the dinosaurs, along with the other land animals, on Day 6 of the Creation Week (Genesis 1:20–25, 31). Adam and Eve were also made on Day 6—so dinosaurs lived at the same time as people, not separated by eons of time.

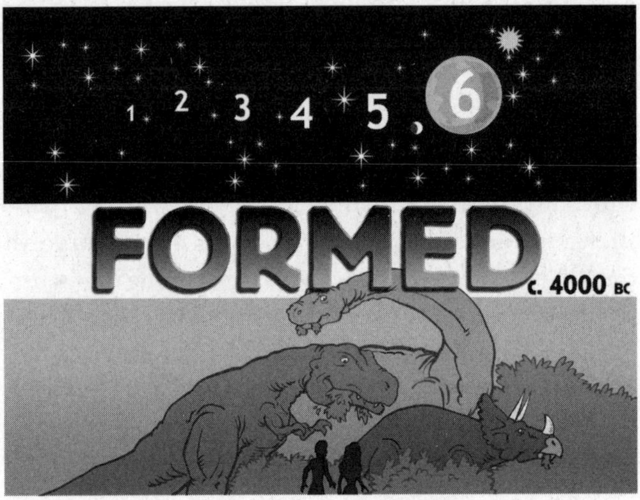

Dinosaurs could not have died out before people appeared because dinosaurs had not previously existed; and death, bloodshed, disease, and suffering are a result of Adam's sin (Genesis 1:29–30; Romans 5:12, 14; 1 Corinthians 15:21–22).

---

2 D. Gish, *Evolution: the Fossils Still Say No!* Institute for Creation Research, El Cajon, California, 1995, 129ff, discusses evolutionists' views from a creationist position; Norell et al., *Discovering Dinosaurs in the American Museum of Natural History,* 2: "Dinosaurs belong to a group called Archosauria … . The living Archosauria are the twenty-one extant crocodiles and alligators, along with the more than ten thousand species of living theropod dinosaurs (birds)."

3 J. Morris, *The Young Earth,* Master Books, Green Forest, Arkansas, 1994; H. Morris, *The Genesis Record,* Baker Book House, Grand Rapids, Michigan, 1976, 42–46. On the biblical chronology, see J. Ussher, *The Annals of the World,* Master Books, Green Forest, Arkansas, 2003; original published in 1658.

Representatives of all the *kinds* of air-breathing land animals, including the dinosaur kinds, went aboard Noah's Ark. All those left outside the Ark died in the cataclysmic circumstances of the Flood, and many of their remains became fossils.

After the Flood, around 4,300 years ago, the remnant of the land animals, including dinosaurs, came off the Ark and lived in the present world, along with people. Because of sin, the judgments of the Curse and the Flood have greatly changed earth. Post-Flood climatic change, lack of food, disease, and man's activities caused many types of animals to become extinct. The dinosaurs, like many other creatures, died out. Why the big mystery about dinosaurs?

Why Such Different Views?

How can there be such totally different explanations for dinosaurs? Whether one is an evolutionist or accepts the Bible's account of history, the evidence for dinosaurs is *the same*. All scientists have the same facts—they have the same world, the same fossils, the same living creatures, the same universe.

If the "facts" are the same, then how can the explanations be so different? The reason is that scientists have only the present—dinosaur fossils exist only in the present—but scientists are trying to connect the fossils in the present to the past. They ask, "What happened in history to bring dinosaurs into existence, wipe them out, and leave many of them fossilized?"[4]

The science that addresses such issues is known as *historical* or *origins science*, and it differs from the *operational science* that gives us computers, inexpensive food, space exploration, electricity, and the like. Origins science deals with the past, which is not accessible to direct experimentation, whereas operational science deals with how the world works in the here and now, which, of course, is open to repeatable experiments. Because of difficulties in reconstructing the past, those who study fossils (paleontologists) have diverse views on dinosaurs.[5] As has been said, "Paleontology (the study of fossils) is much like politics: passions run high, and it's easy to draw very different conclusions from the same set of facts."[6]

[4] M. Benton, *Dinosaurs: An A–Z Guide,* Derrydale Books, New York, 1988, 10–11.

[5] Benton, *Dinosaurs: An A–Z Guide.* See also D. Lambert and the Diagram Group, *The Dinosaur Data Book,* Avon Books, New York, 1990, 10–35; Norell, *et al., Discovering Dinosaurs in the American Museum of Natural History*, 62–69; V. Sharpton and P. Ward, Eds., *Global Catastrophes in Earth History,* The Geological Society of America, Special Paper 247, 1990.

[6] M. Lemonick, Parenthood, dino-style, *Time*, p. 48, January 8, 1996.

A paleontologist who believes the record in the Bible, which claims to be the Word of God,[7] will come to different conclusions than an atheist who rejects the Bible. Willful denial of God's Word (2 Peter 3:3–7) lies at the root of many disputes over historical science.

Many people think the Bible is just a book about religion or salvation. It is much more than this. The Bible is the History Book of the

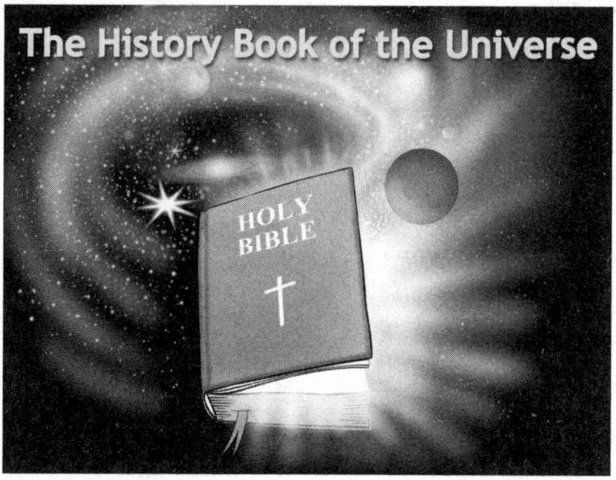

Universe and tells us the future destiny of the universe as well. It gives us an account of when time began, the main events of history, such as the entrance of sin and death into the world, the time when the *whole* surface of the globe was destroyed by water, the giving of different languages at the Tower of Babel, the account of the Son of God coming as a man, His death and Resurrection, and the new heavens and earth to come.

Ultimately, there are only two ways of thinking: starting with the revelation from God (the Bible) as foundational to *all* thinking (including biology, history, and geology), resulting in a *Christian worldview;* or starting with man's beliefs (for example, the evolutionary story) as foundational to all thinking, resulting in a *secular worldview.*

Most Christians have been indoctrinated through the media and education system to think in a secular way. They tend to take secular thinking *to* the Bible, instead of using the Bible to *build* their thinking (Romans 12:1–2; Ephesians 4:20–24).

[7] Psalm 78:5; 2 Timothy 3:14–17; and 2 Peter 1:19–21. God, who inspired the writing, has always existed, is perfect and never lies (Titus 1:2).

The Bible says, "The fear of the Lord is the beginning of knowledge" (Proverbs 1:7) and "the fear of the Lord is the beginning of wisdom" (Proverbs 9:10).

If one begins with an evolutionary view of history (for which there were no witnesses or written record), then this way of thinking will be used to explain the evidence that exists in the present. Thus, we have the evolutionary explanation for dinosaurs above.

But if one begins with the biblical view of history from the written record of an eyewitness (God) to all events of history, then a totally different way of thinking, based on this, will be used to explain the *same* evidence. Thus, we have the biblical explanation given above.

Dinosaur History

Fossil bones of dinosaurs are found around the world. Many of these finds consist of just fragments of bones, but some nearly complete skeletons have been found. Scientists have been able to describe many different types of dinosaurs based on distinctive characteristics, such as the structure of the skull and limbs.[8]

Where Did Dinosaurs Come From?

The Bible tells us that God created different kinds of land animals on Day 6 of Creation Week (Genesis 1:24–25). Because dinosaurs were land animals, this must have included the dinosaur kinds.[9]

8 D. Lambert, *A Field Guide to Dinosaurs,* Avon Books, New York, 1983, 17.

9 If some dinosaurs were aquatic, then these would have been created on Day 5 of Creation Week.

Evolutionists claim that dinosaurs evolved from some reptile that had originally evolved from amphibians. But they cannot point to any clear transitional (in-between) forms to substantiate their argument. Dinosaur family trees in evolutionary books show many distinct types of dinosaurs, but only hypothetical lines join them up to some common ancestor. The lines are dotted because there is *no* fossil evidence. Evolutionists simply cannot prove their belief in a nondinosaur ancestor for dinosaurs.

What Did Dinosaurs Look Like?

Scientists generally do not dig up a dinosaur with all its flesh intact. Even if they found *all* the bones, they still would have less than 40 percent of the animal to work out what it originally looked like. The bones do not tell the color of the animal, for example, although some fossils of skin impressions have been found, indicating the skin texture. As there is some diversity of color among reptiles living today, dinosaurs may have varied greatly in color, skin texture, and so on.

When reconstructing dinosaurs from bony remains, scientists make all kinds of guesses and often disagree. For example, debate has raged about whether dinosaurs were warm- or cold-blooded. It is even difficult to tell whether a dinosaur was male or female from its bones. There is much speculation about such things.

Sometimes scientists, because they're human, make honest mistakes in their reconstructions, which need correction when more bones are found. For instance, the famous *Brontosaurus* was removed for a while from dinosaur dictionaries, because the original "discoverer" put the wrong head on a skeleton of a dinosaur that had already been named Apatosaurus.[10] But now some scientists are arguing that the *Brontosaurus* is a separate species from *Apatosaurus*, thus either name is valid, depending on proper classification based on the minor differences.

Who Discovered Dinosaurs?

Secular books would tell you that the first discovery of what later were called dinosaurs was in 1677 when Dr. Robert Plot found bones so big they were thought to belong to a giant elephant or a giant human.[11]

In 1822, Mary Anne Mantell went for a walk along a country road in

[10] S.West Dinosaur head hunt, *Science News*, 116(18):314–315, 1979.

[11] Benton, *Dinosaurs: An A-Z Guide*, 14.

Sussex, England. According to tradition, she found a stone that glittered in the sunlight and showed it to her fossil-collecting husband. Dr. Mantell, a physician, noticed that the stone contained a tooth similar to, but much larger than, that of modern reptiles. He concluded that it belonged to some extinct giant plant-eating reptile with teeth like an iguana. In 1825 he named the owner of the tooth *Iguanodon* (iguana tooth). It was Dr. Mantell who began to popularize the "age of reptiles."[12]

From a biblical perspective, however, the time of the above discoveries was actually the time when dinosaurs were *rediscovered.* Adam discovered dinosaurs when he first observed them.

When Did Dinosaurs Live?

Evolutionists claim dinosaurs lived millions of years ago. But it is important to realize that when they dig up a dinosaur bone it does not have a label attached showing its date. Evolutionists obtain their dates by *indirect* dating methods that other scientists question, and there is much evidence against the millions of years.[13]

Does God tell us when He made *Tyrannosaurus rex*? Many would say no. But the Bible states that God made all things in six normal days. He made the land animals, including dinosaurs, on Day 6 (Genesis 1:24–25), so they date from around 6,000 years ago—the approximate date of creation obtained by

[12] Lambert et al., *The Dinosaur Data Book*, 279.
[13] Morris, *The Young Earth*, 51–67.

adding up the years in the Bible.[14] So, since *T. rex* was a land animal and God made all the land animals on Day 6, then God made *T. rex* on Day 6.

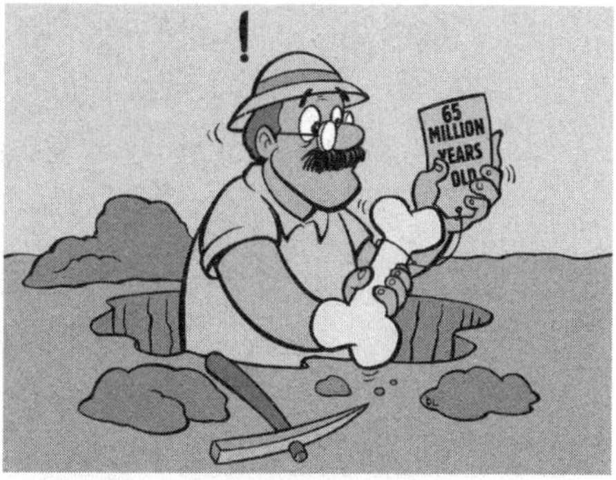

Furthermore, from the Bible we see that there was no death, bloodshed, disease, or suffering before sin.[15] If one approaches Genesis to Revelation consistently, interpreting Scripture with Scripture, then death and bloodshed of man and animals came into the world only *after* Adam sinned. The first death of an animal occurred when God shed an animal's blood in the Garden of Eden and clothed Adam and Eve (Genesis 3:21). This was also a picture of the Atonement—foreshadowing Christ's blood that was to be shed for us. Thus, there could *not* have been bones of dead animals before sin—this would undermine the gospel.

This means that the dinosaurs must have died after sin entered the world, not before. Dinosaur bones could *not* be millions of years old because Adam lived only thousands of years ago.

Does the Bible Mention Dinosaurs?

If people saw dinosaurs, you would think that ancient historical writings, such as the Bible, should mention them. The King James Version was first translated in 1611.[16] Some people think that because the word "dinosaur" is not found in this or other translations, the Bible does not mention dinosaurs.

---

[14] Morris, *The Genesis Record*, 44–46.

[15] J. Stambaugh, Creation, suffering and the problem of evil, *CEN Technical Journal* **10**(3):391–404, 1996.

[16] The KJV most often used today is actually the 1769 revision by Benjamin Blayney of Oxford.

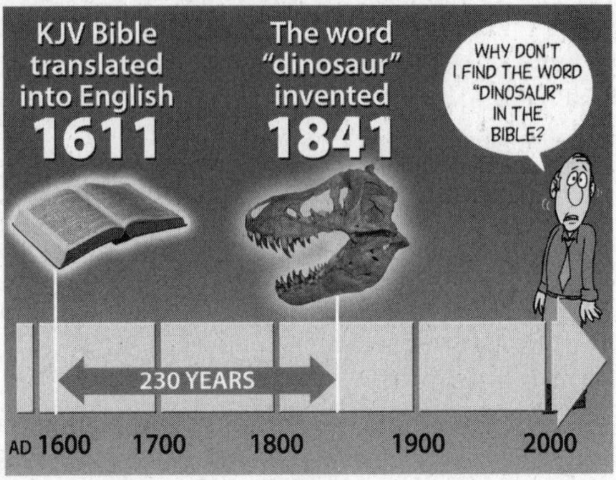

It was not until 1841, however, that the word "dinosaur" was invented.[17] Sir Richard Owen, a famous British anatomist and first superintendent of the British Museum (and a staunch anti-Darwinist), on viewing the bones of *Iguanodon* and *Megalosaurus*, realized these represented a unique group of reptiles that had not yet been classified. He coined the term "dinosaur" from Greek words meaning "terrible lizard."[18]

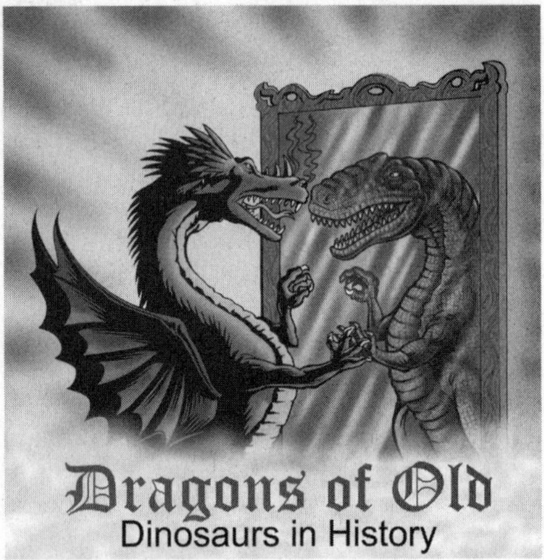

Thus, one would not expect to find the word "dinosaur" in the King James Bible—the word did not exist when the translation was done.

Is there another word for "dinosaur"? There are *dragon* legends from around the world. Many dragon descriptions fit the characteristics of specific dinosaurs. Could these actually be accounts of encounters with what we now call dinosaurs?

[17] D. Dixon et al., *The Macmillan Illustrated Encyclopedia of Dinosaurs and Prehistoric Animals*, Macmillan Publishing Co., New York, 1998, 92.

[18] D. Norman, *The Illustrated Encyclopedia of Dinosaurs,* Salamander Books Limited, London, 1985, 8. The meaning of "terrible lizard" has helped popularize the idea that dinosaurs were all gigantic savage monsters. This is far from the truth. Had Owen known about the *smaller* dinosaurs, he may never have coined the word.

Just as Flood legends are based on a real global Flood (Flood of Noah)—dragon legends are possibly based on actual encounters with real animals that today we call dinosaurs. Many of these land-dragon descriptions do fit with what we know about dinosaurs.

In Genesis 1:21, the Bible says, "And God created the great sea monsters and every living creature that moves, with which the waters swarmed, after their kind." The Hebrew word here for "sea monsters" ("whales" in KJV) is the word translated elsewhere as "dragon" (Hebrew: *tannin*). So, in the first chapter of the first book of the Bible, God may be describing the great sea dragons (sea-dwelling, dinosaur-type animals) that He created.

There are other Bible passages about dragons that lived in the sea: "the dragons in the waters" (Psalm 74:13), "and he shall slay the dragon that is in the sea" (Isaiah 27:1). Though the word "dinosaur" strictly refers to animals that lived on the land, the sea reptiles and flying reptiles are often grouped with the dinosaurs. The sea dragons could have included dinosaur-type animals such as the *Mosasaurus.*[19]

Job 41 describes a great animal that lived in the sea, Leviathan, that even breathed fire. This "dragon" may have been something like the mighty 40 ft. (12 m) *Sarcosuchus imperator* (Super Croc),[20] or the 82 ft. (25 m) *Liopleurodon.*

There is also mention of a flying serpent in the Bible: the "fiery flying serpent" (Isaiah 30:6). This could be a reference to one of the pterodactyls, which are popularly thought of as flying dinosaurs, such as the *Pteranodon, Rhamphorhynchus,* or *Ornithocheirus.*[21]

Not long after the Flood, God was showing a man called Job how great He was as Creator, by reminding Job of the largest land animal He had made:

---

[19] The Hebrew words have a range of meanings, including "sea monster" (Gen. 1:21; Job 7:12; Psa. 148:7; Isa. 27:1; Ezek. 29:3, 32:2) and "serpent" (Exod. 7:9; cf. Exod. 4:3 and Hebrew parallelism of Deut. 32:33). *Tannin/m* are fearsome creatures, inhabiting remote, desolate places (Isa. 34:13, 35:7; Jer. 49:33, 51:37; Mal. 1:8), difficult to kill (Isa. 27:1, 51:9) and/or serpentine (Deut. 32:33; cf. Psa. 91:13) and/or having feet (Ezek. 32:2). However, *tannin* are referred to as suckling their young (Lam. 4:3), which is not a feature of reptiles, but of whales (sea monsters?), for example. The word(s) seems to refer to large, fearsome creatures that dwelled in swampy areas or in the water. The term could include reptiles and mammals. Modern translators often render the words as "jackals," but this seems inappropriate because jackals are not particularly fearsome or difficult to kill and don't live in swamps.

[20] S. Czerkas and S. Czerkas, *Dinosaurs: A Global View,* Barnes and Noble Books, Spain, 1996, 179; P. Booker, A new candidate for Leviathan? *TJ* **19**(2):14–16, 2005.

[21] D. Norman, *The Illustrated Encyclopedia of Dinosaurs,* Gramercy, New York, 1988, 170–172; P. Wellnhofer, *Pterosaurs: The Illustrated Encyclopedia of Prehistoric Flying Reptiles,* Barnes and Noble, New York, 1991, 83–85, 135–136.

> Look now at the behemoth, which I made along with you; he eats grass like an ox. See now, his strength is in his hips, and his power is in his stomach muscles. He moves his tail like a cedar; the sinews of his thighs are tightly knit. His bones are like beams of bronze, his ribs like bars of iron. He is the first of the ways of God; only He who made him can bring near His sword (Job 40:15–19).

The phrase "first of the ways of God" suggests this was the largest land animal God had made. So what kind of animal was "behemoth"?

Bible translators, not being sure what this beast was, often transliterated the Hebrew, and thus the word *behemoth* (e.g., KJV, NKJV, NASB, NIV). However, in many Bible commentaries and Bible footnotes, "behemoth" is said to be "possibly the hippopotamus or elephant."[22] Some Bible versions actually translate "behemoth" this way.[23] Besides the fact that the elephant and hippo were *not* the largest land animals God made (some of the dinosaurs far eclipsed these), this description does not make sense, since the tail of behemoth is compared to the large cedar tree (verse 17).

Now an elephant's tiny tail (or a hippo's tail that looks like a flap of skin) is quite unlike a cedar tree. Clearly, the elephant and the hippo could not possibly be "behemoth."

No *living* creature comes close to this description. However, behemoth is very much like *Brachiosaurus*, one of the large dinosaurs.

[22] E.g., NIV Study Bible, Zondervan, Grand Rapids, Michigan, 1985.

[23] New Living Translation: Holy Bible, Tyndale House Publishers, Wheaton, Illinois, 1996. Job 40:15: "Take a look at the mighty hippopotamus."

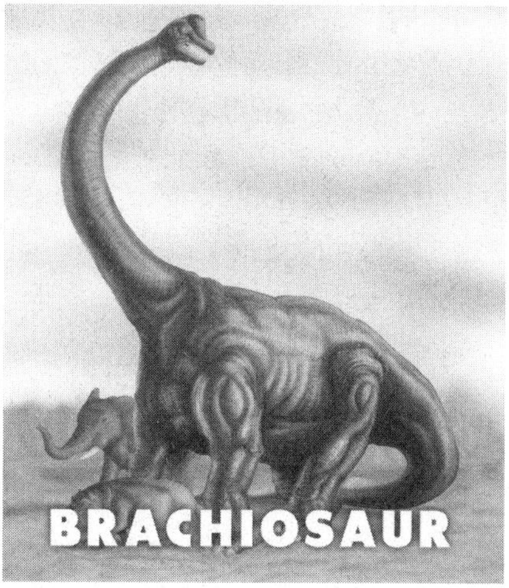

Are There Other Ancient Records of Dinosaurs?

In the film *The Great Dinosaur Mystery*,[24] a number of dragon accounts are presented:

- A Sumerian story dating back to 2000 BC or earlier tells of a hero named Gilgamesh, who, when he went to fell cedars in a remote forest, encountered a huge vicious dragon that he slew, cutting off its head as a trophy.
- When Alexander the Great (c. 330 BC) and his soldiers marched into India, they found that the Indians worshipped huge hissing reptiles that they kept in caves.
- China is renowned for its dragon stories, and dragons are prominent on Chinese pottery, embroidery, and carvings.
- England and several other cultures retains the story of St. George, who slew a dragon that lived in a cave.
- There is the story of a tenth-century Irishman who wrote of his encounter with what appears to have been a *Stegosaurus.*
- In the 1500s, a European scientific book, *Historia Animalium,* listed several living animals that we would call dinosaurs. A well-known naturalist of the time, Ulysses Aldrovandus, recorded an encounter between a peasant

[24] P. Taylor, *The Great Dinosaur Mystery,* Films for Christ, Mesa, Arizona, 1991. See also P. Taylor, *The Great Dinosaur Mystery and the Bible*, Accent Publications, Denver, Colorado, 1989.

named Baptista and a dragon whose description fits that of the small dinosaur *Tanystropheus.* The encounter was on May 13, 1572, near Bologna in Italy, and the peasant killed the dragon.

Saint George (Sankt Goran) and the dragon in Gamla Stan (Old Town) of Stockholm, Sweden

Petroglyphs (drawings carved on rock) of dinosaur-like creatures have also been found.[25]

In summary, people down through the ages have been very familiar with dragons. The descriptions of these animals fit with what we know about dinosaurs. The Bible mentions such creatures, even ones that lived in the sea and flew in the air. There is a tremendous amount of other historical evidence that such creatures have lived beside people.

What Do the Bones Say?

There is also physical evidence that dinosaur bones are not millions of years old. Scientists from the University of Montana found *T. rex* bones that were not totally fossilized. Sections of the bones were like fresh bone and contained what seems to be blood cells and hemoglobin. If these bones really were tens of millions of years old, then the blood cells and hemoglobin would have totally disintegrated. A report by these scientists stated the following:

> A thin slice of *T. rex* bone glowed amber beneath the lens of my microscope … . The lab filled with murmurs of amazement, for I had focused on something inside the vessels that none of us had ever noticed before: tiny round objects, translucent red with a dark center … . Red blood cells? The shape and location suggested them, but blood cells are mostly water and couldn't possibly have stayed preserved in the 65-million-year-old *tyrannosaur* … . The bone sample that had us so excited came from a beautiful, nearly complete specimen of *Tyrannosaurus rex* unearthed in 1990 … . When the team brought the dinosaur into the lab, we noticed that some parts

[25] D. Swift, Messages on stone, *Creation* **19**(2):20–23, 1997.

> deep inside the long bone of the leg had not completely fossilized … . So far, we think that all of this evidence supports the notion that our slices of *T. rex* could contain preserved heme and hemoglobin fragments. But more work needs to be done before we are confident enough to come right out and say, "Yes, this *T. rex* has blood compounds left in its tissues."[26]

Unfossilized duck-billed dinosaur bones have been found on the North Slope in Alaska.[27] The bones could not have survived for the millions of years unmineralized. This is a puzzle to those who believe in an "age of dinosaurs" millions of years ago, but not to someone who builds his thinking on the Bible.

What Did Dinosaurs Eat and How Did They Behave?

Movies like *Jurassic Park* and *The Lost World* portray most dinosaurs as aggressive meat-eaters. But the mere presence of sharp teeth does *not* tell you how an animal behaved or necessarily what food it ate—only what kind of teeth it had (for ripping food and the like). However, by studying fossil dinosaur dung (coprolite), scientists have been able to determine the diet of some dinosaurs.[28]

Originally, before sin, *all* animals, including the dinosaurs, were vegetarian. Genesis 1:30 states, "And to every beast of the earth, and to every bird of the air, and to every thing that creeps upon the earth, which has life, I have given every green herb for food: and it was so."

This means that even *T. rex,* before sin entered the world, ate only plants. Some people object to this by pointing to the big teeth that a large *T. rex* had, insisting they must have been used for attacking animals. However, just because an animal has big, sharp teeth does not mean it eats meat. It just means it has big, sharp teeth![29]

Many animals today have sharp teeth but are basically vegetarian. The giant panda has sharp teeth like a meat-eater, but it eats bamboo. Perhaps

---

26 M. Schweitzer and T. Staedter, The real Jurassic Park, *Earth,* pp. 55–57, June 1997. See report in *Creation* **19**(4):42–43, which describes the careful testing that showed that hemoglobin was present.

27 K. Davies, Duckbill dinosaurs (Hadrosauridae, Ornithischia) from the North Slope of Alaska, *Journal of Paleontology* **61**(1):198–200, 1987.

28 S. Lucas, *Dinosaurs: The Textbook*, Wm. C. Brown Publishers, Dubuque, IA, 1994, 194–196.

29 D. Marrs and V. Kylberg, *Dino Cardz,* 1991. *Estemmenosuchus* was a large mammal-like reptile. "Despite having menacing-looking fangs it apparently was a plant-eater." The authors possibly concluded this from its rear teeth.

the panda's teeth were beautifully designed to eat bamboo. To explain why a giant panda has teeth like a meat-eaters today, yet eats bamboo, evolutionists have to say that the giant panda evolved as a meat eater, and then switched to bamboo.[30]

Different species of bats variously eat fruit, nectar, insects, small animals, and blood, but their teeth do not clearly indicate what they eat.[31] Bears have teeth with carnivore features, but some bears are vegetarian, and many, if not most, are mainly vegetarian.

Before sin, God described the world as "very good" (Genesis 1:31). Some cannot accept this concept of perfect harmony because of the food chain that they observe in today's world. However, one cannot look at the sin-cursed world and the resultant death and struggle, and use this to reject the Genesis account of history. Everything has changed because of sin. That's why Paul describes the present creation as "groaning" (Romans 8:22). One must look through the Bible's "eyes" to understand the world.[32]

Some argue that people or animals would have been hurt even in an ideal world. They contend that even before sin, Adam or an animal could have stood on small creatures or scratched himself on a branch. Now these sorts of situations are true of today's fallen world—the present world is not perfect; it is suffering from the effects of the Curse (Romans 8:22). One cannot look at the Bible through the world's eyes and insist that the world before sin was just like the world we see today. We do not know what a perfect world, continually restored and totally upheld by God's power (Colossians 1:17; Hebrews 1:3), would have been like—we have never experienced perfection (only Adam and Eve did before sin).

We do get little glimpses from Scripture, however; in Deuteronomy 8:4, 29:5 and Nehemiah 9:21, we are told that when the Israelites wandered in the desert for 40 years, their clothes and shoes did not wear out, nor did their feet swell. When God upholds things perfectly, wearing out or being hurt in any way is not even an option.

Think of Shadrach, Meshach, and Abednego (Daniel 3:26–27). They came out of the fire without even the smell of smoke on them. Again, when the Lord upholds perfectly, being hurt is not possible. In a perfect world, before sin and the Curse, God would have upheld everything, but in this cursed world, things run down. Many commentators believe the description in Isaiah 11:6–9 of the

---

30 K. Brandes, *Vanishing Species*, Time-Life Books, New York, 1974, 98.
31 P. Weston, Bats: sophistication in miniature, *Creation* **21**(1):28–31, 1999.
32 Morris, *The Genesis Record*, 78.

wolf and lamb, and the lion that eats straw like an ox, is a picture of the new earth in the future restoration (Acts 3:21) when there will be no more curse or death (Revelation 21:1, 22:3). The animals described are living peacefully as vegetarians (this is also the description of the animal world before sin—Genesis 1:30). Today's world has been changed dramatically because of sin and the Curse. The present food chain and animal behavior (which also changed after the Flood—Genesis 9:2–3) cannot be used as a basis for interpreting the Bible—the Bible explains why the world is the way it is.

In the beginning, God gave Adam and Eve dominion over the animals: "Then God blessed them, and God said to them, 'Be fruitful and multiply; fill the earth and subdue it; have dominion over the fish of the sea, over the birds of the air, and over every living thing that moves on the earth'" (Genesis 1:28). Looking at today's world, we are reminded of Hebrews 2:8: "For in that He put all in subjection under him, He left nothing that is not put under him. But now we do not yet see all things put under him." Man's relationship with all things changed because of sin—they are not "under him" as they were originally.

Most people, including most Christians, tend to observe the world as it is today, with all its death and suffering, and then take that observation to the Bible and interpret it in that light. But we are sinful, fallible human beings, observing a sin-cursed world (Romans 8:22); and thus, we need to start with divine revelation, the Bible, to begin to understand.

So how did fangs and claws come about? Dr. Henry Morris, a founding figure in the modern creation movement, states:

> Whether such structures as fangs and claws were part of their original equipment, or were recessive features which only became dominant due to selection processes later, or were mutational features following the Curse, or exactly what, must await further research.[33]

After sin entered the world, everything changed. Maybe some animals started eating each other at this stage. By the time of Noah, God described what had happened this way: "So God looked upon the earth, and indeed it was corrupt; for all flesh had corrupted their way on the earth" (Genesis 6:12).

Also, after the Flood, God changed the behavior of animals. We read, "And the fear of you and the dread of you shall be on every beast of the earth, on

[33] See chapter 21 for more on the possible origin of defense-attack structures.

every bird of the air, on all that move on the earth, and on all the fish of the sea. They are given into your hand" (Genesis 9:2). Thus, man would find it much more difficult to carry out the dominion mandate given in Genesis 1:28.

Why Do We Find Dinosaur Fossils?

Fossil formation requires a sudden burial. When an animal dies, it usually gets eaten or decays until there is nothing left. To form a fossil, unique conditions are required to preserve the animal and replace it with minerals, etc.

Evolutionists once claimed that the fossil record was formed slowly as animals died and were gradually covered by sediment. But they have acknowledged more recently that the fossil record must involve catastrophic processes.[34] To form the billions of fossils worldwide, in layers sometimes kilometers thick, the organisms, by and large, must have been buried quickly. Many evolutionists now say the fossil record formed quickly, in spurts interspersed by millions of years.

According to the Bible, as time went on, earth became full of wickedness, so God determined that He would send a global Flood "to destroy from under heaven all flesh in which is the breath of life" (Genesis 6:17).

God commanded Noah to build a very large boat into which he would take his family and representatives of every kind of land-dwelling, air-breathing animal (that God Himself would choose and send to Noah, Genesis 6:20). This must have included two of each kind of dinosaur.

How Did Dinosaurs Fit on the Ark?

Many people think of dinosaurs as large creatures that would never have fit into the Ark.

But the average size of a dinosaur, based on the skeletons found over the earth, is about the size of a sheep.[35] Indeed, many dinosaurs were relatively small. For instance, *Struthiomimus* was the size of an ostrich, and *Compsognathus* was no bigger than a rooster. Only a few dinosaurs grew to extremely

[34] For example, D. Ager, *The New Catastrophism,* Cambridge University Press, Cambridge, UK, 1993.

[35] M. Crichton, *The Lost World*, Ballantine Books, New York, 1995, 122. "Dinosaurs were mostly small … . People always think they were huge, but the average dinosaur was the size of a sheep or a small pony." According to Horner and Lessem, *The Complete T. Rex*, 1993, 124: "Most dinosaurs were smaller than bulls."

large sizes (e.g., *Brachiosaurus* and *Apatosaurus),* but even they were not as large as the largest animal in the world today, the blue whale. (Reptiles have the potential to grow as long as they live. Thus, the large dinosaurs were probably very old ones.)

Dinosaurs laid eggs, and the biggest fossil dinosaur egg found is about the size of a football.[36] Even the largest dinosaurs were very small when first hatched. Remember that the animals that came off the boat were to repopulate the earth. Thus, it would have been necessary to choose young adults, which would soon be in the prime of their reproductive life, to go on the Ark. Recent research suggests that dinosaurs underwent rapid adolescent growth spurts.[37] So it is realistic to assume that God would have sent young adults to the Ark, not fully grown creatures.

Some might argue that the 600 or more named species of dinosaurs could not have fit on the Ark. But Genesis 6:20 states that representative *kinds* of land animals boarded the Ark. The question then is, what is a "kind" (Hebrew: *min*)? Biblical creationists have pointed out that there can be many species descended from a kind. For example, there are many types of cats in the world, but all cat species probably came from only a few kinds of cats originally.[38] The cat varieties today have developed by natural and artificial selection acting on the original variation in the information (genes) of the

[36] D. Lambert, *A Field Guide to Dinosaurs*, Avon Books, New York, 1983, 127.

[37] G.M. Erickson, K.C. Rogers, and S.A. Yerby, Dinosaurian growth patterns and rapid avian growth rates, *Nature* **412**(6845):405–408, 429–433, July 26, 2001.

[38] W. Mehlert, On the origin of cats and carnivores, *CEN Technical Journal*, **9**(1):106–120, 1995.

original cats. This has produced different combinations and subsets of information, and thus different types of cats.

Mutations (errors in copying of the genes during reproduction) can also contribute to the variation, but the changes caused by mutations are "downhill," causing loss of the original information.

Even speciation could occur through these processes. This speciation is *not* "evolution," since it is based on the created information *already present* and is thus a limited, downhill process, not involving an upward increase in complexity. Thus, only a few feline pairs would have been needed on Noah's Ark.

Dinosaur names have tended to proliferate, with new names being given to just a few pieces of bone, even if the skeleton looks similar to one that is a different size or found in a different country. There were probably fewer than 50 distinct groups or kinds of dinosaurs that had to be on the Ark.[39]

Also, it must be remembered that Noah's Ark was extremely large and quite capable of carrying the number of animals needed, including dinosaurs.

The land animals that were not on the Ark, including dinosaurs, drowned. Many were preserved in the layers formed by the Flood—thus the millions of fossils. Presumably, many of the dinosaur fossils were buried at this time, around 4,500 years ago. Also, after the Flood, there would have been considerable catastrophism, including such events as the Ice Age, resulting in some post-Flood formation of fossils, too.

The contorted shapes of these animals preserved in the rocks, the massive numbers of them in fossil graveyards, their wide distribution, and some whole

ONLY
ABOUT
50
DINOSAUR
"KINDS"

[39] Norell et al., *Discovering Dinosaurs in the American Museum of Natural History,* figure 56, pp. 86–87. See Czerkas and Czerkas, *Dinosaurs: A Global View,* 151.

skeletons, all provide convincing evidence that they were buried rapidly, testifying to massive flooding.[40]

Why Don't We See Dinosaurs Today?

At the end of the Flood, Noah, his family, and the animals came out of the Ark (Genesis 8:15–17). The dinosaurs thus began a new life in a new world. Along with the other animals, the dinosaurs came out to breed and repopulate the earth. They would have left the landing place of the Ark and spread over the earth's surface. The descendants of these dinosaurs gave rise to the dragon legends.

But the world they came out to repopulate differed from the one they knew before Noah's Flood. The Flood had devastated it. It was now a much more difficult world in which to survive.

After the Flood, God told Noah that from then on, the animals would fear man, and that animal flesh could be food for man (Genesis 9:1–7). Even for man, the world had become a harsher place. To survive, the once easily obtained plant nutrition would now have to be supplemented by animal sources.

Both animals and man would find their ability to survive tested to the utmost. We can see from the fossil record, from the written history of man, and from experience over recent centuries, that many forms of life on this planet have not survived that test.

We need to remember that many plants and air-breathing, land-dwelling animals have become extinct *since* the Flood—either due to man's action or competition with other species, or because of the harsher post-Flood environment. Many groups are still becoming extinct. Dinosaurs seem to be numbered among the extinct groups.

Why then are people so intrigued about dinosaurs and have little interest in the extinction of the fern *Cladophebius,* for example? It's the dinosaurs' appeal as monsters that excites and fascinates people.

Evolutionists have capitalized on this fascination, and the world is awash with evolutionary propaganda centered on dinosaurs. As a result, evolutionary philosophy has permeated modern thinking, even among Christians.

---

[40] For example, reptiles drowned in a flash flood 200 million years ago, according to the interpretation put upon the reptile fossils discovered in Lubbock Quarry, Texas (*The Weekend Australian,* p. 32, November 26–27, 1983).

If you were to ask the zoo why they have endangered species programs, you would probably get an answer something like this: "We've lost lots of animals from this earth. Animals are becoming extinct all the time. Look at all the animals that are gone forever. We need to act to save the animals." If you then asked, "Why are animals becoming extinct?" you might get an answer like this: "It's obvious! People killing them, lack of food, man destroying the environment, diseases, genetic problems, catastrophes like floods—there are lots of reasons."

If you then asked, "Well, what happened to the dinosaurs?" the answer would probably be, "We don't know! Scientists have suggested dozens of possible reasons, but it's a mystery."

Maybe one of the reasons dinosaurs are extinct is that we did not start our endangered species programs early enough. The factors that cause extinction today, which came about because of man's sin—the Curse, the aftermath of the Flood (a judgment), etc.—are the same factors that caused the dinosaurs to become extinct.

Are Dinosaurs Really Extinct?

One cannot prove an organism is extinct without having knowledge of every part of the earth's surface simultaneously. Experts have been embarrassed when, after having declared animals extinct, they were discovered alive and well. For example, in the 1990s explorers found elephants in Nepal that have many features of mammoths.[41]

---

[41] Elephants take mammoth step out of an ancient past, *The Sunday Mail* (Brisbane, Australia), December 17, 1995.

Scientists in Australia found some living trees that they thought had become extinct with the dinosaurs. One scientist said, "It was like finding a 'live dinosaur.'"[42] When scientists find animals or plants that they thought were extinct long ago, they call them "living fossils." There are hundreds of living fossils, a big embarrassment for those who believe in millions of years of earth history.

Explorers and natives in Africa have reported sighting dinosaur-like creatures, even in the twentieth century.[43] These have usually been confined to out-of-the-way places such as lakes deep in the Congo jungles. Descriptions certainly fit those of dinosaurs.

Cave paintings by native Americans seem to depict a dinosaur.[44] Scientists accept the mammoth drawings in the cave, so why not the dinosaur drawings? Evolutionary indoctrination that man did not live at the same time as dinosaurs stops most scientists from even considering that the drawings are of dinosaurs.

It certainly would be no embarrassment to a creationist if someone discovered a dinosaur living in a jungle. However, this should embarrass evolutionists.

And no, we cannot clone a dinosaur, as in the movie *Jurassic Park*, even if we had dinosaur DNA. An egg from a living female dinosaur would also be a must to employ the cloning techniques currently used by scientists to clone a wide variety of animals.

Birdosaurs?

Many evolutionists do not really think dinosaurs are extinct anyway. In 1997, at the entrance to the bird exhibit at the zoo in Cincinnati, Ohio, we read the following on a sign:

> Dinosaurs went extinct millions of years ago—or did they? No, birds are essentially modern short-tailed feathered dinosaurs.

In the mid-1960s, Dr. John Ostrom from Yale University began to popularize the idea that dinosaurs evolved into birds.[45] However, not all

---

[42] See Anon., *Melbourne Sun*, February 6, 1980. More than 40 people claimed to have seen *plesiosaurs* off the Victorian coast (Australia) over recent years.

[43] Anon., Dinosaur hunt, *Science Digest* **89**(5):21, 1981. See H. Regusters, Mokele-mbembe: an investigation into rumors concerning a strange animal in the Republic of Congo, 1981, *Munger Africana Library Notes,* **64:** 2–32, 1982; M. Agmagna, Results of the first Congolese mokele-mbembe expedition, *Cryptozoology* **2:**103, 1983, as cited in *Science Frontiers* **33,** 1983.

[44] D. Swift, Messages on stone, *Creation,* **19**(2):20–23, 1997.

[45] Norell, *Discovering Dinosaurs in the American Museum of Natural History*, 13.

evolutionists agree with this. "It's just a fantasy of theirs," says Alan Feduccia, an ornithologist at the University of North Carolina at Chapel Hill, and a leading critic of the dino-to-bird theory. "They so much want to see living dinosaurs that now they think they can study them vicariously at the backyard bird feeder."[46]

There have been many attempts to indoctrinate the public to believe that modern birds are really dinosaurs. *Time* magazine, on April 26, 1993, had a front page cover of a "birdosaur," now called *Mononykus*, with feathers (a supposed transitional form between dinosaurs and birds) based on a fossil find that had *no* feathers.[47] In the same month, *Science News* had an article suggesting this animal was a digging creature more like a mole.[48]

In 1996, newspapers reported a find in China of a reptile fossil that supposedly had feathers.[49] Some of the media reports claimed that, if it were confirmed, it would be "irrefutable evidence that today's birds evolved from dinosaurs." One scientist stated, "You can't come to any conclusion other than that they're feathers."[50] However, in 1997 the Academy of Natural Sciences in Philadelphia sent four leading scientists to investigate this find. They concluded that they were *not* feathers. The media report stated, concerning one of the scientists, "He said he saw 'hair-like' structures—not hairs—that could have supported a frill, or crest, like those on iguanas."[51]

No sooner had this report appeared than another media report claimed that 20 fragments of bones of a reptile found in South America showed that dinosaurs were related to birds.[52]

Birds are warm-blooded and reptiles are cold-blooded, but evolutionists who believe dinosaurs evolved into birds would like to see dinosaurs as warm-blooded to support their theory. But Dr. Larry Martin, of the University of Kansas, opposes this idea:

> Recent research has shown the microscopic structure of dinosaur bones was "characteristic of cold-blooded animals," Martin said. "So we're back to cold-blooded dinosaurs."[53]

---

46 V. Morell, Origin of birds: the dinosaur debate, *Audubon*, March–April 1997, p. 38.

47 Anon., New "birdosaur" not missing link! *Creation* **15**(3):3, 1993.

48 Anon., "Birdosaur" more like a mole, *Creation* **15**(4):7, 1993.

49 M. Browne, Downy dinosaur reported, *Cincinnati Enquirer*, p. A13, October 19, 1996.

50 Anon., Remains of feathered dinosaur bolster theory on origin of birds, Associated Press, New York, October 18, 1996.

51 B. Stieg, Bones of contention, *Philadelphia Inquirer,* March 31, 1997.

52 P. Recer, Birds linked to dinosaurs, *Cincinnati Enquirer*, p. A9, May 21, 1997.

53 Stieg, Did birds evolve from dinosaurs? *The Philadelphia Inquirer*, March 1997.

Sadly, the secular media have become so blatant in their anti-Christian stand and pro-evolutionary propaganda that they are bold enough to make such ridiculous statements as, "Parrots and hummingbirds are also dinosaurs."[54]

Several more recent reports have fueled the bird/dinosaur debate among evolutionists. One concerns research on the embryonic origins of the "fingers" of birds and dinosaurs, showing that birds could *not* have evolved from dinosaurs.[55] A study of the so-called feathered dinosaur from China revealed that the dinosaur had a distinctively reptilian lung and diaphragm, which is distinctly different from the avian lung.[56] Another report said that the frayed edges that some thought to be "feathers" on the Chinese fossil are similar to the collagen fibers found immediately beneath the skin of sea snakes.[57]

There is *no* credible evidence that dinosaurs evolved into birds.[58] Dinosaurs have always been dinosaurs and birds have always been birds.

What if a dinosaur fossil *was* found with feathers on it? Would that prove that birds evolved from dinosaurs? No, a duck has a duck bill and webbed feet, as does a platypus, but nobody believes that this proves that platypuses evolved from ducks. The belief that reptiles or dinosaurs evolved into birds requires reptilian scales on the way to becoming feathers, that is, transitional scales, not fully formed feathers. A dinosaur-like fossil with feathers would just be another curious mosaic, like the platypus, and part of the pattern of similarities placed in creatures to show the hand of the one true Creator God who made everything.[59]

Why Does It Matter?

Although dinosaurs are fascinating, some readers may say, "Why are dinosaurs such a big deal? Surely there are many more important issues to deal with in today's world, such as abortion, family breakdown, racism, promiscuity, dishonesty, homosexual behavior, euthanasia, suicide, lawlessness,

54 P. Recer, Birds linked to dinosaurs, 1997.

55 A. Burke and A. Feduccia, Developmental patterns and the identification of homologies in the avian hand, *Science* **278:**666–668, 1997; A. Feduccia and J. Nowicki, The hand of birds revealed by early bird embryos, *Naturwissenschaften* **89:**391–393, 2002.

56 J. Ruben et al., Lung structure and ventilation in theropod dinosaurs and early birds, *Science* **278:**1267–1270, 1997.

57 A. Gibbons, Plucking the feathered dinosaur, *Science* **278:**1229, 1997.

58 See chapter 24.

59 For more on the problems with dinosaur-to-bird evolution, see chapter 24.

pornography, and so on. In fact, we should be telling people about the gospel of Jesus Christ, not worrying about side issues like dinosaurs."

Actually, the evolutionary teachings on dinosaurs that pervade society *do* have a great bearing on why many will not listen to the gospel, and thus why social problems abound today. If they don't believe the history in the Bible, why would anyone trust its moral aspects and message of salvation?

The Implications

If we accept the evolutionary teachings on dinosaurs, then we must accept that the Bible's account of history is false. If the Bible is wrong in this area, then it is not the Word of God and we can ignore everything else it says that we find inconvenient.

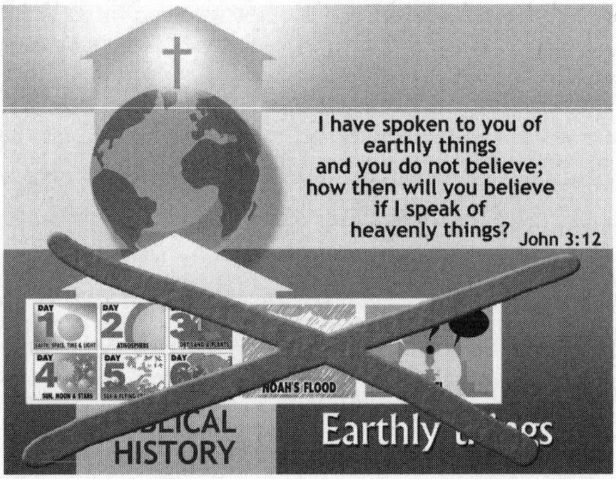
I have spoken to you of
earthly things
and you do not believe;
how then will you believe
if I speak of
heavenly things?
John 3:12
NOAH'S FLOOD
HISTORY

If everything made itself through natural processes—without God—then God does not own us and has no right to tell us how to live. In fact, God does not really exist in this way of thinking, so there is no absolute basis for morality. Without God, anything goes—concepts of right and wrong are just a matter of opinion. And without a basis for morality, there is no such thing as sin. And no sin means that there is no need to fear God's judgment and there is no need for the Savior, Jesus Christ. The history in the Bible is vital for properly understanding why one needs to accept Jesus Christ.

Millions of Years and the Gospel

The teaching that dinosaurs lived and died millions of years before man directly attacks the foundations of the gospel in another way. The fossil record,

I have spoken to you of earthly things and you do not believe; how then will you believe if I speak of heavenly things? John 3:12
NOAH'S FLOOD
BABEL
BIBLICAL HISTORY
Earthly things

of which dinosaurs form a part, documents death, disease, suffering, cruelty, and brutality. It is a very ugly record. Allowing for millions of years in the fossil layers means accepting death, bloodshed, disease, and suffering *before* Adam's sin.

But the Bible makes it clear that death, bloodshed, disease, and suffering are a *consequence of sin*. As part of the Curse, God told Adam in Genesis 3:19 that he would return to the dust from which he was made, showing that the sentence of death was not only spiritual, but physical as well.

After Adam disobeyed God, the Lord clothed Adam and Eve with "coats of skins" (Genesis 3:21). To do this He must have killed and shed the blood of at least one animal. The reason for this can be summed up by Hebrews 9:22:

> And according to the law almost all things are purified with blood, and without shedding of blood there is no remission.

God required the shedding of blood for the forgiveness of sins. What happened in the Garden of Eden was a picture of what was to come in Jesus Christ, who shed His blood on the Cross as "the Lamb of God, who takes away the sin of the world" (John 1:29).

If the shedding of blood occurred before sin, as would have happened if the garden was sitting on a fossil record of dead things millions of years old, then the foundation of the Atonement would be destroyed.

This big picture also fits with Romans 8, which says that the whole creation "groans" because of the effects of the Fall of Adam—it was not "groaning" with death and suffering before Adam sinned. Jesus Christ suffered physical death and shed His blood because death was the penalty for sin. Paul discusses this in detail in Romans 5 and 1 Corinthians 15.

Revelation chapters 21 and 22 make it clear that there will be a "new heaven and a new earth" one day where there will be "no more death" and "no more curse"—just as it was before sin changed everything. Obviously, if there are going to be animals in the new earth, they will not die or eat each other or eat the redeemed people.

Thus, the teaching of millions of years of death, disease, and suffering before Adam sinned is a direct attack on the foundation of the message of the Cross.

Conclusion

If we accept God's Word, beginning with Genesis, as being true and authoritative, then we can explain dinosaurs and make sense of the evidence we observe in the world around us. In doing this, we are helping people see that Genesis is absolutely trustworthy and logically defensible, and is what it claims to be—the true account of the history of the universe and mankind. And what one believes concerning the book of Genesis will ultimately determine what one believes about the rest of the Bible. This, in turn, will affect how a person views himself or herself, fellow human beings, and what life is all about, including their need for salvation.

13

Why Don't We Find Human & Dinosaur Fossils Together?

BODIE HODGE

Biblical creationists believe that man and dinosaurs lived at the same time because God, a perfect eyewitness to history, said that He created man and land animals on Day 6 (Genesis 1:24–31). Dinosaurs are land animals, so logically they were created on Day 6.

In contrast, those who do not believe the plain reading of Genesis, such as many non-Christians and compromised Christians, believe the rock and fossil layers on earth represent millions of years of earth history and that man and dinosaurs did not live at the same time.

Old-earth proponents often argue that if man and dinosaurs lived at the same time, their fossils should be found in the same layers. Since no one has found definitive evidence of human remains in the same layers as dinosaurs (Cretaceous, Jurassic, and Triassic), they say that humans and dinosaurs are separated by millions of years of time and, therefore, didn't live together. So, old-earth proponents ask a very good question: Why don't we find human fossils with dinosaur fossils, if they lived at the same time?

We find human fossils in layers that most creationists consider post-Flood. Most of these were probably buried after the Flood and after the scattering of humans from Babel. So it is true that human and dinosaur fossils have yet to be found in the same layers, but does that mean that long-age believers are correct?

What Do We Find in the Fossil Record?

The first issue to consider is what we actually find in the fossil record.

- ~95% of all fossils are shallow marine organisms, such as corals and shellfish.
- ~95% of the remaining 5% are algae and plants.
- ~95% of the remaining 0.25% are invertebrates, including insects.
- The remaining 0.0125% are vertebrates, mostly fish. (95% of land vertebrates consist of less than one bone, and 95% of mammal fossils are from the Ice Age after the Flood.)[1]

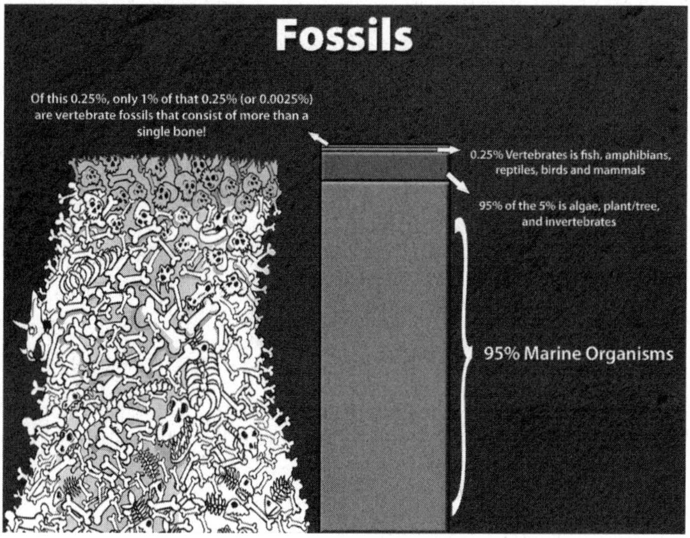

The number of dinosaur fossils is actually relatively small, compared to other types of creatures. Since the Flood was a marine catastrophe, we would expect marine fossils to be dominant in the fossil record. And that is the case.

Vertebrates are not as common as other types of life forms. This makes sense of these percentages and helps us understand why vertebrates, including dinosaurs, are so rare and even overwhelmed by marine organisms in the record.

[1] A. Snelling, Where are all the human fossils? *Creation* **14**(1):28–33, December, 1991; J. Morris, *The Young Earth*, Master Books, Green Forest, Arkansas, 2002, 71.

Yet that still does not explain why there are no fossilized humans *found to date* in Flood sediments.

Were Pre-Flood Humans Completely Obliterated?

In Genesis 6:7 and 7:23, God says He will "blot out" man from the face of the earth using the Flood. Some have suggested that this phrase means to completely obliterate all evidence of man. However, this is not completely accurate. After a lengthy study, Fouts and Wise make it clear that the Hebrew word מחה (māhâ), translated as "blot out" or "destroy," can still leave evidence behind. They say,

> Although māhâ is properly translated "blot out," "wipe," or even "destroy," it is not to be understood to refer to the complete obliteration of something without evidence remaining. In every Biblical use of māhâ where it is possible to determine the fate of the blotted, wiped, or destroyed, the continued existence of something is terminated, but evidence may indeed remain of the previous existence and/or the blotting event itself. Even the theological consideration of the "blotting out" of sin suggests that evidence usually remains (e.g., consequences, scars, sin nature, etc.).[2]

In light of this, it is possible that human fossils from the Flood could still exist but just haven't been found yet.

So, should we find human fossils in layers that contain dinosaur fossils? To answer this further, we need to understand what we actually find in the fossil record, what the likelihood is that humans would have been fossilized, what is unusual about their distribution, and how much Flood sediment there was.

Do Humans Fossilize like Other Creatures?

Fossilization is a rare event, especially of humans who are very mobile. Since the rains of Noah's Flood took weeks to cover the earth, many people could have made it to boats, grabbed on to floating debris, and so on. Some may have made it to higher ground. Although they wouldn't have lasted that long and would have eventually perished, they might not fossilize.

---

[2] D. Fouts and K. Wise, Blotting out and breaking up: miscellaneous Hebrew studies in geocatastrophism, Proceedings of the Fourth International Conference on Creationism, Creation Science Fellowship, Pittsburgh, 1998, 219.

In most cases, dead things decompose or get eaten. They just disappear and nothing is left. The 2004 tsunami in Southeast Asia was a shocking reminder of the speed with which water and other forces can eliminate all trace of bodies, even when we know where to look. According to the United Nation's Office of the Special Envoy for Tsunami Recovery, nearly 43,000 tsunami victims were never found.[3]

Even if rare, it would still be possible to fossilize a human body. In fact, we do find fossils of humans, such as Neanderthals, in the post-Flood sediments. So why don't we find humans in pre-Flood sediments?

One suggestion has been that the human population was relatively small. Let's see how that possibility bears out.

Were Pre-Flood Humans Few in Number?

Estimates for the pre-Flood population are based on very little information, since Genesis 1 doesn't give extensive family size or population growth information. We know that Noah was in the tenth generation of his line, and he lived about 1,650 years after creation. Genesis also indicates that in Noah's lineage children were being born to fathers between the ages of 65 and more than 500 (when Noah bore his three sons).

How many generations were there in other lineages? We don't know. We know that those in the line from Adam to Noah were living upwards of 900 years each, but we can't be certain everyone lived that long. How many total children were born? Again, we don't know. What were the death rates? We simply don't know. Despite this lack of information, estimates have been done. One estimate puts the number as high as 17 billion people.[4] These estimates are based on various population growth rates and numbers of generations. Recall that Noah was in the tenth generation from Adam, however, so these estimates may be too high.

It seems doubtful that there were many hundreds of millions of people before the Flood. If the world was indeed bad enough for God to judge with a Flood, then people were probably blatantly disobedient to God's command to be fruitful and fill the earth. Moreover, the Bible says that violence filled the earth, so death rates may have been extraordinarily high.

---

[3] The Human Toll, www.tsunamispecialenvoy.org/country/humantoll.asp.

[4] T. Pickett, Population of the Pre-Flood World, www.ldolphin.org/pickett.html; H. Morris, *Biblical Cosmology and Modern Science*, Baker Book House, Grand Rapids, Michigan, 1970, 77–78; Morris, *The Young Earth*, 71.

In light of this, the population of humans in the pre-Flood world could have been as low as hundreds of thousands. Even if we make a generous assumption of 200 million people at the time of the Flood, there would be just over one human fossil per cubic mile of sediment laid down by the Flood!

Were Humans Concentrated in High Density Pockets that Have Not Been Discovered?

Today, humans tend to clump together in groups in towns, villages, and cities. In the same way, people were probably not evenly distributed before the Flood. The first city is recorded in Genesis 4:17, long before the Flood. We know that most of the population today lives within 100 miles (160 km) of the coastline. One report states, "Already nearly two-thirds of humanity—some 3.6 billion people—crowd along a coastline, or live with 150 kilometers of one."[5]

This is strong evidence that the pre-Flood civilizations probably were not evenly distributed on the landmass. If man wasn't evenly distributed, then the pockets of human habitation possibly were buried in places that have not yet been discovered.

Not only is fossilization a rare event, but fossils are also difficult to find. Just consider how much sediment was laid down by the Flood, compared to the area that has actually been exposed for us to explore.

John Woodmorappe's studies indicate that there are about 168 million cubic miles (700 million km^3) of Flood sediment.[6] John Morris estimates that there is about 350 million cubic miles of Flood sediment.[7] The latter may be high because the total volume of water on the earth is estimated at about 332.5 million cubic miles, according to the U.S. Geological Survey.[8] But even so, there is a lot of sediment left to sift through. Having such a massive amount of sediment to study is a major reason why we have not found human fossils yet.

So, a small human population and massive amounts of sediment are two prominent factors why we haven't found human fossils in pre-Flood sediments. It also may simply be that we haven't found the sediment where humans were living and were buried.

---

5 D. Hinrichsen, "Coasts in Crisis," www.aaas.org/international/ehn/fisheries/hinrichs.htm.

6 J. Woodmorappe, *Studies in Flood Geology,* Institute for Creation Research, El Cajon, California, 1999, 59. This number actually comes from A.B. Ronov, "The earth's sedimentary shell," *International Geology Review* 24(11):1321–1339, 1982.

7 Morris, *The Young Earth,* 71.

8 "Where is the earth's water located?" U.S. Geological Survey, ga.water.usgs.gov/edu/earthwherewater.html.

Think about It—Would You Want to Live with Dinosaurs?

Often, people believe that if human bones aren't found with dinosaur bones, then they didn't live at the same time. Actually, all we know for sure is that they weren't buried together. It is very easy for creatures to live at the same time on earth, but never even cross paths. Have you ever seen a tiger or a panda in the wild? Just because animals are not found together does not mean they do not live in the same world at the same time.

A great example is the coelacanth. Coelacanth fossils are found in marine deposits below dinosaurs and in other marine layers that "date" about the same age as dinosaurs.[9] It was once thought the coelacanth became extinct about 70 million years ago because their fossils are not found in any deposits higher than this. However, in 1938 living populations were found in the Indian Ocean.[10] It appears that coelacanths were buried with other sea creatures during the Flood—as we would expect. The example of the coelacanth shows that animals are not necessarily buried in the same place as other animals from different environments. We don't find human bones buried with coelacanths, either, but we live together today and people are enjoying them for dinner in some parts of the world.

Coelacanths aren't the only example. We find many examples like this, even with creatures that did not live in the sea. One popular example is the Wollemi Pine, which was fossilized in Jurassic deposits, supposedly 150 million years ago.[11] However, we find these trees living today. Another great living fossil is the Ginkgo tree, which supposedly thrived 240 million years ago, prior to the dinosaurs.[12] Yet, they are not found in layers with dinosaurs or post-Flood humans, even though they exist today. The list of "living fossils" goes on. Because animals and plants aren't buried together, it is no indication that things didn't live together.

In fact, based on human nature, we can assume that humans probably chose not to live in the same place with dinosaurs. So, the real issue is what happened to the local environment where humans lived.

---

[9] L. Dicks, The creatures time forgot, *New Scientist*, **164**(2209): 36–39, October 23, 1999.

[10] R. Driver, Sea monsters...more than a legend? *Creation* **19**(4):38–42, September 1997, www.answersingenesis.org/creation/v19/i4/seamonsters.asp.

[11] www.answersingenesis.org/docs2/4416livingfossil_tree12-25-2000.asp.

[12] www.pbs.org/wgbh/nova/fish/other.html.

What Can We Conclude?

If human and dinosaur bones are ever found in the same layers, it would be a fascinating find to both creationists and evolutionists. Those who hold a biblical view of history wouldn't be surprised but would consider several logical possibilities, such as human parties invading dinosaur lands for sport or for food, or merely humans and dinosaurs being washed up and buried together.

Evolutionists, on the other hand, who believe the geologic layers represent millions of years of time, would have a real challenge. In the old-earth view, man isn't supposed to be the same age as dinosaurs. Yet we can be sure that this finding would not overturn their starting assumptions—they would simply try to develop a hypothesis consistent with their preconceived view of history. For example, they might search for the possibility that the fossils were moved and redeposited.

So, ultimately, the debate is not about the evidence itself—where we find human fossils and dinosaur fossils. Nobody was there to actually observe humans and dinosaurs living together outside of written revelation, which is very limited pre-Flood. We are forced to reconstruct that history based on our existing assumptions about time and history, as well as our limited fossil evidence from the rocks.

As biblical creationists, we don't require that human and dinosaur fossils be found in the same layers. Whether they are found or not, does not affect the biblical view of history.

The fundamental debate is really about the most trustworthy source of information about history. Do we start with the Bible, which God says is true in every detail, including its history, or do we start with the changing theories of imperfect man? God tells Christians to walk by faith and that "without faith it is impossible to please Him" (Hebrews 11:6). But this is not a blind faith. God has filled the world with clear evidences that confirm the truth of His Word and the certainty of the Christian faith. The fossil record itself is an incredible testimony to the truth of God's Word and His promise to "blot out" all land dwelling, air-breathing animals and humans in a worldwide catastrophe.

14

Can Catastrophic Plate Tectonics Explain Flood Geology?

ANDREW A. SNELLING

What Is Plate Tectonics?

The earth's thin rocky outer layer (3–45 mi [5–70 km] thick) is called "the crust." On the continents it consists of sedimentary rock layers—some containing fossils and some folded and contorted—together with an underlying crystalline rocky basement of granites and metamorphosed sedimentary rocks. In places, the crystalline rocks are exposed at the earth's surface, usually as a result of erosion. Beneath the crust is what geologists call the mantle, which consists of dense, warm-to-hot (but solid) rock that extends to a depth of 1,800 mi (2,900 km). Below the mantle lies the earth's core, composed mostly of iron. All but the innermost part of the core is molten (see Figure 1).

Investigations of the earth's surface have revealed that it has been divided globally by past geologic processes into what today is a mosaic of rigid blocks called "plates." Observations indicate that these plates have moved large distances relative to one another in the past and that they are still moving very slowly today. The word "tectonics" has to do with earth movements; so the study of the movements and interactions among these plates is called "plate tectonics." Because almost all the plate motions occurred in the past, plate tectonics is, strictly speaking, an interpretation, model, or theoretical description of what geologists envisage happened to these plates through earth's history.

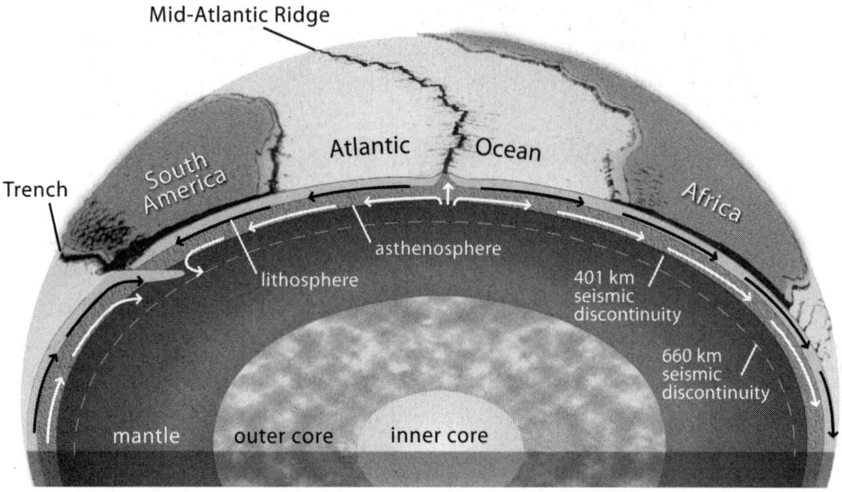

Figure 1. Cross-sectional view through the earth. The two major divisions of the planet are its mantle, made of silicate rock, and its core, comprised mostly of iron. Portions of the surface covered with a low-density layer of continental crust represent the continents. Lithospheric plates at the surface, which include the crust and part of the upper mantle, move laterally over the asthenosphere. The asthenosphere is hot and also weak because of the presence of water within its constituent minerals. Oceanic lithosphere, which lacks the continental crust, is chemically similar on average to the underlying mantle. Because oceanic lithosphere is substantially cooler, its density is higher, and it therefore has an ability to sink into the mantle below. The sliding of an oceanic plate into the mantle is known as "subduction," as shown here beneath South America. As two plates pull apart at a mid-ocean ridge, material from the asthenosphere rises to fill the gap, and some of this material melts to produce basaltic lava to form new oceanic crust on the ocean floor. The continental regions do not participate in the subduction process because of the buoyancy of the continental crust.

The general principles of plate tectonics theory may be stated as follows: deformation occurs at the edges of the plates by three types of horizontal motion—extension (rifting or moving apart), transform faulting (horizontal slippage along a large fault line), and compression, mostly by subduction (one plate plunging beneath another).[1]

Extension occurs where the seafloor is being pulled apart or split along rift zones, such as along the axes of the Mid-Atlantic Ridge and the East Pacific Rise. This is often called "seafloor spreading," which occurs where two

[1] S.E. Nevins and S.A. Austin, Continental drift, plate tectonics, and the Bible; in D.R. Gish and D.H. Rohrer, eds., *Up With Creation!* Creation-Life Publishers, San Diego, California, 1978, 173–180.

oceanic plates move away from each other horizontally, with new molten material from the mantle beneath rising between them to form new oceanic crust. Similar extensional splitting of a continental crustal plate can also occur, such as along the East African Rift Zone.

Transform faulting occurs where one plate is sliding horizontally past another, such as along the well-known San Andreas Fault of California.

Compressional deformation occurs where two plates move toward one another. If an oceanic crustal plate is moving toward an adjacent continental crustal plate, then the former will usually subduct (plunge) beneath the latter. Examples are the Pacific and Cocos Plates that are subducting beneath Japan and South America, respectively. When two continental crustal plates collide, the compressional deformation usually crumples the rock in the collision zone to produce a mountain range. For example, the Indian-Australian Plate has collided with the Eurasian Plate to form the Himalayas.

History of Plate Tectonics

The idea that the continents had drifted apart was first suggested by a creationist, Antonio Snider.[2] He observed from the statement in Genesis 1:9–10 about God's gathering together the seas into one place that at that point in earth history there may have been only a single landmass. He also noticed the close fit of the coastlines of western Africa and eastern South America. So he proposed that the breakup of that supercontinent with subsequent horizontal movements of the new continents to their present positions occurred catastrophically during the Flood.

However, his theory went unnoticed, perhaps because Darwin's book, which was published the same year, drew so much fanfare. The year 1859 was a bad year for attention to be given to any other new scientific theory, especially one that supported a biblical view of earth history. And it also didn't help that Snider published his book in French.

It wasn't until the early twentieth century that the theory of continental drift was acknowledged by the scientific community, through a book by Alfred Wegener, a German meteorologist.[3] However, for almost 50 years the overwhelming majority of geologists spurned the theory, primarily because a handful of seismologists claimed the strength of the mantle rock was too high to allow continents to drift in the manner Wegener had proposed. Their

2 A. Snider, *Le Création et ses Mystères Devoilés*, Franck and Dentu, Paris, 1859.
3 A. Wegener, *Die Entstehung der Kontinente und Ozeane*, 1915.

estimates of mantle rock strength were derived from the way seismic waves behave as they traveled through the earth at that time.

For this half-century the majority of geologists maintained that continents were stationary, and they accused the handful of colleagues who promoted the drift concept of indulging in pseudo-scientific fantasy that violated basic principles of physics. Today that persuasion has been reversed—plate tectonics, incorporating continental drift, is the ruling perspective.

What caused such a dramatic about-face? Between 1962 and 1968 four main lines of independent experiments and measurements brought about the birth of the theory of plate tectonics:[4]

1. Mapping of the topography of the seafloor using echo depth-sounders;
2. Measuring the magnetic field above the seafloor using magnetometers;
3. "Timing" of the north-south reversals of the earth's magnetic field using the magnetic memory of continental rocks and their radioactive "ages;" and
4. Determining very accurately the location of earthquakes using a worldwide network of seismometers.

An important fifth line of evidence was the careful laboratory measurement of how mantle minerals deform under stress. This measurement can convincingly demonstrate that mantle rock can deform by large amounts on timescales longer than the few seconds typical of seismic oscillations.[5]

Additionally, most geologists became rapidly convinced of plate tectonics theory because it elegantly and powerfully explained so many observations and lines of evidence:

1. The jigsaw puzzle fit of the continents (taking into account the continental shelves);
2. The correlation of fossils and fossil-bearing strata across the ocean basins (e.g., the coal beds of North America and Europe);
3. The mirror image zebra-striped pattern of magnetic reversals in the volcanic rocks of the seafloor parallel to the mid-ocean rift zones in the plates on either side of the zone, consistent with a moving apart of the plates (seafloor spreading);

---

[4] A. Co, ed., *Plate Tectonics and Geomagnetic Reversals*, W.H. Freeman and Co., San Francisco, California, 1973.

[5] S.H. Kirby, Rheology of the lithosphere, *Reviews of Geophysics and Space Physics* **25**(1): 219–1244, 1983.

4. The location of most of the world's earthquakes at the boundaries between the plates, consistent with earthquakes being caused by two plates moving relative to one another;

5. The existence of the deep seafloor trenches invariably located where earthquake activity suggests an oceanic plate is plunging into the mantle beneath another plate;

6. The oblique pattern of earthquakes adjacent to these trenches (subduction zones), consistent with an oblique path of motion of a subducting slab into the mantle;

7. The location of volcanic belts (e.g., the Pacific "ring of fire") adjacent to deep sea trenches and above subducting slabs, consistent with subducted sediments on the tops of down-going slabs encountering melting temperatures in the mantle; and

8. The location of mountain belts at or adjacent to convergent plate boundaries (where the plates are colliding).

Slow-and-Gradual or Catastrophic?

Because of the scientific community's commitment to the uniformitarian assumptions and framework for earth history, most geologists take for granted that the movement of the earth's plates has been slow and gradual over long eons. After all, if today's measured rates of plate drift—about 0.5–6 in (2–15 cm) per year—are extrapolated uniformly back into the past, it requires about 100 million years for the ocean basins and mountain ranges to form. And this rate of drift is consistent with the estimated 4.8 mi^3 (20 km^3) of molten magma that currently rises globally each year to create new oceanic crust.[6]

On the other hand, many other observations are incompatible with slow-and-gradual plate tectonics. While the seafloor surface is relatively smooth, zebra-stripe magnetic patterns are obtained when the ship-towed instrument (magnetometer) observations average over mile-sized patches. Drilling into the oceanic crust of the mid-ocean ridges has also revealed that those smooth patterns are not present at depth in the actual rocks.[7] Instead, the magnetic polarity changes rapidly

---

6 J. Cann, Subtle minds and mid-ocean ridges. *Nature* 393:625–627, 1998.

7 J.M. Hall and P.T. Robinson, Deep crustal drilling in the North Atlantic Ocean, *Science* **204:**573–576, 1979.

and erratically down the drill-holes. This is contrary to what would be expected with slow-and-gradual formation of the new oceanic crust accompanied by slow magnetic reversals. But it is just what is expected with extremely rapid formation of new oceanic crust and rapid magnetic reversal during the Flood, when rapid cooling of the new crust occurred in a highly nonuniform manner because of the chaotic interaction with ocean water.

Furthermore, slow-and-gradual subduction should have resulted in the sediments on the floors of the trenches being compressed, deformed, and thrust-faulted, yet the floors of the Peru-Chile and East Aleutian Trenches are covered with soft, flat-lying sediments devoid of compressional structures.[8] These observations are consistent, however, with extremely rapid subduction during the Flood, followed by extremely slow plate velocities as the floodwaters retreated from the continents and filled the trenches with sediment.

If uniformitarian assumptions are discarded, however, and Snider's original biblical proposal for continental "sprint" during the Genesis Flood is adopted, then a catastrophic plate tectonics model explains everything that slow-and-gradual plate tectonics does, plus most everything it can't explain.[9] Also, a 3-D supercomputer model of processes in the earth's mantle has demonstrated that tectonic plate movements can indeed be rapid and catastrophic when a realistic deformation model for mantle rocks is included.[10] And, even

---

[8] D.W. Scholl et al., Peru-Chile trench sediments and seafloor spreading, *Geological Society of America Bulletin* **81**:1339–1360, 1970; R. Von Huene, Structure of the continental margin and tectonism at the Eastern Aleutian Trench. *Geological Society of America Bulletin* **83**:3613–3626, 1972.

[9] S.A. Austin et al., Catastrophic plate tectonics: a global Flood model of earth history; in R.E. Walsh, ed., *Proceedings of the Third International Conference on Creationism*, Creation Science Fellowship, Pittsburgh, Pennsylvania, pp. 609–621, 1994.

[10] J.R. Baumgardner, Numerical simulation of the large-scale tectonic changes accompanying the Flood; in R.E. Walsh, C.L. Brooks, and R.S. Crowell, eds., *Proceedings of the First International Conference on Creationism*, Vol. 2, Pittsburgh, Pennsylvania, pp. 17–30, 1986; J.R. Baumgardner, 3-D finite element simulation of the global tectonic changes accompanying Noah's Flood; in R.E. Walsh, C.L. Brooks, and R.S. Crowell, eds., *Proceedings of the Second International Conference on Creationism*, Vol. 2, Creation Science Fellowship, Pittsburgh, Pennsylvania, pp. 35–45, 1990; J.R. Baumgardner, Computer modeling of the large-scale tectonics associated with the Genesis Flood; in R.E. Walsh, ed., *Proceedings of the Third International Conference on Creationism*, Creation Science Fellowship, Pittsburgh, Pennsylvania, pp. 49–62, 1994; J.R. Baumgardner, Runaway subduction as the driving mechanism for the Genesis Flood, in R.E. Walsh, ed., *Proceedings of the Third International Conference on Creationism*, Creation Science Fellowship, Pittsburgh, Pennsylvania, pp. 63–75, 1994; J.R. Baumgardner, The physics behind the Flood, in R.L. Ivey, Jr., ed., *Proceedings of the Fifth International Conference on Creationism*, Creation Science Fellowship, Pittsburgh, Pennsylvania, pp. 113–126, 2003.

though it was developed by a creation scientist, this supercomputer 3-D plate tectonics modeling is acknowledged as the world's best.[11]

The catastrophic plate tectonics model of Austin et al.[12] begins with a pre-Flood supercontinent surrounded by cold ocean-floor rocks that were denser than the warm mantle rock beneath. To initiate motion in the model, some sudden trigger "cracks" the ocean floors adjacent to the supercontinental crustal block, so that zones of cold ocean-floor rock start penetrating vertically into the upper mantle along the edge of most of the supercontinent.[13]

These vertical segments of ocean-floor rock correspond to the leading edges of oceanic plates. These vertical zones begin to sink in conveyor-belt fashion into the mantle, dragging the rest of the ocean floor with them. The sinking slabs of ocean plates produce stresses in the surrounding mantle rock, and these stresses, in turn, cause the rock to become more deformable and allow the slabs to sink faster. This process causes the stress levels to increase and the rock to become even weaker. These regions of rock weakness expand to encompass the entire mantle and result in a catastrophic runaway of the oceanic slabs to the bottom of the mantle in a matter of a few weeks.[14]

The energy for driving this catastrophe is the gravitational potential energy of the cold, dense rock overlying the less dense mantle beneath it at the beginning of the event. At its peak, this runaway instability allows the subduction rates of the plates to reach amazing speeds of feet-per-second. At the same time the pre-Flood seafloor was being catastrophically subducted into the mantle, the resultant tensional stress tore apart (rifted) the pre-Flood supercontinent (see Figure 2). The key physics responsible for the runaway instability is the fact that mantle rocks weaken under stress, by factors of a billion or more, for the sorts of stress levels that can occur in a planet the size of the earth—a behavior verified by many laboratory experiments over the past forty years.[15]

The rapidly sinking ocean-floor slabs forcibly displace the softer mantle rock into which they are subducted, which causes large-scale convectional flow throughout the entire mantle. The hot mantle rock displaced by these subducting slabs wells up elsewhere to complete the flow cycle, and in particular rises into the seafloor rift zones to form new ocean floor. Reaching the

[11] J. Beard, How a supercontinent went to pieces, *New Scientist* **137**:19, January 16, 1993.
[12] Ref. 9.
[13] Ibid.
[14] Ibid.
[15] Ref. 5.

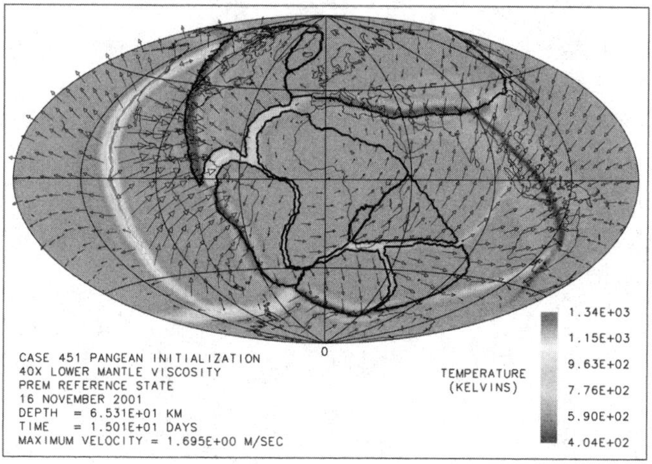

Figure 2(a). Snapshot of 3-D modeling solution after 15 days. The upper plot is an equal area projection of a spherical mantle surface 40 mi (65 km) below the earth's surface in which grayscale denotes absolute temperature. Arrows denote velocities in the plane of the cross-section. The dark lines denote plate boundaries where continental crust is present or boundaries between continent and ocean where both exist on the same plate. The lower plot is an equatorial cross-section in which the grayscale denotes temperature deviation from the average at a given depth.

surface of the ocean floor, this hot mantle material vaporizes huge volumes of ocean water with which it comes into contact to produce a linear curtain of supersonic steam jets along the entire 43,500 miles (70,000 km) of the seafloor rift zones stretching around the globe (perhaps the "fountains of the great deep" of Genesis 7:11 and 8:2). These supersonic steam jets capture large amounts of liquid water as they "shoot" up through the ocean above the seafloor where they form. This water is catapulted high above the earth and then falls back to the surface as intense global rain ("and the floodgates of heaven were opened"). The rain persisted for "40 days and nights" (Genesis 7:11–12) until all the pre-Flood ocean floor had been subducted.

This catastrophic plate tectonics model for earth history[16] is able to explain geologic data that slow-and-gradual plate tectonics over many millions of years cannot. For example, the new rapidly formed ocean floor would have initially been very hot. Thus, being of lower density than the pre-Flood ocean floor, it would have risen some 3,300 ft. (1,000 m) higher than its predecessor, causing a dramatic rise in global sea level. The ocean waters would

[16] Ref. 10.

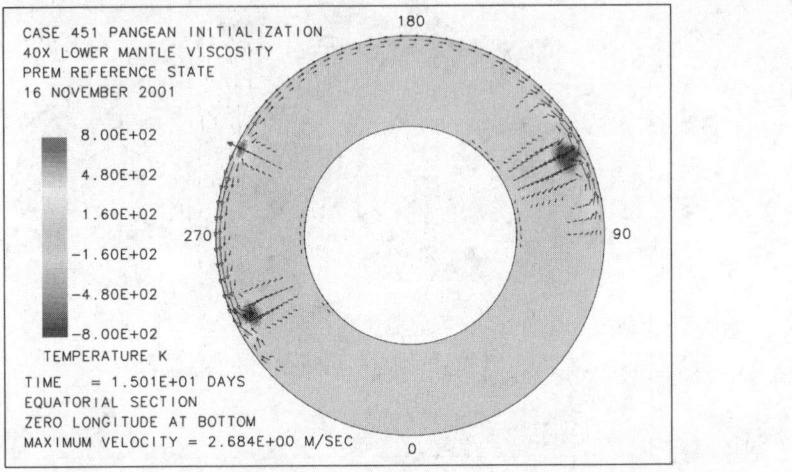

Figure 2(b). Snapshot of the modeling solution after 25 days. Grayscale and arrows denote the same quantities as in Figure 2(a). For a detailed explanation of this calculation, see Baumgardner, 2003.

thus have swept up onto and over the continental land surfaces, carrying vast quantities of sediments and marine organisms with them to form the thick, fossiliferous sedimentary rock layers we now find blanketing large portions of today's continents. This laterally extensive layer-cake sequence of sedimentary rocks is magnificently exposed, for example, in the Grand Canyon region of the southwestern U.S.[17] Slow-and-gradual plate tectonics simply cannot account for such thick, laterally extensive sequences of sedimentary strata containing marine fossils over such vast interior continental areas—areas which are normally well above sea level.

Furthermore, the whole mantle convectional flow resulting from runaway subduction of the cold ocean-floor slabs would have suddenly cooled the mantle temperature at the core-mantle boundary, thus greatly accelerating convection in, and heat loss from, the adjacent outer core. This rapid cooling of the surface of the core would result in rapid reversals of the earth's magnetic field.[18]

[17] S.A. Austin, ed., *Grand Canyon: Monument to Catastrophe*, Institute for Creation Research, Santee, California, 1994.

[18] R.E. Walsh, C.L. Brooks, and R.S. Crowell, eds., *Proceedings of the First International Conference on Creationism*, Vol. 2, Creation Science Fellowship, Pittsburgh, Pensylvania, 1986.

These magnetic reversals would have been expressed at the earth's surface and been recorded in the zebra-shaped magnetic stripes in the new ocean-floor rocks. This magnetization would have been erratic and locally patchy, laterally as well as at depth, unlike the pattern expected in the slow-and-gradual version. It was predicted that similar records of "astonishingly rapid" magnetic reversals ought to be present in thin continental lava flows, and such astonishingly rapid reversals in continental lava flows were subsequently found.[19]

This catastrophic plate tectonics model thus provides a powerful explanation for how the cold, rigid crustal plates could have moved thousands of miles over the mantle while the ocean floor subducted. It predicts relatively little plate movement today because the continental "sprint" rapidly decelerated when all the pre-Flood ocean floor had been subducted.

Also, we would thus expect the trenches adjacent to the subduction zones today to be filled with undisturbed late-Flood and post-Flood sediments. The model provides a mechanism for the retreat of the floodwaters from off the continents into the new ocean basins, when at the close of the Flood, as plate movements almost stopped, the dominant tectonic forces resulted in vertical earth movements (Psalm 104:8). Plate interactions at plate boundaries during the cataclysm generated mountains, while cooling of the new ocean floor increased its density, which caused it to sink and thus deepen the new ocean basins to receive the retreating floodwaters.

Aspects of modeling the phenomenon of runaway behavior in the mantle[20] have been independently duplicated and verified.[21] The same modeling predicts that since runaway subduction of the cold ocean-floor slabs occurred only a few thousand years ago during the Flood, those cold slabs would not have had sufficient time since the catastrophe to be fully "digested" into the surrounding mantle. Evidence for these relatively cold slabs just above the

---

[19] Ibid.; R.S. Coe and M. Prévot, Evidence suggesting extremely rapid field variation during a geomagnetic reversal, *Earth and Planetary Science Letters* **92:**292–298, 1989; A.A. Snelling "Fossil" magnetism reveals rapid reversals of the earth's magnetic field, *Creation* **13**(3):46–50, 1991; R.S. Coe, M. Prévot, and P. Camps, New evidence for extraordinary rapid change of the geomagnetic field during a reversal, *Nature* **374**:687–692, 1995; A.A. Snelling, The "principle of least astonishment"! *TJ* **9**(2):138–139, 1995.

[20] Ref. 9; Ref. 10.

[21] P.J. Tackley et al., Effects of an endothermic phase transition at 670 km depth on spherical mantle convection, *Nature* **361:**699–704, 1993; S.A. Weinstein, Catastrophic overturn of the earth's mantle driven by multiple phase changes and internal heat generation, *Geophysical Research Letters* **20:**101, 104, 1993; L. Moresi and Solomatov, Mantle convection with a brittle lithosphere: thoughts on the global tectonic styles of the earth and Venus, *Geophysical Journal International* **133**:669–682, 1998.

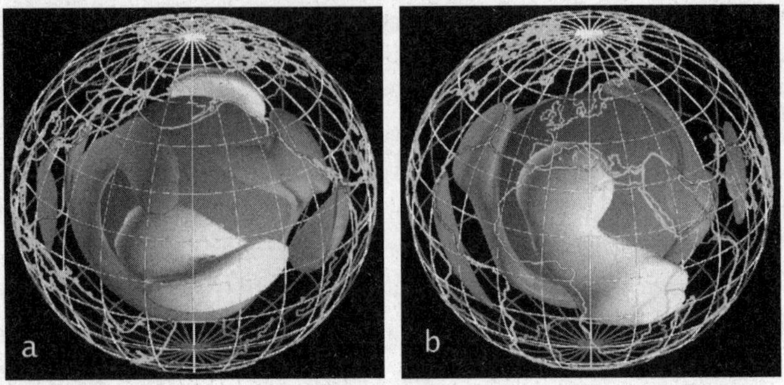

Figure 3. Distribution of hot (light-shaded surfaces) and cold (darker-shaded surfaces) regions in today's lower mantle as determined observationally by seismic tomography (imaging using recordings of seismic waves), viewed from (a) 180° longitude and (b) 0° longitude. The very low temperature inferred for the ring of colder rock implies that it has been subducted quite recently from the earth's surface. The columnar blobs of warmer rock have been squeezed together and pushed upward as the colder and denser rock settled over the core. (Figure courtesy of Alexandro Forte)

core-mantle boundary, to which they would have sunk, therefore should still be evident today, and it is (see Figure 3).[22]

Moreover, whether at the current rate of movement—only 4 in (10 cm) per year—the force and energy of the collision between the Indian-Australian and Eurasian Plates could have been sufficient to push up the Himalayas (like two cars colliding, each only traveling at .04 in/h [1 mm/h]) is questionable. In contrast, if the plate movements were measured as feet-per-second, like two cars each traveling at 62 mph (100 km/h), the resulting catastrophic collision would have rapidly buckled rock strata to push up those high mountains.

Is Catastrophic Plate Tectonics Biblical?

The Bible does not directly mention either continental drift or plate tectonics. However, if the continents were once joined together, as suggested by Genesis 1:9–10, and are now apart, then the only possibility is continental division and "sprint" during the Flood. Some have suggested this continental division

[22] S.P. Grand, Mantle shear structure beneath the Americas and surrounding oceans, *Journal of Geophysical Research* **99**:11591–11621, 1994; J.E. Vidale, A snapshot of whole mantle flow, *Nature* **370**:16–17, 1994.

occurred after the Flood during the days of Peleg when "the earth was divided" (Genesis 10:25). However, this Hebrew expression can be also translated to mean "lands being divided among peoples," which, according to the context, refers to the results of the Tower of Babel judgment. Furthermore, the destruction at the earth's surface, where people and animals were then living during such a rapid continental "sprint," would have been as utterly devastating as the Flood itself.

Therefore, using catastrophic plate tectonics as a model, mechanism, and framework to describe and understand the Genesis Flood event is far more reasonable and is also consistent with the Bible. Early skepticism about the slow-and-gradual plate tectonics model has largely evaporated because it has such vast explanatory power. When applied to the Flood, however, the catastrophic plate tectonics model not only explains those elements in a more consistent way, but it also provides a powerful explanation for the dramatic evidences of massive flooding and catastrophic geologic processes on the continents.

From the late eighteenth century to the present, most scientists, including creationists, rejected the Genesis Flood to explain the fossil-bearing portion of the geological record because it lacked an adequate mechanism to produce such a vast amount of geological change in such a short time. Only now are we beginning to understand at least part of the means God may have used to bring this world-destroying judgment to pass, including catastrophic plate tectonics.

Conclusion

Many creationist geologists now believe the catastrophic plate tectonics concept is very useful as the best explanation for how the Flood event occurred within the biblical framework for earth's history. Even though the Bible does not specifically mention this concept, it is consistent with the biblical account, which implies an original supercontinent that broke up during the Flood, with the resultant continents obviously then having to move rapidly ("sprint") into their present positions.

This concept is still rather new, and of course radical, but its explanatory power makes it compelling. Additional work is now being done to further detail this geologic model for the Flood event, especially to show that it provides a better explanation for the order and distribution of the fossils and strata globally than the failed slow-and-gradual belief. Of course, future discoveries may require adjustments in our thinking and understanding, but such is the nature of the human scientific enterprise. In contrast, "the word of the Lord endures forever" (1 Peter 1:25).

15

Don't Creationists Believe Some "Wacky" Things?

BODIE HODGE

When answering questions about the creation/evolution issue, I have often been accused of believing some strange things. Some accuse me of believing, for example, that the earth is flat, that animals don't change, or that the earth literally sits on several pillars.

When I tell these people I don't believe these things, they are sometimes shocked. I suspect these rumors exist to convince unsuspecting people that the Bible isn't true. With a little research, we can easily debunk some of these myths.

1. Claim: Biblical Creationists Believe the Earth Is Flat.

This charge is often leveled at biblical creationists the moment the Bible is brought up. As far as I'm aware, no biblical creationists believe this. The Bible doesn't teach a flat earth, and this belief was never widespread.[1] In fact, the Bible plainly teaches the earth isn't flat, so it shouldn't be an issue:

> It is He who sits above the *circle of the earth*, and its inhabitants are like grasshoppers, who stretches out the heavens like a curtain and spreads them out like a tent to dwell in (Isaiah 40:22, emphasis added).

> He drew a *circular horizon* on the face of the waters, at the boundary of light and darkness (Job 26:10, emphasis added).

---

[1] Who invented a flat earth? *Creation* **16**(2):48-49, March 1994. Found online at www.answersingenesis.org/creation/v16/i2/flatearth.asp.

The earth is circular as indicated by the Bible, not flat.

Flat-earth beliefs were rather common in ancient Greece before 500 BC. This belief resurfaced in the early AD 300s with Lactantius; few others throughout history, though, have held to it. The humanists later revived this strange belief during the Renaissance and tried to imply that Christians, for the most part, believed this view. However, this simply wasn't the case.[1] Instead, the humanists took some biblical passages out of context. One such example is Revelation 7:1, which prophetically refers to the four corners of the earth. Instead of understanding the figurative nature of the verse, the humanists attempted to impose a strictly literal meaning on the passage. This passage is obviously referring to the directions of North, South, East, and West. Expositor John Gill comments on this verse:

> Four angels are mentioned, in allusion to the four spirits of the heavens, in Zec 6:5; and though the earth is not a plain square with angles, but round and globular, yet it is said to have four corners, with respect to the four points of the heavens; and though there is but one wind, which blows sometimes one way, and sometimes another, yet four are named with regard to the above points, east, west, north, and south, from whence it blows.[2]

Poetic passages, such as Psalm 75:3, which refers to the "pillars" of the earth, were also used to discredit Christians. Commentators such as John Gill[3] and Matthew Henry[4] rightly point out the figurative nature of these passages.

Recommended reading: *Taking Back Astronomy* (Chapter 2)

---

[2] J. Gill, *Exposition of the Old Testament*, Notes on Revelation 7:1, 1748–1763. Found online at https://www.biblestudytools.com/commentaries/gills-exposition-of-the-bible/revelation-7-1.html.

[3] J. Gill, *Exposition of the Old Testament*, Notes on Psalm 75:3, 1748–1763. Found online at https://www.biblestudytools.com/commentaries/gills-exposition-of-the-bible/psalms-75-3.html.

[4] M. Henry, *Matthew Henry Complete Commentary*, Notes on Psalm 75:3. Found online at https://www.biblestudytools.com/commentaries/matthew-henry-complete/psalms/75.html.

2. Claim: Biblical Creationists Don't Believe There Are "Beneficial" Mutations.

Mutations in and of themselves are usually harmful and we would expect this because of the Curse. Most of the other mutations are static, meaning they don't really affect the organism as a whole. However, there are a few cases of *beneficial* mutations that have been observed—these are different from mutations that cause the alleged gain of *new* genetic information. In fact, they should be referred to as mutations with beneficial outcomes—you'll see why in a moment.

A mutation that causes a beetle to lose its wings would be considered beneficial if the beetle lived on a windy island. It would be beneficial because it might keep the beetle from blowing out to sea to die. However, this mutation causes a loss of genetic information since the beetle no longer has the information to make wings. It could also be considered a harmful mutation since it can't get away from predators as easily.

The mutation that causes sickle cell anemia could be considered beneficial because it protects against malaria. However, the person with this mutation has lost the information to make proper, efficient blood cells, and sickled blood cells cause many problems.

Normal blood cell

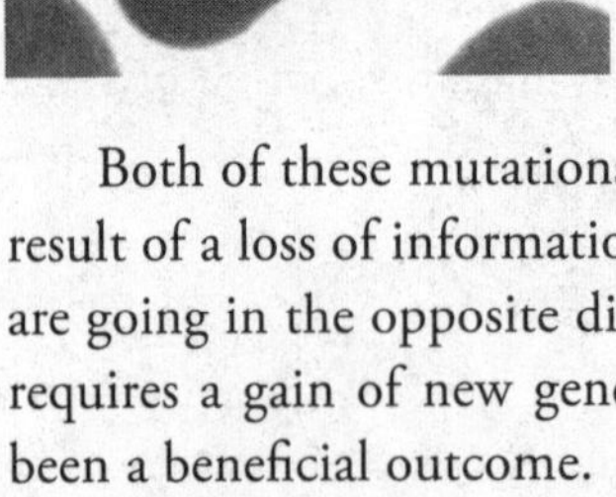

Sickle blood cell

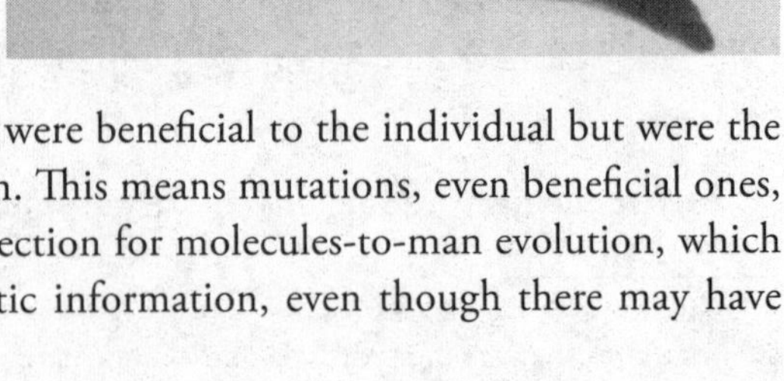

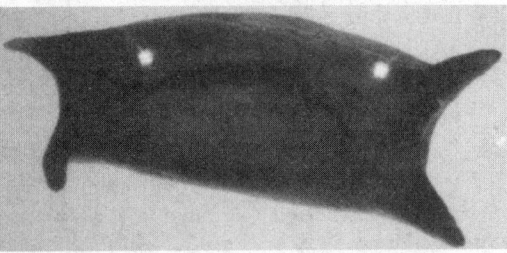

Both of these mutations were beneficial to the individual but were the result of a loss of information. This means mutations, even beneficial ones, are going in the opposite direction for molecules-to-man evolution, which requires a gain of new genetic information, even though there may have been a beneficial outcome.

Consider chickens that lost the information to produce feathers.[5] This can be considered "beneficial" because we no longer have to pluck them! But

[5] E. Young, Featherless chicken creates a flap, May 21, 2002, www.newscientist.com/article.ns?id=dn2307.

the chickens can't fly and have trouble keeping warm. Often, people confuse gains of new information with beneficial mutations, but they are different. For molecules-to-man evolutionary changes, the mutation needs to be beneficial *and* cause a gain of new information.

Recommended reading: *The New Answers Book 2*, chapter 7: Are mutations part of the "engine" of evolution?

3. Claim: You Can't Be a Christian If You Don't Believe in a Young Earth.

Answers in Genesis has continually claimed that one *can* be a Christian regardless of one's stance on the age of the earth or evolution. However, as AiG has also pointed out, these Christians are not being consistent.

Believing in a younger age of the earth (about 6,000 years) is a corollary of trusting the Bible. First, we start with the first five days of creation, then Adam was made on the sixth day, then adding ages given in the genealogies from Adam to Abraham we get about 2,000 years.[6] Both secular historians and Christians place Abraham at about 2,000 BC, so "the beginning" would be about 6,000 years ago. So the earth is about 6,000 years old—which is old—but much younger than the billions of years that are commonly touted.

	Time	Total Time
First 5 days of creation	5 days	5 days
Adam on Day 6 to Abraham	~2000 years	Still ~2000 years
Abraham to Christ	~2000 years	~4000 years
Christ until today	~2000 years	~6000 years

Believing in an approximately 6,000-year-old earth sets a proper foundation for believing Jesus Christ because you are letting God speak through His Word, without taking ideas to the Bible. In the same way, by trusting the Bible first, we realize that sin and death are intrusions into the world that go back to Genesis 3—which is the foundation for the gospel. Jesus came to save us from sin and death. If you give up this foundation of starting with the Bible and you insert evolutionary/millions-of-years ideas for the past history

[6] Bodie Hodge, "Ancient Patriarchs in Genesis," Answers in Genesis, www.answersingenesis.org/articles/2009/01/20/ancient-patriarchs-in-genesis.

of the world over the Bible's teachings in Genesis, it is inconsistent to believe the rest of the Bible—particularly the gospel. Sadly, people do it, and it is wrong, but it won't negate their salvation.

See other chapters in this book:

Chapter 8: Could God Have Created Everything in Six Days?

Chapter 9: Does Radiometric Dating Prove the Earth Is Old?

Chapter 19: Does Distant Starlight Prove the Universe Is Old?

4. Claim: Biblical Creationists Take the Whole Bible Literally.

It is better to say that creationists read and understand the Bible according to the grammatical-historical approach to Scripture. That is, we understand a biblical passage by taking into account its context, author, readership, literary style, etc. In other words, we read and understand the Bible in a plain or straightforward manner. This is usually what people mean when they say "literal interpretation of the Bible." This method helps to eliminate improper interpretations of the Bible.

> But we have renounced the hidden things of shame, not walking in craftiness nor handling the word of God deceitfully, but by manifestation of the truth commending ourselves to every man's conscience in the sight of God (2 Corinthians 4:2).

> All the words of my mouth are with righteousness; nothing crooked or perverse is in them. They are all plain to him who understands, and right to those who find knowledge (Proverbs 8:8–9).

Reading the Bible "plainly" means understanding which passages are written as historical narrative, which are written as poetry, which are written as parable, which are written as prophecy, and so on. The Bible is written in many different literary styles and should be read accordingly. Genesis records actual historical events; it was written as historical narrative, and there is no reason to read it as any other literary style, such as allegory or poetry.

For example, a non-Christian once claimed, "The Bible clearly says 'there is no God' in Psalms 14:1." However, this verse in context says:

> The fool has said in his heart, "There is no God." They are corrupt, they have done abominable works, there is none who does good (Psalm 14:1).

The context helps determine the proper interpretation—that a *fool* claims there is no God.

Someone else claimed, "To interpret the days in Genesis, you need to read 2 Peter 3:8, which indicates the days are each a thousand years." Many people try to use this passage to support the idea that the earth is millions or billions of years old, but let's read it in context:

> But, beloved, do not forget this one thing, that with the Lord one day is as a thousand years, and a thousand years as one day. The Lord is not slack concerning His promise, as some count slackness, but is longsuffering toward us, not willing that any should perish but that all should come to repentance (2 Peter 3:8–9).

This passage employs a literary device called a simile. Here, God compares a day to a thousand years in order to make the point that time doesn't bind Him, in this case specifically regarding His patience. God is eternal and is not limited to the time He created.

Also, this verse does not reference the days in Genesis, so it is not warranted to apply this to the length of the days in Genesis 1. When read plainly, these verses indicate that God is patient when keeping His promises. The gentleman that spoke to me had preconceived beliefs based on man's ideas that the earth was millions of years old. Those beliefs led him to this strange interpretation as opposed to using the historical-grammatical method.

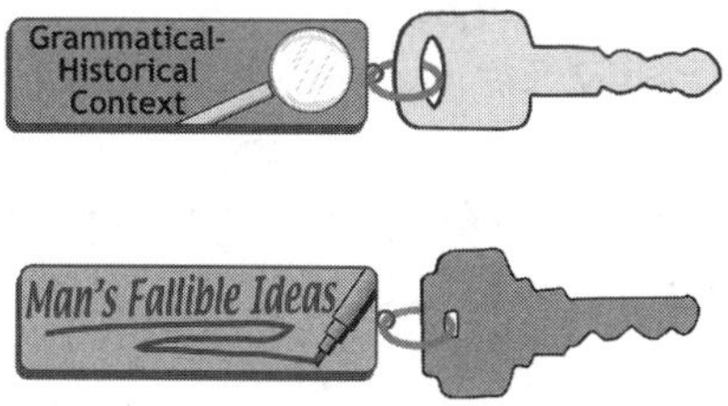

So, biblical Christians read the Bible plainly, or straightforwardly, and in context. Accordingly, we learn from what God says and means, and we don't apply strange literalistic (in the strict sense) meanings on metaphorical or allegorical passages, and vice versa.

> Recommended reading: *The New Answers Book 3*, chapter 8: Did Bible Authors Believe in a Literal Genesis?

5. Claim: Biblical Creationists Don't Have Any Evidence for Their Position.

In fact, we have the same evidence that evolutionists have, whether bones, fossils, or rocks. The difference is the *interpretation* of the evidence. Creationists and evolutionists begin with different starting points when looking at the same evidence, which is why they arrive at different conclusions.

As biblical Christians, we trust as our axiom, or starting point, that God exists and that His Word is truth. From there, we use the Bible to explain the evidence we see in the world around us. Evolutionists commonly use their axiom (naturalism/materialism and a belief that molecules-to-man evolution is true) to interpret evidence. When carefully analyzing the two interpretations, the biblical interpretation is vastly superior—it explains the evidence and is confirmed by operational science.

Recommended reading: *The New Answers Book 2*, chapter 2: What's the Best "Proof" of Creation?

See chapter 1 in this book: Is There Really a God?

6. Claim: Biblical Creationists Believe the Earth Is the Same Now as It Was at the Beginning of Creation.

Biblical creationists believe that significant changes have happened to the earth in its 6,000-year history—two very catastrophic ones: the Fall and the Flood.

The Fall was when Adam and Eve disobeyed God. Prior to this, the earth and all of creation was perfect (Genesis 1:31; Deuteronomy 32:4). Adam was given precious few commands in this perfect world, one of which was to not eat from the fruit of the Tree of the Knowledge of Good and Evil. If he ate, his punishment would be death (Genesis 2:17).

But Adam ate, and he died (Genesis 3:19, 5:5), and now we die because we too sin (disobey God). Death and suffering entered the creation as an intrusion.

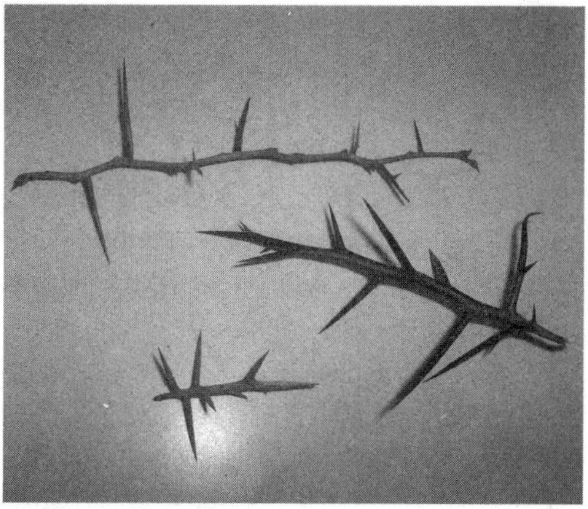

There were also other results of Adam's disobedience (Genesis 3). One was that the ground was cursed. Another was thorns and thistles. There were changes to the animals and humans.

The Fall was a significant event that definitely caused the earth to change (Romans 8:18–22).

The Flood was God's judgment on the people of the world who had turned their back on Him (Genesis 6–8). God said He would destroy them with a Flood, and He did.

This Flood was a global Flood that demolished everything. Many biblical creationists believe there was initially only one continent (Genesis 1:9). This original continent broke apart and was rearranged catastrophically during the Flood and the following years and finally became what we have today.

This massive Flood buried many animals, plants, and marine life, and many became fossils. A vast portion of the sedimentary rock layers we find throughout the world today is a testimony to this global Flood.

The Flood also caused ocean basins to sink down, mountains to be pushed up, etc. Major geological features resulted. Additional after-effects of the Flood were the Ice Age, plate fault lines, etc.

Biblical creationists believe the world has changed. The real question is, in what way? This is an exciting part of creationist research today.

See other chapters in this book:

Chapter 10: Was There Actually a Noah's Ark and Flood?

Chapter 14: Can Catastrophic Plate Tectonics Explain Flood Geology?

Chapter 26: Why Does God's Creation Include Death and Suffering?

7. Claim: Biblical Creationists Are Anti-Science and Anti-Logic.

Biblical creationists love science! In fact, most fields of science were developed by men who believed the Bible, such as Isaac Newton (dynamics, gravitation, calculus), Michael Faraday (electromagnetics, field theory), Robert Boyle (chemistry), Johannes Kepler (astronomy), and Louis Pasteur (bacteriology, immunization). Francis Bacon, a Bible-believing Christian, developed the scientific method.

The reason such fields of science developed was the belief that God created the universe and that He instituted laws that we could investigate. Even today, many great scientists believe the Bible and use good observational science on a daily basis.[7]

Even logic flows naturally from a biblical worldview. Since we are created in the image of a logical God, we would expect to have logical faculties. However, logic is not a material entity, so it becomes a problem for the materialist atheist who denies the immaterial realm. From a materialistic perspective, a logical thought is the same as an illogical thought—merely a chemical reaction in the brain. From a materialistic point of view, then, the perception of logic is due to random processes and has nothing to do with absolute truth, which is also immaterial.

So in a biblical worldview, logic exists and so does truth, both of which are immaterial. But in a purely materialistic worldview, there is no basis for logic or truth to exist, since they are immaterial. And if our brains are the result of random mutations and natural selection, how do we know that our brains have evolved in a way that allows us to think and reason according to truth?

To state that logic *can* yield a truthful result means that absolute truth must exist, hence God. This does not mean that atheists and evolutionists cannot use logic or do science. But when they do, they must borrow from the above Christian principles, an action which is not consistent with their professed worldview.

Recommended reading: *The New Answers Book 2*, chapter 14: Can Creationists Be "Real" Scientists?

See chapter 4 in this book: Don't Creationists Deny the Laws of Nature?

[7] To read about creation scientists and other biographies of interest, see www.answersingenesis.org/go/bios.

16

Where Does the Ice Age Fit?

MICHAEL OARD

If you ask a youngster the question, "Was there really an ice age?" they might say rather quickly that there was. Then they may tell you that there were two of them. Of course, if you listen much longer, they will tell you that they saw both of those movies in the theater.

The ice age is a popular topic that is often discussed in school, at home, or in Hollywood. Sadly, most people hear the secular/uniformitarian view and don't look at this subject from a biblical perspective. This is where it gets interesting, though. The secular view has no good mechanism to cause a single ice age, let alone the many they propose. But the Bible does have a mechanism. Let's take a closer look.

Before I get too deep, let me define a few words you'll need to know to help clarify this chapter:

Glacier: a large mass of ice that has accumulated from snow over the years and is slowly moving from a higher place.

Moraines: stones, boulders, and debris that have been carried and dropped by a glacier.

Uniformitarianism: the belief that rates today are the same as they were in the past, without the possibility of major catastrophes like worldwide floods.

Interglacial: a short period of warming between glacier growth/movement that caused glaciers to melt away.

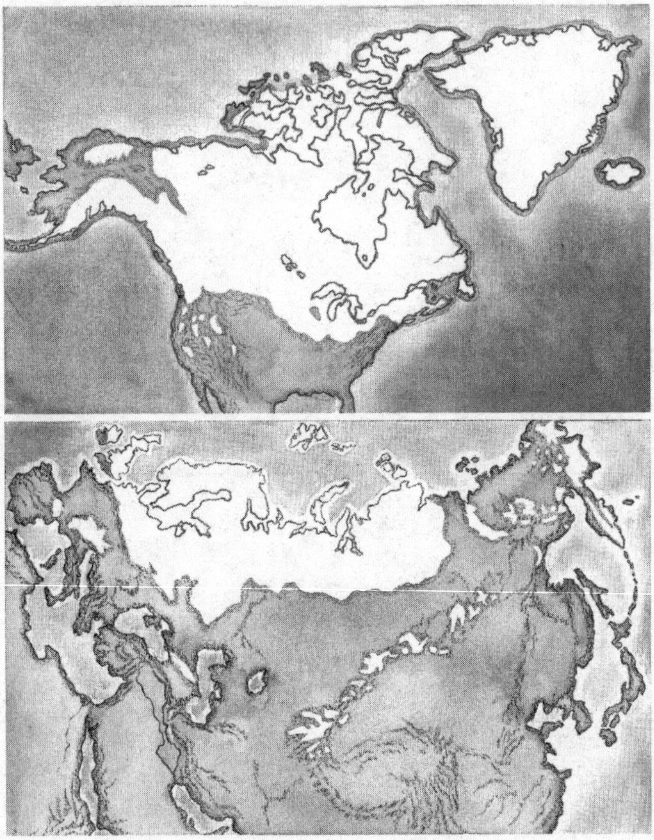

Figures 1 and 2. The extent of the Ice Age over North America and Eurasia.

Ice cores: cores of ice that have been drilled down into a glacier.

Ice Age: when seen in capital letters, refers to the biblical post-Flood Ice Age.

An ice age is defined as a time of extensive glacial activity in which substantially more of the land is covered by ice. During the Ice Age that ended several thousand years ago, 30 percent of the land surface of the earth was covered by ice (Figures 1 and 2). In North America an ice sheet covered almost all of Canada and the northern United States.

We know the extent of the Ice Age in the recent past because similar features, as observed around glaciers today, are also found in formerly glaciated areas, such as lateral and terminal moraines. A *lateral moraine* is a mound of rocks of all sizes deposited on the side of a moving glacier, while a *terminal*, or *end, moraine* is a mound of rocks bulldozed in front of the glacier.

Figure 3 shows a horseshoe-shaped moraine from a glacier that spread out from a valley in the Wallowa Mountains of northeast Oregon. The two

Figure 3. Horseshoe-shaped lateral and end moraines plowed up by a glacier moving out of a valley in the northern Wallowa Mountains of northeast Oregon. Beautiful Wallowa Lake fills the depression within the moraines.

lateral moraines are 600 feet (183 m) high, while the end moraine is 100 feet (30 m) high, enclosing beautiful Wallowa Lake. Scratched bedrock and boulders are telltale signs of previous glaciation (Figures 4 and 5), which are similar to such features found around glaciers today (Figures 6 and 7).

Figures 4 and 5. Striated bedrock and boulders from an ice cap in the northern Rocky Mountains that spread through the Sun River Canyon out onto the high plains, west of Great Falls, Montana.

Figures 6 and 7. Scratched bedrock and boulder from the Athabasca Glacier in the Canadian Rocky Mountains.

Secular/Uniformitarian Belief

Secular/uniformitarian scientists used to believe that there were four ice ages during the past few million years. However, the idea of four ice ages was rejected in the 1970s in favor of thirty or more ice ages separated by interglacials.[1] Such a switch was forced by a paradigm change in glaciology toward belief in the astronomical model of the ice ages (or "Milankovitch mechanism," as it is called). The idea of four ice ages still lingers in public museum displays, though (Figure 8).

The astronomical model postulates regularly repeating ice ages caused by the changing orbital geometry of the earth. Secular glaciologists believe that over the past 800,000 years there were, allegedly, eight ice ages, each lasting about 100,000 years.[2] The glacial phase supposedly dominated for 90,000 years, while the interglacial phase lasted only 10,000 years. Accordingly, the

[1] J. Kennett, *Marine Geology*, Prentice-Hall, Englewood Cliffs, New Jersey, 1982, 747.

[2] D. Paillard, Glacial cycles: toward a new paradigm, *Reviews of Geophysics*, **39**(3):325–346, 2001.

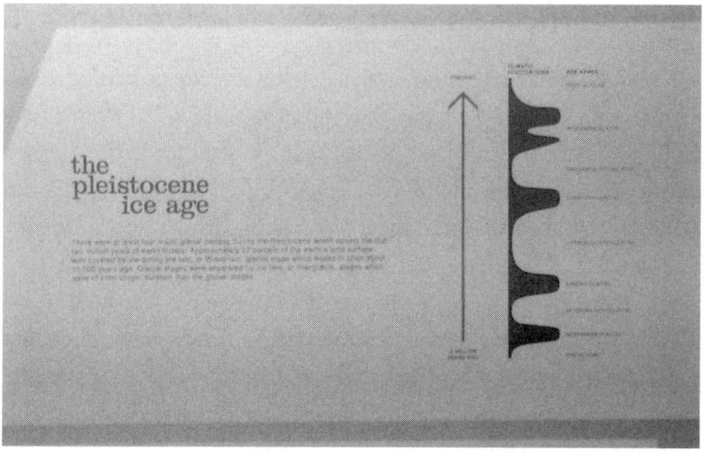

Figure 8. Display of four ice ages at the College of Eastern Utah Prehistoric Museum at Price, Utah, taken in 2006.

story continues that beyond 800,000 years, the ice ages are believed to have cycled every 40,000 years or so.

The secular/uniformitarian model now holds that the Antarctic Ice Sheet developed around 40 million years ago and reached general equilibrium about 15 million years ago.[3] The Greenland Ice Sheet, they say, is younger, having developed only a few million years ago.

Uniformitarian scientists further believe four "ancient ice ages" occurred during geological time (Table 1). These ice ages supposedly occurred hundreds of millions to several billion years ago, with each ice age lasting tens to hundreds of millions of years. Ancient ice ages are deduced from features in the rock that seem to indicate glaciation.

Geological Period	**Secular Approximate Age Range** (million years ago)
Late Paleozoic	256–338
Late Ordovician	429–445
Late Proterozoic	520–950
Early Proterozoic	2200–2400

Table 1. The four main "ancient ice ages" within the uniformitarian paradigm and their inferred age range in millions of years before the present. The age ranges for the earliest "ice ages" are admittedly rough estimates.[4]

[3] M.J. Oard, *The Frozen Record: Examining the Ice Core History of the Greenland and Antarctic Ice Sheets*, Institute for Creation Research, El Cajon, California, 2005, 31–34.

[4] J.C. Crowell, *Pre-Mesozoic Ice Ages: Their Bearing on Understanding the Climate System*, Geological Society of America Memoir 192, Boulder, Colorado, 1999, 3.

Severe Difficulties with Secular/Uniformitarian Beliefs

Secular/uniformitarian scientists have great difficulty explaining any recent ice ages based on rates they observe today. They have proposed dozens of hypotheses, but all have serious flaws. One problem is that the summer temperatures in the northern United States would have to cool more than 50°F (28°C) accompanied by a huge increase in snow. What would trigger or sustain such a dramatic climate change that would persist for thousands of years? David Alt of the University of Montana in Missoula recently admitted, "Although theories abound, no one really knows what causes ice ages."[5]

Ancient ice ages have been somewhat controversial over the years, but recently some uniformitarian scientists have come out with the shocking belief that some Proterozoic ice ages were global.[6] This belief is based on paleomagnetic data that supposedly shows certain rocks, believed to be from ancient ice ages, were marine and equatorial. Because of the reflection of sunlight from a white surface, it is likely that a glaciated earth would never melt. However, advocates of "snowball earth" state not only that such a glaciation completely melted but also that temperatures following glaciation ended up much warmer than today. Such a "freeze-fry" hypothesis indicates that the concept of ancient ice ages is unsound.

Did the Flood Trigger the Ice Age?

If uniformitarian scientists have severe difficulties accounting for ice ages, how would creationists explain an ice age or multiple ice ages? Let's start with the recent ice age.

When attempting to account for ice ages, the uniformitarian scientists do not consider one key element—the Genesis Flood. What if there truly were a worldwide Flood? How would it have affected the climate? A worldwide Flood would have caused major changes in the earth's crust, as well as earth movements and tremendous volcanism. It would have also greatly disturbed the climate.

A shroud of volcanic dust and aerosols (very small particles) would have been trapped in the stratosphere for several years following the Flood. These volcanic effluents would have then reflected some of the sunlight back to space

[5] D. Alt, *Glacial Lake Missoula and its Humongous Floods*, Mountain Press Publishing Company, Missoula, Montana, 2001, 180.

[6] M.J. Oard, Another tropical ice age? *Journal of Creation* **11**(3):259–261, 1997; M.J. Oard, Snowball Earth—a problem for the supposed origin of multicellular animals, *Journal of Creation* **16**(1):6–9, 2002.

and caused cooler summers, mainly over large landmasses of the mid and high latitudes. Volcanoes would have also been active during the Ice Age and gradually declined as the earth settled down. Abundant evidence shows substantial Ice Age volcanism, which would have replenished the dust and aerosols in the stratosphere.[7] The Greenland and Antarctic ice sheets also show abundant volcanic particles and acids in the Ice Age portion of the ice cores.[8]

An ice age also requires huge amounts of precipitation. The Genesis account records the "fountains of the great deep" bursting forth during the Flood. Crustal movements would have released hot water from the earth's crust along with volcanism and large underwater lava flows, which would have added heat to the ocean. Earth movement and rapid Flood currents would have then mixed the warm water, so that after the Flood the oceans would be warm from pole to pole. There would be no sea ice. A warm ocean would have had much higher evaporation than the present cool ocean surface. Most of this evaporation would have occurred at mid and high latitudes, close to the developing ice sheets, dropping the moisture on the cold continent. This is a recipe for powerful and continuous snowstorms that can be estimated using basic meteorology.[9] Therefore, to cause an ice age, rare conditions are required—warm oceans for high precipitation, and cool summers for lack of melting the snow. Only then can it accumulate into an ice sheet.

The principles of atmospheric science can also estimate areas of high oceanic evaporation, the eventual depth of the ice, and even the timing of the Ice Age. Numerical simulations of precipitation in the polar regions using conventional climate models with warm sea surface temperatures have demonstrated that ice sheets thousands of feet thick could have accumulated in less than 500 years.[10]

A Rapid Ice Age

Most creationists agree that there was one major Ice Age following the Flood. The timing of the Ice Age is quite significant, since uniformitarians claim that each ice age over the past 800,000 years lasted about 100,000 years. To estimate the time for a post-Flood Ice Age, we need to know how long the volcanism lasted and the cooling time of the oceans. Once these two

---

[7] M.J. Oard, *An Ice Age Caused by the Genesis Flood*, Institute for Creation Research, El Cajon, California, 1990, 33–38.

[8] Oard, *The Frozen Record.*

[9] Oard, *An Ice Age Caused by the Genesis Flood.*

[10] L. Vardiman, *Climates before and after the Genesis Flood: numerical models and their implications,* Institute for Creation Research, El Cajon, California, 2001.

mechanisms for the Ice Age wane, the ice sheets will reach a maximum and then begin to melt. So, an estimate of the time for the Ice Age can be worked out based on the available moisture for snow and the cooling time of the ocean (the primary mechanism) in a cool post-Flood climate.

I used budget equations for the cooling of the ocean and atmosphere, which are simply based on heat inputs minus heat outputs—the difference causing the change in temperatures. Since there is no way to be precise, I used minimums and maximums for the variables in the equations in order to bracket the time. The best estimate is about 500 years after the Flood to reach glacial maximum with an average ice and snow depth of about 2,300 feet (700 m) in the Northern Hemisphere and 4,000 feet (1,220 m) on Antarctica.[11]

Once the conditions for the Ice Age ended, those ice sheets in unfavorable areas melted rapidly. Antarctica and Greenland, possessing a favorable latitude and altitude, would continue to grow during deglaciation and afterward. To calculate the melting rate for the ice sheets over North America and Eurasia, I used the energy balance over a snow cover, which gives a faster rate than the uniformitarians propose based on their models.

An energy balance equation is a straightforward and more physical method of calculating the melt rate. Using maximum and minimum values for the variable in the melt equation, I obtained a best estimate of the average melt rate along the periphery (a 400-mile [645-km] long strip) of the ice sheet in North America at about 33 feet/year (10 m/year). Such a melting rate compares favorably with current melt rates for the melting zones of Alaskan, Icelandic, and Norwegian glaciers today. At this rate, the periphery of the ice sheets melts in less than 100 years. Interior areas of ice sheets would melt more slowly, but the ice would be gone in about 200 years. The ice sheets melt so fast, catastrophic flooding would be expected, such as with the bursting of glacial Lake Missoula described later in this chapter.

Therefore, the total length of time for a post-Flood Ice Age is about 700 years. It was indeed a rapid Ice Age. This is an example of bringing back the Flood into earth history. As a result, processes that seem too slow at today's rates were much faster in the past. The Flood was never disproved; it was *arbitrarily* rejected in the 1700s and 1800s by secular intellectuals in favor of slow processes over millions of years.

---

11 Oard, *An Ice Age Caused by the Genesis Flood.*

How Many Ice Ages?

Still, there is the claim of many ice ages. Most formerly glaciated areas show evidence for only one ice age, and a substantial amount of information indicates only one ice age.[12] The idea of multiple ice ages is essentially a *uniformitarian assumption*. Today this idea is strongly based on oxygen isotope ratios from seafloor sediments. The paleothermometers developed from these data assume highly questionable statistical comparisons between peaks and valleys in temperature, which are claimed to correspond to orbital changes in the heating of the earth. In a provocative paper concluding that only one ice sheet covered southern and central Alberta late in the uniformitarian timescale, Robert Young and others stated: "Glacial reconstructions commonly assume a multiple-glaciation hypothesis in all areas that contain a till cover."[13]

Areas that appear to have evidence of more than one ice age can be reinterpreted to be the deposits from one ice sheet that advanced and retreated over a short period. The more modern understanding of glacial activity indicates that ice sheets are very dynamic. We do not need 100,000 years for each ice age or 2.5 million years for multiple ice ages.

One of the key assumptions in the multiple glaciation hypothesis is the astronomical model of ice ages. This mechanism is based on cyclical past changes in the geometry of the earth's orbit. Uniformitarian scientists believe that a decrease in solar radiation at about 60° N in summer, resulting from orbital changes, causes repeating ice ages, either every 100,000 years or every 40,000 years. By matching wiggles in variables taken from deep-sea cores, uniformitarian scientists believe they have proven the astronomical mechanism of multiple ice ages.[14] There are many problems with this model and relating deep-sea cores to it; mainly, the decrease in sunshine is too small.[15] Didier Paillard stated,

> Nevertheless, several problems in classical astronomical theory of paleoclimate have indeed been identified: (1) The main cyclicity in the paleoclimate record is close to 100,000 years, but there is [*sic*] no significant orbitally induced changes in the radiative [sunshine] forcing of the Earth in this frequency range (the "100-kyr Problem").[16]

---

[12] Ibid., 135–166.

[13] R.R. Young et. al., A single, late Wisconsin, Laurentide glaciation, Edmonton area and southwestern Alberta, *Geology* **22**:683–686, 1994.

[14] J.D. Hays, J. Imbrie, and N.J. Shackleton, Variations in the earth's orbit: pacemaker of the ice ages, *Science* **194**:1121–1132, 1976.

[15] Oard, *The Frozen Record,* 111–122.

[16] Paillard, Glacial cycles: toward a new paradigm, 325.

Although the main cycle in the astronomical model is 100,000 years, the change in sunshine at high northern latitudes is insignificant for such a dramatic change as an ice age.

Is the Ice Age Biblical?

Since the Flood offers a viable explanation for the Ice Age, one could expect that the Ice Age would be mentioned in the Bible. It is possible that the book of Job, written about 500 years or so after the Flood, may include a reference to the Ice Age in Job 38:29–30, which says, "From whose womb comes the ice? And the frost of heaven, who gives it birth? The waters harden like stone, and the surface of the deep is frozen." However, Job could have observed frost and lake ice during winter in Palestine, especially if temperatures were colder because of the Ice Age. The reason the Ice Age is not directly discussed in the Bible is probably because the Scandinavian ice sheet and mountain ice caps were farther north than the region where the Bible was written. Only an increase in the snow coverage of Mt. Hermon and possibly more frequent snowfalls on the high areas of the Middle East would have been evident to those living in Palestine.

How Are "Ancient Ice Ages" Explained?

The evidence for "ancient ice ages" is found in the hard rocks; these deposits are not on the surface like the deposits from the post-Flood Ice Age. There are substantial difficulties in interpreting these rocks as from ancient ice ages.[17] An alternative mechanism can easily explain these deposits within a biblical framework. This mechanism is gigantic submarine landsides that occurred during the Genesis Flood.

The Mystery of the Woolly Mammoths

Millions of woolly mammoth bones, tusks, and a few carcasses have been found frozen in the surface sediments of Siberia, Alaska, and the Yukon Territory of Canada—a major mystery of uniformitarian paleoclimate. The woolly mammoths were part of a Northern Hemisphere community of animals that lived and died during the post-Flood Ice Age.[18] Woolly

---

[17] M.J. Oard, *Ancient Ice Ages or Gigantic Submarine Landslides?* Creation Research Society Monograph No. 6, Chino Valley, Arizona, 1997.

[18] M.J. Oard, *Frozen In Time: The Woolly Mammoths, the Ice Age, and the Bible*, Master Books, Green Forest, Arkansas, 2004.

Figure 9. Large dust drift to the top of a house during the dust bowl era in the Midwest.

mammoths probably died after the Flood because there are thousands of carcasses scattered across Alaska and Siberia resting above Flood deposits. And there must have been sufficient time for the mammoths to have repopulated these regions after the Flood. The post-Flood Ice Age provides an explanation for the mystery of the woolly mammoths, as well as many other Ice Age mysteries.

The mammoths spread into these northern areas during early and middle Ice Age time because summers were cooler and winters warmer. The areas were unglaciated (just the mountains glaciated) and a rich grassland. However, late in the Ice Age, winter temperatures turned colder and the climate drier with strong wind storms. The mammoths died by the millions and were buried by dust, which later froze, preserving the mammoths. Severe dust storms that produce tall dust drifts (Figure 9) can also explain a number of the secondary mysteries, such as some carcasses that show evidence of suffocation in a generally standing position, and how they become entombed into rock-hard permafrost (for a more complete treatment of this subject, please see my book, *Frozen in Time*).

Is Glacial Lake Missoula Related to the Ice Age?

At the peak of the Ice Age, a finger of the ice sheet in western Canada and the northwest United States filled up the valleys of northern Idaho. A huge lake 2,000 feet (610 m) deep was formed in the valleys of western Montana. This was glacial Lake Missoula (Figure 10). In the course of time, the lake burst and emptied in a few days, causing an immense flood several hundred feet deep that

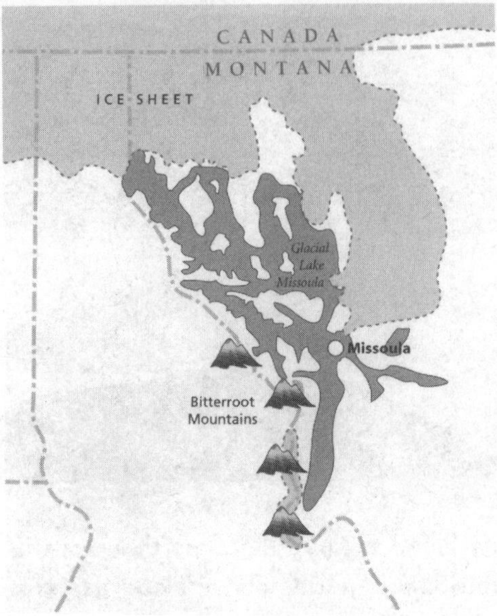

Figure 10. Map of ice sheet and glacial Lake Missoula (drawn by Mark Wolfe)

carved out canyons and produced many flood features from eastern Washington into northwest Oregon (Figure 11).

This flood can help us understand the global Flood. Interestingly, the Lake Missoula flood was rejected for 40 years despite tremendous evidence because of the anti-biblical bias in historical science.[19]

Now this flood is not only accepted, but uniformitarian scientists now believe many more of them occurred. They postulate 40 to 100 at the peak of their last ice age, with perhaps hundreds more from previous ice ages. However, the evidence

Figure 11. The Potholes, remnants of a 400-foot (120 m) high waterfall. The lakes at the bottom are remnant plunge pools.

[19] M.J. Oard, *The Missoula Flood Controversy and the Genesis Flood*, Creation Research Society Monograph No. 13, Chino Valley, AZ, 2004.

is substantial that there was only one gigantic Lake Missoula flood, with possibly several minor floods afterward.[20]

What about Ice Cores?

Uniformitarian scientists claim to be able to count annual layers in the Greenland ice sheet to determine its age, in the same way people can count tree rings. In doing so, they arrive at 110,000 years near the bottom of the Greenland ice sheet. Similar claims for a much greater age are made for the Antarctica ice sheet. These claims are equivocal and are essentially based on the uniformitarian belief that the ice sheets are millions of years old. The data from ice cores can be better explained within the post-Flood Ice Age model, which dramatically reduces the calculated age to well within the biblical limit.[21]

Conclusion

Although a major mystery of uniformitarian history, the Ice Age is readily explained by the climatic consequences of the Genesis Flood—it was a short Ice Age of about 700 years, and there was only one Ice Age.[22] We do not need the hundred thousand years for one ice age, or the few million years for multiple ice ages, as claimed by uniformitarian scientists.

Even their claim of ancient ice ages in the hard rocks can be accounted for by gigantic submarine landslides during the Flood. The post-Flood rapid Ice Age can also account for a number of major mysteries and other interesting phenomena that occurred during the Ice Age, such as the Lake Missoula flood and the life and death of the woolly mammoths in Siberia and elsewhere. When we stick to the Genesis account of the Flood and the short scriptural timescale, major secular/uniformitarian mysteries are readily explained.[23]

---

[20] Ibid.

[21] L. Vardiman, *Ice cores and the Age of the Earth*, Institute for Creation Research, El Cajon, California, 1993; Oard, *The Frozen Record*.

[22] Oard, *An Ice Age Caused by the Genesis Flood*; Oard, *Ancient Ice Ages or Gigantic Submarine Landslides?* M.J. Oard and B. Oard, *Life in the Great Ice Age*, Master Books, Green Forest, Arkansas, 1993.

[23] For more on the Ice Age, see www.answersingenesis.org/go/ice-age.

17

Are There Really Different Races?

KEN HAM

What if a Chinese person were to marry a Polynesian, or an African with black skin were to marry a Japanese, or a person from India were to marry a person from America with white skin—would these marriages be in accord with biblical principles?

A significant number of Christians would claim that such "interracial" marriages directly violate God's principles in the Bible and should not be allowed.

Does the Word of God really condemn the marriages mentioned above? Is there ultimately any such thing as interracial marriage?

To answer these questions, we must first understand what the Bible and science teach about "race."

What Constitutes a "Race"?

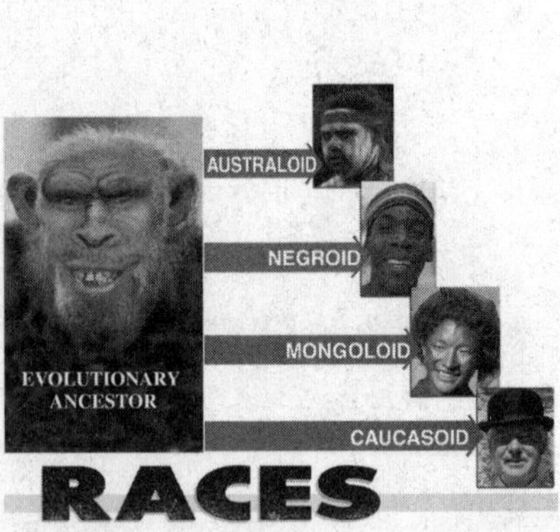

In the 1800s, before Darwinian evolution was popularized, most people, when talking about "races," would be referring to such groups as the "English race," "Irish race," and so on. However, this all changed in 1859 when Charles Darwin published his book *On the Origin of Species by Means*

of Natural Selection or the Preservation of Favoured Races in the Struggle for Life.

Darwinian evolution was (and still is[1]) inherently a racist philosophy, teaching that different groups or "races" of people evolved at different times and rates, so some groups are more like their apelike ancestors than others. Leading evolutionist Stephen Jay Gould claimed, "Biological arguments for racism may have been common before 1859, but they increased by orders of magnitude following the acceptance of evolutionary theory."[2]

The Australian Aborigines, for instance, were considered the missing links between the apelike ancestor and the rest of mankind.[3] This resulted in terrible prejudices and injustices towards the Australian Aborigines.[4]

Ernst Haeckel, famous for popularizing the now-discredited idea that "ontogeny recapitulates phylogeny,"[5] stated:

> At the lowest stage of human mental development are the Australians, some tribes of the Polynesians, and the Bushmen, Hottentots, and some of the Negro tribes. Nothing, however, is perhaps more remarkable in this respect, than that some of the wildest tribes in southern Asia and eastern Africa have no trace whatever of the first foundations of all human civilization, of family life, and marriage. They live together in herds, like apes.[6]

Racist attitudes fueled by evolutionary thinking were largely responsible for an African pygmy being displayed, along with an orangutan, in a cage in the Bronx zoo.[7] Indeed, Congo pygmies were once thought to be "small ape-like, elfish creatures" that "exhibit many ape-like features in their bodies."[8]

---

[1] J.P. Rushton, professor of psychology at the University of Western Ontario, Lond, Ontario, Canada, Race, Evolution and Behavior, www.harbornet.com/folks/theedrich/JP_Rushton/Race.htm.

[2] S.J. Gould, *Ontogeny and Phylogeny*, Belknap-Harvard Press, Cambridge, Massachusetts, 1977, 127–128.

[3] Missing links with mankind in early dawn of history, *New York Tribune*, p. 11, February 10, 1924.

[4] D. Monaghan, The body-snatchers, *The Bulletin*, November 12, 1991, pp. 30–38; Blacks slain for science's white superiority theory, *The Daily Telegraph Mirror*, April 26, 1994.

[5] For more information on the fallacious nature of this idea, see www.answersingenesis.org/go/embryonic.

[6] E. Haeckel, *The History of Creation*, 1876, 363–363.

[7] J. Bergman, Ota Benga: the man who was put on display in the zoo! *Creation* **16**(1):48–50, 1993.

[8] A.H.J. Keane, Anthropological curiosities — the pygmies of the world, *Scientific American Supplement* 64, no. 1650 (August 17, 1907): 99.

As a result of Darwinian evolution, many people started thinking in terms of the different people groups around the world representing different "races," but within the context of evolutionary philosophy. This has resulted in many people today, consciously or unconsciously, having ingrained prejudices against certain other groups of people.[9]

However, *all* human beings in the world today are classified as *Homo sapiens sapiens*. Scientists today admit that, biologically, there really is only one race of humans. For instance, a scientist at the Advancement of Science Convention in Atlanta stated, "Race is a social construct derived mainly from perceptions conditioned by events of recorded history, and it has no basic biological reality." This person went on to say, "Curiously enough, the idea comes very close to being of American manufacture."[10]

Reporting on research conducted on the concept of race, ABC News stated, "More and more scientists find that the differences that set us apart are cultural, not racial. Some even say that the word *race* should be abandoned because it's meaningless." The article went on to say that "we accept the idea of race because it's a convenient way of putting people into broad categories, frequently to suppress them—the most hideous example was provided by Hitler's Germany. And racial prejudice remains common throughout the world."[11]

GET RID OF THIS *EVOLUTIONIZED* TERM!

In an article in the *Journal of Counseling and Development*,[12] researchers argued that the term "race" is basically so meaningless that it should be discarded.

More recently, those working on mapping the human genome announced "that they had put together a draft of the entire sequence of the

---

[9] This is not to say that *evolution* is the cause of racism. *Sin* is the cause of racism. However, Darwinian evolution fueled a particular form of racism.

[10] R.L. Hotz, Race has no basis in biology, researchers say, *Cincinnati Enquirer*, p. A3, February 20, 1997.

[11] We're all the same, ABC News, September 10, 1998, www.abcnews.com/sections/science/DyeHard/dye72.html.

[12] S.C. Cameron and S.M. Wycoff, The destructive nature of the term race: growing beyond a false paradigm, *Journal of Counseling & Development*, 76:277–285, 1998.

human genome, and the researchers had unanimously declared, there is only one race—the human race."[13]

Personally, because of the influences of Darwinian evolution and the resulting prejudices, I believe everyone (and especially Christians) should abandon the term "race(s)." We could refer instead to the different "people groups" around the world.

The Bible and "Race"

The Bible does not even use the word race in reference to people,[14] but it does describe all human beings as being of "one blood" (Acts 17:26). This of course emphasizes that we are all related, as all humans are descendants of the first man, Adam (1 Corinthians 15:45),[15] who was created in the image of God (Genesis 1:26–27).[16] The Last Adam, Jesus Christ (1 Corinthians 15:45) also became a descendant of Adam. Any descendant of Adam can be saved because our mutual relative by blood (Jesus Christ) died and rose again. This is why the gospel can (and should) be preached to all tribes and nations.

CAN THE BIBLE BE USED TO JUSTIFY RACIST ATTITUDES?

The inevitable question arises, "If the Bible teaches all humans are the same, where was the church during the eras of slavery and segregation? Doesn't the Bible actually condone the enslavement of a human being by another?"

---

13 N. Angier, Do races differ? Not really, DNA shows, New York Times web, Aug. 22, 2000.

14 In the original, Ezra 9:2 refers to "seed," Romans 9:3 to "kinsmen according to the flesh."

15 For more on this teaching, see chapter 6, Cain's Wife—Who Was She?

16 Contrary to popular belief, mankind does not share an apelike ancestor with other primates. To find out the truth behind the alleged apemen, visit www.answersingenesis.org/go/anthropology.

Both the Old and New Testaments of the Bible mention slaves and slavery. As with all other biblical passages, these must be understood in their grammatical-historical context.

Dr. Walter Kaiser, former president of Gordon-Conwell Theological Seminary and Old Testament scholar, states:

> The laws concerning slavery in the Old Testament appear to function to moderate a practice that worked as a means of loaning money for Jewish people to one another or for handling the problem of the prisoners of war. Nowhere was the institution of slavery as such condemned; but then, neither did it have anything like the connotations it grew to have during the days of those who traded human life as if it were a mere commodity for sale. … In all cases the institution was closely watched and divine judgment was declared by the prophets and others for all abuses they spotted.[17]

Job recognized that all were equal before God, and all should be treated as image-bearers of the Creator.

> If I have despised the cause of my male or female servant when they complained against me, what then shall I do when God rises up? When He punishes, how shall I answer Him? Did not He who made me in the womb make them? Did not the same One fashion us in the womb? (Job 31:13–15).

In commenting on Paul's remarks to the slaves in his epistles, Peter H. Davids writes:

> The church never adopted a rule that converts had to give up their slaves. Christians were not under law but under grace. Yet we read in the literature of the second century and later of many masters who upon their conversion freed their slaves. The reality stands that it is difficult to call a person a slave during the week and treat them like a brother or sister in the church. Sooner or later the implications of the kingdom they experienced in church seeped into the behavior of the masters during the week. Paul did in the end create a revolution, not one from without, but one from within, in which a changed heart produced changed behavior and through that in the end brought about social change. This change happened wherever the kingdom of God was expressed through the church, so the

---

[17] W.C. Kaiser, Jr. et al., *Hard Sayings of the Bible*, InterVarsity Press, Downers Grove, Illinois, 1996, 150.

world could see that faith in Christ really was a transformation of the whole person.[18]

Those consistently living out their Christian faith realize that the forced enslavement of another human being goes against the biblical teaching that all humans were created in the image of God and are of equal standing before Him (Galatians 3:28; Colossians 3:11). Indeed, the most ardent abolitionists during the past centuries were Bible-believing Christians. John Wesley, Granville Sharp, William Wilberforce, Jonathan Edwards, Jr., and Thomas Clarkson all preached against the evils of slavery and worked to bring about the abolition of the slave trade in England and North America. Harriet Beecher Stowe conveyed this message in her famous novel *Uncle Tom's Cabin*. And of course, who can forget the change in the most famous of slave traders? John Newton, writer of "Amazing Grace," eventually became an abolitionist after his conversion to Christianity, when he embraced the truth of Scripture.

"Racial" Differences

But some people think there must be different races of people because there appear to be major differences between various groups, such as skin color and eye shape.

The truth, though, is that these so-called "racial characteristics" are only minor variations among people groups. If one were to take any two people anywhere in the world, scientists have found that the basic genetic differences between these two people would typically be around 0.2 percent—even if they came from the same people group.[19] But these so-called "racial" characteristics that people think are major differences (skin color, eye shape, etc.) "account for only 0.012 percent of human biological variation."[20]

Dr. Harold Page Freeman, chief executive, president, and director of surgery at North General Hospital in Manhattan, reiterates, "If you ask what percentage of your genes is reflected in your external appearance, the basis by which we talk about race, the answer seems to be in the range of 0.01 percent."[21]

In other words, the so-called "racial" differences are absolutely trivial—overall, there is more variation *within* any group than there is *between* one

[18] Ref. 17, 644.
[19] J.C. Gutin, End of the rainbow, *Discover*, pp. 72–73, November 1994.
[20] Ref. 12.
[21] Ref. 13.

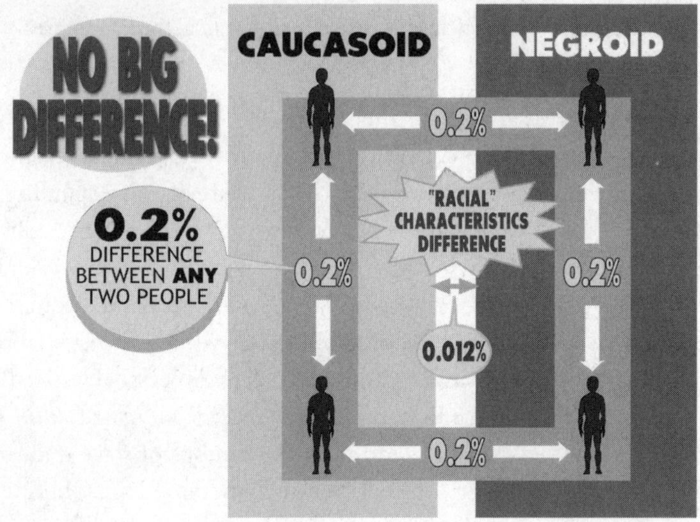

group and another. If a white person is looking for a tissue match for an organ transplant, for instance, the best match may come from a black person, and vice versa. ABC News claims, "What the facts show is that there are differences among us, but they stem from culture, not race."[22]

The only reason many people think these differences are major is because they've been brought up in a culture that has taught them to see the differences this way. Dr. Douglas C. Wallace, professor of molecular genetics at Emory University School of Medicine in Atlanta, stated, "The criteria that people use for race are based entirely on external features that we are programmed to recognize."[23]

If the Bible teaches and science confirms that all are of the same human race and all are related as descendants of Adam, then why are there such seemingly great differences between us (for example, in skin color)? The answer, again, comes with a biblically informed understanding of science.

Skin "Color"

Jesus loves the little children, all the children of the world. Red and yellow, black and white, they are precious in His sight.

When Jesus said, "Let the little children come to Me, and do not forbid them; for of such is the kingdom of heaven" (Matthew 19:14), He did not

22 Ref. 11.
23 Ibid.

distinguish between skin colors. In fact, scientists have discovered that there is one major pigment, called melanin, that produces our skin color. There are two main forms of melanin: eumelanin (brown to black) and pheomelanin (red to yellow). These combine to give us the particular shade of skin that we have.[24]

Melanin is produced by melanocytes, which are cells in the bottom layer of the epidermis. No matter what our shade of skin, we all have approximately the same concentration of melanocytes in our bodies. Melanocytes insert melanin into melanosomes, which transfer the melanin into other skin cells, which are cabaple of dividing (stem cells), primarily in the lowest layer of the epidermis. According to one expert,

> The melanosomes (tiny melanin-packaging units) are slightly larger and more numerous per cell in dark-skinned than light skinned people. They also do no degrade as readily, and disperse into adjacent skin cells to a higher degree.[25]

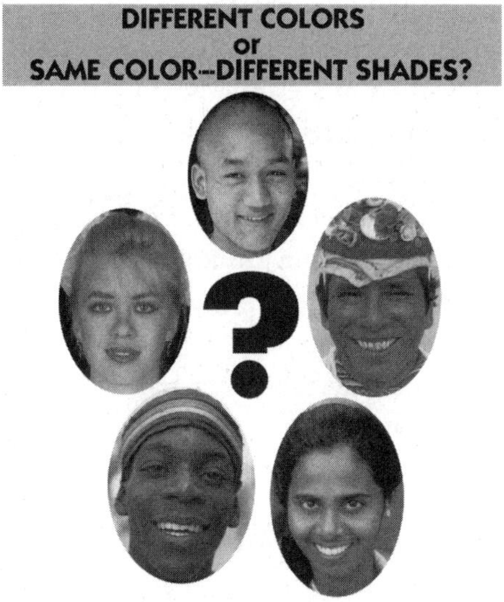
DIFFERENT COLORS
or
SAME COLOR–DIFFERENT SHADES?

In the stem cells, the pigment serves its function as it forms a little dark umbrella over each nucleus. The melanin protects the epidermal cells from being damaged by sunlight. In people with lighter shades of skin, much of the pigment is lost after these cells divide and their daughter cells move up in the epidermis to form the surface dead layer—the stratum corneum.

Geneticists have found that four to six genes, each with

[24] Of course, melanin is not the only factor that determines skin shade: blood vessels close to the skin can produce a reddish tinge, while extra layers of adipose tissue (fat) in the skin yield a yellowish tinge. Exposure to the sun can cause increased melanin production, thus darkening skin, but only to a certain point. Other pigments also affect skin shade but generally have very little bearing on how light or dark the skin will be. The major provider of skin color is melanin.

[25] Ackerman, *Histopathologic Diagnosis of Skin Diseases*, Lea & Febiger, Philadelphia, Pennsylvania, 1978, 44; Lever and Schamberg-Lever, *Histopathology of the Skin*, 7th Ed., J.B. Lippincott, Philadelphia, 1990, 18–20.

multiple alleles (or variations), control the amount and type of melanin produced. Because of this, a wide variety of skin shades exist. In fact, it is quite easy for one couple to produce a wide range of skin shades in just one generation, as will be shown below.

Inheritance

DNA (deoxyribonucleic acid) is the molecule of heredity that is passed from parents to child. In humans, the child inherits 23 chromosomes from each parent (the father donates 23 through his sperm, while the mother donates 23 through her egg). At the moment of conception, these chromosomes unite to form a unique combination of DNA and control much of what makes the child an individual. Each chromosome pair contains hundreds of genes, which regulate the physical development of the child. Note that no new genetic information is generated at conception, but a new *combination* of already-existing genetic information is formed.

To illustrate the basic genetic principles involved in determining skin shade, we'll use a simplified explanation,[26] with just two genes controlling the production of melanin. Let's say that the A and B versions of the genes code for a lot of melanin, while the a and b versions code for a small amount of melanin.

If the father's sperm carried the AB version and the mother's ovum carried the AB, the child would be AABB, with a lot of melanin, and thus very dark skin. Should both parents carry the ab version, the child would be aabb, with very little melanin, and thus very light skin. If the father carries AB (very dark skin) and the mother carries ab (very light skin), the child will be AaBb, with a middle brown

	AB	Ab	aB	ab
AB	AA BB	AA Bb	Aa BB	Aa Bb
Ab	AA Bb	AA bb	Aa Bb	Aa bb
aB	Aa BB	Aa Bb	aa BB	aa Bb
ab	Aa Bb	Aa bb	aa Bb	aa bb

ONLY DARK AABB — ONLY MEDIUM AAbb or aaBB — ONLY LIGHT aabb

[26] The actual genetics involved are much more complicated than this simplified explanation. There are 4 to 6 genes with multiples alleles (versions) of each gene that operate under incomplete dominance, that is, they work together to produce an individual's particular skin shade. However, simplifying the explanation does not take away from the point being made.

shade of skin. In fact, the majority of the world's population has a middle brown skin shade.

A simple exercise with a Punnet Square shows that if each parent has a middle brown shade of skin (AaBb), the combinations that they could produce result in a wide variety of skin shades in just one generation. Based on the skin colors seen today, we can infer that Adam and Eve most likely would have had a middle brown skin color. Their children, and children's children, could have ranged from very light to very dark.

No one really has red, or yellow, or black skin. We all have the same basic color, just different shades of it. We all share the same pigments—our bodies just have different combinations of them.[27]

Melanin also determines eye color. If the iris of the eye has a larger amount of melanin, it will be brown. If the iris has a little melanin, the eye will be blue. (The blue color in blue eyes results from the way light scatters off of the thin layer of brown-colored melanin.)

Hair color is also influenced by the production of melanin. Brown to black hair results from a greater production of melanin, while lighter hair results from less melanin. Those with red hair have a mutation in one gene that causes a greater proportion of the reddish form of melanin (pheomelanin) to be produced.[28]

DNA also controls the basic shape of our eyes. Individuals whose DNA codes for an extra layer of adipose tissue around the eyes have almond-shaped eyes (this is common among Asian people groups). All people groups have adipose tissue around the eyes, some simply have more or less.

Origin of People Groups

Those with darker skin tend to live in warmer climates, while those with lighter skin tend to live in colder climates. Why are certain characteristics more prominent in some areas of the world?

We know that Adam and Eve were the first two people. Their descendants filled the earth. However, the world's population was reduced to eight during the Flood of Noah. From these eight individuals have come all the tribes and nations. It is likely that the skin shade of Noah and his family was middle brown. This would enable his sons and their wives to produce a variety of skin shades in just one generation. Because there was a common language and everybody lived in the same general vicinity, barriers that may have prevented

---

[27] Albinism results from a genetic mutation which prevents the usual production of melanin.

[28] For more information, see www.answersingenesis.org/go/red-hair.

their descendants from freely intermarrying weren't as great as they are today. Thus, distinct differences in features and skin color in the population weren't as prevalent as they are today.

In Genesis 11 we read of the rebellion at the Tower of Babel. God judged this rebellion by giving each family group a different language. This made it impossible for the groups to understand each other, and so they split apart, each extended family going its own way, and finding a different place to live. The result was that the people were scattered over the earth.[29]

Because of the new language and geographic barriers, the groups no longer freely mixed with other groups, and the result was a splitting of the gene pool. Different cultures formed, with certain features becoming predominant within each group. The characteristics of each became more and more prominent as new generations of children were born. If we were to travel back in time to Babel, and mix up the people into completely different family groups, then people groups with completely different characteristics might result. For instance, we might find a fair-skinned group with tight, curly dark hair that has blue, almond-shaped eyes. Or a group with very dark skin, blue eyes, and straight brown hair.[30]

Some of these (skin color, eye shape, and so on) became general characteristics of each particular people group through various selection pressures (environmental, sexual, etc.) and/or mutation.[31] For example, because of the protective factor of melanin, those with darker skin would have been more likely to survive in areas where sunlight is more intense (warmer, tropical areas near the equator), as they are less likely to suffer from diseases such as skin cancer. Those with lighter skin lack the melanin needed to protect them from the harmful UV rays, and so may have been more likely to die before they were able to reproduce. UVA radiation also destroys the B vitamin folate, which is necessary for DNA synthesis in cell division. Low levels of folate in pregnant women can lead to defects in the developing baby. Again, because of this, lighter-skinned individuals may be selected against in areas of intense sunlight.

---

[29] As they went, the family groups took with them the knowledge that had been passed to them about the creation and Flood events. Although these accounts have been changed over time, they reflect the true account found in the Bible. For more information, see www.answersingenesis.org/go/legends.

[30] This assumes that each trait is independently inherited, which may not always be the case. Although there are many instances in which a certain trait shows up in a person of a different ethnic group (e.g., almond-shaped eyes in a woman with very dark skin, or blue eyes in a man with tightly curled brown hair and tan skin).

[31] For more on how selection and mutations operate, see chapter 22 in this book.

On the flip side, melanin works as a natural sunblock, limiting the sunlight's ability to stimulate the liver to produce vitamin D, which helps the body absorb calcium and build strong bones. Since those with darker skin need more sunlight to produce vitamin D, they may not have been as able to survive as well in areas of less sunlight (northern, colder regions) as their lighter-skinned family members, who don't need as much sunlight to produce adequate amounts of vitamin D. Those lacking vitamin D are more likely to develop diseases such as rickets (which is associated with a calcium deficiency), which can cause slowed growth and bone fractures. It is known that when those with darker skin lived in England during the Industrial Revolution, they were quick to develop rickets because of the general lack of sunlight.[32]

Of course, these are generalities. Exceptions occur, such as in the case of the darker-skinned Inuit tribes living in cold northern regions. However, their diet consists of fish, the oil of which is a ready source of vitamin D, which could account for their survival in this area.

Real science in the present fits with the biblical view that all people are rather closely related—there is only one race biologically. Therefore, to return to our original question, there is in essence no such thing as interracial marriage. So we are left with this—is there anything in the Bible that speaks clearly against men and women from different people groups marrying?

The Dispersion at Babel

Note that the context of Genesis 11 makes it clear that the reason for God's scattering the people over the earth was that they had united in rebellion against Him. Some Christians point to this event in an attempt to provide a basis for their arguments against so-called interracial marriage. They believe that this passage implies that God is declaring that people from different people groups can't marry so that the nations are kept apart. However, there is no such indication in this passage that what is called "interracial marriage" is condemned. Besides, there has been so much mixing of people groups over the years, that it would be impossible for every human being today to trace their lineage back to know for certain which group(s) they are descended from.

We need to understand that the sovereign creator God is in charge of the nations of this world. Paul makes this very clear in Acts 17:26. Some people erroneously claim this verse to mean that people from different nations shouldn't marry. However, this passage has nothing to do with marriage. As

[32] en.wikipedia.org/wiki/Melanin.

John Gill makes clear in his classic commentary, the context is that God is in charge of all things—where, how, and for how long any person, tribe, or nation will live, prosper, and perish.[33]

In all of this, God is working to redeem for Himself a people who are one in Christ. The Bible makes clear in Galatians 3:28, Colossians 3:11, and Romans 10:12–13 that in regard to salvation, there is no distinction between male or female or Jew or Greek. In Christ, any separation between people is broken down. As Christians, we are one in Christ and thus have a common purpose—to live for Him who made us. This oneness in Christ is vitally important to understanding marriage.

Purpose of Marriage

Malachi 2:15 informs us that an important purpose of marriage is to produce godly offspring—progeny that are trained in the ways of the Lord. Jesus (in Matthew 19) and Paul (in Ephesians 5) make it clear that when a man and woman marry, they become one flesh (because they were one flesh historically—Eve was made from Adam). Also, the man and woman must be one spiritually so they can fulfill the command to produce godly offspring.

This is why Paul states in 2 Corinthians 6:14, "Do not be unequally yoked together with unbelievers. For what fellowship has righteousness with lawlessness? And what communion has light with darkness?"

According to the Bible then, which of the following marriages in the picture on the right does God counsel against entering into?

The answer is obvious—number 3. According to the Bible, the priority in marriage is that a Christian should marry only a Christian.

---

33 See note on Acts 17:26, in John Gill, D.D., *An exposition of the Old and New Testament*, London: printed for Mathews and Leigh, 18 Strand, by W. Clowes, Northumberland-Court, 1809. Edited, revised, and updated by Larry Pierce, 1994–1995 for Online Bible CD-ROM.

Sadly, there are some Christian homes where the parents are more concerned about their children not marrying someone from another "race" than whether or not they are marrying a Christian. When Christians marry non-Christians, it negates the spiritual (not the physical) oneness in marriage, resulting in negative consequences for the couple and their children.[34]

Roles in Marriage[35]

Of course, every couple needs to understand and embrace the biblical roles prescribed for each family member. Throughout the Scriptures our special roles and responsibilities are revealed. Consider these piercing passages directed to fathers:

> The father shall make known Your truth to the children (Isaiah 38:19).
>
> Fathers, do not provoke your children to wrath, but bring them up in the training and admonition of the Lord (Ephesians 6:4).
>
> For I have known him, in order that he may command his children and his household after him, that they keep the way of the LORD, to do righteousness and justice, that the LORD may bring to Abraham what He has spoken to him (Genesis 18:19).

These are just a few of the many verses that mention *fathers* in regard to training children. Additionally, the writer of Psalm 78 continually admonishes fathers to teach their children so they'll not forget to teach their children, so that they might not forget what God has done and keep His commandments. This includes building within their children a proper biblical worldview and providing them with answers to the questions the world asks about God and the Bible (as this book does). It also includes shepherding and loving his wife as Christ loved the church.

Of course, just as God made the role of the man clear, He has also made His intentions known regarding the role of a godly wife. In the beginning, God fashioned a woman to complete what was lacking in Adam, that she might

[34] It is true that in some exceptional instances when a Christian has married a non-Christian, the non-Christian spouse, by the grace of God, has become a Christian. This is a praise point but it does not negate the fact that Scripture indicates that it should not have been entered into in the first place. This does not mean that the marriage is not actually valid, nor does it dilute the responsibilities of the marital union—see also 1 Corinthians 7:12–14, where the context is of one spouse becoming a Christian after marriage.

[35] For more on this topic, see *Raising Godly Children in an Ungodly World* by K. Ham and S. Ham, Green Forest, Arkansas, Master Books, 2008.

become his helper, that the two of them would truly become one (Genesis 2:15–25). In other Bible passages the woman is encouraged to be a woman of character, integrity, and action (e.g., Proverbs 31:10–31). Certainly mothers should also be involved in teaching their children spiritual truths.

These roles are true for couples in every tribe and nation.

Rahab and Ruth

The examples of Rahab and Ruth help us understand how God views the issue of marriage between those who are from different people groups but trust in the true God.

Rahab was a Canaanite. These Canaanites had an ungodly culture and were descendants of Canaan, the son of Ham. Remember, Canaan was cursed because of his obvious rebellious nature. Sadly, many people state that Ham was cursed—but this is not true.[36] Some have even said that this (non-existent) curse of Ham resulted in the black "races."[37] This is absurd and is the type of false teaching that has reinforced and justified prejudices against people with dark skin.

In the genealogy in Matthew 1, it is traditionally understood that the same Rahab is listed here as being in the line leading to Christ. Thus Rahab, a descendant of Ham, must have married an Israelite (descended from Shem). Since this was clearly a union approved by God, it underlines the fact that the particular "people group" she came from was irrelevant—what mattered was that she trusted in the true God of the Israelites.

The same can be said of Ruth, who as a Moabitess also married an Israelite and is also listed in the genealogy in Matthew 1 that leads to Christ. Prior to her marriage, she had expressed faith in the true God (Ruth 1:16).

---

[36] See Genesis 9:18–27. Canaan, the youngest of Ham's sons, received Noah's curse. Why? The descendants of Canaan were some of the wickedest people on earth. For example, the people of Sodom and Gomorrah were judged for their sexual immorality and rebellion. It may be that Ham's actions toward his father (Genesis 9:22) had sexual connotations, and Noah saw this same sin problem in Canaan and understood that Canaan's descendants would also act in these sinful ways. (The Bible clearly teaches that the unconfessed sin of one generation is often greater in the next generation.) The curse on Canaan has nothing to do with skin color but rather serves as a warning to fathers to train their children in the nurture and admonition of the Lord. We need to deal with our own sin problems and train our children to deal with theirs.

[37] For example: "We know the circumstances under which the posterity of Cain (and later of Ham) were cursed with what we call Negroid racial characteristics" (Bruce McConkie, Apostle of the Mormon Council of 12, *Mormon Doctrine*, p. 554, 1958); "The curse which Noah pronounced upon Canaan was the origin of the black race" (The Golden Age, *The Watchtower* [now called *Awake!*], p. 702, July 24, 1929).

When Rahab and Ruth became children of God, there was no longer any barrier to Israelites marrying them, even though they were from different people groups.

Real Biblical "Interracial" Marriage

If one wants to use the term "interracial," then the real interracial marriage that God says we should not enter into is when a child of the Last Adam (one who is a new creation in Christ—a Christian) marries one who is an unconverted child of the First Adam (one who is dead in trespasses and sin—a non-Christian).[38]

Cross-Cultural Problems

Because many people groups have been separated since the Tower of Babel, they have developed many cultural differences. If two people from very different cultures marry, they can have a number of communication problems, even if both are Christians. Expectations regarding relationships with members of the extended family, for example, can also differ. Even people from different English-speaking countries can have communication problems because words may have different meanings. Counselors should go through this in detail, anticipating the problems and giving specific examples, as some marriages have failed because of such cultural differences. However, such problems have nothing to do with genetics or "race."

Conclusion

1. There is no biblical justification for claiming that people from different so-called races (best described as people groups) should not marry.

2. The biblical basis for marriage makes it clear that a Christian should marry only a Christian.

When Christians legalistically impose nonbiblical ideas, such as no interracial marriage onto their culture, they are helping to perpetuate prejudices that have often arisen from evolutionary influences. If we are really honest, in countries like America, the main reason for Christians being against interracial marriage is, in most instances, really because of skin color.

---

[38] Examples of such "mixed marriages" and their negative consequences can be seen in Nehemiah 9 and 10, and Numbers 25.

The church could greatly relieve the tensions over racism (particularly in countries like America), if only the leaders would teach biblical truths about our shared ancestry: all people are descended from one man and woman; all people are equal before God; all are sinners in need of salvation; all need to build their thinking on God's Word and judge all their cultural aspects accordingly; all need to be one in Christ and put an end to their rebellion against their Creator.

Christians must think about marriage as God thinks about each one of us. When the prophet Samuel went to anoint the next king of Israel, he thought the oldest of Jesse's sons was the obvious choice due to his outward appearance. However, we read in 1 Samuel 16:7, "But the Lord said to Samuel, 'Do not look at his appearance or at his physical stature, because I have refused him. For the Lord does not see as man sees; for man looks at the outward appearance, but the Lord looks at the heart.'" God doesn't look at our outward biological appearance; He looks on our inward spiritual state. And when considering marriage, couples should look on the inside spiritual condition of themselves and each other because it is true that what's on the inside, spiritually, is what really matters.

18

Are ETs & UFOs Real?

JASON LISLE

Are there extraterrestrial life forms out there? The question of life on other planets is a hot topic in our culture today. Science fiction movies and television shows often depict strange creatures from far-away planets. But these ideas are not limited merely to science fiction programming. Many secular scientists believe that one day we will actually discover life on other planets. There are even projects like the Search for Extra-Terrestrial Intelligence (SETI) that scan the heavens with powerful radio telescopes listening for signals from intelligent aliens. Many Christians have bought into the idea of extraterrestrial alien life. But is this idea really biblical? The Christian should constantly examine ideas in light of Scripture and take "every thought into captivity to the obedience of Christ" (2 Corinthians 10:5).

CREATIONWISE
IT'S MIND BOGGLING TO THINK OF THE VASTNESS OF CREATION! THE BIBLE IS ONLY ONE LITTLE BOOK. JUST THINK OF ALL THE INFORMATION GOD HASN'T GIVEN TO US!
YES, BUT WHAT HAVE YOU DONE WITH THE INFORMATION GOD HAS GIVEN TO US?
© AiG 2003
DAN LIETHA

The Evolution Connection

The idea of extraterrestrial life stems largely from a belief in evolution. Recall that in the evolutionary view, the earth is "just another planet"—one where the conditions just happened to be right for life to form and evolve. If there are countless billions of other planets in our galaxy, then surely at least a handful of these worlds have also had the right conditions. Extraterrestrial life is almost inevitable in an evolutionary worldview.

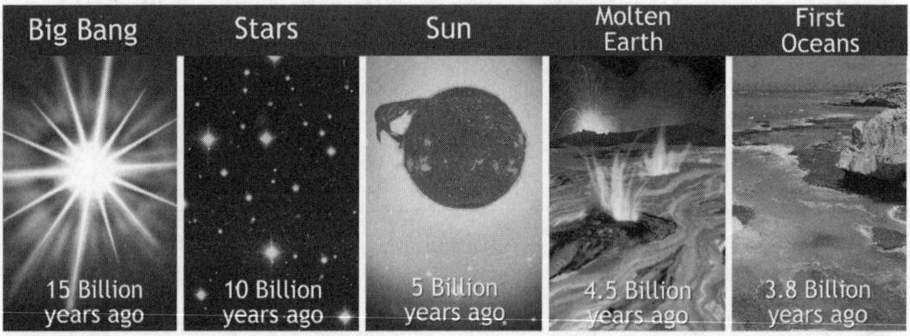

However, the notion of alien life does not square well with Scripture. The earth is unique. God designed the earth for life (Isaiah 45:18). The other planets have an entirely different purpose than does the earth, and thus they are designed differently. In Genesis 1 we read that God created plants on the earth on Day 3, birds to fly in the atmosphere and marine life to swim in the ocean on Day 5, and animals to inhabit the land on Day 6. Human beings were also made on Day 6 and were given dominion over the animals. But where does the Bible discuss the creation of life on the "lights in the expanse of the heavens"? There is no such description because the lights in the expanse were not designed to accommodate life. God gave care of the earth to man, but the heavens are the Lord's (Psalm 115:16). From a biblical perspective, extraterrestrial life does not seem reasonable.

Problems are multiplied when we consider the possibility of *intelligent* alien life. Science fiction programming abounds with races of people who evolved on other worlds. We see examples of Vulcans and Klingons—pseudo-humans similar to us in most respects but different in others. As a plot device, these races allow the exploration of the human condition from the perspective of an outsider. Although very entertaining, such alien races are theologically problematic. Intelligent alien beings cannot be redeemed. God's plan of redemption is for human beings: those descended from Adam. Let us examine the conflict between the salvation message and the notion of alien life.

The Redemption of Mankind

The Bible teaches that the first man, Adam, rebelled against God (Genesis 3). As a result, sin and death entered the world (Romans 5:12). We are all descended from Adam and Eve (Genesis 3:20) and have inherited from them a sin nature (Romans 6:6, 20). This is a problem: sin is a barrier that prevents man from being right with God (Isaiah 59:2). But God loves us despite our sin and provided a plan of redemption—a way to be reconciled with God.

After Adam and Eve sinned, God made coats of skins to cover them (Genesis 3:21). He therefore had to kill at least one animal. This literal action is symbolic of our salvation; an innocent Lamb (Christ—the Lamb of God) would be sacrificed to provide a covering for sin (John 1:29). In the Old Testament, people would sacrifice animals to the Lord as a reminder of their sin (Hebrews 10:3) and as a symbol of the One to come, the Lord Jesus, who would actually pay the penalty for sin.

The animal sacrifices did not actually pay the penalty for sin (Hebrews 10:4, 11). Animals are not related to us; their shed blood cannot count for ours. But the blood of Christ can. Christ is a blood relative of ours since He is descended from Adam as are we; all human beings are of "one blood" (Acts 17:26). Furthermore, since Christ is also God, His life is of infinite value, and thus His death can pay for all the sins of all people. That is why only the Lord Himself could be our Savior (Isaiah 45:21). Therefore, Christ died once for all (Hebrews 10:10).

The Redemption of ET?

When we consider how the salvation plan might apply to any hypothetical extraterrestrial (but otherwise human-like) beings, we are presented with a problem. If there were Vulcans or Klingons out there, how would

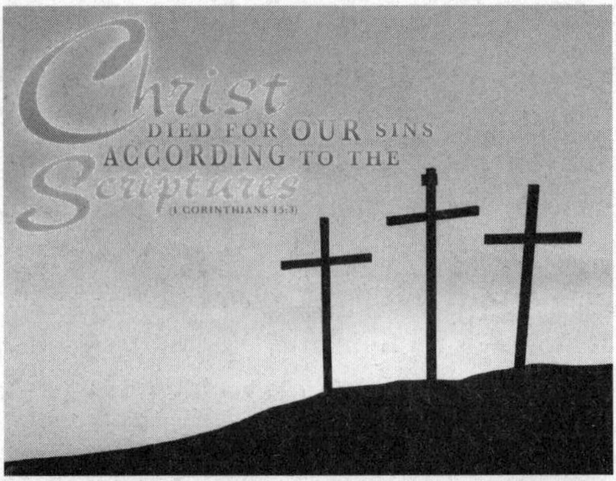

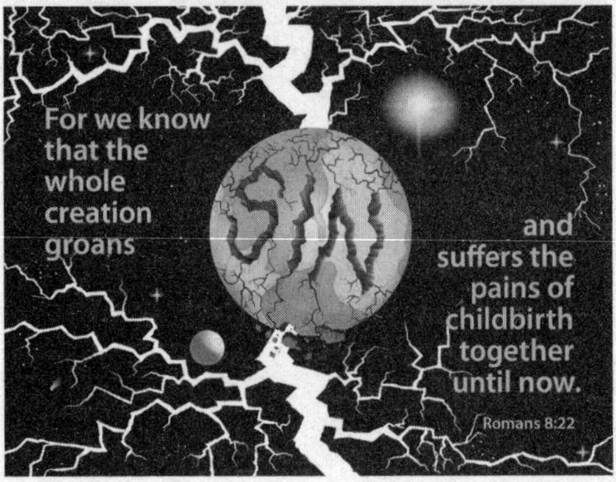

they be saved? They are not blood relatives of Jesus, and so Christ's shed blood cannot pay for their sin. One might at first suppose that Christ also visited their world, lived there, and died there as well, but this is antibiblical. Christ died *once* for *all* (1 Peter 3:18; Hebrews 9:27–28, 10:10). Jesus is now and forever both God and man; but He is *not* an alien.

One might suppose that alien beings have never sinned, in which case they would not need to be redeemed. But then another problem emerges: they suffer the effects of sin, despite having never sinned. Adam's sin has affected all of creation—not just mankind. Romans 8:20–22 makes it clear that the entirety of creation suffers under the bondage of corruption. These kinds of issues highlight the problem of attempting to incorporate an antibiblical notion into the Christian worldview.

Extraterrestrial life is an evolutionary concept; it does not comport with the biblical teachings of the uniqueness of the earth and the distinct spiritual position of human beings. Of all the worlds in the universe, it was the earth that God Himself visited, taking on the additional nature of a human being, dying on a cross, and rising from the dead in order to redeem all who would trust in Him. The biblical worldview sharply contrasts with the secular worldview when it comes to alien life. So, which worldview does the scientific evidence support? Do modern observations support the secular notion that the universe is teeming with life, or the biblical notion that earth is unique?

Where Is Everybody?

So far, no one has discovered life on other planets or detected any radio signals from intelligent aliens. This is certainly what a biblical creationist would expect. Secular astronomers continue to search for life on other worlds, but they have found only rocks and inanimate matter. Their radio searches are met with silence. The real world is the biblical world—a universe designed by God with the earth at the spiritual focal point, not an evolutionary universe teeming with life.

When it comes to extraterrestrial life, science is diametrically opposed to the evolutionary mentality. We currently have *no* evidence of alien life forms. This problem is not lost on the secular scientists. It has been said that the atomic scientist Enrico Fermi was once discussing the topic of extraterrestrial life when he asked the profound question, "Where is everybody?" Since there are quite possibly multiple billions of planets in our galaxy, and since in the secular view these are all accidents, it is almost inevitable that some of these had the right conditions for life to evolve. And if some of these worlds are billions of years older than ours, then at least some of them would have evolved intelligent life eons ago. The universe should therefore have countless numbers of technologically superior civilizations, any one of which could have colonized our galaxy ages ago. Yet, we find no evidence of these civilizations. Where is everybody? This problem has become known as the "Fermi paradox."

This paradox for evolution is a *feature* of creation. We have seen that the earth is designed for life. With its oceans of liquid water, a protective

atmosphere containing abundant free oxygen, and a distance from the sun that is just right for life, earth was certainly designed by God to be inhabited. But the other planets of the universe were not. From the sulfuric acid clouds of Venus to the frozen wasteland of Pluto, the other worlds of our solar system are beautiful and diverse, but they are not designed for life.

What about UFOs?

Sometimes after I speak on the topic of extraterrestrial life, someone will ask me about UFOs. A UFO (unidentified flying object) is just that—an object seen in the sky that is unidentified to the person seeing it. People often want me to explain a sighting of some unknown flying object which they or often a friend have claimed to see. (Sometimes the implication is that if I can't explain it, it somehow proves that it must be an alien spacecraft; but such reasoning is completely vacuous.[1]) These kinds of questions are unreasonable. It is one thing to be asked to interpret evidence that we have, but it is unrealistic to ask someone to interpret undocumented second- or third-hand stories with no actual evidence available for inspection.

There is no doubt that some people sincerely have seen things in the sky that they do not understand. This is hardly surprising since there are lots of things "up there," which can be misunderstood by people not familiar with them. These include Venus, satellites, the international space station, the space shuttle, rockets, Iridium flares, manmade aircraft, internal reflections, meteors, balloons, fireflies, aurorae, birds, ball lightning, lenticular clouds, parhelia, etc. However, a person unfamiliar with these would see a UFO, since the object is "unidentified" to him or her. It is how people interpret what they see that can be questionable.

Remember that we always interpret evidence in light of our worldview. It is therefore crucial to have a correct, biblical worldview. The fallacious worldview of atheism/naturalism may lead someone to draw erroneous conclusions about what they see. From a biblical worldview, we expect to occasionally see things that are not easily explained, since our minds are finite. But UFOs are not alien spacecraft, and of course, there is no tangible evidence to support such a notion.

[1] The argument is that alien spacecraft could not be explained by a natural phenomenon. Therefore, it is suggested that witnessing something that cannot be explained naturally must prove the existence of alien spacecraft. This is a logical fallacy called "affirming the consequent." It's equivalent to saying, "All white dwarf stars are white. Fred is white; therefore Fred is a white dwarf star."

Why the Hype?

In the 1990s the television series *The X-files* entertained millions of fans with stories of aliens, government conspiracies, and one dedicated FBI agent's relentless search for truth. The show's motto, "The truth is out there," is a well-known phrase for sci-fi fans. But why is there such hype surrounding the notion of extraterrestrial life? Why is science fiction programming so popular? Why does SETI spend millions of dollars searching for life in outer space?

The discovery of intelligent extraterrestrial life would certainly be seen as a vindication of evolution; it is an expectation from a naturalistic worldview. But the desire to meet aliens, especially intelligent, technologically advanced ones, seems much more deeply felt than merely to vindicate evolutionary predictions. What is the *real* issue? I've heard a number of different answers from secular astronomers.

In some cases a belief in ETs may stem from a feeling of cosmic loneliness: "If there are aliens, then we would not be alone in the universe." In many cases it comes from an academic desire to learn the mysteries of the universe; a highly developed alien race might have advanced knowledge to pass on to us. Perhaps such knowledge is not merely academic; the hypothetical aliens may know the answers to fundamental questions of existence: "Why am I here? What is the meaning of life?" and so on. An advanced alien race might have medical knowledge far exceeding our own—knowledge which could be used to cure our diseases. Perhaps their medical technology would be so far advanced that they even hold the secret of life and death; with such incredible medical knowledge, perhaps human beings would no longer have to die—*ever.*

In a way, a belief in extraterrestrial life has become a secular replacement for God. God is the one who can heal every disease. God is the one in whom all the treasures of wisdom and knowledge are deposited (Colossians 2:3). God is the one who can answer the fundamental questions of our existence. God alone possesses the gift of eternal life (John 17:3). It is not surprising that the unbelieving scientist would feel a sense of cosmic loneliness, having rejected his Creator. But, we are not alone in the universe; there is God. God created us for fellowship with Him; thus we have an innate need for Him and for purpose. Although human beings have rejected God, in Adam and by our sins as well, our need for fellowship with Him remains.

When I think of the majority of intelligent scientists who have studied God's magnificent creation but have nonetheless rejected Him and have instead chosen to believe in aliens and millions of years of evolution, I am reminded of Romans 1:18–25. God's invisible qualities—His eternal power and divine nature—are clearly revealed in the natural world so that there is no excuse for rejecting God or suppressing the truth about Him. The thinking of man apart from God is nothing more than futile speculations. Exchanging the truth of God, such as creation, for a lie, such as evolution, and turning to a mere creature such as hypothetical aliens for answers is strikingly similar to what is recorded in Romans 1:25.

But when we start from the Bible, the evidence makes sense. The universe is consistent with the biblical teaching that the earth is a special creation. The magnificent beauty and size of a universe, which is apparently devoid of life except for one little world where life abounds, is exactly what we would expect from a biblical worldview. The truth is not "out there;" the truth is *in there*—in the Bible! The Lord Jesus is the truth (John 14:6). So when we base our thinking on what God has said in His Word, we find that the universe makes sense.

19

Does Distant Starlight Prove the Universe Is Old?

JASON LISLE

Critics of biblical creation sometimes use distant starlight as an argument against a young universe. The argument goes something like this: (1) there are galaxies that are so far away, it would take light from their stars billions of years to get from there to here; (2) we can see these galaxies, so their starlight has already arrived here; and (3) the universe must be at least billions of years old—much older than the 6,000 or so years indicated in the Bible.

Many big bang supporters consider this to be an excellent argument against the biblical timescale. But when we examine this argument carefully, we will see that it does not work. The universe is very big and contains galaxies that are very far away, but that does not mean that the universe must be billions of years old.

The distant starlight question has caused some people to question cosmic distances. "Do we really know that galaxies are so far away? Perhaps they are much closer, so the light really doesn't travel very far."[1] However, the techniques that astronomers use to measure cosmic distances are generally logical and scientifically sound. They do not rely on evolutionary assumptions about the past. Moreover, they are a part of *observational* science (as opposed to historical/origins science); they are testable and repeatable in the present. You could repeat the experiment to determine the distance to

[1] See the DVD *Astronomy: What Do We Really Know?* by Dr. Jason Lisle for a more complete treatment of these questions, available at www.answersbookstore.com.

a star or galaxy, and you would get approximately the same answer. So we have good reason to believe that space really is very big. In fact, the amazing size of the universe brings glory to God (Psalm 19:1).

Some Christians have proposed that God created the beams of light from distant stars already on their way to the earth. After all, Adam didn't need any time to grow from a baby because he was made as an adult. Likewise, it is argued that the universe was made mature, and so perhaps the light was created in-transit. Of course, the universe was indeed made to function right from the first week, and many aspects of it were indeed created "mature." The only problem with assuming that the light was created in-transit is that we see things happen in space. For example, we see stars change brightness and move. Sometimes we see stars explode. We see these things because their light has reached us.

But if God created the light beams already on their way, then that means none of the events we see in space (beyond a distance of 6,000 light-years) actually happened. It would mean that those exploding stars never exploded or existed; God merely painted pictures of these fictional events. It seems uncharacteristic of God to make illusions like this. God made our eyes to accurately probe the real universe; so we can trust that the events that we see in space really happened. For this reason, most creation scientists believe that light created in-transit is not the best way to respond to the distant starlight argument. Let me suggest that the answer to distant starlight lies in some of the unstated assumptions that secular astronomers make.

The Assumptions of Light Travel-time Arguments

Any attempt to scientifically estimate the age of something will necessarily involve a number of *assumptions*. These can be assumptions about the starting conditions, constancy of rates, contamination of the system, and many others. If even one of these assumptions is wrong, so is the age estimate. Sometimes an incorrect worldview is to blame when people make faulty assumptions. The distant starlight argument involves several assumptions that are questionable—any one of which makes the argument unsound. Let's examine a few of these assumptions.

The Constancy of the Speed of Light

It is usually assumed that the speed of light is constant with time.[2] At today's rate, it takes light (in a vacuum) about one year to cover a distance of 6 trillion miles. But has this always been so? If we incorrectly assume that the rate has always been today's rate, we would end up estimating an age that is much older than the true age. But some people have proposed that light was much quicker in the past. If so, light could traverse the universe in only a fraction of the time it would take today. Some creation scientists believe that this is the answer to the problem of distant starlight in a young universe.

However, the speed of light is not an "arbitrary" parameter. In other words, changing the speed of light would cause other things to change as well, such as the ratio of energy to mass in any system.[3] Some people have argued that the speed of light can never have been much different than it is today because it is so connected to other constants of nature. In other words, life may not be possible if the speed of light were any different.

This is a legitimate concern. The way in which the universal constants are connected is only partially understood. So, the impact of a changing speed of light on the universe and life on earth is not fully known. Some creation scientists are actively researching questions relating to the speed of light. Other creation scientists feel that the assumption of the constancy of the speed of light is probably reasonable and that the solution to distant starlight lies elsewhere.

The Assumption of Rigidity of Time

Many people assume that time flows at the same rate in all conditions. At first, this seems like a very reasonable assumption. But, in fact, this assumption is false. And there are a few different ways in which the nonrigid nature of time could allow distant starlight to reach earth within the biblical timescale.

Albert Einstein discovered that the rate at which time passes is affected by motion and by gravity. For example, when an object moves very fast, close

---

[2] Many people mistakenly think that Einstein's theory of relativity demands that the speed of light has not changed in time. In reality, this is not so. Relativity only requires that two different observers would measure the same velocity for a beam of light, even if they are moving relative to each other.

[3] This follows from the equation $E=mc^2$, in which c is the speed of light and E is the energy associated with a given amount of mass (m).

to the speed of light, its time is slowed down. This is called "time-dilation." So, if we were able to accelerate a clock to nearly the speed of light, that clock would tick very slowly. If we could somehow reach the speed of light, the clock would stop completely. This isn't a problem with the clock; the effect would happen regardless of the clock's particular construction because it is time itself that is slowed. Likewise, gravity slows the passage of time. A clock at sea-level would tick slower than one on a mountain, since the clock at sea-level is closer to the source of gravity.

It seems hard to believe that velocity or gravity would affect the passage of time since our everyday experience cannot detect this. After all, when we are traveling in a vehicle, time appears to flow at the same rate as when we are standing still. But that's because we move so slowly compared to the speed of light, and the earth's gravity is so weak that the effects of time-dilation are correspondingly tiny. However, the effects of time-dilation have been measured with atomic clocks.

Since time can flow at different rates from different points of view, events that would take a long time as measured by one person will take very little time as measured by another person. This also applies to distant starlight. Light that would take billions of years to reach earth (as measured by clocks in deep space) could reach earth in only thousands of years as measured by clocks on earth. This would happen naturally if the earth is in a *gravitational well,* which we will discuss below.

Many secular astronomers assume that the universe is infinitely big and has an infinite number of galaxies. This has never been proven, nor is there evidence that would lead us naturally to that conclusion. So, it is a leap of "blind" faith on their part. However, if we make a different assumption instead, it leads to a very different conclusion. Suppose that our solar system is located near the center of a finite distribution of galaxies. Although this cannot be proven for certain at present, it is fully consistent with the evidence; so it is a reasonable possibility.

In that case, the earth would be in a *gravitational well.* This term means that it would require energy to pull something away from our position into deeper space. In this gravitational well, we would not "feel" any extra gravity, nonetheless time would flow more slowly on earth (or anywhere in our solar system) than in other places of the universe. This effect is thought to be very small today; however, it may have been much stronger in the past. (If the universe is expanding as most astronomers believe, then physics demands that such effects would have been stronger when the universe was smaller). This being the case, clocks on earth would have ticked much more slowly than clocks in deep space. Thus, light from the most distant galaxies would arrive on earth in only a few thousand years as measured by clocks on earth. This idea is certainly intriguing. And although there are still a number of mathematical details that need to be worked out, the premise certainly is reasonable. Some creation scientists are actively researching this idea.

Assumptions of Synchronization

Another way in which the relativity of time is important concerns the topic of synchronization: how clocks are set so that they read the same time at the same time.[4] Relativity has shown that synchronization is not absolute. In other words, if one person measures two clocks to be synchronized, another person (moving at a different speed) would *not* necessarily measure those two clocks to be synchronized. As with time-dilation, this effect is counter-intuitive because it is too small to measure in most of our everyday experience. Since there is no method by which two clocks (separated by a distance) can be synchronized in an absolute sense, such that all observers would agree regardless of motion, it follows that there is some flexibility in how we choose what constitutes synchronized clocks. The following analogy may be helpful.

Imagine that a plane leaves a certain city at 4:00 p.m. for a two-hour flight. However, when the plane lands, the time is still 4:00. Since the plane arrived at the same time it left, we might call this an instantaneous trip. How is this possible? The answer has to do with time zones. If the plane left Kentucky at 4:00 p.m. local time, it would arrive in Colorado at 4:00 p.m. local time. Of course, an observer on the plane would experience two hours of travel. So, the trip takes two hours as measured by *universal time*. However, as long as the plane is traveling west (and providing it travels fast enough), it will always naturally arrive at the same time it left as measured in *local time*.

There is a cosmic equivalent to local and universal time. Light traveling toward earth is like the plane traveling west; it always remains at the same cosmic local time. Although most astronomers today primarily use cosmic universal time (in which it takes light 100 years to travel 100 light-years), historically cosmic local time has been the standard. And so it may be that the Bible also uses cosmic local time when reporting events.

---

[4] For a discussion on synchrony conventions see W.C. Salmon, The philosophical significance of the one-way speed of light, *Nous* **11**(3):253–292, Symposium on Space and Time, 1977.

Since God created the stars on Day 4, their light would leave the star on Day 4 and reach earth on Day 4 *cosmic local time*. Light from all galaxies would reach earth on Day 4 if we measure it according to cosmic local time. Someone might object that the light itself would experience billions of years (as the passenger on the plane experiences the two hour trip). However, according to Einstein's relativity, light does not experience the passage of time, so the trip would be instantaneous. Now, this idea may or may not be the reason that distant starlight is able to reach earth within the biblical timescale, but so far no one has been able to prove that the Bible does *not* use cosmic local time. So, it is an intriguing possibility.[5]

The Assumption of Naturalism

One of the most overlooked assumptions in most arguments against the Bible is the assumption of *naturalism*. Naturalism is the belief that nature is "all that there is." Proponents of naturalism *assume* that all phenomena can be explained in terms of natural laws. This is not only a blind assumption, but it is also clearly antibiblical. The Bible makes it clear that God is not bound by natural laws (they are, after all, *His* laws). Of course God can use laws of nature to accomplish His will; and He usually does so. In fact, natural laws could be considered a description of the way in which God normally upholds the universe. But God is supernatural and is capable of acting outside natural law.

This would certainly have been the case during Creation Week. God created the universe supernaturally. He created it from nothing, not from previous material (Hebrews 11:3). Today, we do not see God speaking into existence new stars or new kinds of creatures. This is because God ended His work of creation by the seventh day. Today, God sustains the universe in a different way than how He created it. However, the naturalist erroneously assumes that the universe was created by the same processes by which it operates today. Of course it would be absurd to apply this assumption to most other things. A flashlight, for example, operates by converting electricity into light, but the flashlight was not created by this process.

Since the stars were created during Creation Week and since God made them to give light upon the earth, the way in which distant starlight arrived on earth may have been supernatural. We cannot assume that past acts of God are necessarily understandable in terms of a current scientific mechanism, because

---

[5] See Distant Starlight and Genesis, *TJ* **15**(1):80–85, 2001; available online at www.answersingenesis.org/tj/v15/i1/starlight.asp.

science can only probe the way in which God sustains the universe today. It is irrational to argue that a supernatural act cannot be true on the basis that it cannot be explained by natural processes observed today.

It is perfectly acceptable for us to ask, "Did God use natural processes to get the starlight to earth in the biblical timescale? And if so, what is the mechanism?" But if no natural mechanism is apparent, this cannot be used as evidence against *supernatural* creation. So, the unbeliever is engaged in a subtle form of circular reasoning when he uses the assumption of naturalism to argue that distant starlight disproves the biblical timescale.

Light Travel-Time: a Self-Refuting Argument

Many big bang supporters use the above assumptions to argue that the biblical timescale cannot be correct because of the light travel-time issue. But such an argument is self-refuting. It is fatally flawed because the big bang has a light travel-time problem of its own. In the big bang model, light is required to travel a distance much greater than should be possible within the big bang's own timeframe of about 14 billion years. This serious difficulty for the big bang is called the "horizon problem."[6] The following are the details.

In the big bang model, the universe begins in an infinitely small state called a singularity, which then rapidly expands. According to the big bang model, when the universe is still very small, it would develop different temperatures in different locations (Figure 1A). Let's suppose that point A is hot and point B is cold. Today, the universe has expanded (Figure 1B), and points A and B are now widely separated.

However, the universe has an extremely uniform temperature at great distance—beyond the farthest known galaxies. In other words, points A and B have almost exactly the same temperature today. We know this because we see electromagnetic radiation coming from all directions in space in the form of microwaves. This is called the "cosmic microwave background" (CMB). The frequencies of radiation have a characteristic temperature of 2.7 K (-455°F) and are *extremely* uniform in all directions. The temperature deviates by only one part in 10^5.

The problem is this: How did points A and B come to be the same temperature? They can do this only by exchanging energy. This happens in many systems: consider an ice cube placed in hot coffee. The ice heats up and the coffee cools down by exchanging energy. Likewise, point A can give energy to

[6] See www.answersingenesis.org/creation/v25/i4/lighttravel.asp.

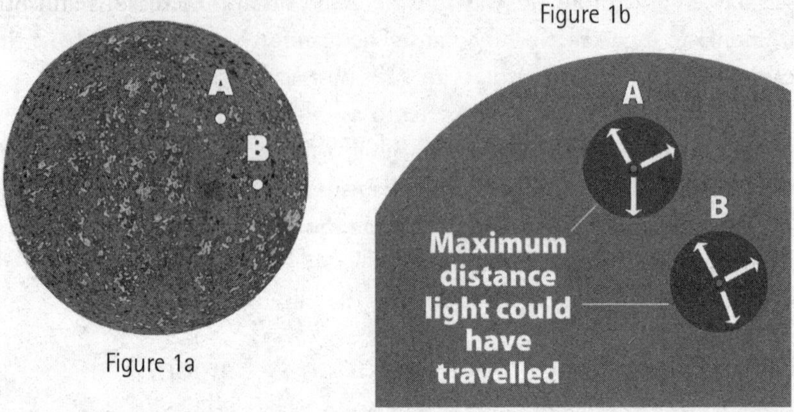

The Horizon Problem

point B in the form of electromagnetic radiation (light), which is the fastest way to transfer energy since nothing can travel faster than light. However, using the big bang supporters' own assumptions, including uniformitarianism and naturalism, there has not been enough time in 14 billion years to get light from A to B; they are too far apart. This is a light travel-time problem—and a very serious one. After all, A and B have almost exactly the same temperature today, and so must have exchanged light multiple times.

Big bang supporters have proposed a number of conjectures which attempt to solve the big bang's light travel-time problem. One of the most popular is called "inflation." In "inflationary" models, the universe has two expansion rates; a normal rate and a fast inflation rate. The universe begins with the normal rate, which is actually quite rapid, but is slow by comparison to the next phase. Then it briefly enters the inflation phase, where the universe expands much more rapidly. At a later time, the universe goes back to the normal rate. This all happens early on, long before stars and galaxies form.

The inflation model allows points A and B to exchange energy (during the first normal expansion) and to then be pushed apart during the inflation phase to the enormous distances at which they are located today. But the inflation model amounts to nothing more than storytelling with no supporting evidence at all. It is merely speculation designed to align the big bang to conflicting observations. Moreover, inflation adds an additional set of problems and difficulties to the big bang model, such as the cause of such inflation and a graceful way to turn it off. An increasing number of secular astrophysicists are rejecting inflation for these reasons and others. Clearly, the horizon problem remains a serious light travel-time problem for the big bang.

The critic may suggest that the big bang is a better explanation of origins than the Bible since biblical creation has a light travel-time problem—distant starlight. But such an argument is not rational since the big bang has a light travel-time problem of its own. If both models have the same problem *in essence*,[7] then that problem cannot be used to support one model over the other. Therefore, distant starlight cannot be used to dismiss the Bible in favor of the big bang.

Conclusions

So, we've seen that the critics of creation must use a number of assumptions in order to use distant starlight as an argument against a young universe. And many of these assumptions are questionable. Do we know that light has always propagated at today's speed? Perhaps this is reasonable, but can we be absolutely certain, particularly during Creation Week when God was acting in a supernatural way? Can we be certain that the Bible is using "cosmic universal time," rather than the more common "cosmic local time" in which light reaches earth instantly?

We know that the rate at which time flows is not rigid. And although secular astronomers are well aware that time is relative, they *assume* that this effect is (and has always been) negligible, but can we be certain that this is so? And since stars were made during Creation Week when God was *supernaturally* creating, how do we know for certain that distant starlight has arrived on earth by entirely *natural* means? Furthermore, when big bang supporters use distant starlight to argue against biblical creation, they are using a self-refuting argument since the big bang has a light travel-time problem of its own. When we consider all of the above, we see that distant starlight has never been a legitimate argument against the biblical timescale of a few thousand years.

As creation scientists research possible solutions to the distant starlight problem, we should also remember the body of evidence that is consistent

[7] The details, of course, differ. The big bang does not have a problem with distant starlight as such. But then again, biblical creation does not have a horizon problem. (The cosmic microwave background does not need to start with different temperatures in a creationist cosmogony.) However, both problems are the same in *essence*: how to get light to travel a greater distance than seems possible in the time allowed.

with the youth of the universe. We see rotating spiral galaxies that cannot last multiple billions of years because they would be twisted-up beyond recognition. We see multitudes of hot blue stars, which even secular astronomers would agree cannot last billions of years.[8] In our own solar system we see disintegrating comets and decaying magnetic fields that cannot last billions of years; and there is evidence that other solar systems have these things as well. Of course, such arguments also involve assumptions about the past. That is why, ultimately, the only way to know about the past *for certain* is to have a reliable historic record written by an eyewitness. That is exactly what we have in the Bible.

8 Secular astronomers believe that blue stars must have formed relatively recently. But there are considerable difficulties in star formation scenarios—problems with magnetic fields and angular momentum to name a couple.

20

Did Jesus Say He Created in Six Literal Days?

KEN HAM

A very important question we must ask is, "What was Jesus' view of the days of creation? Did He say that He created in six literal days?"

When confronted with such a question, most Christians would automatically go to the New Testament to read the recorded words of Jesus to see if such a statement occurs.

Now, when we search the New Testament Scriptures, we certainly find many interesting statements Jesus made that relate to this issue. Mark 10:6 says, "But from the beginning of the creation, God 'made them male and female.'" From this passage, we see that Jesus clearly taught that the creation was young, for Adam and Eve existed "from the beginning," not billions of years after the universe and earth came into existence. Jesus made a similar statement in Mark 13:19 indicating that man's sufferings started very near the beginning of creation. The parallel phrases of "from the foundation of the world" and "from the blood of Abel" in Luke 11:50–51 also indicate that Jesus placed Abel very close to the beginning of creation, not billions of years after the beginning. His Jewish listeners would have assumed this meaning in Jesus' words, for the first-century Jewish historian Josephus indicates that the Jews of his day believed that both the first day of creation and Adam's creation were about 5,000 years before Christ.[1]

[1] See William Whiston, transl., *The Works of Josephus,* Hendrickson, Peabody, Massachusetts, p. 850, 1987, and Paul James-Griffiths, "Creation days and Orthodox Jewish Tradition," *Creation* **26**(2): 53–55, www.answersingenesis.org/creation/v26/i2/tradition.asp.

In John 5:45–47, Jesus says, "Do not think that I shall accuse you to the Father; there is one who accuses you—Moses, in whom you trust. For if you believed Moses, you would believe Me; for he wrote about Me. But if you do not believe his writings, how will you believe My words?" In this passage, Jesus makes it clear that one must believe what Moses wrote. And one of the passages in the writings of Moses in Exodus 20:11 states: "For in six days the LORD made the heavens and the earth, the sea, and all that is in them, and rested the seventh day. Therefore the LORD blessed the Sabbath day and hallowed it." This, of course, is the basis for our seven-day week—six days of work and one day of rest. Obviously, this passage was meant to be taken as speaking of a total of seven literal days based on the Creation Week of six literal days of work and one literal day of rest.

In fact, in Luke 13:14, in his response to Jesus healing a person on the Sabbath, the ruler of the synagogue, who knew the law of Moses, obviously referred to this passage when he said, "There are six days on which men ought to work; therefore come and be healed on them, and not on the Sabbath day." The sabbath day here was considered an ordinary day, and the six days of work were considered ordinary days. This teaching is based on the Law of Moses as recorded in Exodus 20, where we find the Ten Commandments—the six-day Creation Week being the basis for the Fourth Commandment.

We should also note the way Jesus treated as historical fact the accounts in the Old Testament, which religious and atheistic skeptics think are unbelievable mythology. These historical accounts include Adam and Eve as the first married couple (Matthew 19:3–6; Mark 10:3–9), Abel as the first prophet who was killed (Luke 11:50–51), Noah and the Flood (Matthew 24:38–39), Moses and the serpent in the wilderness (John 3:14), Moses and the manna from heaven to feed the Israelites in the wilderness (John 6:32–33, 49), the experiences of Lot and his wife (Luke 17:28–32), the judgment of Sodom and Gomorrah (Matthew 10:15), the miracles of Elijah (Luke 4:25–27), and Jonah and the big fish (Matthew 12:40–41). As New Testament scholar John Wen-

AFTER EDEN
by Dan Lietha
I DON'T BELIEVE THE EARTH WAS CREATED IN 6 DAYS LIKE IT SAYS IN GENESIS.
HOW DO YOU KNOW THAT? WERE YOU THERE?
YES! I WAS THERE!
www.AnswersInGenesis.org
© 2001 AIG
For by Him (Jesus) were all things created, that are in heaven, and that are in earth. Colossians 1:16a

ham has compellingly argued, Jesus did not allegorize these accounts but took them as straightforward history, describing events that actually happened just as the Old Testament describes.[2] Jesus used these accounts to teach His disciples that the events of His death, Resurrection, and Second Coming would likewise certainly happen in time-space reality.

These passages taken together strongly imply that Jesus took Genesis 1 as literal history describing creation in six 24-hour days. But are there any more explicit passages?

I believe there are. However, one has to approach this issue in a slightly different manner. We are not limited to the New Testament when we try to find out if Jesus stated He created in six days; we can also search the Old Testament. After all, Jesus is the Second Person of the Trinity and therefore has always existed.

First, Colossians makes it clear that Jesus Christ, the Son of God, was the one who created all things: "For by Him all things were created that are in heaven and that are on earth, visible and invisible, whether thrones or dominions or principalities or powers. All things were created through Him and for Him. And He is before all things, and in Him all things consist" (Colossians 1:16–17).

We are also told elsewhere in Scripture how Jesus created: "By the word of the LORD the heavens were made, And all the host of them by the breath of His mouth. For He spoke, and it was done; He commanded, and it stood fast" (Psalm 33:6, 9). We see the meaning of this when we consider the miracles of Jesus during His earthly ministry. All the miracles occurred instantly—at His

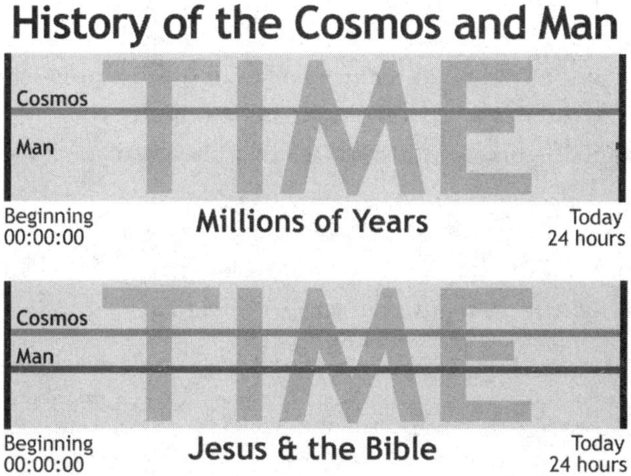

[2] John Wenham, *Christ and the Bible*, IVPress, Downers Grove, Illinois, pp. 11–37, 1973.

Word. He instantly turned water into wine in His very first miracle, which "revealed His glory" as the Creator (John 2:1–11; John 1:1–3, 14, 18). It was the instant calming of the wind and the waves that convinced His disciples that He was no mere man. So it was with all His miracles (Mark 4:35–41). He did not speak and wait for days, weeks, months, or years for things to happen. He spoke and it was done. So, when He said, "Let there be …" in Genesis 1, it did not take long ages for things to come into existence.

We also know that Jesus is in fact called the Word: "In the beginning was the Word, and the Word was with God, and the Word was God. He was in the beginning with God. All things were made through Him, and without Him nothing was made that was made" (John 1:1–3).

Jesus, who is the Word, created everything by simply speaking things into existence.

Now, consider Exodus 20:1: "And God spoke all these words, saying … ." Because Jesus is the Word, this must be a reference to the preincarnate Christ speaking to Moses. As we know, there are a number of appearances of Christ (theophanies) in the Old Testament. John 1:18 states: "No one has seen God at any time. The only begotten Son, who is in the bosom of the Father, He has declared Him." There is no doubt, with rare exception, that the preincarnate Christ did the speaking to Adam, Noah, the patriarchs, Moses, etc. Now, when the Creator God spoke as recorded in Exodus 20:1, what did He (Jesus) say? As we read on, we find this statement: "For in six days the LORD made the heavens and the earth, the sea, and all that is in them, and rested the seventh day" (Exodus 20:11).

Yes, Jesus did explicitly say He created in six days.[3] Not only this, but the one who spoke the words "six days" also wrote them down for Moses: "Then the LORD delivered to me two tablets of stone written with the finger of God, and on them were all the words which the LORD had spoken to you on the mountain from the midst of the fire in the day of the assembly" (Deuteronomy 9:10).

Jesus said clearly that He created in six days. And He even did something He didn't do with most of Scripture—He wrote it down Himself. How clearer and more authoritative can you get than that?

---

[3] Even if someone is convinced that God the Father was the speaker in Exodus 20:11, the Father and Son would never disagree. Jesus said in John 10:30: "I and my Father are one" [neuter—one in the essence of deity, not one in personality]. He also said, "I speak these things as the Father taught me," and "I always do the things that are pleasing to Him" (John 8:28–29).

21

How Did Defense/Attack Structures Come About?

ANDY MCINTOSH & BODIE HODGE

The Relevance of the Issue of DAS (Defense/Attack Structures)

Many people question the goodness of God when they see "nature, red in tooth and claw,"[1] and therefore, they accuse those who believe in the Bible of not seeing reality in nature's fight for survival, which in the view of the secular scientists substantiates evolution.

In the past, many Bible-believers looked to nature as evidence of God's design in nature and attributed the features animals possessed to kill prey or defend themselves as all part of God's original design.

For example, in 1802 William Paley wrote the now-classic book *Natural Theology: or, Evidences of the Existence and Attributes of the Deity, Collected from the Appearances of Nature*. In this work, Paley makes the argument for the design in nature being attributed to a designer—God—and included features that were "red in tooth and claw" as part of this original design.

Darwin, who read Paley's work, realized that organisms have certain design features that make them fit for the environments in which they live. In other words, they were well designed for what they do—even the ability to cause pain, suffering, and death. However, Darwin later

[1] From "In Memoriam" by Alfred Lord Tennyson, 1850.

saw difficulties with Paley's argument concerning design. To Darwin, a creation capable of inflicting pain and death seemed to deny a good and loving Creator God.

Darwin could see that the idea of a benevolent designer did not square with the world that he observed. How could a good God be the author of death and bloodshed? The answer of Darwin and many others was to turn from the God of the Bible to a belief in man's ideas about the past that include millions of years of death and suffering.

A most notable adherent to this view in our present day is David Attenborough. Attenborough is the presenter of many popular nature documentaries produced by the British Broadcasting Corporation. In a similar journey to that of Darwin, he argues strongly for belief in evolution because of the suffering that the natural world exhibits. The quote below is very revealing as to what has moved Attenborough to an evolutionary position.

> When Creationists talk about God creating every individual species as a separate act, they always instance hummingbirds, or orchids, sunflowers and beautiful things. But I tend to think instead of a parasitic worm that is boring through the eye of a boy sitting on the bank of a river in West Africa, [a worm] that's going to make him blind. And [I ask them], "Are you telling me that the God you believe in, who you also say is an all-merciful God, who cares for each one of us individually, are you saying that God created this worm that can live in no other way than in an innocent child's eyeball? Because that doesn't seem to me to coincide with a God who's full of mercy."[2]

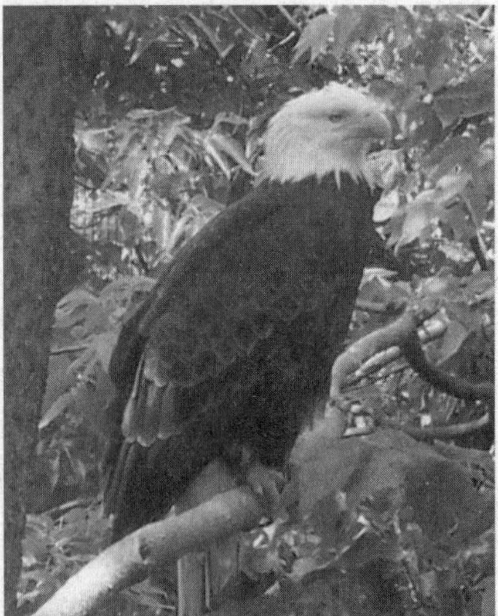

Eagles have pointed claws and sharp beaks.

The examples of Darwin and Attenborough show why the issue of defense/attack structures (DAS) is important, and how it is closely

[2] From M. Buchanan, Wild, wild life, *Sydney Morning Herald*, The Guide, p. 6, March 24, 2003.

related to the existence of suffering and death in the world around us. Defense/attack structures include anything from claws and flesh-tearing beaks on birds of prey or the claws and teeth of cats, to a wasp's stinger or a poison dart frog's toxin.

What Are Some Defense/Attack Structures?

Examples of defense/attack structures are numerous in the world around us, existing in plants as well as animals. Let's look at a few.

PLANT—VENUS FLYTRAP

A great example in plants is the Venus flytrap. This plant snaps two of its lobes on any unsuspecting fly that ventures inside. The mechanism by which the trap snaps shut involves a complex interaction between elasticity, osmotic pressure in the cellular plant material, and growth. When the plant is open, the lobes are convex (bent outwards), but when it is closed, the lobes are concave (forming a cavity). It is stable in both the open and closed positions, but it changes states to close quickly when triggered.[3]

ARACHNID—SPIDER

A good example of DAS is the spider. Spider webs are renowned for their potential to catch flying insects, such as flies and moths. The sophistication of silk production through special glands that keep the polymer soft right up until it is exuded behind the spider is still not understood.[4] Furthermore, the ability of the spider to make some strands sticky and others not, so that the spider itself only walks on the non-sticky parts is clearly a clever design feature. Not all spiders make webs, but they are all capable of producing silk in several varieties. Though the predatory nature of spiders is universal, the actual prey-catching technique of web-building is not the same for each species.

---

[3] Y. Forterre et. al., How the Venus flytrap snaps, *Nature* **433**(7024): 421–5, 2005, found online at www.nature.com/nature/journal/v433/n7024/abs/nature03185.html; How a Venus flytrap snaps up its victims, *New Scientist*, January 29, 2005, found online at www.newscientist.com/channel/life/mg18524845.900.

[4] G. De Luca and A.D. Rey, Biomimetics of spider silk spinning process, pp. 127–136, Design and Nature III: Comparing Design in Nature with Science and Engineering, Vol. 87 of *WIT Transactions on Ecology and the Environment*, C.A. Brebbia, ed., WIT Press, 2006; See also en.wikipedia.org/wiki/Spider_silk.

INSECT—BOMBARDIER BEETLE

Another example in the insect world, and probably the most extraordinary, is the bombardier beetle. This insect possesses a sophisticated defense apparatus, which involves shooting a hot (212°F/100°C) noxious mixture of chemicals out of a special swivel nozzle in its backside, into the face of predators such as rodents, birds, frogs, or other insects.

ANIMALS—CATS AND REPTILES

Of the numerous examples of DAS in the animal world, the meat-eating lion, tiger, and other large cats (cheetah, lynx, etc.) would be the most obvious. It should be noted though that these creatures are not solely dependent on a carnivorous diet because there are known cases of large cats being able to survive on a vegetarian diet when meat has been not available in zoos.[5]

Many animals in the reptile world also give us excellent examples of DAS. Chameleons have the ability to flick their tongues in only fractions of a second to capture their prey. Crocodiles and alligators have powerful jaws, and snakes possess poisonous fangs or deadly coils. The anaconda can kill bulls and tapirs easily with its extremely strong muscles.[6]

Alligator teeth are long and sharp.

These are but a few of the DAS found around the world. If you check the plants and animals in your area, you can probably spot some of these and other defense/attack structures.

Why, Biblically, Is the World Like This?

The biblical response to DAS is that the theology of Darwin and Attenborough has made a major assumption—the world is now what it always

[5] B. Hodge, Unexpectedly vegetarian animals—what does it mean? www.answersingenesis.org/articles/2009/06/02vegetarian-animals.

[6] H. Mayell, Anaconda expert wades barefoot in Venezuela's swamps, National Geographic News, March 13, 2003; found online at news.nationalgeographic.com/news/2002/04/0430_020503_anacondaman.html.

has been. The Bible, as early as Genesis 3, makes it clear that this is not the case.

The world (and indeed the universe) was originally perfect. Six times in Genesis 1 it states that what God had made was "good" and the seventh time that "God saw everything that He had made, and indeed it was very good" (Genesis 1:31). A perfect God would make nothing less. In fact, Moses, who also penned Genesis, declared in Deuteronomy 32:4 that all of God's works are perfect.The original creation was perfect, but we can see by looking at the world around us that there has been a drastic change. The change was a result of the Fall of man—an event which fundamentally altered the world.

The original world had no parasites boring into children's eyes or any other part of nature being "red in tooth and claw." The death and suffering in the past and in the present is a result of man's sin and rebellion against God. When the first man Adam disobeyed his Creator, all of creation was cursed, bringing disease, sickness, pain, suffering, and death into the world.

When God spoke to Adam, He said, "Because you have heeded the voice of your wife, and have eaten from the tree of which I commanded you, saying, 'You shall not eat of it': cursed is the ground for your sake; in toil you shall eat of it all the days of your life. Both thorns and thistles it shall bring forth for you, and you shall eat the herb of the field. In the sweat of your face you shall eat bread till you return to the ground, for out of it you were taken; for dust you are, and to dust you shall return" (Genesis 3:17–19).

God also told Eve, "I will greatly multiply your sorrow and your conception; in pain you shall bring forth children; your desire shall be for your husband, and he shall rule over you" (Genesis 3:16).

And earlier still, the Bible records what God spoke to the serpent: "So the LORD God said to the serpent: 'Because you have done this, you are cursed more than all cattle, and more than every beast of the field; on your belly you shall go, and you shall eat dust all the days of your life'" (Genesis 3:14). So in essence there were several changes at the Fall.

This is not just an Old Testament doctrine. The New Testament picks up on the inseparable connection between the world's state and man's condition. In Romans 8:22–23, Paul states, "For we know that the whole creation groans and labors with birth pangs together until now. Not only that, but we also who have the firstfruits of the Spirit, even we ourselves groan within ourselves, eagerly waiting for the adoption, the redemption of our body."

While the world has been cursed because of man's rebellion in Adam, there is coming a day—a day for the "redemption of our body" (Romans

Verse	Some of the known effects	Said to
Genesis 3:14	1. Serpent cursed more than other animals—specifically mentions crawling on its belly and eating dust. 2. Other animals are cursed; to what extent, we aren't told.	Serpent
Genesis 3:16	1. Increased pain and sorrow in childbearing and raising children. 2. Their desire will be for their husbands.	Woman/ Eve
Genesis 3:17–19	1. Ground is cursed—specifically mentions thorns and thistles and the pain and sorrow associated with working the ground. We aren't told the other effects of the Curse. 2. Death—mankind would return to dust.	Man/ Adam

8:23)—when at the resurrection of God's people, the world will also be liberated from the Curse. In Romans 8, Paul makes it clear that the extent of this Curse encompasses the whole creation.

When we look at defense/attack structures in the animal or plant kingdom, we must look at them in the context of a truly biblical theology. Let's review the clear teachings from Scripture.

1. Man and animals were originally created as vegetarian (Genesis 1:29–30). Throughout Genesis 1 the Lord states repeatedly that the created order was "good" and then in Genesis 1:31, "very good." Thus, "nature, red in tooth and claw" was not part of God's original creation.

2. In verse 30, God explicitly states, "Also, to every beast of the earth, to every bird of the air, and to everything that creeps on the earth, in which there is life, I have given every green herb for food." Literally in the Hebrew, the phrase "in which there is life" is *nephesh chayyah*. This phrase is translated "living soul" and is used in Genesis 1:20–21 and Genesis 2:7 when referring to man and animals. However, this phrase is never used in reference to plants (or invertebrates), thus highlighting the difference between plant life and human and animal life.

3. The Curse in Genesis 3 caused a major change in both animals and plants. The animals were cursed; Genesis 3:14 says, "You are cursed *more than all cattle*, and *more than every beast of the field* [emphasis added]." The plants were also cursed; Genesis 3:17–18 says, "Cursed is the ground for your sake; in toil you shall eat of it all the days of your life. Both thorns and thistles it shall bring forth for you, and you shall eat the herb of the field." (There is evidence that thorns are formed from altered leaves.[7])

4. It was not until after the Flood that God allowed man to eat meat (Genesis 1:29–30, 9:3).

5. Later in Scripture the prophet Isaiah refers to a future time when there will be a reverse of the Curse: "The wolf also shall dwell with the lamb, the leopard shall lie down with the young goat, the calf and the young lion and the fatling together; and a little child shall lead them" (11:6). "The wolf and the lamb shall feed together, the lion shall eat straw like the ox, and dust shall be the serpent's food. They shall not hurt nor destroy in all My holy mountain, says the LORD" (65:25).

6. The book of Revelation speaks of a time when the Curse will be removed (22:3) and there will be no more pain, suffering, or death (21:4).

The Bible provides us with a big picture as we look at defense/attack structures.

Two Major Perspectives to Understand DAS Biblically

Two primary alternatives can easily explain defense/attack structures from a biblical perspective: (1) the present features used in defense and attack were not originally used for that purpose, and (2) the DAS design features were brought in by God *as a result of* the Fall.

The first perspective—that the present features were not originally used for defense/attack purposes—indicates that DAS were used for different functions before the Fall. Another way to clarify this perspective is to say that the design was the same but the function was different.

Let's take sharp teeth as an example. When people see animals with sharp teeth, they most commonly interpret this to mean that the animal is

[7] S. Carlquist, Ontogeny and comparative anatomy of thorns of Hawaiian Lobeliaceae, *American Journal of Botany*, **49**(4): 413–419, April 1962.

a meat-eater. When scientists find fossils of creatures with sharp teeth, they also interpret this to mean that the animal was a meat-eater. But is this a proper interpretation? Not really. Sharp teeth in animals indicate only one thing—the animal has sharp teeth.

Creatures with sharp teeth do not necessarily use them to rip other animals apart today. For example, the giant panda has very sharp teeth, yet it eats entirely bamboo shoots. Also, the fruit bat, which at first might appear to have teeth consistent with a carnivorous diet, eats primarily fruit. The Bible teaches that animals were created to be vegetarian (Genesis 1:30); so, we must be careful not to merely assume what an animal ate based on its teeth.

Other DAS can also be explained in this way. Claws could have been

T. rex originally ate vegetables.

Bears have sharp teeth, but they eat many vegetarian meals.

used to grip vegetarian foods or branches for climbing. And chameleon tongues could have been used to reach out and grab vegetarian foods, etc. This perspective has the advantage of never having to suggest that God designed a structure or system feature to be harmful to another living creature of His creation.

It is evident that for the silk-producing structure in spiders, it is hard to establish an alternative function for these glands, though spiders have been shown to catch and eat pollen.[8] The evidence seems to point to such structures being designed as they are to effectively catch things like insects. However, we may simply not know the original harmless function of these structures.

Consequently, many have suggested the fact that some creatures have continued to eat plants, which actually indicates that predatory habits came due to altered function. Bears commonly eat vegetarian foods. There have been lions and vultures documented to refuse eating meat.[9]

Even viruses (genetic carriers that infect a host with almost always deleterious results) may have originally been used in a different and beneficial role before the Fall. In a similar manner, harmful bacteria may have had a different and better purpose than their current function.

However, this perspective does have some shortcomings, especially when we apply it to the whole of DAS. One such problem is that of thorns. It can

---

[8] *Nature Australia* **26**(7):5, Summer 1999–2000.

[9] B. Hodge, Unexpectedly vegetarian animals—what does it mean? www.answersingenesis.org/articles/2009/06/02vegetarian-animals.

be argued that trees, bushes, etc., use thorns solely as a defense mechanism. But the Bible indicates that thorns and thistles came as a result of the Fall (Genesis 3:17–19). So, something indeed changed at the Curse.

Thorns and Thistles

This first perspective avoids God designing DAS in a perfect world for the purpose of harming something that was alive.

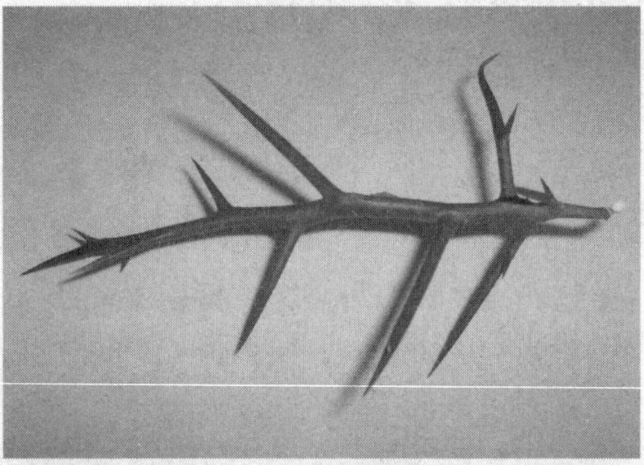

The second perspective—DAS design features were brought in by God *as a result of* the Fall—calls for design alterations after the Fall to allow such attack and defense structures. To clarify, this was the result of man's sin, not God's original design, and the consequences of sin still remain. Such "cursed design" is from God's intelligence as a punishment for the man's, the woman's, and the serpent's disobedience. This second perspective would then better explain some things like sharp teeth, claws, the special glands that make the spider silk, etc.

There is some warrant for this view in Scripture since we know that plants have been made such that now some of them have thorns (physically changed form) and that the serpent changed form to crawl on its belly (physically changed form). Since there was a physical change and this was passed along to offspring, then there had to be genetic alterations. Some of these changes could have been immediate, and others could have been slower in revealing themselves.

Regardless, the genetic blueprint of these systems must have changed such that DAS became evident. Remembering that God knows the future, it is possible that the devices were placed latently in the genetic code of these creatures at creation and were "turned on" at the Fall. Another possibility is that God redesigned the creatures after the Fall to have DAS features in them. Since defense/attack structures are a reminder of a sin-cursed world full of death and suffering, there was more likely a change after the Fall as opposed to these features being simply dormant.

Scripture that gives implied support to this perspective is that after the Fall, man would know pain and hard work and would eventually die (Genesis 3:19). Some biological change is experienced. Pain and sorrow in childbirth are a direct result of the Fall, and the serpent is radically redesigned after his rebellion. So this overall position may be the better of the two, though we wouldn't be dogmatic.

Conclusion

Both biblical perspectives explain the changes that occurred when man sinned and the world fell from a perfect one to an imperfect one, and both positions have merits. But the Bible doesn't specifically say one way or another. In fact, there could be aspects of both perspectives that may have happened. Not all creatures with DAS need to be explained in the same way. For some it may have been that their existing functions adapted, while there seems to be every indication that other mechanisms came in after the Fall.

Regardless, the accusation that a loving and perfect God made the world as we see it today ignores the Bible's teachings about the results of the Curse. A proper understanding of why there are defense/attack structures in the world today should be a reminder that the world is sin-cursed and that we are all sinners in need of a Savior.

After the Fall, God acted justly. He did what was right. But during the curses in Genesis 3, God did something that only a loving God would do—He gave the first prophecy of redemption. He promised a Savior. Genesis 3:15 says, "And I will put enmity between you and the woman, and between your seed and her Seed; He shall bruise your head, and you shall bruise His heel."

The One who would crush the head of the serpent would be born of a virgin, the seed of a woman. This is the first of many prophecies of Jesus Christ coming as the seed of a woman—a virgin birth. It was truly a loving and gracious God who came to earth in the form of a man and died for us and paid the penalty of our sins on the Cross.

DAS should remind us that when God says something, it will come to pass. When one receives Christ as their Savior, they will one day enjoy eternal life in a world that no longer has any curse or death or suffering or pain (Revelation 21:4, 22:3).

> For God so loved the world that He gave His only begotten Son, that whoever believes in Him should not perish but have everlasting life. For God did not send His Son into the world to condemn the world, but that the world through Him might be saved. He who believes in Him is not condemned; but he who does not believe is condemned already, because he has not believed in the name of the only begotten Son of God (John 3:16–18).

22

Is Natural Selection the Same Thing as Evolution?

GEORGIA PURDOM

Let's listen in on a hypothetical conversation between a biblical creationist (C) and an evolutionist (E) as they discuss some recent scientific news headlines:

E: Have you heard about the research findings regarding mouse evolution?

C: Are you referring to the finding of coat color change in beach mice?

E: Yes, isn't it a wonderful example of evolution in action?

C: No, I think it's a good example of natural selection in action, which is merely selecting information that already exists.

E: Well, what about antibiotic resistance in bacteria? Don't you think that's a good example of evolution occurring right before our eyes?

C: No, you seem to be confusing the terms "evolution" and "natural selection."

E: But natural selection is the primary mechanism that drives evolution.

C: Natural selection doesn't drive molecules-to-man evolution; you are giving natural selection a power that it does not have—one that can supposedly add new information to the genome, as molecules-to-man evolution requires. But natural selection simply can't do that because it works with information that already exists.

Natural selection is an observable process that is often purported to be the underlying mechanism of unobservable molecules-to-man evolution. The concepts are indeed different, though some mistakenly interchange the two. So let's take a closer look. There are two major questions to answer:

1. How do biblical creationists rightly view the observable phenomenon of natural selection?
2. Could this process cause the increase in genetic information necessary for molecules-to-man evolution?

What Is Natural Selection?

Below are some definitions evolutionists use to define "natural selection." The problem biblical creationists have with these definitions lies mostly in their misapplication, as noted by the bolded phrases.

> **Evolutionary change** based on the differential reproductive success of individuals within a species.[1]
>
> The process by which genetic traits are passed on to each successive generation. Over time, natural selection helps species become better adapted to their environment. Also known as "survival of the fittest," **natural selection is the driving force behind the process of evolution.**[2]
>
> The process in nature by which, according to **Darwin's theory of evolution,** only the organisms best adapted to their environment tend to survive and transmit their genetic characters in increasing numbers to succeeding generations while those less adapted tend to be eliminated (**also see evolution**).[3]

From a creationist perspective natural selection is a process whereby organisms possessing specific characteristics (reflective of their genetic makeup) survive better than others in a given environment or under a given selective pressure (i.e., antibiotic resistance in bacteria). Those with

[1] Michael A.Park, *Introducing Anthropology: An Integrated Approach*, 2nd Ed., glossary, highered.mcgraw-hill.com/sites/0072549238/student_view0/glossary.html, 2002.
[2] National Geographic's strange days on planet earth, glossary, www.pbs.org/strangedays/glossary/N.html.
[3] Dinosaurs — glossary of terms, http://web.archive.org/web/2016*/www.internal.schools.net.au/edu/lesson_ideas/dinosaurs/glossary.html.

certain characteristics live, and those without them diminish in number or die.

The problem for evolutionists is that natural selection is nondirectional—should the environment change or the selective pressure be removed, those organisms with previously selected for characteristics are typically less able to deal with the changes and may be selected against because their genetic information has decreased—more on this later. Evolution of the molecules-to-man variety, requires directional change. Thus, the term "evolution" cannot be rightly used in the context of describing what natural selection can accomplish.

What Is Evolution?

This term has many definitions just as "natural selection" does. Much of the term's definition depends on the context in which the word "evolution" is used. Below are some recent notable definitions of evolution (note the bold phrases).

> Unfolding in time of a predictable or prepackaged sequence in an inherently **progressive,** or at least **directional manner.**[4]

> The theory that all life forms are **descended** from **one or several common ancestors** that were present on early earth, **three to four billion years** ago.[5]

> The "Big Idea" [referring to evolution] is that living things (species) are related to one another through **common ancestry** from earlier forms that differed from them. Darwin called this **"descent with modification,"** and it is still the best definition of evolution we can use, **especially with members of the general public and with young learners.**[6]

All of these definitions give the same basic idea that evolution is *directional* in producing all the life forms on earth today from one or several ancestral life forms billions of years ago. The last definition is especially intriguing because it indicates that an ambiguous definition of evolution should be used with the public and with children. Most creationists would

---

[4] S.J. Gould, What does the dreaded "E" word *mean*, anyway? *Natural History* 109(1): 28–44, 2000.

[5] D. O'Leary, *By Design or by Chance?* Castle Quay, Kitchener, Ontario, Canada, 7, 2004.

[6] Eugenie C. Scott, Creation or evolution? http://web.archive.org/web/20050406235737/www.ncseweb.org/resources/articles/6261_creation_or_evolution__1_9_2001.asp.

agree partially with the idea of "descent with modification" in that species we have today look different from the original kinds that God created (i.e., the great variety of dogs we have now compared to the original created dog kind). The advantage with using such a broad definition for evolution is that it can include any and all supporting models of evolution (such as traditional Darwinism, neo-Darwinism, punctuated equilibrium, etc.) and can spark the least amount of controversy in the public eye.

Historical Background on the Discovery of Natural Selection

Many people give credit to Charles Darwin for formulating the theory of natural selection as described in his book *On the Origin of Species.* Few realize that Darwin only popularized the idea and actually borrowed it from several other people, especially a creationist by the name of Edward Blyth. Blyth published several articles describing the process of natural selection in *Magazine of Natural History* between 1835 and 1837—a full 22 years before Darwin published his book. It is also known that Darwin had copies of these magazines and that parts of *On The Origin of Species* are nearly verbatim from Blyth's articles.[7]

Blyth, however, differed from Darwin in his starting assumptions. Blyth believed in God as the Creator, rather than the blind forces of nature. He believed that God created original kinds, that all modern species descended from those kinds, and that natural selection acted by conserving rather than originating. Blyth also believed that man was a separate creation from animals. This is especially important since humans are made in the image of God, an attribute that cannot be applied to animals (Genesis 1:27). Blyth seemed to view natural selection as a mechanism designed directly or indirectly by God to allow His creation to survive in a post-Fall, post-Flood world. This is very different from Darwin's view. Darwin wrote, "What a book a devil's chaplain might write

Edward Blyth

[7] J. Foard, The Darwin papers, "Edward Blyth and natural selection," www.thedarwinpapers.com.

on the clumsy, wasteful, blundering low and horridly cruel works of nature."[8]

Is Natural Selection Biblical?

It is important to see natural selection as a mechanism that God used to allow organisms to deal with their changing environments in a sin-cursed world—especially after the Flood. God foreknew that the Fall and the Flood were going to happen, and so He designed organisms with a great amount of genetic diversity that could be selected for or against, resulting in certain characteristics depending on the circumstances. Whether this information was initially part of the original design during Creation Week before the Fall or was added, in part, at the Fall (as a part of the punishment of man and the world by God),[9] we can't be certain. Regardless, the great variety of information in the original created kinds can only be attributed to an intelligence—God.

In addition, natural selection works to preserve the genetic viability of the original created kinds by removing from the population those with severely deleterious/lethal characteristics. Natural selection, acting on genetic information, is the primary mechanism that explains how organisms could have survived after the Fall and Flood when the world changed drastically from God's original creation.

Let me take a moment to clarify an important theological point so there is no confusion. Death entered the world as the result of sin. Death, therefore, is in the world as a punishment for man's disobedience to God, and it should remind us that the world is sin-cursed and needs a Savior. Death is not a good thing but is called an enemy (1 Corinthians 15:26).

But recall that God, in His infinite wisdom, can make good come out of anything, and death is no exception. God is able to make good come out of even death itself. Natural selection, though fueled by death, helps the population by getting rid of genetic defects, etc. In the same way, without death Christ wouldn't have conquered it and been glorified in His Resurrection.

So what can natural selection accomplish and not accomplish? The table on the next page displays some of the main points.

---

[8] Letter from Charles Darwin to Joseph Hooker, Darwin Archives, Cambridge University, July 13, 1856.

[9] See chapter 21 in this book.

Natural Selection Can	Natural Selection Cannot
1. Decrease genetic information.	1. Increase or provide new genetic information.
2. Allow organisms to survive better in a given environment.	2. Allow organisms to evolve from molecules to man.
3. Act as a "selector."	3. Act as an "originator."
4. Support creation's "orchard" of life.	4. Support evolutionary "tree" of life.

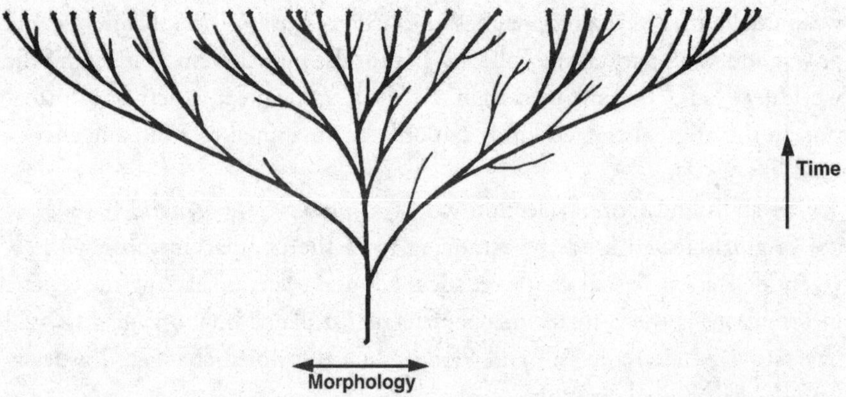

The evolutionary tree, which postulates that all today's species are descended from one common ancestor (which itself evolved from nonliving chemicals).

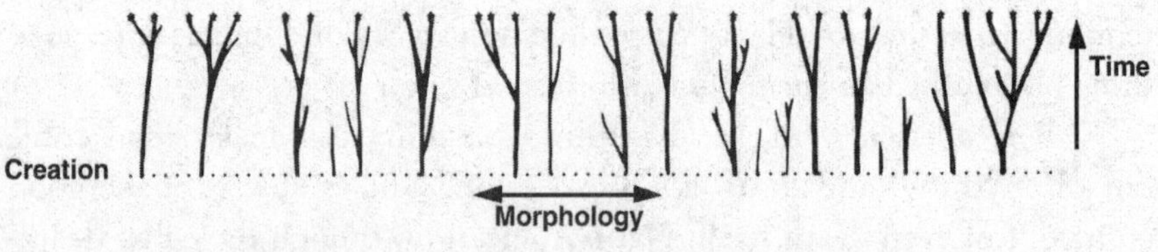

The creationist orchard,[10] which shows that diversity has occurred within the original Genesis kinds over time.[11]

[10] Dr. Kurt Wise developed the "orchard" analogy in the early 90s.

[11] Creationists often refer to each kind as a *baramin,* from Hebrew bara = create and min = kind.

Natural Selection and Dogs

Let's illustrate the possibilities and limitations of natural selection using the example of varying fur length of dogs (designed variation).

There are many different dog species—some with long fur and some with short fur. The original dog kind, most likely resembling today's wolf, had several variants of the gene for fur length. L will be the variant of the gene representing long fur, and S will be the variant of the gene representing short fur.

The original dog kind most likely would have been a mixture of the genes specifying fur length, including both L and S. Because of this makeup, they also most likely had the characteristic of medium fur length. When the original kind (LS dogs) mated, their genetic variability could be seen in their offspring in three ways—LL for long fur, LS for medium fur, and SS for short fur.

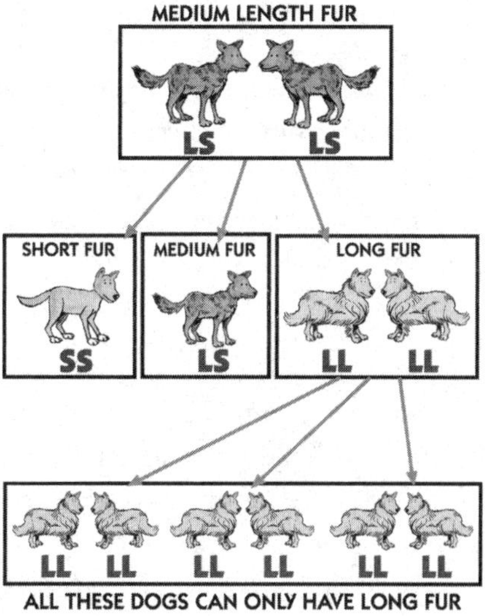

If two long-fur dogs then mated, the only possible outcome for the offspring is LL, long fur. As can be seen in the example below, the long-fur dogs have lost the S gene variant and are thus not capable of producing dogs with short fur or medium fur. This loss may be an advantage if these long-fur dogs live in an area with cold temperatures. The long-fur dogs would then be naturally selected for, as they would survive better in the given environment. Eventually, the majority of this area's dog population would have long fur.

However, the loss of the S variant could be a disadvantage to the long-fur dogs if the climate became warmer or if the dogs moved to a warmer climate. Because of their decreased genetic variety (no S gene), they would be unable to produce dogs with short fur, which would be needed to survive better in a warm environment. In this situation, the long-fur dogs would be naturally selected against and die.

When the two dogs representing the dog kind came off Noah's Ark and began spreading across the globe, we can see how the variation favored some animals and not others.

Using the points from the table for what natural selection can accomplish (seen on previous page), it can be seen that:

1. Through natural selection, genetic information (variety) was lost.
2. The long-fur dogs survive better in a cold environment; they are less able to survive in a warm environment and vice versa.
3. A particular characteristic in the dog population was selected for.
4. Dogs are still dogs since the variation is within the boundaries of "kind."

Natural selection of designed variation within the dog kind is not an example of evolution because it does not lead to the formation of a different kind of animal such as a horse, bear, or human. Instead, it is evidence of God's grace in supplying for His creation in the altered environments of a post-Fall, post-Flood world.

Natural Selection and Bacteria

Another example of natural selection is that of antibiotic resistance in bacteria. Such natural selection is commonly portrayed as evolution in action, but in this case, natural selection works in conjunction with mutation rather than designed variation.

Antibiotics are natural products produced by fungi and bacteria, and the antibiotics we use today are typically derivatives of those. Because of this relationship, it is not surprising that some bacteria would have resistance to certain antibiotics; they must do so to be competitive in their environment. In fact, if you took a sample of soil from outside your home, you would find antibiotic-resistant bacteria.

A bacterium can gain resistance through two primary ways:

1. By losing genetic information, and
2. By using a design feature built in to swap DNA—a bacterium gains resistance from another bacterium that has resistance.

Let's take a look at the first. Antibiotics usually bind a protein in the bacterium and prevent it from functioning properly, killing the bacteria. Antibiotic-resistant bacteria have a mutation in the DNA which codes for that protein. The antibiotic then cannot bind to the protein produced from the mutated DNA, and thus the bacteria live. Although the bacteria can survive well in an environment with antibiotics, it has come at a cost. If the antibiotic-resistant bacteria are grown with the nonmutant bacteria in an environment without antibiotics, the nonmutant bacteria will live and the mutant bacteria will die. This is because the mutant bacteria produce a mutant protein that does not allow them to compete with other bacteria for necessary nutrients.

Let's clarify this some by looking at the bacteria *Helicobacter pylori.* Antibiotic-resistant *H. pylori* have a mutation that results in the loss of information to produce an enzyme. This enzyme normally converts an antibiotic to a poison, which causes death. But when the antibiotics are applied to the mutant *H. pylori*, these bacteria can live while the normal bacteria are killed. So by natural selection the ones that lost information survive and pass this trait along to their offspring.

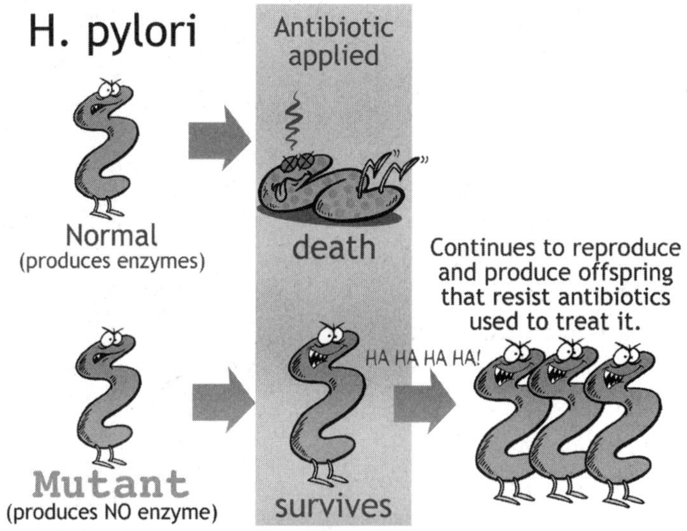

Now let's take a look at the second method. A bacterium can get antibiotic resistance by gaining the aforementioned mutated DNA from another bacterium. Unlike you and me, bacteria can swap DNA. It is important to note that this is still not considered a gain of genetic information since the information already exists and that while the mutated DNA may be new to a particular bacterium, it is not new overall.

Using the points from the table for what natural selection can accomplish, it can be seen that:

1. Through mutation, genetic information was lost.
2. The antibiotic resistant bacteria *only* survive well in an environment with antibiotics; they are less able to survive in the wild. (It is important to keep in mind that the gain of antibiotic resistance is not an example of a beneficial mutation but rather a beneficial outcome of a mutation in a given environment. These types of mutations are rare in other organisms as offspring are more limited in number, therefore, there is a greater need to preserve genetic integrity.)
3. A particular mutation in a bacterial population was selected for.
4. *H. pylori* is still *H. pylori.* No evolution has taken place to change it into something else—it's still the same bacteria with some variation.

Antibiotic resistance in bacteria, rather than being an example of evolution in action, is another example of natural selection seen properly from a biblical/creationist perspective.

Speciation—A Possible Outcome of Natural Selection

A species can be defined as a population of organisms produced by a parent population that has changed so significantly that it can no longer interbreed with the parent population. Using the example of dogs, it is possible that long-fur dogs might change sufficiently (other changes besides fur might also be selected for living in cold environments) to the point that they can no longer mate with short-fur or medium-fur dogs.

Although evolutionists claim that speciation takes long periods of time (millions of years), they are often amazed at how fast species can be observed to form today. Speciation has been observed to

Natural selection at work!

occur in as little as a few years as seen in guppies, lizards, fruit flies, mosquitoes, finches, and mice. This observation does not come as a surprise to creationists as all species alive in the past and today would have had to be produced in fewer than 6,000 years from the original created kinds. In fact, such processes (and perhaps other genetic factors) would have occurred rapidly after the Flood, producing variation within each kind. Such effects are largely responsible for generating the tremendous diversity seen in the living world.[12]

Speciation has never been observed to form an organism of a different kind, such as a dog species producing a cat. Speciation works *only* within a kind. Evolution requires natural selection and speciation to give rise to new kinds from a former kind (e.g., dinosaurs evolving into birds). Speciation, however, leads to a loss of information, not the gain of information required by evolution. Thus, speciation as a possible outcome of natural selection cannot be used as a mechanism for molecules-to-man evolution.

Conclusion

When discussing natural selection as a possible mechanism for evolution, it is important to define terms. Evolutionists and biblical creationists view

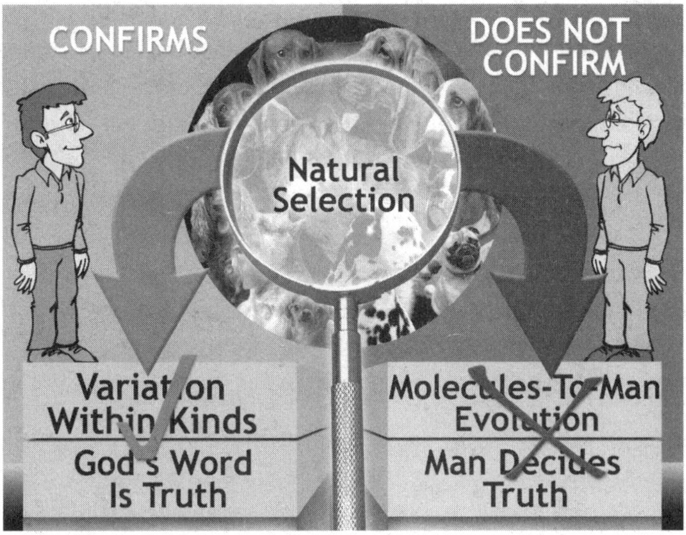

[12] G. Purdom, "Evolution" of finch beaks—again, www.answersingenesis.org/articles/aid/v1/n1/evolution-finch-beaks-again.

these terms differently, but it comes down to how we interpret the evidence in light of our foundation. Do we view natural selection using God's Word as our foundation, or do we use man's truth as our foundation?

The creationist view of natural selection is supported biblically and scientifically. Natural selection is a God-ordained process that allows organisms to survive in a post-Fall, post-Flood world. It is an observable reality that occurs in the present and takes advantage of the variations within the kinds and works to preserve the genetic viability of the kinds.

Simply put, the changes that are observed today show variation within the created kind—a horizontal change. For a molecules-to-man evolutionary model, there must be a change from one kind into another—a vertical change. This is simply not observed. We have never seen a bacterium like *H. pylori* give rise to something like a dog. Instead, we simply observe variations within each created kind.

Evolution requires an increase in information that results in a directional movement from molecules to man. Natural selection cannot be a mechanism for evolution because it results in a decrease in information and is not directional. Speciation may occur as a result of natural selection, but it only occurs within a kind. Therefore, it is also not a mechanism for evolution but rather supports the biblical model.

Natural selection cannot be the driving force for molecules-to-man evolution when it does not have that power, nor should it be confused with molecules-to-man evolution. It is an observable phenomenon that preserves genetic viability and allows limited variation within a kind—nothing more, nothing less. It is a great confirmation of the Bible's history.

23

Hasn't Evolution Been Proven True?

A. J. MONTY WHITE

Anyone who has read Genesis 1–11 realizes that the modern teachings of molecules-to-man evolution are at odds with what God says. So what is the response to evolution from a biblical and scientific perspective? Let's take a closer look.

Evolutionists often say that *evolution* simply means "change." However, in reality it means a certain kind of change. The word is now accepted to mean the change of nonliving chemicals into simple life-forms into more complex life-forms and finally into humans—what might be called *from-goo-to-you-via-the-zoo*. We are informed that this change occurred over millions of years, and the dominant mechanism that is supposed to have driven it is natural selection coupled with mutations.

Furthermore, the word *evolution* has also been applied to nonliving things. Almost everything is said to have evolved—the solar system, stars, the universe, as well as social and legal systems. Everything is said to be the product of evolution. However, the three major forms of evolution are

1. Stellar evolution
2. Chemical evolution
3. Biological evolution.

The story of evolution leaves no room for a supernatural Creator. Evolutionary processes are supposed to be purely naturalistic. This means that even the need for a supernatural Creator disappears because it is argued that the natural world can create new and better or more complex creatures by itself. The implication of this is very revealing: evolution means "no God" and if there is no God, then there are no rules—no commandments, no God-given rules which we must obey. We can therefore live our lives as we please, for according to evolutionary philosophy, there is no God to whom we have to give an account. No wonder molecules-to-man evolution is attractive to so many, for it allows them to live as they please. This is called relative morality.

Does the Bible Teach Evolution?

The simple answer to this question is "No." In Genesis 1 we read the account of the creation (not the evolution) of everything—the universe, the sun, moon, and stars, the planet earth with all its varied plant and animal kinds, including the pinnacle of God's creation—humans. Nowhere in this account do we read about molecules-to-man evolution. Furthermore, there was no time for evolution, for God supernaturally created everything in six literal days (Exodus 20:11, 31:17).

There are those who argue that Genesis 1 is a simplified account of evolution. But such a hypothesis does not stand up to scrutiny. A quick look at the order of the events in Genesis 1 and in evolution shows this (see chart below[1]). The order of events is quite different and the Genesis account of creation bears no relation to the evolutionary account of origins.

Evolution	**Genesis**
Sun before earth	Earth before sun
Dry land before sea	Sea before dry land
Atmosphere before sea	Sea before atmosphere
Sun before light on earth	Light on earth before sun
Stars before earth	Earth before stars
Earth at same time as planets	Earth before other planets
Sea creatures before land plants	Land plants before sea creatures
Earthworms before starfish	Starfish before earthworms

[1] T. Mortenson, Evolution vs. creation: the order of events matters! www.answersingenesis.org/docs2006/0404order.asp

Land animals before trees	Trees before land animals
Death before man	Man before death
Thorns and thistles before man	Man before thorns and thistles
TB pathogens & cancer before man (dinosaurs had TB and cancer)	Man before TB pathogens and cancer
Reptiles before birds	Birds before reptiles
Land mammals before whales	Whales before land animals
Land mammals before bats	Bats before land animals
Dinosaurs before birds	Birds before dinosaurs
Insects before flowering plants	Flowering plants before insects
Sun before plants	Plants before sun
Dinosaurs before dolphins	Dolphins before dinosaurs
Land reptiles before pterosaurs	Pterosaurs before land reptiles

In spite of this, some argue that there is a major difference between "make" and "create" (the Hebrew words are *asah* and *bara*, respectively). They argue that God *created* some things—for example, the heaven and the earth as recorded in Genesis 1:1 and the marine and flying creatures as recorded in Genesis 1:21. They then argue that God *made* other things, perhaps by evolution from pre-existing materials—for example, the sun, moon, and stars as recorded in Genesis 1:16, and the beasts and cattle as recorded in Genesis 1:25. Though these words have slightly different nuances of meaning, they are often used interchangeably, as seen clearly where *asah* (to make) and *bara* (to create) are used in reference to the same act (the creation of man, Genesis 1:26–27). Nothing in Genesis 1 leads to the conclusion that God used evolutionary processes to produce His creation.

There is a further problem with believing that the Genesis account of creation should be interpreted as an evolutionary account. One of the things that drives evolution is *death*. Yet the Bible teaches quite clearly that death was introduced into the perfect world as a result of Adam's sin. Neither human nor animal death existed until this event—both humans and animals were originally vegetarian (Genesis 1:29–30 shows that plants are not living creatures, as land and sea creatures, birds, and people are). The original world that God created was death-free, and so evolution could not have occurred before humans were created.

Stellar Evolution: The Big Bang

The big bang is the most prominent naturalistic view of the origin of the universe in the same way that Neo-Darwinian evolution is the naturalistic view of living systems. The difference between what the Bible teaches about the origin of the universe and what the evolutionists teach can be summed up as follows: the Bible teaches that "in the beginning God created" and the evolutionists teach, in essence, that "in the beginning nothing became something and exploded."

According to the big bang, our universe is supposed to have suddenly popped into existence and rapidly expanded and given rise to the countless billions of galaxies with their countless billions of stars.

In support of the idea that nothing can give rise to the universe, cosmologists argue that quantum mechanics predicts that a vacuum can, under some circumstances, give rise to matter. But the problem with this line of reasoning is that a vacuum is *not* nothing; it is something—it is a vacuum that can be made to appear or disappear, as in the case of the Torricellian vacuum, which

is found at the sealed end of a mercury barometer. All logic predicts that if you have nothing, nothing will happen. It is against all known logic and all laws of science to believe that the universe is the product of nothing. This concept is similar to hoping that an empty bank account will suddenly give rise to billions of dollars all on its own.

However, if we accept that the universe and everything in it came from nothing (and also from *nowhere*) then we have to follow this to its logical conclusion. This means that not only is all the physical material of the universe the product of nothing, but also other things. For example, we are forced to accept that nothing (which has no mind, no morals, and no conscience) created reason and logic; understanding and comprehension; complex ethical codes and legal systems; a sense of right and wrong; art, music, drama, comedy, literature, and dance; and belief systems that include God. These are just a few of the philosophical implications of the big bang hypothesis.

Chemical Evolution: The Origin of Life

It is commonly believed (because it is taught in our schools and colleges) that laboratory experiments have proved conclusively that living organisms evolved from nonliving chemicals. Many people believe that life has been created in the laboratory by scientists who study chemical evolution.

The famous experiment conducted by Stanley Miller in 1953 is often quoted as proof of this. Yet the results of such experiments show nothing of the sort. These experiments, designed as they are by intelligent humans, show that under certain conditions, certain organic compounds can be formed from inorganic compounds.

In fact, what the *intelligent* scientists are actually saying is, "If I can just synthesize life in the laboratory, then I will have proven that no *intelligence* was necessary to form life in the beginning." Their experiments are simply trying to prove the opposite—that an intelligence is required to create life.

If we look carefully at Miller's experiment, we will see that what he did fails to address the evolution of life. He took a mixture of gases (ammonia, hydrogen, methane, and water vapor) and he passed an electric current through them. He did this in order to reproduce the effect of lightning passing through a mixture of gases that he thought might have composed the earth's atmosphere millions of years ago. As a result, he produced a mixture of amino acids. Because amino acids are the building blocks of proteins and proteins are considered to be the building blocks of living systems, Miller's experiment was hailed as proof that life had evolved by chance on the earth millions of years ago.

There are a number of objections to such a conclusion.

1. There is no proof that the earth ever had an atmosphere composed of the gases used by Miller in his experiment.

2. The next problem is that in Miller's experiment he was careful to make sure there was no oxygen present. If oxygen was present, then the amino acids would not form. However, if oxygen was absent from the earth, then there would be no ozone layer, and if there was no ozone layer the ultraviolet radiation would penetrate the atmosphere and would destroy the amino acids as soon as they were formed. So the dilemma facing the evolutionist can be summed up this way: amino acids would not form in an atmosphere *with* oxygen and amino acids would be destroyed in an atmosphere *without* oxygen.

3. The next problem concerns the so-called handedness of the amino acids. Because of the way that carbon atoms join up with other atoms, amino acids exist in two forms—the right-handed form and the left-handed form. Just as your right hand and left hand are identical in all respects except for their handedness, so the two forms of amino acids are identical except for their handedness. In all living systems only left-handed amino acids are found. Yet Miller's experiment produced a mixture of right-handed and left-handed amino acids in identical proportions. As only the left-handed ones are used in living systems, this mixture is useless for the evolution of living systems.

4. Another major problem for the chemical evolutionist is the origin of the information that is found in living systems. There are various claims about the amount of information that is found in the human genome, but it can be conservatively estimated as being equivalent to a few thousand books, each several hundred pages long. Where did this information come from? Chance does not generate information. This observation caused the late Professor Sir Fred Hoyle and his colleague, Professor Chandra Wickramasinghe of Cardiff University, to conclude that the evolutionist is asking us to believe that a tornado can pass through a junk yard and assemble a jumbo jet.

The problems outlined above show that, far from creating life in the laboratory, the chemical evolutionists have not shown that living systems arose by chance from nonliving chemicals. Furthermore, the vast amount of information contained in the nucleus of a living cell shows that living systems could not have evolved from nonliving chemicals. The only explanation for the existence of living systems is that they must have been created.

Biological Evolution: Common Descent?

Comparative anatomy is the name given to the science that deals with the structure of animals. Comparing the anatomy of one kind of animal with another is supposed to prove descent from a common ancestor. This is often put forward as strong evidence for evolution. However, the science of comparative anatomy can just as easily be used as evidence of creation, as we shall see.

The bones of a horse are different from our bones, but there is such a similarity that if we are familiar with the human skeleton, we could easily identify and name the bones of a horse. We could do the same if we studied the skeleton of a salamander, a crocodile, a bird, or a bat. However, not only are the bones similar, but so also are other anatomical structures, such as muscles, the heart, the liver, the kidneys, the eyes, the lungs, the digestive tract, and so on. This is interpreted by the evolutionists as proof that these various animals are all descended from a common ancestor.

One of the classic examples that is often used in biology textbooks to illustrate comparative anatomy is the forelimbs of amphibians, reptiles, humans, birds, bats, and quadrupeds. In the illustration, it can be seen that all the forelimbs of these six different types of creatures have an upper arm bone (the humerus) and two lower arm bones (the radius and the ulna), although in the case of the bat there is only one bone, called the radio-ulna.

Evolutionists teach that these structures are said to be homologous when they are similar in structure and origin, but not necessarily in function. But notice how subtly the notion of origins is introduced into the definition. The bat's wing is considered to be homologous to the forelimb of a salamander because it is similar in structure and believed to have the same origin. However, it is not considered to be homologous to the wing of an insect because, even though it has the same function, it is not considered to have the same origin. However, the fact that the two structures are similar does not necessarily mean that they are derived from a common ancestor.

We have to realize that the entire line of reasoning by evolutionists is based upon a single assumption: that the degree of similarity between organisms indicates the degree of supposed relationship of the said organisms. In other words, it is argued that if animals look alike, then they must be closely related (from an evolutionary point of view), and if they do not look very much alike, then they are more distantly related. But this is just an assumption.

The presence of homologous structures can actually be interpreted as evidence for a common designer. Contrary to the oversimplified claim in this figure, the forelimbs of vertebrates do not form in the same way. Specifically, in frogs the phalanges form as buds that grow outward and in humans they form from a ridge that develops furrows inward. The fact that the bones can be correlated does not mean that they are evidence of a single common ancestor.[2]

Figure 9 Homologous structures **Holt 286**

The forelimbs of vertebrates contain the same kinds of bones, which form in the same way during embryological development.

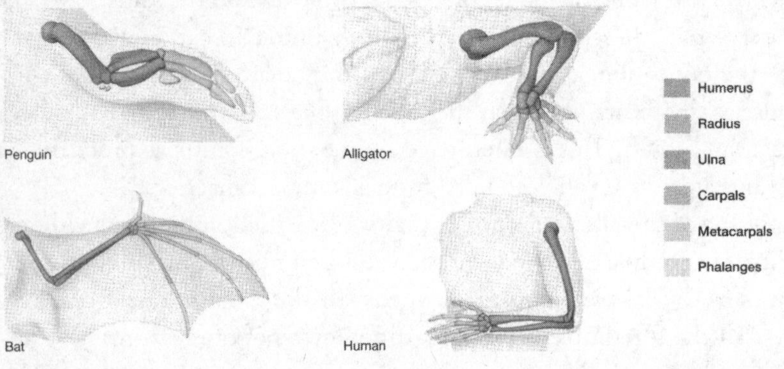

In fact, there is another logical reason why things look alike—creation by an intelligent designer using a common blueprint. This is the reason that Toyota and Ford motor vehicles look so much alike. They are built to a common plan—you only have to look at them to realize this. However, the problem with the living world is that in many cases either explanation (i.e., evolution or creation) appears to be logical and it is often impossible for us to tell which is the more reasonable explanation. This is why it is important for us to understand which worldview we are using to interpret the evidence.

[2] R. Patterson, *Evolution Exposed: Biology*, Answers in Genesis, Petersburg, Kentucky, 2009, 72

There is, however, one discovery that appears to make the evolutionary view of descent from a common ancestor look illogical and flawed. This discovery is that structures that appear homologous often develop under the control of genes that are *not* homologous. If the structures evolved from the same source, you would expect the same genes to make the structures. The fact that these structures are similar (or homologous) is apparent, but the reason is not because of Darwinian evolution. It is more logical and reasonable to believe in a common Creator rather than a common ancestor.

Many evolutionists readily admit that they have failed to find evidence of the evolution of large structures such as bones and muscles, so instead they argue that they have found homology among the complex organic molecules that are found in living systems. One of these is hemoglobin, the protein that carries oxygen in red blood cells. Although this protein is found in nearly all vertebrates, it is also found in some invertebrates (worms, starfish, clams, and insects) and also in some bacteria. Yet there is no evidence of the evolution of this chemical—in all cases, the same kind of molecule is complete and fully functional. If evolution has occurred, it should be possible to map out how hemoglobin evolved, but this cannot be done. To the creationist, however, hemoglobin crops up complete and fully functional wherever the Creator deems it fitting in His plan.

Missing Links

Our English word *fossil* is from the Latin *fossilis,* which means "something dug up." The present-day meaning of the word fossil is a relic or trace of past life preserved in the rocks. This can be a preserved hard part of the plant or animal, such as a stem or a leaf or a shell or a bone or a tooth; it can also be a soft part such as skin or even excrement (called coprolites), or it can be a trace made by the creature when it was alive, such as a footprint. All the fossils that are found in all the sedimentary rocks are regarded together as the fossil record.

Charles Darwin proposed the gradual evolution of life forms over a long period of time. If this has happened, you would expect to find this gradual evolution of one kind of life form into another kind to be recorded in the fossil record. However, this evolutionary account of one kind of life form changing into another kind is *not* recorded in the fossils. There are many instances where variations within a kind are found (for example, different varieties of elephant or dinosaur) but there are no examples of in-between kinds. Both evolutionists and creationists agree that the intermediate transitional forms expected on the basis of slow gradual change of one kind of creature into another kind is not found fossilized in the sedimentary rocks. In other words, the transitional forms are missing—hence the term "missing links."

Charles Darwin himself realized that his theory was not supported by the fossil record, for he wrote in his *Origin of Species*:

> The number of intermediate varieties which have formerly existed on earth must be truly enormous. Why then is not every geological formation and every stratum full of such intermediate links? Geology assuredly does not reveal any such finely graduated organic chain: and this, perhaps, is the most obvious and gravest objection which can be urged against my theory.[3]

When Charles Darwin penned these words, he attributed this absence of transitional forms to what he called the "extreme imperfection" of the fossil record. Since that time, however, literally millions of fossils have been found, but still the transitional forms are absent. The fossil record does not show the continuous development of one kind of creature into another, but it shows different kinds of creatures that are fully functional with no ancestors or descendants which are different kinds of creatures.

It cannot be overemphasized that there are many places in the fossil record where it is expected that plenty of intermediate forms should be found—yet they are not there. All the evolutionists ever point to is a handful of highly debatable transitional forms (e.g., horses), whereas they should be able to show us thousands of incontestable examples. This is very noticeable when looking at the fossil record of some of the more peculiar kinds of animals such as the *cetacean* (whales, dolphins, and porpoises), the *sirenia* (manatees, dugongs, and sea cows), the *pinnipedia* (sea lions, seals, and walruses), kangaroos, bats, dragonflies, and spiders. Their supposed evolutionary origins and descent are represented by missing links and speculations rather than factual evidence.

Even alleged transitional forms in supposed human evolution fall short. In fact, most so-called missing links fall into three categories: extinct ape, living ape, or human. The following chart gives some of the most common scientific names and their classifications.

---

[3] C. Darwin, *The Origin of Species*, Penguin Books, London, 1968, 291.

Name	What is it?*
Australopithecus afarensis, such as "Lucy"	Extinct ape
Australopithecus africanus	Extinct ape
Australopithecus boisei	Extinct ape
Australopithecus robustus	Extinct ape
Pan troglodytes and *Pan paniscus* (chimpanzee)	Living ape
Gorilla gorilla and *Gorilla beringei* (gorilla)	Living ape
Pongo pygmaeus and *Pongo abelii* (orangutan)	Living ape
Ramapithecus	Extinct ape (extinct orangutan)
Homo habilis	Junk category mixing some human and some ape fossils
Homo floresiensis	Human (dwarf, pygmy)
Homo ergaster	Human
Homo erectus, such as "Peking man" and "Java man"	Human**
Homo neanderthalensis (Neanderthals)	Human
Homo heidelbergensis	Human
Homo sapiens (modern & archaic)	Human

* An accurate classification of these kinds of fossils depends on an accurate starting point. Some fossils have been misclassified. The ones labeled as humans (*Homo heidelbergensis, Homo erectus,* etc.), indeed show variation, but they are still human. This is also true of the different ape kinds. Variation, not evolution, is what we would expect from the clear teachings of the Bible.
** For the most part these two classifications are anatomically human. However, a number of finds that are not human but rather apelike have been included as part of the *Homo erectus* category, due to evolutionary beliefs. These apelike finds should be reclassified.[4]

It is obvious that the evolutionists have "faith" in the original existence of the missing transitional forms.

[4] For more on supposed human evolution, see chapter 8, "Did humans really evolve from apelike creatures?" in K. Ham et al., *The New Answers Book 2,* Master Books, Green Forest, Arkansas, 2008.

Evolution of New Kinds?

Charles Darwin visited the Galapagos Islands and brought back samples of the different finches that lived on the different islands. He observed that they had different shaped beaks, which appeared to suit the type of food that the finches ate. From this observation, Darwin concluded that a pair or flock of finches had flown to these islands at some time in the past and that the different beaks on the finches had evolved via natural selection, depending on what island they lived on and consequently what they fed on. From these types of simple observations and conclusions, Darwin developed not only the idea of the evolution of species but also the idea of chemicals-to-chemist evolution!

But let us consider exactly what Darwin actually observed—finches living on different islands feeding on different types of food having different beaks. What did he propose? That these finches had descended from a pair or flock of finches. In other words, he proposed that finches begat finches—that is, they reproduced after their own kind. This is exactly what the Bible teaches in Genesis 1.

It cannot be overemphasized that no one has ever seen one kind of plant or animal changing into another different kind. Darwin did not observe this, even though he proposed that it does happen. There are literally thousands of plant and animal kinds on the earth today, and these verify what the Bible indicates in Genesis 1 about plants and animals reproducing after their own kind.

Plants and animals reproducing after their own kind is what we observe, and it is what Charles Darwin observed in finches on the Galapagos Islands. For example, we see different varieties of *Brassica*—kale, cabbage, cauliflower are all varieties of the wild common mustard *Brassica oleracea*. Furthermore, another perfect example of a kind is the hundreds of different varieties of dogs, including spaniels, terriers, bulldogs, Chihuahuas, Great Danes, German shepherds, Irish wolfhounds, and greyhounds, which are all capable of interbreeding, together with wolves, jackals, dingoes, and coyotes. All are descended from the two representatives of the dog kind that came off Noah's Ark.

Conclusion

We have seen that the Bible does not teach evolution. There is no demonstrable evidence for the big bang, and chemical evolution has failed miserably in spite of evolutionists' attempts to create living systems in the laboratory. Similarities in the structure found in living systems can be interpreted better as evidence for a common design rather than a common ancestry. In spite of billions of fossils being found, there are no unquestionable fossils that show a transition between any of the major life forms.

Natural selection (done in the wild) and artificial selection (as done by breeders) produce enormous varieties *within* the different kinds of plants and animals. It has proved an impossible feat, however, to change one kind of creature into a different kind of plant or animal. The so-called "kind barrier" has never been crossed. Such evolution has never been observed. This has been pointed out by none other than evolutionary Professor Richard Dawkins, who confidently asserted in an interview that evolution has been observed but then added, "It's just that it hasn't been observed while it's happening."[5]

5 www.pbs.org/now/transcript/transcript349_full.html#dawkins.

24

Did Dinosaurs Turn into Birds?

DAVID MENTON

Introduction

According to many evolutionists today, dinosaurs are really not extinct but rather are feeding at our bird feeders even as we speak. For many evolutionists, it would seem, birds simply *are* dinosaurs. With this sort of bias, it is quite easy for evolutionists to find supposed evidence to support the notion that birds evolved from dinosaurs.

But what does the Bible tell us about the origin of birds, and just how good is the scientific evidence that some dinosaurs evolved into birds?

What Does the Bible Say about the Origin of Birds?

BIRDS WERE CREATED ON DAY 5 AND DINOSAURS ON DAY 6.

In the first chapter of Genesis, verse 21, we read that on Day 5 of creation, God created "every winged fowl after its kind." This includes birds that flew above the earth (Genesis 1:20). Man and land animals were created on Day 6 of the Creation Week (Genesis 1:24–31). Were there land birds that didn't fly originally? I would leave open the possibility, but a discussion of this is beyond the scope of this chapter. Most ornithologists say that these birds are *secondarily* flightless (i.e., they lost the ability to fly). This would be due to variance within kind or to mutational losses since creation. So, the best possibility is that bird were created on Day 5 as flyers, and some have lost this ability, but I wouldn't be dogmatic.

The extinct aquatic reptiles, such as the plesiosaurs, and the extinct flying reptiles, such as the pterodactyls, are not classified as dinosaurs, and most evolutionists do not believe that they evolved into birds. Thus, for the Bible-believing Christian, both the fact of creation and the order of creation affirm that birds and dinosaurs originated separately.

BIRDS ARE OF MANY DIFFERENT "KINDS."

Genesis 1:21 says that God created every winged bird after its "kind." The following verse says they were to multiply, or reproduce; so the logical connection is that birds of the same kind can reproduce. The Hebrew word for "kind" in Genesis refers to any group of animals capable of interbreeding and reproducing according to their type. For example, all dogs and dog-like animals, such as wolves and coyotes, are capable of interbreeding and thus would represent one "kind," even though some are classified today as different species.

This does not mean, however, that all birds represent a single created kind and thus share a common ancestry. The Bible tells us that there are many different bird kinds (plural). The Levitical dietary laws (Leviticus 11:13–19), for example, list many different bird kinds as being unclean. This gives further biblical support for multiple created bird kinds.

What Do Evolutionists Claim about the Origin of Birds?

Evolutionists have long speculated that birds evolved from reptiles. At one time or another, virtually every living and extinct class of reptiles has been proposed as the ancestor of birds. The famous Darwinian apologist Thomas Huxley was the first to speculate (in the mid 1800s) that birds evolved from dinosaurs.

While this notion has gone in and out of favor over the years, it is currently a popular view among evolutionists. Indeed, the origin of birds from dinosaurs is touted as irrefutable dogma in our schools, biology textbooks, and the popular media.

While evolutionists now agree that birds are related in some way to dinosaurs, they are divided over whether birds evolved from some early shared ancestor of the dinosaurs within the archosauria (which includes alligators, pterosaurs, plesiosaurs, ichthyosaurs, and thecodonts) or directly from advanced theropod dinosaurs (bipedal meat-eating dinosaurs, such as the well-known *Tyrannosaurus rex*). The latter view has gained in popularity since 1970, when

John Ostrom discovered a rather "bird-like" early Cretaceous theropod dinosaur called *Deinonychus*.

An adult *Deinonychus* measured about 12 feet (3.5 m) long, weighed over 150 pounds (68 kg), and was about 5 feet (1.5 m) tall standing on its two hind legs. Like other theropods (which means "beast foot"), *Deinonychus* had forelimbs much smaller than its hind limbs, with hands bearing three fingers and feet bearing three toes. The most distinctive feature of *Deinonychus* (which means "terrible claw") is a large curved talon on its middle toe.

One of the main reasons that *Deinonychus* and other similar theropod dinosaurs (called dromaeosaurs) seemed to be plausible ancestors to birds is that, like birds, these creatures walked solely on their hind legs and have only three digits on their hands. But as we shall see, there are many problems with transforming any dinosaur, and particularly a theropod, into a bird.

Problems with Dinosaurs Evolving into Birds

WARM-BLOODED VS. COLD-BLOODED

Seemingly forgotten in all the claims that birds are essentially dinosaurs (or at least that they evolved from dinosaurs) is the fact that dinosaurs are reptiles. There are many differences between birds and reptiles, including the fact that (with precious few exceptions) living reptiles are cold-blooded creatures, while birds and mammals are warm-blooded. Indeed, even compared to most mammals, birds have exceptionally *high* body temperatures resulting from a high metabolic rate.

The difference between cold- and warm-blooded animals isn't simply in the relative temperature of the blood but rather in their ability to maintain a constant body core temperature. Thus, warm-blooded animals such as birds and mammals have internal physiological mechanisms to maintain an essentially constant body temperature; they are more properly called "endothermic." In contrast, reptiles have a varying body temperature influenced by their surrounding environment and are called "ectothermic." An ectothermic animal can adjust its body temperature behaviorally (e.g., moving between shade and sun), even achieving higher body temperature than a so-called warm-blooded animal, but this is done by outside factors.

In an effort to make the evolution of dinosaurs into birds seem more plausible, some evolutionists have argued that dinosaurs were also endothermic,[1] but there is no clear evidence for this.[2]

One of the lines of evidence for endothermic dinosaurs is based on the microscopic structure of dinosaur bones. Fossil dinosaur bones have been found containing special microscopic structures called osteons (or Haversian systems). Osteons are complex concentric layers of bone surrounding blood vessels in areas where the bone is dense. This arrangement is assumed by some to be unique to endothermic animals and thus evidence that dinosaurs are endothermic, but such is not the case. Larger vertebrates (whether reptiles, birds, or mammals) may also have this type of bone. Even tuna fish have osteonal bone in their vertebral arches.

Another argument for endothermy in dinosaurs is based on the eggs and assumed brood behavior of dinosaurs, but this speculation too has been challenged.[3] There is in fact no theropod brooding behavior not known to occur in crocodiles and other cold-blooded living reptiles.

Alan Feduccia, an expert on birds and their evolution, has concluded that "there has never been, nor is there now, any evidence that dinosaurs were endothermic."[4] Feduccia says that despite the lack of evidence "many authors have tried to make specimens conform to the hot-blooded theropod dogma."

"BIRD-HIPPED" VS. "LIZARD-HIPPED" DINOSAURS

All dinosaurs are divided into two major groups based on the structure of their hips (pelvic bones): the lizard-hipped dinosaurs (saurischians) and the bird-hipped dinosaurs (ornithiscians). The main difference between the two hip structures is that the pubic bone of the bird-hipped dinosaurs is directed toward the rear (as it is in birds) rather entirely to the front (as it is in mammals and reptiles).

But in most other respects, the bird-hipped dinosaurs, including such huge quadrupedal sauropods as *Brachiosaurus* and *Diplodocus*, are even less

[1] R.T. Bakker, Dinosaur renaissance, *Scientific American* **232**:58–78, 1975.

[2] A. Feduccia, Dinosaurs as reptiles, *Evolution* **27**:166–169,1973; A. Feduccia, *The Origin and Evolution of Birds*, 2nd Ed., Yale University Press, New Haven, Connecticut, 1999.

[3] N.R. Geist and T.D. Jones, Juvenile skeletal structure and the reproduction habits of dinosaurs, *Science* **272**:712–714,1996.

[4] A. Feduccia, T. Lingham-Soliar, and J.R. Hinchliffe, Do feathered dinosaurs exist? Testing the hypothesis on neontological and paleontological evidence, *Journal of Morphology* **266**:125–166, 2005.

bird-like than the lizard-hipped, bipedal dinosaurs such as the theropods. This point is rarely emphasized in popular accounts of dinosaur/bird evolution.

THE THREE-FINGERED HAND

One of the main lines of evidence sighted by evolutionists for the evolution of birds from theropod dinosaurs is the three-fingered "hand" found in both birds and theropods. The problem is that recent studies have shown that there is a digital mismatch between birds and theropods.

Most terrestrial vertebrates have an embryological development based on the five-fingered hand. In the case of birds and theropod dinosaurs, two of the five fingers are lost (or greatly reduced) and three are retained during development of the embryo. If birds evolved from theropods, one would expect the same three fingers to be retained in both birds and theropod dinosaurs, but such is not the case. Evidence shows that the fingers retained in theropod dinosaurs are fingers 1, 2, and 3 (the "thumb" is finger 1) while the fingers retained in birds are 2, 3, and 4.[5]

AVIAN VS. REPTILIAN LUNG

One of the most distinctive features of birds is their lungs. Bird lungs are small in size and nearly rigid, but they are, nevertheless, highly efficient to meet the high metabolic needs of flight. Bird respiration involves a unique "flow-through ventilation" into a set of nine interconnecting flexible air sacs sandwiched between muscles and under the skin. The air sacs contain few blood vessels and do not take part in oxygen exchange, but rather function like bellows to move air through the lungs.

The air sacs permit a unidirectional flow of air through the lungs resulting in higher oxygen content than is possible with the bidirectional air flow through the lungs of reptiles and mammals. The air flow moves through the same tubes at different times both into and out of the lungs of reptiles and mammals, and this results in a mixture of oxygen-rich air with oxygen-depleted air (air that has been in the lungs for awhile). The unidirectional flow

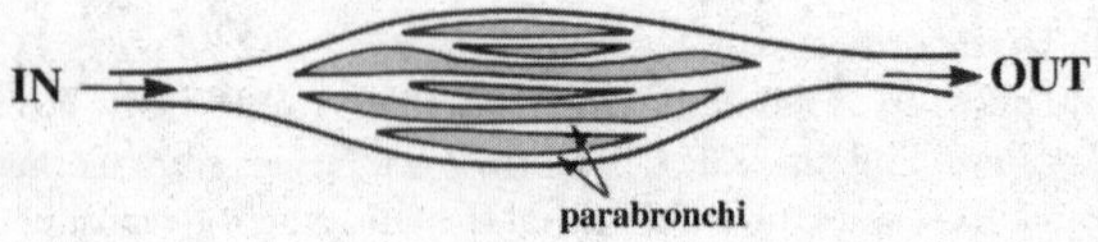

[5] Feduccia et al., 2005.

through bird lungs not only permits more oxygen to diffuse into the blood but also keeps the volume of air in the lungs nearly constant, a requirement for maintaining a level flight path.

If theropod dinosaurs are the ancestors of birds, one might expect to find evidence of an avian-type lung in such dinosaurs. While fossils generally do not preserve soft tissue such as lungs, a very fine theropod dinosaur fossil (*Sinosauropteryx*) has been found in which the outline of the visceral cavity has been well preserved. The evidence clearly indicates that this theropod had a lung and respiratory mechanics similar to that of a crocodile—not a bird.[6] Specifically, there was evidence of a diaphragm-like muscle separating the lung from the liver, much as you see in modern crocodiles (birds lack a diaphragm). These observations suggest that this theropod was similar to an ectothermic reptile, not an endothermic bird.

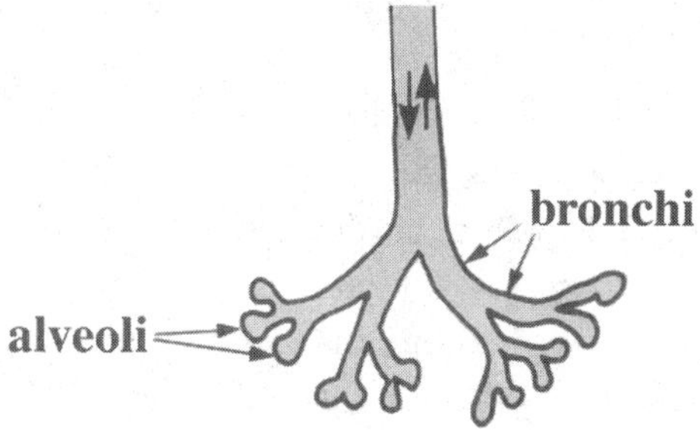

Origin of Feathers

DO FEATHERED DINOSAURS EXIST?

Feathers have long been considered to be unique to birds. Certainly all living birds have feathers of some kind, while no living creature other than birds has been found to have a cutaneous appendage even remotely similar to a feather. Since most evolutionists are certain that birds evolved from dinosaurs (or at least are closely related to them), there has been an intense effort to find dinosaur fossils that show some suggestion of feathers or "protofeathers." With such observer bias, one must be skeptical of recent widely publicized reports of feathered dinosaurs.

Dinosaurs are reptiles, and so it is not surprising that fossil evidence has shown them to have a scaly skin typical of reptiles. For example, a recently discovered well-preserved specimen of *Compsognathus* (a small theropod dinosaur of the type believed to be most closely related to birds) showed unmistakable evidence of scales but alas—no feathers.[7]

---

[6] J.A. Ruben, T.D. Jones, N.R. Geist, and W.J. Hillenius, Lung structure and ventilation in theropod dinosaurs and early birds, *Science* **278**:1267–1270, 1997.
[7] U.B. Gohlich and L.M. Chiappe, A new carnivorous dinosaur from the late Jurassic Solnhofen archipelago, *Nature* **440**:329–332, 2006.

Still, there have been many claims of feathered dinosaurs, particularly from fossils found in Liaoning province in northeastern China.[8] The earliest feathered dinosaur from this source is the very unbird-like dinosaur *Sinosauropteryx,* which lacks any evidence of structures that could be shown to be feather-like.[9]

Structures described as "protofeathers" in the dinosaur fossils *Sinosauropteryx* and *Sinithosaurus* are filamentous and sometimes have interlaced structures bearing no obvious resemblance to feathers. It now appears likely that these filaments (often referred to as "dino-fuzz") are actually connective tissue fibers (collagen) found in the deep dermal layer of the skin. Feduccia laments that "the major and most worrying problem of the feathered dinosaur hypothesis is that the integumental structures have been homologized with avian feathers on the basis of anatomically and paleontologically unsound and misleading information."[10]

Complicating matters even further is the fact that true birds have been found among the Liaoning province fossils in the same layers as their presumed dinosaur ancestors. The obvious bird fossil *Confuciusornis sanctus,* for example, has long slender tail feathers resembling those of a modern scissor-tail flycatcher. Two taxa (*Caudipteryx* and *Protarchaeopteryx*) that were thought to be dinosaurs with true feathers are now generally conceded to be flightless birds.[11]

Thus far, the only obvious dinosaur fossil with obvious feathers that was "found" is *Archaeoraptor liaoningensis*. This so-called definitive feathered dinosaur was reported with much fanfare in the November 1999 issue of *National Geographic* but has since been shown to be a fraud.

What would it prove if features common to one type of animal were found on another? Nothing. Simply put, God uses various designs with various creatures. Take the platypus, for example—a mosaic. It has several design features that are shared with other animals, and yet it is completely distinct. So if a dinosaur (or mammal) is ever found with feathers, it would call into

---

[8] P.J. Chen, Z.M. Dong, and S.N. Zheng, An exceptionally well-preserved theropod dinosaur from the Yixian formation of China, *Nature* 391:147–152,1998; X. Xu, X.Wang, and X. Wu, A dromaeosaurid dinosaur with a filamentous integument from the Yixian formation of China, *Nature* 401:262–266, 1999; P.J. Currie and P.J. Chen, Anatomy of *Sinosauropteryx prima* from Liaoning, northeastern China, *Can. J. Earth Sci.* 38:1705–1727, 2001.

[9] Feduccia et al., 2005.

[10] Feduccia et al., 2005.

[11] Feduccia et al., 2005.

question our human criteria for classification, not biblical veracity. What's needed to support evolution is *not* an unusual mosaic of complete traits, but a trait in transition, such as a "scale-feather," what creationist biologists would call a "sceather."

FEATHERS AND SCALES ARE DISSIMILAR.

If birds evolved from dinosaurs or any other reptile, then feathers must have evolved from reptilian scales. Evolutionists are so confident that feathers evolved from scales that they often claim that feathers are very similar to scales. The popular Encarta computerized encyclopedia (1997) describes feathers as a "horny outgrowth of skin peculiar to the bird but similar in structure and origin to the scales of fish and reptiles."[12]

In actual fact, feathers are profoundly different from scales in both their structure and growth. Feathers grow individually from tube-like follicles similar to hair follicles. Reptilian scales, on the other hand, are not individual follicular structures but rather comprise a continuous sheet on the surface of the body. Thus, while feathers grow and are shed individually (actually in symmetrically matched pairs!), scales grow and are shed as an entire sheet of skin.

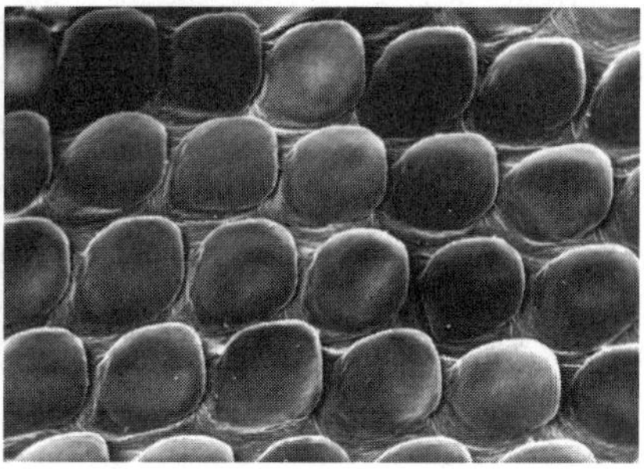

The feather vane is made up of hundreds of barbs, each bearing hundreds of barbules interlocked with tiny hinged hooklets. This incredibly complex structure bears not the slightest resemblance to the relatively simple reptilian scale. Still, evolutionists continue to publish imaginative scenarios of how long-fringed reptile scales evolved by chance into feathers, but evidence of "sceathers" eludes them.

[12] Encarta 98 Encyclopedia. 1993–1997.

Archaeopteryx, a True Bird, Is Older than the "Feathered" Dinosaurs.

A reconstruction of *Archaeopteryx* as displayed in a natural history museum in Stolkholm, Sweden

One of the biggest dilemmas for those who want to believe that dinosaurs evolved into birds is that the so-called feathered dinosaurs found thus far are dated to be about 20 million years more recent than *Archaeopteryx*. This is a problem for evolution because *Archaeopteryx* is now generally recognized to be a true bird.[13] Some specimens of this bird are so perfectly fossilized that even the microscopic detail of its feathers is clearly visible. So, having alleged missing links of dinosaurs changing into birds when birds already exist doesn't help the case for evolution.

For many years *Archaeopteryx* has been touted in biology textbooks and museums as the perfect transitional fossil, presumably being precisely intermediate between reptiles and birds. Much has been made over the fact that *Archaeopteryx* had teeth, fingers on its wings, and a long tail—all supposedly proving its reptilian ancestry. While there are no living birds with teeth, other fossilized birds such as *Hesperornis* also had teeth. Some modern birds, such as the ostrich, have fingers on their wings, and the juvenile hoatzin (a South American bird) has well-developed fingers and toes with which it can climb trees.

Origin of Flight

One of the biggest problems for evolutionists is explaining the origin of flight. To make matters worse, evolutionists believe that the flying birds evolved before the nonflying birds, such as penguins.

The theropod type of dinosaur that is believed to have evolved into flying birds is, to say the least, poorly designed for flight. These dinosaurs have small

---

[13] P.J. Currie et al., eds., *Feathered Dragons: Studies on the Transition from Dinosaurs to Birds*, Indiana University Press, Bloomington, Indiana, 2004.

forelimbs that typically can't even reach their mouths. It is not clear what theropods, such as the well-known *T. rex,* did with its tiny front limbs. It is obvious that they didn't walk, feed, or grasp prey with them, and they surely didn't fly with them!

Another problem is that this bipedal type of dinosaur had a long heavy tail to balance the weight of a long neck and large head. Decorating such a creature with feathers would hardly suffice to get it off the ground or be of much benefit in any other way.

Conclusion

Having a true bird appear before alleged feathered dinosaurs, no mechanism to change scales into feathers, no mechanism to change a reptilian lung into an avian lung, and no legitimate dinosaurs found with feathers are all good indications that dinosaurs didn't turn into birds. The evidence is consistent with what the Bible teaches about birds being unique and created after their kinds.

Genesis is clear that God didn't make birds from pre-existing dinosaurs. In fact, dinosaurs (land animals made on Day 6) came *after* winged creatures made on Day 5, according to the Bible. Both biblically and scientifically, chicken eaters around the world can rest easy—they aren't eating mutant dinosaurs.

25

Does Archaeology Support the Bible?

CLIFFORD WILSON

It is a biblical principle that matters of testimony should be established by the mouths of two or three witnesses. According to Hebrew law, no person could be found guilty of an offence without properly attested evidence from witnesses, even though this law was put aside at the trial of Jesus.

When it comes to the Word of God, a similar principle is demonstrated from the modern science of archaeology. We are told in Psalm 85:11, "Truth shall spring out of the earth," and in Psalm 119:89, "Forever, O Lord, Your word is settled in heaven." God's Word is sure. It outlasts human generations, and in His own time God vindicates its truth. This puts God's Word in a unique category: it is the "other side" of the two-way communication pattern between God and man. Man's speech distinguishes him uniquely from all the animals, and God's written Word distinguishes His special communication to man as immeasurably superior to all other supposed revelations.

According to that biblical principle of "two or three witnesses," we shall now select evidences that support the truth and accuracy of God's Word. In every area, the evidence has been forthcoming: God has vindicated His Word, and His Book is a genuine writing, with prophecies and revelation that must be taken seriously. His Book is unique because it is His Book.

Those inspired men of old wrote down God's message, applicable to themselves in their own times, and also applicable to men and women across the centuries, right down to the present century. The Bible is the "other side" of the Christian's study of the miracle of language. It is God's chosen way of revealing His thoughts—the deep things which are unsearchable except by the revelation of the Holy Spirit.

In the following outline we suggest certain divisions of the Word of God. Then we list three significant evidences from archaeology to confirm that the witness is sufficient to cause the case to be accepted for each section—God's Word is indeed Truth.

Major Evidences Regarding Genesis 1–11

Genesis 1–11 is the "seed-plot of the Bible," an introduction to Abraham and great doctrines, such as God the Creator, Friend, Revealer, Judge, Redeemer, Restorer, and Sustainer. It is actual history, and it is a summary of beginnings.

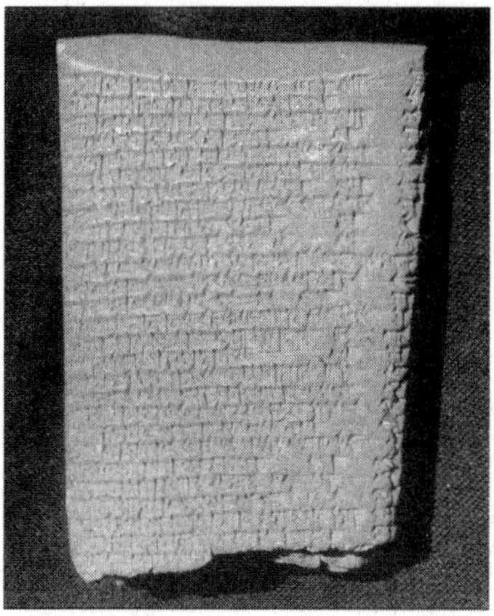

One of the Babylonian Creation Tablets, Enuma Elish

1. Enuma Elish—This is the Babylonian Creation Record. We also have the Ebla Creation Tablet. The Bible record is clearly superior to this as the Enuma Elish has creation from pre-existing matter, which really isn't creation at all. The Bible is the true account of this historical event.

2. The Epic of Gilgamesh includes the Babylonian Flood Story. Again, the biblical record is greatly superior. As Nozomi Osanai wrote in her master's thesis on a comparison between Noah's Flood and the Gilgamish Epic, "According to the specifics, scientific reliability, internal consistency, the correspondence to the secular records, and the existence of common elements among the flood traditions around the world, the Genesis account seems to be more acceptable as an accurate historical record."[1]

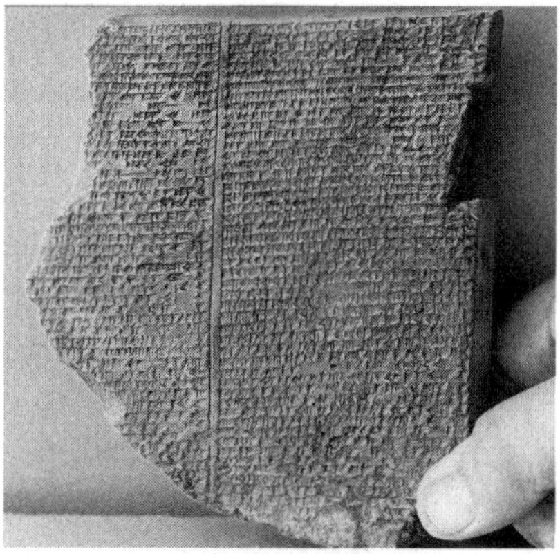

Part of the Gilgamesh Epic

[1] N. Osanai, A comparative study of the flood accounts in the Gilgamesh Epic and Genesis, www.answersingenesis.org/go/gilgamesh.

3. Long-living Kings at Kish (Sumer)—These kings supposedly lived from 10,000 to 64,000 years ago. The Bible's record is conservative and is the true account, while the Babylonian and other traditions have been embellished over time. It was later realized that the Babylonians had two bases for arithmetic calculations, based on either tens or sixties. When the records were retranslated using the system of tens rather than sixties, they came to a total within 200 years of the biblical record.

Major Evidences Regarding Genesis 11–36

This section contains Patriarchal records, with special reference to Abraham, the father of the Hebrews.

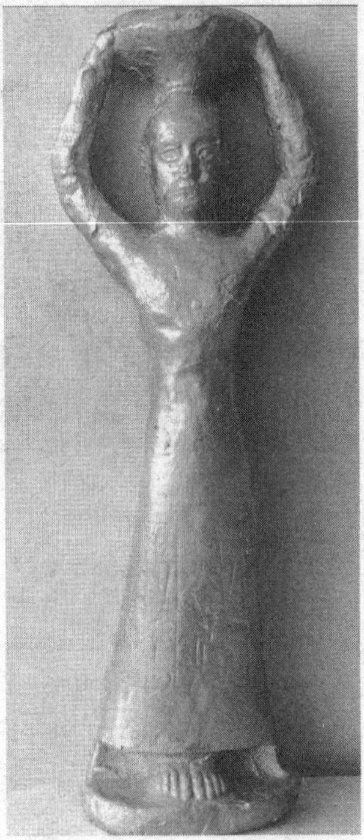

Ur Nammu, the King of Ur who claimed to build a famous tower

1. Abraham's home city of Ur was excavated by Sir Leonard Woolley, with surprising evidence of near-luxury.[2]

2. The customs of Patriarchal times, as described in the Bible, are endorsed by archaeological finds at such places as Ur, Mari, Boghazkoi, and Nineveh. These were written records from that day—not just put down in writing many centuries later. They bear the marks of eyewitness reporting.

Thus, Abraham's relationship with Hagar is seen in a different light by understanding that the woman who could not personally bear a child for her husband should provide him with one of her maidservants. In the Bible record we are told that it was Sarah who made the approach to Abraham, and her maid Hagar was a willing accomplice in having Abraham's child. Thus, she gained economic security and personal prestige. We stress it was not Abraham who made the first approach to Hagar, but Abraham's wife Sarah did in keeping with the customs of the day.

2 There was another Ur to the north, mentioned in the Ebla Tablets. The same name was often used for another city. Woolley's "Ur" in the south was Abraham's city.

The records of the five kings who fought against four kings (Genesis 14) are interesting, in that the names of the people concerned fit the known words and names of the times.

3. Abraham's negotiations with the Hittites (Genesis 23) are accurate and follow the known forms of such Hittite transactions. Neo-Hittites came later, but there were distinct language relationships. The Bible was right in calling the earlier people "sons of Hatti" or "Hittites."

Interestingly, the Hittite word for *retainers*, which means "servants trained in a man's own household" is *hanakim* (Genesis 14:14). This term is used only here in the Bible. Execration texts of the Egyptians (found on fragments of ceramic pots, which seem to have been used in ritual magic cursing of surrounding peoples) gives us the meaning of this term, and it is correctly used in the Bible record in Genesis 14.

Major Evidences Regarding Genesis 37–50

This section tells us the history of Joseph, the son of Jacob and great-grandson of Abraham. His brothers sell him to the Ishmaelites who sell him to an Egyptian eunuch. Joseph becomes successful in Egypt and helps to settle all of Israel there.

1. Known Egyptian titles such as "captain of the guard" (Genesis 39:1), "overseer" (39:4), "chief of the butlers" and "chief of the bakers" (40:2), "father to the Pharaoh" (actually "father to the gods," which to Joseph was blasphemous because he could not accept Pharaoh as a manifestation of Ra the sun god; Joseph Hebraized the title, so that he did not dishonor the Lord), "Lord of Pharaoh's House" (the palace), and "Ruler of all Egypt" (Genesis 45:8) attest to the historicity of this account.

2. Joseph's installation as vizier (chief minister) is very similar to other recorded ceremonies. His new name was Zaphnath-Paaneah, meaning "head of the sacred college" (Genesis 41:41–45). Other Egyptian phrases and other local color are also plentiful throughout the record (e.g., embalming and burial practices [Genesis 50]).

3. The Dead Sea Scrolls make the number of the people of Jacob 75, not 70, in Genesis 46:27, not 70, thus correcting a scribal error and showing that Stephen's figure was right (Acts 7:14).[3]

Major Evidences Regarding Exodus to Deuteronomy

The Eshnunna Law Code dating to c.1900 BC

These are the other four books of the Pentateuch, written by Moses, and probably at times in consultation with Aaron, the chief priest, and Joshua, the military leader.

1. The Law of Moses was written by a man raised in the courts of pharaoh, and it was greatly superior to other law codes, such as those of the Babylonian king Hammurabi, and the Eshnunna code that was found near modern Baghdad.
2. The covenant forms of the writings of Moses follow the same format as those of the Hittites, as endorsed by Professor George Mendenhall. The law code is a unity, dating to about 1500 BC (the time of Moses). These writings come from one source only, and there is no one to fit this requirement at this time except Moses. Ethical concepts of the Law were not too early for Moses, despite earlier hyper-criticism. (Ebla tablets from Syria pre-date Moses and, for example, include penalties against rape.)

At this point it is relevant to comment on two world-famous archaeologists with whom I had the privilege of working as an area supervisor with the American Schools of Oriental Research at the excavation of Gezer in Israel many years ago. Each of them (at two separate excavations) gave wonderful lectures to 140 American college students.

[3] This may not be correcting a scribal error since the 70 figure is referring to the number of Jacob's descendants previously listed in Genesis 46. Thus, it could be excluding Jacob and his two wives and two concubines, which give the number 75 of which Stephen spoke. See Eric Lyons, *Jacob's Journey to Egypt*, Apologetics Press website, http://www.apologeticspress.org/articles/619, 2003.

At the time of his lecture, Professor Nelson Glueck stated, "I have excavated for thirty years with a Bible in one hand and a trowel in the other, and in matters of historical perspective I have never found the Bible to be in error." Being a world-class Jewish scholar, Professor Glueck would have meant the Old Testament when he referred to the Bible, but it is also true that at least on one occasion, to my knowledge, he defended the accuracy of the New Testament writings as well.

The other lecture was given by Professor George Ernest Wright of Harvard University. He spoke on the validity of the writings of Moses, especially the covenant documents in the Pentateuch. He stated that the research of Professor George Mendenhall had led to the conclusion—with which he agreed—that the covenant documents of Moses were a unity and must be dated to approximately 1500 BC.

In further conversation after the lecture, Professor Wright told me that he had lectured for 30 years to graduate students—especially at Harvard—and he had told them that they could forget Moses in the Pentateuch. He now acknowledged that for thirty years he had been wrong, and that Moses really had been personally involved in the actual writing of the Pentateuch.

3. The ten plagues or judgments against the *leading gods* of Egypt (Exodus 12:12) are seen as real judgments, with a leading god of Egypt selected for judgment with each of the plagues.

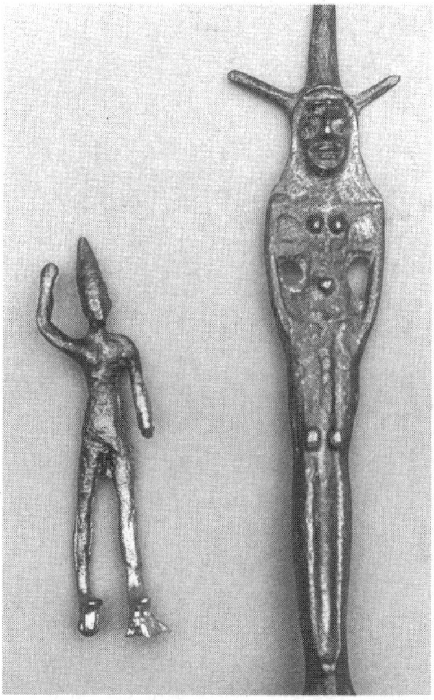

Canaanite deities, Baal and Asherah

Major Evidences Regarding Joshua to Saul

This section includes the conquest, the judges, and the early kingdom.

1. Deities such as Baal, Asherah, and Dagan are properly identified in association with the right people.

2. City-states are also identified (e.g., Hazor as "the head of those kingdoms" [Joshua 11:10]. The excavation of Hazor corroborated its great size).

3. Saul's head and armor were put into two temples at Beth-Shan. Both

Philistine and Canaanite temples were found. The Bible record was endorsed when such an endorsement seemed unlikely (1 Samuel 31:9–10 and 1 Chronicles 10:10).

Major Evidences Regarding David to Solomon

At this time the Kingdom of Israel is established.

1. David's elegy at Saul's death is an accurate reflection of the literary style of his times. Excavations at Ras Shamra (the ancient Ugarit in Syria) clarified various expressions, such as "upsurgings of the deep" instead of "fields of offerings" as in 2 Samuel 1:21.
2. Following the discovery of the Ugaritic library, it has become clear that the Psalms of David should be dated to his times and not to the Maccabean period, 800 years later, as critics claimed. The renowned scholar William Foxwell Albright wrote, "To suggest that the Psalms of David should be dated to the Maccabean period is absurd."[4]
3. Solomonic cities such as Hazor, Megiddo, and Gezer (1 Kings 9:15) have been excavated. Solomon even used similar blueprints for some duplicated buildings.

Major Evidences Regarding the Assyrian Period

This was the time of "The Reign of Terror," not long after Solomon's death.

The entrance to the Solomonic city of Gezer

[4] W.F. Albright, *History, Archaeology, and Christian Humanism*, McGraw-Hill, New York, 1964, 35.

1. Isaiah 20:1 was challenged by critics because they knew of no king named Sargon in lists of Assyrian kings. Now Sargon's palace has been recovered at Khorsabad, including a wall inscription and a library record endorsing the battle against the Philistine city of Ashdod (mentioned in Isaiah 20:1).

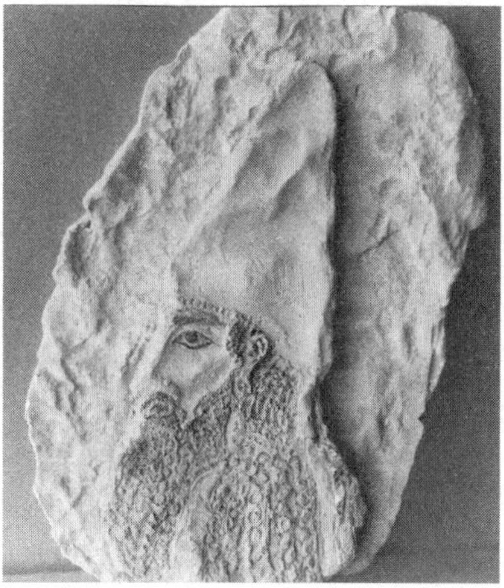
King Sargon of Assyria, mentioned at Isaiah 20:1

2. Assyrian titles such as *tartan* (commander-in-chief), and several others, are used casually yet confidently by Bible writers.

Other Assyrian titles such as *rabmag*, *rabshakeh,* and *tipsarru* were also used by Bible writers. As the Assyrians disappeared from history after the Battle of Carchemish in 605 BC, this retention of "obsolete" words is a strong pointer to the eyewitness nature of the records. Thus it points also to the genuineness of the prophecies because the same men who wrote the historical facts also wrote prophecies.

3. The death of Sennacherib is recorded at Isaiah 37:38 and 2 Kings 19:37 and is confirmed in the records of Sennacherib's son, Esarhaddon. It was later added to by Esarhaddon's son Ashurbani-pal.

Part of a pathway excavated by Dr. Clifford Wilson between Sennacherib's palace and the temple where his sons killed him

Various details about Nineveh and the account of Jonah point to the Bible's historicity. The symbol of Nineveh was a pregnant woman with a fish in her womb.

Adad-Nirari III, who might have been the king of Jonah's time, introduced remarkable reforms—possibly after the message of the prophet

Jonah. Adad-Nirari's palace was virtually alongside the later construction of what is known as "Nebi Yunis" ("the prophet Jonah"). That structure is the supposed site of the tomb of Jonah, and although that is unlikely, the honoring of Jonah is very interesting.

Major Evidences Regarding the Babylonians and Nebuchadnezzar

Nebuchadnezzar sacked Jerusalem and took Judah into captivity.

1. Daniel knew that Nebuchadnezzar was responsible for the splendor of Babylon (Daniel 4:30). This was unknown to modern historians until it was confirmed by the German professor Koldewey, who excavated Babylon approximately 100 years ago.

2. We now know from the Babylonian Chronicle that the date of Nebuchadnezzar's capture of Jerusalem was the night of March 15/16, 597 BC. We also know that Belshazzar really was the king of Babylon at this time because his father Nabonidus, who was undertaking archaeological research, was away from Babylon for about 10 years. He appointed his son Belshazzar as co-regent during that time.

3. Prophecies against Babylon (e.g., Jeremiah 51 and 52) have been literally fulfilled. Nebuchadnezzar wrote that the walls of Babylon would be a perpetual memorial to his name, but Jeremiah said, "The broad walls of

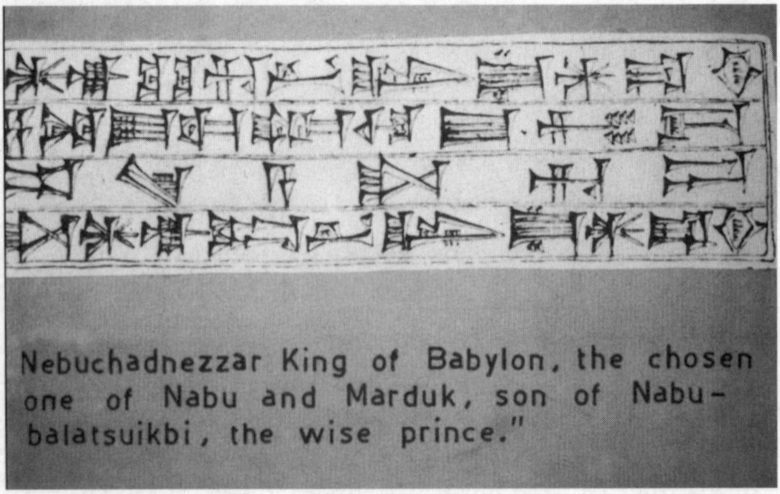
Nebuchadnezzar King of Babylon, the chosen one of Nabu and Marduk, son of Nabu-balatsuikbi, the wise prince."

Critics said 'There was no such king,' but his palace and library were uncovered

Babylon shall be utterly broken" (Jeremiah 51:58). Jeremiah, inspired by God, has been confirmed.

Major Evidences Regarding Cyrus and the Medes and Persians

The Medes and the Persians took over after the Babylonians.

1. Cyrus became king over the Medes and Persians. We read of Cyrus when his name was recorded prophetically in Isaiah 44:28 and 45:1. He issued the famous Cyrus Decree that allowed captive peoples to return to their own lands (2 Chronicles 36:22–23 and Ezra 1:1–4). The tomb of Cyrus has been found.

2. God was in control of His people's history—even using a Gentile king to bring His purposes to pass. The Cyrus Cylinder (a clay cylinder found in 1879 inscribed in Babylonian cuneiform with an account of Cyrus' conquest of Babylon in 539 BC) confirms that Cyrus had a conquest of Babylon.

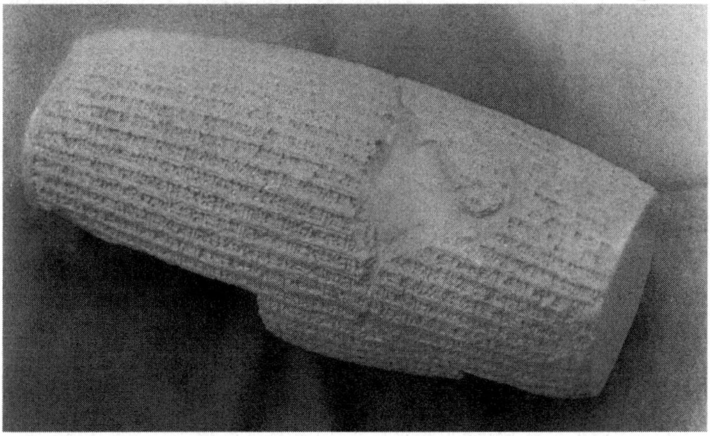

The Cyrus Cylinder–Isaiah referred to him prophetically

3. Some Jews remained in Babylon, as shown in the book of Esther. The type of "unchanging" laws of the Medes and Persians shown therein (Esther 1:19) is endorsed from Aramaic documents recovered from Egypt.

Major Evidences Regarding Ezra and Nehemiah

This was the time of the resettlement in the land after the exile in Babylon.

1. Elephantine papyri, the Dead Sea Scrolls, Targums of Job, etc., show that Aramaic was then in use, as Ezra indicates.

2. Sanballat was, as the Bible says, the Governor of Samaria (Nehemiah 4 and 6), though it was claimed by many writers that Sanballat was much later than Nehemiah. Several Sanballats are now known, and recovered letters even refer to Johanan (Nehemiah 12:13). Geshem the Arab (Nehemiah 6) is also known. Despite longstanding criticisms, Ezra and Nehemiah are accurate records of an actual historical situation.

3. The letters about Sanballat (above) clear up a dating point regarding Nehemiah. Nehemiah's time was with Artaxerxes I who ruled from 465 to 423 BC, not Artaxerxes II. This illustrates the preciseness with which Old Testament dating is very often established by modern research.

Part of the restored wall of Nehemiah

Major Evidences Regarding the Dead Sea Scrolls

The Dead Sea Scrolls

1. After approximately 2,000 years of being buried in caves near the Dead Sea, these scrolls came to light again in AD 1947. The Jews were searching for a Messiah or Messiahs—the king-like David, the great High Priest of the people of Israel, the High Priest after the order of Melchizedek, the prophet like Moses, and possibly the pierced Messiah.

I say "possibly the pierced Messiah" because this refers only to a very small fragment. Also, the future and the imperfect tenses in the Hebrew language are very often the same and can only be determined by the context.

In this case the prophecy could be saying that the expected Messiah will be "pierced" or that "he was pierced." Isaiah 11:4 states, "And with the breath of His lips He shall *slay* the wicked [emphasis added]." And in the NASB, Isaiah 53:5 says, "He was *pierced through* for our transgressions [emphasis added]." Both statements are relevant, for in fact the Messiah *was* pierced, and in a coming judgment those who have rejected the Messiah *will be* pierced.

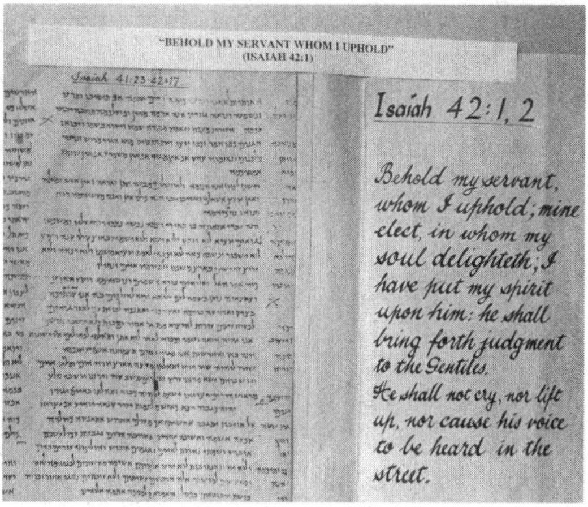
"BEHOLD MY SERVANT WHOM I UPHOLD"
(ISAIAH 42:1)
Isaiah 41:23-42:17
Isaiah 42:1, 2
Behold my servant, whom I uphold; mine elect, in whom my soul delighteth; I have put my spirit upon him: he shall bring forth judgment to the Gentiles.
He shall not cry, nor lift up, nor cause his voice to be heard in the street.
Part of the main Scroll of Isaiah recovered alongside the Dead Sea

2. The Scrolls have provided copies of most of the Old Testament, for fragments of every Old Testament book except Esther have been found in Hebrew, about 1,000 years earlier than previous extant Hebrew copies. (A writing from the book of Esther is found in another scroll.)

3. Considerable light was thrown on New Testament backgrounds and on the Jewish nature of John's Gospel. For example, contrasts such as "light and darkness" are common to John and the "War Scroll," a text that describes the eschatological last battle; and Hebrew was still a living language, not just a priestly language.

The Dead Sea Scroll of Isaiah also shows an old form of the Hebrew letter "tau," which looks like an "X" in the margin of the scroll. It occurs 11 times, at Isaiah 32:1, 42:1, 42:5, 42:19, 44:28, 49:5–7, 55:3–4, 56:1–2, 56:3, 58:13, and 66:5. As already stated, both the records of the Assyrians and the Dead Sea Scrolls (with a near-complete copy of Isaiah) were totally hidden from human eyes for about 2,000 years. Most of the content of these two sources overlapped and thus confirmed the evidence for the genuineness of the prophecies of Isaiah.

An important point about the finding of these scrolls is that they relate to the uncovering of the Assyrian palaces from the 1840s onwards. Isaiah gives a number of historical facts relating to the Assyrians that remarkably confirm the accuracy of Isaiah.

Possibly, the finding of the Dead Sea Scrolls is one of the most wonderful facts regarding the relevance of biblical archaeology and the Bible.[4]

Major Evidences Regarding the Person of Our Lord Jesus

Events surrounding the words and actions of Jesus have been authenticated by archaeological discoveries.

1. Problems about the census at the time of our Lord's birth have been resolved by the findings of important papyrus documents. These documents were found in Egypt inside sacred, embalmed crocodiles. The documents were the Jewish priestly writings that were written immediately before, during, and just after New Testament times.

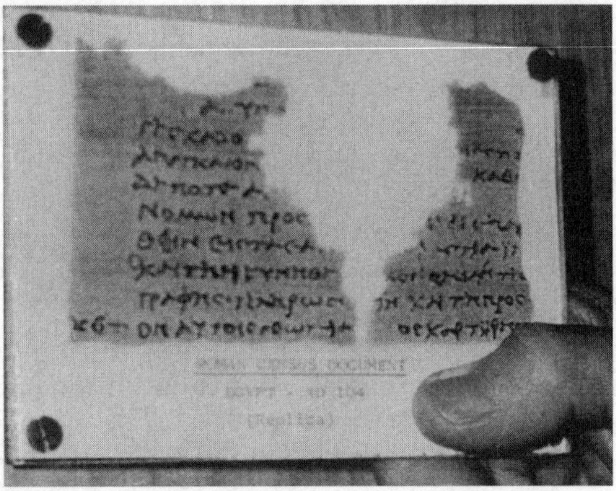

Part of an inscription about enrolling for the poll tax

The excavators Granfell and Hunt reported that their evidence showed that this was the first census (poll tax—enrollment) that took place in the time of Quirinius. (Another inscription has shown that Quirinius was in Syria twice—first as a military leader at a time of civil unrest, and later as Governor of Syria.) The census was probably delayed in Palestine because of that civil unrest.

2. Those papyrus findings have thrown much light on the words our Lord used. It is indeed true that He spoke the language of His time on earth (Mark 12:37).

3. Pilate is now better known because of a recovered inscription at Caesarea. The John Rylands papyrus (AD 125) records part of the trial before Pilate, fragments of which are recorded in John 18:31–33, 37–38.

[4] Many other points of interest from the Dead Sea Scrolls are outlined in the book *The Stones Still Shout* by Clifford Wilson.

Both sides of the Rylands Papyrus

Major Evidences Regarding the New Testament, the Early Church, and the Early Years of Christianity

The documents of the New Testament have been validated as accurate historical documents.

1. The papyrii from those Egyptian "talking crocodiles" have demonstrated that the New Testament documents are remarkable records of the times claimed for them in the language of "everyday" people. Those everyday expressions from Paul's time have also thrown much light on Paul's writings themselves.

2. The findings of Sir William Ramsay and his successors in Asia Minor re-established the veracity of Luke the historian and other New Testament writers.

The three Bible writings most attacked by critics were the Moses' Pentateuch, Ezra/Nehemiah, and Luke. Every one of these has been remarkably confirmed as being accurate and reliable by the research of credible scholars.

3. A flood of evidence shows the continuity between the New Testament documents (e.g., the Rylands Papyrus with parts of John 18:31–33 on one side and 37–38 on the other) and the abundant evidence from the secular Roman writers and the early church fathers.

Does Archaeology Prove the Bible?

Even when excavators are digging to uncover a past time period dealt with in the Bible, it is by no means sure that direct biblical history will be unearthed. Such findings are hoped for, not only by Bible students, but by disinterested archaeologists as well, because they know that they must take Bible records seriously. A link with Bible history is an excellent dating point, always desirable but not possible or achieved. These findings are excellent *confirmations* of God's Word, as opposed to "proving the Bible."

Archaeologists are scholars, usually academics with interest in the Bible as an occasional source book. A substantial number of scholarly archaeologists are committed Christians, but they are a minority. Many people believe that all archaeologists set out to verify biblical history, but that is not the case. Many excavators have virtually no interest in the Bible, but there are notable exceptions.

Superiority Despite Attacks by Critics

We have already said that we do not use the statement: "Archaeology proves the Bible." In fact, such a claim would be putting archaeology above the Bible. What happens when seemingly assured results of archaeology are shown to be wrong after all? Very often archaeology does endorse particular Bible events. And some would say that in this way it "proves the Bible." But such a statement should be taken with reservation because archaeology is the support, not the main foundation.

Thousands of facts in the Bible are not capable of verification because the evidence has long since been lost. However, it is remarkable that where confirmation is possible and has come to light, the Bible survives careful investigation in ways that are unique in all literature. Its superiority to attack, its capacity to withstand criticism, and its amazing facility to be proved right are all staggering by any standards of scholarship. Seemingly assured results "disproving" the Bible have a habit of backfiring.

Over and over again the Bible has been vindicated from Genesis to Revelation. The superiority of Genesis 1–11 has been established, and the patriarchal backgrounds have been endorsed. The writings of Moses *do* date to his time, and the record of the conquest of Canaan under Joshua has many indications of eyewitness recording.

David's Psalms were clearly products of his time, and records about Solomon should no longer be written off as "legendary." Solomon was a literary giant, a commercial magnate, and a powerful ruler—under God. God alone gave Israel their "golden age."

The Assyrian period has given dramatic confirmation to biblical records, with excavations of palace after palace over the last 150 years. Such excavations constantly add to our understanding of the background to Old Testament kings, prophets, peoples, and incidents.

The exile in Babylon is endorsed at various points, and the Cyrus Decree makes it clear that captured people could return to their own lands and worship according to their own beliefs. Ezra and Nehemiah are accurate reflections of that post-exilic period.

Likewise, the New Testament documents have been consistently demonstrated as factual, eyewitness records. Kings, rulers, and officials are named unerringly; titles are used casually but with remarkable accuracy; geographic boundaries are highlighted; and customs are correctly touched on.

It is indeed true that "truth shall spring out of the earth" (Psalm 85:11).

Archaeology as It Relates to the Biblical Record

Our understanding of essential biblical doctrine has never changed because of archaeological findings. It should be acknowledged, however, that at times it has been necessary to look again to see just what the Bible is actually saying. There have been times when new light has been thrown on words used in Scripture in both Old and New Testaments.

We have seen that the titles of officials of Israel's neighbors are now better understood and that many words are better understood because of the records in clay, on papyrus, and on stone.

The Old Testament is an ancient book, not a modern record, and its style is that of the East and not the West. At times it must be interpreted, based on its context, in the symbolic and figurative style of the Jews of ancient times, and not according to the "scientific precision" of our modern materialistic age.

Sometimes the Bible uses "the language of phenomena"—as when it refers to the sun rising. Scientifically speaking, the earth is what "rises." However, though the Bible is not a science textbook, it is yet wonderfully true that where the Bible touches on science it is astonishingly accurate.

The more this new science of archaeology touches the records of the Bible, the more we are convinced that it is a unique record. At many points it is greatly superior to other writings left by neighboring people.

We have not said, "Archaeology proves the Bible," and we do not suggest it. To do so would be quite wrong, even though such a statement is often made by those introducing a lecturer on biblical archaeology. The Bible itself is the absolute; archaeology is not. If archaeology could prove the Bible, archaeology would be greater than the Bible, but it is not. The Bible comes with the authority of almighty God. It is His Word, and He is greater than all else.

Nevertheless, archaeology has done a great deal to restore confidence in the Bible as the revealed Word of God. It has thrown a great deal of light on previously obscure passages and has helped us to understand customs, culture, and background in many ways that seemed most unlikely to our fathers in a previous generation. Archaeology is highly relevant for understanding the Bible today.

The Value of Archaeology for the Bible Student

Archaeology has done a great deal to cause many scholars to take the Bible much more seriously. It has touched the history and culture of Israel and her neighbors at many points and has often surprised researchers by the implicit accuracy of its statements.

If it can be shown (as it can) that the Bible writers lived and gave their message against the backgrounds claimed for them, it becomes clear that their amazing prophetic messages are also genuine, written long before the events they prophesied. Consider five important ways in which archaeology has been of great value for Bible students.

1. Archaeology confirms Bible history, and it often shows that Bible people and incidents are correctly referred to.

One example is that of Sargon, a king named in Isaiah 20:1. Critics at one time said that there was no such king. But then his palace was found at Khorsabad, and there was a description of the very battle referred to by Isaiah. Another illustration is the death of the Assyrian King Sennacherib. His death is recorded in Isaiah 37 and also in the annals of Sennacherib's son Esarhaddon, whom Isaiah says succeeded Sennacherib.

2. Archaeology gives local color, indicating that the background is authentic.

Laws and customs, gods, and religious practices are shown to be associated with times and places mentioned in the Bible. Rachel's stealing her father's clay gods illustrates the correct understanding of customs: she and Leah asked, "Is there yet any portion or inheritance for us in our father's house?" (Genesis 31:14). She knew the teraphim (clay gods) were associated with title deeds, which was a custom of that time.

3. Archaeology provides additional facts.

Archaeological facts help the Bible student understand times and circumstances better than would otherwise be possible. Bible writers tell us the names of such Assyrian kings as Sennacherib and Esarhaddon, and we now know a great deal more about these rulers from records recovered in their palaces and libraries.

4. Archaeology has proved of tremendous value in Bible translations.

The meanings of words and phrases are often illuminated when found in other contexts. 2 Kings 18:17, for example, correctly uses three Assyrian army titles. Those terms are *tartan* (commander-in-chief), *rabshakeh* (chief of the princes), and *rabsaris* (chief eunuch). The meanings of these words were unknown at the time of the production of the King James Version of the Bible in 1611.

Only when Assyrian palaces were excavated was a great deal of light thrown onto their meanings. The fact that these titles are correctly used in the Old Testament is another strong argument for eyewitness recording. People do not know the titles of their enemy without some form of contact.

5. Archaeology has demonstrated the accuracy of many Bible prophecies.

The prophecies against Nineveh, Babylon, and Tyre in Isaiah are typical examples, as are the early records of creation in the Bible. It is also highly important that Isaiah and others so accurately pointed to the coming Messiah. At many points their history has been vindicated, and so have their prophecies about Jesus.

This spiritual application is surely one of the most important aspects of biblical archaeology, reminding us that "holy men of God spoke as they were moved by the Holy Spirit" (2 Peter 1:21).

Archaeology has done much to demonstrate that "the Bible was right after all." Its early records of creation, Eden, the Flood, long-living men, and the dispersal of the nations are not mere legends after all. Other tablets recording the same events have been recovered, but they are often distorted and corrupted.

The Bible record is immensely superior, and quite credible. Those early Bible records can no longer be written off as myth or legend.

"For ever, O LORD, Your word is settled in heaven" (Psalm 119:89).

A Memory Aid Showing the Relevance of Archaeology to the Bible

S **Superiority**—Creation, Flood, Tower of Babel, Laws of Moses, Psalms of David, genuine prophets of Israel, the teachings of Jesus.

C **Customs**—Rachel stealing clay gods; Joseph's story; religious practices; ruthlessness of Assyrians; unchangeable laws of Medes and Persians; enrolling for census when Jesus was born.

A **Additional information**—Moabite Stone; Jehu and the Black Obelisk of Shalmaneser; the assassination of Assyrian King Sennacherib; Belshazzar as co-regent with his father Nabonidus; new light on New Testament backgrounds from the Dead Sea Scrolls and other manuscripts and inscriptions.

L **Language and Languages**—Hebrew, Aramaic, and Greek. Others are touched in passing, including Egyptian, Canaanite, Philistine, Babylonian, Persian, Latin, and Assyrian.

P **Prophecy**—about Bible lands and people, as well as the Lord Jesus Christ. The local color and the integrity of prophecies demonstrate the uniqueness of the Bible.

S **Specific Incidents and People**— Sargon's victory against Ashdod (Isaiah 20:1); the death of Sennacherib (Isaiah 37); Nebuchadnezzar the King of Babylon who campaigned against Jerusalem and Judah; various rulers (such as the Herods) correctly identified (the Gospels and Acts); the census in the time of Caesar Augustus.

Many people have commented that they do not have the knowledge to talk about archaeology and the Bible; this acrostic SCALPS should help.[6]

First Peter 3:15 urges us to "always be ready to give a defense to everyone who asks you a reason for the hope that is in you, with meekness and fear."

That's a command to Christians!

[6] This acrostic may be photocopied and enlarged.

26

Why Does God's Creation Include Death & Suffering?

TOMMY MITCHELL

Why do bad things happen? Through the ages, human beings have sought to reconcile their understanding of an all-powerful, loving God with the seemingly endless suffering around them.

One prominent example of this struggle is the media mogul Ted Turner. Having lost his faith after his sister died of a painful disease, Turner claimed, "I was taught that God was love and God was powerful, and I couldn't understand how someone so innocent should be made or allowed to suffer so."[1]

Is God responsible for human suffering? Is God cruel, capricious, and vindictive, or is He too weak to prevent suffering? If God truly is sovereign, how can He let someone He loves suffer?

A World of Misery and Death

Each day brings new tragedy. A small child is diagnosed with leukemia and undergoes extensive medical treatment only to die in his mother's arms. A newlywed couple is killed by a drunk driver as they leave for their honeymoon. A faithful missionary family is attacked

[1] Associated Press, Ted Turner was suicidal after breakup, www.nytimes.com/aponline/arts/AP-People-Turner.html, April 16, 2001.

and killed by the very people they were ministering to. Thousands are killed in a terrorist attack. Hundreds drown in a tsunami, while scores of others are buried in an earthquake.

How are these things possible if God really loves and cares for us? Is He a God of suffering?

Man's usual response to tragedy is to blame God, as Charles Darwin did after the death of his beloved daughter Annie.

"Annie's cruel death destroyed Charles's tatters of beliefs in a moral, just universe. Later he would say that this period chimed the final death-knell for his Christianity … . Charles now took his stand as an unbeliever."[2]

Is this the proper response? A correct view of history, found in the Bible, provides the answer.

Was God's Creation Really "Very Good"?

In the beginning, about 6,000 years ago, God created the universe and everything in it in six actual days. At the end of His creative acts on the sixth day, God "saw everything that He had made, and indeed it was very good" (Genesis 1:31).

2 A. Desmond and J. Moore, *Darwin: The Life of a Tormented Evolutionist*, W.W. Norton & Company, New York, 1991, 387.

To have been very good, God's creation must have been without blemish, defect, disease, suffering, or death. There was no "survival of the fittest." Animals did not prey on each other, and the first two humans, Adam and Eve, did not kill animals for food. The original creation was a beautiful place, full of life and joy in the presence of the Creator.

Both humans and animals were vegetarians at the time of creation. In Genesis 1:29–30 the Lord said, "See, I have given you every herb that yields seed which is on the face of all the earth, and every tree whose fruit yields seed; to you it shall be for food. Also, to every beast of the earth, to every bird of the air, and to everything that creeps on the earth, in which there is life, I have given every green herb for food."

This passage shows clearly that in God's very good creation, animals did not eat each other (and thus, there was no animal death), as God gave Adam, Eve, and the animals only plants to eat. (It was not until after the worldwide Flood of Noah's Day—1,600 years later—that man was allowed to eat meat, according to Genesis 9:3.)

Because eating a plant can kill it, some people claim that death was part of the original creation. The Bible makes a distinction, though, between plants and animals. This distinction is expressed in the Hebrew word *nephesh*, which describes an aspect of life attributed only to animals and humans. *Nephesh* can be translated "breathing creature" or

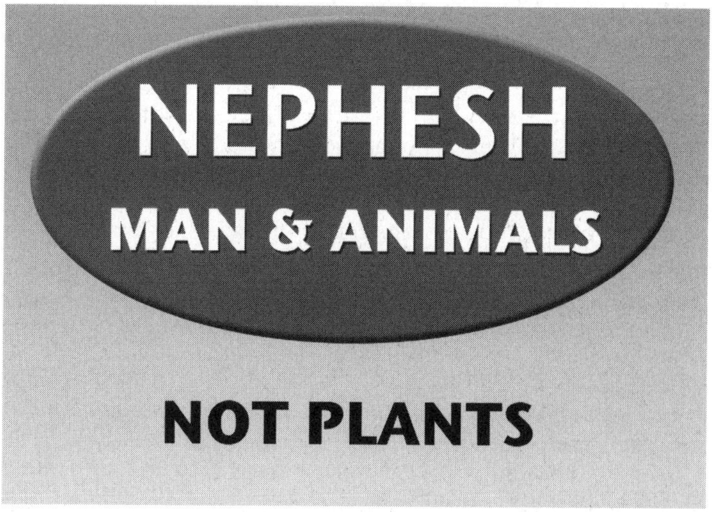

"living creature" (see Genesis 1:20–21, 24). Plants do not possess this *nephesh* quality and so cannot die in the scriptural sense.

The original creation was very good. According to Moses in Deuteronomy 32:4, "His work is perfect." Obviously, things are not like this any longer.

Why Do We Die Now?

If there was no animal or human death when God finished His creation and pronounced it very good, why do we die now? We see death all around us today. Something must have happened to change creation—that something was sin.

God placed Adam and Eve in a perfect paradise. As their Creator, He had authority over them. In His authority, God gave Adam a rule: "But of the tree of the knowledge of good and evil you shall not eat, for in the day that you eat of it you shall surely die" (Genesis 2:17).

Sometime after God declared His completed creation "very good" at the end of the sixth day, one of God's angels, Lucifer, led a rebellion against their Creator.[3] Lucifer then took on the form of a serpent and tempted Eve to eat the fruit God had forbidden. Both Adam and Eve ate it. Their actions resulted in the punishment that God had warned them about. God is holy and cannot tolerate sin in His presence. The just Creator righteously kept His promise that punishment would follow their disobedience. With the rebellious actions of one man, death entered God's creation.

[3] The Bible is not clear when Lucifer rebelled or when Adam and Eve sinned. However, we can surmise that it was not too long after God put Adam and Eve in the Garden of Eden, as He told them to be fruitful and multiply, and they obviously had not had an opportunity to conceive a child before they rebelled.

Ashamed and afraid, Adam and Eve tried to escape the consequences of their sin by making coverings of fig leaves. But by themselves, they could not cover what they had done. They needed something else to provide a covering. According to the writer of Hebrews, "Without shedding of blood, there is no remission [of sin]" (9:22). A blood sacrifice was necessary to cover their guilt before God.

To illustrate the horrible consequences of sin, God killed an animal and made coats of skin (depicted at left) to cover Adam and Eve. We are not told what type of animal was killed, but perhaps it was something like a lamb to symbolize Jesus Christ, the Lamb of God, who would shed His own blood to take away our sins.

Genesis 3 also reveals that the ground was cursed. Thorns and thistles were now part of the world. Animals were cursed, the serpent more than the rest. The world was no longer perfect but sin-cursed. Suffering and death now abounded in that once-perfect creation.

What Does All This Have to Do with Me?

If it was Adam's decision to disobey God that brought sin into the world, why do we all have to suffer punishment?

After Adam and Eve sinned and were banished from the Garden of Eden (Genesis 3:20–24), they began to have children. Each child inherited Adam's sinful nature, and each child rebelled against his or her Creator. Every human is a descendant of Adam and Eve, born with the same problem: a sinful nature.

If we are honest with ourselves, we will realize that Adam is a fair representative for all of us. If a perfect person in a perfect place decided to disobey God's rules, none of us would have done better. The Apostle Paul writes, "Therefore, just as through one man sin entered the world,

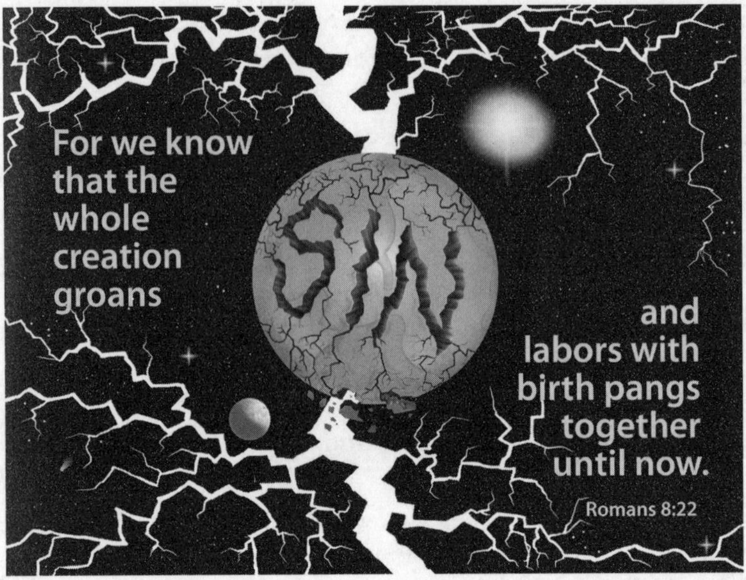
For we know that the whole creation groans
SIN
and labors with birth pangs together until now.
Romans 8:22

and death through sin, and thus death spread to all men, because all sinned" (Romans 5:12).

As children of Adam, we all inherit Adam's sin nature. We have all, at some point, disobeyed a command from the Creator, so we all deserve to die and suffer eternal punishment in hell. We must understand that not one of us is innocent before God. Romans 3:23 says, "For all have sinned and fall short of the glory of God." Not one of us is worthy to stand before the Creator of the universe because we would each bring a sinful, rebellious nature into His presence.

In the beginning, God sustained His creation in its perfect state. The account of the Israelites wandering in the wilderness provides a glimpse of how things might have been in the original creation. The garments of the Israelites did not wear out, nor did their feet swell for the forty years they camped in the desert (Deuteronomy 8:4). God is omnipotent and perfectly capable of sustaining and protecting His creation.

When Adam sinned, however, the Lord cursed the universe. In essence there was a change, and along with that change God began to uphold the creation in a cursed state. Suffering and death entered into His creation. The whole universe now suffers from the effects of sin (Romans 8:22).

The sad things (e.g., the death of a loved one, tsunamis that kill thousands, hurricanes that leave many dead or homeless, etc.) that happen around us and to us are reminders that sin has consequences and that the world needs a Savior.

God took pleasure in all of His creation (Revelation 4:11), but He loved people most of all. He uses the deterioration of the created universe to show us the consequences of our sin. If we did not experience the consequences of our rebellion against the Creator, we would never understand that we need salvation from our sin, and we would never receive His offer of mercy for our sin.

Most people easily recognize that there is a problem in the world. We need to realize that there is One who has overcome this problem of death and suffering—Jesus Christ.

Is There Any Hope?

Sadly, the consequences for our sin are much worse than life in a cursed universe. In addition to living our lives in a sin-filled creation, we must all die physically and then face a punishment much more horrible than anything we have ever known: the second death. The Apostle John tells of a lake of fire called the "second death" that awaits all those whose names are not written in the book of life (Revelation 20:14–15). This second death is the final punishment for our sin.

Even though we rebelled against Him and brought punishment on ourselves, God loves His children and does not want them to spend eternity in hell. Our merciful Creator has provided a way to be reconciled to Him and to escape the terrible eternal punishment for our sin. This way of escape is through the death and resurrection of Jesus Christ.

Jesus Christ, who is God, came to earth as a man, lived a sinless life, and then died to pay the penalty for sin. The Apostle Paul tells us that "as through one man's offense judgment came to all men, resulting in condemnation, even so through one Man's righteous act the free gift came to all men, resulting in justification of life" (Romans 5:18).

God is righteous and justly sentenced man to death, so we received the punishment we deserve. However, God exercised grace because of His love for us and took that punishment upon Himself as the payment for our sin.

Take heart! Christ did not remain in the grave. He showed that He has power over death by rising on the third day after He was buried. Because Christ clearly demonstrated His power over death, those who believe in Him can know that they too will live, and death will have no sting. In fact, the Bible says,

> So when this corruptible has put on incorruption, and this mortal has put on immortality, then shall be brought to pass the saying that is written: "Death is swallowed up in victory. O Death, where is your sting? O Hades, where is your victory?" (1 Corinthians 15:54–55).

In Christ, those who have received the free gift of eternal life can look forward to spending eternity with Him in a perfect, pain-free place (Revelation 21:4). As the Apostle Paul wrote,

> For by grace you have been saved through faith, and that not of yourselves; it is the gift of God, not of works, lest anyone should boast (Ephesians 2:8–9).

Some may suggest that if God really loved us, He would put us in a perfect place where nothing painful can touch us. However, He already did that once, and Adam rebelled. Given the same opportunity, each one of us would do the same thing. God demonstrated His love by dying for the world and rising again. All who receive the free gift of eternal life will spend eternity with Him.

Compared to eternity, the time we spend here in a cursed world is insignificant. God will complete His demonstration of love by placing those who receive His salvation in a perfect place forever.

The Restoration of All Things

The Bible describes death as the last enemy that will be destroyed (1 Corinthians 15:26). Revelation 21:4 says that "God will wipe away every tear from their eyes; there shall be no more death, nor sorrow, nor crying. There shall be no more pain, for the former things have passed away." Those who have received salvation look forward to the time when the Lord will revoke the Curse and restore the universe to a perfect state like the one it had before man sinned (Revelation 22:3).

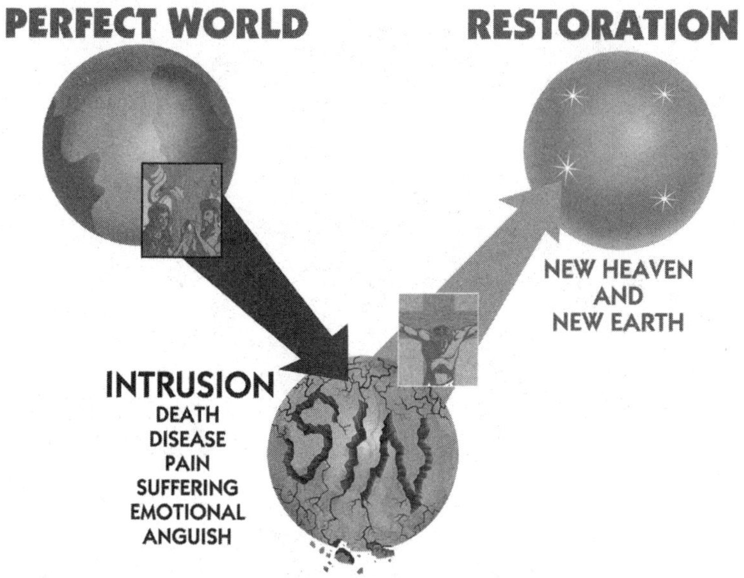

The Lord not only loves His children enough to die for their sin, He also promises to fix the ruined world by creating a new heavens and new earth (Revelation 21:1). And just as the first Adam brought death into the world, Christ, as the "last Adam," brings renewed life into the world.

As Paul wrote,

> And so it is written, "The first man Adam became a living being." The last Adam became a life-giving spirit (1 Corinthians 15:45).

The Alternate View of History

Those who reject the Creator must explain how the world came into existence without God.

Evolutionists and most other "long agers" believe that 13–14 billion years ago, a big bang caused the universe to begin from nothing. Galaxies, stars, and planets formed as matter—scattered across the universe—cooled and coalesced. About five billion years ago, the earth itself began to form. The earth, it is claimed, cooled for a billion years or so, water formed on the surface, and in this primordial ocean, molecules somehow arranged themselves together to form the simplest one-celled life forms.

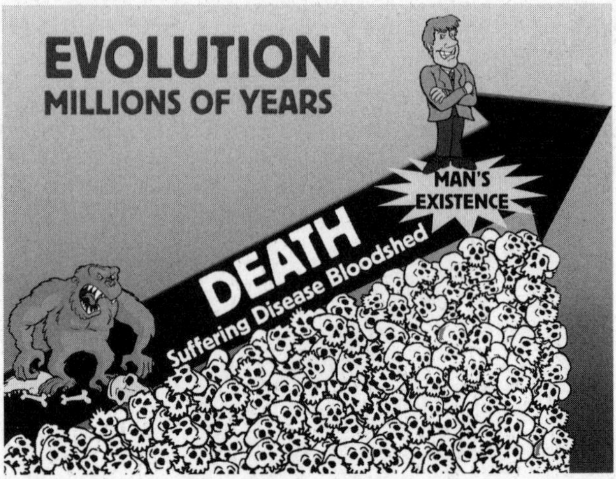

Due to environmental stresses and other forces, directionless mutations, say evolutionists, led to survival advantages for certain organisms. These organisms gradually changed into progressively more complex organisms. The strongest organisms were able to survive and reproduce, and the weaker organisms died off or were killed by the stronger creatures.

This merciless process eventually produced ape-like creatures who evolved into man himself. Thus humans are the ultimate product (so far!) of millions of years of death and suffering.

This naturalistic view of the universe uses the fossil record as proof for the belief that creatures became more advanced over millions of years. This view teaches that the fossil record is a record of millions of years of disease, struggle, and death. The late famous evolutionist Carl Sagan declared that "the secrets of evolution are time and death."[4]

Evolution requires millions of years of struggle and death.

Does This Really Matter?

The Bible says that death came as the result of man's sin. Evolution says that death has always been a part of nature. Can both be true? Obviously not.

[4] C. Sagan, *Cosmos Part 2: One Voice in the Cosmic Fugue,* produced by Public Broadcasting Service, Los Angeles, with affiliate station KCET-TV. First aired in 1980 on PBS stations throughout the US.

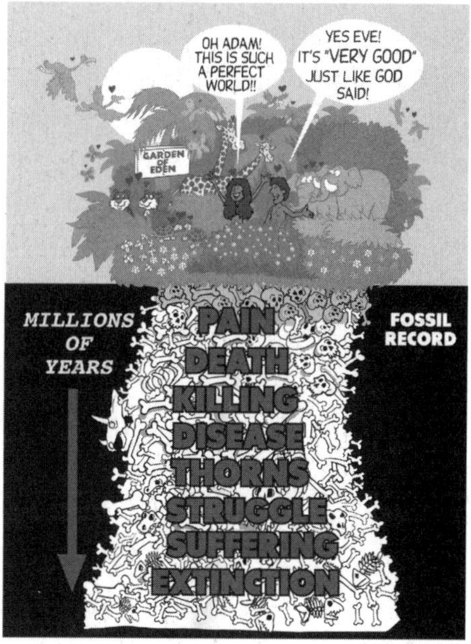

If the fossil record represents millions of years of earth history, there must have been millions of years of death, struggle, and disease before man appeared, contrary to what Genesis teaches.

"Theistic evolution" is an idea that attempts to merge the Genesis account and the concept of millions of years of evolution. Theistic evolution postulates millions of years of death before God stepped into the process, at some point, and created the Garden of Eden. As illustrated below, theistic evolution requires God to call millions of years of death and suffering "very good."

On the other hand, if the fossil record is the product of a catastrophic global Flood in which vast numbers of organisms were suddenly buried

The incorrect view:

The correct view:

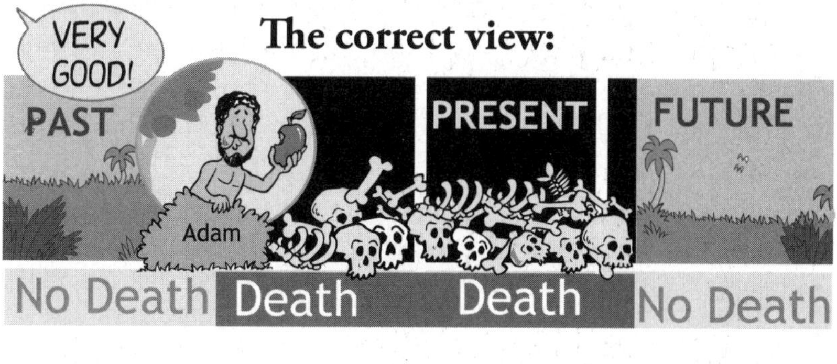

in chemical-rich water and sediment, the need to postulate millions of years of history goes away. God's account of a perfect world ruined by sin and destroyed by a watery judgment (Genesis 6–9) is consistent with the fossil evidence in the world.

God's promise of future restoration, "the restitution of all things" (Acts 3:21), would be nonsensical if evolution really happened. Only an original creation free from death makes God's promise of restoration logical. A perfect creation cannot be the promised future restoration if no perfect creation existed in the past.

Where Do Caring and Mercy Come From?

While many evolutionists cry out that a loving God is inconsistent with this world of cruelty we inhabit, they conveniently overlook other things. For example, how does evolution explain mercy, charity, and caring? If evolution is true, the driving force of nature is "survival of the fittest." Those less able to compete are destined to die. Any attempt to rescue these "less competitive" people would be to work against the most fundamental force of nature. The existence of doctors, hospitals, charitable organizations, and even a police force is contrary to raw evolutionary forces.

The evolutionist has no basis for moral judgments. If man is just the result of millions of years of evolution, our behavior is based on random chemical reactions. There is no ultimate moral code. All morality is relative. So if a person needs money, why is it wrong to rob someone? According to evolution, the stronger person should succeed. Might makes right. So in the evolutionary view, such violence is a natural, and necessary, part of the world.

Those who have a worldview based on the Bible have a consistent basis for acts of kindness, charity, or caring. We are commanded in Scripture to love our neighbors as ourselves, to perform acts of mercy, and to care for the widows and orphans. If we take evolution to its logical conclusion, we will conclude that these widows and orphans should die because they are a drain on the resources of nature.

Only Bible-believers ultimately offer the world a basis to make moral judgments. Those who reject the Bible have no basis for morality.

What about Individual Suffering?

In John 9 Jesus addressed the issue of personal suffering. When His disciples assumed that a man's blindness was the result of the man's sin, Jesus answered, "Neither this man nor his parents sinned, but that the works of God should be revealed in him" (John 9:3). Jesus did not consider the man's suffering to be wasted or capricious, because God would be glorified in the man's life.

The book of Job tells the history of a righteous man who pleased God but nevertheless suffered the loss of his wealth, his ten children, and his health. His friends were sure his sufferings represented judgment for some secret sins, but God denied this accusation. Many people have taken comfort simply in knowing that their personal tragedies did not necessarily represent personal judgments.

Jesus demonstrated that His love for us is not incompatible with personal suffering when Lazarus was sick and about to die. "When Jesus heard that, He said, 'This sickness is not unto death, but for the glory of God, that the Son of God may be glorified through it.' Now Jesus loved Martha and her sister and Lazarus" (John 11:4–5).

Jesus clearly loved Lazarus and his grieving family, but He was able to see a purpose to suffering that they could not see. Christ clearly revealed to them that He had power over death (by raising Lazarus from the dead), even prior to His crucifixion and resurrection.

Jesus commented on the purpose of tragedy after the tower of Siloam collapsed, killing eighteen people. "Or those eighteen on whom the tower in Siloam fell and killed them, do you think that they were worse sinners than all other men who dwelt in Jerusalem? I tell you, no; but unless you repent you will all likewise perish" (Luke 13:4–5).

These examples show that it is not necessarily an individual's sin that leads to suffering, but sin in general already has. God may use suffering as a reminder that sin has consequences—and perhaps for other purposes we do not fully investigate in this chapter. But the presence of suffering does not mean God does not love us. Quite the opposite—Christ came and suffered with us and took that punishment when He didn't have to.

In times of suffering, Christians honor the Lord by trusting Him and knowing that He loves them and has a purpose for their lives. The presence of suffering in the world should remind us all that we are sinners in a sin-cursed world and also prompt us to tell others about the salvation available in Christ—after all, that would be the loving thing to do. We can tell people the truth of how they, too, can be saved from this sin-cursed world and live eternally with a perfect and good God.

> For our light affliction, which is but for a moment, is working for us a far more exceeding and eternal weight of glory, while we do not look at the things which are seen, but at the things which are not seen. For the things which are seen are temporary, but the things which are not seen are eternal (2 Corinthians 4:17–18).

27

How Can I Use This Information to Witness?

KEN HAM

In 1959, I turned eight years old. It was a historic year for my homeland of Australia because a famous American evangelist conducted a series of crusades in the large cities of Melbourne and Sydney.

Some commentators claimed this was the closest Australia ever came to revival.[1]

In the years following, Australia has not seen such an influential crusade. Later crusades did not seem to match the apparent results of 1959.

Today, when such crusades are conducted, whether in Australia, America or other countries, statistics indicate that the small percentage of people who do go forward for first-time commitments seem to fall away or are not incorporated into any church.[2]

Why was it that, even though the entire Australian society "buzzed" as a result of these 1959 crusades, there seemed to be no lasting major impact on the culture itself? And why has Australia's culture (and other Western cultures) been continuously declining in regard to Christian morality, despite numerous evangelistic campaigns?

---

1 S. Piggin, *Evangelical Christianity in Australia: Spirit, Word and World*, Oxford University Press, Melbourne, 1996, 154–171.

2 R. McCune, *Promise Unfulfilled: The Failed Strategy of Modern Evangelicalism*, Ambassador International, Greenville, South Carolina, 2004, 80–82.

It really comes down to understanding the difference between "Jews" and "Greeks" (using the terms as types).

"Crusades" Conducted by Paul and Peter

In 1 Corinthians 1:23 we read the words of the Apostle Paul, "But we preach Christ crucified, to the *Jews a stumbling block*, and to the *Greeks foolishness*" (emphasis added).

In Acts 2, the Apostle Peter preached a bold message that was primarily directed to Jews (or those familiar with the Jewish religion). The main

thrust of his message concerned the death and resurrection of Christ and the need for salvation.

The Scripture records that 3,000 people responded positively to Peter's message. This was a phenomenally successful "crusade."

Now in Acts 17, when Paul preached a similar message concerning the resurrection of Christ to the Greek philosophers, their response indicated that they thought the message was really foolishness.

Why the Difference in Response?

In Acts 2, Peter was preaching to people (Jews) who, at that time, believed in the God of creation as recorded in the Old Testament. They understood the meaning of sin because they knew about the Fall of the first human couple in Genesis 3. They also had the Law of Moses, so they knew exactly what God expected of them and how they fell short. They were not indoctrinated in the evolutionary ideas that the Greeks had developed. (More about that in a moment.) The Word of God had credibility in their eyes and was considered sacred.

The Jews also understood the need for a sacrifice for sin because, after all, according to Acts 2, they were there on that particular day (the day of Pentecost) to sacrifice animals, as they had always done. However, most of the Jews had rejected Jesus as the Messiah, so Peter challenged them concerning who Jesus was and what He had done on the Cross.

Here, then, is an important observation to note: the Jews had the foundational knowledge of creation and sin to understand the message of salvation. Peter didn't have to convince his audience that God was Creator or that man had sinned. He could concentrate on the message of the Cross.

Peter, you see, didn't have to establish the credibility of God's Word or convince the Jews about creation (as opposed to naturalistic explanations of origins or deal with teaching about supposed millions of years—these were not really issues in the Jewish culture at that time).

Evolution in Ancient Times

Now in Acts 17, Paul was preaching to Greek philosophers. In their culture, they did not have any understanding of the God of creation as the Jews understood. They believed in many gods, and that the gods, like humans, had evolved. The Epicureans, for instance, believed man evolved from the dirt (in fact, they were the atheists of the age).

The Greeks had no understanding of sin or what was necessary to atone for sin. God's Word to the Jews had no credibility in this evolution-based culture. Thus when Paul preached the same basic message Peter gave in Acts 2, the Greeks did not understand—it was "foolishness" to them.

As you read on in Acts 17, it's fascinating to see what Paul tried to do in reaching the Greeks with the gospel. He talked to them about the "unknown God" (referred to on one of the Greek altars) and proceeded to define the true God of creation to them.

Paul also explained that all people were of "one blood" (from one man, Adam), thus laying the foundational history necessary to

understand the meaning of the first man Adam's sin and the need for salvation for all of us as Adam's descendants.[3] He countered their evolutionary beliefs, thus challenging their entire way of thinking in a very foundational way.

Having done this, Paul then again preached the message of Christ and the Resurrection. Although some continued to sneer, others were interested to hear more (their hearts were opened) and some were converted to Christ.

Even though Paul didn't see 3,000 people saved as Peter did, Paul was nonetheless very successful (from a human perspective, knowing it is God who opens people's hearts to the truth, as 1 Corinthians 2:14 teaches).

Think about what he had to do: Paul had to first change "Greeks" into "Jews."

In other words, he had to take pagan, evolutionist Greeks and change their whole way of thinking about life and the universe, and then get them to think like Jews concerning the true foundation of history recorded in Genesis.

No wonder only a few were converted at first. Such a change is a dramatic one. Imagine, for example, trying to change an Aborigine from my homeland into an American in regard to his whole way of thinking? Such a change would be extremely difficult, to say the least.

The Culture Change

Now let's go back to 1959. At that time in Australia's history, it was common for public school students to have prayer (even reciting the Lord's Prayer) at an assembly before the start of the day. In elementary schools, it was also not uncommon for students to

[3] To understand what the Bible teaches about the origin of the so-called races around the world, see chapter 17: Are There Really Different "Races"?

be read a section of the Bible or a Bible "story" before they started the day. On the weekends, many children went to Sunday schools. Then through the week, ministers of religion even visited schools and taught students about the Bible.

I suggest to you that generations ago, even in Australia (which did not have a strong Christian heritage that America had), the culture was somewhat like the "Jews." Most people knew the basic concepts of the Christian religion concerning creation, sin, and the message of salvation. So when an evangelist came and preached the message of the Cross, it was sort of like Peter preaching to the Jews in Acts 2. They had the foundational knowledge to understand the message and responded accordingly.

Now, even though much of the Australian society was like the "Jews" in the sense that many people could understand biblical terms—because of the relative familiarity with the Bible "stories"—I believe these "Jews" were already becoming "Greeks" in regard to their thinking about reality. Fifty years ago, Australia's public schools were teaching millions of years and evolutionary ideas. In a low-key way, these schools were subtly undermining the credibility of the Bible's history.

This is one of the reasons that I believe there seems to have been no real lasting Christian impact on the Australian culture, and the culture has progressively become more anti-Christian since then. Underneath it all, Australians still had questions about the validity of the Bible as a whole.

Now today in Australia, saying a prayer during a school assembly or sharing Bible "stories" to start the school day are unheard of. In addition, evolution is also taught as "fact" throughout the education system.

Biblical Illiteracy

The last two generations in Australia have had little or no knowledge of the Bible. By and large they have been thoroughly indoctrinated in an atheistic, evolutionary philosophy. Children don't automatically go to Sunday school or church programs as they used to. Ministers of religion are finding it more and more difficult to conduct programs in schools. And sadly, most church leaders tell their congregations that it's fine to believe in millions of years and/or evolutionary ideas, as long as God is somehow involved.

After years of subtle indoctrination and with an increasing emphasis on rejecting a literal Genesis, Australians basically reject the credibility of the Genesis history, and thus they doubt the reliability of the rest of the Bible.

Whether it's Australia, America, Great Britain or elsewhere, Western societies are no longer made up mainly of "Jews" but are more like the pagan Greeks: increasingly anti-Christian, and holding to a predominantly atheistic, evolutionary secular philosophy.

Indeed, they are probably even worse than Paul's opponents 2,000 years ago. The Greeks at least asked to hear him out; today many secularists try to suppress Christian teachings. In our modern time, there is a remnant of "Jews" who still have an understanding of Christian terminology, but this group is quickly becoming a smaller and smaller minority.

Today's "Greeks" do not have the foundational knowledge to fully understand the gospel. They have been led to believe that the Bible is not a credible book; its history in Genesis (creation in six days and a global Flood) is not seen to be true because so many people have been indoctrinated to believe in millions of years and evolution. Thus when an evangelist today preaches the message of the Cross, like the Greeks in Acts 17, it is foolishness to them.

How Can We Reach Today's "Greeks" Then?

As Paul understood, the "Greeks" need to be turned into "Jews." Their wrong foundation concerning evolution and millions of years

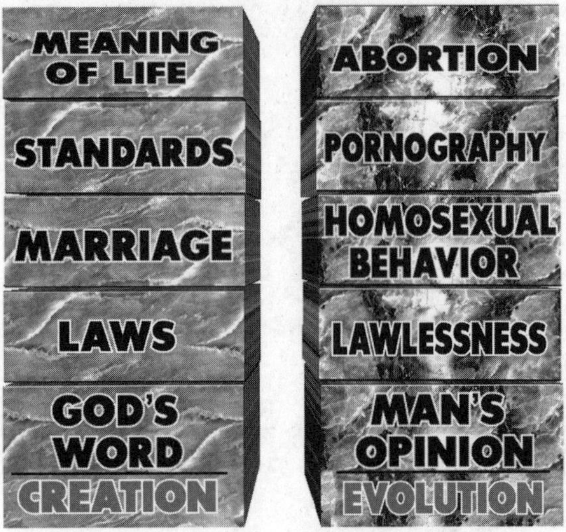

needs to be changed to one of understanding and believing that the Bible's account of creation and the Fall of man is true (i.e., that man is a sinner).

Once they have this different foundation, these "Greeks" can better understand the message of Christ and, we pray, respond accordingly and receive Christ. Sadly, most Christian leaders in recent decades didn't realize this shift. This approach of trying to "turn Greeks into Jews" to make people more open to the gospel message really should have been used even back in 1959 and before.

In the 1900s, the seeds of "Greek" thinking were already infiltrating the minds of people within and without the Christian church. The church by and large was not giving answers in dealing with evolutionary ideas and establishing the credibility of the Bible.

This could have made a real difference in the way people saw the Bible. If they had understood that it wasn't just a book of spiritual and moral issues but a book of history that could really be trusted, then they would have been more likely to trust the gospel based in that history.

When you compare this situation in Australia to the condition of the United States or Great Britain, it is easy to see a similar set of circumstances. Generations ago, their cultures were like the "Jews." The Bible, prayer, creation, etc. were a part of everyday life in public (government-run) schools—so most people were "Jews" in much of their thinking about spiritual matters.

But the seeds of "Greek" thinking were also being laid down subtly through the education system. Even in 1925, public-school students in America were sadly being taught that the Caucasians were the "highest race" and that the earth was millions of years old—by a textbook that also promoted so-called "mercy killing."[4]

Generations in the US (and in other countries) have now come through an education system that is basically devoid of the knowledge of God. In fact, Christianity is often taught *against* or relegated to mere personal belief instead of objective truth about world history. The Bible, prayer, and creation have basically been thrown out of the public education system. Students by and large are taught evolution as fact. The Bible is not a credible book in the eyes of most of these students. They are "Greeks."

If we want to evangelize the once-Christianized Western world today, there needs to be an understanding that the cultures have become Greek-like. The message will not be understood by such people until they can be changed from "Greeks" into "Jews."

The culture today needs the answers from science and the Bible to counter evolutionary and "millions of years" teaching so that the literal history of Genesis 1–11 is established—thus giving credibility to the gospel (in fact, *all* Christian doctrine) that is founded in this history. Indeed, such "creation evangelism" is a part of the beginning of the process of changing "Greeks" into "Jews" so they will better understand the gospel message and respond to it.

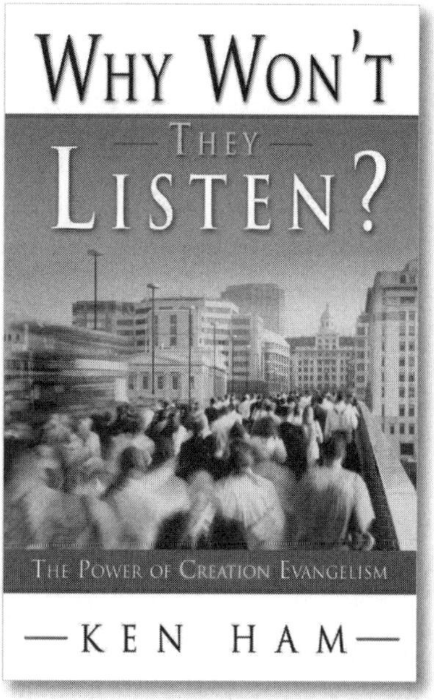

Much more about how to effectively evangelize our secular, "evolutionized" world (including practical advice) can be found in the book *Why Won't They Listen?*[5]

4 G.W. Hunter, *A Civic Biology Presented in Problems*, American Book Company, New York, 1914, 196.

5 K. Ham, *Why Won't They Listen?* Master Books, Green Forest, Arkansas, 2002.

Bonus

How Can We Use Dinosaurs to Spread the Creation Gospel Message?

BUDDY DAVIS

Dinosaurs are some of the most fascinating animals, and children especially are intrigued by them. This is one reason why evolutionists use them, over and over, to teach millions of years and evolution. Christians, however, should use dinosaurs to teach the true history of the universe. When children, young people, and adults are informed about the truth of dinosaurs, they can answer the questions of a skeptical world of and spread the good news of the gospel. When dinosaurs are used to spread the gospel, they become "missionary lizards."

Evidence of Creation

According to evolutionary teaching, dinosaurs roamed the earth millions of years ago and evolved from other types of animals. But what does the Bible say about the origin and history of dinosaurs? From Genesis 1:24–31, we can glean that dinosaurs were created on Day 6, the same day that God made the other land-dwelling, air-breathing animals, along with the first man and woman. Dinosaurs did not evolve from other animals, nor did any other animal evolve from dinosaurs. God created the original kinds of dinosaurs, and they multiplied from there, reproducing according to their kind.

As scientists have studied the fossils of dinosaurs, they have found that dinosaurs can be divided into two main groups: saurischian ("lizard-hipped") and ornithischian ("bird-hipped"). In saurischian dinosaurs, the ischium and pubis bones are forked beneath the ilium. This group of dinosaurs includes the large sauropods, such as *Apatosaurus* and *Diplodocus*. In ornithischian dinosaurs, the pubis and ischium lie side by side underneath the ilium. Ornithischian dinosaurs include *Stegosaurus*, *Triceratops*, and *Hadrosaurus*—the "duck-billed" dinosaur.

These two groups of dinosaurs are still dinosaurs, however, even though they vary in many ways. God created the various dinosaur kinds with great variety. This variety is seen in some of the most popular dinosaurs.

Popular Dinosaurs

Scientists have classified over 300 dinosaur species, but it is reasonable to assume that different sizes, varieties, and sexes of the same kind of dinosaur have ended up with different names. There may have been only 50 or fewer original kinds of dinosaurs that God created on Day 6 of Creation Week.

The following are some of the most well-known and popular dinosaurs with some interesting facts about them.

1) *Stegosaurus*—"Roofed lizard." 30 feet (9 m) long. Weighed 2–4 tons (1.8–3.6 metric tons). Found in North America. Group: Ornithischia.

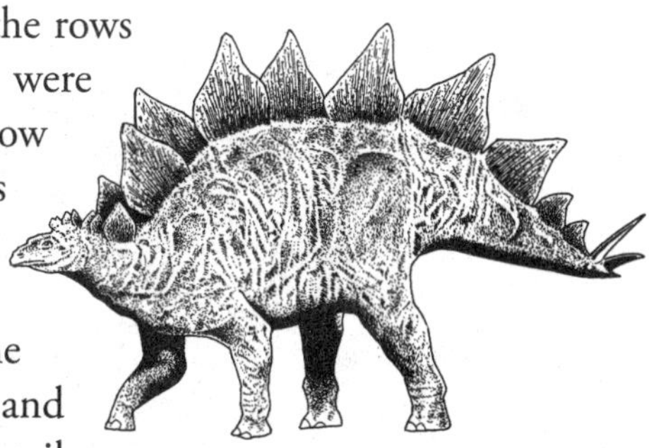

   It was once believed that the rows of plates on *Stegosaurus*'s back were just for protection. Scientists now believe that they were used as solar panels. The plates were thin and full of blood vessels. They were embedded into the skin and not part of the back and tail bone. The neck, hips, and tail of the *Stegosaurus* were further protected by small boney-like studs. The tail was flexible and armed with at least four horns.

2) *Brachiosaurus*—"Arm lizard." 80 feet (24 m) long. Weighed 50 tons (45 metric tons). Found in America, Europe, and Africa. Group: Saurischia.

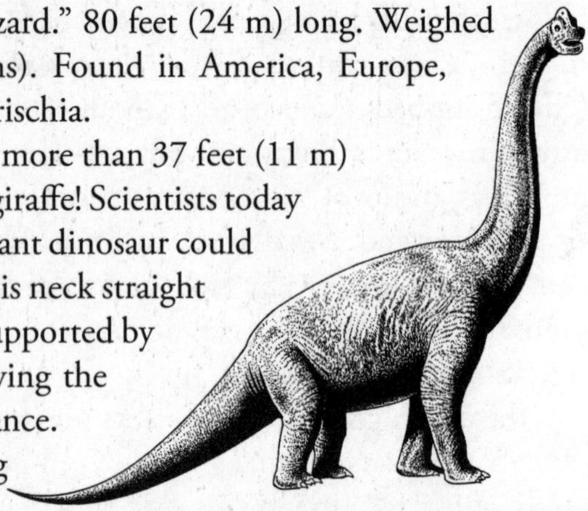

Brachiosaurus stood more than 37 feet (11 m) high, twice the size of a giraffe! Scientists today question whether this giant dinosaur could have been able to raise his neck straight up. *Brachiosaurus* was supported by very long front legs, giving the back a sloping appearance. These giants were among the largest animals to have walked the earth.

3) *Ankylosaurus*—"Fused lizard." 33 feet (10 m) long. Weighed 4 tons (3.6 metric tons). Lived in Asia, North America, Europe, and South America. Group: Ornithischia.

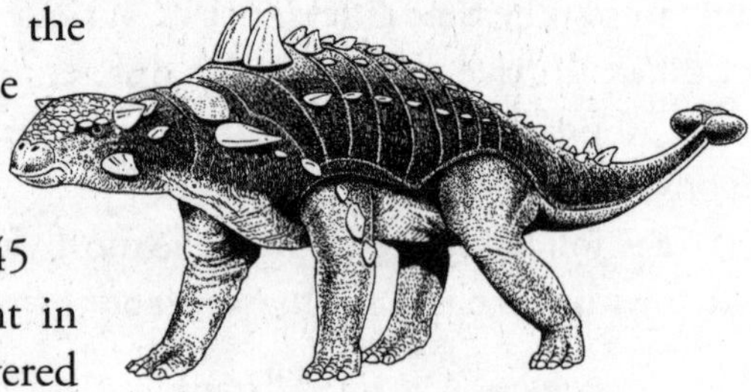

Ankylosaurus is the most popular of the ankylosaurs. The club on its tail weighed over 100 pounds (45 kg). It was like a knight in armor. Its body was covered with a protective armor with rows of horns; even the eyelids were armored.

4) *Triceratops*—"Three-horned face." 30 feet (9 m) long. Weighed 6 tons (5.4 metric tons). Lived in North America. Group: Ornithischia.

Triceratops was a short-filled ceratopsian. It had two brow horns that grew up to 3 feet (1 m) long. The nose horn was much shorter. The frill also had rows of small horn-like bones lining the outside edge. The frill was made of solid bone. *Triceratops* was one of the most massive of all dinosaurs.

5) *Compsognathus*—"Elegant jaw." 3–4 feet (1–1.2 m). Weighed about 5 pounds (2 kg). Found in Western Europe. Group: Saurischia.

Compsognathus was one of the smallest dinosaurs; its body was the size of a chicken. The skull was delicate, and the jaw appeared fragile. The skull and jaw were armed with small, curved teeth. The slender body supported a long tail used for balance.

6) *Tyrannosaurus rex*—"Tyrant lizard." 40 feet (12 m) long. Weighed 7 tons (6.4 metric tons). Found in North America. Group: Saurischia.

This is one of the most famous and well-known dinosaurs. *T. rex* had 50 to 60 teeth with some of its jaw teeth over 7 inches (18 cm) long. The teeth were curved, serrated, and very strong. If a tooth was broken, a new one would replace it. His bottom jaw could flex allowing him to swallow huge chunks of food. The skull of one *T. rex* was 5 feet (1.5 m) long, and scientists who have studied CAT scans of the skull believe that the *T. rex*'s sense of smell and hearing was very good.

T. rex had small arms that were no bigger than a man's arm. They appear to have been well-muscled like a weight-lifter. It is not known exactly how the *T. rex* used them.

7) *Velociraptor*—"Swift hunter." 6.6 feet (2 m) long. Weighed 35 pounds (16 kg). Found in Mongolia, China. Group: Saurischia.

Velociraptor was a member of the dromaeousaur family. Like all raptors, he had a sickle-like claw on his foot and three sharp claws on each hand. This dinosaur also had a mouth full of razor-sharp teeth.

Fossils

Since dinosaurs, as far as we know, are extinct, dinosaur fossils are the only things scientists can study. Dinosaur fossil remains have been found on every continent on earth. Robert Plot described one of the first dinosaur bones in his book *Natural History of Oxford* in 1676. The bone he found has been lost, but it was thought to have been part of a thigh bone of *Megalosaurus*.

One of the first complete fossilized dinosaur skeletons ever found was an *Iguanodon*. Over 30 individual *Iguanodon* skeletons were discovered in a Belgium coal mine in 1878.

One of the first complete skeletons ever assembled for display was a *Hadrosaurus*. It was discovered in 1850 in Haddonfield, New Jersey, and is still on display at the Academy of Natural Sciences in Philadelphia, Pennsylvania.

Since scientists study only the fossils of dinosaurs (not living specimens), and since fossils are the bones of dead things, Christians can use dinosaurs to explain the origin of death. After God created all things, including dinosaurs, He called His creation "very good" (Genesis 1:31). Death was not part of the world until Adam disobeyed God's command not to eat of the Tree of the Knowledge of Good and Evil. Once Adam disobeyed, God cursed all of creation (Genesis 3:14–19). Romans 8:22 tell us, "For we know that the whole creation groans and labors with birth pangs together until now." Creation now groans under the Curse, and death affects everything in creation.

The Flood and the Ice Age

Christians can also use dinosaurs to discuss the global Flood that occurred in Noah's day.

The global Flood may have been one of the reasons dinosaurs went extinct. Before the Flood, dinosaurs freely roamed the earth. But due to man's wickedness, God sent a global Flood that destroyed all life that was not inside the Ark. During the Flood, many of these animals and humans were buried in sediment that later hardened, thus giving us many of the fossils scientists study today.

We also need to remember that dinosaurs were on the Ark. The Bible tells us in Genesis 6:19 and 7:2–9 that two of every land-dwelling, air-breathing animal (and seven of some) were on the Ark with Noah, his wife, his sons, and their wives. So, what happened to these mighty dinosaurs?

After the Flood, these dinosaurs probably went extinct for a variety of reasons, just as animals become extinct today. The Flood greatly changed the earth's habitat, and it may have changed it so much that many of the dinosaurs could not successfully survive the harsher environment. The post-Flood Ice Age also probably contributed to their demise.

Some of the dinosaurs that survived for a while after the Ice Age likely were referred to as "dragons." Most of these eventually died out or were killed. Other reasons for their extinction could be starvation, disease, and hunting pressure.

Conclusion

Dinosaurs and the truths that they share about God's creation, man's sin, death, the Flood, and the Ice Age can be used by Christian young people and adults to share the gospel with unbelievers. These missionary lizards uphold the authority of Scripture, and they can be powerful tools in sharing the salvation message, which should be the ultimate goal of every Christian.

As non-Christians hear the biblical explanation of dinosaurs, many have been, and will be, challenged to listen to the rest of what the Bible states. We rejoice that many have been won to the Lord using the true history of these missionary lizards.

Glossary

abiogenesis: the alleged spontaneous generation of living organisms from non-living matter

adaptation: a physical trait or behavior due to inherited characteristics that gives an organism the ability to survive in a given environment

adaptive radiation: the process of speciation as populations spread and encounter different environments

allele: any of the alternative forms of a gene that occur at a specific spot (locus) in the DNA sequence (genome) of an organism

anthropology: systematic study of the characteristics of humans through history

archaebacteria: the kingdom of prokaryotic cells excluding eubacteria (considered as a separate domain in certain classification schemes) which is alleged to be ancestral to eubacteria by some evolutionists

***Archaeopteryx*:** extinct species of perching bird (known from fossils) with teeth, wing claws, and a bony tail

Archaeoraptor: a fraudulent fossil from China that combined the body of a bird with the tail of a dinosaur

artifact: an item or its remains produced in the past by humans; generally recovered through archaeological exploration

atheism: the belief that God, or any supreme intelligence, does not exist

***Australopithecus*:** genus of extinct apes known from fossils found in Africa, including the infamous "Lucy"

bacteria: a group of unicellular organisms that lack a true nucleus and membrane-bound organelles; including eubacteria and archaebacteria

baramin: (see created kind)

Bible: the collection of 66 books that is the inspired Word of God; used as the authoritative source for determining truth

biblical creation: the supernatural events, occurring over 6 approximately 24-hour days, described in Genesis 1 and 2, by which God caused the formation of heaven and earth and everything in them

biblical creation model: a scientific model based on the biblical account of creation, the curse of nature brought about by Adam's sin, and the global catastrophe of Noah's Flood

big bang model: the cosmological model suggesting the universe began as a single point which expanded to produce the known universe

biology: the systematic study of the characteristics and interactions of living things

beneficial mutation: a mutation which confers a survival advantage to an organism under certain environmental conditions; usually a result of the loss of genetic information (see mutation)

catastrophism: the doctrine that changes in the geologic record are a result of physical processes operating at rates that are dramatically higher than are observed today

cell theory: a theory of biology consisting of three parts: (1) cells are the basic unit of all living things; (2) all living things are composed of one or more cells; and (3) all cells come from preexisting cells

chemistry: the systematic study of the properties and interaction of matter

clone: an organism that is genetically identical to its parent

cloning: producing a new organism using the DNA of an existing organism

compromise: Reinterpreting Scripture based on outside beliefs and developing theology around this belief. Common origins compromise positions accept the secular view of millions of years, as opposed to the global Flood of Noah. Some of these popular views are: Progressive Creation/Day Age Theory, Gap Theory, Framework Hypothesis, and Theistic Evolution.

cosmogony: a belief about the origin of the universe

cosmology: the systematic study of the structure of the universe, including its origin

created kind (baramin): the original organisms (and their descendants) created supernaturally by God as described in Genesis 1; these organisms reproduce only their own kind within the limits of preprogrammed information, but with great variation. **Note:** Since the original creation, organisms of one kind cannot interbreed with a different kind, but individuals within a kind may have lost the ability (information) to interbreed due to the effects of the Curse.

Cro-Magnon man: an extinct people group of Europe and Eastern Asia

Darwinism: a belief that all organisms have a single common ancestor that has produced all living organisms through the process of natural selection; popularized by Charles Darwin in *On the Origin of Species*

day-age theory: a compromise belief that the days of Genesis 1 are actually vast ages of different lengths; based on secular dating methods

deism: a belief in a Creator God that denies His intervention in the history of the universe since its creation

DNA (deoxyribonucleic acid): the basic molecule of hereditary information which serves as a code for the production of proteins and is common to all living organisms

eisegesis: an interpretation of Scripture that incorporates the interpreter's ideas as opposed to the actual meaning of the text (taking ideas to Scripture and reinterpreting it)

endosymbiont hypothesis: the suggestion that mitochondria, chloroplasts, and other organelles originated as bacteria that were ingested and became a part of eukaryotic cells over evolutionary time

entropy (thermodynamics): the measure of the tendency of closed systems to increase in disorder

eubacteria: the kingdom of prokaryotic cells, excluding archaebacteria (considered as a separate domain in certain classification schemes); alleged to be descended from archaebacteria by some evolutionists

evolution: all life on earth has come about through descent with modification from a single common ancestor (a hypothetical, primitive single-celled organism)

exegesis: critical interpretation of Scripture taking into account the writing style, meaning, and context of the passage (learning from what Scripture is saying)

extrapolation: inferring information outside of the range of the actual data based on trends

faith: belief in things that cannot be directly known or observed

Flood (Noah's Flood): the supernatural event described in Genesis 6–10 that covered the entire earth with water, killing all land vertebrates except those aboard the Ark built by Noah

fossil: preserved remains or traces of once living organisms

coprolite: fossilized excrement

included: organisms that are encased in a substance leaving the specimen virtually intact, as in amber

living: organisms that are virtually identical to fossil organisms; often thought to have been extinct and then discovered

mold and cast: a type of replacement fossil which includes the concave or convex impression of an organism; typical of shells and leaves

permineralized: an organism in which the porous parts are filled with mineral deposits leaving the original superstructure intact

replacement (mineralized): organism whose entire structure has been replaced by mineral deposits so that none of the original superstructure remains

trace/track/micro: evidence of the activity of an organism, including tracks, burrows, root traces

fossilization: the process of preserving the remains or traces of an organism, generally by some form of petrification

framework hypothesis: a compromise belief that Genesis 1 is written in a non-literal, non-chronological way; based on secular dating methods

gap theory: a compromise belief that a vast period of time exists between Genesis 1:1 and 1:2 during which time the geologic eras can be fit

gene: a segment of DNA that codes for the production of polypeptides

gene pool: the collection of varying alleles within a population of organisms

genetics: the study of characteristics inherited by the transmission of DNA from parent to offspring

genome: the complete set of genetic material (DNA) of any cell in an organism

geocentric: using the earth as a central frame of reference

geologic column: the layers of rock that compose the crust of the earth

glacier: large mass of ice that has accumulated from snow over the years and is slowly moving from a higher place

half-life: the amount of time required for one half of the atoms of the parent isotope to decay into the daughter isotope

heliocentric: using the sun as a central frame of reference

heredity: acquiring traits by transfer of genes from parent to offspring

historical (origins) science: interpreting evidence from past events based on a presupposed philosophical point of view

hominid: extinct and living members of the family Hominidae, including modern humans and their ancestors

***Homo erectus*:** fossils of extinct human people groups that are misinterpreted as missing links in human evolution

***Homo habilis*:** an invalid category consisting of various ape and human fossil fragments

homologous structure: any feature that shares a common design with a similar feature in another species of organism (alleged to support common ancestry in evolutionary models)

***Homo sapiens*:** the category that includes modern humans, Neandertals, and other extinct human groups

human: any member of the species *Homo sapiens*

humanism: a belief in mankind as the measure of all things; based on relative truth and morality and rejecting any supernatural authority

ice age: the period of glaciation following Noah's Flood during which a significant portion of the earth had a cold climate

Ice Age: when denoted in caps is referring to the biblical post-Flood Ice Age

ice cores: cores of ice that have been drilled down into a glacier

interglacial: short period of warming between glacier growth/movement that caused it to melt away

information: an encoded, symbolically represented message conveying expected action and intended purpose

interpolation: inferring information within the range of the actual data based on trends

Java man: the first fossil specimen of *Homo erectus*

Kennewick man: human remains found in Washington State in 1996

kind (see created kind)

life (biological): anything that: contains genetic information, can reproduce offspring that resemble itself, grow and develop, control cellular organization and conditions including metabolism and homeostasis, and respond to its environment **Note:** The Bible defines life in a different sense, using the Hebrew phrase *nephesh chayyah,* indicating organisms with a life spirit.

local flood: a nonscriptural compromise belief that Noah's Flood was an event confined to the Mesopotamian Valley

logic: systematic application of principles of reasoning to arrive at a conclusion

Lucy: a 40% complete fossil specimen of *Australopithecus afarensis* discovered in Ethiopia in 1974 by Donald Johanson

macroevolution: term used by evolutionists to describe the alleged, unobservable change of one kind of organism to another by natural selection acting on the accumulation of mutations over vast periods of time

mammal: any organism that has fur and nurses young from mammary glands

materialism: a belief claiming that physical matter is the only or fundamental reality and that all organisms, processes, and phenomena can be explained as manifestations or interactions of matter

metamorphic rocks: rocks that have been altered in texture or composition by heat, pressure, or chemical activity after they initially formed

microevolution: term used by evolutionists to describe relatively small changes in genetic variation that can be observed in populations

mineralization: replacement of material from an object, usually organic, with minerals that harden

mitochondrial DNA (mtDNA): small circular loops of DNA found in the mitochondria of eukaryotic cells

mitochondrial Eve: the most recent common ancestor of humans whose lineage can be traced backward through female ancestors; alleged support for the out-of-Africa hypothesis of human evolution

model: physical, mental, or mathematical representations that can be used to explain observed phenomena and make specific, useful predictions

moraines: stones, boulders, and debris that has been carried and dropped by a glacier

Mungo man: fossil human remains from Australia dated by evolutionists to 40,000 years or more

mutation: any change in the sequence of DNA base pairs in the genome of an organism

frameshift: addition or deletion of one or more nucleotide pairs in the coding region of a gene causing the triplet codons to be read in the wrong frame

deletion: removal of one or more nucleotide pairs in the DNA sequence

duplication: large segments of DNA that have been copied and inserted into a new position in the DNA sequence, possibly on different chromosomes

insertion: addition of one or more nucleotide pairs in the DNA sequence

inversion: a section of DNA that has been reversed within the chromosome

neutral: any mutation that does not effect the function of an organism

point: addition, deletion, or substitution of a single nucleotide pair in the DNA sequence

translocation: the movement of a section of a chromosome from one position to another, generally between different chromosomes

natural selection: the process by which individuals possessing a set of traits that confers a survival advantage in a given environment tend to leave more offspring on average that survive to reproduce in the next generation

naturalism: a belief denying that an event or object has a supernatural significance; specifically, the doctrine that scientific laws are adequate to account for all phenomena

Neanderthal/Neandertal: an extinct human people group with relatively thick bones and a distinct culture; disease and nutritional deficiency may be responsible for the bone characteristics

neo-Darwinism: an extension of Darwinism which includes modern genetic concepts to explain the origin of all life on earth from a single common ancestor

Noah's Flood: (see Flood)

old-earth creation: any compromise position that accepts the millions-of-years idea from secular science and attempts to fit that time into the events of Genesis 1–2

operational (observational) science: a systematic approach to understanding that uses observable, testable, repeatable, and falsifiable experimentation to understand how nature commonly behaves

organism: any cell or group of cells that exhibits the properties of life (living things) **(see life)**

paleontology: the systematic study of the history of life on the earth based on the fossil record

permineralization: the filling of cavities of an object, usually organic, with minerals which harden

petrification: processes, including mineralization, permineralization, and inclusion, which change an object, usually organic, into stone or a similar mineral structure

phylogenetic tree: diagrams that show the alleged evolutionary relationships between organisms

Piltdown man: fraudulent "prehuman" fossil consisting of the skull cap of a modern human and the jaw and teeth of an orangutan

plate tectonics: the systematic study of the movement of the plates that make up the earth's crust

uniformitarian model: based on the gradual movement of the plates over hundreds of millions of years

catastrophic model: based on rapid movement of the plates associated with Noah's Flood

polypeptide: a chain of amino acids formed from the DNA template and modified to produce proteins

presupposition: a belief that is accepted as true and is foundational to one's worldview

progressive creation: a compromise belief accepting that God has created organisms in a progressive manner over billions of years to accommodate secular dating methods

punctuated equilibrium: an evolutionary model that suggests evolution occurs in rapid spurts rather than by gradual change

radioactive decay: The breakdown of unstable nuclei of atoms releasing energy and subatomic particles

radiometric dating: using ratios of isotopes produced in radioactive decay to calculate an "age" of the specimen based on assumed rates of decay and other assumptions

parent isotope: original isotope before it has undergone radioactive decay

daughter isotope: isotope resulting from radioactive decay

half-life: the amount of time required for one half of the parent atoms to decay into the daughter atoms

relative dating: estimating the age of a fossil or rock layer by comparing its position to layers of known age

absolute dating: using radiometric dating to test a specimen in an attempt to estimate its age

religion: a cause, principle, or belief system held to with zeal and conviction

RNA (Ribonucleic Acid): a molecule found in all living things that serves various roles in producing proteins from the coded information in the DNA sequence

secular: not from a religious perspective or source

secular humanism: (see humanism)

science: the systematic study of a subject in order to gain information (see also operational science and historical science)

speciation: the process of change in a population that produces distinct populations which rarely naturally interbreed due to geographic isolation or other factors

species: a group of organisms within a genus that naturally reproduce and have fertile offspring

spontaneous generation: the false belief that life can arise from nonliving matter

strata: layers of rock deposited by geologic events

theistic evolution: a compromise belief that suggests God used evolutionary processes to create the universe and life on earth over billions of years

theory: an explanation of a set of facts based on a broad set of observations that is generally accepted within a group of scientists

transitions/transitional forms: species that exhibit traits that may be interpreted as intermediate between two kinds of organisms in an evolutionary framework, e.g., an organism with a fish body and amphibian legs

uniformitarianism: the doctrine that present day processes acting at similar rates as observed today account for the change evident in the geologic record

vestigial organ: any organ that has a demonstrated reduction and/or loss of function **Note:** Vestigial organs include eyes in blind cave-fish but not organs that are assumed to have had a different function in an unknown ancestor.

virus: a nonliving collection of proteins and genetic material that can only reproduce inside of a living cell

Y-chromosome Adam: the most recent common ancestor whose lineage can be traced backward through male ancestors

Yom: one of the Hebrew words for "day" encompassing several definitions such as the daylight portion of a day (12 hours, Genesis 1:5a), a day with one evening and one morning (24 hours, Genesis 1:5b) or a longer period of time (Genesis 2:4). The context reveals which definition is in use.

Index

About the Authors

Ken Ham

Ken is the president and CEO of Answers in Genesis (USA). He has authored several books, including the best-seller *The Lie: Evolution*. He is one of the most in-demand speakers in the U.S. and has a daily radio program called *Answers...with Ken Ham,* which is heard on over 850 stations in the US and over 1,000 worldwide.

Ken has a BS in applied science (with an emphasis in environmental biology) from Queensland Institute of Technology in Australia. He also holds a diploma of education from the University of Queensland (a graduate qualification for science teachers in the public schools in Australia). Ken has been awarded two honorary doctorates: a Doctor of Divinity (1997) from Temple Baptist College in Cincinnati, Ohio, and a Doctor of Literature (2004) from Liberty University in Lynchburg, Virginia.

Jason Lisle

Jason graduated *summa cum laude* from Ohio Wesleyan University, where he double-majored in physics and astronomy, and minored in mathematics. He did graduate work at the University of Colorado, where he earned a master's degree and a PhD in astrophysics.

In graduate school, he specialized in solar astrophysics. While there, Jason used the SOHO spacecraft to investigate motions on the surface of the sun, as well as solar magnetism and subsurface weather.

He has authored papers in both secular and creationist literature, and written several books. Jason is a capable speaker and writer and is currently working as Director of Research at the Institute for Creation Research.

Georgia Purdom

Georgia received her PhD in molecular genetics from Ohio State University in 2000. As an associate professor of biology, she completed five years of teaching and research at Mt. Vernon Nazarene University in Ohio before joining the staff at Answers in Genesis (USA).

Dr. Purdom has published papers in the *Journal of Neuroscience,* the *Journal of Bone and Mineral Research*, and the *Journal of Leukocyte Biology*. She is also a member of the Creation Research Society, American Society for Microbiology, and American Society for Cell Biology.

She is a peer-reviewer for *Creation Research Society Quarterly*. Georgia has a keen interest and keeps a close eye on the Intelligent Design movement.

Andy McIntosh

Andy McIntosh is a professor (the highest teaching/research rank in UK university hierarchy) in combustion theory at Leeds University, UK. His PhD was in aerodynamics. A number of his students later worked for Rolls Royce, designing aircraft engines.

Andy has an extensive work and research background but also has interest in theological matters. His career in mathematics and science has led him to the view that the world and the universe show powerful evidence of design. As a result, he is often asked to speak on the subject of origins both in the UK and abroad.

David Menton

Dr. Menton was an associate professor of anatomy at Washington University School of Medicine from 1966 to 2000 and has since become Associate Professor Emeritus. He was a consulting editor in histology for *Stedman's Medical Dictionary*, a standard medical reference work.

David earned his PhD from Brown University in cell biology. He is a popular speaker and lecturer with Answers in Genesis (USA), showing complex design in anatomy with popular DVDs such as *The Hearing Ear and Seeing Eye* and *Fearfully and Wonderfully Made*. He also has an interest in the famous Scopes Trial, which was a big turning point in the creation/evolution controversy in the USA in 1925.

A.J. Monty White

Now chief executive of Answers in Genesis (UK/Europe), Dr. Monty White joined AiG after leaving the University of Wales in Cardiff where he had been a senior administrator for 28 years. He is a graduate of the University of Wales, obtaining his BS in chemistry in 1967 and his PhD for research in the field of gas kinetics in 1970. Monty spent two years investigating the optical and electrical properties of organic semiconductors before moving to Cardiff, where he joined the administration at the university there.

Monty is well known for his views on creation, having written numerous articles and pamphlets, as well as a number of books dealing with various aspects of creation, evolution, science, and the Bible. Monty has appeared on British television programs and has been interviewed on local and national radio about creation.

Paul F. Taylor

Paul learned to play the piano early in life and was educated at Chetham's School of Music, Manchester, England. However, an interest in science took him to Nottingham University to study chemistry, and there he graduated with a BS in 1982. He then took a year's post-graduate Certificate in Education so that he could become a schoolteacher. Paul taught science in state schools for 17 years, eventually becoming a head of department, and gained a master's degree in science education at Cardiff University.

Paul worked with AiG (UK/Europe) as a writer, speaker, and head of media and publications. He is currently working with the Creation Today ministry in Florida.

Bodie Hodge

Bodie earned a BS and MS in mechanical engineering at Southern Illinois University at Carbondale in 1996 and 1998, respectively. His specialty was in materials science working with advanced ceramic powder processing. He developed a new method of production of submicron titanium diboride.

Bodie accepted a teaching position as visiting instructor at Southern Illinois in 1998 and taught for two years. After this, he took a job working as a test engineer at Caterpillar's Peoria Proving Ground. Bodie currently works at Answers in Genesis (USA) as a speaker, writer, and researcher after working for three years in the Answers Correspondence Department.

Terry Mortenson

Terry earned a BA in math at the University of Minnesota in 1975 and later went on to earn an MDiv in systematic theology at Trinity Evangelical Divinity School in 1992. His studies took him to the U.K. where he earned a PhD in the history of geology at Coventry University.

Terry has done extensive research regarding the beliefs of the nineteenth century Scriptural geologists. An accumulation of this research can be found in his book *The Great Turning Point*. Terry is currently working at Answer in Genesis (USA) as a speaker, writer, and researcher.

Mike Riddle

As a former captain in the Marines, Mike earned a BS in mathematics and MS in education. Mike has been involved in creation apologetics for many years and has been an adjunct lecturer with the Institute for Creation Research. Mike has a passion for teaching and he exhibits a great ability to bring topics down to a lay-audience level in his lectures.

Before becoming a Marine, Mike became a US national champion in the track-and-field version of the pentathlon (in 1976). His best events were the 400 meters, javelin, long jump, and 1,500 meters. In his professional life, Mike worked for many years in the computer field with Microsoft (yes, he has met Bill Gates).

Mike Oard

Retired from the National Weather Service, meteorologist Mike Oard has researched the compelling evidence for Noah's Flood and the resulting Ice Age, and how the incredible wooly mammoth connects to biblical history.

Mike, or "Mr. Ice Age," received his MS in atmospheric science from the University of Washington. He has authored a children's book, *Life in the Great Ice Age*, and a book for teens and adults, *The Weather Book*. Mike recently wrote a semitechnical book called *Frozen in Time* for lay-readers, as well as the technical monograph *An Ice Age Caused by the Genesis Flood*.

Andrew Snelling

Andrew is currently Director of Research with Answers in Genesis (US). He received a BS in applied geology with first-class honors at the University of New South Wales in Sydney, and earned his PhD in geology at the University of Sydney for his thesis entitled "A Geochemical Study of the Koongarra Uranium Deposit, Northern Territory, Australia."

Between studies and since, Andrew worked for six years in the exploration and mining industries in Tasmania, New South Wales, Victoria, Western Australia, and the Northern Territory as a field, mine, and research geologist. Andrew was also a principal investigator in the RATE (Radioisotopes and the Age of The Earth) project hosted by the Institute for Creation Research and the Creation Research Society.

Tommy Mitchell

Tommy graduated with a BA with highest honors from the University of Tennessee–Knoxville in 1980 with a major in cell biology and a minor in biochemistry. He subsequently attended Vanderbilt University School of Medicine in Nashville, where he was granted an MD degree in 1984.

Dr. Mitchell's residency was completed at Vanderbilt University Affiliated Hospitals in 1987. He was board certified in internal medicine, with a medical practice in Gallatin, Tennessee (the city of his birth). In 1991, he was elected to the Fellowship in the American College of Physicians (F.A.C.P.). Tommy became a full-time speaker, researcher, and writer with Answers in Genesis (USA) in 2006.

Clifford Wilson

Cliff has a BA and a MA from Sydney University, a BDiv (which was postgraduate, including Hebrew and Greek) from the Melbourne College of Divinity, and a master of religious education (MRE) from Luther Rice Seminary. His PhD is from the University of South Carolina, and he has done field work in archaeology in association with Hebrew Union College in Jerusalem.

Cliff has carefully studied thousands of archaeological finds for decades and openly declares that they confirm the Bible's history. He was the founding president of Pacific International University in Springfield, Missouri. Cliff and his wife, Barbara, who also holds a PhD, have authored more than 70 books.

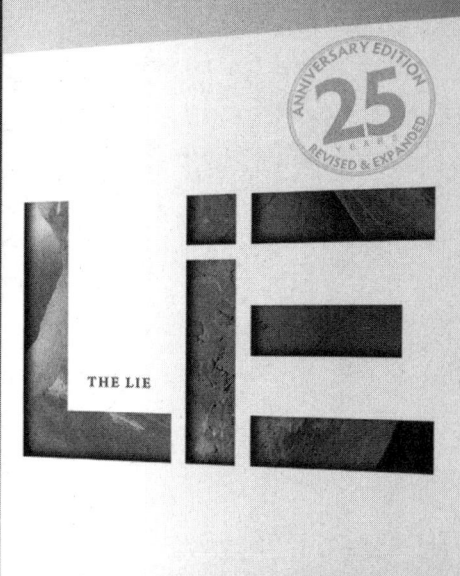

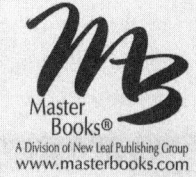

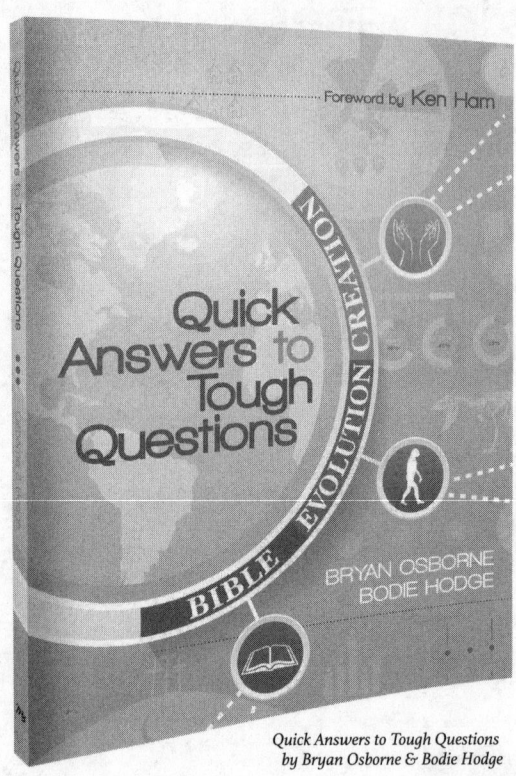

The New Answers Book 2

The New Answers Book 2

Over 30 Questions on

Creation/Evolution and the Bible

Ken Ham General Editor

First printing: May 2008
Eighteenth printing: February 2020

Master Books®, P.O. Box 726, Green Forest, AR 72638.
Master Books® is a division of the New Leaf Publishing Group, Inc.

ISBN: 978-0-89051-537-2
ISBN: 978-1-61458-094-2 (digital)
Library of Congress Number: 2008903202

Unless otherwise noted, the following versions of the Bible were used:
King James Version — introduction, chapters 3, 5, 18, and 20
New King James Version — chapters 4, 13, 14, 17, 19, 22, 25, 28, 29, and conclusion; Scripture taken from the New King James Version. Copyright © 1982 by Thomas Nelson, Inc. Used by permission. All rights reserved.
Modern King James Version — chapters 10, 11, and 12
New International Version — chapter 7

Please consider requesting that a copy of this volume be purchased by your local library system.

Printed in the United States of America

Please visit our website for other great titles:
www.masterbooks.com

For information regarding author interviews,
please contact the publicity department at (870) 438-5288.

Master Books®
A Division of New Leaf Publishing Group
www.masterbooks.com

ACKNOWLEDGMENTS AND SPECIAL THANKS

We are especially grateful to the following people for lending their expertise in reviewing various aspects of this book:

Dr. Bob Compton (DVM), Dr. David Crandall (former international director of Gospel Literature Service), David Down (Egyptologist, editor-in-chief of the bimonthly magazine *Archaeological Diggings)*, Brian Edwards (pastor, apologist, author), Steve Fazekas (theology), Dr. Werner Gitt (engineering; former director and professor at the German Federal Institute of Physics and Technology), Ken Ham (biology, president and CEO of Answers in Genesis-U.S.), Bodie Hodge (engineering, materials), Dave Jolly (biology), Dr. Jason Lisle (astrophysics), Stacia McKeever (biology, psychology), Dr. David Menton (cell biology, retired associate professor of anatomy at Washington University School of Medicine), Dr. Tommy Mitchell (internal medicine), Dr. Terry Mortenson (history of geology), Larry Pierce (chronologist/translator of the *Annals of the World* by James Ussher, and developer of *Online Bible*), Dr. Georgia Purdom (genetics), Dr. Andrew Snelling (geology), Dr. John Whitcomb (theology, president of Whitcomb Ministries), Dr. John Whitmore (geology, associate professor of geology at Cedarville University), Dr. John Reed (geology), David Wright (student of engineering), Gary Vaterlaus (science education).

We are also appreciative of the talents of Dan Lietha, who provided many of the illustrations used in this book. Dan Stelzer did several illustrations as well. All other illustrations that are not AiG images are noted on the illustration, figure, or photograph. Also, thanks to Stacia McKeever for much of the executive editing.

Contents

INTRODUCTION

Why Is the Christian Worldview Collapsing in America?

KEN HAM

Back in the 16th century, William Tyndale was persecuted, imprisoned, strangled, and his body burned at the stake. Why? Because he worked to translate the Scriptures into English and get copies of the Bible to the average person. Influenced by Luther and others, Tyndale was an integral part of the Reformation that spread God's written Word throughout the world — particularly to the Western world.

At that time, many church leaders believed the Bible should not be in the hands of the common person and that only appointed and scholarly church leaders should tell the public what they should believe. But the spread of God's written Word in the 1500s changed all that as it permeated many nations. It resulted in what we called the "Christian West." However, today we see the Christian influence in our Western world waning — Europe (especially the United Kingdom) is nearly dead spiritually. Right here in America, the Christian worldview is collapsing before our very eyes.

So What Is Happening?

First, let me point out that we need to be like the men of Issachar, who had "understanding of the times" (1 Chronicles 12:32). Today we are seeing an undoing of the Reformation, as society is not honoring some great people of God who were martyred for proclaiming the truths of the Bible.

The Reformation was a movement to call people to the authority of the Word of God. Almost 500 years later, we believe the teaching of millions of years and evolution has been the major tool in this era to undo the work of the Reformation.

To understand the times in which we live, we need to know how this sad transformation has come about — including how people view the Bible:

- The majority of church leaders have adopted the secular religion (i.e., millions of years/evolution) of the age and have compromised God's Word — thus undermining its authority to coming generations.
- Statistics are clear that most people in churches do not study their Bibles as they should. Frankly, we have a very biblically illiterate church today. We also observe church academics of our age beginning to impose a similar philosophy to that seen in Tyndale's time — that it is these learned leaders (most of whom have compromised God's Word) who determine what the public should believe. Increasingly, churchgoers are not like the Bereans who "searched the Scriptures daily to find out whether these things were so" (Acts 17:11).

I want to give you two specific examples of this dramatic change — and I believe you will be quite shocked.

The first is of Dr. James F. McGrath, who holds the Clarence L. Goodwin Chair in New Testament Language and Literature at Butler University in Indianapolis. Recently, Dr. McGrath wrote a blog item[1] concerning AiG's stand on a literal Genesis. First, he quoted another writer:

> Some may excuse Mr. Ham on the ground that he has no theological or biblical training (he has a bachelor's degree in applied science). I am not so inclined for one reason: by assuming the pulpit of churches and declaring he intends to interpret the Bible, he de facto sets himself up as a Bible teacher, and should be held accountable to know not only the relevant facts, but the proper way to exegete and teach a passage of scripture.

1. http://blog.echurchwebsites.org.uk/2010/08/03/ken-ham-rachel-held-evans-blogosphere.

> If he does not want to give up seven years of his life and tens of thousands of dollars to get training in the Bible, theology, and the ancient languages (the standard degree program for clergy) then that is perfectly understandable. What is not so understandable is his desire to set himself up as a Bible teacher without getting Bible training.

Then Dr. McGrath followed with his own comments about the above statements:

> Amen! . . . I think that the best course of action is for those who are well-informed about the Bible to debunk, refute and if necessary "refudiate" the statements of those who have no expertise in any field of scholarship related to the Bible, and yet believe that without any real knowledge of the original languages, historical context, and other relevant factors, their pontifications will do anything but harm the souls of believers and the Christian faith itself.

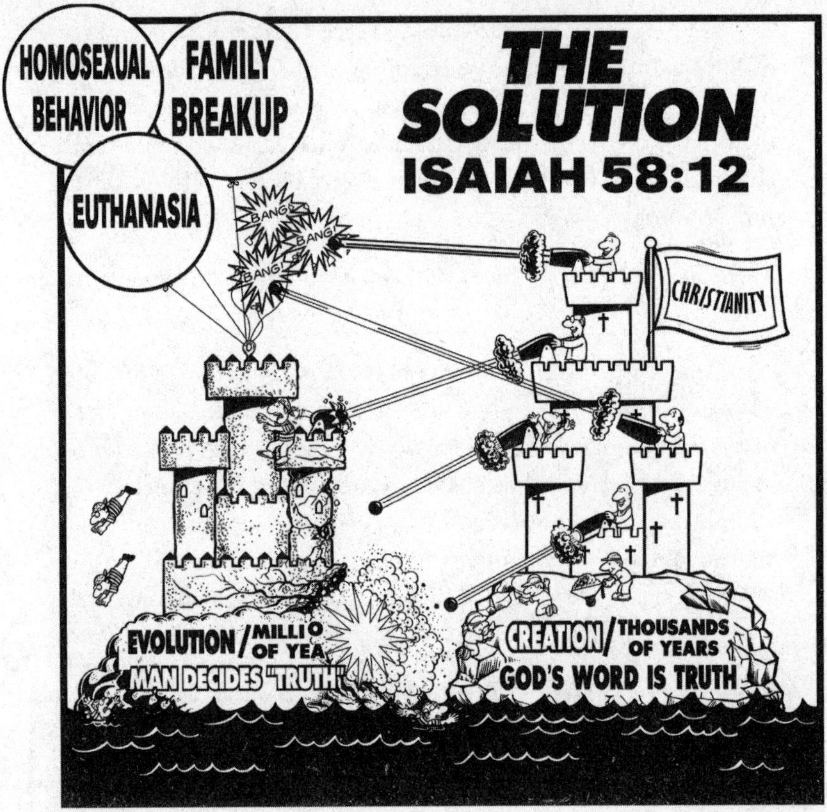

Well, it is true that I personally don't have formal theological training — but there are those at Answers in Genesis who do (e.g., Dr. Terry Mortenson, Steve Fazekas, Tim Chaffey, and some of our board members). And we do have quite a number of other highly qualified theologians whose counsel we seek to ensure we are accurate in handling God's Word.

By the way, I'm so glad I have not been theologically trained in the way Dr. McGrath has (and sadly like many who are now being trained in Bible colleges and seminaries). Otherwise, I might have ended up believing what he wrote below:

> So why am I a Christian? . . . given that I do not espouse Biblical literalism and inerrancy, some might ask whether I am still a Christian. . . . I am a Christian in much the same way that I am an American . . . the tradition that gave birth to my faith and nurtured it is one that has great riches (as well as much else beside. . . . Why am I a Christian? Because I prefer to keep the tradition I

have, rather than discarding it with the bathwater and then trying to make something new from scratch.

The second sad example is from Dr. William Dembski, a professor at what is known as a conservative seminary in the South. What he proposes in his book *The End of Christianity* is an undermining of biblical authority, and it's an unfortunate example of the sort of compromise often being taught to our future pastors. Here are a few excerpts from his book:

> For the theodicy I am proposing to be compatible with evolution, God must not merely introduce existing human-like beings from outside the Garden. In addition, when they enter the Garden, God must transform their consciousness so that they become rational moral agents made in God's image.[2]

Also:

> Moreover, once God breathes the breath of life into them, we may assume that the first humans experienced an amnesia of their former animal life: Operating on a higher plane of consciousness once infused with the breath of life, they would transcend the lower plane of animal consciousness on which they had previously operated — though, after the Fall, they might be tempted to resort to that lower consciousness.[3]

Dr. Dembski also states:

> The young-earth solution to reconciling the order of creation with natural history makes good exegetical and theological sense. Indeed, the overwhelming consensus of theologians up through the Reformation held to this view. I myself would adopt it in a heartbeat except that nature seems to present such a strong evidence against it.[4]

By "nature" he is in essence accepting fallible scientists' interpretations of evidence (such as fossils, geologic layers, and so on). His statement concerning "good exegetical and theological sense" is the point exactly! In other

2. William A. Dembski, *The End of Christianity* (Nashville, TN: B & H Academic, 2009), p. 159.
3. Ibid., p. 154–155.
4. Ibid., p. 55.

words, we know what the clear teaching of Scripture is — and what the great Reformers knew. But Dembski rejects it.

We would say that Dr. Dembski (who may be a fine Christian man) is taking the belief in billions of years (obtained by man's fallible interpretations of the present in an attempt to connect to the past) as infallible, and in reality making God's Word fallible.

A "Genesis 3 Attack"

This is the "Genesis 3 attack" ("Did God Really Say?") in our era — undoing what the Reformation accomplished. We need a new Reformation to call our Church (and culture) back to the authority of the Word of God. This is why the ministry of Answers in Genesis is so vital today — please pray for us!

Thank you for supporting Answers in Genesis . . . and for helping to bring a new and much-needed Reformation to our church and culture. The battle before us is one about authority: is God's Word the authority, or is it man's words?

We will continue (despite the opposition we receive) to hold compromising church leaders accountable, and stand unashamedly and uncompromisingly on the authority of the Word of God. That's what the Answers in Genesis and Creation Museum outreaches are all about.

1

What Is a Biblical Worldview?

STACIA MCKEEVER & KEN HAM

The history as recorded in the Bible has been attacked by our increasingly secular culture. As a result, recent generations have been brought up to see the Bible as a book that contains many interesting stories and religious teaching, but has no connection to reality.

This limited viewpoint helps explain why there are so many questions about how the Bible can explain dinosaurs, fossils, death, suffering, and many other topics that relate to our real world.

This chapter will outline the major events of the past (and even the future) — the "7 Cs of History" — that are foundational to the Bible's important message and demonstrate how the Bible connects to the real world.

Creation

God created the heavens, the earth, and all that is in them in six normal-length days around 6,000 years ago. His completed *creation* was "very good" (Genesis 1:31), and all the original animals (including dinosaurs) and the first two humans (Adam and Eve) ate only plants (Genesis 1:29–30). Life was perfect and not yet affected by the Curse — death, violence, disease, sickness, thorns, and fear had no part in the original creation.

After He was finished creating, God "rested" (or stopped) from His work, although He continues to uphold the creation (Colossians 1:17). His creation of all things in six days and resting on the seventh set a pattern for our week, which He designed for us to follow.

The science of "information theory" confirms that first statement of the Bible, "In the beginning God created. . . ." DNA is the molecule of heredity, part of a staggeringly complex system, more information dense than that in the most efficient supercomputer. Since the information in our DNA can only come from a source of greater information (or intelligence), there must have been something other than matter in the beginning. This other source must have no limit to its intelligence; in fact, it must be an ultimate source of intelligence from which all things have come. The Bible tells us there is such a source — God. Since God has no beginning and no end and knows all (Psalm 147:5), it makes sense that God is the source of the information we see all around us! This fits with real science, just as we would expect.[1]

In Genesis, God created things "after their kinds." And this is what we observe today: great variation within different "kinds" (e.g., dogs, cats, elephants, etc.), but not one kind changing into another, as molecules-to-man evolution requires.[2]

Corruption

After God completed His perfect creation, He told Adam that he could eat from any tree in the Garden of Eden (Genesis 2:8) except one — the Tree of the Knowledge of Good and Evil. He warned Adam that death would be the punishment for disobedience (Genesis 2:17). Instead of listening to the command of his Creator, Adam chose to rebel, eating the fruit from the tree (Genesis 3:6). Because our Holy God must punish sin, He sacrificed animals to make coverings for Adam and Eve, and He sent the first couple from the garden, mercifully denying them access to the Tree of Life so that they would not live forever in their sinful state.

Adam's sin ushered death, sickness, and sorrow into the once-perfect creation (Genesis 3:19; Romans 5:12). God also pronounced a curse on the world (Genesis 3; Romans 8:20–22). As a result, the world that we now live in is a decaying remnant — a *corruption* — of the beautiful, righteous world that Adam and Eve originally called home. We see the results of this corruption all

1. For a more in-depth analysis of the complexity of DNA and information theory, see www.AnswersInGenesis.org/go/information_theory.
2. For more information, see www.AnswersInGenesis.org/go/kinds.

around us in the form of carnivorous animals, mutations, sickness, disease, and death.[3] The good news is that, rather than leave His precious handiwork without hope, God graciously promised to one day send a Redeemer who would buy back His people from the curse of sin (Genesis 3:15).

Catastrophe

As the descendants of Adam and Eve married and filled the earth with offspring, their wickedness was great (Genesis 6:5). God judged their sin by sending a global flood to destroy all men, animals, creatures that moved along the ground, and birds of the air (Genesis 6:7). Those God chose to enter the ark — Noah, his family, and land-dwelling representatives of the animal kingdom (including dinosaurs) — were saved from the watery *catastrophe.*

There was plenty of room in the huge vessel for tens of thousands of animals — even dinosaurs (the average dinosaur was only the size of a sheep, and Noah didn't have to take fully grown adults of the large dinosaurs). Noah actually needed only about 16,000 animals on the ark to represent all the distinct kinds of land-dwelling animals.[4]

This earth-covering event has left its mark even today. From the thousands of feet of sedimentary rock found around the world to the billions of dead things buried in rock layers (fossils), the Flood reminds us even today that our righteous God cannot — and will not — tolerate sin, while the ark reminds us that He provides a way of salvation from sin's punishment. The rainbows we experience today remind us of God's promise never again to destroy the earth with water (Genesis 9:13–15). Incidentally, if the Flood were a local event (rather than global in extent), as some claim, then God has repeatedly broken His promise since we continue to experience local flooding even today.[5]

Confusion

After the Flood, God commanded Noah and his family — the only humans left in the world — and the animals to fill the earth (Genesis 8:17). However, the human race once again disobeyed God's command and

3. For more information, see www.AnswersInGenesis.org/go/curse.
4. See *Noah's Ark: A Feasibility Study* by John Woodmorappe (Santee, CA: Institute for Creation Research, 1996) for a detailed analysis of the capacity of this huge ship to hold all the residents of the ark.
5. For more information, see www.AnswersInGenesis.org/go/flood.

built a tower, which they hoped would keep them together (Genesis 11:3–4). So, around 100 years after the Flood waters had retreated, God brought a *confusion* (a multiplicity) of languages in place of the common language the people shared, causing them to spread out over the earth. The several different languages created suddenly at Babel (Genesis 10–11) could each subsequently give rise to many more. Languages gradually change; so when a group of people breaks up into several groups that no longer interact, after a few centuries they may each speak a different (but related) language. Today, we have thousands of languages but fewer than 20 language "families."[6]

All the tribes and nations in the world today have descended from these various groups. Despite what you may have been led to believe about our seeming superficial differences, we really are all "one blood" (Acts 17:26) — descendants of Adam and Eve through Noah and his family — and all, therefore, are in need of salvation from sin.

God had created Adam and Eve with the ability to produce children with a variety of different characteristics. This ability was passed on through Noah and his family. As the people scattered, they took with them different amounts of genetic information for certain characteristics — e.g., height, the amount of pigment for hair and skin color (by the way, we all have the same pigment, just more or less of it), and so on.

In fact, the recent Human Genome Project supports this biblical teaching that there is only *one* biological race of humans. As one report says, "It is clear that what is called 'race' . . . reflects just a few continuous traits determined by a tiny fraction of our genes."[7] The basic principles of genetics explain various shades of *one* skin color (not different colors) and how the distinct people groups (e.g., American Indians, Australian Aborigines) came about because of the event at the Tower of Babel. The creation and Flood legends of these peoples, from all around the world, also confirm the Bible's anthropology to be true.

Christ

God's perfect creation was corrupted by Adam when he disobeyed God, ushering sin and death into the world. Because of Adam's disobedience and because we have all sinned personally, we are all deserving of the death penalty and need a Savior (Romans 5:12).

6. For more information, see www.AnswersInGenesis.org/go/linguistics.
7. S. Pääbo, "The Human Genome and Our View of Ourselves," *Science* 29, no. 5507 (2001)): 1219–1220.

As mentioned before, God did not leave His precious — but corrupted — creation without hope. He promised to one day send Someone who would take away the penalty for sin, which is death (Genesis 3:15; Ezekiel 18:4; Romans 6:23).

God killed at least one animal in the Garden of Eden because of the sin of Adam; subsequently, Adam's descendants sacrificed animals. Such sacrifices could only cover sin — they pointed toward the time when the One whom God would send (Hebrews 9) would make the ultimate sacrifice.

When God gave Moses the Law, people began to see that they could never measure up to God's standard of perfection (Romans 3:20) — if they broke any part of the Law, the result was the same as breaking all of it (James 2:10). They needed Someone to take away their imperfection and present them faultless before God's throne (Romans 5:9; 1 Peter 3:18).

In line with God's purpose and plan for everything, He sent His promised Savior at just the right time (Galatians 4:4). There was a problem, however. All humans are descended from Adam and therefore, all humans are born with sin. God's chosen One had to be perfect, as well as infinite, to take away the infinite penalty for sin.

God solved this "problem" by sending His Son, Jesus *Christ* — completely human and completely God. Think of it: the Creator of the universe (John 1:1–3, 14) became part of His creation so that He might save His people from their sins!

Jesus fulfilled more than 50 prophecies made about Him centuries before, showing He was the One promised over 4,000 years before by His Father (Genesis 3:15). While He spent over 30 years on earth, He never once sinned — He did nothing wrong. He healed many people, fed huge crowds, and taught thousands of listeners about their Creator God and how to be reconciled to Him. He even confirmed the truth of Genesis by explaining that marriage is between one man and one woman (Matthew 19:3–6, quoting Genesis 1:27 and 2:24).

Cross

Jesus is called the "Last Adam" in 1 Corinthians 15:45. While Adam disobeyed God's command not to eat the forbidden fruit, Jesus fulfilled the Creator's purpose that He die for the sin of the world.

The first Adam brought death into the world through his disobedience; the Last Adam brought eternal life with God through His obedience (1 Corinthians 15:21–22).

Because God is perfectly holy, He must punish sin — either the sinner himself or a substitute to bear His wrath. Jesus bore God's wrath for our sin by dying in our place on the Cross (Isaiah 53:6). The Lamb of God (John 1:29; Revelation 5:12) was sacrificed once for all (Hebrews 7:27), so that all those who believe in Him will be saved from the ultimate penalty for sin (eternal separation from God) and will live with Him forever.

Jesus Christ, the Creator of all things (John 1:1–3; Colossians 1:15–16), was not defeated by death. He rose three days after He was crucified, showing that He has power over all things, including death, the "last enemy" (1 Corinthians 15:26). As Paul wrote, "O death, where is your sting? O grave, where is your victory? . . . But thanks be to God who gives us the victory through our Lord Jesus Christ" (1 Corinthians 15:55–57).

When we believe in Christ and understand what He has done for us, we are passed from death into life (John 5:24). The names of those who receive Him are written in the Lamb's Book of Life (Revelation 13:8; 17:8) — when they die, they will go to be with Him forever (John 3:16).

Just as "science" cannot prove that Jesus rose from the dead, it also cannot prove that God created everything in six days. In fact, "science" can't prove any event from history because it is limited in dealings about the past. Historical events are known to be true because of reliable eyewitness accounts. In fact, there are reliable eyewitness accounts that Jesus' tomb was empty after three days and that He later appeared to as many as 500 people at once (1 Corinthians 15:6). Of course, we know that both the Resurrection and creation in six days are true because God, who cannot lie, states in His Word that these things happened.

While the secular history of millions of years isn't true, and evolutionary geology, biology, anthropology, astronomy, etc., do not stand the test of observational science, the Bible's history, from Genesis 1 onward, *is* true; the Bible's geology, biology, anthropology, astronomy, etc., are confirmed by observational science. Therefore, the fact that the Bible's history is true should challenge people to seriously consider the Bible's message of salvation that is based in this history.

Consummation

Death has been around almost as long as humans have. Romans 8 tells us that the whole of creation is suffering because of Adam's sin. As terrible as things are, however, they are not a permanent part of creation.

God, in His great mercy, has promised not to leave His creation in its sinful state. He has promised to do away with the corruption that Adam brought into the world. He has promised to remove, in the future, the curse He placed on His creation (Revelation 22:3) and to make a new heaven and a new earth (2 Peter 3:13). In this new place there will be no death, crying, or pain (Revelation 21:4).

Those who have repented and believed in what Jesus did for them on the Cross can look forward to the consummation of God's kingdom — this new heaven and earth — knowing they will enjoy God forever in a wonderful place. In the future, God will take away the corruption that was introduced in the Garden of Eden, giving us once again a perfect place to live!

A worldview based on a proper understanding of the history of the world, as revealed in the Bible, is what every Christian needs to combat our society's evolutionary propaganda.

2

What's the Best "Proof" of Creation?

KEN HAM

In the ongoing war between creation and evolution, Christians are always looking for the strongest evidence for creation. They are looking for the "magic bullet" that will prove to their evolutionist friends that creation is true and evolution is false. This craving for evidence has led some Christians to be drawn to what we might call "flaky evidence." Over the past several years, some so-called evidence for creation has been shown not to be reliable. Some of these are

- supposed human and dinosaur footprints found together at the Paluxy River in Texas;
- the small accumulation of moon dust found by the Apollo astronauts;
- a boat-like structure in the Ararat region as evidence of Noah's ark;
- a supposed human handprint found in "dinosaur-age rock";
- a dead "plesiosaur" caught near New Zealand.

Most well-meaning, informed creationists would agree in principle that things which are not carefully documented and researched should not be used. But in practice, many of them are very quick to accept the sorts of facts mentioned here, without asking too many questions. They are less cautious than they might otherwise be, because they are so keen to have "our" facts/evidences

to counter "theirs." What they really don't understand, however, is that it's not a matter of "their facts vs. ours." *All* facts are actually interpreted, and *all* scientists actually have the *same* observations — the same data — available to them.

Evidence

Creationists and evolutionists, Christians and non-Christians, all have the same facts. Think about it: we all have the same earth, the same fossil layers, the same animals and plants, the same stars — the facts are all the same.

The difference is in the way we all *interpret* the facts. And why do we interpret facts differently? Because we start with different *presuppositions*; these are things that are assumed to be true without being able to prove them. These then become the basis for other conclusions. *All* reasoning is based on presuppositions (also called *axioms*). This becomes especially relevant when dealing with past events.

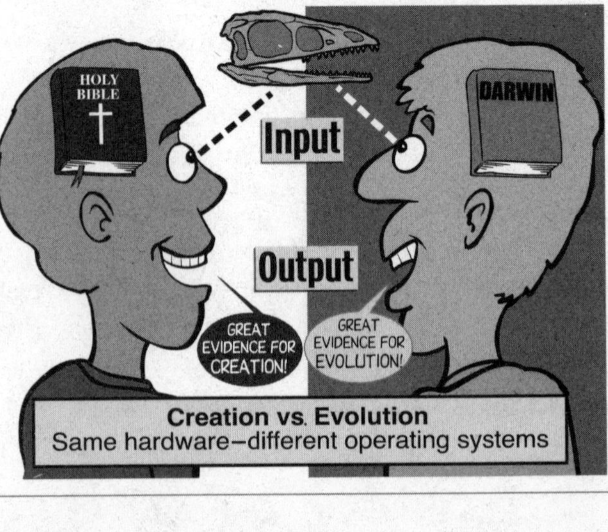

Past and Present

We all exist in the present, and the facts all exist in the present. When one is trying to understand how the evidence came about — Where did the animals come from? How did the fossil layers form? etc. — what we are actually trying to do is to connect the past to the present. However, if we weren't there in the past to observe events, how can we know what happened so that we can explain the present? It would be great to have a time machine so that we could know for sure about past events.

Christians, of course, claim they do have, in a sense, a time machine. They have a book called the Bible, which claims to be the Word of God who has always been there and has revealed to us the major events of the past about which we need to know. On the basis of these events (creation, the Fall, the Flood, Babel, etc.), we have a set of presuppositions to build a way of thinking which enables us to interpret the facts of the present.[1]

Evolutionists have certain beliefs about the past/present that they presuppose (e.g., no God, or at least none who performed acts of special creation), so they build a different way of thinking to interpret the facts of the present.

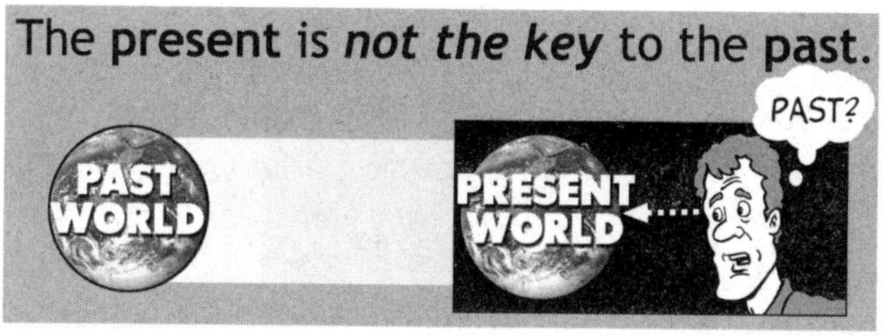

Thus, when Christians and non-Christians argue about the facts, in reality they are arguing about their *interpretations* based on their *presuppositions.*

That's why the argument often turns into something like:

"Can't you see what I'm talking about?"

"No, I can't. Don't you see how wrong you are?"

"No, I'm not wrong. It's obvious that I'm right."

"No, it's not obvious."

And so on.

These two people are arguing about the same facts, but they are looking at the facts through different glasses.

It's not until these two people recognize the argument is really about the presuppositions they have to start with that they will begin to deal with the foundational reasons for their different beliefs. A person will not interpret the facts differently until he or she puts on a different set of glasses — which means to change one's presuppositions.

1. See chapter 1 on "What Is a Biblical Worldview?" for further development of this idea.

A Christian who understands these things can actually put on the evolutionist's glasses (without accepting the presuppositions as true) and understand how he or she looks at facts. However, for a number of reasons, including spiritual ones, a non-Christian usually can't put on the Christian's glasses — unless he or she recognizes the presuppositional nature of the battle and is thus beginning to question his or her own presuppositions.

It is, of course, sometimes possible that just by presenting "evidence" one can convince a person that a particular scientific argument for creation makes sense on "the facts." But usually, if that person then hears a different *interpretation* of the same facts that seems better than the first, that person will swing away from the first argument, thinking he or she has found "stronger facts."

However, if that person had been helped to understand this issue of presuppositions, then he or she would have been better able to recognize this for what it is — a different interpretation based on differing presuppositions (i.e., starting beliefs).

Debate Terms

Often people who don't believe the Bible will say that they aren't interested in hearing about the Bible. They want real proof that there's a God who created. They'll listen to our claims about Christianity, but they want proof *without mentioning the Bible.*

If one agrees to a discussion without using the Bible as these people insist, then we have allowed *them* to set the terms of the debate. In essence these terms are

1. **"Facts" are neutral**. However, there are no such things as "brute facts"; *all* facts are interpreted. Once the Bible is eliminated from the argument, the Christians' presuppositions are gone, leaving them unable to effectively give an alternate interpretation of the facts. Their opponents then have the upper hand as they still have *their* presuppositions.

2. **Truth can/should be determined independently of God.** However, the Bible states: "The fear of the Lord is the beginning of wisdom" (Psalm 111:10); "The fear of the Lord is the beginning of knowledge" (Proverbs 1:7); "But the natural man does not receive the things of the Spirit of God, for they are foolishness to him; neither can he know them, because they are spiritually discerned" (1 Corinthians 2:14).

A Christian cannot divorce the spiritual nature of the battle from the battle itself. A non-Christian is *not* neutral. The Bible makes this very clear: "The one who is not with Me is against Me, and the one who does not gather with Me scatters" (Matthew 12:30); "And this is the condemnation, that the Light has come into the world, and men loved darkness rather than the Light, because their deeds were evil" (John 3:19).

Agreeing to such terms of debate also implicitly accepts the proposition that the Bible's account of the universe's history is irrelevant to understanding that history!

Ultimately, God's Word Convicts

First Peter 3:15 and other passages make it clear we are to use every argument we can to convince people of the truth, and 2 Corinthians 10:4–5 says we are to refute error (as Paul did in his ministry to the Gentiles). Nonetheless, we must never forget Hebrews 4:12: "For the word of God is living and powerful and sharper than any two-edged sword, piercing even to the dividing apart of soul and spirit, and of the joints and marrow, and is a discerner of the thoughts and intents of the heart."

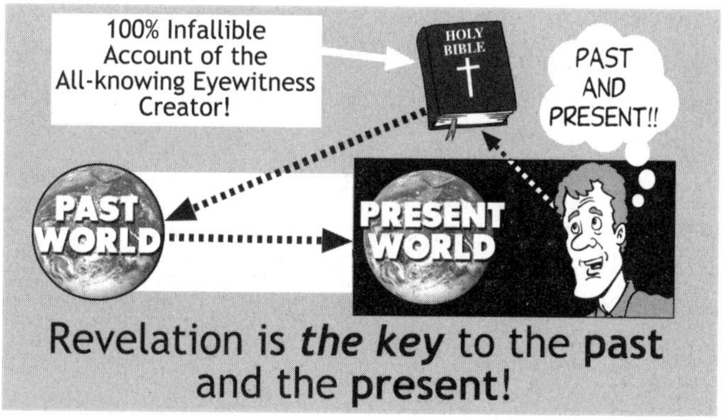

Also, Isaiah 55:11 says, "So shall My word be, which goes out of My mouth; it shall not return to Me void, but it shall accomplish what I please, and it shall certainly do what I sent it to do."

Even though our human arguments may be powerful, ultimately it is God's Word that convicts and opens people to the truth. In all of our arguments, we must not divorce what we are saying from the Word that convicts.

Practical Application

When someone says he wants "proof" or "evidence," not the Bible, one might respond as follows:

> You might not believe the Bible, but I do. And I believe it gives me the right basis to understand this universe and correctly interpret the facts around me. I'm going to give you some examples of how building my thinking on the Bible explains the world and is not contradicted by science.

One can, of course, do this with numerous scientific examples, showing, for example, how the issue of sin and judgment is relevant to geology and fossil evidence; how the fall of man, with the subsequent curse on creation, makes sense of the evidence of harmful mutations, violence, and death; or how the original "kinds" of animals gave rise to the wide variety of animals we see today.

Choose a topic and develop it:

> For instance, the Bible states that God made distinct *kinds* of animals and plants. Let me show you what happens when I build my thinking on this presupposition. I will illustrate how processes such as natural selection, genetic drift, etc., can be explained and interpreted. You will see how the science of genetics makes sense based upon the Bible. Evolutionists believe in natural selection — that is real science, as you observe it happening. Well, creationists also believe in natural selection. Evolutionists accept the science of genetics — well, so do creationists.
>
> However, here is the difference: evolutionists believe that, over millions of years, one kind of animal has changed into a totally different kind. However, creationists, based on the Bible's account of origins, believe that God created separate kinds of animals and plants to reproduce their own kind; therefore, one kind will not turn into a totally different kind.

> Now this can be tested in the present. The scientific observations support the creationist interpretation that the changes we see are not creating new information. The changes are all within the originally created pool of information of that kind — sorting, shuffling, or degrading it. The creationist account of history, based on the Bible, provides the correct basis to interpret the facts of the present; and real science confirms the interpretation.

After this detailed explanation, continue like this:

> Now let me ask you to defend *your* position concerning these matters. Please show me how *your* way of thinking, based on *your* beliefs, makes sense of the same evidence. And I want you to point out where my science and logic are wrong.

In arguing this way, a Christian is

1. using biblical presuppositions to build a way of thinking to interpret the evidence;
2. showing that the Bible and science go hand in hand;
3. challenging the presuppositions of the other person (many are unaware they have these);
4. forcing the debater to logically defend his position consistent with science and his own presuppositions (many will find that they cannot do this), and help this person realize they do have presuppositions that can be challenged;
5. honoring the Word of God that convicts the soul.

If Christians really understood that all facts are actually interpreted on the basis of certain presuppositions, we wouldn't be in the least bit intimidated by the evolutionists' supposed "evidence." We should instead be looking at the evolutionists' (or old-earthers'[2]) *interpretation* of the evidence, and how the same evidence could be interpreted within a biblical framework and confirmed by testable and repeatable science. If more creationists did this, they would be less likely to jump at flaky evidence that seems startling but in reality has been interpreted incorrectly in their rush to find the knockdown, drag-out convincing "evidence" against evolution that they think they desperately need.

The various age-dating methods are also subject to interpretation. All dating methods suffer, in principle, from the same limitations — whether they are used to support a young world or an old world. For instance, the public

2. Those who accept millions of years of history.

reads almost daily in newspapers and magazines that scientists have dated a particular rock at billions of years old. Most just accept this. However, creation scientists have learned to ask questions as to how this date was obtained — what method was used and what *assumptions* were accepted to develop this method? These scientists then question those assumptions (questions) to see whether they are valid and to determine whether the rock's age could be interpreted differently. Then the results are published to help people understand that scientists have not proven that the rock is billions of years old and that the facts can be interpreted in a different way to support a young age.

Consider the research from the creationist group Radioisotopes and the Age of The Earth (RATE) concerning the age of zircon crystals in granite.[3] Using one set of assumptions, these crystals could be interpreted to be around 1.5 billion years old, based on the amount of lead produced from the decay of uranium (which also produces helium). However, if one questions these assumptions, one is motivated to test them. Measurements of the rate at which helium is able to "leak out" of these crystals indicate that if they were much older than about 6,000 years, they would have nowhere near the amount of helium still left in them. Hence, the originally applied assumption of a constant decay rate is flawed; one must assume, instead, that there has been acceleration of the decay rate in the past. Using this revised assumption, the same uranium-lead data can now be interpreted to also give an age of fewer than 6,000 years.

Another example involves red blood cells and traces of hemoglobin that have been found in *T. rex* bones, although these should have long decomposed if they were millions of years old. Yet the reaction of the researchers was a perfect illustration of how evolutionary bias can result in trying to explain away hard facts to fit the preconceived framework of millions of years:

> It was exactly like looking at a slice of modern bone. But, of course, I couldn't believe it. I said to the lab technician: "The bones, after all, are 65 million years old. How could blood cells survive that long?"[4]

Whenever you hear a news report that scientists have found another "missing link" or discovered a fossil "millions of years old," try to think about the

3. R. Humphreys, "Young Helium Diffusion Age of Zircons Supports Accelerated Nuclear Decay," in Larry Vardiman, Andrew Snelling, and Eugene Chaffin, eds., *Radioisotopes and the Age of the Earth*, vol. 2 (El Cajon, CA: Institute for Creation Research; Chino, Valley, AZ: Creation Research Society, 2005), p. 25–100.
4. *Science* 261 (July 9, 1994): 160; see also, "Scientists Recover *T. rex* Soft Tissue: 70-million-year-old Fossil Yields Preserved Blood Vessels," www.msnbc.msn.com/id/7285683/, March 24, 2005.

right questions that need to be asked to challenge the questions these scientists asked to get their interpretations!

All of this should be a lesson for us to take note of the situation when we read the newspaper — we are reading someone's interpretation of the facts of world history — there very well could be a different way of looking at the same "facts." One can see this in practice on television when comparing a news network that's currently considered fairly liberal (CNN) with one that is more conservative (FOX) — one can often see the same "facts" interpreted differently!

The reason so many Christian professors (and Christian leaders in general) have rejected the literal creation position is that they have blindly accepted the interpretation of facts from the secular world, based on man's fallible presuppositions about history. And they have then tried to reinterpret the Bible accordingly. If only they would start with the presupposition that God's Word is true, they would find that they could then correctly interpret the facts of the present and show overwhelmingly that observational science repeatedly confirms such interpretations.

Don't forget, as Christians we need to always build our thinking on the Word of the One who has the answers to all of the questions that could ever be asked — the infinite Creator God. He has revealed the true history of the universe in His Word to enable us to develop the right way of thinking about the present and thus determine the correct interpretations of the evidence of the present. We should follow Proverbs 1:7 and 9:10, which teach that fear of the Lord is the beginning of true wisdom and knowledge.

The Bottom Line

The bottom line is that it's not a matter of who has the better (or the most) "facts on their side." We need to understand that there are no such things as brute facts — *all* facts are interpreted. The next time evolutionists use what seem to be convincing facts for evolution, try to determine the *presuppositions* they have used to interpret these facts. Then, beginning with the big picture of history from the Bible, look at the same facts through these biblical glasses and interpret them differently. Next, using the real science of the present that an evolutionist also uses, see if that science, when properly understood, confirms (by being consistent with) the interpretation based on the Bible. You will find over and over again that the Bible is confirmed by real science.

But remember that, like Job (42:2–6), we need to understand that compared to God we know next to nothing. We won't have all the answers. However, so many answers have come to light now that a Christian can give a

credible defense of the Book of Genesis and show it is the correct foundation for thinking about, and interpreting, every aspect of reality.

Therefore, let's not jump in a blind-faith way at the startling facts we think we need to "prove" creation — trying to counter "their facts" with "our facts." (Jesus himself rose from the dead in the most startling possible demonstration of the truth of God's Word. But many still wouldn't believe — see Luke 16:27–31.) Instead, let's not let apparent facts for evolution intimidate us, but let's understand the right way to think about facts. We can then deal with *the same facts the evolutionists use*, to show they have the wrong framework of interpretation — and that the facts of the real world really do conform to, and confirm, the Bible. In this way we can do battle for a biblical worldview.

Remember, it's no good convincing people to believe in creation, without also leading them to believe and trust in the Creator and Redeemer, Jesus Christ. God honors those who honor His Word. We need to use God-honoring ways of reaching people with the truth of what life is all about.

3

Are Biblical Creationists Divisive?

BODIE HODGE

Biblical creationists[1] are often accused of causing division in the Church. It is claimed that their insistence on accepting Genesis as narrative history introduces dissension by majoring on a "minor" doctrine. However, as will be shown, quite the opposite it true.

Who Is Really Being Divisive?

Far too often, people have the wrong impression about what it means to be divisive. Those who are divisive are those who are against the clear teachings of the Bible. Paul made this clear in his letter to the Christians in Rome.

> Now I urge you, brethren, note those who cause divisions and offenses, contrary to the doctrine which you learned, and avoid them. For those who are such do not serve our Lord Jesus Christ, but their own belly, and by smooth words and flattering speech deceive the hearts of the simple (Romans 16:17–18).

Jude also confirmed that unbiblical beliefs cause divisions:

1. Biblical creationists are often termed *young-earth creationists* (YEC). They adopt a "plain" or "straightforward" reading of the Bible; thus Genesis, which is written as historical narrative, is literal history and the days in Genesis 1 are ordinary days. A corollary to this is that the earth is young. See chapter 4, "How Old Is the Earth?" for more information.

> But you, beloved, remember the words which were spoken before by the apostles of our Lord Jesus Christ: how they told you that there would be mockers in the last time who would walk according to their own ungodly lusts. These are sensual persons, who cause divisions, not having the Spirit (Jude 1:17–19).

Jude wrote that these divisions are caused by sensual, or worldly minded, beliefs. This should serve as a warning to those who accept man-made ideas that are opposed to the clear teachings of Scripture.

What are some of those clear teachings of Scripture?

- Sin entered the world through one man, and death through sin (Genesis 2:17, 3:17; Romans 5:12).
- Man and animals were originally vegetarian (Genesis 1:29–30).
- The week is composed of seven normal-length days (Genesis 1:1–2:4; Exodus 20:11).
- All people are descendants of Adam and Eve (Genesis 1:26–28, 3:20).
- People began to wear clothing after sin entered the world (Genesis 3:7, 21).
- Thorns and thistles resulted from the curse God placed on His creation after sin entered the world (Genesis 3:18).
- The flood of Noah's day was global in extent (Genesis 6–8).

These are not new doctrines — Paul, the other apostles, and Christ himself accepted these teachings. They (and biblical creationists today) understood that Genesis is a record of actual historical events. As a corollary of this, biblical creationists accept, based on careful study of the Bible, that the earth is thousands of years old (not billions).

The questioning of these teachings by many in the Church began in earnest around two hundred years ago. This was not due to a reexamination of Scripture, but rather because the culture had begun to teach an earth history of "millions of years."[2] The acceptance of the culture's ideas about the past has led to the reinterpretation of Genesis to fit with these man-made ideas. Some

2. Dr Terry Mortenson's book *The Great Turning Point* (Green Forest, AR: Master Books, 2004) discusses this in detail.

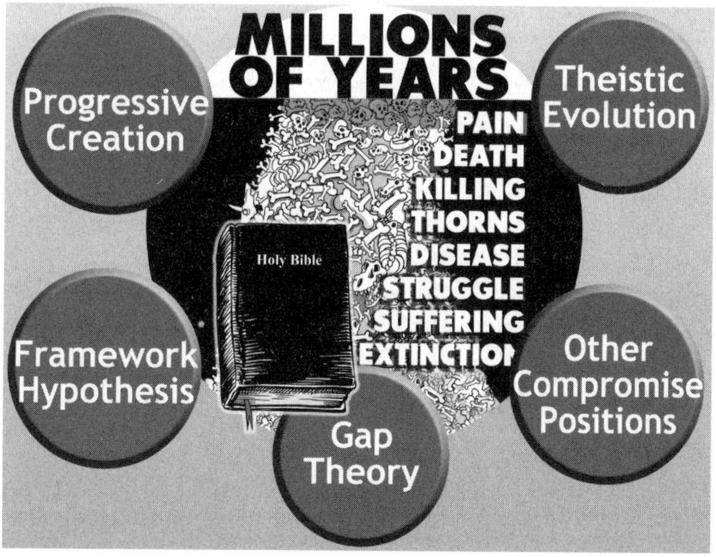

of these reinterpretations are the *framework hypotheses, gap theory, progressive creation*, and *theistic evolution*.

Each of these views attempts to combine the secular/evolutionary view of millions of years with the biblical view of history. In doing so, these views reject one or more of the clear teachings listed above. For example, each view rejects that the Genesis flood was a global event. Whenever one combines a man-made view with the Bible, something has to give. In most cases, this something is the Scripture. When one mixes the Word of God with another belief system, the result is doctrines that deviate from the Bible's clear teachings.

Sadly, these compromising beliefs have infiltrated many churches, Christian colleges, and seminaries. When a biblical creationist teaches people what the Bible plainly says, he is often told by adherents of these compromising views that *he* is being divisive. However, according to Paul, the ones causing division are those who deny the

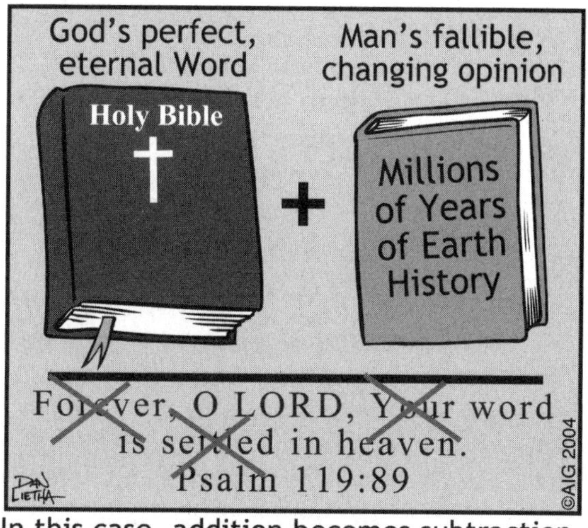

In this case, addition becomes subtraction.

doctrines clearly taught in Scripture. The divisive ones are those who mix the Bible with secular views and refuse to heed the call to return to the clear teachings in the Bible.

Wasn't Jesus Divisive?

Some have claimed, based on the following passages, that Jesus was divisive.

> For from now on five in one house will be *divided*: three against two, and two against three. Father will be *divided* against son and son against father, mother against daughter and daughter against mother, mother-in-law against her daughter-in-law and daughter-in-law against her mother-in-law (Luke 12:52–53; italics added).

> So there was a *division* among the people because of Him (John 7:43; italics added).

> Do not think that I came to bring peace on earth. I did not come to bring peace but a sword. For I have come to "*set a man against his father, a daughter against her mother, and a daughter-in-law against her mother-in-law*"; and "*a man's enemies will be those of his own household.*" He who loves father or mother more than Me is not worthy of Me. And he who loves son or daughter more than Me is not worthy of Me. And he who does not take his cross and follow after Me is not worthy of Me (Matthew 10:34–38; italics in original).

Carefully reread these passages and note that Jesus was not divisive! The *people* were divided because of the message. Jesus' message conformed to the doctrines laid down in the Old Testament; He came to fulfill the Law, not abolish it (Matthew 5:17)! As the perfect God, He was the One who inspired the writings of the Old Testament in the first place. *He* wasn't divisive; rather, those disagreeing with Him were causing divisions. Christ knew that His message would cause division among the people because many wouldn't believe and wouldn't adhere to the doctrines previously established.

The passages in Luke, John, and Matthew teach that the people were divided. There were those who received what Jesus taught (which is what the Scriptures taught, and thus was *not* divisive), and there were those who didn't. The ones who were divisive were those not adhering to what Jesus taught.

How Should I Deal with Those Who Are Divisive?

Paul and Barnabas's message divided the Jews. Some followed the apostle's teachings and others didn't. Remember, those being divisive were the ones opposed to the scriptural teachings.

> Therefore they stayed there a long time, speaking boldly in the Lord, who was bearing witness to the word of His grace, granting signs and wonders to be done by their hands. But the multitude of the city was divided: part sided with the Jews, and part with the apostles. And when a violent attempt was made by both the Gentiles and Jews, with their rulers, to abuse and stone them (Acts 14:3–5).

But notice what happened — those who were divisive found like-minded Gentiles (nonbelievers) to oppose Paul! Did Paul compromise like they did? No. Paul continued teaching the same message.

This is similar to what is happening in today's Church. Many readily adhere to secular millions-of-years teachings over the Bible's teachings. They are opposing the Scriptures. They are being divisive. Biblical creationists will continue to defend the authority of the Bible in all areas, just like Paul did.

Paul instructs us regarding divisive people:

> Reject a divisive man after the first and second admonition, knowing that such a person is warped and sinning, being self-condemned (Titus 3:10–11).

We are to confront the divisive twice or answer them twice. If they refuse to heed the words of correction, we are to have nothing more to do with them. This confirms what Jesus taught to the disciples when they were ministering;

they were to shake the dust from their feet as a testimony against those who refused to listen (Luke 9:5). They weren't to get wrapped up in an argument for extended periods of time but were to continue preaching the truth.

This is an important message for us today. We need to be careful that we don't get caught up in discussions with a divisive person for long periods of time (via e-mail, message boards, letters, phone calls, etc.). Instead, we need seek the millions waiting eagerly to hear the message that the Bible's history is true and the message of the gospel is likewise true.

The harvest is plentiful but the workers are few (Matthew 9:37). And there are fewer still when the harvesters get caught up trying to harvest wheat from a thistle when ten heads of wheat are waiting. Answer a divisive person twice. If that person continues to be divisive, have nothing more to do with him/her. If that person is genuinely willing to learn, continue to answer him/her with gentleness and respect (1 Peter 3:15).

Unity Comes by Uniting around What the Bible Clearly Teaches

Paul, as well as Jesus (John 17:22–23), makes it clear that there shouldn't be divisions but unity.

> Now I plead with you, brethren, by the name of our Lord Jesus Christ, that you all speak the same thing, and that there be no divisions among you, but that you be perfectly joined together in the same mind and in the same judgment (1 Corinthians 1:10).

This unity should not come at the expense of compromising the Scriptures, but should come by adhering to what the Scriptures say. This is why Paul exhorted the Roman Christians to take note of those causing divisions and avoid them (Romans 16:17). In other words, don't learn from those causing divisions (those who have accepted fallible man's ideas), but learn from those who adhere to the doctrines that have been laid down by Scripture.

> See to it that no one takes you captive through hollow and deceptive philosophy, which depends on human tradition and the basic principles of this world rather than on Christ (Colossians 2:8; NIV).

Before Adam and Eve sinned, they were in complete unity with each other and with their Creator. After they sinned, disunity became the norm. Restoring that unity comes at a cost. Christ has paid the price. This is a call for all Christians to return to what the Bible clearly teaches, and obey Christ's Word — starting in Genesis.

The following points provide some practical ways that we can encourage unity among our Christian brethren:

- Pray that the Lord would bring about unity among His people. Pray that He would turn the hearts of His children to the clear teachings in His Word and would keep them from being influenced by fallible man's ideas (Ephesians 4:13).
- Respond to those who are divisive (going against Scripture) twice, with gentleness and respect (1 Peter 3:15). If they are willing to learn, continue to help them. If they continue to be divisive, have nothing more to do with them.
- Avoid those who are openly divisive (going against Scripture). Encourage others to refrain from following their divisive example (Romans 16:17–18).
- Drop any pride of your own (Proverbs 16:18). Read and study the Word of God. Allow it to teach you, and be aware of bringing man-made ideas to it. Learn to love God's Word, and ask the Lord to show you where you are being divisive. No one is perfect, and all are subject to the teachings of the Bible. When we make mistakes, we need to return to the authority of God's Word with humility and a teachable spirit.

4

How Old Is the Earth?

BODIE HODGE

The question of the age of the earth has produced heated discussions on Internet debate boards, TV, radio, in classrooms, and in many churches, Christian colleges, and seminaries. The primary sides are

- Young-earth proponents (biblical age of the earth and universe of about 6,000 years)[1]
- Old-earth proponents (secular age of the earth of about 4.5 billion years and a universe about 14 billion years old)[2]

The difference is immense! Let's give a little history of where these two basic calculations came from and which worldview is more reasonable.

Where Did a Young-earth Worldview Come From?

Simply put, it came from the Bible. Of course, the Bible doesn't say explicitly anywhere, "The earth is 6,000 years old." Good thing it doesn't; otherwise it would be out of date the following year. But we wouldn't expect an all-knowing God to make that kind of a mistake.

1. Not all young-earth creationists agree on this age. Some believe that there may be small gaps in the genealogies of Genesis 5 and 11 and put the maximum age of the earth at about 10,000–12,000 years. However, see chapter 5, "Are There Gaps in the Genesis Geologies?"
2. Some of these old-earth proponents accept molecules-to-man biological evolution and so are called theistic evolutionists. Others reject neo-Darwinian evolution but accept the evolutionary timescale for stellar and geological evolution, and hence agree with the evolutionary order of events in history.

God gave us something better. In essence, He gave us a "birth certificate." For example, using a personal birth certificate, a person can calculate how old he is at any point. It is similar with the earth. Genesis 1 says that the earth was created on the first day of creation (Genesis 1:1–5). From there, we can begin to calculate the age of the earth.

Let's do a rough calculation to show how this works. The age of the earth can be estimated by taking the first five days of creation (from earth's creation to Adam), then following the genealogies from Adam to Abraham in Genesis 5 and 11, then adding in the time from Abraham to today.

Adam was created on day 6, so there were five days before him. If we add up the dates from Adam to Abraham, we get about 2,000 years, using the Masoretic Hebrew text of Genesis 5 and 11.[3] Whether Christian or secular, most scholars would agree that Abraham lived about 2,000 B.C. (4,000 years ago).

So a simple calculation is:

5 days
+ ~2,000 years
+ ~4,000 years
———
~6,000 years

At this point, the first five days are negligible. Quite a few people have done this calculation using the Masoretic text (which is what most English translations are based on) and with careful attention to the biblical details, they have arrived at the same time frame of about 6,000 years, or about 4000 B.C. Two of the most popular, and perhaps best, are a recent work by Dr. Floyd Jones[4] and a much earlier book by Archbishop James Ussher[5] (1581–1656). See table 1.

Table 1. Jones and Ussher

Name	Age Calculated	Reference and Date
Archbishop James Ussher	4004 B.C.	*The Annals of the World*, A.D. 1658
Dr. Floyd Nolan Jones	4004 B.C.	*The Chronology of the Old Testament*, A.D. 1993

3. Bodie Hodge, "Ancient Patriarchs in Genesis," Answers in Genesis, www.answersingenesis.org/articles/2009/01/20/ancient-patriarchs-in-genesis.
4. Floyd Nolan Jones, *Chronology of the Old Testament* (Green Forest, AR: Master Books, 2005).
5. James Ussher, *The Annals of the World,* transl. Larry and Marion Pierce (Green Forest, AR: Master Books, 2003).

The misconception exists that Ussher and Jones were the only ones to arrive at a date of 4000 B.C.; however, this is not the case at all. Jones[6] lists several chronologists who have undertaken the task of calculating the age of the earth based on the Bible, and their calculations range from 5501 to 3836 B.C. A few are listed in table 2.

Table 2. Chronologists' Calculations According to Dr. Jones

	Chronologist	When Calculated?	Date B.C.
1	Julius Africanus	c. 240	5501
2	George Syncellus	c. 810	5492
3	John Jackson	1752	5426
4	Dr William Hales	c. 1830	5411
5	Eusebius	c. 330	5199
6	Marianus Scotus	c. 1070	4192
7	L. Condomanus	n/a	4141
8	Thomas Lydiat	c. 1600	4103
9	M. Michael Maestlinus	c. 1600	4079
10	J. Ricciolus	n/a	4062
11	Jacob Salianus	c. 1600	4053
12	H. Spondanus	c. 1600	4051
13	Martin Anstey	1913	4042
14	W. Lange	n/a	4041
15	E. Reinholt	n/a	4021
16	J. Cappellus	c. 1600	4005
17	E. Greswell	1830	4004
18	E. Faulstich	1986	4001
19	D. Petavius	c. 1627	3983
20	Frank Klassen	1975	3975
21	Becke	n/a	3974
22	Krentzeim	n/a	3971
23	W. Dolen	2003	3971
24	E. Reusnerus	n/a	3970
25	J. Claverius	n/a	3968
26	C. Longomontanus	c. 1600	3966
27	P. Melanchthon	c. 1550	3964
28	J. Haynlinus	n/a	3963
29	A. Salmeron	d. 1585	3958
30	J. Scaliger	d. 1609	3949
31	M. Beroaldus	c. 1575	3927
32	A. Helwigius	c. 1630	3836

6. Jones, *Chronology of the Old Testament*, p. 26.

As you will likely note from table 2, the dates are not all 4004 B.C. There are several reasons chronologists have different dates,[7] but two primary reasons:

1. Some used the Septuagint or another early translation instead of the Hebrew Masoretic text. The Septuagint is a Greek translation of the Hebrew Old Testament, done about 250 B.C. by about 70 Jewish scholars (hence it is often cited as the LXX, which is the Roman numeral for 70). It is good in most places, but appears to have a number of inaccuracies. For example, one relates to the Genesis chronologies where the LXX indicates that Methuselah would have lived past the Flood, without being on the ark!

2. Several points in the biblical time-line are not straightforward to calculate. They require very careful study of more than one passage. These include exactly how much time the Israelites were in Egypt and what Terah's age was when Abraham was born. (See Jones's and Ussher's books for a detailed discussion of these difficulties.)

The first four in table 2 (bolded) are calculated from the Septuagint, which gives ages for the patriarchs' firstborn much higher than the Masoretic text or the Samarian Pentateuch (a version of the Old Testament from the Jews in Samaria just before Christ). Because of this, the Septuagint adds in extra time. Though the Samarian and Masoretic texts are much closer, they still have a few differences. See table 3.

Using data from table 2 (excluding the Septuagint calculations and including Jones and Ussher), the average date of the creation of the earth is 4045 B.C. This still yields an average of about 6,000 years for the age of the earth.

Extra-biblical Calculations for the Age of the Earth

Cultures throughout the world have kept track of history as well. From a biblical perspective, we would expect the dates given for creation of the earth to align more closely to the biblical date than billions of years.

This is expected since everyone was descended from Noah and scattered from the Tower of Babel. Another expectation is that there should be some discrepancies

7. Others would include gaps in the chronology based on the presences of an extra Cainan in Luke 3:36. But there are good reasons this should be left out. See chapter 5, "Are There Gaps in the Genesis Genealogies?" and chapter 27, "Isn't the Bible Full of Contradictions?"

Table 3. Septuagint, Masoretic, and Samarian Early Patriarchal Ages at the Birth of the Following Son

Name	Masoretic	Samarian Pentateuch	Septuagint
Adam	130	130	230
Seth	105	105	205
Enosh	90	90	190
Cainan	70	70	170
Mahalaleel	65	65	165
Jared	162	62	162
Enoch	65	65	165
Methuselah	187	67	167
Lamech	182	53	188
Noah	500	500	500

about the age of the earth among people as they scattered throughout the world, taking their uninspired records or oral history to different parts of the globe.

Under the entry "creation," *Young's Analytical Concordance of the Bible*[8] lists William Hales's accumulation of dates of creation from many cultures, and in most cases Hales says which authority gave the date. See table 4.

Historian Bill Cooper's research in *After the Flood* provides intriguing dates from several ancient cultures.[9] The first is that of the Anglo-Saxons, whose history has 5,200 years from creation to Christ, according to the Laud and Parker Chronicles. Cooper's research also indicated that Nennius's record of the ancient British history has 5,228 years from creation to Christ. The Irish chronology has a date of about 4000 B.C. for creation, which is surprisingly close to Ussher and Jones! Even the Mayans had a date for the Flood of 3113 B.C.

This meticulous work of many historians should not be ignored. Their dates of only thousands of years are good support for the biblical date of about 6,000 years, but not for billions of years.

The Origin of the Old-earth Worldview

Prior to the 1700s, few believed in an old earth. The approximate 6,000-year age for the earth was challenged only rather recently, beginning in the late 18th

8. Robert Young, *Young's Analytical Concordance to the Bible* (Peadoby, MA: Hendrickson, 1996), referring to William Hales, *A New Analysis of Chronology and Geography, History and Prophecy,* vol. 1 (1830), p. 210.
9. Bill Cooper, *After the Flood* (UK: New Wine Press, 1995), p. 122–129.

Table 4. Selected Dates for the Age of the Earth by Various Cultures

Culture	Age, B.C.	Authority listed by Hales
Spain by Alfonso X	6984	Muller
Spain by Alfonso X	6484	Strauchius
India	6204	Gentil
India	6174	Arab records
Babylon	6158	Bailly
Chinese	6157	Bailly
Greece by Diogenes Laertius	6138	Playfair
Egypt	6081	Bailly
Persia	5507	Bailly
Israel/Judea by Josephus	5555	Playfair
Israel/Judea by Josephus	5481	Jackson
Israel/Judea by Josephus	5402	Hales
Israel/Judea by Josephus	4698	University history
India	5369	Megasthenes
Babylon (Talmud)	5344	Petrus Alliacens
Vatican (Catholic using the Septuagint)	5270	N/A
Samaria	4427	Scaliger
German, Holy Roman Empire by Johannes Kepler*	3993	Playfair
German, reformer by Martin Luther*	3961	N/A
Israel/Judea by computation	3760	Strauchius
Israel/Judea by Rabbi Lipman*	3616	University history

* Luther, Kepler, Lipman, and the Jewish computation likely used biblical texts to determine the date.

century. These opponents of the biblical chronology essentially left God out of the picture. Three of the old-earth advocates included Comte de Buffon, who thought the earth was at least 75,000 years old. Pièrre LaPlace imagined an indefinite but very long history. And Jean Lamarck also proposed long ages.[10]

10. Terry Mortenson, "The Origin of Old-earth Geology and its Ramifications for Life in the 21st Century," *TJ* 18, no. 1 (2004): 22–26, online at www.answersingenesis.org/tj/v18/i1/oldearth.asp.

However, the idea of millions of years really took hold in geology when men like Abraham Werner, James Hutton, William Smith, Georges Cuvier, and Charles Lyell used their interpretations of geology as the standard, rather than the Bible. Werner estimated the age of the earth at about one million years. Smith and Cuvier believed untold ages were needed for the formation of rock layers. Hutton said he could see no geological evidence of a beginning of the earth; and building on Hutton's thinking, Lyell advocated "millions of years."

From these men and others came the consensus view that the geologic layers were laid down slowly over long periods of time based on the rates at which we see them accumulating today. Hutton said:

> The past history of our globe must be explained by what can be seen to be happening now. . . . No powers are to be employed that are not natural to the globe, no action to be admitted except those of which we know the principle.[11]

This viewpoint is called naturalistic uniformitarianism, and it excludes any major catastrophes such as Noah's flood. Though some, such as Cuvier and Smith, believed in multiple catastrophes separated by long periods of time, the uniformitarian concept became the ruling dogma in geology.

11. James Hutton, *Theory of the Earth* (Trans. of Roy. Soc. of Edinburgh, 1785); quoted in A. Holmes, *Principles of Physical Geology* (UK: Thomas Nelson & Sons Ltd., 1965), p. 43–44.

Thinking biblically, we can see that the global flood in Genesis 6–8 would wipe away the concept of millions of years, for this Flood would explain massive amounts of fossil layers. Most Christians fail to realize that a global flood could rip up many of the previous rock layers and redeposit them elsewhere, destroying the previous fragile contents. This would destroy any evidence of alleged millions of years anyway. So the rock layers can theoretically represent the evidence of either millions of years or a global flood, but not both. Sadly, by about 1840, even most of the Church had accepted the dogmatic claims of the secular geologists and rejected the global flood and the biblical age of the earth.

After Lyell, in 1899, Lord Kelvin (William Thomson) calculated the age of the earth, based on the cooling rate of a molten sphere, at a maximum of about 20–40 million years (this was revised from his earlier calculation of 100 million years in 1862).[12] With the development of radiometric dating in the early 20th century, the age of the earth expanded radically. In 1913, Arthur Holmes's book, *The Age of the Earth,* gave an age of 1.6 billion years.[13] Since then, the supposed age of the earth has expanded to its present estimate of about 4.5 billion years (and about 14 billion years for the universe).

Table 5. Summary of the Old-earth Proponents for Long Ages

Who?	Age of the Earth	When Was This?
Comte de Buffon	78 thousand years old	1779
Abraham Werner	1 million years	1786
James Hutton	Perhaps eternal, long ages	1795
Pièrre LaPlace	Long ages	1796
Jean Lamarck	Long ages	1809
William Smith	Long ages	1835
Georges Cuvier	Long ages	1812
Charles Lyell	Millions of years	1830–1833
Lord Kelvin	20–100 million years	1862–1899
Arthur Holmes	1.6 billion years	1913
Clair Patterson	4.5 billion years	1956

12. Mark McCartney, "William Thompson: King of Victorian Physics," *Physics World,* December 2002, physicsweb.org/articles/world/15/12/6.
13. Terry Mortenson, "The History of the Development of the Geological Column," in *The Geologic Column,* eds. Michael Oard and John Reed (Chino Valley, AZ: Creation Research Society, 2006).

But there is growing scientific evidence that radiometric dating methods are completely unreliable.[14]

Christians who have felt compelled to accept the millions of years as fact and try to fit them into the Bible need to become aware of this evidence. It confirms that the Bible's history is giving us the true age of the creation.

Today, secular geologists will allow some catastrophic events into their thinking as an explanation for what they see in the rocks. But uniformitarian thinking is still widespread, and secular geologists will seemingly never entertain the idea of the global, catastrophic flood of Noah's day.

The age of the earth debate ultimately comes down to this foundational question: Are we trusting man's imperfect and changing ideas and assumptions about the past? Or are we trusting God's perfectly accurate eyewitness account of the past, including the creation of the world, Noah's global flood, and the age of the earth?

Other Uniformitarian Methods for Dating the Age of the Earth

Radiometric dating was the culminating factor that led to the belief in billions of years for earth history. However, radiometric dating methods are not the only uniformitarian methods. Any radiometric dating model or other uniformitarian dating method can and does have problems, as referenced before. All uniformitarian dating methods require assumptions for extrapolating present-day processes back into the past. The assumptions related to radiometric dating can be seen in these questions:

- Initial amounts?
- Was any parent amount added?
- Was any daughter amount added?
- Was any parent amount removed?
- Was any daughter amount removed?
- Has the rate of decay changed?

If the assumptions are truly accurate, then uniformitarian dates should agree with radiometric dating across the board for the same event. However,

14. For articles at the layman's level, see www.answersingenesis.org/home/area/faq/dating.asp. For a technical discussion, see Larry Vardiman, Andrew Snelling, and Eugene Chaffin, eds., *Radioisotopes and the Age of the Earth*, vol. 1 and 2 (El Cajon, CA: Institute for Creation Research; Chino Valley, AZ: Creation Research Society, 2000 and 2005). See also "Half-Life Heresy," *New Scientist*, October, 21 2006, pp. 36–39, abstract online at www.newscientist.com/channel/fundamentals/mg19225741.100-halflife-heresy-accelerating-radioactive-decay.html.

radiometric dates often disagree with one another and with dates obtained from other uniformitarian dating methods for the age of the earth, such as the influx of salts into the ocean, the rate of decay of the earth's magnetic field, and the growth rate of human population.[15]

The late Dr. Henry Morris compiled a list of 68 uniformitarian estimates for the age of the earth by Christian and secular sources.[16] The current accepted age of the earth is about 4.54 billion years based on radiometric dating of a group of meteorites,[17] so keep this in mind when viewing table 6.

Table 6. Uniformitarian Estimates Other than Radiometric Dating Estimates for Earth's Age Compiled by Morris

	0 – 10,000 years	>10,000 – 100,000 years	>100,000 – 1 million years	>1 million – 500 million years	>500 million – 4 billion years	>4 billion – 5 billion years
Number of uniformitarian methods*	23	10	11	23	0	0

* When a range of ages is given, the maximum age was used to be generous to the evolutionists. In one case, the date was uncertain so it was not used in this tally, so the total estimates used were 67. A few on the list had reference to Saturn, the sun, etc., but since biblically the earth is older than these, dates related to them were used.

As you can see from table 6, uniformitarian maximum ages for the earth obtained from other methods are nowhere near the 4.5 billion years estimated by radiometric dating; of the other methods, only two calculated dates were as much as 500 million years.

The results from some radiometric dating methods completely undermine those from the other radiometric methods. One such example is carbon-14 (^{14}C) dating. As long as an organism is alive, it takes in ^{14}C and ^{12}C from the atmosphere; however, when it dies, the carbon intake stops. Since ^{14}C is radioactive (decays into ^{14}N), the amount of ^{14}C in a dead organism gets less and less

15. For many more examples see www.answersingenesis.org/go/young.
16. Henry M. Morris, *The New Defender's Study Bible* (Nashville, TN: World Publishing, 2006), p. 2076–2079.
17. C.C. Patterson, "Age of Meteorites and the Age of the Earth," *Geochemica et Cosmochemica Acta*, 10 (1956): 230–237.

over time. Carbon-14 dates are determined from the measured ratio of radioactive carbon-14 to normal carbon-12 ($^{14}C/^{12}C$). Used on samples that were once alive, such as wood or bone, the measured $^{14}C/^{12}C$ ratio is compared with the ratio in living things today.

Now, ^{14}C has a derived half-life of 5,730 years, so the ^{14}C in organic material supposedly 100,000 years old should all essentially have decayed into nitrogen.[18] Some things, such as wood trapped in lava flows, said to be millions of years old by other radiometric dating methods, still have ^{14}C in them.[19] If the items were really millions of years old, then they shouldn't have any traces of ^{14}C. Coal and diamonds, which are found in or sandwiched between rock layers allegedly millions of years old, have been shown to have ^{14}C ages of only tens of thousands of years.[20] So which date, if any, is correct? The diamonds or coal can't be millions of years old if they have any traces of ^{14}C still in them. This shows that these dating methods are completely unreliable and indicates that the presumed assumptions in the methods are erroneous.

Similar kinds of problems are seen in the case of potassium-argon dating, which has been considered one of the most reliable methods. Dr. Andrew Snelling, a geologist, points out several of these problems with potassium-argon, as seen in table 7.[21]

These and other examples raise a critical question. If radiometric dating fails to give an accurate date on something of which we *do* know the true age, then how can it be trusted to give us the correct age for rocks that had no human observers to record when they formed? If the methods don't work on rocks of known age, it is most unreasonable to trust that they work on rocks of unknown age. It is far more rational to trust the Word of the God who created the world, knows its history perfectly, and has revealed sufficient information in the Bible for us to understand that history and the age of the creation.

---

18. This does not mean that a ^{14}C date of 50,000 or 100,000 would be entirely trustworthy. I am only using this to highlight the mistaken assumptions behind uniformitarian dating methods.
19. Andrew Snelling, "Conflicting 'Ages' of Tertiary Basalt and Contained Fossilized Wood, Crinum, Central Queensland Australia," *Technical Journal* 14, no. 2 (2005): p. 99–122.
20. John Baumgardner, "^{14}C Evidence for a Recent Global Flood and a Young Earth," in *Radioisotopes and the Age of the Earth: Results of a Young-Earth Creationist Research Initiative*, ed. Vardiman *et al.* (Santee, CA: Institute for Creation Research; Chino Valley, AZ: Creation Research Society, 2005), p. 587–630.
21. Andrew Snelling, "Excess Argon: The 'Achilles' Heel' of Potassium-Argon and Argon-Argon Dating of Volcanic Rocks," *Impact*, January 1999, online at www.icr.org/article/436.

Table 7. Potassium-argon (K-Ar) Dates in Error

Volcanic eruption	When the rock formed	Date by (K-Ar) radiometric dating
Mt. Etna basalt, Sicily	122 B.C.	170,000–330,000 years old
Mt. Etna basalt, Sicily	A.D. 1972	210,000–490,000 years old
Mount St. Helens, Washington	A.D. 1986	Up to 2.8 million years old
Hualalai basalt, Hawaii	A.D. 1800–1801	1.32–1.76 million years old
Mt. Ngauruhoe, New Zealand	A.D. 1954	Up to 3.5 million years old
Kilauea Iki basalt, Hawaii	A.D. 1959	1.7–15.3 million years old

Conclusion

When we start our thinking with God's Word, we see that the world is about 6,000 years old. When we rely on man's fallible (and often demonstrably false) dating methods, we can get a confusing range of ages from a few thousand to billions of years, though the vast majority of methods do not give dates even close to billions.

Cultures around the world give an age of the earth that confirms what the Bible teaches. Radiometric dates, on the other hand, have been shown to be wildly in error.

The age of the earth ultimately comes down to a matter of trust — it's a worldview issue. Will you trust what an all-knowing God says on the subject or will you trust imperfect man's assumptions and imaginations about the past that regularly are changing?

> Thus says the Lord: "Heaven is My throne, and earth is My footstool. Where is the house that you will build Me? And where is the place of My rest? For all those things My hand has made, and all those things exist," says the Lord. "But on this one will I look: On him who is poor and of a contrite spirit, and who trembles at My word" (Isaiah 66:1–2).

5

Are There Gaps in the Genesis Genealogies?

LARRY PIERCE & KEN HAM

Most of us love to read portions of Scripture that give accounts of victories, miracles, and drama. We enjoy far less the Scriptures that outline a certain person begat a son or daughter, who in turn begat a son, thus beginning a long list of begats. Most people believe the genealogies contain only dull details, but those of us who keep in mind that "every word is given by inspiration of God" see that even these so-called dull passages contain vital truth that can be trusted.

Genesis 5 and 11 contain two such genealogies. It may be hard to believe, but Genesis 5 and 11 are actually two of the more controversial chapters in the Bible, even in Christian circles.

Because so many Christians and Christian leaders have accepted the secular dates for the origin of man and the universe, they must work out ways that such dates can somehow be incorporated into the Bible's historical account. In other words, they must convince people that the Bible's genealogical records do not present an unbroken line of chronology. If such an unbroken line exists, then we should be able to calculate dates concerning the creation of man and the universe.

To fit the idea of billions of years into Scripture, many Christian leaders, since the early 19th century, have reinterpreted the days of creation to mean long ages. Biblical creationist literature has meticulously addressed this topic

many times, showing clearly that the word *day*, as used in Genesis 1 for each of the six days of creation, means an ordinary, approximately 24-hour day.[1]

A straightforward addition of the chronogenealogies yields a date for the beginning near 4000 B.C. Chronologists working from the Bible consistently get 2,000 years between Adam and Abraham. Few would dispute that Abraham lived around 2000 B.C. Many Christian leaders, though, claim there are gaps in the Genesis genealogies. One of their arguments is that the word *begat*, as used in the time-line from the first man Adam to Abraham in Genesis 5 and 11, can skip generations. If this argument were true, the date for creation using the biblical time-line of history cannot be worked out.

In a recent debate,[2] a well-known progressive creationist[3] stated that he believed a person could date Adam back 100,000 years from the present. Since most modern scholars place the date of Abraham around 2000 B.C. (Ussher's date for Abraham's birth is 1996 B.C.), the remaining 96,000 years must fit into the Genesis 5 and 11 genealogies, between Adam and Abraham.

1. See, for example, www.answersingenesis.org/go/days-of-creation.
2. Ken Ham, Jason Lisle, Hugh Ross, Walt Kaiser, *The Great Debate: Young Earth vs. Old Earth*, DVD (Kentucky: Answers in Genesis, 2006), program 10, bonus 2.
3. Most progressive creationists believe that the six days of creation were actually long periods of time, not 24-hour days.

4004 BC	2348 BC	2242 BC	1996-1821 BC
Creation	Flood	Tower of Babel	Abraham

By accepting these three points (right), we can determine the dates for key events (above) going back to the beginning of time:*

*All are based on the Masoretic text.

1. The word for *day* (Hebrew: *yom*) in Genesis 1 for the days of creation are ordinary days (of approximately 24 hours).
2. The word for *begat* (*yalad*) in the genealogies in Genesis 5 and 11 does not skip any generations.
3. The date of Abraham was 1996 BC.

Now, if we estimate that 40 years equals one generation, which is fairly generous,[4] this means that 2,500 generations are missing from these genealogies. But this makes the genealogies ridiculously meaningless.

Two Keys to Consider

Those who claim that there are gaps in these genealogies need to demonstrate this from the biblical text and not simply say that gaps exist. However, consider the following:

1. Although in the Hebrew way of thinking, the construction "X is the son of Y" does not always mean a literal father/son relationship, additional biographical information in Genesis 5 and 11 strongly supports the view that there are no gaps in these chapters. So we know for certain that the following are literal father/son relationships: Adam/Seth, Seth/Enosh, Lamech/Noah, Noah/Shem, Eber/Peleg, and Terah/Abram. Nothing in these chapters indicates that the "X *begat* Y" means something other than a literal father/son relationship.

2. Nowhere in the Old Testament is the Hebrew word for *begat* (*yalad*) used in any other way than to mean a single-generation (e.g., father/son or mother/daughter) relationship. The Hebrew word *ben* can mean *son* or *grandson*, but the word *yalad* never skips generations.

Six Arguments Refuted

In the recent debate (mentioned previously), various biblical references were given as proofs that the Hebrew word *yalad* does not always point to the

4. Jonathan Sarfati, *Refuting Compromise* (Green Forest, AR: Master Books, 2004), p. 295.

very next generation. However, when analyzed carefully, these arguments actually confirm what we are asserting concerning the word *begat.*

Argument 1

Genesis 46:15 says, "These be the sons of Leah, which she bare unto Jacob in Padanaram, with his daughter Dinah: all the souls of his sons and his daughters were thirty and three" (KJV). The word *bare* here is the Hebrew word *yalad*, which is also translated *begat.* It is claimed by some that because there are sons of various wives, grandsons, daughters, etc., in this list of "thirty and three," the word *begat* is referring to all these and can't be interpreted as we assert.

Is Argument 1 Relevant?

A person needs to read the quoted verse carefully to correctly understand its meaning. The *begat* (*bare*) refers to the sons born in Padanaram. Genesis 35:23 lists the six sons born in Padanaram (those whom Leah begat), who are listed as part of the total group of 33 children in Genesis 46:15. Thus, this passage confirms that *begat* points to the generation immediately following — a literal parent/child relationship.

Argument 2

Matthew 1:8 omits Ahaziah, Joash, and Amaziah, going directly from Joram to Uzziah. Matthew 1:11 skips Jehoiakim between Josiah and Jeconiah. These passages prove that the word *begat* skips generations.

Is Argument 2 Relevant?

Here, the Greek word for *begat* is *gennao*, which shows flexibility not found in the Hebrew word and does allow for the possibility that a generation or more may be skipped. The only way we would know that a generation has been skipped is by checking the Hebrew passages. However, it is linguistically deceptive to use the Greek word for *begat* to define the Hebrew word for *begat.* Also, Matthew 1 is intentionally incomplete when reading Matthew 1:1 and Matthew 1:17, merely giving 14 generations between key figures of Abraham, David, and Jesus.

Argument 3

Genesis 46:18, 22, and 25 says, "These are the sons of Zilpah, whom Laban gave to Leah his daughter, and these she bare unto Jacob, even sixteen souls. . . . These are the sons of Rachel, which were born to Jacob: all the souls were fourteen. . . . These are the sons of Bilhah, which Laban gave to Rachel his daughter, and she bare these unto Jacob: all the souls were seven" (KJV). In verse 18, the Hebrew word *yalad* (*begat* or *bore*) implies a grandson, as well as a son; so the word *begat* cannot be used to show a direct relationship.

Is Argument 3 Relevant?

The word *bare* in verse 18 refers to Zilpah's actual sons, referenced in verses 16 (Gad) and 17 (Asher). Note the pattern in this chapter. In verse 15 we are given the total number of Leah's offspring (33), in verse 18 the total of Zilpah's offspring (16), in verse 22 the total of Rachel's offspring (14), and in verse 25 the total of Bilhah's offspring (7). This makes a total of 70. But nowhere is it stated that these four wives physically bore the total number of sons listed for each.

What this passage shows, as stated earlier, is that the Hebrew word for *son* (*ben*) may include grandsons. In the case of Zilpah, her two sons are clearly listed, as well as the children of Gad and Asher. To insist that in this case only (and not the cases of Leah, Rachel, and Bilhah) the summary total given at the end of verse 18 implies that all these were begotten of Zilpah is not justified by the context, and therefore, is not sound hermeneutics. The context makes it very clear that Zilpah had only two sons, and this passage does not show that the Hebrew word *yalad* (*begat* or *bore*) implies a grandson, as well as a son.

Argument 4

An example of where the word *begat* omits generations is 1 Chronicles 7:23–27. It is clear from this passage that there are ten generations from Ephraim to Joshua, whereas Genesis 15:16 says there were only four generations from the time the children of Israel entered Egypt to the time they left. Therefore, the Hebrew word for *begat* does not always mean the next generation.

Is Argument 4 Relevant?

This argument seems logically airtight except for two minor points. The Hebrew word *yalad* for *begat* is not used in the 1 Chronicles passage, and Genesis 15:16 is misquoted. Genesis states that "in the fourth generation" the children of Israel would leave Egypt — not that there would be a maximum of four generations. For this prophecy in Genesis to be fulfilled, some of the fourth generation would be in the exodus from Egypt — and they were. Exodus 6 lists the generations from Levi to Moses, showing that Moses and Aaron were in the fourth generation. Therefore the passage in 1 Chronicles cannot be used to prove that the Hebrew word for *begat* can skip a generation.

It is quite helpful, however, to explain how the Israelites became so numerous during their stay in Egypt. The descendants of Joshua appear to have had a new generation about every 20 years, whereas the descendants of Moses and Aaron had a new generation about every 50 years.

Argument 5

In Luke 3:36, the name Cainan is listed, which is not listed in the Old Testament chronologies.

Is Argument 5 Relevant?

The present copies of the Septuagint (ancient Greek translation of the Old Testament) incorrectly have the name Cainan inserted in the Old Testament genealogies. The great Baptist Hebrew scholar John Gill (c. A.D. 1760), in his exposition on this verse, wrote:

> This Cainan is not mentioned by Moses in Genesis 11:12 nor has he ever appeared in any Hebrew copy of the Old Testament, nor in the Samaritan version, nor in the Targum; nor is he mentioned by Josephus, nor in 1 Chronicles 1:24 where the genealogy is repeated; nor is it in Beza's most ancient Greek copy of Luke: it indeed stands in the present copies of the Septuagint, but was not originally there; and therefore could not be taken by Luke from there, but seems to be owing to some early negligent transcriber of Luke's Gospel, and since put into the Septuagint to give it authority: I say early, because it is in many Greek copies, and in the Vulgate Latin, and all the Oriental versions, even in the Syriac, the oldest of them; but ought not to stand neither in the text, nor in any version: for certain it is, there never was such a Cainan, the son of Arphaxad, for Salah was his son; and with him the next words should be connected.[5]

Since Gill's commentary was written, the oldest manuscript we have of Luke, the *P75*, was found. It dates to the late second century A.D. and does not include Cainan in the genealogy. This verse in Luke should not be used to prove that the genealogies in Genesis have gaps, because it has poor textual authority.

Argument 6

Author and radio host Harold Camping argues for a unique interpretation of the chronologies in Genesis 5 and 11. According to his interpretation, Adam was created in 11,013 B.C. The chronological statements in these two chapters are of the following form.

When X was A years old he begat Y. He lived B years after he begat Y and died at the age of C years. So A + B = C.

5. Note on Luke 3:36 in: John Gill, D.D., *An Exposition of the Old and New Testament; The Whole Illustrated with Notes, Taken from the Most Ancient Jewish Writings* (London: printed for Mathews and Leigh, 18 Strand, by W. Clowes, Northumberland-Court, 1809). Edited, revised, and updated by Larry Pierce, 1994–1995 for The Word CD-ROM. See also chapter 27, "Isn't the Bible Full of Contradictions?"

Camping interprets this statement as follows:

> When X was A years old he begat a progenitor of Y. He lived B years after he begat a progenitor of Y and died at age C, which was the same year that Y was born.

Is Argument 6 Relevant?

We must give Mr. Camping credit for originality and ingenuity, for we are not aware of anyone who interpreted these verses as such before him. As proof for this interpretation, Mr. Camping cites Matthew 1:8 that the word *begat* does not mean a father/son relationship. We have already discussed this line of reasoning in argument 2 and refuted it, thus exploding Mr. Camping's argument.

While claiming to honor the text of the Bible, Mr. Camping demonstrates a profound misunderstanding of the Hebrew verb forms for *begat* found in chapter 5 and 11 of Genesis. These verbs use the *hiphil* form of the verb. Most Hebrew verbs use the *qal* form, which corresponds to the active indicative tense in English. *Hiphil* usually expresses the causative action of *qal.*

he eats	he causes to eat
he comes	he causes to come, he brings
he reigned	he made king, he crowned

The *hiphil* has no exact English equivalent and is difficult to capture the meaning in English. Some modern English translations use the word *fathered* instead of the word *begat,* thus removing the ambiguity. To make it absolutely clear, the verb could be translated *X himself fathered Y,* but that is awkward English. It is difficult to father a remote descendant without committing incest! When the Hebrew verb form is honored in English, it precludes the interpretation Mr. Camping places on it. God chose this form to make it absolutely clear that we understand that there are no missing generations in chapters 5 and 11 of Genesis. Any other Hebrew verb form would not have been nearly as emphatic as the *hiphil* form.

In his latest book *Time Has an End*, Mr. Camping sets out a complete chronology for the Bible using his defective understanding of the chronologies in Genesis 5 and 11, which includes the following mistakes.

- Israel's time in Egypt was 430 years.
- The date for the Exodus is wrong.
- The chronology for the time of the judges is confused.

- The chronology of the divided kingdom is partially based on Dr. Edwin Thiele's work *The Mysterious Numbers of the Hebrew Kings*, which contradicts the Bible in many places.
- The end of the world in 2011. (His earlier prediction of 1994 had to be reinterpreted.)

Rather than refute these incorrect ideas, we recommend the *Chronology of the Old Testament* (Master Books, 2005) by Dr. Floyd Jones for a more accurate, biblically based chronology that is devoid of the speculations of Mr. Camping and refutes most of Camping's chronology.

Missing Generations?

Many creationists believe the earth is about 10,000 years old in an attempt to make the biblical record conform to modern archaeological ideas. According to these ideas, Egypt began around 3500 B.C. and Babylon in 4000 B.C. Since these nations speak different languages, their founding must have been after the Tower of Babel, which occurred after the Flood. So some creationists place the Flood around 5000 B.C. and the creation around 10,000 B.C. It is curious that, having rejected the evidence for long ages, these creationists are inadvertently and blindly trusting man's fallible dating methods for archaeological data, which rests on just as flimsy a foundation as does the evidence for long ages.[6]

Assuming these creationists are correct, how many generations are missing from Genesis 5 and 11? We will use the Hebrew text for these calculations; using other versions such as the Septuagint (LXX) makes the matter even more improbable.

According to the Hebrew text, there were 1,656 years between creation and the Flood and 1,556 years between creation and Noah's first son, or 10 generations. Assuming the average generation (from father to son) was 156 years (divide 1,556 by 10), how many extra generations are needed to get 5,000 years from the creation to Noah's first son? Divide 5,000 by 156 and you get about 32 generations. On the average, then, for every generation listed in Genesis 5, two are missing! However, let's examine Genesis 5 more closely:

1. There are no missing generations between Adam and Seth, since Seth is a direct replacement for Abel, whom Cain murdered (Genesis 4:25).

6. See Larry Vardiman, Andrew Snelling, and Eugene Chaffin, eds., *Radioisotopes and the Age of the Earth*, vol. 2 (El Cajon, CA: Institute for Creation Research; Chino Valley, AZ: Creation Research Society, 2005).

2. There are no missing generations between Seth and Enosh, since Seth named him (Genesis 4:25).
3. Jude says Enoch was the seventh from Adam (Jude 14), so there are no missing generations between Adam and Enoch.
4. Lamech named Noah, so there are no missing generations there (Genesis 5:29).
5. Some Hebrew scholars believe that the name *Methuselah* means "when he dies it is sent," referring to the Flood. Assuming no gaps in the chronology, Methuselah died the same year the Flood began. Some Jews believed that God gave Noah time to mourn the death of Methuselah, whom they believe died a week before the Flood began (Genesis 7:4). If this is so, then no missing generations can be inserted here. If this were not the case, then this is the only place in Genesis 5 one might attempt to shoehorn the missing 22 generations! Would you trust a chronologist who was so careful to record names and ages yet omit 22 generations in his tabulation in one place? It simply doesn't follow.

As we have seen, careful exegesis of the Bible simply does not allow for an extra 22 generations.

A similar analysis can be done for Genesis 11, which features 10 generations over 355 years, therefore averaging 36 years per generation. Those who hold to a creation occurring in 10,000 B.C. and the Flood happening in 5,000 B.C. have expanded this time period from 355 years to over 2,600 years. Assuming each generation lasts 36 years, then there would be 72 generations, such that for every generation listed, six are missing. If the writer of Genesis was so careless as to omit over 85 percent of the generations in Genesis 11, why did he waste time giving us the information in the first place? What purpose would it serve, since it would be so inaccurate?

These examples show the folly of accepting a creation event as distant as 10,000 B.C. Those who accept even longer ages have a worse problem; they must insert 10 to 100 times as many "missing generations" in Genesis 5 and 11 as those who hold to a creation of about 10,000 B.C. Interestingly, both camps loathe explaining where these missing generations are to be inserted. All they know for sure is that they are missing! Those who hold to the inerrancy of the Scriptures should reject all attempts to make the earth older than the Hebrew text warrants, which is about 4000 B.C.

Conclusion

The Scriptures themselves attest to the fact that the secular dates given for the age of the universe, man's existence on the earth, and so on, are not correct, because they are based on the fallible assumptions of fallible humans. Nothing in observational science contradicts the time-line of history as recorded in the Bible.

But there are two more reasons that these genealogies are vital. First, they are given in Scripture to show clearly that the Bible is real history and that we are all descendants of a real man, Adam; thus all human beings are related.

Second, the Son of God stepped into this history to fulfill the promise of Genesis 3:15, the promise of a Savior. This Savior died and rose again to provide a free gift of salvation to the descendants of Adam — all of whom are sinners and are separated from their Creator.

Without the genealogies, how can it be proven that Jesus is the One who would fulfill this promise? Indeed, perhaps the primary purpose of the genealogies is to show that Jesus fulfilled the promise of God the Father.

We can trust these genealogies because they are a part of the infallible, inerrant Word of God.

6

Can Natural Processes Explain the Origin of Life?

MIKE RIDDLE

When considering how life began, there are only two options. Either life was created by an intelligent source (God) or it began by natural processes. The common perception presented in many textbooks and in the media is that life arose from nonlife in a pool of chemicals about 3.8 billion years ago. The claim by evolutionists is that this formation of life was the result of time, chance, and natural processes. One widely used example of how life could have formed by natural processes is the Miller-Urey experiment, performed in the early 1950s.

Miller's objective was not to create life but to simulate how life's basic building structures (amino acids[1]) might have formed in the early earth. In the experiment, Miller attempted to simulate the early atmosphere of earth by using certain gases, which he thought might produce organic compounds necessary for life. Since the gases he included (water, methane, ammonia, and hydrogen) do not react with each other under natural conditions, he generated electrical currents to simulate some form of energy input (such as lightning) that was needed to drive the chemical reactions. The result was

1. The basic building blocks of all living systems are proteins, which consist of only 20 different types of amino acids. The average number of amino acids in a biological protein is over 300. These amino acids must be arranged in a very specific sequence for each protein.

production of amino acids. Many textbooks promote this experiment as the first step in explaining how life could have originated. But there is more to this experiment than what is commonly represented in textbooks.

The Rest of the Story – Some Critical Thinking

When we examine the purpose, assumptions, and results of the Miller experiment, there are three critical thinking questions that can be raised:

1. How much of the experiment was left to chance processes or how much involved intelligent design?
2. How did Miller know what earth's early atmosphere (billions of years ago) was like?
3. Did Miller produce the right type of amino acids used in life?

The Method Used

In the experiment, Miller was attempting to illustrate how life's building blocks (amino acids) could have formed by natural processes. However, throughout the experiment Miller relied on years of intelligent research in chemistry. He purposely chose which gases to include and which to exclude. Next, he had to isolate the biochemicals (amino acids) from the

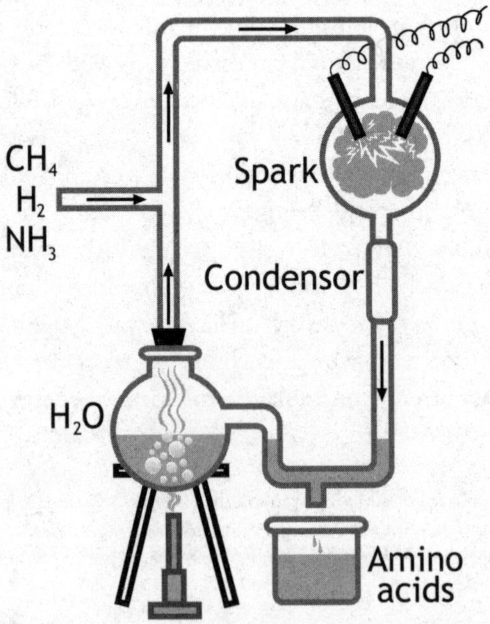

environment he had created them in because it would have destroyed them. No such system would have existed on the so-called primitive earth. It appears Miller used intelligent design throughout the experiment rather than chance processes.

The Starting Ingredients

How did Miller know what the atmosphere was like billions of years ago? Miller assumed that the early earth's atmosphere was very different from today. He based his starting chemical mixture on the assumption that the early earth had a reducing atmosphere (an atmosphere that contains no free oxygen). Why did Miller and many other evolutionists assume there was no free oxygen in earth's early atmosphere? As attested below, it is well known that biological molecules (specifically amino acid bonds) are destroyed in the presence of oxygen, making it impossible for life to evolve.

> Oxygen is a poisonous gas that oxidizes organic and inorganic materials on a planetary surface; it is quite lethal to organisms that have not evolved protection against it.[2]

> In the atmosphere and in the various water basins of the primitive earth, many destructive interactions would have so vastly diminished, if not altogether consumed, essential precursor chemicals, that chemical evolution rates would have been negligible.[3]

Therefore, in order to avoid this problem, evolutionists propose that earth's first atmosphere did not contain any freestanding oxygen. We must ask ourselves, "Is there any evidence to support this claim, or is it based on the assumption that evolution must be true?" As it turns out, the existence of a reducing atmosphere is merely an assumption not supported by the physical evidence. The evidence points to the fact that the earth has always had oxygen in the atmosphere.

> There is no scientific proof that Earth ever had a non-oxygen atmosphere such as evolutionists require. Earth's oldest rocks contain evidence of being formed in an oxygen atmosphere.[4]

---

2. P. Ward and D. Brownlee, *Rare Earth* (New York: Copernicus, 2000), p. 245.
3. C. Thaxton, W. Bradley, and R. Olsen, *The Mystery of Life's Origin: Reassessing Current Theories* (New York: Philosophical Library, 1984), p. 66.
4. H. Clemmey and N. Badham, "Oxygen in the Atmosphere: An Evaluation of the Geological Evidence," *Geology* 10 (1982): 141.

> The only trend in the recent literature is the suggestion of far more oxygen in the early atmosphere than anyone imagined.[5]

If we were to grant the evolutionists' assumption of no oxygen in the original atmosphere, another fatal problem arises. Since the ozone is made of oxygen, it would not exist; and the ultraviolet rays from the sun would destroy any biological molecules. This presents a no-win situation for the evolution model. If there was oxygen, life could not start. If there was no oxygen, life could not start. Michael Denton notes:

> What we have is sort of a "Catch 22" situation. If we have oxygen we have no organic compounds, but if we don't have oxygen we have none either.[6]

Because life could not have originated on land, some evolutionists propose that life started in the oceans. The problem with life starting in the oceans, however, is that as organic molecules formed, the water would have immediately destroyed them through a process called *hydrolysis*. Hydrolysis, which means "water splitting," is the addition of a water molecule between two bonded molecules (two amino acids in this case), which causes them to split apart. Many scientists have noted this problem.

> Besides breaking up polypeptides, hydrolysis would have destroyed many amino acids.[7]

> In general the half-lives of these polymers in contact with water are on the order of days and months — time spans which are surely geologically insignificant.[8]

> Furthermore, water tends to break chains of amino acids apart. If any proteins had formed in the oceans 3.5 billion years ago, they would have quickly disintegrated.[9]

Scientifically, there is no known solution for how life could have chemically evolved on the earth.

5. Thaxton, Bradley, and Olsen, *The Mystery of Life's Origin*, p. 80.
6. M. Denton, *Evolution: A Theory in Crisis* (Bethesda, MD: Adler & Adler, 1985), p. 261.
7. *Encyclopedia of Science and Technology*, Vol. 1, 1982: p. 411–412.
8. K. Dose, *The Origin of Life and Evolutionary Biochemistry* (New York: Plenum Press, 1974), p. 69.
9. R. Morris, *The Big Questions* (New York: Times Books/Henry Holt, 2002), p. 167.

On the Other Hand . . .

Because the scientific evidence contradicts the origin of life by natural processes, Miller resorted to unrealistic initial conditions to develop amino acids in his experiment (no oxygen and excessive energy input). However, there is more to the story. Producing amino acids is not the hard part. The difficult part is getting the right type and organization of amino acids. There are over 2,000 types of amino acids, but only 20 are used in life. Furthermore, the atoms that make up each amino acid are assembled in two basic shapes. These are known as *left-handed* and *right-handed.* Compare them to human hands. Each hand has the same components (four fingers and a thumb), yet they are different. The thumb of one hand is on the left, and the thumb of the other is on the right. They are mirror images of each other. Like our hands, amino acids come in two shapes. They are composed of the same atoms (components) but are mirror images of each other, called left-handed amino acids and right-handed amino acids. Objects that have handedness are said to be *chiral* (pronounced "ky-rul"), which is from the Greek for *hand.*

Handedness is an important concept because all amino acids that make up proteins in living things are 100 percent left-handed. Right-handed amino acids are never found in proteins. If a protein were assembled with just one right-handed amino acid, the protein's function would be totally lost. As one PhD chemist has said:

> Many of life's chemicals come in two forms, "left-handed" and "right-handed." Life requires polymers with all building blocks having the same "handedness" (*homochirality*) — proteins have only

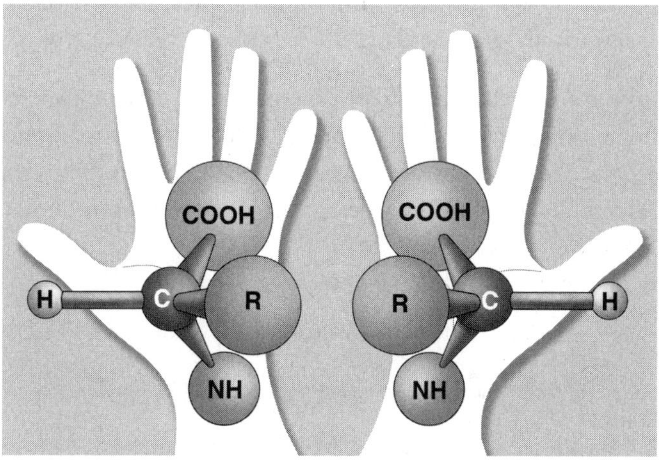

> "left-handed" amino acids. . . . But ordinary undirected chemistry, as is the hypothetical primordial soup, would produce equal mixtures of left- and right-handed molecules, called *racemates*.[10]

A basic chemistry textbook admits:

> This is a very puzzling fact. . . . All the proteins that have been investigated, obtained from animals and from plants from higher organisms and from very simple organisms — bacteria, molds, even viruses — are found to have been made of L-amino [left-handed] acids.[11]

The common perception left by many textbooks and journals is that Miller and other scientists were successful in producing the amino acids necessary for life. However, the textbooks and media fail to mention that what they had actually produced was a mixture of left- and right-handed amino acids, which is detrimental to life. The natural tendency is for left- and right-handed amino acids to bond together. Scientists still do not know why biological proteins use only left-handed amino acids.

> The reason for this choice [only left-handed amino acids] is again a mystery, and a subject of continuous dispute.[12]

Jonathan Wells, a developmental biologist, writes:

> So we remain profoundly ignorant of how life originated. Yet the Miller-Urey experiment continues to be used as an icon of evolution, because nothing better has turned up. Instead of being told the truth, we are given the misleading impression that scientists have empirically demonstrated the first step in the origin of life.[13]

Despite the fact that the Miller experiment did not succeed in creating the building blocks of life (only left-handed amino acids), textbooks continue to promote the idea that life could have originated by natural processes. For example, the following statement from a biology textbook misleads students into thinking Miller succeeded:

---

10. John Ashton, ed., *In Six Days*, (Green Forest, AR: Master Books, 2000), p. 82.
11. Linus Pauling, *General Chemistry*, 3rd ed. (San Francisco, CA: W.H. Freeman & Co., 1970), p. 774.
12. Robert Shapiro, *Origins: A Skeptic's Guide to the Creation of Life on Earth* (New York: Summit Books, 1986), p. 86.
13. J. Wells, *Icons of Evolution* (Washington, DC: Regnery Pub., 2000), p. 24.

> By re-creating the early atmosphere (ammonia, water, hydrogen and methane) and passing an electric spark (lightning) through the mixture, Miller and Urey proved that organic matter such as amino acids could have formed spontaneously.[14]

First, note the word *proved.* Miller and Urey proved nothing except that life's building blocks could *not* form in such conditions. Second, the textbook completely ignores other evidence, which shows that the atmosphere always contained oxygen. Third, the textbook ignores the fact that Miller got the wrong type of amino acids — a mixture of left- and right-handed.

The Miller experiment (and all experiments since then) failed to produce even a single biological protein by purely naturalistic processes. Only God could have begun life.

Information

Another important component of life is information. The common factor in all living organisms is the information contained in their cells. Where and how did all this coded information arise? Proteins are amazingly versatile and carry out many biochemical functions, but they are incapable of assembling themselves without the assistance of DNA. The function of DNA is to store information and pass it on (transcribe) to RNA, while the function of RNA is to read, decode, and use the information received from DNA to make proteins. Each of the thousands of genes on a DNA molecule contains instructions necessary to make a specific protein that, in turn, is needed for a specific biological function.

Any hypothesis or model meant to explain how all life evolved from lifeless chemicals into a complex cell consisting of vast amounts of information also has to explain the source of information and how this information was encoded into the genome. All evolutionary explanations are unable to answer this question. Dr. Werner Gitt, former physics professor and director of information processing at the Institute of Physics and Technology in Braunschweig, Germany, and Dr. Lee Spetner both agree that information cannot arise by naturalistic processes:

14. Kenneth Miller and Joseph Levine, *Biology,* 5th ed. (Upper Saddle River, NJ: Pearson Prentice Hall, 2000).

> There is no known law of nature, no known process and no known sequence of events which can cause information to originate by itself in matter.[15]

> Not even one mutation has been observed that adds a little information to the genome. This surely shows that there are not the millions upon millions of potential mutations the theory [evolution] demands.[16]

The DNA code within all plant and animal cells is vastly more compact than any computer chip ever made. DNA is so compact that a one-square-inch chip of DNA could encode the information in over seven billion Bibles. Since the density and complexity of the genetic code is millions of times greater than man's present technology, we can conclude that the originator of the information must be supremely intelligent.

Two biologists have noted:

> DNA is an information code. . . . The overwhelming conclusion is that information does not and cannot arise spontaneously by mechanistic processes. Intelligence is a necessity in the origin of any informational code, including the genetic code, no matter how much time is given.[17]

God, in His Word, tells us that His creation is a witness to himself and that we do not have an excuse for not believing (Romans 1:19–20). The fact that the information encoded in DNA ultimately needs to have come from an infinite source of information testifies to a Creator. And as we saw above, the only known way to link together left-handed amino acids is through purposeful design.

Since no human was present to assemble the first living cell, it is further testimony to an all-wise Creator God.

Given Enough Time . . .

Nobel prize-winning scientist George Wald once wrote:

15. W. Gitt, *In the Beginning Was Information* (Green Forest, AR: Master Books, 2006).
16. L. Spetner, *Not by Chance* (New York: Judaica Press, 1997), p. 160.
17. L. Lester and R. Bohlin, *The Natural Limits to Biological Change*, (Dallas, TX: Probe Books, 1989), p. 157.

> However improbable we regard this event [evolution], or any of the steps which it involves, given enough time it will almost certainly happen at least once. . . . Time is in fact the hero of the plot. . . . Given so much time, the "impossible" becomes possible, the possible probable, the probable virtually certain. One has only to wait; time itself performs the miracles.[18]

In the case of protein formation, the statement "given enough time" is not valid. When we look at the mathematical probabilities of even a small protein (100 amino acids) assembling by random chance, it is beyond anything that has ever been observed.

What is the probability of ever getting one small protein of 100 left-handed amino acids? (An average protein has at least 300 amino acids in it — all left-handed.) To assemble just 100 left-handed amino acids (far shorter than the average protein) would be the same probability as getting 100 heads in a row when flipping a coin. In order to get 100 heads in a row, we would have to flip a coin 10^{30} times (this is 10 x 10, 30 times). This is such an astounding improbability that there would not be enough time in the whole history of the universe (even according to evolutionary time frames) for this to happen.

The ability of complex structures to form by naturalistic processes is essential for the evolution model to work. However, the complexity of life

18. George Wald, "The Origin of Life," *Scientific American* 191 no. 2 (1954): 48.

appears to preclude this from happening. According to the laws of probability, if the chance of an event occurring is smaller than 1 in 10^{-50}, then the event will never occur (this is equal to 1 divided by 10^{50} and is a very small number).[19]

What have scientists calculated the probability to be of an average-size protein occurring naturally? Walter Bradley, PhD, materials science, and Charles Thaxton, PhD, chemistry,[20] calculated that the probability of amino acids forming into a protein is:

$$4.9 \times 10^{-191}$$

This is well beyond the laws of probability (1×10^{-50}), and a protein is not even close to becoming a complete living cell. Sir Fred Hoyle, PhD, astronomy, and Chandra Wickramasinghe, professor of applied math and astronomy, calculated that the probability of getting a cell by naturalistic processes is:

$$1 \times 10^{-40,000}$$

> No matter how large the environment one considers, life cannot have had a random beginning. . . . There are about two thousand enzymes, and the chance of obtaining them all in a random trial is only one part in $(10^{20})^{2000} = 10^{40,000}$, an outrageously small probability that could not be faced even if the whole universe consisted of organic soup.[21]

Conclusion

As we have seen, the scientific evidence confirms that "in the beginning, God created." Life cannot come from nonlife; only God can create life. True science and the Bible will always agree. Whether in biology, astronomy, geology, or any other field of study, we can trust God's Word to be accurate when it speaks about these topics. Let us stand up for the truth of Genesis and take back our culture.

19. Probability expert Emile Borel wrote, "Events whose probabilities are extremely small never occur. . . . We may be led to set at 10^{-50} the value of negligible probabilities on the cosmic scale." (E. Borel, *Probabilities and Life*, [New York: Dover Publications, 1962], p. 28.)
20. J.P. Moreland, editor, *The Creation Hypothesis*, "Information and the Origin of Life," by Walter Bradley and Charles Thaxton (Downers Grove, IL: Inter-Varsity Press, 1994), p. 190.
21 F. Hoyle and C. Wickramasinghe, *Evolution from Space* (New York: Simon and Schuster, 1984), p. 176.

7

Are Mutations Part of the "Engine" of Evolution?

BODIE HODGE

In the evolutionary model, mutations are hailed as a dominant mechanism for pond-scum-to-people evolution and provide "proof" that the Bible's history about creation is wrong. But are we to trust the ideas of imperfect, fallible men about how we came into existence, or should we believe the account of a perfect God who was an eyewitness to His creation? Let's look at mutations in more detail and see if they provide the information necessary to support pond-scum-to-people evolution, or if they confirm God's Word in Genesis.

Mutations are primarily permanent changes in the DNA strand. DNA (deoxyribonucleic acid) is the information storage unit for all organisms, including humans, cats, and dogs. In humans, the DNA consists of about three billion base pairs. The DNA is made of two strands and forms a double helix. In sexual reproduction, one set of chromosomes (large segments of DNA) comes from the mother and one set from the father. In asexual reproduction, the DNA is copied whole and then passed along when the organism splits.

The double helix is made up of four types of nitrogen bases called *nucleotides*. These types are guanine, cytosine, adenine, and thymine. They are represented by the letters G, C, A, and T. Each of these base pairs, or "letters," is part of a code that stores information for hair color, height, eye shape, etc. The bases pair up as follows: adenine to thymine and guanine to cytosine.

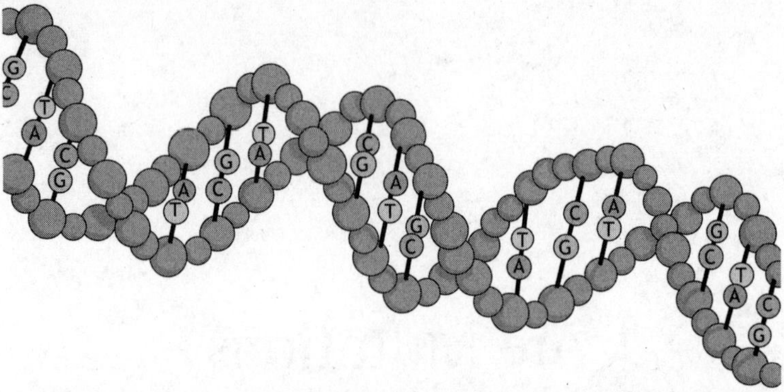

Think of it like Morse code. Morse code is a system in which letters are represented by dashes and dots (if audible, then it is a long sound and short sound). When you combine different dots and dashes, you can spell out letters and words. Here is a copy of Morse code:

A •-	N -•	0 -----
B – •••	O ---	1 •----
C -•-•	P •--•	2 ••---
D -••	Q --•-	3 •••--
E •	R •-•	4 ••••-
F ••-•	S •••	5 •••••
G -- •	T -	6 -••••
H ••••	U ••-	7 --•••
I ••	V •••-	8 ---••
J •---	W •--	9 ----•
K -•-	X -••-	Fullstop •-•-•-
L •-••	Y -•--	Comma --••--
M --	Z --••	Query ••--••

If someone wanted to call for help using Morse code, for instance, he or she would send the letters SOS (which is the international distress signal). Morse code for SOS is:

S is *dot dot dot* [• • •] or three short sounds.
O is *dash dash dash* [– – –] or three long sounds.
S is *dot dot dot* [• • •] or three short sounds.

Therefore, it would be [• • • – – – • • •], or three short sounds followed by three long sounds, followed by three short sounds.

A mutation would be like changing a dot to a dash in Morse code. If we tried to spell SOS in Morse code, but changed the first dot to a dash, it would accidentally read:

[– • • – – – • • •]

Dash dot dot is the sequence for D, not S; so it would now read:

D [– • •]
O [– – –]
S [• • •]

So, because of the mistake (mutation), we now read DOS, instead of SOS. If you sent this, no one would think you needed help. This mutation was significant because it did two things to your message:

1. The original word was lost.
2. The intent/meaning was lost.

The DNA strand is similar to, but much more complicated than, Morse code. It uses four letters (G, A, T, C) instead of dashes and dots to make words and phrases. And like Morse code, mutations can affect the DNA strand and cause problems for the organism. These DNA mistakes are called *genetic* mutations.

Theoretically, genetic mutations (that are not static) can cause one of two things:

1. Loss of information[1]
2. Gain of new information

Virtually all observed mutations are in the category of *loss of information.* This is different from loss or gain of *function.* Some mutations can cause an organism to lose genetic information and yet gain some type of function. This is rare but has happened. These types of mutations have a *beneficial* outcome. For example, if a beetle loses the information to make a wing on a windy island, the mutation is beneficial because the beetle doesn't get blown out to

1. For a definition of information that is based on the laws of science, see W. Gitt, *In the Beginning Was Information* (Green Forest, AR: Master Books, 2006).

sea and killed. Genetically, the mutation caused a loss of information but was helpful to the beetle. Thus, it was a beneficial outcome.

Besides mutations that cause information loss, in theory there could also be mutations that cause a *gain of new information*. There are only a few alleged cases of such mutations. However, if a mutated DNA strand were built up with a group of base pairs that didn't do anything, this strand wouldn't be useful. Therefore, to be useful to an organism, a mutation that has a gain of new information must also cause a gain of new function.

Types of Genetic Mutations

The DNA strand contains instructions on how to make proteins. Every three "letters" code for a specific amino acid, such as TGC, ATC, GAT, TAG, and CTC. Many amino acids together compose a protein. For simplicity's sake, to illustrate concepts with the DNA strand, we will use examples in English. Here is a segment illustrating DNA in three-letter words:

The car was red. The red car had one key.
The key has one eye and one tip.

Point Mutations

Point mutations are mutations where one letter changes on the DNA sequence. A point mutation in our example could cause "car" in the second sentence to be read "cat":

The car was red. The red **cat** had one key.
The key has one eye and one tip.

With this point mutation, we lost the information for one word (car) as well as changed the meaning of the sentence. We did gain one word (cat), but we lost one word (car) and lost the meaning of one phrase. So the overall result was a loss of information.

But many times, point mutations won't produce another word. Take for instance another point mutation, which changes "car" not to "cat" but to "caa":

The car was red. The red **caa** had one key.
The key has one eye and one tip.

With this point mutation, we lost the information for one word (car) as well as the meaning. We did not gain any new words, and we lost one word and lost the meaning of one phrase. So again, the overall result of this point mutation was a loss of information, but even more so this time.

Point mutations can be very devastating. There is a children's disease called Hutchinson-Gilford progeria syndrome (HGPS), or simply progeria. It was recently linked to a single point mutation. It is a mutation that causes children's skin to age, their head to go bald at a very early age (pre-kindergarten), their bones to develop problems usually associated with the elderly, and their body size to remain very short (about one-half to two-thirds of normal height). Their body parts, including organs, age rapidly, which usually causes death at the average age of 13 years.[2]

Not all point mutations are as devastating, yet they still result in a loss of information. According to biophysicist Lee Spetner, "All point mutations that have been studied on the molecular level turn out to reduce the genetic information and not to increase it."[3]

Inversion Mutations

An inversion mutation is a strand of DNA in a particular segment that reverses itself. An inversion mutation would be like taking the second sentence of our example and spelling it backwards:

The car was red. **Yek eno dah rac der eht.**
The key has one eye and one tip.

With inversion mutations, we can lose quite a bit of information. We lost several words from, and the meaning of, the second sentence. These mutations can cause serious problems to the organism. The bleeding disorder hemophilia A is caused by an inversion in the Factor VIII (F8) gene.

Insertion Mutations

An insertion mutation is a segment of DNA, whether a single base pair or an extensive length, that is inserted into the DNA strand. For this example, let's copy a word from the second sentence and insert it into the third sentence:

The car was red. The red car had one key.
Had the key has one eye and one tip.

2. B. Hodge, "One Tiny Flaw and 50 Years Lost," *Creation* 27(1) (2004): 33.
3. L. Spetner, *Not by Chance* (New York: Judaica Press, 1997), p. 138.

This insertion really didn't help anything. In fact, the insertion is detrimental to the third sentence in that it makes the third sentence meaningless. So this copied word in the third sentence destroyed the combined meanings of the eight words in the third sentence. Insertions generally result in a protein that loses function.[4]

Deletion Mutations

A deletion mutation is a segment of DNA, whether a single base pair or an extensive length, that is deleted from the strand. This will be an obvious loss. In this instance, the second sentence will be deleted.

The car was red. The key has one eye and one tip.

The entire second sentence has been lost. Thus, we have lost its meaning as well as the words that were in the sentence. Some disorders from deletion mutations are facioscapulohumeral muscular dystrophy (FSHD) and spinal muscular atrophy.[5]

Frame Shift Mutations

There are two basic types of frame shift mutations: frame shift due to an insertion and frame shift due to a deletion. These mutations can be caused by an insertion or deletion of one or more letters not divisible by three, which causes an offset in the reading of the "letters" of the DNA.

If a mutation occurs where one or more letters are inserted, then the entire sentence can be off. If a **t** were inserted at the beginning of the second sentence, it would read like this:

The car was red. T**t**h **ere** dca rha **don eke** yth
eke yha **son** eey ean **don** eti p.

Four new words were produced (two of them twice): *ere*, *don*, *eke* and *son*. These 4 words were not part of the original phrase. However, we lost 14 words. Not only did we lose these words, but we also lost the meaning behind the words. We lost 14 words while gaining only 4 new ones.

Therefore, even though the DNA strand became longer and produced 4 words via a single insertion, it lost 14 other words. The overall effect was a loss of information.

4. DNA Direct website, www.dnadirect.com/resource/genetics_101/GH_DNA_mutations.jsp.
5. Athena Diagnostics website, www.athenadiagnostics.com/site/content/diagnostic_ed/genetics_primer/part_2.asp.

A frame shift mutation can also occur by the deletion of one or more "letters." If the first **t** in the second sentence is deleted, the letters shift to the left, and we get:

The car was red. **Her** edc arh **ado** nek eyt hek
eyh aso **nee yea** ndo **net** ip.

Five new words were produced: *her*, *ado*, *nee*, *yea*, and *net*. However, once again, we lost 14 words. So again, the overall effect was a loss of information, and the DNA strand became smaller due to this mutation.

Frame shift mutations are usually detrimental to the organism by causing the resulting protein to be nonfunctional.

This is just the basics of mutations at a genetic level.[6]

What Does Evolution Teach about Mutations?

Pond-scum-to-people evolution teaches that, over time, by natural causes, nonliving chemicals gave rise to a living cell. Then, this single-celled life form gave rise to more advanced life forms. In essence, over millions of years, increases in information caused by mutations plus natural selection developed all the life forms we see on earth today.

For molecules-to-man evolution to happen, there needs to be a gain in *new* information within the organism's genetic material. For instance, for a single-celled organism, such as an amoeba, to evolve into something like a cow, *new* information (not random base pairs, but complex and ordered DNA) would need to develop over time that would code for ears, lungs, brain, legs, etc.

6. For more on specific mutations and more complex examples, please visit www.answersingenesis.org/go/mutations.

If an amoeba were to make a change like this, the DNA would need to mutate *new* information. (Currently, an amoeba has limited genetic information, such as the information for protoplasm.) This increase of new information would need to continue in order for a heart, kidneys, etc., to develop. If a DNA strand gets larger due to a mutation, but the sequence doesn't code for anything (e.g., it doesn't contain information for working lungs, heart, etc.), then the amount of DNA added is useless and would be more of a hindrance than a help.

There have been a few arguable cases of information-gaining mutations, but for evolution to be true, there would need to be *billions* of them. The fact is, we don't observe this in nature, but rather we see the opposite — organisms losing information. Organisms are changing, but the change is in the wrong direction! How can losses of information add up to a gain?

What Does the Bible Teach?

From a biblical perspective, we know that Adam and Eve had perfect DNA because God declared all that He had made "very good" (Genesis 1:31). This goes for the original animal and plant kinds as well. They originally had perfect DNA strands with no mistakes or mutations.

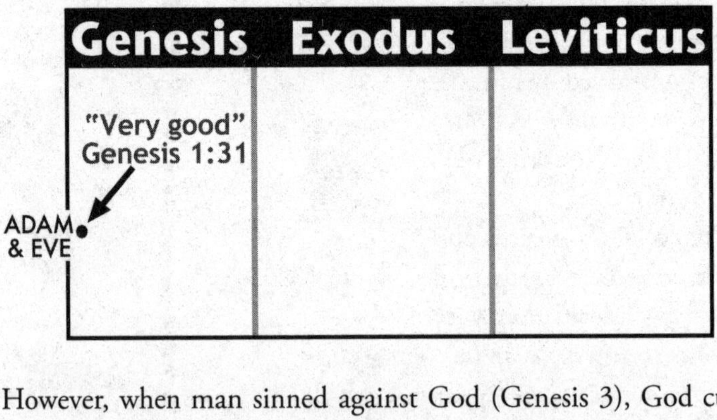

However, when man sinned against God (Genesis 3), God cursed the ground and the animals, and He sentenced man to die (Genesis 2:17; 3:19). At this time, God seemed to withdraw some of His sustaining power to no longer completely uphold everything in a perfect state.

Since then, we would expect mutations to occur and DNA flaws to accumulate. The incredible amount of information that was originally in the DNA has been filtered out, and in many cases lost, due to mutations and natural selection.

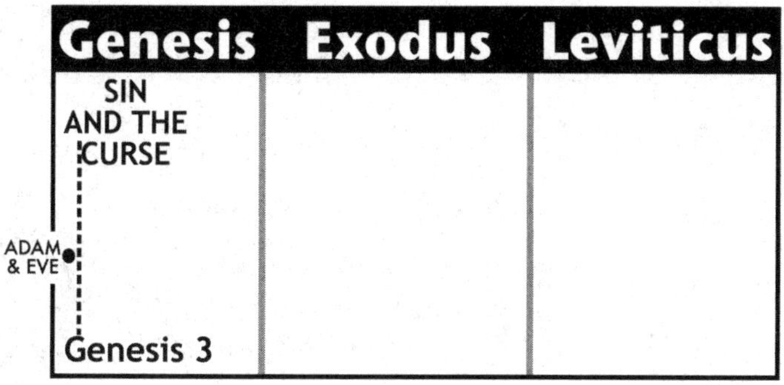

At the time of Noah's flood, there was a genetic bottleneck where information was lost among many land animals and humans. The only genetic information that survived came from the representatives of the kinds of land-dwelling, air-breathing animals and humans that were on the ark.

Over time, as people increased on the earth, God knew that mutations were rising within the human population and declared that people should no longer intermarry with close relatives (Leviticus 18). Why did He do this? Intermarriage with close relatives results in the possibility of similar genetic mutations appearing in a child due to inheriting a common mutation from both the father and mother. If both parents inherited the same mutated gene from a common ancestor (e.g., a grandparent), this would increase the possibility of both parents passing this mutated gene along to their child.

Marrying someone who is not a close relative reduces the chances that both would have the same mutated gene. If the segment of DNA from the mother had a mutation, it would be masked by the father's unmutated gene. If the segment of DNA from the father had a mutation, it would be masked by the mother's unmutated gene. If the genes from both parents were mutated,

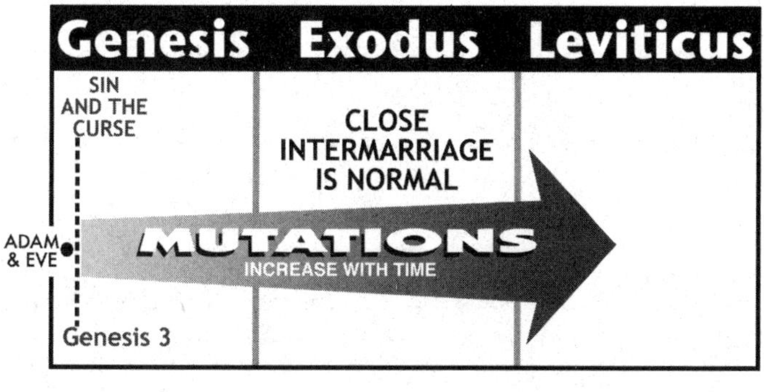

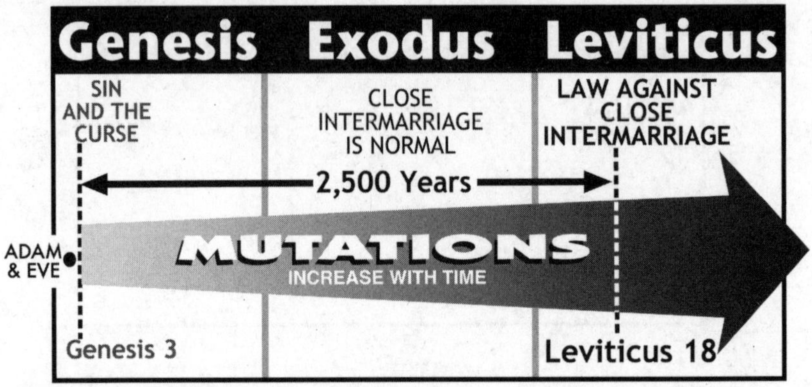

then the mutation would show in the child.[7] Our all-knowing God obviously knew this would happen and gave the command in Leviticus not to marry close relations.

Conclusion

The biblical perspective on change within living things doesn't require that new information be added to the genome as pond-scum-to-people evolution does. In fact, we expect to see the opposite (loss of genetic information) due to the curse in Genesis 3. Biblically, we would expect mutations to produce defects in the genome and would not expect mutations to be adding much, if any, new information.

Observations confirm that mutations overwhelmingly cause a loss of information, not a net gain, as evolution requires.

Mutations, when properly understood, are an excellent example of science confirming the Bible. When one sees the devastating effects of mutations, one can't help but be reminded of the curse in Genesis 3. The accumulation of mutations from generation to generation is due to man's sin. But those who have placed their faith in Christ, our Creator, look forward to a new heaven and earth where there will be no more pain, death, or disease.

7. This is only true for recessive mutations like the one that causes cystic fibrosis. There are some dominant mutations that will appear in the child regardless of having a normal copy of the gene from one parent.

8

Did Humans Really Evolve from Apelike Creatures?

DR. DAVID MENTON

Perhaps the most bitter pill to swallow for any Christian who attempts to "make peace" with Darwin is the presumed ape ancestry of man. Even many Christians who uncritically accept evolution as "God's way of creating" try to somehow elevate the origin of man, or at least his soul, above that of the beasts. Evolutionists attempt to soften the blow by assuring us that man didn't exactly evolve from apes (tailless monkeys) but rather from *apelike* creatures. This is mere semantics, however, as many of the presumed apelike ancestors of man are apes and have scientific names, which include the word *pithecus* (derived from the Greek meaning "ape"). The much-touted "human ancestor" commonly known as "Lucy," for example, has the scientific name *Australopithecus afarensis* (meaning "southern *ape* from the Afar triangle of Ethiopia"). But what does the Bible say about the origin of man, and what exactly is the scientific evidence that evolutionists claim for our ape ancestry?

Biblical Starting Assumptions

God tells us that on the same day He made all animals that walk on the earth (the sixth day), He created man separately in His own image with the intent that man would have dominion over every other living thing on earth (Genesis 1:26–28). From this it is clear that there is no animal that is man's equal, and certainly none his ancestor.

Thus, when God paraded the animals by Adam for him to name, He observed that "for Adam there was not found an help meet for him" (Genesis 2:20). Jesus confirmed this uniqueness of men and women when He declared that marriage is to be between a man and a woman because "from the beginning of the creation God made them male and female" (Mark 10:6). This leaves no room for prehumans or for billions of years of cosmic evolution prior to man's appearance on the earth. Adam chose the very name "Eve" for his wife because he recognized that she would be "the mother of all living" (Genesis 3:20). The apostle Paul stated clearly that man is not an animal: "All flesh is not the same flesh: but there is one kind of flesh of men, another flesh of beasts, another of fishes, and another of birds" (1 Corinthians 15:39).

Evolutionary Starting Assumptions

While Bible-believing Christians begin with the assumption that God's Word is true and that man's ancestry goes back only to a fully human Adam and Eve, evolutionists begin with the assumption that man has, in fact, evolved from apes. No paleoanthropologists (those who study the fossil evidence for man's origin) would dare to seriously raise the question, "*Did* man evolve from apes?" The only permissible question is, "From *which* apes did man evolve?"

Since evolutionists generally do not believe that man evolved from any ape that is now living, they look to fossils of humans and apes to provide them with their desired evidence. Specifically, they look for any anatomical feature that looks "intermediate" (between that of apes and man). Fossil apes having such features are declared to be ancestral to man (or at least collateral relatives) and are called *hominids*. Living apes, on the other hand, are not considered to be hominids, but rather are called *hominoids* because they are only similar to humans but did not evolve into them. Nonetheless, evolutionists are willing to accept mere similarities between the fossilized bones of extinct apes and the bones of living men as "proof" of our ape ancestry.

What Is the Evidence for Human Evolution?

Though many similarities may be cited between living apes and humans, the only historical evidence that could support the ape ancestry of man must come from fossils. Unfortunately, the fossil record of man and apes is very sparse. Approximately 95 percent of all known fossils are marine invertebrates, about 4.7 percent are algae and plants, about 0.2 percent are insects and other invertebrates, and only about 0.1 percent are vertebrates (animals with bones). Finally, only the smallest imaginable fraction of vertebrate fossils consists of primates (humans, apes, monkeys, and lemurs).

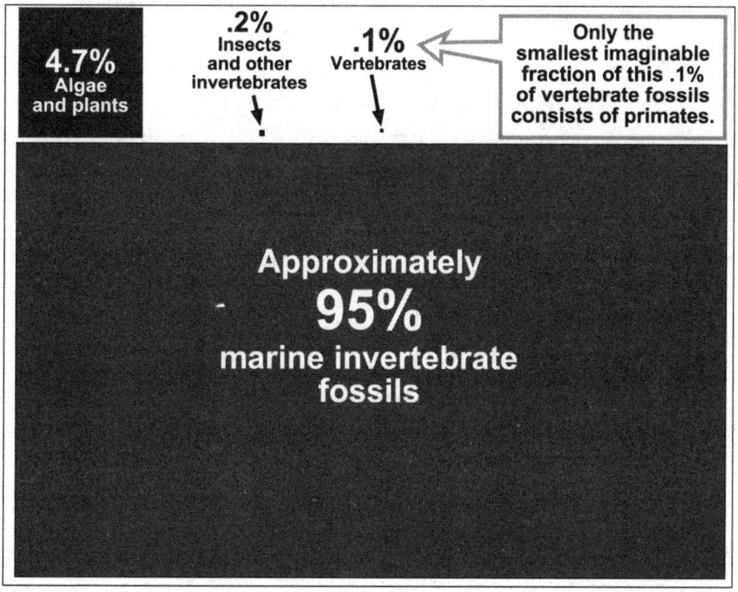

Because of the rarity of fossil hominids, even many of those who specialize in the evolution of man have never actually seen an original hominid fossil, and far fewer have ever had the opportunity to handle or study one. Most scientific papers on human evolution are based on casts of original specimens (or even on published photos, measurements, and descriptions of them). Access to original fossil hominids is strictly limited by those who discovered them and is often confined to a few favored evolutionists who agree with the discoverers' interpretation of the fossil.

Since there is much more prestige in finding an ancestor of man than an ancestor of living apes (or worse yet, merely an extinct ape), there is immense pressure on paleoanthropologists to declare almost any ape fossil to be a "hominid." As a result, the living apes have pretty much been left to find their own ancestors.

Many students in our schools are taught human evolution (often in the social studies class!) by teachers having little knowledge of human anatomy, to say nothing of ape anatomy. But it is useless to consider the fossil evidence for the evolution of man from apes without first understanding the basic anatomical and functional differences between human and ape skeletons.

Jaws and Teeth

Because of their relative hardness, teeth and jaw fragments are the most frequently found primate fossils. Thus, much of the evidence for the ape ancestry of man is based on similarities of teeth and jaws.

In contrast to man, apes tend to have incisor and canine teeth that are relatively larger than their molars. Ape teeth usually have thin enamel (the hardest surface layer of the tooth), while humans generally have thicker enamel. Finally, the jaws tend to be more U-shaped in apes and more parabolic in man.

The problem in declaring a fossil ape to be a human ancestor (i.e., a hominid) on the basis of certain humanlike features of the teeth is that some living apes have these same features and they are not considered to be ancestors of man. Some species of modern baboons, for example, have relatively small canines and incisors and relatively large molars. While most apes do have thin enamel, some apes, such as the orangutans, have relatively thick enamel. Clearly, teeth tell us more about an animal's diet and feeding habits than its supposed evolution. Nonetheless, thick enamel is one of the most commonly cited criteria for declaring an ape fossil to be a hominid.

Artistic imagination has been used to illustrate entire "apemen" from nothing more than a single tooth. In the early 1920s, the "apeman" *Hesperopithecus* (which consisted of a single tooth) was pictured in the *London Illustrated News* complete with the tooth's wife, children, domestic animals, and cave! Experts used this tooth, known as "Nebraska man," as proof for human evolution during the Scopes trial in 1925. In 1927, parts of the skeleton were discovered together with the teeth, and Nebraska man was found to really be an extinct peccary (wild pig)!

Skulls

Skulls are perhaps the most interesting primate fossils because they house the brain and give us an opportunity, with the help of imaginative artists, to look our presumed ancestors in the face. The human skull is easily distinguished from all living apes, though there are, of course, similarities.

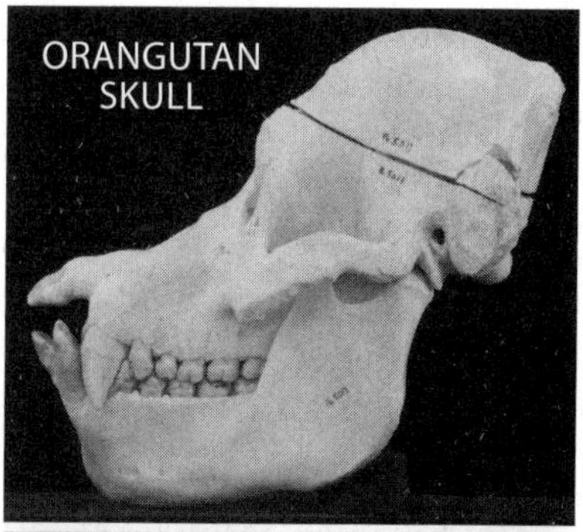

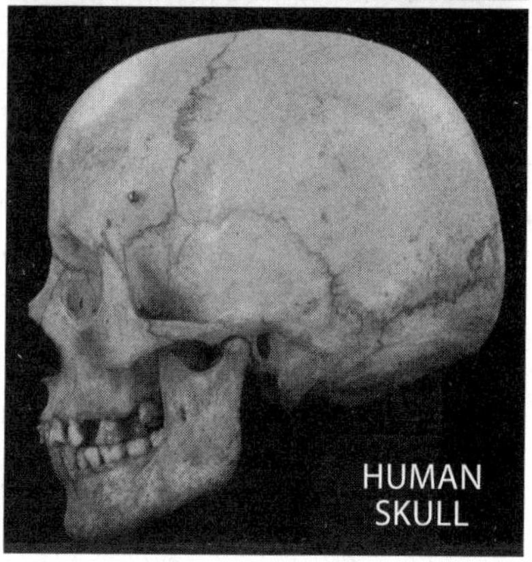

The vault of the skull is large in humans because of their relatively large brain compared to apes. Even so, the size of the normal adult human brain varies over nearly a threefold range. These differences in size in the human brain do not correlate with intelligence. Adult apes have brains that are generally smaller than even the smallest of adult human brains, and of course they are not even remotely comparable in intelligence.

Perhaps the best way to distinguish an ape skull from a human skull is to examine it from a side view. From this perspective, the face of the human is nearly vertical, while that of the ape slopes forward from its upper face to its chin.

From a side view, the bony socket of the eye (the *orbit*) of an ape is obscured by its broad, flat upper face. Humans, on the other hand, have a more curved upper face and forehead, clearly revealing the orbit of the eye from a side view.

Another distinctive feature of the human skull is the nose bone that our glasses rest on. Apes do not have protruding nasal bones and would have great difficulty wearing glasses.

Leg Bones

The most eagerly sought-after evidence in fossil hominids is any anatomical feature that might suggest *bipedality* (the ability to walk on two legs). Since humans walk on two legs, any evidence of bipedality in fossil apes is considered by evolutionists to be compelling evidence for human ancestry. But we should bear in mind that the way an ape walks on two legs is entirely different from the way man walks on two legs. The distinctive human gait requires the complex integration of many skeletal and muscular features in our hips, legs, and feet. Thus, evolutionists closely examine the hipbones (*pelvis*), thighbones (*femur*), leg bones (*tibia* and *fibula*), and foot bones of fossil apes in an effort to detect any anatomical features that might suggest bipedality.

Evolutionists are particularly interested in the angle at which the femur and the tibia meet at the knee (called the *carrying angle*). Humans are able to keep their weight over their feet while walking because their femurs converge toward the knees, forming a carrying angle of approximately nine degrees with the tibia (in other words, we're sort of knock-kneed). In contrast, chimps and gorillas have widely separated, straight legs with a carrying angle of essentially zero degrees. These animals manage to keep their weight over their feet when walking by swinging their body from side to side in the familiar "ape walk."

Evolutionists assume that fossil apes with a high carrying angle (human-like) were bipedal and thus evolved into man. Certain *australopithecines* (ape-like creatures) are considered to have walked like us and thus to be our ancestors largely because they had a high carrying angle. But high carrying angles are not confined to humans — they are also found on some modern apes that walk gracefully on tree limbs and only clumsily on the ground.

Living apes with a high carrying angle (values comparable to man) include such apes as the orangutan and spider monkey — both adept tree climbers and capable of only an apelike bipedal gait on the ground. The point is that there are *living* tree-dwelling apes and monkeys with some of the same anatomical features that evolutionists consider to be definitive evidence for bipedality, yet none of these animals walks like man and no one suggests they are our ancestors or descendants.

Foot Bones

The human foot is unique and not even close to the appearance or function of the ape foot. The big toe of the human foot is in-line with the foot and does not jut out to the side like an ape's. Human toe bones are relatively straight, rather than curved and grasping like ape toes.

While walking, the heel of the human foot hits the ground first and then the weight distribution spreads from the heel along the outer margin of the foot up to the base of the little toe. From the little toe it spreads inward across the base of the toes and finally pushes off from the big toe. No ape has a foot or push-off like that of a human, and thus, no ape is capable of walking with our distinctive human stride or making human footprints.

Hipbones

The pelvis (hipbones) plays a critically important role in walking, and the characteristic human gait requires a pelvis that is distinctly different from that of the apes. Indeed, one only has to examine the pelvis to determine if an ape has the ability to walk like a man.

The part of the hipbones that we can feel just under our belt is called the iliac blade. Viewed from above, these blades are curved forward like the handles of a steering yolk on an airplane. The iliac blades of the ape, in contrast, project straight out to the side like the handlebars of a scooter. It is simply not possible to walk like a human with an apelike pelvis. On this feature alone one can easily distinguish apes from humans.

Only Three Ways to Make an "Apeman"

Knowing from Scripture that God didn't create any apemen, there are only three ways for the evolutionist to create one:

1 Combine ape fossil bones with human fossil bones and declare the two to be one individual — a real "apeman."
2 Emphasize certain humanlike qualities of fossilized ape bones, and with imagination upgrade apes to be more humanlike.
3 Emphasize certain apelike qualities of fossilized human bones, and with imagination downgrade humans to be more apelike.

These three approaches account for *all* of the attempts by evolutionists to fill the unbridgeable gap between apes and men with fossil apemen.

Combining Men and Apes

The most famous example of an apeman proven to be a combination of ape and human bones is Piltdown man. In 1912, Charles Dawson, a medical doctor and an amateur paleontologist, discovered a mandible (lower jawbone) and part of a skull in a gravel pit near Piltdown, England. The jawbone was apelike, but had teeth that showed wear similar to the human pattern. The skull, on the other hand, was very humanlike. These two specimens were combined to form what was called "Dawn man," which was calculated to be 500,000 years old.

The whole thing turned out to be an elaborate hoax. The skull was indeed human (about 500 years old), while the jaw was that of a modern female orangutan whose teeth had been obviously filed to crudely resemble the human wear pattern. Indeed, the long ape canine tooth was filed down so far that it exposed the pulp chamber, which was then filled in to hide the mischief. It would seem that any competent scientist examining this tooth would have concluded that it was either a hoax or the world's first root canal! The success of this hoax for over 50 years, in spite of the careful scrutiny of the best authorities in the world, led the human evolutionist Sir Solly Zuckerman to declare: "It is doubtful if there is any science at all in the search for man's fossil ancestry."[1]

Making Man out of Apes

Many apemen are merely apes that evolutionists have attempted to upscale to fill the gap between apes and men. These include all the australopithecines, as well as a host of other extinct apes such as *Ardipithecus*, *Orrorin*, *Sahelanthropus*, and *Kenyanthropus*. All have obviously ape skulls, ape pelvises, and ape hands and feet. Nevertheless, australopithecines (especially *Australopithecus afarensis*) are often portrayed as having hands and feet identical to modern man; a ramrod-straight, upright posture; and a human gait.

The best-known specimen of *A. afarensis* is the fossil commonly known as "Lucy." A life-like mannequin of "Lucy" in the *Living World* exhibit at the St. Louis Zoo shows a hairy, humanlike female body with human hands and feet but with an obviously apelike head. The three-foot-tall Lucy stands erect in a deeply pensive pose with her right forefinger curled under her chin, her eyes gazing off into the distance as if she were contemplating the mind of Newton.

Few visitors are aware that this is a gross misrepresentation of what is known about the fossil ape *Australopithecus afarensis*. These apes are known

1. S. Zuckerman, *Beyond the Ivory Tower* (London: Weidenfeld & Nicolson, 1970), p. 64.

to be long-armed knuckle-walkers with locking wrists. Both the hands and feet of this creature are clearly apelike. Paleoanthropologists Jack Stern and Randall Sussman[2] have reported that the hands of this species are "surprisingly similar to hands found in the small end of the pygmy chimpanzee–common chimpanzee range." They report that the feet, like the hands, are "long, curved and heavily muscled" much like those of living tree-dwelling primates. The authors conclude that no living primate has such hands and feet "for any purpose other than to meet the demands of full or part-time arboreal (tree-dwelling) life."

Despite evidence to the contrary, evolutionists and museums continue to portray Lucy (*A. afarensis*) with virtually human feet (though some are finally showing the hands with long, curved fingers).

Making Apes out of Man

In an effort to fill the gap between apes and men, certain fossil *men* have been declared to be "apelike" and thus, ancestral to at least "modern" man. You might say this latter effort seeks to make a "monkey" out of man! Human fossils that are claimed to be "apemen" are generally classified under the genus *Homo* (meaning "man"). These include *Homo erectus*, *Homo heidelbergensis,* and *Homo neanderthalensis.*

The best-known human fossils are of Cro-Magnon man (whose marvelous paintings are found on the walls of caves in France) and Neandertal man. Both are clearly human and have long been classified as *Homo sapiens*. In recent years, however, Neandertal man has been downgraded to a different species — *Homo neanderthalensis*. The story of how Neandertal man was demoted to an apeman provides much insight into the methods of evolutionists.

Neandertal man was first discovered in 1856 by workmen digging in a limestone cave in the Neander valley near Dusseldorf, Germany. The fossil bones were examined by an anatomist (Professor Schaafhausen) who concluded that they were human.

At first, not much attention was given to these finds, but with the publication of Darwin's *Origin of Species* in 1859, the search began for the imagined "apelike ancestors" of man. Darwinians argued that Neandertal man was an apelike creature, while many critical of Darwin (like the great anatomist Rudolph Virchow) argued that Neandertals were human in every respect, though some appeared to be suffering from rickets or arthritis.

2. *American Journal of Physical Anthropology* 60 (1983): 279–317.

Over 300 Neandertal specimens have now been found scattered throughout most of the world, including Belgium, China, Central and North Africa, Iraq, the Czech Republic, Hungary, Greece, northwestern Europe, and the Middle East. This group of people was characterized by prominent eyebrow ridges (like modern Australian Aborigines), a low forehead, a long, narrow skull, a protruding upper jaw, and a strong lower jaw with a short chin. They were deep-chested, large-boned individuals with a powerful build. It should be emphasized, however, that none of these features fall outside the range of normal human anatomy. Interestingly, the brain size (based on cranial capacity) of Neandertal man was actually *larger* than average for that of modern man, though this is rarely emphasized.

Most of the misconceptions about Neandertal man resulted from the claims of the Frenchman Marcelin Boule who, in 1908, studied two Neandertal skeletons that were found in France (LeMoustier and La Chapelle-aux-Saints). Boule declared Neandertal men to be anatomically and intellectually inferior brutes who were more closely related to apes than humans. He asserted that they had a slumped posture, a "monkey-like" arrangement of certain spinal vertebrae, and he even claimed that their feet were of a "grasping type" (like those of gorillas and chimpanzees). Boule concluded that Neandertal man could not have walked erectly, but rather must have walked in a clumsy fashion. These highly biased and inaccurate views prevailed and were even expanded by many other evolutionists up to the mid-1950s.

In 1957, the anatomists William Straus and A.J. Cave examined one of the French Neandertals (La Chapelle-aux-Saints) and determined that the individual suffered from severe arthritis (as suggested by Virchow nearly 100 years earlier), which had affected the vertebrae and bent the posture. The jaw also had been affected. These observations are consistent with the Ice Age climate in which Neandertals had lived. They may well have sought shelter in caves, and this, together with poor diet and lack of sunlight, could easily have led to diseases that affect the bones, such as rickets.

In addition to anatomical evidence, there is a growing body of cultural evidence for the fully human status of Neandertals. They buried their dead and had elaborate funeral customs that included arranging the body and covering it with flowers. They made a variety of stone tools and worked with skins and leather. A wood flute was recently discovered among Neandertal remains. There is even evidence that suggests that Neandertals engaged in medical care. Some Neandertal specimens show evidence of survival to old age despite numerous wounds, broken bones, blindness, and disease. This suggests that these individuals were cared for and nurtured by others who showed human compassion.

Still, efforts continue to be made to somehow dehumanize Neandertal man. Many evolutionists now even insist that Neandertal man is not even directly related to modern man because of some differences in a small fragment of DNA! There is, in fact, nothing about Neandertals that is in any way inferior to modern man. One of the world's foremost authorities on Neandertal man, Erik Trinkaus, concludes: "Detailed comparisons of Neandertal skeletal remains with those of modern humans have shown that there is nothing in Neandertal anatomy that conclusively indicates locomotor, manipulative, intellectual, or linguistic abilities inferior to those of modern humans."[3]

Conclusion

Why then are there continued efforts to make apes out of man and man out of apes? In one of the most remarkably frank and candid assessments of the whole subject and the methodology of paleoanthropology, Dr. David Pilbeam (a distinguished professor of anthropology) suggested the following:

> Perhaps generations of students of human evolution, including myself, have been flailing about in the dark; that our data base is too sparse, too slippery, for it to be able to mold our theories. Rather the theories are more statements about us and ideology than about the past. Paleoanthropology reveals more about how humans view themselves than it does about how humans came about. But that is heresy.[4]

Oh, that these heretical words were printed as a warning on every textbook, magazine, newspaper article, and statue that presumes to deal with the bestial origin of man!

No, we are not descended from apes. Rather, God created man as the crown of His creation on day 6. We are a special creation of God, made in His image, to bring Him glory. What a revolution this truth would make if our evolutionized culture truly understood it!

3. *Natural History* 87 (1978):10.
4. *American Scientist* 66 (1978):379.

9

Does the Bible Say Anything about Astronomy?

DR. JASON LISLE

The Bible is the history book of the universe. It tells us how the universe began and how it came to be the way it is today.

The Bible is much more than just a history book, however; it was written by inspiration of God. The Lord certainly understands how this universe works; after all, He made it. So His Word, the Bible, gives us the foundation for understanding the universe.

It has been said that the Bible is not a science textbook. This is true, of course, and it's actually a good thing. After all, our science textbooks are based on the ideas of human beings who do not know everything and who often make mistakes. That's why science textbooks change from time to time, as people discover new evidence and realize that they were wrong about certain things.

The Bible, though, never changes because it never needs to. God got it right the first time! The Bible is the infallible Word of God. So when it touches on a particular topic, it's right. When the Bible talks about geology, it's correct. When Scripture addresses biology or anthropology, it's also right.

What does the Bible teach about astronomy? Let's take a look at some of the things the Bible has to say about the universe. We will see that the Bible is absolutely correct when it deals with astronomy.

The Earth Is Round

The Bible indicates that the earth is round. One verse we can look at is Isaiah 40:22, where it mentions the "circle of the earth." From space, the earth always appears as a circle since it is round. This matches perfectly with the Bible.

Another verse to consider is Job 26:10, where it teaches that God has "inscribed" a circle on the surface of the waters at the boundary of light and darkness. This boundary between light and darkness is where evening and morning occur. The boundary is a circle since the earth is round.

The Earth Floats in Space

A very interesting verse to consider is Job 26:7, which states that God "hangs the earth on nothing." This might make you think of God hanging the earth like a Christmas tree ornament, but hanging it on empty space. Although this verse is written in a poetic way, it certainly seems to suggest that the earth floats in space; and indeed the earth does float in space. We now have pictures of the earth taken from space that show it floating in the cosmic void. The earth literally hangs upon nothing, just as the Bible suggests.

- The Hindus believe the earth to be supported on the backs of four elephants, which stand on the shell of a gigantic tortoise floating on the surface of the world's waters.
- The earth of the Vedic priests was set on 12 solid pillars; its upper side was its only habitable side.
- The Altaic people of Northern Siberia affirm that their mighty Ulgen created the earth on the waters and placed under it three great fish to support it.
- The Tartars and many of the other tribes of Eurasia believe the earth to be supported by a great bull.

The Expansion of the Universe

The Bible indicates in several places that the universe has been "stretched out" or expanded. For example, Isaiah 40:22 teaches that God stretches out the heavens like a curtain and spreads them out like a tent to dwell in. This would suggest that the universe has actually increased in size since its creation. God is stretching it out, causing it to expand.

Now, this verse must have seemed very strange when it was first written. The universe certainly doesn't *look* as if it is expanding. After all, if you look at the night sky tonight, it will appear about the same size as it did the previous night, and the night before that.

In fact, secular scientists once believed that the universe was eternal and unchanging. The idea of an expanding universe would have been considered nonsense to most scientists of the past. So it must have been tempting for Christians to reject what the Bible teaches about the expansion of the universe.

I wonder if any Christians tried to "reinterpret" Isaiah 40:22 to read it in an unnatural way so that they wouldn't have to believe in an expanding universe. When the secular world believes one thing and the Bible teaches another, it is always tempting to think that God got the details wrong. But God is never wrong.

Most astronomers today believe that the universe is indeed expanding. In the 1920s, astronomers discovered that virtually all clusters of galaxies appear to be moving away from all other clusters; this indicates that the entire universe is expanding.

You can think of this like points on a balloon. As the balloon is inflated, all the points move farther away from each other. If the entire universe was being stretched out, the galaxies would all be moving away; and that is what they actually appear to be doing.

It is fascinating that the Bible recorded the idea of an expanding universe thousands of years before secular science came to accept the idea.

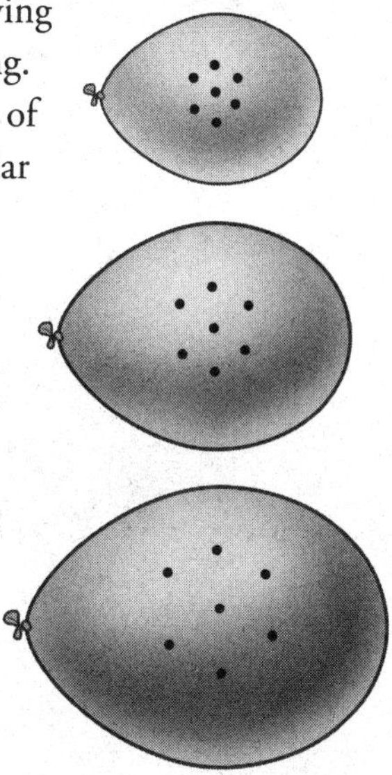

The Age of the Universe

Scripture also addresses the age of the universe. The Bible teaches that the entire universe was created in six days (Exodus 20:11). We know from the genealogies and other events recorded in Scripture that this creation happened about 6,000 years ago.

Yet, this is quite different from what most schools teach. Most secular scientists believe that the universe is many billions of years old, and they usually hold to the big-bang theory. The big bang is a secular speculation about the origin of the universe; it is an alternative to the Bible's teaching. The big bang attempts to explain the origin of the universe without God (see the next chapter, "Does the Big Bang Fit with the Bible?").

People who believe in the big bang usually interpret the evidence according to their already-existing belief in the big bang. In other words, they just assume that the big bang is true; they interpret the evidence to match their beliefs. Of course, the Bible can also be used to interpret the evidence. And since the Bible records the true history of the universe, we see that it makes a lot more sense of the evidence than the big bang does.

Now let's look at some facts about the universe regarding its age. We will see that the evidence is consistent with 6,000 years but doesn't make sense if we hold to the big bang.

Of course, big bang supporters can always reinterpret the evidence by adding extra assumptions. So the following facts are not intended to "prove" that the Bible is right about the age of the universe. The Bible is right in all matters because it is the Word of God. However, when we understand the scientific evidence, we will find that it agrees with what the Bible teaches. The evidence is certainly consistent with a young universe.

Recession of the Moon

The moon is slowly moving away from the earth. As the moon orbits the earth, its gravity pulls on the earth's oceans, which causes tides. The tides actually "pull forward" on the moon, causing the moon to gradually spiral outward. So the moon moves about an inch and a half away from the earth every year. That means that the moon would have been closer to the earth in the past.

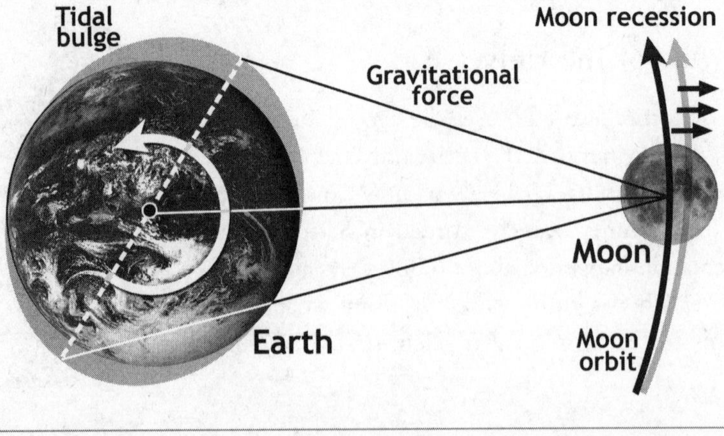

For example, 6,000 years ago, the moon would have been about 800 feet closer to the earth (which is not much of a change, considering the moon is a quarter of a million miles away). So this "spiraling away" of the moon is not a problem over the biblical time scale of 6,000 years. But if the earth and moon were over four billion years old (as evolutionists teach), then we would have big problems. In this case, the moon would have been so close that it would actually have been touching the earth only 1.4 billion years ago. This problem suggests that the moon can't possibly be as old as secular astronomers claim.

Secular astronomers who assume that the big bang is true must use other explanations to get around this. For example, they might assume that the rate at which the moon was receding was actually smaller in the past. But this is an extra assumption needed to make their billions-of-years model work. The simplest explanation is that the moon hasn't been around for that long. The recession of the moon is a problem for a belief in billions of years but is perfectly consistent with a young age.[1]

Magnetic Fields of the Planets

Many of the planets of the solar system have strong magnetic fields. These fields are caused by electrical currents that decay with time. We can even measure this decay of the earth's magnetic field: it gets weaker and weaker every year. If the planets were really billions of years old (as evolutionists believe), then their magnetic fields should be extremely weak by now. Yet they are not. The outer planets of the solar system, in particular, have quite strong magnetic fields. A reasonable explanation for this is that these planets are only a few thousand years old, as the Bible teaches.

Spiral Galaxies

A galaxy is an enormous assembly of stars, interstellar gas, and dust. The galaxy in which we live is called the Milky Way; it has over 100 billion stars. Some galaxies are round or elliptical. Others have an irregular shape, but some of the most beautiful galaxies are spiral in nature, such as our own. Spiral galaxies slowly rotate, but the inner regions of the spiral rotate faster than the outer regions. This means that a spiral galaxy is constantly becoming more and more twisted up as the spiral becomes tighter. After a few hundred million years, the galaxy would be wound so tightly that the spiral structure would no longer be recognizable. According to the big-bang scenario, galaxies are supposed to be many billions of years old. Yet we do see spiral galaxies

1. www.answersingenesis.org/home/area/feedback/2006/0811.asp/

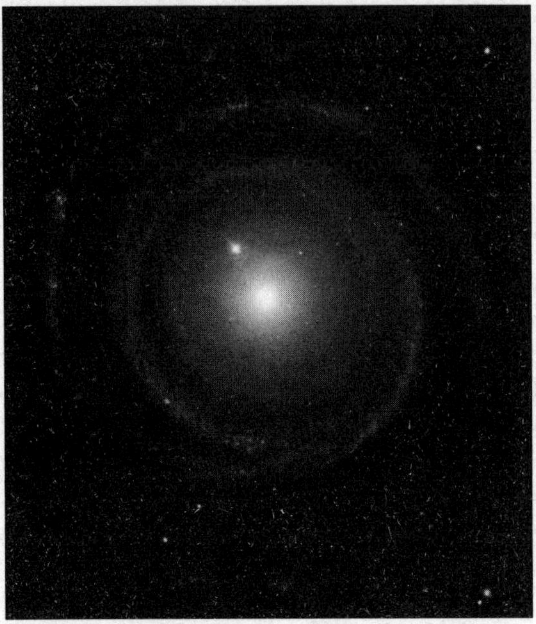

— and lots of them. This suggests that they are not nearly as old as the big bang requires. Spiral galaxies are consistent with the biblical age of the universe but are problematic for a belief in billions of years.

Comets

Comets are balls of ice and dirt. Many of them orbit the sun in elliptical paths. They spend most of their time far away from the sun, but occasionally they come very close to it. Every time a comet comes near the sun, some of its icy material is blasted away by the solar radiation. As a result, comets can orbit the sun for only so long (perhaps about 100,000 years at most) before they completely run out of material. Since we still have a lot of comets, this suggests that the solar system is much younger than 100,000 years; this agrees perfectly with the Bible's history.

Yet, secular astronomers believe the solar system is 4.5 billion years old. Since comets can't last that long, secular astronomers must assume that new comets are created to replace those that are gone. So they've invented the idea of an "Oort cloud." This is supposed to be a vast reservoir of icy masses orbiting far away from the sun. The idea is that occasionally an icy mass falls into the inner solar system to become a

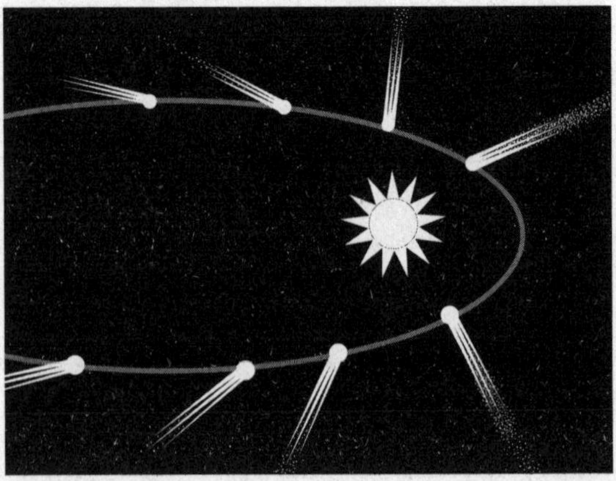

"new" comet. It is interesting that there is currently no evidence of an Oort cloud. And there's no reason to believe in one if we accept the creation account in Genesis. Comets are consistent with the fact that the solar system is young.

Supernatural Creation

Aside from age, there are other indications that the universe was supernaturally created as the Bible teaches. These evidences show God's creativity — not a big bang. For example, astronomers have discovered "extrasolar" planets. These are planets that orbit distant stars, not our sun. These planets have not been directly observed. Instead, they have been detected indirectly, usually by the gravitational "tug" they produce on the star they orbit. But the principles being used here are all good "operational science," the kind of testable, repeatable science that can be done in a laboratory. So we have every reason to believe that these are indeed real planets that God created.

These extrasolar planets are actually a problem for big-bang evolutionary models of solar system formation. Secular astronomers had expected that other solar systems would resemble ours, with small planets forming very closely to their star, and large planets (like Jupiter and Saturn) forming farther away. But many of these extrasolar planets are just the opposite; they are large, Jupiter-sized planets orbiting very closely to their star. This is inconsistent with evolutionary models of solar system formation, but it's not a problem for biblical creation. God can create many different varieties of solar systems, and apparently He has done just that.

Conclusion

We have seen that when the Bible addresses the topic of astronomy, it is accurate in every aspect. This shouldn't be surprising, because the Bible, which teaches that the heavens declare the glory and handiwork of God (Psalm 19:1), is the written Word of the Creator. God understands every aspect of the universe He has created, and He never makes mistakes.

In addition, the Word of God provides the correct foundation for understanding the scientific evidence. At the same time, the Bible provides more than just information on the physical universe. It also answers the most profound questions of life. Why are we here? How should we live? What happens when we die? The Word of God even answers the question of why there is death and suffering in the world.[2]

2. See www.AnswersInGenesis.org/go/curse.

We can have confidence that what the Bible says about our need for salvation is true, because the Bible has demonstrated itself to be accurate time after time. Showing our children how true science confirms the Bible will help them answer the evolutionary attacks they encounter at schools and in the media.

10

Does the Big Bang Fit with the Bible?

DR. JASON LISLE

The "big bang" is a story about how the universe came into existence. It proposes that billions of years ago the universe began in a tiny, infinitely hot and dense point called a *singularity*. This singularity supposedly contained not only all the mass and energy that would become everything we see today, but also "space" itself. According to the story, the singularity rapidly expanded, spreading out the energy and space.

It is supposed that over vast periods of time, the energy from the big bang cooled down as the universe expanded. Some of it turned into matter — hydrogen and helium gas. These gases collapsed to form stars and galaxies of stars. Some of the stars created the heavier elements in their core and then exploded, distributing these elements into space. Some of the heavier elements allegedly began to stick together and formed the earth and other planets.

This story of origins is entirely fiction. But sadly, many people claim to believe the big-bang model. It is particularly distressing that many professing Christians have been taken in by the big bang, perhaps without realizing its atheistic underpinnings. They have chosen to reinterpret the plain teachings of Scripture in an attempt to make it mesh with secular beliefs about origins.

Secular Compromises

There are several reasons why we cannot just add the big bang to the Bible. Ultimately, the big bang is a *secular* story of origins. When first proposed, it was an attempt to explain how the universe could have been created without God. Really, it is an *alternative* to the Bible, so it makes no sense to try to "add" it to the Bible. Let us examine some of the profound differences between the Bible and the secular big-bang view of origins.

The Bible teaches that God created the universe in six days (Genesis 1; Exodus 20:11). It is clear from the context in Genesis that these were days in the ordinary sense (i.e., 24-hour days) since they are bounded by evening and morning and occur in an ordered list (second day, third day, etc.). Conversely, the big bang teaches the universe has evolved over billions of years.

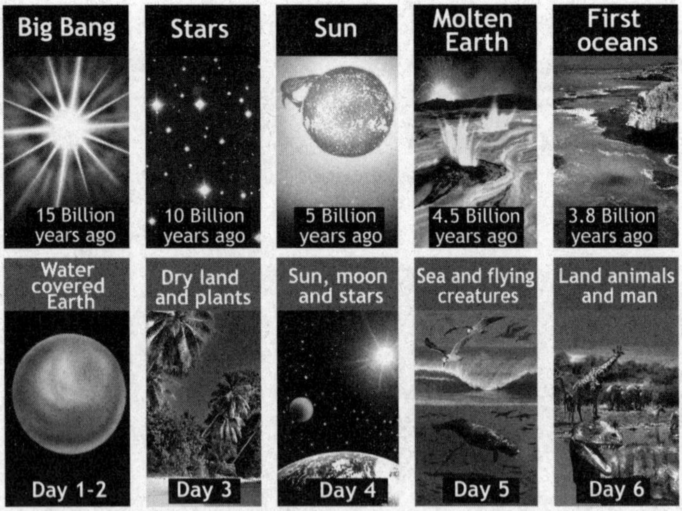

The Bible says that earth was created before the stars and that trees were created before the sun.[1] However, the big-bang view teaches the exact

1. The sun and stars were made on day 4 (Genesis 1:14–19). The earth was made on day 1 (Genesis 1:1–5). Trees were made on day 3 (Genesis 1:11–13).

opposite. The Bible tells us that the earth was created as a paradise; the secular model teaches it was created as a molten blob. The big bang and the Bible certainly do not agree about the past.

Many people don't realize that the big bang is a story not only about the past but also about the future. The most popular version of the big bang teaches that the universe will expand forever and eventually run out of usable energy. According to the story, it will remain that way forever in a state that astronomers call "heat death."[2] But the Bible teaches that the world will be judged and remade. Paradise will be restored. The big bang denies this crucial biblical teaching.

Scientific Problems with the Big Bang

The big bang also has a number of scientific problems. Big-bang supporters are forced to accept on "blind faith" a number of notions that are completely *inconsistent* with real observational science. Let's explore some of the inconsistencies between the big-bang story and the real universe.

Missing Monopoles

Most people know something about magnets — like the kind found in a compass or the kind that sticks to a refrigerator. We often say that magnets

2. Despite the name *heat death*, the universe would actually be exceedingly cold.

have two "poles" — a north pole and a south pole. Poles that are alike will repel each other, while opposites attract. A "monopole" is a hypothetical massive particle that is just like a magnet but has only one pole. So a monopole would have either a north pole or a south pole, but not both.

Particle physicists claim that many magnetic monopoles should have been created in the high temperature conditions of the big bang. Since monopoles are stable, they should have lasted to this day. Yet, despite considerable search efforts, monopoles have not been found. Where are the monopoles? The fact that we don't find any monopoles suggests that the universe never was that hot. This indicates that there never was a big bang, but it is perfectly consistent with the Bible's account of creation, since the universe did not start infinitely hot.

The Flatness Problem

Another serious challenge to the big-bang model is called the flatness problem. The expansion rate of the universe appears to be very finely balanced with the force of gravity; this condition is known as flat. If the universe were the accidental by-product of a big bang, it is difficult to imagine how such a fantastic coincidence could occur. Big-bang cosmology cannot explain why the matter density in the universe isn't greater, causing it to collapse upon itself (closed universe), or less, causing the universe to rapidly fly apart (open universe).

The problem is even more severe when we extrapolate into the past. Since any deviation from perfect flatness tends to increase as time moves forward, it logically follows that the universe must have been *even more* precisely balanced in the past than it is today. Thus, at the moment of the big bang, the universe would have been virtually flat to an extremely high precision. This must have been the case (assuming the big bang), despite the fact that the laws of physics allow for an *infinite* range of values. This is a coincidence that stretches

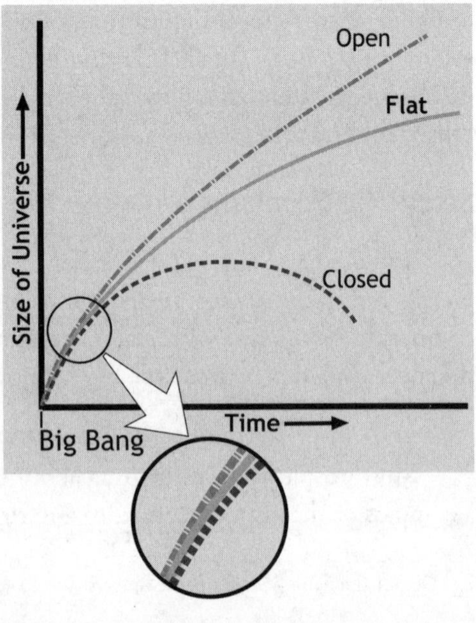

credulity to the breaking point. Of course, in the creation model, "balance" is expected since the Lord has fine-tuned the universe for life.

Inflating the Complexities

Many secular astronomers have come up with an idea called "inflation" in an attempt to address the flatness and monopole problems (as well as other problems not addressed in detail here, such as the horizon problem). Inflation proposes that the universe temporarily went through a period of accelerated expansion. Amazingly, there is no real supporting evidence for inflation; it appears to be nothing more than an unsubstantiated conjecture — much like the big bang itself. Moreover, the inflation idea has difficulties of its own, such as what would start it and how it would stop smoothly. In addition, other problems with the big bang are not solved, even if inflation were true. These are examined below.

Where Is the Antimatter?

Consider the "baryon number problem." Recall that the big bang supposes that matter (hydrogen and helium gas) was created from energy as the universe expanded. However, experimental physics tells us that whenever matter is created from energy, such a reaction also produces *antimatter*. Antimatter has similar properties to matter, except the charges of the particles are reversed. (So whereas a proton has a positive charge, an antiproton has a negative charge.) Any reaction where energy is transformed into matter produces an exactly equal amount of antimatter; there are no known exceptions.

The big bang (which has no matter to begin with, only energy) should have produced exactly equal amounts of matter and antimatter, and that should be what we see today. But we do not. The visible universe is comprised almost entirely of matter — with only trace amounts of antimatter anywhere.

This devastating problem for the big bang is actually consistent with biblical creation; it is a design feature. God created the universe to be essentially matter only — and it's a good thing He did. When matter and antimatter come together, they violently destroy each other. If the universe had equal amounts of matter and antimatter (as the big bang requires), life would not be possible.

Missing Population III Stars

The big-bang model by itself can only account for the existence of the three lightest elements (hydrogen, helium, and trace amounts of lithium).

This leaves about 90 or so of the other naturally occurring elements to be explained. Since the conditions in the big bang are not right to form these heavier elements (as big-bang supporters readily concede), secular astronomers believe that stars have produced the remaining elements by nuclear fusion in the core. This is thought to occur in the final stages of a massive star as it explodes (a supernova). The explosion then distributes the heavier elements into space. Second- and third-generation stars are thus "contaminated" with small amounts of these heavier elements.

If this story were true, then the *first* stars would have been comprised of only the three lightest elements (since these would have been the only elements in existence initially). Some such stars[3] should still be around today since their potential life span is calculated to exceed the (big bang) age of the universe. Such stars would be called "Population III" stars.[4] Amazingly (to those who believe in the big bang), Population III stars have not been found anywhere. All known stars have at least trace amounts of heavy elements in them. It is amazing to think that our galaxy alone is estimated to have over 100 billion stars in it, yet not one star has been discovered that is comprised of *only* the three lightest elements.

The Collapse of the Big Bang

With all the problems listed above, as well as many others too numerous to include, it is not surprising that quite a few secular astronomers are beginning to abandon the big bang. Although it is still the dominant model at present, increasing numbers of physicists and astronomers are realizing that the big bang simply is not a good explanation of how the universe began. In the May 22, 2004, issue of *New Scientist*, there appeared an open letter to the scientific community written primarily by *secular* scientists[5] who challenge the big bang. These scientists pointed out that the copious

3. Small (red main sequence) stars do not use up their fuel quickly. These stars theoretically have enough fuel to last significantly longer than the estimated age of the (big bang) universe.
4. If a star has a very small amount of heavy elements, it is called a "Population II" star. Population II stars exist primarily in the central bulge of spiral galaxies, in globular star clusters, and in elliptical galaxies. If a star has a relatively large amount of heavy elements (like the sun), it is called "Population I." These stars exist primarily in the arms of spiral galaxies. The (hypothetical) Population III star would have no heavy elements at all.
5. The alternatives to the big bang that these scientists had suggested are equally unbiblical. These included a steady-state theory and plasma cosmology.

arbitrary assumptions and the lack of successful big-bang predictions challenge the legitimacy of the model. Among other things, they state:

> The big bang today relies on a growing number of hypothetical entities, things that we have never observed — inflation, dark matter and dark energy are the most prominent examples. Without them, there would be a fatal contradiction between the observations made by astronomers and the predictions of the big bang theory. In no other field of physics would this continual recourse to new hypothetical objects be accepted as a way of bridging the gap between theory and observation. It would, at the least, raise serious questions about the validity of the underlying theory.[6]

This statement has since been signed by hundreds of other scientists and professors at various institutions. The big bang seems to be losing considerable popularity. Secular scientists are increasingly rejecting the big bang in favor of other models. If the big bang is abandoned, what will happen to all the Christians who compromised and claimed that the Bible is compatible with the big bang? What will they say? Will they claim that the Bible actually does not teach the big bang, but instead that it teaches the latest secular model? Secular models come and go, but God's Word does not need to be changed because God got it exactly right the first time.

Conclusion

The big bang has many scientific problems. These problems are symptomatic of the underlying incorrect worldview. The big bang erroneously assumes that the universe was *not* supernaturally created, but that it came about by natural processes billions of years ago. However, reality does not line up with this notion. Biblical

6. E. Lerner et al., An open letter to the scientific community, *New Scientist* 182(2448):20, May 22, 2004. Available online at www.cosmologystatement.org.

creation explains the evidence in a more straightforward way without the ubiquitous speculations prevalent in secular models. But ultimately, the best reason to reject the big bang is that it goes against what the Creator of the universe himself has taught: "In the beginning God created the heaven and the earth" (Genesis 1:1).

11

Where Did the Idea of "Millions of Years" Come From?

DR. TERRY MORTENSON

Today, most people in the world, including most people in the Church, take for granted that the earth and universe are millions and millions (even billions) of years old. Our public schools, from kindergarten on up, teach these vast ages, and one is scoffed at if he questions them. But it has not always been that way, and it is important to understand how this change took place and why.

Geology's Early Beginnings

Geology, as a separate field of science with systematic field studies, collection and classification of rocks and fossils, and development of theoretical reconstructions of the historical events that formed those rock layers and fossils, is only about 200 years old. Prior to this, back to ancient Greek times, people had noticed fossils in the rocks. Many believed that the fossils were the remains of former living things turned to stone, and many early Christians (including Tertullian, Chrysostom, and Augustine) attributed them to Noah's flood. But others rejected these ideas and regarded fossils as either jokes of nature, the products of rocks endowed with life in some sense, the creative works of God, or perhaps even the deceptions of Satan. The debate was finally settled when Robert Hooke (1635–1703) confirmed by microscopic analysis of fossil wood that fossils were the mineralized remains of former living creatures.

Prior to 1750, one of the most important geological thinkers was Niels Steensen (1638–1686), or Steno, a Danish anatomist and geologist. He established the principle of *superposition*, namely that sedimentary rock layers are deposited in a successive, essentially horizontal fashion, so that a lower stratum was deposited before the one above it. In his book *Forerunner* (1669), he expressed belief in a roughly 6,000-year-old earth and that fossil-bearing rock strata were deposited by Noah's flood. Over the next century, several authors, including the English geologist John Woodward (1665–1722) and the German geologist Johann Lehmann (1719–1767), wrote books essentially reinforcing that view.

In the latter decades of the 18th century, some French and Italian geologists rejected the biblical account of the Flood and attributed the rock record to natural processes occurring over a long period of time. Several prominent Frenchmen also contributed to the idea of millions of years. The widely respected scientist Comte de Buffon (1707–1788) imagined in his book *Epochs of Nature* (1779) that the earth was once like a hot molten ball that had cooled to reach its present state over about 75,000 years (though his unpublished manuscript says about 3,000,000 years). The astronomer Pierre Laplace (1749–1827) proposed the *nebular hypothesis* in his *Exposition of the System of the Universe* (1796). This theory said that the solar system was once a hot, spinning gas cloud, that over long ages gradually cooled and condensed to form the planets. Jean Lamarck, a specialist in shell creatures, advocated a theory of biological evolution over long ages in his *Philosophy of Zoology* (1809).

Abraham Werner (1749–1817) was a popular mineralogy professor in Germany. He believed that most of the crust of the earth had been precipitated chemically or mechanically by a slowly receding global ocean over the course of about a million years. It was an elegantly simple theory, but Werner failed to take into account the fossils in the rocks. This was a serious mistake, since the fossils tell much about when and how quickly the sediments were deposited and transformed into stone. Many of the greatest geologists of the 19th century were Werner's students, who were impacted by his idea of a very long history for the earth.

In Scotland, James Hutton (1726–1797) was developing a different theory of earth history. He studied medicine at the university. After his studies, he took over the family farm for a while. But he soon discovered his real love: the study of the earth. In 1788 he published a journal article and in 1795 a book, both by the title *Theory of the Earth*. He proposed that the continents were being slowly eroded into the oceans. Those sediments were gradually hardened by the internal heat of the earth and then raised by convulsions to become new landmasses that would later be eroded into the oceans, hardened, and elevated. So in his view, earth history was cyclical, and he stated that he could find no evidence of a beginning in the rock record, making earth history indefinitely long.

Catastrophist – Uniformitarian Debate

Neither Werner nor Hutton paid much attention to the fossils. However, in the early 1800s, Georges Cuvier (1768–1832), the famous French comparative anatomist and vertebrate paleontologist, developed his *catastrophist* theory of earth history. It was expressed most clearly in his *Discourse on the Revolutions of the Surface of the Globe* (1812). Cuvier believed that over the course of long, untold ages of earth history, many catastrophic floods of regional or nearly global extent had destroyed and buried creatures in sediments. All but one of these catastrophes occurred before the creation of man.

William Smith (1769–1839) was a drainage engineer and surveyor who in the course of his work around Great Britain became fascinated with the strata and fossils. Like Cuvier, he had an old-earth catastrophist view of earth history. In three works published from 1815 to 1817, he presented the first geological map of England and Wales and explained an order and relative chronology of the rock formations as defined by certain characteristic (index) fossils. He became known as the "Father of English Stratigraphy" because he developed the method of giving relative dates to the rock layers on the basis of the fossils found in them.

Georges Cuvier (1768–1832)

A massive blow to catastrophism came during the years 1830 to 1833, when Charles Lyell (1797–1875), a lawyer and former student of Buckland, published his influen-

Charles Lyell
(1797–1875)

tial three-volume work *Principles of Geology*. Reviving and augmenting the ideas of Hutton, Lyell's *Principles* set forth the principles by which he thought geological interpretations should be made. His theory was a radical *uniformitarianism* in which he insisted that only present-day processes of geological change at *present-day rates of intensity and magnitude* should be used to interpret the rock record of past geological activity. In other words, geological processes of change have been uniform throughout earth history. No continental or global catastrophic floods have ever occurred, insisted Lyell.

Lyell is often given too much credit (or blame) for destroying faith in the Genesis flood and the biblical time scale. But we must realize that many Christians (geologists and theologians) contributed to this undermining of biblical teaching before Lyell's book appeared. Although the catastrophist theory had greatly reduced the geological significance of Noah's flood and expanded earth history well beyond the traditional biblical view, Lyell's work was the final blow for belief in the Flood. By explaining the whole rock record by slow gradual processes, he thereby reduced the Flood to a geological non-event. Catastrophism did not die out immediately, although by the late 1830s only a few catastrophists remained, and they believed Noah's flood was geologically insignificant.

By the end of the 19th century, the age of the earth was considered by all geologists to be in the hundreds of millions of years. Radiometric dating methods began to be developed in 1903, and over the course of the 20th century that age of the earth expanded to 4.5 billion years.

Christian Responses to Old-earth Geology

During the first half of the 19th century, the Church responded in various ways to these old-earth theories of the catastrophists and uniformitarians. A number of writers in Great Britain (and a few in America), who became known as "scriptural geologists," raised biblical, geological, and philosophical arguments against the old-earth theories. Some of them were scientists and some were clergy. Some were both ordained and scientifically well informed, as was common in those days. Many of them were very geologically competent by the standards of their day, both by reading and by their own careful observations of rocks and fossils. They believed that the biblical account of creation and Noah's flood explained the rock record far better than the old-earth theories.[1]

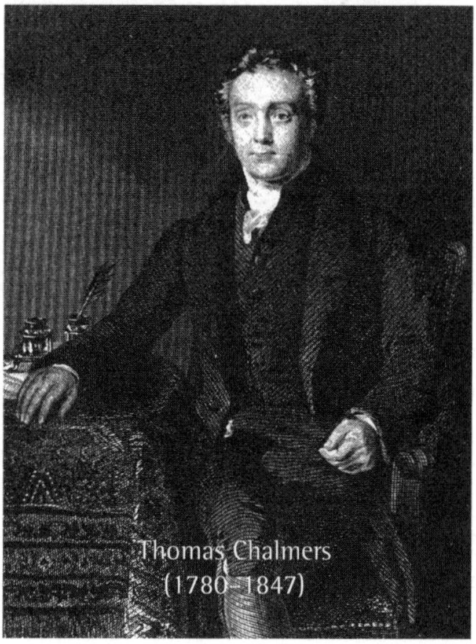
Thomas Chalmers (1780–1847)

Other Christians in the early 1800s quickly accepted the idea of millions of years and tried to fit all this time into Genesis, even though the uniformitarians and catastrophists were still debating and geology was in its infancy as a science. In 1804, Thomas Chalmers (1780–1847), a young Presbyterian pastor, began to preach that Christians should accept the millions of years; and in an 1814 review of Cuvier's book, he proposed that all the time could fit between Genesis 1:1 and 1:2. By that time, Chalmers was becoming a highly influential evangelical leader and, consequently, this *gap theory* became very popular. In 1823, the respected Anglican theologian George Stanley Faber (1773–1854) began to advocate the *day-age view*, namely that the days of creation were not literal but figurative for long ages.

To accept these geological ages, Christians also had to reinterpret the Flood. In the 1820s, John Fleming (1785–1857), a Presbyterian minister, contended that Noah's flood was so peaceful that it left no lasting geological evidence. John Pye Smith (1774–1851), a Congregational theologian, preferred to see it as a localized inundation in the Mesopotamian valley (modern-day Iraq).

1. See T. Mortenson, *The Great Turning Point: The Church's Catastrophic Mistake on Geology—Before Darwin* (Green Forest, AR: Master Books, 2004) for a full discussion of these men and the battle they fought against these developing old-earth theories and Christian compromises.

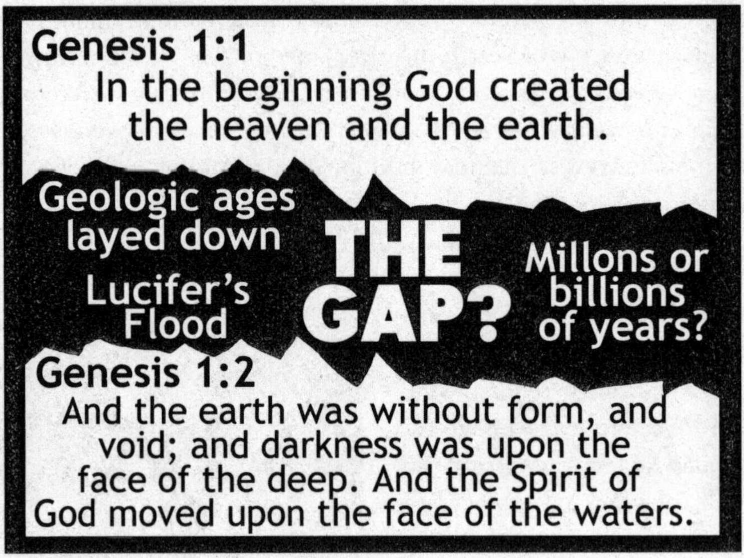
Genesis 1:1
In the beginning God created the heaven and the earth.
Geologic ages layed down
Lucifer's Flood
THE GAP?
Millons or billions of years?
Genesis 1:2
And the earth was without form, and void; and darkness was upon the face of the deep. And the Spirit of God moved upon the face of the waters.

Liberal theology, which by the early 1800s was dominating the Church in Europe, was beginning to make inroads into Britain and North America in the 1820s. The liberals considered Genesis 1–11 to be as historically unreliable and unscientific as the creation and flood myths of the ancient Babylonians, Sumerians, and Egyptians.

In spite of the efforts of the scriptural geologists, these various old-earth reinterpretations of Genesis prevailed, so that by 1845 all the commentaries on Genesis had abandoned the biblical chronology and the global flood; and by the time of Darwin's *Origin of Species* (1859), the young-earth view had essentially disappeared within the Church. From that time onward, most Christian leaders and scholars of the Church accepted the millions of years and insisted that the age of the earth was not important. Many godly men soon accepted evolution as well. Space allows only mention of a few examples.

The Baptist "prince of preachers" Charles Spurgeon (1834–1892) uncritically accepted the old-earth geological theory (though he never explained how to fit the long ages into the Bible). In an 1855 sermon he said:

> Can any man tell me when the beginning was? Years ago we thought the beginning of this world was when Adam came upon it; but we have discovered that thousands of years before that God was preparing chaotic matter to make it a fit abode for

> man, putting races of creatures upon it, who might die and leave behind the marks of his handiwork and marvelous skill, before he tried his hand on man.[2]

The great Presbyterian theologian at Princeton Seminary, Charles Hodge (1779–1878), insisted that the age of the earth was not important. He favored the gap theory initially and switched to the day-age view later in life. His compromise contributed to the eventual victory of liberal theology at Princeton about 50 years after his death.[3]

C.I. Scofield put the gap theory in notes on Genesis 1:2 in his Scofield Reference Bible, which was used by millions of Christians around the world. More recently, a respected Old Testament scholar reasoned:

> From a superficial reading of Genesis 1, the impression would seem to be that the entire creative process took place in six

2. C.H. Spurgeon, "Election," *The New Park Street Pulpit* 1 (1990): 318.
3. See J. Pipa and D. Hall, eds., *Did God Create in Six Days?* (Whitehall, WV: Tolle Lege Press, 2005), p. 7–16, for some of the documentation of this sad slide into apostasy.

> twenty-four-hour days. If this was the true intent of the Hebrew author . . . this seems to run counter to modern scientific research, which indicates that the planet Earth was created several billion years ago. . . .[4]

Numerous similar statements from Christian scholars and leaders in the last few decades could be quoted to show that their interpretation of Genesis is controlled by the fact that they assume that geologists have proven millions of years. As a result, most seminaries and Christian colleges around the world are compromised.

Compromise Unnecessary

The sad irony of all this compromise is that in the last half century, the truth of Genesis 1–11 has been increasingly vindicated, often unintentionally, by the work of evolutionists. Lyell's uniformitarian *Principles* dominated geology until about the 1970s, when Derek Ager (1923–1993), a prominent British geologist, and others increasingly challenged Lyell's assumptions and argued that much of the rock record shows evidence of rapid catastrophic erosion or sedimentation, drastically reducing the time involved in the formation of many geological deposits. Ager, an atheist to his death (as far as one can tell from his writings), explained the influence of Lyell on geology this way:

> My excuse for this lengthy and amateur digression into history is that I have been trying to show how I think geology got into the hands of the theoreticians [uniformitarians] who were conditioned by the social and political history of their day more than by observations in the field. . . . In other words, we have allowed ourselves to be brain-washed into avoiding any interpretation of the past that involves extreme and what might be termed "catastrophic" processes.[5]

These "neocatastrophist" reinterpretations of the rocks have developed contemporaneously with a resurgence of "Flood geology," a view of earth history very similar to that of the 19th-century scriptural geologists and a key ingredient of young-earth creationism, which was essentially launched into the world by the publication of *The Genesis Flood* (1961) by Drs. John Whitcomb

4. G. Archer, *A Survey of Old Testament Introduction* (Chicago, IL: Moody Press, 1985), p. 187.
5. D. Ager, *The Nature of the Stratigraphical Record* (New York: Wiley, 1981), p. 46–47.

and Henry Morris. This movement is now worldwide in scope, and the scientific sophistication of the scientific model is rapidly increasing with time.

Many Christians today are arguing that we need to contend against Darwinism with "intelligent design" arguments and leave Genesis out of the public discussion. But this strategy was tried in the early 19th century with many writings on natural theology, culminating in the famous eight volumes of the 1830s that collectively became known as the *Bridgewater Treatises*. These books were "preaching to the choir" and did nothing to retard the slide in the culture toward atheism and deism. In fact, by compromising on the age of the earth and ignoring Scripture in their defense of Christianity, they actually contributed to the weakening of the Church. The same is happening today.

The renowned atheist evolutionist and Harvard University biologist Ernst Mayr said this:

> The [Darwinian] revolution began when it became obvious that the earth was very ancient rather than having been created only 6,000 years ago. This finding was the snowball that started the whole avalanche.[6]

Mayr was right about the age of the earth (not Darwin's theory) being the beginning of the avalanche of unbelief. He was wrong that the idea of millions of years was a "finding" of scientific research. Rather, it was the fruit of antibiblical philosophical assumptions used to interpret the rocks and fossils. Historical research has shown that Laplace was an open atheist, that Buffon, Lamarck, Werner, and Hutton were deists or atheists, and that Cuvier, William Smith, and Lyell were deists or vague theists. These men (who influenced the thinking of compromised Christians) were NOT unbiased, objective pursuers of truth.

Typical of what Lyell, Buffon, and others wrote is Hutton's statement. He insisted, "The past history of our globe must be explained by what can be seen to be happening now. . . . No powers are to be employed that are not natural to the globe, no action to be admitted except those of which we know the principle."[7] By insisting that geologists must reason only from known, present-day natural processes, he ruled out supernatural creation and the unique global flood, as described in Genesis, before he ever looked at the rocks.

6. E. Mayr, "The Nature of the Darwinian Revolution," *Science* 176 (1972): 988.
7. J. Hutton, "Theory of the Earth," Trans. of the Royal Society of Edinburgh, 1788, quoted in A. Holmes, *Principles of Physical Geology* (New York: Ronald Press Co., 1965), p. 43–44.

It is no wonder that Hutton could not see the overwhelming geological evidence confirming the biblical teaching about creation, the Flood, and the age of the earth. And no wonder all the geology students who have been brainwashed with the same presuppositions for the last 200 years haven't been able to see it either. We should not be surprised that most Christian leaders and scholars are ignorant of the evidence. They, too, have been brainwashed, as many young-earth creationists once were also.

Disastrous Consequences of Compromise

The scriptural geologists of the early 19th century opposed old-earth geological theories not only because the theories reflected erroneous scientific reasoning and were contrary to Scripture, but also because they believed that Christian compromise with such theories would eventually have a catastrophic effect on the health of the Church and her witness to a lost world. Henry Cole, an Anglican minister, wrote:

> Many reverend geologists, however, would evince their reverence for the divine Revelation by making a distinction between its *historical* and its *moral* portions; and maintaining, that the latter only is inspired and absolute Truth; but that the former is not so; and therefore is open to any latitude of philosophic and scientific interpretation, modification or denial! According to these impious and infidel modifiers and separators, there is not one third of the Word of God that *is* inspired; for not more, nor perhaps so much, of that Word, is occupied in abstract moral revelation, instruction, and precept. The other two thirds, therefore, are open to any scientific modification and interpretation; or, (if scientifically required), to a total denial! It may however be safely asserted, that whoever professedly, before men, disbelieves the inspiration of any part of Revelation, disbelieves, in the sight of God, its inspiration altogether. . . . What the consequences of such things must be to a revelation-possessing land, time will rapidly and awfully unfold in its opening pages of national skepticism, infidelity, and apostasy, and of God's righteous vengeance on the same![8]

8. H. Cole, *Popular Geology Subversive of Divine Revelation* (London: Hatchard and Son, 1834), p. ix–x, 44–45 footnote.

Cole and other opponents of the old-earth theories rightly understood that the historical portions of the Bible (including Genesis 1–11) are foundational to the theological and moral teachings of Scripture. Destroy the credibility of the former and sooner or later you will see rejection of the latter, both inside and outside the Church. If the scriptural geologists were alive today and saw the castle diagram shown below, they would say, "That picture's exactly what we were concerned about!" The history of the once-Christian nations in Europe and North America has confirmed the scriptural geologists' worst fears about the Church and society.

It is time for the Church, especially her leaders and scholars, to stop ignoring the age of the earth and the scientific evidence that increasingly vindicates the Word of God. Christians must repent of their compromise with millions of years and once again believe and preach the literal truth of Genesis 1–11. It is time to take our culture back.

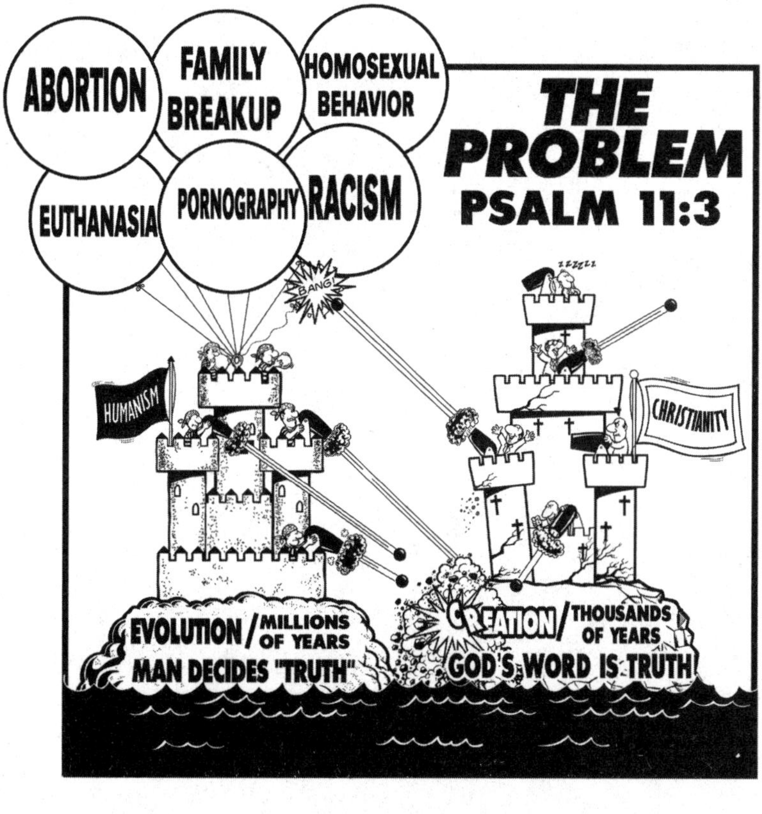

12

What's Wrong with Progressive Creation?

KEN HAM & DR. TERRY MORTENSON

One result of compromising with our evolutionary culture is the view of creation called the "day-age" theory or "progressive creation." This view, while not a new one, has received wide publicity in the past several years. Much of this publicity is due to the publications and lectures of astronomer Dr. Hugh Ross — probably the world's leading progressive creationist. Dr. Ross's views on how to interpret the Book of Genesis won early endorsements from many well-known Christian leaders, churches, seminaries, and Christian colleges. The teachings of Dr. Ross seemingly allowed Christians to use the term "creationist" but still gave them supposed academic respectability in the eyes of the world by rejecting six literal days of creation and maintaining billions of years. However, after his views became more fully understood, many who had previously embraced progressive creation realized how bankrupt those views are and removed their endorsement.

In this chapter, some of the teachings of progressive creation will be examined in light of Scripture and good science.[1]

1. For a more complete analysis, see Tim Chaffey and Jason Lisle, *Old-Earth Creationism on Trial* (Green Forest, AR: Master Books, 2008); Mark Van Bebber and Paul S. Taylor, *Creation and Time: A Report on the Progressive Creation Book by Hugh Ross* (Gilbert, AZ: Eden Publications, 1994); www.answersingenesis.org/go/compromise.

In Summary, Progressive Creation Teaches:

- The big-bang origin of the universe occurred about 13–15 billion years ago.
- The days of creation were overlapping periods of millions and billions of years.
- Over millions of years, God created new species as others kept going extinct.
- The record of nature is just as reliable as the Word of God.
- Death, bloodshed, and disease existed before Adam and Eve.
- Manlike creatures that looked and behaved much like us (and painted on cave walls) existed before Adam and Eve but did not have a spirit that was made in the image of God, and thus had no hope of salvation.
- The Genesis flood was a local event.

The Big Bang Origin of the Universe

Progressive creation teaches that the modern big-bang theory of the origin of the universe is true and has been proven by scientific inquiry and observation. For Hugh Ross and others like him, big-bang cosmology becomes the basis by which the Bible is interpreted. This includes belief that the universe and the earth are billions of years old. Dr. Ross even goes so far as to state that life would not be possible on earth without billions of years of earth history:

> It only works in a cosmos of a hundred-billion trillion stars that's precisely sixteen-billion-years old. This is the narrow window of time in which life is possible.[2]

> Life is only possible when the universe is between 12 and 17 billion years.[3]

This, of course, ignores the fact that God is omnipotent — He could make a fully functional universe ready for life right from the beginning, for with God nothing is impossible (Matthew 19:26).[4]

2. Dallas Theological Seminary chapel service, September 13, 1996.
3. Toccoa Falls Christian College, Staley Lecture Series, March 1997.
4. For an evaluation of the big-bang model, see chapter 10, "Does the Big Bang Fit with the Bible?"

The Days of Creation in Genesis 1

Progressive creationists claim that the days of creation in Genesis 1 represent long periods of time. In fact, Dr. Ross believes day 3 of creation week lasted more than 3 billion years![5] This assertion is made in order to allow for the billions of years that evolutionists claim are represented in the rock layers of earth. This position, however, has problems, both biblically and scientifically.

The text of Genesis 1 clearly states that God supernaturally created all that is in six actual days. If we are prepared to let the words of the text speak to us in accord with the context and their normal definitions, without influence from outside ideas, then the word for "day" in Genesis 1 obviously means an ordinary day of about 24 hours. It is qualified by a number, the phrase "evening and morning," and for day 1, the words "light and darkness."[6]

Dr. James Barr, Regius Professor of Hebrew at Oxford University, who himself does not believe Genesis is true history, admitted the following, as far as the language of Genesis 1 is concerned:

> So far as I know, there is no professor of Hebrew or Old Testament at any world-class university who does not believe that the writer(s) of Gen. 1–11 intended to convey to their readers the ideas that (a) creation took place in a series of six days which were the same as the days of 24 hours we now experience, (b) the figures contained in the Genesis genealogies provided by simple addition a chronology from the beginning of the world up to later stages in the biblical story, (c) Noah's Flood was understood to be world-wide and extinguish all human and animal life except for those in the ark.[7]

Besides the textual problems, progressive creationists have scientific dilemmas as well. They accept modern scientific measurements for the age of the earth, even though these measurements are based on evolutionary, atheistic assumptions. Dr. Ross often speaks of the "facts of nature" and the "facts of science" when referring to the big bang and billions of years. This demonstrates his fundamental misunderstanding of evidence. The scientific "facts" that evolutionists claim as proof of millions of years are really *interpretations* of selected observations that have been made with *antibiblical and usually atheistic, philosophical assumptions*. We all have the same facts: the same living

5. Reasons to Believe, "Creation Timeline," www.reasons.org/creation-timeline.
6. See *The New Answers Book 1*, chapter 8 by Ken Ham, for a more detailed defense of literal days in Genesis 1 (Green Forest, AR: Master Books, 2006), p. 88–112.
7. Letter to David C.C. Watson, April 23, 1984.

creatures, the same DNA molecules, the same fossils, the same rock layers, the same Grand Canyon, the same moon, the same planets, the same starlight from distant stars and galaxies, etc. These are the facts; how old they are and how they formed are the *interpretations* of the facts. And what one believes about history will affect how one interprets these facts. History is littered with so-called "scientific facts" that supposedly had proven the Bible wrong, but which were shown years or decades later to be not facts but erroneously interpreted observations because of the antibiblical assumptions used.[8]

The Order of Creation

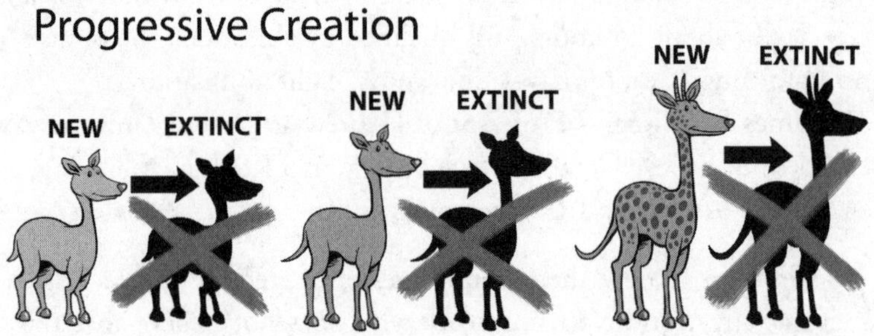

As their name indicates, progressive creationists believe that God progressively created species on earth over billions of years, with new species replacing extinct ones, starting with simple organisms and culminating in the creation of Adam and Eve. They accept the evolutionary order for the development of life on earth, even though this contradicts the order given in the Genesis account of creation.[9] Evolutionary theory holds that the first life forms were marine organisms, while the Bible says that God created land plants first. Reptiles are supposed to have predated birds, while Genesis says that birds came first. Evolutionists believe that land mammals came before whales, while the Bible teaches that God created whales first.

Dr. Davis Young, emeritus geology professor at Calvin College, recognized this dilemma and abandoned the "day-age" theory. Here is part of his explanation as to why he discarded it:

> The biblical text, for example, has vegetation appearing on the third day and animals on the fifth day. Geology, however, had long

8. See chapter 2, "What's the Best 'Proof' of Creation?" for more on how our presuppositions influence our interpretations.
9. Dr. Terry Mortenson, "Evolution vs. Creation: The Order of Events Matters!" Answers in Genesis, www.answersingenesis.org/docs2006/0404order.asp.

> realized that invertebrate animals were swarming in the seas long before vegetation gained a foothold on the land. . . . Worse yet, the text states that on the fourth day God made the heavenly bodies after the earth was already in existence. Here is a blatant confrontation with science. Astronomy insists that the sun is older than the earth.[10]

The Sixty-seventh Book of the Bible

Dr. Ross has stated that he believes nature to be "just as perfect" as the Bible. Here is the full quote:

> Not everyone has been exposed to the sixty-six books of the Bible, but everyone on planet Earth has been exposed to the sixty-seventh book — the book that God has written upon the heavens for everyone to read.
>
> And the Bible tells us it's impossible for God to lie, so the record of nature must be just as perfect, and reliable and truthful as the sixty-six books of the Bible that is part of the Word of God. . . . And so when astronomers tell us [their attempts to measure distance in space] . . . it's part of the truth that God has revealed to us. It actually encompasses part of the Word of God.[11]

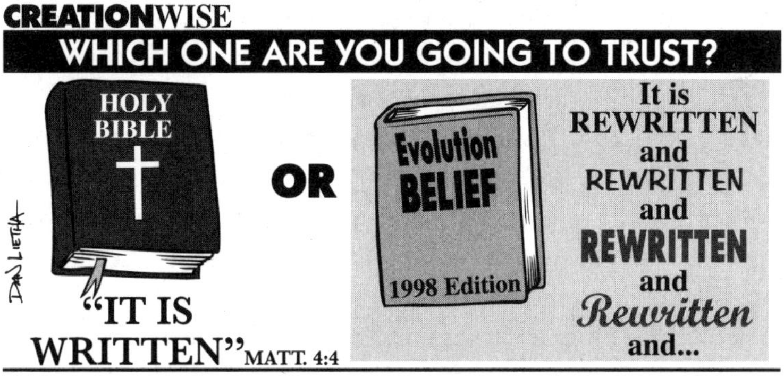

CREATIONWISE
WHICH ONE ARE YOU GOING TO TRUST?
HOLY BIBLE
OR
Evolution BELIEF
1998 Edition
It is REWRITTEN and REWRITTEN and REWRITTEN and Rewritten and...
"IT IS WRITTEN" MATT. 4:4

Dr. Ross is right that God cannot lie, and God tells us in Romans 8:22 that "the whole creation groans and labors with birth pangs" because of sin. And not only was the universe cursed, but man himself has been affected by the Fall. So how can sinful, fallible human beings in a sin-cursed universe say that their interpretation of the evidence is as perfect as God's written revelation?

10. D. Young, *The Harmonization of Scripture and Science*, science symposium at Wheaton College, March 23, 1990.
11. Toccoa Falls Christian College, Staley Lecture Series, March 1997.

Scientific assertions must use *fallible* assumptions and *fallen* reasoning — how can this be the Word of God?

The respected systematic theologian Louis Berkhof said:

> Since the entrance of sin into the world, man can gather true knowledge about God from His general revelation only if he studies it in the light of Scripture, in which the elements of God's original self-revelation, which were obscured and perverted by the blight of sin, are republished, corrected, and interpreted. . . . Some are inclined to speak of God's general revelation as a second source; but this is hardly correct in view of the fact that nature can come into consideration here only as interpreted in the light of Scripture.[12]

In other words, Christians should build their thinking on the Bible, not on fallible interruptations of scientific observations about the past.

Death and Disease before Adam

Progressive creationists believe the fossil record was formed from the millions of animals that lived and died before Adam and Eve were created. They accept the idea that there was death, bloodshed, and disease (including cancer) before sin, which goes directly against the teaching of the Bible and dishonors the character of God.

God created a perfect world at the beginning. When He was finished, God stated that His creation was "very good." The Bible makes it clear that man and all the animals were vegetarians before the Fall (Genesis 1:29-30). Plants were given to them for food (plants do not have a *nephesh* [life spirit] as man and animals do and thus eating them would not constitute "death" in the biblical sense[13]).

Concerning the entrance of sin into the world, Dr. Ross writes, "The groaning of creation in anticipation of release from sin has lasted fifteen billion years and affected a hundred billion trillion stars."[14]

However, the Bible teaches something quite different. In the context of human death, the apostle Paul states, "Through one man sin entered the world, and death through sin" (Romans 5:12). It is clear that there was no sin in the world before Adam sinned, and thus no death.

12. L. Berkhof, Introductory volume to *Systematic Theology* (Grand Rapids, MI: Wm. B. Eerdmans Publ. Co., 1946), p. 60, 96.
13. See *The New Answers Book 1*, chapter 21 by Andy McIntosh and Bodie Hodge, p. 259–270, for more details.
14. Hugh Ross, "The Physics of Sin," *Facts for Faith*, Issue 8, 2002, www.reasons.org/resources/fff/2002issue08/index.shtml.

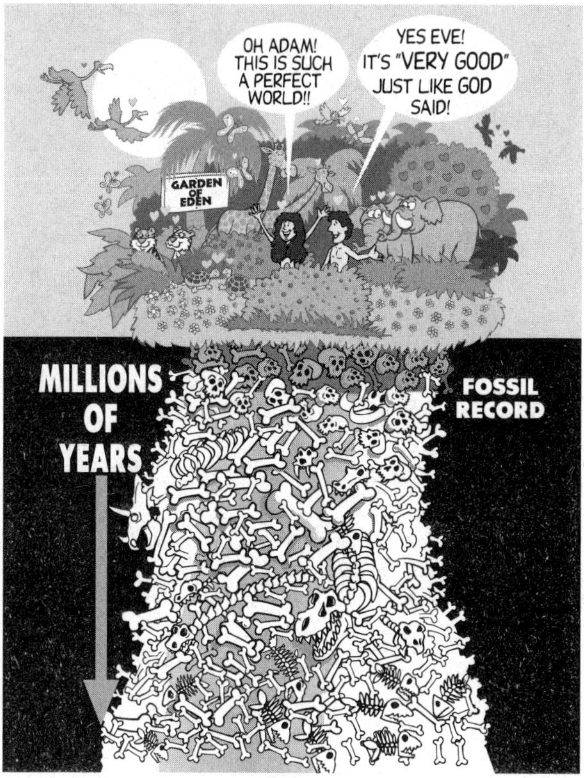

God killed the first animal in the Garden and shed blood because of sin. If there were death, bloodshed, disease, and suffering before sin, then the basis for the atonement is *destroyed.* Christ suffered death because death was the penalty for sin. There will be no death or suffering in the perfect "restoration" — so why can't we accept the same in a perfect ("very good") creation before sin?

God must be quite incompetent and cruel to make things in the way that evolutionists imagine the universe and earth to have evolved, as most creatures that ever existed died cruel deaths. Progressive creation denigrates the wisdom and goodness of God by suggesting that this was God's method of creation. This view attacks His truthfulness as well. If God really created over the course of billions of years, then He has misled most believers for 4,000 years into believing that He did it in six days.[15]

Spiritless Hominids before Adam

Since evolutionary radiometric dating methods have dated certain humanlike fossils as older than Ross's date for modern humans (approx. 40,000 years), he and other progressive creationists insist that these are fossils of pre-Adamic creatures that had no spirit, and thus no salvation.

Dr. Ross accepts and defends these evolutionary dating methods, so he must redefine all evidence of humans (descendants of Noah) if they are given

15. Dr. Terry Mortenson, "Genesis According to Evolution," *Creation* 26(4) September 2004: 50–51.

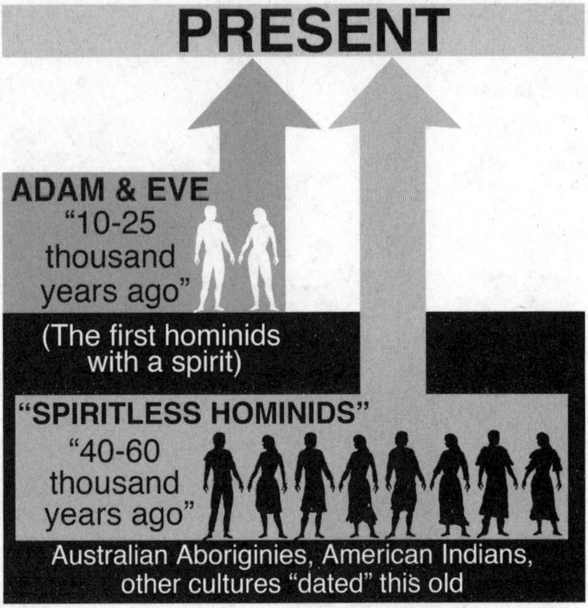

evolutionary dates of more than about 40,000 years (e.g., the Neandertal cave sites) as related to spiritless "hominids," which the Bible does not mention. However, these same methods have been used to "date" the Australian Aborigines back at least 60,000 years (some have claimed much older) and fossils of "anatomically modern humans" to over 100,000 years.[16] By Ross's reasoning, none of these (including the Australian Aborigines) could be descendants of Adam and Eve. However, Acts 17:26 says, "And He has made from one blood every nation of men to dwell on all the face of the earth, and has determined their preappointed times and the boundaries of their dwellings" (NKJV). All people on earth are descendants of Adam.

In addition, the fossil record cannot, by its very nature, conclusively reveal if a creature had a spirit or not, since spirits are not fossilized. But there is clear evidence that creatures, which Ross (following the evolutionists) places before Adam, had art and clever technology and that they buried their dead in a way that many of Adam's descendants have.[17] Therefore, we have strong reason to believe that they were fully human and actually descendants of Adam, and that they lived only a few thousand years ago.

16. T. White et al., "Pleistocene *Homo sapiens* from Middle Awash, Ethiopia," *Nature* 423 (\ June 12, 2003): 742–747. Dr. Ross will permit up to 60,000 years, but this is extreme for this position.
17. Marvin Lubelow, *Bones of Contention*, revised and updated (Grand Rapids, MI: Baker Books, 2004).

The Genesis Flood

One important tenet of progressive creation is that the flood of Noah's day was a local flood, limited to the Mesopotamian region. Progressive creationists believe that the rock layers and fossils found around the world are the result of billions of years of evolutionary earth history, rather than from the biblical flood.

Dr. Ross often says that he believes in a "universal" or "worldwide" flood, but in reality he does not believe that the Flood covered the whole earth. He argues that the text of Genesis 7 doesn't really say that the Flood covered the whole earth. But read it for yourself:

> 19 They [the flood waters] rose greatly on the earth, and *all* the high mountains under the *entire* heavens were covered.
>
> 21 *Every* living thing that moved on the earth perished — birds, livestock, wild animals, *all* the creatures that swarm over the earth, and *all* mankind.
>
> 22 *Everything* on dry land that had the breath of life in its nostrils died.
>
> 23 *Every* living thing on the face of the earth was wiped out; men and animals and the creatures that move along the ground and the birds of the air were wiped from the earth. *Only* Noah was left, and those with him in the ark [emphasis added].

Also, many questions remain for those who teach that the Genesis flood was only local:

- If the Flood was local, why did Noah have to build an ark? He could have walked to the other side of the mountains and missed it.
- If the Flood was local, why did God send the animals to the ark so they could escape death? There would have been other

animals to reproduce that kind if these particular ones had died.

- If the Flood was local, why was the ark big enough to hold all the different kinds of vertebrate land animals? If only Mesopotamian animals were aboard, the ark could have been much smaller.[18]
- If the Flood was local, why would birds have been sent on board? These could simply have winged across to a nearby mountain range.
- If the Flood was local, how could the waters rise to 15 cubits (8 meters) above the mountains (Genesis 7:20)? Water seeks its own level. It couldn't rise to cover the local mountains while leaving the rest of the world untouched.
- If the Flood was local, people who did not happen to be living in the vicinity would not be affected by it. They would have escaped God's judgment on sin. If this had happened, what did Christ mean when He likened the coming judgment of all men to the judgment of "all" men in the days of Noah (Matthew 24:37–39)? A partial judgment in Noah's day means a partial judgment to come.
- If the Flood was local, God would have repeatedly broken His promise never to send such a flood again.

18. See John Woodmorappe, *Noah's Ark: A Feasibility Study* (El Cajon, CA: Institute for Creation Research, 1996).

Conclusion

It is true that whether one believes in six literal days does not ultimately affect one's salvation, if one is truly born again. However, we need to stand back and look at the "big picture." In many nations, the Word of God was once widely respected and taken seriously. But once the door of compromise is unlocked and Christian leaders concede that we shouldn't take the Bible as written in Genesis, why should the world take heed of it in *any* area? Because the Church has told the world that one can use man's interpretation of the world (such as billions of years) to reinterpret the Bible, it is seen as an outdated, scientifically incorrect "holy book," not intended to be taken seriously.

As each subsequent generation has pushed this door of compromise open farther and farther, increasingly they are not accepting the morality or salvation of the Bible either. After all, if the history in Genesis is not correct as written, how can one be sure the rest can be taken as written? Jesus said, "If I have told you earthly things and you do not believe, how will you believe if I tell you heavenly things?" (John 3:12; NKJV).

It would not be exaggerating to claim that the majority of Christian leaders and laypeople within the church today do not believe in six literal days. Sadly, being influenced by the world has led to the Church no longer powerfully influencing the world.

The "war of the worldviews" is not ultimately one of young earth versus old earth, or billions of years versus six days, or creation versus evolution — the real battle is the authority of the Word of God versus man's fallible theories.

Belief in a historical Genesis is important because progressive creation and its belief in millions of years (1) contradicts the clear teaching of Scripture, (2) assaults the character of God, (3) severely damages and distorts the Bible's teaching on death, and (4) undermines the gospel by undermining the clear teaching of Genesis, which gives the whole basis for Christ's atonement and our need for a Redeemer. So ultimately, the issue of a literal Genesis is about the authority of the Word of God versus the authority of the words of sinful men.

Why do Christians believe in the bodily resurrection of Jesus Christ? Because of the *words of Scripture* ("according to the Scriptures").

And why should Christians believe in six literal days of creation? Because of the *words of Scripture* ("In six days the Lord made . . .").

The real issue is one of authority — let us unashamedly stand upon God's Word as our sole authority!

13

Is the Intelligent Design Movement Christian?

DR. GEORGIA PURDOM

One player in the "war of the worldviews" is the intelligent design movement. ID has gained increasing recognition and publicity over the last several years at both local and national levels. It is especially well known in educational circles, where it has been heralded as an alternative to Darwinism/naturalism.

Intelligent design can be defined as a theory that holds that "certain features" of living things were designed by an "intelligent cause" as opposed to being formed through purely natural means.[1] The ID theory does not name the intelligent cause, and it does not claim that everything is designed, thus allowing for evolution/natural causes to play a role.

The historical roots of the ID movement lie in the natural theology movement of the 18th and 19th centuries. William Paley (1743–1805) reasoned that if one walked across a field and came upon a watch, the assumption would be that there had to be a watchmaker — the complexity and purpose of the watch points to the fact that it is not the result of undirected, unintelligent causes, but the product of a designer.[2] Natural theology sought to support the existence of God through nature (general revelation) apart from the Bible (special revelation), since the Bible was facing much criticism at that time. The scientific

1. Discovery Institute Center for Science and Culture, www.discovery.org/csc/topQuestions.php, September 13, 2005.
2. W. Paley, *Paley's Watchmaker,* edited by Bill Cooper (West Sussex, England: New Wine Press, 1997, first published in 1802), p. 29–31.

knowledge of that time was grossly deficient, and it was thought that natural causes were sufficient to bring everything into existence.

In the last 100 years or so, there has been an explosion of knowledge about the complexity of cells, DNA, and microorganisms. Thus, the need for a designer has become even greater. The current ID movement has more than just philosophical arguments for a designer; it uses scientific evidence drawn from biology, chemistry, and physics.

Irreducible Complexity

The ID concept affirms that living things are designed and exhibit *irreducible complexity*. Some examples are the biochemistry of vision and the mammalian blood-clotting pathway. These biological pathways consist of many factors, and *all* the factors are necessary for the pathway to function properly. Thus, evolution (which works via the mechanism of small, gradual steps that keep only that which is immediately functional) could not have formed these pathways. For example, if only three of the blood-clotting factors (there are many factors in the complete pathway) were formed in an organism, blood would not clot, and thus the factors would not be kept because they are not currently useful to the organism. Evolutionary processes do not allow the organism to keep the three factors in the hopes that one day the rest of the blood-clotting factors will form. Evolution is goalless and purposeless; therefore, it does not keep the leftovers.

The question of whether a feature of a living organism displays design can be answered by using what is called an explanatory filter. The filter has three levels of explanation:

1. Necessity — did it have to happen?
2. Chance — did it happen by accident?
3. Design — did an intelligent agent cause it to happen?

This is a very logical, common-sense approach used by individuals every day to deduce cause and effect. For example, consider the scenario of a woman falling:

1. Did she have to fall? No, but she did.
2. Was it an accident?
3. Or was she pushed?

If we apply this explanatory filter to living organisms, a feature must be designed if the first two answers are no.

Let us evaluate the blood-clotting pathway with respect to these three questions:

1. The blood-clotting pathway is compatible with, but not required by, the natural laws of biology and chemistry; so it is not a necessity specified by natural phenomena.
2. It is complex because it is composed of many factors, thus the remote probability that it happened by chance. (Note that complex structures fall into two categories: ordered complexity and specified complexity. A snowflake, although complex structurally, has little information and thus is considered an example of ordered complexity. It is the direct result of natural phenomena rather than intelligent design[3]).
3. The blood-clotting pathway does show design, referred to as specified complexity, because it is complex and has a high amount of information. It is the direct result of an intelligent agent. All the factors must be present and interact with each other in a specified manner in order for the pathway to be functional. Thus, the blood-clotting pathway meets all the requirements for irreducible complexity, and so must be designed.

What the ID Movement Is and Is Not

William Dembski states, "ID is three things: a scientific research program that investigates the effects of intelligent causes; an intellectual movement that challenges Darwinism and its naturalistic legacy; and a way of understanding divine action."[4] The ID theory focuses on what is designed rather than answering the questions of who, when, why, and how. Those within the movement believe this promotes scientific endeavor by looking for function and purpose in those things that are designed, whereas an evolutionary mindset presupposes waste and purposelessness and aborts further scientific thinking. Although it may be a way of understanding divine action outside of a biblical framework, there are some serious implications for the Creator, which we will discuss later.

The ID movement does not speak to the optimality of design because it does not attempt to explain all designs. Remember, only "certain features" are designed, and evolutionary processes are not ruled out. The ID movement also claims not to be religiously motivated. It focuses not on the whom but on the

3. See www.intelligentdesign.org/menu/complex/complex3.htm for a more detailed discussion.
4. W. Dembski, "Science and Design," *First Things* 86 (1998): 21–27.

what. This may sound very appealing at first glance. Some biblical creationists believe that the ID movement's tolerance and acceptance of a wide range of beliefs about the supernatural could be useful in reaching a larger audience. Since the movement is very careful not to associate itself with Christianity or any formal religion, some think it will stand a better chance of gaining acceptance as an alternative to Darwinism in the schools, because it does not violate the so-called separation of church and state.

The ID movement does have several positives. The movement has produced many resources, including books and multimedia, that support the biblical creationist viewpoint. It makes clear that Darwinism/naturalism is based on the presupposition that the supernatural does not exist, thus affecting the way one interprets the scientific evidence. ID is based on the presupposition that the supernatural does exist.

ID may serve as a useful tool in *preliminary* discussions about God and creation to gain an audience that might be turned off at the mention of the Bible. However, in further discussions, the Bible as the biblical creationists' foundation should be primary.

However, the central problem with the ID movement is a divorce of the Creator from creation. The Creator and His creation cannot be separated; they reflect on each other. All other problems within the movement stem from this one.

Those within the ID movement claim their science is neutral. However, science is not neutral because it works with hypotheses based on beliefs or presuppositions. It is ironic that ID adherents refuse to see this about their own science, considering that they claim the problem with Darwinism is the presupposition that nothing supernatural exists. All scientists approach their work with presuppositions. The question is whether those beliefs are rooted in man's fallible ideas about the past or rooted in the infallible Word of God, the Bible.

The natural theology movement of the 1800s failed because it did not answer the next logical question: if it is designed, then who designed it? Although most within this movement claimed that design pointed to the God of the Bible, by divorcing general revelation (nature) from special revelation (the Bible), they opened the door to other conclusions. Deism (another movement of the same period) took the idea of excluding the Bible to the extreme and said God can only be known through nature and human reason, and that faith and revelation do not exist.

In today's culture, many are attracted to the ID movement because they can decide for themselves who the creator is — a Great Spirit, Brahman, Allah, God, etc. The current movement does not have unity on the naming of the

Intelligent design
Biblical creation
THERE HAS TO BE A "MAKER"
GOD IS MY "MAKER," AND SAVIOR!
©Answers In Genesis 2005
Romans 1:20
DAN LIETHA
As he spake by the mouth of his holy prophets, which have been since the world began: ... To give knowledge of salvation unto his people by the remission of their sins. Luke 1:70, 77

creator and focuses more on what is designed. Thus, adherents do not oppose an old age for the earth and allow evolution to play a vital role once the designer formed the basics of life. They fail to understand that a belief in long ages for the earth formed the foundation of Darwinism. If God's Word is not true concerning the age of the earth, then maybe it's not true concerning other events of the creation week, and maybe God was not a necessary part of the equation for life after all.

The ID movement's belief in evolution also allows them to distance themselves from the problem of evil in the natural world. Examples of this include pathogenic microbes, carnivorous animals, disease, and death.

Without the framework of the Bible and the understanding that evil entered the world through man's actions (Genesis 3), God appears sloppy and incompetent, if not downright vicious. People ask why God is unable to prevent evil from thwarting His plans, resulting in such poor design, instead of understanding that because of the Fall there is now a "cursed" design. In addition, because the ID movement does not acknowledge God as Redeemer, there seems to be no final solution for the evil in this world, and by all appearances evil will continue to reign supreme. However, when we trust the Bible, we read that Jesus clearly conquered death by His Resurrection (Romans 6:3–10) and one day death will no longer reign (Revelation 21:4). Again, the Creator and His creation cannot be separated.

The attributes of God are very important when resolving apparent discrepancies in His creation. For example, according to the Bible, the earth is around 6,000 years old. However, starlight can be seen from stars millions of light years away. Also, according to the Bible, God does not lie. Therefore, we must lack some information that would resolve this apparent discrepancy. (Some good research has been done on this issue, and there are several plausible solutions.[5])

Our Creator and Redeemer

Romans 1:20 states that all men know about God through His creation. However, just recognizing that there is a designer is only the first step. Colossians 1:15–20 and 2 Peter 3:3–6 point to the inexorable link between God's role as Creator *and* Redeemer. In Colossians, Paul talks about God as Creator and moves seamlessly to His role as Redeemer. Paul sees creation as a foundation for redemption. In 1 Peter, Peter states that people started disbelieving in the second coming of Christ because they started doubting God's role as Creator.

5. See *The New Answers Book 1*, chapter 19 by Jason Lisle, (Green Forest, AR: Master Books, 2006), p. 245–254.

Again, God's role as Creator becomes foundational to His role as Redeemer. Recognizing a designer is not enough to be saved; submitting to the Redeemer is also necessary. While some might consider ID to be a noble attempt to counter the evolutionary indoctrination of our culture, it falls far short of a thoroughly biblical response.

We must not separate the creation from its Creator; knowledge of God must come through both general revelation (nature) and special revelation (the Bible). The theologian Louis Berkhof said, "Since the entrance of sin into the world, man can gather true knowledge about God from His general revelation only if he studies it in the light of Scripture."[6] It is only then that the *entire* truth about God and what is seen around us can be fully understood and used to help people understand the bad news in Genesis and the good news of Jesus Christ.

AFTER EDEN
by Dan Lietha
GOD, IF YOU'RE REAL, SHOW ME A SIGN!
© 2003 AiG
DAN LIETHA
I'M WAITING!
For since the creation of the world God's invisible attributes, His eternal power and divine nature, have been clearly seen, being understood through what has been made, so that they are without excuse. Romans 1:20

6. L. Berkhof, Introductory volume to *Systematic Theology* (Grand Rapids, MI: Wm. B. Eerdmans Publ. Co., 1946), p. 60.

14

Can Creationists Be "Real" Scientists?

DR. JASON LISLE

Some evolutionists have stated that creationists cannot be real scientists. Several years ago, the National Academy of Sciences published a guidebook entitled *Teaching about Evolution and the Nature of Science*. This guidebook states that biological evolution is "the most important concept in modern biology, a concept essential to understanding key aspects of living things." Famous geneticist Theodosius Dobzhansky stated that "nothing in biology makes sense except in the light of evolution."[1]

But is a belief in particles-to-people evolution really necessary to understand biology and other sciences? Is it even helpful? Have any technological advances been made because of a belief in evolution?

Although evolutionists interpret the evidence in light of their belief in evolution, science works perfectly well without any connection to evolution. Think about it this way: is a belief in molecules-to-man evolution necessary to understand how planets orbit the sun, how telescopes operate, or how plants and animals function? Has any biological or medical research benefited from a belief in evolution? Not at all. In fact, the PhD cell biologist (and creationist) Dr. David Menton has stated, "The fact is that though widely believed, evolution contributes nothing to our understanding of empirical science and thus

1. *The American Biology Teacher* 35:125–129.

plays no essential role in biomedical research or education."[2] And creationists are not the only ones who understand this. Dr. Philip Skell, Emeritus Evan Pugh Professor of Chemistry, Penn State University, wrote:

> I recently asked more than 70 eminent researchers if they would have done their work differently if they had thought Darwin's theory was wrong. The responses were all the same: No.
>
> I also examined the outstanding biodiscoveries of the past century: the discovery of the double helix; the characterization of the ribosome; the mapping of genomes; research on medications and drug reactions; improvements in food production and sanitation; the development of new surgeries; and others. I even queried biologists working in areas where one would expect the Darwinian paradigm to have most benefited research, such as the emergence of resistance to antibiotics and pesticides. Here, as elsewhere, I found that Darwin's theory had provided no discernible guidance, but was brought in, after the breakthroughs, as an interesting narrative gloss. . . . From my conversations with leading researchers it had became [*sic*] clear that modern experimental biology gains its strength from the availability of new instruments and methodologies, not from an immersion in historical biology.[3]

The rise of technology is not due to a belief in evolution, either. Computers, cellular phones, and DVD players all operate based on the laws of physics, which God created. It is because God created a logical, orderly universe and gave us the ability to reason and to be creative that technology is possible. How can a belief in evolution (that complex biological machines do *not* require an intelligent designer) aid in the development of complex machines, which are clearly intelligently designed? Technology has shown us that sophisticated machines require intelligent designers — not random chance. Science and technology are perfectly consistent with the Bible, but not with evolution.

Differing Assumptions

The main difference between scientists who are creationists and those who are evolutionists is their starting assumptions. Creationists and evolutionists have a different view of history, but the way they do science in the present is the same. Both creationists and evolutionists use observation and experimentation

2. David Menton, "A Testimony to the Power of God's Word," Answers in Genesis, www.answersingenesis.org/docs2003/0612menton_testimony.asp.
3. P. Skell, "Why Do We Invoke Darwin?" *The Scientist* 16:10.

to draw conclusions about nature. This is the nature of observational science. It involves repeatable experimentation and observations in the present. Since observational scientific theories are capable of being tested in the present, creationists and evolutionists are generally in agreement on these models. They agree on the nature of gravity, the composition of stars, the speed of light in a vacuum, the size of the solar system, the principles of electricity, etc. These things can be checked and tested in the present.

But historical events cannot be checked scientifically in the present. This is because we do not have access to the past; it is gone. All that we have is the circumstantial evidence (relics) of past events. Although we can make educated guesses about the past and can make inferences from things like fossils and rocks, we cannot directly test our conclusions because we cannot repeat the past. Furthermore, since creationists and evolutionists have very different views of history, it is not surprising that they reconstruct past events very differently. We all have the same evidence; but in order to draw conclusions about what the evidence means, we use our worldview — our most basic beliefs about the nature of reality. Since they have different starting assumptions, creationists and evolutionists interpret the same evidence to mean very different things.

Ultimately, biblical creationists accept the recorded history of the Bible as their starting point. Evolutionists reject recorded history, and have effectively made up their own pseudohistory, which they use as a starting point for interpreting evidence. Both are using their beliefs about the past to interpret the evidence in the present. When we look at the scientific evidence today, we find that it is very consistent with biblical history and not as consistent with millions of years of evolution. We've seen in this book that the scientific evidence is consistent with biblical creation. We've seen that the geological evidence is consistent with a global flood — not millions of years of gradual deposition. We've seen that the changes in DNA are consistent with the loss of information we would expect as a result of the Curse described in Genesis 3, not the hypothetical gain of massive quantities of genetic information required by molecules-to-man evolution. Real science confirms the Bible.

Real Scientists

It shouldn't be surprising that there have been many *real* scientists who believed in biblical creation. Consider Isaac Newton (1642–1727), who codiscovered calculus, formulated the laws of motion and gravity, computed the nature of planetary orbits, invented the reflecting telescope, and made a number of discoveries in optics. Newton had profound knowledge of, and faith in, the Bible. Carl Linnaeus (1707–1778), the Swedish botanist who developed the

Sir Isaac Newton

double-Latin-name system for taxonomic classification of plants and animals, also believed the Genesis creation account. So also did the Dutch geologist Nicolaus Steno (1631–1686), who developed the basic principles of stratigraphy.

Even in the early 19th century when the idea of millions of years was developed, there were prominent Bible-believing English scientists, such as chemists Andrew Ure (1778–1857) and John Murray (1786?–1851), entomologist William Kirby (1759–1850), and geologist George Young (1777–1848). James Clerk Maxwell (1831–1879) discovered the four fundamental equations that light and all forms of electromagnetic radiation obey. Indeed, Maxwell's equations are what make radio transmissions possible. He was a deep student of Scripture and was firmly opposed to evolution. These and many other great scientists have believed the Bible as the infallible Word of God, and it was their Christian faith that was the driving motivation and intellectual foundation of their excellent scientific work.

Today there are many other PhD scientists who reject evolution and believe that God created in six days, a few thousand years ago, just as recorded in Scripture. Russ Humphreys, a PhD physicist, has developed (among many other things) a model to compute the present strength of planetary magnetic fields,[4] which enabled him to accurately predict the field strengths of the outer

4. Russell Humphreys, "The Creation of Planetary Magnetic Fields," Creation Research Society, www.creationresearch.org/crsq/articles/21/21_3/21_3.html.

planets. Did a belief in the Bible hinder his research? Not at all. On the contrary, Dr. Humphreys was able to make these predictions precisely because he started from the principles of Scripture. John Baumgardner, a PhD geophysicist and biblical creationist, has a sophisticated computer model of catastrophic plate tectonics, which was reported in the journal *Nature*; the assumptions for this model are based on the global flood recorded in Genesis. Additionally, think of all the people who have benefited from a magnetic resonance imaging (MRI) scan. The MRI scanner was developed by the creationist Dr. Raymond Damadian.[5]

Dr. John Baumgardner

Consider the biblical creationists Georgia Purdom and Andrew Snelling (both authors in this book), who work in molecular genetics and geology, respectively. They certainly understand their fields, and yet are convinced that they do not support evolutionary biology and geology.[6] On the contrary, they confirm biblical creation.

I have a PhD from a secular university and have done extensive research in solar astrophysics. In my PhD research, I made a number of discoveries about the nature of near-surface solar flows, including the detection of a never-before-seen polar alignment of supergranules, as well as patterns indicative of giant overturning cells. Was I hindered in my research by the conviction that the early chapters of Genesis are literally true? No, it's just the reverse. It is because a logical God created and ordered the universe that I, and other creationists, expect to be able to understand aspects of that universe through logic, careful observation, and experimentation.

Clearly, creationists can indeed be real scientists. And this shouldn't be surprising since the very basis for scientific research is biblical creation. This is not to say that noncreationists cannot be scientists. But, in a way, an evolutionist is being inconsistent when he or she does science. The big-bang supporter claims the universe is a random chance event, and yet he or she studies it as if it were logical and orderly. The evolutionist is thus forced to borrow certain creationist principles in order to do science. The universe is logical and orderly

5. Answers in Genesis, "Super-scientist Slams Society's Spiritual Sickness!" www.answersingenesis.org/creation/v16/i3/science.asp.
6. See various articles at www.answersingenesis.org/go/evolution and www.answersingenesis.org/go/geology.

because its Creator is logical and has imposed order on the universe. God created our minds and gave us the ability and curiosity to study the universe. Furthermore, we can trust that the universe will obey the same physics tomorrow as it does today because God is consistent. This is why science is possible. On the other hand, if the universe is just an accidental product of a big bang, why should it be orderly? Why should there be laws of nature if there is no lawgiver? If our brains are the by-products of random chance, why should we trust that their conclusions are accurate? But if our minds have been designed, and if the universe has been constructed by God, as the Bible teaches, then of course we should be able to study nature. Science is possible because the Bible is true.

15

How Should a Christian Respond to "Gay Marriage"?

KEN HAM

Most people have heard of the account of Adam and Eve. According to the first book of the Bible, Genesis, these two people were the first humans from whom all others in the human race descended. Genesis also records the names of three of Adam and Eve's many children — Cain, Abel, and Seth. Christians claim that this account of human history is accurate, because the Bible itself claims that it is the authoritative Word of the Creator God, without error.

To challenge Christians' faith in the Bible as an infallible revelation from God to humans, many skeptics have challenged the Bible's trustworthiness as a historical document by asking questions like, "Where did Cain find his wife?" (Don't worry — this will become highly relevant to the topic of gay marriage shortly!) This question of Cain's wife is one of the most-asked questions about the Christian faith and the Bible's reliability. In short, Genesis 5:4 states that Adam had "other sons and daughters"; thus, originally, brothers had to marry sisters.[1]

1. For more information on this topic, see Ken Ham, *The New Answers Book 1*, chapter 6, "Cain's Wife — Who Was She?" (Green Forest, AR: Master Books, 2006), p. 64–76.

An Atheist on a Talk Show

This background is helpful in offering the context of a conversation I had with a caller on a radio talk show. The conversation went something like this:

Caller: "I'm an atheist, and I want to tell you Christians that if you believe Cain married his sister, then that's immoral."

AiG: "If you're an atheist, then that means you don't believe in any personal God, right?"

Caller: "Correct!"

AiG: "Then if you don't believe in God, you don't believe there's such a thing as an absolute authority. Therefore, you believe everyone has a right to their own opinions — to make their own rules about life if they can get away with it, correct?"

Caller: "Yes, you're right."

AiG: "Then, sir, you can't call me immoral; after all, you're an atheist, who doesn't believe in any absolute authority."

AiG: "Do you believe all humans evolved from apelike ancestors?"

Caller: "Yes, I certainly believe evolution is fact."

AiG: "Then, sir, from your perspective on life, if man is just some sort of animal who evolved, and if there's no absolute authority, then marriage is whatever you want to define it to be — if you can get away with it in the culture you live in.

"It could be two men, two women or one man and ten women; in fact, it doesn't even have to be a man with another human — it could be a man with an animal.[2]

"I'm sorry, sir, that you think Christians have a problem. I think it's you who has the problem. Without an absolute authority, marriage, or any other aspect of how to live in society, is determined on the basis of opinion and ultimately could be anything one decides

2. See "Man Marries Dog for Luck — Then Dies," www.theage.com.au/articles/2004/02/04/1075853937098.html?from=storyrhs; and M. Bates, "Marriage in the New Millennium: Love, Honor and Scratch between the Ears," *Oak Lawn (Illinois) Reporter*, April 5, 2001, as referenced at www.freerepublic.com/forum/a3ac9e00d0a87.htm. There are many articles online that discuss the possibility of a man marrying his dog if the sanctity of marriage is not upheld; search for words like *marriage*, *man*, and *dog*.

— if the culture as a whole will allow you to get away with this. You have the problem, not me."

It was a fascinating — and revealing — exchange.

So the questions, then, that could be posed to this caller and other skeptics are: "Who has the right to determine what is good or bad, or what is morally right or wrong in the culture? Who determines whether marriage as an institution should be adhered to, and if so, what the rules should be?"

The "Pragmatics" Aspect of Opposing Gay Marriage – Some Cautions

Some who defend marriage as a union between one man and one woman claim that it can be shown that cultures that have not adhered to this doctrine have reaped all sorts of problems (whether the spread of diseases or other issues). Thus, they claim, on this basis, it's obvious that marriage should be between one man and one woman only.

Even though such problems as the spread of HIV might be shown to be a sound argument in this issue, ultimately it's not a good basis for stating that one man for one woman must be the rule. It may be a sound argument based on the pragmatics of wanting to maintain a healthy physical body, but why should one or more human beings have the right to dictate to others what they can or can't do in sexual relationships? After all, another person might decide that the relationship between one man and woman in marriage might cause psychological problems and use that as the basis for the argument. So which one is correct?

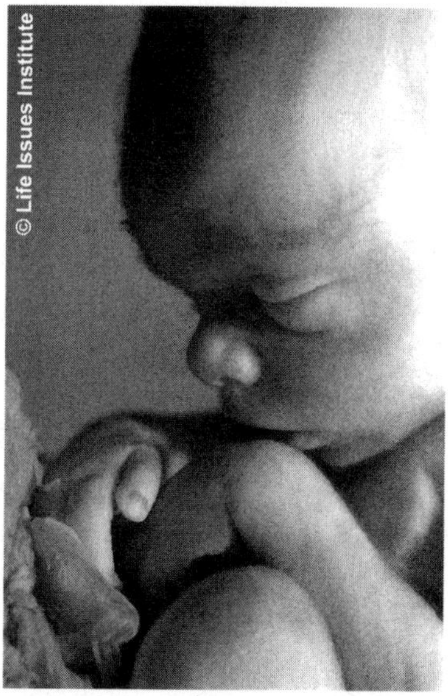

Say that a person used the argument that research has shown, for example, that the children of gay parents have a higher incidence of depression. Or the argument that since HIV kills people, it is vital that marriage is between a man and a woman. But note how such arguments have also been tried in the case of abortion and *rejected* by the culture.

Let us illustrate. Some researchers claim to have shown a high incidence of depression in people who have had an abortion. The culture, however, has rejected such pragmatic "we shouldn't hurt people" arguments, claiming that it is more important that others have the "right to choose." The argument that abortion kills people is an important one because most people still accept the basic biblical prohibition against taking innocent human life. So we should ensure that people know that the baby is really human. But is it going to be enough in the long term, as even this prohibition cannot be absolute without the Bible?

Allowing the Killing of a Newborn?

A slowly increasing minority of people, like Professor Peter Singer, the Ira W. DeCamp Professor of Bioethics at Princeton University,[3] are quite content to accept the obvious fact that abortion kills human beings, but this does not affect their view of abortion in the slightest. In fact, consistent with the fact that he rejects the Bible and the view that man was made in the image of God, Singer has argued that society should consider having a period after birth in which a baby is still allowed to be killed if socially desirable (e.g., if it has an unacceptable handicap).

Ultimately, it comes down to this: How does a culture determine what is right and what is wrong? If the majority agrees on a set of standards, what happens when that majority is replaced by a different majority?

After all, the majority in power in many of our Western nations once believed abortion was wrong — but now the majority in power doesn't believe this, so the rules have been changed.

The majority in power in many of our Western societies once believed the institution of marriage should be one man for one woman. But this has changed. Many are now allowing "gay marriage." So how long before polygamous or pedophiliac relationships are allowed, which some people are starting to advocate?[4] Who is to say they are wrong, if the majority agrees with them?

Before the Hitler era, nobody would have believed that the majority in a progressive, industrialized Western nation such as Germany could have agreed that it was ethically proper to mass murder the mentally retarded and those with incurable long-term illnesses. Yet the majority of Germans were convinced by their society to see euthanasia as ethically acceptable, even kindhearted.

Some might say that there is no way Western culture would allow pedophilia. Fifty years ago, however, most people probably would not have dreamed

3. www.answersingenesis.org/docs/1186.asp.
4. B. Sorotzkin, "The Denial of Child Abuse: The Rind, *et al.* Controversy," NARTH.com; L. Nicolosi, "The Pedophilia Debate Continues — and DSM Is Changed Again," NARTH.com; and "Russian Region Wants to Allow Men Up to Four Wives," CNN.com, July 21, 1999.

that America or Britain would ever allow gay marriage. Where does one draw the line? And who determines who draws that line? What's the answer?

Does the Church Have the Answer?

The gay marriage issue has been headline news across North America and on other continents. Even the acceptance of gay clergy has been widely noted in both secular and Christian media outlets.

- In November 2003, a part of the Episcopal Church voted to ordain a gay bishop. Thus, the world saw part of the Church now condoning homosexual behavior.[5]
- On March 18, 2004, the Pacific Northwest Conference of the United Methodist Church in America supported a lesbian pastor. Once again, the world looked on as a large denomination legitimized homosexual behavior.[6]

As part of the public debate on the gay marriage issue, many Church leaders have been interviewed on national TV programs and asked to share their position on this topic. While the majority of Church leaders have been speaking against gay unions and have been defending marriage as being between one man and one woman, many of these same Church leaders have not been able to adequately defend their position.

One Christian leader was interviewed on MSNBC-TV and was asked about the gay marriage issue. The interview went something like this:

TV host: "Did Jesus deal directly with the gay marriage issue?"

Christian leader: "No, but then Jesus didn't deal directly with the abortion issue or many other issues. . . ."

This is such a disappointing response. A proper response could have been such a powerful witness — not only to the interviewer but to the potential millions of viewers watching the news program, so people could understand why this Christian leader opposed gay marriage.

The same Christian leader appeared on CNN-TV doing an interview that, in part, went something like the following:

Interviewer: "Why are you against gay marriage?"

5. "Episcopal Church Consecrates Openly Gay Bishop," CNN.com, November 3, 2003.
6. Read the church proceedings for and against the Rev. Karen Dammann at www.pnwumc.org/Dammann.htm.

Christian leader: "Because down through the ages, culture after culture has taught that marriage is between a man and a woman."

We believe this kind of answer actually opens the door to gay marriage! How? Because it basically says that marriage is determined by law or opinion.

So, why is it that we don't see many Christian leaders giving the right sorts of answers? I think it's because the majority of them have compromised with the idea of millions of years of history, as well as evolutionary beliefs in astronomy, geology, and so on. As a result, the Bible's authority has been undermined, and it's no longer understood to be the absolute authority.[7]

Gay Marriage – Is Evolution the Cause?

After reading explanations from *Answers in Genesis* such as those above, some critics have concluded that we are saying that belief in millions of years or other evolutionary ideas is the cause of social ills like gay marriage. This is not true at all.

It is accurate to say that the increasing acceptance of homosexual behavior and gay marriage has gone hand in hand with the popularity and acceptance of millions of years and evolutionary ideas. But this does not mean that every person who believes in millions of years/evolution accepts gay marriage or condones homosexual behavior.

But the more people (whether Christian or not) believe in man's ideas concerning the history of the universe, regardless of what God's Word appears to be plainly teaching, the more man's fallible ideas are used as a basis for determining "truth" and overriding the Bible's authority.

People need to understand that homosexual behavior and the gay marriage controversy are ultimately not the problems in our culture, but are the *symptoms* of a much deeper problem. Even though it's obvious from the Bible that homosexual behavior and gay marriage are an abomination (Romans 1 and other passages make this very clear), there is a foundational reason as to why there is an increasing acceptance of these ills in America and societies like it.

> What does the Bible says about homosexual behavior and gay marriage? Study the following verses: Genesis 2:18–25; Leviticus 18:22; Mark 10:6; Romans 1:26–27; 1 Corinthians 6:9–10; 1 Timothy 1:9–10

7. For more information on this important point, see chapter 11, "Where Did the Idea of 'Millions of Years' Come From?"

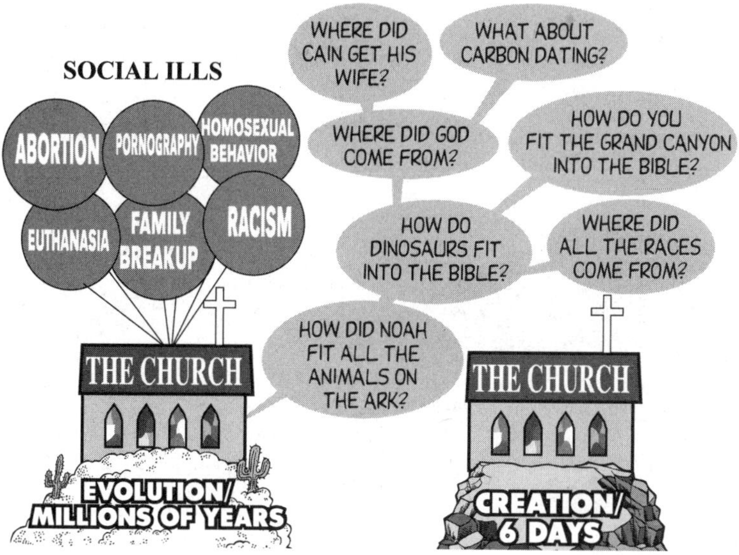

Cultures in the West were once pervaded by a primarily Christian worldview because the majority of people at least respected the Bible as the authority on morality. It needs to be clearly understood that over the past 200 years the Bible's authority has been increasingly undermined, as much of the Church has compromised with the idea of millions of years (this began before Darwin) and has thus begun reinterpreting Genesis. When those outside the Church saw Church leaders rejecting Genesis as literal history, one can understand why they would have quickly lost respect for all of the Bible. If the Church doesn't even believe this Book to be true, then why should the world build its morality on a fallible work that modern science supposedly has shown to be inaccurate in its science and history?

The Bible has lost respect in people's eyes (both within and without the Church) to the extent that the culture as a whole now does not take the Bible's morality seriously at all. The increasing acceptance of homosexual behavior and gay marriage is a symptom of the loss of biblical authority, and is primarily due to the compromise the Church has made with the secular world's teaching on origins.

Mocking the Bible

For example, consider the following. A New Orleans newspaper printed a commentary entitled "In Gay Rights Debate, Genesis Is Losing."[8] The column pointed out (correctly) that God intended marriage to be between one man and

8. J. Gill, *Times-Picayune*, New Orleans, Louisiana, March 5, 2004.

one woman. The writer even quoted Genesis 2:24, which declares, "Therefore shall a man leave his father and his mother and shall cleave to his wife and they shall be one flesh."

The author then, mockingly, wrote, "Ah, Genesis. Heaven and earth created in six days, a serpent that talks, and a 600-year-old man building an ark. Just the guide we need to set rational policy."

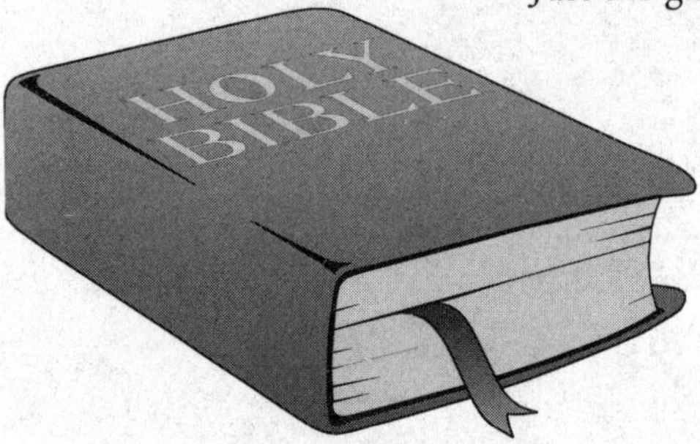

This secular writer recognized that the literal history of Genesis was the basis for the belief that marriage is one man for one woman. However, by mocking the Genesis account (just as many church leaders effectively do when they reinterpret Genesis 1–11 on the basis of man's fallible ideas), the writer removed the foundations upon which the institution of marriage stands. This opens the door to gay marriage or anything else one might determine about marriage.

Were Homosexuals Created That Way?

Human sexuality is very complex, and the arguments will long rage as to the causes of homosexual behavior. In this fallen world, most behaviors are a complex mix of one's personal choices superimposed on a platform of predisposition. This can come both from one's genetic makeup and one's environment (for example, one's upbringing). Few students of human nature would doubt the proposition that some personalities are much more predisposed to alcoholism and/or wife beating, for instance. But would anyone argue that this would make wife beating acceptable?

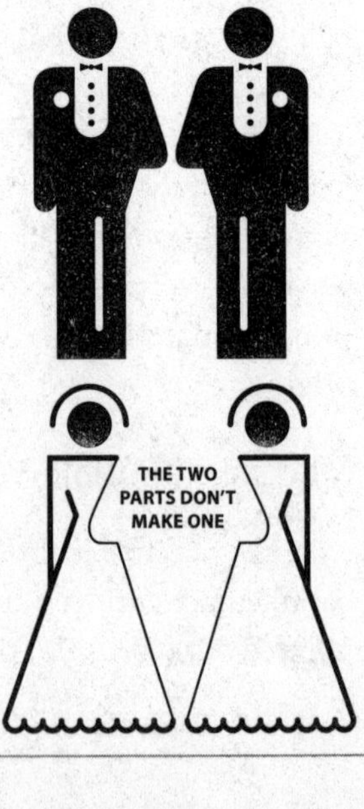

The case for a "homosexual gene" has evaporated, but let's say that researchers really were able to identify such a gene. After all, mutations in a cursed, fallen world can cause all sorts of abnormalities and malfunctions. For one thing, that would be a result of the Curse, not creation. And would knowledge of such a gene make right what Scripture clearly says is wrong? Absolute right and wrong exist independent of any secondary causative agencies.

In fact, it is quite possible that a contributing factor to at least some cases of homosexuality is a dysfunctional upbringing right at the time when the child is gaining crucial environmental input regarding his or her own sexual identity. (Notice the importance the Bible places on bringing up children, the family unit, and so on.) But if anything, this highlights one of the huge risks of "married" gay people bringing up adopted children, namely the vulnerability of the children to confused messages about their own sexual identity. To put it simply, if one's environment contributes to homosexuality, gay marriage will tend to increase the likelihood of the next generation being gay.[9]

Gay Marriage – What Is the Answer?

In the Bible in Judges 17:6, we read this statement: "In those days there was no king in Israel; every man did what was right in his own eyes" (NAS95). In other words, when there is no absolute authority to decide right and wrong, everyone has his or her own opinion about what to do.

So how could the Christian leader whose interviews were quoted earlier in this chapter have responded differently? Well, consider this answer:

> First of all, Jesus (who created us and therefore owns us and has the authority to determine right and wrong), as the God-man, *did* deal directly with the gay marriage issue, in the Bible's New Testament, in Matthew 19:4–6: "And He answered and said to them, 'Have you not read that He who made them at the beginning "made them male and female," and said, "For this cause a man shall leave father and mother and shall cling to his wife, and the two of them shall be one flesh?" So then, they are no longer two but one flesh. Therefore what God has joined together, let not man separate.' "

He could have continued:

> Christ quoted directly from the book of Genesis (and its account of the creation of Adam and Eve as the first man and woman — the first marriage) as literal history, to explain the doctrine of marriage as

9. Two things to note in this section: (1) The idea is already with us that gay "couples" should be freely able to donate their sperm to surrogate mothers or to clone their DNA to perpetuate their own genes. So if there is any genetic basis to homosexuality (i.e., "made that way"), then this too will increase the frequency of homosexuality in future generations. (2) Regarding the capacity of an individual to stop his or her homosexual behavior, we wish to observe that even with what sin has done in this fallen world, the Bible promises that we will not be tested beyond what we can endure (1 Corinthians 10:13) because the power of God is available to all believers.

being one man for one woman. Thus marriage cannot be a man and a man, or a woman and a woman.

Because Genesis is real history (as can be confirmed by observational science, incidentally), Jesus dealt quite directly with the gay marriage issue when he explained the doctrine of marriage.

Not only this, but in John 1 we read: "In the beginning was the Word, and the Word was with God, and the Word was God. The same was in the beginning with God. All things were made by him; and without him was not any thing made that was made" (KJV).

Jesus, the Creator, is the Word. The Bible is the written Word. Every word in the Bible is really the Word of the Creator—Jesus Christ.[10]

Therefore, in Leviticus 18:22, Jesus deals directly with the homosexual issue, and thus the gay marriage issue. This is also true of Romans 1:26–27 and 1 Timothy 1:9–10.

Because Jesus in a real sense wrote all of the Bible, whenever Scripture deals with marriage and/or the homosexual issue, Jesus himself is directly dealing with these issues.

Even in a secular context, the only answer a Christian should offer is this:

> The Bible is the Word of our Creator, and Genesis is literal history. Its science and history can be trusted. Therefore, we have an absolute authority that determines marriage.
>
> God made the first man and woman — the first marriage. Thus, marriage can only be a man and a woman because we are accountable to the One who made marriage in the first place.
>
> And don't forget — according to Scripture, one of the primary reasons for marriage is to produce godly offspring.[11] Adam and Eve were told to be fruitful and multiply, but there's no way a gay marriage can fulfill this command!

The battle against gay marriage will ultimately be lost (like the battle against abortion) *unless* the church and the culture return to the absolute authority beginning in Genesis. Then and only then will there be a true foundation for the correct doctrine of marriage — one man for one woman for life.

---

10. See Colossians 1:15–20 as well.
11. Malachi 2:15: "Has not the Lord made them one? In flesh and spirit they are his. And why one? Because he was seeking godly offspring. So guard yourself in your spirit, and do not break faith with the wife of your youth."

16

Did People Like Adam and Noah Really Live Over 900 Years of Age?

DR. DAVID MENTON & DR. GEORGIA PURDOM

"Methuseleh lived 900 years . . . but these stories you're liable to read in the Bible, they ain't necessarily so."[1]

Along with American composer George Gershwin, many people find it difficult to believe that Methuselah lived to be 969 years old. Nevertheless, the Bible teaches quite plainly that the early patriarchs often lived to be nearly 1,000 years old and even had children when they were several hundred years old! Similar claims of long life spans are found in the secular literature of several ancient cultures (including the Babylonians, Greeks, Romans, Indians, and Chinese). But even a life span of nearly 1,000 years is sadly abbreviated when we consider that God initially created us to live *forever.*

According to the Bible, God created the first humans — Adam and Eve — without sin and with the ability to live forever. God gave the first human couple everything they needed for their eternal health and happiness in the Garden of Eden; but He warned them not to eat fruit from the Tree of the Knowledge of Good and Evil or they would die, as indeed would all their descendants after them (Genesis 2:16–17). When Satan's deception prompted

1. George Gershwin, "It Ain't Necessarily So," *Porgy & Bess*, 1934.

Eve to disobey this command and then Adam willfully disobeyed, their minds and bodies profoundly changed (Genesis 3). Not only did they become subject to death, but their firstborn child (Cain) became the world's first murderer. Truly, the wages of sin is death, physically and spiritually. It is sobering to think that the Bible would have been only a few pages long — from creation to the fall into sin — were it not for the undeserved love of God who both promised and sent the Messiah to save us from sin and death (Genesis 3:15; Isaiah 25:8; Psalm 49:14–15; 1 John 5:13).

For 1,500 years after creation, men lived such long lives that most were either contemporaries of the first man, Adam, or personally knew someone who was! The ten patriarchs (excluding Enoch) who preceded the Great Flood lived an average of 912 years. Lamech died the youngest at the age of 777, and Methuselah lived to be the oldest at 969. See table 1.

Table 1. Ages of the Patriarchs from Adam to Noah

	Patriarch	Age	Bible Reference
1	Adam	930	Genesis 5:4
2	Seth	912	Genesis 5:8
3	Enosh	905	Genesis 5:11
4	Cainan	910	Genesis 5:14
5	Mahalalel	895	Genesis 5:17
6	Jared	962	Genesis 5:20
7	Enoch	365 (translated)	Genesis 5:23
8	Methuselah	969	Genesis 5:27
9	Lamech	777	Genesis 5:31
10	Noah	950	Genesis 9:29

During the 1,000 years following the Flood, however, the Bible records a progressive decline in the life span of the patriarchs, from Noah who lived to be 950 years old until Abraham at 175 (see figure 1 and table 2). In fact, Moses was unusually old for his time (120 years) because, when he reflected on the brevity of life, he said: "The days of our lives are seventy years; and if by reason of strength they are eighty years, yet their boast is only labor and sorrow; for it is soon cut off, and we fly away" (Psalm 90:10).

Table 2. Ages of the Patriarchs after Noah to Abraham

	Patriarch	Age	Bible Reference
11	Shem	600	Genesis 11:10–11
12	Arphaxad	438	Genesis 11:12–13
13	Shelah	433	Genesis 11:14–15
14	Eber	464	Genesis 11:16–17
15	Peleg	239	Genesis 11:18–19
16	Reu	239	Genesis 11:20–21
17	Serug	230	Genesis 11:22–23
18	Nahor	148	Genesis 11:24–25
19	Terah	205	Genesis 11:32
20	Abraham	175	Genesis 25:7

Extrabiblical evidence to support the long life spans of the people in Genesis is found in the Sumerian King List. This list mentions a flood and gives the length of the reigns of kings before and after a flood. There are many striking parallels between the Sumerian King List and Genesis, such as a flood event, numerical parallels between the pre-Flood biblical patriarchs and the antediluvial kings, and a substantial decrease in life span of people following the flood.[2] One author on this subject concludes, "It is highly unlikely that the biblical account was derived from the Sumerian in view of the differences of the two accounts, and the obvious superiority of the Genesis record both in numerical precision, realism, completion, and moral and spiritual qualities."[3] It is more likely that the Sumerian King List was composed using Genesis for numerical information. Obviously, the Book of Genesis would only be used if the person writing the list believed it to be a true historical account containing accurate information.

Today, man's maximum life span is about 120 years,[4] and our average life expectancy is still only 70–80 years — just as it was when the 90th Psalm was

2. Raul Lopez, "The Antediluvian Patriarchs and the Sumerian King List," *CEN Technical Journal* 12, no. 3 (1998): 347–357.
3. Ibid.
4. It should be noted that Genesis 6:3 does not refer to God mandating a maximum life span for people of 120 years. If this is the case, then the Bible is in error as many people have been recorded as living longer than 120 years. Rather, it refers to the amount of time from when God determined to destroy mankind to when God sent a global flood.

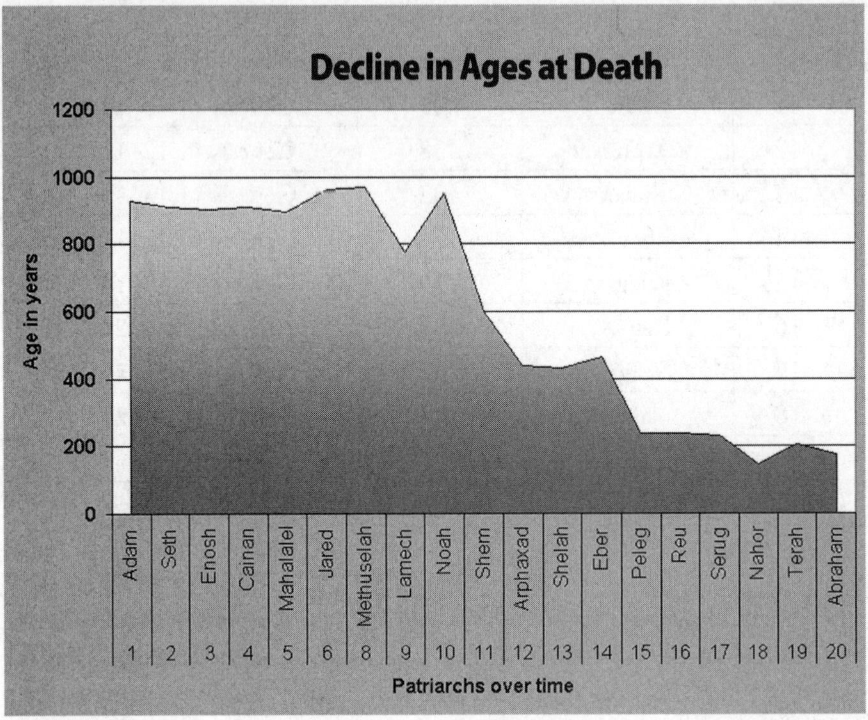

Figure 1

written 3,400 years ago! The precipitous plunge in life spans after the Flood suggests that something changed at the time of the Flood, or shortly thereafter, that was responsible for this decline. A line graph of this decline reveals an exponential curve (see figure 1). An exponential decay rate is often called a "natural" decay rate because it is so often observed in nature. For example, this is the decay curve we see when living organisms are exposed to lethal doses of toxic substances or radiations. Since it is unlikely that people living in pre-Flood times were familiar with exponential decay curves, it is thus unlikely that these dates were fabricated.

The fossil record reveals that prior to the Flood, most of the earth appears to have had a tropical type of environment. Following the Flood, there was clearly an environmental change resulting in an ice age that covered nearly 30 percent of the earth with ice (primarily in the northern latitudes). This, together with other changes following the Flood, could have adversely affected life spans.

Biological Causes of Aging

What exactly causes this process of aging in our body? Although the mechanism of aging (and its prevention) has long been an object of biomedical research, science still has no definitive answer to this question. Around the turn of the century, it was believed that aging didn't directly involve the living cells of our body but rather was an extracellular phenomenon. It was believed that our normal living cells, if properly nourished, could grow and divide indefinitely outside our body. In 1961, this idea was refuted by Leonard Hayflick, who grew human cells outside the body in covered glass dishes containing the necessary nutrients. Hayflick discovered that cells cultured in this way normally died after about 50 cell divisions (*Hayflick's limit*). This suggests that even the individual cells of our body are mortal, apart from any other bodily influence.

Genetic Determinants

Both aging and life span are processes that have genetic determinants that are overlapping and unique. Approximately 20–30 percent of factors affecting life span are thought to be heritable and thus genetic.[5] Life span varies greatly among individuals, indicating that while aging plays a role, other factors are also involved.

Mutations and Genetic Bottlenecks

A mutation is any change in the sequence of DNA.[6] All known mutations cause a loss of information. The rate at which all types of mutations occur per generation has been suggested to be greater than 1,000.[7] We inherit mutations from our parents and also develop mutations of our own; subsequently, we pass a proportion of those on to our children. So it is conceivable in the many generations between Adam and Moses that a large number of mutations would have been present in any given individual.

Genetic bottlenecks (or population bottlenecks) occur when significant proportions of the population dies or proportions become isolated. Such a bottleneck occurred at the time of Noah's flood when the human population was reduced to eight people (Genesis 6–9). Other smaller bottlenecks occurred following the Tower of Babel dispersion (Genesis 11). These events would have resulted in a major reduction of genetic variety.

For every gene there are two or more versions called alleles. This is analogous to the color red (gene) but different shades of red — light and dark (alleles). It is

5. T. Perls and D. Terry, "Genetics of Exceptional Longevity," *Experimental Gerontology* 38 (2003): 725–730.
6. See chapter 7, which provides an overview of mutations.
7. J. Sanford, *Genetic Entropy and the Mystery of the Genome* (New York: Ivan Press, 2005), p. 37.

possible for "good" (unmutated) alleles to mask or hide "bad" (mutated) alleles. However, in a smaller population with less allelic variation, this becomes more difficult to accomplish, and thus mutated alleles have a greater effect.

Although Noah lived 950 years, his father, Lamech, lived only 777 years (granted we do not know if he died from old age). In addition, we do not know how long Noah's wife lived, but Noah's son Shem only lived 600 years. Considering that the longest recorded life span of someone born after the Flood was Eber at 464 years, it would appear that both mutations and genetic bottlenecks had severe effects on aging and life span.

Examples of Genetic Determinants Affecting Aging and Life Span

Although many genetic factors are suggested to affect aging and life span, these processes largely remain a mystery. Aging can be thought of as increased susceptibility to internal (i.e., agents that damage DNA) and external (i.e., disease-causing bacteria) stressors because of a decrease in the maintenance, repair, and defensive systems of the body.

For example, DNA repair systems are needed to protect the genome (all our DNA) from mutation. *Xeroderma pigmentosum* (XP) is a genetic disorder caused by a deficient (due to mutations) DNA repair system that normally repairs mutations caused by ultraviolet light. Individuals with this disease must severely limit their exposure to sunlight. Outer surfaces of the body such as skin and lips commonly show signs of premature aging.[8] While this is an extreme example, any mutation that decreases the efficiency of our maintenance, repair, and defensive systems will likely lead to more rapid aging and decreased life span.

Telomeres, long, repetitive sequences of DNA at the ends of human chromosomes, are also thought to play an important role in aging. With each division of the cell, telomeres shorten due to the inability of the enzyme that copies the DNA to go all the way to the end of the chromosome.[9] When telomeres have become too short, the cell stops dividing. This limitation plausibly serves as a quality control mechanism. Older cells will have accumulated many mutations in their DNA, and their continued division may lead to diseases like cancer. Most body cells cannot replicate indefinitely, leading to aging and eventually death. Thus, telomeres are important in determining the life span of cell types that directly affect aging.

8. DermNet NZ, "Xeroderma Pigmentosum," www.dermnetnz.org/systemic/xeroderma-pigmentosum.html.
9. P. Monaghan and M. Haussmann, "Do Telomere Dynamics Link Lifestyle and Lifespan?" *TRENDS in Ecology and Evolution* 21 (2006): 47–53.

Job hunting in the days before Noah's flood.

Genetic determinants of life span or longevity are difficult to pinpoint. Even if the genes are determined to be associated with people who live for many years, their actual role in increasing life span is unknown. Genetic studies of centarians (people who have lived more than 100 years) have produced several possible candidate longevity genes. The gene for apolipoprotein E (APOE), important in the regulation of cholesterol, has certain alleles that are more common among centarians.[10] This is also true for certain alleles of insulin-like growth factor 1 (IGF1), important in cell proliferation and cell death, and superoxide dismutases (SOD), important in the breakdown of agents that damage DNA.[11] Possibly the alleles associated with the centarians more closely reflect the genetic makeup of individuals with a long life span 6,000 years ago. Still, these alleles show the effects of the curse if the highest achievable age today is around 120 years!

10. K. Christensen et al., "The Quest for Genetic Determinants of Human Longevity: Challenges and Insights," *Nature Reviews Genetics* 7 (2006): 436–448.
11. Ibid.

Evolution and the Genetics of Aging and Life Span

Evolution has a difficult time explaining aging and life span. Aging is often viewed as a default. Genes are selected on the basis of how they benefit an individual in their young reproductive years, or the " 'warranty period' [which] is the time required to fulfill the Darwinian purpose of life in terms of successful reproduction for the continuation of generations."[12] However, these same genes may be harmful overall, leading to aging and eventually death.

The problem for evolution is that longevity genes are selected for. To deal with this seeming dichotomy, some evolutionists have suggested that selection of longevity genes serves a purpose in that long-lived individuals can care for more of their descendants, known as the "grandmother effect."[13] The problem is that any theory that is so flexible it can account for everything isn't a very good theory.

Genes associated with aging and life span have been affected as a result of the Fall either directly through mutations or indirectly through genetic bottlenecks. Modern medicine and anti-aging therapies may slow the process of aging and extend our life span, but they will never eradicate the ultimate end — death. Only Jesus Christ, who was victorious over death, can promise eternal life with Him to all who believe (Romans 6:23, 10:9).

Physiological Determinants

In one sense, most of the substance of our body really doesn't continue to get *older* during our life: a great many of our body's parts are constantly repairing and replacing themselves. The epidermal cells that cover the entire surface of our skin, for example, never get older than one month. New cells are continually produced (by cell division) deep in the epidermis, while the older ones continually slough off at the surface. Similarly, the cells lining our intestines completely replace themselves every 4 days; our red blood cells are entirely replaced about every 90 days; and our white blood cells are replaced about every week.

Even cells that never (or rarely) divide, such as cardiac muscle cells and brain cells, turn over molecule by molecule. It is believed that little or nothing in our body is more than about 10 years old. Thus, thanks to cell turnover and replacement, most of the organs in the body of a 90-year-old man are perhaps

12. S. Rattan, "Theories of Biological Aging: Genes, Proteins, and Free Radicals," *Free Radical Research* 41 (2006): 1230–1238.
13. W. Browner et al., "The Genetics of Human Longevity," *The American Journal of Medicine* 11 (2004): 851–860.

no older than those of a child. Indeed, you might say our body never actually grows older.

It's rather like the story about "grandpa's ax." It seems a man had an old ax that hung over his fireplace and which he claimed had been passed down in his family for five generations. When asked how old the ax was, he said he wasn't sure because although his great-great-great-great grandfather bought the ax about 300 years ago, he also understood that over the years, the ax had 6 new heads and 12 new handles. Our bodies are something like grandpa's ax in that we too are constantly replacing "heads and handles," and in a sense we never get older.

At this point we might be inclined to ask, why did Methuselah die so *young*? How, indeed, is it even possible for anyone to age and die if the body constantly repairs and replaces its parts? Surely, if our automobile could do this, we would expect it to last forever. Part of the answer may be that certain key parts of our body *fail* to repair or replace themselves. Our critically important heart muscle cells, for example, fail to multiply, repair, or replace themselves after birth (although, like all muscle cells, they can increase in size). This is why any disruption in the blood supply to the heart muscle during a heart attack leads to permanent death of that part of the heart. Most nerve cells of our brain — including those of our eye and inner ear — also fail to multiply or repair themselves. From the time of our birth to the end of our life, we lose thousands of nerve cells a minute from our central nervous system, and we can never replace them. As we get older, this causes a progressive loss of our ability to hear, see, smell, taste, and . . . ahh . . . something else, but I just can't remember it!

The important point is that science offers no hope for eternal life, or even for the significant lengthening of life. It has been estimated that if complete cures, or preventions, were found for the three major killers (cancer, stroke, and coronary artery disease), the maximum life span of man would still not increase (although more people would approach this maximum). And such long-lived people would still become progressively weaker with age, as critical components of their body continue to deteriorate.

We may conclude that God's Word, not science, has the complete solution to the problem of aging and death. The solution has been "revealed by the appearing of our Savior Jesus Christ, who has abolished death and brought life and immortality to light through the gospel" (2 Timothy 1:10).

17

Why 66?

BRIAN H. EDWARDS

How can we be sure that we have the correct 66 books in our Bible? The Bible is a unique volume. It is composed of 66 books by 40 different writers over 1,500 years. But what makes it unique is that it has one consistent storyline running all the way through, and it has just one ultimate author — God. The story is about God's plan to rescue men and women from the devastating results of the Fall, a plan that was conceived in eternity, revealed through the prophets, and carried out by the Son of God, Jesus Christ.

Each writer of the Bible books wrote in his own language and style, using his own mind, and in some cases research, yet each was so overruled by the Holy Spirit that error was not allowed to creep into his work. For this reason, the Bible is understood by Christians to be a book without error.[1]

This collection of 66 books is known as the "canon" of Scripture. That word comes from the Hebrew *kaneh* (a rod), and the Greek *kanon* (a reed). Among other things, the words referred equally to the measuring rod of the carpenter and the ruler of the scribe. It became a common word for anything that was the measure by which others were to be judged (see Galatians 6:16,

1. For a more full discussion of the inspiration of the Bible, see Brian Edwards, *Nothing But the Truth* (Darlington, UK: Evangelical Press, 2006), p.116–143. In this, the following definition can be found: "The Holy Spirit moved men to write. He allowed them to use their own style, culture, gifts and character, to use the results of their own study and research, to write of their own experiences and to express what was in their mind. At the same time, the Holy Spirit did not allow error to influence their writings; he overruled in the expression of thought and in the choice of words. Thus they recorded accurately all that God wanted them to say and exactly how he wanted them to say it, in their own character, style and language."

for example). After the apostles, church leaders used it to refer to the body of Christian doctrine accepted by the churches. Clement and Origen of Alexandria, in the third century, were possibly the first to employ the word to refer to the Scriptures (the Old Testament).[2] From then on, it became more common in Christian use with reference to a collection of books that are fixed in their number, divine in their origin, and universal in their authority.

In the earliest centuries, there was little *debate* among Christians over which books belonged in the Bible; certainly by the time of the church leader Athanasius in the fourth century, the number of books had long been fixed. He set out the books of the New Testament just as we know them and added:

> These are the fountains of salvation, that whoever thirsts may be satisfied by the eloquence which is in them. In them alone is set forth the doctrine of piety. Let no one add to them, nor take anything from them.[3]

Today, however, there are attempts to undermine the clear witness of history; a host of publications, from the novel to the (supposedly) academic challenge the long-held convictions of Christians and the clear evidence of the past. Dan Brown in *The Da Vinci Code* claimed, "More than eighty gospels were considered for the New Testament, and yet only relatively few were chosen for inclusion — Matthew, Mark, Luke and John among them."[4] Richard Dawkins, professor of popular science at Oxford, England, has made similar comments.[5]

So, what is the evidence for our collection of 66 books? How certain can we be that these are the correct books to make up our Bible — no more and no less?

The Canon of the Old Testament

The Jews had a clearly defined body of Scriptures that collectively could be summarized as the Torah, or Law. This was fixed early in the life of Israel, and there was no doubt as to which books belonged and which did not. They did

2. Clement of Alexandria, *The Miscellanies* bk. VI.15. He comments, "The ecclesiastical rule (canon) is the concord and harmony of the Law and the Prophets." B.F. Westcott, referring to Origen's commentary on Matthew 28, wrote: "No one should use for the proof of doctrine books not included among the canonized Scriptures." (*The Canon of the New Testament During the First Four Centuries* [Cambridge: Macmillan & Co.,1855], p. 548).
3. From the Festal Epistle of Athanasius XXXIX. Translated in *Nicene and Post-Nicene Fathers*, vol. IV., p. 551–552.
4. Dan Brown, *The Da Vinci Code* (London: Bantam Press, 2003), p. 231.
5. Richard Dawkins, *The God Delusion* (London: Bantam Press, 2006), p. 237.

not order them in the same way as our Old Testament, but the same books were there. *The Law* was the first five books, known as the Pentateuch, which means "five rolls" — referring to the parchment scrolls on which they were normally written. *The Prophets* consisted of the Former Prophets (unusually for us these included Joshua, Judges, Samuel, and Kings) and the Latter Prophets (Isaiah, Jeremiah which included Lamentations, and the 12 smaller prophetic books). *The Writings* gathered up the rest. The total amounted generally to 24 books because many books, such as 1 and 2 Samuel and Ezra and Nehemiah, were counted as one.

When was the canon of the Old Testament settled? The simple response is that if we accept the reasonable position that each of the books was written at the time of its history — the first five at the time of Moses, the historical records close to the period they record, the psalms of David during his lifetime, and the prophets written at the time they were given — then the successive stages of acceptance into the canon of Scripture is not hard to fix. Certainly, the Jews generally held this view.

There is a lot of internal evidence that the books of the Old Testament were written close to the time they record. For example, in 2 Chronicles 10:19, we have a record from the time of Rehoboam that "Israel has been in rebellion against the house of David to this day." Clearly, therefore, that must have been recorded prior to 721 B.C., when the Assyrians finally crushed Israel and the cream of the population was taken away into captivity — or at the very latest before 588 B.C., when Jerusalem suffered the same fate. We know also that the words of the prophets were written down in their own lifetime; Jeremiah had a secretary called Baruch for this very purpose (Jeremiah 36:4).

Josephus, the Jewish historian writing around A.D. 90, clearly stated in his defense of Judaism that, unlike the Greeks, the Jews did not have many books:

> For we have not an innumerable multitude of books among us, disagreeing from and contradicting one another [as the Greeks have] but only twenty-two books, which contain the records of all the past times; which are justly believed to be divine.[6]

The Council of Jamnia

Between A.D. 90 and 100, a group of Jewish scholars met at Jamnia in Israel to consider matters relating to the Hebrew Scriptures. It has been suggested that the canon of the Jewish Scriptures was agreed here; the reality is

6. Josephus, *Against Apion*, trans. William Whiston (London: Ward, Lock & Co.), bk. 1, ch. 8. His 22 books consisted of exactly the same as our 39 for the reasons given in the text.

that there is no contemporary record of the deliberations at Jamnia and our knowledge is therefore left to the comments of later rabbis. The idea that there was no clear canon of the Hebrew Scriptures before A.D. 100 is not only in conflict with the testimony of Josephus and others, but has also been seriously challenged more recently. It is now generally accepted that Jamnia was not a council nor did it pronounce on the Jewish canon; rather it was an assembly that examined and discussed the Hebrew Scriptures. The purpose of Jamnia was not to decide which books should be included among the sacred writings, but to examine those that were already accepted.[7]

The Apocrypha and the Septuagint

There is a cluster of about 14 books, known as the Apocrypha, which were written some time between the close of the Old Testament (after 400 B.C.) and the beginning of the New. They were never considered as part of the Hebrew Scriptures, and the Jews themselves clearly ruled them out by the confession that there was, throughout that period, no voice of the prophets in the land.[8] They looked forward to a day when "a faithful prophet" should appear.[9]

The Old Testament had been translated into Greek during the third century B.C., and this translation is known as the Septuagint, a word meaning *70,* after the supposedly 70 men involved in the translation. It was the Greek Septuagint that the disciples of Jesus frequently used since Greek was the common language of the day.

Whether or not the Septuagint also contained the Apocrypha is impossible to say for certain, since although the earliest copies of the Septuagint available today do include the Apocrypha — placed at the end — these are dated in the fifth century and therefore cannot be relied upon to tell us what was common half a millennium earlier. Significantly, neither Jesus nor any of the apostles ever quoted from the Apocrypha, even though they were obviously using the Greek Septuagint. Josephus was familiar with the Septuagint and made use of it, but he never considered the Apocrypha part of the Scriptures.[10]

7. This is a widespread view. See for example R. Beckwith, *The Old Testament Canon of the New Testament Church* (London: SPCK, 1985), p. 276. Also, A. Bentzen, *Introduction to the Old Testament*, vol. 1 (Copenhagen: G.E.C. Gad, 1948), p. 31; Bruce Metzger, *The Canon of the New Testament* (Oxford: Oxford University Press, 1987), p. 110; John Wenham, *Christ and the Bible* (London: Tyndale Press, 1972), p.138–139.
8. The Apocrypha. 1 Maccabees 9:27 at the time of revolt against Syrian occupation in the mid second century B.C. by Judas Maccabeas: "There was a great affliction in Israel, the like whereof was not since the time that a prophet was not seen among them."
9. The Apocrypha. 1 Maccabees 14:41.
10. It should be noted that the Roman Catholic and Eastern Orthodox churches do accept some of the Apocryphal books as Scripture because they support, for example, praying for the dead.

The Dead Sea Scrolls

The collection of scrolls that has become available since the discovery of the first texts in 1947 near Wadi Qumran, close by the Dead Sea, does not provide scholars with a definitive list of Old Testament books, but even if it did, it would not necessarily tell us what mainstream orthodox Judaism believed. After all, the Samaritans used only their own version of the Pentateuch, but they did not represent mainstream Judaism.

What can be said for certain, however, is that all Old Testament books are represented among the Qumran collection with the exception of Esther, and they are quoted frequently as Scripture. Nothing else, certainly not the Apocrypha, is given the same status.

In spite of suggestions by critical scholars to the contrary, there is no evidence, not even from the Dead Sea Scrolls, that there were other books contending for a place within the Old Testament canon.

For the Jews, therefore, Scripture as a revelation from God through the prophets ended around 450 B.C. with the close of the book of Malachi. This was the Bible of Jesus and His disciples, and it was precisely the same in content as our Old Testament.

The New Testament scholar John Wenham concludes: "There is no reason to doubt that the canon of the Old Testament is substantially Ezra's canon, just as the Pentateuch was substantially Moses' canon."[11]

Jesus, His Disciples, and the Early Church Leaders

For their part, the Christian community both in the days of Jesus and in the centuries following had no doubt that there was a body of books that made up the records of the old covenant. Since there are literally hundreds of direct quotations or clear allusions to Old Testament passages by Jesus and the apostles, it is evident what the early Christians thought of the Hebrew Scriptures. The New Testament writers rarely quote from other books and never with the same authority. The Apocrypha is entirely absent in their writing.

While it is true that some of the early church leaders quoted from the Apocrypha — though very rarely compared to their use of the Old Testament books — there is no evidence that they recognized these books as equal to the Old Testament.[12]

The conviction that there was a canon of old covenant books that could not be added to or subtracted from doubtless led the early Christians to expect

11. John Wenham, *Christ and the Bible* (London: Tyndale Press, 1972), p.134.
12. This is a point made firmly by John Wenham in *Christ and the Bible*, p. 146–147.

the same divine order for the story of Jesus, the record of the early church, and the letters of the apostles.

The Canon of the New Testament

The earliest available list of New Testament books is known as the Muratorian Canon and is dated around A.D. 150. It includes the four Gospels, Acts, thirteen letters of Paul, Jude, two (perhaps all three) letters of John, and the Revelation of John. It claims that these were accepted by the "universal church." This leaves out 1 and 2 Peter, James, and Hebrews. However, 1 Peter was widely accepted by this time and may be an oversight by the compiler (or the later copyist). No other books are present except the Wisdom of Solomon, but this must be an error since that book belongs in the Apocrypha and no one ever added it to the New Testament!

By A.D. 240, Origen from Alexandria was using all our 27 books as "Scripture," and no others, and referred to them as the "New Testament."[13] He believed them to be "inspired by the Spirit."[14] But it was not until A.D. 367 that Athanasius, also from Alexandria, provided us with an actual *list* of New Testament books identical with ours.[15]

However, long before we have that list, the evidence shows that the 27 books, and only those, were widely accepted as Scripture.

Why Did It Take So Long?

The New Testament was not all neatly printed and bound by the Macedonian Pub. Co. at Thessalonica shortly after Paul's death and sent out by the pallet load into all the bookstores and kiosks of the Roman Empire. Here are six reasons why it took time for the books of the New Testament to be gathered together.

---

13. *Origen De Principiis (Concerning Principles)*, pref. 4. He used the title "New Testament" six times in *De Principiis*.
14. *Origen De Principiis*, pref. 4, ch. 3:1.
15. From the Festal Epistle of Athanasius XXXIX. Translated in *Nicene and Post-Nicene Fathers*, vol. IV. p. 551–552. This is what he wrote: "As the heretics are quoting apocryphal writings, an evil which was rife even as early as when St. Luke wrote his gospel, therefore I have thought good to set forth clearly what books have been received by us through tradition as belonging to the Canon, and which we believe to be divine. [Then follows the books of the Old Testament with the unusual addition of the Epistle of Baruch.] Of the New Testament these are the books . . . [then follows the 27 books of our New Testament, and no more]. These are the fountains of salvation, that whoever thirsts, may be satisfied by the eloquence which is in them. In them alone is set forth the doctrine of piety. Let no one add to them, nor take anything from them."

1. The originals were scattered across the whole empire. The Roman Empire reached from Britain to Persia, and it would have taken time for any church even to learn about all the letters Paul had written, let alone gather copies of them.
2. No scroll could easily contain more than one or two books. It would be impossible to fit more than one Gospel onto a scroll, and even when codices (books) were used, the entire New Testament would be extremely bulky and very expensive to produce. It was therefore far more convenient for New Testament books to be copied singly or in small groups.
3. The first-century Christians expected the immediate return of Christ. Because of this, they didn't plan for the long-term future of the Church.
4. No one church or leader bossed all the others. There were strong and respected leaders among the churches, but Christianity had no supreme bishop who dictated to all the others which books belonged to the canon and which did not.
5. The early leaders assumed the authority of the Gospels and the apostles. It was considered sufficient to quote the Gospels and apostles, since their authority was self-evident. They did not need a list — inconvenient for us, but not significant for them.
6. Only when the heretics attacked the truth was the importance of a canon appreciated. It was not until the mid-second century that the Gnostics and others began writing their own *pseudepigrapha* (false writing); this prompted orthodox leaders to become alert to the need for stating which books had been recognized across the churches.

In the light of all this, the marvel is not how long it took before the majority of the churches acknowledged a completed canon of the New Testament, but how soon after their writing each book was accepted as authoritative.

Facts about the New Testament Canon

- There were only ever the four Gospels used by the churches for the life and ministry of Jesus. Other pseudo-gospels were written but these were immediately rejected by the churches across the empire as spurious.

- The Acts of the Apostles and 13 letters of Paul were all accepted without question or hesitation from the earliest records.
- Apart from James, Jude, 2 and 3 John, 2 Peter, Hebrews, and Revelation, all other New Testament books had been universally accepted by A.D. 180. Only a few churches hesitated over these seven.
- Well before the close of the first century, Clement of Rome quoted from or referred to more than half the New Testament and claimed that Paul wrote "in the Spirit" and that his letters were "Scriptures."
- Polycarp, who was martyred in A.D. 155, quoted from 16 NT books and referred to them as "Sacred Scriptures."
- Irenaeus of Lyons, one of the most able defenders of the faith, around A.D. 180 quoted over 1,000 passages from all but four or five New Testament books, and called them "the Scriptures" given by the Holy Spirit.
- Tertullian of Carthage, around A.D. 200, was the first serious expositor and used almost all the NT books. They were equated with the Old Testament, and he referred to "the majesty of our Scriptures." He clearly possessed a canon almost, if not wholly, identical to ours.
- By A.D. 240, Origen of Alexandria was using all our 27 books, and only those, as Scripture alongside the Old Testament books.

And these are just examples of many of the church leaders at this time.

What Made a Book "Scripture"?

At first, the churches had no need to define what made a book special and equal to the Old Testament Scriptures. If the letter came from Paul or Peter, that was sufficient. However, it was not long before others began writing additional letters and gospels either to fill the gaps or to propagate their own ideas. Some tests became necessary, and during the first 200 years, five tests were used at various times.

1. Apostolic — does it come from an apostle?

   The first Christians asked, "Was it written by an apostle or under the direction of an apostle?" They expected this just as the Jews had expected theirs to be underwritten by the prophets. Paul was insistent that his readers should be reassured that the letters they received actually came from his pen (e.g., 2 Thessalonians 3:17).

2. Authentic — does it have the ring of truth?

The authoritative voice of the prophets, "This is what the Lord says," is matched by the apostles' claim to write not the words of men but the words of God (1 Thessalonians 2:13). It was the internal witness of the texts themselves that was strong evidence of canonicity.

3. Ancient — has it been used from the earliest times?

Most of the false writings were rejected simply because they were too new to be apostolic. Early in the fourth century, Athanasius listed the New Testament canon as we know it today and claimed that these were the books "received by us through tradition as belonging to the Canon."[16]

4. Accepted — are most of the churches using it?

Since, as we have seen, it took time for letters to circulate among the churches, it is all the more significant that 23 of the 27 books were almost universally accepted well before the middle of the second century.

When tradition carries the weight of the overwhelming majority of churches throughout the widely scattered Christian communities across the vast Roman Empire, with no one church controlling the beliefs of all the others, it has to be taken seriously.

5. Accurate — does it conform to the orthodox teaching of the churches?

There was widespread agreement among the churches across the empire as to the content of the Christian message. Irenaeus asked the question whether a particular writing was consistent with what the churches taught.[17] This is what ruled out so much of the heretical material immediately.

Providence

Our final appeal is not to man, not even to the early church leaders, but to God, who by His Holy Spirit has put His seal upon the New Testament. By

16. Athanasius, *Festal Epistle* XXXIX.
17. Irenaeus, *Against Heresies,* bk. III, ch. 3:3. "This is most abundant proof that there is one and the same vivifying faith, which has been preserved in the Church from the apostles until now, and handed down in truth."

their spiritual content and by the claim of their human writers, the 27 books of our New Testament form part of the "God breathed" Scripture. It is perfectly correct to allow this divine intervention to guard the process by which eventually all the canonical books — and no others — were accepted. The idea of the final canon being an accident, and that any number of books could have ended up in the Bible, ignores the evident unity and provable accuracy of the whole collection of 27 books.

Bruce Metzger expressed it well: "There are, in fact, no historical data that prevent one from acquiescing in the conviction held by the Church Universal that, despite the very human factors . . . in the production, preservation, and collection of the books of the New Testament, the whole process can also be rightly characterized as the result of divine overruling."[18]

A belief in the authority and inerrancy of Scripture is bound to a belief in the divine preservation of the canon. The God who "breathed out" (2 Timothy 3:16) His word into the minds of the writers ensured that those books, and no others, formed part of the completed canon of the Bible.

18. Metzger, *The Canon of the New Testament*, p. 285.

18

What Was the Christmas Star?

DR. JASON LISLE

The apostle Matthew records that the birth of Jesus was accompanied by an extraordinary celestial event: a star that led the magi[1] (the "wise men") to Jesus. This star "went before them, till it came and stood over where the young child was" (Matthew 2:9). What was this star? And how did it lead the magi to the Lord? There have been many speculations.

Common Explanations

The star mentioned in Matthew is not necessarily what we normally think of as a star. That is, it was not necessarily an enormous mass of hydrogen and helium gas powered by nuclear fusion. The Greek word translated *star* is *aster* (αστηρ), which is where we get the word *astronomy*. In the biblical conception of the word, a star is any luminous point of light in our night sky. This would certainly include our modern definition of a star, but it would also include the planets, supernovae, comets, or anything else that resembles a point of light. But which of these explanations best describes the Christmas star?

A supernova (an exploding star) fits the popular Christmas card conception of the star. When a star in our galaxy explodes, it shines very brightly for

1. Magi (pronounced mā'jī') were scholars of the ancient world, possibly a class of Zoroastrian priests from Media or Persia. It is commonly assumed that three magi came on the journey to visit Christ since they brought three gifts. However, the Bible does not actually give the number of magi.

several months. These beautiful events are quite rare and outshine all the other stars in the galaxy. It seems fitting that such a spectacular event would announce the birth of the King of kings — the God-man who would outshine all others. However, a supernova does not fit the biblical text. The Christmas star must not have been so obvious, for it went unnoticed by Israel's King Herod (Matthew 2:7). He had to ask the magi when the star had appeared, but everyone would have seen a bright supernova.

Nor could the Christmas star have been a bright comet. Like a supernova, everyone would have noticed a comet. Comets were often considered to be omens of change in the ancient world. Herod would not have needed to ask the magi when a comet had appeared. Moreover, neither a comet nor a supernova moves in such a way as to come and stand over a location on earth as the Christmas star did (Matthew 2:9). Perhaps the Christmas star was something more subtle: a sign that would amaze the magi but would not be noticed by Herod.

A Conjunction?

This leads us to the theory that the Christmas star was a *conjunction* of planets. A conjunction is when a planet passes closely by a star or by another planet. Such an event would have been very meaningful to the magi, who were knowledgeable of ancient astronomy, but would likely have gone unnoticed by others. There were several interesting conjunctions around the time of Christ's birth. Two of these were triple conjunctions; this is when a planet passes a star (or another planet), then backs up, passes it again, then reverses direction and passes the star/planet a third time. Such events are quite rare.

Nonetheless, there was a triple conjunction of Jupiter and Saturn beginning in the year 7 B.C. Also, there was a triple conjunction of Jupiter and the bright star Regulus beginning in the year 3 B.C. Of course, we do not know the exact year of Christ's birth, but both of these events are close to the estimated time. Advocates of such conjunction theories point out that the planets and stars involved had important religious significance in the ancient world. Jupiter was often considered the king of the gods, and Regulus was considered the "king star." Did such a conjunction announce the birth of the King of kings? However, the Bible describes the Christmas star as a single star — not a conjunction of two or more stars. Neither of the above conjunctions was close enough to appear as a single star.

But there was one (and *only* one) extraordinary conjunction around the time of Christ's birth that could be called a "star." In the year 2 B.C., Jupiter and Venus moved so close to each other that they briefly appeared to merge into a single bright star. Such an event is extremely rare and may have been perceived as highly significant to the magi. Although this event would have been really spectacular, it does not fully match the description of the Christmas star. A careful reading of the biblical text indicates that the magi saw the star on at least two occasions: when they arrived at Jerusalem (Matthew 2:2) and after meeting with Herod (Matthew 2:9). But the merging of Jupiter and Venus happened only once — on the evening of June 17.

Although each of the above events is truly spectacular and may have been fitting to announce the birth of the King of kings, none of them seems to fully satisfy the details of the straightforward reading of Matthew 2. None of the above speculations fully explain how the star "went ahead of" the magi nor how it "stood over where the child was." Indeed, no known natural phenomenon would be able to stand over Bethlehem since all natural stars continually move due to the rotation of the earth.[2] They appear to rise in the east and set in the west, or circle around the celestial poles. However, the Bible does not say that this star was a *natural* phenomenon.

Natural Law

Of course, God can use natural law to accomplish His will. In fact, the laws of nature are really just descriptions of the way that God normally upholds the universe and accomplishes His will. But God is not bound by natural law; He is free to act in other ways if He so chooses. The Bible records a number of occasions where God has acted in a seemingly unusual way to accomplish an extraordinary purpose.

The Virgin Birth itself was a supernatural event; it cannot be explained within the context of known natural laws. For that matter, God has previously used apparently supernatural signs in the heavens as a guide. In Exodus 13:21, we find that God guided the Israelites by a cloud by day and a pillar of fire by night. It should not be surprising that a supernatural sign in the heavens would accompany the birth of the Son of God. The star that led the magi seems to be one of those incredible acts of God — specially designed and created for a unique purpose.[3] Let us examine what this star did according to Matthew 2.

2. The star that moves the least is the North Star because it is almost directly in line with the earth's North Pole. However, this would not have been the case at the time of Christ's birth, due to a celestial phenomenon called "precession." There was no "North Star" during Christ's earthly ministry.
3. Although this star seems to break all the rules, it is perhaps even more amazing that essentially all the other stars do not. The fact that all the stars in our night sky obey orderly logical laws of nature is consistent with biblical creation and inconsistent with secular notions. For more information on the laws of nature, see www.answersingenesis.org/articles/am/v1/n2/god-natural-law.

Purpose of the Star

First, the star alerted the magi to the birth of Christ, prompting them to make the long trek to Jerusalem. These magi were "from the East," according to verse 1; they are generally thought to be from Persia, which is east of Jerusalem. If so, they may have had some knowledge of the Scriptures since the prophet Daniel had also lived in that region centuries earlier. Perhaps the magi were expecting a new star to announce the birth of Christ from reading Numbers 24:17, which describes a star coming from Jacob and a King ("Scepter")[4] from Israel.[5]

Curiously, the magi seem to have been the only ones who saw the star — or at least the only ones who understood its meaning. Recall that King Herod had to ask the magi when the star had appeared (Matthew 2:7). If the magi alone saw the star, this further supports the notion that the Christmas star was a supernatural manifestation from God rather than a common star, which would have been visible to all. The fact that the magi referred to it as "His star" further supports the unique nature of the star.[6]

The position of the star when the magi first saw it is disputed. The Bible says that they "saw His star in the east" (Matthew 2:2). Does this mean that the *star* was in the eastward heavens when they first saw it, or does it mean that the *magi* were "in the East" (i.e., Persia) when they saw the star?[7] If the star was in the East, why did the magi travel west? Recall that the Bible does not say that the star guided the magi to Jerusalem (though it may have); we only know for certain that it went before them on the journey from Jerusalem to the house of Christ. It is possible that the star initially acted only as a sign, rather than as a guide. The magi may have headed to Jerusalem only because this would have seemed a logical place to begin their search for the King of the Jews.

But there is another interesting possibility. The Greek phrase translated *in the East* (εν ανατολη) can also be translated *at its rising*. The expression can be used to refer to the east since all normal stars rise in the east (due to earth's rotation). But the Christmas star may have been a supernatural exception — rising

4. This verse makes use of synecdoche — the part represents the whole. In this case, the scepter represents a scepter bearer (i.e., a king). This is clear from the synonymous parallelism (see the next note).
5. This verse is written in synonymous parallelism, which is a form of Hebrew poetry in which a statement is made followed by a very similar statement with the same basic meaning. "A star shall come forth from Jacob, and a Scepter shall rise from Israel." Both statements poetically indicate the coming of a future king (Christ). Star and Scepter (bearer) both indicate a king, and Israel and Jacob are two names for the same person who is the ancestor of the coming king.
6. Granted, all stars were created by God and therefore belong to Him. But the Christmas star is specially designated as "His" (Christ's), suggesting its unusual nature.
7. The latter view is indicated by John Gill in his commentary.

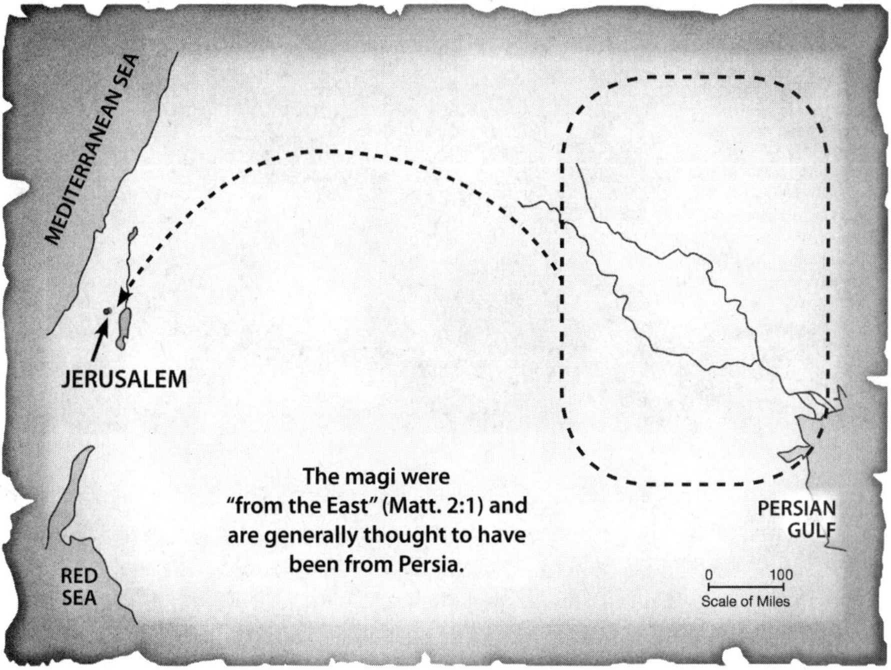

in the *west* over Bethlehem (which from the distance of Persia would have been indistinguishable from Jerusalem). The wise men would have recognized such a unique rising. Perhaps they took it as a sign that the prophecy of Numbers 24:17 was fulfilled since the star quite literally rose from Israel.

Clearing Up Misconceptions

Contrary to what is commonly believed, the magi did not arrive at the manger on the night of Christ's birth; rather, they found the young Jesus and His mother living in a house (Matthew 2:11). This could have been nearly two years after Christ's birth, since Herod — afraid that his own position as king was threatened — tried to have Jesus eliminated by killing all male children under the age of two (Matthew 2:16).

It seems that the star was not visible at the time the magi reached Jerusalem but then reappeared when they began their (much shorter) journey from Jerusalem to the Bethlehem region, approximately 6 miles (10 km) away. This view is supported by the fact that first, the magi had to ask King Herod where the King of the Jews was born, which means the star wasn't guiding them at that time (Matthew 2:2). And second, they rejoiced exceedingly when they saw the star (again) as they began their journey to Bethlehem (Matthew 2:10).

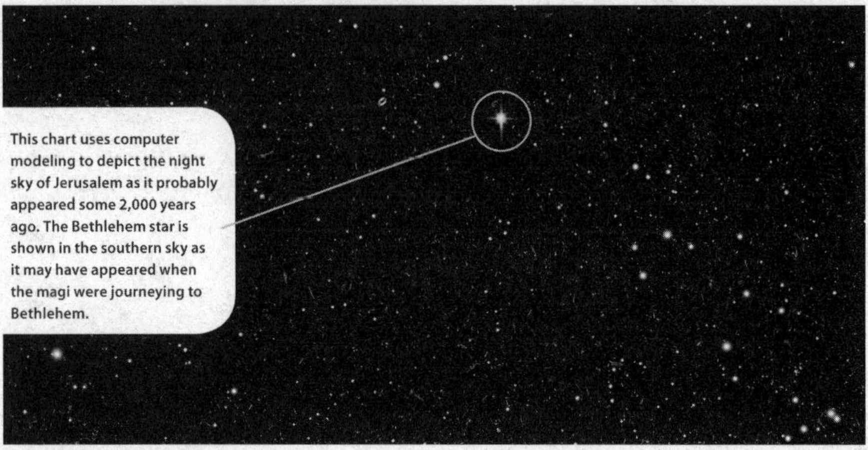

This chart uses computer modeling to depict the night sky of Jerusalem as it probably appeared some 2,000 years ago. The Bethlehem star is shown in the southern sky as it may have appeared when the magi were journeying to Bethlehem.

After the magi had met with Herod, the star went on before them to the Bethlehem region[8] and stood over the location of Jesus. It seems to have led them to the very house that Jesus was in — not just the city. The magi already knew that Christ was in the Bethlehem region. This they had learned from Herod, who had learned it from the priests and scribes (Matthew 2:4–5, 8). For a normal star, it would be impossible to determine which house is directly beneath it. The star over Christ may have been relatively near the surface of earth (an "atmospheric" manifestation of God's power) so that the magi could discern the precise location of the Child.

Whatever the exact mechanism, the fact that the star led the magi to Christ is evidence that God uniquely designed the star for a very special purpose. God can use extraordinary means for extraordinary purposes. Certainly the birth of our Lord was deserving of honor in the heavens. It is fitting that God used a celestial object to announce the birth of Christ since "the heavens declare the glory of God" (Psalm 19:1).

8. Although we know Christ was born in the town of Bethlehem, there is no reason to suppose that He remained there for the entire time of the magi's journey. We know that Christ's family brought Him to the Temple in Jerusalem after the days of purification (Luke 2:22); it is possible that they went directly to Nazareth after that (Luke 2:39) and then returned to Bethlehem sometime later. The wise men apparently did meet Christ in the Bethlehem region, however. We know this because as soon as they departed, God warned Joseph (Matthew 2:13) that Herod was about to kill all the male children in Bethlehem and its environs (Matthew 2:16), necessitating the escape to Egypt.

19

Is Jesus God?

DR. RON RHODES

Is Jesus really God? There are many cults and false religions today that deny it. The Jehovah's Witnesses, for example, believe Jesus was created by the Father billions of years ago as the Archangel Michael and is hence a "lesser god" than the Father. The Mormons say Jesus was born as the first and greatest spirit child of the Heavenly Father and heavenly mother, and was the spirit-brother of Lucifer. New Agers claim Jesus was an enlightened master. Unitarian Universalists say Jesus was just a good moral teacher. What is the truth about Jesus Christ? We turn to the Scriptures for the answer.

Jesus Truly Is God

There are numerous evidences for the absolute deity of Jesus Christ in the Bible. The following is a summary of the more important evidences.

Jesus Has the Names of God

Jesus Christ possesses divine names — names that can *only* be used of God. For example:

Jesus is Yahweh. Yahweh is a very common Hebrew name for God in the Old Testament, occurring over 5,300 times. It is translated *LORD* (all capitals) in many English translations of the Bible.

We first learn of this name in Exodus 3, where Moses asked God by what name He should be called. God replied to him, "I AM WHO I AM. . . .Thus you shall say to the children of Israel, 'I AM has sent me to you' " (verse 14). *Yahweh* is basically a shortened form of "I AM WHO I AM" (verse 15). The name conveys the idea of eternal self-existence. Yahweh never came into being at a point in time for He has always existed.

Jesus implicitly ascribed this divine name to himself during a confrontation He had with a group of hostile Jews. He said, "I say to you, before Abraham was, I AM" (John 8:58). Jesus deliberately contrasted the created origin of Abraham — whom the Jews venerated — with His own eternal, uncreated nature as God.

Jesus is Kurios. The New Testament Greek equivalent of the Old Testament Hebrew name Yahweh is *Kurios*. Used of God, *Kurios* carries the idea of a sovereign being who exercises absolute authority. The word is translated *Lord* in English translations of the Bible.

To an early Christian accustomed to reading the Old Testament, the word *Lord*, when used of Jesus, would point to His identification with the God of the Old Testament (*Yahweh*). Hence, the affirmation that "Jesus is Lord" (*Kurios*) in the New Testament constitutes a clear affirmation that Jesus is Yahweh, as is the case in passages like Romans 10:9, 1 Corinthians 12:3, and Philippians 2:5–11.

Jesus is Elohim. Elohim is a Hebrew name that is used of God 2,570 times in the Old Testament. The name literally means "strong one," and its plural ending (*im* in Hebrew) indicates fullness of power. Elohim is portrayed in the Old Testament as the powerful and sovereign governor of the universe, ruling over the affairs of humankind.

Jesus is recognized as both Yahweh *and* Elohim in the prophecy in Isaiah 40:3: "Prepare the way of the Lord [*Yahweh*]; make straight in the desert a highway for our God [*Elohim*]." This verse was written in reference to John the Baptist preparing for the coming of Christ (as confirmed in John 1:23) and represents one of the strongest affirmations of Christ's deity in the Old Testament. In Isaiah 9:6, we likewise read a prophecy of Christ with a singular variant (*El*) of *Elohim*: "And His name will be called Wonderful, Counselor, Mighty God [*El*], Everlasting Father, Prince of Peace."

Jesus is Theos. The New Testament Greek word for God, *Theos*, is the corresponding parallel to the Old Testament Hebrew term *Elohim*. A well-known example of Christ being addressed as God (*Theos*) is found in the story of "doubting Thomas" in John 20. In this passage, Thomas witnesses the resurrected Christ and worshipfully responds: "My Lord and my God [*Theos*]" (John 20:28).

Jesus is called *Theos* throughout the rest of the New Testament. For example, when a jailer asked Paul and Silas how to be saved, they responded: "Believe on the Lord Jesus, and you will be saved, you and your household" (Acts 16:31).

After the jailer believed and became saved, he "rejoiced, having believed in God [*Theos*] with all his household" (verse 34). Believing *in Christ* and believing *in God* are seen as identical acts.

Jesus Possesses the Attributes of God

Jesus possesses attributes that belong only to God.

Jesus is eternal. John 1:1 affirms: "In the beginning was the Word, and the Word was with God, and the Word was God." The word *was* in this verse is an imperfect tense, indicating continuous, ongoing existence. When the time-space universe came into being, Christ already existed (Hebrews 1:8–11).

Jesus is self-existent. As the Creator of all things (John 1:3; Colossians 1:16; Hebrews 1:2), Christ himself must be *un*created. Colossians 1:17 tells us that Christ is "before all things, and in Him all things consist."

Jesus is everywhere-present. Christ promised His disciples, "Where two or three are gathered together in My name, I am there in the midst of them" (Matthew 18:20). Since people all over the world gather in Christ's name, the only way He could be present with them all is if He is truly omnipresent (see Matthew 28:20; Ephesians 1:23, 4:10; Colossians 3:11).

Jesus is all-knowing. Jesus knew where the fish were in the water (Luke 5:4, 6; John 21:6–11), and He knew just which fish contained the coin (Matthew 17:27). He knew the future (John 11:11, 18:4), specific details that would be encountered (Matthew 21:2–4), and knew from a distance that Lazarus had died (John 11:14). He also knows the Father as the Father knows Him (Matthew 11:27; John 7:29, 8:55, 10:15, 17:25).

Jesus is all-powerful. Christ created the entire universe (John 1:3; Colossians 1:16; Hebrews 1:2) and sustains the universe by His own power (Colossians 1:17; Hebrews 1:3). During His earthly ministry, He exercised power over nature (Luke 8:25), physical diseases (Mark 1:29–31), demonic spirits (Mark 1:32–34), and even death (John 11:1–44).

Jesus is sovereign. Christ presently sits at the right hand of God the Father, "angels and authorities and powers having been made subject to Him" (1 Peter 3:22). When Christ comes again in glory, He will be adorned with a majestic robe, and on the thigh section of the robe will be the words, "KING OF KINGS AND LORD OF LORDS" (Revelation 19:16).

Jesus is sinless. Jesus challenged Jewish leaders: "Which of you convicts Me of sin?" (John 8:46). The apostle Paul referred to Jesus as "Him who knew no sin" (2 Corinthians 5:21). Jesus is one who "loved righteousness and hated lawlessness" (Hebrews 1:9), was "without sin" (Hebrews 4:15), and was "holy, harmless, [and] undefiled" (Hebrews 7:26).

Jesus Possesses the Authority of God

Jesus always spoke in His own divine authority. He never said, "Thus saith the Lord" as did the prophets; He always said, "Verily, verily, I say unto you. . . ." He never retracted anything He said, never guessed or spoke with uncertainty, never made revisions, never contradicted himself, and never apologized for what He said. He even asserted, "Heaven and earth will pass away, but My words will by no means pass away" (Mark 13:31), hence elevating His words directly to the realm of heaven.

Jesus Performs the Works of God

Jesus' deity is also proved by His miracles. His miracles are often called "signs" in the New Testament. Signs always *signify* something — in this case, that *Jesus is the divine Messiah*.

Some of Jesus' more notable miracles include turning water into wine (John 2:7–8); walking on the sea (Matthew 14:25; Mark 6:48; John 6:19); calming a stormy sea (Matthew 8:26; Mark 4:39; Luke 8:24); feeding 5,000 men and their families (Matthew 14:19; Mark 6:41; Luke 9:16; John 6:11); raising Lazarus from the dead (John 11:43–44); and causing the disciples to catch a great number of fish (Luke 5:5–6).

Jesus Is Worshiped as God

Jesus was worshiped on many occasions in the New Testament. He accepted worship from Thomas (John 20:28), the angels (Hebrews 1:6), some wise men (Matthew 2:11), a leper (Matthew 8:2), a ruler (Matthew 9:18), a blind man (John 9:38), an anonymous woman (Matthew 15:25), Mary Magdalene (Matthew 28:9), and the disciples (Matthew 28:17).

Scripture is emphatic that only God can be worshiped (Exodus 34:14; Deuteronomy 6:13; Matthew 4:10). In view of this, the fact that both humans and angels worshiped Jesus on numerous occasions shows He is God.

Old Testament Parallels Prove Jesus Is God

A comparison of the Old and New Testaments provides powerful testimony to Jesus's identity as God. For example, a study of the Old Testament indicates that it is *only* God who saves. In Isaiah 43:11, God asserts: "I, even I, am the Lord, and besides Me there is no savior." This verse indicates that (1) a claim to be Savior is, in itself, a claim to deity; and (2) there is only one Savior — the Lord God. It is thus highly revealing of Christ's divine nature that the New Testament refers to Jesus as "our great God and Savior" (Titus 2:13).

Likewise, God asserted in Isaiah 44:24: "I *am* the Lord, who makes all things, who stretches out the heavens *all alone*, who spreads abroad the earth by *Myself*" (emphasis added). The fact that God *alone* "makes all things" (Isaiah 44:24) — and the *accompanying* fact that Christ is claimed to be the Creator of "all things" (John 1:3; Colossians 1:16; Hebrews 1:2) — proves that Christ is truly God.

Preincarnate Appearances of Christ

Many theologians believe that appearances of the "angel of the Lord" (or, more literally, "angel of Yahweh") in Old Testament times were preincarnate appearances of Jesus Christ. (The word *preincarnate* means "before becoming a human being.") There are a number of evidences for this view:

1. The angel of Yahweh appeared to Moses in the burning bush and *claimed to be God* (Exodus 3:6).
2. Yet, the angel of Yahweh was *sent* into the world *by* Yahweh (Judges 13:8–9), just as Jesus was sent into the world in New Testament times by the Father (John 3:17).
3. The angel of Yahweh prayed *to* Yahweh on behalf of the people of God (Zechariah 1:12), just as Jesus prays to the Father for the people of God today (Hebrews 7:25; 1 John 2:1–2).
4. It would seem that appearances of this "angel" could not be the Father or the Holy Spirit. After all, the Father is One "whom no one has seen or can see" (1 Timothy 6:16, NIV; see also John 1:18, 5:37). Moreover, the Holy Spirit cannot be physically seen (John 14:17). That leaves only Jesus.
5. The angel of Yahweh and Jesus engaged in amazingly similar ministries — such as delivering the enslaved (Exodus 3; Galatians 1:4; 1 Thessalonians 1:10; 2 Timothy 4:18; Hebrews 2:14–15) and comforting the downcast (Genesis 16:7–13; 1 Kings 19:4–8; Matthew 14:14, 15:32–39).

These evidences suggest that appearances of the angel of Yahweh in Old Testament times were preincarnate appearances of Christ. Assuming this is correct, the word "angel" is used of Christ in these verses in accordance with its Hebrew root, which means "messenger, one who is sent, envoy." Christ, as the angel of Yahweh, was acting on behalf of the Father, just as He did in New Testament times.

The Biblical Basis for the Trinity

The deity of Christ is intimately connected to the doctrine of the Trinity. This doctrine affirms that there is only one God and that in the unity of the one godhead there are three coequal and coeternal persons — the Father, the Son, and the Holy Spirit. Let us briefly consider the evidence for this doctrine.

There Is One God

In the course of God's self-disclosure to humankind, He revealed His nature in progressive stages. First, God revealed that He is the *only true God*. This was a necessary starting point for God's self-revelation. Throughout history, Israel was surrounded by pagan nations deeply engulfed in the belief that there are many gods. Through the prophets, God communicated to Israel that there is only one true God (Deuteronomy 6:4, 32:39; Psalm 86:10; Isaiah 44:6). Even at this early juncture, however, we find preliminary indications of the Trinity (Genesis 1:26, 11:7; Isaiah 6:8, 48:16). God's oneness is also emphasized in the New Testament (Romans 3:29–30; 1 Corinthians 8:4; Galatians 3:20; 1 Thessalonians 1:9; 1 Timothy 1:17, 2:5; James 2:19; Jude 25).

The Father Is God

As history unfolded, God progressively revealed more about himself. It eventually became clear that while there is only one God, there are three distinct persons within the one godhead, each individually recognized as God (Matthew 28:19).

The Father, for example, is explicitly called God (John 6:27; Romans 1:7; Galatians 1:1; 1 Peter 1:2). He is also portrayed as having all the attributes of deity — such as being everywhere-present (Matthew 19:26), all-knowing (Romans 11:33), all-powerful (1 Peter 1:5), holy (Revelation 15:4), and eternal (Psalm 90:2).

The Son Is God

Jesus is also explicitly called "God" in Scripture (Titus 2:13; Hebrews 1:8). And He, too, has all the attributes of deity — including being everywhere-present (Matthew 28:20), all-knowing (Matthew 9:4), all-powerful (Matthew 28:18), holy (Acts 3:14), and eternal (Revelation 1:8, 17).

The Holy Spirit Is God

The Holy Spirit is also recognized as God (Acts 5:3–4). He, too, possesses the attributes of deity, including being everywhere-present (Psalm 139:7–9), all-knowing (1 Corinthians 2:10–11), all-powerful (Romans 15:19), holy (John 16:7–14), and eternal (Hebrews 9:14).

Three-in-Oneness in the Godhead

Scripture also indicates there is three-in-oneness in the godhead. In Matthew 28:19, the resurrected Jesus instructed the disciples, "Go therefore and make disciples of all the nations, baptizing them in the name of the Father and of the Son and of the Holy Spirit" (Matthew 28:19). The word *name* is singular in the Greek, thereby indicating God's oneness. However, the definite articles in front of Father, Son, and Holy Spirit (in the original Greek) indicate they are distinct personalities, even though there is just one God.

These distinct personalities relate to each other. The Father and Son, for example, *know* each other (Matthew 11:27), *love* each other (John 3:35), and *speak* to each other (John 11:41–42). The Holy Spirit *descended upon* Jesus at His baptism (Luke 3:22), is called *another* comforter (John 14:16), was *sent* by the Father and Jesus (John 15:26), and seeks to *glorify* Jesus (John 16:13–14).

An Analogy

A helpful analogy of the Trinity is that God is like a triangle that is one figure yet has three different sides (or corners) at the same time. So there is a simultaneous *threeness* and *oneness*. Of course, no analogy is perfect since in every analogy there is a similarity and a difference. For example, water can exist simultaneously in three different states as ice, water, and steam; that is, as a solid, liquid, and a gas at pressure of 4 Torr and temperature of 273K. One substance but three totally different personalities.

Answering Objections

Cults and false religions often raise objections against both the deity of Christ and the doctrine of the Trinity. In what follows, key objections will be briefly summarized and answered.

Jesus Is the Son of God

Some claim that because Jesus is the Son of God, He must be a lesser God than God the Father. Among the ancients, however, an important meaning of *Son of* is "one who has the same nature as." Jesus, as the Son of God, has the very *nature of God* (John 5:18, 10:30, 19:7). He is thus not a lesser God.

The Father Is "Greater" Than Jesus

Some cults argue that because Jesus said the Father is "greater" than Him (John 14:28), this must mean Jesus is a lesser God. Biblically, however, Jesus is equal with the Father in His divine nature (John 10:30). He was *positionally* lower than the Father from the standpoint of His becoming a servant by taking on human likeness (Philippians 2:6–11). Positionally, then, the Father was "greater" than Jesus.

Jesus Is the Firstborn

Some cults argue that because Jesus is the "firstborn of creation" (Colossians 1:15), He is a created being and hence cannot be truly God. Biblically, however, Christ was not created but is the Creator (Colossians 1:16; John 1:3). The term *firstborn*, defined biblically, means Christ is "first in rank" and "preeminent" over the creation He brought into being.

Jesus Is Not All-Knowing

Some cults argue that because Jesus said no one knows the day or hour of His return except the Father (Mark 13:32), Jesus must not be all-knowing, and hence He must not be truly God. In response, Jesus in the Gospels sometimes spoke from the perspective of His divinity and at other times from the perspective of His humanity. In Mark 13:32, Jesus was speaking from the limited perspective of His humanity (see Philippians 2:5–11). Had he been speaking from His divinity, He would not have said He did not know the day or hour. Other verses show that Christ, *as God*, knows all things (Matthew 17:27; Luke 5:4–6; John 2:25, 16:30, 21:17).

Jesus Prayed

Some cults argue that because Jesus prayed to the Father, He could not truly be God. Biblically, however, it was in His humanity that Christ prayed to the Father. Since Christ came as a man — and since one of the proper duties of man is to worship, pray to, and adore God — it was perfectly proper for Jesus to address the Father in prayer. Positionally speaking as a man, as a Jew, and as our High Priest — "in all things He had to be made like His brethren" (Hebrews 2:17) — Jesus could pray to the Father. But this in no way detracts from His intrinsic deity.

The Trinity Is Illogical

Some cults claim the Trinity is illogical ("three in one"). In response, the Trinity may be *beyond* reason, but it is not *against* reason. The Trinity does not entail three gods in one God, or three persons in one person. Such claims would be nonsensical. There is nothing contradictory, however, in affirming three persons in one God (or three *whos* in one *what*).

The Trinity Is Pagan

Some cults have claimed the doctrine of the Trinity is rooted in ancient paganism in Babylon and Assyria. In response, the Babylonians and Assyrians believed in triads of gods who headed up a pantheon of many other gods. These triads constituted three separate gods (polytheism), which is utterly different from the doctrine of the Trinity that maintains that there is only one God (monotheism) with three persons within the one godhead.

Our God Is an Awesome God

We have seen that Jesus must be viewed as God by virtue of the facts that He has the *names* of God, the *attributes* of God, and the *authority* of God; He does the *works* of God; and He is *worshiped* as God. We have also seen persuasive scriptural evidences for the doctrine of the Trinity. *Our triune God is an awesome God!*

20

Information: Evidence for a Creator?

MIKE RIDDLE

The battle of the ages began when Satan deceived himself into thinking he could overthrow the sovereign rule of God. Since then, Satan has opposed God and has become known as the adversary or great deceiver. Two opposing kingdoms are in conflict. The kingdom of Satan attacked the kingdom of God with the goal of destroying it. Both God and Satan have a purpose for history; but since God is God, and Satan is His created creature, God's purpose is the ultimate one.

With the birth of the Church, Satan had a new enemy to contend with. The Church's preaching of the gospel poses a serious threat to his kingdom. Every time the gospel is preached to nonbelievers, Satan is in danger of them believing it and leaving his kingdom. Thus, in order to prevent losing members in his kingdom, Satan must attack the Church and its message. Throughout the history of the Church, Satan has used various tactics from physical persecution to deceiving the Church into believing wrong ideas and compromising God's Word. Satan has launched these attacks from both outside and inside the Church. People have burned the Bible, banned it, changed it, or considered it irrelevant, especially in this modern scientific age. One of Satan's major strategies against the church has been and is the philosophy of *materialism.*

Materialism is the assumption that all that exists is mass and energy (matter); there are no supernatural forces, nothing exists that is nonmaterial, and no God. Materialism is the foundational presupposition for atheism, humanism, and evolutionism.

> The Cosmos is all that is or ever was or ever will be.[1]

> We atheists . . . try to find some basis of rational thinking on which we can base our actions and our beliefs, and we have it. . . . We accept the technical philosophy of materialism. *It is a valid philosophy which cannot be discredited.* Essentially, materialism's philosophy holds that nothing exists but natural phenomena. . . . There are no supernatural forces, no supernatural entities such as gods, or heavens, or hells, or life after death (emphasis added).[2]

The challenge by materialists is that the Church cannot defend against the philosophy of materialism. The materialists do not believe the Church can demonstrate the existence of God. Further, they know that if materialism is true, then evolution must also be true. But what if the assumption of materialism is false? What if it could be shown through empirical science that the universe consists of more than just mass and energy?

Good News

For Darwinian (molecules-to-man) evolution to actually work, new genetic information is required each step of the way. In order for a fish to grow legs, new information must be encoded into the DNA. For a reptile to grow feathers, new information must be encoded into the DNA. For an apelike creature to evolve into a human, new information must be encoded into the DNA. This new information must add to or replace old information with new instructions to grow legs, or feathers, or human characteristics. But what is information and where does it come from?

Follow me in this illustration: Imagine for a moment that it is your mother's birthday and you want to wish her a happy birthday, but you are stuck in an area without power. You know your friend a couple of miles away has power and knows Morse code. So you build a fire and begin using smoke signals to spell out Morse code for your friend to call your brother to have him send an e-mail on your behalf to your mother for her birthday.

Information went from you to the smoke signals directly to your friend's eyes and from your friend's mouth through sound waves to the phone receiver then through electronic signals in the phone to your brother and back into sound waves for your brother to hear it. Then the information went through

1. Carl Sagan, *Cosmos* (New York: Random House,1980), p. 4.
2. Madalyn O'Hair, *What on Earth Is an Atheist!* (Austin, TX: American Atheist Press, 1972), p. 39–40.

his fingers and was transferred into code on the computer and again through electronic means to your mother who received the information on her computer screen as an understandable concept — *Happy Birthday*. Nothing material actually transferred from you to your mother, but information did, which shows that everything isn't material.

This is the good news! Why is this good news? Because the foundation for materialism (atheism, humanism, evolution) is that the universe consists of only two entities[3]: mass and energy. Therefore, if a third entity can be shown to exist, then materialism and all philosophies based on it must also be false. Information is this third fundamental entity.

What Is Information?

There are several definitions of information currently in use; however, each of these definitions are generally too broad. For example, one definition of information includes symbols with or without meaning, and another includes everything in its definition of information. Imagine sending random symbols as smoke signals to your friend — would *Happy Birthday* ever get sent to your mother on her birthday? Imagine sending a bunch of smoke signal dots in the air to your friend — would *Happy Birthday* ever get sent to your mother?

In July 2006, a team of scientists representing various scientific disciplines met to evaluate a definition of information proposed by information scientist Dr. Werner Gitt,[4] which is precise and corresponds very well to human languages and machine languages. The team proposed that this definition be called Universal Definition of Information (UDI) and agreed that there are four essential attributes that define it:

1. **Code** (syntax): Information within all communications systems contains a code. A code contains a set of symbols and rules for using letters, words, phrases, or symbols to represent something else. One reason for coding is to enable communication. Examples of codes would be the English alphabet,

3. Entity: The state of having existence; something with distinct and independent existence.
4. Team members included Werner Gitt, PhD, engineering/information; Jason Lisle, PhD, astrophysics; John Sanford, PhD, genetics; Bob Compton, PhD, physiology, DVM; Georgia Purdom, PhD, molecular genetics; Royal Truman, PhD, chemistry; Kevin Anderson, PhD, microbiology; John Oller, PhD, linguistics; Andy McIntosh, PhD, combustion theory/thermodynamics; Mike Riddle, BS, mathematics/ MA, education; Dave Mateer, BS, mathematics and computer science.

words, and syntax; hieroglyphics; or codes used in computers (for example, C, Fortran, or Cobol).

2. **Meaning** (semantics): Meaning enables communication by representing real objects or concepts with specific symbols, words, or phrases. For example, the word *chair* is not the physical chair but represents it. Likewise, the name "Bob" is not the physical person but represents the real person. When words are associated with real objects or concepts, it gives the word meaning.

   For example, *aichr* and *Bbo* do not have meaning because they do not represent any real object or concept. However, if in the future one of these character strings were to represent a real object or concept, it would have meaning. Prior to the computer Internet age, the word *blog* had no meaning; today it is associated with a web page that serves as a personal log (derived from *web log*) of thoughts or activities. It can also mean a discussion community about personal issues. Another new word with meaning is *simplistic*. New words are continually being designated with meaning.

3. **Expected Action** (pragmatics): Expected action conveys an implicit or explicit request or command to perform a given task. For example, in the statement, "Go to the grocery store and buy some chocolate chips," the expected action is that someone will go to the store. This does not mean the action will actually happen, but it is expected to happen.

4. **Intended Purpose** (apobetics): Intended purpose is the anticipated goal that can be achieved by the performance of the expected action(s). For example, in the statement, "Go to the grocery store and buy some chocolate chips," the intended purpose might be to bake and eat chocolate chip cookies.

These four essential attributes specify the definition domain for information. A definition of information (Universal Definition of Information) was formulated by using these four attributes:

> An encoded, symbolically represented message conveying expected action and intended purpose.

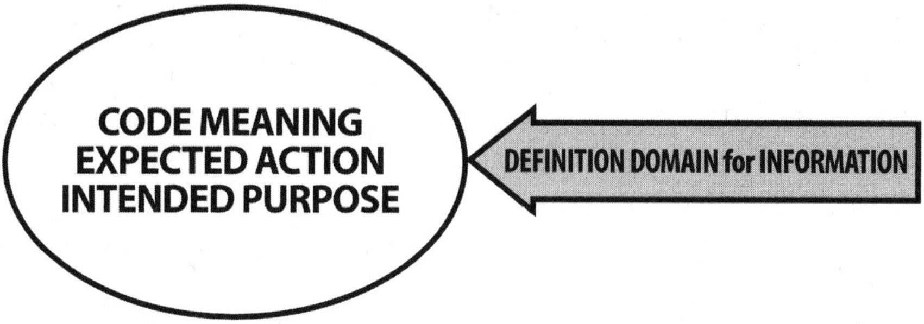

Encoded	Code
Symbolically represented message	Meaning
Expected action	Action
Intended purpose	Purpose

Anything not containing *all four* attributes is not considered information by this Universal Definition of Information (UDI).

Examples of entities that do contain Information [UDI]:

- The Bible
- Newspaper
- Hieroglyphics
- Sheet music
- Mathematical formulas

Examples of entities that do *not* contain Information [UDI] (one or more of the attributes are missing):

- **A physical star:** Lacks a code and lacks meaning because it does not represent something else; it is the physical object that the word *star* represents.
- **A physical snowflake:** Lacks a code and lacks meaning because it does not represent something else; it is the physical object.
- **Random sequence of letters:** Has a symbol set, but lacks rules for words or grammar (no code). Since it is random, it has no meaning to any sequence of letters.
- **A physical piano:** Lacks meaning because it does not represent something else; it is the physical object.

Investigating Information [UDI] Scientifically

The lowest level of operational science begins with ideas originated and formulated by man. These include models, hypotheses, theories, assumptions, speculations, etc. This is the lowest level of scientific certainty because man's understanding of reality is incomplete, faulty, and constantly changing. A very large gap exists between this level of science and the highest level. This highest level contains scientific laws.

Scientific laws are precise statements formulated from discoveries made through observations and experiments that have been repeatedly verified and never contradicted. There are scientific laws about matter (Newton's law of gravity, laws of thermodynamics, laws of electricity, and laws of magnetism). There is Pasteur's law about life (law of biogenesis). Each of these laws is universal with no known exceptions. Scientific evidence that supports or refutes a scientific concept determines its level of certainty.

The information team evaluated scientific laws about information formulated by Dr. Werner Gitt that determine the nature and origin of information [UDI].

Fundamental Law 1 (FL1)
A purely material entity, such as physicochemical processes, cannot create a nonmaterial entity. (Something material cannot create something nonmaterial.)

Physical entities include mass and energy (matter). Examples of something that is not material (nonmaterial entity) include thought, spirit, and volition (will).

Fundamental Law 2 (FL2)
Information is a nonmaterial fundamental entity and *not* a property of matter.

The information recorded on a CD is nonmaterial. If you weigh a modern blank CD, fill it with information, and weigh it again, the two weights will be the same. Likewise, erasing the information on the CD has no effect on the weight.

The same information can be transmitted on a CD, a book, a whiteboard, or using smoke signals. This means the information is independent of the material source. A material object is required to store information, but the information is not part of the material object. Much like people in an airplane are being stored and transferred in the plane, they are not part of the physical plane.

The first law of thermodynamics makes it clear that mass and energy (matter) can neither be created nor destroyed. All mass and energy in the universe is being conserved (the total sum is constant). However, someone can write a new complicated formula on a whiteboard and then erase the formula. This is a case of creating and destroying information.

White Board

New Math Formula

Since the first law of thermodynamics states that mass and energy (matter) cannot be created or destroyed, and information (UDI) can be created and destroyed, information (UDI) must be nonmaterial.

> The genetic information system is the software of life and, like the symbols in a computer, is purely symbolic and independent of its environment. Of course, the genetic message, when expressed as a sequence of symbols, is nonmaterial but must be recorded in matter and energy.[5]

> Indeed, Einstein pointed to the nature and origin of symbolic information as one of the profound questions about the world as we know it. He could identify no means by which matter could bestow meaning to symbols. The clear implication is that symbolic information, or language, represents a category of reality distinct from matter and energy.[6]

5. Hubert Yockey, *Information Theory, Evolution, and the Origin of Life* (New York: Cambridge University Press, 2005), p. 7.
6. John Baumgardner, "Highlights of the Los Alamos Origins Debate," www.globalflood.org.

First Law of Information (LI1)
Information cannot originate in statistical processes. (Chance plus time cannot create information no matter how many chances or how much time is available.)

There is no known law of nature, no known process, and no known sequence of events which can cause information to originate by itself in matter.[7]

Second Law of Information (LI2)
Information can only originate from an intelligent sender

Corollary 1[8]
All codes result from an intentional choice and agreement between sender and recipient.

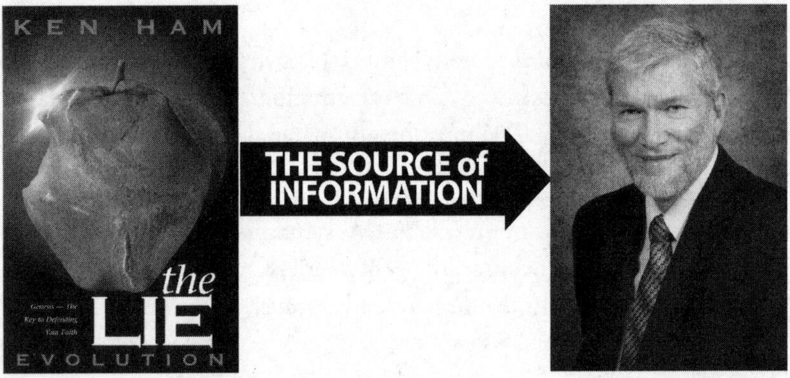

This is not the source... **Author: Ken Ham**

We observe daily a continual input of new information (UDI) from an intelligent source (human beings). At present, on earth, the only new information we have detected being created is from human beings. Careful examination of other systems will determine if there are any other intelligent sources of new UDI.

7. Werner Gitt, *In the Beginning Was Information* (Green Forest, AR: Master Books, 1997), p. 106.
8. A corollary is a logical inference that follows directly from the proof of another proposition.

Corollary 2
Any given chain of information can be traced backward to an intelligent source.

For two people to effectively communicate, there must be some agreement on the language or code that is used.

Law of Matter about Machines (LM1)
When information (UDI) is utilized in a material domain, it always requires a machine.

Definition of a machine: A machine is a material device that uses energy to perform a specific task.

Corollary 1 to LM1
Information is required for the design and construction of machines.

What does this mean? Both information (UDI) and matter are necessary for the development of a machine. It is the information that determines and directs the assembly of the material system into the necessary configuration, thereby creating a machine. This means that tracing backward to the manufacture and design of any machine capable of performing useful work will lead to the discovery or necessity of information and ultimately to its intelligent source.

Testing UDI Universally (Living Systems)

Does the code in all living systems (DNA) exhibit all four attributes of Universal Definition of Information (UDI)?

Since all living systems contain DNA and DNA information contains all four attributes, it meets the UDI definition of information. Furthermore, the capacity and density of the information encoded in DNA surpasses anything mankind has accomplished.

> There is no information system designed by man that can even begin to compare to it [DNA].[9]

9. John Sanford, *Genetic Entropy and the Mystery of the Genome* (Lima, NY: Elim Publishing, 2005), p. 1.

Code	The decoded portion of DNA contains 4 letters (ATCG) that make up three-letter words (codon). These codons are arranged linearly in a various sequence (syntax).
Meaning	Each three-letter word represents 1 of the 20 specific amino acids used in life. The sequence (syntax) of the DNA words designates the specific sequence of the amino acids in protein formation.
Expected Action	Cellular proteins are biomachines essential for construction, function, maintenance, and reproduction of the entire organism
Intended Purpose	Existence of life

The information encoded in DNA is billions of times more compact than a modern PC hard drive.

How long would it take using naturalistic processes to type out such a code?

A billion universes each populated by billions of typing monkeys could not type out a single gene of this genome.[10]

But a purposeful, all-knowing, all-powerful Creator could create complex codes in less than a day.

> Ah Lord God! behold, thou hast made the heaven and the earth by thy great power and stretched out arm, and there is nothing too hard for thee (Jeremiah 32:17).

The information team agreed upon a precise definition of information (UDI) that is consistent with the information found in human natural languages and in machine languages. Additionally, scientific laws that govern the UDI definition domain were established. It was agreed that the information encoded within the DNA belongs to the UDI domain.

10. Johnjoe McFadden, *Quantum Evolution* (New York: W.W. Norton & Company, 2002), p. 84.

Seven Conclusions

If we apply these laws governing UDI to DNA information, we can make logically sound arguments (conclusions).

1. Since the DNA code of all life forms is clearly within the UDI definition domain of information, **we conclude there must be a sender** (LI 1, 2).

2. Since the density and complexity of the DNA encoded information is billions of times greater than man's present technology, **we conclude the sender must be supremely intelligent** (LI 2, plus corollaries).

3. Since the sender must have
   - encoded (stored) the information into the DNA molecules
   - constructed the molecular biomachines required for the encoding, decoding, and synthesizing processes
   - designed all the features for the original life forms

   we conclude the sender must be purposeful and supremely powerful (LM 1, plus corollary).

4. Since information is a nonmaterial fundamental entity and cannot originate from purely material quantities, **we conclude the sender must have a nonmaterial component (Spirit). God is Spirit** (FL1, 2; LI 2, plus corollaries)!

5. Since information is a nonmaterial fundamental entity and cannot originate from purely material quantities, and since information also originates from man, **we conclude man's nature must have a nonmaterial component (spirit). Man has a spirit** (FL 1, 2; LI 2, plus corollaries)!

6. Since information is nonmaterial and the third fundamental entity, **we conclude that the assumption "the universe is composed solely of mass and energy" is false** (FL 1, 2).

   The philosophy of materialism is false!

7. Since all theories of chemical and biological evolution require that information must originate solely from mass and energy alone (no sender), **we conclude all theories of**

chemical and biological evolution are false (Fl 1, 2; LI 1, 2, plus corollaries).

The evolution of life is false!

Therefore, the scientific laws governing the UDI domain have

- Refuted the presupposition of atheism, humanism, and the like, including the theories of chemical and biological evolution.
- Confirmed the existence of an eternal, all-knowing, all-powerful being (God).

Summary

The importance of information to the creation/evolution debate is founded in the presuppositions of each model. The presupposition of the evolutionary model is materialism, which is the idea that everything in the universe is solely comprised of matter (mass and energy). From this foundational assumption, evolutionists logically conclude that cosmic evolution, chemical evolution, and biological evolution are all true. The presupposition of materialism has been shown scientifically to be false.

The presupposition of the Bible is that there is a God who created the universe, the earth, and all organisms living on earth. This has been shown to be consistent with scientific discoveries that there is a nonmaterial third fundamental entity called information that originates only from an intelligent source. The universe consists of more than just mass and energy, and the information found within the DNA system of all life originated from an all-knowing, all-powerful Creator God.

The Challenge

Anyone who disagrees with these laws and conclusions must falsify them by demonstrating the initial origin of information from purely material sources. This challenge has never been scientifically achieved.

21

Is Evolution a Religion?

DR. TOMMY MITCHELL & DR. A.J. MONTY WHITE

We are sure that many people will find the question posed as the title of this chapter a little strange. Surely, evolution is about the origin and development of life forms on earth — what has this got to do with religion? Evolution is science, isn't it? And we are told that it has got to be separate from religious belief — at least in the classroom! Well, let's see if evolution fits the bill as a true science as opposed to a religious belief. In order to do so, we must define some terms.

What Is Science?

Creationists are often accused of being unscientific or pseudoscientific, while at the same time those who promote evolution assume the mantle of "real scientist." But what is science anyway? According to *The American Heritage Dictionary*, science is "the observation, identification, description, experimental investigation, and theoretical explanation of phenomena."[1] Or put more simply, science involves observing things in the real world and trying to explain how they work. The key word here is *observation.*

You see, creationists do, indeed, believe in real "observational science," sometimes called "operational science." We enjoy the benefits of observational science every day. Whether flying in an airplane, having our illness cured by the wonders of modern medicine, or writing this book on a space-age laptop computer, we are benefiting from the technology that applies genuine observational

1. *The American Heritage Dictionary of the English Language*, 1996, s.v. "Science."

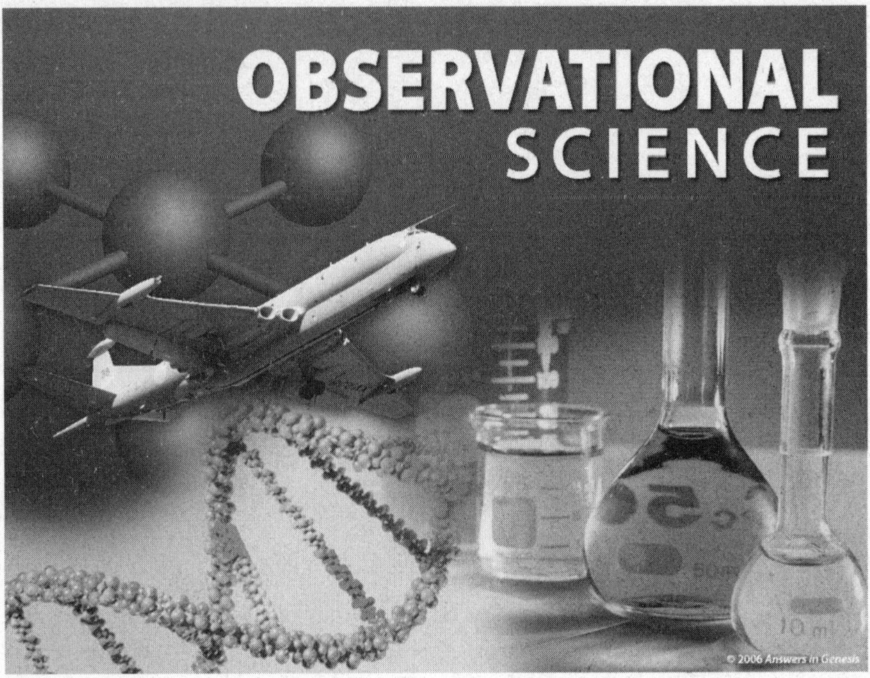

science to real-world needs. These triumphs of science exist in the present and can therefore be the subjects of examination and investigation.

Another type of science is known as "historical science," sometimes called "origins science." Historical science is the process of using the methods of science in the present to determine what happened in the past. Since the physical world exists in the present, all the evidence a scientist has available to examine the physical world also exists in the present. The scientist has no method to examine directly the past, thus he must make *assumptions* in order to come to conclusions. However, assumptions are unproven, and generally unprovable, beliefs. Assumptions are no more than untestable guesses.

Things that happened in the past are just that, past. They cannot be observed or tested in the present. They cannot be repeated or verified in the present. Then, you ask, how do we *know* so much about the past?

Understanding the Past

Perhaps an example here would help illustrate this issue. If you were to ask a roomful of people, "Do you think George Washington was a real person?" what would you expect the response to be? Of course, everyone would say that he or she believed George Washington actually existed.

George Washington

Now ask this question: "Can you give me a way to prove his existence scientifically, that is, by some experimental procedure?" The usual responses are "Test his DNA," or "Dig up his bones." But actually, these methods won't work. First of all, DNA testing would only work if you already had a valid sample of his DNA to use as a comparison. If you dug up his bones, you still could not prove they were his. In order to make any conclusions, you would have to make some assumptions based on things you could not actually test.

Well then, if there is no scientific method to prove he lived, how do we know George Washington existed? It's easy! We have abundant historical documentation of his life. These documents were held to be valid by the people who lived in that day and are not disputed. Thus, we have reliable evidence that he actually walked the earth. (Whether or not he actually chopped down a cherry tree is still a matter of debate!)

What Does This Have to Do with Evolution?

Molecules-to-man evolution is based on the premise that, through mutation and natural selection, organisms have, over the past three billion or so years, become more complex. These organisms have then progressed into an ever-increasing array of creatures until, ultimately, humans arrived on the scene.

When asked if anyone has ever seen one type of creature change into another, the answer is always no. Confronted with this, the evolutionists will usually counter that it happens *too slowly to be seen*. The claim is that it takes millions of years for these painfully slow processes to occur. Well then, if the process is too slow to be seen, how do we know it happened at all? After all, no one was there to observe all these organisms slowly changing into more complex forms. Also, there is no way in the present to test or repeat what happened in the past. Any conclusions about things that are not testable in the present must be based on improvable assumptions about the untestable past.

Ernst Mayr, who is considered by many to be one of the 20th century's most influential evolutionists, put it this way:

> Evolutionary biology, in contrast with physics and chemistry, is an *historical science* — the evolutionist attempts to explain events and processes that have already taken place. Laws and experiments are *inappropriate techniques* for the explication of such events and processes. Instead one *constructs a historical narrative*, consisting of a tentative reconstruction of the particular scenario that led to the events one is trying to explain[2] (emphasis added).

He then amazingly concludes, "No educated person any longer questions the validity of the so-called theory of evolution, which we know now to be *a simple fact*"[3] (emphasis added).

So-called Evidence for Evolution

What is so obvious in our world that Mayr can call goo-to-you evolution "a simple fact," which according to him no educated person would question? There are many supposed evidences for evolution. We will now consider two of these supposed evidences here and will examine them in the light of observational, rather than historical, science.

Evolutionists often claim that the theory of evolution has nothing to do with origin of life. They argue that evolution only deals with issues of the changes in organisms over time. They contend that life has progressed through purely naturalistic means, without any supernatural intervention. However, if they argue that life progresses by purely naturalistic mechanisms, then they must also delineate a natural process by which life came into being.

One supposed evidence for evolution is that life began spontaneously in the earth's vast oceans approximately three billion years ago.[4] Textbooks, magazines, and television documentaries constantly bombard us with this so-called fact. Just what is the evidence for the evolution of life from nonliving molecules? There

2. Ernst Mayr, "Darwin's Influence on Modern Thought," *Scientific American*, July 2000, p. 80.
3. Ibid., p. 83.
4. For the sake of this discussion, we will not consider the proposal made by some who claim that primitive life was brought to earth by aliens in the distant past.

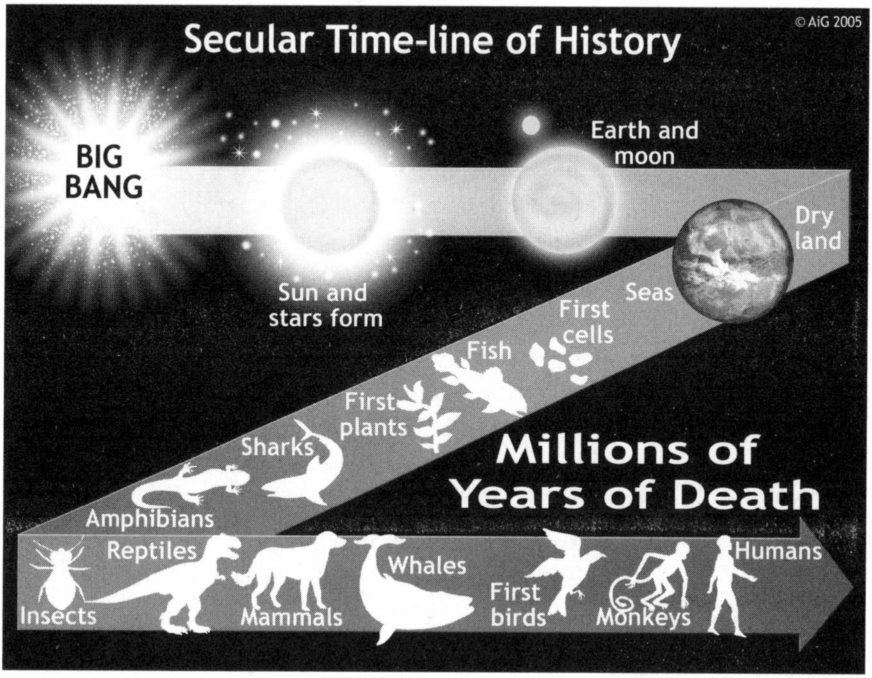

isn't any! There is no method to determine what the earth's "ancient atmosphere" was like or the composition of the oceans at that time.[5] No one was there to test or examine that environment. No one can say with certainty what the chemical makeup of the primordial oceans was. So how can it be claimed that simple proteins and nucleic acids arose spontaneously?

Based on our knowledge of these molecules using observational science in the present, it is difficult to imagine these processes happening by naturalistic processes. No scientific observation has ever shown how these complex molecules could arise spontaneously, let alone evolve simultaneously and assemble themselves in such a way as to become alive. One prominent evolutionist, Leslie Orgel, notes, "And so, at first glance, one might have to conclude that life could never, in fact, have originated by chemical means."[6]

One of the primary evidences used to support the theory of evolution is the fossil record. Evolutionists have long proposed that the fossilized remains of dead organisms, both plant and animal, found in the rock layers prove that life has evolved on the earth over millions of years. Using observational science, how

5. The authors accept the biblical view of history, not the millions-of-years view. They do not accept the evolutionary time scale; this is presented here merely for the sake of this discussion.
6. Leslie E. Orgel, "The Origin of Life on Earth," *Scientific American*, October 1994, p. 78.

can this conclusion be reached? There are only the fossils themselves to examine. These fossils only exist in the present. There is no method to determine directly what happened to these creatures; neither to determine how they died, nor how they were buried in the sediment, nor how long it took for them to fossilize. Although it is possible to make up a story to explain the fossil record, this contrived story does not meet the criteria for true scientific investigation. A story about the past cannot be scientifically tested in the present.

The creationist looking at the fossil record reaches a far different conclusion from the evolutionist. To the creationist, the fossils in the rocks represent the result of a global cataclysm with massive sedimentation rapidly burying millions upon millions of creatures. This catastrophic event would account not only for the fossil record but also for the rock layers themselves. (Deposition of sediment in layers would have resulted from sorting in the turbulent Flood and post-Flood waters.) So which viewpoint is correct? Neither the creationist's nor the evolutionist's explanation can be tested in the present.

But in this regard the creationist does have evidence. Evidence is found in a book called the Bible. The Bible claims to be the Word of God. It is a record of what God did and when He did it. In the Bible we learn how life began — God created it. The Bible helps us understand the fossil record — much of it is the result of a worldwide flood as described in Genesis 6–8. Like the historical documents that establish George Washington existed, we have a reliable historical document called the Bible to give us answers about our origin and about our world.

An evolutionist has no historical documentation for his viewpoint. He relies on the assumptions of historical science for support. Herein lies a fundamental misunderstanding of the purpose and potential of science. Scientific inquiry properly involves the investigation of processes that are observable, testable, and repeatable. The origin and development of life on earth cannot be observed, tested, or repeated because it happened in the past.

So, is evolution observable science? No, evolution falls under the realm of historical science; it is a belief system about the past. How can an evolutionist believe these things without rigorous scientific proof? The answer is that he *wants to*. Evolutionists are quite sincere in their beliefs, but ultimately these beliefs are based on their view that the world originated by itself through totally naturalistic processes. There is a term for this type of belief system — that term is *religion.* Religion is "a cause, a principle, or an activity pursued with zeal or conscientious devotion."[7] It should be pointed out that religion does not necessarily involve the concept of God.

Perhaps a few observations from some of the world's leading evolutionists will now put the question posed in the title of this chapter into perspective.

Evolution as a Religion

Dr. Michael Ruse, from the Department of Philosophy at the University of Guelph in Ontario, is a philosopher of science, particularly of the evolutionary sciences. He is the author of several books on Darwinism and evolutionary theory and in an article in the *National Post* he wrote:[8]

> Evolution is promoted by its practitioners as more than mere science. Evolution is promulgated as an ideology, a secular religion — a full-fledged alternative to Christianity, with meaning and morality. . . . Evolution is a religion. This was true of evolution in the beginning, and it is true of evolution still today.

This is an incredible admission: the study of the origin and development of life-forms on earth is not "mere science," but "a secular religion."

However, this is also the view of William Provine, the Charles A. Alexander Professor of Biological Sciences at the Department of Ecology and Evolutionary Biology at Cornell University. Writing in *Origins Research,* he tells us, "Let me summarize my views on what modern evolutionary biology tells us loud and clear."[9] Now you would expect this leading professor of biology to say that modern evolutionary biology tells us something about the origin of life or something about natural selection or something about the origin of species or something about genetics. But, no! According to this leading evolutionary biologist, modern evolutionary biology tells us loud and clear that:

> There are no gods, no purposes, no goal-directed forces of any kind. There is no life after death. When I die, I am absolutely certain that I am going to be dead. That's the end for me. There is no ultimate foundation for ethics, no ultimate meaning to life, and no free will for humans, either.[10]

7. *The American Heritage Dictionary of the English Language*. 1996, s.v., "Religion."
8. Michael Ruse, "Saving Darwinism from the Darwinians," *National Post,* May 13, 2000, p. B-3.
9. William B Provine, *Origins Research* 16, no. 1 (1994): 9.
10. Ibid.

It is obvious that these two influential biologists in the United States believe that evolution *is* a religion — a religion of atheism where there are no end products and where evolution reigns supreme.

Religion of Atheism

Writing a superb article about the rise of the Darwinian fundamentalism in *The Spectator*, the journalist Paul Johnson sums up the belief system of atheistic evolutionists with great insightfulness.

> Nature does not distinguish between a range of mountains, like the Alps, or a stone or a clever scientist like Professor Dawkins, because it is sightless, senseless and mindless, being a mere process operating according to rules which have not been designed but simply are.[11]

Although Paul Johnson uses the word *nature*, he actually is referring to evolution. By this he means chance random processes honed by natural selection over eons of time. This is the process by which *everything* has been created, according to the evolutionists. The *everything* can be an inanimate object like a range of mountains, or it can be incredibly complex creatures like you and the authors of this book.

This belief in molecules-to-man evolution can and does cause people to become atheists as admitted by leading atheist Dr. Richard Dawkins, the Charles Simonyi Professor of the Public Understanding of Science at Oxford University. In answer to the question "Is atheism the logical extension of believing evolution?" Dawkins replied, "My personal feeling is that understanding evolution led me to atheism."[12]

Evolution Contrasted with Christianity

The only true real religion is Christianity, and this can be used as *the* template to explain what a religion is. A religion will therefore give an explanation for

- *A holy book* — Christianity teaches that the Bible is the Word of God and that this book teaches us what to believe concerning God and what God requires of us. The holy book of the evolutionists is Darwin's *Origin of Species*. The evolutionists

11. Paul Johnson, "Where the Darwinian Fundamentalists Are Leading Us," *The Spectator*, April 23, 2005, p. 32.
12. Laura Sheahen and Dr. Richard Dawkins, "The Problem with God: Interview with Richard Dawkins," www.beliefnet.com/story/178/story_17889.html.

believe that this book gives an explanation for the origin and development of life on earth[13] without the need of any God or supernatural agent.

- *The origin of everything* — Christianity teaches that in the beginning God created everything (that is, the entire universe with all its stars and planets, all plant life and all animal and human life) out of nothing over a period of six literal days. In comparison, evolution teaches that in the beginning nothing exploded and gradually evolved over billions of years into the universe that we see today.
- *The origin of death and suffering* — Christianity teaches that when God created everything, it was perfect. As a result of the sin of the first man, Adam, death, disease, and suffering entered the scene of time. Evolution does not recognize the word *sin* but teaches that fish-to-philosopher evolution can only proceed via death. Hence death, disease, and suffering are the necessary driving forces of evolution; from this concept, we get the phrase *survival of the fittest.*
- *The reason why humans are here* — Christianity teaches that humans are the pinnacle of God's creation and that they were made in God's image and likeness. In contrast, amoeba-to-architect evolution teaches that humans have evolved from some apelike ancestor, which in turn evolved from another sort of animal.
- *The future of humans* — Christianity teaches that one day the Lord Jesus Christ will return to this earth and that He will create a new heavens and earth where those people who trusted Him as their Lord and Savior in this life will live with God forever. Evolution, on the other hand, teaches that humans are not the end product of evolution; evolution will continue and humans will either become extinct or evolve into some other species of creature that will definitely not be human.
- *The future of the universe* — Christianity teaches that the present universe will be burned up by God, and He will then

13. I (MW) once knew a professor of biology who told me that he believed that Darwin's writings were inspired and that he read from the *Origin of Species* for at least 20 minutes every night before retiring to bed!

create a new heavens and earth. Evolution, on the other hand, teaches that one day the universe will reach what is called a *heat death,* although it is in effect a *cold* death, for the temperature of the universe will be just a fraction of a degree above absolute zero. This will happen when all the energy that is available to do work will have been used up, and then nothing will happen — the universe will just "be." The time period for the universe to reach this state is almost unimaginable. It is thought that it will take about a thousand billion years for all the stars to use up all their fuel and fizzle out. By then, of course, there will be no life in the universe; every single life-form, including humans, will have become extinct billions of years previously. There will still be, however, occasional flashes of starlight in the dark universe as very large stars collapse in on themselves to form black holes. For the next 10^{122} (that is the figure 1 followed by 122 zeros!) years, this so-called Hawking radiation will be the only thing happening in the universe. Then, when all the black holes have evaporated, there will be darkness for 10^{26} years, during which time the universe will simply "be" and nothing will happen.

Evolution – an Attractive Religion

At first sight, believing in evolution may not seem an attractive proposition. However, what makes it attractive is that there is no God to whom you have to give an account of your actions. This is borne out by the following quote from an atheist:

> We no longer feel ourselves to be guests in someone else's home and therefore obliged to make our behavior conform with a set of pre-existing cosmic rules. It is our creation now. We make the rules. We establish the parameters of reality. We create the world, and because we do, we no longer feel beholden to outside forces. We no longer have to justify our behavior, for we are now the architects of the universe. We are responsible to nothing outside ourselves, for we are the kingdom, the power, and the glory forever and ever. [14]

Evolution therefore leads to the teaching that you can do as you please. You can live your life just to please yourself. Many people today live such a life.

14. Jeremy Rifkin, *Algeny* (New York: Viking Press, 1983), p. 244.

They have abandoned the faith of their forefathers and have embraced the doctrines of evolution with its atheism. No wonder we are living in a "me, me, me" hedonistic society where everything that you do is to try to please and bring pleasure to yourself. This is more than "selfish ambition"; it is totally decadent and is in total contrast to what Christianity teaches about what our ambition should be — our chief end is to glorify God (not oneself) and to enjoy Him (not oneself) forever.

22

Is the Bible Enough?

PAUL TAYLOR

For so many people today, it would appear that the Bible is not enough. This is the case even (or perhaps especially) among people who have not actually read it. Witness the current popularity of those who would add extra books to the canon of Scripture. Or witness the claims that certain ancient documents are supposedly *more reliable* than the books of the Bible but were kept out of the canon because of petty jealousies.

The last few years have seen the publication of books such as *Holy Grail, Holy Blood*; *The Da Vinci Code*; and *The Gospel of Judas*. What such works proclaim, along with myriad TV documentaries, is that our Bible is suspect, allegedly having been compiled some three centuries after Christ by the winners of an intense theological/political debate. Are such claims true? Are there really other books that should be viewed as Scripture?

Other chapters in this book lay to rest the myth that the Bible was compiled three centuries after Christ. It is the purpose of this chapter to show that the books that allegedly "didn't quite make it" are not inspired and have no merit compared with the books that are part of the canon of Scripture.

Canon

We have become quite used to the word *canon* these days. The word is frequently used of a body of literature. For example, one can refer to the complete works of Shakespeare as the *Shakespearian* canon. More bizarrely, I recently read a discussion about whether certain novels about *Doctor Who* could be considered to be part of the *Doctor Who* canon. Strangely, this last usage was closer to the correct use of the word *canon*, as applied to Scripture. The argument went that the novels introduced concepts and ideas that were later contradicted or not found to be in harmony with events reported in the recent revised TV series. Presumably, the writer of the article felt that these *Doctor Who* novels were not following an accepted rule or pattern.

The word *canon*, in the context of literature, comes from a Greek word meaning "rule." We see the word used in Galatians 6:16.

> And as many as walk according to this rule, peace and mercy be upon them, and upon the Israel of God.

The Strong's number[1] for the word *rule* is 2583 and catalogues the Greek word from which we derive the word *canon*. The word is not referring to a law, but rather a way of doing things — a pattern of behavior. In the context of biblical literature, the word implies that the Bible is self-authenticating — that it is not merely complete, but that it is also internally self-consistent.

Another chapter in this book deals with the subject of alleged discrepancies in the Bible. In that chapter, we see that it is possible to interpret different passages of the Bible as if they contradict each other, but that if one approaches the Bible acknowledging that it is internally self-consistent, then the alleged discrepancies all easily disappear. That is why the apostle Peter describes the people who twist Scripture in this way as "untaught and unstable" (2 Peter 3:16). In our present study, we will see that the extrabiblical writings — and the so-called missing gospels — do not pass the test of self-consistency with the rest of Scripture and are therefore easy to dismiss as not being part of the consistent whole pattern of the Bible — the *canon*.

Apocrypha

The existence in the English language of names such as Toby (from Tobit) and Judith testify to the fact that the so-called Apocrypha was once influential in English society. The word *apocrypha* comes from the Greek word meaning "hidden." However, it popularly refers to a group of books considered by the Roman Catholic Church as part of the Old Testament.

1. Dr. James Strong (1822–1894) published his *Exhaustive Concordance of the Bible* in 1890. One invaluable feature of the work was that he assigned numbers to root words from the original Hebrew or Greek. These numbers have frequently been used by other Bible concordances and, more recently, by Bible software. The numbers enable students of the Bible to recognize where the same original words have been used, even if they do not know Hebrew or Greek.

Traditionally, Protestant churches have dismissed the apocryphal books. For example, Article VI of the Church of England's Thirty-Nine Articles lists first the canonical books of the Old Testament, and then lists the apocryphal books prefaced with this warning:

> And the other Books (as Hierome saith) the Church doth read for example of life and instruction of manners; but yet doth it not apply them to establish any doctrine; such are these following:
>
> The Third Book of Esdras, The rest of the Book of Esther, The Fourth Book of Esdras, The Book of Wisdom, The Book of Tobias, Jesus the Son of Sirach, The Book of Judith, Baruch the Prophet, The Song of the Three Children, The Prayer of Manasses, The Story of Susanna, The First Book of Maccabees, Of Bel and the Dragon, The Second Book of Maccabees.

The Hierome referred to in the Articles is Jerome. Jerome lived c. 347 to c. 420. He translated the Bible into Latin — the well-known *Vulgate* or common version. Originally, he used the Septuagint as the source of his Old Testament translation. The Septuagint (usually abbreviated to LXX) is a translation of the Old Testament into Greek. Many LXX manuscripts contain the apocryphal books. However, Jerome later revised the Vulgate, going back to Hebrew manuscripts for the Old Testament. It was at this point that he expressed dissatisfaction with the apocrypha, making the comment the Church of England used in its Articles above.

This illustrates that it was not merely a Protestant Reformation decision to remove the Apocrypha. In fact, the Apocrypha was never originally part of the OT canon and was added later. Interestingly, the apocryphal books themselves do not actually claim to be canonical. For example, in 1 Maccabees 9:27, the writer states: "So there was a great affliction in Israel, unlike anything since the time *a prophet had ceased to be seen among them*" (emphasis mine). Moreover, New Testament writers do not quote from apocryphal books, even though they are prepared to quote from other extrabiblical books (e.g., Paul quoted from Greek poets in Acts 17, and Jude quoted from the *Book of Enoch*).

The apocryphal books fail the internal self-consistency test. For example, 2 Maccabees 12:42 contains this exhortation to pray for the dead.

> And they turned to prayer, beseeching that the sin which had been committed [by the dead] might be wholly blotted out (Revised Oxford Apocrypha).

This sentiment is contrary to what is found in the rest of Scripture, both Old and New Testaments, such as Deuteronomy 18:11 and Hebrews 9:27. Similarly, inconsistencies and inaccuracies can be found between other apocryphal books and the correct canon of Scripture.

Da Vinci Decoded

Much of the modern preoccupation with extrabiblical writings has come from the publication of Dan Brown's novel *The Da Vinci Code*, and the earlier "serious" treatise on the subject, *Holy Blood, Holy Grail* by Richard Leigh and Michael Baigent. These, and other sensational books and TV documentaries, tend to focus on opposing biblical truth by stating the following:

- Jesus did not die on the cross.
- Jesus married, or had a close and sexual relationship with, Mary Magdalene.
- Mary Magdalene was supposed to be the leader of the new "church," but misogynist disciples usurped her position.
- These "truths" have been kept secret from the general public over the centuries and are known only to special initiates.

The "initiates" who have this secret knowledge are reputed to be found in many of the traditional "secret" organizations, such as Freemasons or the Knights Templar. At the heart of the so-called secret knowledge are the various doctrines and practices collectively known as *Gnosticism*. Before one even notes the way in which Gnosticism diverges from biblical truth, it is worth reflecting that the Bible makes claim that it should be understood mostly by plain reading. Gnostics, on the other hand, always have codes or secret knowledge required to interpret what God has said. Perhaps it was Gnostics that the apostle Paul had in mind when he warned Timothy thus:

> O Timothy! Guard what was committed to your trust, avoiding the profane and idle babblings and contradictions of what is falsely called knowledge — by professing it some have strayed concerning the faith. Grace be with you (1 Timothy 6:20–21).

The Strong's number for *knowledge* in this passage is 1108 and indicates the Greek word *gnosis*, meaning *knowledge*. In the Authorized Version, the word is translated as *science*. Certainly, Paul's criticism of the requirement for special knowledge is pertinent even if he didn't actually have the people we know as Gnostics in mind.

In his book *The Missing Gospels*, Darrell Bock shows that the documents and people labeled as *Gnostic* in fact hold to quite a wide variety of views and doctrines. There are, however, some common traits:

> An essential aspect of Gnosticism was its view of deity, namely, the distinction between and relationship of the transcendent God to the Creator God. This is important because this view of God produced the orthodox reaction against those texts.[2]

Bock observes five characteristics by which Gnostic writings differ from the Bible:

1. Dualism. Gnostics see a distinction between the transcendent God and the Creator God.
2. Cosmogony. This leads to a different view of the universe. Gnostics see an eternal battle between good and evil and do not view God as necessarily being more powerful than the devil.
3. Soteriology. Gnosticism's mode of salvation is by gaining the higher levels of secret knowledge.
4. Eschatology. In common with their view that matter is suspect, Gnostics are not usually looking forward to a bodily resurrection.
5. Cult. Gnostic groups perform various rituals. One of those described in *The Da Vinci Code* involved one of the characters taking part in a naked dance in the forest.

Bock goes on to place the rise of Gnosticism as clearly later than the writing of biblical texts, though there may be reference to Gnostic principles in the passage quoted above. Bock shows Gnosticism to be an unbiblical aberration, rather than being able to live up to the claim that it is the correct teaching of Christ — and that all the other scholars down the centuries have it wrong.

Are These Books Really Scripture?

Brian Edwards has produced a useful little summary of Gnostic ideas as presented in *The Da Vinci Code*.[3] Some of his thoughts are further summarized in the following.

2. Darrell Bock, *The Missing Gospels* (Nashville, TN: Thomas Nelson, 2006), p. 21.
3. Brian Edwards, *Da Vinci: A Broken Code* (Leominster, UK: Day One Publications, 2006).

The *Gospel of Thomas* does not contain a life story. Instead, it is a collection of 114 alleged sayings of Jesus. Some of these are contrary to the rest of Scripture. Not one serious scholar believes that the document was written by the apostle Thomas.

The *Gospel of Philip* contains a lot of Gnostic teaching. Some of the teachings are obscure, in a mystical kind of way.

> Light and darkness, life and death, right and left, are brothers of one another. They are inseparable. Because of this neither are the good good, nor evil evil, nor is life life nor death death.

Other teachings are aberrant, such as the idea that God made a mistake in creation.

> For he who created it wanted to create it imperishable and immortal. He fell short of attaining his desire.

The teaching given here is that the world is imperfect because God made a mistake. The Bible makes clear that God did indeed make the world perfect, but it is imperfect today because of our sin. In other words, by this teaching, Gnosticism is seeking to remove the responsibility from the human race and hand it to God.

The *Gospel of Mary* purports to be by Mary Magdalene. It certainly attempts to boost her position. It is an article of faith in Dan Brown's novel that Mary Magdalene was actually Jesus's chosen successor and wife — and father of his child.

> Peter said to Mary, "Sister we know that the Savior loved you more than the rest of woman. Tell us the words of the Savior which you remember, which you know, but we do not, nor have we heard them." Mary answered and said, "What is hidden from you I will proclaim to you."

The legends put forward in the books by Brown and Baigent and Leigh are not new. The legend is that, after the crucifixion, Mary fled, as she was pregnant with Jesus's son. She eventually arrived in what is today called France. The Merovingian dynasty claimed to be descended from her, as did Joan of Arc, as did the Stuart dynasty in Scotland and England. They claim that the Holy Grail was actually Mary's womb, and now represents the so-called holy bloodline of descendants of Jesus.

One thinks immediately of Isaiah 53, where the prophet makes clear that the Messiah, the Suffering Servant, will have no descendants.

> He was taken from prison and from judgment, and who will declare His generation? For He was cut off from the land of the living; for the transgressions of My people He was stricken (Isaiah 53:8).

The only people who can really have any claim of "descent" from Jesus are those of us who are saved by repentance and faith in Him.

> When You make His soul an offering for sin, He shall see His seed, He shall prolong His days, and the pleasure of the Lord shall prosper in His hand. He shall see the labor of His soul, and be satisfied. By His knowledge My righteous Servant shall justify many, for He shall bear their iniquities (Isaiah 53:10–11).

The concept of a married Jesus runs counter to the whole theme of the Bible. Passages in both Old and New Testaments compare our relationship with the Savior as individuals, but more specifically as the Church to a marriage. See, for example, Song of Songs, Psalm 45, and Revelation 19. If Jesus had a real, earthly wife, then this analogy would be inappropriate.

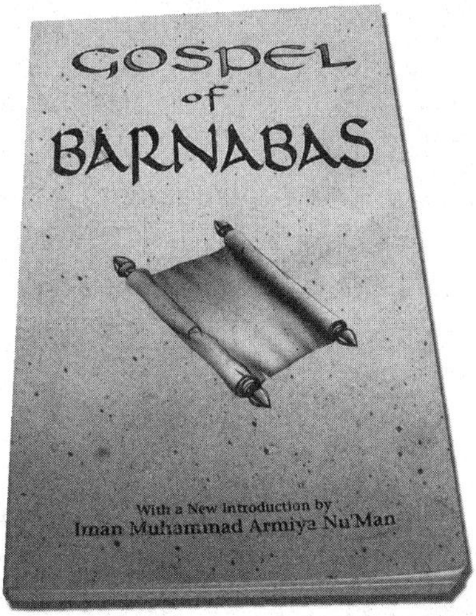

In the *Gospel of Barnabas*, it is claimed that Judas took on the appearance of Jesus and was mistakenly crucified in Jesus's place. The gospel also claims that Jesus told His mother and disciples that He had not been crucified.

It is noteworthy that the *Gospel of Barnabas* claims that the Messiah was to be descended, not from Isaac, but from Ishmael. The document is therefore much quoted by Muslims wanting to prove Islam to be the true faith. It has since been found that it was written in medieval times long after Christ.[4]

The *Gospel of Judas*, an extraordinary document written by Gnostics, claims that Jesus taught one message to 11 of His disciples, but a special, true, secret message to Judas. As part of the secret plan, Jesus persuaded Judas to "betray" Him, thus taking on the highest service for Jesus. This rehabilitation of Judas is remarkable, but as with other Gnostic

4. Answering Islam, "The Gospel of Barnabas," www.answering-islam.org/Nehls/Answer/barnabas.html.

writings, the authenticity of authorship is dubious, plus it still suffers from being entirely contrary to what is taught in actual biblical books.

Other Publications

The *Book of Enoch* falls into a different category from the pseudipigraphal or apocryphal works listed above. Although it is an intertestamental book, it is not part of the official Apocrypha. No books from the official Apocrypha are quoted in the New Testament, but there is a quote from the *Book of Enoch*; Jude quotes a prophecy of Enoch (see verses 14–15), taken from Enoch 1:9. It should be noted that the inclusion of such a quotation in a canonical work does not qualify the rest of the *Book of Enoch* to be part of the canon of Scripture. A similar example is that Paul quotes Greek poets in his address at Mars Hill in Athens (Acts 17). Clearly, the inclusion of this particular prophecy of Enoch proves this individual prophecy to be inspired, but it is not possible therefore to assume inspiration of any of the rest of the book.

A similar claim of authority is sometimes made for the *Book of Jasher*. This book is mentioned in the Bible twice. It is referred to in Joshua 10:13 and again in 2 Samuel 1:18. The title literally means "the book of the upright one." This book is, however, lost, and this loss would itself seem to underline that it is not an inspired, canonical book. Once again, the mention in the Bible of extrabiblical literature does not in itself add any authenticity to that literature. Numerous manuscripts have been published claiming to be the actual *Book of Jasher*. The most well known of these was published by the Church of Jesus Christ of Latter Day Saints. Another example of their literature is discussed below.

The popular name for the Latter Day Saints' Church is *Mormonism*. This name derives from their main "holy" book, the *Book of Mormon*. Many Christians have written detailed criticism of this work, so this paragraph can do no more than scratch the surface.[5] Suffice it to say that there are many reasons why the *Book of Mormon* cannot be accepted as genuine Scripture. The teenaged "prophet" Joseph Smith supposedly translated it

5. I would personally recommend J. Ankerberg and J. Weldon, *Behind the Mask of Mormonism* (Eugene, OR: Harvest House, 1996).

from gold plates. These plates have conveniently vanished. It is remarkable, therefore, that some passages of this book quote word for word not just from the Bible, but from a specific translation of the Bible — the KJV. If the book were genuinely inspired, one might expect it to include the same material. But for the wording to be identical to a specific English translation, when the OT was in Hebrew and the *Book of Mormon* supposedly in some other language, is beyond coincidence — for example, compare Isaiah 53:5 from the KJV with Mosiah 14:5. Even the (noninspired) verse divisions are identical, proving that the *Book of Mormon*, far from translating God's words from gold plates, is, in fact, just made up while using direct copies from books such as the KJV Bible.

The Watchtower Bible and Tract Society, or Jehovah's Witnesses, have published a number of magazines (*Watchtower, Awake,* etc.) and books, without which, they claim, it is impossible to interpret the Bible correctly. Although they claim to believe only the Bible, in practice, their religion has added to God's words. Not only that, but it has changed God's Word to suit its own ends. For example, their *New World Translation* of the Bible famously renders John 1:1 as, "In the beginning was the Word, and the Word was with God and the Word was *a god*" (emphasis mine). This use of the term *a god* is in contradiction to all accepted translations, and indeed is contrary to the rabbinical concept of the *Mamre* (or Word of God), to which John, under inspiration, was alluding. As with Mormon literature above, there is a great deal more to be said on the subject of Watchtower literature.[6]

Conclusion

From Edwards and Bock we have seen that the Gnostic documents are of dubious authenticity, not having been written by the authors claimed for them. Secondly, we have seen that their teaching fails the internal self-consistency test, as the documents contain teaching that is counter to what is taught in the accepted canon of Scripture.

6. For more information, see Ron Rhodes, *Reasoning from the Scriptures with the Jehovah's Witnesses* (Eugene, OR: Harvest House Publishers, 1993).

At Answers in Genesis, we understand that the Bible is under severe attack in today's world. Most of that attack seems to be centered on the Book of Genesis, but this is not an exclusive attack. What better way to undermine our belief in Scripture than to produce extra books, outside of the Bible, claiming that their omission from the Bible was merely due to fourth-century politics.

Neither the Old Testament Apocrypha nor the so-called missing gospels have any right to be treated as Scripture. Their authorship is dubious, their quotability negligible, and their agreement with the rest of Scripture nonexistent. Moreover, the argument about the listing of the canon not occurring until the third or fourth centuries is fallacious. As early as A.D. 90, verses from New Testament books were being quoted and referred to as Scripture.

The reader can be sure to have confidence in God's Word. It is all true — all 66 books of the accepted canon. For those who would disbelieve parts of the Bible, there is a warning. For those who would like to study all these other possible ways to God, the same warning applies:

> Every word of God is pure; He is a shield to those who put their trust in Him. Do not add to His words, lest He rebuke you, and you be found a liar (Proverbs 30:5–6).

23

Aren't Millions of Years Required for Geological Processes?

DR. JOHN WHITMORE

Geology became established as a science in the middle to late 1700s. While some early geologists viewed the fossil-bearing rock layers as products of the Genesis flood, one of the common ways in which most early geologists interpreted the earth was to look at present rates and processes and assume these rates and processes had acted over millions of years to produce the rocks they saw. For example, they might observe a river carrying sand to the ocean. They could measure how fast the sand was accumulating in the ocean and then apply these rates to a sandstone, roughly calculating how long it took sandstone to form.

Similar ideas could be applied to rates of erosion to determine how long it might take a canyon to form or a mountain range to be leveled. This type of thinking became known as *uniformitarianism* (the present is the key to the past) and was promoted by early geologists like James Hutton and Charles Lyell.

These early geologists were very influential in shaping the thinking of later biologists. For example, Charles Darwin, a good friend of Lyell, applied slow and gradual uniformitarian processes to biology and developed the theory of naturalistic evolution, which he published in the *Origin of Species* in 1859. Together, these early geologists and biologists used uniformitarian theory as an atheistic explanation of the earth's rocks and biology, adding millions of years

Figure 1. Layered sedimentary rocks from the Grand Canyon, Arizona. Photo by John Whitmore.

to earth history. The earlier biblical ideas of creation, catastrophism, and short ages were put aside in favor of slow and gradual processes and evolution over millions of years.

This chapter will document that geological processes that are usually assumed to be slow and gradual can happen quickly. It will document that millions of years are not required to explain the earth's rocks, as Hutton, Lyell, Darwin, and so many others have assumed.

Rapid Lithification of Sedimentary Rock

Long periods of time are not required to harden rock. Sedimentary rock generally consists of sediment (mud, sand, or gravel) that has been turned into rock. Sedimentary rocks include sandstones, shales, and limestones. Sedimentary rock is usually formed under water and is easy to recognize because of its many layers. A familiar example would be the layered rocks of the Grand Canyon (figure 1).

Layers in sedimentary rocks can be seen at small scales too, like the finely laminated beds from the Green River Formation in Wyoming (figure 2). When sediment turns into rock, or becomes hard, we say the sediment has become *lithified*. Lithification occurs during sediment compaction (which drives out water) and cementation, or gluing together of the sedimentary grains. The process of lithification is not time dependent, but rather dependent upon whether the rock becomes compacted or not and whether a source of cement is present (usually a mineral like quartz or calcite). If these conditions are met, sediment can be turned rapidly into rock.

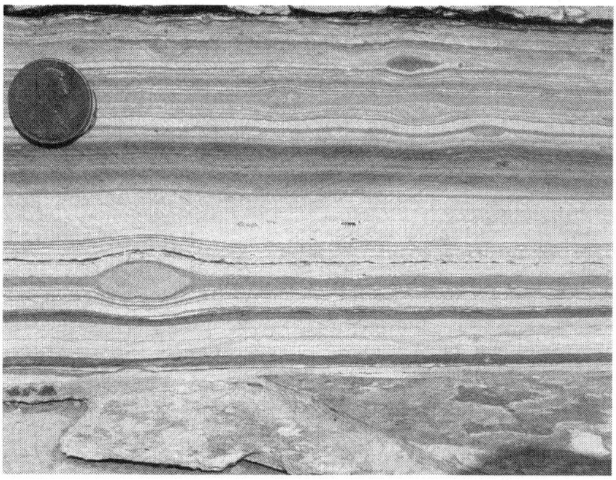

Figure 2. Finely laminated sedimentary layers from the Green River Formation, Wyoming. The U.S. penny is for scale (1.9 cm diameter). The dark oblong-shaped objects between the laminations are fish coprolites (feces). As many as ten laminations per mm can occur in these rocks. Photo by John Whitmore.

Many examples of rock forming rapidly have been reported in the creationist literature: a clock (figure 3), a sparkplug, and keys have all been found in cemented sedimentary rock. Also a hat and a bag of flour[1] have been found petrified. Examples of bolts, anchors, and bricks found in beach rock have also been reported.[2] All of these examples show that sediment and other materials can be hardened within a relatively short time span. In many of these examples, rock probably formed as microbes (microscopic bacteria and other small organisms) precipitated calcite cement, which in turn bound sediments together and/or filled pore

Figure 3. Remains of a clock encased in sedimentary rock. It was found on a beach along the coast of Washington state by Dolores Testerman.

1. Tas Walker, "Petrified Flour," *Creation*, December 2000, p. 17.
2. K.A. Rasmussen, I.G. Macintyre, L. Prufert, and V.V. Romanovsky, "Late Quaternary Coastal Microbialites and Beachrocks of Lake Issyk-Kul, Kyrgyzstan; Geologic, Hydrographic, and Climatic Significance," *Geological Society of America Abstracts with Programs* 28, no. 7 (1996): 304.

spaces. Examples of rapid lithification of this type include limestones that have been cemented together on the ocean floor.[3]

Rapid Formation of Thin, Delicate Rock Layers

Thin, delicate rock layers don't necessarily represent quiet, docile sedimentary processes; thin layers of rock can be formed catastrophically. On May 18, 1980, Mount St. Helens violently erupted. It was one of the most well-studied and scientifically documented volcanic eruptions in earth history, both by conventional scientists[4] and creationists.[5]

The volcano remained geologically active during the months and years following the 1980 eruption. Fresh lava is still oozing out of the volcano today. During the violent eruptions of the volcano, pyroclastic material (hot volcanic ash and rock) was thrown from the volcano with hurricane force velocity. One of the most fascinating discoveries following the eruption was that some of these pyroclastic deposits, those that contained fine volcanic ash particles, were thinly laminated.[6] When geologists see thin layers like this (figure 4), they usually assume that slow, delicate processes formed the layers (like mud settling to a lake bottom). However, in this case, the layers were formed during a catastrophic volcanic eruption.

Figure 4. Finely laminated beds produced during a violent pyroclastic flow from Mount St. Helens on June 12, 1980. Photo by Steve Austin, copyright 1989, Institute for Creation Research; used by permission.

3. J.A.M. Kenter, P.G. Della, and P.M. Harris, "Steep Microbial Boundstone-dominated Platform Margins; Examples and Implications," *Sedimentary Geology* 178, no. 1-2 (2005): 5–30.
4. For example, see P.W. Lipman and D.R. Mullineaux, eds., *The 1980 Eruptions of Mount St. Helens, Washington*, U.S. Geological Survey Professional Paper 1250 (Washington, D.C.: United States Government Printing Office, 1981).
5. S.A. Austin, ed., *Grand Canyon: Monument to Catastrophe* (Santee, CA: Institute for Creation Research, 1994), p. 284; H.G. Coffin, "Erect Floating Stumps in Spirit Lake, Washington," *Geology* 11 (1983): 298–299.
6. S.A. Austin, "Mt. St. Helens and Catastrophism," *Impact*, July 1986, online at www.icr.org/article/261/. Laminations are thin layers of sediment, usually a few millimeters or less.

Other types of thin, delicate rock layers can also form quickly too. Fossil fish are very abundant in the thin, laminated mudstones of the Green River Formation of Wyoming (figure 2). After death, fish rot very quickly. Scales and flesh can slough off within a matter of days, and fish can completely disappear within a week or two.[7] In order for the Green River fish to be preserved as well as they are, it would have been necessary for a thin layer of calcite mud to cover the fish immediately after death (figure 5).

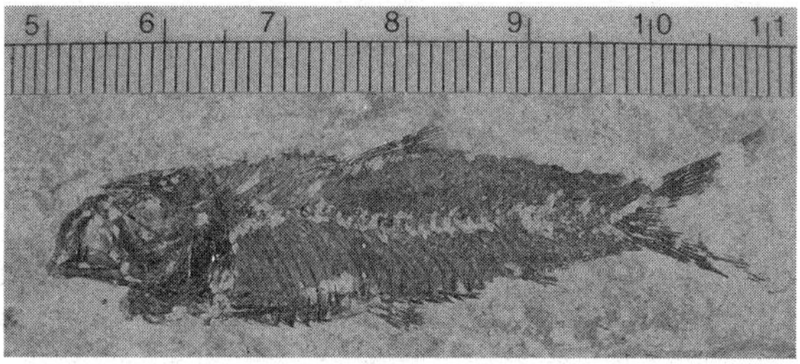

Figure 5. A well-preserved fish *(Knightia)* from the Green River Formation, Wyoming. In order for fish to be preserved like this, before major decay ensues, the fish must be buried within days of death. Scale in cm. Photo by John Whitmore.

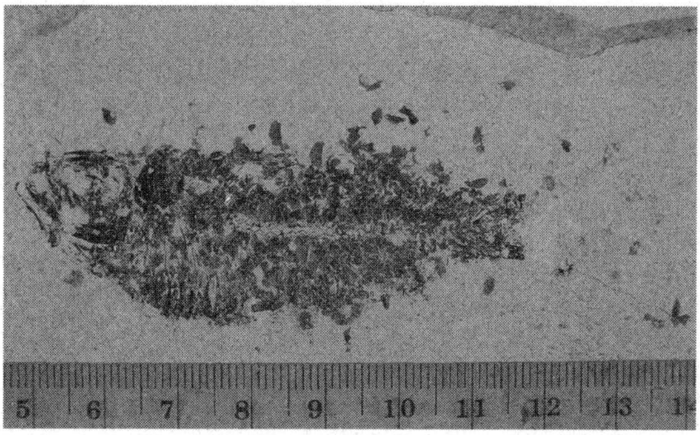

Figure 6. A fish *(Diplomystus)* that decayed for several days before burial, Green River Formation, Wyoming. Note the sloughed scales. Burial a few days after death, by a thin layer of calcite mud, arrested the decay and prevented the fish from complete destruction. Scale in cm. Photo by John Whitmore.

These thin layers of mud are what make up the thin, laminated layers of the Green River Formation. If a fish is not covered immediately, but several days after its death, scales will slough off and scatter around the fish carcass (figure 6). Because many of the layers in the Green River Formation contain well-preserved fish, we can conclude that many of layers were formed within a day or two. A study of fish coprolites (feces) also concluded that the thin layers must have formed quickly.[8] The Green River Formation was probably made in a post-Flood lake setting where sediments were accumulating rapidly.[9] These few examples of thin layers being made quickly

7. J.H. Whitmore, "Experimental Fish Taphonomy with a Comparison to Fossil Fishes," (PhD diss., Loma Linda, CA: Loma Linda University, 2003).
8. D.A. Woolley, "Fish Preservation, Fish Coprolites and the Green River Formation," *TJ* 15, no. 1 (2001): 105–111.
9. J.H. Whitmore, "The Green River Formation: A Large Post-Flood Lake System," *Journal of Creation* 20, no. 1 (2006): 55–63.

does not mean that all thinly laminated rock layers have formed quickly; it shows that some thinly laminated layers can form quickly.

Rapid Erosion

Erosion can happen catastrophically, at scales that are difficult for us to imagine. When standing along the edge of a canyon and seeing a river in the bottom, one is inclined to imagine that the very river in the bottom of the gorge has cut the canyon over long periods of time. However, geologists are realizing that many canyons have been cut by processes other than rivers that currently occupy canyons.

Massive erosion during catastrophic flooding occurs by several processes. This includes abrasion,[10] hydraulic action,[11] and cavitation.[12] The "Little Grand Canyon" of the Toutle River was cut by a mudflow on March 19, 1982, that originated from the crater of Mount St. Helens. The abrasive mudflow cut through rockslide and pumice deposits from the 1980 eruptions. Parts of the new canyon system are up to 140 feet deep.

Engineer's Canyon was also cut by the mudflow and is 100 feet deep. There is a small stream in the bottom of Engineer's Canyon (figure 7). One would be inclined to think that this stream was responsible for cutting the canyon over long periods of time if one did not know the canyon was cut catastrophically by a mudflow. In this case, the canyon is responsible for the stream; the stream is *not* responsible for the canyon.

Other large canyons and valleys are known to have been cut catastrophically as well. One of the most famous examples is the formation of the Channeled

10. Abrasion is wearing away of bedrock by particles that are being carried in the water or along the stream bottom. As rocks and sand are being carried along, they grind away the bedrock on the stream bottom. The process is similar to smoothing a piece of wood with sandpaper.
11. Hydraulic action is erosion of bedrock by the shee r force or energy of the water. Water moving at great speeds can work its way into cracks and force rocks apart, can slam boulders against a cliff face, causing rocks to crumble, and can pluck large pieces of bedrock from the stream bottom.
12. Cavitation is erosion by exceedingly rapidly moving water that creates vacuum bubbles as it flows across imperfections or depressions in a bedrock surface. As the vacuum bubbles implode (collapse violently in on themselves), they can destroy the bedrock below them, acting like sledgehammer blows. Cavitation has been known to quickly deteriorate bedrock, cement, and even steel. For example, rapidly rotating submarine propellers can create vacuum bubbles that destroy the propeller and the rudder behind it, removing large chunks of steel. A concrete spillway tunnel was damaged by cavitation in the Glen Canyon Dam in 1983. Cavitation produced a 32- x 40- x 150-foot hole in the bottom of a 40-foot diameter, 3-foot thick, steel-reinforced concrete spillway (Austin, *Grand Canyon: Monument to Catastrophe*, p. 104–107).

Figure 7. Engineer's Canyon, Mount St. Helens, Washington. The canyon was cut by a mudflow originating from the crater of the volcano on March 19, 1982. The cliff on the left is about 100 feet high. Note the small stream in the bottom of the canyon. In this case, *the stream did not form the canyon,* the canyon came first and *is responsible for the stream* being there! Photo by Steve Austin, copyright 1989, Institute for Creation Research; used by permission.

Scabland[13] of eastern Washington state. The catastrophic explanation of the enigmatic topography is now well accepted, but when it was first proposed in the 1920s by J Harland Bretz,[14] it was radical. The idea was not well accepted until nearly 50 years later, in 1969.

Bretz was trying to explain a whole series of deep, abandoned canyons (cut in hard, basaltic bedrock), dry waterfalls, deep plunge pools, hanging valleys, large stream ripples, gravel bars, and large exotic boulders. The Scabland formed as a glacier blocked the Clark Fork River in Idaho during the Ice Age. The glacially dammed river caused water to back up and form a huge lake (Lake Missoula) in western Montana, in places 2,000 feet deep!

13. The Scablands are a whole series of deep, abandoned canyons, hundreds of feet deep, cut into hard, basaltic bedrock.
14. See the following papers by JH. Bretz all in the *Journal of Geology*, "The Channeled Scablands of the Columbia Plateau," 31 (1923): 617–649; "Alternative Hypothesis for Channeled Scabland I," 36 (1928): 193–223; "Alternative Hypothesis for Channeled Scabland II," 36 (1928): 312–341; "The Lake Missoula Floods and the Channeled Scabland," 77 (1969): 505–543.

Figure 8. Dry Falls, near Coulee City, Washington. This is part of Grand Coulee, a canyon that is 50 miles long and as much as 900 feet deep, cut during the catastrophic Missoula Flood. The flood water poured over the lip of this 350-foot escarpment in the center of the photo, at five times the width of Niagara Falls. The lakes are filled plunge pools (300 feet deep) cut by water cascading over the cliff. Photo by John Whitmore.

Eventually, the ice dam burst, releasing water equivalent in volume to Lakes Erie and Ontario combined. The water rushed through Idaho and into eastern Washington, carving the Scabland topography. Hard basaltic bedrock was rapidly cut by abrasion, hydraulic action, and cavitation (figure 8). As the water drained into the Pacific Ocean, it created a delta more than 200 mi^2 in size. It took Lake Missoula about two weeks to drain. It has been estimated that at peak volume, the flood represented about 15 times the combined flow of all the rivers in the world![15] Catastrophic floods of this magnitude were unthinkable at the height of uniformitarian geology in the early 1900s. Today, they are becoming more widely accepted as explanations of large parts of the earth's topography.[16]

The origin of the Grand Canyon has been a topic of much speculation. Conventional geologists have not reached any consensus on its origin. Dr. Steve Austin, of the Institute for Creation Research, published in 1994 that the Grand Canyon was cut by a catastrophic flood that originated from post-Flood lakes ponded behind the Kaibab Upwarp.[17] In 2000, a symposium was

15. An excellent creationist summary on the formation of the Channeled Scabland region can be found in M.J. Oard, "Evidence for Only One Gigantic Lake Missoula Flood," *Proceedings of the Fifth International Conference on Creationism*, ed. R.L. Ivey Jr. (Pittsburgh, PA: Creation Science Fellowship, 2003), p. 219–231.
16. For example, see I.P. Martini, V.R. Baker, and G. Garzen, eds., *Flood and Megaflood Processes and Deposits: Recent and Ancient Examples*, Special Publication 32, International Association of Sedimentologists (Oxford: Blackwell Science, 2002).
17. Austin, *Grand Canyon: Monument to Catastrophe*, p. 83–110. Whitmore and Austin discussed and independently originated this idea in 1985, while Whitmore was a graduate student at ICR. The first person to originate this idea may have been E. Blackwelder in 1934 (GSA Bulletin, v. 45, p. 551–566).

convened in Grand Canyon National Park to discuss the canyon's origin. One paper[18] was published that was similar to Austin's idea, although the authors gave him no credit. Evidence in favor of the lake failure hypothesis for the catastrophic carving of the Grand Canyon is growing.

Recent work from the Anza Borrego Desert of California also supports this theory.[19] Austin believes that several lakes ponded behind the Kaibab Upwarp, containing a volume of about 3,000 mi^3 of water, about three times the volume of Lake Michigan,[20] or about six times the volume of Lake Missoula. Austin proposed that the lakes drained because the limestones of the Kaibab Upwarp, which held back the ponded water and developed caves (through solution by carbonic acid), catastrophically piping the water out of the lakes, cutting the canyon.

Rapid Fossil Formation[21]

When an organism is turned into stone (i.e., fossilized), the process usually must happen rapidly, or the organism will be lost to decay. *Taphonomy* is a relatively new branch of geology that studies everything that happens from the death of an organism to its inclusion in the fossil record. Many experiments have been performed to see what happens to all types of animal carcasses in all types of settings including marine, freshwater, and terrestrial settings.

The goal of many of these experiments is to make actualistic taphonomic observations so the fossil record can be better understood. One common theme throughout many of these experiments is rapid disintegration of soft animal tissue. In the absence of scavengers, bacteria and other microbes can rapidly digest animal carcasses in nearly all types of environments. For example, I have documented that fish can completely disintegrate in time frames from days to weeks in both natural and laboratory settings under all types of variable

18. N. Meek and J. Douglass, "Lake Overflow: An Alternative Hypothesis for Grand Canyon Incision and Development of the Colorado River" in *Colorado River Origin and Evolution*, eds. R.A. Young and E.E. Spamer, Proceedings of a Symposium held at Grand Canyon National Park in June 2000 (Grand Canyon, AZ: Grand Canyon Association, 2001), p. 199–204.
19. R.J. Dorsey, A. Fluette, K. McDougall, B.A. Housen, S.U. Janecke, G J. Axen and C.R. Shirvell, "Chronology of Miocene-Pliocene Deposits at Split Mountain Gorge, Southern California: A Record of Regional Tectonics and Colorado River Evolution," *Geology* 35, no. 1 (2007): 57–60.
20. Austin, *Grand Canyon Monument to Catastrophe*, p. 104.
21. An expanded version of this section can be found in Whitmore, "Fossil Preservation," in *Rock Solid Answers: Responses to Popular Objections to Flood Geology*, eds. M.J. Oard and J.K. Reed (Green Forest, AR: Master Books, in press).

conditions (temperature, depth, oxygen levels, salinity, and species).[22] The taphonomic literature has shown this is generally true for many other types of organisms as well.[23]

Simply put, in order for an animal carcass to be turned into a fossil, it must be sequestered from decay very soon after death. The most common way for this to happen is via deep rapid burial so the organism can be protected from scavengers that may churn through the sediment in search of nutrients. Many fossil deposits around the world are considered to be *Lagerstätten* deposits (like the Green River Formation), or deposits that contain abundant fossils with exceptional preservation. It is widely recognized that most of these deposits were formed by catastrophic, rapid burial of animal carcasses.[24]

Common experience tells us that soft tissues disappear quickly if something doesn't happen to prevent their decay. However, what about the hard parts of organisms, like clam or snail shells? Shouldn't they be able to last almost indefinitely without being buried? Numerous experiments have been completed, watching what happens to shells on the ocean floor over time.[25] Not surprisingly, these experiments have shown that thick, durable shells last longer than thin, fragile shells.

If the fossil record has accumulated by slow gradual processes, like those that are occurring in today's oceans, then the fossil record should be biased toward thick, durable shells and against thin, fragile shells. This was exactly the hypothesis that a recent paper tested.[26] The authors used the online Paleobiology Data Base, consisting of extensive fossil data from all over the world and throughout geologic time. Contrary to their expectations, they found thin, fragile material is just as likely to be found in the fossil record as thick, durable material. A reasonable interpretation of this finding (which the authors did not consider) is that much of the fossil record was produced catastrophically! This finding supports the hypothesis that much of the record was produced rapidly, during the Flood.

---

22. See reference 11.
23. For good reviews of the literature see S.M. Kidwell and K.W. Flessa, "The Quality of the Fossil Record: Populations, Species, and Communities," *Annual Reviews of Ecology and Systematics* 26 (1995): 269–299; or P.A. Allison and D.E.G. Briggs, eds., *Taphonomy: Releasing the Data Locked in the Fossil Record* (New York: Plenum Press, 1991).
24. C.E. Brett and A. Seilacher, "Fossil Lagerstätten: A Taphonomic Consequence of Event Stratification," in *Cycles and Events in Stratigraphy*, eds. G. Einsele, W. Ricken, and A. Seilacher (Berlin: Springer-Verlag, 1991), p. 283–297.
25. See reference 27.
26. A.K. Behrensmeyer, F.T. Fursich, R.A. Gastaldo, S.M. Kidwell, M.A. Kosnik, M. Kowalewski, R.E. Plotnick, R.R. Rogers, and J. Alroy, "Are the Most Durable Shelly Taxa also the Most Common in the Marine Fossil Record?" *Paleobiology* 31 (2005): 607–623.

Rapid Coal Formation

Coal does not take long periods of time to form. Coal forms from peat, which is highly degraded wood and plant material. Peat looks much like coffee grounds or peat moss. During the Flood, large quantities of peat were likely produced and buried as a result of pre-Flood vegetation being ripped up and destroyed.

The extensive coal beds we find throughout the world may have also been the result of pre-Flood floating forests that were destroyed and buried.[27] Coal has been produced experimentally in the laboratory from wood and peat.[28] Most of these experiments have used reasonable geologic conditions of temperature (212–390°F, 100–200° C) and pressure (to simulate depth of burial). These experiments have succeeded in producing coal in just weeks of time. It appears time is probably not a significant factor in coal formation. The most important factors appear to be the quality of the organic material (peat), heat, and pressure (depth of burial).

Rapid Formation of Salt Deposits

Salt deposits can form in other places and in other ways besides large salt lakes that evaporate over long periods of time (like the Great Salt Lake in Utah or the Dead Sea in Israel). Geologists have traditionally interpreted thick salt deposits as *evaporites.* In other words, they picture a large basin of seawater (like the Mediterranean Sea) being enclosed and sealed off from the surrounding ocean. The confined salt water evaporates, forming a thick deposit of salt on the bottom of the basin.

Conventional scientists have recognized that this model is fraught with many paradoxes and unresolved problems.[29] Recently, a new theory of salt formation

27. K.P. Wise, "The Pre-Flood Floating Forest: A Study in Paleontological Pattern Recognition," in *Proceedings of the Fifth International Conference on Creationism*, ed. R.L. Ivey Jr. (Pittsburgh, PA: Creation Science Fellowship, 2003), p. 371–381.
28. Many experiments in "artificial coalification" have been carried out. A few examples are: W.H. Orem, S.G. Neuzil, H.E. Lerch and C.B. Cecil, "Experimental Early-stage Coalification of a Peat Sample and a Peatified Wood Sample from Indonesia," *Organic Geochemistry* 24, no. 2 (1996): 111–125; A.D. Cohen and A.M. Bailey, "Petrographic Changes Induced by Artificial Coalification of Peat: Comparison of Two Planar Facies (Rhizophora and Cladium) from the Everglades-mangrove Complex of Florida and a Domed Facies (Cyrilla) from the Okefenokee Swamp of Georgia," *International Journal of Coal Geology* 34 (1997): 163–194; S. Yao, C. Xue, W. Hu, J. Cao, C. Zhang, "A Comparative Study of Experimental Maturation of Peat, Brown Coal and Subbituminous Coal: Implications for Coalification," *International Journal of Coal Geology* 66 (2006): 108–118.
29. J.K. Warren, *Evaporites: Their Evolution and Economics* (Oxford: Blackwell Science, 1999).

has been proposed that overcomes some of these difficulties.[30] This theory points out that salt is not very soluble[31] at high temperatures and pressures. These situations are common near deep-sea hydrothermal vents. The authors cite examples from the Red Sea and Lake Asale (Ethiopia) where these situations exist and are associated with abundant salts. Several times throughout the paper, the authors cite that rapid deposition of the salt with accompanying rapid sedimentation rates are necessary conditions for the salt to be preserved. If the salt is not rapidly covered, it will dissolve back into the seawater when the conditions change.

Rapid Coral Reef Formation[32]

Under certain conditions, coral reefs can grow rapidly. Modern coral reefs are often small accumulations of corals, coralline algae, and other organisms that secrete calcium carbonate (calcite, the main ingredient of limestone) exoskeletons. However, some can be massive and thick, like the Great Barrier Reef (thickness of 180 feet [55 m])[33] off the coast of Australia or Eniwetok Atoll[34] (thickness of 4,590 feet [1,400 m])[35] in the Marshall Islands of the Pacific. Some have argued that because of the slow growth rate of corals, large reefs need tens of thousands of years to grow.[36] Corals, which build coral reefs, have been reported to grow as much as 4 to 17 inches (99–432 mm) per year.[37]

Large coral accumulations have been found on sunken World War II ships after only several decades.[38] *Acropora* colonies have reached 23–31 inches (60–80

30. M. Hovland, H.G. Rueslåtten, H.K. Johnsen, B. Kvamme and T. Kuznetsova, "Salt Formation Associated with Sub-surface Boiling and Supercritical Water," *Marine and Petroleum Geology* 23 (2006): 855–869.
31. If something is not very soluble, it means that it can't dissolve easily, or it will come out of solution easily and form a solid precipitate.
32. An expanded version of this section can be found in: Whitmore, "Modern and Ancient Reefs," in *Rock Solid Answers: Responses to Popular Objections to Flood Geology*, eds. M.J. Oard and J.K. Reed (Green Forest, AR: Master Books, in press).
33. P. Read and A. Snelling, "How Old Is Australia's Great Barrier Reef?" *Creation Ex Nihilo*, November 1985, p. 6–9.
34. An atoll is a circular reef with a central lagoon that rises from the deep ocean floor, not the continental shelf like the Great Barrier Reef of Australia. It has been documented that most atolls sit on volcanic pedestals.
35. H.S. Ladd and S.O. Schlanger, "Drilling Operations on Eniwetok Atoll," *U.S. Geological Survey Professional Paper* 260-Y (1960): 863–903.
36. D.E. Wonderly, *God's Time-Records in Ancient Sediments* (Hatfield, PA: Interdisciplinary Biblical Research Institute, 1977, reprinted in 1999 with minor corrections).
37. A.A. Roth, *Origins*, (Hagerstown, MD: Review and Herald Publishing Association, 1998), p. 237.
38. S.A. Earle, "Life Springs from Death in Truk Lagoon," *National Geographic* 149, no. 5 (1976): 578–603.

cm) in diameter in just 4.5 years in some experimental rehabilitation studies.[39] At the highest known growth rates, the Eniwetok Atoll (the thickest known reef at 4,590 feet [1,400 m]) would have taken about 3,240 years to rise from the ocean floor. However, *coral growth rate* is not equal to *reef growth rate*; it is usually much less. Reef growth is a balance between constructive and destructive processes and has proved particularly difficult to measure. Reefs are constructed by coral growth and sediment, which settles and becomes cemented between reef organisms.

Modern reefs are destroyed by a number of processes, including active bioeroders (parrotfish, sea urchins), chemical dissolution, boring organisms (sponges, clams, and various worms), tsunamis, and storm waves. Reef growth occurs by the addition of mass, particularly from corals. Reef volume increases as living animals and their dead remains become cemented together with sediments to form the reef. Reef growth slows or even stops as the reef reaches sea level, because the reef organisms need to be submerged in water. Hence, the growth rate of a reef is slower than that of fast-growing corals.

So how might a thick reef, like the Eniwetok Atoll, have grown from the ocean floor since the time of the Flood? The Eniwetok Atoll is not made completely of corals that have grown on top of each other. Drilling operations into the atoll have shown that a significant amount of the material (up to 70 percent of the bore hole) was "soft, fine, white chalky limestone,"[40] *not* well-cemented reef limestone. It may be significant that this atoll, along with many of the other atolls in the western Pacific, ultimately rise from volcanic pedestals. It is known that heat coming from these volcanoes draws cold, nutrient-rich water into the cavernous atoll framework and circulates it upward, through the atoll via convection. This process is called geothermal endo-upwellling[41] and helps provide nutrients to the reef organisms near sea level.

Here is a possible scenario of how the Eniwetok Atoll may have become so thick in the few thousand years since the Flood (figure 9). The reef began as a volcanic platform. Carbonates (limestones) began to accumulate on the platform as the result of bacteria and other organisms that can precipitate calcite, especially in volcanically warmed water. This produced much of the "soft, fine,

39. H.E. Fox, "Rapid Coral Growth on Reef Rehabilitation Treatments in Komodo National Park, Indonesia," *Coral Reefs* 24 (2005): 263.

40. See reference 38.

41. F. Rougerie and J.A. Fagerstrom, "Cretaceous History of Pacific Basin Guyot Reefs: A Reappraisal Based on Geothermal Endo-upwelling," *Palaeogeography, Palaeoclimatology, Palaeoecology* 112 (1994): 239–260; A.H. Saller and R.B. Koepnick, "Eocene to Early Miocene Growth of Eniwetok Atoll: Insight from Strontium-isotope Data," *Geological Society of America Bulletin* 102, no. 3 (1990): 381–390.

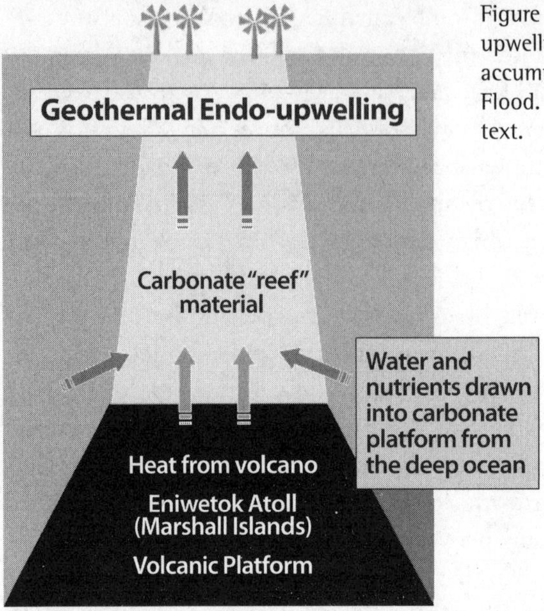

Figure 9. How geothermal endo-upwelling might explain thick "reef" accumulations since the time of the Flood. The process is explained in the text.

chalky limestone" found within the reef. Carbonate-producing organisms (like corals) were brought to the platform as small larval forms, transported by ocean currents. This explains the occasional occurrence of various corals and mollusks found within the deeper parts of the drill core. The volcanic heat source allowed the carbonate mound to grow, deep below sea level, and the process of geothermal endo-upwelling to begin. The combination of nutrient supply and heat may have allowed the carbonate mound to grow much faster than observed coral reef growth rates today. As the carbonate mound approached sea level, shallow water reef corals were permanently established and thrived as a result of the upwelling process.

Concluding Remarks

Many modern geologists realize that most rocks contain evidence of rapid accumulation. However, the idea that the earth is millions of years old is still a common belief. So if the time is not within the rocks, where is it? Many believe the time is within the "cracks" or "hiatuses" between the rocks (see figure 10). Derek Ager, who was not friendly toward creationist ideas, explained it like this: "The history of any one part of the earth, like the life of a soldier, consists of long periods of boredom and short periods of terror."[42] He viewed most of the physical rock record as accumulating quickly (i.e., "the short periods of terror") and the breaks in between rock layers representing long periods of time (i.e., "the long periods of boredom"). In other words, the "breaks" or "cracks" are

42. D.V. Ager, *The Nature of the Stratigraphical Record*, 2nd edition, (London: MacMillan Press, 1981), p. 106–107.

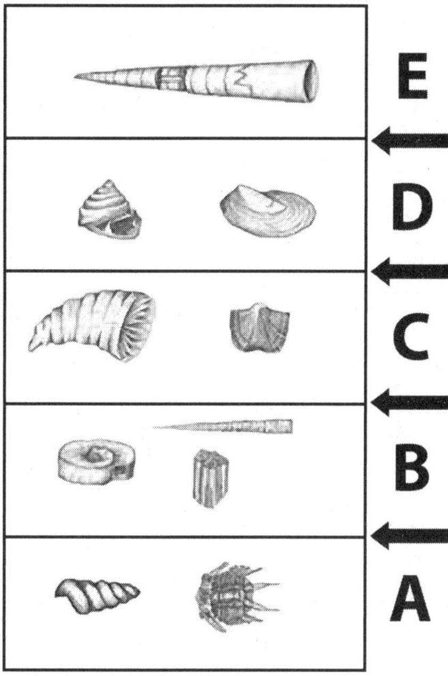

Figure 10. Today, conventional geologists still believe that the earth is millions of years old. However, they believe that individual rock layers may represent short periods of time, or "events." So where do they put all of the time? The time is placed in between the layers (at the arrows). Each event (A, B, C, D, E) represents a short period of time, but each arrow represents a long period of time, or "hiatus." During the hiatus, either perfectly flat erosion levels the surface before the next event (removing accumulated deposits), or nondeposition occurs over millions of years.

where most of the time is placed. The belief then is that these surfaces represent either long periods of nondeposition or surfaces of perfectly flat erosion. But both of these propositions have problems. For example, if a surface is exposed for long periods of time, why don't organisms churning through the mud extensively disturb the sediments below the surface? In observational studies, it is estimated that bottom-dwelling organisms can rework the annual sediment accumulation several times over![43]

This chapter has only examined a few processes in geology that are assumed to take long periods of time. There are many more issues that could be addressed. Today, ideas of uniformitarianism are fading quickly in geology. In fact, many conventional geologists would like to abandon the idea of uniformitarianism altogether, although they are careful not to advocate biblical catastrophism.[44]

Conventional geologists are recognizing that catastrophic processes can form many parts of the geologic record, and this is being widely reported in the literature.[45] The eventual nemesis will be time. Time will continue to be placed

43. D.C. Rhoads, "Rates of Sediment Reworking by Yoldia limatula in Buzzards Bay, Massachusetts, and Long Island Sound," *Journal of Sedimentary Petrology* 33, no. 3 (1963): 723–727.
44. K.J. Hsü, "Actualistic Catastrophism," *Sedimentology* 30 (1983): 3–9; P.D. Krynine, "Uniformitarianism Is a Dangerous Doctrine," *Journal of Sedimentary Petrology* 26, no. 2 (1956): 184; J.H. Shea, "Twelve Fallacies of Uniformitarianism," *Geology* 10 (1982): 455–460.
45. W.A. Berggren and J.A. Van Couvering, eds., *Catastrophes and Earth History; the New Uniformitarianism* (Princeton, NJ: Princeton University Press, 1984). This book is a

in between the rocks, not because there is evidence for it, but that is the only place left for it.[46] Conventional geological paradigms demand long periods of time be accounted for, whether there is evidence for it or not.

collection of 18 essays. Note especially the essays by S.J. Gould (chapter 1) and D.V. Ager (chapter 4).

46. One example of an attempt to place time in between rock layers is carbonate hardgrounds. These are hardened cement-like surfaces that occur on the ocean floor. It is often claimed these surfaces occur in the rock record, too, and represent surfaces where long periods of time passed. Creationists have recently begun to address hardgrounds at a classic site in Ohio: J. Woodmorappe, and J.H. Whitmore, "Field Study of Purported Hardgrounds of the Cincinnatian," *TJ* 18, no. 3 (2004): 82–92, 2004; J. Woodmorappe, "Hardgrounds and the Flood: The Need for a Re-evaluation," *Journal of Creation* 20, no. 3 (2006): 104–110.

24

Doesn't Egyptian Chronology Prove That the Bible Is Unreliable?

DR. ELIZABETH MITCHELL

Egyptology, originally expected to support the history recorded in the Old Testament, has produced a chronology that contradicts the Bible. This so-called traditional Egyptian chronology would have the pyramids predate the flood of Noah's day; such cannot be the case, for pyramids could never withstand a worldwide flood. And when traditional Egyptian chronology is used to evaluate archaeological findings, landmark events such as the mass exodus of Hebrew people from Egypt appear to have left no evidence. Such discrepancies between traditional Egyptian chronology and the Bible are used to attack the Bible's historical accuracy. Instead of simply assuming the accuracy of traditional Egyptian chronology and modifying the Bible, people should carefully examine traditional chronology to see if it is as reliable as some claim it to be.

Traditional Egyptian Chronology

Though traditional Egyptian chronology dominates modern understanding of ancient history, traditional chronology is inconsistent with the Bible. When there is a discrepancy between traditional chronology and the Bible's chronology, scholars usually ignore the Bible. Though many claim that traditional chronology is indisputable, a close look at this chronology reveals its shaky foundation. Dr. Rene Grognard of the University of Sydney says, "It is important to show the weaknesses or errors in our understanding of a theory in order to leave our minds free to think of a more acceptable alternative."[1] Before exploring an acceptable alternative to traditional Egyptian chronology, this chapter will show some of the errors it is built on.

Traditional Egyptian chronology is built on Manetho's history and the Sothic theory. In the third century B.C., Manetho compiled a list of pharaohs and the lengths of their reigns. The Sothic cycle theory assigns familiar calendar dates to those reigns. However, both Manetho's history and the Sothic theory have flaws that make them an unreliable foundation for chronology.

Manetho's History

Ptolomy II commissioned a priest named Manetho to compile a history of Egypt. Traditional Egyptian chronology bases its outlines of Egyptian dynasties on Manetho's history (see chart). However, Manetho's writings are unsuitable for establishing a reliable Egyptian chronology because Manetho's history:

- was never intended to be a chronological account of Egyptian history,
- is inconsistent with contemporary Egyptian sources.

Traditional Egyptian Chronology (simplified overview)[2]

Old Kingdom	Dynasties 1–6	2920–2770 B.C.
Great Pyramids of Giza	4th Dynasty	2600–2500 B.C.
First Intermediate Period	Dynasties 7–11	2150–1986 B.C.
Middle Kingdom	Dynasties 12–13	1986–1759 B.C.
Second Intermediate Period	Dynasties 14–17	1759–1525 B.C.
New Kingdom	Dynasties 18–20	1525–1069 B.C.
Third Intermediate Period	Dynasties 21–25	1069–664 B.C.
Late Period (Persian)	Dynasties 26–31	664–332 B.C.
Alexander the Great		332–323 B.C.
Ptolemaic Period		323–30 B.C.
Roman Period		began 30 B.C.

1. D. Mackey, "Sothic Star Dating: The Sothic Star Theory of the Egyptian Calendar," abridged thesis, Sydney, Australia, 1995; available at www.specialtyinterests.net/sothic_star.html.
2. D. Rohl, *Pharaohs and Kings: A Biblical Quest* (New York: Crown Publishers, 1995), p. 24. Dynasties are grouped in sets called Old Kingdom, Middle Kingdom, and New Kingdom. After each set is an Intermediate Period whose history is less clear. Duration of dynasties comes from Manetho. Dates come from various interpretations of the Sothic cycle. Note: Meyer, Breasted, and many others give even earlier dates.

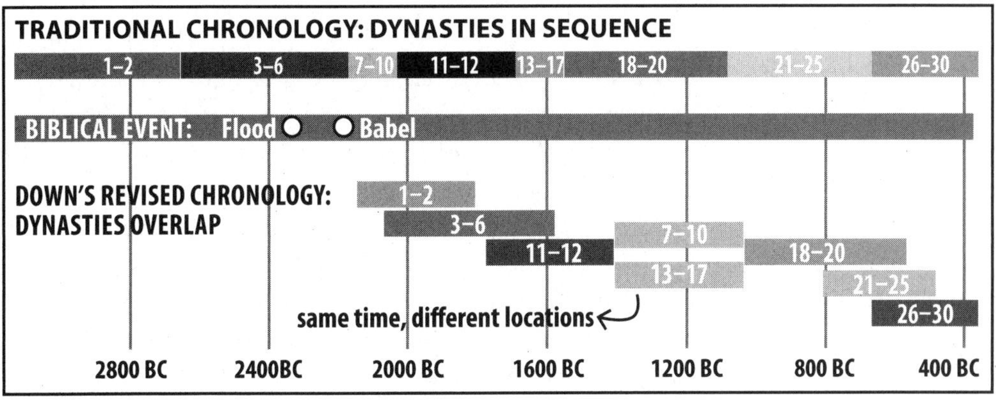

Several Egyptian pharaohs may have ruled at the same time in different regions of the land, as archaeologist David Down suggests in his revised chronology.

Manetho, whose writings only survive as a partially preserved "garbled abridgement,"[3] did not intend for his history to be a chronological account of Egyptian history. Like everyone else in the ancient world, Manetho measured time in regnal years ("in the fifth year of King So-and-So"). Eusebius, the fourth-century historian who quoted Manetho extensively, did not believe that Manetho intended for his regnal years to be added up consecutively. Eusebius says, "Several Egyptian kings ruled at the same time. . . . It was not a succession of kings occupying the throne one after the other, but several kings reigning at the same time in different regions."[4] Because Manetho's history lists the reigns of kings who ruled simultaneously, historians should not add the years of the kings' reigns together as if the kings ruled one after another.

Manetho's history is also inconsistent with contemporary Egyptian sources. Professor J. H. Breasted, author of *History of Egypt,* calls Manetho's history "a late, careless and uncritical compilation, which can be proven wrong from the contemporary monuments in the vast majority of cases, where such documents have survived."[5] Manetho's interpretation of each variation in spelling as a different king creates numerous nonexistent generations. Because Manetho's history contradicts actual Egyptian records from the time of the pharaohs, historians should not consider Manetho's history authoritative.

3. A. Gardiner, *Egypt of the Pharaohs* (Oxford: Oxford University Press, 1961), p. 46, quoted in D. Mackey's thesis. Manetho is quoted by Josephus, Eusebius, Africanus, and Syncellus.
4. J. Ashton and D. Down, *Unwrapping the Pharaohs* (Green Forest, AR: Master Books, 2006), p. 73.
5. D. Mackey, "Sothic Star Dating."

The Sothic Cycle

Eduard Meyer created the Sothic cycle in 1904 to give Egypt a unified calendar[6] that aligns Egyptian regnal years with modern historians' B.C. dates. Historians combine the Sothic cycle dates with Manetho's history to get traditional Egyptian dates. Meyer proposed that the Egyptian calendar, having no leap year, fell steadily behind until it corrected itself during the year of the "rising of Sothis." The theory says the Egyptians knew that 1,460 years were necessary for the calendar to correct itself because the annual sunrise appearance of the star Sirius corresponded to the first day of Egypt's flood season only once every 1,460 years.[7] Sothic theory claims that the Egyptian calendar was correct only once every 1,460 years (like a broken watch that is correct twice a day) and that the Egyptians dated important events from this Great Sothic Year. In reality, there is no evidence for this Sothic cycle in ancient Egypt.

The Sothic cycle is not reliable because it

- is based on contradictory starting points,
- has little historical support.

Meyer had to depend on later non-Egyptian writers to establish a starting point for his calculations, and those sources are contradictory. Censorinius, a third-century Roman writer, and Theon, a fourth-century Alexandrian astronomer, give different starting points. According to Censorinius, the Great Sothic Year occurred in A.D. 140, but according to Theon, it occurred in 26 B.C. Meyer subtracted multiples of 1,460 years from A.D. 140 and proposed 4240 B.C. as a totally certain date for the establishment of Egypt's civil calendar.[8]

The Sothic cycle finds little historical support. History gives no hint that the Egyptians regularly dated important events from the rising of Sothis. The second-century astronomer Claudius Ptolemy never mentions the rising of Sothis.[9] Furthermore, whenever Egyptian writings mention the rising of Sothis in connection with a regnal year, the pharaoh is unnamed,[10] or the reference is ambiguous.[11] For these reasons, many Egyptologists have consistently rejected Sothic-cycle-based chronology.

---

6. D. Mackey, "Fall of the Sothic Theory: Egyptian Chronology Revisited," *TJ* 17 no. 3 (2003): 70–73, available at www.answersingenesis.org/tj/v17/i3/sothic_theory.asp.
7. Rohl, *Pharaohs and Kings: A Biblical Quest,* p. 129–130.
8. Mackey, "Fall of the Sothic Theory: Egyptian Chronology Revisited."
9. Ibid.
10. Ibid.
11. Rohl, *Pharaohs and Kings: A Biblical Quest,* p. 134–135. The famous Ebers Papyrus allegedly confirms a 1517 B.C. date for the ninth year of Amenhotep I. However, this document refers to a *monthly* rising of Sothis, an astronomical impossibility.

Discrepancies

Whenever two chronologies disagree, at least one must be wrong. Traditional Egyptian chronology disputes the Hebrew chronology recorded in the Bible as well as secular data from neighboring nations. As Damien Mackey summarized in his thesis:

> The value of any one nation's absolute chronology must ultimately depend on its ability to *integrate with all known data from other regions as well.* It would be useless to establish a complete system of chronology that can exist only in isolation, but that cannot stand up to scrutiny by comparison with other systems. For the Sothic scheme [of Egyptian chronology] to be valid — just as for Mesopotamian, Palestinian, Greek or Anatolian chronologies to be valid — *it is necessary for each period of Egyptian history to be capable of perfect alignment with any relevant period of history of one or another ancient nation.* This is most especially true in the case of Egyptian history because . . . *the historians of other nations tend to look to Egyptian chronology as the rule according to which they estimate and adjust their own chronologies*[12] (emphasis added).

Biblical Discrepancies

Traditional dates for Egyptian pyramids predate Noah's flood (see chart). Since the pyramids could not have survived a global flood, some people question the reliability of the Bible's chronology. Others use the traditional dates for the pyramids to support the idea that Noah's flood was a local flood that did not affect Egypt.[13] The pyramids do not come with labels declaring their dates, and the traditional dates used for them create an irreconcilable discrepancy with the Bible.

Bible Time-line (B.C.)

4004	2348	1491	586	4
Creation	Noah's flood	Exodus	Temple destroyed	Christ's birth

Traditional Egyptian Dates (B.C.)

3150[14] to 2920	2600 to 2500	1290
Zoser's pyramid	Great Pyramid	Exodus

12. D. Mackey, "Sothic Star Dating: The Sothic Star Theory of the Egyptian Calendar," abridged thesis, Sydney, Australia, 1995.
13. The inconsistency of the local flood idea with both science and the rest of the Bible is discussed in chapter 10 of *The New Answers Book 1* (Green Forest, AR: Master Books, 2006).
14. Earlier date comes from W. Durant, *Our Oriental Heritage* (New York: Simon and Schuster, 1954), p. 147.

Traditional dates for the Old Testament stories involving Egypt remain unconfirmed by archaeology and actually contradict Scripture. The characters of the Bible stories left no archaeological evidence of their existence in the times traditionally assigned to them. Bible-believing Egyptologists assigned these dates in error. The early Egyptologists, hoping to find the Bible confirmed in Egypt, contributed to the errors in traditional chronology by incorrectly applying the Bible in two instances. They incorrectly:

- assumed that Ramses the Great was the pharaoh of the oppression,
- identified Shoshenq as Shishak of the Bible.

The first error assigned an Exodus date inconsistent with the rest of Scripture. The second error provided support for the excessive antiquity of traditional dating. Both errors caused scholars to assign inconsistent, unsupported dates to the Bible accounts.

Scholars routinely disregard the biblical date for the Exodus.[15] As Gleason Archer says, "But notwithstanding . . . consistent testimony of Scripture to the 1445 date (or an approximation thereof), the preponderance of scholarly opinion today is in favor of a considerably later date, the most favored one at present being 1290 B.C., or about ten years after Ramses II began to reign."[16] The traditional date for Ramses II "the Great," a 19th dynasty king, is nearly two centuries after the Exodus. Because Exodus 1:11 says that the Hebrew slaves built the city Ramses, early Egyptologists assumed that Ramses II was the pharaoh who oppressed the Israelites. On that basis, most scholars assign Ramses' traditional date to the Exodus and ignore the Bible's testimony.

The name *Ramses* should not restrict the oppression to the 19th dynasty because this name is not unique to the 19th dynasty. *Ramses,* which means "son of Ra — the sun god," was a name commonly used to honor pharaohs. For instance, Ahmose, the founder of the 18th dynasty, was also called Ramses, as was a later 18th dynasty king, Amenhotep III.[17] Archaeology of the 18th and 19th dynasties shows no evidence of enslaved Israelites because the Hebrews

15. Conservative Bible scholars calculate the Exodus to have occurred sometime between 1491–1445 B.C. Solomon began to build the temple in the fourth year of his reign, in the 480th year after the Exodus from Egypt, according to I Kings 6:1. Accepted dates for the beginning of Solomon's reign, as calculated from the lengths of the reigns of Old Testament kings, range from 1015 to 970 B.C. From this data, the Exodus occurred around 1491 to 1445 B.C. The dates are confirmed by additional Scriptures. See Dr. Jones's *Chronology of the Old Testament* for a full discussion.
16. G. Archer, *A Survey of Old Testament Introduction* (Chicago, IL: Moody Press, 1994), p. 241.
17. F.N. Jones, *Chronology of the Old Testament* (Green Forest, AR: Master Books, 2004), p. 50–51.

had left Egypt centuries before. Scholars should neither assume that Ramses II was the pharaoh of the oppression nor assign his date to the Exodus.

Jean Champollion,[18] the father of Egyptology, unwittingly gave support to biblically inconsistent chronology when he erroneously identified pharaoh Shoshenq as the Shishak of the Bible. Champollion found an inscription about Shoshenq, founder of the 22nd dynasty, at the temple of Karnak. Because the names sound similar, Champollion assumed that Shoshenq was the Shishak who plundered Jerusalem in the fifth year of King Rehoboam.[19] Using the biblical date for Rehoboam as a starting point, chronologists used Manetho's list to outline the next three centuries of Egyptian history.

The two problems with Shoshenq's identification involve military strategy and phonics. According to the inscriptions, Shoshenq attacked the northern part of Israel, not Rehoboam's Jerusalem or Judah. During Rehoboam's time, Jeroboam ruled the northern kingdom. Jeroboam was Shishak's ally.[20] If Shoshenq were Shishak, then Shoshenq attacked his ally and ignored his enemy. Furthermore, the phonetics of these two pharaohs' names only sound similar in their transliterated forms, not in the original languages.[21] Because of this faulty identification of Shoshenq with Shishak, Egyptologists ignore the rest of the biblical facts relating to the geography and characters involved. Because the dates constructed from this biblical misinterpretation actually coincide with the traditional dating of the third intermediate period, many Bible scholars trust the traditional chronology even when it disputes the Old Testament.

Secular Discrepancies

Traditional Egyptian chronology disputes not only biblical chronology but also information from nonbiblical sources. Egypt's traditional dates clash with secular data in at least two areas:

- The Hittite connection with Assyrian chronology
- Carbon dating

The Hittites built a powerful empire based in Asia Minor, but scholars have to depend on dates from other ancient nations to determine Hittite chronology. *Synchronisms* are events shared by two cultures, and Egypt shares many

18. Jean Champollion translated the famous Rosetta stone, unlocking the secret of Egyptian hieroglyphics.
19. Rohl, *Pharaohs and Kings: A Biblical Quest,* p. 120–121. See 1 Kings 14:25.
20. Rohl, *Pharaohs and Kings: A Biblical Quest,* p. 122–127 and 1 Kings 11:40. Jeroboam had fled to Shishak during Solomon's lifetime.
21. Ashton and Down, *Unwrapping the Pharaohs,* p. 185.

synchronisms with the Hittites. Therefore, Egypt's erroneous dates have been assigned to the Hittites. For instance, the traditional date of 1353 B.C. for pharaoh Akhenaten's accession[22] to the throne is assigned to Hittite king Supiluliumas because Supiluliumas sent to a letter of congratulations to Akhenaten.[23] The date 1275 B.C. for the battle of Kadesh,[24] at which both Ramses II and Hittite king Muwatalli II claimed victory, comes from the traditional dates for Ramses the Great. (His dates derive from Sothic theory and Manetho's history.) Finally, when Ramses III recorded his traditionally dated 1180 B.C.[25] victory over sea people, he said that the sea people had already annihilated the Hittites. According to these Egyptian dates, the Hittites became extinct about 1200 B.C. (see chart).

Traditional Time-line (B.C.)

3150[26]	2600	1290	1275	1200
Zoser's Pyramid	Great Pyramid	Exodus	Kadesh	Hittites extinct

The Egyptian version of Hittite chronology falls apart, however, when compared to more recent Assyrian archaeological discoveries. Assyrian inscriptions record wars with the Hittites during the eighth and ninth centuries B.C., centuries after the Hittites supposedly ceased to exist. These inscriptions describe wars during the reigns of Assyrian kings Shalmaneser III and Sennacherib and even name the same Hittite kings as the Egyptian records[27] (see chart). The Assyrian time-line is consistent with well-established dates such as Nebuchadnezzar's conquest of Jerusalem. Traditional Egyptian dates must be wrong.

Problems Time-line (B.C.)

2600	2348	1275	1200	800s–700s
(trad.)	(bib.)	(trad.)	(trad.)	(Assyr.)
Great Pyramid	flood	Kadesh	Hittites extinct	Hittite/Assyrian wars

Acceptance of the biblical account of Hittite history could have prevented the incorrect dating of the Hittites even before the discovery of the Assyrian monumental inscriptions. According to 2 Kings 7:6, during Elisha's lifetime the

22. Rohl, *Pharaohs and Kings: A Biblical Quest,* p. 20.
23. Ashton and Down, *Unwrapping the Pharaohs*, p. 75.
24. *Anatolia: Cauldron of Cultures* (Alexandria, VA: Time-Life Books, 1995), p. 64.
25. Ibid, p. 69.
26. W. Durant, *Our Oriental Heritage* (New York: Simon and Schuster, 1954), p. 147.
27. Ashton and Down, *Unwrapping the Pharaohs*, p. 75–76.

Hittites were as formidable as Egypt. One explorer, Irish missionary William Wright, correctly evaluated the hieroglyphics he found in Asia Minor because he accepted the Bible's history. In 1872, despite scholarship that insisted the Hittites and the Bible were unhistorical, Wright believed that the inscriptions he had found "would show that a great people, called Hittites in the Bible, but never referred to in classic history, had once formed a mighty empire in that region."[28]

Carbon dating[29] also disputes traditional chronology. According to the *Cambridge Encyclopedia on Archaeology*:

> When the radiocarbon method was first tested, good agreement was found between radiocarbon dates and historical dates for samples of known age. . . . As measurements became more precise, however, it gradually became apparent that there were systematic discrepancies between the dates that were being obtained and those that could be expected from historical evidence [i.e., the traditional dates]. These differences were most marked in the period before about the mid-first millennium B.C., in which *radiocarbon dates appear too recent, by up to several hundred years, by comparison with historical dates.* Dates for the earliest comparative material available, reeds used as bonding between mud brick courses of tombs of Egyptians Dynasty I, about 3,100 B.C., *appeared to be as much as 600 years, or about 12% too young*[30] (emphasis added).

Just as carbon dating is more consistent with a young earth than most people realize, carbon dating is consistent with a much younger Egyptian civilization than traditional chronology claims.

Revised Chronologies

In *Centuries of Darkness,* Peter James calls traditional chronology a "gigantic academic blunder."[31] David Rohl writes, "The only real solution to the archaeological problems which have been created is to pull down the whole structure and start again, reconstructing from the foundations upward."[32] Revised chronology reflects the relationships between ancient nations more accurately

28. *Anatolia*, p. 41.
29. Carbon dating is discussed in chapter 7 of *The New Answers Book 1* (Green Forest, AR: Master Books, 2006).
30. D. Downs, "The Chronology of Egypt and Israel," from *Diggings*, available at www.biblicalstudies.qldwide.net.au/chronology_of_egypt_and_israel.html.
31. P. James, *Centuries of Darkness*, 320, quoted in Ashton and Down, *Unwrapping the Pharaohs*, p. 184.
32. Rohl, *Pharaohs and Kings: A Biblical Quest*, p. 9.

and reveals "remarkable agreement between the histories of Egypt and Israel."[33] Revised chronology bolsters the Christian's trust in the Bible and equips him with answers for a skeptical world.

Efforts to assign familiar dates to events of antiquity require a starting point, a known date. Four starting points provide secure anchors for the chronology of the Middle East. By counting both backward and forward from these four dates, the chronologist can assign familiar dates from creation to Christ[34] and combine the annals of the ancient nations to build a consistent chronology. These four anchor points are summarized on the "Starting Points" chart.

Starting Points

664 B.C. Thebes sacked	621 B.C. Lunar eclipse	605 B.C. Battle of Carchemish	586 B.C. Temple destroyed	A.D. 26 15th year of Tiberias
Taharka Dies	Nabopolassar's 5th year	Nebuchadnezzar's 1st year (*sole rex*)	Nebuchadnezzar's 19th year	Christ's 30th year

Space does not permit analysis of all the revised chronologies. A number of scholars, including Peter James, David Rohl, D.A. Courville, and David Down, have produced fine work in this area. Some begin with the Bible, while others begin with starting points such as the battle of Thebes. The Christian should only accept revised chronology that is consistent with the Bible. New evidence may someday shed new light on the identity of a pharaoh, but nothing should ever rock the Christian's faith in the trustworthiness of God's Word.

David Down, in *Unwrapping the Pharaohs*, has synthesized the work of many experts into a cohesive narrative consistent with the Bible. He points out many synchronisms between the histories of Israel and Egypt, providing a highly plausible identification for many of the characters in the Old Testament. Furthermore, his work is consistent with the history of surrounding nations and allows the Hittites to slip into their proper niche in the context of their Assyrian and Egyptian neighbors.

33. Down, "The Chronology of Egypt and Israel."

34. Jones, *Chronology of the Old Testament,* p. 23, 123, and 309. Claudius Ptolemy documented a lunar eclipse that occurred on April 15, 621 B.C. (Gregorian calendar), during the fifth year of Nabopolassar, Nebuchadnezzar's father. Counting forward gives the 605 B.C. and 586 B.C. dates. Ashurbanipal's sacking of Thebes in 664 B.C. comes from several independent ancient sources. (See Rohl, *Pharaohs and Kings: A Biblical Quest,* p. 119.) Contemporary Roman writers confirm the Tiberius date. (See Jones, *Chronology of the Old Testament,* p. 218.)

The Revision Compared to the Absolute Authority – the Bible

Synchronisms between Old Testament characters and Egypt include the following:

- Pre-Dynastic and Old Kingdom
  Mizraim, Abram
- Middle Kingdom
  Joseph, Moses
- New Kingdom
  Solomon, Rehoboam, Asa, Ahab
- Third Intermediate and Late Periods
  Hezekiah, Josiah, Jeremiah

Predynastic Egypt and Old Kingdom – the Post-Flood World

Most histories begin with the unsubstantiated notion that primitive people slowly developed civilization from rudimentary beginnings. Archaeology around the world has instead revealed advanced ancient technology without discernible periods of evolution.[35] This sudden appearance of cultures possessing advanced technology approximately 4,000 years ago is consistent with the Bible's account of the Flood, the proliferation of intelligent people on the plains of Shinar, and their subsequent scattering from the Tower of Babel.[36]

1. Mizraim's Family

Each group leaving Babel took with it whatever skills its members possessed.

Mizraim, Noah's grandson, founded Egypt around 2188 B.C., a date consistent with both biblical and secular records.[37] The Egyptians, the Sumerians, and the Mayans all retained the technology to build pyramids. Imhotep designed Egypt's first pyramid for third dynasty pharaoh Zoser. The Great Pyramid of

35. D. Chittick, *Puzzle of Ancient Man* (Newberg, OR: Creation Compass, 2006), p. 8–15.
36. Archbishop Ussher calculated the date for the Tower of Babel 2242 B.C. from Genesis and from Manetho's statement that the confusion occurred in the fifth year of Peleg's life. L. Pierce, "In the Days of Peleg," *Creation* 22 no. 1 (1999): p. 46–49, available at www.answersingenesis.org/creation/v22/i1/peleg.asp.
37. Ibid. The 12th-century historian Constantinus Manasses wrote that Egypt endured for 1,663 years. Egypt lost her independence around 526 B.C. with the Persian conquest. Hence, 2188 B.C. is a reasonable date for Egypt's founding and is consistent with a 2242 B.C. date for the Tower of Babel.

Giza, built for pharaoh Khufu of the fourth dynasty, is "the largest and most accurately constructed building in the world."[38] This pyramid required advanced optical, surveying, mathematical, and construction techniques, an impressive leap beyond the technology demonstrated in earlier pyramids.

2. Abram and Khufu's Pyramid

Abram's visit to Egypt may explain Egypt's sudden advance. Abram grew up in the advanced but idolatrous culture of Ur about three centuries after the Flood. Josephus wrote that Abram "communicated to them arithmetic, and delivered to them the science of astronomy; for before Abram came into Egypt they were unacquainted with those parts of learning; for that science came from the Chaldeans into Egypt."[39] Based on Josephus's statement, Abram's visit to Egypt may well have occurred during the fourth dynasty.

Middle Kingdom – Joseph and Moses

In contrast to the lack of evidence for an Israelite population in Egypt during the New Kingdom of Ramses' time, there is significant evidence of the Israelite presence during the Middle Kingdom. The 12th and 13th dynasties provide the backdrop for the stories of Joseph, the oppression of the Israelites, Moses, and the Exodus. The biblical dates for these events can provide dates for these dynasties (see chart).

1. Joseph as Vizier

Sesostris I of the 12th dynasty had a powerful vizier named Mentuhotep. Mentuhotep held the office of chief treasurer and wielded authority "like the declaration of the king's power."[40] "Mentuhotep . . . appears as the alter ego of the king. When he arrived, the great personages bowed down before him at the outer door of the royal palace."[41]

Compare Mentuhotep to Joseph in Genesis 41:40, 43. Furthermore, Ameni, a provincial governor under Sesostris I, had the following inscribed on his tomb: "No one was unhappy in my days, not even in the years of famine, for I had tilled all the fields of the Nome of Mah, up to its southern and northern frontiers. Thus I prolonged the life of its inhabitants and preserved the food which it produced."[42]

---

38. Ibid, p. 106.
39. Josephus, *The Works of Josephus: New Updated Edition*, book 1, chapter 8, as translated by William Whiston (Peabody, MA: Hendrickson Publishers, 1987), p. 39.
40. Ashton and Down, *Unwrapping the Pharaohs*, p. 83, quoting from James Henry Breasted's *History of Egypt*.
41. Ibid, quoting from Emille Brugsch's *Egypt Under the Pharaohs*.
42. Ibid, p. 83–84.

Ameni sounds like a man with the inside track on the agricultural forecast! Ameni's employer, vizier Mentuhotep, may have been Jacob's son Joseph.

2. Israelite Slavery

The late 12th dynasty reveals evidence for Israelite slavery. Sesostris III, the fifth king of the 12th dynasty, built cities in the delta including Bubastis, Qantir, and Ramses. The building material of choice in the Middle Kingdom was no longer stones but rather bricks composed of mud and straw.[43] A large Semitic slave population lived in the villages of Kahun and Gurob during the latter half of the 12th dynasty. On one papyrus slave list, 48 of the 77 legible names are typical of a "Semitic group from the northwest,"[44] many listed beside the Egyptian name assigned by the owner.[45] The presence of Semitic slaves in Egypt during this time is consistent with the biblical account of the oppression of the Israelites.

3. Moses' Adoption

Traditional chronology has tried to fit Moses into the 18th or 19th dynasty where there is no evidence of Semitic slavery on a large scale, but Moses' unusual adoption does fit into the late 12th dynasty. Amenemhet III, the dynasty's sixth king, had two daughters but no sons. Josephus describes a childless daughter of pharaoh finding a child in the river and telling her father, "As I have received him [Moses] from the bounty of the river, in a wonderful manner, I thought proper to adopt him for my son and the heir of thy kingdom."[46] Amenemhet III's daughter Sobekneferu was childless and eventually ruled briefly as pharaoh herself, making Sobekneferu a likely candidate for Moses' foster mother.[47]

4. Testimony of the Dead

Examinations of cemeteries at Tell ed-Daba and Kahun, areas with high Semitic slave populations, have been particularly supportive of the biblical narrative. Graves at ed-Daba reveal that 65 percent of the dead were infants.[48] This extraordinarily high figure is consistent with the slaughter of Israelite infants ordered by Pharaoh. Also consistent with the prescribed slaughter are "wooden boxes . . . discovered underneath the floors of many houses at

43. Ibid, p. 79.
44. Ibid, p. 92, quoting from Dr. Rosalie David's *The Pyramid Builders of Ancient Egypt.*
45. Rohl, *Pharaohs and Kings: A Biblical Quest,* p. 275–276.
46. W. Whiston, transl., book 2, chapter 9, section 7, *The Works of Josephus* (Peabody, MA: Hendrickson Publishers, 1987), p. 68.
47. Ashton and Down, *Unwrapping the Pharaohs,* p. 92.
48. Rohl, *Pharaohs and Kings: A Biblical Quest,* p. 271.

Kahun. They contained babies, sometimes buried two or three to a box, and aged only a few months at death."[49]

Examination of graves in a more recent section, datable to the late 13th dynasty, reveals shallow mass graves without the customary grave goods. These disorganized, crowded burials suggest the need for rapid burial of large numbers of people.[50] The death of the firstborn in the tenth plague would have created just such a situation.

5. The Exodus

In the 13th dynasty, during the reign of Neferhotep I, the Semitic slaves suddenly departed from Tel ed-Daba[51] and Kahun.

> Completion of the king's pyramid was not the reason why Kahun's inhabitants eventually deserted [Kahun], abandoning their tools and other possessions in the shops and houses. . . . The quantity, range, and type of articles of everyday use which were left behind suggest that the departure was sudden and unpremeditated.[52]

Furthermore, Neferhotep I's mummy has never been found, and his son Wahneferhotep did not ever reign, Neferhotep being succeeded by his brother Sobkhotpe IV.[53] The sudden departure of the Semitic slave population fits the biblical account of the Hebrew slaves' sudden exodus from Egypt after the tenth plague. The pharaoh's mummy is missing because he died in the Red Sea with his army when he pursued the slaves, and his son never ruled because he died in the tenth plague.

6. The Hyksos

Just a few years after the Exodus, the 13th dynasty ended, and the Second Intermediate Period, the time of Hyksos rule, began. The Hyksos have puzzled scholars, and everyone has a pet theory as to the Hyksos's identity. Manetho reported:

> Men of ignoble birth out of the eastern parts . . . had boldness enough to make an expedition into our country and with ease subdue

49. D. Down, "Searching for Moses," *TJ* 15 no. 1 (2001): 53-57, available at www.answersingenesis.org/tj/v15/i1/moses.asp.
50. Rohl, *Pharaohs and Kings: A Biblical Quest,* p. 279.
51. Ibid., reporting findings by Professor Manfred Bietak of Austrian Institute for Egyptology.
52. Ashton and Down, *Unwrapping the Pharaohs,* p. 100, quoting Dr. Rosalie David's *The Pyramid Builders of Ancient Egypt.*
53. Ibid., p. 103.

> it by force, yet *without our hazarding a battle with them. . . .* This whole nation was styled Hycsos[54] (emphasis added).

Manetho places this conquest at the end of the 13th dynasty.[55]

Since no evidence of chariots had been found in pre-Hyksos Egypt, tradition has held that the Hyksos were able to defeat Egypt because they possessed chariots. Therefore, since Exodus 14 describes Pharaoh's pursuit with chariots, many have thought that the Exodus occurred after the Hyksos conquest. However, discoveries in recent years have confirmed the use of horses and chariots in the 12th and the 13th dynasties, prior to the Hyksos invasion. For example, an engraving from the 13th dynasty shows Khonsuemmwaset, a pharaoh's son and army commander, with a pair of gloves, the symbol for charioteer, under his seat.[56]

The drowning of the Egyptian army in the Red Sea explains the conquest of the powerful nation of Egypt without a battle. Some have hypothesized that the Hyksos were Amalekites.[57] Whoever the Hyksos were, they ruled Egypt from Avaris in the delta as the 15th and 16th dynasties, while their puppets in the 17th dynasty ruled from Thebes nearly 500 miles to the south. The 17th dynasty overthrew the Hyksos[58] and began the New Kingdom.

New Kingdom – Israel's Early Monarchy

1. David and Tahpenes's Husband

During David's reign, a young Edomite named Hadad found refuge in Pharaoh's house and married Queen Tahpenes's sister.[59] Hadad and the queen's sister had a son named Genubath. Genubath eventually became king of Edom. Records of the 18th dynasty's founder, Ahmose, refer to a name that resembles Tahpenes.[60] Later in the 18th dynasty, Thutmosis III received tribute from the land of Genubatye.[61]

54. Ibid.,, p. 102, quoting Josephus.
55. Rohl, *Pharaohs and Kings: A Biblical Quest,* p. 280–281.
56. Rohl, *Pharaohs and Kings: A Biblical Quest,* p. 285.
57. Ashton and Down, *Unwrapping the Pharaohs,* p. 103, referencing Courville's *The Exodus Problem and Its Ramifications.*
58. Ashton and Down, *Unwrapping the Pharaohs,* p. 106. Rebellion arose after the Hyksos king picked a fight with the Theban king Seqenenre by claiming the hippopotamus noise from the new canal in Thebes was keeping him awake at night.
59. 1 Kings 11:15–20
60. Phonetic similarity is certainly no guarantee of identity, as the case of Shishak's misidentification has shown. However, the occurrence of both of these names in the time sequence consistent with the times of David's and Solomon's reigns is at least a strong suggestion of synchronism.
61. "Contemporary Personalities and Affairs of the Early Israelite and 18th Dynasty Egyptian Kings," from The California Institute for Ancient Studies, www.specialtyinterests.net/solsen.html.

2. Solomon and the Egyptian Princess

Thutmosis I of the 18th dynasty had two daughters, Hatshepsut and Nefrubity. Nefrubity dropped out of the Egyptian records and may have been the Egyptian princess that Solomon married to seal his 1 Kings 3:1 treaty with Egypt.[62]

3. Queen of Sheba and Hatshepsut

Another mysterious Bible character emerges from the 18th dynasty. The female pharaoh Hatshepsut's trip to the land of Punt is famous, but the identity of Punt has remained a mystery despite engravings commemorating the treasures she brought home. First Kings 10 says the queen of Sheba visited Solomon, giving and receiving great gifts. Josephus identified this queen of Sheba as "queen of Egypt and Ethiopia."[63] In Matthew 12:42 the Lord Jesus refers to the queen of Sheba as "the queen of the south." "The south" is a biblical designation for Egypt.[64] Thus, Hatshepsut was probably the queen of Sheba.

4. Rehoboam and Shishak

When Thutmosis III became pharaoh, he conquered much of Palestine, ultimately taking away the treasures in Rehoboam's Jerusalem without a battle. He listed these treasures on the wall of the temple at Karnak. His list mirrors the Bible's account from 1 Kings 6:32, 10:17, and 14:25–26, including the 300 gold shields and doors overlaid with gold.[65] Thutmosis III was Shishak.

5. Asa and Zerah the Ethiopian

Asa, Rehoboam's grandson, had an encounter with Egypt. Second Chronicles 14 describes God's miraculous defense against an overwhelming attack by Zerah the Ethiopian. Ethiopia (Kush) refers to southern Egypt or Sudan. The 18th dynasty's headquarters was in southern Egypt, so this reference likely refers to another 18th dynasty pharaoh, possibly Amenhotep II.[66]

6. Ahab and Akhenaton

Late in the 18th dynasty, one of Egypt's most famous families set the stage for both biblical and Hittite synchronisms. Clay tablets found in Akhenaton's archives at Tel el-Amarna in 1887 included 60 letters from the king of Sumur, likely the Egyptian name for *Samaria.* The city of Samaria, according to 1 Kings 22:26, had a governor named Amon (an Egyptian name). The Amarna letters

62. Ashton and Down, *Unwrapping the Pharaohs,* p. 111. See 1 Kings 3:1.
63. Ibid., p. 121.
64. Daniel 11:5 and 8–9.
65. Ashton and Down, *Unwrapping the Pharaohs,* p. 126–128.
66. Ibid., p. 134.

call this governor Aman-appa and describe a severe famine that is consistent with the famine in the days of Ahab and Elijah.[67]

7. The Hittites and Tutankhamen

Akhenaton's son, the famous King Tutankhamen, died young, leaving no heir and a widowed queen called Ankhesenamen. According to the *Deeds of Suppiluliuma as told by his son Mursili II* in the Hittite archives, Tut's widow wrote to the powerful Hittite king Supililiumas, pleading, "Give me one son of yours . . . he would become my husband. . . . In Egypt he will be king"[68] Had Supililiumas's son Zannanza survived his trip to Egypt, the balance of power would have shifted against Assyria in favor of a Hittite-Egyptian coalition. Zannanza was assassinated, and Tut's general, Harmheb, assumed power. Upon Harmheb's death, his vizier, Ramses I the Great, took the throne as the first pharaoh of the 19th dynasty.

The dates for Ramses the Great's reign[69] and his battle of Kadesh with the Hittites are uncertain, because historians have no biblical parallels and no way to assess the preceding dynasty's duration. The rest of the revised chronology shifts the 19th dynasty dates three to five centuries later than the traditional dates. Ramses III, of the 20th dynasty, reported the annihilation of the Hittites during his reign. Revised chronology allows the Hittites to still exist at the time the Assyrians claimed to be at war with them.

8. "Israel Is Laid Waste"

The real 19th dynasty was concerned with the power of Assyria, not the plagues of Moses. Merneptah, the son of Ramses the Great, recorded the change in the region's power structure by listing many places Assyria had seized. His monument states, "Israel is laid waste, his seed is not."[70] This inscription not only places the latter part of the 19th dynasty in the 8th century B.C.; it also documents that Israel was an actual nation by the time of the 19th dynasty.

Third Intermediate and Late Periods – Judah's Late Monarchy and Captivity

The Third Intermediate Period contains dynasties 21–25, but some of these dynasties were concurrent, not sequential as assumed in the traditional

67. Ibid., p. 154.
68. G. Johnson, "Queen Ankhesenamen and the Hittite Prince," 1999, available at www.guardians.net/egypt/georgejohnson/queenankhesenamen.htm.
69. Rohl, *Pharaohs and Kings: A Biblical Quest,* places him in 900s B.C. (p. 175); Down, *Unwrapping the Pharaohs,* in 700s B.C. (p. 209) depending on uncertain 18th dynasty co-regencies.
70. Ashton and Down, *Unwrapping the Pharaohs,* p. 178.

chronology. In fact, the Royal Cache at Luxor contained a labeled 21st dynasty mummy wrapped in 22nd dynasty linen![71] The linen label names Sheshonq, the same pharaoh earlier mistaken for Shishak.

1. Hezekiah and Taharka

The biblical synchronism in this period involves Hezekiah. The imminent arrival of Assyria's enemy Taharka,[72] the last pharaoh of the 25th dynasty, helped Hezekiah by putting Sennacherib to flight in 709 B.C. Taharka later rebelled against the Assyrian domination of Egypt, dying in 664 B.C. when Ashurbanipal sacked Thebes.[73]

2. Josiah and Necho

After Ninevah's destruction, Pharaoh Necho II of the 26th dynasty marched to Carchemish, where the Assyrian remnant was making its last stand. On the way, according to 2 Chronicles 35, Necho killed Judah's king Josiah at Megiddo. Returning from his 605 B.C. defeat at Carchemish, Necho took Jehoahaz as a hostage and placed Jehoiakim on the throne of Judah.

3. Jeremiah and Hophra

One final biblical synchronism occurs in connection with the fate of 26th dynasty pharaoh, Hophra. Following a coup, Hophra fled to Babylon. There, he acquired an army and returned to reclaim his throne. Jeremiah predicted his defeat, and the prophecy recorded in Jeremiah 44:30 was fulfilled.

Table of Biblical and Egyptian Synchronisms[74]

Date B.C.	Bible	Egyptians	Dynasty
4004	Adam		
2348	Noah's flood		
post-Babel	Mizraim		
late 1900s	Abraham	Khufu	4
1706	Joseph; Jacob to Egypt	Sesostris I	12
1635	Joseph dies		
after 1635	enslavement	Sesostris III	12
1571	Moses born	Amenemhet III	12

71. Rohl, *Pharaohs and Kings: A Biblical Quest,* pp. 75–76.
72. 2 Kings 19:9, referred to as Tirhakah king of Ethiopia.
73. Rohl, *Pharaohs and Kings: A Biblical Quest,* p. 22.
74. Dates for biblical events are from Dr. Floyd-Nolen Jones's *Chronology of the Old Testament,* chosen for its careful analysis and internal consistency with regard to Scripture.

1491	Exodus	Neferhotep I	13
	Judges	Hyksos	15-17
late 1000s	David (1 Kings 11:19)	Ahmosis or Amenhotep I	18
1012	Solomon starts temple	Thutmosis I	18
	Queen of Sheba	Hatshepsut	18
971	Rehoboam; Shishak invades	Thutmosis III	18
late 900s	Asa; Zerah the Ethiopian	Amenhotep II	18
late 900s	Ahab; Elijah	Akhenaton	18
uncertain		Raamses II	19
722	Assyria destroys Israel	Merneptah	19
709	Hezekiah; Assyrian invasion	Taharka	25
664	Manasseh	Taharka dies	25
609	Josiah dies	Necho	26
605		Necho; Carchemish	26
589	Jeremiah	Hophra	26
586	Temple destroyed		
525		Cambyses of Persia	

Conclusion

Isaiah warned against going down to Egypt for help (Isaiah 31:1). This phrase has come to symbolize a warning not to go to the world for truth. God determines truth. Historians examine fragmentary clues and fill in the gaps based on their presuppositions. Those presuppositions may be biblical or traditional. Accepting traditional Egyptian chronology necessitates rejection of biblical truth. Accepting biblical chronology allows a reconstruction of ancient chronology on a foundation of truth. Viewing the evidence from a biblical framework makes the histories of Egypt and the Old Testament fit together like two sides of a zipper.

Since the original publication of this chapter, Isaac Newton's work on revised chronologies has become available in English. *Newton's Revised History of Ancient Kingdoms* makes available much additional information and insight about the history of ancient Egypt as well as the history of other ancient kingdoms. For further studies of revised chronologies, because the Bible is the ultimate standard, I suggest consulting Dr. Floyd Jones' book *The Chronology of the Old Testament.*

25

What about Satan and the Origin of Evil?

BODIE HODGE

Christians are often asked questions about Satan: Who is he? Was he created? When was he created?

These and similar questions are valid questions to ask. To answer them, we need to carefully consider what the Bible says, since it is the only completely reliable source of information about Satan. The Bible doesn't give much information about Satan or the angels, but it does give enough to answer some of these questions.

God's Word is infallible and the absolute authority, and we need to be leery of conclusions drawn from sources outside the Bible, such as man's ideas or traditions. Let's consider what the Bible says related to these questions.

Who Is Satan and Was He Always Called Satan?

The first use of the name *Satan* is found in 1 Chronicles 21:1; chronologically, Job, which was written much earlier, surpasses this. *Satan* is found throughout Job 1 and 2. Satan literally means "adversary" in Hebrew.

Another name appears in the Old Testament in the King James Version:

> How art thou fallen from heaven, O Lucifer, son of the morning! How art thou cut down to the ground, which didst weaken the nations! (Isaiah 14:12; KJV).

This is the only passage that uses the name *Lucifer* to refer to Satan. This name doesn't come from Hebrew but Latin. Perhaps this translation into English was influenced by the Latin Vulgate, which uses this name. In Latin, *Lucifer* means "light bringer."

The Hebrew is *heylel* and means "light bearer," "shining one," or "morning star." Many modern translations translate this as *star of the morning* or *morning star*. In this passage, *heylel* refers to the king of Babylon and Satan figuratively. Of course, Jesus lays claim to this title in Revelation 22:16. Though the passage in Revelation is in Greek while the passage in Isaiah is Hebrew, both are translated similarly.

Some believe that Lucifer was a heavenly or angelic name that was taken from Satan when he rebelled. The Bible doesn't explicitly state this, though Satan is nowhere else referred to as Lucifer but instead is called other names like the devil, Satan, etc. This tradition may hold some truth, although the idea seems to miss that this verse is referring to him *during* and *after* his fall — not before. Since other scriptural passages refer to him as Satan, Lucifer wasn't necessarily his pre-Fall name any more than Satan would be.

Even though Satan is first mentioned by name in Job, previous historical accounts record his actions (see Genesis 3, when Satan influenced the serpent, and Genesis 4 where Cain belonged to him [1 John 3:12]).

In the New Testament, other names reveal more about Satan's current nature. *Devil* (*diabolos*) means "false accuser, Satan, slanderer" in Greek and is the word from which the English word *diabolical* is formed. Satan is called a dragon in Revelation 12:9 and 20:2, as well as the "evil one" in several places. Revelation 12:9 calls him "that ancient serpent" or "serpent of old," and Matthew 4:3 calls him the "tempter." Other names for Satan include *Abaddon* (destruction), *Apollyon* (destroyer, Revelation 9:11), *Beelzebub* or *Beelzebul* (Matthew 12:27) and *Belial* (2 Corinthians 6:15). Satan is also referred to as the god of this world/age (2 Corinthians 4:4), prince of this world (John 12:31), and father of lies (John 8:44).

Was Satan Originally a Fallen Angel from Heaven?

Satan is mentioned in conjunction with angels (Matthew 25:41; Revelation 12:9) and the "sons of God" (Job 1:6, 2:1), which many believe to be angels. Although no Bible verse actually states that he was originally an angel, he is called a cherub in Ezekiel 28:16. The meaning of *cherub* is uncertain, though it is usually thought of as an angelic or heavenly being. (Ezekiel 28 is discussed in more detail later.)

In 2 Corinthians 11:14, we find that Satan masquerades as an angel of light — another allusion to his angel-like status:

> And no wonder! For Satan himself transforms himself into an angel of light.

Although it is possible that Satan was an angel, it may be better to say that he was originally a "heavenly host" (which would include angels), since we know that he came from heaven, but don't know with certainty that he was an actual angel. Recall Isaiah 14:12:

> How you are fallen from heaven, O Lucifer, son of the morning! How you are cut down to the ground, you who weakened the nations!

When Satan, the great dragon in Revelation (12:9), fell, it appears that he took a third of the heavenly host with him (a "third of the stars" were taken to earth with him by his tail, Revelation 12:4). We know that angels who fell have nothing good to look forward to:

> Then He will also say to those on the left hand, "Depart from Me, you cursed, into the everlasting fire prepared for the devil and his angels" (Matthew 25:41).

> For if God did not spare the angels who sinned, but cast them down to hell and delivered them into chains of darkness, to be reserved for judgment (2 Peter 2:4).

What these passages *don't* say is who and where the angels and Satan were originally.

> And it grew up to the host of heaven; and it cast down some of the host and some of the stars to the ground, and trampled them (Daniel 8:10).

Daniel is speaking of heavenly hosts and angels, which were often spoken of as stars or luminaries (see Judges 5:20; Daniel 8:10; Jude 13; Revelation 1:20). It is unlikely that this passage refers to physical stars, as such would destroy the earth. The Hebrew word for stars (*kowkab*) also includes planets, meteors, and comets. Were these stars comets and meteors? Likely not, since the context refers to heavenly beings, which would be trampled on. This is further confirmation that Satan (and perhaps some other heavenly host) and his angels sinned and fell.

Another key passage to this is Ezekiel 28:15–17 (discussed in more detail later). The passage indicates that Satan was indeed perfect before his fall. He was in heaven and was cast to the earth.

Were the Heaven of Heavens, Satan, and His Angels Created?

The Bible doesn't give an *exact* time of Satan's creation or of his fall but does give some clues. Paul says in Colossians that *God/Christ created all things*:

> For by Him all things were created that are in heaven and that are on earth, visible and invisible, whether thrones or dominions or principalities or powers. All things were created through Him and for Him (Colossians 1:16).

So logically, Satan was created, as was the "heaven of heavens." We already found that Satan was originally in heaven prior to his fall. So the question becomes, when was the heaven of heavens created? The Bible uses the word *heaven* in several ways. The first mention is Genesis 1:1:

> In the beginning God created the heavens and the earth.

The Hebrew word for *heavens* is plural (dual form): *shamayim,* dual of an unused singular *shameh*. The word itself means "heaven, heavens, sky, visible heavens, abode of stars, universe, atmosphere," and "the abode of God." The context helps determine the meaning of a particular word; *heavens* is properly plural, and many Bible scholars and translators have rightly translated it as such.

Therefore, it seems safe to assume that the "heaven of heavens" was created along with the physical heavens (the space-time continuum, i.e., the physical universe, where the stars, sun, and moon would abide after they were created on day 4) during creation week.

The definition of the Greek word for *heaven(s)* (*ouranos*) is similar: "the vaulted expanse of the sky with all things visible in it; the universe, the world; the aerial heavens or sky, the region where the clouds and the tempests gather, and where thunder and lightning are produced; the sidereal or starry heavens; the region above the sidereal heavens, the seat of order of things eternal and consummately perfect where God dwells and other heavenly beings."

By usage, this could include the heaven of heavens. However, other biblical passages also help to answer whether the heaven of heavens was created.

> You alone are the Lord; You have made heaven, the heaven of heavens, with all their host, the earth and everything on it, the seas and all that is in them, and You preserve them all. The host of heaven worships You (Nehemiah 9:6).

A clear distinction is made between at least two heavens — the physical heavens and the heaven of heavens. The physical heavens include the expanse made on day 2, the place where the stars were placed on day 4, and the atmosphere (birds are referred to as "of the air" and "of the heavens," e.g., 1 Kings 14:11; Job 12:7; Psalm 104:12). The heaven of heavens is the residing place of the heavenly host, angels, and so on. This would seem to be the third heaven, which Paul mentions:

> I know a man in Christ who fourteen years ago — whether in the body I do not know, or whether out of the body I do not know, God knows — such a one was caught up to the third heaven (2 Corinthians 12:2).

The passage in Nehemiah indicates that God made the heavens; they are not infinite as God is. So the question now becomes, when?

Since the heaven of heavens is referred to with the earth, seas, and physical heaven, we can safely assume that they were all created during the same time frame — during creation week. The creation of the heaven of heavens did not take place on day 7, as God rested on that day from all of His work of creating. So it must have happened sometime during the six prior days.

> Then God saw everything that He had made, and indeed it was very good. So the evening and the morning were the sixth day. Thus the heavens and the earth, and all the host of them, were finished (Genesis 1:31–2:1).

Everything that God made, whether on earth, sky, seas, or heaven, was "very good." Did this include the heaven of heavens and Satan and the angels? Absolutely! Satan is spoken to in Ezekiel 28:15:

> You were perfect in your ways from the day you were created, till iniquity was found in you.

This passage says that Satan was blameless, hence he was *very good* originally. It would make sense then that the heaven of heavens was also a recipient of this blessed saying, since Satan was. In fact, this is what we would expect from an all-good God: a very good creation. Deuteronomy 32:4 says every work of God is perfect. So the heaven of heavens, Satan, and the angels were originally very good.

Ezekiel 28:15 says "from the *day*" (emphasis added) Satan was created. Obviously, then, Satan had a beginning; he is not infinite as God is. Thus, Satan has some sort of binding to time. Other Scriptures also reveal the relationship between Satan and time.

> For this reason, rejoice, O heavens and you who dwell in them. Woe to the earth and the sea, because the devil has come down to you, having great wrath, knowing that *he has only a short time* (Revelation 12:12; NASB, emphasis added).

> When the devil had finished every temptation, he departed from Him until an *opportune time* (Luke 4:13; NASB, emphasis added).

As a created being with a beginning, Satan is bound by time. He is not omnipresent as God is, nor is he omniscient. God has declared the end from the beginning (Isaiah 46:10); Satan cannot.

We can be certain that Satan, the heaven of heavens, and all that is in them had a beginning.

When Were the Angels and Satan Created?

The Bible doesn't give the exact timing of the creation of Satan and the angels; however, we can make several deductions from Scripture concerning the timing. Let's begin by examining Ezekiel 28:11–19:

> 11 Moreover the word of the LORD came to me, saying,
>
> 12 "Son of man, take up a lamentation for the king of Tyre, and say to him, 'Thus says the Lord GOD: "You were the seal of perfection, full of wisdom and perfect in beauty.
>
> 13 You were in Eden, the garden of God; every precious stone was your covering: the sardius, topaz, and diamond, beryl, onyx, and jasper, sapphire, turquoise, and emerald with gold. The workmanship of your timbrels and pipes was prepared for you on the day you were created.
>
> 14 You were the anointed cherub who covers; I established you; you were on the holy mountain of God; you walked back and forth in the midst of fiery stones.
>
> 15 You were perfect in your ways from the day you were created, till iniquity was found in you.
>
> 16 By the abundance of your trading you became filled with violence within, and you sinned; therefore I cast you as a profane thing out of the mountain of God; and I destroyed you, O covering cherub, from the midst of the fiery stones.
>
> 17 Your heart was lifted up because of your beauty; you corrupted your wisdom for the sake of your splendor; I cast you to the ground, I laid you before kings, that they might gaze at you.

> 18 You defiled your sanctuaries by the multitude of your iniquities, by the iniquity of your trading; therefore I brought fire from your midst; it devoured you, and I turned you to ashes upon the earth in the sight of all who saw you.
>
> 19 All who knew you among the peoples are astonished at you; you have become a horror, and shall be no more forever." ' "

In the sections prior to this, the word of the Lord was to Tyre itself (Ezekiel 27:2) and to the ruler of Tyre (Ezekiel 28:2). Beginning in Ezekiel 28:11, a lament (expression of grief or mourning for past events) is expressed to the king of Tyre; or more specifically, to the one *influencing* the king of Tyre. Note well that the king of Tyre was never a model of perfection (verse 12), nor was he on the mount of God (verse 14), nor was he in the Garden of Eden (verse 13; note that the Flood has destroyed the Garden of Eden several hundred years prior to this time period).

God easily sees Satan's influence and speaks directly to him. Elsewhere the Lord spoke to the serpent in Genesis 3: Genesis 3:14 is said to the serpent; Genesis 3:15 is said to Satan who influenced the serpent. Jesus rebuked Peter and then spoke to Satan (Mark 8:33). In Isaiah 14, the passage speaks to the king of Babylon and some parts to Satan, who was influencing him.

In the Ezekiel passage we note that Satan was originally perfect (blameless) from the *day* he was created until he sinned (wickedness was found in him). Thus, we can deduce that Satan was created during creation week; since he was blameless, he was under God's "very good" proclamation (Genesis 1:31) at the end of day 6.

In Job 38:4–7, God spoke to Job:

> Where were you when I laid the foundations of the earth? Tell Me, if you have understanding. Who determined its measurements? Surely you know! Or who stretched the line upon it? To what were its foundations fastened? Or who laid its cornerstone, when the morning stars sang together, and all the sons of God shouted for joy?

Although a poetic passage, it may tell us that some of God's creative work was eyewitnessed by angels and that morning stars sang. Are morning stars symbolic of heavenly host or other angelic beings? It is possible — recall stars are often equated with angelic or heavenly beings, and most commentators suggest this refers to angels.

If so, the creation of the angels was prior to day 3 during creation week. From Genesis 1, God created the foundations of the earth on either day 1 (earth created) or day 3 (land and water separated). The logical inference is that the angels were created on either day 1 or at least by day 3.

If not, then the physical stars (created on day 4) were present while the angels shouted for joy. If this was the case, then morning stars and angels did their singing and shouting after the stars were created.

It seems most likely that *morning stars* symbolize heavenly host. Satan, a heavenly host, was called a morning star; therefore, Satan and the angels were created sometime prior to day 3 (or early on day 3), possibly on day 1.

When Did Satan Fall?

Satan sinned when pride overtook him and he fell from perfection (Ezekiel 28:15–17). When was this? The Bible doesn't give an exact answer either, but deductions can again be made from the Scriptures.

> How you are fallen from heaven, O Lucifer, son of the morning! How you are cut down to the ground, you who weakened the nations! For you have said in your heart: "I will ascend into heaven, I will exalt my throne above the stars of God; I will also sit on the mount of the congregation on the farthest sides of the north; I will ascend above the heights of the clouds, I will be like the Most High" (Isaiah 14:12–14).

When he sinned, he was cast from heaven (Isaiah 14:12). This must have been after day 6 of creation week because God pronounced everything very good (Genesis 1:31). Otherwise, God would have pronounced Satan's rebellion very good; yet throughout Scripture, God is absolute that sin is detestable in His eyes.

God sanctified the seventh day. It seems unlikely that God would have sanctified a day in which a great rebellion occurred. In Genesis 1:28, God commanded Adam and Eve to be fruitful and multiply. Had they waited very long to have sexual relations, they would have been sinning against God by not being fruitful. So, it couldn't have been long after day 7 that Satan tempted the woman through the serpent.

Archbishop Ussher, the great 17th-century Bible scholar, placed Satan's fall on the tenth day of the first year, which is the Day of Atonement. The Day of Atonement seems to reflect back to the first sacrifice when God made coverings for Adam and Eve from the coats of animal skins (Genesis 3:21). It may be that the generations to come (from Abel to Noah to Abraham to the Israelites) followed this pattern of sacrificing for sins on the Day of Atonement.

Regardless, the fall of Satan would likely have been soon after day 7.

How Could Satan, Who Was Created Good, Become Evil?

The answer to this question delves deep into the "sovereignty of God vs. man's responsibility" debate over which the Church has battled for ages.

From what we can tell from studying the Bible, Satan was the first to sin. He sinned before the woman sinned, and before Adam sinned. Some claim that we sin because Satan enters us and causes us to sin, but the Bible doesn't teach this. We sin whether Satan enters us or not. Satan was influencing the serpent when the woman sinned and when Adam sinned; they sinned on their own accord without being able to claim, "Satan made me do it."

But what causes this initial sin; why did Satan sin in the first place?

> Let no one say when he is tempted, "I am tempted by God"; for God cannot be tempted by evil, nor does He Himself tempt anyone. But each one is tempted when he is drawn away by his own desires and enticed. Then, when desire has conceived, it gives birth to sin; and sin, when it is full-grown, brings forth death (James 1:13–15).

Death is the punishment for sin. Sin originates in desire — one's own desire. James (1:14) hints that evil comes from one's own desire. It was by Satan's own desire that his pride in his own beauty and abilities overtook him.

In the "very good" original creation, it seems likely that Satan and mankind had the power of contrite choice.[1] In the Garden of Eden, the woman was convinced by her own *desire* (the tree was *desirable* to make one wise — Genesis 3:6). Satan had not entered her; she was enticed by her own desire.

God is not the author of sin; our desires are. God did not trick or deceive Satan into becoming full of pride. God hates pride (Proverbs 8:13), and it would not be in His character to cause one to become prideful. Nor was He the one who deceived Eve. Deception and lies go hand in hand (Psalm 78:36; Proverbs 12:17), yet God does not lie or deceive (Titus 1:2; Hebrews 6:18).

Note that since Satan's *own desires* caused his pride, the blame for evil's entrance into creation cannot be God's. To clarify, this doesn't mean God was unaware this would happen, but God permitted it to happen. God is sovereign and acted justly by casting Satan out of heaven after he rebelled against the Creator.

Therefore, when God incarnate came to destroy evil and the work of the devil (1 John 3:8), it was truly an act of love, not a gimmick to correct what He "messed up." He was glorified in His plan for redemption.

1. Whether mankind had this power after the Fall is not the topic of discussion in this section.

Some have asked why God didn't send Satan to hell instead of casting him to earth, assuming this would have prevented death, suffering, or curses for mankind. But God is love, and this shows that God was patient with him as God is patient with us. Perhaps Satan would have had a possibility of salvation had he not continued in his rebellion and sealed his fate, although Genesis 3:15 revealed that Satan's head would be crushed (after his continued sin and deception of the woman).

A related question is: was Satan required for man to sin? Satan's temptation of the woman instigated her to look at the fruit of the tree of the knowledge of good and evil, but it was she who *desired* it and sinned. Can we really say with certainty that on another day, without Satan, the woman and/or Adam would not have desired the fruit and sinned? However, in the words of Aslan, the lion in C.S. Lewis's *Chronicles of Narnia*, "There are no what-ifs."

In reality, we suffer death and the Curse because Adam sinned (Genesis 3) and we sinned in Adam (Hebrews 7:9–10), and we continue to sin (Romans 5:12). Adam did his part, but we must take responsibility for our part in committing high treason against the Creator of the universe. It is faulty to think that death and suffering are the result of Satan's rebellion. Man had dominion over the world, not Satan. When Satan rebelled, the world wasn't cursed; when Adam sinned, the ground was cursed, death entered the world, and so on. This is why we needed a last Adam (1 Corinthians 15:45), not a last Eve or a last Satan. This is why Christ came. The good news is that for those in Christ, the punishment for sin (death) will have no sting (1 Corinthians 15:55).

Why Would God, Who Is Not Evil, Allow Evil to Continue to Exist?

As with the other questions in this chapter, great theologians have struggled over how to effectively answer this. Paul, in his book to the Christians in Rome, offers some insight into the overarching perspective that we should have:

> And we know that all things work together for good to those who love God, to those who are the called according to His purpose (Romans 8:28).

All things, including the evil in this world, have a purpose. God is glorified through the plan of salvation that He worked out from the beginning. From the first Adam to the Last Adam, God planned a glorious way to redeem a people for himself through the promise of a Savior who would conquer both sin and death.

Jesus was glorified when He conquered Satan, sin, and death through His death and resurrection (see John 7:39, 11:4, 12:16, 12:23; 1 Peter 1:21; Acts 3:13). Both God the Son and God the Father were glorified through the Resurrection (see John 11:4, 13:31–32). Everything that happens is for the glory of God, even when we can't see how God can be glorified from our limited perspective.

Those who have received the gift of eternal life look forward to the time when we join God in heaven — a place there will be no evil (Revelation 21:27). This 6,000-year-old cursed world is only a blip compared to eternity. This relatively brief time on earth is all the time that evil will be permitted.

What Will Become of Satan?

Satan's days are numbered, and he will be condemned eternally.

> Therefore rejoice, O heavens, and you who dwell in them! Woe to the inhabitants of the earth and the sea! For the devil has come down to you, having great wrath, because he knows that he has a short time (Revelation 12:12).

> And he cast him into the bottomless pit, and shut him up, and set a seal on him, so that he should deceive the nations no more till the thousand years were finished. But after these things he must be released for a little while (Revelation 20:3).

We should have no fear of Satan or his minions, since God has power over him and has already decreed what his outcome will be — a second death — an eternal punishment called hell.

> Then He will also say to those on the left hand, "Depart from Me, you cursed, into the everlasting fire prepared for the devil and his angels" (Matthew 25:41).

> The devil, who deceived them, was cast into the lake of fire and brimstone where the beast and the false prophet are. And they will be tormented day and night forever and ever (Revelation 20:10).

> Then Death and Hades were cast into the lake of fire. This is the second death (Revelation 20:14).

Some people may claim that they want to "rule with Satan in hell," rather than go to heaven with and enjoy the infinite goodness of God. Sadly, these people fail to realize that Satan has *no* power in hell, nor will they. Satan is not the "ruler" in hell but a captive just as they will be if they don't receive the free gift of eternal life by repenting of their sins and believing in the finished work of Jesus Christ on the cross.

We trust those reading this book will realize that the only way of salvation is found through a personal relationship with Jesus Christ. God has provided a way of salvation, a right relationship with Him, and a means of forgiveness; have you received Christ as your Savior?

26

Why Is the Scopes Trial Significant?

BY KEN HAM & DR. DAVID MENTON

In recent years, removing the Ten Commandments from public spaces has been big news. In fact, Christian morality on the whole seems to be rapidly declining in America and the western hemisphere: abortion is on the rise, divorce rates are climbing, gay marriage issues are increasing. But did you know there is a connection between these events and the 1925 Scopes trial?

In 2003, news reports featured many people demonstrating in front of the Alabama court building after the decision to remove the Ten Commandments monument as a public display. Some were lying prostrate on the ground, crying out to the Lord to stop this from happening. But how many of these people really understood the foundational nature of this battle?

If we asked the demonstrators, "Do you believe in millions of years for the age of the earth — and what about the days of creation in Genesis 1?" — well, our long experience in creation ministry indicates that the answer would most likely be something like "What? They're taking the Ten Commandments out — why are you asking me irrelevant questions?"

Or if asked, "Where did Cain get his wife?" they might say, "Can't you see what's happening? They're taking the Ten Commandments out of a courthouse — don't waste my time asking a question that has nothing to do with this!"

In fact, these questions do relate to the real reason the culture is acting this way. During the Scopes trial similar questions were asked; the answers given still resonate today. Let us explain.

The Scopes Trial

The Scopes trial[1] took place during a hot July in 1925 in the little town of Dayton, nestled in the Cumberland Mountains of Tennessee. In a time when modern court trials can drag on for months or even years, it is amazing to consider that the Scopes trial lasted only 12 days (July 10–21) — including the selection of the jury!

The leadership of the American Civil Liberties Union (ACLU) in New York City initiated the Scopes trial. The ACLU became alarmed over "anti-evolution" bills that were being introduced in the legislatures of 20 states in the early 1920s. These bills were all very similar and forbade public schools to teach the evolution of man but generally ignored the evolution of anything else.

The ACLU hoped that a test case might overthrow these bills or at least make them unenforceable. They chose to pursue their case in Tennessee, where the state legislature had unanimously passed the Butler Act. This act declared that it shall be "unlawful for any teacher in any of the Universities, Normals, and all other public schools of the state which are supported in whole or in part by the public school funds of the State, to teach any theory that denies the story of the Divine Creation of man as taught in the Bible, and to teach instead that man has descended from a lower order of animals."

The ACLU placed advertisements in Tennessee newspapers that read in part: "We are looking for a Tennessee teacher who is willing to accept our services in testing this law in the courts." George Rappleyea, a mine operator in Dayton, read the ACLU ad in a Chattanooga newspaper and decided that he would like to see such a trial held in Dayton. Rappleyea's interest was neither scientific nor educational, but rather he hoped that hosting the trial would bring national attention to the town of Dayton and encourage investments in his mining operations.

John Scopes

Rappleyea approached a young friend named John Scopes who had taught math and coached the football team for one year at the local Rhea County high school. Scopes had no background in science and had little interest or understanding of evolution. Indeed, the only qualification Scopes had as a science teacher was that he filled in for an ill biology teacher the last two weeks of the school year. Nonetheless, Rappleyea talked a reluctant Scopes into participating in the ACLU's test case.

1. Many believe the movie *Inherit the Wind* to be a factual account of the Scope trial. It's not. To find out how the real trial differs from the Hollywood version portrayed in the movie, visit www.answersingenesis.org/creation/v19/i1/scopes.asp.

John Scopes

Although Scopes never taught evolution during his two weeks as a biology teacher, and thus really didn't violate the Butler Act, it was considered sufficient that the class textbook, Hunter's *Civic Biology*, did cover the evolution of man. For example, the Hunter textbook speculated that in his early history, "Man must have been little better than one of the lower animals" and concluded, "At the present time there exist upon the earth five races or varieties of man . . . the highest type of all, the Caucasians, represented by the civilized white inhabitants of Europe and America." Sadly, this sort of blatant racism in the name of evolution was enthusiastically endorsed by most of the academic world as well as by many Christian groups.

After the ACLU agreed to accept John Scopes for their test case and pay all expenses, he was arrested for teaching the evolution of man and immediately released on a $1,000 bond. The Dayton lawyer who served the warrant for Scopes' arrest was Sue Hicks (the subject of the Johnny Cash hit song "A Boy Named Sue," by the way). It was also Hicks who came up with the idea of calling upon the popular Christian lawyer/politician William Jennings Bryan to serve as head of the prosecution of John Scopes. When the ACLU chose the famous criminal lawyer and outspoken atheist/agnostic Clarence Darrow to head the defense team for John Scopes, a high visibility trial was virtually guaranteed.

William Jennings Bryan

William Jennings Bryan

Bryan had been the leader of the Democratic Party for 25 years and had run three times unsuccessfully for president of the United States. While considered a conservative Christian, his political views were very liberal for his time; indeed even the archliberal Clarence Darrow supported him in his first two attempts for the presidency. Bryan served as secretary of state under President Woodrow Wilson.

Bryan was well informed about the creation/evolution controversy and regularly corresponded with scientists of his time, such as Henry Fairfield Osborn, on the evidence for and against evolution. While Bryan was a staunch creationist and a strong critic of biological evolution, he accepted geological evolution and an old age for the earth. In his autobiography, *The Memoirs of William Jennings Bryan*, Bryan said that his objectives in the Scopes trial were to "establish the right of taxpayers to control what is taught in their schools" and to "draw a line between teaching evolution as a fact and teaching it as a theory."

Clarence Darrow

Clarence Darrow was an immensely successful criminal lawyer who specialized in defending unpopular people and radical causes, often winning seemingly impossible cases. His agnostic convictions led him to believe that man's actions were ultimately just the result of body chemistry, and that concepts of good and evil were essentially meaningless. In his autobiography, *The Story of My Life*, Darrow explained his purpose for participating in the Scopes trial: "My object and my only object, was to focus the attention of the country on the program of Mr. Bryan and the other Fundamentalists in America."

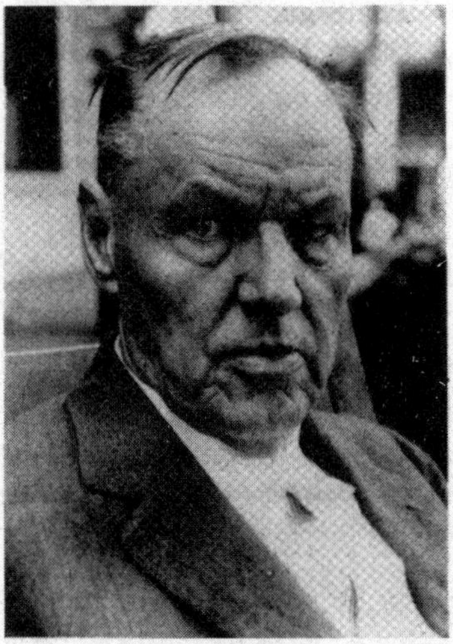
Clarence Darrow

The Trial

Technically, the only legal issue in the Scopes trial was: did John Scopes violate the Butler Act by teaching that man descended from a lower order of animals? For both Bryan and Darrow, however, the real issue wasn't Scopes's guilt or innocence, but rather should evolution be taught as fact in the public schools? Darrow had hoped to have a number of evolutionist scientists testify in the court to the "fact" of evolution, but this wasn't permitted by the judge because the evidence for evolution was technically not at issue in the trial, and Darrow refused to allow his evolutionists to be cross-examined by the prosecution. As a result, most of the testimony by the scientists at the trial was written and filed into record — none was heard by the jury.

Anyone taking the time to read the transcript of the Scopes trial (*The World's Most Famous Court Trial*, Bryan College) will note that Darrow and his defense team of lawyers knew little about evolution and failed in their efforts to establish why it was necessary to teach evolution in the classroom. They lamely attempted to justify its reality and importance by equating evolution with human embryology.

For example, the development of the embryo from a single cell (the fertilized egg) was often cited as evidence that all life came (evolved?) from a single cell. Even the evolutionary expert Dr. Maynard Metcalf of Johns Hopkins University confused evolution with human embryonic development and the aging process!

Much of Darrow's effort at the trial amounted to a caustic diatribe against the Bible and Christianity. His anti-Christian hostility was so intense that there was fear on the part of liberal theologians and organizations that supported his evolutionary views that he might turn popular opinion against them. Darrow even turned his anger and hostility against Judge John T. Raulston by repeatedly interrupting and insulting him, for which he was cited for contempt of court.

After a self-serving apology from Darrow, Judge Raulston forgave Darrow for his contempt with these words: "The Man that I believe came into the world to save man from sin, the Man that died on the cross that man might be redeemed, taught that it was godly to forgive and were it not for the forgiving nature of himself I would fear for man. The Savior died on the Cross pleading with God for the men who crucified Him. I believe in that Christ. I believe in these principles. I accept Col. Darrow's apology." It's difficult to imagine a judge saying such a thing in our "enlightened" day, but not difficult to imagine what would happen to one who did.

Bryan Takes the Witness Stand

On the seventh day of the trial, Darrow challenged Bryan to take the witness stand as an expert on the Bible. Going against the advice of his co-counsel, Bryan foolishly agreed to this outrageous and unprecedented arrangement, with the agreement that Darrow would in turn take his turn at the witness stand to be questioned on his agnostic and evolutionary views.

In his questioning, Darrow sarcastically and often inaccurately recounted several miracles of the Old Testament such as Eve and the serpent, Jonah and the whale, Joshua's long day, Noah's flood, confusion of tongues at the Tower of Babel, and biblical inspiration. Darrow ridiculed Bryan for his belief and defense of these miracles, but Bryan steadfastly stuck with the clear words of Scripture, forcing Darrow to openly deny the Word of God.

Then came the turning point. Darrow raised the matter of a six-day creation. Bryan denied that the Bible says God created everything in six ordinary days of approximately 24 hours. When Darrow asked, "Does the statement 'the morning and the evening were the first day,' and 'the morning and the evening were the second day' mean anything to you?" Bryan replied, "I do not see that there is any necessity for constructing the words, 'the evening and the morning,' as meaning necessarily a 24-hour day."

When Darrow asked, "Creation might have been going on for a very long time?" Bryan replied, "It might have continued for millions of years." With the help of Bryan's compromise on the days of creation, Darrow achieved his goal of making the Bible subject to reinterpretation consistent with the ever-changing scientific and philosophical speculations of man.

The Significance

At the time of the trial, some probably thought, *What have the age of the earth, the days of creation, and Cain's wife got to do with this trial?* But actually, Darrow understood the connection — the same connection that these questions have to the Ten Commandments controversy (and general loss of Christian morality) today.

While in the witness box, Bryan, who stood for Christianity, couldn't answer the question about Cain's wife, and admitted he didn't believe in six literal creation days but accepted the millions of years for the earth's age.

That's when Darrow knew he had won, because he had managed to get the Christian to admit, in front of a worldwide audience, that he couldn't defend the Bible's history (e.g., Cain's wife), and didn't take the Bible as written (the days of creation), and instead accepted the world's teaching (millions of years). Thus, Bryan (unwittingly) had undermined biblical authority and paved the way for secular philosophy to pervade the culture and education system.

Sadly, most Christians today have, like Bryan, accepted the world's teaching and rejected the plain words of the Bible regarding history. Thus, they have helped the world teach generations of children that the Bible cannot be trusted in Genesis. After years of such indoctrination, a generation has now arisen that is also (logically) rejecting the morality based on the Bible. Today, with, for example, the removal of the Ten Commandments from public places, we are seeing the increasing elimination of the Christian foundational structure in the nation.

This is a major reason why the influence of Christianity has been so weakened in our Western world — the Church is giving the message that we need to trust in man's theories — not the Word of God. The answer isn't to merely protest such removals — or to simply protest other anti-Christian actions (e.g., abortion, euthanasia, gay marriage) — but to teach people *why* they can believe the Bible is true in every area it touches on. We need to provide Bible-based answers to the questions the world asks about the Christian faith (Who was Cain's wife? Isn't the earth millions of years old? Weren't the days in Genesis 1 long periods of time?). As we do this, people will begin to see that they can trust the Bible when it speaks of "earthly" things, and thus, when it speaks of "heavenly" things (salvation, absolute moral standards, etc.), as Jesus teaches in John 3:12.

27

Isn't the Bible Full of Contradictions?

PAUL F. TAYLOR

A Christian talk radio show in America frequently broadcasts an advertisement for a product. In this ad, a young lady explains her take on Scripture: "The Bible was written a long time ago, and there wasn't a lot of knowledge back then. I think that if you read between the lines, it kinda contradicts itself." The show's host replies, "Oh no, it doesn't!" but nevertheless her view is a common view among many people.

Some years ago, I was participating in an Internet forum discussion on this topic. Another participant kept insisting that the Bible couldn't be true because it contradicts itself. Eventually, I challenged him to post two or three contradictions, and I would answer them for him. He posted over 40 alleged contradictions. I spent four hours researching each one of those points and then posted a reply to every single one. Within 30 seconds, he had replied that my answers were nonsense. Obviously, he had not read my answers. He was not interested in the answers. He already had an *a priori* commitment to believing the Bible was false and full of contradictions. It is instructive to note that after a quick Google search, I discovered that his list of supposed Bible contradictions had been copied and pasted directly from a website.

This anecdote shows that, for many people, the belief that the Bible contains contradictions and inaccuracies is an excuse for not believing. Many such people have not actually read the Bible for themselves. Still fewer have analyzed any of the alleged contradictions. It has been my experience that, after a little research, all the alleged contradictions and inaccuracies are explainable.

If you, the reader, are prepared to look at these answers with an open mind, then you will discover that the excuse of supposed inaccuracies does not hold water. If, however, you have already convinced yourself that such an old book as the Bible just has to contain errors, then you may as well skip this chapter. Like my Internet forum opponent, nothing (apart from the work of the Holy Spirit) is going to convince you that the Bible is 100 percent reliable — especially not the facts!

On Giants' Shoulders

In attempting to explain some of the Bible's alleged errors, I am standing on the shoulders of giants. I will not be able to address every alleged error for reason of space; others have done the job before me. In my opinion, chief among these is John W. Haley, who wrote the definitive work on the subject, *Alleged Discrepancies of the Bible.*[1] Haley tackles a comprehensive list of alleged discrepancies under the headings "doctrinal," "ethical," and "historical." This chapter uses a similar thematic approach because it will be possible to examine only a representative sample of alleged discrepancies. Readers are referred to Haley's work for a more exhaustive analysis of the subject.

Law of Noncontradiction

$$A \neq \text{Non-}A$$

One of our own presuppositions could be labeled as the "law of noncontradiction." This stems directly from the belief that the Bible is the inspired, inerrant, and authoritative word of God. Although the 66 books of the Bible were written by diverse human authors in differing styles over a long period of time, it is our contention that the Bible really has only one author — God. The law of noncontradiction has been defined by theologian James Montgomery Boice as follows: "If the Bible is truly from God, and if God is a God of truth

1. For explanations of more supposed contradictions, see www.answersingenesis.org/go/contradictions; John W. Haley, *Alleged Discrepancies of the Bible* (Grand Rapids, MI: Baker, 1988). The book was originally published in 1874.

(as He is), then . . . if two parts seem to be in opposition or in contradiction to each other, our interpretation of one or both of these parts must be in error."[2] Wayne Grudem makes the same point thus:

> When the psalmist says, "The sum of your word is truth; and every one of your righteous ordinances endures for ever" (Ps 119:160), he implies that God's words are not only true individually but also viewed together as a whole. Viewed collectively, their "sum" is also "truth." Ultimately, there is no internal contradiction either in Scripture or in God's own thoughts.[3]

Boice proceeds to describe two people who are attempting to understand why we no longer perform animal sacrifices. One sees the issue as consistent with the evolution of religion. Another emphasizes the biblical concept of Jesus' ultimate and perfect fulfilment and completion of the sacrificial system. Boice says:

> The only difference is that one approaches Scripture looking for contradiction and development. The other approaches Scripture as if God has written it and therefore looks for unity, allowing one passage to throw light on another.[4]

Our presupposition that the Bible will not contain error is justified by the Bible itself. In Titus 1:2, Paul refers to God "who cannot lie," and the writer to the Hebrews, in 6:17–18, shows that by His counsel and His oath "it is impossible for God to lie." However, if a Bible student is determined to find error in the Bible, he will find it. It is a self-fulfilling prophecy. Yet, the error is not really there.

Inerrancy Only for Original Manuscripts

Historical evangelical statements of faith claim inerrancy for the Scriptures for the original manuscripts. Apparently, this is a problem for some and leads to claims of inconsistency. The argument goes that there have

2. James M. Boice, *Foundations of the Christian Faith* (Downers Grove, IL: InterVarsity Press, 1986), p. 91.
3. Wayne Grudem, *Systematic Theology* (Grand Rapids, MI: Zondervan, 1994), p. 35.
4. Boice, *Foundations of the Christian Faith,* p. 93.

been many translators and copyists since the Bible times and that these translators and copyists must have made errors. Therefore, it is said, we cannot trust current translations of the Bible to be accurate. Boice asks if an appeal to an inerrant Bible is meaningless.

> It would be if two things were true: (1) if the number of apparent errors remained constant as one moved back through the copies toward the original writing and (2) if believers in infallibility appealed to an original that differed substantially from the best manuscript copies in existence. But neither is the case.[5]

In fact, recent discoveries of biblical texts show that the Bible is substantially the same as when it was written. What few discrepancies might still remain are due to mistranslations or misunderstandings. These issues are all known to biblical scholars and are easily explained.

Presuppositional Discrepancies

A number of alleged Bible discrepancies could be described as *presuppositional discrepancies*. What I mean by the term is that there are a number of alleged discrepancies that are only discrepancies because of the presuppositions of the one making the allegations. Many such alleged discrepancies involve scientific argument and are covered in detail in other literature, including elsewhere in this book. Such discrepancies disappear immediately if the reader decides to interpret them in the light of a belief in the truth of the Bible.

> *The Bible says the world is only 6,000 years old and was created in six days, but science has proved that the earth is millions of years old.*

This sort of alleged discrepancy is very common. The supposed inaccuracy of the early chapters of Genesis is very often used as a reason to state that the whole Bible is not true. Many articles on the Answers in Genesis website (www.answersingenesis.org) and in *Answers* magazine tackle such issues, so it is not relevant to repeat the arguments again here. Readers are referred to the chapter "Did Jesus Say He Created in Six Literal Days?" in the *New Answers Book 1*[6] or to my detailed analysis in the *Six Days of Genesis*.[7]

Answers in Genesis endeavors show that a belief in the truth of Scripture from the very first verse is a reasonable and rational position to take. Once that

5. Boice, *Foundations of the Christian Faith,* p. 75.
6. Ken Ham, ed., *The New Answers Book 1* (Green Forest, AR: Master Books, 2007).
7. Paul F. Taylor, *The Six Days of Genesis* (Green Forest, AR: Master Books, 2007).

point is understood, many of these pseudoscientific objections to Scripture fade away.

Let us briefly comment on another such presuppositional discrepancy.

> *Genesis 6–8 suggest that the whole world was once covered by water. There is no evidence for this.*

Detailed answers to this allegation can, once again, be found in much of our literature. For example, see the relevant chapter in *The New Answers Book 1.*[8]

It cannot be emphasized too strongly that creationists and evolutionists do not have different scientific evidence. We have the same scientific evidence; the *interpretation* of this evidence is different.

Thus, if one starts from the assumption that the fossil record was laid down over millions of years before human beings evolved, then the fossils do not provide evidence for the Flood. However, if one starts with the presupposition that the Bible's account is true, then we see the fossil record itself as evidence for a worldwide flood and there is no evidence of millions of years! As Ken Ham has often said, "If there really was a worldwide flood, what would you expect to see? Billions of dead things, buried in rock layers laid down by water all over the earth." This is exactly what we see.

Incorrect Context

Strongly related to the presuppositional discrepancies are the supposed errors caused by taking verses out of context. For example, a passage in the Bible states, "There is no God." However, the meaning of the phrase is very clear when we read the context: "The fool has said in his heart, 'There is no God.'" (Psalm 14:1). The words "There is no God" are consequently found on the lips of someone the Bible describes as a *fool.*[9]

This discrepancy might seem trivial, but there are more sophisticated examples of the same problem. These often arise by comparing two separate passages, which are referring to slightly different circumstances. For example, consider the following:

> *Ecclesiastes says that we are upright, while Psalms says that we are sinners.*

8. Ham, *The New Answers Book 1,* Ken Ham and Tim Lovett, "Was There Really a Noah's Ark & Flood?" p. 125–140.
9. Unless otherwise stated, Bible passages quoted in this chapter are from the New King James Version (NKJV). Other translations are indicated by standard letters, such as KJV (King James Version), NIV (New International Version), and Tyndale (William Tyndale's translation).

The verses to which this statement alludes are these:

> God made man upright (Ecclesiastes 7:29).
> Behold, I was brought forth in iniquity (Psalm 51:5).

Looking at the contexts of both verses removes the discrepancy. In Ecclesiastes 7:29, the writer is talking about Adam and Eve, stating that we were *originally* created upright. In Psalm 51, David is speaking of his personal situation as a sinner, especially in the light of his sinful adultery with Bathsheba and his causing the death of Uriah. Thus, there is no contradiction between these passages.

Translational Errors

A common allegation against the Bible is that it is likely to have been mistranslated. When one actually analyzes possible mistranslations, however, it is found that there are actually very few real mistranslations. All of these have been studied and documented and can be found in Haley's book. As we have a number of good English translations today, it is often helpful to compare a couple of these. Once this comparison has been made, many of the so-called translational errors disappear.

> *There are two creation accounts: Genesis 1 and 2 give different accounts. In chapter 1, man and woman are created at the same time after the creation of the animals. In chapter 2, the animals are created after people.*

This apparent contradiction is best illustrated by looking at Genesis 2:19.

> Out of the ground the LORD God formed every beast of the field and every bird of the air, and brought them to Adam to see what he would call them (NKJV).

The language appears to suggest that God made the animals after making Adam and then He brought the animals to Adam. However, in Genesis 1, we have an account of God creating animals *and then* creating men and women.

The difficulty with Genesis 2:19 lies with the use of the word *formed*. The same style is read in the KJV.

> And out of the ground the LORD God formed every beast of the field, and every fowl of the air; and brought them unto Adam to see what he would call them.

The NIV has a subtly different rendition.

> Now the LORD God had formed out of the ground all the beasts of the field and all the birds of the air. He brought them to the man to see what he would name them.

The NIV suggests a different way of viewing the first two chapters of Genesis. Genesis 2 does not suggest a chronology. That is why the NIV suggests using the style "the LORD God *had formed* out of the ground all the beasts of the fields." Therefore, the animals being brought to Adam had already been made and were not being brought to him immediately after their creation. Interestingly, Tyndale agrees with the NIV — and Tyndale's translation predates the KJV.

> The Lord God had made of the earth all manner of beasts of the field and all manner fowls of the air.

Tyndale and the NIV are correct on this verse because the verb in the sentence can be translated as *pluperfect* rather than *perfect*. The pluperfect tense can be considered as the past of the past — that is to say, in a narration set in the past, the event to which the narration refers is already further in the past. Once the pluperfect is taken into account, the perceived contradiction completely disappears.

In the Book of Leviticus, bats are described as birds.

The passage to which the allegation refers is Leviticus 11:13–20.

> 13 And these you shall regard as an abomination among the birds; they shall not be eaten, they are an abomination: the eagle, the vulture, the buzzard,
> 14 the kite, and the falcon after its kind;
> 15 every raven after its kind,
> 16 the ostrich, the short–eared owl, the sea gull, and the hawk after its kind;
> 17 the little owl, the fisher owl, and the screech owl;
> 18 the white owl, the jackdaw, and the carrion vulture;
> 19 the stork, the heron after its kind, the hoopoe, and the bat.
> 20 All flying insects that creep on all fours shall be an abomination to you (NKJV).

> 13 And these are they which ye shall have in abomination among the **fowls**; they shall not be eaten, they are an abomination: the eagle, and the ossifrage, and the ospray,
> 14 And the vulture, and the kite after his kind;
> 15 Every raven after his kind;
> 16 And the owl, and the night hawk, and the cuckow, and the hawk after his kind,
> 17 And the little owl, and the cormorant, and the great owl,
> 18 And the swan, and the pelican, and the gier eagle,
> 19 And the stork, the heron after her kind, and the lapwing, and the bat.
> 20 All **fowls** that creep, going upon all four, shall be an abomination unto you (KJV).

Bible critics point out that, in their view, the writer of Leviticus is ignorant. He must have thought bats were birds, whereas we now classify them as mammals. Many Bible critics might also go on to discuss the supposed evolutionary origin of bats and birds.

A look at the KJV sheds some light on what the passage actually means. The KJV uses the word *fowls* instead of *birds*. Today, we would not see a significant difference, but notice that the KJV also describes insects as *fowls* in verse 20. The actual Hebrew word is *owph* (Strong's 05775). Although *bird* is usually a good translation of *owph*, it more accurately means *has a wing*. It is therefore completely in order for the word to be used of birds, flying insects, and bats. It could presumably also be used of the pteradons and other flying reptiles.

This translation of *owph* is supported by noting its use in Genesis 1:20.

> Then God said, "Let the waters abound with an abundance of living creatures, and let birds fly above the earth across the face of the firmament of the heavens" (NKJV).

> *How could the young Samuel have been sleeping in the Temple when the Temple was not built until much later?*

There are two allegations referred to 1 Samuel 3:3. The verse is quoted below from the KJV, the NIV, and the NKJV.

> And ere the lamp of God went out in the temple of the Lord, where the ark of God was, and Samuel was laid down to sleep (KJV).

> The lamp of God had not yet gone out, and Samuel was lying down in the temple of the Lord, where the ark of God was (NIV).

> And before the lamp of God went out in the tabernacle of the Lord where the ark of God was, and while Samuel was lying down (NKJV).

The translation used by the NKJV gives a clue as to where the first misunderstanding comes from. The Hebrew word is *hēkäl.* This word is used of the temple, but the word is literally a large building or edifice. Commentators[10] have suggested that before the building of the temple the word was often applied to the sacred tabernacle. Therefore, it is perfectly possible for Samuel to have been asleep in this tabernacle. This alleged discrepancy is not so much a mistranslation as a misunderstanding.

The other alleged discrepancy with this verse is that Samuel was sleeping in the sacred portion of this tabernacle, the holy of holies, where the ark of God was. The NKJV gets it correct by pointing out that light went out where the holy of holies was while Samuel was lying down, not that he was lying down in this very holy place. This shows the difficulty of translating Hebrew into English when not careful. This brings us to our next section, where we find alleged discrepancies due to use of language.

Use of Language

Some alleged discrepancies occur because of the way that language has changed. It is interesting that while Hebrew has changed very little over the

10. See, for example, Haley, *Alleged Discrepancies of the Bible*, p. 396.

centuries, English is a language undergoing constant major change. The study of how English has altered is fascinating, though outside the scope of this chapter. As an aside, we can easily see how different strands of English have developed in different ways. The best example of this is the divergence between British and American English — a source of tremendous scope for misunderstanding, one-upmanship, and humor (or is it humour?).

Many of the biblical misunderstandings caused by change of language are found in the KJV, which was first translated in 1611. The English language has changed much since 1611, on both sides of the Atlantic. For example, we know that few people today refer to each other as *thee* and *thou,* except some of the older generation in the counties of Lancashire and Yorkshire in Northern England. The KJV uses this terminology to address God, and we can mistakenly think that this is a term of respect. In fact, the use of *thou* is much more specific. It is used to refer to a close friend or relative. In a society that uses the word *thou*, it would never be used in reference to someone to whom one was being especially polite. For example, in his youth my Lancastrian father would refer to his school friends as *thee* but to his teacher as *you*. Therefore, to refer to God as *thou*, while certainly not being disrespectful, implies a degree of intimacy usually associated with families or close friends.

Genesis 1 must contain a gap, because God commanded people to "replenish" the earth. You cannot replenish something, unless it was once previously full.

Genesis 1:28 contains the following command: "Be fruitful, and multiply, and replenish the earth, and subdue it" (KJV). Most other translations use the word *fill* rather than *replenish.* In fact, the Tyndale Bible, which predates the KJV, uses the word *fill.* So did the translators of the KJV get it wrong?

On the contrary. The word *replenish* was a very suitable word to choose in 1611 because at that time the word meant *to fill completely*, refuting any alleged gap. It therefore carries a slightly stronger emphasis than simply the word *fill*, and the Hebrew word has this emphasis. The word *replenish* did not imply doing something again as many words beginning with *re* do. Its etymology is common with the word *replete*, which still today carries no connotation of a repeated action. However, over the centuries the meaning of *replenish* has altered, so that if we now, for example, suggest replenishing the stock cupboard, we are suggesting that we refill a cupboard, which is now less full than it once was.

There are many other examples of misunderstandings caused by these changes in the English language. None of these misunderstandings were caused by errors on the part of the KJV translators. In fact, they chose the best English words at the time. The problems are caused simply because of the way that English has changed.

Another example of this is to ask why the Psalmist seems to be trying to prevent God from doing something in Psalm 88.

> But unto thee have I cried, O Lord; and in the morning shall my prayer *prevent* thee (Psalm 88:13, KJV, emphasis mine).

The NKJV renders the same verse as follows:

> But to You I have cried out, O Lord, And in the morning my prayer *comes before* You (Psalm 88:13, NKJV, emphasis mine).

Which translation is correct? The answer is that they both are. In 1611, the word *prevent* meant *to come before.* Compare the French verb *venir* (to come) with *prevenir* (to come before). However, in the following centuries, the word *prevent* has altered its meaning in English.

Some problems with use of language exist because of the sort of idioms used in the original languages, which would have been familiar to the original readers but sometimes pass us by. For example:

Moses says insects have four legs, whereas we know they have six.

I have come across this alleged discrepancy frequently. I sometimes wonder if those using this allegation have really thought it through. Do they honestly believe that Moses was so thick that he couldn't count the legs on an insect correctly?

The passage concerned is Leviticus 11:20–23.

> All flying insects that creep on all fours shall be an abomination to you. Yet these you may eat of every flying insect that creeps on all fours: those which have jointed legs above their feet with which to leap on the earth. These you may eat: the locust after its kind, the destroying locust after its kind, the cricket after its kind, and the grasshopper after its kind (NKJV).

In fact, we use the phrase *on all fours* in a similar manner to Hebrew. The phrase is colloquial. It is referring to the actions of the creature (i.e., walking around) rather than being a complete inventory of the creature's feet. Also, when the Bible is referring to locusts and similar insects, it is actually being very precise. Such insects do indeed have four legs with which to "creep" and another two legs with which to "leap," which Moses points out (*those which have jointed legs above their feet with which to leap*). Once again, we find that the allegation of biblical discrepancy does not show up under the light of common sense.

> *If Jesus was to be in the grave three days and nights, how do we fit those between Good Friday and Easter Sunday?*

There are several solutions to this problem. Some have suggested that a special Sabbath might have occurred, so that Jesus was actually crucified on a Thursday. However, a solution, which seems to me to be more convincing, is that Jesus was indeed crucified on a Friday but that the Jewish method of counting days was not the same as ours.

In Esther 4:16, we find Esther exhorting Mordecai to persuade the Jews to fast. "Neither eat nor drink for three days, night or day" (NKJV). This was clearly in preparation for her highly risky attempt to see the king. Yet just two verses later, in Esther 5:1, we read: "Now it happened on the third day that Esther put on her royal robes and stood in the inner court of the king's palace." If three days and nights were counted in the same way as we count them today, then Esther could not have seen the king until the fourth day. This is completely analogous to the situation with Jesus's crucifixion and resurrection.

> For as Jonah was three days and three nights in the belly of the great fish, so will the Son of Man be three days and three nights in the heart of the earth (Matthew 12:40; NKJV).

> Now after the Sabbath, as the first day of the week began to dawn, Mary Magdalene and the other Mary came to see the tomb (Matthew 28:1; NKJV).
>
> Then, as they were afraid and bowed their faces to the earth, they said to them, "Why do you seek the living among the dead? He is not here, but is risen! Remember how He spoke to you when He was still in Galilee, saying, 'The Son of Man must be delivered into the hands of sinful men, and be crucified, and the third day rise again'" (Luke 24:5–7; NKJV).

If the three days and nights were counted the way we count them, then Jesus would have to rise on the fourth day. But, by comparing these passages, we can see that in the minds of people in Bible times, "the third day" *is equivalent to* "after three days."

In fact, the way they counted was this: part of a day would be counted as one day. The following table, reproduced from the Christian Apologetics and Research Ministry (CARM) website, shows how the counting works.[11]

Day One		**Day Two**		**Day Three**	
FRI starts at sundown on Thursday	**FRI** ends at sundown	**SAT** starts at sundown on Friday	**SAT** ends at sundown	**SUN** starts at sundown on Saturday	**SUN** ends at sundown
Night	Day	Night	Day	Night	Day
Crucifixion		Sabbath		Resurrection	

This table indicates that Jesus died on Good Friday; that was day one. In total, day one includes the day and the previous night, even though Jesus died in the day. So, although only part of Friday was left, that was the first day and night to be counted. Saturday was day two. Jesus rose in the morning of the Sunday. That was day three. Thus, by Jewish counting, we have three days and nights, yet Jesus rose on the third day.

It should not be a surprise to us that a different culture used a different method of counting days. As soon as we adopt this method of counting, all the supposed biblical problems with counting the days disappear.

11. Christian Apologetics and Research Ministry, "How Long Was Jesus Dead in the Tomb?" www.carm.org/diff/Matt12_40.htm.

Copyist Error

It does not undermine our belief in the inerrancy of Scripture to suppose that there may be a small number of copyist errors. With a little logical analysis, this sort of error is not too difficult to spot.

> *There must be an error in Luke 3:36. The genealogy gives an extra Cainan not found in similar genealogies, such as Genesis 11:12.*

Expositor Dr. John Gill gives ample reasons why this was a copyist error.[12] Gill says:

> This Cainan is not mentioned by Moses in #Ge 11:12 nor has he ever appeared in any Hebrew copy of the Old Testament, nor in the Samaritan version, nor in the Targum; nor is he mentioned by Josephus, nor in #1Ch 1:24 where the genealogy is repeated; nor is it in Beza's most ancient Greek copy of Luke: it indeed stands in the present copies of the Septuagint, but was not originally there; and therefore could not be taken by Luke from thence, but seems to be owing to some early negligent transcriber of Luke's Gospel, and since put into the Septuagint to give it authority: I say "early," because it is in many Greek copies, and in the Vulgate Latin, and all the Oriental versions, even in the Syriac, the oldest of them; but ought not to stand neither in the text, nor in any version: for certain it is, there never was such a Cainan, the son of Arphaxad, for Salah was his son; and with him the next words should be connected.

If the first Cainan was not present in the original, then the Greek may have read in a manner similar to the following. Remember that NT Greek had no spaces, punctuation, or lower case letters.

ΤΟΥΣΑΡΟΥΧΤΟΥΡΑΓΑΥΤΟΥΦΑΛΕΓΤΟΥΕΒΕΡΤΟΥΣΑΛΑ
ΤΟΥΑΡΦΑΞΑΔΤΟΥΣΗΜΤΟΥΝΩΕΤΟΥΛΑΜΕΧ
ΤΟΥΜΑΘΟΥΣΑΛΑΤΟΥΕΝΩΧΤΟΥΙΑΡΕΔΤΟΥΜΑΛΕΛΕΗΛΤΟΥΚΑΙΝΑΝ
ΤΟΥΕΝΩΣΤΟΥΣΗΘΤΟΥΑΔΑΜΤΟΥΘΕΟΥ

12. Note on Luke 3:36, in: John Gill, D.D., *An Exposition of the Old and New Testament; The Whole Illustrated with Notes, Taken from the Most Ancient Jewish Writings* (London: printed for Mathews and Leigh, 18 Strand, by W. Clowes, Northumberland-Court, 1809), edited, revised, and updated by Larry Pierce, 1994–1995 for The Word CD-ROM. Available online at eword.gospelcom.net/comments/luke/gill/luke3.htm. See also chapter 5, "Are There Gaps in the Genesis Genealogies?"

If an early copyist glanced at the third line, while copying the first line, it is conceivable that the phrase TOYKAINAN (son of Cainan) may have been copied there.

ΤΟΥΣΑΡΟΥΧΤΟΥΡΑΓΑΥΤΟΥΦΑΛΕΓΤΟΥΕΒΕΡΤΟΥΣΑΛΑΤΟΥΚΑΙΝΑΝ
ΤΟΥΑΡΦΑΞΑΔΤΟΥΣΗΜΤΟΥΝΩΕΤΟΥΛΑΜΕΧ
ΤΟΥΜΑΘΟΥΣΑΛΑΤΟΥΕΝΩΧΤΟΥΙΑΡΕΔΤΟΥΜΑΛΕΛΕΗΛΤΟΥΚΑΙΝΑΝ
ΤΟΥΕΝΩΣΤΟΥΣΗΘΤΟΥΑΔΑΜΤΟΥΘΕΟΥ

There is some circumstantial evidence for this theory. The Septuagint (LXX) is a Greek translation of the Old Testament said to be translated by about 72 rabbis. Early copies of LXX do not have the extra Cainan in Genesis 11, but later copies postdating Luke's gospel do have the extra Cainan.

It might seem odd to suggest that there could be a copyist error in our translations of the Bible. What is even more remarkable to me, however, is that such possible copyist errors are so extremely rare. Paradoxically, the possible existence of such an error merely reinforces how God has preserved His Word through the centuries.

Conclusion

This chapter has discussed only some of the many alleged Bible contradictions and discrepancies. However, the methods of disposing of the supposed discrepancies used here can also be used on other alleged errors. There is one matter on which the reader should be very confident — the supposed Bible errors are well known to Bible scholars and have all been addressed and found not to be errors after all. In every case, there is a logical explanation for the supposed error. The Bible is a book we can trust — no, more than that — it is the *only* book we can fully trust.

28

Was the Dispersion at Babel a Real Event?

BODIE HODGE

When did the events at the Tower of Babel happen? What did the tower look like? Are there any records of Noah's descendants found throughout the world after they left Babel? What about different languages? Are Noah and his sons found in any ancient genealogies? In this chapter, we'll examine the fascinating answers to questions about what happened on the plain of Shinar. For background to this chapter, please read Genesis 10–11.

When Did the Event at Babel Occur?

Renowned chronologist Archbishop James Ussher[1] placed the time of Babel at 106 years after the Flood, when Peleg was born.[2]

> To Eber were born two sons: the name of one *was* Peleg, for in his days the earth was divided; and his brother's name *was* Joktan (Genesis 10:25).

Although this may not be the exact date, it is in range because Peleg was in the fourth generation after the Flood.

1. James Ussher, *The Annals of the World,* trans. Larry and Marion Pierce (Green Forest, AR: Master Books, 2003), p. 22.
2. The use of Ussher's dates are not an across-the-board endorsement of his work. We recognize that any human work contains errors; however, Ussher meticulously researched biblical and ancient history, and we are comfortable with using many of the dates he proposed.

Some have suggested that this division refers to a geophysical splitting of the continents; however, this is associated with the flood of Noah's time — not the events at Babel. The massive amounts of water and the crustal breakup indicated in Genesis 7:11 (the fountains of the great deep burst forth) were substantial enough to cause catastrophic movements of plates. Continental collision formations, such as high mountains, were already in place prior to Peleg's day. For example, we know the mountains of Ararat had formed by the end of the Flood because the ark landed there. These mountains are caused by a collision with the Arabian plate and the Eurasian plate. So these would have already moved by the time the Flood had ended.

Continental splitting during the day of Peleg would have caused another global flood! Instead, the division mentioned here refers to the linguistic division that happened when God confused the language at Babel. Even the Jewish historian Josephus (who lived near the time of Christ) stated:

> He was called Peleg, because he was born at the dispersion of the nations to their various countries. . . .[3]

Prominent modern theologians such as John Whitcomb reaffirm this as well.[4] According to Archbishop Ussher, the date of Babel would have been near 2242 B.C.[5] See table 1 for a comparison to other events according to Ussher.

Table 1. Major Dates According to Ussher

Major event	Date (According to Ussher)
Creation	4004 B.C.
Global Flood	2348 B.C.
Tower of Babel	2242 B.C.
Call of Abraham	1921 B.C.
Time of the Judges (Moses was first)	1491 B.C. (God appeared to Moses in the burning bush)
Time of the Kings (Saul was the first)	1095 B.C.
Split Kingdom	975 B.C.
Christ Was Born	5 B.C.

3. William Whiston, *The Works of Josephus Complete and Unabridged* (Peabody, MA: Hendrickson Publishers, 1987), p. 37.
4. John Whitcomb, "Babel," *Creation,* June 2002, p. 31–33, online at www.answersingenesis.org/creation/v24/i3/babel.asp.
5. Ussher, *The Annals of the World,* p. 22.

It was during the days of Peleg that the family groups left the plain of Shinar and traveled to different parts of the world, taking with them their own language that other families couldn't understand. Not long after this, Babylon (2234 B.C.), Egypt (2188 B.C.), and Greece (2089 B.C.) began.[6] Civilizations that were closer to Babel (e.g., those in the Middle East) were established prior to civilizations farther from Babel (e.g., those in Australia or the Americas).

Even more fascinating is that as people went around the world, they left evidence of this event! Let's take a look.

Ziggurats throughout the World

The Tower of Babel has traditionally been depicted as a type of ziggurat, although the Bible doesn't give specific dimensions. The Hebrew word for *tower* used in Genesis 11, referring to the Tower of Babel, is *migdal*: a tower; by analogy, a rostrum; figuratively, a (pyramidal) bed of flowers.

Interestingly, this word means *tower* but figuratively reflects a flowerbed that yields a *pyramidal* shape. This gives a little support to the idea that the Tower of Babel may have been pyramidal or ziggurat shaped.

In what is now Iraq, Robert Koldewey excavated a structure some think to be the foundation of the original Tower of Babel. It underlays a later ziggurat that was thought to be built by Hammurabi in the 19th century B.C.[7]

When people were scattered from the Tower of Babel in the time of Peleg, they likely took this building concept with them to places all over the world. It makes sense that

6. Larry Pierce, "In the Days of Peleg," *Creation*, December 1999, p. 46–49.
7. David Down, "Ziggurats in the News," *Archaeological Diggings*, March–April 2007, p. 3–7.

many of the families that were scattered from Babel took varying ideas of the tower to their new lands and began building projects of their own.

Ziggurats, pyramids, mounds, and the like have been found in many parts of the world — from Mesopotamia to Egypt to South America. The ancient Chinese built pyramids and the Mississippian culture built mounds. Pyramids are classed slightly differently from ziggurats, as are mounds, but the similarities are striking.

Why did the people at Shinar build a tower? Some suspect that they were afraid of another flood, similar to the one that Noah and his sons had informed them about. However, Dr. John Gill casts doubt on this idea.

> It is generally thought what led them to it was to secure them from another flood, they might be in fear of; but this seems not likely, since they had the covenant and oath of God, that the earth should never be destroyed by water any more; and besides, had this been the thing in view, they would not have chosen a plain to build on, a plain that lay between two of the greatest rivers, Tigris, and Euphrates, but rather one of the highest mountains and hills they could have found: nor could a building of brick be a sufficient defense against such a force of water, as the waters of the flood were; and besides, but few at most could be preserved at the top of the tower, to which, in such a case, they would have betook themselves.[8]

The Bible records that the people said among themselves:

> Come, let us build ourselves a city, and a tower whose top is in the heavens; let us make a name for ourselves, lest we be scattered abroad over the face of the whole earth. (Genesis 11:4)

It seems that the tower was to be a special place to keep people together, rather than filling the earth as God had commanded them to (Genesis 9:1). It is possible that the tower was built under the guise that it was a place for sacrifice unto God. This would have prevented people from going too far since they would have to come back to offer sacrifices at Babel.

A recurring theme in Scripture is that people seek to do things they think will honor God but end up disobeying God. One example is when Saul offered a sacrifice when he wasn't supposed to (1 Samuel 13:8–13). It is better to obey

8. Note on Genesis 11:4 in: John Gill, D.D., *An Exposition of the Old and New Testament; The Whole Illustrated with Notes, Taken from the Most Ancient Jewish Writings* (London: printed for Mathews and Leigh, 18 Strand, by W. Clowes, Northumberland-Court, 1809). Edited, revised, and updated by Larry Pierce, 1994–1995 for The Word CD-ROM.

than sacrifice. In fact, many ziggurats and pyramids around the world were used for sacrifice or other sacred religious events, such as burying people (e.g., pharaohs of Egypt). Perhaps the concept of sacred sacrifice and religious festivities with ziggurats was a carryover from Babel.

Regardless, ziggurats and pyramids all over the world are an excellent confirmation of the original recorded in God's Word — the Tower of Babel.

Noah in Royal Genealogies of Europe

The Bible in Genesis 10 gives an outline of family groups that left Babel (see table 2).

These people moved throughout the world and populated virtually every continent. (Was Antarctica ever settled in the past? At this point I am unaware.) Historians have commented on genealogical records in the past and other ancient documents on the origins of various peoples.[9]

These genealogies seem to connect prominent modern houses and royal lines with the Table of Nations listed in the Bible. In these genealogies, Noah is found on the top of the lists on many of these documents, some of which feature variant spellings such as *Noe*, *Noa*, and *Noah*.

One historian discovered a relationship between the ancient name of *Sceaf* (*Seskef*, *Scef*) and the biblical *Japheth*.[10] This seems reasonable, as Japheth has

9. Nennius, *Historia Brittonium*, edited in the 10th century by Mark the Hermit, with English version by the Rev. W. Gunn, rector of Irstead, Norfolk, printed in London, 1819; Flavius Josephus, *The Complete Works of Flavius Josephus the Jewish Historian* (~100 A.D.), translated by William Whiston (~1850 A.D.) (Green Forest, AR: Master Books, 2008).
10. Bill Cooper, *After the Flood* (Chichester, England: New Wine Press, 1995), p. 92–96.

traditionally been seen as the ancestor of the European nations. Most of the European genealogies researched have a variant of *Sceaf* with the exception of

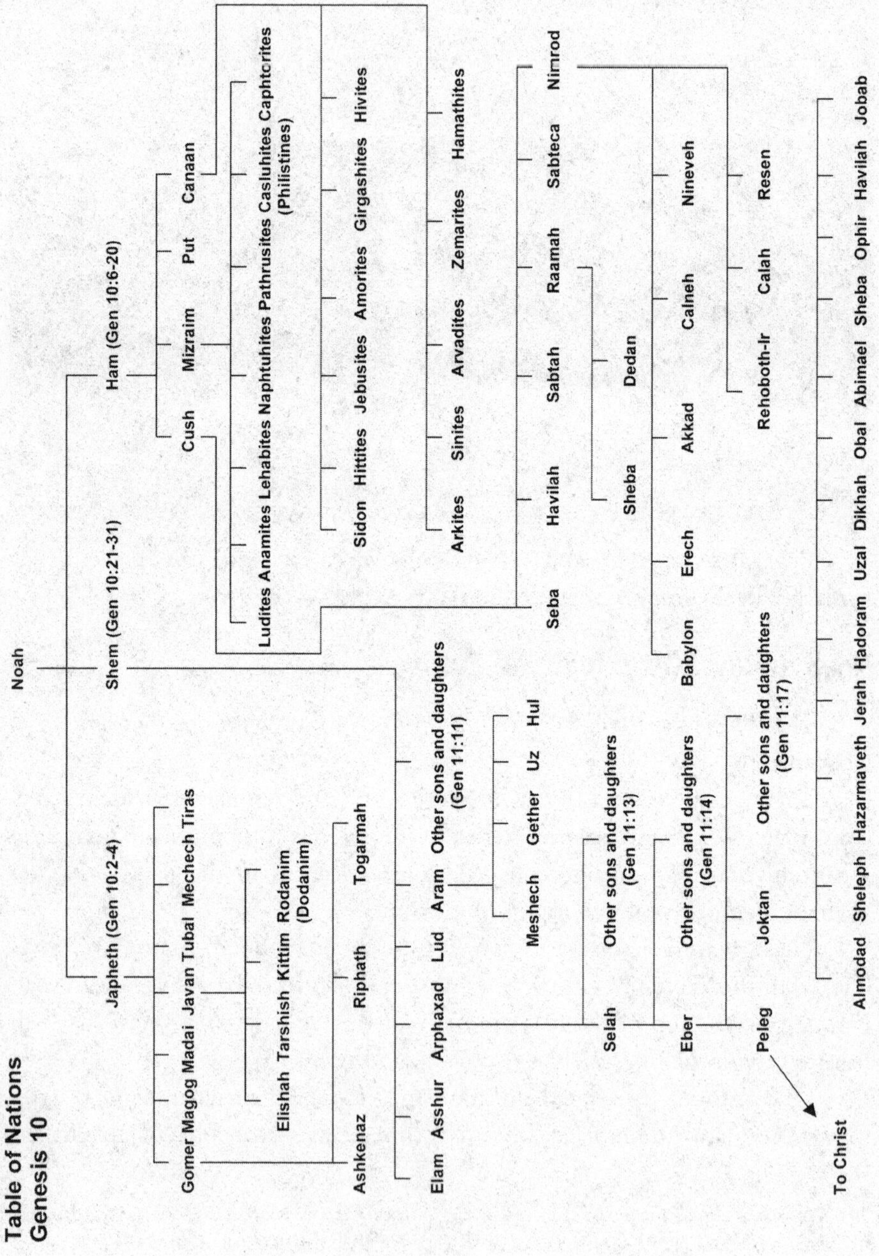

Table 2. Biblical Table of Nations

Irish genealogies, which still used the name *Japheth*. The Irish genealogical chart is reprinted in table 3.[11]

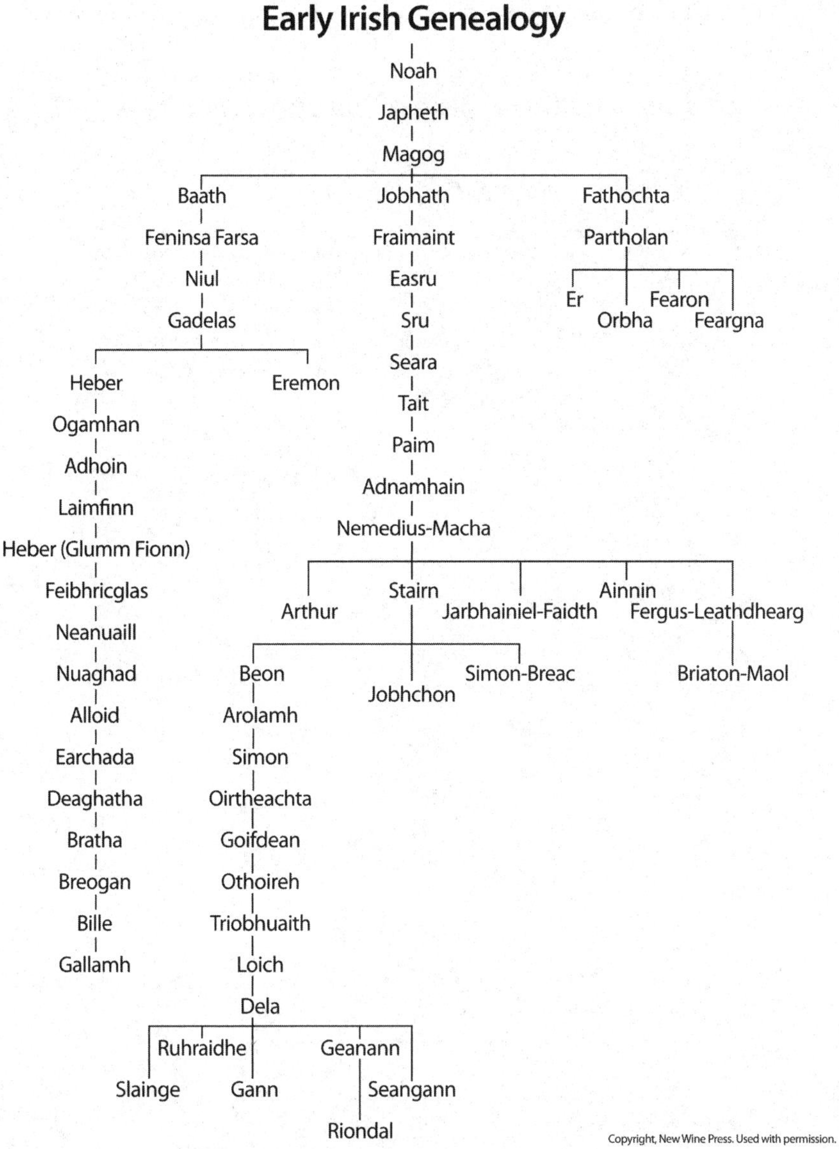

Table 3. Irish Genealogies

Permission for use granted by New Wine Press

11. Ibid., p. 108.

Anglo-Saxon chronologies feature six royal houses.[12] An eighth century Welsh monk living in Powys, Nennius developed a table of nations of the lineages of many of the European people groups from Noah's son Japheth: Gauls, Goths, Bavarians, Saxons, and Romans. Nennius's table of nations is reproduced in table 4.[13]

Nennius and the Table of European Nations

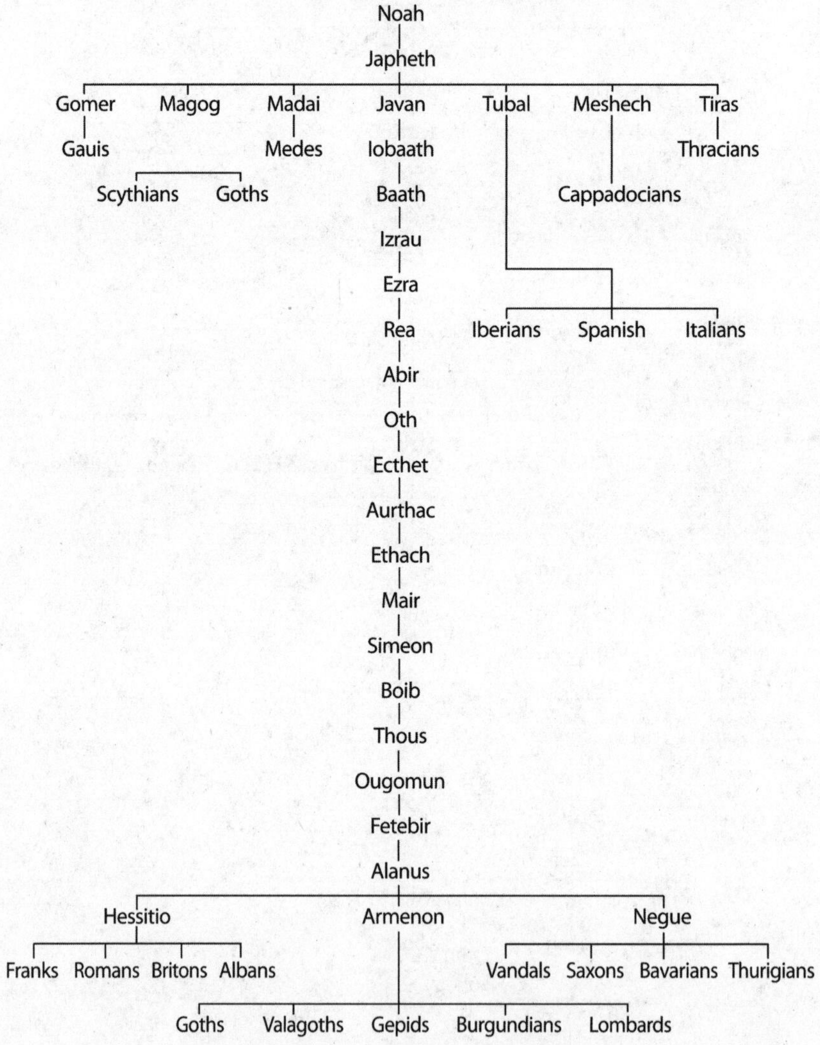

Table 4. Nennius's Table of Nations

Permission for use granted by New Wine Press

12. Ibid., p. 84–86.
13. Ibid., p. 49.

Though it repeats the Goths in two different areas, Nennius's chart bears strong similarities to the history that Josephus recorded,[14] as well as the Bible's Table of Nations. However, there are clearly enough differences to show that it was neither a copy from the biblical text nor from the Jewish historian Josephus.[15]

Chinese records also describe *Nuah* with three sons, *Lo Han, Lo Shen,* and *Jahphu,* according to the Miautso people of China.[16] Although original documents of ancient sources sometimes no longer exist and one has to rely on quotes from other ancient books, it is interesting how in many places we find similarities to the Table of Nations given in the Bible.

Noah's Grandsons' Names Are Everywhere!

History abounds with names that are reused. Names of places become names of people; names of people become names of places. After the Flood, several of Noah's descendants were named for places prior to the Flood. See table 5 for a list.

Table 5. A Few Pre-Flood and Post-Flood References

Name	Bible Reference Pre-Flood	Bible Reference Post-Flood	Person
Havilah	Genesis 2:11	Genesis 10:7, Genesis 10:29	Noah's grandson through Ham; Noah's great, great, great, great grandson through Shem.
Cush	Genesis 2:13	Genesis 10:6	Noah's grandson through Ham
Asshur	Genesis 2:14	Genesis 10:22	Noah's grandson through Shem

Names may vary throughout history. For example, Pennsylvania was named for William Penn; St. Petersburg in Russia was named for Peter the Great, who was ultimately named for Peter who penned two books of the Bible. Names can undergo many changes such as variations in spelling, differences in symbols, and alterations in pronunciation.

14. Whiston, *The Works of Josephus Complete and Unabridged*, p. 36–37.
15. Cooper, *After the Flood*, chapter 3.
16. Edgar Traux, "Genesis According to the Miao People," *Impact*, April 1991, online at www.icr.org/article/341/.

Despite any changes, however, the names of post-Flood regions, cities, rivers, or languages should bear similarity to the names of those leaving Babel. One would be surprised how often these names appear. Table 6 lists some of these.

Table 6. Noah's Descendants' Names Reflected Around the World[17]

Name	Descendant of Noah	What Is It?
Aramaic	Aram	Language that came out of Babel and still survives, likely with changes down the ages. Some short parts of the Bible are written in Aramaic. Jesus spoke it on the cross when He said: "ELOI, ELOI, LAMA SABACHTHANI?" (Mark 15:34).
Cush	Cush	Ancient name of Ethiopia. In fact, people of Ethiopia still call themselves Cushites.
Medes	Madai	People group often associated with the Persians.
Ashkenaz	Ashkenaz	Still the Hebrew name for Germany.
Galacia, Gaul, and Galicia	Gomer	These regions are the old names for an area in modern Turkey, France ,and Northwestern Spain, respectively, where Gomer was said to have lived. His family lines continued to spread across southern Europe. The Book of Galatians by Paul was written to the church at Galatia.
Gomeraeg	Gomer	This is the old name for the Welsh language on the British Isles from their ancestor, Gomer, whose ancestors began to populate the Isle from the mainland.

17. Information in this table comes from the following sources: Whiston, *The Works of Josephus Complete and Unabridged*, p. 36–37; Cooper, *After the Flood*, p. 170–208.

Javan	Javan	This is still the Hebrew name for Greece. His sons, Elishah, Tarshish, Kittim (Chittim), and Dodanim still have reference to places in Greece. For example, Paul, the author who penned much of the New Testament, was from the region of Tarshish (Acts 21:39) and a city called Tarsus. Jeremiah mentions Kittim in Jeremiah 2:10 and is modern-day Cyprus (and other nearby ancient regions that now had varied names such as Cethim, Citius, Cethima Cilicia). The Greeks worshiped Jupiter Dodanaeus from Japheth/ Dodanim. The Elysians, were ancient Greek people.
Meshech/ Moscow	Mechech	Mechech is the old name for Moscow, Russia, and one region called the Mechech Lowland still holds the original name today.
Canaan	Canaan	The region of Palestine that God removed from the Canaanites for their sin and gave as an inheritance to the Israelites beginning with the conquest of Joshua. It is often termed the Holy Land and is where modern-day Israel resides.
Elamites	Elam	This was the old name for the Persians prior to Cyrus.
Assyria	Asshur	Asshur is still the Hebrew name for Assyria.
Hebrew	Eber	This people group and language was named for Eber. Abraham was a Hebrew, and the bulk of the Old Testament is written in Hebrew.
Taurus/ Toros	Tarshish	A mountain range in Turkey.
Tanais	Tarshish	The old name of the Don River flowing into the Black Sea.
Mizraim	Mizraim	This is still the Hebrew name for Egypt.

We Don't Speak the Same Language Anymore!

The Tower of Babel explains why everyone doesn't speak the same language today.

There are over 6,900 spoken languages in the world today.[18] Yet the number of languages emerging from Babel at the time of the dispersion would have been much less than this — likely less than 100 different original language families.

So where did all these languages come from? Linguists recognize that most languages have similarities to other languages. Related languages belong to what are called *language families*. These original language families (probably less than 100) resulted from God's confusion of the language at Babel. Since that time, the original language families have grown and changed into the abundant number of languages today.

Noah's great-great-grandson Eber fathered Peleg when the events at Babel took place. The modern language of Hebrew is named after Eber. Noah's grandson Aram was the progenitor of Aramaic. The Bible lists Noah's grandsons, great-grandsons, great-great-grandsons, and great-great-great-grandsons who received a language at Babel in Genesis 10. Eber and Aram were but two!

From Japheth (Genesis 10:2–5) came at least 14 language families; from Ham (Genesis 10:6–20), 39; from Shem (Genesis 10:22–31), at least 25 (excluding Peleg and other children who may have just been born). The total number of languages that may have come out of Babel according to Genesis 10 may have been at least 78, assuming Noah, Ham, Shem, Japheth, and Peleg didn't receive a new language. This excludes some descendants of Shem who are given slight mention in Genesis 11:11–17; they may have also received a language.

Both *Vistawide World Languages and Cultures*[19] and *Ethnologue*,[20] companies that provide statistics on language, agree that only 94 languages *families* have been so far ascertained. With further study in years to come, this may change, but this figure is well within the range of families that dispersed from Babel (Genesis 10).

Is it feasible for 7,000 languages to develop from less than 100 in 4,000 years? The languages that came out of the confusion at Babel were "root languages" or language families. Over time, those root languages have varied by borrowing from other languages, developing new terms and phrases, and losing previous words and phrases.

18. Vistawide, "World Language Families," www.vistawide.com/languages/language_families_statistics1.htm.
19. Ibid.
20. Ethnologue, "Statistical Summaries," www.ethnologue.com/ethno_docs/distribution.asp?by=family.

Let's look at changes in the English language, as an example. English has changed so much over the course of 1,000 years that early speakers would hardly recognize it today. Table 7 provides a look at the changes in Matthew 6:9.

Table 7[21]

Beginning of Matthew 6:9	**Date**
Our Father who art in heaven and/or Our Father who is in heaven	Late Modern English (1700s)
Our father which art in heauen	Early Modern English (1500–1700) (KJV 1611)
Oure fader that art in heuenis	Middle English (1100–1500)
Fæder ure þu þe eart on heofonum	Old English (c. A.D. 1000)

Just as English has changed significantly over the past 1,000 years, it becomes easy to see how the original languages at Babel could have rapidly changed in the 4,000 years since that time, whether spoken or written.

In conclusion, there exist a great many confirmations of the Bible's account of the Tower of Babel and what happened as a result. Even stories about a tower and sudden language changes appear in ancient histories from Sumerian, Grecian, Polynesian, Mexican, and Native American sources.[22] This is what we would expect since the Tower of Babel was a real event. Language changes, ziggurats, names of Noah found throughout the world, and tower legends are excellent confirmations of the events at Babel.

21. Comparison of Matthew 6:9, Mansfield University, www.faculty.mansfield.edu/bholtman/holtman/101/GmcVaterunser.pdf.
22. Pam Sheppard, "Tongue-Twisting Tales," *Answers*, April–June 2008, p.56–57.

29

When Does Life Begin?

DR. TOMMY MITCHELL

When does human life begin? This question has confounded individuals and divided our society. Opinions have come from the right and the left, from prolife advocates and those in favor of abortion on demand, from physicians and lawyers, from the pulpit and the courtroom.

When did I begin to be me? Is this a scientific question or a theological one?[1] Would this question be best left to scientists or to preachers and philosophers? Information and viewpoints from secular scientific sources and from theologians will be examined in this chapter, but the ultimate answer can have no authority unless that answer is based squarely on the Word of God. The Bible, because it is true, will not disagree with genuine science. Furthermore, the Bible is the only valid and consistent basis for making moral judgments, since it comes from the Creator of the whole world and all people in it. Any other basis for judgment would be a useless clamor of divergent, man-made opinions.

Who Is More Human?

Life is a continuum. From the season of growing in the womb to being born, from playing as a child to growing older, each stage of life seems to blend gracefully (or not so gracefully in my case) into the next. Life progresses

1. The answer to the question "What is life?" is beyond the scope of this article. There are several excellent resources dealing with this topic: see James Stambaugh, " 'Life' According to the Bible, and the Scientific Evidence," *TJ* 6, no. 2 (1992): 98–121, online at www.answersingenesis.org/tj/v6/i2/life.asp.

and time passes, culminating in death. Death, a very visible end point, is more easily defined than the point at which the continuum of human life begins.

Where is the starting point? If life is indeed a continual process, can we not just work backward to its beginning? There are a variety of opinions about life's beginnings. Many say life begins at conception. Others argue strongly that life does not start until implantation in the womb. Still others say that human life begins only when the umbilical cord is cut, making the newborn child an independent agent. How is fact separated from opinion?

Perhaps another way to ask the question is, when do we become human? Certainly a child sitting on grandpa's knee or a fully grown adult would be considered human. Is the adult *more* human than the child? Of course not. No reasonable person would consider the child to be less human. At what point along the journey did this child become human? Was it at conception, somewhere during his development, or at birth?

The Process

The initial event along the road of human development is fertilization. Twenty-three chromosomes from the mother and 23 chromosomes from the father are combined at the time of fertilization. At this point, the genetic makeup of the individual is determined. At this time, a unique individual, known as a *zygote*, begins to exist. But is this zygote human?

This zygote then divides again and again. Some cells develop into the placenta and are essential for implantation. Other cells develop into the anatomical parts of the baby.[2] The number of cells increases rapidly, and the name changes as the number increases. By the time this rapidly dividing ball of cells arrives in the uterus, it is called a *blastocyst*. Implantation in the uterine wall normally occurs about six days after fertilization.[3]

For reasons unclear to medical science, the mass of cells sometimes splits to produce identical twins. These twins are called identical because their sets of chromosomes are identical. Depending upon the stage of development when the split occurs, the twins may share certain placental parts, but the twins produced are distinct individuals. If the split occurs between the 13th and 15th days, the twins will actually share body parts, a condition known as *conjoined*

2. This process, called *differentiation,* is the process by which the dividing cells gradually become different from one another.
3. The name of the rapidly diving ball of cells continues to change as size and shape changes, with the name *embryo* being assigned at about three weeks after fertilization. The term *fetus* is used from about the eighth week of development.

twins, commonly called *Siamese twins.* (After that time, development and differentiation are too far along to allow successful splitting.)

Even though the names arbitrarily change throughout this process and certain milestones in development are evident, the process set in motion at the moment of conception is a continuous chain of events. In this sequence, groups of cells multiply and develop into specific body parts with amazing precision and a remarkably low rate of error, considering the complexity of changes that must occur. However, at no time in this process is there a scientific point at which the developing individual clearly "becomes a person," any more than a baby becomes more human when it walks, talks, or is weaned. These milestones in zygote, blastocyst, embryonic, and fetal development are simply descriptions of anatomy, not hurdles met in the test of humanness. From a scientific point of view, the words are arbitrary and purely descriptive.

Can Science Help?

Scientists have studied the marvelous process previously described for decades. The changes in the form of the embryo through each stage are well documented. The question still remains, at what point does human life begin? There are numerous positions on this. Some of these will be reviewed here.

A Genetic Position

The simplest view is based on genetics. Those who hold this position argue that since a genetically unique individual is created at the time of fertilization, each human life begins at fertilization. The zygote formed at fertilization is different from all others and, if it survives, will grow into a person with his or her own unique set of genes. In this view, the terms *fertilization* and *conception* are interchangeable. Thus, in this view, life would be said to begin at conception.

The phenomenon of twinning is sometimes used to argue against this position. Until about day 14, there is the possibility that the zygote will split, producing twins. Those who oppose a genetic view say that there is no uniqueness to the zygote, no humanness or personhood, until the potential for twinning has passed. They ask, if the zygote is an individual "person" at fertilization, then what is the nature of that "personhood" if the zygote should split into two individuals?

Another objection to this view is the fact the many fertilized eggs never successfully implant. An estimated 20–50 percent of fertilizations die or are spontaneously aborted.[4] Thus, those who raise this objection hold that, since

4. Christian Answers Net, "Does Life Begin Only When the Embryo Implants?" www.christiananswers.net/q-sum/q-life014.html.

there are such a large number of zygotes that never fully develop, those zygotes are not truly human.

However, neither of the objections can be so easily supported. The twinning objection falls short when one considers the problem presented by the existence of so-called Siamese twins. In these cases, the zygote does not completely split, and the children are born joined together, often sharing certain body organs. Nonetheless, both twins have distinct personalities and are distinct individuals. Here the "personhood" obviously could not be granted after twinning since the process was never completed.

The second objection, the high loss rate of zygotes, is also not logical. The occurrence of spontaneous abortions does not mean that the lost were not fully human, any more than the development of some deadly disease in a child makes the child suddenly nonhuman.

The Implantation View

An increasingly heard viewpoint today is related to the implantation of the blastocyst into the uterine lining. This implantation process begins on day six following fertilization and can continue until around day nine. Some now suggest that it is not until this time that the zygote can be called human life. However, achieving implantation does not make the individual more human; rather, implantation makes the individual more likely to survive.

Interestingly enough, the popularity of this view has led to some changes in how some define conception. Until recently, *conception* was synonymous with *fertilization*. In fact, in the 26th edition of *Stedman's Medical Dictionary*, conception was defined as the "act of conceiving, or becoming pregnant; fertilization of the oocyte (ovum) by a spermatozoon to form a viable zygote."[5] Conception was defined as the time of fertilization.

However, something interesting happened in the next five years. In the 27th edition of *Stedman's Medical Dictionary*, conception is defined as follows: "Act of conceiving; the implantation of the blastocyte in the endometrium."[6] Note here that *implantation* is now the defining point in conception. The scientific community arbitrarily, without any scientific justification, redefined the starting point of life.

According to the redefined view, a zygote less than nine or so days old, having not yet completed implantation, would not be considered alive. If it is not alive, it certainly cannot be human. This change was completely arbitrary,

5. *Stedman's Medical Dictionary*, 26th edition (Baltimore: Williams & Wilkins, 1995), p. 377.
6. *Stedman's Medical Dictionary*, 27th edition (Baltimore: Williams & Wilkins, 2000), p. 394.

for there was no basic change in the understanding of the developmental process that would make this redefinition necessary.

The new definition would, however, have great implications in the political, ethical, and moral arenas. Personal and governmental decision-making on such issues as embryonic stem cell research, cloning, and the so-called "morning after pill" directly depends on the validity of this definition. If preimplantation blastocysts were not really alive, they could be guiltlessly harvested or destroyed prior to the six-to-nine day mark because "conception" had not yet occurred.

The Embryological View

The embryological view holds that human life begins 12–14 days after fertilization, the time period after which identical twins would not occur. (*Embryo* can refer to the developing baby at two to three weeks after fertilization or more loosely to all the stages from zygote to fetus.) No individuality and therefore no humanness is considered to exist until it is not possible for twinning to happen. Here, the initial zygote is not human and possesses no aspect of "personhood." As stated previously, this line of reasoning fails because of the shortcoming of the twinning argument itself. Specifically, the fact that conjoined (Siamese) twins are distinct persons is undeniable; their humanity is not obviated by the fact that they share body parts.

The Neurologic View

In this view, human life begins when the brain of the fetus has developed enough to generate a recognizable pattern on an electroencephalogram (EEG). Here, it is proposed that humanness is attained when the brain has matured to the point that the appropriate neural pathways have developed.[7] This point is reached at about 26 weeks after fertilization. After this level of maturation has been achieved, the fetus is presumably able to engage in mental activity consistent with being human.

Others take a different view of neurological maturation and propose that human life begins at around 20 weeks gestation. This is the time when the thalamus, a portion of the brain that is centrally located, is formed. The thalamus is involved in processing information before the information reaches the cerebral cortex and also is a part of a complex system of neural connections that play a role in consciousness.

7. H.J. Morowitz and J.S. Trefil, *The Facts of Life: Science and the Abortion Controversy* (New York: Oxford University Press, 1992).

These distinctions are arbitrary. The developing brain does display some electrical activity before the 26-week mark. It could just as easily be argued that any brain activity would constitute humanness.

The Ecological View

Proponents of the ecological view hold that the fetus is human when it reaches a level of maturation when it can exist outside the mother's womb.[8] In other words, a fetus is human when it can live separated from its mother. Here the limiting factor is usually not neurological development, but rather the degree of maturation of the lungs.

This view of humanness presents a very interesting problem. The problem is that, over the last century, we have been becoming human earlier and earlier. Here the issue is not the actual stage of development of the fetus. The limiting factor rather is the current state of medical technology. For example, some 20 years ago the age of viability of a prematurely born fetus was about 28 weeks; today it is around 24 weeks. Thus, in this view, man himself, through his advances in technology, can grant humanness where it did not previously exist!

The Birthday View

Some hold the position that human life begins only at the point when the baby is born. Here the baby is human when the umbilical cord is cut, and the child survives based on the adequate functioning of its own lungs, circulatory system, etc.

The shortcoming of this reasoning is that even after birth, the child is not truly independent of its mother. Without care from someone, an infant would die very shortly after birth. This supposed "independence" is very much an arbitrary concept.

Other Views

There are still other points of view as to the question of when human life begins. Some suggest that a fetus is human when the mother can feel it move in the womb. Others say that humanness begins when the child takes its first breath on its own. Francis Crick, one of the co-discoverers of the structure of DNA, says that a child should not be declared "human" until three days after birth.[9]

---

8. Scott Gilbert, "When Does Human Life Begin?" Developmental Biology 8e Online, www.8e.devbio.com/article.php?ch=21&id=7.
9. Mark Blocher, *Vital Signs* (Chicago, IL: Moody Press, 1992), p. 91.

There are clearly significant differences in the way that the scientific community views the beginning of life. There is no obvious consensus among scientists about when human life begins. So, can science really help us answer this question? Perhaps science, by its nature, is not capable of dealing directly with this problem. Scott Gilbert, PhD, professor of biology at Swarthmore College, notes, "If one does not believe in a 'soul,' then one need not believe in a moment of ensoulment. The moments of fertilization, gastrulation, neurulation, and birth, are then milestones in the gradual acquisition of what it is to be human. While one may have a particular belief in when the embryo becomes human, it is difficult to justify such a belief solely by science."[10]

If Not Science, Then What?

If science cannot give us the answer, then is there another place we can turn? As Christians, we should turn to the Bible, God's Word, to see if there is a solution to this dilemma.

Psalm 139:13–16

Perhaps the most often quoted portion of Scripture on this subject is Psalm 139:13–16.

> For You formed my inward parts:
> You covered me in my mother's womb.
> I will praise You, for I am fearfully and wonderfully made;
> Marvelous are Your works,
> And that my soul knows very well.
> My frame was not hidden from You,
> When I was made in secret,
> And skillfully wrought in the lowest parts of the earth.
> Your eyes saw my substance, being yet unformed.
> And in Your book they all were written,
> The days fashioned for me,
> When as yet there were none of them.

Here we read about God knowing the Psalmist while he was "yet unformed," while he was being "made in secret," in a place invisible to human eyes. The uses of the personal pronouns in these verses indicate that there was, indeed, a person present before birth. R.C. Sproul notes, "Scripture does assume a continuity of life from before the time of birth to after the time of

10. Gilbert, "When Does Human Life Begin?"

birth. The same language and the same personal pronouns are used indiscriminately for both stages."[11]

Jeremiah 1:4–5

> Then the word of the LORD came to me, saying:
> "Before I formed you in the womb I knew you;
> Before you were born I sanctified you;
> I ordained you a prophet to the nations."

Here God tells Jeremiah that he was set apart before he was born. This would indicate that there was personhood present before Jeremiah's birth. The verse even indicates that God considered Jeremiah a person and that he was known before he was formed. Sproul indicates, "Even those who do not agree that life begins before birth grant that there is continuity between a child that is conceived and a child that is born. Every child has a past before birth. The issue is this: Was that past personal, or was it impersonal with personhood beginning only at birth?"[12]

Psalm 51:5

This verse is frequently used to make the case for human life beginning at conception. It reads:

> Behold, I was brought forth in iniquity,
> And in sin my mother conceived me.

The most often heard interpretation of this passage is that the author, David, sees that he was sinful even at the time he was conceived. If he was not a person, then it follows that he could not have a sinful human nature at that time. A prehuman mass of cells could not have any basis for morality. Only the "humanness" occurring at the time of conception would allow David to possess a sinful nature at that time.

Life before Birth

These Scriptures reveal that there is personhood before birth. The personal nature of the references in the Bible shows how God views the unborn child. Another text frequently used to prove the humanness of the fetus is found in the first chapter of Luke:

---

11. R.C. Sproul, *Abortion: A Rational Look at an Emotional Issue* (Colorado Springs, CO: NavPress, 1990), p. 53–54.
12. Ibid., p. 55.

> Now Mary arose in those days and went into the hill country with haste, to a city of Judah, and entered the house of Zacharias and greeted Elizabeth. And it happened, when Elizabeth heard the greeting of Mary, that the babe leaped in her womb; and Elizabeth was filled with the Holy Spirit. Then she spoke out with a loud voice and said, "Blessed are you among women, and blessed is the fruit of your womb! But why is this granted to me, that the mother of my Lord should come to me? For indeed, as soon as the voice of your greeting sounded in my ears, the babe leaped in my womb for joy" (Luke 1:39-44).

We read in this passage of a meeting between Mary the mother of Jesus and Elizabeth, her cousin, the mother of John the Baptist. Here Elizabeth describes the life in her womb as "the babe." God's inspired Word reports Elizabeth's assessment that John "leaped" in the womb because of the presence of Jesus. Some try to discount this episode as a miracle, claiming it does not relate to the personhood of the unborn. Nonetheless, God's Word describes this unborn child as capable of exhibiting joy in the presence of his Savior.

Are the Unborn of Less Worth?

Exodus 21 has been put forth by some to suggest the God himself holds that the life of an unborn is less valuable than the life of an adult.

> If men fight, and hurt a woman with child, so that she gives birth prematurely, yet no harm follows, he shall surely be punished accordingly as the woman's husband imposes on him; and he shall pay as the judges *determine.* But if *any* harm follows, then you shall give life for life, eye for eye, tooth for tooth, hand for hand, foot for foot . . . (Exodus 21:22–24).

This verse gives directions for dealing with a situation in which two men are fighting and they accidentally harm a pregnant woman. Two circumstances are noted here. The first situation is when the woman gives birth prematurely and "no harm follows." The common interpretation states that here the child is lost due to a premature birth, and the woman herself does not suffer a serious injury. Here the penalty is a fine of some type to compensate for the loss of the child.

The second circumstance is "if any harm follows." Here the common interpretation is that is the woman gives birth prematurely, the child dies, and the woman herself dies. Here the penalty is life for life. It is argued that since there is only a fine imposed in the first circumstance for the loss of only the premature child while the death penalty is imposed for the loss of the mother, the unborn is less valuable than an adult. Thus, the unborn need not be considered to have achieved full humanness before birth.

However, upon closer examination, this type of interpretation may not be valid. The "harm" indicated in these verses may refer to the child and not to the mother. In the first circumstance, the injured mother gives birth prematurely and no "harm" comes to the child. In other words, the premature child lives. Thus, a fine is levied for causing the premature birth and the potential danger involved. In the second situation, there is a premature birth and the "harm" that follows is the death of the child. Here the penalty is life for life. Therefore, the Bible does not hold that the life of the unborn is less valuable than the life of an adult.

John Frame, in the book *Medical Ethics,* says this, "There is *nothing* in Scripture that even remotely suggests that the unborn child is anything less *than a human person from the moment of conception*"[13] (emphasis his). Here, conception is meant to imply the time of fertilization.

So Where Are We?

A purely scientific examination of human development from the moment of fertilization until birth provides no experimental method that can gauge humanness. Stages of maturation have been described and cataloged. Chemical processes and changes in size and shape have been analyzed. Electrical activity has been monitored. However, even with this vast amount of knowledge, there is no consensus among scientists as to where along this marvelous chain of events an embryo (or zygote or fetus or baby, depending upon who is being asked) becomes human.

Science has, however, revealed the intricate developmental continuum from fertilization, through maturation, to the birth of the child. Each stage flows seamlessly into the next with a myriad of detailed embryological changes followed by organ growth and finely tuned development choreographed with precision. The more we learn about the process, the more amazingly complex we find it to be.

Life Begins at Conception

Although science has shown us the wonderful continuity of the development of life throughout all its stages, science has been unable to define the

13. John Frame, *Medical Ethics* (Phillipsburg, NJ: Presbyterian and Reformed Publishing Company, 1988), p. 95.

onset of humanness. However, there is ample information in Scripture for us to determine the answer to this problem.

The Bible contains numerous references to the unborn.[14] Each time the Bible speaks of the unborn, there is reference to an actual person, a living human being already in existence. These Scriptures, taken in context, all indicate that God considers the unborn to be people. The language of the text continually describes them in personal terms.

Since the Bible treats those persons yet unborn as real persons, and since the development of a person is a continuum with a definite beginning at the moment of fertilization, the logical point at which a person begins to be human is at that beginning. The answer is that life begins at conception (using the now older definition of the term, here to be synonymous with fertilization). Frankly, no other conclusion is possible from Scripture or science.

What are the implications of this conclusion? Why is this important? Quite simply, the status of the zygote/embryo/fetus is central to many issues facing our society. The most obvious issue in this regard is abortion. If the zygote is a human life, then abortion is murder. The same can be said of issues surrounding the embryonic stem cell debate. If the embryo is human, then destroying it is murder, no matter what supposedly altruistic reason is given as justification. The ethics of cloning require consideration of the concept of humanness and the timing of its onset. A person's acceptance or rejection of the controversial morning after pill is based upon the determination of when human life begins.[15]

Complex issues may not have simple solutions, but when examined objectively in light of God's Word, without biases introduced by other motivations, God's truth will reveal the correct answers. Science can give us better understanding of the world God created, and what we see in God's world will agree with the truth we read in God's Word. We dare not play word games with human life to justify personal agendas. Scripture provides no real loopholes or escape clauses to excuse us from the principle that God created human beings in His own image, designed them to reproduce after their kind, and sent Jesus Christ into the world as a human being to die for us all, thus demonstrating the inestimable love our Creator has for each human life.

14. See also Genesis 25:21–23; Isaiah 45.
15. David Menton, "Plan B: Over-the-Counter Abortion?" Answers in Genesis, www.answersingenesis.org/articles/am/v2/n1/plan-b.

30

Do Creationists Believe in "Weird" Physics like Relativity, Quantum Mechanics, and String Theory?

DR. DANNY FAULKNER

Science is the study of the natural world using the five senses. Because people use their senses every day, people have always done some sort of science. However, good science requires a systematic approach. While ancient Greek science did rely upon some empirical evidence, it was heavily dominated by deductive reasoning. Science as we know it began in the 17th century. The father of the scientific method is Sir Francis Bacon (1561–1626), who clearly defined the scientific method in his *Novum Organum* (1620). Bacon also introduced inductive reasoning, which is the foundation of the scientific method.

The first step in the scientific method is to define clearly a problem or question about how some aspect of the natural world operates. Some preliminary investigation of the problem can lead one to form a *hypothesis*. A hypothesis is an educated guess about an underlying principle that will explain the phenomenon that we are trying to explain. A good hypothesis can be tested. That is, a hypothesis ought to make predictions about certain observable phenomena, and we can devise an experiment or observation to test those predictions. If we conduct the experiment or observation and find that the predictions match the results, then we say that we have confirmed our hypothesis, and we have some confidence that our hypothesis is correct. On the other hand, if our predictions are not borne out, then we say that our hypothesis is disproved, and we can either alter our hypothesis or develop a new one and repeat the process of testing. After repeated testing with positive results, we say that the hypothesis is confirmed, and we have confidence that our hypothesis is correct.

Notice that we did not "prove" the hypothesis, but that we merely confirmed it. This is a big difference between deductive and inductive reasoning. If we have a true premise, then properly applied deductive reasoning will lead to a true conclusion. However, properly applied inductive reasoning does not necessarily lead to a true conclusion. How can this be? Our hypothesis may be one of several different hypotheses that produce the same experimental or observational results. It is very easy to assume that our hypothesis, when confirmed, is the end of the matter. However, our hypothesis may make other predictions that future, different tests may not confirm. If this happens, then we must further modify or abandon our hypothesis to explain the new data. The history of science is filled with examples of this process, and we ought to expect that this will continue.

This puts the scientist in a peculiar position. While we can definitely disprove a number of propositions, we can never be entirely sure that what we believe to be true is indeed true. Thus, science is a very changing thing. History shows that scientific "truth" changes over time. The uncertainty is the reason why continued testing of our ideas is so important in science. Once we test a hypothesis many times, we gain enough confidence that it is correct, and we eventually begin to call our hypothesis a *theory*. So a theory is a grown-up, well-developed hypothesis.

At one time, scientists conferred the title of *law* to well-established theories. This use of the word "law" probably stemmed from the idea that God had imposed some order (law) onto the universe, and our description of how the world operates is a statement of this fact. However, with a less Christian understanding of the world, scientists have departed from using the word *law*. Scientists continue to refer to older ideas, such as Newton's law of gravity or laws of motion as law, but no one has termed any new ideas in science as law for a very long time.

In 1687, Sir Isaac Newton (1643–1727) published his *Principia*, which detailed work that he had done about two decades earlier. In the *Principia*, Newton presented his law of gravity and laws of motion, which are the foundation of the branch of physics known as mechanics. Because he required a mathematical framework to present his ideas, Newton invented calculus. His great breakthrough was to hypothesize that the force that held us to the earth was the same force that kept the moon orbiting around the earth each month. From knowledge of the moon's distance from the earth and orbital period, Newton used his laws of motion to conclude that the moon is accelerated toward the earth 1/3600 of the measured acceleration of gravity at the surface of the earth. The fact that we on the earth's surface are 60 times closer to the earth's center than the moon allowed Newton to devise his inverse square law for gravity ($60^2 = 3{,}600$).

Isaac Newton (1643–1727)

This unity of gravity on the earth and the force between the earth and moon was a good hypothesis, but could Newton test it? Yes. When Newton applied his laws of gravity and motion to the then-known planets orbiting the sun (Mercury, Venus, Earth, Mars, Jupiter, and Saturn), he was able to predict several things:

1. The planets orbit the sun in elliptical orbits with the sun at one focus of the ellipses.
2. The line between the sun and a planet sweeps out equal areas in equal intervals of time.
3. The square of a planet's orbital period is proportional to the third power of the planet's mean distance from the sun.

These three statements are known as Kepler's three laws of planetary motion, because the German mathematician Johannes Kepler (1571–1630) had found them in a slightly different form several decades before Newton. Kepler empirically found his three laws by studying data on planetary motions taken by the Danish astronomer Tycho Brahe (1546–1601) over a period of 20 years in the latter part of the 16th century. Kepler arrived at his result by laborious trial and error for over two decades, but he had no explanation of why the planets behaved the way that they did. Newton easily showed (or predicted) that the planets must follow Kepler's law as a consequence of his law of gravity.

Johannes Kepler (1571–1630)

Many other predictions of Newton's new physics followed. Besides Earth, Jupiter and Saturn had satellites that obeyed Newton's formulation of Kepler's three laws. Newton's good friend who privately funded the publication of the *Principia*, Sir Edmond Halley (1656–1742), applied Newton's work to the observed motions of comets. He found that comets also followed the laws, but that their orbits were much more elliptical and inclined than the orbits of planets. In his study, Halley noticed that one comet that he observed had an orbit identical to one seen about 75 years before and that both comets had a 75-year orbital period. Of course, when the comet returned once again, Halley was long dead, but this comet bears his name.

In 1704, Newton first published his other seminal work in physics, *Optics*. In this book, he presented his theory of the wave nature of light. Together, his *Principia* and *Optics* laid the foundation of physics as we know it. Over the next two centuries, scientists applied Newtonian physics to all sorts of situations, and in each case the predictions of the theory were borne out by experiment and observation. For instance, William Herschel stumbled upon the planet Uranus in 1781, and its orbit followed Kepler's three laws as well. However, by 1840, astronomers found that there were slight discrepancies between the predicted and observed motion of Uranus. Two mathematicians independently hypothesized that there was an additional planet beyond Uranus whose gravity was tugging on Uranus. This led to the discovery of Neptune in 1846. These successes gave scientists a tremendous confidence in Newtonian physics, and thus Newtonian physics is one of the most well-established theories in history. However, by the end of the 19th century, experimental results began to conflict with Newtonian physics.

Quantum Mechanics

Near the end of the 19th century, physicists turned their attention to how hot objects radiate, with one practical application being the improvement of efficiency of the filament of the recently invented light bulb. Noting that at low temperatures good absorbers and emitters of radiation appear black, they dubbed a perfect absorber and emitter of radiation a *black body*. Physicists experimentally determined that a black body of a certain temperature emitted the greatest amount of energy at a certain frequency and that the amount of energy that it radiated diminished toward zero at higher and lower frequencies. Attempts to explain this behavior with classical, or Newtonian, physics worked very well at most frequencies but failed miserably at higher frequencies. In fact, at very high frequencies, classical physics required that the energy emitted increase toward infinity.

In 1901, the German physicist Max Planck (1858–1947) proposed a solution. He suggested that the energy radiated from a black body was not exactly in waves as Newton had shown, but was instead carried away by tiny particles (later called photons). The energy of each photon was proportional to its frequency. This was a radical departure from classical physics, but this new theory did exactly explain the spectra of black bodies.

Max Planck (1858–1947)

In 1905, the German-born physicist Albert Einstein (1879–1955) used Planck's theory to explain the photoelectric effect. What is the photoelectric effect? A few years earlier, physicists had discovered that when light shone on a metal to which an electric potential was applied, electrons were emitted. Attempts to explain the details of this phenomenon with classical physics had failed, but Einstein's application of Planck's theory explained it very well.

Other problems with classical physics had mounted. Physicists found that excited gas in a discharge tube emitted energy at certain discrete wavelengths or frequencies. The exact wavelengths of emission depended upon the composition of the gas, with hydrogen gas having the simplest spectrum. Several physicists investigated the problem, with the Swedish scientist Johannes Rydberg (1854–1919) offering the most general description of the hydrogen spectrum in 1888. However, Ryberg did not offer a physical explanation. Indeed, there was no classical physics explanation for the spectral behavior of hydrogen gas until 1913, when the Danish physicist Niels Bohr (1885–1962) published his model of the hydrogen atom that did explain hydrogen's spectrum.

In the Bohr model, the electron orbits the proton only at certain discrete distances from the proton, whereas in classical physics the electron can orbit at any distance from the proton. In classical physics the electron must continually emit radiation as it orbits, but in Bohr's model the electron emits energy only when it leaps from one possible orbit to another. Bohr's explanation of the hydrogen atom worked so well that scientists assumed that it must work for other atoms as well. The hydrogen atom is very simple, because it consists of only two particles, a proton and an electron. Other atoms have increasing numbers of particles (more electrons orbiting the nucleus, which contains more protons as well as neutrons) which makes their solutions much more difficult, but the Bohr model worked for them as well. The Bohr model is essentially the model that most of us learned in school.

While Bohr's model was obviously successful, it seemed to pull some new principles out of the air, and those principles contradicted principles of classical physics. Physicists began to search for a set of underlying unifying principles to explain the model and other aspects of the emerging new physics. We will omit the details, but by the mid-1920s, those new principles were in place. The basis of this new physics is that in very small systems, as within atoms, energy can exist in only certain small, discrete amounts with gaps between adjacent values. This is radically different from classical physics, where energy can assume any value. We say that energy is quantized because it can have only certain discrete values, or quanta. The mathematical theory that explains the energies of small systems is called quantum mechanics.

Quantum mechanics is a very successful theory. Since its introduction in the 1920s, physicists have used it to correctly predict the behavior and characteristics of elementary particles, nuclei of atoms, atoms, and molecules. Many facets of modern electronics are best understood in terms of quantum mechanics. Physicists have developed many details and applications of the theory, and they have built other theories upon it.

Quantum mechanics is a very successful theory, yet a few people do not accept it. Why? There are several reasons. One reason for rejection is that the postulates of quantum mechanics just do not feel right. They violate our everyday understanding of how the physical world works. However, the problem is that very small particles, such as electrons, do not behave the same way that everyday objects do. We invented quantum mechanics to explain small things such as electrons because our everyday understanding of the world fails to explain them. The peculiarities of quantum mechanics disappear as we apply quantum mechanics to larger systems. As we increase the size and scope of small systems, we find that the oddities of quantum mechanics tend to smear out and assume properties more like our common-sense perceptions. That is, the peculiarities of quantum mechanics disappear in larger, macroscopic systems.

Another problem that people have with quantum mechanics is certain interpretations applied to quantum mechanics. For instance, one of the important postulates of quantum mechanics is the Schrödinger wave equation. When we apply the Schrödinger equation to a particle such as an electron, we get a mathematical wave as a description of the particle. What does this wave mean? Early on, physicists realized that the wave represented a probability distribution. Where the wave had a large value, the probability was large of finding the particle in that location, but where the wave had low value, there was little probability of finding the particle there. This is strange. Newtonian physics had led to determinism — the absolute knowledge of where a particle was at a particular time from the forces and other information involved. Yet, the probability function does accurately predict the behavior of small particles such as electrons. Even Albert Einstein, whose early work led to much of quantum mechanics, never liked this probability. He once famously remarked, "God does not play dice with the universe." Erwin Schrödinger (1887–1961), who had formulated his famous Schrödinger equation stated in 1926, "If we are going to stick to this ****** quantum-jumping, then I regret that I ever had anything to do with quantum theory."

Note that with the probability distribution we cannot know precisely where a particle is located. A statement of this is the Heisenberg Uncertainty Principle (named for Werner Heisenberg, 1901–1976). We explain this by acknowledging that particles such as electrons have a wave nature as well as a particle nature. For that matter, we also believe that waves (such as light and sound) also have a particle nature. This wave-particle duality is a bit strange to us, because we do not sense it in everyday experience, but it is borne out by numerous experimental results.

For instance, let us consider a double slit experiment. If we send a wave toward an obstruction with two slits in it, the wave will pass through both slits and produce a distinctive interference pattern behind the slits. This is because the wave passes through both slits. If we send a large number of electrons toward a similar apparatus, the electrons will also produce an interference pattern behind the slits, suggesting that the electrons (or their wave functions) went through both slits. However, if we send one electron at a time toward the slits and look for the emergence of each electron behind the slits, we will find that each electron will emerge through one slit or the other, but not both. How can this be? Indeed, this is perplexing. The most common resolution is the Copenhagen interpretation, named for the city where it was developed. This interpretation posits that an individual electron does not go through either slit, but instead exists in some sort of meta-stable state between the two states until we observe (detect) the electrons. At the point of observation, the electron's wave equation collapses, allowing the electron to assume one state or the other. Now, this is weird, but most alternate explanations are even weirder, so you might understand why some people may have a problem with quantum mechanics.

Is there a way out of this dilemma? Yes. Why do we need an interpretation to quantum mechanics? No one demanded any such interpretation of Newtonian physics. No one asked, "What does it mean?" There is no meaning, other than the fact that Newtonian physics does a good job of describing what we see in the macroscopic world. The same ought to be true for quantum mechanics. It does a good job of describing the microscopic world. Whereas classical physics introduced determinism, quantum mechanics introduced indeterminism. This indeterminism is fundamental in the sense that uncertainty in outcome will still exist even if we have all knowledge of the relevant input parameters. Newtonian determinism fit well with the concept of God's sovereignty, but the fundamental uncertainty of quantum mechanics appears to rob God of that attribute. However, this assumes that quantum mechanics is a complete theory, that is, that quantum mechanics is an ultimate theory. There are limits to the applications of quantum mechanics, such as the fact that there is no theory of quantum gravity. If the history of science is any teacher, we can expect that quantum mechanics will one day be replaced by some other theory. This other theory probably will include quantum mechanics as a special case of the better theory. That theory may clear up the uncertainty question.

As an aside, we perhaps ought to mention that the determinism derived from Newtonian physics also produces a conclusion unpalatable to many Christians. If determinism is true, then all future events are predetermined from the initial conditions of the universe. Just as the Copenhagen interpretation of quantum mechanics led to even God not being able to know the outcome of an experiment, many people applying determinism concluded that God was unable to alter the outcome of an experiment. That is, God was bound by the physics that rules the universe. This quickly led to deism. Most, if not all, people today who reject quantum mechanics refuse to accept this extreme interpretation of Newtonian physics. They ought to recognize that just as determinism is a perversion of Newtonian physics, the Copenhagen interpretation is a perversion of quantum mechanics.

The important point is that just as classical mechanics does a good job in describing the macroscopic world, quantum mechanics does a good job in describing the microscopic world. We ought not expect any more of a theory. Consequently, most physicists who believe the biblical account of creation have no problem with quantum mechanics.

Relativity

There are two theories of relativity, the special and general theories. We will briefly describe the special theory of relativity first. Even before Newton, Galileo (1564–1642) had conducted experiments with moving bodies. He realized that if we move toward or away from a moving object, the relative speed that we measure for that object depends upon that object's motion and our motion. This Galilean relativity is a part of Newtonian mechanics. The same behavior is true for the speed of waves. For instance, if we ride in a boat moving through water with waves, the speed of the waves that we measure will depend upon our motion and on the motion of the waves. In 1881, Albert A. Michelson (1852–1931) conducted a famous experiment that he refined and repeated in 1887 with Edward W. Morley (1838–1923). In this experiment, they measured the speed of light parallel and perpendicular to our annual motion around the sun. Much to their surprise, they found that the speed of light was the same regardless of the direction they measured it. This null result baffled physicists, for if taken at face value, it suggested that the earth did not orbit the sun, while there is other evidence that the earth does indeed orbit the sun.

In 1905, Albert Einstein took the invariance of the speed of light as a postulate and worked out its consequences. He made three predictions concerning an object as its speed approaches the speed of light:

1. The length of the object as it passes will appear to shorten toward zero.
2. The object's mass will increase without bound.
3. The passage of time as measured by the object will approach zero.

These behaviors are strange and do not conform to what we might expect from everyday experience, but keep in mind that in everyday experience we do not encounter objects moving at any speed close to that of light.

Eventually, these predictions were confirmed in experiments. For instance, particle accelerators accelerate small particles to very high speeds. We can measure the masses of the particles as we accelerate them, and their masses increase in the manner predicted by the theory. In other experiments, very fast-moving, short-lived particles exist longer than they do when moving very slowly. The rate of time dilation is consistent with the predictions of the theory. Length contraction is a little more difficult to directly test, but we have tested it as well.

Einstein's theory of special relativity applies to particles moving at a constant rate but does not address their acceleration. Einstein addressed that problem with his general theory in 1916, but he also treated the acceleration due to gravity. In general relativity, space and time are physical things that have a structure in some ways similar to a fabric. Einstein treated time as a fourth dimension in addition to the normal three dimensions of space. We sometimes call this four-dimensional entity *space-time* or simply *space*. The presence of a large amount of matter or energy (Einstein previously had shown their equivalence) alters space. Mathematically, the alteration of space is like a curvature, so we say that matter or energy bends space. The curvature of space telegraphs the presence of matter and energy to other matter and energy in space, and this more deeply

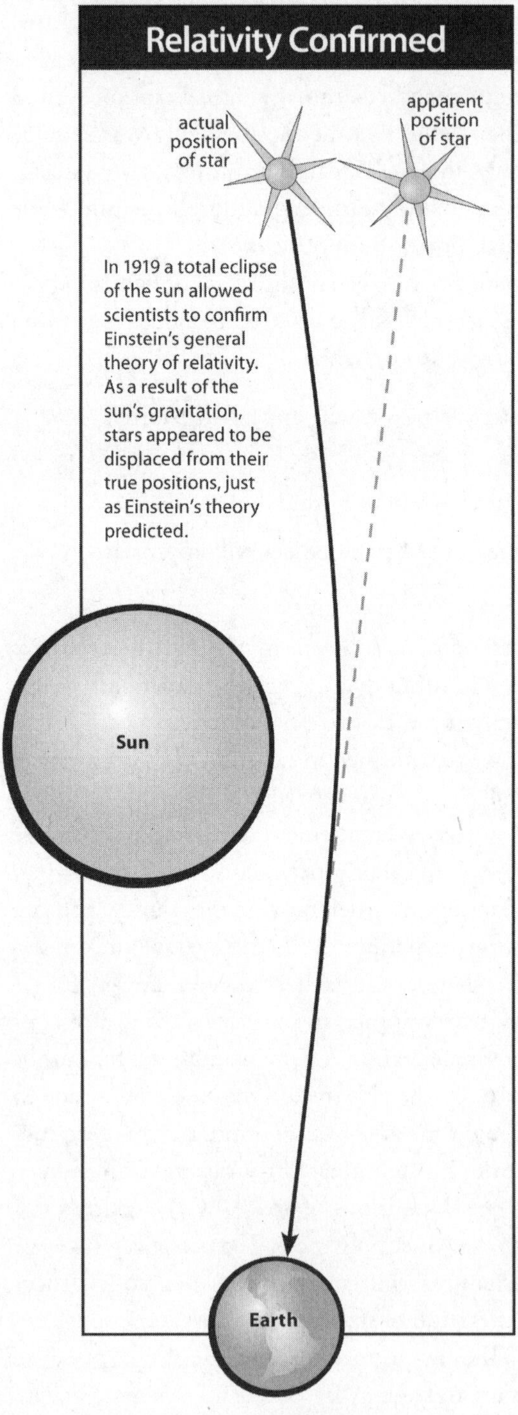
Relativity Confirmed
actual position of star
apparent position of star
In 1919 a total eclipse of the sun allowed scientists to confirm Einstein's general theory of relativity. As a result of the sun's gravitation, stars appeared to be displaced from their true positions, just as Einstein's theory predicted.
Sun
Earth

answered a question about gravity. Newton had hypothesized that gravity operated through empty space, but his theory could not explain at all how the information about an object's mass and distance was transmitted through space. In general relativity, an object must move through a straight line in space-time, but the curvature of space-time induced by nearby mass causes that straight-line motion to appear to us as acceleration.

Einstein's new theory made several predictions. The first opportunity to test the theory happened during a total solar eclipse in 1919. During the eclipse, astronomers were able to photograph stars around the edge of the sun. The light from those stars had to pass very close to the sun to get to the earth. As the stars' light passed near the sun, the sun attracted the light via the curvature of space-time. This caused the stars to appear closer to the sun than they would have otherwise. Newtonian gravity also predicts a deflection of starlight toward the sun, but the deflection is less than with general relativity. The observed amount of deflection was consistent with the predictions of general relativity. Astronomers have repeated the experiment many times since 1919 with ever-improving accuracy.

For many years, radio astronomers have measured with great precision the locations of distant-point radio sources as the sun passed by, and those results beautifully agree with the predictions. Another early confirmation was the explanation of a small anomaly in the orbit of the planet Mercury that Newtonian gravity could not explain. Many other experiments of various types have repeatedly confirmed general relativity. Some experiments today even allow us to test for slight variations of Einstein's theory.

We can apply general relativity to the universe as a whole. Indeed, when we do this, we discover that it predicts that the universe is either expanding or contracting; it is a matter of observation to determine which the universe actually is doing. In 1928, Edwin Hubble (1889–1953) showed that the universe is expanding. Most people today think that the expansion began with the big bang, the supposed sudden appearance of the universe 13.7 billion years ago. However, there are many other possibilities. For instance, the creation physicist Russell Humphreys proposed his *white hole cosmology,* assuming that general relativity is the correct theory of gravity (see his book *Starlight and Time*[1]). It is interesting to note that universal expansion is consistent with certain Old Testament passages (e.g., Psalm 104:2) that mention the stretching of the heavens.

Seeing that there is so much evidence to support Einstein's theory of general relativity, why do some creationists oppose the theory? There are at least three reasons. One reason is that, as with quantum mechanics, modern relativity theory appears to violate certain common-sense views of the way that the world works. For instance, in everyday experience, we don't see mass change and time appear to slow. Indeed, general relativity forces us to abandon the concept of simultaneity of time. *Simultaneity* means that time progresses at the same rate for all observers, regardless of where they are. As we previously stated, in special relativity, time slows with greater speed. However, with general relativity, the rate at which time passes depends not only upon speed but also on one's location in a gravitational field. The deeper one is in a gravitational field, the slower that time passes. For example, a clock at sea level will record the passage of time more slowly than a clock at mile-high Denver. Admittedly, this is weird. However, the discrepancy between the clocks at these two locations is so miniscule as to not appear on most clocks, save the most accurate atomic clocks. This sort of thing has been measured several times, and the discrepancies between the clocks involved always are the same as those predicted by theory. Thus, while our perception is that time flows uniformly everywhere, the reality is that the passage

1. D. Russell Humhreys, *Starlight and Time* (Green Forest, AR: Master Books, 1994).

of time does depend upon one's location, but the differences are so small in the situations encountered on the earth that we cannot perceive them. That is, the predictions of general relativity on earth are consistent with our ability to perceive time. However, there are conditions beyond the earth that the loss of simultaneity would be very obvious if we could experience them.

A second reason why some creationists oppose modern relativity theory is the misappropriation of modern relativity theory to support moral relativism. Unfortunately, modern relativity theory arose at precisely the time that moral relativism became popular. Moral relativists proclaim that "all things are equal," and they were very eager to snatch some of the triumph of relativity theory to support their cause. There are at least two problems with this misappropriation. First, it does not follow that a principle that works in the natural world automatically operates in the world of morality. The physical world is material, but the world of morality is immaterial. Second, the moral relativists either did not understand relativity or they intentionally misused it. Despite the common misconception, modern relativity theory does not tell us that everything is relative. There are absolutes in modern theory of relativity. The speed of light is a constant. While the passage of time may vary, general relativity provides an absolute way in which to compare the passage of time in two reference frames. The modern theory of relativity in no way supports moral relativism.

The third reason why some creationists reject modern relativity theory is that they think that general relativity inevitably leads to the big-bang model. However, the big-bang model is just one possible origin scenario for the universe; there are many other possibilities. We have already mentioned Russ Humphreys's white hole cosmology, and there are other possible recent creation models based upon general relativity. True — if general relativity is not correct, then the big-bang model would be in trouble. However, if general relativity is correct, then the shortcut attempt to undermine the big-bang model will doom us from ever finding the correct cosmology.

String Theory

With the establishment of quantum mechanics in the 1920s, the development of the science of particle physics soon followed. At first, only a few particles were known: the electron, proton, and neutron. These particles all had mass and were thought at the time to be the fundamental building blocks of matter. Quantum mechanics introduced the concept that material particles could be described by waves, and conversely that waves could be described by particles. That led to the concept of particles that had no mass, such as photons, the particles that make up light. Eventually, physicists saw the need for other particles, such as neutrinos and antiparticles. Evidence for these odd particles soon followed. Experimental results suggested the existence of other particles, such as the meson, muon, and tau particles, as well as their antiparticles. Many of these new particles were very short-lived, but they were particles nevertheless.

Physicists began to see patterns in the growing zoo of particles. They could group particles according to certain properties. For instance, elementary particles possess angular momentum, a property normally associated with spinning objects, so physicists say that elementary particles have "spin." Imagining elementary particles as small spinning spheres is useful, but modern theories view this as a bit naive. Spin comes in a quantum amount. Some particles have whole integer values of quantum spin. That is, they have integer multiples (0, ±1, ±2, etc.) of the basic unit of spin. Physicists call these particles Bosons. Other particles have half integer (±1/2, ±3/2, etc.) amounts of spin, and are known as fermions. Bosons and fermions have very different properties. Physicists also noticed that elementary particles tended to have certain mathematical relationships between one another. Physicists eventually began to use group theory, a concept from abstract algebra, to classify and study elementary particles.

By the 1960s, physicists began to suspect that many elementary particles, such as protons and neutrons, were not so elementary after all, but consisted of even more elementary particles. Physicists called these more elementary particles *quarks*, after an enigmatic word in a James Joyce poem. According to the theory, there are six types of quarks. Many particles, such as protons and neutrons, consist of the combination of two quarks. The different combinations of quarks lead to different particles. Some of those combinations of quarks ought to produce particles that no one had yet seen, so these combinations amounted to predictions of new particles. Particles physicists were able to create these particles in experiments in particle accelerators, so the successful search for those predicted particles was confirmation of the underlying theory. Therefore, quark theory now is well established.

In recent years, particle physicists have in similar fashion developed string theory. Physicists have noticed that certain patterns among elementary particles can be explained easily if particles behave as tiny vibrating strings. These strings would require the existence of at least six additional dimensions of space. We already know that the universe has three normal spatial dimensions as well as the dimension of time, so these six extra dimensions bring the total number of dimensions to ten. The reason why we do not normally see the other six dimensions is that they are tightly curled up and hidden within the tiny particles themselves. At extremely high energies, the extra dimensions ought to manifest themselves. Therefore, particle physicists can predict what kind of behavior

The Size of Strings

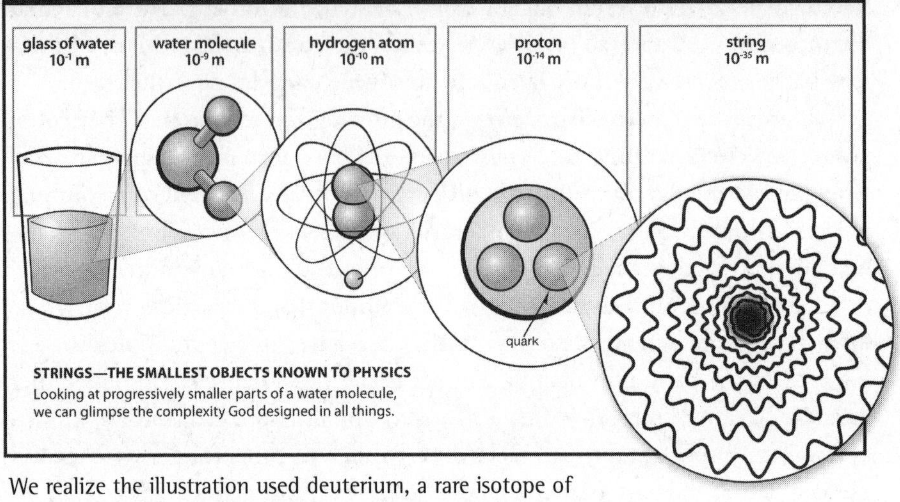

STRINGS—THE SMALLEST OBJECTS KNOWN TO PHYSICS
Looking at progressively smaller parts of a water molecule, we can glimpse the complexity God designed in all things.

We realize the illustration used deuterium, a rare isotope of hydrogen, to help convey the point.

strings ought to exhibit when they accelerate particles to extremely high energies. The problem is that current particle accelerators are not nearly powerful enough to produce these effects. As theoretical physicists refine their theories and we build new, powerful particle accelerators, physicists expect that one day we can test whether string theory is true, but for now there is no experimental evidence for string theory.

Currently, most physicists think that string theory is a very promising idea. Assuming that string theory is true, there still remains the question as to which particular version of string theory is the correct one. You see, string theory is not a single theory but instead is a broad outline of a number of possible theories. Once we confirm string theory, we can constrain which version properly describes our world. If true, string theory could lead to new technologies. Furthermore, a proper view of elementary particles is important in many cosmological models, such as the big bang. This is because in the big-bang model, the early universe was hot enough to reveal the effects of string theory.

Conclusion

Modern physics is a product of the 20th century and relies upon twin pillars: quantum mechanics and general relativity. Both theories have tremendous experimental support. Christians ought not to view these theories with such great suspicion. True, some people have perverted or hijacked these theories to support some nonbiblical principles, but some wicked people have even perverted Scripture to support nonbiblical things. We ought to recognize that modern physics is a very robust, powerful theory that explains much. At the same time, the theory is very incomplete in some respects. In time, we ought to expect that some new theories will come along that will better explain the world than these theories do. However, we know that God's Word does not change.

String theory has emerged in the 21st century as the next great idea in physics. Time will tell if string theory will live up to our expectations. What ought to be the reaction of Christians to this? We must be vigilant to investigate the amount of nonbiblical influences that may have crept into modern thinking, particularly in the interpretation of string theory (as with modern physics). However, we must be careful not to throw out the baby with the bath water. That is, can we reject the anti-Christian thinking that many have brought to the discussion? The answer is certainly yes. As with the question of origins, we must strive to interpret these things on our terms, guided by the Bible. Do the new theories adequately describe the world? Can we see the hand of the Creator in our new physics? Can we find meaning in our studies that brings glory to God? If we can answer yes to each of these questions, then these new theories ought not to be a problem for the Christian.

31

Doesn't the Order of Fossils in the Rock Record Favor Long Ages?

DR. ANDREW A. SNELLING

Fossils are the remains, traces, or imprints of plants or animals that have been preserved in the earth's near-surface rock layers at some time in the past.[1] In other words, fossils are the remains of *dead* animals and plants that were buried in sedimentary layers that later hardened to rock strata. So the fossil record is hardly "the record of life in the geologic past" that so many scientists incorrectly espouse,[2] assuming a long prehistory for the earth and life on it. Instead, it is a record of the *deaths* of countless billions of animals and plants.

The Fossil Record

For many people, the fossil record is still believed to be "exhibit A" for evolution. Why? Because most geologists insist the sedimentary rock layers were deposited gradually over vast eons of time during which animals lived, died, and then were occasionally buried and fossilized. So when these fossilized animals (and plants) are found in the earth's rock sequences in a particular order

1. K.K.E. Neuendorf, J.P. Mehl Jr., and J.A. Jackson, eds., *Glossary of Geology* (Alexandria, VA: American Geological Institute, 2005), p. 251.
2. Ibid.

of first appearance, such as animals without backbones (invertebrates) in lower layers followed progressively upward by fish, then amphibians, reptiles, birds, and finally mammals (e.g., in the Colorado Plateau region of the United States), it is concluded, and thus almost universally taught, that this must have been the order in which these animals evolved during those vast eons of time.

However, in reality, it can only be dogmatically asserted that the fossil record is the record of the order in which animals and plants were buried and fossilized. Furthermore, the vast eons of time are unproven and unproveable, being based on assumptions about how quickly sedimentary rock layers were deposited in the unobserved past. Instead, there is overwhelming evidence that most of the sedimentary rock layers were deposited rapidly. Indeed, the impeccable state of preservation of most fossils requires the animals and plants to have been very rapidly buried, virtually alive, by vast amounts of sediments before decay could destroy delicate details of their appearance and anatomy. Thus, if most sedimentary rock layers were deposited rapidly over a radically short period of time, say in a catastrophic global flood, then the animals and plants buried and fossilized in those rock layers may well have all lived at about the same time and then have been rapidly buried progressively and sequentially.

Furthermore, the one thing we can be absolutely certain of is that when we find animals and plants fossilized together, they didn't necessarily live together in the same environment or even die together, but they certainly were buried together, because that's how we observe them today! This observational certainty is crucial to our understanding of the many claimed mass extinction events in the fossil record. Nevertheless, there is also evidence in some instances that the fossils found buried together may represent animals and plants that did once live together (see later).

Mass Extinctions

In the present world, when all remaining living members of a particular type of animal die, that animal (or plant) is said to have become extinct. Most scientists (incorrectly) regard the fossil record as a record of life in the geologic past. So, when in the upward progression of strata the fossils of a particular type of animal or plant stop occurring in the record and there are no more fossils of that animal or plant in the strata above, or any living representatives of that animal or plant, most scientists say that this particular creature went extinct many years ago. Sadly, there are many animals and plants that are extinct, and we only know they once existed because of their fossilized remains in the geologic record. Perhaps the most obvious and famous example is the dinosaurs.

There are distinctive levels in the fossil record where vast numbers of animals (and plants) are believed to have become extinct. Evolutionists claim that all these animals (and plants) must have died, been buried, and become extinct all at the same time. Since this pattern is seen in the geologic record all around the globe, they call these distinctive levels in the fossil record mass extinctions. Furthermore, because something must have happened globally to wipe out all those animals (and plants), the formation of these distinctive levels in the fossil record are called mass extinction events. However, in the context of catastrophic deposition of the strata containing these fossils, this pattern would be a preserved consequence of the Flood.

Now geologists have divided the geologic record into time periods, according to their belief in billions of years of elapsed time during which the sedimentary strata were deposited. Thus, those sedimentary strata that were supposedly deposited during a particular time period are so grouped and named accordingly. This is the origin of names such as Cambrian, Ordovician, Silurian, Devonian, Carboniferous, Permian, Triassic, Jurassic, Cretaceous, and more.

There are some 17 mass extinction events in the fossil record recognized by geologists, from in the late Precambrian up until the late Neogene, "just before the dawn of written human history." However, only eight of those are classed as major mass extinction events — end-Ordovician, late-Devonian, end-Permian, end-Triassic, early-Jurassic, end-Jurassic, middle-Cretaceous, and end-Cretaceous. Most people have probably heard about the end-Cretaceous mass extinction event, because that's when the dinosaurs are supposed to have been wiped out, along with about a quarter of all the known families of animals. However, the end-Permian mass extinction event was even more catastrophic, because 75 percent of amphibian families and 80 percent of reptile families were supposedly wiped out then, along with 75 to 90 percent of all pre-existing species in the oceans.

Asteroid Impacts and Volcanic Eruptions

So what caused these mass extinction events? Evolutionary geologists are still debating the answer. The popularized explanation for the end-Cretaceous mass extinction event is that an asteroid hit the earth, generating choking dust clouds and giant tsunamis (so-called tidal waves) that decimated the globe and its climate, supposedly for a few million years. A layer of clay containing a chemical signature of an asteroid is pointed to in several places around the globe as one piece of evidence, and the 124-mile (200 km) wide Chicxulub impact crater in Mexico is regarded as "the scene of the crime."

However, at the same level in the geologic record are the massive remains of catastrophic outpourings of staggering quantities of volcanic lavas over much of India, totally unlike any volcanic eruptions experienced in recent human history. The Pinatubo eruption in the Philippines in 1991 blasted enough dust into the atmosphere to circle the globe and cool the following summer by 1–2°C, as well as gases which caused acid rain. Yet that eruption was only a tiny firecracker compared to the massive, catastrophic Indian eruption. Furthermore, volcanic dust has a similar chemical signature to that of an asteroid. Interestingly, even more enormous quantities of volcanic lavas are found in Siberia and coincide with the end-Permian mass extinction event.

The Biblical Perspective

What then should Bible-believing Christians make of these interpretations of the fossil and geologic evidence? Of course, we first need to recognize that both creationists and evolutionists start with presupposed assumptions, which they then use to interpret the presently observed evidence. So this difference of interpretations cannot be "religion vs. science," as it is so often portrayed.

Furthermore, it needs to be noted that in the geologic record there are very thick sequences of rock layers, found below the main strata record containing prolific fossils, which are either totally devoid of fossils or only contain very rare fossils of microorganisms and minor invertebrates. In the biblical framework of earth history, these strata would be classified as creation week and pre-Flood. Also, a few fossils may also have been formed since the Flood due to localized, residual catastrophic depositional events, so Flood geologists do not claim all fossils were formed during the Flood.

As already noted, the only dogmatic claim which can be made is that the geologic strata record the order in which animals and plants were buried and fossilized. However, it is clear from Genesis 1–3, Romans 5:12, 8:20–22, and 1 Corinthians 15:21–22 that God created a good world which was severely marred by death as a result of Adam's sin. Because the animals were created as vegetarians (Genesis 1:29–30) and the whole creation was subsequently impacted with corruption and death due to the Fall, there could have been no animal fossils in Eden's rocks. Indeed, fossilization under present-day conditions is exceedingly rare, so evolutionary geologists applying "the present is the key to the past" have a real problem in explaining how the vast numbers of fossils in the geologic record could have formed. Thus, the global destruction of all the pre-Flood animals and plants by the year-long Flood cataclysm alone makes sense of this fossil and geologic evidence (though as noted above, a small percent of the geological and fossil evidence is from post-Flood residual catastrophism).

Indeed, not only did the animals and plants have to be buried rapidly by huge masses of water-transported sediments to be fossilized, but the general vertical order of burial is also consistent with the biblical flood. The first fossils in the record are of marine animals exclusively, and it is only higher in the strata that fossils of land animals are found, because the Flood began in the ocean basins ("the fountains of the great deep burst open") and the ocean waters then flooded over the continents. How else would there be marine fossils in sedimentary layers stretching over large areas of the continents? Added to this, "the floodgates of heaven" were simultaneously opened, and both volcanism and earth movements accompanied these upheavals.

In a global watery cataclysm, therefore, there would be simultaneous wholesale destruction of animals and plants across the globe. The tearing apart of the earth's crust would release stupendous outpourings of volcanic lavas on the continental scale found in the geologic record. The resultant "waves" of destruction are thus easily misinterpreted as mass extinction events, when these were just stages of the single, year-long, catastrophic global flood.

It is also significant that some fossilized animals and plants once thought to be extinct have in fact been found still alive, thus demonstrating the total unreliability of the evolutionary time scale. The last fossilized coelacanth (a fish) is supposedly 65 million years old, but coelacanths are still here, so where did they "hide" for 65 million years? The Wollemi pine's last fossil is supposedly 150 million years old, but identical living trees were found in 1994. The recent burial and fossilization of these animals and plants, and the extinction of many other animals and plants, during the single biblical flood thus makes better sense of all the fossil and geologic evidence.

Accounting for the Order of Fossils in the Rock Record

Even though the order of strata and the fossils contained in them (sometimes extrapolated and interpolated) has been made the basis of the accepted millions-of-years system of geochronology and historical geology, the physical reality of the strata order and the contained fossils is generally not in dispute. Details of local strata sequences have been carefully compiled by physical observations during field work and via drill-holes. Careful correlations of strata of the same rock types have then been made between local areas and from region to region, often by physical means, so that the robustness of the overall fossil order and strata sequence of the geologic record has been clearly established.

Indeed, it is now well recognized that there are at least six thick sequences of fossil-bearing sedimentary strata, known as megasequences, which can be traced right across the North American continent and beyond to other continents.[3] Such global-scale deposition of sediment layers (e.g., chalk and coal beds) is, of course, totally inexplicable to uniformitarian (long-ages) geologists by the application of only today's slow-and-gradual geologic processes that only operate over local to regional scales. But it is powerful evidence of catastrophic deposition during the global Genesis flood. Thus, it is not the recognized order of the strata in the geologic record that is in dispute, but rather the millions-of-years interpretation for the deposition of the sedimentary strata and their contained fossils.

It is true that the complete geologic record is hardly ever, if at all, found in any one place on the earth's surface. Usually several or many of the strata in local sequences are missing compared to the overall geologic record, but usually over a given region there is more complete preservation of the record via correlation and integration. However, quite commonly there is little or no physical or physiographic evidence of the intervening period of erosion or non-deposition of the missing strata systems, suggesting that at such localities neither erosion nor deposition ever occurred there. Yet this is exactly what would be expected based on the biblical account of the Genesis flood and its implications. In some areas one sequence of sedimentary strata with their contained fossil assemblages would be deposited, and in other areas entirely different strata sequences would be deposited, depending on the source areas and directions of the water currents transporting the sediments. Some strata units would have been deposited over wider areas than others, with erosion in some areas but continuous deposition in others, even when intervening strata units were deposited elsewhere. Thus, as a result of the complex interplay of currents, waves, and transported sediments with their entombed organisms, a variety of different types of sedimentary rocks and strata sequences would have been laid down directly on the pre-Flood strata sequences and the crystalline basement that probably dates back to the creation week itself. Thus the pattern of deposition of the strata sequences and their contained fossils is entirely consistent with the strata record the Flood might be expected to have produced. In contrast, by using the present to interpret the past, evolutionary geologists have no more true scientific certainty of their version of the unobservable, unique historic events which they claim produced the geologic record.

Nevertheless, if the general order of the strata and their contained fossil assemblages is not generally in dispute, then that order in the strata sequences still must reflect the geological processes and their timing responsible for the formation of the strata and their order. If, as is assiduously maintained here, the

3. L.L. Sloss, "Sequences in the cratonic interior of North America," *Geological Society of America Bulletin* 74 (1963): 93–114.

order in the fossil record does not represent the sequence of the evolutionary development of life, then the fossil record must be explainable within the context of the tempo of geologic processes burying these organisms in the sediment layers during the global flood cataclysm. Indeed, both the order of the strata and their contained fossils could well provide us with information about the pre-Flood world, and evidence of the progress of different geological processes during the Flood event. There are a number of factors that have been suggested to explain the order in the fossil record in terms of the Flood processes, rather than over the claimed long ages.

Pre-Flood Biogeography

If we look at today's living biology, we find that across mountains such as the Sierra Nevada of California, or in a trip from the South Rim of the Grand Canyon down to the Colorado River, there are distinct plant and animal communities in different life or ecology zones that are characteristic of the climates at different elevations. Thus, we observe cacti growing in desert zones and pines growing in alpine zones rather than growing together. Therefore, just as these life/ecology zones today can be correlated globally (all deserts around the world have similar plants and animals), so too some fossil zones and fossil communities may be correlated globally within the geologic record of the Flood.

Thus it has been suggested that there could well have been distinct biological communities and ecological zones in the pre-Flood world that were spatially and geographically separated from one another and that that were then sequentially inundated, swept away, and buried as the Flood waters rose. This ecological zonation model for the order of fossils in the geologic record[4] would argue that the lower fossiliferous layers in the strata record must therefore represent the fossilization of biological communities at lower elevations and warmer climates, while higher layers in the geologic record must represent fossilization of biological communities that lived at higher elevations and thus cooler temperatures.

Based on the vertical and horizontal distribution of certain fossil assemblages in the strata record, it has been concluded that the pre-Flood biogeography consisted of distinct and unique ecosystems which were destroyed by the Flood and did not recover to become re-established in the post-Flood world of today. These include a floating-forest ecosystem consisting of unique trees called lycopods of various sizes that contained large, hollow cavities in their trunks and branches and hollow root-like rhizomes, with associated similar plants. It also includes some unique animals, mainly amphibians, that lived in these forests

4. H. Clark, *The New Diluvialism* (Angwin, CA: Science Publications, 1946).

that floated on the surface of the pre-Flood ocean.[5] Spatially and geographically separated and isolated from this floating-forest ecosystem were stromatolite reefs adjacent to hydrothermal springs in the shallow waters of continental shelves making up a hydrothermal-stromatolite reef ecosystem.[6]

In the warmer climates of the lowland areas of the pre-Flood land surfaces, dinosaurs lived where gymnosperm vegetation (naked seed plants) was abundant, while at high elevations inland in the hills and mountains where the climate was cooler, mammals and humans lived among vegetation dominated by angiosperms (flowering plants).[7] Thus these gymnosperm-dinosaur and angiosperm-mammal-man ecosystems (or biomes) were spatially and geographically separated from one another on the pre-Flood land surfaces. In Genesis chapter 2, the river coming out of the Garden of Eden is described as dividing into four rivers, which may imply the Garden of Eden (with its fruit trees and other angiosperms, mammals, and man) was at a high point geographically, the rivers flowing downhill to the lowland swampy plains bordering the shorelines where the gymnosperms grew and the dinosaurs lived. This would explain why we don't find human and dinosaur fossil remains together in the geologic record, dinosaurs and gymnosperms only fossilized together, and angiosperms only fossilized with mammals and man higher in the record separate from the dinosaurs and gymnosperms.

It can therefore be argued that in a very general way the order of fossil "succession" in the geologic record would reflect the successive burial of these pre-Flood biological communities as the Flood waters rose up onto the continents. The Flood began with the breaking up of the fountains of the great deep (the breaking up of the pre-Flood ocean floor), so there would have been a sudden surge of strong ocean currents and tsunamis picking up sediments from the ocean floor and moving landward that would first of all have overwhelmed the stromatolite reefs in the shallow seas fringing the shorelines. This destruction of the protected lagoons between the stromatolite reefs and the shorelines by these severe storms would have then caused the strange animals that probably were unique to these stromatolite reefs to be buried and thus preserved in the lowermost Flood strata directly overlaying the burial of the stromatolites.

5. K.P. Wise, "The Pre-Flood Floating Forest: A Study in Paleontological Pattern Recognition," in *Proceedings of the Fifth International Conference on Creationism*, ed. Robert L. Ivey, Jr., (Pittsburgh, PA: Creation Science Fellowship, 2003), p. 371–381.
6. K.P. Wise, "The Hydrothermal Biome: A Pre-Flood Environment," in *Proceedings of the Fifth International Conference on Creationism*, ed. Robert L. Ivey Jr. (Pittsburgh, PA: Creation Science Fellowship, 2003), p. 359–370.
7. K.P. Wise, *Faith, Form, and Time* (Nashville, TN: Broadman & Holman, 2002), p. 170–175.

Increasing storms, tidal surges, and tsunamis generated by earth movements, earthquakes, and volcanism on the ocean floor would have resulted in the progressive breaking up of the floating-forest ecosystem on the ocean surface, and thus huge rafts of vegetation would have been swept landward to be beached with the sediment load on the land surfaces being inundated. Thus, the floating-forest vegetation would have been buried higher in the strata record of the Flood, well above the stromatolites and the strange animals that lived with them. Only later, in the first 150 days of the Flood, as the waters rose higher across the land surface, would the gymnosperm-dinosaurs ecosystem be first swept away and buried, followed later by the angiosperm-mammal-man ecosystem that lived at higher elevations. People would have continued to move to the highest ground to escape the rising Flood waters, and so would not necessarily have been buried with the angiosperms and mammals. Thus the existence of these geographically separated distinct ecosystems in the pre-Flood world could well explain this spatial separation and order of fossilization in the geologic record of the Flood.

Early Burial of Marine Creatures

The vast majority by number of fossils preserved in the strata record of the Flood are the remains of shallow-water marine invertebrates (brachiopods, bivalves, gastropods, corals, graptolites, echinoderms, crustaceans, etc.).[8] In the lowermost fossiliferous strata (Cambrian, Ordovician, Silurian, and Devonian), the contained fossils are almost exclusively shallow-water marine invertebrates, with fish and amphibian fossils only appearing in progressively greater numbers in the higher strata.[9] The first fish fossils are found in Ordovician strata, and in Devonian strata are found amphibians and the first evidence of continental-type flora. It is not until the Carboniferous (Mississippian and Pennsylvanian) and Permian strata higher in the geologic record that the first traces of land animals are encountered.

Because the Flood began in the ocean basins with the breaking up of the fountains of the great deep, strong and destructive ocean currents were generated by the upheavals and moved swiftly landward, scouring the sediments on the ocean floor and carrying them and the organisms living in, on, and near them. These currents and sediments reached the shallower continental shelves, where the shallow-water marine invertebrates lived in all their prolific diversity.

8. K.P. Wise, quoted in J.D. Morris, *The Young Earth* (Green Forest, AR: Master Books, 1994), p. 70.
9. S.M. Stanley, *Earth and Life Through Time*, second edition (New York: W.H. Freeman and Company, 1989); R. Cowen, *History of Life*, third edition (Oxford, England: Blackwell Scientific Publications, 2000).

Unable to escape, these organisms would have been swept away and buried in the sediment layers as they were dumped where the waters crashed onto the land surfaces being progressively inundated farther inland. As well as burying these shallow-water marine invertebrates, the sediments washed shoreward from the ocean basins would have progressively buried fish, then amphibians and reptiles living in lowland, swampy habitats, before eventually sweeping away the dinosaurs and burying them next, and finally at the highest elevations destroying and burying birds, mammals, and angiosperms.

Hydrodynamic Selectivity of Moving Water

Moving water hydrodynamically selects and sorts particles of similar sizes and shapes. Together with the effect of the specific gravities of the respective organisms, this would have ensured deposition of the supposedly simple marine invertebrates in the first-deposited strata that are now deep in the geologic record of the Flood. The well-established "impact law" states that the settling velocity of large particles is independent of fluid viscosity, being directly proportional to the square root of particle diameter, directly proportional to particle sphericity, and directly proportional to the difference between particle and fluid density divided by fluid density.[10] Moving water, or moving particles in still water, exerts "drag" forces on immersed bodies which depend on the above factors. Particles in motion will tend to settle out in proportion mainly to their specific gravity (or density) and sphericity.

It is significant that the marine organisms fossilized in the earliest Flood strata, such as the trilobites, brachiopods, etc., are very "streamlined" and quite dense. The shells of these and most other marine invertebrates are largely composed of calcium carbonate, calcium phosphate, and similar minerals which are quite heavy (heavier than quartz, for example, the most common constituent of many sands and gravels). This factor alone would have exerted a highly selective sorting action, not only tending to deposit the simpler (that is, the most spherical and undifferentiated) organisms first in the sediments as they were being deposited, but also tending to segregate particles of similar sizes and shapes. These could have thus formed distinct faunal "stratigraphic horizons," with the complexity of structure of deposited organisms, even of similar kinds, increasing progressively upward in the accumulating sediments.

It is quite possible that this could have been a major process responsible for giving the fossil assemblages within the strata sequences a superficial appearance

10. W.C. Krumbein, and L.L. Sloss, *Stratigraphy and Sedimentation*, second edition (San Francisco, CA: W.H. Freeman and Company, 1963), p. 198.

of "evolution" of similar organisms in the progressive succession upward in the geologic record. Generally, the sorting action of flowing water is quite efficient, and would definitely have separated the shells and other fossils in just the fashion in which they are found, with certain fossils predominant in certain stratigraphic horizons, and the supposed complexity of such distinctive, so-called "index" fossils increasing in at least a general way in a progressive sequence upward through the strata of the geologic record of the Flood.

Of course, these very pronounced "sorting" powers of hydraulic action are really only valid generally, rather than universally. Furthermore, local variations and peculiarities of turbulence, environment, sediment composition, etc., would be expected to cause local variations in the fossil assemblages, with even occasional heterogeneous combinations of sediments and fossils of a wide variety of shapes and sizes, just as we find in the complex geological record.

In any case, it needs to be emphasized that the so-called "transitional" fossil forms that are true "intermediates" in the strata sequences between supposed ancestors and supposed descendants according to the evolutionary model are exceedingly rare, and are not found at all among the groups with the best fossil records (shallow-marine invertebrates like mollusks and brachiopods).[11] Indeed, even evolutionary researchers have found that the successive fossil assemblages in the strata record invariably only show trivial differences between fossil organisms, the different fossil groups with their distinctive body plans appearing abruptly in the record, and then essentially staying the same (stasis) in the record.[12]

Behavior and Higher Mobility of the Vertebrates

There is another reason why it is totally reasonable to expect that vertebrates would be found fossilized higher in the geologic record than the first invertebrates. Indeed, if vertebrates were to be ranked according to their likelihood of being buried early in the fossil record, then we would expect oceanic fish to be buried first, since they live at the lowest elevation.[13] However, in the ocean, the fish live in the water column and have great mobility, unlike the

11. Wise, *Faith, Form, and Time*, p. 197–199.
12. N. Eldridge and S.J. Gould, "Punctuated Equilibria: An Alternative to Phyletic Gradualism," in *Mammals in Paleobiology*, ed. T.J.M. Schopf (San Francisco, CA: Freeman, Cooper and Company, 1972), p. 82–115; S.J. Gould and N. Eldridge, "Punctuated Equilibria: The Tempo and Mode of Evolution Reconsidered," *Paleobiology* 3 (2007): 115–151; S.J. Gould and N. Eldridge, "Punctuated Equilibrium Comes of Age," *Nature* 366 (1993): 223–227.
13. L. Brand, *Faith, Reason, and Earth History* (Berrien Springs, MI: Andrews University Press, 1997), p. 282–283.

invertebrates that live on the ocean floor and have more restricted mobility, or are even attached to a substrate. Therefore, we would expect the fish to only be buried and fossilized subsequent to the first marine invertebrates.

Of course, fish would have inhabited water at all different elevations in the pre-Flood world, even up in mountain streams, as well as the lowland, swampy habitats, but their ranking is based on where the first representatives of fish are likely to be buried. Thus it is hardly surprising to find that the first vertebrates to be found in the fossil record, and then only sparingly, are in Ordovician strata. Subsequently, fish fossils are found in profusion higher up in the Devonian strata, often in great "fossil graveyards," indicating their violent burial.

A second factor in the ranking of the likelihood of vertebrates being buried is how animals would react to the Flood. The behavior of some animals is very rigid and stereotyped, so they prefer to stay where they are used to living, and thus would have had little chance of escape. Adaptable animals would have recognized something was wrong, and thus made an effort to escape. Fish are the least adaptable in their behavior, while amphibians come next, and then are followed by reptiles, birds, and lastly, the mammals.

The third factor to be considered is the mobility of land vertebrates. Once they become aware of the need to escape, how capable would they then have been of running, swimming, flying, or even riding on floating debris? Amphibians would have been the least mobile, with reptiles performing somewhat better, but not being equal to the mammals' mobility, due largely to their low metabolic rates. However, birds, with their ability to fly, would have had the best expected mobility, even being able to find temporary refuge on floating debris.

These three factors would tend to support each other. If they had worked against each other, then the order of vertebrates in the fossil record would be more difficult to explain. However, since they all do work together, it is realistic to suggest that the combination of these factors could have contributed significantly to producing the general sequence we now observe in the fossil record.

In general, therefore, the land animals and plants would be expected to have been caught somewhat later in the period of rising Flood waters and buried in the sediments in much the same order as that found in the geologic record, as conventionally depicted in the standard geologic column. Thus, generally speaking, sediment beds burying marine vertebrates would be overlain by beds containing fossilized amphibians, then beds with reptile fossils, and, finally, beds containing fossils of birds and mammals. This is essentially in the order:

1. Increasing mobility, and therefore increasing ability to postpone inundation and burial;

2. Decreasing density and other hydrodynamic factors, which would tend to promote later burial; and
3. Increasing elevation of habitat and therefore time required for the Flood waters to rise and advance to overtake them.

This order is essentially consistent with the implications of the biblical account of the Flood, and therefore it provides further circumstantial evidence of the veracity of that account. Of course, there would have been many exceptions to this expected general order, both in terms of omissions and inversions, as the water currents waxed and waned, and their directions changed due to obstacles and obstructions as the land became increasingly submerged and more and more amphibians, reptiles, and mammals were overtaken by the waters.

Other factors must have been significant in influencing the time when many groups of organisms met their demise. As the catastrophic destruction progressed, there would have been changes in the chemistry of seas and lakes from the mixing of fresh and salt water, and from contamination by leaching of other chemicals. Each species of aquatic organism would have had its own physiological tolerance to these changes. Thus, there would have been a sequence of mass mortalities of different groups as the water quality changed. Changes in the turbidity of the waters, pollution of the air by volcanic ash, and/or changes in air temperatures, would likely have had similar effects. So whereas ecological zonation of the pre-Flood world is a useful concept in explaining how the catastrophic processes during the Flood would have produced the order of fossils now seen in the geologic record, the reality was undoubtedly much more complex, due to many other factors.

Conclusions

In no sense is it necessary to capitulate to the vociferous claim that the order in the fossil record is evidence of the progressive organic evolution to today's plants and animals through various transitional intermediary stages over millions of years from common ancestors. While there are underlying thick strata sequences which are devoid of fossils and were therefore formed during creation week and the pre-Flood era, most of the fossil record is a record of death and burial of animals and plants during the Flood, as described in the biblical account, rather than being the order of a living succession that suffered the occasional mass extinction.

Asteroid impacts and volcanic eruptions accompanied other geological processes that catastrophically destroyed plants, animals, and people, and reshaped the earth's surface during the Flood event. Rather than requiring long ages, the order of fossils in the rock record can be accounted for by the year-long Flood, as a result of the pre-Flood biogeography and ecological zonation, the early burial of marine creatures, the hydrodynamic selectivity of moving water, and the behavior and higher mobility of the vertebrates. Thus, the order of the fossils in the rock record doesn't favor long ages, but is consistent with the global, catastrophic, year-long Genesis flood cataclysm, followed by localized residual catastrophism.

CONCLUSION

The Biggest Question of All

DR. DAVID R. CRANDALL

Hopelessness Abroad

My body trembled as I watched a young student from the University of Rome take a suicide plunge from the top of the Roman Coliseum. I was only 19 years old and visiting my first foreign country. I witnessed firsthand the hopelessness of a world without the Lord Jesus Christ. That young man had asked the question "Is life worth living?" Obviously, his answer was no.

That experience changed my life. I stood on Via Cavour in Rome, Italy, and promised God that I would spend my life telling others the truth about the loving Creator God so that people all over the world could have hope. For the last 40 years I have ministered on every habitable continent and have preached in 86 countries. But to this day, the image of the student in Rome still motivates me to be involved in spreading the good news of the gospel that gives people a reason for living and a plan for life and eternity.

The Biggest Question of All

Late in His life and ministry, Jesus wanted His disciples to articulate in His presence their beliefs about Him. So He asked them the biggest question of all time: "Who do you say that I am?" (Matthew 16:15).

In this book, you have read answers to many questions. As important as these questions are, they pale in significance compared to Jesus' question. You can be wrong about your answers to many questions in this book, but you dare not be wrong about your answer to this question. You see, your answer to Jesus' question will determine your eternal fate; it will determine where you will be living 200 years from now.

A Divine Answer

The disciples were a bit stunned by the question, and only the apostle Peter attempted an answer. His answer needs to be your answer: "You are the Christ, the Son of the living God" (Matthew 16:16).

The amazing thing about this answer is that Peter did not come up with it on his own. It had come from the Heavenly Father (Matthew 16:17). God himself helped Peter mold the correct answer to this all-important question. So, when you read this answer, realize that it is a divinely given response! "You are the Christ, the Son of the living God."

Three Views of Christ

"Who do you say that I am?" Former atheist-turned-Christian C.S. Lewis tackled this question by coming up with three possible responses: Jesus Christ is the Lord (God); Jesus was a liar; or Jesus was a lunatic. As we look at the historical Jesus, we find His uniqueness in His birth (conceived by the Holy Spirit, miraculously born of a virgin), His death upon the cross, and His resurrection from the grave.

John Duncan, quoted in *Colloquia Peripatetica* (1870), put it best: "Christ either deceived mankind by conscious fraud, or He was Himself deluded and self-deceived, or He was Divine." Who do you believe Jesus Christ is?

Your Answer

This chapter is the last chapter in this book, but someday you will face the last chapter of your life. Are you prepared for "The End"? About 6,000 years ago, a young couple by the name of Adam and Eve lived in a beautiful garden that their Creator had made especially for them. They were told by the Creator that they could enjoy this home fully with one exception: they were not to eat from the tree of the knowledge of good and evil, otherwise they would die (Genesis 2:17).

Sadly, they disobeyed God, and sin and death came into the world. As a result, everyone is born in sin, and we all are under a death sentence because we sin, too (Romans 5:12). But God is a God of grace and mercy and He did the unthinkable. He took that punishment upon himself due to His love for each one of us. Jesus Christ came to earth and paid the penalty for sin. He offers himself to us as Savior (1 Corinthians 15:22).

The Bible is clear on the subject of salvation: "Whoever believes in Him [Jesus] should not perish but have eternal life" (John 3:15). As you consider the world's most important question, consider taking these action steps:

- Admit that you are a sinner (Romans 3:23; 1 John 1:8; Galatians 5:19–21).
- Repent of your sins before a Holy God and turn away from them (Mark 1:15; 2 Corinthians 7:10; Acts 20:21).
- Receive Jesus Christ as Lord of your life (John 1:12).
- Realize that eternal life is a gift from God (Romans 6:23).
- Receive God's gift by faith — by taking God at His Word (Romans 10:8–11).
- Read and believe what the Bible says: "For by grace you have been saved through faith, and that not of yourselves; it is the gift of God, not of works, lest anyone should boast" (Ephesians 2:8–9).
- Express this to God in prayer. Although there is no one prayer that should be prayed, you may want to say something like: "Dear God, thank you for sending Your Son, the Lord Jesus Christ, to pay for my sins on the cross. Thank you that He died for me. I acknowledge that I am a sinner and that I cannot save myself. I repent of my sins and I receive Your gift of salvation by faith. Thank you for loving me enough to save me. In Jesus' name, amen."

It's a Sure Thing

The Bible says "that if you confess with your mouth the Lord Jesus and believe in your heart that God has raised Him from the dead, you will be saved" (Romans 10:9).

How do you know that you are saved? The Bible says you can know! "These things I have written to you who believe in the name of the Son of God, that you may know that you have eternal life" (1 John 5:13).

Saved from Death

During the 2000 Olympic games, I had the joy of taking 65 Americans to Sydney, Australia, for personal evangelism. Because of their grandeur and beauty, the Blue Mountains outside Sydney became a magnet for the tourists. Teams from our group were sent to the mountains to talk to visitors about the Lord. Among those visitors was an Aussie from Melbourne named Paul. One of our team members engaged Paul in conversation for over 45 minutes, explaining about the God of the Bible and His love for Paul.

Paul responded with enthusiasm and prayed to receive Christ as his personal Savior. When he was asked where he was going from there, Paul shared that he had no place to go. He told us a story of family rejection and bad decisions on his part. We stood speechless as he told us that he had come to the Blue Mountains that day to commit suicide "Because no one in this world cares about me."

Once again, like in Rome, my body trembled; but this time it trembled with delight. Paul had found hope for what appeared to be a hopeless life. Paul had found love, forgiveness, and acceptance from God. Paul had found friendship from a bunch of Americans who lived halfway around the world.

Three months after his salvation on the Blue Mountains, Paul suffered insulin shock and died. But we know that Paul is in heaven with his Lord! That, my friend, is real *hope*!

The Christian's Global Assignment

If you have trusted Christ as your personal Savior from sin, God has given you a new mandate. It is called the Great Commission: "'Go therefore and make disciples of all the nations, baptizing them in the name of the Father and of the Son and of the Holy Spirit, teaching them to observe all things that I have commanded you; and lo, I am with you always, even to the end of the age.' Amen." (Matthew 28:19–20). The word translated *nations* in verse 19 in Greek is *ethnos,* which speaks to us of *all the ethnic groups* in the world. Our Lord wants us to reach all the ethnic groups in the world. He did not give us this assignment knowing it would be impossible for us to reach; rather, He gave us this assignment expecting us to fulfill it.

Reaching the World, Closing the Knowledge Gap

Answers in Genesis is called to proclaim the life-changing message of the gospel, beginning in the Book of Genesis. One of the most thrilling developments in recent years is a method of evangelism called "creation evangelism," in which the Bible is taught chronologically. People hear about a loving Creator God who made them in His image and is the Creator of the universe. This God sent His Son, Jesus, to die on a cross in Jerusalem to pay the penalty of our sins. This form of evangelism answers modern mankind's most searching questions and gives every reason for hope.

Because of our mission, AiG WorldWide is translating a massive amount of creation literature, DVDs, radio programs, and web articles into the languages of the world. After the translation teams have completed a project, we will print and distribute the material, preferably without cost, to mission field leaders all over the globe. Here are three ways we plan to carry out our vision:

- AiG libraries will provide literature for Christian Bible schools and mission organizations to give answers to the next generation of Christian leaders.
- Christian pastors and leaders will be brought to AiG for training on how they can become creation spokesmen in their own countries.
- New and innovative programs will be initiated to help provide answers for believers and hope for the lost. We want to create a massive creation movement worldwide.

To Every People Group

As I travel globally, I still tremble with raw "Roman emotion" when I see the masses of unsaved people without hope. I have watched them light incense, bow before statues, chant memorized prayers, beat themselves, and worship multitudes of gods. And with the world's population edging closer to seven billion people, I see greater opportunities for missions today than ever before. Jesus commands us to get the Word out. So we prayerfully invite you to join us! Together we can dispel the hopelessness abroad with the hope of the glorious gospel of Jesus Christ.

Contributors

Dr. David Crandall

David has been involved in full-time ministry for nearly 40 years. For the last 12 years he has served as international director of Gospel Literature Services. He led this ministry to publish and translate Christian literature into 117 different languages, and he has ministered cross-culturally in 68 different countries. In 2006, he joined Answers in Genesis as the director of AiG World-Wide. Dr. Crandall currently serves on the executive board of the Association of Baptists for World Evangelism.

Brian H. Edwards

Brian was pastor of an evangelical church in a southwest London suburb for 29 years, and then president of the Fellowship of Independent Evangelical Churches from 1995–1998. He is the author of 16 books, and continues a ministry of writing and itinerant preaching and lecturing. His wife, Barbara, died in 1998; he has two sons and three granddaughters.

Dr. Danny R. Faulkner

Danny has a BS (math), MS (physics), MA and PhD (astronomy, Indiana University). He is full professor at the University of South Carolina–Lancaster, where he teaches physics and astronomy. He has published about two dozen papers in various astronomy and astrophysics journals.

Ken Ham

Ken is the president and CEO of Answers in Genesis (USA). He has authored several books, including the best-seller *The Lie: Evolution*. He is one of the most in-demand speakers in the United States and has a daily radio program called *Answers . . . with Ken Ham,* which is heard on over 850 stations in the United States and over 1,000 worldwide.

Ken has a BS in applied science (with an emphasis in environmental biology) from Queensland Institute of Technology in Australia. He also holds a diploma of education from the University of Queensland (a graduate qualification for science teachers in the public schools in Australia). Ken has been awarded two honorary doctorates: a Doctor of Divinity (1997) from Temple Baptist College in Cincinnati, Ohio, and a Doctor of Literature (2004) from Liberty University in Lynchburg, Virginia.

Bodie Hodge

Bodie earned a BS and MS in mechanical engineering at Southern Illinois University at Carbondale in 1996 and 1998, respectively. His specialty was in materials science, working with advanced ceramic powder processing. He developed a new method of production of submicron titanium diboride.

Bodie accepted a teaching position as visiting instructor at Southern Illinois in 1998 and taught for two years. After this, he took a job working as a test engineer at Caterpillar's Peoria Proving Ground. Bodie currently works at Answers in Genesis (USA) as a speaker, writer, and researcher after working for three years in the Answers Correspondence Department.

Carl Kerby

Carl, who is one of AiG's most dynamic lecturers on the Book of Genesis, is a founding board member of AiG. In addition to being AiG's vice president for ministry relations, Carl conducts a number of faith-building AiG meetings each year. His passion is to proclaim the authority and accuracy of the Bible, and Carl does so in a highly effective way for all audiences. He is much in demand as a speaker among both young people and adults. A former air traffic controller at Chicago's busy O'Hare International Airport, Carl's thrilling testimony has been shared in churches throughout America.

Carl has several DVDs, including *Genesis: The Bottom Strip of the Christian Faith; Genesis — Today's Answer to Racism;* and *What Is the Best Evidence that God Created?*

Dr. Jason Lisle

Dr. Lisle received his PhD in astrophysics from the University of Colorado at Boulder. He specializes in solar astrophysics and has interests in the physics of relativity and biblical models of cosmology. Dr. Lisle has published a number of books, including *Taking Back Astronomy* and *The Ultimate Proof of Creation*, plus articles in both secular and creationist literature. He is a speaker, researcher, and writer for Answers in Genesis–USA.

Stacia McKeever

Stacia worked for Answers in Genesis for 12 years until leaving full-time work to raise her son She is currently involved in the Answers in Genesis VBS programs and other writing projects. She coauthored the "Answers for Kids" section in *Creation* magazine for several years and has written or coauthored a number of articles for *Creation* and the AiG website. She has a BS in biology and a BA in psychology from Clearwater Christian College.

Stacia has conducted hands-on workshops for young children around the United States for several years and has written curricula (*Beginnings*, *The Seven C's of History*) and workbooks for elementary-aged children and adults. Stacia has written for *The Godly Business Woman* and *Evangelizing Today's Child* and has researched and written copy for several Bible-themed calendars.

Dr. David Menton

Dr. Menton was an associate professor of anatomy at Washington University School of Medicine from 1966 to 2000 and has since become associate professor emeritus. He was a consulting editor in histology for *Stedman's Medical Dictionary*, a standard medical reference work.

David earned his PhD from Brown University in cell biology. He is a popular speaker and lecturer with Answers in Genesis (USA), showing complex design in anatomy with popular DVDs such as *The Hearing Ear and Seeing Eye* and *Fearfully and Wonderfully Made*. He also has an interest in the famous Scopes trial, which was a big turning point in the creation/evolution controversy in the USA in 1925.

Dr. Elizabeth Mitchell

Elizabeth earned her MD from Vanderbilt University School of Medicine and practiced medicine for seven years until she retired to be a stay-at-home mom. Her interest in ancient history strengthened when she began to homeschool her daughters. She desires to make history come alive and to correlate it with biblical history.

Dr. Tommy Mitchell

Tommy graduated with a BA with highest honors from the University of Tennessee–Knoxville in 1980 with a major in cell biology and a minor in biochemistry. He subsequently attended Vanderbilt University School of Medicine in Nashville, where he was granted an MD in 1984.

Dr. Mitchell's residency was completed at Vanderbilt University Affiliated Hospitals in 1987. He was board certified in internal medicine, with a medical practice in Gallatin, Tennessee (the city of his birth). In 1991, he was elected a Fellow of the American College of Physicians (FACP). Tommy became a full-time speaker, researcher, and writer with Answers in Genesis (USA) in 2006.

Dr. Terry Mortenson

Terry earned a BA in math at the University of Minnesota in 1975 and later went on to earn an MDiv in systematic theology at Trinity Evangelical Divinity School in 1992. His studies took him to the United Kingdom, where he earned a PhD in the history of geology at Coventry University.

Terry has done extensive research regarding the beliefs of the 19th century scriptural geologists. An accumulation of this research can be found in his book *The Great Turning Point*. Terry is currently working at Answer in Genesis (USA) as a speaker, writer, and researcher.

Larry Pierce

Larry is a retired programmer from Canada who did his undergraduate and graduate work at the University of Waterloo in mathematics. He greatly enjoys ancient history. This passion led him to spend five years translating *The Annals of the World* from Latin into English. He is also the creator of a sophisticated and powerful Bible program, The Online Bible.

Dr. Georgia Purdom

Georgia received her PhD in molecular genetics from Ohio State University in 2000. As an associate professor of biology, she completed five years of teaching and research at Mt. Vernon Nazarene University in Ohio before joining the staff at Answers in Genesis (USA).

Dr. Purdom has published papers in the *Journal of Neuroscience*, the *Journal of Bone and Mineral Research*, and the *Journal of Leukocyte Biology*. She is also a member of the Creation Research Society, American Society for Microbiology, and American Society for Cell Biology.

She is a peer reviewer for *Creation Research Society Quarterly*. Georgia has a keen interest in and keeps a close eye on the Intelligent Design movement.

Dr. Ron Rhodes

Ron has a ThM and ThD in systematic theology from Dallas Theological Seminary and is the president of Reasoning from the Scriptures Ministries. He has authored 35 books and is a popular conference speaker across the United States. During his schedule, he finds time to teach cult apologetics at several well-known seminaries.

Mike Riddle

As a former captain in the Marines, Mike earned a BS in mathematics and an MS in education. He has been involved in creation apologetics for many years and has been an adjunct lecturer with the Institute for Creation Research. Mike has a passion for teaching and he exhibits a great ability to bring topics down to a lay-audience level in his lectures.

Before becoming a Marine, Mike became a U.S. national champion in the track-and-field version of the pentathlon (in 1976). His best events were the 400 meters, javelin, long jump, and 1,500 meters. In his professional life, Mike worked for many years in the computer field with Microsoft (yes, he has met Bill Gates).

Dr. Andrew A. Snelling

Dr. Andrew Snelling received his PhD (geology) from the University of Sydney (Australia). After research experience in the mineral exploration industry, he was founding editor of the *Creation Ex Nihilo Technical Journal* (Australia). He also served as a professor of geology at the Institute for Creation Research. In 2007 he joined Answers in Genesis–USA as director of research. A member of several professional geological societies, Dr. Snelling has written numerous technical papers in geological journals and creationist publications.

Paul F. Taylor

Paul F. Taylor is the senior speaker for Answers in Genesis (UK/Europe). He holds a BSc in chemistry from Nottingham University and a masters in science education from Cardiff University. He is the author of several books, including *Cain and Abel*, *In the Beginning*, and *The Six Days of Genesis*. Paul and his wife, Geri, have five children between them.

Dr. A.J. Monty White

Formerly the chief executive of Answers in Genesis (UK/Europe), Dr. Monty White joined AiG after leaving the University of Wales in Cardiff where he had been a senior administrator for 28 years. He is a graduate of the University of Wales, obtaining his BS in chemistry in 1967 and his PhD for research in the field of gas kinetics in 1970. Monty spent two years investigating the optical and electrical properties of organic semiconductors before moving to Cardiff, where he joined the administration at the university there.

Monty is well known for his views on creation, having written numerous articles and pamphlets, as well as a number of books dealing with various aspects of creation, evolution, science, and the Bible. Monty has appeared on British television programs and has been interviewed on local and national radio about creation.

Dr. John Whitmore

John received a BS in geology from Kent State University, an MS in geology from the Institute for Creation Research, and a PhD in biology with paleontology emphasis from Loma Linda University. Currently an associate professor of geology, he is active in teaching and research at Cedarville University. Dr. Whitmore serves on the board of Creation Research Science Education Foundation located in Columbus, Ohio, and he is also a member of the Creation Research Society and the Geological Society of America.

Index

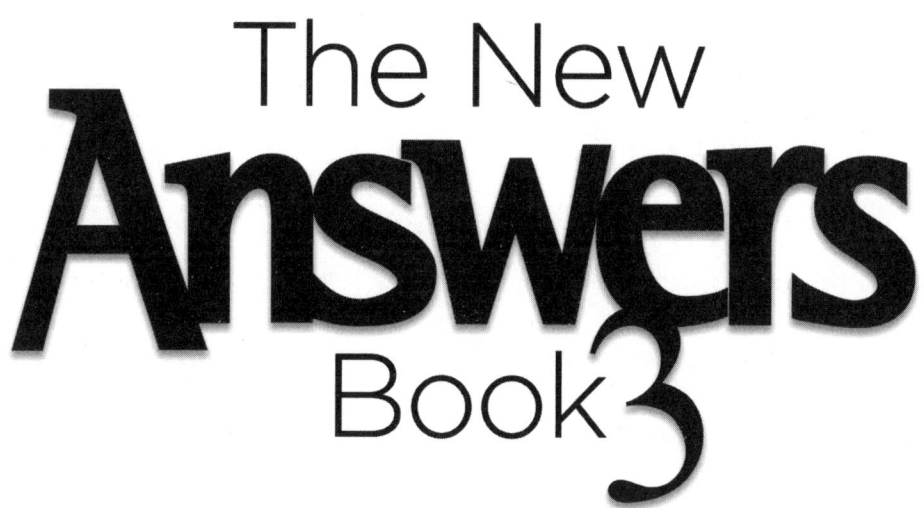

The New Answers Book 3

Over 35 Questions on

Creation/Evolution and the Bible

Ken Ham General Editor

First printing: February 2010
Fourteenth printing: March 2021

Master Books®, P.O. Box 726, Green Forest, AR 72638
Master Books® is a division of the New Leaf Publishing Group, Inc.

ISBN: 978-0-89051-579-2
ISBN: 978-1-61458-108-6 (digital)
Library of Congress Number: 2008903202

Printed in the United States of America

Master
Books®
A Division of New Leaf Publishing Group
www.masterbooks.com

ACKNOWLEDGMENTS AND SPECIAL THANKS

Acknowledgments and special thanks for reviewing or editing chapters:

Steve Fazekas (theology, AiG), Frost Smith (biology, editor, AiG), Mike Matthews (editor, AiG), Gary Vaterlaus (science education, editor, AiG), Tim Chaffey (theology, Midwest Apologetics), Dr. John Whitcomb (theology, president of Whitcomb Ministries), Dr. Larry Vardiman (atmospheric science, chairman of the department of astro-geophysics at the Institute for Creation Research), Ken Ham (biology, president and CEO of Answers in Genesis), Donna O'Daniel (biology, AiG), Dr. Tim Clarey (geology), Christine Fidler (CEO of Image in the UK), Mark Looy (editor, AiG), Dr. Terry Mortenson (history of geology, AiG), John Upchurch (editor, AiG), Dr. Jason Lisle (astrophysics, AiG), Dr. John Morris (geological engineering, president of the Institute for Creation Research), Dr. Andrew Snelling (geology, director of research at AiG), Dr. David Menton (retired, cell biology, former associate professor of anatomy at Washington University School of Medicine, now AiG), Dr. Tommy Mitchell (internal medicine, AiG), Dr. Georgia Purdom (genetics, AiG), Roger Patterson (biology, editor, AiG), Bodie Hodge (engineering, materials, AiG), Dr. John Whitmore (geology, associate professor of geology, Cedarville University), Lori Jaworski (editor, AiG), Dr. Danny Faulkner (astronomy, physics, chair of the division of math, science, nursing & public health at the University of South Carolina, Lancaster), Stacia McKeever (biology, psychology, AiG), David Wright (student of engineering, AiG), and Dr. Elizabeth Mitchell (MD).

Special thanks to Dan Lietha for many of the illustrations used in this book, and to Laura Strobl, Dan Stelzer, and Jon Seest for much of the graphic design layout and illustrations, unless otherwise supplied.

We would also like to thank the authors for their contributions to the Answers Book Series and all the research and hard work they put into these chapters.

Contents

INTRODUCTION

Evolving Tactics

KEN HAM

Over the past 30 years of my personal, intimate experience in the biblical creation ministry, I have observed "evolving" (in the sense of "changing") tactics used by prominent secularists to respond to arguments from creationist scholars and researchers. Based on my experience, I would divide the interactions of biblical creationists and outspoken secularists into four basic eras.

The Debate Era of the 1970s

When I first became aware of the U.S. creation movement in the 1970s (while I was a teacher in Australia), I learned that Duane Gish (Ph.D. in biochemistry from the University of California, Berkeley) of the Institute for Creation Research was actively debating evolutionary scientists from various academic institutions.

At that time, creationist arguments against evolution consisted of arguments against so-called ape-men, and arguments that the Cambrian Explosion and lack of transitional forms illustrated that Darwinian evolution did not happen.

Evolutionists argued back with supposed counters to these arguments. For instance, they claimed that *Archaeopteryx* was a transitional form between reptiles and birds (since refuted), that the "mammal-like reptiles" were transitional forms, and so on. However, in the long run, such "evidences" were just interpreted differently by both sides according to their starting points — creation or evolution!

Evolutionists still use this fossil to support the transition of one kind of animal to another. Creationists interpret the same evidence in light of the Bible and come to different conclusions.

The Rise of Creationist Media in the Early 1980s

Although secular educational institutions and secular journals, by and large, taught evolution as fact, I noticed more deliberate attempts to increase public indoctrination about evolution and earth history occurring over millions of years. For this reason, I might also call this the *Intensified Evolutionary Indoctrination* era.

At the same time, evolutionists increasingly refused to debate creationists. In today's world (early 21st century), such debates are rare.

Nonetheless, the biblical creation movement began publishing more and more books, videos, and other materials; and the "creation versus evolution" issue rose to greater prominence in the culture and secular media. Secularist opposition to the creation movement intensified, with many articles in print. Although they included some ridicule, many articles tried to outline the supposed scientific reasons why creationists were wrong.

The Public School Controversies of the 1980s and 1990s

Frustrated by how evolutionary teaching had taken over much of the secular education system, and seeing that creation was basically outlawed from the classroom, many Christians tried (unsuccessfully) in the courts to force public school teachers to teach creation in their classes, or at least to allow critiques of evolution.

This era eventually sparked the rise of the non-Christian "intelligent design" movement[1] —which many Christians thought might be the answer to the education problem — but soon found it was not.

Secularists fought hard to falsely accuse creationists of being anti-science. They typically labeled belief in the Genesis account of history — or even the simple belief that God created — as just a "religious" view, while belief in Darwinian evolution was a "scientific" view.

The Name-Calling Era of the Early 21st Century

Some secularists have reverted to name-calling in a desperate attempt to discredit biblical creationists. In the early part of the 21st century, articles against the creation movement became more scathing, sarcastic, and mocking, with increasing name-calling. Rather than attempting to use logical arguments to dissuade people, evolutionists mocked not just the Genesis account of creation but also belief in any unnamed intelligence behind the universe.

No longer satisfied to argue that creationists could not be real scientists and that belief in creation is anti-scientific, secularists began accusing

1. The intelligent design movement does not claim to be Christian. It is a movement (with both Christians and non-Christians) that is against naturalism, teaching that an unnamed intelligence is behind the universe.

creationists of being anti-technology. I began to see the argument appear that people who believe in creation are inconsistent if they use modern technology, such as computers and airplanes, which are products of man's scientific ingenuity.

Increased name-calling against creationists, in an attempt to defame their integrity, began to appear, not just in newspaper articles but in various evolutionist books and reputable science magazines. Biblical creationists were equated with terrorists, as secular writers used words like *fundamentalists* to describe both Christians and terrorists. All of this name-calling by unscrupulous secularists is part of a deliberate attempt to smear Christians and use fear tactics to brainwash people into a false understanding of what Christians believe.

This era also saw the rise of the "New Atheists," who began overtly attacking Christianity and preaching atheism around the world. This radical atheist movement is spearheaded by Dr. Richard Dawkins of Oxford University, summed up in a quote from his best-selling book *The God Delusion*, in which he vehemently attacks the Christian God:

> The God of the Old Testament is arguably the most unpleasant character in all fiction: jealous and proud of it; a petty, unjust, unforgiving control-freak; a vindictive, bloodthirsty ethnic cleanser; a misogynistic, homophobic, racist, infanticidal, genocidal, filicidal, pestilential, megalomaniacal, sadomasochistic, capriciously malevolent bully. Those of us schooled from infancy in his ways can become desensitized to their horror.[2]

In June of 2008, Paul Myers, associate professor of biology at the University of Minnesota–Morris, decided to oppose me on his blog by beginning a name-calling exercise.

> Millions of people, including some of the most knowledgeable biologists in the world, think just about every day that you are . . . [and then he launched into a long list of names, from airhead to birdbrain, blockhead, bonehead, and bozo to sap, scam artist, sham, simpleton, a snake oil salesman, wacko] and much, much worse. You're a clueless schmuck who knows nothing about science and has arrogantly built a big fat fake museum to promote medieval [expletive] — you should not be surprised to learn that you are held in very low esteem by the community of scholars and scientists, and by

2. Richard Dawkins, *The God Delusion* (Boston, MA: Houghton Mifflin Co., 2006), p. 31.

the even larger community of lay people who have made the effort to learn more about science than you have (admittedly, though, you have set the bar very, very low on that, and there are 5 year old children who have a better grasp of the principles of science as well as more mastery of details of evolution than you do).[3]

More troublesome is the accusation, which I now observe from different sources, that creationists and Christians are "child abusers." Such an emotionally charged term is really meant to marginalize Christians in the culture. If the secular elite had total control of the culture, they could prosecute this in the courts.

Richard Dawkins agrees that this term is appropriate for Christians who teach about the doctrine of hell: "I am persuaded that the phrase 'child abuse' is no exaggeration when used to describe what teachers and priests are doing to children whom they encourage to believe in something like the punishment of unshriven mortal sins in an eternal hell."[4]

3. Posted on blog of P.Z. Myers on 6/21/2008.
4. Dawkins, *The God Delusion*, p. 318.

In chapter 16 of the best-selling book *God Is Not Great*, entitled "Is Religion Child Abuse?" another New Atheist, Christopher Hitchens, answers the chapter title in the affirmative, claiming that all related customs, such as circumcision, are child abuse. He even equates teaching children about religion to indoctrination and child abuse.

When the Creation Museum opened near the Cincinnati International Airport on Memorial Day weekend 2007, secular scientists and an atheist group demonstrated outside the museum with signs simply mocking my name, such as "Behold, the curse of Ham," rather than using logical scientific arguments to argue their case.

Resorting to such name-calling not only shows that this issue strikes at deep spiritual problems, but that those who can't prove their position by logic or science are driven by emotion. We can expect such name-calling to increase as secularists become more frustrated in not being able to refute the powerful truth that the Creator is clearly seen (see Romans 1:18–20) and "in the beginning God created the heavens and the earth" (Genesis 1:1).

We need to remember what God said in Proverbs 21:24: "A proud and haughty man — 'Scoffer' is his name; he acts with arrogant pride." In contrast, God expects His people to take the higher ground, to earn a reputation for kind and gentle words, as we speak "the truth in love" (Ephesians 4:15). The theme verse of my life and Answers in Genesis includes every Christian's duty to give answers "with meekness":

> But sanctify the Lord God in your hearts, and always be ready to give a defense to everyone who asks you a reason for the hope that is in you, with meekness and fear (1 Peter 3:15).

1

Where Was the Garden of Eden Located?

KEN HAM

Most Bible commentaries state that the site of the Garden of Eden was in the Middle East, situated somewhere near where the Tigris and Euphrates Rivers are today. This is based on the description given in Genesis 2:8–14:

> The LORD God planted a garden eastward in Eden. . . . Now a river went out of Eden to water the garden, and from there it parted and became four riverheads. The name of the first is Pishon. . . . The name of the second river is Gihon. . . . The name of the third river is Hiddekel [Tigris]. . . . The fourth river is the Euphrates.

Even the great theologian John Calvin struggled over the exact location of the Garden of Eden. In his commentary on Genesis he states:

> Moses says that one river flowed to water the garden, which afterwards would divide itself into four heads. It is sufficiently agreed among all, that two of these heads are the Euphrates and the Tigris; for no one disputes that . . . (Hiddekel) is the Tigris. But there is a great controversy respecting the other two. Many think, that Pison and Gihon are the Ganges and the Nile; the error, however, of these men is abundantly refuted by the distance of the positions of these rivers. Persons are not wanting who fly across even to the Danube; as if indeed the habitation of one man stretched itself from the most remote part of Asia to the extremity of Europe. But since many other

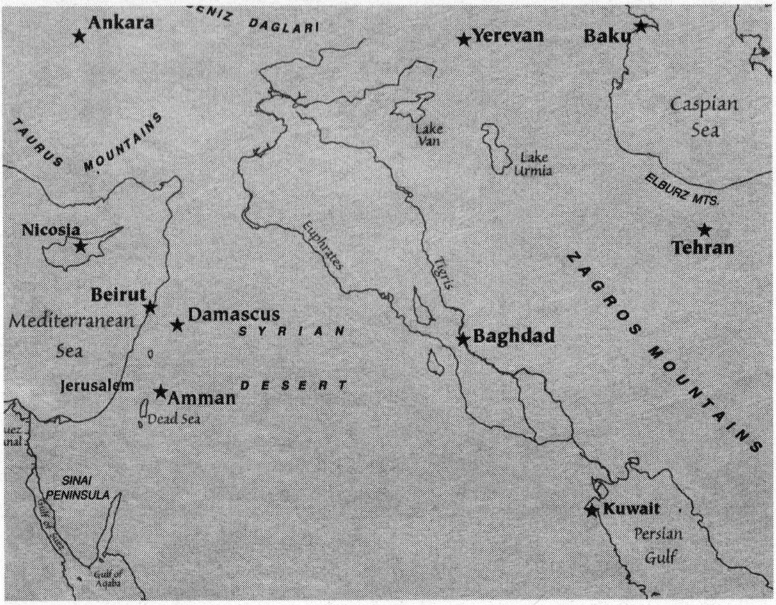

Many wrongly conclude that the Garden of Eden was somewhere in the Middle East based on the names of the rivers in Genesis 2.

> celebrated rivers flow by the region of which we are speaking, there is greater probability in the opinion of those who believe that two of these rivers are pointed out, although their names are now obsolete. Be this as it may, the difficulty is not yet solved. For Moses divides the one river which flowed by the garden into four heads. Yet it appears, that the fountains of the Euphrates and the Tigris were far distant from each other.[1]

Calvin recognized that the description given in Genesis 2 concerning the location of the Garden of Eden does not fit with what is observed regarding the present Tigris and Euphrates Rivers. God's Word makes it clear that the Garden of Eden was located where there were four rivers coming from one head. No matter how one tries to fit this location in the Middle East today, it just can't be done.

Interestingly, Calvin goes on to say:

> From this difficulty, some would free themselves by saying that the surface of the globe may have been changed by the deluge. . . .[2]

1. John Calvin, *Commentary on Genesis*, Volume 1, online at: www.ccel.org/ccel/calvin/calcom01.viii.i.html.
2. Ibid.

This is a major consideration that needs to be taken into account. The worldwide, catastrophic Flood of Noah's day would have destroyed the surface of the earth. If most of the sedimentary strata over the earth's surface (many thousands of feet thick in places) is the result of this global catastrophe as creationists believe, then we would have no idea where the Garden of Eden was originally located — the earth's surface totally changed as a result of the Flood.

Not only this, but underneath the region where the present Tigris and Euphrates Rivers are located there exists hundreds of feet of sedimentary strata — a significant amount of which is fossiliferous. Such fossil-bearing strata had to be laid down at the time of the Flood.

Therefore, no one can logically suggest that the area where the present Tigris and Euphrates Rivers are today is the location of the Garden of Eden, for this area is sitting on Flood strata containing billions of dead things (fossils). The perfect Garden of Eden can't be sitting on billions of dead things before sin entered the world!

This being the case, the question then is why are there rivers named Tigris and Euphrates in the Middle East today?

In my native country of Australia, one will recognize many names that are also used in England (e.g., Newcastle). The reason is that when the settlers came out from England to Australia, they used names they were familiar with in England to name new places/towns in Australia.

Another example is the names given to many rivers in the United States. There is the Thames River in Connecticut, the Severn River in Maryland, and the Trent River in North Carolina — all named for prominent rivers in the UK.

In a similar way, when Noah and his family came out of the ark after it landed in the area we today call the Middle East (the region of the Mountains of Ararat), it would not have been surprising for them to use names they were familiar with from the pre-Flood world (e.g., Tigris and Euphrates), to name places and rivers, etc., in the world after the Flood.

Ultimately, we don't know where the Garden of Eden was located. To insist that the Garden was located in the area around the present Tigris and Euphrates Rivers is to deny the catastrophic effects of the global Flood of Noah's day, and to allow for death before sin.

2

What Did Noah's Ark Look Like?

TIM LOVETT (WITH BODIE HODGE)

Most of us have seen various depictions of Noah's ark — from the large, box-like vessel to the one in children's nurseries with the giraffes' heads sticking out the top. But what did the ark really look like? Can we really know for sure?

Depicting the Ark — A Sign of the Times?

Noah's ark has been a popular subject for artists throughout the centuries. However, it is not easy to adequately depict this vessel because the description in Genesis 6 is very brief. To paint a complete picture, the artist must assume some important details.

As the invention of Gutenberg's movable-type printing press in the 1400s made rapid and widespread distribution of the Holy Scriptures possible, Noah's ark quickly became the subject of lavish illustrations. Many designs were pictured, and some were more biblical than others. Often, artists distorted the biblical specifications to match the ships of the day. For instance, the picture shown in figure 1 has the hull of a caravel, which was similar to two of the small sailing vessels used by Christopher Columbus in 1492.

Unlike most other artists, Athanasius Kircher (a Jesuit scientist, 1602–1680) was committed to accurately depicting the massive ark specified in Genesis. He has been compared to Leonardo da Vinci for his inventiveness and his works' breadth and depth. This early "creation scientist" calculated the number of animals that could fit in the ark, allowing space for provisions and Noah's family. His realistic designs (figure 2) set the standard for generations of artists.

For the next two centuries, Bible artists stopped taking Noah's ark quite so seriously, and ignored the explicit biblical dimensions in their illustrations. These artists simply reflected the scholars of the day, who had rejected the Bible's history of the world. Few Christians living in 1960 had ever seen a biblically based rendering of Noah's ark. Cute bathtub shapes and smiling cartoonish animals illustrated the pervasive belief that Noah's ark was nothing more than a tool for character-building through fictionalized storytelling.

Figure 1. Artist's depiction of the construction of Noah's ark, from H. Schedel's *Nuremburg Chronicle* of 1493.

Figure 2. Athanasius Kircher (1600s) was careful to follow the Bible's instructions and used a rectilinear hull, based on the dimensions in Genesis 6:15, including three decks, a door in the side, and a window of one cubit.

Then in 1961 Dr. John Whitcomb and Dr. Henry Morris published *The Genesis Flood*, which made sense of a global cataclysm and a real, shiplike Noah's ark. This book was a huge thrust to help begin the modern creationist movement.

The primary focus in *The Genesis Flood* was the size of the ark and its animal-carrying capacity. A block-shaped ark was ideal for this, easily suggesting that the ark had plenty of volume. Later studies confirmed that a ship with a rectangular cross-section 50 cubits wide and 30 cubits high was stable. Images

Figure 3. This 1985 painting by Elfred Lee was completed after multiple interviews in the early 1970s with George Hagopian, an "eyewitness" of a box-shaped ark. (Image used with permission from Elfred Lee.)

of a rectangular ark strikingly similar to Kircher's design rendered centuries earlier began to appear in publications (see figure 3).

The next few decades saw another popular phenomenon — the search for Noah's ark. Documentary movies and books claimed Noah's ark was hidden on Mt. Ararat, and prime-time television broadcast some mysterious photos of dark objects jutting out from the snow. George Hagopian was one of the first modern "eyewitnesses" who purported to have seen a box-shaped ark. And so it happened — Noah's ark was illustrated worldwide as a box.

When looking at history, artists in each generation have defined Noah's ark according to the cultural setting and what they knew at the time. While we used to see variety in the shape of the ark, more recent depictions have seemingly locked into the box shape. But new insights — in keeping with the biblical specifications of the ark and conditions during the Flood — suggest that it's time we start thinking "outside the box."

Thinking Outside the Box

While the Bible gives us essential details on many things, including the size and proportions of Noah's ark, it does not necessarily specify the precise shape of this vessel. It is important to understand, however, that this lack of physical description is consistent with other historical accounts in Scripture.[1] So how should we illustrate what the ark looked like? The two main options include a default rectangular shape reflecting the lack of specific detail, and a more fleshed-out design that incorporates principles of ship design from maritime science, while remaining consistent with the Bible's size and proportions.

Genesis describes the ark in three verses, which require careful examination:

1. Other objects spoken of in Scripture lack physical details that have been discovered (through archaeology and other research) later (e.g., the walls of Jericho were actually double and situated on a hillside — one higher than the other with a significant space of several feet between them).

> 6:14—Make yourself an ark [*tebah*] of gopherwood; make rooms [*qinniym*] in the ark, and cover it inside and outside with pitch [*kofer*].
>
> 6:15—And this is how you shall make it: The length of the ark shall be three hundred cubits, its width fifty cubits, and its height thirty cubits.
>
> 6:16—You shall make a window [*tsohar*] for the ark, and you shall finish it to a cubit from above; and set the door of the ark in its side. You shall make it with lower, second, and third decks.

Most Bibles make some unusual translation choices for certain key words. Elsewhere in the Bible, the Hebrew word translated here as "rooms" is usually rendered "nests"; "pitch" would normally be called "covering"; and "window" would be "noon light." Using these more typical meanings, the ark would be something like the following.

The *tebah* (ark) was made from gopher wood, it had nests inside, and it was covered with a pitch-like substance inside and out. It was 300 cubits long, 50 cubits wide, and 30 cubits high. It had a noon light that ended a cubit upward and above, it had a door in the side, and there were three decks. (For the meaning of "upward and above," see the section "2. A cubit upward and above" on the following pages.)

As divine specifications go, Moses offered more elaborate details about the construction of the tabernacle, which suggests this might be the abridged version of Noah's complete directions. On the other hand, consider how wise Noah must have been after having lived several centuries. The instructions that we have recorded in Genesis may be all he needed to be told. But in any case, 300 cubits is a big ship, not some whimsical houseboat with giraffe necks sticking out the top.

Scripture gives no clue about the shape of Noah's ark beyond its proportions — length, breadth, and depth. Ships have long been described like this without ever implying a block-shaped hull.

The scale of the ark is huge yet remarkably realistic when compared to the largest wooden ships in history. The proportions are even more amazing — they are just like a modern cargo ship. In fact, a 1993 Korean study was unable to find fault with the specifications.

All this makes nonsense of the claim that Genesis was written only a few centuries before Christ, as a mere retelling of earlier Babylonian flood legends such as the *Epic of Gilgamesh*. The *Epic of Gilgamesh* story describes a cube-shaped ark, which would have given a dangerously rough ride. This is neither accurate nor scientific. Noah's ark is the original, while the Gilgamesh Epic is a later distortion.

What about the Shape?

For many years, biblical creationists have simply depicted the ark as a rectangular box. This helped emphasize its size. It was easy to explain capacity and illustrate how easily the ark could have handled the payload. With the rectangular shape, the ark's stability against rolling could even be demonstrated by simple calculations.

Yet the Bible does not say the ark must be a rectangular box. In fact, Scripture does not elaborate about the shape of Noah's ark beyond those superb, overall proportions.

Scientific Study Endorses Seaworthiness of Ark

Noah's ark was the focus of a major 1993 scientific study headed by Dr. Seon Hong at the world-class ship research center KRISO, based in Daejeon, South Korea. Dr. Hong's team compared 12 hulls of different proportions to discover which design was most practical. No hull shape was found to significantly outperform the 4,300-year-old biblical design. In fact, the ark's careful balance is easily lost if the proportions are modified, rendering the vessel either unstable, prone to fracture, or dangerously uncomfortable.

The research team found that the proportions of Noah's ark carefully balanced the conflicting demands of stability (resistance to capsizing), comfort (seakeeping), and strength. In fact, the ark has the same proportions as a modern cargo ship.

The study also confirmed that the ark could handle waves as high as 100 feet (30 m). Dr. Hong is now director general of the facility and claims "life came from the sea," obviously not the words of a creationist on a mission to promote the worldwide Flood. Endorsing the seaworthiness of Noah's ark obviously did not damage Dr. Hong's credibility.

The word *ark* in Hebrew is the obscure term *tebah*, a word that appears only one other time when it describes

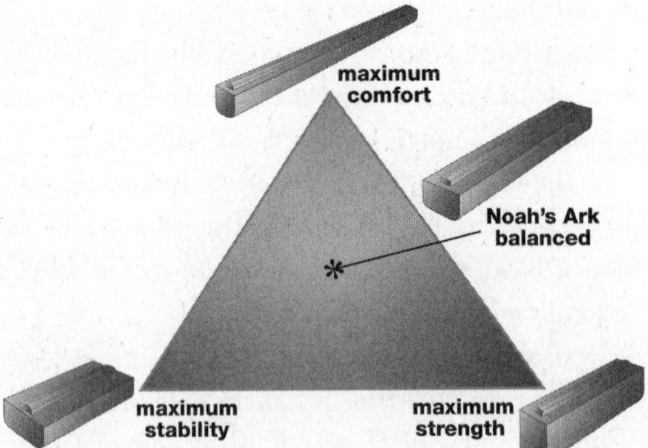

Figure 4. The proportions of the ark were found to carefully balance the conflicting demands of stability, comfort, and strength.

the basket that carried baby Moses (Exodus 2:3). One was a huge, wooden ship and the other a tiny, wicker basket. Both floated, both preserved life, and both were covered; but the similarity ends there. If the word implied anything about shape, it would be "an Egyptian basket-like shape," typically rounded. More likely, however, *tebah* means something else, like "lifeboat."[2]

The Bible leaves the details regarding the shape of the ark wide open — anything from a rectangular box with hard right angles and no curvature at all, to a shiplike form. Box-like has the largest carrying capacity, but a ship-like design would be safer and more comfortable in heavy seas. Such discussion is irrelevant if God intended to sustain the ark no matter how well designed and executed.

Clues from the Bible

Some people question whether the ark was actually built to handle rough seas, but the Bible gives some clues about the sea conditions during the Flood:

> The ark had the proportions of a seagoing vessel built for waves (Genesis 6:15).
>
> Logically, a mountain-covering, global flood would not be dead calm (Genesis 7:19).
>
> The ark moved about on the surface of the waters (Genesis 7:18).
>
> God made a wind to pass over the earth (Genesis 8:1).
>
> The Hebrew word for the Flood (*mabbul*) could imply being carried along.

The 1993 Korean study showed that some shorter hulls slightly outperformed the ark model with biblical proportions. The study assumed waves came from every direction, favoring shorter hulls like that of a modern lifeboat. So why was Noah's ark so long if it didn't need to be streamlined for moving through the water?

The answer lies in ride comfort (seakeeping). This requires a longer hull, at the cost of strength and stability, not to mention more wood. The ark's high priority for comfort suggests that the anticipated waves must have been substantial.

1. Something to Catch the Wind

Wind-driven waves would cause a drifting vessel to turn dangerously side-on to the weather. However, such waves could be safely navigated by making the ark

2. C. Cohen, "Hebrew TBH: Proposed Etymologies," *The Journal of the Ancient Near Eastern Society* (JANES), April 1, 1972, p. 36–51. (The journal was at that time called The Journal of the Ancient Near Eastern Society of Columbia University.)

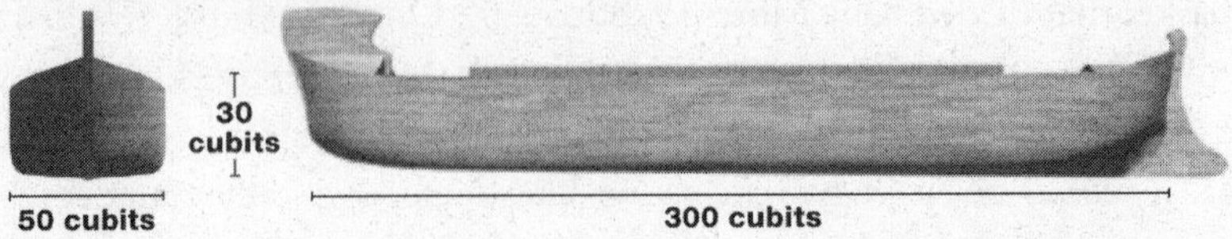

Figure 5. Scripture gives no clue about the shape of Noah's ark beyond its proportions that are given in Genesis 6:15, which reads: "And this is how you shall make it: The length of the ark shall be three hundred cubits, its width fifty cubits, and its height thirty cubits."

steer itself with a wind-catching obstruction on the bow. To be effective, this obstruction must be large enough to overcome the turning effect of the waves. While many designs could work, the possibility shown here reflects the high stems which were a hallmark of ancient ships.

2. A Cubit Upward and Above

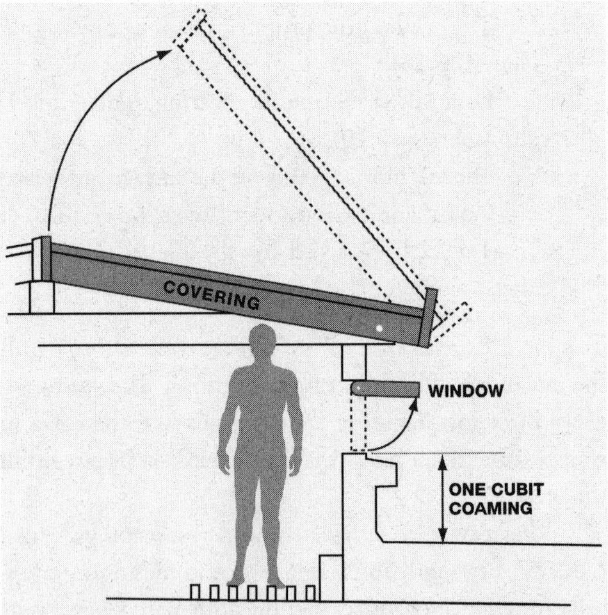

Any opening on the deck of a ship needs a wall (coaming) to prevent water from flowing in, especially when the ship rolls. In this illustration, the window "ends a cubit upward and above," as described in Genesis 6:16. The central position of the skylight is chosen to reflect the idea of a "noon light." This also means that the window does not need to be exactly one cubit. Perhaps the skylight had a transparent roof (even more a "noon light"), or the skylight roof could be opened (which might correspond to when "Noah removed the covering of the ark"). While variations are possible, a window without coaming is not the most logical solution.

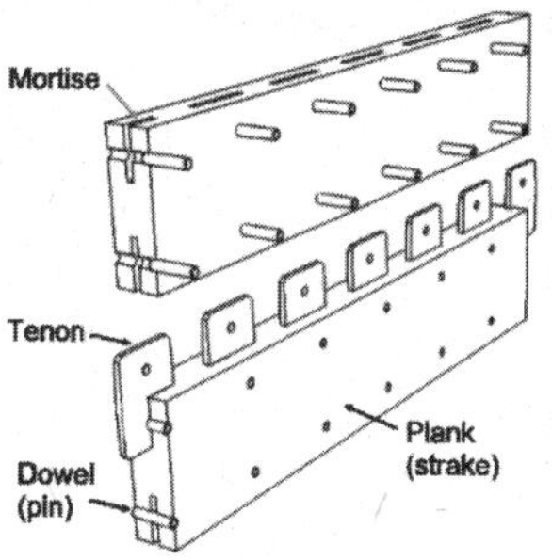

3. Mortise and Tenon Planking

Ancient shipbuilders usually began with a shell of planks (strakes) and then built internal framing (ribs) to fit inside. This is the complete reverse of the familiar European method where planking was added to the frame. In shell-first construction, the planks must be attached to each other somehow. Some used overlapping (clinker) planks that were dowelled or nailed, while others used rope to sew the planks together. The ancient Greeks used a sophisticated system where the planks were interlocked with thousands of precise mortise and tenon joints. The resulting hull was strong enough to ram another ship, yet light enough to be hauled onto a beach by the crew. If this is what the Greeks could do centuries before Christ, what could Noah do centuries after Tubal-Cain invented forged metal tools?

4. Ramps

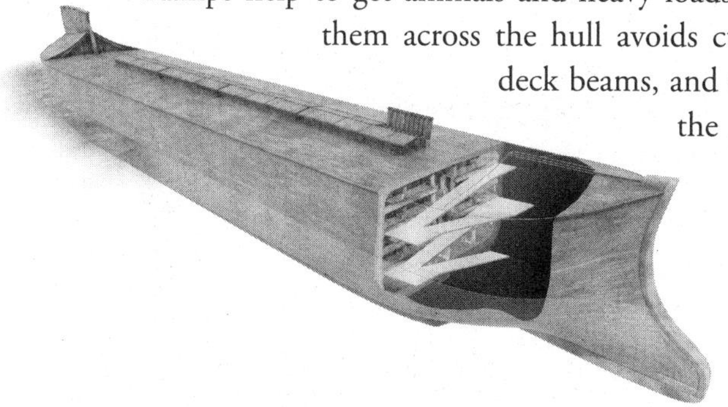

Ramps help to get animals and heavy loads between decks. Running them across the hull avoids cutting through important deck beams, and this location is away from the middle of the hull where bending stresses are highest. (This placement also better utilizes the irregular space at bow and stern.)

5. Something to Catch the Water

To assist in turning the ark to point with the wind, the stern should resist being pushed sideways. This is the same as a fixed rudder or skeg that provides directional control. There are many ways this could be done, but here we are reflecting the "mysterious" stern extensions seen on the earliest large ships of the Mediterranean.

How Long Was the Original Cubit?

Do we really know the size of Noah's ark (Genesis 6:15), the ark of the covenant (Exodus 25:10), the altar (Exodus 38:1), Goliath (1 Samuel 17:4), and Solomon's Temple (1 Kings 6:2)? While the Bible tells us that the length of Noah's ark was 300 cubits, its width 50 cubits, and its height 30 cubits, we must first ask, "How long is a cubit?" The answer, however, is not certain because ancient people groups assigned different lengths to the term "cubit" (Hebrew word *ammah*), the primary unit of measure in the Old Testament.

Table 1. The length of a cubit was based on the distance from the elbow to the fingertips, so it varied between different ancient groups of people. Here are some samples from Egypt, Babylon, and ancient Israel:

Culture	Inches (centimeters)
Hebrew (short)	17.5 (44.5)
Egyptian	17.6 (44.7)
Common (short)	18 (45.7)
Babylonian (long)	19.8 (50.3)
Hebrew (long)	20.4 (51.8)
Egyptian (long)	20.6 (52.3)

But when Noah came off the ark, only one cubit measurement existed — the one he had used to construct the ark. Unfortunately, the exact length of this cubit is unknown. After the nations were divided years later at the Tower of Babel, different cultures (people groups) adopted different cubits. So it requires some logical guesswork to reconstruct the most likely length of the original cubit.

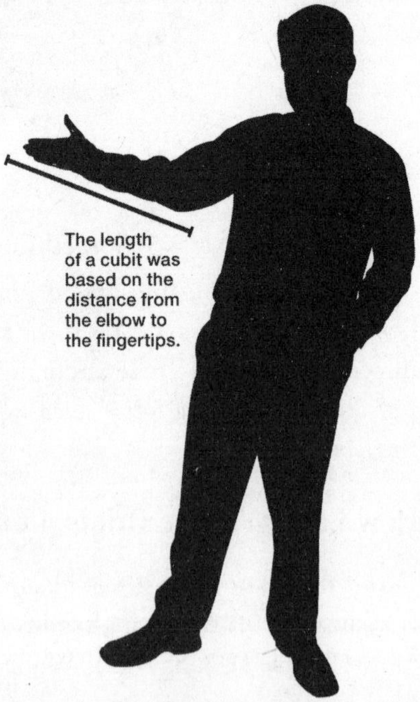
The length
of a cubit was
based on the
distance from
the elbow to
the fingertips.

Since the Babel dispersion was so soon after the Flood, it is reasonable to assume that builders of that time were still using the cubit that Noah used. Moreover, we would expect that the people who settled near Babel would have retained or remained close to the original cubit. Yet cubits from that region (the ancient Near East) are generally either a common (short) or a long cubit. Which one is most likely to have come from Noah?

In large-scale construction projects, ancient civilizations typically used the long cubit (about 19.8–20.6 inches [52 cm]). The Bible offers some input in 2 Chronicles 3:3, which reveals that Solomon used an older (long) cubit in construction of the Temple.

Most archaeological finds in Israel are not as ancient as Solomon, and these more modern finds consistently reveal the use of a short cubit, such as confirmed by measuring Hezekiah's tunnel. However, in Ezekiel's vision, an angel used "a cubit plus a handbreadth," an unmistakable definition for the long cubit (Ezekiel 43:13). The long cubit appears to be God's preferred standard of measurement. Perhaps this matter did not escape Solomon's notice, either.

Though the original cubit length is uncertain, it was most likely one of the long cubits (about 19.8–20.6 inches). If so, the ark was actually bigger than the size described in most books today, which usually use the short cubit. Using the 20.4-inch cubit gives an ark that is 510 feet long, 85 feet wide, and 51 feet high (not counting the extensions of the bow fin and the skeg).

Was Noah's Ark the Biggest Wooden Ship Ever Built?

Few wooden ships have ever come close to the size of Noah's ark. One possible challenge comes from the Chinese treasure ships of Yung He in the 1400s. An older contender is the ancient Greek trireme *Tessarakonteres.*

At first, historians dismissed ancient Greek claims that the *Tessarakonteres* was 425 feet (130 m) long. But as more information was learned, the reputation of these early shipbuilders grew markedly. One of the greatest challenges to the construction of large wooden ships is finding a way to lay planks around the outside in a way that will ensure little or no leaking, which is caused when there is too much movement between the planks. Apparently, the Greeks had

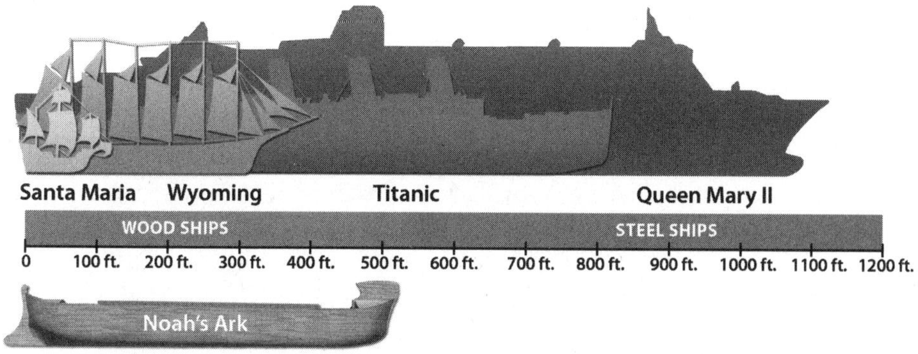

Figure 6. The ark is near the maximum size that is known to be possible for a wooden vessel. How big was the ark? To get the 510 feet (155 m) given here, we used a cubit of 20.4 inches (51.8 cm).

access to an extraordinary method of planking that was lost for centuries, and only recently brought to light by marine archaeology.

It is not known when or where this technique originated. Perhaps they used a method that began with the ark. After all, if the Greeks could do it, why not Noah?

Designed for Tsunamis?

Was the ark designed for tsunamis? Not really. Tsunamis devastate coastlines, but when a tsunami travels in deep water, it is almost imperceptible to a ship. During the Flood, the water was probably very deep — there is enough water in today's oceans to cover a relatively flat terrain to a consistent depth of over two miles (3.2 km). The Bible states that the ark rose "high above the earth" (Genesis 7:17) and was stranded early (Genesis 8:4), before mountaintops were seen. If the launch was a mirror of the landing — the ark being the last thing to float — it would have been a deep-water voyage from start to finish.

The worst waves may have been caused by wind, just like today. After several months at sea, God made a wind to pass over the earth. This suggests a large-scale weather pattern likely to produce waves with a dominant direction. It is an established fact that such waves would cause any drifting vessel to be driven sideways (broaching). A long vessel like the ark would remain locked in this sideways position, an uncomfortable and even dangerous situation in heavy weather.

However, broaching can be avoided if the vessel catches the wind at one end and is "rooted" in the water at the other — turning like a weather vane into the wind. Once the ark points into the waves, the long proportions create a more comfortable and controlled voyage. It had no need for speed, but the ark did "move about on the surface of the waters."

The box-like ark is not entirely disqualified as a safe option, but sharp edges are more vulnerable to damage during launch and landing. Blunt ends would also produce a rougher ride and allow the vessel to be more easily thrown around (but, of course, God could have miraculously kept the ship's precious cargo safe, regardless of the comfort factor). Since the Bible gives proportions consistent with those of a true cargo ship, it makes sense that it should look and act like a ship, too.

Coincidentally, certain aspects of this design appear in some of the earliest large ships depicted in pottery from Mesopotamia, not long after the Flood. It makes sense that shipwrights, who are conservative as a rule, would continue to include elements of the only ship to survive the global Flood — Noah's ark.

Scripture does not record direction-keeping features attached to the ark. They might have been obvious to a 500 year old, or perhaps they were common among ships in Noah's day as they were afterward. At the same time, the brief specifications in Genesis make no mention of other important details, such as storage of drinking water, disposal of excrement, or the way to get out of the ark. Obviously, Noah needed to know how many animals were coming, but this is not recorded either.

The Bible gives clear instruction for the construction of a number of things, but it does not specify many aspects of the ark's construction. Nothing in this newly depicted ark contradicts Scripture, even though it may be different from more accepted designs. But this design, in fact, shows us just how reasonable Scripture is as it depicts a stable, comfortable, and seaworthy vessel that was capable of fulfilling all the requirements stated in Scripture.

3

Should Christians Be Pushing to Have Creation Taught in Government Schools?

KEN HAM AND ROGER PATTERSON

Although this item specifically targets public schools in the United States, the principles can be applied to any school system in any country.

There have been a number of recent, highly controversial instances involving school boards discussing the topic of creation/evolution in the government-run school classroom, in science textbook disclaimers, and so on.

On the one hand, it's encouraging to see the increasing interest from citizens to put pressure on school boards deciding what is taught in the classrooms. The humanist elites are livid that this is even a topic for discussion. They want a monopoly on the teaching of molecules-to-man evolution in the public school science classrooms. On the other hand, if creation were taught in the science classrooms, would it be taught accurately and respectfully by a qualified individual?

The Issue

Public school teachers know that they can critically discuss different theories in regard to just about every issue — but not evolution. Even if a school board simply wants evolution to be critically analyzed (a good teaching technique, after all) without even mentioning creation or the Bible, the American Civil

Liberties Union and other humanists are immediately up in arms. There are the usual accusations of trying to get "religion" into schools and that it's a front for what they label as "fundamentalist Christianity."

By the way, when the public school system threw out prayer, Bible readings, creation, and the Ten Commandments, they didn't throw out religion. They replaced the Christian worldview influence with an atheistic one. The public schools, by and large, now teach that everything a student learns about science, history, etc., has nothing to do with God — it can all be explained without any supernatural reference. This is a religious view — an anti-Christian view with which students are being indoctrinated. Humanists know that naturalistic evolution is foundational to their religion — their worldview that everything can be explained without God. That is why they are so emotional when it comes to the topic of creation/evolution.

We are certainly encouraged at Answers in Genesis that there are moves in different places to stop the censorship of the anti-Christian propagandists in the public schools and allow students to, at the very minimum, question evolution. We are sure this is in part due to the influence of the creation ministries in society and the plethora of creationist and anti-evolutionist materials now available to parents and students. On the other hand, Christians have to understand that fighting the evolution issue in public schools is actually the same battle as fighting abortion, homosexual behavior, pornography, etc.

In other words, just as these issues are *symptoms* of the foundational change in our culture (i.e., from believing that God's Word is the absolute authority to that of man's opinions being the authority), so the evolution issue is also a symptom of this same foundational change.

Evolutionism as a Religion

If you were to ask the average person if evolution is a religion, he would probably say no. However, evolution is actually one of the cornerstones of the religion of humanism. (Now keep in mind that evolutionists do use real observational science such as natural selection, speciation, genetic studies, etc., as part of their overall argument. However, evolutionism in the sense of the belief aspects of evolution [life arising by natural processes, etc.] is a belief system — a religion.) Despite the vigorous objections of many humanists, humanism is a religion. Even a cursory reading of the "Humanist Manifesto I" penned in 1933 reveals that it is a religious document:

> FIRST: Religious humanists regard the universe as self-existing and not created.
>
> SECOND: Humanism believes that man is a part of nature and that he has emerged as a result of a continuous process.
>
> SEVENTH: Religion consists of those actions, purposes, and experiences which are humanly significant. Nothing human is alien to the religious. It includes labor, art, science, philosophy, love, friendship, recreation — all that is in its degree expressive of intelligently satisfying human living. The distinction between the sacred and the secular can no longer be maintained.[1]

Many other points in the document point to humanism as a religion that is to replace "the old attitudes" of traditional religions. John Dewey, considered the father of the modern American public school systems, was a signatory on the document. His application of his religious ideals to the education system cannot be denied. As a result, the public school system in America, and much of the world, is dominated by humanist philosophies.

Later versions of the manifesto also include the idea that humans have evolved as part of nature with no supernatural intervention at all.[2] Also presented are the beliefs that we can only know about the world around us by observation and experimentation — no biblical revelation is accepted — and that man is the measure of all things. All of these ideas are solidly anti-Christian in their sentiments.

Notable signatories of the "Humanist Manifesto III" include Eugenie Scott, executive director of the National Center for Science Education, and Richard Dawkins. Both of these individuals work hard to have their religious views presented to the students in classrooms across the world. Ultimately, we

1. American Humanist Association, "Humanist Manifesto I," www.americanhumanist.org/who_we_are/about_humanism/Humanist_Manifesto_I.
2. American Humanist Association, "Humanism and Its Aspirations: Humanist Manifesto III," www.americanhumanist.org/Who_We_Are/About_Humanism/Humanist_Manifesto_III. This article includes the tenet: "Humans are an integral part of nature, the result of unguided evolutionary change." The same idea is presented in the Humanist Manifesto II.

should think of their efforts to promote evolutionary teaching in schools as support for their respective religious organizations.

Many humanists would call themselves secular humanists in order to avoid the connection to the word "religion." They have adopted a similar manifesto founded on the same basic principles but avoiding the religious phrasing.

> **Separation of Church and State:** Because of their commitment to freedom, secular humanists believe in the principle of the separation of church and state. The lessons of history are clear: wherever one religion or ideology is established and given a dominant position in the state, minority opinions are in jeopardy. A pluralistic, open democratic society allows all points of view to be heard. Any effort to impose an exclusive conception of Truth, Piety, Virtue, or Justice upon the whole of society is a violation of free inquiry.[3]

Then, in the section on evolution we read:

> Today the theory of evolution is again under heavy attack by religious fundamentalists. Although the theory of evolution cannot be said to have reached its final formulation, or to be an infallible principle of science, it is nonetheless supported impressively by the findings of many sciences. There may be some significant differences among scientists concerning the mechanics of evolution; yet the evolution of the species is supported so strongly by the weight of evidence that it is difficult to reject it. Accordingly, we deplore the efforts by fundamentalists (especially in the United States) to invade the science classrooms, requiring that creationist theory be taught to students and requiring that it be included in biology textbooks. This is a serious threat both to academic freedom and to the integrity of the educational process. We believe that creationists surely should have the freedom to express their viewpoint in society. Moreover, we do not deny the value of examining theories of creation in educational courses on religion and the history of ideas; but it is a sham to mask an article of religious faith as a scientific truth and to inflict that doctrine on the scientific curriculum. If successful, creationists may seriously undermine the credibility of science itself.[4]

3. Council for Secular Humanism, "A Secular Humanist Declaration," www.secularhumanism.org/index.php?section=main&page=declaration.
4. Ibid.

The secular humanists basically believe we should not "impose an exclusive conception of truth" unless it involves suppressing religious ideas (including creation) — it is mandatory that the truth of evolution can have exclusive reign in the science classrooms. What they fail to realize is that they are simply substituting one "article of religious faith" for another in an arbitrary way that fits their agenda; Christians could assert the opposite claim.

If the documents from the humanists are not enough to be convincing about whether humanism (with the belief in naturalistic evolution as its foundation) is a religion that attempts to explain the meaning of life, the U.S. Supreme Court has also recognized humanism as a religion. In the 1961 case *Torcaso* v. *Watkins* regarding the legality of requiring a religious test for public office, the rationale for the finding includes the view that "religions in this country which do not teach what would generally be considered a belief in the existence of God, are Buddhism, Taoism, Ethical Culture, Secular Humanism, and others."[5]

Humanism, whether secular or religious, is a religion, albeit a non-theistic one for most of its adherents. One of humanism's fundamental tenets — evolution by natural processes alone — is the sole view allowed to be taught in public school science classrooms. This demonstrates that the public school systems are indeed promoting one religious view over another in the science classrooms. Again, religion was not removed from schools; Christian views were simply replaced by humanistic views. There is indeed a state religion in the American government school system — secular humanism![6]

Despite the assertion by humanists that evolution is an undeniable fact, is it really a scientific idea?

Science is generally limited to those things that are observable, testable, and repeatable. Language in the humanist documents mentioned above would affirm this notion. When we are discussing operational science, conducting experiments, and building technology based on those principles, creationists and humanists have no disagreement. It is only when we look to explain the past that the disagreements occur.

Everyone has the same evidence to examine, but we all look at the evidence in light of our pre-existing worldview. Evolutionists believe that life has evolved by natural processes, so they interpret the evidence in light of that belief. Creationists do the same, using God's Word, the Bible, as the standard. Since events of the past cannot be observed, tested, or repeated, we cannot ultimately call

5. *Torcaso* v. *Watkins*, 81 S. Ct. 1681 (1961).
6. The same basic case can be made for other humanist ideas such as moral relativism, situational ethics, the rejection of the supernatural, the value of human life, etc. Humanism has become the *de facto* religion in the public schools.

our understanding of those events scientific.[7] Christians should trust what God has revealed in Scripture and build their thinking, in every area, on that foundation.

What Are Christians Doing?

Some Christians who are teaching in the government schools sometimes find themselves in a situation where they can openly teach creation in the science classrooms. Teachers should understand what is allowed according to their state and local laws and statutes, and take advantage of those opportunities. However, there are often political implications to consider and a teacher who even legally teaches biblical creation may face other repercussions.

Some teachers choose to avoid teaching evolutionary ideas in the biology classroom. While on the surface this might sound like a wise idea, it may present some problems. Many standardized tests that students may have to take include information on evolution. Not teaching the basic concepts may lead to these students doing poorly on these exams. Also, if the curriculum requires the teaching of evolutionary ideas, teachers could be violating their contract by intentionally eliminating this subject. Avoiding the issue is not the best strategy, as it will likely lead to problems on many different levels.

What Should Christians Be Doing?

Whenever permissible, evolutionary ideas should be taught — but warts and all. There are many inconsistencies within the evolutionary framework and many disagreements about how to interpret the evidence. When appropriate, point out that many scientists, both creationists and evolutionists, do not believe that Darwinian evolution is adequate for explaining the existence of life on earth.

Also, many states have allowances for students to be released from school for special religious instruction. Consider supporting or starting a ministry that

7. The unscientific, even anti-scientific, nature of evolution is not the focus of this article. For more information on this topic, please see "Science or the Bible?" available at www.answersingenesis.org/articles/am/v2/n3/science-or-the-bible, and "Evolution: The Anti-science" available at www.answersingenesis.org/articles/aid/v3/n1/evolution-anti-science.

uses this time to teach students the true history of the earth from the Bible. Providing Christian students with this instruction will equip them to share this truth with teachers and other students. Additionally, these students should be equipped to share the gospel with their fellow students and teachers. Salvation is the ultimate goal for Christians in such a ministry, not just converting evolutionists into creationists.

We need bold Christians who will become active in their communities, school boards, and other organizations who will be prompting these changes from the bottom up. In these settings, Christians can start asking challenging questions about the exclusion of Christianity from schools, the acceptance of the religion of humanism, the absence of critical thinking when it comes to teaching evolution, etc. Based on the U.S. Constitution, no single religion should be endorsed in a government-run school. If no one stands up to challenge these ideas, the schools will continue to indoctrinate students with the religious beliefs of humanists.

AiG's True Position on Teaching Creation in Public Schools

Answers in Genesis is often misrepresented as trying to get creationist teaching into the public schools.[8] AiG does not lobby any government agencies to include the teaching of biblical creation in the public schools. As we have stated many times, we do not believe that creation should be mandated in public school science classrooms. If teaching creation were mandated, it would likely be taught poorly (and possibly mockingly) by a teacher who does not understand what the Bible teaches and who believes in evolution.

At the same time, it is not right that the tenets of secular humanism can be taught at the exclusion of Christian ideas. This type of exclusivity does not promote the critical thinking skills of students demanded by most science education standards. Teachers should be allowed, at the very least, the academic freedom to present various models of the history of life on earth and teach the strengths and weaknesses of those models. Recognizing that in the current political climate we can only expect to see evolution taught, it is only reasonable to include teaching the shortcomings of evolutionary ideas.

Advice to Christian Teachers and Administrators

Many Christians in the public school system view themselves as missionaries in a hostile environment. Their presence there is undoubtedly valuable and

8. For example, the prominent U.S. newspaper *Star-Tribune* of Minneapolis-St. Paul, Minnesota (August 14, 2009) falsely stated that AiG is on a "mission to get creationism into science classrooms nationwide."

provides an opportunity to be salt and light in their communities. If you are a teacher or administrator in the public schools, we encourage you to be both wise and bold as you prayerfully consider your role in this controversial creation/evolution issue.

If your state or country does not even permit the questioning of evolution or discussing creation, there are other options that will help keep a teacher from getting fired. There are many strategies that take the responsibility for any creation teaching away from the school and its administration:

1. Offer an optional course after school that is free for students (perhaps once or twice a month) to refute some of the evolution and long-age teachings. You or someone you know who enjoys teaching creation can use many of the great biblically based, creationist resources. Show the students that they are not getting all of the information from the textbooks. The books *Evolution Exposed: Biology* and *Earth Science* are designed to counter the unbiblical notions taught in the science textbooks. Get these books into the hands of the students so that they can understand both sides of the origins debate.
2. Offer to be an adult sponsor for students who wish to start a Creation Club in the school. This student-led alternative can be very effective at getting outside speakers into the school to address the club and present information. Clubs can meet at lunch or after school, just like a Chess Club or any other.
3. Have your local church youth group provide a short course to counter evolutionary claims. Students in the youth group can invite other students to attend and learn more about the issue. Use a DVD-based curriculum such as *Demolishing Strongholds* or *The Answers Academy* to communicate these ideas to students.
4. As stated above, many states offer the option of "released time." Support or start a ministry in your community that would provide biblical instruction for public school students.
5. Support local ministries like the Gideons, Fellowship of Christian Athletes, Young Life, and many others that seek to share the gospel and God's Word with students in public schools. (Be discerning, as not all groups will have a literal view of Genesis and local chapters often hold varying views.)
6. Understand the limits a teacher has to discuss ideas with students outside of the classroom. The political climate in your

district should be taken into consideration, as well as recognizing that you may face persecution.

7. Have a local church or group of churches offer to bring students to the Creation Museum where they can be presented with biblical truths about God as the Creator. Students can help pass the word around at public schools.

Conclusions

As much as we want to see students know that true science confirms the creation account in Genesis and that molecules-to-man evolution is a blind-faith belief that flies in the face of much scientific evidence, in the long run the school battle will not be successful unless society as a whole (and the Church) returns to the Bible as *the* authority. That's why at AiG, we spend so much energy to equip the Church to restore biblical authority beginning with Genesis. Then, and only then, will the secular worldview of society be successfully challenged. More important, recognize that spreading the glorious gospel of Jesus Christ is the ultimate goal.

If you are not directly involved in public schools in any way, pray for those who are and support Christian teachers and administrators who are trying to make a difference. Support and pray for families and students in the public schools. Volunteer to be a mentor or to assist in the public schools or teach a Sunday school class to help the students understand the origins issue.

4

What Are "Kinds" in Genesis?

DR. GEORGIA PURDOM AND BODIE HODGE

Zonkeys, Ligers, and Wholphins, Oh My!" Although not exactly the same mantra that the travelers in the classic *Wizard of Oz* repeat, these names represent real life animals just the same. In fact, two of these strange-sounding animals, a zonkey and a zorse, can now be seen at the new Creation Museum petting zoo. But what exactly are these animals and how did they come to be? Are they new species? Can the Bible explain such a thing?

What Is a "Kind"?

The first thing that needs to be addressed is: "What is a kind?" Often, people are confused into thinking that a "species" is a "kind." But this isn't necessarily so. A *species* is a man-made term used in the modern classification system. And frankly, the word *species* is difficult to define, whether one is a creationist or not! There is more on this word

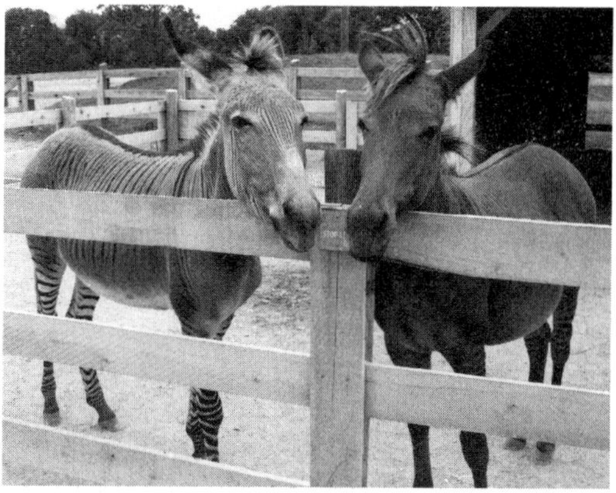

Figure 1. Zonkey and zorse at the Creation Museum

and its definition and relationship to "kinds" later in this chapter. The Bible uses the term "kind." The Bible's first use of this word (Hebrew: *min*) is found in Genesis 1 when God creates plants and animals "according to their kinds." It is used again in Genesis 6 and 8 when God instructs Noah to take two of every kind of land-dwelling, air-breathing animal onto the ark and also in God's command for the animals to reproduce after the Flood. A plain reading of the text infers that plants and animals were created to reproduce within the boundaries of their kind. Evidence to support this concept is clearly seen (or rather not seen) in our world today, as there are no reports of dats (dog + cat) or hows (horse + cow)! So a good rule of thumb is that if two things can breed together, then they are of the same created kind. It is a bit more complicated than this, but for the time being, this is a quick measure of a "kind."

As an example, dogs can easily breed with one another, whether wolves, dingoes, coyotes, or domestic dogs. When dogs breed together, you get dogs; so there is a dog kind. It works the same with chickens. There are several breeds of chickens, but chickens breed with each other and you still get chickens. So there is a chicken kind. The concept is fairly easy to understand.

But in today's culture, where evolution and millions of years are taught as fact, many people have been led to believe that animals and plants (that are

Figure 2. Domestic dogs all belong to the same dog kind.

Figure 3. The amazing variety of chicken breeds all belong to the same kind.

classed as a specific "species") have been like this for tens of thousands of years and perhaps millions of years. So when they see things like lions or zebras, they think they have been like this for an extremely long time.

From a biblical perspective, though, land animals like wolves, zebras, sheep, lions, and so on have at least two ancestors that lived on Noah's ark, only about 4,300 years ago. These animals have undergone many changes since that time. But dogs are still part of the dog kind, cats are still part of the cat kind, and so on. God placed variety within the original kinds, and other variation has occurred since the Fall due to genetic alterations.

Variety within a "Kind"

Creation scientists use the word baramin to refer to created kinds (Hebrew: *bara* = created, *min* = kind). Because none of the original ancestors survive today, creationists have been trying to figure out what descendants belong to each baramin in their varied forms. Baramin is commonly believed to be at the level of family and possibly order for some plants/animals (according to the common classification scheme of kingdom, phylum, class, order, family, genus, species). On rare occasions, a kind may be equivalent to the genus or species levels.

Baraminology is a field of study that attempts to classify fossil and living organisms into baramins. This is done based on many criteria, such as physical

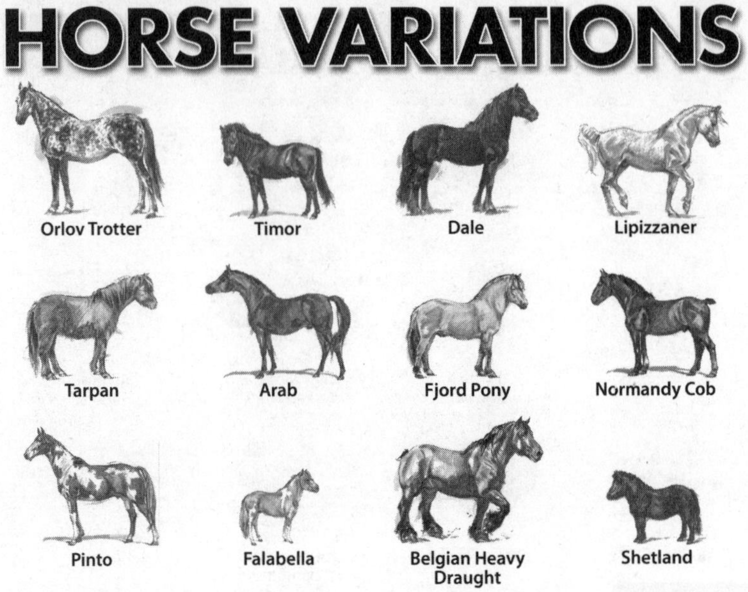

Figure 4. Horses of all shapes and sizes are of the same kind.

characteristics and DNA sequences. For living organisms, hybridization is a key criterion. If two animals can produce a hybrid, then they are considered to be of the same kind.[1] However, the inability to produce offspring does not necessarily rule out that the animals are of the same kind, since this may be the result of mutations (since the Fall).

Zonkeys (from a male zebra bred with a female donkey), zorses (male zebra and female horse), and hebras (male horse and female zebra) are all examples of hybrid animals. Hybrid animals are the result of the mating of two animals of the same "kind." Perhaps one of the most popular hybrids of the past has been the mule, the mating of a horse and donkey. So seeing something like a zorse or zonkey shouldn't really surprise anyone, since donkeys, zebras, and horses all belong to the horse kind.

The concept of kind is important for understanding how Noah fit all the animals on the ark. If kind is at the level of family/order, there would have been plenty of room on the ark to take two of every kind and seven of some. For example, even though many different dinosaurs have been identified, creation scientists think there are only about 50 "kinds" of dinosaurs. Even though breeding studies are impossible with dinosaurs, by studying fossils one can

1. Some might argue that if the hybrid offspring are infertile, then this indicates that the parent animals are of separate created kinds. However, fertility of the offspring has no bearing on the kind designation. Hybridization is the key.

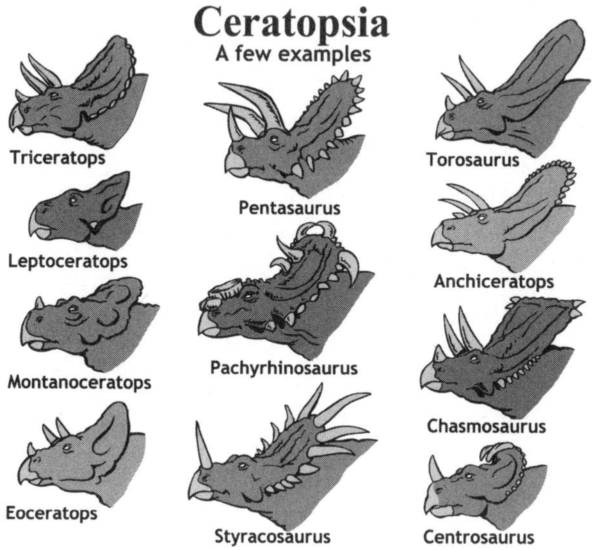

Figure 5. Using fossil evidence, we can identify kinds within the dinosaurs.

ascertain that there was likely one Ceratopsian kind with variation in that kind and so on.

After the Flood, the animals were told to "be fruitful and multiply on the earth" (Genesis 8:17). As they did this, natural selection, mutation, and other mechanisms allowed speciation within the kinds to occur. Speciation was necessary for the animals to survive in a very different post-Flood world. This is especially well illustrated in the dog kind in which current members (e.g., coyotes, dingoes, and domestic dogs) are confirmed to be descended from an ancestral type of wolf.[2]

Hybrid animals are usually the result of parent animals that have similar chromosome numbers. Many times the hybrids are infertile due to an uneven chromosome number that affects the production of eggs and sperm. However, this is not always the case, as even some mules (horse + donkey) have been known to reproduce. Consider some of the following amazing animal hybrids.

Zonkeys, Zorses, and Mules

These hybrids are the result of mating within the family Equidae. As we've said before, zonkeys are the result of mating a male zebra and a female donkey; zorses are the result of mating a male zebra and a female horse; and mules are the result of mating a male donkey and a female horse. But reverse matings (such as hinnies produced from a male horse and female donkey) are rare, although still possible. All are considered "infertile" due to uneven chromosome numbers, but fertility has been observed in some cases. Zonkeys and zorses have a mixture of their parents' traits, including the beautiful striping patterns of the zebra parents.

2. Savolainen et al., "Genetic Evidence for an East Asian origin of Domestic Dogs," *Science* 298 (2002): 1610–1613.

Ligers, Tigons, and Other Cats

These hybrids are the result of mating within the family Felidae. Ligers are the result of mating a male lion and a female tiger. Ligers are the largest cats in the world, weighing in at over 1,000 pounds (450 kg). Tigons are the result of mating a female lion and a male tiger. These matings only occur in captivity, since lions live in Africa, tigers live in Asia, and the two are enemies in the wild. Female hybrids are typically fertile while male hybrids are not.

Other hybrids in this family include bobcats that mate with domestic cats and bobcats with lynx (Blynx and Lynxcat). There have been mixes of the cougar and the ocelot, as well as many others. This shows that large, midsize, and small cats can ultimately interbreed, and therefore suggests that there is only one cat kind.

Wolphin

Turning to the ocean, this hybrid is the result of mating within the family Delphinidae. The wholphin is the result of mating a false killer whale (genus *Pseudorca*) and bottlenose dolphin (genus *Tursiops*). Such a mating occurred in captivity at Hawaii's Sea Life Park in 1985.[3] The wholphin is fertile. This hybrid shows the difficulty of determining the species designation, since a major criterion is the ability to interbreed and produce fertile offspring. Even though the whale and dolphin are considered separate genera, they may, in fact, belong to the same species. This shows how difficult it is to define the term *species*. Of course, from a biblical perspective it is easy to say they are both the same kind!

"Fixity of Species" and Changing Definitions

So what is the relationship between the *kinds* and *species* anyway? If one were to ask around to see what kind of definitions people have of the word *species* [or *genus*], most would respond by saying they have something to do with classification. In today's society, the words *genus* and *species* are synonymous with the Linnaean taxonomy system.

In the early 1700s, if someone said something about a "species" or "genus," it would have had nothing to do with classification systems. So why is this important today and what can we learn from it? The word *species*, and its changing definition, were partly responsible for the compromise of the Church in late 1800s. In fact, the Church is still struggling over this change. Let's do a brief history review.

3. Stephen Adams, "Dolphin and Whale Mate to Create a 'Wholphin,' " Telegraph.co.uk website news, April 2, 2008, Telegraph Media Group Limited, www.telegraph.co.uk/news/uknews/1582973/Dolphin-and-whale-mate-to-create-a-wolphin.html.

Species: Origin and Meaning

The English word *species* comes directly from Latin. For example, the Latin Vulgate (early Latin Bible translation), by Jerome around A.D. 400, says of Genesis 1:21:

> creavitque Deus cete grandia et omnem animam viventem atque motabilem quam produxerant aquae in *species* suas et omne volatile secundum *genus* suum et vidit Deus quod esset bonum [emphasis added].

Species is also found in the Latin version in Genesis 1:24, 25 as well. The Latin basically meant the biblical "kind." In fact, this word carried over into English (and other languages that have some Latin influence). It means a "kind, form, or sort." Another word that was commonly used for a kind in the Latin Vulgate was *genus*. This is evident in Genesis 1:11, 12, and 21. In both cases, these two words (*species* and *genus*) were used for the Hebrew word *min* or kind.

It made sense that Carl Linnaeus, a Swedish Christian, began using Latin terms for his new classification system. It was logical to use these common terms, which were a part of the commercial language throughout Europe (much in the way that English, for example, is seen as a universal language in the world today for communication and so on). Linnaeus even wrote his large treatise, *Systema Natvrae*, and other findings, in Latin in the mid to late 1700s.

Early commentators recognized that species originally meant the biblical kinds, as even John Calvin, prominent reformer in the 1500s, stated in his notes on Genesis 1:24:

> I say, moreover, it is sufficient for the purpose of signifying the same thing, (1) that Moses declares animals were created "according to their species": for this distribution carried with it something stable. It may even hence be inferred that the offspring of animals was included. For to what purpose do distinct species exist, unless that individuals, by their several kinds, may be multiplied?

Of course, Calvin originally wrote in Latin, but this early English translation by Thomas Tymme in 1578 still shows the point that the word *species* was used to mean the biblical kind. Calvin is even pointing out stability or fixity (i.e., biblical kinds). Dr. John Gill, about the same time as Linnaeus, equates species and kinds in his note under Genesis 1:22 by saying:

> With a power to procreate their kind, and continue their species, as it is interpreted in the next clause; *saying, be fruitful, and multiply, and fill the waters in the seas.*

Others, such as Basil, prior to the Latin Vulgate, discussed *species* as the biblical kind in the fourth century in his *Homilies* on Genesis 1. Matthew Henry, in the late 1600s and early 1700s, used *species* as kinds in his notes on Genesis 2:3, saying there would be no new "species" created after creation week had completed. The list could continue. The point is that *species* originally meant the biblical kind.

Species: A Change

After Linnaeus, both of these words (*species* and *genus*) were commonly used in modern biological classification systems with slightly different definitions. In the mid-to-late 1700s, *species* began taking on a new, more specific definition in scientific circles as a biological term (that definition is still being debated even today). But, by and large, the definition had changed so that, instead of there being a dog *species* (or dog kind), there were many dog *species*.

In the common and Church sense, the word *species* was still viewed as the biblical "kind." But as the scientific term gained popularity, this led to a problem. When theologians and members of the Church said "fixity of species" (meaning fixity of the biblical kinds) people readily saw that there were variations among the *species* (by the new definition). They thought, *But* species *do change!* Of course, no one ever showed something like a dog changing into something like a cat. Dogs were still dogs, cats were still cats, and so on.

However, a bait-and-switch fallacy had taken place. Christians were teaching fixity of species (kinds), but the definition of *species* changed out from under them. So Christians looked ignorant when people began observing that species — by the *new* definition — do change. Of course, in reality, this was merely variation within the created kinds. For example, dogs could be observed changing into something different — still dogs, but not looking like other "species" (by the new definition) of dogs. So it appeared that the created kinds were becoming new species (new definition), even though the animals did not change into a different kind of animal. It *appeared* that the Church was wrong.

Perhaps the most influential critique of fixity of species came from Charles Darwin, whose book *On the Origin of Species* tackled the misunderstood idea of *fixity of species* (though it never used the term "fixity"). Mr. Darwin studied many creatures during his travels and realized there was variation and not *fixity of species* (by the new definition).

The Implications

The Dog Species

Figure 6. Original definition of species: all dogs were one species.

The results of this were devastating to the Church. And people began doubting the Word of God as a result, walking away from Christianity, and embracing an evolutionary philosophy. George Bentham, writing May 30, 1882, to Francis Darwin regarding his father Charles's ideas, said:

> I have been throughout one of his most sincere admirers, and fully adopted his theories and conclusions, notwithstanding the severe pain and disappointment they at first occasioned me. On the day that his celebrated paper was read at the Linnean Society, July 1st, 1858, a long paper of mine had been set down for reading, in which, in commenting on the British Flora, I had collected a number of observations and facts illustrating what I then believed to be a fixity in species, however difficult it might be to assign their limits, and showing a tendency of abnormal forms produced by cultivation or otherwise, to withdraw within those original limits when left to themselves. Most fortunately my paper had to give way to Mr. Darwin's and when once that was read, I felt bound to defer mine for reconsideration; I began to entertain doubts on the subject, and on the appearance of the "Origin of Species," I was forced, however reluctantly, to give up my long-cherished convictions, the results of much labour and study, and I cancelled all that part of my paper which urged original fixity, and published only portions of the remainder in another form, chiefly in the "Natural History Review."[4]

Several Dog Species

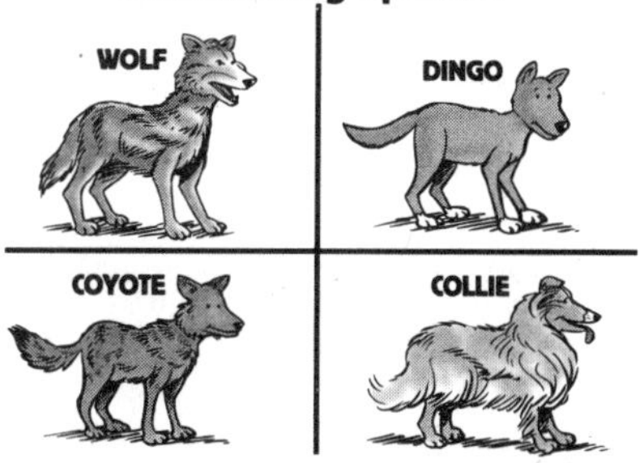

Figure 7. New definition of species: several wolf species, several coyote species, etc.

4. Francis Darwin, ed., *The Life and Letters of Charles Darwin Including an Autobiographical Chapter*, Volume 2, as produced by Classic Literature Library, Free Public Book Domain, originally published in 1897, www.charles-darwin.classic-literature.co.uk/the-life-and-letters-of-charles-darwin-volume-ii/ebook-page-41.asp

Even today, an objection commonly leveled at the Bible is that it claims that species are fixed. A good response would be: "To which definition of *species* are you referring?" By the old definition (as a kind), creationists would agree, but would probably better state it in modern English as fixity of the created kinds so as not to confuse the issue. The idea of one kind changing into another can be argued against based on the fact that no such change has ever been observed.

After Darwin's book, many churches gave up *fixity of species* (by either definition) and began taking compromised positions such as theistic evolution (basically giving up Genesis for molecules-to-man evolution and then picking up with Abraham). Realizing that the Church had been duped by a bait-and-switch fallacy provides a valuable learning tool. When people fail to understand history, they often repeat it.

A Great Place for Creation Research

All of these animals' ancestors that we have discussed above — horses, donkey, zebras, tigers, lions, whales, and dolphins — were created with genetic diversity within their various kinds (or by the older definition of *species*). Through time, the processes of natural selection, mutation, and other mechanisms have altered that original information (decreased or degenerated) to give us even more variation within a kind.

Great variety can be observed in the offspring of animals of the same kind, just as the same cake recipe can be used to make many different cakes with various flavors and colors. Hybrids have a portion of the same genetic information as their parents but combined in a unique way to give a very unique-looking animal. What an amazing diversity of life God has created for us to enjoy!

The study of created kinds is an exciting area of research, and our hope is to help encourage others to get involved. Whether studying the duck-goose kind, elephant-mammoth kind, camel-llama kind, apple-pear kind, or others, the field of baraminology is a great place for biologists, botanists, geneticists, and paleontologists (for extinct kinds) to get immersed in creation research.

5

How Could All the Animals Fit on the Ark and Eight People Care for Them?

MICHAEL BELKNAP AND TIM CHAFFEY

One of the most important issues relating to the Bible's Flood account is the topic of animals on the ark. The estimated numbers, sizes, and types of ark animals affect nearly every aspect of the vessel's interior operations, including time and labor expenditures, food and water needs, space and waste management, and enclosure design.[1]

The subject of fitting and caring for all the required animals on the ark is a significant point of contention between biblical creationists and skeptics. However, properly addressing these concerns is more complicated than a mere compilation of data about different animal species. First, we have to answer some fundamental questions.

How Large Was the Ark?

According to Genesis 6:15, the ark was 300 cubits long, 50 cubits wide, and 30 cubits high — the proportions of a genuine shipping vessel. A cubit is typically considered to be the length from a man's farthest fingertip to his elbow. While various cubit lengths have been used throughout history, the

1. Laura Welch, ed., *Inside Noah's Ark: Why It Worked* (Green Forest, AR: Master Books, 2016).

The proposed skylight roof could be opened. This might be the covering when "Noah removed the covering of the ark" (Genesis 8:13).

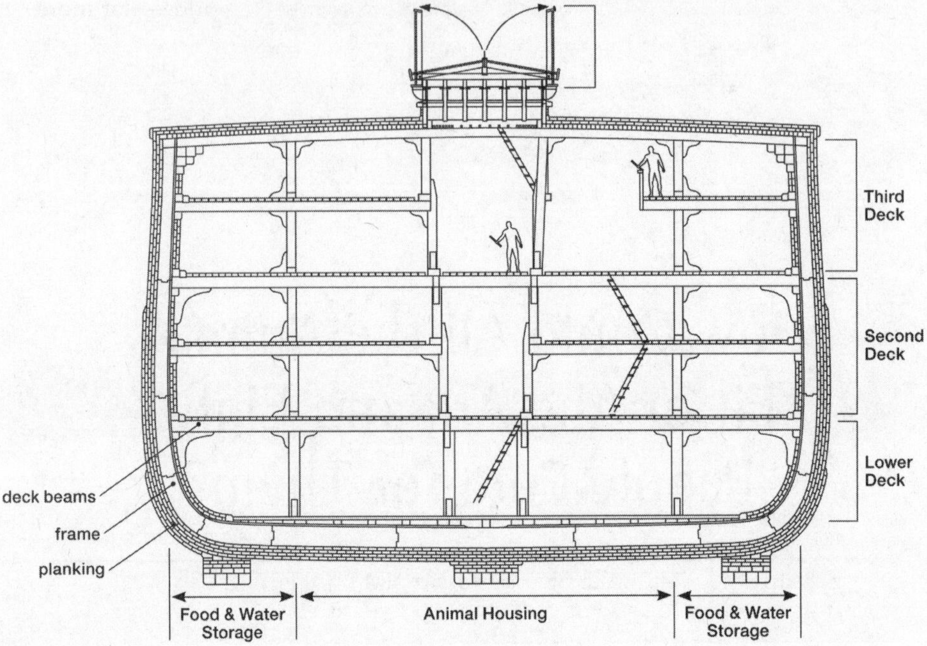

Figure 1. This is a cross-section view of a possible design of the interior of the ark.

Ark Encounter calculated the size of the ark based on a 20.4-inch (52 cm) cubit. The result is a vessel 510 feet (155 m) long, 85 feet (26 m) wide, and 51 feet (16 m) high. Accounting for a 15 percent reduction in volume due to the curvature of the hull, an ark this size could contain the equivalent of 450 semi-trailers of cargo or about 1.88 million cubic feet (53,200 m^3) — a truly massive ship.

Which Animals Were Required on the Ark?

The Bible informs us that the ark housed representatives of every land-dependent, air-breathing animal — ones that could not otherwise survive the Flood (Genesis 7:21–23). Conversely, Noah did not care for marine animals, and he probably did not need to bring insects, with the possible exception of delicate insects like butterflies and moths, since most insects could survive outside the ark. Also, they take in oxygen through spiracles in their skin, and the Bible specifies that those creatures killed outside of the ark were those "in whose nostrils was the breath of the spirit of life."

How Many Species Are in the World Today?

Skeptics often assert that there are millions of species in the world — far more than the number that could fit on the ark. However, according to estimates published in 2014, there are fewer than 1.8 million documented species of organisms in the world. Consider also that over 98 percent of those species are fish, invertebrates, and non-animals (like plants and bacteria). This means that there are fewer than 34,000 species of known, land-dependent vertebrates in the world today.[2]

Species or Kinds?

Though wild animals today are often considered according to their *species*, the Bible deals with animals according to their *min* — a common Hebrew word usually translated as "kind." We can infer from Scripture that God created plants and animals to reproduce after their kinds (Genesis 1:11–25), and it is clear from various texts that a kind is often a broader category than the current concept of a species. This means that a kind may contain many different species. Since Noah was only sent select representatives from relevant kinds, all land-dwelling vertebrate species not present on the ark were wiped out. Therefore, if we see an ark kind represented today by different species — e.g., horses, zebras, and donkeys of the equid kind — those species have developed *since* the time of the Flood. Therefore, species are simply varying expressions of a particular kind.

What Is an Animal Kind?

There are numerous approaches to defining a kind, but one of the simplest is: *a created distinct type of organism and all of the related descendants.* Kinds are often referred to as baramins (from the Hebrew words for "created" and "kind"), and the study of created kinds is called baraminology.

What Are the Criteria for Identifying Kinds?

In 2011, Ark Encounter researchers began in-depth animal studies with the goal of identifying the maximum number of ark kinds. The researchers applied three primary criteria in estimating the ark kinds: hybridization, cognitum, and statistical baraminology.[3]

2. IUCN 2014. IUCN Red List of Threatened Species. Version 2014.2. <www.iucnredlist.org>. Downloaded on November 9, 2016.
3. These last two methods are admittedly subjective and play off the assumption of common design equating with common ancestry. Even though these methods are utilized, the results are not seen as absolute. There are examples where these methods have failed to be accurate. But due to the limited amount of hybridization data, we are left with no other choice.

Hybrid data is the most favored method in identifying kinds. Researchers believe that only closely related animals can successfully produce offspring, and this is consistent with the Bible's emphasis on the relationship between reproduction and created kinds. Since only animals in the same kind are related, hybrids positively identify which animals are part of the same kind. The usefulness of hybrid data is limited, however, in that not all potential crosses have been tested or reliably documented. Some organisms have even gone extinct. Hybridization is also strictly an *inclusive criterion,* as not even all related animals can produce offspring together (i.e., they have lost the ability to reproduce with certain others of their kind).

The cognitum approach estimates animal kinds using the human senses of perception. This method assumes that animal kinds have maintained their core distinctiveness even as they have diversified over time (*overall* design is equated with common ancestry). Presently, extinct animals are most often classified using this approach. For example, woolly mammoths are extinct, and there are no hybrid data connecting them with elephants. However, their extreme similarity to elephants has resulted in their placement in the elephant kind.

In statistical analyses, continuities and discontinuities of animals are identified by comparing physical traits using statistical tests called baraminic distance correlation (BDC). Like the cognitum approach, this method assumes that the physical similarities and dissimilarities identified in the tests are reliable indictors of relatedness (*specific* designs or differences equate with common ancestry). It also assumes that the traits selected for comparison are baraminologically significant.

What Are Some Safeguards against Underestimating the Number of Ark Animals?

The Ark Encounter researchers put several safeguards in place to avoid underestimating the number of animals on the ark. These include a tendency to "split" rather than "lump" animal groups. Also, all "clean" and all flying creatures — not just "clean" ones — were multiplied by 14 instead of 7 animals.

What Are Splitting and Lumping?

Estimating the number of animals on the ark depends upon several factors. Near the top of that list is the decision to split or lump the animals that may or may not be related as a kind. Coyotes, wolves, dingoes, and domestic dogs can generally interbreed. Thus, they can be lumped into the same kind. So, Noah just needed two members of the dog kind on the ark.

On the other hand, there are approximately two dozen known families of bats, living and extinct. Based on anatomy and other features, many of these families probably belong to the same kind (e.g., if we used the cognitum criteria). In fact, it is possible that every bat belongs to the same kind. However, since breeding studies have not yet confirmed this idea, the Ark Encounter researchers split the bats into their various families. So, instead of including as few as 14 bats on the ark, the information depicts over 300 of them (14 from each family). In keeping with the worst-case approach to estimating the number of animals on the ark, the animals were split into separate kinds whenever the data was insufficient to support lumping them into a single kind.

Why 14 Instead of 7?

Some Bible translations indicate that Noah was to bring 7 of each flying creature and clean animal. Yet other Bibles state that 7 pairs of these creatures were on the ark. The Hebrew text literally reads, "seven seven — a male and his female" (Genesis 7:2). Does this unique phrasing mean 7 or 14?

Fossil groups	**Kinds (est.)**	**Per kind**	**Total animals (est.)**
Amphibians	54	2	108
Reptiles	211	Flying 24 x 14, Flightless 187 x 2	710
Non-mammal synapsids	78	2	156
Mammals	314	Clean/flying 13 x 14, Unclean 301 x 2	784
Birds	89	Flying 68 x 14, Flightless 21 x 2	994
Non-fossil groups	**Kinds (est.)**	**Per kind**	**Total animals (est.)**
Amphibians	194	2	388
Reptiles	101	2	202
Mammals	137	Clean/flying 31 x 14, Unclean 106 x 2	646
Birds	195	Flying 190 x 14, Flightless 5 x 2	2,670
Total	**1,373**		**6,658**

7 of each kind:	7 pairs of each kind:
KJV*	NLT
NKJV	ESV*
NASB*	HCSB
NET*	NRSV
NIV (1984)*	NIV (2011)

* Asterisks indicate that a textual note appears in these Bibles that mentions the possibility of the other view.

In favor of the "seven" view is that Genesis 8:20 states that Noah sacrificed clean animals and birds after the Flood. While it does not say that Noah sacrificed just one animal of each clean kind, those who hold to the "seven" view could point to the common "six and one" pattern seen in the Old Testament. For example, God created the world in six days and rested for one (Genesis 1; Exodus 20:11). Perhaps six of each clean animal were for man's use and one was dedicated to the Lord.

In favor of the "seven pairs" view is the text's mention that there would be a male and "his female" for the clean animals. If an odd number was brought to Noah, then plenty of animals did not have a mate. Furthermore, the Hebrew text does not use similar wording with the unclean animals in verse two. That is, readers can know that one pair of unclean animals was in view, but the text does not say "two two, a male and his female" — it just has the word for two.

Since Hebrew language scholars do not agree about this issue, it seems wise to be tentative about which view is accurate. Since a worst-case approach is being used in regard to the animals, these calculations are based on the "seven pairs" position.

What Is Meant by a "Worst-Case Scenario"?

The Ark Encounter depicts a worst-case approach when estimating the number of animal kinds. Some people believe Noah brought 2 of every unclean animal and 7 of every clean animal. The text seems to indicate that Noah cared for more animals than this (Genesis 7:2–3), particularly when it comes to the clean animals and flying creatures. The Lord may have sent 7 pairs of the clean animals and 7 pairs of all the flying creatures (not just the clean varieties).

Although this worst-case approach more than doubles the estimated number of animals on the ark, this model shows that even a high-end estimate of total animals would have fit on board. Obviously, if the Lord sent just 7 of each clean animal and just 7 of each of the clean flying creatures, the ark would have had plenty of space to accommodate this lower total.

How Many Animal Kinds Were on the Ark?

Based on initial projections, the Ark Encounter team estimates that there were fewer than 1,400 animal kinds on the ark. It is anticipated that future research may reduce that number even further.

How Many Individual Animals Were on the Ark?

The Ark Encounter team projects that there were fewer than 6,700 animals on board the ark. The wide discrepancy between the number of ark kinds and ark individuals is due to the relatively large number of flying and "clean" kinds — each estimated at 14 animals apiece.

How Big Were the Ark Animals?

People often wonder how all of the animals could have fit on the ark, particularly when considering the massive dinosaurs. We see so many illustrations of large creatures packed tightly into a little boat. But this image is inaccurate. Noah's ark was much larger than it is usually depicted, and many of the animals were probably smaller than shown in popular pictures.

It makes more sense to think that God would have sent to Noah juveniles or smaller varieties within the same kind. Consider the following advantages to bringing juveniles or smaller versions of a creature:

1. They take up less space.
2. They eat less.
3. They create less waste.
4. They are often easier to manage.
5. They are generally more durable.
6. In the case of juveniles, they would have more time to reproduce after the Flood.

Indeed, even when the giant dinosaurs and elephant-sized creatures are factored in, the ark animals were probably much smaller than is frequently assumed. According to Ark Encounter estimates, it is projected only 15 percent of ark animals would have achieved an average adult mass over 22 pounds (10 kg). This means that the vast majority of ark animals were smaller than a beagle, with most of those being much smaller. Starting with a mass category of 0.035–0.35 oz. (1–10 g), the animal groups were distributed into eight logarithmically increasing size classes. Amazingly, the size range with the greatest projected number of ark animals was 0.35–3.5 oz. (10–100 g).

Did the Ark Animals Hibernate?

One common explanation for how Noah and his family could have cared for so many animals is that the animals hibernated during the voyage. While this would have certainly been convenient for the family, we are not told from Scripture that this was the case. So, when the Ark Encounter design team considered food, water, waste management needs, etc., they once again applied their "worst-case scenario" approach and assumed that all the animals were awake and active.

Now, was this actually the case inside the real ark? Dark, closed-in spaces do tend to induce inactivity in many animals — particularly reptiles, who will often fall into a state of inactivity called torpor. So, while many of the ark animals may have hibernated or been largely inactive for part of the Flood year, it is likely that Noah and his family could not count on this behavior being the norm and would have planned accordingly. After all, God told Noah, "And you shall take for yourself of all food that is eaten, and you shall gather it to yourself; and it shall be food for you and for them" (Genesis 6:21) — that is, the animals. The foregone conclusion is that the animals would be awake to eat these provisions.

Were the Animals Caged?

Some people assume that the ark animals were free to roam the ark, but there are problems with this idea. First, it would not be safe for the animals on a vessel that surely rocked and pitched in the stormy seas. Second, there is no guarantee that all of the ark's animals were vegetarian. Finally, and most importantly, the Lord told Noah to "make rooms in the ark" (Genesis 6:14). Some Bibles use "nests" instead of "rooms." Essentially, Noah was to make enclosures for the ark's animals. A cage or pen system was the easiest way to ensure every animal remained safe and received care.

How Did They Care for Creatures with Specialized Needs?

Many animals today require special care in order for them to survive well in captivity. Hippopotamuses are frequently cited as an example of this, as they have skin that must be kept wet much of the time. Is it unrealistic to believe that Noah and his family could have kept a pair of hippos alive on the ark?

The ark had plenty of water, so it is possible that Noah's family developed a system to regularly deliver water to keep the hippos moist. On the other hand,

Figure 2. The ark would have contained animal cages in which the animals were fed and watered per their needs.

these creatures may not have been as difficult to tend as many people imagine. There are two species of the hippo kind in the world today.

The animal most people think of is the common hippopotamus, but there is also the pygmy hippo. The pygmy hippo is more terrestrial than the common hippo (albeit still semiaquatic), and they are most similar to the fossil hippos found in early post-Flood rock layers. So, the ark's hippos were likely smaller and more terrestrial — and therefore easier to care for — than the large common hippos of today.

How Much Food Would Need to Be Gathered?

Once a number of important data points are logged (e.g., the number of animals, average animal masses and metabolic rates per kind, etc.), a simple formula can be used to estimate the food requirements on the ark. Based on information collected by Ark Encounter researchers, the minimum bulk food volume needed on the ark was 17,125 cu. ft. (485 m^3), assuming 80 percent dry matter — that equals over 400 tons (363 tonnes).[4]

Scripture does not record for us if Noah was given all this information, so how could he know how much food to store? It is possible that the family simply loaded the ark with as much food as it could take, allowing space for other necessities, such as the animal enclosures, working spaces, and living

4. Noah's family and the animals were on the ark for about a year. They exited the ark a year and ten days after the Flood began, but we are not certain what calendar is referenced in Genesis 7–8, so we cannot know how many days were in that year. This calculation is based on a 360-day year. Eleven days were added to account for the additional 10 days into the second year and the first and last days in the ark.

quarters. While constructing the "Half Ark Model" for an exhibit in the Ark Encounter, researchers discovered that everything fit nicely.

What Types of Food Would Need to Be Gathered?

After making everything, the Lord stated that people and animals were to eat vegetation (Genesis 1:29–30). It was not until after the Flood that God permitted man to eat meat (Genesis 9:3). After sin, we cannot be sure when certain animals began to eat meat (it may have been immediate or more gradual), although the fossil record provides strong evidence that carnivory occurred prior to the Flood. If carnivorous activity was prevalent in the pre-Flood world as is implied in Genesis 6:12, it is still possible that the individual animals the Lord sent did not eat meat.

However, if some of the ark's animals did eat meat, there are several methods of preserving or supplying their food. Meat can be preserved through drying, pickling, salting, or smoking. Certain fish can pack themselves in mud and survive for years without water — these could have been stored on the ark. Mealworms and other insects can be bred for both carnivores and insectivores.

In any case, most of the animals were vegetarian or could survive as vegetarians for a period of time. To provide food for all of these animals, Noah's family could have grown or purchased vast stores of grains, grasses, seeds, and nuts. Certain vegetables with a long shelf life may have been brought aboard, as could dried varieties of some fruits and vegetables.

What Did They Feed Picky Eaters?

When you look at the animals in the world today, many of them cannot simply be classified as either an herbivore (plant eater) or carnivore (meat eater). Also, some animals today have special needs that require specific handling. How could Noah's family provide for these types of animals? Remember that today's animals are descendants of the ark representatives and many modern animals are likely more specialized in their dietary and habitat needs.

Take the modern koala for example. Koalas can eat the leaves of some other trees, such as the wattle, paperbark, and tea trees, but they prefer certain eucalyptus leaves. Like other animals, the koala's ark ancestors were less specialized, and probably heartier than modern representatives. As such, they may have eaten a much wider variety of food, including grains, fruits, and vegetables. Indeed, certain extinct koalas (e.g., *Litokoala*) apparently possessed just these sorts of generalizations.

Devouring up to 35,000 insects per day, anteaters also sound difficult to keep. However, they are often fed fruit and eggs in captivity. In the wild, giant

Figure 3. Some animals on the ark may have had less specialized diets than some of their descendants.

anteaters also consume fruit that has dropped to the ground while other anteater species climb trees to get fruit.

So, when looking at the animal kinds we know today, it is important to keep in mind that the variations of animal kinds expressed in our world today may be more specialized or adapted to specific environments than the ark's animals were at the time. We cannot assume these limitations for the animals on the ark.

How Could So Much Food Be Stored?

Placing food in large silos, bags, or other corruptible containers could have resulted in an unacceptable degree of spoilage and waste. Alternatively, using sealed earthen vessels — perhaps stored in shelving units or bundled together with ropes — would have provided better moisture stability and reduced the likelihood of infestation.

And while Noah's family could have used other storage methods (e.g., barrels and crates), earthenware makes a lot of sense given the simplicity of production and requirements for keeping goods free of contaminants. Strategic placement of food would minimize the effort needed to retrieve and distribute it.

How Could the Water Be Collected and Stored?

Water has always been at the heart of any civilization. On the ark it would have been needed for many things as well, such as drinking, cleaning animal stalls, bathing, and washing clothes and dishes.

For this journey, Noah faced a number of challenges related to water management. How much drinking water would be needed? How would they maintain sanitary conditions on board? Would the ark be able to carry enough water for the duration of the Flood? Could water be collected and used along the way?

Unlike ancient sailors, who often relied on staying close to shore or planned stops at islands during long ocean voyages, Noah faced an entire world covered by water. There are numerous examples of early civilizations boiling water or using sand as a filter to acquire safe water. Developing a large-scale method of water treatment for the ark would have been a monumental feat.

Assuming Noah did not develop such a process, and that the Lord did not miraculously filter the water for them, there are other solutions to ensure the ark had enough fresh water.

In the history of ships and ocean voyages, water collection using runoff from the sails or through use of barrels on deck has been documented. Neither of these techniques are directly applicable to the ark, but the basic ideas behind them could be relevant. In order to reduce the occurrence of contamination, water could have been collected beforehand into numerous cisterns, earthen vessels, or other waterproofed containers. In addition to this, using the ark's roof surface as a massive rainwater collection device would combine elements of sustainability, redundancy, and efficiency — especially if the water could be channeled into overhead cisterns for storage and distribution as needed.

Watering Systems

Methods for distributing water in the various areas throughout the ark — e.g., amphibian containers, small animal cages, large animal enclosures, and human living spaces — may have varied. It would have been unnecessarily laborious and time-consuming for the family members to carry large containers of water around all day. Instead, utilizing a simple system of fixed pipes and spigots would permit easy access to water from the animal pens, living quarters, and other areas.

Bamboo is a practical natural material for this task, being strong, lightweight, easily cultivated, and resistant to degradation. But we can leave open the option of using more advanced materials as well — recall the pre-Flood world was not primitive.

Aside from the water collection and distribution systems, each animal cage could have been equipped with simple, appropriately sized, vacuum-fed water containers — similar to those still in use today, though possibly crafted as specialized earthen vessels. Such a design would accomplish significant time and labor savings, particularly for the larger animals. Though working in two-person teams is often the most efficient arrangement, utilizing partially automated systems would mean that single tasks would not always require the attention of both individuals. This is crucial since the ark contained only eight laborers.

Figure 4. Water distribution to areas of the ark by bamboo pipes and vacuum-fed water containers in the animal cages are among simple yet functional solutions.

Feeding Systems

The feeding systems on the ark would have been much easier to develop than the watering systems. For large animal enclosures, bamboo or wood chutes leading to a food dish could have been filled from an overhead catwalk. External chutes leading to interior food trays could have also been used in the small animal cages, greatly accelerating the feeding process.

Wastewater Systems

It is a simple but unpleasant fact of life — both humans and animals produce liquid and solid waste. Without an effective management system for removal of this waste from living areas, people and animals can sicken and eventually die. Put a large number of animals with eight people in a closed environment like the ark for about a year during the Flood, and it is a huge challenge that had to be addressed before the journey began. It is inevitable that there would have been a solution on board the ship for a number of reasons.

1. The design of the vessel was not meant for either crew or animals to be walking about on the roof of the ark, at least not while it was afloat. During the Flood event, the only decks that could be walked on safely on a regular basis were interior ones.
2. While there is a door noted for the ark, it likely could not be opened during transit.
3. There was an opening at the top of the ark, but nothing hints at this being a site that waste products could be efficiently tossed out of without landing on areas of the roof and causing sanitation problems. More importantly, if Noah's family collected rainwater from the roof for their water supply, as discussed previously, they would not want to pollute it with sewage.
4. The amount of labor it would take to remove the waste using various types of manual labor alone would have been difficult but manageable. The system solutions for human waste and animal waste could have been completely different, but they may have had a common collection point and labor-reducing method of removal from the ship.

Animal Waste

Factoring in the sizes, number, and estimated metabolisms of the projected 6,658 ark animals, it is likely that the daily solid waste production on the ark reached a few tons. Human occupant contributions would have been negligible.

The Ark Encounter designs show Noah's family using carts and small wagons to move the solid animal waste. While this sounds like a lot of work,

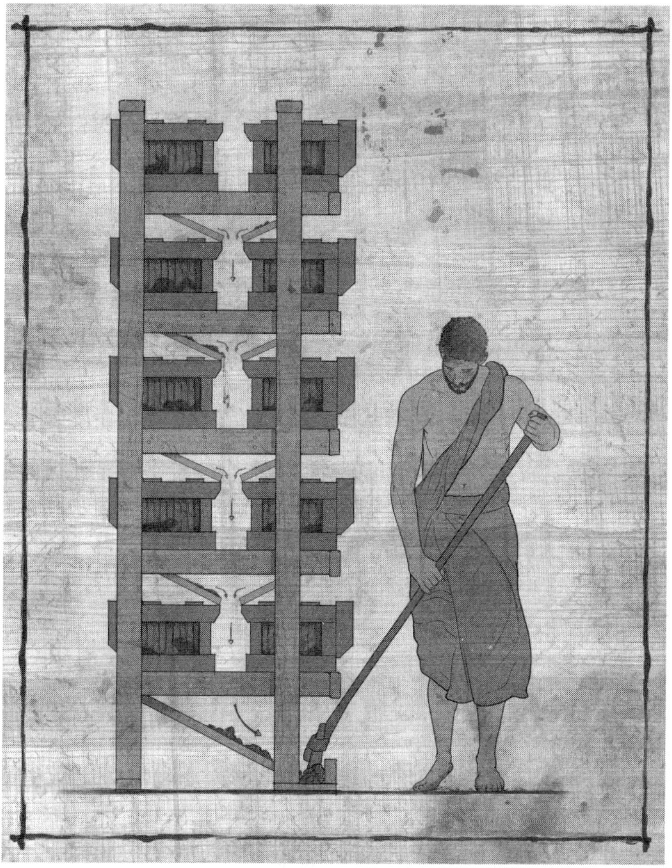

Figure 5. Cage design could have minimized some of the labor in caring for the animals, such as this example where waste is collected from a central point at the bottom.

it would have been manageable. Some manual cleaning would be expected even with solutions built into the cage or enclosure designs. The design of the enclosures could have made the waste removal task much simpler. Sloped floors or designs that incorporated slatted floors could have been used — the latter would have permitted waste to slip through and collect below. Large animal enclosures, on the other hand, are typically designed with flat and solid floors, since slotted floors can result in leg or foot injuries.

Concerning liquid waste, collection points funneling into bamboo pipes could have moved urine and excess water away from the enclosures. Each enclosure complex could have been connected to a central waste-water collection tank. While the inclusion of bedding often increases animal comfort, storing and appropriately distributing it takes a lot of space and effort. Thus, it may have been deemed unnecessary for the year-long stay aboard the ark.

Animal Transfer System

In the absence of bedding, each stall must eventually be cleaned out by hand. Regardless of animal diet or behavior, such confined spaces with large animals on a moving ark was a genuinely dangerous proposition — in fact, the most dangerous creatures on the ark may have been the large herbivores. Enclosures could have implemented simple animal transfer systems, permitting safe and efficient cleaning.

Close Proximity

While you may find it hard to imagine waste management with people and animals in such close proximity, it is important to remember that humans have lived close to their livestock since ancient times — both to protect the animals and to oversee their care. Life on the ark would have been a natural extension of this and animal care standard practices seen throughout history.

A Hole in The Ship?

One significant design feature proposed as a potential component of the ark is a moonpool, which is essentially a large cavity running from the bottom of the ship to the upper deck or roof. Even though the moonpool is open to the ocean, the water is confined within its interior, moving up and down like a piston of an engine. The Ark Encounter designers proposed two moonpools in the stern, straddling the keel. One moonpool they designed for ventilation, as the in-and-out movement of the water acts like a massive bellows, circulating fresh air throughout the ark. The other was designed as an integral part of the waste removal system, exploiting the moonpool's secure access to the waters outside the ark.

Light Source Solutions

While it may not receive as much attention as other solutions, lighting surely played a key role in life on the ark. Whether providing an energy source for growing supplemental plants or making it easier to complete chores in the depths of the ship, it was essential that methods be found for all the lighting needs on the ark. We are not sure what the "covering" was that Noah opened (Genesis 8:13), but if it was a roof that was translucent or could be drawn back, then this could have allowed light to fill the ark. Also, oil lamps (or other technologies) could have been used to light the interior. Roof panels could be raised and lowered so that natural light would be utilized on the ark. Windows are another potential source of natural light.

Is This How They Actually Did It?

It is easy to let your imagination wander when it comes to the possibilities of how the ark worked. However, it is always important to come back to what we know from the Scriptures and consider practical solutions that have worked well in the past. The point of the ship was not simply to save just eight members of one family – that would have certainly been easier! Noah's task also included the care of all the animals the Lord brought to him.

The possible solutions provided in this chapter really just scratch the surface of what Noah and his family could have designed and built for the real ark. One thing is for certain, though. We know that in the end, "Thus Noah did; according to all that God commanded him, so he did" (Genesis 6:22). Whichever solutions Noah developed, we can certainly give thanks for the faithfulness of our brilliant ancestor.

6

Was the Flood of Noah Global or Local in Extent?

KEN HAM AND DR. ANDREW A. SNELLING

Many Christians and their leaders believe that it is not relevant whether the Flood of Noah described in Genesis 6–8 was global or localized (in the Mesopotamian Valley of the Tigris and Euphrates Rivers). After all, they say, it's not relevant to a Christian's salvation, and the gospel message to be preached is all about Jesus.

Besides, matters about rocks and the earth's history are the domain of the geologists, because the Bible isn't a science textbook. So if the geologists say there never was a global Flood, then that settles it! Thus, Christians who advocate an old earth agree with the secular geologists, and therefore they oppose any notion that the Flood of Noah was global.

However, whether the Flood of Noah was global or local in extent *is* a crucial question. This is because ultimately what is at stake is the authority of *all* of God's Word. Indeed, if the text of Scripture in Genesis 6–8 clearly teaches that the Flood was global and we reject that teaching, then we undermine the reliability and authority of other parts of Scripture, including John 3:16. God's Word must be trustworthy and authoritative in all that it affirms.

Millions of Years or a Global Flood?

Secular geologists have interpreted the fossil-bearing sedimentary layers, such as those exposed in the walls of the Grand Canyon, as having taken millions of years to form. Countless sea creatures lived on shallow seafloors, for example, and were slowly buried, to be replaced by new sea creatures growing on the seafloors. The various sedimentary rock layers that we now see stacked up on top of one another thus supposedly slowly accumulated as sea creatures were progressively buried.

The guiding principle used by secular geologists to interpret the rock record is "the present is the key to the past," which means that the geologic processes we see operating today, at the rates they operate today, are all that are necessary to explain the rock layers (figure 1). While catastrophes such as local flooding and volcanic eruptions are allowable because they do occur today, any suggestion of a global catastrophic Flood as described in the Bible is totally ruled out before the geological evidence is even examined.

On the other hand, the description of the Flood in Genesis 6–8 is not hard to understand. We are told that the "fountains of the great deep" burst open and poured water out onto the earth's surface for 150 days (five months). Simultaneously, and for the same length of time, the "floodgates of heaven" were open, producing torrential global rainfall.[1]

Figure 1. Two views of the rock layers: the world teaches that the vast majority of the rock layers were laid down slowly over millions of years; but in light of a global Flood in Genesis 6–9, it makes more sense that the bulk of the rock layers that contain fossils were laid down during this catastrophe only thousands of years ago.

The combined result was that the waters destructively rose across the face of the earth to eventually cover "all the high hills under the whole heaven." The mountains also were eventually covered, so that every creature "in whose nostrils is the breath of life" perished. Only Noah, his family, and all the air-breathing, land-dwelling creatures he took on board the ark were saved.

Based on that clear description of this real historical event, it is very rational to conclude that we should expect to find evidence today of billions of dead

1. The reference to 40 days and 40 nights (Genesis 7:12, 17) appears to be telling us how long it was before the ark started to float, for the windows of heaven were closed on the same day (150th) as the fountains of the deep were (Genesis 7:24–8:3). For a detailed argument based on the Hebrew text see William Barrick, "Noah's Flood and its Geological Implications," in Terry Mortenson and Thane H. Ury, eds., *Coming to Grips with Genesis* (Green Forest, AR: Master Books, 2008), p. 251–282.

animals and plants buried in rock layers composed of water-deposited sand, lime, and mud all around the earth. And indeed, that's exactly what we do find — billions of fossils of animals and plants buried in sedimentary rock layers stretching across every continent all around the globe.[2] So instead of taking millions of years to form, most of the fossil-bearing sedimentary rock layers, as seen in the walls of the Grand Canyon and elsewhere, could have formed rapidly during the year of this global catastrophic Flood of Noah.[3]

It should immediately be obvious that these two interpretations of the evidence are mutually exclusive! Most of these rock layers are either the sobering testimony to Noah's Flood or the record of millions of years of history on this earth. One must be true and the other must be false. We can't consistently or logically believe in both, because the millions of years can't be fitted into the 370-day length of the global cataclysmic Flood of Noah described in Genesis 6–8. That is ultimately the fundamental reason why many old-earth advocates in the Christian community oppose the clear teaching of Scripture that the Flood was global. Only a relatively insignificant local flood would fit with the secular geological interpretation of millions of years of slow and gradual geologic processes for most of the fossil record.

Biblical Problems

In order to relegate Noah's Flood to being only local in extent, and/or to being a myth, the Hebrew text of Genesis 6–8 and also the larger context have to be virtually ignored.

The Book of Genesis is clearly divided into two main sections. Chapters 1–11 deal with *universal* origins (the material universe, the plant and animal kingdoms, humans, marriage, sin, death, redemption, the nations of the earth, etc.). Chapters 12–50, on the other hand, concentrate on the *particular* origin of the Hebrew nation and its tribes, mentioning other nations only insofar as they came in contact with Abraham and his descendants.[4]

The realization of this fact of the context of the Flood account within the section of Genesis on universal origins sheds important light on the question of the magnitude of the Flood. Furthermore, the biblical account of the Flood catastrophe occupies more than 3 chapters of these 11 chapters on universal

2. See chapter 28 in this volume: Andrew A. Snelling, "What Are Some of the Best Flood Evidences?"
3. Some localized fossil-bearing deposits may have formed after the Fall of Adam and Eve in sin and before Noah's Flood, and some of the localized fossiliferous rock layers at the top of the geological record were formed in post-Flood events. But creationist geologists are in general agreement that most of the fossil-bearing sedimentary rock record is a result of Noah's Flood.
4. W.H. Griffith Thomas, *Genesis: A Devotional Commentary* (Grand Rapids, MI: Eerdmans, 1946), p. 18–19.

origins, while only 2 chapters are devoted to the creation of all things! How important, therefore, must the Flood account be! Yet nobody denies that the account in Genesis 1–2 of the creation of all things is referring to the scale of the whole earth, and indeed the whole universe. Thus the context of Genesis 6–8 demands that the scriptural narrative be understood to be describing a watery catastrophe of global proportions.

But when we read the Flood account itself, we see this conclusion confirmed. We are immediately struck with prolific usage of universal terms such as "all," "every," "under heaven," and "in whose nostrils was the breath of life." For example, Genesis 6:7–13 tells us why God sent the Flood judgment:

> The Lord said, "I will blot out man whom I have created from the face of the land, from man to animals to creeping things and to birds of the sky; for I am sorry that I have made them." . . . God looked on the earth, and, behold, it was corrupt; for all flesh had corrupted their way upon the earth. Then God said to Noah, "The end of all flesh has come before Me; for the earth is filled with violence because of them; and, behold, I am about to destroy them with the earth" (NASB).

Note in particular God's emphasis on "all flesh" and "the earth," not just some flesh or part of the earth. Also, note that the Flood came to destroy animals and birds, not just sinful humans. The Apostle Paul tells us in Romans 8:19–23 that the whole non-human creation was subjected to the Curse because of man's sin, and thus the whole of creation suffers death and decay. So also in the Flood, the non-human creation suffered, regardless of whether animals or birds had come into close contact with sinful man or not.

Then when the Flood began, we are told in Genesis 7:11–12 that "all the fountains of the great deep (were) broken up," and "the rain was upon the earth." Again, the words "all" and "the earth" are clearly intended to imply global extent. Indeed, this usage of universal terms is prolific as the Flood account reaches a crescendo in Genesis 7:18–24:

> The waters prevailed, and greatly increased on the earth. . . . And the waters prevailed exceedingly on the earth, and all the high hills under the whole heaven were covered. . . . and the mountains were covered. And all flesh died that moved upon the earth . . . every creeping thing . . . and every man: All in whose nostrils was the breath of the spirit of life, all that was on the dry land, died. So He destroyed all living things which were on the face of the ground. . . . They were

destroyed from the earth. . . . And the waters prevailed on the earth one hundred and fifty days.

Figure 2. A flood that covered the highest hills by a significant amount, yet was local does not make sense!

So frequent is this use of universal terms, and so powerful are the points of comparison ("high hills," "whole heaven," and "mountains"), that it is extremely difficult to imagine what more could have been written under the direction of the Holy Spirit to express the concept of a global Flood! In the words of a leading Hebrew scholar of the 19th century, who strongly opposed those who tried to tone down the universal terms of the Genesis Flood account:

> They have distorted the spirit of the language, and disregarded the dictates of common sense. It is impossible to read the narrative of our chapter (Genesis 7) without being irresistibly impressed that the whole earth was destined for destruction. This is so evident throughout the whole of the description, that it is unnecessary to adduce single instances. . . . In our case the universality does not lie in the words merely, but in the tenor of the whole narrative.[5]

Something else in the Flood account is irreconcilable with the Flood being localized in the Mesopotamian Valley. In Genesis 7:20 we are told that "the mountains were covered." Because water always seeks its own level, how could the mountains only be covered in one local area without also covering the mountains in all adjoining areas and even on the other side of the planet (figure 2)? This clear statement in God's Word only makes physical and scientific sense if the Flood were global in extent.

Even the renowned and theologically liberal Hebrew scholar James Barr, then Oriel Professor of the Interpretation of Holy Scripture at Oxford University in England, was prepared to admit in a letter to David C.C. Watson dated April 23, 1984:

5. M.M. Kalisch, *Historical and Critical Commentary on the Old Testament* (London: Longman, Brown, Green, et al., 1858), p. 143–144.

> . . . so far as I know, there is no Professor of Hebrew or Old Testament at any world-class university who does not believe that the writer(s) of Genesis 1–11 intended to convey to their readers the ideas that . . . Noah's Flood was understood to be world-wide and extinguish all human and animal life except for those in the Ark. Or to put it negatively, the apologetic arguments which suppose . . . the flood to be a merely local Mesopotamian flood are not taken seriously by any such Professors, as far as I know.[6]

Theological Problems

If the Flood were only a relatively recent local event of no geologic significance, then the fossil-bearing sedimentary layers that were supposedly laid down over millions of years must have preceded the appearance of man on the earth, including Adam. After all, man only appears very recently in the fossil record. For a Christian who accepts the millions of years, this would mean that animals were living, dying, suffering disease, eating each other, and being buried and fossilized prior to Adam's appearance in the Garden of Eden. In the geologic record we find the fossilized remains of fish eating other fish, animals eating other animals, animals with diseases like cancer, and much more, which indicates that these fossils are a record of disease, violence, and death.

However, theologically there is a big problem here. In Genesis 1:30–31 we are told that when God created all the animals they all were vegetarians, and that God was pleased with everything that He had created because it was "very good." This means that all of creation was perfect when measured against the goodness of God — the only standard God uses (Matthew 19:17).

Furthermore, it is not until *after* God pronounced the Curse on all of creation because of Adam and Eve's disobedience that we are told that the ground would bring forth thorns and thistles (Genesis 3:17–18). But the evolutionary geologists tell us that there are fossilized thorns in Canadian sedimentary layers that are supposedly 400 million years old.[7] The Bible-believing Christian cannot accept this age-claim however.

If the plain statements of God's Word have any authority, then these fossilized thorns could only have grown after the Curse, after Adam was created by God. So the geologic record in which these fossilized thorns are found could only have been deposited after the Curse. However, the only event after the Curse that could have been responsible for burying and fossilizing these thorns,

6. Copy of this letter on file.
7. W.N. Stewart and G.W. Rothwell, *Paleobotany and the Evolution of Plants* (Cambridge, UK: Cambridge University Press, 1993), p. 172–176.

and the billions of other plants and animals we see in the vast rock layers of the earth, is the year-long Genesis Flood. This then rules out the millions of years.

Another theological problem arises when we come to Genesis 9:11–15. God made a promise to Noah and his descendants that "never again shall there be a flood to destroy the earth." In other words, God was promising never to send another event like the one Noah experienced, where we are told specifically in Genesis 7:21 that "all flesh died."

Obviously, if the Flood of Noah were only local in extent, then because we have seen lots of local floods since the time of Noah, that have destroyed both man and animals, God has broken His promise many times over! To the contrary, this rainbow covenant God made with Noah and his descendants could only have been kept by God if the Flood were global in extent, because never since in human history has a global flood been experienced.

The Views of Jesus and the New Testament Authors

The Lord Jesus Christ, God's living Word (John 1:1–3), made special reference to Noah and the Flood in Luke 17:26–30, where He said that "the Flood came and destroyed them all."

There is no biblical or logical reason to assume that all of pre-Flood humanity was living in the Mesopotamian Valley. Genesis 4 indicates that early man built cities, had nomadic herds of animals, invented things, and explored the earth (v. 17–22). So if all the ungodly globally on the earth will be judged when He comes again, when Jesus by way of comparison describes the Flood and all the ungodly being destroyed by it, then He was saying that the Flood also was global.

Similarly, the Apostle Peter in 2 Peter 3:3–7 warned of last-days scoffers who would wilfully forget that after the earth was created by God, it perished, "being flooded with water," and that the present earth is "reserved for fire until the day of judgment." There are three events he is thus referring to: the creation of the world (Greek *kosmos*), the destruction of that world (Greek *kosmos*) by a watery cataclysm (the Flood), and the coming destruction of the heavens and the earth by fire in the future.

In context, it is clear that Peter had to be teaching that the Flood was global, because the creation of the world was global, and the future judgment by fire will also be global. Indeed, the use of the Greek term *kosmos* for both the world that was created and the world that was flooded leaves us no doubt that the Apostle Peter, under the inspiration of the Holy Spirit, was teaching that the Flood was global in extent.

Scientific Problems

If the Flood were only local in extent, why did Noah have to take birds on board the ark (Genesis 7:8), when the birds in that local flooded area could simply have flown away to safe unflooded areas? Similarly, why would Noah need to take animals on board the ark from his local area, when other representatives of those same animal kinds would surely have survived in other, unflooded areas?

Indeed, why would Noah have had to build the ark to the scale specified by God (Genesis 6:15) — 300 cubits long, 50 cubits wide, and 30 cubits high, or approximately 510 feet long, 85 feet wide, and 51 feet high? With these dimensions, the total volume of the ark would have been approximately 1.88 million cubic feet, and with three decks it would have had a total deck area of approximately 110,000 square feet, equivalent to slightly more than the area of 23 standard basketball courts! (Depending on the thickness of the hull, the space would be slightly smaller.) The gross tonnage of the ark would have been about 15,600 tons, well within the category of large metal ocean-going vessels today.[8]

Quite obviously, an ark of such dimensions would only be required if the Flood were global in extent, designed by God to destroy all animals and birds around the world, except for those preserved on that ark. Indeed, because the Bible implies that Noah was warned 120 years before the Flood came (Genesis 6:3), God could have simply told Noah and his family to migrate with any required animals and birds out of the area that was going to be flooded.

In Genesis 1:28 we are told that God commanded Adam and Eve to fill the earth. Adam and his descendants' life-spans were hundreds of years, in which they would have had ample time to produce many children. The chronological framework from Adam to the Flood based on the genealogies given in Genesis 5 indicates a period of 1,656 years for the human population to grow and spread around the earth in obedience to God's command.

8. For fuller details regarding the size and construction of the ark, see Tim Lovett, *Noah's Ark: Thinking Outside the Box* (Green Forest, AR: Master Books, 2008).

Depending on the assumptions used for the number of children in each family, one could easily calculate, using a standard population growth equation, that the human population at the time of the Flood could have been up to a billion or more people. If so, there is no question that they would have spread beyond some localized area, and thus have required a global Flood to destroy them all. God gave a similar command to Noah and his descendants after the Flood to fill the earth (Genesis 9:1, 7), and in a matter of about 150 years God judged them for not obeying that command. Clearly, in the 1,656 years between Adam and the Flood, with the number of people in the pre-Flood population, the earth would have been filled, which is confirmed by God's assessment in Genesis 6:13 that because the earth was filled with violence through man's sin- fulness He would destroy them "with the earth," obviously necessitating that the Flood judgment was of global extent.

Conclusions

This has only been a brief survey of the problems associated with the local Flood view designed to accommodate the supposed millions of years of earth history. The Lord Jesus Christ and the Apostle Peter clearly taught that the Flood of Noah was global in extent, and both the context and the descriptive words used in Genesis 6–8 quite plainly describe the Flood as global in extent.

It wasn't until popularization of the belief in geology that only slow and gradual geological processes formed the geologic record over millions of years that the local Flood compromise became increasingly popular. Yet the Scriptures are clear that the deaths of animals and man only came into the world as a result of the Curse. So the fossils must have been produced after that tragic event. The subsequent global Flood could have produced most of the fossil-bearing sedimentary layers, including the fossilized thorns we find.

And Noah would not have needed to build an ark or take animals on board if the Flood were only local, as there was plenty of warning to escape to another region. These and many more biblical, theological, and scientific considerations make the local Flood compromise totally untenable. This is all ultimately about the authority of all of God's Word, which plainly teaches that the Flood of Noah was global in extent.

7

Is Man the Cause of Global Warming?

MICHAEL OARD

Global warming is big news. The media, environmentalists, and politicians, such as Al Gore,[1] continue to pound away that global warming is real, it is man-caused, and great harm will come to our world because of it. Some even say that global warming is the most significant threat to ever affect man. Bjorn Lomborg quotes respected scientist James Lovelock as saying: "Before this century is over, billions of us will die and the few breeding pairs of people that survive will be in the Arctic where the climate remains tolerable."[2] Intense storms of various sorts, drought, and heat waves will devastate the earth.[3]

Is all this true? Is global warming real? Is it all caused by man? Should we as Christians care about global warming, and if we do care, what should we do about it?

Man Is a Steward of God's Creation

We should be concerned with global warming, as well as other environmental issues, simply because God created the universe, the world, and everything in it (Exodus 20:11). It is His creation; He created it directly with a purpose and with man in mind. It did not evolve over billions of years. Man was told right

1. Al Gore, *An Inconvenient Truth: The Planetary Emergency of Global Warming and What We Can Do About It* (New York, NY: Rodale Press, 2006).
2. B. Lomborg, *Cool It: The Skeptical Environmentalist's Guide to Global Warming* (New York, NY: Alfred A. Knopf, 2007), p. 41.
3. N. Shute, "The Weather Turns Wild," *U.S. News & World Report*, February 5, 2001, p. 44–52.

away in the Garden of Eden to take dominion over the earth (Genesis 1:26–28), which means that we are to be stewards of His creation. We are to cultivate and take care of our surroundings, which at that time was the Garden of Eden: "Then the LORD God took the man and put him in the garden of Eden to tend and keep it" (Genesis 2:15).

What Should Christians Do about Global Warming?

Amidst all the hype, Christians need to first apply 1 Thessalonians 5:21: "Test all things; hold fast what is good." We are to hold fast to God's Word, the Bible, and Jesus as our Lord and Savior. Then we need to examine the evidence *carefully*. As stewards of God's creation, it will take time and energy to find out the facts. It is too easy to accept a superficial analysis of a controversial subject, in which case we might learn just enough to get into trouble. No, we need to dig deeper than the superficial level.

It is no different than evaluating the creation/evolution issue. At the superficial level, evolutionists can paint a pretty picture. It is only when you dig below the surface that you find out that evolution is unsubstantiated.

Since creationists are used to separating data from interpretation (the battle between creation and evolution is not over the data but the interpretation of the data), it is relatively easy to apply the same principles to the global warming issue. So we need to check the real data *first*. We need to be as objective as possible when examining the data, realizing that bias for man-made global warming and its harms is rampant.[4]

Evaluating the Data

When we examine the data, what can we say about global warming? This section will evaluate the facts, while the next main section will delve into additional evidence. We will then be able to evaluate global warming.

Carbon Dioxide and Other Greenhouse Gases Have Increased

First, carbon dioxide and other greenhouse gases, such as methane, have increased significantly over the past 100 years (figures 1a, b). These have been measured continuously since the middle of the 20th century and inferred from proxy indicators before that.

4. J. Pena and R. Vogel, eds., *Global Warming: A Scientific and Biblical Exposé of Climate Change* (DVD), Coral Ridge Ministries and Answers in Genesis, 2008; L. Vardiman, *Some Like It Hot* (Dallas, TX: Institute for Creation Research, 2009); M.J. Oard, "Global Warming: Examine the Issue Carefully," *Answers*, October–December 2006, p. 24–26.

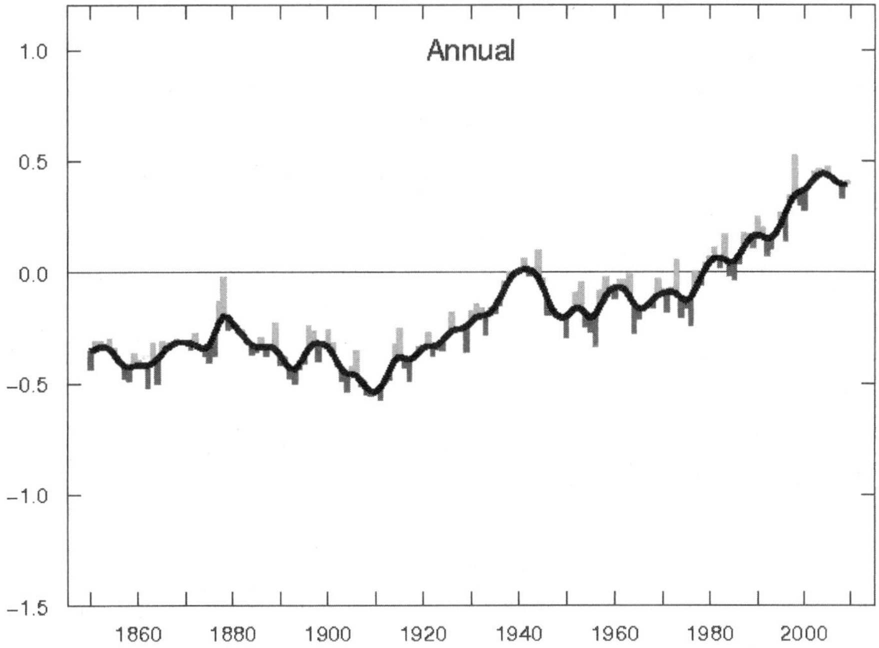

Figure 1a: Annual global temperature from 1850 to July 2009, from the U.K. Met Office Hadley Centre and the Climate Research Unit at the University of East Anglia. Note that temperatures have been cooling since about 2002.

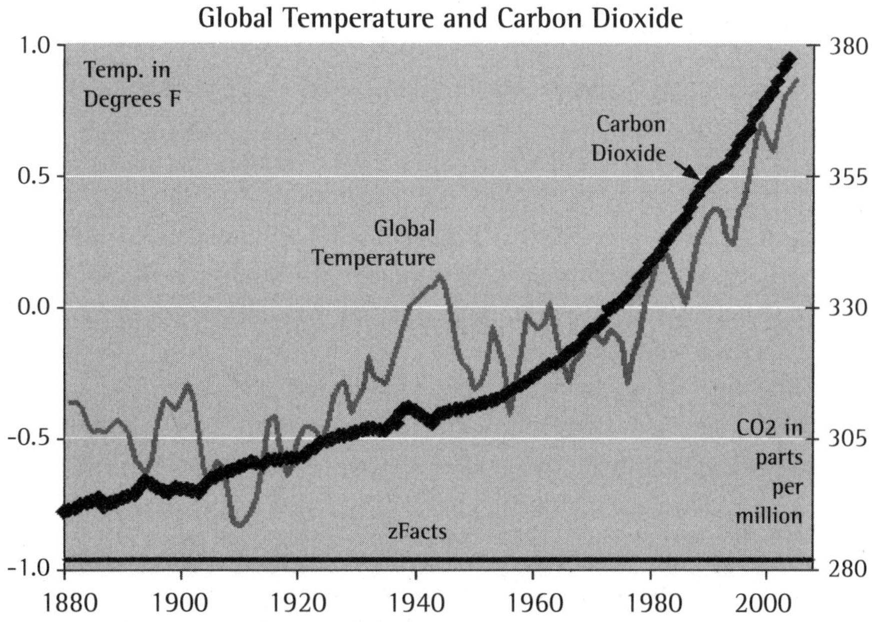

Figure 1b: The increase in carbon dioxide since 1880

Carbon dioxide has been continuously measured since 1959 and been inferred mainly from tree rings and ice cores before 1959. Note that carbon dioxide has increased more after about 1960 than before. Despite the title of the article from which this graphic is taken, the correlation of CO_2 and temperature does not demonstrate a cause-effect relationship.[5]

Carbon dioxide has been added to the air primarily because of the burning of fossil fuels since the industrial revolution. A secondary source for carbon dioxide is believed to be tropical deforestation. As trees are cut down, they rot and the carbon in the wood is oxidized to carbon dioxide. It is true that forests are being cut down in the tropics, especially in Brazil. However, forests grow back. So, it is not deforestation that counts, but the total amount of forest. When we consider the total amount of forest, the trend is unclear; we cannot be certain if it is increasing or decreasing.[6] So the rotting of tropical trees likely is not a significant source of carbon dioxide for the atmosphere.

Carbon dioxide is actually a minor gas in the greenhouse effect. The major greenhouse gas is water vapor, which accounts for 95 percent of greenhouse warming.[7] The greenhouse effect is actually good. Without these greenhouse gases the earth would be about 60°F cooler, and we would likely all freeze to death.

It is theoretically true that the increase in carbon dioxide and other greenhouse gases should cause warmer temperatures. The main question is how much.

There Are Natural Causes of Climate Change

A second fact is that there are natural causes of climate change. There are short-period natural processes that change the temperature by about a degree over several years. Two of these are a strong volcanic eruption that causes cooler global temperatures and an El Nińo that causes warmer global temperatures. Volcanism causes cooling by the reflection of sunlight back to space from particles trapped in the stratosphere. The amount of carbon dioxide and water vapor given off during volcanism is insignificant over the space and time periods significant to climate change.

There are also long-period temperature changes caused by effects on the sun that can be correlated to the number of sunspots: the more sunspots, the warmer the temperature on earth and vice versa. Since sunspots are cool spots,

5. ZFacts, "Evidence that CO_2 Is Cause," www.zfacts.com/p/226.html.
6. A. Granger, "Difficulties in Tracking the Long-term Global Trend in Tropical Forest Area," *Proceedings of the National Academy of Science* 105 no. 2 (2008): 818–823.
7. P.J. Michaels and R.C. Balling Jr., *The Satanic Gases: Clearing the Air About Global Warming* (Washington, D.: Cato Institute, 2000), p. 25–28.

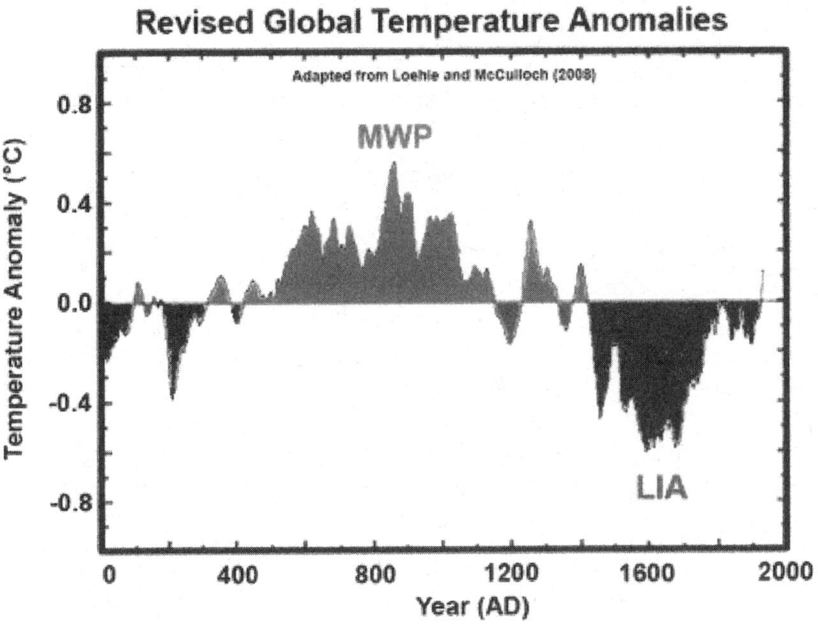

Figure 2. Average global temperature for the past 2,000 years showing the Medieval Warm Period (MWP) and the Little Ice Age (LIA). Before about the middle 1800s, there was little change in carbon dioxide to cause these fluctuations.

heating on earth seems counter intuitive. But when there are many sunspots, there are also many hot spots, called faculae, that more than make up for the cool spots. Two long period temperature changes recognized are the Medieval Warm Period (MWP) from about 800 to 1200 and the Little Ice Age (LIA) from about 1400 to 1880 (figure 2). These have been based on historical records and are well correlated to the number of sunspots using proxy data.[8] Variations in carbon dioxide levels were not responsible for these changes.

The climatic effect of natural processes is also seen during the 20th century by comparing the increase in carbon dioxide with the temperature change (figure 1b). Carbon dioxide increased slowly until after World War II and then accelerated. But the global temperature rose strongly from 1910 to 1940 (remember the dust bowl years in the 1930s), dropped a little between 1940 and 1975 (remember the coming ice age scare), and rose strongly again from 1975 to about 2002. The temperature has generally been cooling from 2002 to 2009 while carbon dioxide in the atmosphere continues to increase

8. C. Loehle and J.H. McCulloch, "Correction to a 2000-year Global Temperature Reconstruction Based on Non-tree Ring Proxies," *Energy & Environment* 19 no. 1 (2008): 93–100.

substantially. Nicola Scafetta and Bruce West stated that natural cycles from the sun account for at least 50 percent of the 20th century global warming.[9]

The increase in the amount of sunshine reaching the earth with a large number of sunspots is small. This is why many man-made global warming advocates discount the significance of the sun. It is known that higher sunspot numbers, which cause a stronger solar magnetic field, better shield the earth from cosmic rays. It is possible that fewer cosmic rays result in fewer low clouds that cause warmer surface temperatures and vice versa.[10] This hypothesis has been seriously challenged, so only time will tell if the hypothesis stands.

There Is No Consensus of Scientists

Third, although it is commonly claimed that there is a consensus of scientists that blame man for global warming, in actuality there is no consensus at all. Many prominent scientists disagree. Dr. Art Robinson has maintained a website since 1998, signed by around 20,000 scientists, saying that, as of 2009, there is no convincing scientific evidence that greenhouse gases are causing or will cause catastrophic heating of the earth's atmosphere and disruption of the earth's climate.[11] Of these, over 2,500 are physicists, geophysicists, climatologists, meteorologists, oceanographers, and environmental scientists, who are particularly qualified to evaluate global warming.

Climate Simulations Exaggerate Carbon Dioxide Warming

Fourth, dozens of computer climate simulations have attempted to quantify the relationship between increased carbon dioxide and temperature change. In the simulations, the scientists double carbon dioxide and leave every other variable alone. The resulting temperature changes range from 3°F to 11° F warming, usually by the year 2100.

But these simulations are crude, since the computer models cannot accurately simulate the many types of clouds and their effects, solar and infrared radiation processes, ocean processes, ice processes, etc.[12] The strengths and weaknesses of computer models need to be understood, but it seems that those who want runaway global warming believe these models *without question.*

9. N. Scafetta and B. West, "Is Climate Sensitive to Solar Variability?" *Physics Today* 61 no. 3 (2008): 50–51.
10. H. Svensmark, "Cosmoclimatology: A New Theory Emerges," *Astronomy and Geophysics* 48 no. 1 (2007): 18–24; L. Vardiman, "A New Theory of Climate Change," *Acts & Facts* 37 no. 11 (2008): 10–12.
11. Global Warming Petition Project, www.petitionproject.org.
12. Michaels and Balling, *The Satanic Gases: Clearing the Air About Global Warming*, p. 55–73.

It is interesting that nature has partially run the experiment for us. Carbon dioxide has increased a little more than 30 percent since the industrial revolution. Other greenhouse gases, not including water vapor, also have increased about 30 percent in "carbon dioxide equivalency units," for a total increase of about 60 percent in "carbon dioxide."[13] Global warming is claimed to be 1.2°F (yes, you heard right, the warming has been very small so far), but at least half is from natural causes. So if a 60 percent increase in "carbon dioxide" causes only a 0.6 degree Fahrenheit warming (man's share), a doubling of carbon dioxide should cause a 1°F warming. The models are therefore 3 to 11 times too sensitive to a doubling of carbon dioxide and should not be believed.

Some Benefits of Global Warming

Fifth, despite all the well-publicized harms, there are benefits to global warming. The media typically exaggerate the harms. Take for example the supposed decreasing polar bear populations as a result of less sea ice in the Northern Hemisphere. This was the theme behind the popular movie *Arctic Tale*.[14] Lomborg documents that the polar bear populations actually have increased.[15]

Some net benefits are that global warming will save the lives of more people, since many more people die of the cold than die of the heat. For instance, in Europe, seven times as many people die of the cold than die of the heat.[16] Other benefits include more plant growth due to higher carbon dioxide levels, aiding farming and ranching; crops able to be grown at higher latitudes; and shipping through ice-free areas of the Arctic Ocean, which will save much fuel and money. At this point it is difficult to tell whether there will be a net benefit or a net harm. Only objective research will determine this.

The Cost to "Fight" Global Warming Is Horrendous

Sixth, if certain environmentalists and politicians get their way, the cost to "fight" global warming will be horrendous, if the attempt is successful and doesn't produce even worse side effects. Lomborg estimates the cost of fighting global warming at many trillions of dollars.[17] Although Lomborg actually believes in the temperature rises suggested by the computer models, he makes a strong case that this money is best spent elsewhere, and that the earth will adapt to warming.

13. Ibid., p. 27.
14. M.J. Oard, movie review: "Arctic Tale — Exaggerating the Effects of Global Warming," www.answersingenesis.org/articles/aid/v2/n1/arctic-tale.
15. Lomborg, *Cool It: The Skeptical Environmentalist's Guide to Global Warming*, p. 3–9.
16. Ibid., p. 17.
17. Ibid, p. 32–38.

Figure 3: Athabasca Glacier, Canadian Rockies, was near the sign in 1890. It has since melted back to its current location due to global warming.

Additional Evidence

In any analysis of such a controversial subject, there is bound to be uncertainty in some variables. Only four of the most important will be evaluated.

Global Warming Is Real

First, global warming is real. Although some claim there is no global warming or we cannot measure it, the evidence for global warming is compelling. The claimed warming since 1880 is only 1.2°F. But we see the effects of the warming in that practically all glaciers have

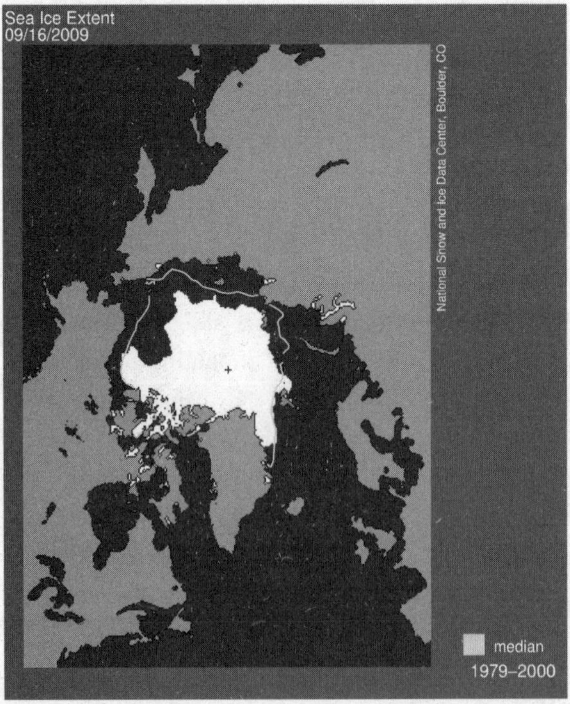

Figure 4: Minimum sea ice extent mid-September 2009, compared with the 1979 to 2000 average minimum (courtesy of the National Snow and Ice Data Center). However, the amount of sea ice has recovered 15 percent over 2008, which was about 10 percent greater than the record low in 2007, possibly due to global cooling as shown in figure 1a.

receded since about 1880 (figure 3), and the sea ice in the Arctic Ocean has decreased (figure 4).

The recent global warming is not caused by the earth breaking out of the Ice Age[18] about 4,000 years ago, because the atmosphere responds very fast to changes that affect climate. The change in seasons is one example of how fast the temperature can change when the angle of the sun changes. The earth has been more or less in steady state since the end of the Ice Age.

However, there is a question of whether the amount of claimed warming is accurate, since there are many biases (mostly favoring warming) in the long-term temperature records. Although those who analyze global temperatures have mostly purged the record of these errors, the claimed warming likely is too warm. Professor Robert Balling has studied these biases for a long time and has concluded:

> But as this chapter makes clear, major problems remain. First, the temperature records are far from perfect and contain contaminations from urbanization, distribution of measurement stations, instrument changes, time of observation biases, assorted problems in measuring near-surface temperatures in ocean areas, and on and on. This could introduce a total bias of 0.2°C to 0.3°C, or about one-third of the observed warming.[19]

This means that global warming since 1880 may be only about 0.8°F, which is closer to the satellite and weather balloon data of the lower atmosphere.

The Lower Atmosphere is Warming Less than 1.2°F

Second, satellites have been measuring the amount of temperature change in the atmosphere since 1979. Weather balloons have been probing the atmosphere for longer than that. At first, it was thought that the satellite temperatures showed a cooling trend. However, there were some errors in the measurements. Now, it appears that the satellite and weather balloon data both show less warming in the lower atmosphere than the claimed 1.2°F surface warming.[20]

18. M.J. Oard, *Frozen In Time: The Woolly Mammoth, the Ice Age, and the Biblical Key to Their Secrets* (Green Forest, AR: Master Books, 2004).
19. R.C. Balling Jr., "Observational Surface Temperature Records Versus Model Predictions," in P.J. Michaels, ed., *Shattered Consensus: The True State of Global Warming* (New York, NY: Rowman & Littlefield Publishers, Inc, 2005), p. 67.
20. J. Christy, "Temperature Changes in the Bulk of the Atmosphere," in P.J. Michaels, ed., *Shattered Consensus: The True State of Global Warming* (New York, NY: Rowman & Littlefield Publishers, Inc, 2005), p. 72–105.

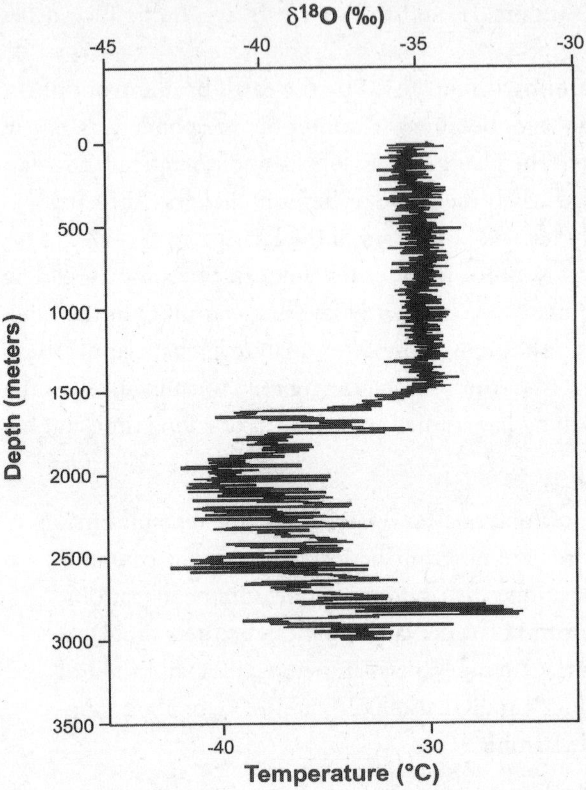

Figure 5: The oxygen isotope ratio from bedrock to the top of the GISP2 ice core at Summit on the Greenland Ice Sheet (plot courtesy of Dr. Larry Vardiman). The oxygen isotope ratio is generally proportional to temperature with cooler temperatures to the left.

The Climate Likely Cannot Jump to a More Catastrophic Mode

Third, some scientists have concluded that the atmosphere goes through thresholds to different climatic states. They believe that although our climate has been steady, global warming may bring the temperature up to the "tipping point" that will cause a shift to a much different climate, possibly leading to an ice age. The threshold idea is based on Greenland ice cores showing large, abrupt changes in temperature during the Ice Age portion (figure 5).

Some suggest that global warming will halt the Gulf Stream that transports warm water into the high North Atlantic Ocean. Temperatures then plummet in Europe and an ice age can occur. This is the basis of the movie *The Day After Tomorrow*,[21] taken from the preposterous book *The Coming Global Superstorm*.[22] Despite Hollywood fantasy, some scientists believe that such a scenario is possible over a time frame of a decade or two.

21. M.J. Oard, "The Greenhouse Warming Hype of the Movie The Day After Tomorrow," *Acts & Facts Impact*, 373 (Dallas, TX: Institute for Creation Research, 2004).
22. A. Bell and W. Strieber, *The Coming Global Superstorm* (New York, NY: Pocket Books, 2000).

The idea of an abrupt climate change after the temperature passes a "threshold" is where the worldview issue between creation and evolution is crucial. Evolutionary scientists date the ice cores at hundreds of thousands of years old, and the ice sheets are believed to have been more or less the same thickness for millions of years. But these abrupt temperature changes in the ice cores are due to a rapid, post-Flood Ice Age and are only related to changes during a unique Ice Age.[23]

Storms and Droughts Likely Unchanged

Fourth, it seems like every large storm, drought, or heat wave that occurs in the world is blamed on man-made global warming. But these things have been happening for millennia. The problem is that most people have short memories about past events. Furthermore, damage is greater now because more people and property lie in harm's way. But overall, there do not seem to be any long-term trends in any of these weather events.[24]

Summary

In the face of claims that man is causing disastrous global warming, an objective look at the facts and additional evidence show otherwise. Natural processes on the sun account for over 50 percent of the claimed 1.2°F global warming, which is likely too warm. Since the climate simulations greatly exaggerate the temperature rise from an increase in carbon dioxide, these models cannot be trusted. Thus, man is likely responsible for only about 0.5°F warming — miniscule and likely impossible to mitigate.

What is really needed is unbiased research in climate change. Climate disaster is not just around the corner; we have sufficient time for careful research.

Acknowledgment

I thank Drs. Larry Vardiman and Jason Lisle for reviewing the manuscript and offering valuable improvements.

23. M.J. Oard, *The Frozen Record: Examining the Ice Core History of the Greenland and Antarctic Ice Sheets* (Dallas, TX: Institute for Creation Research, 2005).
24. P.J. Michaels, *The Predictable Distortion of Global Warming by Scientists, Politicians, and the Media* (Washington DC: CATO Institute, 2004).

8

Did Bible Authors Believe in a Literal Genesis?

DR. TERRY MORTENSON

Anyone who has read the Bible very much will recognize that there are different kinds of literature in the Old and New Testaments. There are parables, poetry, prophetic visions, dreams, epistles, proverbs, and historical narrative, with the majority being the latter. So, how should we interpret Genesis 1–11? Is it history? Is it mythology? Is it symbolic poetry? Is it allegory? Is it a parable? Is it a prophetic vision? Is it a mixture of these kinds of literature or some kind of unique genre? And does it really matter anyway?

We will come back to the last question later, but suffice it to say here that the correct conclusion on genre of literature is foundational to the question of the correct interpretation. If we interpret something literally that the author intended to be understood figuratively, then we will misunderstand the text. When Jesus said "I am the door" (John 10:9), He did not mean that He was made of wood with hinges attached to His side. Conversely, if we interpret something figuratively that the author intended to be taken literally, we will err. When Jesus said, "The Son of Man is about to be betrayed into the hands of men, and they will kill Him, and the third day He will be raised up" (Matthew 17:22–23), He clearly meant it just as literally as if I said to my wife, "Margie, I'm going to fill up the gas tank with gas and will be back in a few minutes."

There are many lines of evidence we could consider to determine the genre of Genesis 1–11, such as the internal evidence within the Book of Genesis and how the Church has viewed these chapters throughout church history. But in this chapter we want to answer the question, "How did the other biblical authors (besides Moses, who wrote Genesis[1]) and Jesus interpret them?" From my reading and experience it appears that most people who consider the question of how to interpret the early chapters of Genesis have never asked, much less answered, that question.

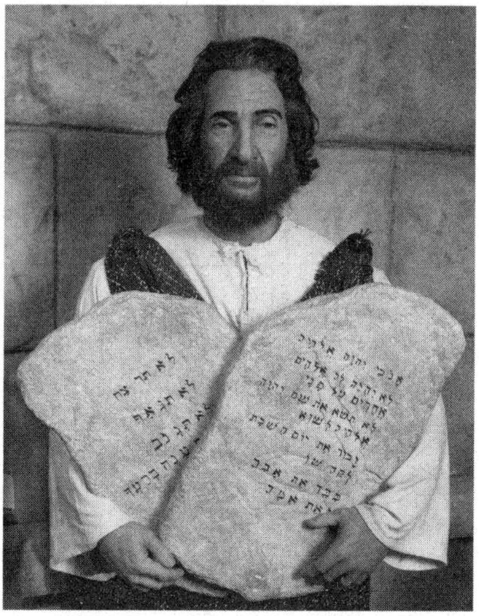
Moses as depicted in the Creation Museum's biblical authority room.

To begin, consider what God says about the way He spoke to Moses in contrast to the way He spoke to other prophets. In Numbers 12:6–8 we read:

> Then He said, "Hear now My words: if there is a prophet among you, I, the LORD, make Myself known to him in a vision; I speak to him in a dream. Not so with My servant Moses; he is faithful in all My house. I speak with him face to face, even plainly, and not in dark sayings; And he sees the form of the LORD. Why then were you not afraid to speak against My servant Moses?"

So God says that He spoke "plainly" to Moses, not in "dark sayings," that is, not in obscure language. That strongly suggests that we should not be looking for mysterious, hard-to-understand meanings in what Moses wrote. Rather, we should read Genesis as the straightforward history that it appears to be. An examination of how the rest of the Bible interprets Genesis confirms this.

1. That Moses was the author of the first five books (called the Pentateuch) of the Old Testament is clear from Scripture itself. The Pentateuch explicitly claims this in Exodus 17:14, 24:4, 34:27; Numbers 33:1–2; Deuteronomy 31:9–11. Other OT books affirm that Moses wrote these books, which by the time of Joshua became known collectively as "the Law," "the book of the Law," or "the Law of Moses": Joshua 1:8, 8:31–32; 1 Kings 2:3; 2 Kings 14:6 (quoting Deuteronomy 24:16), 21:8; Ezra 6:18; Nehemiah 13:1; Daniel 9:11–13; Malachi 4:4. The New Testament agrees in Matthew 19:8; John 5:46–47, 7:19; Acts 3:22 (quoting from Deuteronomy 18:15); Romans 10:5 (quoting from Leviticus 18:5), and Mark 12:26 (referring to Exodus 3:6). Jewish tradition also ascribes the Pentateuch to Moses. Also, the theories of liberal theologians who deny the Mosaic authorship of these books are fraught with false assumptions and illogical reasoning. See Gleason L. Archer Jr., *A Survey of Old Testament Introduction* (Chicago, IL: Moody Press, 1985), p. 109–113.

Old Testament Authors and Their Use of Genesis

When we turn to other Old Testament authors, there are only a few references to Genesis 1–11. But they all treat those chapters as literal history.

The Jews were very careful about genealogies. For example, in Nehemiah 7:61–64 the people who wanted to serve in the rebuilt temple needed to prove that they were descended from the priestly line of Aaron. Those who could not prove this could not serve as priests. First Chronicles 1–8 gives a long series of genealogies all the way back to Adam. Chapter 1 (verses 1–28) has no missing or added names in the genealogical links from Adam to Abraham, compared to Genesis 5 and 11. The author(s) of 1 Chronicles obviously took these genealogies as historically accurate.

Outside of Genesis 6–11, Psalm 29:10 contains the only other use of the Hebrew word *mabbul* (translated "flood").[2] God literally sat as King at the global Flood of Noah. If that event was not historical, the statement in this verse would have no real force and the promise of verse 11 will give little comfort to God's people.

David, the writer of many of the psalms, from a Creation Museum display.

Psalm 33:6–9 affirms that God created supernaturally by His Word, just as Genesis 1 says repeatedly. Creatures came into existence instantly when God said, "Let there be. . . ." God did not have to wait for millions or thousands of years for light or dry land or plants and animals or Adam to appear. "He spoke and it was done; He commanded and it stood fast" (Psalm 33:9).

Psalm 104:5 and 19 speak of events during creation week.[3] But verses 6–9 in this psalm give additional information to that provided in Genesis 8, which describes how the waters receded off the earth at the end of the Flood.[4] The Psalmist is clearly describing historical events.

2. There are four other Hebrew words that are used in the OT to describe lesser, localized floods.
3. Most of this psalm is referring to aspects of God's creation as it existed at the time the Psalmist was writing. Contrary to what some old-earth creationists assert, Psalm 104 is not a "creation account."
4. That these verses do not refer to creation week is evident from the promise reflected in verse 9, which echoes the promise of Genesis 9:11–17. God made no such promise on the third day of creation week when He made dry land appear.

In beautiful poetic form, Psalm 136 recounts many of God's mighty acts in history, beginning with statements about some of His creative works in Genesis 1.

Isaiah recorded God's Word, not mythical tales.

In Isaiah 54:9 God says (echoing the promise of Psalm 104:9) to Israel, "For this is like the waters of Noah to Me; for as I have sworn that the waters of Noah would no longer cover the earth, so have I sworn that I would not be angry with you, nor rebuke you." The promise of God would have no force if the account of Noah's Flood was not historically true. No one would believe in the Second Coming of Christ if the promise of it (as recorded in Matt. 24:37–39) was given as, "Just as Santa Claus comes from the North Pole in his sleigh pulled by reindeer on Christmas Eve and puts presents for the whole family under the Christmas tree in each home, so Jesus is coming again as the King of kings and Lord of lords." In fact, the analogy would convince people that the Second Coming is a myth.

In Ezekiel 14:14–20 God refers repeatedly to Noah, Daniel, and Job and clearly indicates that they were all equally historical and righteous men. There is no reason to doubt that God meant that everything the Bible says about these men is historically accurate.

New Testament Authors' View of Genesis

The New Testament has many more explicit references to the early chapters of Genesis.

The genealogies of Jesus presented in Matthew 1:1–17 and Luke 3:23–38 show that Genesis 1–11 is historical narrative. These genealogies must all be equally historical or else we must conclude that Jesus was descended from a

Paul relied heavily on Genesis as plainly written.

myth and therefore He would not have been a real human being and therefore not our Savior and Lord.[5]

Paul built his doctrine of sin and salvation on the fact that sin and death entered the world through Adam. Jesus, as the Last Adam, came into the world to bring righteousness and life to people and to undo the damaging work of the first Adam (Romans 5:12–19; 1 Corinthians 15:21–22, 45–47). Paul affirmed that the serpent deceived Eve, not Adam (2 Corinthians 11:3; 1 Timothy 2:13–14). He took Genesis 1–2 literally by affirming that Adam was created first and Eve was made from the body of Adam (1 Corinthians 11:8–9). In Romans 1:20, Paul indicated that people have seen the evidence of God's existence and some of His attributes since the creation of the world.[6] This means that Paul believed that man was right there at the beginning of history, not billions of years after the beginning.

Peter similarly based some of his teachings on the literal history of Genesis 1–11. In 1 Peter 3:20, 2 Peter 2:4–9, and 2 Peter 3:3–7, he referred to the Flood. He considered the account of Noah and the Flood just as historical

5. In Matthew 1:1–17, Matthew has clearly left out some names in his genealogy (for a literary purpose), as seen by comparing it to the Old Testament history. But all the people are equally historical all the way back to Abraham, who is first mentioned in Genesis 11. Luke 3:23–38 traces the lineage of Jesus back to Adam. There is no reason to think there are any missing names in Luke's genealogy, because 1) he was concerned about giving us the exact truth (Luke 1:4), and 2) his genealogy from Adam to Abraham matches 1 Chronicles 1:1–28 and Genesis 5 and 11, and there is no good reason for concluding that Genesis has missing names. See Ken Ham and Larry Pierce, "Who Begat Whom? Closing the Gap in Genesis Genealogies," www.answersingenesis.org/articles/am/v1/n2/who-begat-whom, and Travis R. Freeman, "Do the Genesis 5 and 11 Genealogies Contain Gaps?" in Terry Mortenson and Thane H. Ury, eds., *Coming to Grips with Genesis* (Green Forest, AR: Master Books, 2008), p. 283–314.
6. The New King James and the King James Version translate the Greek in this verse as "from the creation of the world." The word "from" in English has a similar range of meanings as the Greek word (*apo*) that it translates here. There are a number of reasons to take it in a temporal sense, meaning "since" as the NAS, NIV, and ESV translate it. For a fuller discussion of this important verse, see Ron Minton, "Apostolic Witness to Genesis Creation and the Flood," in Terry Mortenson and Thane H. Ury, eds., *Coming to Grips with Genesis* (Green Forest, AR: Master Books, 2008), p. 351–354.

as the account of the judgment of Sodom and Gomorrah (Genesis 19). He affirmed that only eight people were saved and that the Flood was global, just as the future judgment at the Second Coming of Christ will be. He argued that scoffers will deny the Second Coming because they deny the supernatural creation and Noah's Flood. And Peter told his readers that scoffers will do this because they are reasoning on the basis of the philosophical assumption that today we call uniformitarian naturalism: "all things continue as they were from the beginning of creation" (2 Peter 3:4).[7]

The words of John and Peter demonstrate their trust in the historicity of the Genesis accounts.

It has been objected that the apostles did not know the difference between truth and myth. But this is also false. In 1 Corinthians 10:1–11 Paul refers to a number of passages from the Pentateuch where miracles are described and he emphasizes in verses 6 and 11 that "these things happened." In 2 Timothy 4:3–4 Paul wrote:

> For the time will come when they will not endure sound doctrine, but according to their own desires, because they have itching ears, they will heap up for themselves teachers; and they will turn their ears away from the truth, and be turned aside to fables.

The Greek word translated here as "fables" is *muthos*, from which we get our English word "myth." In contrast to "truth" or "sound doctrine," the same Greek word is used in 1 Timothy 1:4, 4:7; Titus 1:14; and 2 Peter 1:16. In a first-century world filled with Greek, Roman, and Jewish myths, the apostles clearly knew the difference between truth and myth. And they constantly affirmed that the Word of God contains truth, not myth.

7. For more discussion of this, see Terry Mortenson, "Philosophical Naturalism and the Age of the Earth: Are They Related?" *The Master's Seminary Journal* 15 no. 1 (2004): 71–92, online at www.answersingenesis.org/docs2004/naturalismChurch.asp.

Christ and His Use of Genesis

In John 10:34–35 Jesus defended His claim to deity by quoting from Psalm 82:6 and then asserting that "Scripture cannot be broken." That is, the Bible is faithful, reliable, and truthful. The Scriptures cannot be contradicted or confounded. In Luke 24:25–27 Jesus rebuked His disciples for not believing all that the prophets have spoken (which He equates with "all the Scriptures"). So in Jesus's view, all Scripture is trustworthy and should be believed.

Another way that Jesus revealed His complete trust in the Scriptures was by treating as historical fact the accounts in the Old Testament, which most contemporary people think are unbelievable mythology. These historical accounts include Adam and Eve as the first married couple (Matthew 19:3–6, Mark 10:3–9), Abel as the first prophet who was martyred (Luke 11:50–51), Noah and the Flood (Matthew 24:38–39), the experiences of Lot and his wife (Luke 17:28–32), the judgment of Sodom and Gomorrah (Matthew 10:15), Moses and the serpent in the wilderness wanderings after the exodus from Egypt (John 3:14), Moses and the manna from heaven (John 6:32–33, 49), the miracles of Elijah (Luke 4:25–27), and Jonah in the big fish (Matthew 12:40–41). As Wenham has compellingly argued,[8] Jesus did not allegorize these accounts but took them as straightforward history, describing events that actually happened, just as the Old Testament describes. Jesus used these accounts to teach His disciples that the events of His own death, resurrection, and Second Coming would likewise certainly happen in time-space reality. Jesus also indicated that the Scriptures are essentially perspicuous (or clear): 11 times the gospel writers record Him saying, "Have you not read. . . ?"[9] And 30 times He defended His teaching by saying "It is written."[10] He rebuked His listeners for not understanding and believing what the text plainly says.

Besides the above-mentioned evidence that Jesus took Genesis 1–11 as straightforward, reliable history, the gospel writers record three important statements that reveal Jesus' worldview. Careful analysis of these verses (Mark 10:6; Mark 13:19–20; Luke 11:50–51) shows that Jesus believed that Adam and Eve were in existence essentially at the same time that God created everything else (and Abel was very close to that time), not millions or billions of years after God

8. John Wenham, *Christ and the Bible* (Downers Grove, IL: InterVarsity Press, 1973), p. 11–37.
9. In these instances Jesus referred to Genesis 1–2, Exodus 3–6, 1 Samuel 21:6, Psalm 8:2 and 118:22, and to unspecified Levitical law — in other words, to passages from the historical narrative, the Law, and the poetry of Scripture.
10. Passages He specifically cited were from all five books of the Pentateuch, Psalms, Isaiah, Jeremiah, Zechariah, and Malachi. Interestingly, in the temptation of Jesus, Satan used Scripture literally and, in response, Jesus did not imply that the literal interpretation of Satan was wrong. Rather, He corrected Satan's misapplication of the text's literal meaning by quoting another text, which He took literally (see Matthew 4:6–7).

made the other things.[11] This shows that Jesus took the creation days as literal 24-hour days. So everything Jesus said shows that we can justifiably call Him a young-earth creationist.

It has been objected that in these statements Jesus was just accommodating the cultural beliefs of His day. But this is false for four reasons. First, Jesus was the truth (John 14:6), and therefore He always spoke the truth. No deceitful or misleading words ever came from His mouth (1 Peter 2:22). Even his enemies said, "Teacher, we know that You are truthful and defer to no one; for You are not partial to any, but teach the way of God in truth" (Mark 12:14; NASB). Second, Jesus taught with authority on the basis of God's Word, which He called "truth" (John 17:17), not as the scribes and Pharisees taught based on their traditions (Matthew 7:28–29). Third, Jesus repeatedly and boldly confronted all kinds of wrong thinking and behavior in his listeners' lives, in spite of the threat of persecution for doing so (Matthew 22:29; John 2:15–16, 3:10, 4:3–4, 9; Mark 7:9–13). And finally, Jesus emphasized the foundational importance of believing what Moses wrote in a straightforward way (John 5:45; Luke 16:31, 24:25–27, 24:44–45; John 3:12, Matthew 17:5).

Why Is This Important?

We should take Genesis 1–11 as straightforward, accurate, literal history because Jesus, the Apostles, and all the other biblical writers did so. There is absolutely no biblical basis for taking these chapters as any kind of non-literal, figurative genre of literature. That should be reason enough for us to interpret Genesis 1–11 in the same literal way. But there are some other important reasons to do so.

Only a literal, historical approach to Genesis 1–11 gives a proper foundation for the gospel and the future hope of the gospel. Jesus came into the world to solve the problem of sin that started in real, time-space history in the real Garden of Eden with two real people called Adam and Eve and a real serpent that spoke to Eve.[12] The sin of Adam and Eve resulted in spiritual and physical

11. See Terry Mortenson, "Jesus' View of the Age of the Earth," in Terry Mortenson and Thane H. Ury, eds., *Coming to Grips with Genesis* (Green Forest, AR: Master Books, 2008), p. 315–346.
12. Why Christians have trouble believing Genesis 3 when it speaks of a talking serpent is a mystery to me. We have talking parrots today, which involves no miracles. And if the Christian believes in any miracles of the Bible, then he must believe that Balaam's donkey was used by God to speak to the false prophet (Numbers 22:28). Since Satan is a supernatural being who can do supernatural things (e.g., 2 Corinthians 11:11–13; Matthew 4:1–11; 2 Thessalonians 2:8–9), it is not difficult at all to understand or believe that he could speak through a serpent to deceive Eve (cf. 2 Corinthians 11:3; Revelation 12:9).

death for them, but also a divine curse on all of the once "very good" creation (see Genesis 1:31 and 3:14–19). Jesus is coming again to liberate all Christians and the creation itself from that bondage to corruption (Romans 8:18–25). Then there will be a new heaven and a new earth, where righteousness dwells and where sin, death, and natural evils will be no more. A non-literal reading of Genesis destroys this message of the Bible and ultimately is an assault on the character of God.[13]

Genesis is also foundational to many other important doctrines in the rest of the Bible, such as male, loving headship in the home and the church.

Conclusion

The Bible is crystal clear. We must believe Genesis 1–11 as literal history because Jesus, the New Testament Apostles, and the Old Testament prophets did, and because these opening chapters of Genesis are foundational to the rest of the Bible.

As we and many other creationists have always said, a person doesn't have to believe that Genesis 1–11 is literally true to be saved. We are saved when we repent of our sins and trust solely in the death and Resurrection of Jesus Christ for our salvation (John 3:16, Romans 10:9–10). But if we trust in Christ and yet disbelieve Genesis 1–11, we are being inconsistent and are not faithful followers of our Lord.

God said through the prophet Isaiah (66:1–2):

> Thus says the LORD: "Heaven is My throne, and earth is My footstool. Where is the house that you will build Me? And where is the place of My rest? For all those things My hand has made, and all those things exist, says the LORD. But on this one will I look: on him who is poor and of a contrite spirit, and who trembles at My word."

Will you be one who trembles at the words of God, rather than believing the fallible and erroneous words of evolutionists who develop hypotheses and myths that deny God's Word? Ultimately, this question of the proper interpretation of Genesis 1–11 is a question of the authority of God's Word.

13. See James Stambaugh, "Whence Cometh Death? A Biblical Theology of Physical Death and Natural Evil," and Thane H. Ury, "Luther, Calvin, and Wesley on the Genesis of Natural Evil: Recovering Lost Rubrics for Defending a Very Good Creation," in Terry Mortenson and Thane H. Ury, eds., *Coming to Grips with Genesis* (Green Forest, AR: Master Books, 2008), p. 373–398 and 399–424, respectively.

9

Do Fossils Show Signs of Rapid Burial?

DR. JOHN D. MORRIS

Uniformitarianism (or gradualism) is the secular belief that rock layers were laid down slowly over millions of years. This view was prominently taught by Charles Lyell through much of the 19th century and strongly influenced Charles Darwin, as well as much of modern geology.

But is uniformitarianism really a true understanding of the rock and fossil records? Did it really take long ages to lay down all these rock layers? Today, that view is being seriously questioned, and rightfully so! Consider a modern geology professor's comments:

> Furthermore, much of Lyell's uniformitarianism, specifically his ideas on identity of ancient and modern causes, gradualism, and constancy of rate, has been explicitly refuted by the definitive modern sources, as well as by an overwhelming preponderance of evidence that, as substantive theories, his ideas on these matters were simply wrong.[1]

When we look at the geologic record in light of the Bible, however, a whole new way of understanding the formation of rock layers and their contained fossils opens up. Earth history as described in the Bible was dominated by several great, world-changing events. First, the earth resulted totally from the six-day creation event (Genesis 1). It was subsequently altered by the Curse on all creation due to Adam's rebellion (Genesis 3). This was soon followed by the great

1. James H. Shea, "Twelve Fallacies of Uniformitarianism," *Geology* 10 no. 9 (1982): 456.

Flood of Noah's day (Genesis 6–9). The Flood is described as nothing less than a tectonic and hydraulic restructuring of the planet, particularly its surface layers. No place on earth escaped its terror. All land-dwelling, air-breathing life not on the ark was drowned by the Flood waters (Genesis 7:22).

What Would a Major, Catastrophic, Global Flood Do?

A global Flood would have done what major floods do. Such a Flood would have eroded and dissolved both soil and rock. Fragments would have been transported and redeposited elsewhere as sediments full of dead plants and animals, the creatures that died in Noah's Flood. Now we observe those sediments hardened into sedimentary rock layers, while the dead things have hardened into fossils.

We can be certain the great majority of earth's sedimentary rock layers and their contained fossils are the result of that great Flood. Evolutionists often wrongly use rocks and fossils to support long ages of evolutionary change, but since Noah's Flood really occurred, it must have laid down the rock layers and fossils. Take rocks and fossils away from evolutionists and evolution's story, and they have no evidence remaining!

Don't think of the Flood as a time when things were merely carried along and then settled out of moving water. Rather, the sediments were deposited in dynamic episodes, one following the other until thick sequences of layers had accumulated, triggered by a combination of consecutive tidal waves (tsunamis), tides, pulses of gravity-driven underwater mud flows, and other processes. The whole sediment package amassed quickly, within the Flood year, not over the hundreds of millions of years claimed by evolutionists. The fossils are evidence of this rapid accumulation.

The conventional secular idea about sediment and fossil deposition involves long ages of slow and gradual accumulation in calm and placid seas. However, fossils are almost never found today in the sea. Life abounds in the sea, but fossils of sea creatures do not. Fossils are hardly ever preserved in an oceanic context. Great deposits of fossils are found in marine sediments, but always on the land! They show evidence of dynamic marine forces destroying life on the continents. What can we make of the myriads of marine fossils found in Kansas, but none in the south Pacific?

How to Make a Fossil

An oft-repeated series of textbook illustrations shows a hypothetical animal dying alongside a stream. Before nature's degradative influences have full sway, the stream overflows, burying the carcass in mud, protecting it from ruin. Over the years, the mud accumulates around the remains, and eventually the entire region subsides, allowing even greater thicknesses of lake bottom or ocean bottom mud to blanket the area, mineralizing the bones and consolidating the mud into rock. Eventually, the region rises again, and erosion exposes the now-fossilized remains.

This scenario would, no doubt, be applicable in rare cases, but it ignores significant advances in sedimentation theory made in recent decades. Geologists now recognize that most rock units are the result of widespread, high-intensity processes, accomplishing in minutes what has traditionally been attributed to slow and gradual processes.

Clams

Most animal fossils are of marine invertebrates, especially shellfish — animals with a hard outer shell, such as clams. Clam fossils are found by the millions, perhaps billions. Clams are surprisingly agile creatures, able to burrow in the sand in their search for food and shelter. The muscle that connects the clam's two halves relaxes at death. The dead clam opens up, and scavengers eat the insides. But often the fossils retain both shell halves, tightly closed — all "clammed" up (figure 1). This is how a clam protects itself from danger.

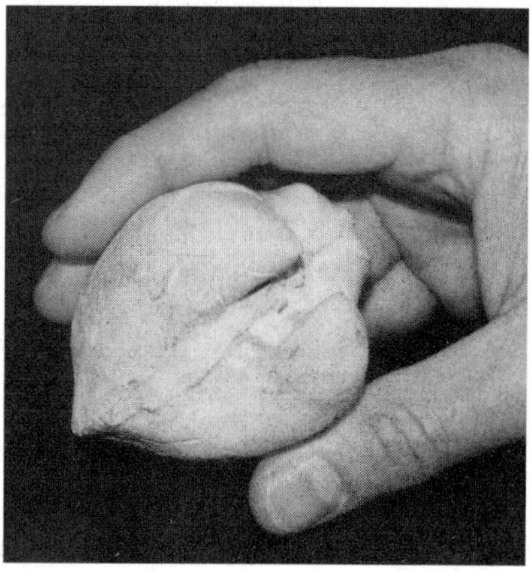

Figure 1. A clam fossilized in the closed position

Usually when we find clam fossils they are jammed together in great numbers, not at all how they live in their life zones today. Thus, we discern the clams felt themselves in danger as they were transported and deposited along with other clams of roughly the same density and shape with many others, buried so deeply they couldn't burrow out. They speak of a rapid depositional process, requiring only a short time.

Fish

Sometimes the fossilized animals appear to have been caught suddenly and buried in life poses — true "action shots." For instance, occasionally a fish fossil is found in the process of eating another fish! How long does it take for a fish to

Figure 2 (left). A fish fossilized in the process of eating another fish.

Figure 3 (below). Rapid fossils such as an ichthyosaur trapped in sediment at the moment of giving birth

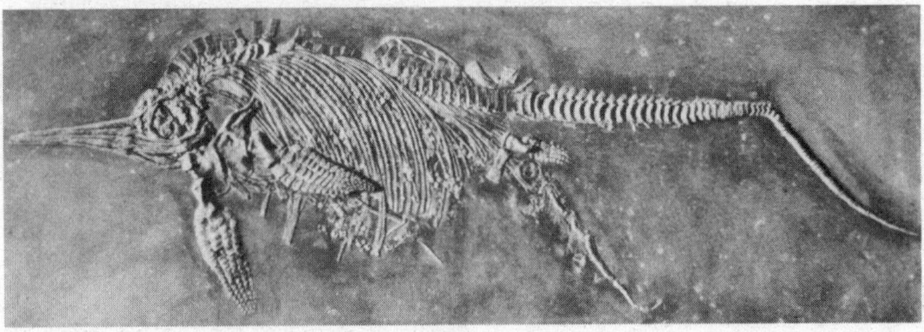

swallow his lunch? It might take a few seconds, but in that brief interval it was trapped and buried (figure 2). Sometimes we see animals such as the ichthyosaur (an extinct marine reptile) pictured in the process of giving birth (figure 3). No great time required here — only a mighty and rapid process.

Jellyfish

Another remarkable fossil is the jellyfish (figure 4). Jellyfish are made mostly of water, and when they get washed up on shore, they quickly dry out. Within a very short time there is nothing left. Yet huge jellyfish fossil graveyards have been found, requiring rapid deposition, burial, and fossilization.[2] Fossil jellyfish graveyards refute the favorite evolutionist excuse

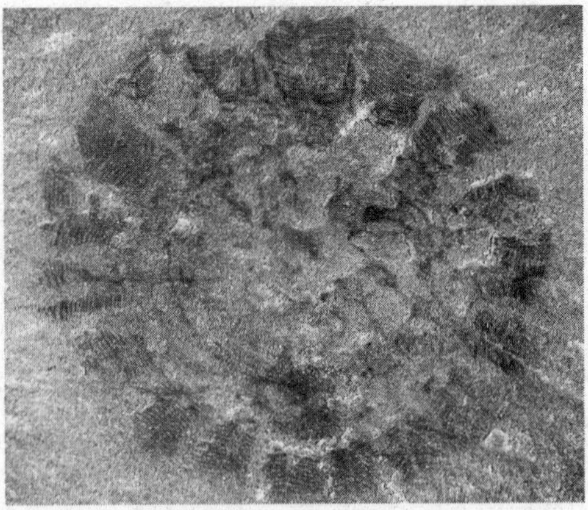

Figure 4. Fossilized jellyfish are only possible if they were buried catastrophically.

2. Reginald Sprigg, "Early Cambrian (?) Jellyfishes from the Flinders Ranges, South Australia," *Transactions of the Royal Society of South Australia* 71 no. 2 (1947): 212–224.

for the lack of transitional fossils, with the claim that only hard parts of fossils are preserved. Instead, we don't expect to find transitional fossils, because none existed. Under the right conditions, any fossil can be preserved.

Right Conditions

Just what are the "right conditions"? Obviously, animals or plants must be quickly buried to avoid the action of scavengers. Many animals are specifically designed to clean the environment of dead and rotting carcasses, and they do a marvelous job. Our world would be a stinking garbage dump without the action of ants, termites, and dung beetles, as well as hyenas, etc.

To become a fossil, a living thing must be out of the reach of other creatures and processes which would destroy it. This includes not only scavengers, but also decomposers, like bacteria. Where can you hide from microscopic bacteria? Likewise, the dead body must be kept from oxidation. Only by undergoing rapid burial, away from scavengers, bacteria, and oxygen, can an organism be fossilized. Yet we find fossils in almost every rock type. Surely catastrophic processes are displayed in the fossils.

Polystrate Fossils

Usually, fossils are found in only one particular layer, but sometimes fossils are discovered straddling two or more geologic layers, each thought to have required long ages to accumulate in conventional thinking. For instance, in the coal regions of Kentucky, trees are often found standing upright in growth position, with their base in one layer, but extending up through several more layers, including, in some cases, layers of coal.

Figure 5. Polystrate fossils, like this tree trunk, cross several geologic layers and cannot be explained by processes that require millions of years of deposition.

Geologists are taught that coal is the metamorphosed remains of plant material, which slowly accumulated as peat in peat swamps. Eventually, as the story goes, the layers of peat were submerged under the ocean and great thicknesses of sediments were deposited on top of them. "Millions of years" of heat and pressure altered the peat into coal. Later, the entire areas emerged from the water to receive more peat and the cycle repeated. But if so, how could one tree stand upright through whole sequences of layers, especially under the sea, while awaiting several cycles of deposition of overlying sediment layers and the necessary heat and pressure? "Polystrate" trees are a good example of rapid deposition (figure 5).

The same argument goes for a fossilized animal whose body thickness extends from one layer into the next. For instance, a whale fossil was found in California that spanned several layers. The entire rock unit could not have required more time to accumulate than is required for a whale carcass to decay.[3]

How Long Does it Take to Fossilize Something?

Figure 6. This hat was turned to stone after being left in an abandoned mine.

It does not take long ages for buried creatures and plants to petrify. Much has been made of a miner's hat found after having been lost for several years. When it was re-located, it had completely petrified — a real "hard" hat (figure 6). Similarly, wood can petrify quickly. A farmer laid a fence in the mid 1800s, and over the years the portion above ground rotted away. But around the year 2000, a fence line of stumps was found totally petrified.[4] It doesn't take a long time to petrify something; it just takes the right conditions. Those right conditions would have often been available during the great Flood of Noah's day.

And how about animal tracks? These are found in many places and many different types of geologic layers. Sometimes the deep trails of large animals like dinosaurs were "fossilized," but often the animal was a small lizard or salamander. Worm trails and burrows were often fossilized.

When an animal makes a track, the sediment layer must be in a soft, unconsolidated condition. Later, as the sediment hardens, the track's shape is preserved. But while it was still soft, the track was fragile and subject to

3. Andrew A. Snelling, "The Whale Fossil in Diatomite, Lompoc, California," *Technical Journal* 9 no. 2 (1995): 244–258.
4. John D. Morris, "Are Human Artifacts Ever Petrified?" Institute for Creation Research, www.icr.org/article/are-human-artifacts-ever-petrified.

erosion. The question must then be raised, how long does it take for sediments to harden into sedimentary rock? Not long at all. A concrete sidewalk is essentially a man-made rock. The presence of a proper "cementing agent" is necessary, but when present, the soft cement can rapidly harden into solid "rock." Many examples of rapid solidification could be cited. It doesn't take a long time, but it does take the right conditions.

Figure 7. This fossil footprint provides trace evidence of the animal that made it.

And that's the point. Things don't necessarily take a long time to fossilize; they just take the right conditions. The conditions for rapid burial would have occurred globally across the continents at the time of the great Flood of Noah's day. Continual erosion provided the sediment to bury organisms. The proper cementing agents would also be present in the waters that transported the sediments that buried the organisms. It doesn't take long for sediments to harden if the cement is provided. The Flood also provided lots of heat, which spurs on some types of hardening. The fact that the fossils are found in profusion as they are is evidence that such conditions were often met.

Conclusion

This brief look at fossilization and these few examples are a great confirmation of the Scriptures, specifically the Flood of Noah's day. These examples can also be problematic for uniformitarianism (gradualism), which sadly, many today are taught as fact. We can have confidence in Scripture. Not only does it speak with authority about spiritual things, but when it speaks of scientific things, even fossils, it can be trusted.

> Thy word is true from the beginning: and every one of thy righteous judgments endureth for ever (Psalm 119:160; KJV).

10

What about the Similarity Between Human and Chimp DNA?

DR. DAVID A. DEWITT

The first thing I want to do is clear up a common misconception — especially among many within the Church. Many falsely believe that in an evolutionary worldview humans evolved from chimpanzees. And so they ask, "If humans came from chimps, then why are there still chimps?" However, this is not a good question to ask because an evolutionary worldview does not teach this. The evolutionists commonly teach that humans and chimpanzees are both basically "cousins" and have a common ancestor in our past. If you go back far enough, *all* life likely has a single common ancestor in the evolutionary view. This, of course, does not mesh with Genesis 1–2.

Evolutionists frequently assert that the similarity in DNA sequences provides evidence that all organisms (especially humans and chimps) are descended from a common ancestor. However, DNA similarity could just as easily be explained as the result of a common Creator.

Human designers frequently reuse the same elements and features, albeit with modifications. Since all living things share the same world, it should be expected that there would be similarities in DNA as the organisms would have similar needs. Indeed, it would be quite surprising if every living thing had completely different sequences for each protein — especially ones that carried out the same function. Organisms that have highly similar functionality and physiological needs would be expected to have a degree of DNA similarity.

What Is DNA?

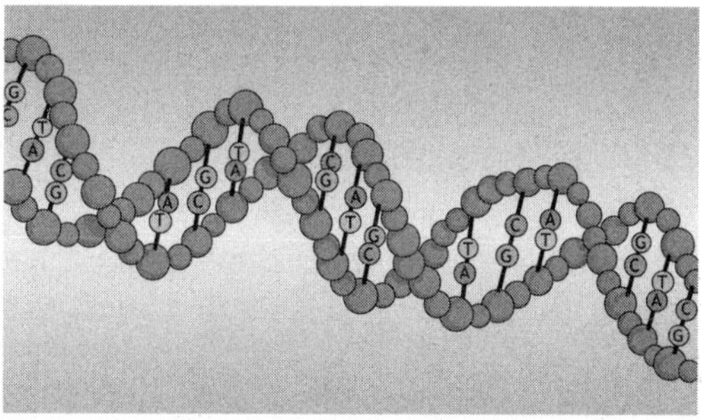

Figure 1. The double-stranded DNA molecule forms with an A opposite a T and a G opposite a C. This sequence determines the structure of proteins.

Every living cell contains DNA (deoxyribonucleic acid), which provides the hereditary instructions for living things to survive, grow, and reproduce. The DNA is comprised of chemicals called bases, which are paired and put together in double-stranded chains. There are four different bases, which are represented by the letters A, T, C, and G. Because A is always paired with T and C is always paired with G, one strand of DNA can serve as a template for producing the other strand.

The DNA is transcribed into a single chain of nucleotides called RNA (ribonucleic acid), which is then translated into the amino acid sequence of a protein. In this way, the sequence of bases in DNA determines the sequence of amino acids in a protein which in turn determines the protein structure and function.

In the human genome (total genetic information in the nucleus of the cell), there are roughly three billion base pairs of DNA with about 20,000 genes (regions that code for proteins). Surprisingly, only about 1 percent of the DNA actually codes for proteins. The rest is non-coding DNA. Some of this DNA comprises control areas — segments of DNA responsible for turning genes on and off, controlling the amount and timing of protein production. There are also portions of DNA that play structural roles. Still other regions of DNA have as yet unknown functions.

What Is the Real Percent Similarity between Humans and Chimpanzees?

Ever since the time of Darwin, evolutionary scientists have noted the anatomical (physical/visible) similarities between humans and the great apes, including chimpanzees, gorillas, and orangutans. Over the last few decades, molecular biologists have joined the fray, pointing out the similarities in DNA sequences. Previous estimates of genetic similarity between humans and chimpanzees suggested they were 98.5–99.4 percent identical.[1]

1. For example:, D.E. Wildman et al., "Implications of Natural Selection in Shaping 99.4% Nonsynonymous DNA Identity between Humans and Chimpanzees: Enlarging Genus Homo," *Proc. Natl. Acad. Sci.* 100 no. 12 (2003): 7181–7188.

Figures 2 and 3: Evolutionists believe that the similarity in the DNA sequence of gorillas, chimpanzees, and humans is proof that they all share a common ancestor (Photos: Shutterstock)

Because of this similarity, evolutionists have viewed the chimpanzee as "our closest living relative." Most early comparative studies were carried out only on genes (such as the sequence of the cytochrome c protein), which constituted only a very tiny fraction of the roughly three billion DNA base pairs that comprise our genetic blueprint. Although the full human genome sequence has been available since 2001, the whole chimpanzee genome has not. Thus, much of the previous work was based on only a fraction of the total DNA.

In the fall of 2005, in a special issue of *Nature* devoted to chimpanzees, researchers reported the draft sequence of the chimpanzee genome.[2] At the time, some researchers called it "the most dramatic confirmation yet"[3] of Darwin's theory that man shared a common ancestor with the apes. One headline read: "Charles Darwin Was Right and Chimp Gene Map Proves It."[4]

Figure 4: The journal *Nature* often trumpets the common ancestry of humans and chimps.

So what is this great and overwhelming "proof" of chimp-human

2. The Chimpanzee Sequencing and Analysis Consortium 2005, "Initial Sequence of the Chimpanzee Genome and Comparison with the Human Genome," *Nature* 437 (2005): 69–87.
3. Alan Boyle, "Chimp Genetic Code Opens Human Frontiers," MSNBC, www.msnbc.msn.com/id/9136200.
4. The Medical News, "Charles Darwin Was Right and Chimp Gene Map Proves It," www.news-medical.net/news/2005/08/31/12840.aspx.

common ancestry? Researchers found 96 percent genetic similarity and a difference between us of 4 percent. This is a very strange kind of proof because it is actually *double* the percent difference that evolutionists have claimed for years![5] Even so, no matter what the actual percent difference turned out to be, whether 2, 4, or 10 percent, they still would have claimed that Darwin was right to support their worldview.

Further, the use of percentages obscures the magnitude of the differences. For example, 1.23 percent of the differences are single base pair substitutions (figure 5).[6] This doesn't sound like much until you realize that it represents about 35 million differences! But that is only the beginning. There are 40–45 million bases present in humans that are missing from chimps and about the same number present in chimps that are absent from man. These extra DNA nucleotides are called "insertions" or "deletions" because they are thought to have been added to or lost from the original sequence. (Substitutions and insertions are compared in figure 5.) This puts the total number of DNA differences at about 125 million. However, since the insertions can be more than one nucleotide long, there are about 40 million total separate mutation events that would separate the two species in the evolutionary view.

To put this number into perspective, a typical 8½ x 11-inch page of text might have 4,000 letters and spaces. It would take 10,000 such pages full of text to equal 40 million letters! So the difference between humans and chimpanzees includes about 35 million DNA bases that are different, about 45 million in the human that are absent from the chimp, *and* about 45 million in the chimp that are absent from the human.

Creationists believe that God made Adam directly from the dust of the earth just as the Bible says in Genesis 2. Therefore, man and the apes have never had an ancestor in common. Assuming they did, for the sake of analyzing the argument, then 40 million separate mutation events would have had to take place and become fixed in the population in only 300,000 generations. This is an average of 133 mutations locked into the genome every generation. Locking

5. Studies of chimp-human similarity have typically ignored insertions and deletions although these account for most of the differences. A study by Roy Britten included these insertions and deletions and obtained a figure that is close to the 4 percent reported for the full sequence. See Roy J. Britten, "Divergence Between Samples of Chimpanzee and Human DNA Sequence Is 5% Counting Indels," *Proc. Nat. Acad. Sci.* 99 no. 21 (2002): 13633–13635.
6. Individuals within a population are variable and some chimps will have more or fewer nucleotide differences with humans. This variation accounts for a portion of the differences. 1.06 percent are believed to be fixed differences. Fixed differences represent those that are universal. In other words, all chimpanzees have a given nucleotide and all humans have a different one at the same position.

A	G	T	C	G	T	A	C	C
|	|	|	|		|	|	|	|
A	G	T	C	A	T	A	C	C

Substitution

A	G	T	C	G	T	A	C	C
|	|	|	|		|	|	|	|
A	G	T	C		T	A	C	C

Insertion/deletion

Figure 5: Comparison between a base substitution and an insertion/deletion. Two DNA sequences can be compared. If there is a difference in the nucleotides (e.g., an A instead of a G) at a given position, this is a substitution. In contrast, if there is a nucleotide base that is missing it is considered an insertion/deletion. It is assumed that a nucleotide has been inserted into one of the sequences or one has been deleted from the other. It is often too difficult to determine whether the difference is a result of an insertion or a deletion and thus it is called an "indel." Indels can be of virtually any length.

in such a staggering number of mutations in a relatively small number of generations is a problem referred to as "Haldane's dilemma."[7]

The Differences Make the Difference

There are many other differences between chimpanzee and human genomes that are not quantifiable as percentages.[8] Specific examples of these differences include:

> At the end of each chromosome is a string of repeating DNA sequences called telomeres. Chimpanzees and other apes have about 23,000 base pairs of DNA at their telomeres. Humans are unique among primates with much shorter telomeres only 10,000 long.[9]

While 18 pairs of chromosomes are virtually identical, chromosomes 4, 9, and 12 show evidence of being "remodeled."[10] In other words, the genes and markers on these chromosomes are not in the same order in the human and

7. Walter J. ReMine, "Cost Theory and the Cost of Substitution — A Clarification," *TJ* 19 no. 1 (2005): 113–125. Note also: This problem is exacerbated because most of the differences between the two organisms are likely due to neutral or random genetic drift. That refers to change in which natural selection is not operating. Without a selective advantage, it is difficult to explain how this huge number of mutations could become fixed in both populations. Instead, many of these may actually be intrinsic sequence differences present from the beginning of creation.
8. Discussed in D.A. DeWitt, "Greater than 98% Chimp/Human DNA Similarity? Not Any More," *TJ* 17 no. 1 (2003): 8–10.
9. S. Kakuo, K. Asaoka, and T. Ide, "Human Is a Unique Species Among Primates in Terms of Telomere Length," *Biochem. Biophys. Res. Commun.* 263 (1999): 308–314.
10. Ann Gibbons, "Which of Our Genes Make Us Human?" Science 281 (1998): 1432–1434.

chimpanzee. Instead of being "remodeled," as the evolutionists suggest, these could also be intrinsic differences as each was a separate creation.

Even with genetic similarity, there can be differences in the amount of specific proteins produced. Just because DNA sequences are similar does not mean that the same amounts of the proteins are produced. Such differences in protein expression can yield vastly different responses in cells. Roughly 10 percent of genes examined showed significant differences in expression levels between chimpanzees and humans.[11]

Gene families are groups of genes that have similar sequences and also similar functions. Scientists comparing the number of genes in gene families have revealed significant differences between humans and chimpanzees. Humans have 689 genes that chimps lack and chimps have 86 genes that humans lack. Such differences mean that 6 percent of the gene complement is different between humans and chimpanzees, irrespective of the individual DNA base pairs.[12]

Thus, the percentage of matching DNA is only one measure of how similar two organisms are, and not really a good one at that. There are other factors besides DNA sequence that determine an organism's phenotype (how traits are physically expressed). Indeed, even though identical twins have the same DNA sequence, as they grow older, twins show differences in protein expression.[13] Therefore, there must be some interaction between the genes and the environment.

Importantly, not all of the data support chimp-human common ancestry as nicely as evolutionists typically suggest. In particular, when scientists made a careful comparison between human, chimpanzee, and gorilla genomes, they found a significant number of genetic markers where humans matched gorillas more closely than chimpanzees! Indeed, at 18–29 percent of the genetic markers, either humans and gorillas or chimpanzees and gorillas had a closer match to each other than chimpanzees and humans.[14]

These results are certainly not what one would expect according to standard evolutionary theory. Chimpanzees and humans are supposed to share a more recent common ancestor with each other than either have with the gorilla. Trying to account for the unexpected distribution of common markers that would otherwise conflict with evolutionary predictions, the authors

11. Y. Gilad et al., "Expression Profiling in Primates Reveals a Rapid Evolution of Human Transcription Factors," *Nature* 440 (2006): 242–245.
12. J.P. Demuth et al., "The Evolution of Mammalian Gene Families," PLoS ONE 1 no. 1 (2006): e85, www.plosone.org/article/info:doi%2F10.1371%2Fjournal.pone.0000085.
13. M.F. Fraga et al., "Epigenetic Differences Arise During the Lifetime of Monozygotic Twins," *Proc. Natl. Acad. Sci.* 102 no. 30 (2005): 10,604–10,609.
14. N. Patterson et al., "Genetic Evidence for Complex Speciation of Humans and Chimpanzees," *Nature* 441 (2006): 315–321.

of this study made the bizarre suggestion: Perhaps chimpanzees and humans split off from a common ancestor, but later descendants of each reproduced to form chimp-human hybrids. Such an "explanation" appears to be an attempt to rescue the concept of chimp-human common ancestry rather than to provide the data to confirm this hypothesis.

All Similarities Are Not Equal

A high degree of sequence similarity does not equate to proteins having exactly the same function or role. For example, the FOXP2 protein, which has been shown to be involved in language, has only 2 out of about 700 amino acids which are different between chimpanzees and humans.[15] This means they are 99.7 percent identical. While this might seem like a trivial difference, consider exactly what those differences are. In the FOXP2 protein, humans have the amino acid asparagine instead of threonine at position 303 and then a serine that is in place of an asparagine at 325. Although apparently a minor alteration, the second change can make a significant difference in the way the protein functions and is regulated.[16] Thus, a very high degree of sequence similarity can be irrelevant if the amino acid that is different plays a crucial role. Indeed, many genetic defects are the result of a single change in an amino acid. For example, sickle cell anemia results from a valine replacing glutamic acid in the hemoglobin protein. It does not matter that every other amino acid is exactly the same.

Usually people think that differences in amino acid sequence only alter the three-dimensional shape of a protein. FOXP2 demonstrates how a difference in one amino acid can yield a protein that is regulated differently or has altered functions. Therefore, we should not be too quick to trivialize even very small differences in gene sequences. Further, slight differences in regions that don't code for proteins can impact how protein levels are regulated. This alteration can change the amount of protein that is produced or when it is produced. In such cases, the high degree of similarity is meaningless because of the significant functional differences that result from altered protein levels.

What about Similar "Junk DNA" in Human and Chimp DNA?

Evolutionists have suggested that there are "plagiarized mistakes" between the human and chimpanzee genome and that these are best explained by a common

15, W. Enard et al., "Molecular Evolution of FOXP2, a Gene Involved in Speech and Language," *Nature* 418 (2002): 869–872.

16. This difference in amino acid sequence opens up a potential phosphorylation site for protein kinase C. Phosphorylation is a major mechanism for regulating the activity of enzymes as well as transcription factors.

ancestor. A teacher who found identical errors on two students' papers would be rightly inclined to believe that the students cheated. The best explanation for two papers with an identical error is that they are both from the same original source. In the same way, some evolutionists have suggested that differences or deactivated genes shared by humans and chimps are best explained by common ancestry. They claim that the only alternative is a Creator who put the same error in two different organisms — a claim they would call incredible.

Evolutionists may consider something to be an error when there is a perfectly good reason that is yet unexplained. They conclude that the error is the result of an ancient mutation based on evolutionary assumptions. Further, when it comes to DNA, there may be genetic hotspots that are prone to the same mutation. For example, humans and guinea pigs share alleged mistakes in the vitamin C pseudogene without sharing a recent common ancestor.[17]

Examples of the alleged "plagiarized mistakes" are endogenous retroviruses (ERVs) — part of the so-called "junk DNA." ERVs are stretches of DNA that can be spliced (cut out), copied, and inserted into other locations within the genome. There are many different types of these mobile pieces of DNA.[18]

The ERVs are not always consistent with evolutionary expectations. For example, scientists analyzed the complement component C4 genes (an aspect of the immune system) in a variety of primates.[19] Both chimpanzees and gorillas had short C4 genes. The human gene was long because of an ERV. Interestingly, orangutans and green monkeys had the same ERV inserted at exactly the same point. This is especially significant because humans are supposed to have a more recent common ancestor with both chimpanzees and gorillas and only more distantly with orangutans. Yet the same ERV in exactly the same position would imply that humans and orangutans had the more recent common

17. Y. Inai, Y. Ohta, and M. Nishikimi, "The Whole Structure of the Human Nonfunctional L-Gulono-Gamma-Lactone Oxidase Gene — the Gene Responsible for Scurvy — and the Evolution of Repetitive Sequences Theron," *J Nutr Sci Vitimol* 49 (2003): 315–319.
18. Humans have many more short interspersed elements (SINEs) than chimps, but chimps have two novel families of retroviral elements, which are absent from man. Comparing endogenous "retroviral elements" yielded 73 human-specific insertions and 45 chimpanzee-specific insertions. Humans have two SINE (Alu) families that the chimpanzees lack and humans have significantly more copies (approximately 7,000 human-specific copies versus approx. 2,300 chimpanzee-specific ones). There are also approximately 2,000 lineage specific L1 elements. All of these lineage specific changes would be required to take place sometime between the last chimp/human common ancestor and the most recent common ancestor for all people on the planet. Importantly, these are modifications for which there is no known selective advantage.
19. A.W. Dangel et al., "Complement Component C4 Gene Intron 9 Has a Phylogenetic Marker for Primates: Long Terminal Repeats of the Endogenous Retrovirus ERV-K(C4) Are a Molecular Clock of Evolution," *Immunogenetics* 42 no. 1 (1995): 41–52.

ancestor. Here is a good case where ERVs do not line up with the expected evolutionary progression. Nonetheless, they are still held up as evidence for common ancestry.

Additional evidence has suggested that ERVs may in fact have functions.[20] One very important function has to do with implantation during pregnancy.[21]

What about the Alleged Fusion of Human Chromosome 2?

Humans normally have 23 pairs of chromosomes while chimpanzees have 24. Evolutionary scientists believe that human chromosome 2 has been formed through the fusion of two small chromosomes in an ape-like ancestor in the human lineage instead of an intrinsic difference resulting from a separate creation. While this may account for the difference in chromosome number, a clear and practical mechanism for how a chromosomal abnormality becomes universal in such a large population is lacking. The fusion would have occurred once in a single individual. Every single human being on earth would have to be a descendant of that one individual. Because there is no selective advantage to a fused chromosome, this becomes even more difficult for evolutionists to explain since natural selection would not be a factor.

Evolution proponents who insist that the chromosome 2 fusion event proves that humans and chimpanzees shared a common ancestor are employing a logical fallacy known as affirming the consequent. Affirming the consequent follows the pattern:

> If P, then Q
> Q
> Therefore, P

In other words,

> If humans and chimpanzees share a common ancestor, then there will be evidence of chromosome fusion.
>
> There is evidence of chromosome fusion.
>
> Therefore, humans and chimpanzees share a common ancestor.

Here is why it is a logical fallacy: For the sake of the argument, let us assume that humans are descended from ancestors that had 48 chromosomes just like the apes, and that there was a common ancestor five million years ago. The

20. Georgia Purdom, "Human Endogenous Retroviruses (HERVs) —Evolutionary "Junk" or God's Tools?" www.answersingenesis.org/docs2006/1219herv.asp.
21. K.A. Dunlap et al., "Endogneous Retroviruses Regulate Periimplantation Placental Growth and Differentiation," Proc. Nat. Acad. Sci. 103 no. 29 (2006): 14,390–14,395.

alleged chromosome 2 fusion would have occurred after the human line split from that of chimpanzees and been passed to all humans on the planet. Even in an evolutionary scenario, the chromosome fusion does not provide evidence for continuity between humans and chimps because it only links those individuals that share the fusion.[22]

In other words, there is no extra evidence for humans having an ancestor in common with chimpanzees provided by the fusion of chromosome 2. It is no more compelling than it would be if humans and chimpanzees had the same number — 48. One could even argue that common ancestry with chimpanzees is less compelling because of the alleged fusion on chromosome 2.

Conclusion

The similarity between human and chimpanzee DNA is really in the eye of the beholder. If you look for similarities, you can find them. But if you look for differences, you can find those as well. There are significant differences between the human and chimpanzee genomes that are not easily accounted for in an evolutionary scenario.

Creationists expect both similarities and differences, and that is exactly what we find. The fact that many humans, chimps, and other creatures share genes should be no surprise to the Christian. The differences are significant. Many in the evolutionary world like to discuss the similarities while brushing the differences aside. Emphasis on percent DNA similarity misses the point because it ignores both the magnitude of the actual differences as well as the significance of the role that single amino acid changes can play.

Please consider the implications of the worldviews that are in conflict regarding the origin of mankind. The Bible teaches that man was uniquely formed and made in the image of God (Genesis 1 and 2). The Lord directly fashioned the first man Adam from dust and the first woman Eve from Adam's side. He was intimately involved from the beginning and is still intimately involved. Keep in mind that the Lord Jesus Christ stepped into history to become a man — not a chimp — and now offers the free gift of salvation to those who receive Him.

22. There is debate among creationists as to whether the evidence for a chromosome 2 fusion event in humans is compelling. Some believe it is an intrinsic difference; others are open to it occurring early in human history, perhaps shortly before Noah. In both cases, evidence linking humans to chimpanzees based on chromosome fusion is lacking.

11

Was There Death Before Adam Sinned?

KEN HAM

Annie's cruel death destroyed Charles's tatters of beliefs in a moral, just universe. Later he would say that this period chimed the final death-knell for his Christianity. . . . Charles [Darwin] now took his stand as an unbeliever."[1]

When Charles Darwin wrote his famous book *On the Origin of Species*, he was in essence writing a history concerning death. In the conclusion of the chapter entitled "On the Imperfections of the Geological Record," Darwin wrote, "Thus, from the war of nature, from famine and death, the most exalted object which we are capable of conceiving, namely, the production of the higher animals, directly follows."[2]

From his evolutionary perspective on the origin of life, Darwin recognized that death had to be a permanent part of the world. Undoubtedly, he struggled with this issue as he sought to reconcile some sort of belief in God with the death and suffering he observed all around him, and which he believed had gone on for millions of years.

This struggle came to a climax with the death of his daughter Annie — said to be "the final death-knell for his Christianity."

Belief in evolution and/or millions of years necessitates that death has been a part of history since life first appeared on this planet. The fossil layers (con-

1. A. Desmond and J. Moore, *Darwin: The Life of a Tormented Evolutionist* (New York, NY: W.W. Norton & Company, 1991), p. 387.
2. C. Darwin, *On the Origin of Species* (Cambridge, MA: Harvard University Press, 1964), p. 490.

taining billions of dead things) supposedly represent the history of life over millions of years. As Carl Sagan is reported to have said, "The secrets of evolution are time and death."[3]

Time and Death

This phrase sums up the history of death according to those who believe in evolution and/or millions of years. In this system of belief:

- death, suffering, and disease over millions of years led up to man's emergence;
- death, suffering, and disease exist in this present world; and
- death, suffering, and disease will continue on into the unknown future. Death is a permanent part of history.

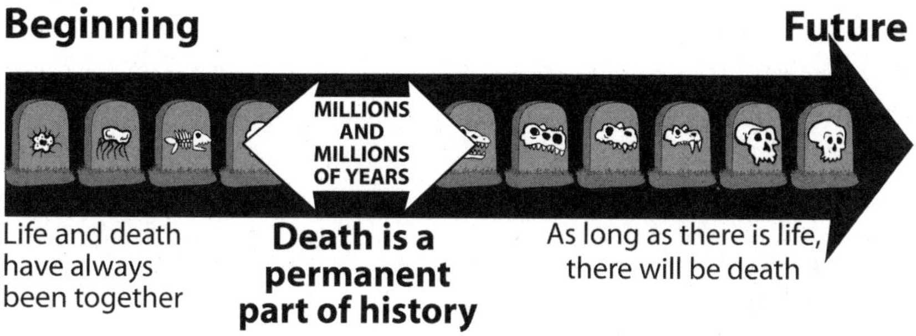
Beginning
Future
MILLIONS AND MILLIONS OF YEARS
Life and death have always been together
Death is a permanent part of history
As long as there is life, there will be death

Sin and Death

Rather than "time and death," the phrase "sin and death" sums up the history of death according to the Bible. From a perspective of the literal history of the

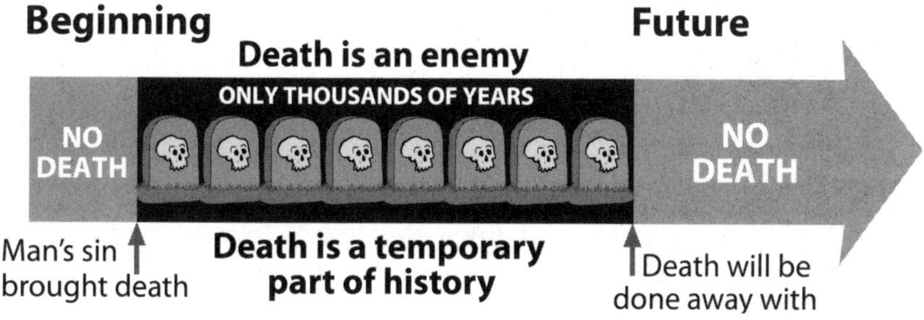
Beginning
Future
Death is an enemy
ONLY THOUSANDS OF YEARS
NO DEATH
NO DEATH
Man's sin brought death
Death is a temporary part of history
Death will be done away with

3. C. Sagan, *Cosmos, Part 2: One Voice in the Cosmic Fugue*, produced by Public Broadcasting Company, Los Angeles, with affiliate station KCET, and first aired in 1980 on PBS stations throughout the United States.

Book of Genesis, there was a perfect world to start with — described by God as "very good" (Genesis 1:31) — but it was marred because of Adam's rebellion. Sin and its consequence of death entered the world that was once a paradise (Romans 5:12 ff., 8:20–22; 1 Corinthians 15:21–22).

In 1 Corinthians 15:26, Paul describes death as the "last enemy." And that's the point — death is an enemy — it's an intrusion. The death of man and the animals was not part of the original creation. And even though death reigns in this present world, one day in the future there will be no more death: "And God shall wipe away all tears from their eyes; and there shall be no more death, neither sorrow, nor crying, neither shall there be any more pain: for the former things are passed away" (Revelation 21:4; KJV).

The idea of millions of years came from the belief that most of the fossil-bearing layers were laid down millions of years before man existed. Those Christians who accept the idea of millions of years and try to fit it into the Bible also must accept death of animals, disease, suffering, thorns, and animals eating each other before sin. But all of this flies in the face of the clear teaching of Scripture that such things could not have existed until after sin.

Consider the following biblical truths in support of that conclusion.

Human Death

Scripture makes it very clear there could not have been human death (physical death) before Adam sinned. For example, Romans 5:12 states:

> Therefore, just as through one man sin entered the world, and death through sin, and thus death spread to all men, because all sinned. . . .

This "death" referred to in Romans 5 cannot have just been "spiritual" death, but also included physical death. The context confirms this. In verses 6–11 the Apostle Paul speaks repeatedly of Christ dying for us, and of someone dying for a good man. Christ did not merely die spiritually on the cross, but physically. When we go back to Genesis we find that after Adam sinned God said:

> Because you have heeded the voice of your wife, and have eaten from the tree of which I commanded you, saying, "You shall not eat of it": Cursed is the ground for your sake; in toil you shall eat of it all the days of your life. . . . In the sweat of your face you shall eat bread till you return to the ground, for out of it you were taken; for dust you are, and to dust you shall return (Genesis 3:17, 19).

God decreed that our bodies would return to dust (physical death) as a result of sin. There is no doubt there could not have been human death before sin.

Animal Death

Unlike the case of Romans 5:12, there is no verse of Scripture that specifically teaches that there was no animal death before sin. However, there are passages of Scripture that, when taken together, lead us to conclude this.

First, it should be noted that the Bible is not about animals — it is about man and his relationship with God. Thus, we would not expect as much specific teaching concerning animals as there is about man. However, consider the following passages:

> A. Genesis 1:29–30 — And God said [to Adam and Eve], "See, I have given you every herb that yields seed which is on the face of all the earth, and every tree whose fruit yields seed; to you it shall be for food. Also, to every beast of the earth, to every bird of the air, and to everything that creeps on the earth, in which there is life, I have given every green herb for food"; and it was so.

These verses seem to clearly teach that man and the animals were to be vegetarian originally. This is confirmed by the fact that in Genesis 9:3 after the Flood, concerning the diet of man, God said to Noah, "Every moving thing that lives shall be food for you. I have given you all things, even as the green herbs."

In other words, it is clear that originally man was to be vegetarian, but now God changed that diet so that man could eat the flesh of animals. As Genesis 1:30 concerns the diet of animals, and it is connected to Genesis 1:29, it is a strong indication that the animals were to be vegetarian originally (before sin).

Problem: For those Christians who believe in millions of years, the fossil record that is claimed by secularists to be millions of years old has in it numerous examples of animals having eaten other animals — supposedly millions of years before man! This is contrary to the Bible's clear teaching that animals were vegetarian originally (before sin).

> B. Genesis 1:31 — Then God saw everything that He had made, and indeed it was very good. So the evening and the morning were the sixth day.

At the end of the sixth day of creation, God described the entire creation as "very good." However, in the fossil record, there are many examples of diseases in the bones of animals (e.g., tumors [cancer]; arthritis; abscesses etc). Such diseases could not be described as "very good," when in the rest of the Bible diseases are always viewed as bad and a result of sin and the curse. These diseases simply could not have existed before sin, if the Bible is true (and it is).

Problem: Those Christians who believe in millions of years for most of the fossil layers to form must accept that diseases like cancer were in the bones of animals before sin, and that God described such diseases as "very good."

C. Romans 8:20–22 — For the creation was subjected to futility, not willingly, but because of Him who subjected it in hope; because the creation itself also will be delivered from the bondage of corruption into the glorious liberty of the children of God. For we know that the whole creation groans and labors with birth pangs together until now.

Paul makes it clear in Romans 8 that the "whole creation" groans because of sin. Most commentators on this passage in the history of the Church have interpreted this "whole creation" to refer to the whole non-human creation (including the animals).[4] That is the only interpretation that makes sense. First, Paul already established the connection between sin and human death in Romans 5. Second, the reference to "birth pangs" (8:22) seems to be an allusion to the judgment on Eve in Genesis 3:16. Also, the groaning of the creation is linked in this passage (Romans 8:18–25) to the groaning of believers in this sinful world. Furthermore, the liberation of the whole creation will happen with the future final redemption of believers (when they get their resurrection bodies) at the Second Coming of Jesus Christ (Romans 8:23–25). Since the Christian's and the creation's liberation are linked, it is most reasonable theologically to conclude that they came into bondage to corruption at the same time also.

Finally, if we reject that conclusion and imagine that the whole creation was in bondage to corruption as soon as God created it (Genesis 1), then what kind of God would He be to call that corruption "very good"? So the whole originally perfect creation was put into bondage to corruption by God's curse recorded in Genesis 3.

Problem: Christians who believe in millions of years have to accept animals eating each other, diseases like cancer, and animals dying and going extinct over the course of millions of years before man, and then on into the present. This would mean that the Fall of man didn't change anything, and that God described all this death and disease as "very good." In this case, the creation is not "groaning" because of sin. But as we have seen, Paul makes it

4. Douglass Moo, *The Epistle to the Romans* (Grand Rapids, MI: Eerdmans, 1996), p. 514; Thane H. Ury, "Luther, Calvin, and Wesley on the Genesis of Natural Evil: Recovering Lost Rubrics for Defending a *Very Good* Creation," in Terry Mortenson and Thane H. Ury, eds., *Coming to Grips with Genesis* (Green Forest, AR: Master Books, 2008), p. 399–423.

clear the creation is groaning because of sin. This only makes sense if a "very good" creation (perfect creation — no death, disease, suffering, etc.) was subject to "futility" and now "groans" because sin changed everything.

D. Acts 3:21 — . . . whom heaven must receive until the times of restoration of all things, which God has spoken by the mouth of all His holy prophets since the world began.

The Bible teaches that there will one day be a "restoration" of "all things." This is because something happened (sin) to cause a problem (the whole creation groans). We look forward to a new heaven and new earth where there will be no death (which the Bible describes as an enemy) or suffering, because there will be no more Curse (Revelation 21:3–5, 22:3). It will be a perfect place — just as everything was once perfect before sin.

Problem: For those Christians who believe in millions of years, and thus must accept death, disease, and suffering of animals before sin, what will this restoration look like? More death and suffering and disease for millions of years or forever? That would be a horrible prospect. No, the restoration will look like things were before sin — all was "very good." And that indeed is something to look forward to!

From the above and other passages of Scripture (including Colossians 1:15–20, which speaks of Jesus Christ as the Creator and Redeemer of "all things"), we have good reasons to believe that animals could not have eaten other animals and died of diseases before sin. The only other ways animals could have died would be from old age (wearing out) or accidents (catastrophes, etc.) — but these would not fit with everything originally being "very good," and would not fit with Paul's teaching in Romans 8 that the whole creation groans now because of sin.

We can therefore conclude with confidence there was no animal death before sin.

Plant Death

Some people argue that there was death before sin, because plants were given for food for man and the animals (Genesis 1:29–30), thus plants died before sin.

However, this objection fails to note carefully what the Bible says about life and death. Biblically speaking, plants do not have a life, as animals and man do. At the end of Genesis 1:30 we see that humans and animals have "life," but plants do not. The word "life" is a translation of two Hebrew words there: *nephesh chayyah*. *Nephesh* is the word usually translated "soul" or "creature" depending on context, and *chayyah* is the noun form of the verb "to live." *Nephesh* or *nephesh chayyah* is never used to describe plants in the Old Testament. They only describe people and animals. Just as plants are not "alive" in the same sense as animals and man are, so also they do not "die" in the same sense. In only one place does the Old Testament use the Hebrew word for "die" (*mut*) when referring to plants, and in that passage (Job 14:7–12) it is very clear that the death of a plant (tree) is categorically different from the death of a man. So when animals and people ate plants in the world before sin, it did not involve death, because plants do not "die" in the sense that man and animals do.

Implications

If the Bible makes it clear there was no animal death and disease and no carnivorous animals before sin, then we cannot add millions of years into the Bible — to do so undermines the authority of Scripture, and comes with severe implications.

In reality, the battle between creation and evolution, between young-earth and old-earth views, is in fact a battle between two totally different histories of death.

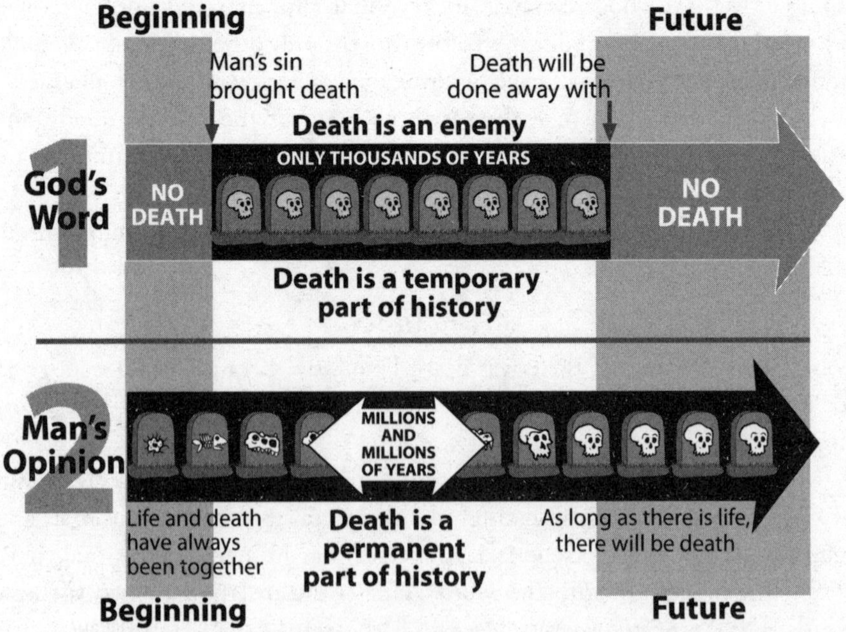
Beginning
Future
Man's sin brought death
Death will be done away with
Death is an enemy
ONLY THOUSANDS OF YEARS
1
God's Word
NO DEATH
NO DEATH
Death is a temporary part of history
2
Man's Opinion
MILLIONS AND MILLIONS OF YEARS
Life and death have always been together
Death is a permanent part of history
As long as there is life, there will be death
Beginning
Future

For the Christian, which history of death you accept has major theological implications.

1. If a Christian accepts the history of death over millions of years, then when God stated in Genesis 1:31 that everything He had made was "very good," this would mean that death, suffering, violence, and diseases like cancer (as represented in the fossil record) were also "very good." This situation is represented in the following diagram:

This view of history, if consistently applied, would lead to the situation summed up by the heretical Bishop John Shelby Spong:

> But Charles Darwin says that there was no perfect creation because it is not yet finished. It is still unfolding. And there was no perfect human life which then corrupted itself and fell into sin. . . . And so the story of Jesus who comes to rescue us from the fall becomes a nonsensical story. So how can we tell the Jesus story with integrity and with power, against the background of a humanity that is not fallen but is simply unfinished?[5]

Bishop Spong accepts the history of death over millions of years. As a result of this, he cannot accept a perfect creation that was marred by sin. Thus, the groaning (death and suffering, etc.) we observe today has continued for millions of years. This is also true of all "long-age creationists." These are those who accept the secular belief in an old world, while opposing evolution in favor of "progressive creation" or "intelligent design."

2. However, if a Christian accepts the history of death as given by a literal reading of the Genesis account, then this history can be represented by the following diagram:

5. Australian Broadcasting Corporation TV "Compass" interview with Bishop John Shelby Spong, by Geraldine Doogue, in front of a live audience at the Eugene Groosen Hall, ABC Studios, Ultimo, Sydney, July 8, 2001. From a transcript at www.abc.net.au/compass/intervs/spong2001.htm.

The perfect creation with no death, disease, or suffering is described as "very good." The Bible makes it clear that God does not delight in death. We read in Ezekiel 33:11, "Say to them, 'As I live,' says the Lord GOD, 'I have no pleasure in the death of the wicked, but that the wicked turn from his way and live. Turn, turn you from your evil ways! For why should you die, O house of Israel?' " God takes no pleasure in the afflictions and calamities (death, etc.) of people.

The Bible makes it obvious that death is the penalty for our sin. In other words, it is really our fault that the world is the way it is — God is a loving, merciful God. When we sinned in Adam, we effectively said that we wanted life without God. All of us also sin individually (Romans 3:23). God had to judge sin, as He warned Adam He would (Genesis 2:17, cf. 3:19). In doing so, God has given us a taste of life without Him — a world that is running down — a world full of death and suffering. As Romans 8:22 says, "The whole creation groans and labors with birth pangs." Man, in essence, forfeited his right to live.

However, even though we are sinners, those who have turned from their sin and trusted Christ for forgiveness will spend eternity with their Creator in a place where righteousness dwells — and there will be no more crying, suffering, or death.

The true history of death, as understood from a literal Genesis, enables us to recognize a loving Creator who hates death, the enemy that will one day be thrown into the lake of fire (Revelation 20:14).

Which history of death do you accept? Is it one that makes God an ogre responsible for millions of years of death, disease, and suffering? Or is it one that correctly places the blame on our sin, and correctly represents our Creator God as a loving, merciful Savior who wept at the tomb of dead Lazarus (John 11:35)?

12

Abortion: Is It Really a Matter of Life and Death?

PAUL F. TAYLOR

The whole subject of abortion[1] produces very strong emotions on both sides of the argument. The two primary sides are:

Pro-life: The pro-life position is that life begins at fertilization, and that all human life is precious and made in the image of God.

Pro-choice (or more so "anti-life"): The pro-choice position is that it is the woman's right to choose whether or not to have an abortion, because an unborn child is considered to be a part of the woman's body. Under this definition, the unborn child is not considered to be fully human.

You would think that pro-choice meant that someone would allow the baby to choose whether he or she should live or die (miscarry), but that is not the case. Even while the baby is choosing to live and continuing to develop, some do not respect *that* choice. And that has brought us to the heat of a debate that rages around the world.

1. If you need help and support as a result of the issues raised in this chapter, please contact one of the following organizations: UK: CARE (Christian Action Research and Education) (www.care.org.uk), IMAGE (www.imagenet.org.uk), Society for the Protection of the Unborn Child (www.spuc.org.uk), LIFE (www.lifeuk.org), US: Heartbeat International (www.heartbeatinternational.org), Care Net (www.care-net.org), National Right to Life (www.nrlc.org). Or feel free to contact one of these organizations to find something closer to home as well.

Such emotions are understandable and can cloud the debate, hiding the truth of what the Bible teaches. However, as this chapter will hopefully make clear, emotional responses to the subject of abortion are not necessarily inappropriate — indeed, such responses may be the most appropriate. Also, an acknowledgment that emotional issues cloud both sides of the debate should not be taken to imply that this chapter will steer a "middle ground" between the two positions. It will not — because the Bible does not do so.

The emotional arguments against abortion include a disgust at the nature of the procedure being discussed. Emotional arguments in favor of abortion focus on an anger that suggests that no one has the right to undermine a woman's right to choose what she does with her own body.

Although this essay is not designed to steer a middle way, it will be necessary to examine some issues dispassionately. This is not because I believe the subject does not demand one's emotions, but because I want to start by cutting through the emotional charge and examining the issues from a "first-principles" biblical perspective. Only when this foundation is laid can we return to the issue of which emotional responses may be appropriate.

Life Before Birth

Of crucial importance to the debate is the status of the embryo, fetus, or baby before birth. Please forgive the coldness of the question — but what exactly is it? Should we refer to it as it, or is it a he or she?

The Bible does not directly refer to abortion. There are many other issues about which the Bible does not give specific comment. However, in many cases, it is clear what the biblical position is. And the Bible does have a great deal to say about the status of life before birth. In Jeremiah 1:4–5 we read:

> Then the word of the Lord came to me, saying:
> "Before I formed you in the womb I knew you;
> Before you were born I sanctified you;
> I ordained you a prophet to the nations."

The Lord is giving a number of pieces of information to the prophet. First, God says that He knew Jeremiah when he was in the womb. Second, He makes clear that He knew Jeremiah even before He was formed in the womb. Third, He tells Jeremiah that his growth in the womb was as a result of being "formed" by God Himself.

Today, we have a great deal of knowledge of how a baby develops in his or her mother's womb. In this passage, God is making clear that this is not an arbitrary process. It is a direct act of formation by God. The Hebrew word that

AFTER EDEN
by Dan Lietha
ISN'T IT AMAZING THAT THE GOD WHO SPOKE THE UNIVERSE INTO EXISTENCE, KNOWS ABOUT AND CARES ABOUT EACH ONE OF US?!
IN FACT, HE KNOWS EACH ONE OF US BY NAME!
WOW, I'M IMPRESSED! MY PARENTS DON'T EVEN KNOW WHAT MY NAME IS YET!
Names For Your Baby
© 2002 AIG
Psalm 139:13-18

is translated as formed is *yatsar*, and refers to being formed or shaped in the same sort of manner that a potter shapes clay. This analogy is interesting, because the image of God as a potter is closely associated with the Book of Jeremiah. Jeremiah 18 is the famous chapter that talks about the potter and the clay. It is significant that a similar image is being used of an unborn child in Jeremiah 1:5.

The passage implies that there is a personhood associated with the unborn Jeremiah. Therefore, the unborn child should be considered as a full human being, with all the implications that the fact entails. We need to examine whether other passages of Scripture make a similar assumption of personhood for other characters, and, hence, whether we can determine if the Bible counts unborn babies as human beings.

Jesus and John

Scripture makes clear that both Jesus and John the Baptist were human before their birth. Jesus was given a name, and His birth was foretold to Mary, at the time of His conception, as recorded in Luke 1:26–38. Some might want to argue, however, that Jesus was a special case. However, no special case argument can be made to apply to John, the account of whose birth is closely wound up with the account of Jesus' birth.

In Luke 1:41 we note that Elizabeth was "filled with the Holy Spirit." She was immediately able to ascertain that Mary was pregnant with the Messiah.

> Why is this granted to me, that the mother of my Lord should come to me? (Luke 1:43).

What is interesting about this passage is that the unborn John joins in the celebration.

> For indeed, as soon as the voice of your greeting sounded in my ears, the babe leaped in my womb for joy (Luke 1:44).

John does not just leap — he leaps for joy! Under the inspiration of the Holy Spirit, this Scripture has been recorded in order to emphasize that John's prophetic work in "preparing the way of the Lord" was beginning before his birth. Therefore, John must have been fully human before his birth.

Mosaic Law

There is an interesting account in the Mosaic Law about the various penalties for different types of murders.

> If men fight, and hurt a woman with child, so that she gives birth prematurely, yet no harm follows, he shall surely be punished accordingly as the woman's husband imposes on him; and he shall pay as the judges determine. But if any harm follows, then you shall give life for life, eye for eye, tooth for tooth, hand for hand, foot for foot, burn for burn, wound for wound, stripe for stripe (Exodus 21:22–25).

When examining Mosaic Law, it is important to remember that the penalties prescribed do not necessarily apply to today, because these laws were civil laws for the children of Israel. For example, because the Church does not hold the sword of the state today, we are not entitled to legislate stoning for adultery. Nevertheless, the fact that stoning is the punishment prescribed for adultery in the Mosaic theocracy illustrates to us how seriously God views that particular sin.

So when we analyze the passage from Exodus 21 quoted above, we see that there are differing sanctions, based on differing circumstances. In the first case analyzed, we have a pregnant woman who is hurt and gives birth prematurely. In this case, however, the baby is not harmed. So the offense is treated in the same manner as it would if the woman had not been pregnant.

The situation changes notably if harm comes to the baby. On this occasion, there is to be recompense of the "eye for an eye" model. This is not to suppose that we are entitled to use the same sanctions today. Nevertheless, the concept of "life for life" illustrates that God considers the death of the unborn to be equivalent to the death of the living. Accordingly, a society should reflect this value in its laws, even if the sanction prescribed is different.

What we have seen from this analysis of Bible passages is that the Bible considers the unborn baby to be human and to have personality, and that God views the value of the life of the unborn, when it is prematurely harmed, to be of equal value to that of any other human being.

Amazingly, this passage has actually been used by some to attempt to condone abortion. This is because of a mistranslation in certain modern versions of the Bible. For example, the Message Bible has:

> When there's a fight and in the fight a pregnant woman is hit so that she *miscarries* but is not otherwise hurt . . . [emphasis added]

The Message Bible puts the emphasis on the harm to the woman, whereas other editions emphasize the harm to both mother and baby. The Hebrew term translated either as premature birth or miscarriage is *yatsa*. This word, which means "to come out," is used many times in the Old Testament, and in each case always refers to a whole birth. It usually refers to a live birth, though one passage refers to a still birth. In no other place, however, is the term used for a miscarriage.[2]

Fearfully and Wonderfully Made

The most famous passage referring to the life of the unborn must be from Psalm 139.

> For You formed my inward parts; You covered me in my mother's womb. I will praise You, for I am fearfully and wonderfully made; Marvelous are Your works, And that my soul knows very well. My frame was not hidden from You, When I was made in secret, And skillfully wrought in the lowest parts of the earth. Your eyes saw my substance, being yet unformed. And in Your book they all were written, The days fashioned for me, When as yet there were none of them (Psalm 139:13–16).

This is a pictorial account of the development of an unborn baby. It refers to the formation of flesh (covering), internal organs (inward parts), and bones (frame). None of these developments was hidden from God, though they were "secret" from people, indicating that we cannot directly see the formation of the unborn. The concept of the "lowest parts of the earth" is a euphemism for the female reproductive system. Even in this unborn state, it is clear that the baby is human, as God has already determined "the days fashioned" for the baby.

What these passages from Scripture show us is that the unborn baby has personality and sensitivity before birth. It is therefore human, and subject to all the protections of the moral laws that protect other humans. If the unborn baby was an integral part of the woman's body, then it would not have the separate actions and reactions outlined in these scriptural passages. Viewing the evidence that shows that unborn babies can react to external stimuli, such as light and sound, is a further confirmation of their unique life apart from the mother.

Caring for the Mother

An argument frequently used in favor of abortion is that we need to have concern for the mother. Abortion was supposedly legalized in the UK and the United States to alleviate the suffering of women undergoing crisis pregnancies.

2. G. Butner, "Exodus 21:22–25: Translations and Mistranslations", www.errantskeptics.org/Exodus2122.htm.

Such crises in pregnancies are very real. Women can be in very real distress during times of pregnancy, particularly if the pregnancy is not planned, or is going wrong because of illness, etc.

Nevertheless, a lot of the difficult cases become clearer once we have determined from Scripture, as above, that the unborn baby is human. Both the UK's Abortion Act of 1967, and the famous U.S. case of Roe v. Wade, were supposed to eliminate dangerous backstreet abortions, and reduce difficult cases, without being used as a general abortion-on-demand measure. Nevertheless, the practical outworking of these laws on both sides of the Atlantic has been startling.

David Reardon has suggested that many women get abortions because they feel under pressure to do so.[3] Some such pressures he identifies as circumstantial — women concerned about how they might cope, financially, emotionally, etc. But many more pressures come from other people. He particularly notes that the pressures frequently come from men — husbands, boyfriends, fathers, etc. Women are often coming under pressure to "do the right thing," even if they have severe doubts. This is one of the factors, Reardon notes, which has made Post-Abortion Trauma such a major psychological illness among women in the last 20 years or so. Reardon's studies suggested that 53 percent of women felt coerced into abortion by other people, and 65 percent by circumstances (obviously some overlap here). Only 33 percent had felt that their abortion was a "free" choice.

In the case of coercion by others, it can be seen that abortion is frequently not even an answer to this coercion. Many women have had abortions because of pressure from male partners in the hope of saving their relationships, only to find that the partner leaves anyway.

In the case of coercion by circumstances, it is my belief that pro-life Christians need to be pro-active in providing help and care for mothers undergoing crisis pregnancies. Is the proposed abortion happening because the mother cannot afford baby equipment and care? Then Christians should be providing that equipment and care. Will the mother be thrown out of her home if she proceeds with the pregnancy? Then Christians must provide emergency refuge and shelter.

Reardon's study, which examined women whose abortions had been about ten years previous to the study, also noted that adolescent women (aged 20 or under) were frequently likely to leave abortions to later in gestation, due to reduced ability to make decisions. This immaturity among younger women led to a greater likelihood of post-abortion trauma, and also physical issues, such as a high rate of subsequent infertility. The work of Christian post-abortion counselors, such as Image (see reference 1), has shown that women can be most

3. D.C. Reardon, "Women at Risk of Post-Abortion Trauma," www.afterabortion.org/women_a.html.

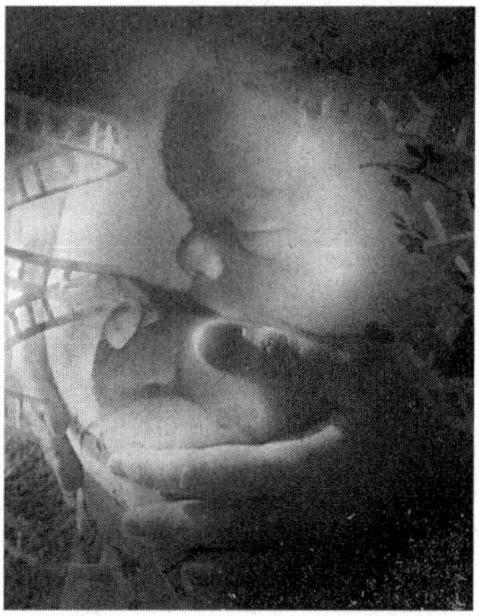

helped through the application of God's forgiveness, when the woman repents.

Reasons for Abortion

In 2007, 205,598 abortions were carried out in England and Wales,[4] and 13,703 in Scotland.[5] This UK total[6] of 219,301 compares with 23,641 in 1968. There are currently more than 600 abortions performed per day in the UK. Of these figures, 82 percent were performed on single women. About 1 percent of abortions were performed because of suspected handicap in the unborn child. One in five pregnancies in the UK ends in abortion. Abortion law was further liberalized under the 1990 Human Fertilization and Embryology Bill, with the result that, in certain cases, abortion can be carried out up to full term. Statistics like these seem to run counter to the generally held mythology that legalized abortions are not carried out for social reasons. Indeed, one top surgeon has recently criticized the "cavalier" way that young surgeons carry out abortions, complaining, "I know of no case where the Department of Health has questioned the legality of abortions."[7]

Social justifications for abortion would seem to be of secondary importance, if the unborn baby is defined as human. Yet the overwhelming majority of abortions carried out in the UK are for "social reasons" — government statistics suggest that 98 percent of all abortions are for social reasons.[8] The earlier sections have shown that abortions are not even in the interest of the mother, when one considers the violence that can be done to the body, the risk for young adolescent pregnant women, and the dangers of post-abortion trauma. However, many difficult cases continue to be cited, so it is worth examining the practical outcome of a couple of these.

Anecdotal evidence suggests that abortions are often offered to mothers when Down's Syndrome is suspected. Indeed, in the UK, nine out of ten babies

4. It is common in the UK for England and Wales statistics to be grouped, with the other states' figures quoted separately.
5. Abortion in Britain, Image (an evangelical, pro-life organization), www.imagenet.org.uk/pages/abortionfactsheet.php.
6. Abortion is illegal in Northern Ireland, though 1,343 women traveled to the mainland for abortions in 2007.
7. "Top Surgeon Tells Court that Junior Doctors Are 'Cavalier' Over Abortions," Christian Concern For Our Nation, www.ccfon.org/view.php?id=751.
8. Abortion Statistics, England and Wales 2006, Dept of Health June 2007, para 4.2.2.

suffering from Down's Syndrome are aborted.[9] The attitude frequently seems to be that it is "kinder" in some way for such a child not to live, because of its "quality of life." But the people concerned — the "sufferers" of Down's Syndrome — may have very different opinions about their quality of life. The issue of "quality of life" is an evolutionary concept and has no place in a biblical worldview, which sees all human life as being in the image of God.

Anya Souza — a Down Syndrome sufferer — was allowed to address the 2003 International Down Syndrome Screening Conference in London. She said:

> I can't get rid of my Down's Syndrome, but you can't get rid of my happiness. You can't get rid of the happiness I give others either. It's doctors like you that want to test pregnant women and stop people like me being born. Together with my family and friends I have fought to prevent my separation from normal society. I have fought for my rights. . . . I may have Down's syndrome but I am a person first.[10]

Another set of difficult cases often cited in support of abortion "rights" is what to do about pregnancies resulting from incest or rape. In these cases, it is clear that a crime has taken place — and that crime could well have been a very violent crime. The woman concerned has been violated, and is clearly already going to be suffering as a result of what has happened to her.

Abortion itself is an act of violence on the unborn baby (and the mother). It is not clear that the difficulties of undergoing an abortion could be in any way a comfort to the woman who has suffered the crimes of incest or rape. Moreover, the unborn baby is an innocent party to the event. It does not make sense to end the life of the innocent party because of another act of violence. Add to this the dangers that the mothers themselves may suffer, as stated above — such as infertility and post-abortion trauma.

The Life of the Mother

All human life is valuable. The unborn baby's life is precious — and so is the mother's. There are certainly a precious few occasions when, tragically, there is a choice between the life of the baby and the life of the mother. It may be necessary, under these extreme conditions, to consider saving the life of the mother or the child. These tragic situations arise because we live in a fallen world.

One example of the above would be an ectopic pregnancy, where the unborn baby has started to develop in the fallopian tube, rather than in the

9. D. Mutton et al., "Trends in Prenatal Screening for, and Diagnosis of, Down's Syndrome: England and Wales, 1989–97," British Medical Journal, October 3, 1998.
10. "Ability and Disability or Eugenic Abortion," Society for the Protection of the Unborn Child, www.spuc.org.uk/students/abortion/disability.

uterus. It may not be possible to move the baby, and the baby would, in any case, die in such circumstances, as would the mother. Sadly, it may be necessary for the baby to be removed surgically, which will result in his death. With this situation though, it is a matter of trying to save a life or two, as opposed to forcing death on one or both of them.

Other circumstances can be more complicated. When there is a tragic choice between saving either the baby or the mother, but it is impossible to do both, then individual families will need, prayerfully, to come to their own decisions on this matter, and no one would be able to criticize their painful choice. It is fortunate that such events are very rare — about 0.004 percent of all cases involve the possible death of the mother.[11]

Language of Abortion

The issue of the personhood of Ms. Souza leads us to examine the use of terminology in the abortion debate. The terminology is important, because language that denies the humanity of the unborn child makes it easier for abortionists to make their case.

The unborn baby is often referred to using two terms. *Embryo* indicates the fertilized product of conception from implantation to eight weeks. *Fetus* (or *foetus*) indicates the baby from the eighth week to birth. Such terms are often easier to use, if the baby is to be terminated, as they do not sound human. The etymology of the latter term is interesting — *fetus* means "little one."

There is also the word *conception*. It always has been, and to most people still is, the combination of the sperm and egg — or fertilization. But the 27th edition of *Stedman's Medical Dictionary* now defines *conception* as implantation of the combined sperm and egg (that must be over 4–5 days old [blastocyte]) when it attaches to the lining of the uterus (endometrium). This has now led to people aborting children with "morning after pills," cloning of humans, and embryonic stem cell research all the while declaring that it is "before conception."

The Real Issue

As with so many cases, we find that abortion is not the real problem. The real problem is much deeper, and abortion is a symptom of the deeper problem. A society that permits abortion does not do so by chance. It is a society that has neglected the fundamentals of God's law. The basis for our objection to abortion has been the biblical position that the unborn baby is human. However, in an evolutionary view, why should any human be accorded special status,

11. Dr. Michael Jarmulowicz, cited in "The Physical and Psycho-Social effects of Abortion on Women: A Report by the Commission of Inquiry into the Operation and Consequences of The Abortion Act," June 1994, p. 5.

compared with, for instance, the welfare of animals? To put it crudely, if an animal is sick or injured, we will often take it to the vet to be "put down."

The difference between the welfare of humans and animals stems back to Genesis. Humans were not made *ex nihilo* in the way that animals were. The first man was fashioned out of the dust, and God breathed into him the breath of life (Genesis 2:7). The first chapter of the Bible reminds us that God made us in His image (Genesis 1:26). This statement was not made of any other animal.

Evolutionary beliefs have influenced us to think that we are simply evolved animals — that we share a common ancestor with the apes — indeed, further back, we are supposed to share a common ancestor with all mammals. As one modern and rather base pop song puts it — "you and me baby ain't nothin' but mammals." If that is the case, then the arguments against abortion become hollow. Even if the unborn baby is human, such humans are dispensable if we are just mammals. The dignity of human life means nothing if humans have evolved by millions of years of death, disease, and bloodshed.

The Bible's position is vastly different. We did not evolve by millions of years of death, disease, and bloodshed; we are not just animals. We are special because we are made in the image of God. We are fallen from that image, certainly, but that image still sets us apart from the animals. Our certainty of the truth of Genesis provides us with the assurance that we are human, and that our humanity began at the moment of conception. It is for that reason that we oppose abortion, because it is a denial of the humanity of the unborn baby.

Further Study

Because this short chapter can only cover so much, I want to encourage you to do further study. I suggest the following resources:

www.answersingenesis.org/get-answers/topic/abortion-euthanasia

David Menton, *Fearfully and Wonderfully Made*, DVD, Answers in Genesis, 2005.

Mike Riddle, *Cloning, Stem Cells, and the Value of Life*, DVD, Answers in Genesis, 2007.

Tommy Mitchell, "When Does Life Begin," *The New Answers Book 2* (Green Forest, AR: Master Books, 2008), p. 313–323.

13

Is the Christian Worldview Logical?

DR. JASON LISLE

Many people have the impression that Christians live in two "worlds" — the world of faith and the world of reason. The world of faith is the realm that Christians live in on Sunday morning, or the world to which they refer when asked about spiritual or moral matters. However, it would seem that Christians live in the world of reason throughout the rest of the week, when dealing with practical, everyday matters. After all, do we really need to believe in the Bible to put gasoline in the car, or to balance our checkbook?

Misconceptions of Faith

The notion of "faith versus reason" is an example of a *false dichotomy*. Faith is not antagonistic to reason. On the contrary, biblical faith and reason go well together. The problem lies in the fact that many people have a misunderstanding of *faith*. Faith is not a belief in the absurd, nor is it a belief in something simply for the sake of believing it. Rather, faith is having confidence in something that we have not perceived with the senses. This is the biblical definition of faith, and follows from Hebrews 11:1. Whenever we have confidence in something that we cannot see, hear, taste, smell, or touch, we are acting upon a type of faith. All people have faith, even if it is not a saving faith in God.

For example, people believe in laws of logic. However, laws of logic are not material. They are abstract and cannot be experienced by the senses. We can write down a law of logic such as the law of non-contradiction ("It is impossible to have **A** and **not A** at the same time and in the same relationship."), but the sentence is only a physical representation of the law, not the law itself.[1] When people use laws of logic, they have confidence in something they cannot actually observe with the senses; this is a type of faith.

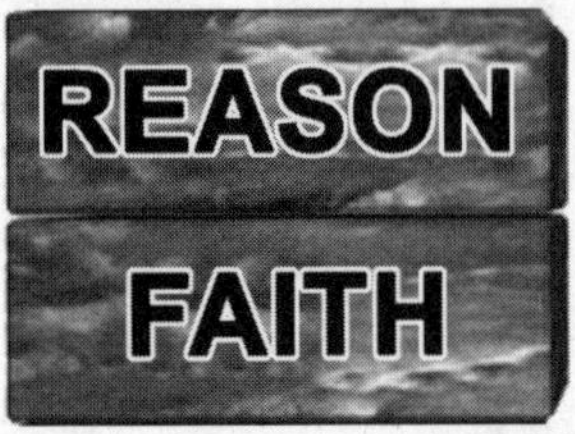

When we have confidence that the universe will operate in the future as it has in the past, we are acting on faith. For example, we all presume that gravity will work the same next Friday as it does today. But no one has actually observed the future. So we all believe in something that goes beyond sensory experience. From a Christian perspective, this is a very reasonable belief. God (who is beyond time) has promised us that He will uphold the universe in a consistent way (e.g., Genesis 8:22). So we have a good reason for our faith in the uniformity of nature. For the consistent Christian, reason and faith go well together.

It is appropriate and biblical to have a good reason for our faith (1 Peter 3:15). Indeed, God encourages us to *reason* (Isaiah 1:18). The apostle Paul *reasoned* with those in the synagogue and those in the marketplace (Acts 17:17). According to the Scriptures, the Christian faith is not a "blind faith." It is a faith that is rationally defensible. It is logical and self-consistent. It can make sense of what we experience in the world. Moreover, the Christian has a moral obligation to think rationally. We are to be imitators of God (Ephesians 5:1), patterning our thinking after His revelation (Isaiah 55:7–8; Psalm 36:9).

The Mark of Rationality

There are those who would challenge the rationality of the biblical worldview. Some say that the Christian worldview is illogical on the face of it. After all, the Bible speaks of floating ax heads, the sun apparently going backward, a universe created in six days, an earth that has pillars and corners, people walking on water, light before the sun, a talking serpent, a talking donkey, dragons, and a senior citizen taking two of every land animal on a big boat! The critic suggests that no rational person can possibly believe in such things in our modern age of scientific enlightenment. He claims that to believe in such things would be *illogical*.

The Bible does make some extraordinary claims. But are such claims truly *illogical*? Do they actually violate any laws of logic? Although the above biblical examples go beyond our ordinary, everyday experiences, none of them are *contradictory*. They do not violate any laws of logic. Some biblical criticisms

1. Otherwise, when you erase the sentence, the law would cease to exist!

involve a misuse of language: taking figures of speech (e.g., "pillars of the earth") as though these were literal, when this is clearly not the case. This is an error on the part of the critic, not an error in the text. Poetic sections of the Bible, such as the psalms, and figures of speech should be taken as such. To do otherwise is academically dishonest.

Most of the criticisms against the Bible's legitimacy turn out to be nothing more than a subjective opinion of what is possible. The critic arbitrarily asserts that it is not possible for the sun to go backward in the sky, or for the solar system to be created in six days. But what is his evidence for this? He might argue that such things cannot happen based on known natural laws. With this we agree. But who said that natural laws are the limit of what is possible? The biblical God is not bound by natural laws. Since the Bible is indeed correct about the nature of God, then there is no problem at all in God reversing the direction of the planets, or creating the solar system in six days. An infinitely powerful, all-knowing God can do anything that is rationally possible.

Non-Christian circles of reasoning are ultimately self-defeating. They do not pass their own test.

When the critic simply dismisses those claims of the Bible that do not appeal to his personal, unargued sense of what is possible, he is being irrational. He is committing the logical fallacy known as "begging the question." Namely, he has decided in advance that such things as miracles are impossible, thereby tacitly assuming that the Bible is not true because it contains miracles. But this is the very assumption with which he began his reasoning. The critic is reasoning in a vicious circle. He has decided in advance that there is not an all-powerful God who is capable of doing the things recorded in Scripture, and then argues on this basis against the biblical God. Such reasoning is not cogent at all. So, when the critic accuses the Bible of being illogical because it goes against his subjective assessment of what is possible, it turns out that it is the critic — not the Bible — who is being illogical.

When people argue that something in the Bible seems strange or unreasonable, we must always ask, "strange or unreasonable *by what standard*?" If it is merely the critic's personal, arbitrary opinion, then we must politely point out that this has no logical merit whatsoever. Personal feelings are not the limit of what is true or possible. In fact, since all the treasures of knowledge are in Christ

(Colossians 2:3), it turns out that God Himself is the limit of what is possible. His Word is therefore the standard of what is reasonable, and we have no independent (and non-arbitrary) standard by which we can judge the Word of God.

The Laws of Logic

The extraordinary claims of Scripture cannot be dismissed merely on the basis that they are extraordinary. If indeed the biblical God exists, and if indeed He has the characteristics attributed to Him by the Bible (all-knowing, all-powerful, beyond time, etc.) then the critic has no basis whatsoever for denying that the miraculous is possible. Clearly, an all-powerful God can make a donkey talk, can create the universe in six days, can bring two of every animal to Noah, etc. These are simply not problems in the biblical worldview. When the critic dismisses the miraculous solely on the basis that it is miraculous, he is simply begging the question.

However, sometimes the critic asserts that the Bible has actually violated a law of logic; he claims that two passages in the Scriptures are contradictory. This is a more serious challenge, because two contradictory statements cannot both be true — even in principle. If the Bible actually endorsed two contradictory statements, then necessarily one of them would have to be false, and the Bible could not be totally inerrant. In reality, most alleged contradictions turn out to be nothing of the kind. They simply reveal that the critic does not truly understand what a contradiction is. A contradiction is "**A** and **not A** at the same time and in the same relationship" where **A** is any proposition. To contradict is to both affirm and deny the same proposition. And this is not the nature of most alleged biblical contradictions. (See the contradictions series on the Answers in Genesis website for more information on this.) Here's an example:

Everyone has an ultimate standard, whether he realizes it or not. If it is not the Bible, it will be something else.

The fact that Christ has two genealogies is not contradictory. Indeed, all people have (at least) two genealogies — one through their dad, and one through their mom. Some people have more than two because their biological father may not be their legal father. The fact that Jesus was born in Bethlehem, but is nonetheless "of Nazareth" is no contradiction since Jesus did grow up in Nazareth. The fact that Matthew (8:28) mentions two demon-possessed men does not contradict the fact that Mark (5:2) and Luke (8:27) chose to mention only one of the two. Perhaps one was much more violent than the other; in any case, there is no contradiction.

Alleged Contradictions Demonstrate That the Bible Is True!

Amazingly, when the critic asserts that the Bible contains contradictions, he has unwittingly refuted his own position, and has demonstrated that the Bible is true. The reason is this: the truth of the Bible is the only cogent reason to believe in the law of non-contradiction. Virtually everyone believes in the law of non-contradiction. We all "know" that two contradictory statements cannot both be true. But have you ever thought about *why* this is?

The law of non-contradiction stems from the nature of the biblical God. God does not deny Himself (2 Timothy 2:13), and all knowledge is in God (Colossians 2:3), thus true knowledge will not contradict itself. The law of non-contradiction (as with all laws of logic) is a universal, invariant law because God Himself upholds the entire universe (Hebrews 1:3), and God does not change with time (Hebrews 13:8). We know these things because God has revealed them in His Word. Thus, the Bible is the only objective basis for knowing that the law of non-contradiction is universally and invariantly true in all situations.

Therefore, when the unbeliever applies the law of non-contradiction, he is implicitly standing upon the Christian worldview. Even when he argues against the Bible, the critic must use God's standard of reasoning in order to do it. The fact that the critic is able to argue at all demonstrates that he is wrong. God alone is the correct standard for reasoning because all truth is in Him. We must therefore start with God as revealed in His Word in order to have genuine knowledge (Proverbs 1:7), whether we admit this truth or suppress it (Romans 1:18). So while it may seem at first that we do not need to believe the Bible in order to put gasoline into the car or to balance our checkbook, implicitly we must indeed rely upon the Bible. Without God as revealed in the Bible, there would be no rational basis for the laws of logic upon which we depend in order to function in our everyday life.

Since rationality itself stems from the nature of the biblical God, it follows that the Christian worldview is necessarily rational. This isn't to say that all Christians are rational all the time. We do not always follow God's standard in practice, even though God has saved us by His grace. Nonetheless, the Christian worldview as articulated in the Scriptures is fully logical and without error. This must be the case since the Bible is the inspired Word of the infallible God. It also follows that non-biblical worldviews are inherently illogical; they deny implicitly or explicitly the revelation of the biblical God in whom are deposited all the treasures of wisdom and knowledge (Colossians 2:3).

Although non-biblical worldviews may have "pockets" of rationality within them, they must ultimately appeal to Scripture as a basis for laws of logic, which they then deny as the one and only inspired Word of God. So not only is the Christian worldview logical, it is the *only* worldview that is ultimately, consistently logical. The Christian has faith — he believes in things (such as the accounts of Scripture) that he has not personally observed by sensory experience. But he has a very good reason to believe in the Scriptures; the biblical God alone makes reason possible. So a good reason for my faith is that my faith makes reason possible.

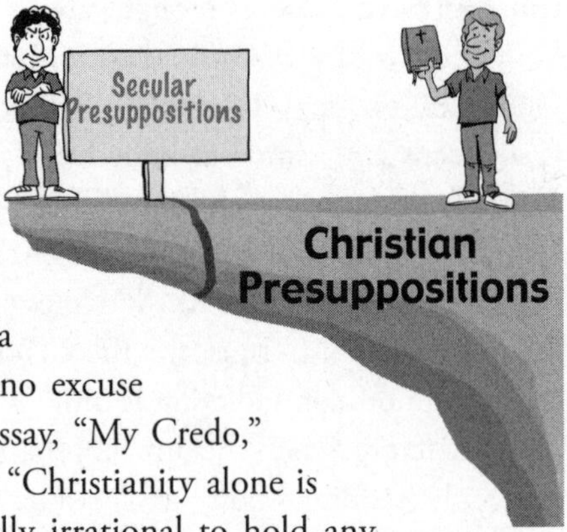

The unbeliever must use Christian principles to argue against the Bible. The fact that he is able to argue at all proves that he is wrong.

The non-Christian does not have a good reason for his beliefs. He has a type of faith, too, but his faith is "blind." He is without an apologetic (a defense of his faith) such that he has no excuse for his beliefs (Romans 1:20). In the essay, "My Credo," Cornelius Van Til cogently argued that "Christianity alone is reasonable for men to hold. It is wholly irrational to hold any other position than that of Christianity. Christianity alone does not slay reason on the altar of 'chance.' "

Yes, the Christian worldview is logical. But what's more, only the Christian worldview is logical. Competing systems of thought cannot account for laws of logic and their properties, the ability of the human mind to access and use laws of logic, or the moral obligation to reason logically. Such truths are entirely contingent upon Almighty God as objectively revealed in the Bible.

14

What about Cloning and Stem Cells?

DR. TOMMY MITCHELL AND DR. GEORGIA PURDOM

There are few issues in our society that raise as many emotional and ethical concerns as cloning and stem cells. Scientists, journalists, special interest groups, and even patients themselves regularly bombard us with their particular views on this issue. How are we to know what to think regarding these issues? How do we separate fact from fiction? Since cloning and stem cells are two separate (but related) issues, we will deal with them individually.

What Is Cloning?

Cloning is a process by which a genetically identical copy of a gene, an entire cell, or even an organism is produced. For this chapter, we will confine the discussion primarily to the cloning of an entire organism. This is a topic about which there is much misinformation. It is also a subject that raises some very serious ethical issues.

Cloning as usually understood is an artificial process, meaning it is carried out in a laboratory setting. It can and does, however, occur regularly in nature. There are organisms (e.g., bacteria, protists, and some plants) that typically reproduce by asexual reproduction. Here a genetically identical copy of the parent is produced by the splitting of a single cell (the parent cell).

Identical twins are also clones. In fact, identical twins have been called "natural clones" since splitting of a fertilized egg causes this, producing two copies of the same organism.

It is the issue of artificial cloning that has captured the interest of so many in our society. This process has garnered much attention in recent years, especially with the birth of the famous sheep, Dolly. Actually, many different types of animals have been cloned including tadpoles, mice, cats, sheep, cattle, a horse, and others.

Dolly was the result of cloning a mammary cell from a mature sheep, but at what cost?

How Is a Clone Made?

The simplest method for making a clone is to remove the nucleus (containing the organism's DNA) from a somatic (body) cell in the animal you want to clone. You then take an egg cell (from the same type of animal) that has had its own nucleus removed, and you place the donor nucleus into the egg cell. This is called somatic cell nuclear transfer (SCNT). This egg is grown briefly in a test tube and then implanted into the womb of an adult animal. If there are no complications, at the end of gestation an animal is born with an identical genetic makeup of the donor animal. As one might imagine, the process is technically quite difficult. Let's use the aforementioned Dolly as an example. It took 277 eggs that ultimately produced 29 embryos and only one living sheep to create Dolly.[1] This is consistent with the failure rate for other animals. As can be seen, many embryos are wasted in these attempts.

It should also be noted that Dolly died at age six. She apparently died of a respiratory infection. Some have suggested that she exhibited signs of premature aging, but others have disputed these reports. The strongest speculation is that her early death was due to shortening of telomeres. Telomeres are segments of DNA that exist on the ends of chromosomes. They progressively shorten with age due to repeated cell division until they reach a point that no further replication of the chromosome can occur. Since Dolly was cloned from a 6-year-old sheep, it could be said that Dolly's DNA when she died was actually 12 years old. The telomere issue remains a significant problem for those involved in cloning research.

How Can Cloning Be Used?

There are two main purposes for cloning: to produce an identical organism (reproductive cloning) or to produce a cloned embryo for the purpose of obtaining embryonic stem cells (therapeutic cloning). There are those who promote reproductive cloning in many different areas. For example, cloning of certain

1. www.en.wikipedia.org/wiki/Cloning

animals used for food or for animals used in specific work environments has been suggested. This has also been proposed as a possible solution for the rescue of many endangered species. Although it is far beyond our present technology, some have theorized that the extinct woolly mammoth or even dinosaurs might in the future be produced through reproductive cloning!

Therapeutic cloning is aimed at producing cloned embryos from which embryonic stem cells may be obtained. This is done ostensibly to use the stem cells to treat disease or illness. While this is laudable in one sense, there are serious ethical issues that arise (see the following stem cell section).

The obvious next step would be to consider cloning a human being. There are those who advocate therapeutic cloning of humans to provide an adequate supply of embryonic stem cells. It has even been suggested by some that humans should be reproductively cloned in order to provide a ready reserve of tissues and organs should they ever be needed. The clone would simply be "spare parts," to be used at the discretion of the "parent" human.

So What's the Problem?

If man is just another animal, just a higher form of pond scum, there really is no problem. Cloning a person is totally justifiable. Just make copies of ourselves and chop them up as we please. People are nothing special.

But those of us who trust in God's Word know there is a problem here. We are not just a higher form of pond scum. We are not just animals. We are made in the image of the Creator.

> And God said, "Let us make man in our image, after our likeness . . ." (Genesis 1:26; ASV).

Therefore, humans are not to be created at man's whim. Rather, we are a special creation of our Father in heaven.

What Is a Stem Cell?

Simply put, a stem cell is a cell in the body (or in an embryo) that has the capability of turning into many specialized cell types. At the time of conception, when the sperm and egg unite, we consist of only one cell. Ultimately, as this cell divides into two cells, then four, then eight, and so on, the roughly 200 different cell types in the body must be produced. This process can occur because of stem cells.

The very earliest cells produced after fertilization are called *totipotent* because they have the capability of turning into any other cell type in the body as well as extra-embryonic cell types such as those which form the placenta. As cells divide and begin to specialize (a process called differentiation), the stem cells along these pathways lose the ability to produce certain types of cells. At this point they are called *pluripotent* in that they can still develop into all the tissue types of the body but not the extra-embryonic tissue types. After further differentiation, the cells become *multipotent*, meaning the number of potential cell types that can be derived from them has been reduced. This process continues until cells are only able to produce cells of one type.

Along the way, some stem cells stop differentiating and merely reproduce themselves, thereby giving the body a reservoir of stem cells. These cells then provide a source of new cells for tissue replacement and repair.

Why Are Stem Cells Important?

Medical researchers are interested in stem cells for their potential to treat various diseases. For example, stem cells that could be induced to change into insulin-secreting cells could help cure those with diabetes. Patients with damage to the spinal cord could benefit from new nerve tissue generated from stem cells. Those suffering from heart muscle damage after a heart attack might be able to have new heart muscle derived from stem cells. Think about Parkinson's disease, multiple sclerosis, Alzheimer's disease . . . the list of potential interventions based on stem cell therapy seems almost endless.

This research is very important. It is certainly one of the most exciting medical advances in our lifetime. Physicians strive to relieve the suffering of their patients. Who among them would not want to have available a means to cure many horrible diseases? This research has so much potential. However, that potential comes with a grave concern. That concern revolves around how we obtain these stem cells.

How Are Stem Cells Obtained?

In order to understand the basis of the debate over stem cell research, one must understand that there are two basic types of stem cells: embryonic stem cells (ESC) and adult stem cells (ASC).

Embryonic stem cells are, as you would expect from their name, derived from embryos. Four to five days after fertilization, the embryo consists of a hollow ball of cells called a blastocyst. It is from this ball of cells that all the body's tissues are ultimately derived. To harvest embryonic stem cells, the embryo is disrupted (killed) and the cells collected. As this cell harvest occurs very early in development, very little differentiation of the stem cells would have taken place. These cells would be considered pluripotent.

Adult stem cells, on the other hand, are not necessarily derived from "adults." This is somewhat confusing to many as adult stem cells can be obtained from any fully formed person, whether newborn, infant, child, or adult. These cells can be found in many tissues in the body: bone marrow, skin, teeth, liver, brain, intestines, blood vessels, skeletal muscle, among others.

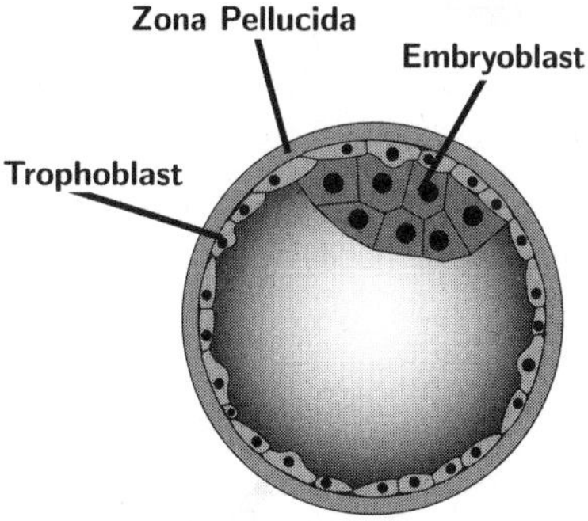

While an embryo must be killed in order to harvest embryonic stem cells, harvesting adult stem cells does not lead to the death of the donor.

The concern regarding adult stem cells is that these cells could have limitations on the types of tissues that could be obtained from them. In theory, these stem cells are further down the path of differentiation (they would be multipotent rather than pluripotent), thus limiting their potential usefulness in medical research and intervention.

How Are Stem Cells Used?

After being isolated, stem cells are then grown in laboratory culture. They are placed in dishes containing a special culture medium. The cells divide and multiply. The initial phase of the process would be designed to grow an adequate supply of the stem cells themselves.

So how are different tissue types generated from stem cells? There are many different methods used to cause a stem cell to differentiate into a specific cell type. Manipulation of the culture medium can guide this process. Other research techniques include hormonal stimulation or genetic modification of the stem cells. This is still an area of intensive investigation, with new techniques becoming available seemingly every few months.

As an example, let's select a patient who has suffered a heart attack. A portion of the heart muscle has been damaged or killed as a result of this event. It would certainly be to the patient's benefit to be able to repair the heart muscle. In this situation, the patient's medical team might choose to intervene with stem cell therapy. Stem cells could be induced to differentiate into heart muscle cells. These cells could then be administered to the patient in hopes of improving the function of the damaged heart. This type of intervention is, in fact, taking place at this time.[2] The results have been very promising thus far.

2. Medical News Today, "First Human Receives Cardiac Stem Cells in Clinical Trial to Heal Damage Caused By Heart Attacks," July 1, 2009, www.medicalnewstoday.com/articles/155915.php.

Depending on the particular situation, the stem cells might be given intravenously,[3] by injection or direct deposit of the cells into the target site, given by intracoronary injection (for cardiac intervention), or injection into the spinal fluid (for neurologic problems). The hope for the future is that by using stem cells, entire organs might be grown for transplant. Again, research in this area is promising.

Are There Problems with Stem Cell Treatments?

As with any new research endeavor, there are pitfalls associated with stem cell research. While we tout the successes, we should also be aware of the problems and limitations.

Several major issues have limited embryonic stem cell therapy. First of all, in laboratory animals, embryonic stems cell have shown a tendency to form tumors. The reasons for this are unclear, although some have speculated that ESC can form tumors due to their tendency to associate with each other rather than with the target tissue.[4] Obviously, this is an area of intense investigation at present.

The other major issue is that of tissue rejection. As with any transplant, foreign tissue is recognized by the body as "non self." So an embryonic stem cell transplant from a random donor would be no different than a heart or a kidney transplant. After all, these embryonic stem cells would come from another person. These cells would be seen as foreign tissue by the body. Thus, anti-rejection drugs would be needed to prevent rejection of the new tissue. However, it should be noted that therapeutic cloning using a person's own cells would avoid this problem.

Adult stem cells apparently do not have the problem of tumor formation. Therapy with ASC would also not have the problem of tissue rejection as long as the stem cells are harvested from the patients themselves.[5]

That is not to say that intervention with adult stem cells is without problems. The main problem is that even though ASC can be found in many body tissues, they occur in very small numbers and can be quite difficult to isolate. Thus, obtaining an adequate supply of ASC for a given therapeutic intervention can be difficult. The most often claimed problem with adult stem cells is the supposed limitation on the number of tissue types that can be derived from them. Since embryonic stem cells have undergone less initial differentiation, there is the potential to derive all needed cell types from ESC. Thus, it

3. This has the problem of the so-called "first pass effect" where the cells given are filtered out of the circulation by the lungs.
4. Joseph Panno, *Stem Cell Research* (New York, NY: Facts on File, 2005), p. 9.
5. ASC therapy using cells from another person would encounter that same rejection potential as ESC.

would stand to reason that ASC are more limited.

However, actual research does not bear out these claims. Adult stem cells have, to date, been used to generate almost every different cell type in the body.[6] It has been shown that adult stem cells of one cell lineage can be induced to produce cells in another. For example, blood-forming stem cells in bone marrow can differentiate into cardiac muscle cells, etc.[7] Some multipotent adult stem cells have been induced to revert to an apparent pluripotent state, and have subsequently produced many more cell types than would have been predicted. So while this may have seemed a problem in theory, it has not been a problem in practice.

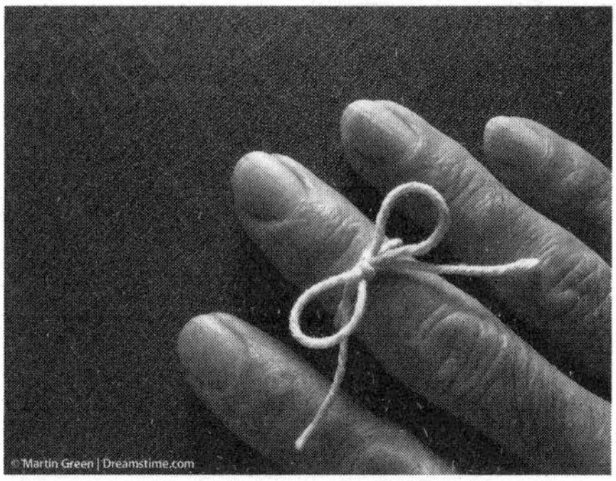

Although adult stem cells have their problems, they have produced far more success than embryonic trials. Alzheimer's and other degenerative diseases are prime candidates for such research.

Should Both ESC and ASC Research and Treatments Be Pursued?

So should we not pursue any avenue we can to help the sick and the dying? Do we not want stem cell therapy to succeed? Certainly we want to help the sick. We want medical science to progress. But we must also examine the facts and ask the question, "At what cost?"

First of all, what can be said for adult stem cell research? Simply this: adult stem cell therapy has been used to treat over 70 diseases to date. Some stem cell therapies have been used for over 40 years.[8] ASC have a proven track record with the hope of greater successes to come. Thus far the only significant clinical interventions available are from using adult stem cells. For some reason the media, when reporting on these issues, has consistently downplayed these successes and even implied that these successes are the result of embryonic stem cell research rather than adult stem cell research.

Embryonic stem cells, on the other hand, despite the regularly reported theoretical benefits, have yet to achieve any significant clinical success. Those in favor of ESC argue that given time these advances will come. Perhaps this is so, but again we need to ask, "At what cost?"

6. "Adult Stem Cell Pluripotency," www.stemcellresearch.org/facts/ASCRPlasticity.pdf.
7. "Stem Cell Basics," National Institutes of Health, Stem Cell Information, www.stemcells.nih.gov/info/basics/basics4.asp.
8. Specifically those involving blood-forming stem cells for bone marrow diseases.

What Is the Cost of Embryonic Stem Cell Research?

Although one could point to the lack of success of embryonic stem cell research, in spite of the years and countless dollars invested in it, and say, "It hasn't been worth the cost," the biggest cost is yet to be counted. The cost is that of human life.

As has been noted, embryonic stem cells are obtained by the destruction of an embryo. An embryo is fully human. So in order to get these stem cells to help one person, another person must be killed. This is simply morally unacceptable.

Scripture tells us that life begins at conception (here defined as the moment of fertilization). In His Word, God tells us:

> Then the word of the LORD came to me, saying: " Before I formed you in the womb I knew you; Before you were born I sanctified you; I ordained you a prophet to the nations" (Jeremiah 1:4–5).

> Behold, I was brought forth in iniquity, And in sin my mother conceived me (Psalm 51:5).

> For You formed my inward parts; You covered me in my mother's womb. I will praise You, for I am fearfully and wonderfully made; Marvelous are Your works, And that my soul knows very well. My frame was not hidden from You, When I was made in secret, And skillfully wrought in the lowest parts of the earth (Psalm 139:13–15).

We are told that the Lord knew us before we were conceived. We have a sin nature in the womb. How could this be if we are not fully human at the time of fertilization?

What's the Ethical Solution?

As has been shown, stem cell therapy has the potential to alleviate much suffering. It is an avenue of medical research that should be pursued in hopes of building on the successes already achieved. However, in our haste to help the sick, we must not neglect those who cannot speak for themselves. Adult stem cell therapy can allow us to fight disease without the destruction of human life.

Although everyone wants to see such devastating diseases come to an end, we all must realize our work will only lead to a temporary alleviation. Jesus Christ, the true conqueror of disease and death, will create a new heaven and a new earth where the effects of sin have been removed. That is the cure we eagerly await.

15

How Old Does the Earth Look?

DR. ANDREW A. SNELLING

Insisting that the earth and the universe are young, only 6,000 years old or so, does not make the biblical view popular in today's enlightened "scientific" culture. It would be so easy just to go along with the view believed and followed by the overwhelming majority of scientists — and taught in nearly all universities and museums around the world — that the universe is 13–14 billion years old and the earth 4.5 billion years old.

After all, many Christians and most scientists who are Christians believe in such a vast antiquity for the earth and universe. Consequently, they even insist the days in Genesis 1 were not literal days, but were countless millions of years long. Also, they claim the Genesis account of creation by God is just poetic and/or figurative, so it is not meant to be read as history.

Why a Young Age for the Earth?

Of course, the reason for insisting on a young earth and universe is because other biblical authors took Genesis as literal history and an eyewitness account provided and guaranteed accurate by the Creator Himself (2 Timothy 3:16a; 2 Peter 1:21). Jesus also took Genesis as literal history (Mark 10:6–9; Matthew 19:4–5; Luke 17:27). So, the outcome of letting Scripture interpret Scripture is a young earth and universe.

The Hebrew language and context used in Genesis 1 can only mean literal (24 hour) days.[1] Furthermore, as history, the genealogies in Genesis 5 and 11 provide an accurate chronology, so that from the creation of the first man, Adam, to the present day is only about 6,000 years. Since the earth was only created five literal days before Adam, then on the authority of God's Word, the earth is only about 6,000 years old.

Does the Earth Look Old?

Nevertheless, most people, including Christians, would still claim dogmatically that the earth *looks* old. But why does the earth supposedly look old? And how old does the earth really look? If we rightly ask such questions, then we are likely to get closer to the right answers.

The use of the word *looks* gives us the necessary clue to finding the answers. Looking at an object and making a judgment about it requires two operations by the observer. There is first the observation of the object with one's eyes. Light impulses then go from the eyes to be processed by one's brain. How one's brain interprets what has been seen through one's eyes is dependent on what information is already stored in the brain. Such information has been progressively acquired and stored in our brains since birth. So, for example, as a child we learn what a rock is by being shown a rock.

We observe that a sandstone is made of sand cemented together, and we see a trilobite fossil inside the sandstone (figure 1), so we wonder how the trilobite came to be fossilized in the sandstone and how both the sandstone and the trilobite fossil formed. However, we never actually observed either the trilobite being buried by sand and fossilized or the deposition of the sand and its cementation into sandstone. Therefore, we don't really know how and when the trilobite fossil and the

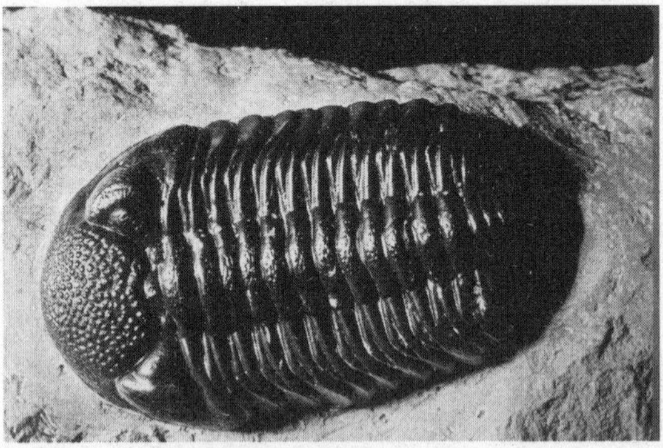

Figure 1. A trilobite fossil in a piece of sandstone

1. S.W. Boyd, "Statistical Determination of Genre in Biblical Hebrew: Evidence for an Historical Reading of Genesis 1:1–2:3," *Radioisotopes and the Age of the Earth: Results of a Young-Earth Creationist Research Initiative*, L. Vardiman, A.A. Snelling, and E.F. Chaffin, eds. (El Cajon, CA: Institute for Creation Research and Chino Valley, AZ: Creation Research Society, 2005), p. 631–734.

sandstone formed — so just by looking at them we really don't know how old they are.

How, then, can we work out how old they might be and how they formed? Because we can't go back to the past, it *seems* logical to think in terms of what we see happening around us today — in the present. Today, rivers slowly erode land surfaces and gradually transport the sand downstream to their mouths, where they build deltas. The sediments also are eventually spread gradually out on the seafloor, where bottom-dwelling creatures like trilobites could perhaps be occasionally buried and then fossilized.

So with this apparently logical scenario in our minds, based on our everyday experience, when we look at that piece of sandstone with the trilobite fossil in it, it seems totally reasonable to conclude that, because it took such a long time to erode and transport the sand and then deposit it to bury and fossilize the trilobite, the sandstone and trilobite fossil must be very old. Perhaps they may even be millions of years old. However, it needs to be remembered that there are no particular intrinsic features of the sandstone and the trilobite fossil that are incontestably diagnostic of any supposed great age. The conclusion that they must be old wasn't because they actually look old, but because it was assumed they took a long time to form based on present-day experience.

Long Age Reasoning Questioned

Now let's extend this reasoning to the earth itself. Why is it that most people think the earth looks old? Isn't it because they *assume* it took a long time to form based on their present-day experience of geological processes? After all, volcanic eruptions only occur sporadically today, so the vast, thick lava flows stacked on top of one another — for example, in the USA's Pacific Northwest — *must* have taken a long time to accumulate. However, this reasoning is wrong for three very valid reasons:

First, it ignores the fact that we *cannot* go back to the past to actually verify by direct observations that vast, thick stacks of lava flows — and sandstones with trilobite fossils — took a long time to form millions of years ago. The inference that the present is the key to the past is only an assumption, not a fact.

Second, that assumption deliberately ignores the fact that we do have direct eyewitnesses from the past who have told us what did happen to the earth and how old it really is. The Bible claims to be the communication to us of the Creator God who has always existed. Its authenticity is overwhelmingly verified by countless exactly fulfilled predictions, archeological and scientific evidences, corroborating eyewitness accounts, and the changed lives and testimonies of Bible-believing Christians. In Genesis 1–11, it is revealed how to calculate the age of the earth, and how rock layers and fossils were rapidly and recently formed in the year-long, global, catastrophic Flood.

And third, there is now abundant scientific evidence that rock layers and fossils can only form rapidly due to catastrophic geological processes not usually seen today, and not on the scale they must have occurred at in the past.[2]

Catastrophism Today

Geologists are always studying present-day geological processes, including rare catastrophic events, such as floods, earthquakes, and violent volcanic eruptions. Such processes have been observed to produce and change geological features very rapidly; so geologists have learned not to ignore such currently rare catastrophic events when interpreting how the earth's features were produced in the past.

Further examples of why most people think the earth looks old are river valleys and canyons. Because the rivers in most valleys and canyons today seem to only slowly and imperceptibly erode their channels, even during occasional floods, most people assume it must have taken millions of years to erode valleys and canyons.

However, the observational realities are more instructive than such an erroneous assumption. For example, since the Colorado River today does not erode its channel, the only truly viable explanation for the carving of Grand Canyon is rapid catastrophic erosion on an enormous scale by dammed waters left over from the global Genesis Flood.[3] Such rapid catastrophic erosion carving canyons has even been observed. As a result of the 1980 and subsequent eruptions at Mount St. Helens, up to 600 feet of rock layers rapidly accumulated nearby. A mudflow on March 18, 1982, eroded a canyon system over 100 feet deep in these sediment layers, resulting in a one-fortieth scale model of the real Grand Canyon (figure 2).[4]

---

2. S.A. Austin, "Interpreting Strata of Grand Canyon," in *Grand Canyon: Monument to Catastrophe*, S. A. Austin, ed. (Santee, CA: Institute for Creation Research, 1994), p. 21–56; A.A. Snelling, "The World's a Graveyard," *Answers*, April–June, p. 76–79; J.H. Whitmore, "Aren't Millions of Years Required for Geological Processes?" *The New Answers Book 2*, Ken Ham, ed. (Green Forest, AR: Master Books, 2008), p. 229–244.
3. S.A. Austin, "How was Grand Canyon Eroded?" in *Grand Canyon: Monument to Catastrophe*, S.A. Austin, ed. (Santee, CA: Institute for Creation Research, 1994), p. 83–110. See also chapter 18 in this volume, "When and How Did the Grand Canyon Form?"
4. S.A. Austin, "Mount St. Helens and Catastrophism," *Proceedings of the First International Conference on Creationism*, vol. 1 (Pittsburgh, PA: Creation Science Fellowship, 1986), p. 3–9.

Figure 2. This canyon system, with 100-foot high cliffs, was eroded adjacent to Mount St. Helens in less than a day!

Uniformitarianism Predicted

In 2 Peter 3, we read a prediction that Peter made around A.D. 62 that scoffers would arise who would challenge and deny that God created the earth and subsequently destroyed the earth by the cataclysmic global Flood. Peter says they would be "willingly ignorant" and deliberately reject the evidence for a created earth and the year-long global Flood. They would claim instead that the present is the key to the past, that present-day geological processes have always operated at today's snail's pace, and that they alone are necessary to explain how rock layers and fossils formed and how old the earth is.

This prediction was actually fulfilled about 200 years ago — about 1,750 years after the prediction was made. James Hutton, a doctor and farmer-turned-geologist, claimed in his 1785 Royal Society of Edinburgh paper and 1795 book *Theory of the Earth* that he saw "no vestige of a beginning" for the earth because present-day geological processes have slowly recycled rock materials over vast eons of time. This was a deliberate rejection of the biblical account of the recent, catastrophic global Flood, up until that time accepted by most scholars to be the explanation for fossil-bearing rock layers. Indeed, Hutton insisted that "the past history of our globe *must* be explained by what can be seen happening now"[5] (emphasis added).

It was Charles Lyell, a lawyer-turned-geologist, with his two-volume *Principles of Geology* (1830–33), who eventually convinced the geological establishment to abandon the biblical Flood in favor of this "principle" he called uniformitarianism. Lyell openly declared that he wanted to remove the influence of Moses (the human author of Genesis) from geology, revealing his motivation was spiritual, *not* scientific.[6] He insisted on the uniformity through time of natural processes only at today's rates — a belief that was later encapsulated in the phrase "the present is the key to the past."

This is the belief that now underpins virtually all modern geological explanations about the earth and its rock layers. And it is a *belief* because it cannot be proved that *only* today's geological processes can explain the earth's history and determine its age. No one has ever observed past geological processes, except for God — and Noah and his family — during the Flood when these processes were definitely catastrophic on a global scale. Yet most people today, even Christians, have unwittingly imbibed this uniformitarian belief, having been brainwashed by the constant barrage of teaching over many decades by the world's education systems (schools, colleges, and universities), museums, and media (newspapers, magazines, television, and even Hollywood). Indeed, most people automatically see the earth as old because they have accepted it is a proven scientific fact that it is old!

Using the Right Glasses

However, based on the authority of God's Word, we can dogmatically say they are absolutely wrong. Looking at the world through "glasses" that are based on human reasoning alone (man's word) makes people wrongly think the earth looks really old. On the other hand, when we as Christians see the world through the biblical "glasses" provided by God's inerrant Word — so that we

5. A. Holmes, *Principles of Physical Geology*, second ed. (London: Thomas Nelson and Sons, 1965), p. 43–44, 163.
6. R.S. Porter, "Charles Lyell and the Principles of the History of Geology," *British Journal for the History of Science*, IX, 32 no. 2 (1976): 91–103.

see the world as God sees it — we can assert unashamedly that the earth does not really look that old at all, being only about 6,000 years old (which, of course, is young). Indeed, the earth we see today is the way it looks because it is the destroyed remains of the original earth God created, still marred by the subsequent Curse.

Furthermore, not only should we understand that the Bible provides the true history of the earth, but that history tells us the earth only looks the way it does today because of what happened in the past. In other words, the past is the key to the present!

Conclusion

Paul, in 2 Corinthians 11:3, warns us about the way Satan subtly beguiled the mind of Eve in the Garden of Eden by questioning and twisting God's Word. Today, Satan has subtly beguiled so many people, including Christians, by twisting the clear testimony of God's Word that "the past is the key to the present" into "the present is the key to the past." And just as he used the appealing look of the fruit on that tree to entice Eve, so he uses the snail's pace of geological processes today to make people doubt or deny what God has told us about the young age of the earth and His eyewitness account of the formation of the rock layers and fossils.

It also must be emphasized that even though we must trust God and His Word by faith alone (Hebrews 11:3), it is neither an unreasonable nor a subjective faith. This is because God is not a man that He should lie, so the evidence we see in God's world will always ultimately be consistent with what we read in God's Word. Thus, when we put on our biblical "glasses," we should be able to immediately see and recognize the overwhelming evidence that the earth looks (and is) young and that the earth's fossil-bearing rock layers are a product of the global, catastrophic Flood.

After all, if the Genesis Flood really did occur, what evidence would we look for? Genesis 7 says all the high hills and mountains under the whole heaven were covered by the water from the fountains of the great deep and the global torrential rainfall so that all land-dwelling, air-breathing creatures not on the ark perished. Wouldn't we, therefore, expect to find the remains of billions of plants and creatures buried in rock layers rapidly laid down by water all around the earth? Yes, of course! And that's exactly what we find — billions of rapidly buried fossils in rock layers up on the continents, rapidly deposited by the ocean waters rising up and over the continents all around the earth. This confirms that the rocks and fossils aren't millions of years old — and neither is the earth.

So how old does the earth look? If we look at the earth through the "glasses" of human reasoning — that only snail-paced present geological processes can explain the past — then the earth does indeed look old. However, that autonomous human reasoning blatantly denies what God's Word clearly tells us about the true age of the God-created earth and about what happened in the recent past during the global, cataclysmic Flood, which is the key to understanding why the earth looks the way it does today.

16

Does Evolution Have a ... *Chance?*

MIKE RIDDLE

> One has only to contemplate the magnitude of this task to concede that the spontaneous generation of a living organism is impossible. Yet we are here — as a result, I believe, of spontaneous generation.[1] — George Wald, Nobel Laureate

In today's culture, molecules-to-man evolution is being taught as a fact, even though it is known to "go against the odds." But few realize the odds they are up against! And they are immense!

The Bible teaches that God is the Creator of all things (Genesis 1; Colossians 1:16; John 1:1–3; Revelation 4:11). While these passages rule out any possibility of Darwinian evolution, they do allow for variation within a created kind. But there is much opposition to what the Bible teaches. People holding to evolution would argue that random chance events, natural selection, and billions of years are sufficient to account for the universe and all life forms.

Do You Believe in "Magic"?

Most people recognize "magic" as an illusionary feat or trickery by sleight of hand. But how far are you willing to go to believe something can happen by

1. George Wald [biochemist and winner of Noble Prize in Physiology or Medicine, 1967], "The Origin of Life," *Scientific American* 191 no. 48 (1954): 46.

"dumb luck" or chance? For example, if I were to roll a die and have it come up six three times in a row, would you consider that lucky? How about if I rolled six ten times in a row? Now you might suspect that I am using some trickery or that the die is weighted.

How far are we willing to go to accept something as a chance occurrence or before we recognize that it was just an illusion? We can test this by measuring our *credulity factor*. Credulity is the willingness to believe something on little evidence.

Measuring Our Credulity Factor against Evolution

Evolutionists state that life originated by natural processes about 3.8 billion years ago. Is there any evidence for this happening? Freeman Dyson, theoretical physicist, mathematician, and member of the U.S. National Academy of Sciences states:

> Concerning the origin of life itself, the watershed between chemistry and biology, the transition between lifeless chemical activity and organized biological metabolism, there is no direct evidence at all. The crucial transition from disorder to order left behind no observable traces.[2]

Since the origin of life has never been observed, this is a major hurdle! We are left with the question, "Is the origin of life by naturalistic processes possible?" This can, in part, be tested by examining two areas:

1. The success of scientists in creating life or the components of a living cell.
2. The probability that such an event could occur.

The Structural Unit of Living Organisms – The Cell

Cells are made up of thousands of components. One of these components is protein. Proteins are large molecules made up of a chain of amino acids. In order to get a protein useful for life, the correct amino acids must be linked together in the right order. How easy is this and does it happen naturally? It turns out that this is not an easy process. There are large hurdles that evolutionary processes must overcome in order to build a biological protein.

2. Freeman Dyson, *Origins of Life* (New York, NY: Cambridge University Press, 1999), p. 36.

Protein molecules contain very specific arrangements of amino acids. Even one missing or incorrect amino acid can lead to problems with the protein's function.

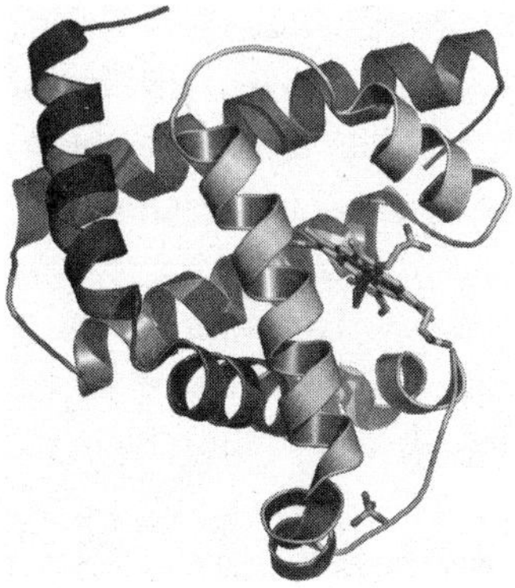

Making Mathematics Painless

Before applying mathematics and probability to the origin of life, we need to consider seven parameters that will affect the formation of a single protein.

First, there are over 300 different types of amino acids. However, only 20 different amino acids are used in life. This means that in order to have life, the selection process for building proteins must be very discriminating.

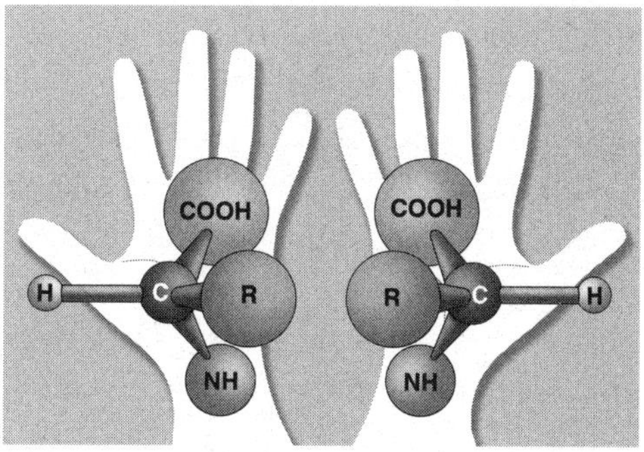

Second, each type of amino acid molecule comes in two shapes commonly referred to as left-handed and right-handed forms. Only left-handed amino acids are used in biological proteins; however, the natural tendency is for left- and right-handed amino acid molecules to bond indiscriminately.

Third, the various left-handed amino acids must bond in the correct order or the protein will not function properly.[3]

Fourth, if there was a pond of chemicals ("primordial soup"), it would have been diluted with many of the wrong types of amino acids and other chemicals available for bonding, making the proper amino acids no longer usable. This means there would have been fewer of the required amino acids used to build a biological protein.

Fifth, amino acids require an energy source for bonding.[4] Raw energy from the sun needs to be captured and converted into usable energy. Where did the energy converter come from? It would require energy to build this biological

3. "The order of the amino acids in a protein determines its function and whether indeed it will have a function at all." Lee Spetner, *Not By Chance* (New York, NY: Judaica Press, 1997), p. 31.
4. "The important fact that amino acids do not combine spontaneously, but require an input of energy, is a special problem." Charles Thaxton, Walter Bradley, and Roger Olsen, *The Mystery of Life's Origin* (Dallas, TX: Lewis and Stanley, 1992), p. 55.

machine. However, before this energy converter can capture raw energy, it needs an energy source to build it — a catch-22 situation.[5]

Sixth, proteins without the protection of the cell membrane would disintegrate in water (hydrolysis), disintegrate in an atmosphere containing oxygen, and disintegrate due to the ultraviolet rays of the sun if there was no oxygen present to form the protective ozone layer.[6]

Seventh, natural selection cannot be invoked at the pre-biotic level. The first living cell must be in place before natural selection can function.

Considering all seven of these hurdles, how probable is it that a single protein could have evolved from a pool of chemicals? Probability outcomes are measured with a value ranging from zero through one. The less likely an event will happen, the smaller the value (closer to zero). The more likely an event will occur, the larger the value (closer to one).

Let's practice this using a coin. What are the chances of getting a heads when we flip a penny? The answer is 50 percent, or one chance in two (written 1/2). What is the chance of getting two heads in a row? Since each toss is 1/2 we can multiply each occurrence to get the final probability. This would be 1/2 x 1/2 which would equal 1/4 (or one chance in four). Now let's use some bigger numbers.

When we flip a coin we have two possible outcomes, heads or tails. In this problem, we want to calculate the probability of getting all heads every time we flip a coin. We can use this exercise to test our credulity factor. How many heads in a row are we willing to accept as a chance occurrence? At what point would we suspect an illusion or some form of magic (trickery)?

The objective of using probabilities is to demonstrate the probability or chance of getting a certain result. On average, how many times and how often will we need to flip the coin to achieve 100 heads in a row? *Over 300 million times a second for over one quadrillion years!*

The chances of getting all heads 100 times in a row is similar to the chance of getting 100 left-handed amino acids to form a biological protein. Proteins range in size from about 50 to over 30,000 amino acids. To get a small protein of 100 left-handed amino acids from an equal mixture of left- and right-handed amino acids, the probability would then be 10^{30} or 1 followed by 30 zeros (1,000,000,000,000,000,000,000,000,000,000). How believable (credulity factor)

5. "A source of energy alone is not sufficient, however, to explain the origin or maintenance of living systems. The additional crucial factor is a means of converting this energy into the necessary useful work to build and maintain complex living systems." Thaxton, Bradley, and Olsen, *The Mystery of Life's Origin*, p. 124.
6. "What we have then is a sort of 'catch 22' situation. If we have oxygen we have no organic compounds, but if we don't have oxygen we have none either." Michael Denton, *Evolution: A Theory in Crisis* (Bethesda, MD: Adler and Adler, 1985), p. 262.

Number of desired heads in a row	Probability	Number of flips	Credulity factor (chance)
1	1/2 have	2	Yes / No
2	1/4 ($1/2^2$)	4	Yes / No
3	1/8 ($1/2^3$)	8	Yes / No
4	1/16 ($1/2^4$)	16	Yes / No
5	1/32 ($1/2^5$)	32	Yes / No
8	1/256 ($1/2^8$)	256	Yes / No
10	1/1024 ($1/2^{10}$)	1024	Yes / No
20	1/1,048,576 ($1/2^{20}$)	1,048,576	Yes / No
100	$1/10^{30}$ ($1/2^{100}$)	1 followed by 30 zeros	Yes / No

is it that this could happen by random chance? Also, consider that this has never been observed! Chance protein formation has always been accepted as a matter of faith by evolutionists.

But wait, there is more! This number, 10^{30}, only measures the possibility of getting all left-handed amino acids. It does not say anything about their order. In our example, we have a chain of 100 amino acids. Each position can be occupied by any 1 of 20 different amino acids common to living things, and these must be in a specific order to form a functional protein. What is the probability that the correct amino acid will be placed in position number 1 of the chain? It will be 1/20. What is the probability that the first two positions will be correct? This can be calculated by multiplying the two probabilities together ($1/20 \times 1/20 = 1/20^2$). Therefore, the probability of getting all 100 amino acids in the correct position would be 1/20 multiplied by itself 100 times or $1/20^{100}$ (this equates to $1/10^{130}$). This is 1 followed by 130 zeros!

Large numbers can be hard to visualize or even comprehend. To put this in picture format we can use a smaller number 10^{21} (1 followed by 21 zeros). If we were to take 10^{21} silver dollars and lay them on the face of the earth; they would cover the entire land surface to a depth of 120 feet.[7]

Are there upper limits for which we can logically expect an event will not occur by random chance? The mathematician Emile Borel proposed $1/10^{50}$ as

7. Peter Stoner, *Science Speaks* (Wheaton, IL: Van Kampen Press, 1952), p. 75.

a universal probability bound. This means that any specified event beyond this value would be improbable and could not be attributed to chance.[8]

As we can see, the probability of getting a single small protein ($1/10^{130}$) far exceeds this limit. Even if the protein can interchange amino acids at various positions (such as in the case of the protein cytochrome a),[9] the resulting probability still exceeds the limit of $1/10^{50}$. So far we have only looked at the probability of getting a single small protein by random chance. What are the chances of getting all the proteins necessary for life?

> No matter how large the environment one considers, life cannot have had a random beginning . . . there are about two thousand enzymes, and the chance of obtaining them all in a random trial is only one part in $(10^{20})^{2000} = 10^{40,000}$, an outrageously small probability that could not be faced even if the whole universe consisted of organic soup.[10]

This number is so large (1 followed by 40,000 zeros) that it staggers the imagination how life could have evolved by natural, random processes. Yet, people continue to hold onto their belief that life did evolve by random chance (high credulity factor).

> Time is in fact the hero of the plot. . . . What we regard as impossible on the basis of human experience is meaningless here. Given so much time, the "impossible" becomes possible, the possible probable, and the probable virtually certain. One has only to wait: time itself performs the miracles.[11]

This statement attributes supernatural qualities to *time*! It also allows for anything to happen. This means we are no longer bound by the laws of science or any other natural limits. The statement thus becomes meaningless.

8. Emile Borel, *Probabilities and Life* (New York, NY: Dover, 1962), p. 28.
9. A transport protein involved in the transfer of energy (electrons) within cells.
10. Sir Fred Hoyle and Chandra Wickramasinghe, *Evolution from Space* (London: Dent, 1981), p. 148, 24.
11. George Wald, "The Origin of Life," p.48.

Tricks of the Trade

Since scientists have been unable to create life, they are forced to speculate through research and sometimes "sleight of hand" how it might have arrived on earth. Below are some of the tricks of the trade used to avoid the obvious — that God is the Creator of all things (Colossians 1:16).

1. It happens naturally

"The formation of biological polymers from monomers is a function of the laws of chemistry and biochemistry, and these are decidedly *not* random."[12]

Explanation

This is an incorrect statement. If it happens naturally, then why can't scientists duplicate this in the lab? Amino acids do not spontaneously bond together to make proteins. First, it takes a source of energy to do this. Second, the natural tendency is to bond left- and right-handed amino acids, but life requires all left-handed amino acids. Third, they must be in the correct order or the protein will not function properly. Fourth, it requires the instructions of DNA to get the right amino acids. Where did DNA come from? Fifth, protein molecules tend to break down in the presence of oxygen or water.

2. The deck of 52 cards

In a deck of 52 playing cards there are almost 10^{68} possible orderings of the cards. If we shuffle the deck we can conclude that the possible ordering of the cards having occurred in the order we got is 1 chance in 10^{68}. This is certainly highly improbable, but we did come up with this exact order of cards. Therefore, no matter how low the probability, events can still occur and evolution is not mathematically impossible.

Explanation

In this example the math is correct but the interpretation is wrong. If the arrangement had been specified beforehand, then the actual outcome would be surprising. By shuffling the cards, the probability is one that a sequence will occur. The fallacy is that the order is predicted after the fact.

3. All the people

We are in a room of 100 people. What is the probability that all 100 people would be here in this room at this exact time? The probability is enormous, but yet we are all here.

12. Ian Musgrave, "Lies, Damned Lies, Statistics, and Probability of Abiogenesis Calculations," TalkOrigins, www.talkorigins.org/faqs/abioprob/abioprob.html.

Explanation

Two things are wrong with this reasoning. First, the people were not pre-specified. This is another example of an after-the-fact prediction. Second, each person made a decision to attend; therefore, this is not a chance gathering. This turns out to be a misunderstanding between a chance event and intelligent choice.

4. Probability is not involved

Probability has nothing to do with evolution because evolution has no goal or objective.

Explanation

This statement disagrees with modern biology textbooks.

When there is more than one possible outcome and the outcome is not predetermined, probability can become a factor. In the case of evolution there is no pre-assigned chemical arrangement of amino acids to form a protein. Therefore, the formation of a biological protein is based on random chance. Scientists know today that it is only because of the instructions (information) in DNA that only left-handed amino acids are linked in the proper order.

> Cells link amino acids together into proteins, but only according to instructions encoded in DNA and carried in RNA.[13]

Both creationists and evolutionists agree that DNA is essential for linking the correct amino acids in a chain to form a protein. The unanswered question is, "Where and how did DNA acquire the enormous amount of information (instructions) to form a protein?" There is no known natural explanation that can adequately explain the origin of life, or even a single protein. The evolutionists are then left to rely on the odds (chance) that such a tremendous, improbable event occurred. Molecular biologist Michel Denton puts the event in perspective:

> Is it really credible that random processes could have constructed a reality, the smallest element of which — a functional protein or gene — is complex beyond our own creative capacities, a reality which is the very antithesis of chance, which excels in every sense anything produced by the intelligence of man?[14]

But wait, there is still more!

13. G.B. Johnson, Biology: *Visualizing Life* (Austin, TX: Holt, Rinehart, and Winston, 1998), p. 193.
14. Denton, *Evolution: A Theory in Crisis*, p. 342.

The Human Body, Time, and Evolution

It is estimated that the human body is made up of 60 trillion cells (60,000,000,000,000).[15] How long would it take to just assemble this many cells, one at a time and in no particular order at the rate of:

One per second	1.9 million years
One per minute	114 million years
One per hour	6.8 billion years

These ages assume no mistakes! However, the evolutionary mechanism is based upon random errors (mistakes) in the DNA. Also included in assembling all the 60 trillion cells is that they have to make the right organs which all have to interact.

The human body contains more than 40 billion capillaries extending over 25,000 miles, a heart that pumps over 100,000 times a day, red blood cells that transport oxygen to tissues, white blood cells that rush to identify enemy agents in the body and mark them for destruction, eyes and ears that are more complex than any man-made machine, a brain that contains over 100 trillion interconnections, plus many other parts such as the nervous system, skeleton, liver, lungs, skin, stomach, and kidneys.

The complexity and dimensions of the human body are staggering. The probability of assembling 60 trillion cells that form specific organs that all work together to form a single human being in the evolutionary time scale of 3.8 billion years is a giant leap of faith. However, an all-knowing, all-powerful Creator has told us in His Word that He is the designer.

> The hearing ear and the seeing eye, The LORD has made them both (Proverbs 20:12).

Every human body is a testimony to a purposeful Creator. As Malcolm Muggeridge said:

> One of the peculiar sins of the twentieth century which we've developed to a very high level is the sin of credulity. It has been said that when human beings stop believing in God they believe in nothing. The truth is much worse: they believe in anything.[16]

15. Boyce Rensberger, *Life Itself* (New York, NY: Oxford University Press, 1996), p. 11.
16. Malcolm Muggeridge, "An Eighth Deadly Sin," *Woman's Hour* radio broadcast, March 23, 1966. Quoted in Malcolm Muggeridge and Christopher Ralling, *Muggeridge Through the Microphone: B.B.C. Radio and Television* (London: British Broadcasting Corporation, 1967).

Conclusion

Probability arguments can present a strong argument for the existence of a Creator God. However, even when such evidence is presented to an evolutionist there is no guarantee that he or she will be persuaded. The real issue is not about evidence; it is a heart issue. As Christians we are called to have ready answers and break down strongholds that act as stumbling blocks to the unbeliever. It is the Holy Spirit that changes lives.

> But sanctify the Lord God in your hearts, and always be ready to give a defense to everyone who asks you a reason for the hope that is in you, with meekness and fear (1 Peter 3:15).

> For the weapons of our warfare are not carnal but mighty in God for pulling down strongholds, casting down arguments and every high thing that exalts itself against the knowledge of God, bringing every thought into captivity to the obedience of Christ (2 Corinthians 10:4–5).

17

What about Eugenics and Planned Parenthood?

DR. GEORGIA PURDOM

In 1915 a baby boy was born to Anna Bollinger. The baby had obvious deformities, and medical doctor Harry Haiselden decided the baby was not worth saving.[1] The baby was denied treatment and died. The story became national news and the cruelty of eugenic practices became public knowledge.

The year 1915 seems far removed from our modern times, but the concept of eugenics is alive and well. In 2005, two doctors from the Netherlands published "The Groningen Protocol — Euthanasia in Severely Ill Newborns."[2] This protocol was published to help doctors decide whether or not a newborn should be actively killed based on the newborn's disease and perceived quality of life.[3]

In this chapter we will explore historical and modern perspectives of eugenics, how Planned Parenthood has played a role in furthering the cause of

1. "A friend of Anna's asked the doctor, 'If the poor little darling has one chance in a thousand won't you operate and save it?' The doctor laughed and replied, 'I'm afraid it might get well.' " Edwin Black, *War against the Weak* (New York, NY: Four Walls Eight Windows, 2003), p. 252.
2. Eduard Verhagen and Pieter J. J. Sauer, "The Groningen Protocol — Euthanasia in Severely Ill Newborns," *New England Journal of Medicine* 352 no. 10 (2005): 959–962.
3. The doctors analyzed 22 cases of newborns with severe spina bifida that had been euthanized. What is the typical outcome for individuals with spina bifida? The March of Dimes web page on spina bifida states, "With treatment, children with spina bifida [all forms] usually can become active individuals. Most live normal or near-normal life spans." March of Dimes, "Spina Bifida," www.marchofdimes.com/pnhec/4439_1224.asp. And yet these children were considered by the doctors to not have a life worth living.

eugenics in the past and present, and what the proper biblical perspective on these issues should be.

What Is Eugenics?

The term *eugenics* was first coined in 1883 by Francis Galton, father of eugenics and cousin of Charles Darwin. The term comes from the Greek roots *eu* (good) and *genics* (in birth) to communicate the idea of being well-born.

The ultimate goal of eugenics was to create a superior race of humans.[4] Many adherents believed in evolution by natural selection, but that natural selection was moving too slowly in favoring the best and eliminating the worst.[5] They also believed that charity in the form of taking care of the poor and sick was prohibiting natural selection from working properly and thus the need to intervene with artificial selection.[6]

Artificial selection was accomplished through two types of eugenics — positive and negative. Positive eugenics focused on increasing the "fit" through promoting marriages among the well-born and promoting those fit couples to have multiple children. Negative eugenics focused on decreasing the number of the "unfit" through prohibiting birth (birth control and sterilization) and segregation (e.g., institutionalization of the unfit, marriage restriction laws, and immigration restriction).

---

4. Dr. John Harvey Kellogg, founder of the Race Betterment Foundation, stated, "We have wonderful new races of horses, cows, and pigs. Why should we not have a new and improved race of men?" Black, *War against the Weak*, p. 88.
5. Leading eugenicist Paul Popenoe in his 1915 paper entitled, "Natural Selection in Man," stated, "Science knows no way to make good breeding stock out of bad, and the future of the race is determined by the kind of children which are born and survive to become parents in each generation. There are only two ways to improve the germinal character of the race, to better it in a fundamental and enduring manner. One is to kill off the weaklings born in each generation. That is Nature's way, the old method of natural selection which we all agreed must be supplanted. When we abandon that, we have but one conceivable alternative, and that is to adopt some means by which fewer weaklings will be born in each generation. The only hope for permanent race betterment under social control is to substitute a selective birth-rate for Nature's selective death-rate. That means — eugenics." Steven Selden, *Inheriting Shame: The Story of Eugenics and Racism in America* (New York, NY: Teachers College Press, 1999), p. 11.
6. In her 1922 book *Pivot of Civilization*, Margaret Sanger, founder of Planned Parenthood, stated, "Organized charity itself is the symptom of a malignant social disease. Those vast, complex interrelated organizations aiming to control and to diminish the spread of misery and destitution and all the menacing evils that spring out of this sinisterly fertile soil, are the surest sign that our civilization has bred, is breeding and is perpetuating constantly increasing numbers of defectives, delinquents and dependents. My criticism, therefore, is not directed at the 'failure' of philanthropy, but rather at its success." Black, *War Against the Weak*, p. 129

History of Eugenics

Although many people associate eugenics with the late 1800s and early 1900s, it is an ancient idea that was in practice long before it was called eugenics. The Law of the Twelve Tables (449 B.C.), which served as the foundation of Roman Law, states "*Cito necatus insignis ad deformitatem puer esto,*" which means, "An obviously deformed child must be put to death."[7] Both Plato and Aristotle supported this practice[8] and it was not uncommon for infants to be exposed or left outside the home for a period of time to determine if they were fit enough to survive. The Romans wanted only the most fit for their future warriors.

Francis Galton, upon reading his cousin Charles's book *Origin of Species,* [9] decided to apply the mechanisms of natural and artificial selection to man. He stated, "Could not the undesirables be got rid of and the desirables multiplied?"[10] Galton promoted the ideas that human intelligence and other hard-to-measure traits such as behaviors were greatly influenced by heredity (not the environment, which was the popular mindset of the day).[11] He advocated for a program of positive eugenics. His book *Hereditary Genius* (1869) was well liked by Charles[12] and had a great influence on the ideas presented in his book *Descent of Man* (1871).[13]

7. Wikipedia, "Twelve Tables," www.en.wikipedia.org/wiki/Twelve_Tables.
8. Christian Medical and Dental Association, "A History of Eugenics," www.cmda.org/AM/Template.cfm?Section=Home&CONTENTID=4214&TEMPLATE=/CM/ContentDisplay.cfm.
9. Galton writing to Darwin stated, "I have laid [*Origin of Species*] down in the full enjoyment of a feeling that one rarely experiences after boyish days, of having been initiated into an entirely new province of knowledge, which, nevertheless, connects itself with other things in a thousand ways." Correspondence between Charles Darwin and Francis Galton, Letter 82, www.galton.org/letters/darwin/correspondence.htm.
10. Black, *War against the Weak*, p. 16.
11. Galton wrote, "I have not patience with the hypothesis occasionally expressed, and often implied, especially in tales written to teach children to be good, that babies are born pretty much alike and that the sole agencies in creating differences between boy and boy, and man and man, are steady application and moral effort. It is in the most unqualified manner that I object to pretensions of natural equality." Donald DeMarco and Benjamin Wiker, *Architects of the Culture of Death* (San Francisco, CA: Ignatius Press, 2004), p. 94.
12. Darwin, writing to Galton, stated, "Exhale myself [*sic*], else something will go wrong with my inside. I do not think I ever in all my life read [*Hereditary Genius*] anything more interesting and original — and how well and clearly you put every point!" DeMarco and Wiker, *Architects of the Culture of Death*, p. 92.
13. "But some remarks on the action of natural selection on civilised nations may be here worth adding. This subject has been ably discussed by Mr. W.R. Greg, and previously by Mr. Wallace and Mr. Galton. Most of my remarks are taken from these three authors. With savages, the weak in body or mind are soon eliminated; and those that survive commonly exhibit a vigorous state of health. We civilised men, on the other hand, do our utmost to

In the early 1900s the eugenics movement became well established in the United States. The movement was well-funded by men like Carnegie, Rockefeller, and Kellogg. Eugenic societies, conferences, research institutions, and journals gave a façade of real science to the study of eugenics. This was further promoted by eugenic departments and courses at the university level.

Francis Galton, Darwin's cousin, promoted eugenic beliefs.

The American eugenics movement focused heavily on negative eugenics.[14] Ten classes of social misfits were determined upon which programs of negative eugenics were applied.

> First, the feebleminded; second, the pauper class; third, the inebriate class or alcoholics; fourth, criminals of all descriptions including petty criminals and those jailed for nonpayment of fines; fifth, epileptics; sixth, the insane; seventh, the constitutionally weak class; eighth, those predisposed to specific diseases; ninth, the deformed; tenth, those with defective sense organs, that is, the deaf, blind, and mute.[15]

All of these traits were thought to be inheritable.[16] Ten percent of the American population was thought to fit into these broad, ill-defined categories (sometimes known as the "submerged tenth").[17] Many of those people were forcibly institutionalized in asylums for the "feebleminded and epileptic." Although not stated in

---

check the process of elimination; we build asylums for the imbecile, the maimed, and the sick; we institute poor-laws; and our medical men exert their utmost skill to save the life of every one to the last moment. There is reason to believe that vaccination has preserved thousands, who from a weak constitution would formerly have succumbed to small-pox. Thus the weak members of civilised societies propagate their kind. No one who has attended to the breeding of domestic animals will doubt that this must be highly injurious to the race of man. It is surprising how soon a want of care, or care wrongly directed, leads to the degeneration of a domestic race; but excepting in the case of man himself, hardly any one is so ignorant as to allow his worst animals to breed." Charles Darwin, *The Descent of Man, and Selection in Relation to Sex*, 1st edition (London: John Murray, 1871), p.168–169.

14. Positive eugenics was also encouraged but to a lesser degree. Fitter family contests were held at many county fairs to disseminate information about eugenics and to encourage with prizes and recognition of the "fittest" families to reproduce.
15. Black, *War against the Weak*, p. 58.
16. Leading eugenicist Charles Davenport stated, "When we look among our acquaintances we are struck by their diversity in physical, mental, and moral traits . . . they may be selfish or altruistic, conscientious or liable to shirk . . . for these characteristics are inheritable." Ibid., p. 105–106.
17. Ibid., p. 52.

Logo of the Second International Congress of Eugenics, 1921

the list, those of "races" other than the Caucasian "race" would also, by the mere fact of ethnic background, be placed into one or more of these categories. Unfortunately, the eugenics movement in the United States heavily influenced Hitler and his scientists and, in return, many eugenicists and eugenic publications supported the horrifying practices of Hitler's Nazi regime. Negative eugenic practices were even sanctioned by the American government.

Forced Sterilization

In 1907, Indiana enacted the first forced sterilization law. The law would be applied to "mentally impaired patients, poorhouse residents, and prisoners."[18] Over 30 states enacted sterilization laws, and between 60,000 and 70,000 people were forcibly sterilized between 1900 and 1970.[19] Most forced sterilizations were performed after 1927. In 1927 the Supreme Court ruled in favor of the forced sterilization of Carrie Buck[20] (in *Buck* v. *Bell*) with justice Oliver Wendell Holmes stating, "It is better for all the world, if instead of waiting to execute degenerate offspring for crime . . . society can prevent those who are manifestly unfit from continuing their kind. . . . Three generations of imbeciles are enough."[21]

Immigration Restriction

The Immigration Act of 1924 set quotas on the number of people allowed into the United States from other countries. Lawmakers were heavily influenced by

18. Ibid., p. 67.
19. Joan Rothschild, *The Dream of the Perfect Child* (Bloomington, IN: Indiana University Press, 2005), p. 45; Black, *War against the Weak*, p. 398.
20. Carrie's widowed mother, Emma, was considered feebleminded and institutionalized at the Colony for Epileptics and Feebleminded in Virginia. Carrie was raped as a teenager and subsequently institutionalized at the same colony. Carrie's baby, Vivian, who was eight months old when evaluated, was said to not look quite right. Thus, "three generations of imbeciles" as declared by Holmes. Black, *War against the Weak*, p.108–123.
21. Robert Marshall and Charles Donovan, *Blessed are the Barren: The Social Policy of Planned Parenthood* (San Francisco: Ignatius Press, 1991), p. 277.

"scientific data" presented to them by high-ranking members of the eugenics movement.[22]

Marriage Restriction Laws

These laws (which varied by state) were designed to keep the Caucasian "race" pure. The laws prohibited "mixed race" marriages (i.e., Negro and Caucasian) but also marriages with those considered defective (e.g., blind).

What Was the Christian Response to Eugenics?

The Christian response to eugenics was mixed. The Christian apologist G.K. Chesterton condemned eugenics in his 1922 book *Eugenics and Other Evils*. He saw how eugenics was being used in Germany to support Nazi ideals.[23]

However, some pastors used their pulpits to promote eugenics. The American Eugenics Society sponsored a sermon contest in 1926. Of the five sermons I read online, all were filled with popular rhetoric from the eugenics movement with little scriptural support given for eugenics. The pastors seemed to have accepted the "science" of eugenics without analyzing it in light of the Bible.[24] This is very similar to the modern situation in which many Christian pastors accept the "science" of evolution, promote the idea in their churches, and don't analyze the conflicts between evolution and Scripture.

History of Planned Parenthood and Its Relationship to Eugenics

The name most commonly associated with Planned Parenthood is that of its founder Margaret Sanger. Margaret was born in 1879, the 6th of 11 children in a poor family, in New York.[25] She was initially quite committed to the Catholic faith but eventually became very cynical in part due to the influence of her

22. Black, *War against the Weak*, p. 202.
23. George Grant, *Grand Illusions: The Legacy of Planned Parenthood* (Franklin, TN: Adroit Press, 1992), p. 94.
24. Scriptural supports for eugenics were often verses taken out of context. For example: "Of a certain moral weakling Jesus said: 'It would be better for him if he had not been born' [referring to Judas Iscariot, Mark 14:21; NIV]. The same thing might be said of millions of weaklings today. . . . And if these millions might be prevented from reproduction so that succeeding generations might appear without their handicaps what a great step would be taken toward the realization of a better order of society of which Jesus dreamed! . . . And the Christian eugenicist believes that in the spirit and purpose of his work he would have the unqualified approval of Jesus." "Eugenics," Sermon #36 excerpt, American Eugenic Society Sermon Contest, 1926, www.eugenicsarchive.org/eugenics/topics_fs.pl?theme=32&search=&matches. However, Paul wrote, "for all have sinned and fall short of the glory of God" (Romans 3:23). Nothing we do on earth can bring about the perfect "new heaven and new earth" (Revelation 21:1) that will someday be brought into existence by God Himself.
25. Grant, *Grand Illusions: The Legacy of Planned Parenthood*, p. 47.

"free thinking" father.[26] Margaret married into money and eventually became an active member of the Socialist Party. She was attracted to the party's fight for "women's suffrage, sexual liberation, feminism, and birth control."[27] Sanger also became a fan of the concepts promoted by Thomas Malthus (who also heavily influenced Charles Darwin in the development of the concept of evolution by natural selection). Malthus was concerned that the human population was growing too rapidly (especially the poor, diseased, and racially inferior) and would outgrow natural resources. The solution proposed by his followers, like Sanger, was to decrease and eliminate the "inferior" population through birth control (including sterilization and abortion).[28] Sanger stated, "The most merciful thing a large family can do to one of its infant members is to kill it."[29]

Margaret Sanger, founder of Planned Parenthood, promoted birth control as a means of controlling the "unfit" in society.

Sanger became one of the foremost champions of birth control and not just for the benign reason of helping poor women who could not afford large families, but also for "the liberation of sexual desire and the new science of eugenics."[30] In 1921 she organized the American Birth Control League. In 1922 she published the book *The Pivot of Civilization* which "unashamedly called for the elimination of 'human weeds,' for the cessation of charity, for the segregation of 'morons, misfits, and the maladjusted' and for the sterilization of 'genetically inferior races.' "[31] Sanger stated:

> The emergency problem of segregation and sterilization must be faced immediately. Every feeble-minded girl or woman of the hereditary type, especially of the moron class, should be segregated

26. Ibid., p. 48.
27. Ibid., p. 50.
28. Ibid., p. 56.
29. Ibid., p. 63.
30. DeMarco and Wiker, *Architects of the Culture of Death*, p. 291.
31. Grant, *Grand Illusions: The Legacy of Planned Parenthood*, p.59. Sanger, in *Pivot of Civilization*, stated, "Birth control, which has been criticized as negative and destructive, is really the greatest and most truly eugenic method, and its adoption as part of the program of Eugenics would immediately give a concrete and realistic power to their science. As a matter of fact, Birth Control has been accepted by the most clear thinking and far seeing of the Eugenicists themselves as the most constructive and necessary of the means to racial health." Black, *War against the Weak*, p. 129.

> during the reproductive period. Otherwise, she is almost certain to bear imbecile children, who in turn are just as certain to breed other defectives. . . . Moreover, when we realize that each feeble-minded person is a potential source of an endless progeny of defect, we prefer the policy of immediate sterilization, of making sure that parenthood is absolutely prohibited to the feeble-minded.[32]

Her magazine, *The Birth Control Review*, contained many articles authored by leading eugenicists of her day. Sanger openly endorsed the concepts and methods of race purification carried out by the Nazis.[33] Sanger believed sex was an evolutionary force that should not be prohibited because of its ability to create genius.[34] In 1942, the American Birth Control League became the Planned Parenthood Federation of America (PPFA).

Modern Perspectives on Eugenics and Planned Parenthood

Eugenics became associated with the horrors of the Nazi regime in the 1940s and so its popularity in the public arena began to fade. In addition, much of the so-called "science" of eugenics was shown to be false by increased knowledge in the field of genetics. It became almost laughable to think that the eugenic-defined trait of "sense of humor" (no pun intended!) could be associated with a particular gene and/or somehow quantified.

However, eugenic concepts and the eugenic ideals of PPFA didn't die. Edwin Black states, "While human genetics was becoming established in America, eugenics did not die out. It became quiet and careful."[35] The eugenic agenda today is not different in principle or goal but only in name and methods. Eugenicist Frederick Osborn in 1965 stated, "The term medical genetics has taken the

32. Ibid., p.131.
33. Grant, *Grand Illusions: The Legacy of Planned Parenthood*, p. 61.
34. DeMarco and Wiker, *Architects of the Culture of Death*, p. 295. Sanger, in *Pivot of Civilization*, stated, "Modern science is teaching us that genius is not some mysterious gift of the gods. . . . Nor is it. . . the result of a pathological and degenerate condition. . . . Rather it is due to the removal of physiological and psychological inhibitions and constraints which makes possible the release and channeling of the primordial inner energies of man into full and divine expression." Ibid. Sanger, in *Pivot of Civilization*, stated, "Slowly but surely we are breaking down the taboos that surround sex; but we are breaking them down out of sheer necessity. The codes that have surrounded sexual behavior in the so-called Christian communities, the teachings of the churches concerning chastity and sexual purity, the prohibitions of the laws, and the hypocritical conventions of society, have all demonstrated their failure as safeguards against the chaos produced and the havoc wrought by the failure to recognize sex as a driving force in human nature — as great as, if indeed no greater than, hunger. Its dynamic energy is indestructible." Ibid., p. 295–296.
35. Black, *War against the Weak*, p.421.

place of the term negative eugenics."[36] Genetic databases filled with individual genetic identities could now generate precise family genetic profiles as opposed to the subjective determination of non-measurable traits by self or other family members stored on millions of index cards that filled eugenic institutions in the early 20th century. In recent years, many feared the adverse use of genetic identities and profiles when applying for jobs and insurance.[37]

James Watson, co-discoverer of the structure of DNA, stated in 2003, "If you are really stupid, I would call that a disease. The lower 10 percent who really have difficulty, even in elementary school, what's the cause of it? A lot of people would like to say, 'Well, poverty, things like that,' It probably isn't. So I'd like to get rid of that, to help lower the 10 percent."[38] The idea of the "submerged tenth" is still alive and well in the 21st century.

Preimplantation genetic diagnosis (PGD) allows parents who have embryos created for use in in vitro fertilization (IVF) to check for genetic disorders and chromosomal abnormalities before the embryos are implanted. The "defective" embryos are destroyed. PGD is also being used for sex selection (only babies of the desired sex are used for IVF), disability selection (e.g., deafness), and predisposition or late-onset disease selection (i.e., predispositions to cancer and late-onset diseases like Alzheimer's).[39] Embryos are destroyed if they are not the desired sex, will have a disability, or may have cancer or disease later in life. PPFA endorses prenatal diagnosis procedures and genetic counseling.[40] Eugenic concepts of prohibiting the birth of the "unfits" is still popular in the 21st century.

Planned Parenthood still endorses many eugenic ideas. This should not be surprising as the PPFA website "History and Successes" page clearly states, "Margaret Sanger, the founder of Planned Parenthood, is one of the movement's great heroes. Sanger's early efforts remain the hallmark of Planned Parenthood's mission. . . ."[41] Sanger's efforts advocated sterilization, abortion, and infanticide of "defectives" in the name of eugenics. Further indicative of the promotion of eugenics, PPFA endorses abortion of deformed babies:

36. Ibid., p. 424.
37. The Anti-Genetic Discrimination Bill was passed into U.S. law in 2008. The law states that genetic information cannot be used against an individual for insurance or job purposes. Many countries have no such law.
38. Black, *War Against the Weak*, p. 442.
39. Susannah Baruch, David Kaufman, and Kathy L. Hudson, "Genetic Testing of Embryos: Practices and Perspectives of US in vitro Fertilization Clinics," *Fertility and Sterility* 89 no. 5 (2008): 1053–1058.
40. DeMarco and Wiker, *Architects of the Culture of Death*, p. 301.
41. Planned Parenthood, "History and Successes," www.plannedparenthood.org/about-us/who-we-are/history-and-successes.htm.

> From 1956 to 1962, hundreds of women in the U.S. and Europe who took the drug thalidomide while pregnant give birth to children missing arms and legs. Sherri Finkbine, an American mother of four who used thalidomide, is refused an abortion. More than 60 percent of Americans disapprove of the refusal. Mrs. Finkbine flees to Sweden for a safe, legal abortion. (The fetus is gravely deformed.) Her case and others involving women who have taken thalidomide convince many Americans that anti-abortion laws need reform.[42]

Thus, those infants who are "gravely deformed" should have been permitted to be eliminated according to PPFA. According to the American Life League, in 2006 PPFA was directly responsible (through its clinics) for 289,750 abortions.[43] Thus, PPFA was responsible for almost 25 percent of the abortions estimated to have occurred in the U.S. in 2006.[44]

PPFA also still advocates for sexual liberation by encouraging the concept that sex and sexual desire is part of a normal, healthy lifestyle.[45] These concepts are in line with Sanger's view of sex, which she wrote about in a letter to her 16-year-old granddaughter: "Kissing, petting, and even intercourse are alright as long as they are sincere."[46] Alan Guttmacher, former president of PPFA stated, "We are merely walking down the path that Mrs. Sanger carved out for us."[47] How true!

Biblical Perspectives on Eugenics and Planned Parenthood

When we start with the truth of God's Word, we see that eugenics and the ideas promoted by Planned Parenthood do not align with the Bible.

The Bible shows that God considers all people equal.

> There is neither Jew nor Greek, there is neither slave nor free, there is neither male nor female; for you are all one in Christ Jesus (Galatians 3:28).

42. Ibid.
43. American Life League, "Abortion and Planned Parenthood Statistics," www.all.org/article.php?id=10123.
44. The estimated number of abortions that occurred in the US in 2006 is 1,206,200. National Right to Life, "Abortion in the United States: Statistics and Trends," www.nrlc.org/ABORTION/facts/abortionstats.html.
45. "A basic understanding of sex and sexuality can help us sort out myth from fact and help us all enjoy our lives more." and ". . . the more we know about sex and sexuality, the better we are able to take charge of our sex lives and our sexual health." Planned Parenthood, "Sex and Sexuality," www.plannedparenthood.org/health-topics/sexuality-4323.htm.
46. DeMarco and Wiker, *Architects of the Culture of Death*, p. 294.
47. Grant, *Grand Illusions: The Legacy of Planned Parenthood*, p. 63.

> And He has made from one blood every nation of men to dwell on all the face of the earth (Acts 17:26a).

God doesn't care whether people have dark brown skin or light brown skin, whether they are deaf or have perfect hearing — God does not show partiality.

The Bible shows that life is precious to God.

> Then God said, "Let Us make man in Our image, according to Our likeness; let them have dominion over the fish of the sea, over the birds of the air, and over the cattle, over all the earth and over every creeping thing that creeps on the earth." So God created man in His own image; in the image of God He created him; male and female He created them (Genesis 1:26–27).

> For You formed my inward parts;
> You covered me in my mother's womb.
> I will praise You, for I am fearfully and wonderfully made;
> Marvelous are Your works,
> And that my soul knows very well.
> My frame was not hidden from You,
> When I was made in secret,
> And skillfully wrought in the lowest parts of the earth.
> Your eyes saw my substance, being yet unformed.
> And in Your book they all were written,
> The days fashioned for me,
> When as yet there were none of them (Psalm 139:13–16).

> For God so loved the world that He gave His only begotten Son, that whoever believes in Him should not perish but have everlasting life (John 3:16).

God created each of us individually and we are His image-bearers on earth. He loved us so much that He sent His Son Jesus to die for us so that we might have eternal life.

The Bible shows the importance of caring for the needy.

> You shall neither mistreat a stranger nor oppress him, for you were strangers in the land of Egypt. You shall not afflict any widow or fatherless child. If you afflict them in any way, and they cry at all to Me, I will surely hear their cry (Exodus 22:21–23).

> Then the King will say to those on His right hand, "Come, you blessed of My Father, inherit the kingdom prepared for you from the foundation of the world: for I was hungry and you gave Me food; I was thirsty and you gave Me drink; I was a stranger and you took Me in; I was naked and you clothed Me; I was sick and you visited Me; I was in prison and you came to Me" (Matthew 25:34–36).

God commands us to care for people no matter what their affliction.

Conclusion

My friends John and Tina were told after 19 years of marriage that they were going to have a baby.[48] They were very excited and then the news came that the baby might have a chromosomal abnormality. Tina shared with me:

> Our doctor advised us multiple times to abort our baby because she was considered high risk for chromosomal issues. We were never swayed because we knew that this surprise little bundle was a gift from God. We experienced sheer ecstasy when our eyes beheld Eden Lanay for the first time. Our seven days with her will no doubt be the highlight of our entire lives [Eden was born with Trisomy 18, Edward's Syndrome]. We are so grateful to God for blessing us beyond measure with our beautiful baby girl.[49]

John said:

> As difficult as Eden's death was, we cherish our time with her. My heart breaks for those who lose their child before birth due to miscarriage or abortion. They have missed out on a marvelous experience with a new life.
>
> The seven days we had with Eden were more glorious than I can describe. I will hold on to those precious memories for the rest of my life.[50]

Life is precious — no matter how short or how impaired that life may be. Contrary to the ideas supported by eugenics and Planned Parenthood, all human life has value because it comes from the Life Giver.

48. To read more about their amazing journey and testimony, see their blog, "Baby Graves," www.babygravesdownunder.blogspot.com.
49. John and Tina Graves, email message to author, August 26, 2009.
50. Ibid.

18

When and How Did the Grand Canyon Form?

DR. ANDREW A. SNELLING AND TOM VAIL

The Grand Canyon is one of the world's most awesome erosional features. It is 277 miles (446 km) long, including the 60 miles (96 km) of Marble Canyon upstream. The depth of the main segment of the Grand Canyon varies between 3,000 and 6,000 feet (900 and 1,800 m), with the rim-to-rim width between 4 and 18 miles (6 and 29 km). Its origin has plagued geologists since the time of John Wesley Powell's first courageous voyage down the Colorado River in 1869. Despite an increase in knowledge about its geology, evolutionary geologists have yet been unable to explain the canyon.[1]

Into What Was the Grand Canyon Carved?

Before discussing when and how the Grand Canyon was formed, it is first important to understand where and through what geologic feature it was carved. Located in northern Arizona, the Grand Canyon has been eroded through the southern end of the Colorado Plateau. Carved through sedimentary layers of sandstone, limestone, and shale and into the basement formations of mostly metamorphic schists and igneous granites, the Grand Canyon is a testimony to the erosive power of water.

1. J.W. Powell, *Grand Canyon: Solving Earth's Grandest Puzzle* (New York, NY: PI Press, 2005); W. Ranney, *Carving Grand Canyon: Evidence, Theories and Mystery* (Grand Canyon, AZ: Grand Canyon Association, 2005); R. Young and E. Spamer, eds., *Colorado River Origin and Evolution: Proceedings of a Symposium held at Grand Canyon National Park in June 2000* (Grand Canyon, AZ: Grand Canyon Association, 2001).

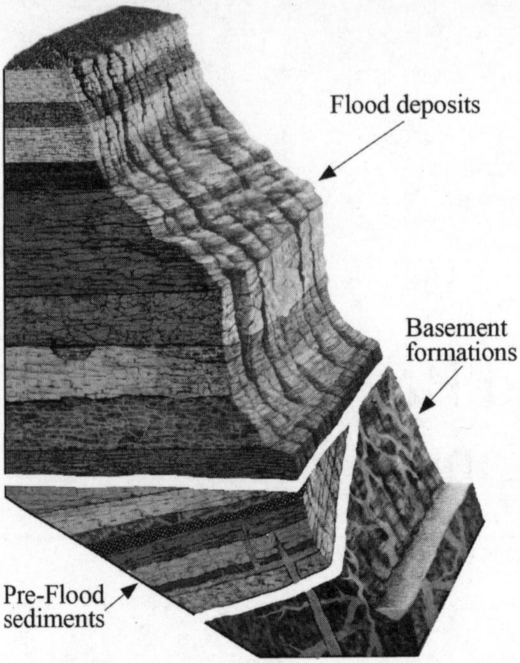

Figure 1. Grand Canyon strata diagram

But how did these rock layers first form? They can be divided into three groups as shown in figure 1. The crystalline basement formations are believed by most creation geologists to have been set in place on day 3 of the creation week. The tilted pre-Flood sediment layers are up to 14,000 feet (4,260 m) in thickness, but are only exposed in the eastern canyon and in a few other areas. The upper layers — the horizontal Flood deposits — cover the entire plateau and, in some cases, the vast majority of the North American continent.[2]

Three Undisputed Observations

The Enormous Scale of Erosion

A simple calculation of the volume of the Grand Canyon reveals almost 1,000 cubic miles (4,000 cubic km) of material have been removed from northern Arizona to produce just the topographic shape of the canyon itself. However, that is not all the erosion which occurred. The Grand Canyon has been carved into a broad elevated area known as the Colorado Plateau (figure 2). The Colorado Plateau covers an area of about 250,000 square miles (647,000 square km) and consists of several smaller plateaus, which today stand at slightly varying elevations. The Kaibab Plateau, which reached more than 9,000 feet (2,740 m), forms part of the North Rim of the Grand Canyon. The sequence of sedimentary rock layers that forms these plateaus consists of many more layers than those exposed in the walls of the Grand Canyon today. In addition, to the north of the canyon there is a sequence of ascending cliffs called the Grand Staircase in which a further 10,000 feet (3,000 m) of sedimentary layers are exposed (figure 3). However, in the Grand Canyon region, most of these layers have been eroded away leaving just a few remnants, such as Red Butte (figure

2. L.L. Sloss, "Sequences in the Cratonic Interior of North America," *Geological Society of America Bulletin* 74 (1963): 93–114.

Figure 2. The extent of the Colorado Plateau

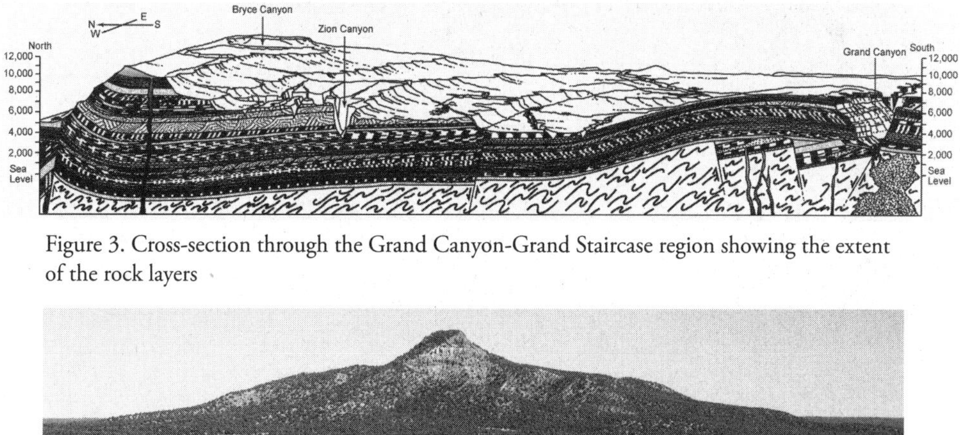

Figure 3. Cross-section through the Grand Canyon-Grand Staircase region showing the extent of the rock layers

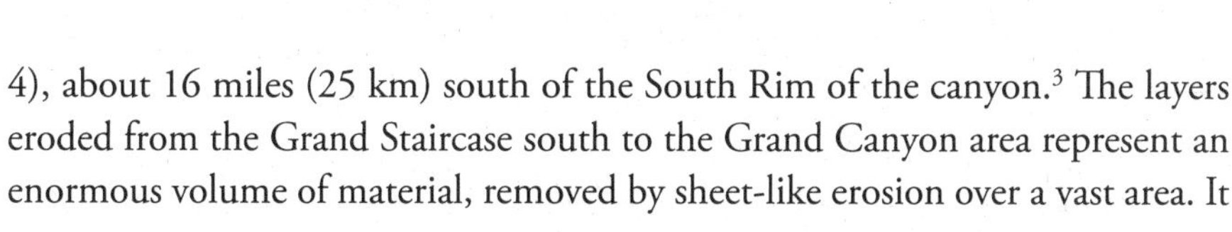

Figure 4. Red Butte

4), about 16 miles (25 km) south of the South Rim of the canyon.[3] The layers eroded from the Grand Staircase south to the Grand Canyon area represent an enormous volume of material, removed by sheet-like erosion over a vast area. It

3. T. Vail, M.J. Oard, J. Hergenrather, and D. Bokovoy, *Your Guide to the Grand Canyon: A Different Perspective* (Green Forest, AR: Master Books, 2008), p. 54.

has been estimated that this volume of sediments eroded from the plateau was around 100,000 cubic miles (400,000 cubic km)![4]

The Grand Canyon Was Cut Through the Plateau

Perhaps the most baffling observation, even to evolutionary geologists, is that the Grand Canyon cuts through, not around, a great plateau. Ranney, in his 2005 book *Carving Grand Canyon: Evidence, Theories and Mystery*, said:

> Oddly enough, the Grand Canyon is located in a place where it seemingly shouldn't be. Some twenty miles east of Grand Canyon Village, the Colorado River turned sharply ninety degrees, from a southern course to a western one and into the heart of the uplifted Kaibab Plateau. . . . It appears to cut right through this uplifted wall of rock, which lies three thousand feet above the adjacent Marble Platform to the east.[5]

Indeed, the headwaters of the Colorado River are at a lower elevation than the top of the Kaibab Plateau through which the Grand Canyon has been cut (figure 5).

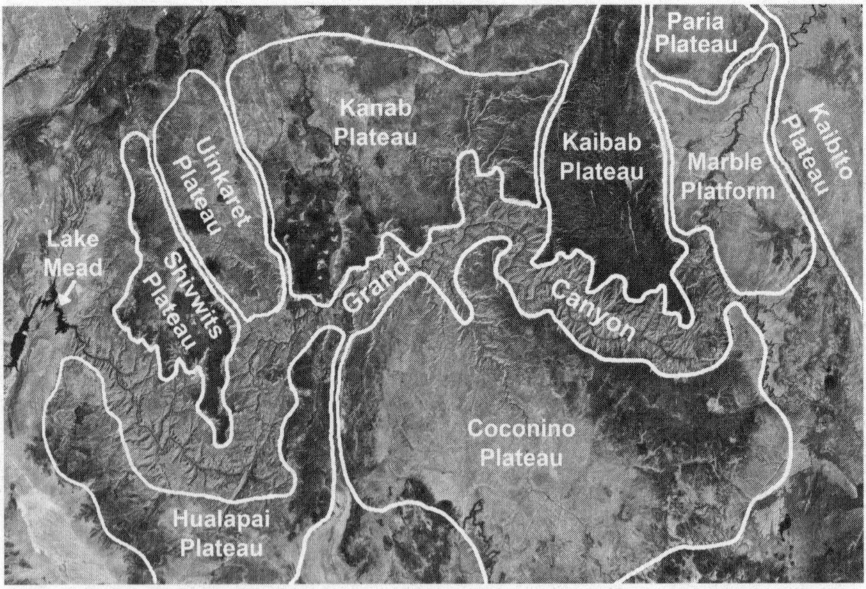

Figure 5. NASA satellite image of the Grand Canyon area, with outline of the different plateaus through which the canyon cuts.

4. M.J. Oard, T. Vail, J. Hergenrather, and D. Bokovoy, "Formation of Rock Layers in the Grand Staircase," in *Your Guide to Zion and Bryce Canyon National Parks: A Different Perspective* (Green Forest, AR: Master Books, 2010), p. 140.
5. Ranney, *Carving Grand Canyon: Evidence, Theories and Mystery*, p. 20.

Uplift of This Plateau Occurred Before Erosion of the Grand Canyon

This third observation also has profound implications concerning the origin of the Grand Canyon. At the eastern edge of the Kaibab Plateau, the sedimentary rock layers were bent, or as a geologist would say "folded," along the East Kaibab Monocline at the time the plateau was uplifted. The uppermost folded layers have been beveled by erosion and overlaid by the flat-lying Wasatch Formation, which is younger.[6] Furthermore, gravel deposits (from the Paleocene and Eocene epochs and thus younger than the folded Cretaceous layers) occur within channels eroded into the surface of the Kaibab Plateau, indicating the major uplift of the plateau and the accompanying erosion of its surface coincided with the uplift of the whole Colorado Plateau.[7] Therefore, in evolutionary thinking the plateau is geologically "old," and most evolutionary geologists believe its uplift occurred before erosion of the canyon into and through the plateau. But that leaves the headwaters of the Colorado River at a lower elevation than the top of that plateau, which indicates the Colorado River could not have carved the Grand Canyon!

The Secular Controversy Over When the Grand Canyon Was Eroded

Over the last 30 years, the time frame for the carving of the Grand Canyon has gone full circle. Thirty years ago, most evolutionists believed the canyon was about 70 million years old. But that estimate changed as radioisotope dating was utilized to show the plateau to be much older than the canyon itself. Basalts found on the North Rim near the western end of the canyon were estimated to be only 6 million years old, but these same basalts are also found on the South Rim![8] This means these lavas had to flow across from one rim to the other, a process which could not have occurred had the canyon been in place at the time. The age of at least the western Grand Canyon was thus reduced to 6 million years, but many continued to believe the central and eastern canyon was 70 million years old, based on the stream-capturing theory outlined below.

6. P.L. Babenroth and A.N. Strahler, "Geomorphology and Structure of the East Kaibab Monocline, Arizona and Utah," *Geological Society of America Bulletin* 56 (1945): 107–150.
7. D.P. Elston, R.A. Young, E.M. McKee, and M.L. Dennis, "Paleontology, Clast Ages, and Paleomagnetism of Upper Paleocene and Eocene Gravel and Limestone Deposits, Colorado Plateau and Transition Zone, Northern and Central Arizona," in *Geology of Grand Canyon, Northern Arizona (with Colorado River Guides)*, D.P. Elston, G.H. Billingsley, and R.A. Young, eds. (Washington, DC: American Geophysical Union, 1989), p. 155–173.
8. I. Lucchitta, "History of the Grand Canyon and of the Colorado River in Arizona," in *Grand Canyon Geology*, second edition, S.S. Beus, and M. Morales, eds. (New York, NY: Oxford University Press, 2003), p. 270–272.

Subsequently, the 70-million-year date was gradually reduced to 17 million years, based on several pieces of the puzzle indicating a younger canyon.[9]

New findings continue to question the age of the canyon. Some scientists still suggest 70 million years as the correct age, while others place it at less than 6 million years.[10] The debate goes on, with none of the accepted dating methods providing a clear-cut answer to the age of the Grand Canyon.[11]

The Secular Claims About How the Canyon Was Eroded

John Wesley Powell was the first to attempt an explanation of how the Grand Canyon was formed. Known as the "antecedent river" theory, Powell theorized an ancient river eroded down into the Colorado Plateau at the same rate the plateau was being uplifted.[12] Although this slow, gradual process fit nicely into the ruling uniformitarian thinking, over the next 50 to 75 years it was rejected by most geologists. The fatal blow against it came with radioisotope dating of the rim rocks.

The antecedent river theory was replaced by the idea of "stream capturing." Stream capturing suggests that through a process called headward erosion, the Grand Canyon was cut from the west through the plateau to "capture" the river, which ran a different direction at the time.[13] This is the theory many evolutionary geologists hold today, but it has seen significant changes over the last 30 years.

The initial stream capturing model had the ancestral Colorado River running through Marble Canyon to the Little Colorado River drainage, where the river then took a southeasterly direction, draining east into the Rio Grande River (figure 6). Another drainage existed to the west of the plateau cutting back through the plateau. However, its headward erosion then cut eastward through about 200 miles (320 km) of the Colorado Plateau and captured the ancestral Colorado River, which then changed its flow to a westerly direction. Subsequent to this capture, the area to the southeast was uplifted so the Little Colorado River now flows *into* the Colorado River. This idea met its demise in

9. V. Polyak, C. Hill, and Y. Asmerom, "Age and Evolution of the Grand Canyon Revealed by U-Pb Dating of Water Table-type Speleotherms," *Science* 319 (2008): 1377–1380.
10. K.E. Karlstrom et al., "$^{40}Ar/^{39}Ar$ and Field Studies of Quaternary Basalts in Grand Canyon and Model for Carving Grand Canyon: Quantifying the Interaction of River Incision and Normal Faulting Across the Western Edge of the Colorado Plateau," *Geological Society of America Bulletin* 119 (2007): 1283–1312; K.E. Karlstrom et al., "Model for Tectonically Driven Incision of the Younger than 6 Ma Grand Canyon," *Geology* 36 (2008): 835–838.
11. A.A. Snelling, "Radiometric Dating: Problems with the Assumptions," *Answers*, October–December 2009, p. 70–73.
12. J.W. Powell, "Exploration of the Colorado River of the West and its Tributaries," *Smithsonian Institution Annual Report*, 1875.
13. E.D. McKee et al., "Evolution of the Colorado River in Arizona," *Museum of Northern Arizona Bulletin* 44, 1967.

Figure 6. Ancestral Colorado River drainage flowing southeastward toward the Rio Grande

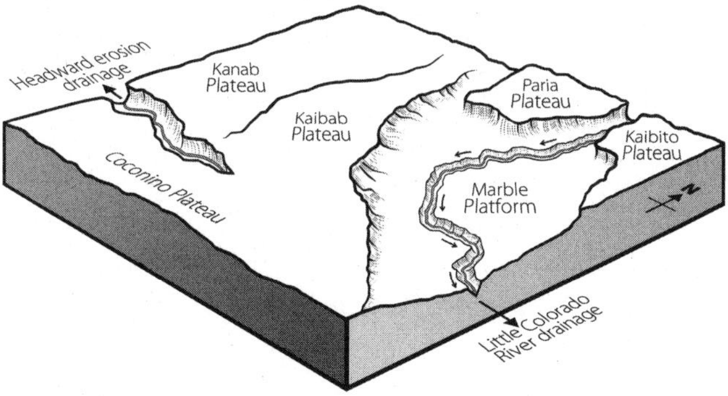

part because the necessary erosional debris could not be found anywhere east of the canyon.

Still having the problem of the basalts on both rims of the western plateau, the theory was modified. The now widely accepted theory has the stream capturing taking place at one of the northwest-tending drainages believed to have existed prior to the plateau uplift.[14] The ancestral Colorado River was thought to have taken a turn to the north, draining into the Great Salt Lake region (figure 7). Again, once the capture took place, the plateau was uplifted, causing the northwest-tending drainages to flow *into* the Colorado River. This modified model seems to be the predominant theory among evolutionary geologists today.

Figure 7. Ancestral Colorado River drainage flowing northwestward toward the Great Salt Lake

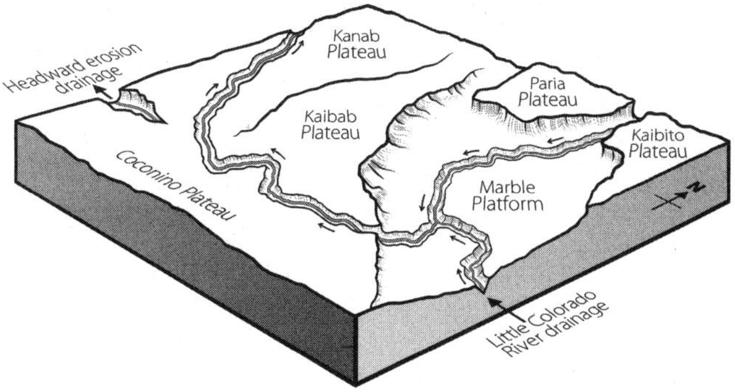

Evidences that Canyon Erosion Was Recent and Rapid

There are several pieces of evidence which suggest the Grand Canyon is a recent or "young" canyon. When considered individually, they are significant challenges to the uniformitarian (long-age) model; when taken as a whole, they become catastrophic. Following is a brief outline of some of those challenges.

14. Lucchitta, "History of the Grand Canyon and of the Colorado River in Arizona," in *Grand Canyon Geology*, p. 263.

Debris Not in the Present River Delta

Almost 1,000 cubic miles (4,000 cubic km) of material has been eroded to form the Grand Canyon. Where did it go? If the canyon was eroded by the Colorado River, an enormous delta should be found at the mouth of the river where it empties into the Gulf of California. But the delta contains only a small fraction of this eroded material.[15] This same problem is found with most river deltas; they only contain enough material to represent thousands, not millions, of years of erosion.

Stable Cliffs

One of the most striking features of the Grand Canyon is the massive sheer cliffs of sedimentary rocks. It is the difference in the rocks' makeup that gives the canyon its color and progressive stair-stepped profile of cliffs above broad slopes. The cliffs are made mostly of limestone and sandstone, with some formations reaching 500 feet (150 m) in thickness. The dark, almost black, color of large sections of the sheer cliffs is due to a coating of desert varnish, which develops slowly over many years[16] and is indicative of their stability. Where recent rockfalls occur, the desert varnish is missing. The fact that the cliffs maintain their desert varnish color indicates they are rarely experiencing even minor rockfalls; thus they are very stable. This is only consistent with their formation by recent catastrophic erosion, not millions of years of slow erosion.

No Talus

The lack of debris, or talus, at the base of the cliffs is also a challenge to the evolutionary model. Over millions of years of erosion, one would expect to find large amounts of talus at the base of the cliffs within the Grand Canyon.[17] The most obvious areas of this lack of talus is within the side canyons ending in broad U-shaped amphitheaters. Some of these amphitheaters are hundreds of feet deep and extend back as much as a mile (1.6 km) from the river. The majority have no water source to remove material, yet the bases of most of these cliffs are relatively "clean," with very little talus. Within the evolutionary model, there is no mechanism for the removal of this material.

---

15. P. Lonsdale, "Geology and Tectonic History of the Gulf of California," in E.L. Winteren, D.M. Hussong, and R.W. Decker, eds., *The Eastern Pacific Ocean and Hawaii, The Geology of North America*, vol. N (Boulder, CO: Geological Society of America, 1989), p. 499–521.
16. T. Liu and W.S. Broecker, "How Fast Does Rock Varnish Form?" *Geology* 28 no. 2 (2000): 183–186.
17. E.W. Holroyd III, "Missing Talus," *Creation Research Society Quarterly* 24 (1987): 15–16.

Relict Landforms

The stability of the Grand Canyon cliffs and the lack of talus at their bases are indicative of the canyon being a relict landform. In other words, the Grand Canyon has changed very little since it was carved. It is a relatively unchanged remnant or relict of the event that eroded it, which therefore could not have been today's slow river processes extrapolated back into the past.

There are several remnants, or relict landforms, of the material that now makes up the Grand Staircase to the north of the Grand Canyon. The two most noticeable ones are Red Butte, 16 miles (25 km) south of the South Rim (see figure 4), and Cedar Mountain just east of Desert View Overlook on the South Rim. These remnants, and others like them, are mostly capped with volcanic basalt, which has protected the sedimentary layers from being eroded away. These same sedimentary layers also form the base of the San Francisco Peaks just north of Flagstaff, Arizona.

These relicts testify to a massive erosional event, which in the biblical model is explained by the receding waters of the catastrophic global Genesis Flood.

Examples of Catastrophic Erosion

Catastrophic geologic events are not generally part of the uniformitarian geologist's thinking, but rather include events that are local or regional in size. One example of a regional event would be the 15,000 square miles (39,000 square km) of the Channeled Scablands in eastern Washington. Initially thought to be the product of slow gradual processes, this first came into question in 1923 when J. Harlen Bretz presented a paper to the Geological Society of America suggesting the Scablands were eroded catastrophically.[18] For the next 30 years Bretz was ridiculed for his theory, but in 1956 additional information was presented supporting the idea. Over the next 20 years, the evidence was pieced together to show the Scablands were, in fact, catastrophically eroded by the "Spokane Flood."[19] This Spokane flood was the result of the breaching of an ice dam that had created glacial Lake Missoula. Today, the United States Geological Survey estimates the flood released 500 cubic miles (2,000 cubic km) of water, which drained in as little as 48 hours, gouging out millions of tons of solid rock.

A more recent example of the power of catastrophic processes was observed at Mount St. Helens in 1980. Two hundred million cubic yards (153 million cubic meters) of material was catastrophically deposited by volcanic flows at

18. J.H. Bretz, "Glacial Drainage of the Columbia Plateau," *Geological Society of America Bulletin* 34 (1923): 573–608.
19. J.E. Allen, M. Burns, and S.C. Sargent, *Cataclysms of the Columbia* (Portland, OR: Timber Press, 1986).

the base of the mountain in just a matter of hours. Less than two years later, a minor eruption caused a mudflow, which carved channels through the recently deposited material.[20] These channels, which are 1/40th the size of the Grand Canyon, exposed flat contacts between the catastrophically deposited layers, contacts similar to those seen between the layers exposed in the walls of the Grand Canyon.

Both these events were relatively minor compared to a global flood. For example, the eruption of Mount St. Helens contained only 0.27 cubic miles (1.1 cubic km) of material compared to other eruptions, which have been as much as 950 cubic miles (3,960 cubic km). That is over 2,000 times the size of Mount St. Helens!

If Noah's Flood laid down the layers rapidly, one on top of another as was observed at Mount St. Helens, the boundaries between the layers would be flat and smooth, just as they are so magnificently displayed in the Grand Canyon. And the Channeled Scablands present a clear example of how the layers of the Grand Canyon could have easily been eroded catastrophically, possibly in a matter of just a few days.

An example of how quickly water can erode through the formations of the Grand Canyon region took place on June 28, 1983, when the pending overflow of Lake Powell required the use of the Glen Canyon Dam's 40-foot (12-m) diameter spillway tunnels for the first time. As the volume of water increased, the entire dam started to vibrate and large boulders spewed from one of the spillways. The spillway was immediately shut down and an inspection revealed catastrophic erosion had cut through the three-foot-thick reinforced concrete walls and eroded a hole 40 feet (12 m) wide, 32 feet (10 m) deep, and 150 feet (46 m) long in the sandstone beneath the dam.[21]

Catastrophic erosion such as this often starts when vacuum bubbles form and implode with jackhammer-like power, eating away anything in their way. This is called cavitation.[22] As volumes increase, whirlpool-like vortexes form, sucking material from the bottom in a process called kolking. That material then enters the flow and acts as projectiles, removing even more material. The erosive power of these forces continues almost exponentially as the volume of water increases. These same forces would have had a major role in the formation of the Grand Canyon.

20. S.A. Austin, "Rapid Erosion at Mount St. Helens," *Origins* 11 (1984): 90–98.
21. *Challenge at Glen Canyon Dam*, VHS, directed by W.L. Rusho (Denver, CO: U. S. Department of Interior, Bureau of Reclamation, 1983).
22. H.L. Barnes, "Cavitation as a Geological Agent," *American Journal of Science* 254 (1956): 493–505.

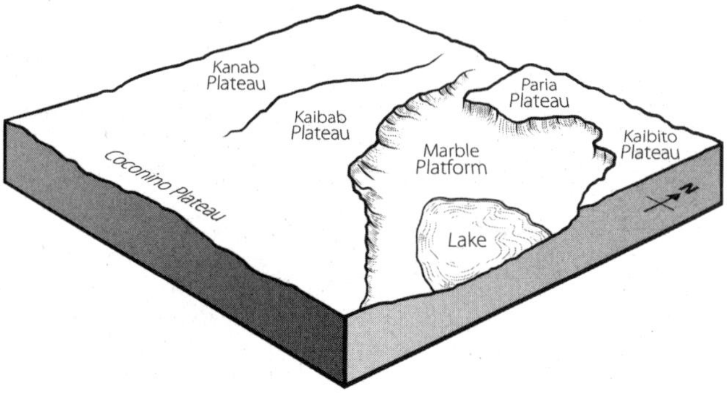

Figure 8. Natural dams trap receding Flood waters creating large lake(s).

Erosion of Grand Canyon Within the Biblical Account of Earth History

Not long after all the fossil-bearing sedimentary layers of the Colorado Plateau had been deposited by the rising Flood waters, those same waters began to recede. We are told in Psalm 104:8 that at the end of the Flood, the mountains rose and the valleys sank down, causing the waters to drain off the continents back into new ocean basins. Massive sheet erosion occurred across the plateau while it was being uplifted, carving the Grand Staircase and leaving behind the colored cliffs, canyons like Zion Canyon, and isolated remnants like Red Butte. As the Flood receded, water would have become trapped behind natural dams north and east of what is now the Grand Canyon area. Some estimate these lakes could have contained as much as 3,000 cubic miles (12,500 cubic km) of water (about three times the volume of today's Lake Michigan).[23] Figure 8 shows where one of these lakes may have been, with additional lake(s) potentially north of the Paria-Kaibito Plateau.

The warming of the oceans caused by the opening of the fountains of the great deep during the Flood would also have resulted in increased rainfall in this region immediately after the Flood. Storms potentially dumped as much as 100 inches (2.5 m) of rain at a time in the area just north of the canyon.[24] This rainfall would have increased the water level in the impounded lakes and would have been a powerful erosional force of its own.

23. S.A. Austin, "How Was Grand Canyon Eroded?" in *Grand Canyon: Monument to Catastrophe*, S.A. Austin, ed. (Santee, CA: Institute for Creation Research, 1994), p. 83–110; W. Brown, *In the Beginning: Compelling Evidence for Creation and the Flood*, sixth edition (Phoenix, AZ: Center for Scientific Creation, 1995), p. 92–95, 102–105.
24. L. Vardiman, "Hypercanes Following the Genesis Flood," in *Proceedings of the Fifth International Conference on Creationism*, R.L. Ivey, Jr., ed. (Pittsburgh, PA: Creation Science Fellowship, 2003), p. 17–28.

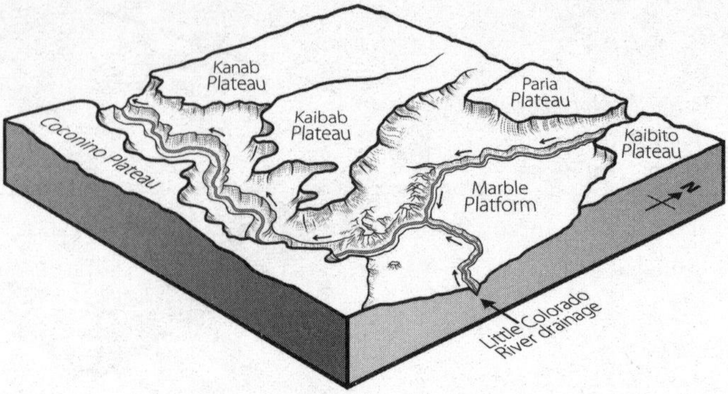

Figure 9. Current drainage of the Colorado and Little Colorado Rivers

As the Flood waters continued to recede, the sheet erosion across the rising Colorado Plateau would have diminished and the water would have started to channelize. This channelization would have then cut the initial path of the canyon.

The Kaibab Plateau now stands some 3,000 feet (900 m) above the adjacent Marble Platform, both part of the Colorado Plateau (figure 5). But the lack of erosional cliffs on the north and eastern sides of the Kaibab Plateau suggests that the southern end of the plateau continued to be uplifted after the rest of the region had stabilized. If this uplifting occurred just prior to, or even during, the channelization phase of the receding Flood waters, it would account for the lack of cliffs. It would also account for the direction of the side canyons eroded into the Kaibab Plateau. For example, some of the side canyons carved into the Marble Platform that join to form Marble Canyon, drain to the northeast, which seems to be the wrong direction. But that would have been the direction in which the receding waters flowed as the Kaibab Plateau was uplifted. Since the Kaibab Plateau is higher at its southern rim, this would also account for the longer and deeper side canyons carved into the North Rim of the Grand Canyon, which also follows along that southern edge of the plateau. Thus the South Rim of the canyon follows the northern edge of the Coconino Plateau (figure 5).

Within the uplifted Colorado Plateau are several limestone layers susceptible to being dissolved by surface and ground waters, as evidenced today by all the caves in the Redwall Limestone, from many of which streams flow. Because of all the volcanic activity during the Flood, the waters could have been slightly acidic, increasing their ability to dissolve limestone. So no sooner had these leftover Flood waters been dammed than they would have begun to find and exploit weaknesses in the limestone and other layers making up the plateau.

Whether it happened as the Flood year ended, or soon thereafter, the lakes would have soon breached their dams, washing over the plateau and exploiting any channels already there, rapidly carving through the plateau resulting in a deep canyon very similar to what we see today (figure 9).

A Few Perplexing Questions

As creationists, we do not have all the answers. In fact, there are many unanswered questions when it comes to the formation of the Grand Canyon. For example, exactly when the Kaibab Plateau was uplifted during the formation of the Grand Canyon is uncertain. Another question relates to the erosional evidence associated with the breaching of the natural dams. It is unclear as to why the waters would have eroded the course they appear to have taken, and why the remaining landscape has some of the features shown today. Also, unknown is what effect the increased rainfall in the region had on carving the canyon.

Some creationists attribute the formation of the canyon almost solely to the breaching of the dams, while others see the receding of the Flood waters to be the main carving mechanism. It is suggested here that combining the strengths of both models best explains the evidence and what we see in the Grand Canyon today.

These issues, however, do not weaken the evidence for the catastrophic carving of the Grand Canyon and its relationship to the Flood. It only shows there is still research to be done in order to better understand the canyon's formation.

Conclusion

Although we cannot be certain of the sequence and timing of these events, the evidence shows the Grand Canyon was formed rapidly, as were the layers into which it is carved. Thus, rather than slow and gradual erosion by the Colorado River over eons of time, the Grand Canyon was carved rapidly by a lot of water in a little bit of time! The reason the Colorado River exists today is because the Grand Canyon was eroded first, soon after the end of the Genesis Flood.

19

Does Astronomy Confirm a Young Universe?

DR. DON B. DEYOUNG AND DR. JASON LISLE

One of the common objections to biblical creation is that scientists have supposedly demonstrated that the universe is much older than the Bible teaches. The first chapter of Genesis clearly teaches that God created all things in six days ("ordinary" days as defined by an evening and morning) and that human beings were created on the sixth day. This is confirmed and clarified in the other Scriptures as well (e.g., Exodus 20:8–11; Mark 10:6). And since the Bible records about four thousand years between Adam and Christ (Genesis 5:3–32), the biblical age of the universe is about 6,000 years. This stands in stark contrast with the generally accepted secular age estimate of 4.6 billion years for the earth, and three times longer still, 13.7 billion years, for the universe beyond.

This fundamental time discrepancy is no small matter. It is obvious that if the secular age estimate is correct, then the Bible is in error and cannot be trusted. Conversely, if the Bible really is what it claims to be, the authoritative Word of God (2 Timothy 3:16), then something is seriously wrong with the secular estimates for the age of the universe. Since the secular time scale challenges the authority of Scripture, Christians must be ready to give an answer — a defense of the biblical time scale (1 Peter 3:15).

The Assumptions of Age Estimates

Why such a difference? What is really going on here? It turns out that all secular age estimates are based on two fundamental (and questionable) assumptions.

These are *naturalism* (the belief that nature is all there is),[1] and *uniformitarianism* (the belief that present rates and conditions are generally representative of past rates and conditions).

In order to estimate the age of something (whose age is not known historically), we must have information about how the thing came to be, and how it has changed over time. Secular scientists assume that the earth and universe were *not* created supernaturally (the assumption of naturalism), and that they generally change in the slow-and-gradual way that we see today (the assumption of uniformitarianism).[2] If these starting assumptions are not correct, then there is no reason to trust the resulting age estimates.

But notice something about the assumptions of naturalism and uniformitarianism: they are *anti-biblical* assumptions. The Bible indicates that the universe was created *supernaturally* by God (Genesis 1:1) and that present rates are *not* always indicative of past rates (such as the global Flood described in Genesis 7–8). So, by assuming naturalism and uniformitarianism, the secular scientist has already assumed that the Bible is wrong. He then estimates that the universe is very, very old, and concludes that the Bible must be wrong. But this is what he assumed at the start. His argument is circular. It's the logical fallacy called "begging the question." But all old-earth (and old-universe) arguments assume naturalism and uniformitarianism. Therefore, they are all fallacious circular arguments. That's right — all of them.

Refuting an Old Earth and Universe

A much better way to argue for the age of the universe is to hypothetically assume the opposite of what you are trying to prove, and then show that such an assumption leads to inconsistencies. In other words, we temporarily assume naturalism and uniformitarianism for the sake of argument, and then show that even when we use those assumptions, the universe appears to be much younger than secular scientists claim. This technique is called a *reductio ad absurdum* (reduction to absurdity). So the secular worldview is unreasonable since it is inconsistent with itself. In the following arguments, we will temporarily assume

1. Some scientists hold to a softer form of naturalism called "methodological naturalism." This is the concept that a supernatural realm may indeed exist, but should not be considered when doing scientific study. For all intents and purposes, the naturalist does not accept that there is anything beyond nature — at least when he or she is doing science.
2. Uniformitarianism is a matter of degree. Some secular scientists are willing to accept that catastrophes play a major role in the shaping of the earth's features. However, virtually all of them deny the worldwide Flood, which would have been the most significant geological event in earth's history since its creation. In this sense, virtually all secular scientists embrace uniformitarianism to a large extent.

(for the sake of argument) that naturalism and uniformitarianism are true, and then show that the evidence still indicates a solar system much younger than the secular estimate of 4.6 billion years, and a universe much younger than 13.7 billion years.

Moon Recession

Our nearest neighbor, the moon, has much to contribute to the recent creation worldview. A parade of lunar origin theories has passed by over the decades. These include fission of the moon from the earth (1960s), capture of the moon by earth's gravity from elsewhere in space (1970s), and formation of the moon from the collapse of a dust cloud or nebula (1980s). The currently popular model calls for lunar origin by an ancient collision of the earth with a Mars-size space object. All such natural origin theories are unconvincing and temporary; a recent supernatural creation remains the only credible explanation. Inquiry into origins need not be limited to natural science alone, as often assumed. The historical definition of science is the search for truth. If God is indeed the Creator, then scientists should not arbitrarily dismiss this fact. Many feel that modern science has been impoverished by its artificial limitation to naturalism, or secularism.

The moon reveals multiple design features. Lunar tides keep our oceans healthy, protecting marine life. The moon's (roughly circular) orbit stabilizes the earth's tilt and seasons. The moon also provides us with a night light, compass, clock, and calendar. The extent to which the moon controls the biorhythms of plants and animals, both on land and in the sea, is not well understood but is surely essential to life.

The moon also instructs us concerning the age of the earth. Consider the gravitational tide force between the earth and moon. This interaction also results in a very gradually receding moon, and slowing of the earth's rotation. These changes are highly dependent on the earth-moon separation, and are in direct conflict with the evolutionary time scale. Figure 1 shows the spinning earth and orbiting moon. A slight delay in the earth's high tides (the dark bumps) results in a forward pull on the moon, causing it to slowly spiral outward from the earth. In turn, the moon's gravity pulls back on the earth, slightly decreasing its spin.

Currently, the moon is moving outward from the earth by 3.82 cm/yr (1.5 in/yr). However, this recession is highly nonlinear and would have been greater in the past. If one assumes unlimited extrapolation back in time, gravity theory shows the moon in direct physical contact with earth about 1.55 billion years ago.[3] This is not to say that the moon was ever this near or this old.

3. Don B. DeYoung, "Tides and the Creation Worldview," *Creation Research Society Quarterly*, 45 no. 2 (2008): 100–108.

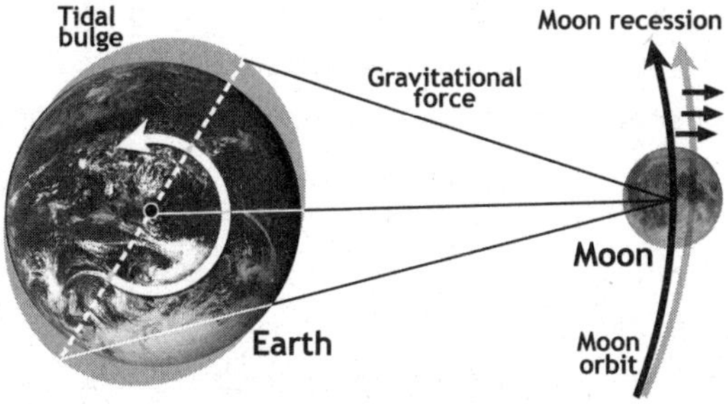

Figure 1. The moon is slowly drifting away from the earth, but the rate of recession would have been much faster in the past.

In fact, a moon located anywhere in the vicinity of the earth would be fragmented, resulting in a Saturn-like ring of debris encircling the earth. This follows because the earth's gravity force would overcome the moon's own cohesive force. The tides lead to a limited time scale for the moon, far less than 1.55 billion years. However, evolutionists assume that the moon and solar system are 4.6 billion years old. Also, life is said to have originated on earth about 3.5 billion years ago. The fundamental problem with the evolutionary time scale is obvious.

On a much shorter time scale, 6,000 years, the moon has moved outward by only about 755 feet (230 m) since its creation. Therefore, the creationist suggestion is that the moon was placed in orbit close to its present earth distance. Due to the earth's rotational slowing, the length of a day 6,000 years ago is calculated to be just 0.12 seconds shorter than at present.

Comets

Comets silently orbit the sun and put on occasional majestic displays in our night sky. Each year, dozens of comets loop the sun. About one-half of them have been named and studied on previous orbits. These comets don't last forever. Sooner or later they may be ejected from the solar system, may collide with the sun or planets, or they may break into fragments like a poorly packed snowball. There are clouds of dusty debris in the solar system, ghosts of disintegrated comets from the past. When the earth happens to pass through such a cloud, it sweeps up some of this comet dust. Then we see "shooting stars," an echo of the comet's original light show. In a spectacular 1994 display, comet Shoemaker-Levi was destroyed when it collided with Jupiter. The gravity of the massive outer planets protects the earth from similar comet collisions.

The question arises, why do comets still exist in the solar system? On a time scale of multiple billions of years, should they not all be long gone, either by escape, collision, or disintegration? The average number of solar revolutions before a comet dissipates is estimated to be about 40 trips. Comet Halley has already been observed through at least 28 orbits, dating back to 240 B.C. Its remaining years are numbered.

Astronomers recognize two comet varieties with respectively short and long revolving periods. The short-period comets have orbit times less than about 200 years. Halley's Comet is such an example with a period of about 76 years. Meanwhile, the long-period comets may require thousands of years for each solar pass. The origin of both kinds of comets remains a mystery to secular astronomers. Based on the rate at which comets are destroyed today, it is surprising (from an old-universe perspective) that either long-period or short-period comets are still present. The supply should have been depleted billions of years ago. How then do secular astronomers explain these apparently "young" comets in a solar system that they believe to be billions of years old?

To account for this paradox, secular astronomers have proposed that myriads of icy, comet-sized objects formed early in the solar system and continue to orbit at a tremendous distance from the sun where they remain permanently frozen for billions of years. It is suggested that every now and then one of these objects is dislodged from its distant orbit and injected into the inner solar system to become a new comet. According to this idea, as old comets are destroyed, new ones replace them.

Two present-day comet reservoirs are suggested by astronomers: one to supply short-period comets, the other to account for long-period comets. The Kuiper belt is thought to exist on the outer fringe of the known solar system, named for astronomer Gerald Kuiper (1905–1973). More than one hundred large, icy objects have been observed beyond planet Neptune, and multitudes more are assumed. It is thought that these trans-Neptunian objects (TNOs) are the largest members of the Kuiper belt. It is assumed that the unseen smaller members of the Kuiper belt occasionally fall inward toward the sun to become short-period comets. Hundreds of times further outward from Neptune is an assumed, vast Oort cloud of icy masses, named for Jan Oort (1900–1992). It is further assumed that a passing star may disturb this remote cloud from time to time, deflecting some of these icy objects toward the inner solar system, thereby replenishing the supply of long-period comets.

So far, the only objects detected at these great distances are much larger than any known comet. The existence of vast Kuiper and Oort clouds of actual comet-sized objects is not verifiable with current technology. The simplest explanation would appear to line up with the biblical time scale: the presence of comets may be evidence that the solar system is not nearly as old as is often assumed. Comets teach us two valuable lessons. First, their eventual loss is a reminder of the temporary nature of the solar system and universe. As Psalm 102:25–26 describes it:

> . . . the heavens are the work of Your hands. They will perish, but You will endure; Yes, they will all grow old like a garment.

As a second lesson, the exact motions of comets, planets, and stars are elegant evidence of God's controlling presence throughout the physical universe.

Faint Young Sun Paradox

Astronomers use the term *stellar evolution* for the aging process of stars. Our sun is assumed to be in its midlife stage, 4.6 billion years of age, as it gradually converts its hydrogen to helium via nuclear fusion reactions in its core. However, a basic time problem arises. Computer modeling of the sun on an evolutionary time scale predicts that the sun must gradually brighten. If true, the sun would be 30 percent dimmer during the period 3.8–2.5 billion years ago. The early earth would have been locked in a global ice age, with the crust and seas frozen solid. This in turn precludes the development of early life on earth.

In conflict with the icy prediction of solar models, geologic evidence points to an earth that was warmer in the past (irrespective of the time scale). This means that there is a fundamental problem with the unlimited extrapolation back in time of solar energy output. The creationist alternative is that the sun was placed in the heavens, on day 4 of the creation week, with a temperature very close to that of the present day.

Rapid Star Aging

Stellar evolution might better be called star decay or degeneration. Current models predict very gradual changes in the nature of stars. The sun, for example, is predicted to pass through several stages in coming ages. At present it is called a "main sequence" star. In the distant future, it is predicted to expand in size and grow cooler as it becomes a red giant star. Following this, the sun reverts to a small, hot white dwarf star. Each stage is assumed to last for millions of years.

Observations suggest that some stars may age much more rapidly than generally believed. For example, consider Sirius, the brightest nighttime star. At a distance of 8.6 light years from earth, it is known as the Dog Star, prominent in the Canis Major constellation. Sirius has a dwarf companion star, and there is intriguing evidence that this dwarf may have formed from a red giant in just the past 1,000 years. Historical records, including those of Ptolemy, describe Sirius as red or pink in color. The suggestion is that the red giant companion dominated the pair at this early time. Today, Sirius is a brilliant blue-white color and its dwarf companion is basically invisible. Other stars also occasionally show unexpected color changes, indicating possible rapid aging processes. Such events call into question the fundamental time scale of current stellar evolution models.

Spiral Galaxies

Spiral galaxies also pose a problem for the secular time scale. Spiral galaxies contain blue stars in their arms. But blue stars are very luminous and expend their fuel quickly. They cannot last billions of years. Secular astronomers realize this and so they simply assume that new blue stars form continuously (from collapsing clouds of gas) to replenish the supply. However, star formation is riddled with theoretical problems. It has never been observed, nor could it truly be observed since the process is supposed to take hundreds of thousands of years. Gas in space is very resistant to being compressed into a star. Compression of gas causes an increase in magnetic field strength, gas pressure, and angular momentum, which would all tend to prevent any further compression into a star. Although these problems may not be insurmountable, we should be very skeptical of star formation — especially given the lack of observational support.

Perhaps even more compelling is the fact that spiral arms cannot last billions of years. The spiral arms of galaxies rotate differentially — meaning the inner portions rotate faster than the outer portions. Every spiral galaxy is essentially twisting itself up — becoming tighter and tighter with time. In far less than one billion years, the galaxy should be twisted to the point where the arms are no longer recognizable. Many galaxies are supposed to be ten billion years old in the secular view, yet their spiral arms are easily recognizable. The spiral structure of galaxies strongly suggests that they are much younger than generally accepted.

There is a common misunderstanding here because people sometimes confuse *linear* velocity with *angular* velocity. Many people have heard or read that spiral galaxies have a nearly "flat" rotation curve — meaning that stars near the edge have about the same linear speed as stars near the core. This is true — but it doesn't alleviate the problem. In fact it is the *cause*. A star near the core makes a very small circle when it orbits, whereas a star near the edge makes a very large circle — which takes much longer if the star travels at the same speed. So in physics terminology we say that the stars have the same speed, but the inner star has a greater angular velocity because it completes an orbit in far less time than the outer star. This is why spiral galaxies rotate differentially.

Additionally, some people are under the mistaken impression that dark matter was hypothesized to alleviate the spiral wind-up problem. But this is not so. Dark matter explains (possibly) why the stars have a flat rotation curve to begin with. It does not explain how a spiral structure could last billions of years.

To get around the spiral galaxy wind-up problem, secular astronomers have proposed the "spiral density wave hypothesis." In this model, as the spiral arms become twisted and homogenized, new spiral arms are formed to replace the old ones. The new arms are supposed to form by a pressure wave that travels around the galaxy, triggering star formation. If this idea were true, then galaxies could be ten billion years old, whereas their arms are constantly being merged and reformed.

However, the spiral density wave hypothesis may create more problems than it solves. There are difficulties in creating such a pressure wave in the first place. The spiral density wave hypothesis cannot easily explain why galactic magnetic fields are aligned with the spiral arms (since magnetic fields move with the material — not with pressure waves); nor can it easily account for the tight spiral structure near the core of some galaxies such as M51. Perhaps most significantly, the spiral density wave hypothesis presupposes that star formation is possible. We have already seen that this is a dubious assumption at best. The simplest, most straightforward explanation for spiral galaxies is the biblical one: God created them thousands of years ago.

Conclusion

Many more such evidences for a young earth, solar system, and universe could be listed. Space does not permit us to discuss in detail how planetary magnetic fields decay far too quickly to last billions of years, or how the internal heat of the giant planets suggests they are not as old as is claimed. In all cases, the age estimates are far too young to be compatible with an old universe. It should be noted that all these age estimates are an upper limit — they denote the *maximum possible age*, not the actual age. So they are all compatible with the biblical time scale, but challenge the notion of an old universe.

It should also be noted that in all cases we have (for argument's sake) based the estimate on the assumptions of our critics. That is, we have assumed hypothetically that both naturalism and uniformitarianism are true, and yet we still find that the estimated ages come out far younger than the old-universe view requires. This shows that the old-universe view is internally inconsistent. It does not comport with its own assumptions. However, the biblical view is self-consistent. As with other fields of science, the evidence from astronomy confirms that the Bible is true. The answer to the title of this chapter is a resounding *yes* — the heavens declare a recent, supernatural creation!

References and Resources for Further Study

Don B. DeYoung, *Astronomy and Creation* (Winona Lake, IN: BMH Books, 2010).
Danny Faulkner, *Universe by Design* (Green Forest, AR: Master Books, 2004).
Jason Lisle, *Taking Back Astronomy* (Green Forest, AR: Master Books, 2006).
Jason Lisle, *The Ultimate Proof of Creation* (Green Forest, AR: Master Books, 2009).

20

How Could Fish Survive the Genesis Flood?

DR. ANDREW A. SNELLING

Some skeptics and long-age Christians lampoon the biblical account of the global Genesis Flood cataclysm by insisting that it was impossible for Noah to have a giant aquarium aboard the ark to preserve all the marine creatures, including trilobites.[1] However, this accusation is of course easily dismissed, because a careful reading of the relevant biblical text (Genesis 7:13–16, 21–23) clearly shows that God only brought to the ark representatives of all the created kinds of air-breathing, land-dwelling creatures. After all, the water-dwelling creatures would surely have been able to survive in the Flood waters.

Obviously, the air-breathing, land-dwelling creatures could not have lived through the earth-covering global Flood, but one would think the aquatic animals would have been right at home in all that water. Perhaps not, however, if during the Flood there was mixing of fresh and salt waters. Yet even that is uncertain, because we don't know how much mixing of fresh and salt waters would have occurred during the Flood. What we do know is that many of today's fish species, for example, are specialized, so they do not survive in water of radically different saltiness from their usual habitats. So how did freshwater and saltwater fish survive the Flood?

1. Ian R. Plimer, *Telling Lies for God: Reason vs. Creationism* (Sydney, Australia: Random House, 1994), p. 111; H. Ross, *A Matter of Days* (Colorado Springs, CO: NavPress, 2004), p. 123.

Saltiness of the Pre-Flood Ocean

To begin with, we do not know how salty the oceans were before the Flood, although early in the fossil record of the Flood we find echinoderms that could have only lived in a salty pre-Flood ocean. What we do know is that if at creation the oceans originally were totally freshwater, then at the current estimated rate of salt build-up in the oceans, all the salt in the oceans would have accumulated in only about 62 million years.[2] Of course, this assumes that the salt accumulation has always been at today's rate.

However, in the biblical account of earth history we are told that the Flood was initiated by the breaking up of the "fountains of the great deep" (Genesis 7:11), which likely were huge outpourings of hot water and steam that burst from inside the earth, associated with cataclysmic volcanic eruptions.[3] Such waters today are very salty, because of dissolved minerals in them. Furthermore, toward the end of the Flood there was massive erosion of the new continental land surfaces as the flood waters drained back into the new ocean basins, thereby carrying a lot more salt with them.

So the oceans before the Flood were a lot less salty than they are now. And since salt has not been added to the oceans uniformly through earth history at today's estimated rate, their current saltiness accumulated in far less than 62 million years.

However, this is still assuming freshwater oceans to begin with! We cannot, of course, be sure, because the Bible is silent about the salinity of the ocean waters at the conclusion of the creation week. We are told that when God created the earth on day 1, it was covered in water, which He divided on day 2. It may be safe to assume this was all freshwater because Genesis 1:2 reveals this water was formless and "empty" (perhaps meaning void or pure).

However, on day 3 God raised the land, and the covering waters were gathered together to form the seas.[4] Thus the earth's land surface was shaped by erosion by these retreating waters, no doubt carrying salts with them. So it's

2. S.A. Austin and D.R. Humphreys, "The Sea's Missing Salt: A Dilemma for Evolutionists," in *Proceedings of the Second International Conference on Creationism*, R.E. Walsh and C.L. Brooks, eds. (Pittsburgh, PA: Creation Science Fellowship, 1990), p. 17–33.
3. S.A. Austin et al., "Catastrophic Plate Tectonics: A Global Flood Model of Earth History," in *Proceedings of the Third International Conference on Creationism,* R.E. Walsh, ed. (Pittsburgh, PA: Creation Science Fellowship, 1994), p. 609–621.
4. There are three major possibilities regarding Genesis 1:9. One possibility has the ocean basins dropping, a second possibility has the continents lifted up through the waters, and the other possibility leaves open the miraculous — that the waters were instantly gathered into one place and dry land merely appeared. However, the focus of this chapter is not to debate these possibilities, but instead to show that that some possibilities would help add salt to the oceans.

possible a lot of salt may have been introduced to the pre-Flood oceans by this means.

Then God created marine creatures on day 5 to live and thrive in those ocean waters, so they must have been created with the ability to tolerate the salty oceans, just as marine creatures are able to today. Thus salt tolerance was not an outcome of the biological changes we are told occurred as a result of the Curse, being instead an ability given marine creatures at their creation by the Creator. Indeed, it was much more likely that God created animals suitable for mild salinity, but with the information available to survive in both extremes (freshwater and even more saline water).

Water Conditions in Which Fish Survive

Living in water requires specific physiological and ecological capabilities, different to those of terrestrial organisms.[5] Thus, for example, freshwater fish tend to absorb water because the saltiness of their body fluids draws water into their bodies (by osmosis), whereas saltwater fish tend to lose water from their bodies because the surrounding water is saltier than their body fluids.

The global scale of the Flood cataclysm produced gigantic problems affecting the very survival of many species. Indeed, the fossil record contains many groups of aquatic organisms that became extinct during the Flood deposition of the sedimentary rock layers.[6] Some organisms would have simply succumbed to the trauma of the turbulence, being swept away and effectively buried alive.[7]

Others would have found their suitable living spaces destroyed, and hence died for lack of appropriate habitats. Too much freshwater for marine-dependent organisms or vice versa would have killed those unable to adapt. However, not only are there such salt versus freshwater problems for aquatic organisms, but also problems of temperature, light, oxygen, contaminants, and nutritional conditions.

To simplify this discussion, only the three main factors affecting survival will be highlighted, primarily with respect to fish — salinity, temperature, and turbidity.

---

5. M.M. Ellis, "Detection and Measurement of Stream Pollution," in *Biology of Water Pollution*, L.E. Keup, W.M. Ingram, and K.M. Mackenthun, eds., US Department of Interior, Federal Water Pollution Control Administration, p. 129–155, 1967.
6. S.M. Stanley, *Extinction* (New York, NY: Scientific American Books, 1987); J.C. Briggs, "A Cretaceous-Tertiary Mass Extinction?" *BioScience* 41 (1991) 619–724; D.J. Bottjer et al., eds., *Exceptional Fossil Preservation: A Unique View on the Evolution of Marine Life* (New York, NY: Columbia University Press, 2002).
7. A.A. Snelling, "The World's a Graveyard," *Answers*, April–June 2008, p. 76–79.

Salinity

Many of today's marine organisms are able to survive large salinity changes, especially estuarine and tidal pool organisms. For example, starfish can tolerate indefinitely seawater with salt concentrations as low as 16–18 percent of the normal level.[8] Barnacles can withstand exposure to less than 10 percent the usual salt concentration of seawater.

Fish, as with all other marine organisms, however, have a problem balancing the fluids outside their bodies with those inside. Freshwater fish are constantly adding too much fresh water to their bodies from food, drinking water, and tissue transfer. On the other hand, marine fish get too little fresh water to maintain their fluid balance, due to the large salt input in their drinking water and the constant osmotic pressure to draw fresh water out of their tissues into the surrounding sea water.[9]

The kidneys and gills are used by fish to manage this balance. If a freshwater fish takes in too much water, then its kidneys secrete as much water as possible, while retaining the circulating salts. Marine bony fish get rid of excess salts largely through their gills, and conserve internal water through resorption. Saltwater sharks have high concentrations of urea in their blood to retain water in the saltwater environment, whereas freshwater sharks have low concentrations of urea to avoid accumulating water. When sawfish move from saltwater to freshwater they increase their urine output 20-fold, and their blood urea concentration decreases to less than one-third.[10]

There are migratory fish that travel between salt and freshwater. For example, salmon, striped bass, sea-run trout, and Atlantic sturgeon move from seawater to freshwater to spawn, but they return to seawater to mature. Eels do just the opposite, reproducing in saltwater but growing to maturity in freshwater streams and lakes. Obviously, all these fish are able to reverse their removal of water and salt by osmotic regulation according to the amount of salt in their environment. On the other hand, sunfishes and cod remain in freshwater and seawater, respectively, for their whole life cycles. Such fish have very narrow limits of salt tolerance, beyond which the environmental conditions are lethal to them.[11]

Within many families of fish there is much evidence of hybridization, suggesting that these families may represent the biblical created "kinds." In most families

8. D.J. Batten, "How Did Fresh- and Saltwater Fish Survive the Flood?" in *The Answers Book: Updated and Expanded*, D.J. Batten, K.A. Ham, J. Sarfati, and C. Wieland, eds. (Brisbane, Australia: Answers in Genesis, 1999), p. 175–178.
9. E. Florey, *An Introduction to General and Comparative Animal Physiology* (Philadelphia, PA: W.B. Saunders, 1966), p. 97–110.
10. Batten, "How Did Fresh- and Saltwater Fish Survive the Flood?" *The Answers Book*.
11. E.P. Odum, *Fundamentals of Ecology* (Philadelphia, PA: W.B. Saunders, 1971), p. 328, 354.

of fish alive today there are also both freshwater and saltwater varieties[12] ("species" in the man-made classification system) — for example, toadfish, gar-pike, bowfin, sturgeon, herring/anchovy, salmon/trout/pike, catfish, clingfish, stickleback, scorpion-fish, and flatfish. This suggests that the ability to tolerate and adjust to large changes in water salinity was probably present in most fish at the time of the Flood.

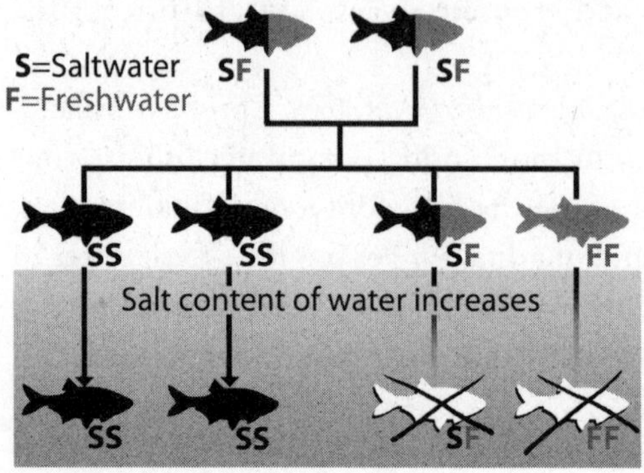

We have to also remember that there has been some post-Flood specialization in some fish "kinds." For example, the Atlantic sturgeon is a migratory salt and freshwater species, but the Siberian sturgeon (a different species in the same "kind") lives only in freshwater. Natural selection has probably resulted in the loss of the ability to tolerate saltwater.

Furthermore, hybrids of freshwater trout and migratory salmon are known, suggesting the differences between freshwater and marine fish may be quite minor. Indeed, the physiological differences may only be largely differences in degree rather than in kind. Many of today's fish species have the capacity to adapt to both fresh and salt water within their own lifetimes. This is why major aquariums are able to house freshwater and saltwater fish together, by using this ability of fish to adapt to water of different salinity from their normal habitats.

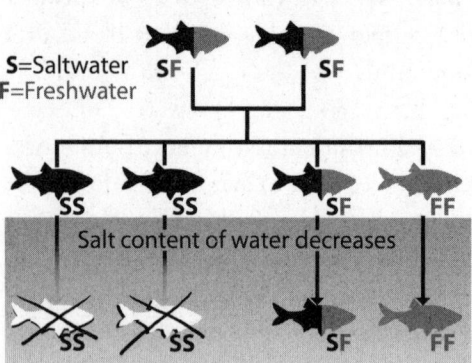

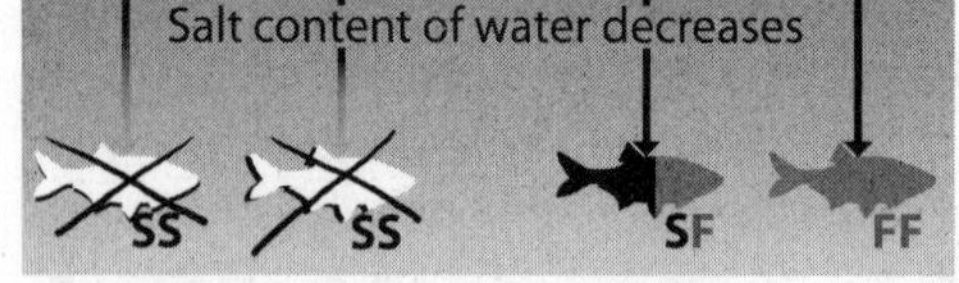

Temperature

The range of temperatures tolerated by fish varies from species to species and their habitats. Some fish have a very narrow range of temperature tolerance in cold, warm, or hot water. Other fish tolerate a wide range of temperatures, from freezing to hot waters (0–32°C, 32–90°F). Stages in the development of juvenile fish are frequently limited by the same narrow range of temperatures required by the adult fish.

12. Batten, "How Did Fresh- and Saltwater Fish Survive the Flood?" *The Answers Book*.

Most fish species, including cold-water types, can tolerate at least brief exposure to warm water at 24°C (75°F) and colder water approaching 2°C (36°F), as long as there are prolonged acclimation periods (several days to weeks). The preferred temperature ranges for some representative adult fish are: trout 16–21°C (61–70°F), sunfish 16–28°C (61–82°F), catfish 21–29°C (70–84°F), eel 16–28°C (61–82°F), and codfish 12–16°C (54–61°F).[13]

It should be emphasized that these abilities pertain to fish today. These fish species have probably been naturally selected within their kinds since the Flood and may have lost much of their original ability to survive in more extreme temperature ranges.[14] It makes more sense to postulate that God created fish to survive in moderate temperatures, with the genetic information available to subsequently select for survival in various more extreme environments.

Turbidity

Organic particles, dust and fine silt, bacteria, and plankton that are usually in suspension in natural waters are measured photoelectrically as turbidity. Such materials adversely affect fish by sinking to the seafloor, lake floor, etc., and covering it with a smothering layer that adversely affects spawning sites and kills organisms that the fish eat. Additionally, the abrasiveness of silt particles damages the gills of fish.

Turbidity also screens out light, decreasing the photic zone where photosynthesis can occur, and thus reduces the available oxygen for fish. The turbidity ranges can be described as clear (less than 10 parts per million, ppm or mg/l of particles in the water), turbid (10–250 ppm), and very turbid (greater than 250 ppm). It has been found that many fish species can survive in water with turbidities of 100,000 ppm for a week or more.[15]

Survival Strategies During the Flood

The heavy rainfall over the land would have quickly filled river basins with torrential flows. Such flooded rivers would have emptied these torrential flows

13. A. Calhorn, *Inland Fisheries Management* (The Resources Agency of California, Department of Fish and Game, 1966), p. 194, 375, 348; W.A. Anikouchine and R.W. Sternberg, *The World Ocean: An Introduction to Oceanography* (Englewood Cliffs, NJ: Prentice-Hall, 1973), p. 215, 233.
14. Batten, "How Did Fresh- and Saltwater Fish Survive the Flood?" *The Answers Book*; G. Purdom, "Is Natural Selection the Same Thing as Evolution?" in *The New Answers Book*, K.A. Ham, ed. (Green Forest, AR: Master Books, 2006), p. 271–282.
15. I.E. Wallen, "The Direct Effect of Turbidity on Fishes," *Oklahoma Agriculture and Mechanics College Bulletin* 48 (1951): 18–24.

out into the oceans as freshwater blankets. Such massive freshwater outflows from the continents would combine with the rainfall over the oceans to form freshwater layers sitting on top of the salty ocean waters, technically known as haloclines that are stable for extended time periods. In such highly stratified, strong density gradients or salt-wedge estuary situations,[16] fish flushed out from land aquatic systems could have continued to survive in freshwater environment pockets. In similar situations today, both marine and freshwater organisms are found living in the same water column, but within their preferred water conditions.

Different levels of salinity
in pockets of water during the flood

Stratification of water layers like this might even have survived strong winds if the depths of the freshwater layers were great enough to prevent internal current mixing. Turbulence may also have been sufficiently low at high latitudes for such layering to persist. Thus, situations are quite likely to have occurred during the Flood where freshwater and marine fish could have survived in water suited to them, in spite of being temporarily displaced from their normal habitats.

On the other hand, very turbid water carrying silt and sediment particles, and water flows with enormous sediment bedloads, would have also moved off the continents out into the oceans. There the silt and sediment particles would have settled in the deeper water, "raining" down on the seafloor, across which ground-hugging slurries and debris flows traversed. Heavier sediment particles would have fallen out in the slower-moving coastal waters to be deposited near the landward-advancing coastlines as the sea level rose, whereas the mudflows and debris flows would deposit their loads out over the deeper seafloors.

Although there would have obviously been turbulence at the interfaces between the freshwater and saltwater layers, the silt and sediment particles would probably have settled without appreciable mixing of the waters, especially given the predominance of the powerful horizontal currents during the Flood. With the range of tolerance already cited above, many fish would have been able to survive the extended exposure to high water turbidities.

As already noted above, the hybridization within many fish kinds today suggests that the ability to tolerate and adjust to large changes in water salinity

16. Odum, *Fundamentals of Ecology*.

and turbidity was probably present in most fish at the time of the Flood. If fish were thus capable of such hybridization during the Flood, then they definitely had the ability to cope with the wide fluctuations and ranges of temperatures and turbidities of the Flood waters. Perhaps what fish we have today are more extreme examples of selection and thus are less apt to survive now compared to the fish during the Flood.

Another possibility is that the eggs of marine organisms survived the Flood to then develop into the adults that re-populated the post-Flood ocean waters. These might have done better than full-grown fish, for example, at surviving the harsh water conditions during the Flood, because the "skin" of the eggs would maintain the necessary conditions within the eggs for embryo survival.

Another Lesson from Mount St. Helens

The recovery of animals and plants at Mount St. Helens after the May 18, 1980, eruption both demonstrates and documents rapid and widely ranging restoration after a geologic catastrophe.[17] Obviously, the Flood was several orders of magnitude greater than a catastrophe, but such an eruption event does show us how the biosphere recovers and re-establishes itself.

With regard to the three key water properties of interest, significant changes were recorded in the affected areas. Salinity increased from 0.01 ppm (mg/l) before the eruption to 150.5 ppm after it. Similarly, the surface water temperatures increased from 4°C (39°F) to 22.4°C (72°F), and turbidity increased from 0.75 ppm (mg/l) to 24.6 ppm.[18]

A little more than a month after the eruption (June 30), the lake most exposed to the catastrophic event, Spirit Lake, had tolerable salinity, ambient temperature, and low turbidity. All endemic fish had obviously been killed by the catastrophe, and probably could not have survived if re-introduced in those waters at that time, due to the large demand for oxygen from the water for the decaying tree debris, and the seeps of methane and sulfur dioxide. But within ten years this lake was able to support fish, with many other aquatic species back and well established.

Perhaps the most significant post-eruption observation, though, was that a variety of habitats within and adjacent to the blast zone survived the

17. K.B. Cumming, "How Could Fish Survive the Genesis Flood?" *Impact* #222, (Dallas, TX: Institute for Creation Research, 1991).
18. R.C. Wissmar et al., "Chemical Changes of Lakes Within the Mount St. Helens Blast Zone," *Science* 216 (1982): 175–178; R.C. Wissmar et al., "Biological Responses of Lakes in the Mount St. Helens Blast Zone," *Science* 216 (1982): 178–181.

catastrophe with minimal impact on continuity of the ecosystems. Meta Lake, within the blast zone, for example, had an ice cover at the time of the searing blast, which protected the dormant ecosystem underneath from experiencing much disruption from the heat, oxygen depletion, and air-fall volcanic ash. Fish and support systems picked up where they left off before the onset of the winter season.

Similar observations were made in Swift Reservoir, in spite of massive mud and debris flows into the lake. Fish were displaced into adjacent unaffected watersheds or downstream into lower reservoirs. However, within two years massive plankton blooms had occurred and ecosystem recovery was well underway with migrant "recruits."

Such a confined catastrophe (500 square miles around Mount St. Helens) does enable projection of expectations to a major catastrophe such as the global Flood. First, in spite of the Flood's enormous magnitude there would have been refuges for survival even in close proximity to the most damaging action spots. Second, biological recovery can be incredibly fast — from one month to ten years. Third, recruitment into the recovery zones from nearby minimally affected zones can occur with normal migratory behavior of organisms. Thus, even though some animal and plant populations, or even species, might be annihilated in catastrophic events, remnant individuals can re-establish new populations.

Conclusion

Many aquatic creatures were killed in the Flood because of the turbidity of the water and changes in salinities and temperatures. Indeed, the geologic record testifies to the massive destruction of marine life, with shallow-water marine invertebrates alone accounting for an estimated 95 percent by number of the fossil record.[19]

Many marine creatures, such as trilobites and ichthyosaurs, probably became extinct as a result of the Flood. However, many fish must have survived in the Flood waters, as they were not taken aboard the ark, and yet they are in today's oceans, lakes, and rivers. As discussed here, there are many simple, plausible explanations for how freshwater and saltwater fish could have survived in spite of the water conditions during the Flood.

Furthermore, if the hybridization within many fish kinds today suggests that the ability to tolerate and adjust to large changes in water salinity and

19. K.P. Wise, in a recorded lecture, c.1992, as quoted in J.D. Morris, *The Young Earth: The Real History of the Earth-Past, Present, and Future*, second edition (Green Forest, AR: Master Books,2007), p. 74.

turbidity was probably present in most fish at the time of and during the global Flood, then they definitely had the ability to cope with the wide fluctuations and ranges of temperatures and turbidities of the Flood waters. Indeed, there are more species of fish today than any other group of vertebrates, which possibly attests to their ability to hybridize and diversify. Thus, there is no reason to doubt the reality of a global Flood as described in God's Word.

21

What about Cosmology?

DR. DANNY FAULKNER

Since the late 1960s, the dominant cosmology has been the big-bang model. The big bang is a hypothetical event in which the universe suddenly appeared 13.71 billion years ago. Initially much smaller, denser, and hotter, the universe expanded and cooled to the one that we see today. The big-bang theory is a radical departure from more than two millennia of thinking on cosmology, for since ancient times many Western scientists and philosophers had assumed that the universe was eternal. It ought to be obvious that an eternal universe does not square with Genesis 1:1, which declares that "in the beginning God created the heavens and the earth," collectively referring to the creation of all that exists in the physical world.

Many Christians have embraced the big-bang cosmology, distilling the theory down to the fact that the big bang represents a beginning of the universe, apparently in some concordance with Genesis 1:1. However, closer examination reveals that the big bang does not agree with the details of the biblical creation account at all.

Big Bang Background

Before delving into that, we ought to mention a little background on the big-bang model. The big-bang model relies upon the expansion of the universe, first confirmed by Edwin Hubble in 1928. The expansion of the universe had been predicted by Einstein's theory of gravity, general relativity, more than a decade earlier. Einstein had realized the implication of an expanding universe possibly requiring a beginning, so he introduced the cosmological constant into his solution of the universe to produce a cosmology that was not expanding, and hence could exist eternally.

This static universe was no longer tenable once Hubble showed that the universe was expanding. However, proponents of an eternal universe did not give in. Prior to 1965, the most popular cosmology was the steady state theory. The steady state model acknowledges that expansion occurs, but hypothesizes that as the universe expands, more matter spontaneously comes into existence to preserve a constant density. This steady state model is eternal — without beginning and without end.

The steady state model had tremendous philosophical appeal well into the 1960s as people had difficulty thinking about the universe in any other way than being eternal. All this changed with the 1965 discovery of the cosmic microwave background (CMB), a low-level, nearly uniform radiation permeating the universe from all directions. Since the CMB had been predicted by the big-bang model as early as 1948, and the steady state theory could not account for the CMB, most scientists adopted the big-bang model shortly after 1965.

Variations on the steady state model have not entirely gone away — there are a few adherents around today. A notable variation of the steady state theory today is plasma cosmology. However, the big-bang theory is so dominant today that these steady state variations are virtually irrelevant, so we will concentrate on the big-bang model.

Genesis and the Big Bang

As previously mentioned, some Christians adopt the big-bang model as part of their apologetic. If the universe had a beginning in agreement with the creation account, then they reason that this offers a tool to somehow prove that the Bible is true. Since the modern big-bang model hypothesizes that the universe began 13.71 billion years ago, belief in the big bang requires belief in a universe much older than the few thousand years calculated from the Old Testament chronologies. How do people reconcile this colossal difference?

The gap theory, day-age theories, allegory, and the framework hypothesis are all different ways that people have attempted to reconcile the discrepancy. All of these attempts have been discussed elsewhere, but suffice it to say here that each poorly handles Scripture. And these solutions raise thorny questions as well. What is the origin of sin? What is the penalty of sin? Did death precede the Fall of mankind? While many old-age creationists attempt to hold onto some semblance of orthodoxy on these questions, regrettably many do not, opting to allegorize much of the first few chapters of Genesis.

Other Issues

But there are other problems as well. For instance, the big-bang model posits that stars and galaxies formed within a billion years of the beginning of the universe, and that star formation continues today. Since the solar system is supposedly 4.6 billion years old, this would mean at least eight billion years of star formation preceded the creation of the earth. However, the Genesis 1 text tells us that earth preceded the stars; the earth was made on day 1 and the stars on day 4. Furthermore, in no way could one say that the creation has been completed if it is an ongoing process now.

The naturalistic origin of stars itself is a problem. Astronomers have theories of how stars form, but each one requires that stars first exist. Current cosmological theories suggest that star formation in the early universe was very intense, with stars forming at a rate many orders of magnitude faster than today. How this came about is unknown. Astronomers generally agree that some unknown mechanism triggered star formation at a prodigious rate. Is this science?

With the largest telescopes we can see distant galaxies that presumably formed in the early universe, and hence are very young. We see that these distant galaxies have the same structure that nearby, supposedly much older, galaxies have. However, physics would seem to dictate that galaxies change with age, but this does not appear to be the case in a big-bang universe.

Additionally, the big-bang model requires that the universe begin with all matter in the universe in the form of hydrogen and helium (and a trifling bit of lithium). All other heavier elements, such as oxygen, calcium, and iron, were gradually forged through nucleosynthesis in later generations of stars. However, the spectra of distant (and supposedly younger) galaxies are rich in heavier elements. This would seem to violate a basic tenet of the big-bang model, instead allowing for heavier elements to exist from the beginning.

An Evolving Model

The big-bang model itself has undergone quite an evolution since its wide-spread acceptance. It is interesting to compare the big-bang model at the time of the writing of this chapter, 2009, to the big-bang model of 1984, just 25 years earlier. From about 1960 to the early 1990s, the best measure of the expansion rate of the universe (the Hubble constant) was about 50 km/s/Mpc (kilometers per second per megaparsec). In the early 1990s, that rate was increased to nearly 80 km/s/Mpc. Around 1980 some astronomers had attempted to raise the Hubble constant to nearly 100 km/s/Mpc, but their work was largely rejected at the time as most astronomers thought that the lower value was firmly established.

Of course, now astronomers think that the higher value is firmly established. With a faster expansion rate, the inferred age of the universe has diminished. With the earlier value of the Hubble constant, the age was solidly thought to be 16–18 billion years old. Now cosmologists think that the universe is 13.71 billion years, give or take 1 percent. Notice that the error bars on those two figures do not overlap, so they both cannot be correct.

Several physical effects now taken as a given in big-bang models were not largely accepted in 1984. One example is inflation, first proposed by Alan Guth in 1980. Inflation was invoked to explain several problems with the big bang, such as the flatness and horizon problems. While inflationary big-bang models were being developed in 1984, these models were not widely accepted until a few years later. Other effects include string theory, dark matter, and dark energy.

Among theoretical physicists, string theory is the current explanation of how elementary particles of matter work. String theory is the idea that elementary particles are a sort of vibration in at least six additional dimensions of space. These six dimensions occur in two sets of three each. Today, these additional dimensions are rolled up beyond our ability to detect them, but in the hot cauldron of the early big bang the high temperatures would have made these extra dimensions manifest. So any serious big-bang model now must incorporate them, though in 1984 few, if any, models did. Dark matter is a mysterious substance that reveals itself by its gravitational influence. While dark matter was first proposed in the 1930s, good data to support it began to arise in the 1970s. Yet dark matter generally was not included in big-bang models of 1984.

Finally in 1999, an extensive study combined data from several different programs to produce what was then the definitive description of some parameters of the big bang. Much to the amazement of all, the data showed what appeared to be an increase in the rate of expansion of the universe. Normally, gravity ought to be sufficient to rein in expansion, but this effect was as if space were repelling itself. This effect is very similar to Einstein's cosmological constant, though the re-christened "dark energy" was intentionally named to underscore the more modern approach to how such a thing might happen. We ought to add that some of these effects, such as string theory and inflation, have no data to support them — they are included only because theoretical physicists and cosmologists think that they describe the way the world works.

So let us compare the big-bang model of today and 25 years earlier. Then, the expansion rate and hence the inferred age of the universe were remarkably different from the rate and age today. Then, there was no inflation, while today one would not think of leaving inflation out of a big-bang model. Then, string theory had not yet been developed, but now it must be included in a big-bang model. Then, dark matter, though known, was not included in cosmological models, but today it is a must to include it. Einstein's cosmological constant was thrown out by 1930, but it came roaring back 70 years later to be included in today's model.

In short, the big-bang model of 25 years ago bears almost no resemblance to the big-bang model of today. How confident of the big-bang model were cosmologists 25 years ago? They had complete confidence. How confident are today's cosmologists of the current big-bang model? They have complete confidence. If cosmologists were right then, they cannot be right today; if cosmologists are right today, they could not have been right 25 years ago. We have no idea what the big-bang model will be like 25 years hence, but we can be certain of two things: the model will be very different then from now, and cosmologists will have complete confidence in that model.

Some of these changes to the big-bang model were driven by changes in theoretical physics, as with string theory. Others were driven by new data, such as dark energy and a revised expansion rate and age. However, some, such as inflation, were invoked merely to salvage the big-bang model. This reveals a deep philosophical problem with the big-bang model. The model has become very plastic. That is, any unexpected new observation or problem can be solved by the appropriate addition of some new effect or some new field.

Some view this as constraining the model and providing physical rigor, but at some point one has to question whether the big-bang model is falsifiable. That is, is there some new result or data that could disprove the model? It would appear that with proper corrections to the model allowed, this will never happen. If this indeed is the situation, then is the big-bang model a scientific model in any way?

Nearly two millennia ago, Claudius Ptolemy published his famous geocentric model of the solar system with planets moving along epicycles that in turn orbited about the earth. In terms of longevity, the Ptolemaic model is the most successful scientific theory of all time, lasting 15 centuries. Throughout the middle ages, scientists found that when the theory did not match observations, they could fix the problem by adding additional epicycles. Unlimited modification allowed for the model to explain everything and anything that happened to arise.

Ultimately, most people realized that the Ptolemaic model became far too complex for its own good, and it collapsed under its own weight in favor of the much simpler heliocentric model. A model that can explain anything and everything is not a good theory. The big-bang model has already demonstrated that it, too, can endure modification *ad infinitum*. At some point we must question whether the big-bang model really is a good theory in the sense that it could be falsified by some new hypothetical result.

Conclusion

The history of science is filled with examples of theories once thought to be unassailable but later discarded. If the history of science is any teacher, then we would expect that the big bang will also be discarded. If we have wedded our apologetic to the big-bang model, then the rejection of the big-bang model will logically lead to the rejection of our apologetic. Many in the Roman Catholic Church four centuries ago embraced the Ptolemaic model, attempting to make it part of Christianity, and it brought discredit to them and their church when that model fell. Those who wish to make the big bang part of the biblical creation model ought to take this lesson to heart. More importantly, the big bang (in any of its versions) is not compatible with a natural reading of the Bible. The Christian should have confidence that God's Word is reliable, regardless of whether it is fashionable for fallible men to agree.

22

Did Life Come from Outer Space?

DR. GEORGIA PURDOM

The simple answer is NO! The Bible states that God created all living things on earth by His spoken word on days 3, 5, and 6 of the creation week. However, the concept that life originated in outer space and was then transferred to earth is popular in today's society. Some believe that bacteria (considered "primitive" life) or organic molecules necessary for life came from other planets, meteors, or comets. Some even suggest that intelligent extraterrestrial aliens sent life to earth. Many people are eager to believe in any ideas concerning the origin of life as long as they exclude the Creator God and the truth of His Word.

Why Life from Outer Space?

Why do scientists want to push the origin of life into outer space rather than believe that life originated on earth? The answer: complexity and time.

Complexity

Life on earth is very complex. Bacteria are considered to be the simplest life form. However, several examples from the bacterial world make it clear that the word "simple" is a relative term. Some of the "simplest" are endosymbionts — organisms that live entirely within other organisms. *Candidatus Carsonella rudii*, a bacterium that lives within the cells of the psyllid insect *Pachypsylla venusta*, is considered to have the smallest genome of any endosymbiotic bacteria.[1] It has 159,662 base pairs (DNA), which encode approximately 182 genes. The genes encode proteins for amino acid (components of protein) biosynthesis, which the host insect cannot get from its diet. The host insect provides necessary proteins that are not encoded by the bacterial genome.

Nanoarchaeum equitans, an archaeal (single-celled microorganism similar to bacteria) symbiont of the archaea *Ignicoccus*, has 490,885 base pairs, which encode approximately 552 genes. Although many of the gene functions are currently unknown, the authors of the paper that sequenced the genome stated that "the *complexity* of its information processing systems and the simplicity of its metabolic apparatus suggests an unanticipated world of organisms to be discovered" (emphasis mine).[2]

Mycoplasma genitalium has 580,076 base pairs, which encode approximately 521 genes.[3] Because of its small genome size, *M. genitalium* was the bacteria of choice for determination of the minimal genome (or minimum number of genes) needed to sustain life. However, determination of the minimal genome has been hampered by the finding that many bacterial genomes encode backup or alternative pathways, which are used when the main pathway is removed. Scientists have stated that this may lead up to a 45 percent *underestimation* of the minimal genes needed to sustain life.[4] As can be seen from these examples, life in even its "simplest" forms is very complex!

Time

According to secular timelines, the earth is 4.5 billion years old. Other parts of outer space are much older (up to 15 billion years old according to big-bang models). Since evolution works by random chance and even the simplest bacteria isn't very simple, a lot of time would be required for life to evolve. Many secular scientists suggest the earth is simply not old enough to allow for the evolution of living organisms. Thus, many scientists push the origin of life into outer space to gain the time needed for life to evolve.

1. Atsushi Nakabachi et al., "The 160-Kilobase Genome of the Bacterial Endosymbiont Carsonella," *Science* 314 no. 5797 (2006): 267.
2. Elizabeth Waters et al., "The Genome of Nanoarchaeum equitans: Insights Into Early Archaeal Evolution and Derived Parasitism," *Proceedings of the National Academy of Sciences USA* 100 no. 22 (2003): 12984–12988.
3. JCVI Comprehensive Microbial Resource, "*Mycoplasma genitalium* G-37 Genome Page," www.cmr.jcvi.org/cgi-bin/CMR/GenomePage.cgi?org=gmg.
4. Csaba Pál et al., "Chance and Necessity in the Evolution of Minimal Metabolic Networks," *Nature* 440 no. 7084 (2006): 667–670.

Does Life Exist in Outer Space?

If life came to earth from outer space, then many scientists suggest that we should be able to find evidence for living things on nearby planets, meteors, and comets. Although billions of dollars have been spent in the search for extraterrestrial life, none has been found.

Mars

Several unmanned exploration probes, rovers, and landers have been sent to Mars to determine if our closest rocky neighbor supports life or may have harbored it in the past. NASA's Phoenix lander identified water in a sample of martian soil.[5] Another NASA space probe identified specific minerals that suggested liquid water had been present on the martian surface for a longer period of time then previously estimated.[6] Scott Murchie of Johns Hopkins University stated, "This is an exciting discovery because it extends the time range for liquid water on Mars, and the places where it might have supported life."[7] Although water is certainly needed for life to exist, water alone does not result in life.

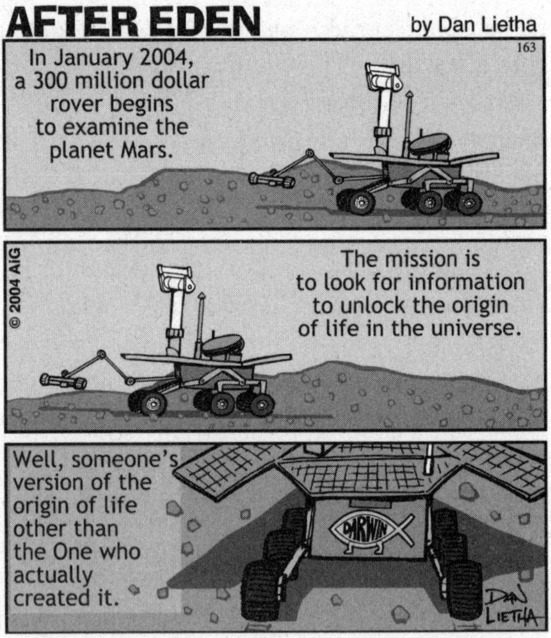
AFTER EDEN
by Dan Lietha
163
In January 2004, a 300 million dollar rover begins to examine the planet Mars.
© 2004 AiG
The mission is to look for information to unlock the origin of life in the universe.
Well, someone's version of the origin of life other than the One who actually created it.
DARWIN
DAN LIETHA

Other components of martian rocks and soil make the likelihood of finding life very unlikely. NASA's Opportunity rover produced evidence that rocks had once been in an environment that was very salty and acidic.[8] Dr. Andrew Knoll, biologist at Harvard University, stated, "It was really salty — in fact, it was salty enough that only a

5. BBC News, "Nasa's Lander Samples Mars Water," www.news.bbc.co.uk/2/hi/science/nature/7536123.stm.
6. BBC News, "New Minerals Point to Wetter Mars," www.news.bbc.co.uk/2/hi/science/nature/7696669.stm.
7. Ibid.
8. BBC News, "Early Mars 'Too Salty' for Life," www.news.bbc.co.uk/2/hi/science/nature/7248062.stm.

handful of known terrestrial organisms would have a ghost of a chance of surviving there when conditions were at their best."[9]

Methane, a gas associated with biological activity (think belching cows!), has been found in the martian atmosphere. Colin Pillinger, planetary scientist at the Open University (UK), stated, "The most obvious source of methane is organisms. So if you find methane in an atmosphere, you can suspect there is life. It's not proof, but it makes it worth a much closer look."[10] However, Nick Pope, formerly associated with the British Government's UFO project at the Ministry of Defense, thinks methane is proof, calling this discovery "the most important discovery of all time," and saying further, "We've really only scratched the surface — it's an absolute certainty that there is life out there and we are not alone."[11]

The biological source of methane is believed to be bacteria living deep underground. However, it could also be due to volcanism or an unknown geological process on Mars since "plumes" of methane were identified in 2003 and the distribution of methane was found to be patchy.[12] If the methane is of geological origin then it would actually make the martian surface very inhospitable for life.[13]

Moons of Jovian Planets

Several moons of Jupiter and Saturn, including Europa, Titan, and Enceladus, are thought to be possible sources of extraterrestrial life. All are thought to have interior oceans that might harbor bacterial life. Plumes containing water vapor erupting from Enceladus have been shown to contain organic molecules such as methane, formaldehyde, ethanol, and other hydrocarbons.[14]

Europa's underground oceans are predicted to be violent.[15] The waves generated in these oceans are postulated to provide an energy source necessary for life. Robert Tyler, an oceanographer at the University of Washington, stated, "The big thing is to have liquid water — and to the extent that this new paper [on violent

9. Ibid.
10. Fox News, "Clouds of Methane May Mean Life on Mars," www.foxnews.com/story/0,2933,479997,00.html.
11. Ibid.
12. Judith Burns, "Martian Methane Mystery Deepens," BBC News, www.news.bbc.co.uk/2/hi/science/nature/8186314.stm.
13. Ibid.
14. Lori Stiles, "Evidence for Ocean on Enceladus: Tiny Saturn Moon Could Be Targeted in Search for Extraterrestrial Life," PhysOrg, www.physorg.com/news167498118.html.
15. Anne Minard, "Jupiter Moon Has Violent, Hidden Oceans, Study Suggests," National Geographic News, www.news.nationalgeographic.com/news/2008/12/081210-europa-oceans.html.

oceans in Europa] adds an energy source, all the better for life's prospects."[16] But water plus organic molecules plus energy does not equal life. Life requires information (DNA), and information requires an intelligent source (God).

Comets

Scientists have made calculations (based on cosmological time frames of billions of years) that in the past, comets had liquid water interiors.[17] NASA's Stardust spacecraft collected samples from the dust of comet Wild 2 and found the amino acid glycine (the simplest of all amino acids).[18] Carl Pilcher, director of the NASA Astrobiology Institute, stated, "The discovery of glycine in a comet supports the idea that the fundamental building blocks are prevalent in space, and strengthens the argument that life in the universe may be common rather than rare."[19] This seems to be an overstatement since only 1 amino acid of the 20 required for life was found and other components for life such as DNA, fats, and sugars have not been found. Again, the formula of water plus amino acids (or other organic molecules) does not equal life.

Life has not yet been found in outer space and it is unlikely to exist because conditions appear too hostile for even the hardiest forms of life to exist. Even if the ingredients necessary for life (organic molecules like amino acids) were transported to earth and added to water and an energy source, life would not miraculously emerge. Life only comes from life, and life only from the Life-Giver.

If Life Did Exist in Outer Space, Could it Have Been Transferred to Earth?

Panspermia is the common name given to the concept that life originated in outer space and then migrated or was transported to earth. Panspermia is not a new idea. Lord Kelvin in 1871 suggested that life came to earth on meteors. Svante Arrhenius coined the term in 1908 and is considered the father of panspermia. We will look at the three categories mentioned in the previous section and determine if transfer of life from these sources to earth is plausible.

Mars

Several meteors of suspected martian origin have been discovered on earth. It is estimated that 5–10 percent of martian ejecta (derived from impacts by

16. Ibid.
17. PhysOrg, "Evidence of Liquid Water in Comets Reveals Possible Origin of Life," www.physorg.com/news168179623.html.
18. Bill Steigerwald, "First Discovery of Life's Building Block in Comet Made," PhysOrg, www.physorg.com/news169736472.html.
19. Ibid.

comets or asteroids) would reach earth in 100 million years (with the minimum amount of time being seven months).[20] Small ejecta (> 1 cm) could arrive on earth as meteorites with a burnt outer area but an inner cool area where bacteria could presumably survive.[21] But of the known martian meteorites, M.J. Burchell of the Centre for Astrophysics and Planetary Sciences at the University of Kent (UK), says that "given their size and transfer times (estimated from exposure to radiation in space), all will have received a sterilizing radiation dose during their transit to earth."[22]

Moons of Jovian Planets

Impacts of these moons by comets or asteroids are also thought to generate ejecta that could then travel to other locations (but not directly to earth).[23] The ejecta are postulated to travel farther into space and possibly be transferred to comets or asteroids.[24] The bacteria would presumably survive in the icy interior of the comets/asteroids.[25] The comets/asteroids could then travel to earth and so indirectly bring life from the Jovian moons.

Comets

During their travel close to a planet, comets could leave behind dust grains that would fall into planetary atmospheres.[26] If life existed in the dust grains and could survive travel through the atmosphere, then presumably a comet could transfer life to earth.

The transfer of material from Mars, Jovian moons, and comets is plausible and in some cases has been documented. However, dust and rocks are not affected by the extreme cold and radiation of outer space, whereas life would be and would probably not survive the journey to earth. Since life has not been found to exist in outer space it is doubtful life was transferred to earth from these locations.[27]

20. M.J. Burchell, "Panspermia Today," International Journal of Astrobiology 3 no. 2 (2004): 73–80.
21. Ibid.
22. Ibid.
23. Ibid.
24. Ibid.
25. Ibid.
26. Ibid.
27. If bacterial life is ever discovered on Mars or anywhere else in outer space, we must recognize that it might be contamination from our own space exploration. Just as it is possible for material from outer space to be transferred to earth, the reverse transfer (from earth to outer space) is also possible. Precautions are taken to ensure that spacecrafts are sterile, but none are completely sterile.

Could Life Have Been Brought to Earth by Intelligent Extraterrestrial Aliens?

The concept that aliens brought life to earth is called directed panspermia. The term was first coined by the co-discoverer of the structure of DNA, Francis Crick, and Leslie Orgel in 1973.[28] They postulated that since earth is relatively young compared to the rest of the universe that it was conceivable that a technologically advanced society in outer space developed even before earth existed (since it only took 4.5 billions years for a technological society to form on earth). Crick and Orgel believe that this alien society then seeded or "infected" other parts of outer space including earth with primitive forms of life (like bacteria). In their 1973 paper they propose the spaceship payload, the mechanisms needed to protect the bacteria for their long trip to earth, and possible motivations by the alien society for seeding life in outer space.

One of their main evidences to support this possibility comes from the similarity of the genetic code in all living things. They stated, "The universality of the genetic code follows naturally from an 'infective' theory of the origins of life. Life on earth would represent a clone derived from a single extraterrestrial organism."[29] The universality of the genetic code only follows "naturally" from their theory because of their presuppositions or starting point that their ideas about the past are supreme to God's Word concerning the history of the origin of life on earth. When we begin with God's Word we see that the universality of the genetic code follows naturally from a common Designer who created all living things by His Word.

The concept of directed panspermia is still advocated by many scientists today. In the movie *Expelled*, Ben Stein asked Richard Dawkins, a very prominent evolutionary biologist, the question, "What do you think is the possibility that . . . intelligent design might turn out to be the answer to some issues in genetics or in evolution?"[30]

Dawkins's reply:

> Well it could come about in the following way: it could be that at some earlier time somewhere in the universe a civilization evolved by, probably by, some kind of Darwinian means to a very, very high level of technology and designed a form of life that they seeded onto perhaps this this [*sic*] planet. Now that is a possibility and an intriguing possibility. And I suppose it's possible that you might find

28. F.H.C. Crick and L.E. Orgel, "Directed Panspermia," *Icarus* 19 (1973): 341–346.
29. Ibid.
30. *Expelled*, DVD, directed by Nathan Frankowski (Premise Media, 2008).

> evidence for that if you look at the, at the detail . . . details of our chemistry molecular biology you might find a signature of some sort of designer.

Burchell stated, "At present, Panspermia can neither be proved nor disproved. Nevertheless, Panspermia is an intellectual idea which holds strong attraction."[31] Sadly, this is true for many who want to exclude God and the history presented in His Word in deference to their own ideas about the past — no matter how outlandish.

Could God Have Created Life on Planets Other than Earth?

Yes, but why? Remember that God spent the vast majority of the creation week preparing the earth for the crowning glory of His creation — man. Everything God created was for man's benefit and enjoyment. Even those things which we don't often consider, like bacteria, were created to benefit man. Bacteria can accomplish this directly through symbiotic relationships in our guts, which help us digest food, and indirectly through cycling of nutrients and chemicals in the environment.[32] Man would seem to gain no benefit or enjoyment from bacteria that exist in outer space. Although we can't rule out that some form of non-intelligent life, such as bacteria, was created on another planet, it seems unlikely knowing the purposes of living organisms and their relationship to man on earth set forth by the Creator God.

31. Burchell, "Panspermia Today."
32. Joseph Francis, "The Organosubstrate of Life," Answers in Genesis, www.answersingenesis.org/articles/aid/v4/n1/organosubstrate-of-life, originally published in *Proceedings of the Fifth International Conference on Creationism*, Robert L. Ivey Jr., ed. (Pittsburgh, PA: Creation Science Fellowship, 2003), p. 433–444.

23

Did the Continents Split Apart in the Days of Peleg?

DR. ANDREW A. SNELLING AND BODIE HODGE

In Genesis chapter 10, two-thirds of the way through the genealogies of the post-Flood patriarchs, we read in verse 25:

> To Eber were born two sons: the name of one was Peleg, for in his days the earth was divided; and his brother's name was Joktan.

The same phrase, "for in his days the earth was divided," also appears in the repetition of this genealogical entry in 1 Chronicles 1:19.

Many find these genealogical lists very boring to read. So they skip over the details and often miss this phrase. However, there are some Christians who get excited about this phrase, and latch on to it, suggesting that maybe this is where continental drift, which secular scientists have proposed, fits into the Bible!

It seems odd that this little "nugget" should appear in this genealogy of Noah's three sons and their descendants after the Flood. But, does this phrase, "for in his days the earth was divided," suggest that continents drifted apart in the days of Peleg as a result of God dividing and separating the continents?

Have the Continents Shifted?

In today's secular society, people have been taught as fact that the continents were once joined together in a supercontinent that spilt apart and then the

Figure 1. The Mid-Atlantic Ridge is strong support for the concept of plate movement.

resultant continents drifted over millions of years into their present positions. One primary piece of conclusive evidence usually presented to support this idea is the jigsaw-puzzle fit of Europe and Africa matching closely with North and South America, respectively. If the North and South Atlantic Ocean basins are closed, these continents fit together at approximately the Mid-Atlantic Ridge, a range of mountains on the ocean floor centrally located in the Atlantic Ocean basins (figure 1).

It was this reconstruction that led to the idea of the earlier supercontinent called Pangea. The secular concept for this continental drift, now known as plate tectonics, goes further and suggests that there were supercontinents even earlier than Pangea, including Pannotia, and before that Rodinia, encompassing the earth's proposed multi-billion year geologic history.

Creation scientists and Flood geologists do not deny that these continents may have been connected to one another in the past as a single supercontinent in light of Genesis 1:9. Actually, it was a Christian geologist named Antonio Snider in 1859 who was the first person to publicly comment on this jigsaw puzzle fit of all the continents, except that he believed the spreading apart and separation of the continents occurred catastrophically during the Genesis Flood.

Creation scientists believe, along with their secular colleagues, that there is good observational evidence that is consistent with an original supercontinent in the past that was split apart, and that today's continents moved to their present positions on the earth's surface. However, the main difference is the timing!

Whereas our secular colleagues believe these processes were slow and gradual over millions of years, creation scientists insist it all took place by catastrophic means, involving continental *sprint* rather than continental *drift*. However, many Christians who see the specific mention of the earth being divided in the days of Peleg, as quoted from Genesis 10:25, appeal to this particular time for the biblical explanation for continental shifting.

What Happened in the Days of Peleg?

The Context of Genesis 10

A careful search of the context of Genesis 10:25 clearly reveals that the division of the earth refers to the dividing up of the post-Flood people on the basis of languages and families, and moving them into different geographical locations. In fact, all of Genesis 10 is dedicated to dividing up Noah's family into its three major divisions based on Noah's three sons and their families, and then to further list the sub-family groups.

Because these genealogical lists encompassed all the people on the post-Flood earth, the division referred to in verse 25 must have affected the entire post-Flood human population. Several other verses dotted throughout Genesis chapter 10 indicate that it was these people who were being split up by language and moved across the earth to different geographical locations or lands:

> ". . . separated into their land . . ." (Genesis 10:5).
>
> ". . . the families of the Canaanites were dispersed" (Genesis 10:18).
>
> ". . . according to their families, according to their languages, in their lands and in their nations" (Genesis 10:20).

Even the culminating verse to the chapter states:

> These were the families of the sons of Noah, according to their generations, in their nations; and from these the nations were divided on the earth after the Flood (Genesis 10:32).

The chronological sequence of events on what happened at the Tower of Babel is given in Genesis chapter 11, where we are told in verse 8 that "the Lord scattered them abroad from there over the face of all the earth. . . ." But it's actually in Genesis chapter 10 where we are told about the different groups of people who were divided up into their families with different languages as a result of the Tower of Babel judgement. It is appropriate to compare Scripture with Scripture in the context it is written. There are four verses listed above in chapter 10 (verses 5, 18, 20, and 32) which explain the statement (in verse 25) that "in his days the earth was divided," as the division of family groupings according to the languages God gave them into different lands across the face of the earth. Verse 25 was not referring to an actual physical division of the earth from one supercontinent into today's many continents.

Another Flood!

Let's stop and consider for a moment what would be the effect of the break-up of a supercontinent followed by the sprinting of the new continents into their present positions. In late 2004, there was an earthquake in the eastern Indian Ocean, resulting in movements of up to 15 or more feet along faults. The energy of the earthquake was transmitted through the water above, producing an enormous tsunami that devastated coastlines all around the Indian Ocean basin, killing more than 220,000 people. And even a few feet of movement on the San Andreas Fault in Southern California causes the ground to shake for many miles, often resulting in the collapse of freeways and other structures.

If the continents did indeed split apart in the days of Peleg, moving thousands of miles into their current positions in a catastrophic manner, the resulting devastation would have utterly destroyed the face of the earth and everything living on it. The ocean waters would have flooded over the continents in huge tsunamis, creating a second worldwide flood event!

In Genesis chapters 10 and 11, we see no written description of such an event. To the contrary, at the end of the Flood in Noah's day God made specific statements that He would never allow another worldwide Flood to ravage the earth's surface and its inhabitants (Genesis 8:21–22; Genesis 9:11). God specifically stated that He had set the boundaries around the land beyond which the waters would never again flood the earth (Psalm 104:8–9).

Of course, in order to shore up their belief that Genesis 10:25 is a reference to continental break-up during the days of Peleg, some may respond that God somehow miraculously held back the ocean waters to keep another flood from happening while this land division occurred. However, there is absolutely no indication in Scripture, not even a hint, that this was the case. Quite clearly, it is far better to err on the side of caution with regard to these Scriptures. This is particularly necessary when the context of Genesis chapter 10 has four other verses that confirm the meaning of verse 25 as referring to the division of people according to their languages into lands of their own across the face of the earth. Furthermore, this is in keeping with God's command to Noah and his family after the Flood to be fruitful and multiply, and to fill the earth (Genesis 9:1).

A Major Geographical Problem

There is also a major geographical flaw with the claim that the continents split apart in the days of Peleg. The description of the Flood of Noah's day in Genesis 8 says that on day 150 of that global, year-long event the ark ran aground in the mountains of Ararat. We read in verses 3-4:

> And the waters receded continually from the earth. At the end of the hundred and fifty days the waters decreased. Then the ark rested in the seventh month, the seventeenth day of the month, on the mountains of Ararat.

Why is this so significant? The mountains of Ararat should not be confused merely with the post-Flood volcano in Turkey called Mt. Ararat. As far as we can tell, the biblical reference to "the mountains of Ararat" speaks of mountains located in the region of eastern Turkey and eastward toward the Caspian Sea.

The buckling of the rock layers within these mountains indicate that they were formed by continental collisions. Thus, if a supercontinent such as Pangea broke apart in the days of Peleg to arrive at their present positions, then these mountains of Ararat would only have formed in the days of Peleg. Thus, they would not have existed on day 150 of the Flood for the ark to run aground on them!

The mountains of Ararat appear to have been caused by the collision of the Eurasian Plate with the Arabian and African Plates, perhaps influenced by the concurrent collision of the Indian Plate with the Eurasian Plate (figure 2). Thus, it would seem that most of the continental shifting between Europe, Asia, Africa, Arabia, and India most likely would have largely been completed by day 150 of the Genesis Flood.

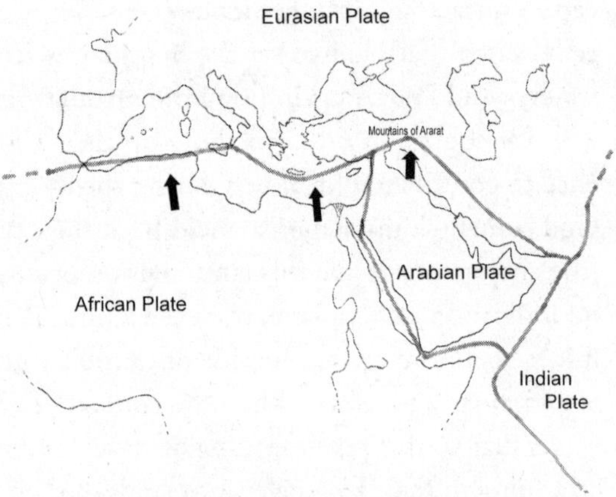

Figure 2: Plate movement resulted in the formation of the Mountains of Ararat.

Naturally, there still could have been comparatively minor adjustments after this point in the Flood, as the mountains of Ararat could still have been rising as further mountain building occurred after the ark ran aground. Again, there appear to be no hints in the biblical narrative of the Flood in Genesis 6–9 that there was any major continental shifting across the earth's surface after day 150, at least in the region of the mountains of Ararat. It's possible, however, that there still could have been some minor continental movement on the other side of the globe, with respect to North and South America, Australia, Antarctica, etc., and so this can't be ruled out entirely. However, according to Genesis 8:2, the fountains or springs of the great deep were stopped, and the windows of heaven were closed, on day 150, implying that the Flood waters possibly had reached their zenith at that point.

Furthermore, if the springs were associated with the rifting of the earth's crust and the seafloor spreading, and subduction and mantle convection that had moved the continental plates apart catastrophically, then the closure of these springs or fountains would seem to imply that the processes allowing for major movement of the continents would have stopped at this point, or at the very least, began to start decelerating, and eventually reach their present snail's pace.

All the primary geologic processes responsible for forcing the catastrophic continental movements during the Flood appear to have likewise begun to rapidly decelerate on and after day 150. Also, Genesis 8:3 indicates that the Flood waters began to steadily decrease and therefore recede from this time point onward, which would seem to indicate that the Flood waters were now subject to new land surfaces and topography rising and valleys sinking as a result of vertical earth movements. This is in stark contrast to the large horizontal movement and associated mountain-building that shifted the continents apart in the first portion of the Flood year.

Further Scientific Support for Continental Shift During the Flood

There are numerous evidences that support the contention that the pre-Flood supercontinent split and the resultant continents shifted apart catastrophically during the Flood. Several of these are highlighted below.

Folded Fossil-bearing Sediment

The fossil-bearing sedimentary layers produced by the Flood include the massive amount of plants buried and fossilized in coal beds. Many of these coal beds in the eastern United States can be traced right across Europe as far as Russia, a testimony to their global-scale formation during the Flood. There are also folded coal seams in the Appalachian Mountains and in the Ural Mountains of Russia. These deformed rock units formed by collisions of continents as a result of rapid continental movement. They were formed before other fossil-bearing sedimentary layers were deposited above the coal beds as a result of continued erosion elsewhere during the Flood. In order to fold and later bury these coal deposits, the continental division and shifting responsible for these mountain-building collisions had to have occurred during the Flood. This is also the same movement of continents that then subsequently separated what is now North and South America from Europe and Africa to form the Atlantic Ocean basins, all during the Flood.[1]

1. J.F. Dewey and J.M. Bird, "Mountain Belts and the New Global Tectonics," *Journal of Geophysical Research* 75 no. 14 (1970): 2625–2647; R.S. Dietz, "Geosynclines, Mountains and Continent-building," *Scientific American* 226 no. 3, as reproduced in *Continents Adrift and Continents Aground* (San Francisco, CA: W.H. Freeman and Company, 1976), p. 103–111.

Basalts

Huge areas consisting of thick volcanic rock layers stacked on top of one another are found in a number of places on today's continents. Even secular geologists recognize these as catastrophic outpourings of huge volumes of lavas, so they call them flood basalts.[2] The two largest examples are the Siberian Traps and the Deccan Traps of India. A smaller example is the Columbia River Basalt of the U.S. Pacific Northwest. Flood basalts are found on every continent, usually with fossil-bearing sedimentary layers beneath them, and further fossil-bearing sedimentary layers above them, indicating they are also the result of catastrophic volcanic eruptions during the Genesis Flood.

The only way huge volumes of basalt lavas could be supplied for such catastrophic eruptions on such a grand scale was via huge mantle upwellings, called plumes. These plumes likely formed as a result of mantle-wide convection during catastrophic continental break-up and shifting. Since these flood basalts had to be produced during the Genesis Flood, then the rapid continental *sprint* and associated mantle processes also had to have occurred during the Flood.

Apes Buried First

In the post-Flood sediments of Africa (including some volcanic layers), we find fossilized remains of apes, and, by and large, in sediment layers *on top* of them are found fossilized human remains and other evidences of human occupation. This fossil sequence on the African continent is thus trumpeted by evolutionists as evidence of apes and then humans having progressively "evolved" from a common ancestor.

However, in the biblical model the apes started migrating from the Ararat area to Africa as soon as they left the ark, arriving early in the post-Flood period.[2] Noah's descendants stayed together at Babel and didn't migrate into Africa until after the confusion of languages. Thus, humans would have arrived in Africa long after the apes had arrived, and expectedly, we find apes buried and fossilized in the localized post-Flood sediment layers before humans. This logical fossil explanation requires that Africa was in its present position before Babel and the days of Peleg as a result of continental division and shifting having occurred during the Flood when the apes were still initially on board the ark.[3]

2. S.A. Austin et al., "Catastrophic Plate Tectonics: A Global Flood Model of Earth History," in *Proceedings of the Third International Conference on Creationism*, R.E. Walsh, ed. (Pittsburgh, PA: Creation Science Fellowship, 1994), p. 609–621.
3. K.P. Wise, "Lucy Was Buried First: Babel Helps Explain the Sequence of Ape and Human Fossils," *Answers*, April–June, 2008, p. 66–68.

Conclusion

Though continental division and shifting in the days of Peleg appears feasible from a superficial reading of Genesis 10:25 in isolation, this concept has some major problems associated with it for the following reasons:

- When Genesis 10:25 is read within the context of the whole chapter, the four other verses (5, 18, 20, and 32) speaking of the division clearly emphasize that this was a linguistic and family division of all post-Flood people into different lands (geographical locations).
- Had the division of continents occurred during the days of Peleg, then the associated catastrophism would have resulted in another worldwide Flood, in violation of God's specific promise to Noah.
- Had the division of continents occurred during the days of Peleg, then the ark (with Peleg's ancestors) would have had no place to land, as the mountains of Ararat produced by continental collisions would not have yet existed.
- There is tremendous fossil and geologic evidence for continental division having occurred only during the Flood.

We, therefore, gently and lovingly encourage our brothers and sisters in Christ to refrain from claiming the division of continents occurred during the days of Peleg. The phrase "for in his days the earth was divided" in Genesis 10:25 needs to be kept and read within its context of Genesis 10 to give it its correct meaning. On the other hand, we also want to encourage people to realize that the Flood is the only major catastrophic event and the only logical mechanism for splitting apart the continents.

Answering a Few Objections[4*]

1. Objection: "To start with, the Hebrew word for 'earth' in Genesis 10:25, 10:32, and 11:1 is Hebrew #776 (*erets*) in the Strong's Concordance, which says this word means earth, field, ground, land and world. . . . In fact, the clear meaning of this Hebrew word for 'earth' . . . is a very strong indication that the Peleg reference has to do with actual breakup of the land mass."

Answer: There are other uses of this word, as Hebrew lexicons readily point out, meaning particular nations or inhabitants. For those not fluent in Hebrew, one needs to consult reputable Hebrew lexicons such as *The Brown-*

4. These are actual comments made to Answers in Genesis regarding the issue of an alleged continental split during the days of Peleg.

Driver-Briggs Hebrew and English Lexicon (BDB) or *The Hebrew and Aramaic Lexicon of the Old Testament* by Koehler and Baumgartner (KB) to see how each Hebrew word, in its context, should be used. Naturally, these do not hit every instance but representative instances (keep in mind *erets* is used well over 2,000 times in the Old Testament). BDB uses as one of its representative examples that Genesis 11:1 is in reference to inhabitants of earth.[5] However, the two brought up in Genesis 10 are not mentioned.

Though KB does not use any of the three pointed out, it does use 2 Chronicles 12:8, where *erets* is used for nations/kingdoms as well as several others to indicate countries and regions of people.[6]

So to exclude this definition may not be wise. And considering that BDB used *erets* specifically in Genesis 11:1 to refer to people confirms the point. According to leading Hebrew lexicons that utilize the context, these would be referring to the people being divided by language. Also, keep in mind that if one wants to argue for *erets* to mean "continent(s)," this is not even listed as a definition among the lexicons.

2. Objection: "You are incorrect that the Peleg reference comes in the middle of an account of the division of languages. It comes in the middle of a genealogy. The story of the division of languages comes afterwards, separately."

Answer: Genesis 10 is a breakdown of the language divisions that are discussed in more detail, with the chronological account in Genesis 11. Even Genesis 10 points out after each genealogy of Japheth, Ham, and Shem that it was a *linguistic division* in accordance with their family group to their nations. Consider the phrases in Genesis 10 that summarize and signify the context of language in these verses:

> Genesis 10:5 — From these the coastland peoples of the Gentiles were separated into their lands, everyone according to his *language*, according to their families, into their nations (emphasis added).
>
> Genesis 10:20 — These were the sons of Ham, according to their families, according to their *languages*, in their lands and in their nations (emphasis added).
>
> Genesis 10:31 — These were the sons of Shem, according to their families, according to their *languages*, in their lands, according to their nations (emphasis added).

5. *Brown-Driver-Briggs Hebrew and English Lexicon* (Peabody, MA: Hendrickson Publishers, 2005), p. 76.
6. L. Koehler and W. Baumgartner, *Hebrew and Aramaic Lexicon of the Old Testament*, Volume 1 (Boston, MA: Brill Publishers, 2001), p. 90–91.

The context of Genesis 10 is indeed referring to linguistic divisions from which the nations were being divided. Even a prominent Jewish historian understood this to mean a division of nations. Consider Josephus's comments here:

> Heber begat Joctan and Phaleg: he was called Phaleg, because he was born at the dispersion of the nations to their several countries; for Phaleg, among the Hebrews, signifies *division.*[7]

3. Objection: "Have you carefully looked at the word for 'divided' in each reference? They are two different Hebrew words: vs. 25 *palag* vs. 32 *parad.* The former can mean to split or cleave and the latter to scatter. . . . What is being divided appears different since the Hebrew verb is different in both verses."

Answer: The name of Peleg [Strong's Concordance #06389] in verse 25 is a variant of [#06388] *peleg*, which in turn is a derivation of [#06385] *palag*. This same root word for Peleg's name is also used in Genesis 10:25. It makes sense why this was used in direct reference to Peleg's name. But this is still different from verse 32 where [#06504] *parad* is used. However, they each appear in the same context.

Parad

Working backward, the *Theological Wordbook of the Old Testament* points out that *parad* is in reference to the scattering of peoples under comment 1806 (discussing *parad*). They in turn reference A. Wieder, "Ugaritic-Hebrew Lexicographical Notes," JBL 84:160–64, esp. p. 163–64. In fact, *parad* is also the Hebrew word used in Genesis 10:5 where it states: "From these the coastland *peoples* of the Gentiles were separated [*parad*] into their lands, everyone according to his language, according to their families, into their nations." Later Mosaic writings in Deuteronomy 32:8 also use *parad* in reference to the split of nations.

Palag

This Hebrew word *palag* is used only three times in Scripture outside of Genesis 10. In 1 Chronicles 1:19 it repeats Genesis 10:25. In one case it refers to a splitting of a water channel when it overflows in poetic Job 38:25. The other usage is in Psalm 55:9 where it refers to splitting of languages. David was speaking of his enemies and was asking the Lord to judge them with the splitting of their tongues. Obviously, David was conjuring thoughts of the Tower of Babel and tongue-shifting there.

Peleg's name was a direct derivation of *palag*, and considering the context of Genesis 10, it makes sense this Hebrew name was indeed referring to the linguistic division. So there would be no reason to distance from this plain interpretation.

7. *The Works of Josephus, Complete and Unabridged*, trans. by William Whiston (Peabody, MA: Hendrickson Publishers, 1988), p. 37.

24

Vestigial Organs — Evidence for Evolution?

DR. DAVID N. MENTON

Vestigial organs have long been one of the classic arguments used as evidence for evolution. The argument goes like this: living organisms, including man, contain organs that were once functional in our evolutionary past, but that are now useless or have reduced function. This is considered by many to be compelling evidence for evolution. More importantly, vestigial organs are considered by some evolutionists to be evidence against creation because they reason a perfect Creator would not make useless organs.

The word *vestige* is derived from the Latin word *vestigium*, which literally means a "footprint." The Merriam-Webster's Dictionary defines a biological vestige as a "a bodily part or organ that is small and degenerate or imperfectly developed in comparison to one more fully developed in an earlier stage of the individual, in a past generation, or in closely related forms."

Darwin on "Rudimentary Organs"

Charles Darwin was perhaps the first to claim vestigial organs as evidence for evolution. In chapter 13 of his *Origin of Species*, Darwin discussed what he called "rudimentary, atrophied and aborted organs." He described these organs as "bearing the plain stamp of inutility [uselessness]" and said that they are "extremely common or even general throughout nature." Darwin speculated that these rudimentary organs once served a function necessary for survival, but over time that function became either diminished or nonexistent.

In Darwin's book *The Descent of Man*, he claimed about a dozen of man's anatomical features to be useless including the muscles of the ear, wisdom teeth, the appendix, the coccyx (tailbone), body hair, and the semilunar fold in the corner of the eye. To Darwin, this was strong evidence that man had evolved from primitive ancestors.

The List of "Vestigial Organs" Grows

In 1893 the German anatomist Robert Wiedersheim expanded Darwin's list of "useless organs" to 86. Listed among Wiedersheim's "vestigial" organs were such organs as the parathyroid, pineal and pituitary glands, as well as the thymus, tonsils, adenoids, appendix, third molars, and valves in veins.[1] All of these organs have been subsequently shown to have useful functions and indeed some have functions essential for life.

Wiedersheim's vestigial organs were presented as one of the so-called "proofs" of evolution in the famous Scopes "Monkey Trial" of 1925. Horatio Hackett Newman, a zoologist from the University of Chicago, stated on the witness stand that "there are, according to Robert Wiedersheim, no less than 180 [*sic*] vestigial structures in the human body, sufficient to make a man a veritable walking museum of antiquities."[2]

Vestigial Organs Still Used as Evidence for Evolution

For over 100 years, evolutionists have continued to use vestigial organs as evidence for evolution. In 1971 the *Encyclopedia Britannica* claimed there were more than 100 vestigial organs in man, and even as recently as 1981, some biology textbook authors were claiming as many as 100 vestigial organs in the human body.[3] One of the most popular current biology textbooks declares that "many species of animals have vestigial organs." Examples cited in humans include the appendix, "tailbone," and muscles that move the ear.[4]

In addition to textbooks, countless popular science magazines, evolution blogs, and websites continue to promote vestigial organs as evidence for evolution.

1. R. Wiedersheim, *The Structure of Man: An Index to His Past History* (London: Macmillan and Co., 1895).
2. *The World's Most Famous Court Trial* (Dayton, TN: Bryan College, 1990). This book is a word-for-word transcript of the famous court test of the Tennessee Anti-Evolution Act, at Dayton, July 10 to 21, 1925, including speeches and arguments of attorneys, testimony of noted scientists, and Bryan's last speech.
3. S.R. Scadding, "Do Vestigial Organs Provide Evidence for Evolution?" *Evolutionary Theory* 5 (1981): 173–176.
4. K.R. Miller and J. Levine, *Biology: Teachers Edition* (Upper Saddle River, NJ: Pearson Prentice Hall, 2006), p. 384.

A website sponsored by the Discovery Channel, for example, assures us that "the human body has something akin to its own junk drawer," and that this junk drawer "is full of vestigial organs, or souvenirs of our evolutionary past."[5]

Problems with Vestigial Organs as Evidence for Evolution

Why Do Useless Organs Persist?

Darwin himself pointed out a flaw in the vestigial organ argument. He wondered how once an organ is rendered useless, it can continue to be further reduced in size until the merest vestige is left. In chapter 14 of *Origin of Species* he declared, "It is scarcely possible that disuse can go on producing any further effect after the organ has once been rendered functionless. Some additional explanation is here requisite which I cannot give." Why, indeed, would useless organs continue to exist for millions of years after they ceased to have any selective advantage?

The Loss of Useful Organs Doesn't Explain Their Origin

A problem for using vestigial organs as evidence for "amoeba to man" evolution is that the chief burden of the macro evolutionary explanation is to account for the spontaneous origin of new functional organs — not the loss of functional organs. While evolution might require the loss of functional organs, it is the acquisition of fundamentally new organs that remains unexplained by random mutations and natural selection.

How Can We Be Certain an Organ Is Useless?

The problem with declaring any organ to be without function is discriminating between truly functionless organs and those that have functions that are simply unknown. Indeed, over the years nearly all of the organs once thought to be useless have been found to be functional. When we have no evidence for the function of an organ, it is well to bear in mind that absence of evidence is not evidence of absence.

Declaring Useful Organs to Be Useless Can Be Dangerous

Once an organ is considered to be useless, it may be ignored by most scientists, or even worse, surgically removed by physicians as a useless evolutionary leftover. The oft repeated claim that the human appendix is useless is a case in point. The evolutionist Alfred Romer in his book *The Vertebrate Body* said of the human appendix: "Its major importance would appear to be financial support of the surgical profession."[6] We can only wonder how many normal appendices

5. www.health.howstuffworks.com/vestigial-organ.htm/printable.
6. A.S. Romer and T.S. Parsons, *The Vertebrate Body* (Philadelphia: Saunders College Publishers, 1986), p. 389.

have been removed by surgeons since Darwin first claimed them to be a useless vestige. Even more frightening would be the surgical removal of a "useless" parathyroid or pituitary gland.

The Definition of Vestigial Organs Has Been Changed

As the list of "functionless" organs has grown smaller and smaller with advancing knowledge, the definition of vestigial organs has been modified to include those whose functions are claimed to have "changed" to serve different functions. But such a definition removes the burden of proof that vestigial organs are a vestige of evolution. Thus, the evolutionist might concede that the human coccyx ("tail bone") does indeed serve an important function in anchoring the pelvic diaphragm — but still insist, without evidence, that it was once used by our ancestors as a tail.

Circular Reasoning

The most conspicuous logical flaw in the use of vestigial organs as evidence for evolution is circular reasoning. Evolutionists first declare vestigial organs to be a result of evolution, and then they turn around and argue that their existence is evidence for evolution. This kind of argument would hardly stand up in a court of law.

There Are Other Explanations for Vestigial Organs

Vestiges of Embryology

Evolutionists insist on explaining vestigial organs only in terms of evolution, but other explanations are more plausible and even provable. For example, the human body does have many organs and structures that are clearly vestiges of our embryological development. While it is quite easy to prove that an organ or structure is a vestige of embryology, there can be no empirical evidence to support the speculation that an organ is a vestige of evolution.

There are several structures that function during the development of the embryo and fetus that appear to be no longer used after birth. Remnants of these once-functional structures persist throughout life. Such structures perfectly fit the definition of a vestige, but they are not vestiges of evolution. The following are a few examples of embryological vestiges.

Ligamentum arteriosum — obliterated remnant of the ductus arteriosus, an artery that shunted blood from the pulmonary trunk to the descending aorta, thus bypassing the lung during fetal development. In certain cases of congenital heart defects, the ductus arteriosus actually continues to function for some time after birth to keep the baby alive.

Ligamentum teres hepatis — obliterated remnant of the umbilical vein that shunted much of the oxygenated blood away from the liver to the inferior vena cava during fetal development.

Median umbilical ligament — an obliterated vestige of the allantois, a pouch extending off of the embryonic cloaca. The allantois disappears very early in gestation after functioning as a scaffolding to help construct the umbilical cord; this remnant is seen as a ligament extending from the bladder to the umbilicus (bellybutton).

Sexual Dimorphism

In most primates there are striking anatomical differences between males and females of the same species. These differences between the sexes are refered to as *sexual dimorphism*. The skulls of a male and female gorilla, for example, might not be recognized as from the same species if one had never seen them in the flesh. The difference between the sexes is not as dramatic in the case of humans, though they are dimorphic. The bodies of human males and females differ mostly in the organs related to reproduction.

Up until the end of the sixth week of embryological development, the reproductive organs of males and females are indistinguishable. After this time, the genital organs of both sexes develop from the same common starting tissues under the control of sex chromosomes (XX in the female and XY in the male) and various hormones. As a result of their embryological development from the same primordia, each sex contains vestigial components of the other sex.

Almost every organ of the female reproductive system can be found in a different or vestigial form in the male reproductive system (and vice versa). For example, in the male, the prostatic utricle (an out pouching of the prostatic urethra having no known function) is a remnant of the paramesonephric duct that develops into the uterus and oviducts of the female. Clearly, the vestigial organs of reproduction are not a result of evolution but rather embryological development.

Homology

Many vestigial organs are examples of homology but not necessarily of evolution. Homology is an underlying similarity between different kinds of animals recognized by both evolutionists and creationists. All terrestrial vertebrates, for example, share a widespread similarity (homology) of body parts. Evolutionists insist that this similarity is the result of evolution from a common ancestor. Creationists, on the other hand, argue that this similarity reflects the theme of a common Creator and the need to meet similar biological requirements.[7]

7. G.E. Parker, *Creation: Facts of Life* (Green Forest, AR: Master Books, 2006), p. 43–53.

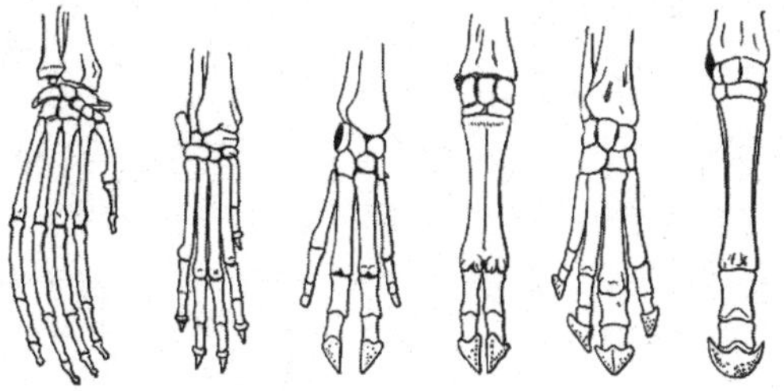

Homology in vertebrate limbs does not prove they came from a common ancestor.

For example, all vertebrates with true limbs (amphibians, reptiles, birds, and mammals) have the same basic limb structure at least during their embryological development. This standard vertebrate limb consists of an upper limb comprising one bone, a lower limb comprising two bones, and a hand or foot bearing five digits (fingers and toes). Thus, the limbs of all limbed vertebrates share fundamental similarities, with each being specialized to meet the needs of each species.

Horses have five digits while developing as an embryo, but generally all but one (the third digit) is absorbed before birth. Vestiges of the second and third metacarpal (and metatarsal) bones are visible in the modern horse as the splint bones. Some fossil horses, however, had three toes, but both three-toed and one-toed horses have been found together in the fossil record. In *National Geographic* magazine, for example, there is a picture of the feet of both a three-toed horse (Pliohippus) and a one-toed horse (Equus) that were found at the same volcanic site in Nebraska.[8]

Human hair is an example of a homologous structure declared to be vestigial by evolutionists. All mammals have hair. Hair may vary from the compacted hairs of a rhinoceros horn to the quills of a porcupine. To declare the unique hairs of one mammal to be vestigial to those of another is biological nonsense.

Evaluating Currently Claimed "Vestigial" Organs

It may prove useful for the reader to use the forgoing discussion of vestigial organs to evaluate some current claims for such structures. The website LiveScience lists what it regards as the top ten "vestigial" organs.[9] Five of these are found in humans, and are discussed below in order of their perceived importance by LiveScience.

The Appendix

Ever since Darwin, the appendix has been the prime example of a "useless" organ. LiveScience says of the appendix that "it is a vestigial organ left behind

8. M.R. Voorhies, "Ancient Ashfall Creates a Pompeii of Prehistoric Animals," *National Geographic*, January 1981, p. 74.
9. www.livescience.com/animals/top10_vestigial_organs.html.

from a plant-eating ancestor." In the middle of the 20th century, surgeons often removed the appendix electively during abdominal surgery, assuming it had no function. According to most evolutionists, the appendix is a vestige of the caecum (an expanded area at the beginning of the large intestine) left over from our plant-eating ancestors. But since humans have a well-developed caecum as well as an appendix, the appendix can hardly be considered a vestigial caecum. In his book *The Vertebrate Body*, evolutionist Alfred Romer said that the appendix is "frequently cited as a vestigial organ supposedly proving something about evolution. This is not the case. . . ."[10]

The important point is that the presence or absence of an appendix (or a caecum) reveals no evolutionary pattern whatever. An appendix is not found in any invertebrate, amphibian, reptile, or bird. Only a few diverse mammals have an appendix. The appendix is found, for example, in rabbits and some marsupials such as the wombat, but is not found in dogs, cats, horses, or ruminants. Both Old World and New World monkeys lack an appendix, while anthropoid apes and man have an appendix.[11]

The appendix is a complex, highly specialized organ with a rich blood supply — not what one would expect from a vestigial organ. The appendix is part of the gut associated lymphoid tissue (GALT), and has long been suspected of playing an immunological role much like that of the tonsils and adenoids (also once considered to be vestigial).

Recent evidence suggests that the appendix is well suited to serve as a "safe house" for commensal (mutually beneficial) bacteria in the large intestine. Specifically, the appendix is believed to provide support for beneficial bacterial growth by facilitating re-inoculation of the colon with essential bacteria in the event that the contents of the intestinal tract are purged following exposure to a pathogen.[12]

Male Breast Tissue and Nipples

It is surprising that evolutionists still continue to bring up the matter of the male breast (mammary gland) as a vestigial organ. Are they proposing that the males once nursed the young early in their evolution but no longer do so? Of course not. So how then does the evolutionist explain the male's vestigial mammary gland if it is not a consequence of evolution?

Vestigial mammary glands in males can only be understood in terms of embryology — not evolution. Mammary glands begin to develop in both males

10. Romer and Parsons, *The Vertebrate Body*, p. 358.
11. J.W. Glover, "The Human Vermiform Appendix: A General Surgeon's Reflections," *Technical Journal* 3 no. 1 (1988): 31–38.
12. R.R. Bollinger et al., "Biofilms in the Large Bowel Suggest an Apparent Function of the Human Vermiform Appendix," *Journal of Theoretical Biology* 249 no. 4 (2007): 826–831.

and females in the sixth week of gestation. At the time of birth, the rudimentary mammary glands of males and females are identical. In fact, both male and female mammary glands may be slightly enlarged at birth and secrete a fluid that is commonly known as "witches milk." This results from hormones that induce milk production in the mother being passed through the placenta to the fetal circulation.[13]

The male mammary gland is clearly a rudimentary or vestigial structure, but even the mammary gland of the nonlactating female might be considered vestigial. Female mammary glands are never fully developed and functional except during times of breast feeding the young. Should the evolutionist then consider the nonlactating female mammary gland to also be a vestige of evolution? The old evolutionist axiom that "nothing in biology makes sense except in the light of evolution" might better say that nothing in biology makes sense in the light of evolution.

Wisdom Teeth

Darwin was the first to popularize the notion that wisdom teeth are vestigial leftovers from our ape-like ancestors. The inherent racism of Darwinism is apparent when in his *Descent of Man*, Darwin declared that wisdom teeth are often lacking in "the more civilized races of man" in contrast to the "melanin (black) races where the wisdom teeth are furnished with three separate fangs, and are generally sound."[14]

Wisdom teeth, properly known as third molars, generally appear between the ages of 15 and 27 in both the upper and lower jaws of man. Many evolutionists consider them to be vestigial because unlike apes, third molars often fail to develop properly in man due to lack of space in the jaw. They argue that apes with their sloping face have longer jaws than man, and that when ape-like creatures evolved into humans with a vertical face and shorter jaws, there was no longer room for third molars.

Third molars are hardly useless vestiges. When there is adequate room for their development, they are fully functional molars and are used in chewing much as the first and second molars. Thinking them to be vestigial, many dentists in the past routinely removed third molars whether or not they were causing problems. It has been estimated that in America, only 20 percent of all young people with otherwise healthy teeth develop impacted third molars that require medical attention, while in the past, nearly nine

13. K.L. Moore, *The Developing Human* (Philadelphia, PA: W.B. Saunders Company, 1988), p. 427.
14. C. Darwin, *The Descent of Man and Selection in Relation to Sex* (New York, NY: D. Appleton and Company, 1896), p. 20.

out of ten American teenagers with dental insurance had their third molars extracted.[15]

The "Tailbone" (Coccyx)

The so-called "tailbone" is perhaps the most commonly touted example in man of a "useless" evolutionary vestige. According to evolutionary dogma, the tailbone, properly called the coccyx (because of its similarity to the shape of a cuckoo's beak), is a vestigial tail left over from our tailed monkey-like ancestors. Once again, many in the medical profession have been taken in by evolutionary speculation but mercifully, they have refrained from surgically removing the normal coccyx.

Even human abnormalities that have nothing to do with the coccyx have been declared to be "human tails." In a report in *The New England Journal of Medicine*, titled "Evolution and the Human Tail," Ledley described a two-inch long fleshy growth on the back of a baby, which he claimed to be a "human tail," though he conceded that it showed none of the distinctive biological characteristics of a tail! In fact, the "tail" was merely a fatty outgrowth of skin that wasn't even located in the right place on the back to be a tail! Still, Ledley declared that "even those of us who are familiar with the literature that defined our place in nature (Darwinism) — are rarely confronted with the relation between human beings and their primitive ancestors on a daily basis. The caudal appendage brings this reality to the fore and makes it tangible and inescapable."[16]

The human coccyx is a group of four or five small vertebrae fused into one bone at the lower end of our vertebral column. The coccyx is commonly called the "tailbone" because of its superficial similarity to a tail. The coccyx does occupy the same relative position at the end of our vertebral column as does the tail in tailed primates, but then, where else would it be? The vertebral column is a linear row of bones that supports the head at one end and the other must end somewhere. Wherever it ends, evolutionists will be sure to call it a vestigial tail.

Many modern biology textbooks give the erroneous impression that the human coccyx has no real function other than to remind us of our evolutionary ancestry. In fact, the coccyx has some very important functions. Six muscles converge from the ring-like bones of the pelvic brim to anchor on the coccyx, forming a bowl-shaped muscular floor of the pelvis called the pelvic diaphragm.

15. A.J. MacGregor, *The Impacted Lower Wisdom Tooth* (New York, NY: Oxford University Press, 1985); a good review of wisdom teeth and the consequences of considering them to be evolutionary vestiges may be found at www.answersingenesis.org/tjv12/i3/wisdomteeth.asp
16. F.D. Ledley, "Evolution and the Human Tail: A Case Report." *N Engl J Med* 306 no. 20 (1982): 1212–1215.

The incurved coccyx with its attached pelvic diaphragm supports the organs in our abdominal and pelvic cavities such as the urinary bladder, uterus, prostate, rectum, and anus. Without this critical muscular support, these organs could be easily herniated. The urethra, vagina, and anal canal pass through the muscular pelvic diaphragm, and thus the diaphragm serves as a sphincter for these structures.

Erector Pili and Body Hair

Evolutionists have long insisted that human body hair, and the small muscles (erector pili) attached to these hairs, are useless vestiges from our hairy ancestors. But human hair is as fully functional as that of any other mammal.

The body of man, like that of most mammals, is covered with hairs except for the palms and soles. But man, unlike other mammals, has mostly tiny colorless hairs called vellus hairs covering the seemingly "unhaired" parts of his body. This gives humans the appearance of being "hairless" with the exception of such areas as the scalp, axilla, chest, and genital regions. But in fact, if we count the tiny vellus hairs, humans have about as many hairs per square inch on their nose and forehead as they do on the top of their head. Indeed, hair density per square inch is approximately the same on the human body as it is for most primates.

Hair grows from tube-like structures in the skin called *hair follicles*. Most hair follicles are capable of making more than one type of hair depending in part on age, location, and hormonal stimulation. The first hairs to grow from the follicles of the developing baby are long silky hairs called *lanugo hairs*. These hairs, which cover most of the body, are usually shed before birth and are replaced with tiny vellus hairs. Thus, the newborn baby may appear to be mostly hairless, but in fact is covered with vellus hairs.

The long pigmented hairs on our scalp and elsewhere on our body are called *terminal hairs*. Terminal hairs grow from follicles that once produced lanugo and vellus hairs and with age may be replaced once again with vellus hairs. For example, after a boy reaches sexual maturity he may begin to lose terminal scalp hairs, which are replaced with vellus hairs, giving the appearance of baldness. Conversely, some vellus hairs on the face may be replaced with terminal hairs, producing a beard.

Evolutionists argue that human body hairs are vestigial (useless) because there are so few long terminal hairs compared to tiny vellus hairs. Hair serves as thermal insulation in most mammals, which is important because most animals are incapable of regulating their body temperature by sweating. Man, on the other hand, is a profuse sweater and can maintain body temperature over a much wider range of ambient temperature than nearly all other mammals. Long body hair of the type seen on most mammals would interfere with the evaporative water loss necessary for human thermoregulation by sweating.

In most mammals, hair serves as an important barrier to ultraviolet radiation from the sun. While human scalp hair serves a similar function on the typically exposed top of our head, our primary defense against UV damage is tanning and wearing clothes.

An important function of hair is its sensory function. All hair follicles, regardless of size, are supplied with sensory nerves so that they may be considered to be mechanoreceptors. Our hairs are like small levers that, when moved by any physical stimulus including air, send sensory signals to our brain. This is true of both the tiny vellus hairs and the long terminal hairs. This sensory function of hair can hardly be considered vestigial.

Another important function of hair follicles is the restoring of the epidermal skin surface following cuts and deep abrasions. Human hair follicles, regardless of size, serve as an important source of epidermal cells for recovering the skin's surface (reepithelialization) when broad areas of the epidermis are lost. If it were not for man's abundant hair follicles and sweat ducts, even routine skin abrasions might require a skin transplant.

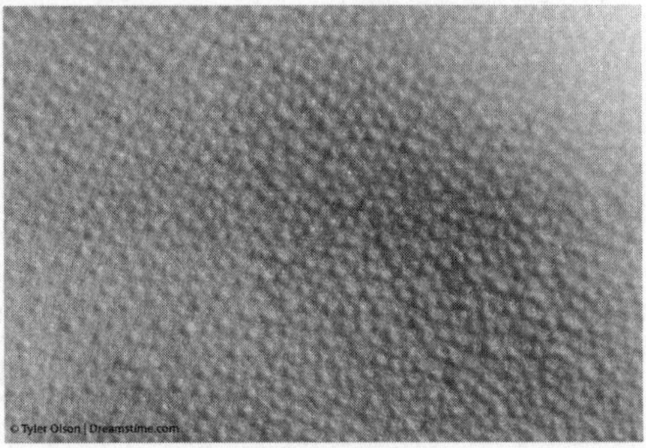

Goose bumps are not remnants of an evolutionary past, but serve several functions for humans.

All hairs are associated with muscles, and most have a muscle called the erector pili, which serves to move the hair from its normal inclined position to a more erect position. In the case of the vellus hairs of man this produces what is commonly called "goose bumps." This muscle is in a position to help squeeze oil from the sebaceous glands, which are also attached to the hair follicle. Erector pili muscles are supplied with nerves of the sympathetic nervous system, which is often associated with our response to "flight and fright" stimuli. Thus, when we are frightened we may get goose bumps. We also get goose bumps when we are chilled. Contraction of the erector pili muscles produces heat, and if this response is inadequate to warm the body, shivering may follow, which involves repeated contractions of the large body muscles.

Is the Argument for Vestigial Organs Vestigial?

Over the years, advancement in our understanding of biological science has raised serious doubts about vestigial organs as evidence for evolution. Creationists have subjected the evolutionary interpretation of vestigial organs to strong criticism.[17] Even some evolutionists are now urging that vestigial organs be downplayed or even abandoned as evidence for evolution. The evolutionist S.R. Scadding, for example, has critically examined vestigial organs as evidence for evolution. He concluded: "Since it is not possible to unambiguously identify useless structures, and since the structure of the argument used is not scientifically valid, I conclude that 'vestigial organs' provide no special evidence for the theory of evolution."[18] But like the long discredited recapitulation myth (that embryos pass through stages of their evolutionary history), vestigial organs continue to be used as evidence for evolution.

LET THERE BE TRUTH

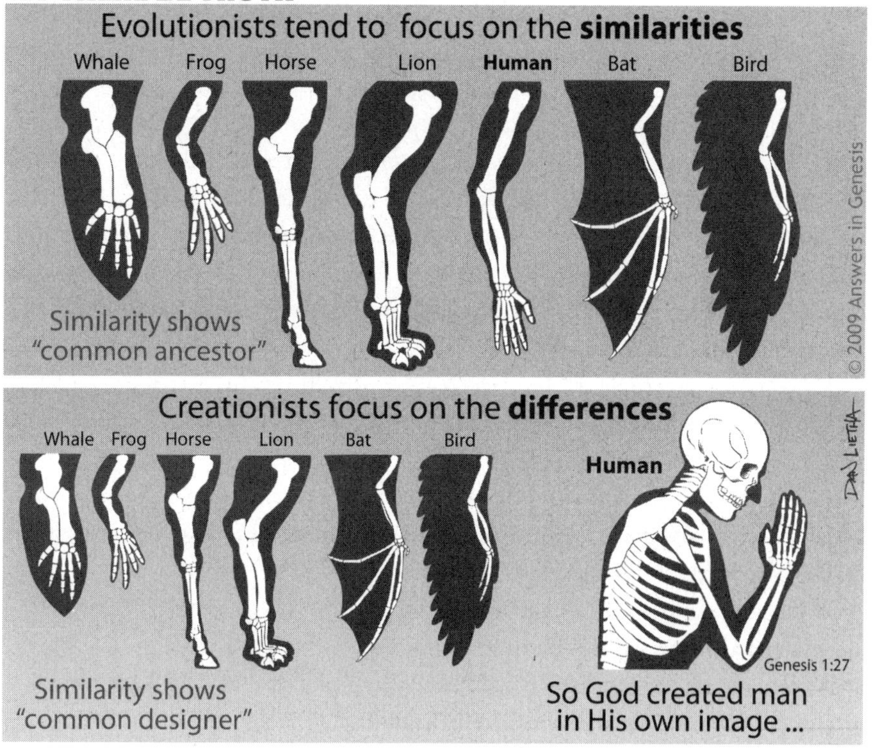

17. J. Bergman and G. Howe, *Vestigial Organs are Fully Functional* (Terre Haute, IN: Creation Research Society Books, 1990).
18. S.R. Scadding, "Do Vestigial Organs Provide Evidence for Evolution?" *Evolutionary Theory* 5 (1981): 173.

25

Is *Tiktaalik* Evolution's Greatest Missing Link?

DR. DAVID N. MENTON

In both the print and broadcast media in 2006 and 2007, reports of the discovery of the fossil fish known as *Tiktaalik* were hyped as convincing proof that, through a random chance process of evolution, fish sprouted legs and walked out onto the land, where they turned into amphibians, reptiles, mammals, and, ultimately people. But the media's excitement seems to stem not so much from being able to report a real scientific discovery as in being able to discredit the biblical account of creation.

A front page article in the *New York Times*,[1] for example, hailed *Tiktaalik* "as a powerful rebuttal to religious creationists, who hold a literal biblical view on the origins and development of life."

The whole idea of walking fish has come to be symbolic of the evolutionary worldview and its opposition to biblical Christianity. Many evolutionists display the familiar "Darwin fish" symbol on their automobiles, T-shirts, and office doors as a public declaration of their allegiance to evolution. The "Darwin fish" is a desecration of the fish symbol used by early Christians as a means of mutual identification during a time of persecution. Christians chose the

The walking fish has become symbolic of evolutionism.

1. John Noble Wilford, "Fossil Called Missing Link From Sea to Land Animals," *New York Times*, Late Edition — Final, Section A, Page 1, Column 5, April 6, 2006.

fish symbol because the individual letters of the Greek word *ichthys* (for "fish") served as an anagram for "Jesus Christ Son of God, Savior." Evolutionists have substituted the word "Darwin" for "*ichthys*" and have placed walking legs with feet on the fish. Thus, the Darwin fish reflects the fact that many evolutionists have indeed replaced Christianity with Darwinism. As for the legs on the Darwin fish, we will see that there are no known fish with true "legs" (and certainly no feet), and none capable of actually "walking" — except in the most trivial sense of the word.

We Must Be Cautious of Evolutionary Claims

In the next months and years, there will doubtless be further claims in the popular media of "irrefutable proofs" for evolution and, more importantly, "proofs" against the biblical account of creation. The popular media — as with tax-supported zoos, science museums, and public schools — are often zealous supporters of the quasi-scientific religion of materialism.

However, few reporters, teachers, or laymen have ever read the original scientific reports upon which grandiose evolutionary claims are based. Moreover, these reports are often convoluted, conflicting, and couched in unprovable assumptions that make evolutionary claims difficult to evaluate even for those who do examine the original scientific papers.

To evaluate the claims that there are fossil fish with legs that walked out of water to take up permanent residence on the land, one needs to understand something about fish, tetrapods (limbed vertebrates including humans), legs, and what is required anatomically to walk and swim. So let us begin by looking at the wide world of fish, and see which ones are supposed to be the "walkers."

There Are Lots of Fish!

The first thing to consider is that there are a *lot* of fish — both living and fossilized. Approximately 25,000 species of currently living fish have been identified, with 200–300 new species discovered — not evolved — every year. Indeed, fish comprise fully half of all known vertebrates!

It is not clear how many different fish species have been found as fossils, but some experts claim that there were once nearly a million species of fish! It appears that over time we have lost a lot of species of fish — and retained relatively fewer. But losing thousands of species of fish is hardly evolution — it's extinction. The question is, have we really gained any fundamentally *new* fish (to say nothing of fish that evolved true legs and walked out onto the land as permanent residents)?

Classification of Fish

Fish come in a bewildering variety of forms that defy consistent classification. As a result, there are competing classification schemes based on the particular bias of the classifier.

Basically, all species of fish have been divided into two main types — the jawless fish (hagfish and lampreys) and the jawed fish (all the rest). The jawed fish are, in turn, divided into two groups: the cartilaginous fish (such as the sharks and rays that have a skeleton made of flexible cartilage) and the much more numerous bony fish, which have hard bony skeletons.

Many of the so-called transitional forms have been greatly disputed, discovered (e.g., coelacanth), or dismissed, and *Tiktaalik* has recently been propped up as the "savior" of the evolutionary paradigm. How soon will it be before *Tiktaalik* is abandoned also?

Evolutionists believe that it took about 100 million years for invertebrates (animals with no bones) to evolve into vertebrates (animals with backbones). However, no compelling fossil evidence documents this purported major and unambiguous transition. While evolutionists believe that fish were the first true vertebrates, they're not sure which evolved first — cartilaginous or bony fish.

During the embryological development of vertebrates, most bones develop first as cartilage models that are later replaced by bone (called endochondral bone). Following the dictates of the embryonic recapitulation myth, it would be attractive for evolutionists to propose that cartilaginous fish evolved into bony fish, but most evolutionists consider the cartilaginous fish to be far too specialized to have been the ancestors of the bony fish.

Tiktaalik

The Bony Fish (*Osteicthyii*)

Bony fish are by far the most numerous of all fish, comprising about 24,000 living species, and they come in an amazing variety of forms and sizes (ranging from a half-inch-long sea horse weighing a fraction of an ounce to a 1,000-pound blue marlin). The purported evolutionary relationship of all these fish is at best highly speculative.

All bony fish have gills for breathing and fins for swimming. Starting from front to back, the most important fins for swimming are the paired pectoral fins (which are typically attached to the posterior margin of the skull), the generally smaller paired pelvic fins (that occupy a position near the anus), and the caudal fin (tail fin).

Bony fish are divided into two groups, the lobe-finned fish, known mostly from fossils, and the vastly more numerous ray-finned fish. Both have fins made up of bony rays, but the lobe-fins have fin rays mounted on a short, fleshy stalk supported by successive segments of bone. It is the superficial resemblance of these bony fins to tetrapod legs that has led evolutionists to speculate that the lobe-fin fish are the ancestors of tetrapods in the late Devonian (approximately 380 million years ago). So let's focus our investigation on the lobe-fins.

The Lobe-fin Fish (*Sarcopterygii*)

The lobe-finned fish have been divided into two rather dissimilar groups, the *Dipnoi* (lungfish) and the *Crossopterygii* (coelacanths and fossil relatives).

Lungfish (*Dipnoi*)

There are only three surviving types of lungfish. They are all eel-like in appearance, and have long and slender fleshy pectoral and pelvic fins, which are highly mobile. This group derives its name from the fact that these fish have air sacks ("lungs") that function at least partially in breathing (though all, at least in their immature state, have functional gills as well). The fact that these fish can breathe air, survive out of water for long periods of time, and have the ability to pull themselves along on their bellies (i.e., "walk") across mud flats with the aid of their fins, has caught the imagination of some evolutionists who consider them to be ancestral to tetrapods.

Many Living Fish Are Air-breathers and "Walkers"

But air-breathing fish are not uncommon among living fish species. For example, many popular aquarium fish (such as the paradise fish, betta, and gourami) are surface air-breathers that can actually drown if kept under water! Evolutionists are not even in agreement on whether lungs evolved before gills (as proposed by the famous vertebrate evolutionist Alfred Romer), or gills evolved before lungs.

Even the sort of "walking" that lungfish engage in is not uncommon among living fish species. Many fish are known to pull themselves along on their bellies, with the help of their pectoral fins, across large expanses of mud flats and even dry land. For example, the northern snakehead (*Channa argus*) and the walking catfish (*Clarias batrachus*) are air-breathing fish that can travel overland for considerable distances. The mudskippers are fish that breathe oxygen through their skin and "skip" along on land with the aid of their fleshy fins — indeed some of the larger species are said to skip faster than the average person can run! The climbing perch (*Anabas testudineus*) not only breathes air and "walks" on land but is even said to be capable of climbing trees! Yet *none* of these curious fish are considered by evolutionists to be ancestors of tetrapods — they are simply interesting and specialized fish. In fact there are even "flying fish" (with specialized fins that permit them to fly or glide in the air for hundreds of yards over water), but evolutionists have never considered them to be ancestors of birds.

Crossopterygians

Most evolutionists now look to fossil *Crossopterygians* for the ancestors of tetrapods — even though none of them are known to be capable of either walking or breathing out of water.

The distinguishing features of these fish are the division of the skull into anterior and posterior units (considered similar to embryonic tetrapod skulls); and fleshy pectoral fins containing bony elements (considered similar to tetrapod legs). These similarities have prompted evolutionists to confidently declare that *Crossopterygians* evolved into tetrapods.

Snakehead fish

According to evolutionists, the *Crossopterygians* flourished during the middle to late Devonian (extending from 385 million years ago to 365 million years ago) and all were once believed to have become extinct about 80 million years ago (even before the extinction of the dinosaurs).[2]

The Coelacanth — One of Many "Living Fossils"

However, in 1938 a fishing trawler netted a strange large blue fish in the Indian Ocean off the coast of Madagascar. This distinctive fish was soon identified as a *Crossopterygian* fish previously known only from the fossil record as the coelacanth.

Coelacanths are distinctly different from all other living fishes. They have an extra lobe on their tails (compared to other lobe-finned fish) and are the only living animal to have a fully functional joint in their cranium, which allows the front part of the head to be lifted when the fish is feeding.

The discovery of a coelacanth came as a surprise to evolutionists. (It was comparable to finding a living dinosaur, because these fish were believed to have become extinct 80 million years ago when they disappeared from the fossil record.) However, since 1938, dozens of living coelacanths have been found and studied, some as far as 7,000 miles away from the location of the first sightings![3]

Understandably, evolutionists are puzzled by how coelacanths could disappear for over "80 million years" and then turn up alive and well in the 20th century. They speculate that the fossilized coelacanths lived in environments favoring fossilization, whereas modern coelacanths live at great depths (over 600 feet) in caves and overhangs of steep marine reefs that don't favor fossil formation. This, however, is special pleading, since essentially no modern marine environment favors the formation of fossils and, indeed, none are being formed, as this would require rapid burial, which is not observed under normal conditions.

More importantly the coelacanth (and many other "living fossils") show that evolutionists can never assume that a plant or animal did not live during any particular period of assumed geologic time simply because it does not appear in the fossil record of this period. If 200-pound coelacanths can "hide" for "80 million years," it would seem anything can hide.

Another reason finding a living coelacanth caused so much surprise at the time of its discovery was that coelacanths were widely believed to be the

2. "New Fossils Fill the Evolutionary Gap Between Fish and Land Animals," www.nsf.gov/news/news_summ.jsp?cntn_id=106807, 2006.
3. Another was recently caught near Indonesia. See www.news.bbc.co.uk/2/hi/science/nature/6925784.stm.

Contrary to early suggestions of walking behavior, coelacanths have only been observed using their fins to swim.

ancestors of the tetrapods. Indeed, many evolutionists assumed that the very reason the coelacanths disappeared from the fossil record was because they evolved into land-dwelling tetrapods; yet here they were very much alive — and swimming!

Coelacanths Don't Walk

At the very least, evolutionists expected to observe some hint of walking behavior in the coelacanth, but the fish have done nothing to accommodate them. Although living coelacanths have often been observed swimming in their natural habitat, they have never been observed walking. Indeed, coelacanths have been observed swimming backward, upside-down, and even standing on their head! Alas — they absolutely refuse to walk on land or in the sea.

Evolutionists Look to Other Lobe-fins

Since living lobe-fin fish have not met expectations, evolutionists have turned to other fossilized lobe-fins for the ancestors of tetrapods. (After all, one can speculate endlessly about fossils without fear of contradiction — until they turn up alive.)

Currently, the three most popular Crossopterygian candidates for ancestors of tetrapods are *Eusthenopteron*, *Panderichthys*, and the recently discovered *Tiktaalik*.

Eusthenopteron

For several years, the evolutionist's "gold standard" of fish with "legs" has been the fossil fish *Eusthenopteron* (which, like the coelacanth, has fleshy pectoral fins with bones). If you have seen an artist's illustration in a textbook showing a fish walking out of the water, most likely it was *Eusthenopteron*.

Like most other jawed fish, *Eusthenopteron* has its pectoral fin girdle (bones that anchor the pectoral fins) attached to the back of its skull by means of a dermal bone called the *cleithrum*. Dermal bones develop directly from connective tissue cells under the skin, rather than from cartilage models as is the case for endochondral bones. (Fish scales, by the way, are dermal bones as well, and reside just under the superficial layer of the skin.)

Panderichthys

Panderichthys is yet another fossil *Crossopterygian* fish that has been declared to be an ancestor of tetrapods. *Panderichthys* lacks dorsal and ventral fins and has a relatively small tail fin (thus looking less obviously fish-like than *Eusthenopteron*).

Like the other *Crossopterygian* fish, *Panderichthys* has thick bony pectoral fins. Evolutionists argue that the shape of these fins and their pectoral girdle look more like that of tetrapods than *Eusthenopteron.* But Daeschler, Shubin, and Jenkins — the discoverers of *Tiktaalik* — claim that "*Panderichthys* possesses relatively few tetrapod synapomorphies, and provides only partial insight into the origin of major features of the skull, limbs, and axial skeleton of early tetrapods." As a result, they insist that "our understanding of major transformations at the fish-tetrapod transition has remained limited."[4]

Tiktaalik to the Rescue?

In the April 2006 issue of *Nature*, Daeschler et al. reported the discovery of several fossilized specimens of a *Crossopterygian* fish named *Tiktaalik roseae.* These well-preserved specimens were found in sedimentary layers of siltstone — cross-bedded with sandstones — in Arctic Canada.[4]

Like the other lobe-fin fish, *Tiktaalik* was declared to be late Devonian (between 385–359 million years old) by means of a "dating" method known as *palynomorph biostratigraphy*. This method presumes to date sedimentary rock layers on the basis of the assumed evolutionary age of pollen and spores contained in the rock. Most importantly, the discoverers of *Tiktaalik* claim that it "represents an intermediate between fish with fins and tetrapods with limbs."

Tiktaalik Is a Fish

Whatever else we might say about *Tiktaalik*, it *is a fish*. In a review article on *Tiktaalik* (appearing in the same issue of the scientific journal *Nature* that reported the discovery of *Tiktaalik*), fish evolution experts Ahlberg and Clack concede that "in some respects *Tiktaalik* and *Panderichthys* are straightforward fishes: they have small pelvic fins, retain fin rays in their paired appendages and have well-developed gill arches, suggesting that both animals remained mostly aquatic."[5]

In other respects, however, Ahlberg and Clack argue that *Tiktaalik* is more tetrapod-like than *Panderichthys* because "the bony gill cover has disappeared, and the skull has a longer snout." The authors weakly suggest that

4. Edward B. Daeschler, Neil H. Shubin, and Farish A. Jenkins, "A Devonian Tetrapod-like Fish and the Evolution of the Tetrapod Body Plan," *Nature* 440 no. 6 (2006): 757–763.
5. P.E. Ahlberg and J.A. Clack, News and Views, *Nature* 440 no. 6 (2006): 747–749.

the significance of all this is that "a longer snout suggests a shift from sucking towards snapping up prey, whereas the loss of gill cover bones probably correlates with reduced water flow through the gill chamber. The ribs also seem larger in *Tiktaalik*, which may mean it was better able to support its body out of water."

Without the author's evolutionary bias, of course, there is no reason to assume that *Tiktaalik* was anything other than exclusively aquatic. And how do we know that *Tiktaalik* lost its gill cover as opposed to never having one? The longer snout and lack of bony gill covers (found in many other exclusively aquatic living fish) are interpreted as indicating a reduced flow of water through the gills, which, in turn, is declared to be suggestive of partial air-breathing — but this is quite a stretch. Finally, what does any of this have to do with fish evolving into land-dwelling tetrapods?

Are the Pectoral Fins of *Tiktaalik* Really Legs?

Before we get into *Tiktaalik*'s "legs," it might be instructive to consider an old trick question. If we call our arms "legs," then how many legs would we have? The answer, of course, is *two legs* — just because we call our arms "legs" doesn't make them legs. The same might be said of the bony fins of *Crossopterygian* fish — we may call them "legs" but that doesn't necessarily make them legs.

Shubin et al. make much of the claim that *Tiktaalik*'s bony fins show a reduction in dermal bone and an increase in endochondral bone.[6] This is important to them because the limb bones of tetrapods are entirely endochondral. They further claim that the *cleithrum* (a dermal bone to which the pectoral fin is attached in fish) is detached from the skull, resembling the position of the scapula (shoulder blade) of a tetrapod. They also claim that the endochondral bones of the fin are more similar to those of a tetrapod in terms of structure and range of motion. However, none of this, if true, proves that *Tiktaalik*'s fins supported its weight out of water, or that it was capable of a true walking motion. (It certainly doesn't prove that these fish evolved into tetrapods.)

The Limbs of Tetrapods

The limbs of tetrapods share similar characteristic features. These unique features meet the special demands of walking on land. In the case of the forelimbs there is one bone nearest the body (proximal) called the *humerus* that articulates (flexibly joins) with two bones, the *radius* and *ulna,* farther away

6. Neil H. Shubin, Edward B. Daeschler, and Farish A. Jenkins, "The Pectoral Fin of Tiktaalik roseae and the Origin of the Tetrapod Limb," *Nature* 440 no. 6 (2006): 764–771.

from the body (distal). These in turn articulate with multiple wrist bones, which finally articulate with typically five digits. The hind limbs similarly consist of one proximal bone, the *femur*, which articulates with two distal bones, the *tibia* and *fibula*, which in turn articulate with ankle bones; and finally with typically five digits. In order to support the weight of the body on land, and permit walking, the most proximal bones of the limbs must be securely attached to the rest of the body. The humerus of the forelimb articulates with the pectoral girdle, which includes the *scapula* (shoulder blade) and the *clavicle* (collar bone). The only bony attachment of the pectoral girdle to the body is the clavicle.

The femur of the hind limb articulates with the pelvic girdle, which consists of fused bones collectively called the *pelvis* (hip bone). It is this hind limb — with its robust pelvic girdle securely attached to the vertebral column — that differs radically from that of any fish. (The tetrapod arrangement is important for bearing the weight of the animal on land.)

All tetrapod limb bones and their attachment girdles are endochondral bones. In the case of all fish, including *Tiktaalik*, the cleithrum and fin rays are dermal bones.

It is significant that the "earliest" true tetrapods recognized by evolutionists (such as *Acanthostega* and *Ichthyostega*) have all of the distinguishing features of tetrapod limbs (and their attachment girdles) and were clearly capable of walking and breathing on land. The structural differences between the tetrapod leg and the fish fin is easily understood when we realize that the buoyant density of water is about a thousand times greater than that of air. A fish has no need to support much of its weight in water where it is essentially weightless.

The Fins of Fish (including *Tiktaalik*)

Essentially all fish (including *Tiktaalik*) have small pelvic fins relative to their pectoral fins. The legs of tetrapods are *just the opposite*: the hind limbs attached to the pelvic girdle are almost always more robust than the forelimbs attached to the pectoral girdle. (This is particularly obvious in animals such as kangaroos and theropod dinosaurs.) Not only are the pelvic fins of all fish small, but they're *not even attached to the axial skeleton* (vertebral column) and thus can't bear weight on land.

While the endochondral bones in the pectoral fins of *Crossopterygians* have some similarity to bones in the forelimbs of tetrapods, there are significant differences. For example, there is nothing even remotely comparable to the digits in any fish. The bony rays of fish fins are dermal bones that are not related in any way to digits in their structure, function, or mode of development.

Clearly, fin rays are relatively fragile and unsuitable for actual walking and weight bearing.

Even the smaller endochondral bones in the distal fin of *Tiktaalik* are not related to digits. Ahlberg and Clack point out that "although these small distal bones bear some resemblance to tetrapod digits in terms of their function and range of movement, they are still very much components of a fin. There remains a large morphological gap between them and digits as seen in, for example *Acanthostega*: if the digits evolved from these distal bones, the process must have involved considerable developmental rearranging."[7]

So Is *Tiktaalik* a Missing Link?

Finally, what about the popular claim that *Tiktaalik* is the "missing link" between fish and tetrapods?

In their review article on *Tiktaalik*, Ahlberg and Clack tell us that "the concept of 'missing links' has a powerful grasp on the imagination: the rare transitional fossils that apparently capture the origins of major groups of organisms are uniquely evocative." The authors concede that the whole concept of "missing links" has been loaded with "unfounded notions of evolutionary 'progress' and with a mistaken emphasis on the single intermediate fossil as the key to understanding evolutionary transition."

Sadly, "unfounded notions" of this kind continue to be uncritically taught and accepted in the popular media and in our schools. Even more sadly, these unfounded notions have been used to undermine the authority of Holy Scripture.

7. Ahlberg and Clack, News and Views.

26

Why Is Mount St. Helens Important to the Origins Controversy?

DR. STEVEN A. AUSTIN

On May 18, 1980, a catastrophic geologic event occurred that not only shocked the world because of its explosive power, but challenged the foundation of evolutionary theory. That event was the eruption of Mount St. Helens in the state of Washington. The eruption of Mount St. Helens is regarded by many as the most significant geologic event of the 20th century, excelling all others in its extraordinary documentation and scientific study. Undeniable facts confront us. Although not the most powerful explosion of the last century, Mount St. Helens provided a significant learning experience within a natural laboratory for the understanding of catastrophic geologic processes.

On May 18, and also during later eruptions, certain critical energy thresholds were exceeded by potent geologic processes. These were able to accomplish significant changes in short order to the landscape (figure 1), providing us a rare, user-friendly opportunity to observe and understand the effects of catastrophic geologic processes.

What would 20 megatons of steam-blast energy do to a landscape? How would mudflows and giant water waves modify the earth? Geologists, who were accustomed to thinking about slow evolutionary processes forming geologic features, were astounded to witness many of these same features form rapidly at Mount St. Helens. Ultimately, the events and processes at the

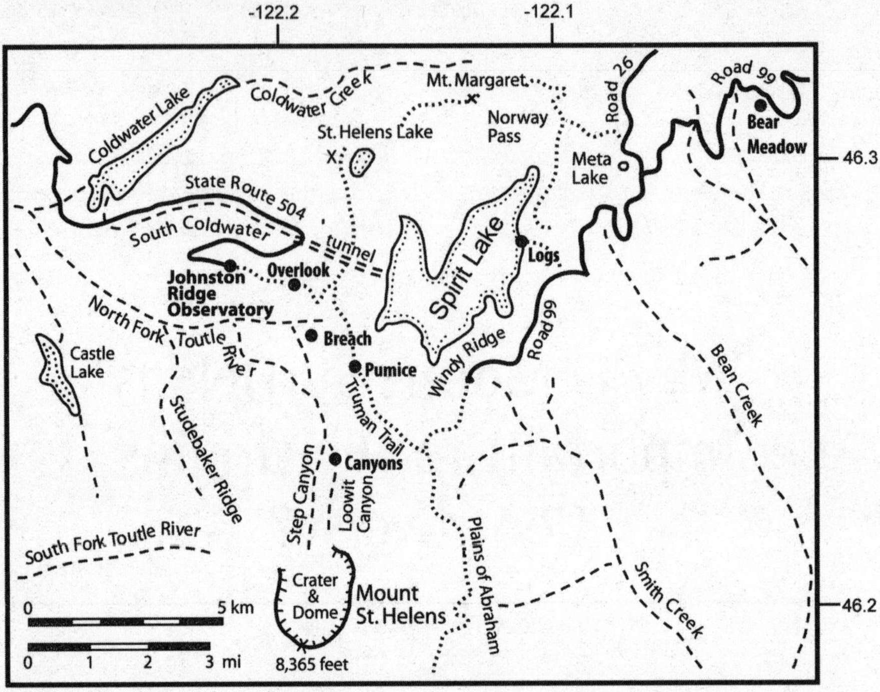

Figure 1. Map of the north flank of the volcano showing areas of special interest to understanding catastrophic geologic processes within Mount St. Helens National Volcanic Monument. "Overlook" is an observation post one mile east of Johnston Ridge Observatory on the ridge showing the eroded landscape on the North Fork of the Toutle River. "Breach" is the March 19, 1982, erosion feature with the "Little Grand Canyon" that is currently closed to off-trail activity without special use permit. "Pumice" is the Pumice Plain deposit consisting of various laminated pyroclastic flow deposits. "Canyons" displays the mudflow-eroded bedrock channels up to 600 feet deep on the north flank of the volcano. "Logs" is the shore observation locality on the Harmony Trail on the east side of Spirit Lake. "Bear Meadow" is the northeastern observation location outside the blast zone.

volcano challenge our way of thinking about how the earth works. Does the earth change piecemeal by slow and gradual processes, which accumulate small changes over immense periods of time? Or have rapid processes accomplished significant geologic changes in very short periods of time? What we have seen at Mount St. Helens has application to many geologic features. Can processes at Mount St. Helens explain the origin of finely laminated strata? Do canyons in hard rock bear evidence of catastrophic erosion? Have the Yellowstone "petrified forests" and coal deposits accumulated by catastrophic sedimentary processes? Was there a global flood on the earth?

Rapid Formation of Stratification

Up to 600 feet thickness of new strata have formed since 1980 at Mount St. Helens. These deposits accumulated from primary air blast, landslide, water wave on Spirit Lake, pyroclastic flows, mudflows, air fall, and stream water. Perhaps the most surprising accumulations are the pyroclastic flow deposits amassed from ground-hugging, fluidized slurries of fine volcanic debris that moved at high velocities off the north flank of the volcano. These deposits include fine pumice ash laminae beds from one millimeter thick to greater than one meter thick, each representing just a few seconds to several minutes of accumulation (see "Pumice" in figure 1).

Figure 2 shows 25 feet of the stratified deposit accumulated within three hours during the evening of June 12, 1980. It was deposited from pyroclastic flows generated by collapse of the eruption plume of debris over the volcano. The strata are very extensive and even contain thin laminae and crossbedding. Within the pyroclastic flow deposits are very thin laminae. It staggers the mind to think how the finest stratification has formed in an event of the

Figure 2. Deposits exposed by mudflow erosion on the North Fork of the Toutle River. The laminated and bedded pyroclastic flow deposit of June 12, 1980, is 25 feet thick in the middle of the cliff. That three-hour deposit is underlain by the pyroclastic flow deposit of May 18, 1980, and overlain by the mudflow deposit of March 19, 1982. (Photo by Steven A. Austin)

violence of a hurricane. Coarse and fine sediment were separated into distinct strata by the catastrophic flow process from a slurry moving at freeway speed. Conventionally, sedimentary laminae and beds are assumed to represent longer seasonal variations — or annual changes — as the layers accumulated very slowly. That is the typical uniformitarian interpretation. Furthermore, our natural way of thinking about catastrophic sedimentary process is that it homogenizes materials depositing coarse and fine together without obvious stratification. Mount St. Helens teaches us that stratification does form very rapidly by flow processes.

Rapid Erosion

If we reason from our everyday experience concerning the way rivers and creeks erode, we might assume that great time periods are needed to form deep canyons. At Mount St. Helens, however, very rapid erosion has occurred since the 1980 eruptions. These erosion features challenge our way of thinking about how landscapes form. What is exceptional at Mount St. Helens is the variety of new erosion features and their concentration within a limited and intensely studied area. There is no place in the blast zone at Mount St. Helens where the effects of recent erosion cannot be seen. That is what makes Mount St. Helens extraordinary! Scientists discovered that the kinds of processes causing erosion were as varied as the different features formed. The major agents of erosion unleashed at Mount St. Helens are listed in summary here.

1. Direct blast — the 20-megaton TNT equivalent, northward-directed steam blast of May 18 caused hot gas and rock fragments to abrade slopes around the mountain.
2. Pyroclastic flows — explosive blasts on and after May 18 created superheated, erosive "rivers" of ground-hugging volcanic ash and steam.
3. Debris avalanche — the movement of great masses of rock, ice, and debris over the earth's surface next to the volcano caused significant abrasion.
4. Mudflows — viscous streams of mud gouged out soft volcanic ash deposits and, to our astonishment, even the hardest rocks to form new canyons.
5. Water in channels — overland flow of floodwater caused extraordinary rill and gully patterns to appear, even in nearly level slopes.

6. Water waves — enormous water waves generated in Spirit Lake by the avalanche on May 18 inflicted severe erosion on slopes adjacent to the lake.
7. Jetting steam — eruptions of steam from buried glacier ice reamed holes through hot volcanic ash deposits, forming distinctive explosion pits.
8. Mass wasting — gravitational collapse induced significant changes to unstable slopes, especially those areas sculptured by other agents, leaving behind a varied landscape.

Two-thirds cubic mile of landslide and eruption debris from May 18, 1980, occupies 23 square miles of the North Fork of the Toutle River north and west of the crater (see figure 1). It was the largest debris avalanche observed in human history! This debris was deposited across the entire width of the valley along the uppermost 16 miles of the North Fork of the Toutle River. These deposits average 150 feet in thickness and form a hummocky surface that blocks the pre-1980 channel. Before the May 18 eruption, Spirit Lake had an outlet river draining westward into the Pacific Ocean.

From May 18, 1980, to March 19, 1982, the upper drainage area of the debris avalanche deposit was not connected to the Pacific Ocean, and water from Spirit Lake basin and the crater of the volcano did not connect to the Toutle River, due to debris blocking the valley. Because of this debris, there has been no natural outlet formed for Spirit Lake.

An explosive eruption of Mount St. Helens on March 19, 1982, melted a thick snowpack in the crater, creating a destructive, sheet-like flood of water, which became a mudflow. Breaching the deposits on the upper North Fork of the Toutle River (see "Breach" in figure 1). The most significant erosion occurred in the biggest steam explosion pit. The mudflow filled the big steam explosion pit with mud, which then overflowed the west rim of the pit as a deep ravine was cut into the 1980 deposits to the west.

The flow formed channels over much of the hummocky rockslide debris, allowing cataracts to erode headward, and established for the first time since 1980 a dendritic integration of channels on the Toutle's North Fork drainage. Erosion has occurred intermittently since then, but most of the streams were established in their present channel locations on March 19, 1982. Figure 2 is within the breach formed by the big mudflow that day. Bedrock was eroded up to 600 feet deep to form Step Canyon and Loowit Canyon on the north flank of the volcano (see "Canyons" in figure 1).

Individual canyons on the debris avalanche deposit have a depth of up to 140 feet and are cut through landslide debris and pumice from pyroclastic flows (see "Breach" in figure 1). That erosion left elevated plateaus north and south of a great breach, resembling the north and south rims of the Colorado River. Also, gully-headed side canyons and amphitheater-headed side canyons in the breach resemble side canyons in the Grand Canyon. The breach did not occur straight through the obstruction but has a meandering path, which reminds us of the meandering path of the Grand Canyon through the high plateaus of northern Arizona. The "Little Grand Canyon of the Toutle River" is a 1/40th-scale model of the real Grand Canyon of Arizona.

Small creeks that flow through the headwaters of the Toutle River today might seem, by present appearances, to have carved these canyons very slowly over a very long time period, except for the fact that the erosion was observed to have occurred rapidly! Geologists should learn that because the long time scale they have been trained to assign to landform development would lead to obvious error at Mount St. Helens, it also may be useless or misleading elsewhere.

Rapid Formation of Fossil Deposits

One million logs floated on Spirit Lake on the late afternoon of May 18, 1980, after they were uprooted and washed into the basin by the 860-foot-high water waves. Careful observation of the floating conifer logs in the lake indicates that such logs show a strong tendency to float upright, best seen from the eastern shore of the lake (see "Logs" in figure 1). Many upright deposited logs possess roots attached to the log, but many have no root ball, and those without roots also show strong tendency to float upright. It appears that the root end of these logs is denser wood and perhaps floods with water more easily, allowing the root ends to sink before the top of the log. All six of the common conifer species were observed to float in an upright position.

Hundreds of upright, fully submerged logs were located by sidescan sonar, and scuba divers verified that they were indeed trunks of trees that the sonar detected. It was estimated that 20,000 upright stumps existed on the floor of the lake in August 1985. It would appear that about ten percent of the deposited logs were in an upright position. If Spirit Lake were drained, the bottom would look like a forest of trees. These, however, did not grow where they are now, but have been replanted!

Scuba investigation of the upright-deposited logs shows that some are already solidly buried by sedimentation with more than three feet of sediment around their bases. Others, however, have none. This proves that the upright logs were deposited at different times, with their roots buried at different levels. If found buried in the rock strata, logs such as the ones in Spirit Lake might be interpreted as multiple forests that grew on different levels over periods of many thousands of years. The Spirit Lake upright-deposited logs, therefore, have considerable implications for interpreting "petrified forests" in the strata record. Direct application of the Spirit Lake logs may be made to the Yellowstone petrified forests. There at Yellowstone, at Specimen Ridge, geologists have commonly attributed the petrified upright logs to many thousands of years of forest growth, but the upright logs in Spirit Lake call that interpretation into question.

Rapid Formation of Peat Layer

The enormous log mat floating on Spirit Lake has lost its bark and branches by the abrasive action of wind and waves. Scuba investigations of the lake bottom showed that water-saturated sheets of conifer bark are especially abundant intermingled with volcanic sediment added from the lake shore, forming a layer of peat many inches thick. The peat shows coarse texture. The primary component is sheets of tree bark, which comprise about 25 percent, by volume, of the peat. Scuba divers recovered sheets of tree bark having lengths of greater than eight feet from the peat bed. Together with broken branch and root material, bark sheets impart the peat's noteworthy coarse texture and dominantly layered appearance.

The "Spirit Lake peat" contrasts strongly with peats that have accumulated in swamps. Typical swamp peats are very finely macerated by organic degradation processes. They are "coffee grounds to mashed potatoes" in general texture. Furthermore, swamp peats possess a homogeneous appearance because of the intense penetration of roots which dominate swamps. Root material is the dominant coarse component of modern swamp peats while bark sheets are extremely rare.

The Spirit Lake peat resembles, both compositionally and texturally, certain coal beds of the eastern United States, which also are dominated by tree bark and appear to have accumulated beneath floating log mats. Conventionally, coal is supposed to have accumulated from organic material built up in swamps by growth in place of plants. Because the accumulation of peat in swamps is a slow process, geologists have supposed that coal beds required about one thousand years to form each inch of coal. The peat layer in Spirit Lake reveals that "floating mat peat" can accumulate very rapidly and possesses textures resembling coal. Swamp peats, however, possess very rare bark-sheet material, because the intrusive action of tree roots disintegrates and homogenizes the peat. The Spirit Lake peat, in contrast, is texturally very similar to coal. Thus, at Spirit Lake, we may have seen the first stage in the formation of coal.

Conclusion

Mount St. Helens provides a rare opportunity to study transient geologic processes which, produced within a few months, changes what geologists might otherwise assume required many thousands of years. The volcano challenges our way of thinking about how the earth works, how it changes, and the time scale attached. These processes and their effects allow Mount St. Helens to serve as a miniature laboratory for catastrophism.

Catastrophism is documented as a viable theory of geologic change and may have far-reaching implications on other scientific disciplines and philosophical inquiries. Many scientists recognize that Darwin's theory (which assumed slow evolutionary change) may be in error. Darwin built his theory of the evolution of living things on the notion that earth has slowly evolved. With catastrophism, we have tools to interpret the stratigraphic record including the geologic evidence of the Genesis Flood. Mount St. Helens "speaks" directly to issues of our day.

Creationists have been intensely interested in the geologic formations at Mount St. Helens because they provide a very graphic and real explanation for features which are often supposed to support evolutionary theory and uniformitarian speculation. Mount St. Helens can also be used as a steppingstone to help us imagine what the Genesis Flood was like.

Bibliography

Arct, M.A., and A.V. Chadwick, "Dendrochronology in the Yellowstone Fossil Forest," *Geological Society of America, Abstracts with Programs* 15 (1983): 5.

Austin, S.A., "Depositional Environment of the Kentucky No. 12 Coal Bed (Middle Pennsylvanian) of Western Kentucky, with Special Reference to the Origin of Coal Lithotypes," Ph.D. dissertation, Pennsylvania State University, 1979, p. 411.

Austin, S.A., "Uniformitarianism — A Doctrine That Needs Rethinking," *Compass* 56 no. 2 (1979): 29–45.

Austin, S.A., "Rapid Erosion at Mount St. Helens," *Origins* 11 no. 2 (1984): 90–98.

Austin, S.A., *Catastrophes in Earth History: A Source Book of Geologic Evidence, Speculation, and Theory*, Technical Monograph No. 13 (El Cajon, CA: Institute for Creation Research, 1984).

Austin, S.A., "Floating Logs and Log Deposits of Spirit Lake, Mount St. Helens Volcano National Monument, Washington," *Geological Society of America, Abstracts with Programs* 23 (1991).

Austin, S.A., "The Dynamic Landscape on the North Flank of Mount St. Helens" in J.E. O'Connor, R.J. Dorsey, and I.P. Madlin, eds., "Volcanoes to Vineyards: Geologic Field Trips through the Dynamic Landscape of the Pacific Northwest," *Geological Society of America Field Guide* 15 (2009).

Chadwick, A.D., and T. Yamoto, "A Paleoecological Analysis of the Petrified Trees in the Specimen Creek Area of Yellowstone National Park, Montana, U.S.A.," *Paleogeography, Palaeoclimatology, Palaeoecology* 45 (1984): 39–48.

Coffin, H.G., "Mount St. Helens and Spirit Lake," *Origins* 10 (1983): 9–17.

Coffin, H.G., "Erect Floating Stumps in Spirit Lake, Washington," *Geology* 11 (1983): 298–299.

Coffin, H.G., "Sonar and Scuba Survey of a Submerged Allochthonous Forest in Spirit Lake, Washington," *Palaios* 2 (1987): 178–180.

Criswell, C.W., "Chronology and Pyroclastic Stratigraphy of the May 18, 1980, Eruption of Mount St. Helens, Washington," *Journal of Geophysical Research* 92 (1987): 10,237–10,266.

Foxworthy, B.L., and M. Hill, "Volcanic Eruptions of 1980 at Mount St. Helens — The First 100 Days," *United States Geological Survey Professional Paper 1249*: 1982.

Fritz, W.J., "Reinterpretation of the Depositional Environment of the Yellowstone Fossil Forests." *Geology* 8 (1980): 309–313.

Glicken, H., "Study of the Rockslide-Debris Avalanche of May 18, 1980, Mount St. Helens Volcano," Ph.D. dissertation., University of California, Santa Barbara, 1986.

Hickson, C.J., "The May 18, 1980, Eruption of Mount St. Helens, Washington State: A Synopsis of Events and Review of Phase 1 from an Eyewitness Perspective," *Geoscience Canada* 17 no. 3 (1990): 127–130.

Karowe, A.L., and T.M. Jefferson, "Burial of Trees by Eruptions of Mount St. Helens, Washington: Implications for the Interpretation of Fossil Forests," *Geological Magazine* 124 no. 3 (1987): 191–204.

Lipman, P.O., and D.R. Mullineaux, eds., "The 1980 Eruptions of Mount St. Helens, Washington," *U.S. Geological Survey Professional Paper* 1250, 1981.

Malone, S.D., "Mount St. Helens, the 1980 Re-awakening and Continuing Seismic Activity," *Geoscience Canada* 17 no. 3 (1990): 163–166.

Meyer, D.F., and H.A. Martinson, "Rates and Processes of Channel Development and Recovery Following the 1980 Eruption of Mount St. Helens, Washington," *Hydrological Sciences Journal* 34 (1989): 115–127.

Morris, J.D., and S.A. Austin, *Footprints in the Ash: The Explosive Story of Mount St. Helens* (Green Forest, AR: Master Books, 2003), p. 128.

Peterson, D.W., "Overview of the Effects and Influence of the Activity of Mount St. Helens in the 1980s," *Geoscience Canada* 17 (1990): 163–166.

Rosenfeld, C.L., and G.L. Beach, "Evolution of a Drainage Network: Remote Sensing Analysis of the North Fork Toutle River, Mount St. Helens, Washington," Corvallis, Oregon State University Water Resources Research Institute, WRRI-88, 1983.

Rowley, P.D., et al., "Proximal Bedded Deposits Related to Pyroclastic Flows of May 18, 1980, Mount St. Helens, Washington," *Geological Society of America Bulletin* 96 (1985): 1373–1383.

Scott, K.M., "Magnitude and Frequency of Lahars and Lahar-runout Flows in the Toutle-Cowlitz River System," *United States Geological Survey Professional Paper 1447-B*, 1989.

Waitt, R.B. Jr., et al., "Eruption-Triggered Avalanche, Flood, and Lahar at Mount St. Helens — Effects of Winter Snowpack," *Science* 221 (1983): 1394–1397.

Weaver, C.S., and S.D. Malone, "Overview of the Tectonic Setting and Recent Studies of Eruptions of Mount St. Helens, Washington," *Journal of Geophysical Research* 92 (1987): 10,149–10,154.

27

What Is the Best Argument for the Existence of God?

DR. JASON LISLE

There are a number of common arguments for the existence of God. But most of these arguments are not as effective as many Christians would like to think. Let's consider a hypothetical conversation between a Christian and an atheist.

Christian: "Everything with a beginning requires a cause. The universe has a beginning and therefore requires a cause. That cause is God."

Atheist: "Even if it were true that everything with a beginning requires a cause, how do you know that the cause of the universe is God? Why not a big bang? Maybe this universe sprang from another universe, as some physicists now believe."

Christian: "The living creatures of this world clearly exhibit design. Therefore, they must have a designer. And that designer is God."

Atheist: "The living creatures only *appear* to be designed. Natural selection can account for this apparent design. Poorly adapted organisms tend to die off, and do not pass on their genes."

Christian: "But living creatures have irreducible complexity. All their essential parts must be in place at the same time, or the organism dies. So God must have created these parts all at the same time. A gradual evolutionary path simply will not work."

Atheist: "Just because you cannot imagine a gradual stepwise way of constructing an organism does not mean there isn't one."

Christian: "DNA has information in it — the instructions to form a living being. And information never comes about by chance; it always comes from a mind. So DNA proves that God created the first creatures."

Atheist: "There could be an undiscovered mechanism that generates information in the DNA. Give us time, and we will eventually discover it. And even if DNA did come from intelligence, why would you think that intelligence is God? Maybe aliens seeded life on earth."

Christian: "The Resurrection of Jesus proves the existence of God. Only God can raise the dead."

Atheist: "You don't really have any proof that Jesus rose from the dead. This section of the Bible is simply an embellished story. And even if it were true, it proves nothing. Perhaps under certain rare chemical conditions, a dead organism can come back to life. It certainly doesn't mean that there is a God."

Christian: "The Bible claims that God exists, and that it is His Word to us. Furthermore, what the Bible says must be true, since God cannot lie."

Atheist: "That is a circular argument. Only if we knew in advance that God existed would it be reasonable to even consider the possibility that the Bible is His Word. If God does not exist — as I contend — then there is no reason to trust the Bible."

Christian: "Predictive prophecy shows that the Bible really must be inspired by God. All of the Old Testament prophecies concerning Christ, for example, were fulfilled. The odds of that happening by chance are very low."

Atheist: "A low probability isn't the same as zero. People do win the lottery. Besides, maybe the Gospels have embellished what Jesus did, so that it would agree with the Old Testament prophecies. Perhaps some so-called prophetic books were actually written after the events they 'predict.' Maybe certain gifted individuals have abilities not yet understood by science and can occasionally predict the future. It certainly doesn't prove the Bible is inspired by God."

Christian: "I have personally experienced God, and so have many other Christians. He has saved us and transformed our lives. We know that He exists from experience."

Atheist: "Unfortunately, your personal experiences are not open to investigation; I have only your word for it. And second, how do you know that such subjective feelings are really the result of God? The right drug might produce similar feelings."

Not Conclusive

It should be noted that all the facts used by the Christian in the above hypothetical conversation are *true*. Yes, God is the first cause, the designer of life, the resurrected Christ, the Author of Scripture, and the Savior of Christians. Yet the way these facts are used is not decisive. That is, none of the above arguments really prove that God exists.

Some of the above arguments are very weak: appeals to personal experience, vicious circular reasoning, and appeals to a first cause. While the facts are true, the arguments do not come close to proving the existence of the biblical God. Some of the arguments seem stronger; I happen to think that irreducible complexity and information in DNA are strong confirmations of biblical creation. And predictive prophecy does confirm the inspiration of Scripture. Nonetheless, for each one of these arguments, the atheist was able to invent a "rescuing device." He was able to propose an explanation for this evidence that is compatible with his belief that God does not exist.

Moreover, most of the atheist's explanations are actually pretty reasonable, given his view of the world. He's not being illogical. He is being consistent with his position. Christians and atheists have different worldviews — different philosophies of life. And we must learn to argue on the level of worldviews if we are to argue in a cogent and effective fashion.

The Christian in the above hypothetical conversation did not have a correct approach to apologetics. He was arguing on the basis of specific evidences with someone who had a totally different professed worldview than his own. This approach is never conclusive, because the critic can always invoke a rescuing device to protect his worldview.[1] Thus, if we are to be effective, we must use an argument that deals with worldviews, and not simply isolated facts. The best argument for the existence of God will be a "big-picture" kind of argument.

God Doesn't Believe in Atheists

The Bible teaches that atheists are not *really* atheists. That is, those who profess to be atheists do ultimately believe in God in their heart-of-hearts. The Bible

1. Of course, sometimes people are persuaded by such arguments. But that doesn't mean the argument is cogent. After all, people can be persuaded by very bad arguments.

teaches that everyone knows God, because God has revealed Himself to all (Romans 1:19). In fact, the Bible tells us that God's existence is so obvious that anyone who suppresses this truth is "without excuse" (Romans 1:20). The atheist denies with his lips what he knows in his heart. But if they know God, then why do atheists claim that they do not believe in God?

The answer may be found in Romans 1:18. God is angry at unbelievers for their wickedness. And an all-powerful, all-knowing God who is angry at you is a terrifying prospect. So even though many atheists might claim that they are neutral, objective observers, and that their disbelief in God is purely rational, in reality, they are strongly motivated to reject the biblical God who is rightly angry with them. So they suppress that truth in unrighteousness. They convince themselves that they do not believe in God.[2] The atheist is intellectually schizophrenic — believing in God, but believing that he does not believe in God.[3]

Therefore, we do not really need to give the atheist any more specific evidences for God's existence. He already knows in his heart-of-hearts that God exists, but he doesn't want to believe it. Our goal is to expose the atheist's suppressed knowledge of God.[4] With gentleness and respect, we can show the atheist that he already knows about God, but is suppressing what he knows to be true.

Exposing the Inconsistency

Because an atheist does believe in God, but does not believe that he believes in God, he is simply a walking bundle of inconsistencies. One type to watch for is a *behavioral inconsistency*; this is where a person's behavior does not comport with what he claims to believe. For example, consider the atheist university professor who teaches that human beings are simply chemical accidents — the end result of a long and purposeless chain of biological evolution. But then he goes home and kisses his wife and hugs his children, as if they were *not* simply chemical accidents, but valuable, irreplaceable persons deserving of respect and worthy of love.

Consider the atheist who is outraged at seeing a violent murder on the ten o'clock news. He is very upset and hopes that the murderer will be punished for

2. This is called an "iterated belief" — a belief about a belief.
3. Self-deception is quite common. People frequently attempt to convince themselves of what they want to believe. The Bible tells us that those who hear God's Word but do not act on it are self-deceived (James 1:22).
4. In some cases, we can use scientific evidence to expose such inconsistency. Consider the evolutionist who admits that the probability of a cell forming by chance is infinitesimal. He is going against the odds. Yet, he decides to carry an umbrella with him when there is a 90 percent chance of rain.

his wicked actions. But in his view of the world, why should he be angry? In an atheistic, evolutionary universe where people are just animals, murder is no different than a lion killing an antelope. But we don't punish the lion! If people are just chemical accidents, then why punish one for killing another? We wouldn't get upset at baking soda for reacting with vinegar; that's just what chemicals do. The concepts that human beings are valuable, are not simply animals, are not simply chemicals, have genuine freedom to make choices, are responsible for their actions, and are bound by a universal objective moral code all stem from a Christian worldview. Such things simply do not make sense in an atheistic view of life.

Many atheists behave morally and expect others to behave morally as well. But absolute morality simply does not comport with atheism. Why should there be an absolute, objective standard of behavior that all people should obey if the universe and the people within it are simply accidents of nature? Of course, people can assert that there is a moral code. But who is to say what that moral code should be? Some people think it is okay to be racist; others think it is okay to kill babies, and others think we should kill people of other religions or ethnicities, etc. Who is to say which position should be followed? Any standard of our own creation would necessarily be subjective and arbitrary.

Now, some atheists might respond, "That's right! Morality is subjective. We each have the right to create our own moral code. And therefore, you cannot impose your personal morality on other people!" But of course, this statement is self-refuting, because when they say, "you cannot impose your personal morality on other people" they are imposing their personal moral code on other people. When push comes to shove, no one really believes that morality is merely a subjective, personal choice.

Logical Inconsistency

Another inconsistency occurs when atheists attempt to be rational. Rationality involves the use of laws of logic. Laws of logic prescribe the correct chain of reasoning between truth claims. For example, consider the argument: "If it is snowing outside, then it must be cold out. It is snowing. Therefore, it is cold out." This argument is correct because it uses a law of logic called *modus ponens*. Laws of logic, like *modus ponens*, are immaterial, universal, invariant, abstract entities. They are immaterial because you can't touch them or stub your toe on one. They are universal and invariant because they apply in all places and at all times (*modus ponens* works just as well in Africa as it does in the United States, and just as well on Friday as it does on Monday). And they are abstract because they deal with concepts.

Laws of logic stem from God's sovereign nature; they are a reflection of the way He thinks. They are immaterial, universal, invariant, abstract entities, because God is an immaterial (Spirit), omnipresent, unchanging God who has all knowledge (Colossians 2:3). Thus, all true statements will be governed by God's thinking — they will be logical. The law of non-contradiction, for example, stems from the fact that God does not deny Himself (2 Timothy 2:13). The Christian can account for laws of logic; they are the correct standard for reasoning because God is sovereign over all truth. We can know some of God's thoughts because God has revealed Himself to us through the words of Scripture and the person of Jesus Christ.

However, the atheist cannot account for laws of logic. He cannot make sense of them within his own worldview. How could there be immaterial, universal, invariant, abstract laws in a chance universe formed by a big bang? Why should there be an absolute standard of reasoning if everything is simply "molecules in motion"? Most atheists have a materialistic outlook — meaning they believe that everything that exists is material, or explained by material processes. But laws of logic are not material! You cannot pull a law of logic out of the refrigerator! If atheistic materialism is true, then there could be no laws of logic, since they are immaterial. Thus, logical reasoning would be impossible!

No one is denying that atheists are able to reason and use laws of logic. The point is that if atheism were true, the atheist would not be able to reason or use laws of logic because such things would not be meaningful. The fact that the atheist is able to reason demonstrates that he is wrong. By using that which makes no sense given his worldview, the atheist is being horribly inconsistent. He is using God's laws of logic, while denying the biblical God that makes such laws possible.

How could there be laws at all without a lawgiver? The atheist cannot account for (1) the existence of laws of logic, (2) why they are immaterial, (3) why they are universal, (4) why they do not change with time, and (5) how human beings can possibly know about them or their properties. But of course, all these things make perfect sense on the Christian system.

YOU JUST BROKE A LAW OF LOGIC!
NON-CON
TRADICTION

Laws of logic owe their existence to the biblical God. Yet they are required to reason rationally, to prove things. So the biblical God must exist in order for reasoning to be possible. Therefore, *the best proof of God's existence is that without Him we couldn't prove anything at all!* The existence of the biblical God is the prerequisite for knowledge and rationality. This is called the "transcendental argument for God" or TAG for short. It is a devastating and conclusive argument, one that only a few people have even attempted to refute (and none of them successfully).[5]

Proof Versus Persuasion

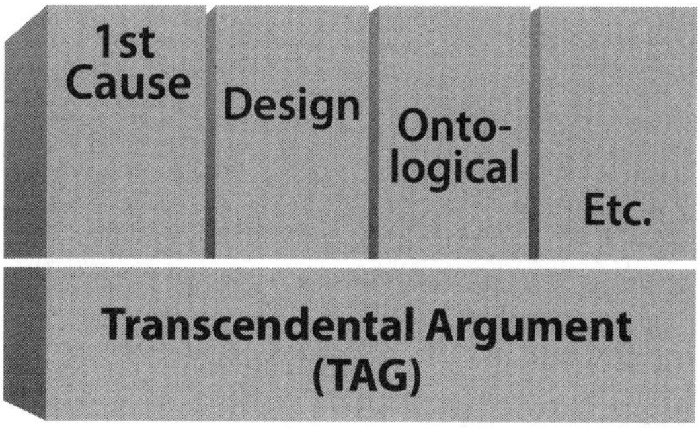

Other arguments for the existence of God (First Cause, Design, etc.) actually assume the transcendental argument before they can even make their argument.

Though the transcendental argument for God is deductively sound, not all atheists will be convinced upon hearing it. It may take time for them to even understand the argument in the first place. As I write this chapter, I am in the midst of an electronic exchange with an atheist who has not yet fully grasped the argument. Real-life discussions on this issue take time. But even if the atheist fully understands the argument, he may not be convinced. We must remember that there is a difference between proof and persuasion. Proof is objective, but persuasion is subjective. The transcendental argument does indeed objectively prove that God exists. However, that does not mean that the atheists will necessarily cry "uncle." Atheists are strongly motivated to not believe in the biblical God — a God who is rightly angry at them for their treason against Him.

But the atheist's denial of God is an emotional reaction, not a logical one. We might imagine a disobedient child who is about to be punished by his father. He might cover his eyes with his hands and say of his father, "You don't exist!" but that would hardly be rational. Atheists deny (with their lips) the biblical God, not for logical reasons, but for psychological reasons. We must also keep in mind that the unbeliever's problem is not simply an emotional

5. Perhaps most significantly, philosopher Michael Martin has attempted to rebut TAG indirectly by making a transcendental-style argument for the non-existence of God (TANG). Martin's argument has been refuted by John Frame, and independently by Michael Butler.

issue, but a deep spiritual problem (1 Corinthians 2:14). It is the Holy Spirit that must give him the ability to repent (1 Corinthians 12:3; 2 Timothy 2:25).

So we must keep in mind that it is not our job to convert people — nor can we. Our job is to give a defense of the faith in a way that is faithful to the Scriptures (1 Peter 3:15). It is the Holy Spirit that brings conversion. But God can use our arguments as part of the process by which He draws people to Himself.

28

Do Evolutionists Believe Darwin's Ideas about Evolution?

ROGER PATTERSON AND DR. TERRY MORTENSON

Charles Darwin first published his ideas on evolution over 150 years ago. In those 150 years we have come to understand the complexity of life, and many new scientific fields have shed light on the question of the validity of Darwin's evolutionary hypothesis. Few people have actually read the works of Darwin, and if they did they might be shocked to read some of Darwin's ideas. In this chapter we will take a look at what Darwin and other early evolutionists believed and how those ideas have changed over time.

Darwin was wrong on many points, and there would be few who would disagree with this claim. But if Darwin was wrong on some points, does that mean that the entire hypothesis of evolution is proven wrong?

What Is Evolution?

Like many words, *evolution* has many different uses depending on its context. The general concept of the word is "change over time." In that sense, one might say that a butterfly evolves from an egg to a caterpillar to a winged butterfly and a child evolves into an adult. There is no disputing that individual organisms change over time. However, using the word in this way is quite misleading for the origins debate. Darwin's hypothesis involves a very different concept.

As *evolution* is used in this chapter and in all science textbooks, natural history museums, and science programs on television, it refers to the biological idea that all life on earth has descended from a single common ancestor. There are many different variations on this theme as well as several explanations of how the first organism came into existence from non-living matter. Examining some of the historical evolutionary positions and comparing them to the ideas that are popular in scientific circles today shows how much those concepts have changed. In general, evolution will be used to refer to the concept of molecules turning into men over time. This concept of evolution is in direct opposition to the biblical account of creation presented in the Book of Genesis.[1]

Evolution – an Ancient Idea

The concept of molecules-to-man evolution is certainly not a new idea. Several Greek philosophers before the time of Christ wrote on the topic. For example, Lucretius and Empedocles promoted a form of natural selection that did not rely on any type of purpose. In *De Rerum Natura (On the Nature of Things)* Lucretius writes:

> And many species of animals must have perished at that time, unable by procreation to forge out the chain of posterity: for whatever you see feeding on the breath of life, either cunning or courage or at least quickness must have guarded and kept that kind from its earliest existence. . . . But those to which nature gave no such qualities, so that they could neither live by themselves at their own will, nor give us some usefulness for which we might suffer to feed them under our protection and be safe, these certainly lay at the mercy of others for prey and profit, being all hampered by their own fateful chains, until nature brought that race to destruction.[2]

This stands in opposition to the thinking of Aristotle, who promoted the idea of purpose in nature. Aristotle also imagined forms of life advancing through history, but he believed nature had the aim of producing beauty.[3] This idea of purpose in nature, or teleology, is later seen in the works of Thomas Aquinas and other Christian philosophers.

1. For an explanation of some of the contradictions between the biblical creation account and the widely held evolution story, see the article "Evolution vs. Creation: The Order of Events Matters!" at www.answersingenesis.org/docs2006/0404order.asp.
2. Sharon Kaye, "Was There No Evolutionary Thought in the Middle Ages? The Case of William of Ockham," *British Journal for the History of Philosophy* 14 no. 2 (2006): 225–244.
3. Henry Fairfield Osborn, *From the Greeks to Darwin* (London: Macmillan, 1913), p. 43–56.

The concept of evolution was not lost from Western thinking until Darwin rediscovered it — it was always present in various forms. Because much of the thinking was dominated by Aristotelian ideas, the idea of a purposeless evolutionary process was not popular. Most saw a purpose in nature and the interactions between living things. The dominance of the Roman Catholic Church in Europe (where modern science was born) and its adherence to Aristotelian philosophies also played a role in limiting the promotion of evolution and other contrary ideas as these would have been seen as heresy. As the Enlightenment took hold in Europe in the 17th and 18th centuries, explanations that looked beyond a directed cause became more popular.

Erasmus Darwin

Coming to the mid-to-late 18th century, Kant, Liebnitz, Buffon, and others began to talk openly of a natural force that has driven the change of organisms from simple to complex over time. The idea of evolution was well established in the literature, but there seemed to be no legitimate mechanism to adequately explain this idea in scientific terms. Following the spirit of the Greek poets Lucretius and Empedocles, Erasmus Darwin, the atheist grandfather of Charles, wrote some of his ideas in poetic verse. Brushing up against the idea of survival of the fittest, Erasmus spoke of the struggle for existence between different animals and even plants. This struggle is a part of the evolutionary process he outlines in his *Temple of Nature* (1803) in the section titled "Production of Life":

> Hence without parent by spontaneous birth
> Rise the first specks of animated earth;
> From Nature's womb the plant or insect swims,
> And buds or breathes, with microscopic limbs.[4]

And he continues:

> Organic Life beneath the shoreless waves
> Was born and nursed in Ocean's pearly caves;
> First forms minute, unseen by spheric glass,
> Move on the mud, or pierce the watery mass;
> These, as successive generations bloom,
> New powers acquire, and larger limbs assume;
> Whence countless groups of vegetation spring,
> And breathing realms of fin, and feet, and wing.[5]

4. Erasmus Darwin, *The Temple of Nature* (London: Jones & Company, 1825), p. 13.
5. Ibid., p. 14–15.

Starting with spontaneous generation from inanimate matter, Erasmus imagined life evolving into more complex forms over time. He did not identify any mechanisms that may have caused the change, other than general references to nature and a vague driving force.

In the introduction to this work, Erasmus Darwin states that it is not intended to instruct but rather to amuse, and he then includes many notes describing his ideas. Despite his claimed-to-be-innocent intentions, this poem lays out the gradual, simple-to-complex progression of matter to living creature — a view very consciously different from the biblical account of creation which the vast majority of his contemporaries knew and believed. He traces the development of life in the seas to life on land with the four-footed creatures eventually culminating in humans and the creation of society. There is no doubt that when Charles began his studies, the idea of evolution apart from the supernatural was present in Western thought (even in his own extended family). The arguments in support of special creation were certainly prominent, but evolutionary ideas were being pressed into mainstream thinking in the era of modernism.[6]

To underscore the early acceptance of evolution, the following passage from *Zoonomia* (3 vol., 1794–1796) illustrates Erasmus Darwin's belief that all life had come from a common "filament" of life.

> From thus meditating on the great similarity of the structure of the warm-blooded animals . . . would it be too bold to imagine that, in the great length of time since the earth began to exist, perhaps millions of ages before the commencement of the history of mankind would it be too bold to imagine that all warm-blooded animals have arisen from one living filament?[7]

Lamarckian Evolution or Use and Disuse

In France, and at the same time as Erasmus, Jean Baptiste Lamarck developed his theories of the origin and evolution of life. Initially, he had argued for the

6. Modernism was the dominant philosophy in Western culture from the late 18th to the late 20th centuries. This philosophy placed science as the supreme authority for determining truth. Science was viewed as the "savior" of mankind — eventually finding cures for all diseases, ending war, famine, etc. Though it has been largely replaced by post-modernism, this modernist thinking is still very prominent among scientists and many others in our culture. Post-modernism, on the other hand, is a radical skepticism about anyone's ability to know truth. Post-modernists argue that truth and morality are relative — there are no absolutes. It also reflects disenchantment with the promises made by modernist philosophers and scientists. Both philosophies reject Scripture as authoritative truth and are based on evolutionary thinking.
7. Erasmus Darwin, *Zoonomia*, volume 1 (Philadelphia, PA: Edward Earle, 1818), p. 397.

immutability of species, but in his later works he laid out a clear alternative to the special creation of plants and animals. Lamarck believed that the geology of the earth was the result of gradual processes acting over vast periods of time — a view later to be known as uniformitarianism. Lamarck developed four laws of evolution and put them forward in his *Philosophie Zoologique* published in 1809. Lamarck proposed that an internal force and the need for new organs caused creatures to develop new characteristics. Once developed, the use or disuse of the organs would determine how they would be passed on to a creature's offspring. This idea of the transmission or inheritance of acquired characteristics is the hallmark of this model of evolution.

Lamarck's mechanism of use and disuse of characters was widely rejected in his lifetime, especially by the prominent French naturalist Georges Cuvier, and was never supported by observations. Lamarck did attempt to explain how the characteristics were inherited, but there was still no clear biological mechanism of inheritance that would support his claims. Lamarck also proposed a tree of life with various branching structures that showed how life evolved from simple to complex forms. Much of what Lamarck proposed seems unreasonable to us today with a modern understanding of genetics. A husband and wife who are both bodybuilders will not have an extraordinarily muscular child — that acquired trait does not have any affect on the genetic information in the germ cells of the parents' bodies. However, recent research has revealed instances of bacterial inheritance that appear to be very Lamarckian in nature. Future research in this area may reveal that Lamarck was correct to some degree. But

In Lamarckian evolution, animals change due to environmental factors and the use or disuse of a feature. For example, a giraffe's neck will get longer over time as it continually stretches it to reach higher leaves on trees.

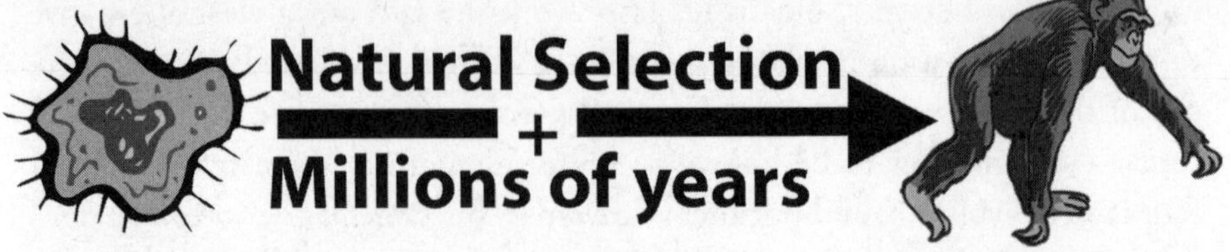

Darwin originally proposed that natural selection would be the primary mechanism acting to change organisms over millions of years. He was not aware of the role of mutations in heredity.

there are many good reasons to expect that this would provide no support for the idea of molecules-to-man evolution.[8]

Darwinian Evolution

Charles Darwin was at least familiar with all of these different views, and their influence can be found throughout his writings. Darwin often referred to the effects of natural selection along with the use or disuse of the parts. The legs and wings of the ostrich, the absence of feet and wings in beetles, and the absence of eyes in moles and cave-dwelling animals are all mentioned by Darwin as a result of use or disuse alongside natural selection.[9] Exactly how this process happened was a mystery to Darwin. He proposed the idea of "pangenesis" as the mechanism of passing traits from parent to offspring. This idea is not significantly different from Lamarck's, for it relies on the use and disuse of organs and structures that are passed on to offspring through pangenes over vast ages.

In his work *The Variation of Animals and Plants under Domestication*, Darwin suggested that gemmules are shed by body cells, and that the combination of these gemmules would determine the appearance and constitution of

8. Even if Lamarckian mechanisms are uncovered, the fossil record would not support the evolution story. See Duane Gish, *Evolution: The Fossils Still Say No*, (Santee, CA: Institute for Creation Research, 1996); Carl Werner, *Evolution: The Grand Experiment,* vol. 1 (Green Forest, AR: New Leaf Press, 2007); and *Living Fossils*, vol. 2 (Green Forest, AR: New Leaf Press, 2008). Natural selection can only "select" from existing genetic information (it cannot create new information), and mutations cause a loss or reshuffling of existing genetic information. See Terry Mortenson's DVD *Origin of the Species: Was Darwin Right?* (Answers in Genesis, 2007) and John Sanford, *Genetic Entropy and the Mystery of the Genome* (Lima NY: Elim Publishing, 2005). Also, what bacteria can do should not be directly applied to other forms of life because bacteria are categorically and significantly different. This is explained in Georgia Purdom's DVD *All Creatures Great and Small: Microbes and Creation* (Petersburg, KY: Answers in Genesis, 2009).
9. Charles Darwin, *The Origin of Species* (New York, NY: The Modern Library, 1993), p. 175–181.

the offspring. If the parent had a long neck, then more gemmules for a long neck would be passed to the offspring. In Darwin's defense, he was not aware of the work of his contemporary, Gregor Mendel. In his garden in the Czech lands, Mendel was studying the heredity of pea plants. Neither man knew of the existence of genes, or the DNA genes are composed of, but both of them understood there was a factor involved in transmitting characteristics from one generation to the next. Despite evidence from experiments conducted by his cousin Francis Galton, Darwin clung to his pangenesis hypothesis and defended it in his later work *Descent of Man*.

Darwin believed that all organisms had evolved by natural processes over vast expanses of time. In the introduction to *Origin of Species* he wrote:

> As many more individuals of each species are born than can possibly survive; and as, consequently, there is a frequently recurring struggle for existence, it follows that any being, if it vary however slightly in any manner profitable to itself, under the complex and sometimes varying conditions of life, will have a better chance of surviving, and thus be *naturally selected*. From the strong principle of inheritance, any selected variety will tend to propagate its new and modified form.[10]

Darwin's belief that slight modifications were selected to produce big changes in organisms over the course of millions of years was the foundation of his model for the evolution of life on earth. We know today that Darwin's notion of gemmules and pangenes leading to new features or the development of enhanced characteristics is a false notion. However, that does not mean, by itself, that Darwin's conclusion is wrong — just that his reasoning was faulty.

Neo-Darwinian Evolution and the Modern Synthesis

The discovery of DNA and the rediscovery of Mendel's work on heredity in pea plants have shown that Darwin's hereditary mechanism does not work. But his conclusion of molecules-to-man transformation over millions of years is still held as true by proponents of evolution. In the early 20th century, Mendelian genetics was rediscovered and it came to be understood that DNA was responsible for the transmission and storage of hereditary information. The scientific majority was still fixed on a naturalistic explanation for the evolution of organisms. That evolution happened was never a question — finding the mechanism was the goal of these naturalistic scientists.

10. Ibid., p. 21.

Mutation of genetic information came to be viewed as the likely mechanism for providing the raw material for natural selection to act on. Combining genetic studies of creatures in the lab and in the wild, models of speciation and change over time were developed and used to explain what was seen in the present. These small changes that resulted from mutations were believed to provide the genetic diversity that would lead to new forms over eons of time. This small change was referred to as "microevolution" since it involved small changes over a short amount of time. The evolutionists claim that the small changes add up to big changes over millions of years, leading to new kinds of life. Thus, microevolution leads to "macroevolution" in the evolutionary view. However, the acceptance of these terms just leads to confusion, and they should be avoided.

This is not fundamentally different from what Charles Darwin taught; it simply uses a different mechanism to explain the process. The problem is that the change in speciation and adaptation is heading in the opposite direction needed for macroevolution. The small changes seen in species as they adapt to their environments and form new species through mutation are the result of losses of information. Darwinian evolution requires the addition of traits (such as forelimbs changing into wings, and scales turning into feathers in dinosaur-to-bird evolution), which requires the addition of new information. Selecting from information that is already present in the genome and that was damaged through copying mistakes in the genes cannot be the process that adds new information to the genome.

Today, evolution has been combined with the study of embryology, genetics, the fossil record, molecular structures, plate tectonics, radiometric dating, anthropology, forensics, population studies, psychology, brain chemistry, etc. This leads to the intertwining of so many different ideas that the modern view of evolution can explain anything. It has become so plastic that it can be molded to explain any evidence, no matter how inconsistent the explanations may become. Even Darwin was willing to admit that there may be evidence that would invalidate his hypothesis. That is no longer the view held by the vast majority of evolutionists today — evolution has become a fact, even a scientific law (on par with the law of gravity), in the minds of many.

To help us see this more clearly, let us take a look at the idea of different races. Darwin published his views on the different races in *Descent of Man*. Though Darwin spoke against slavery, he clearly believed that the different people groups around the world were the result of various levels of evolutionary development. Darwin wrote:

> At some future period, not very distant as measured by centuries, the civilized races of man will almost certainly exterminate and

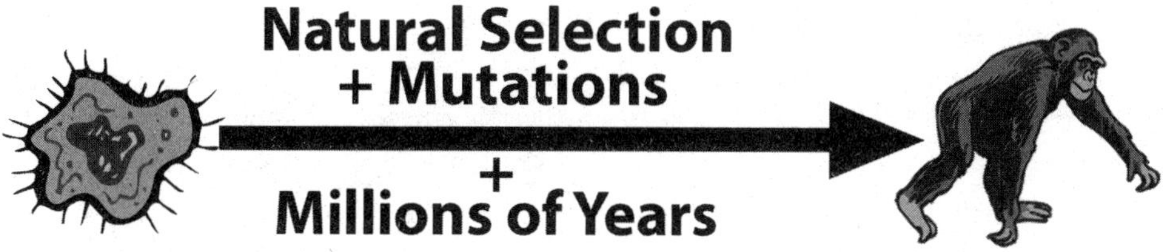

After the discovery of DNA and its role in inheritance, evolutionists pointed to mutations in the DNA as the source for new traits. These accidental mutations provide differences in the offspring that can be selected for. This selection is believed to lead to new kinds of life.

> replace the savage races throughout the world. At the same time the anthropomorphous apes . . . will no doubt be exterminated. The break between man and his nearest allies will then be wider, for it will intervene between man in a more civilized state, as we may hope, even than the Caucasian, and some ape as low as a baboon, instead of as now between the negro or Australian [Aborigine] and the gorilla.[11]

This is the conclusion Darwin came to — that different rates of evolution would lead to different classes of humans. He often refers to the distinction between the civilized Europeans and the savages of various areas of the world. He concludes that some of these savages are so closely related to apes that there is no clear dividing line in human history "where the term 'man' ought to be used."[12] Consistent with his naturalistic view of the world, Darwin saw various groups of humans, whether they are distinct species or not, as less advanced than others. This naturally leads to racist attitudes and, as Dr. Stephen J. Gould noted, biological arguments for racism "increased by orders of magnitude following the acceptance of evolutionary theory,"[13] though this was likely only an excuse to act on underlying social prejudices.

Dr. James Watson (co-discoverer of the double-helix structure of the DNA molecule and a leading atheistic evolutionist) was caught in a storm of evolutionary racism in 2007. The *Times* of London reported in an interview:

> He says that he is "inherently gloomy about the prospect of Africa" because "all our social policies are based on the fact that their intelligence is the same as ours — whereas all the testing says

11. Charles Darwin, *The Origin of Species and The Descent of Man* (New York, NY: The Modern Library, 1936), p. 521.
12. Ibid., p. 541.
13. Stephen Jay Gould, *Ontogeny and Phylogeny* (Cambridge, MA: Belknap Press of Harvard University Press, 1977), p. 127.

> not really," and I know that this "hot potato" is going to be difficult to address. His hope is that everyone is equal, but he counters that "people who have to deal with black employees find this not true." He says that you should not discriminate on the basis of colour, because "there are many people of colour who are very talented, but don't promote them when they haven't succeeded at the lower level." He writes, "there is no firm reason to anticipate that the intellectual capacities of peoples geographically separated in their evolution should prove to have evolved identically. Our wanting to reserve equal powers of reason as some universal heritage of humanity will not be enough to make it so."[14]

Though he later stated that he did not intend to imply that black Africans are genetically inferior, he is being consistent with his evolutionary beliefs. His remarks were considered offensive, even by those who endorse evolution. This exposes an inconsistency in the thinking of many evolutionists today — if we evolved by random chance, we are nothing special. If humans evolved, it is only reasonable to conclude that different groups have evolved at different rates and with different abilities, and mental ability could be higher in one group than another. If the data supported this claim, in the evolutionary framework, then it should be embraced. Those who would suggest that evolution can explain why all humans have value must battle against those evolutionists who would disagree.

This exposes the inconsistent and plastic nature of evolution as an overarching framework — who gets to decide what evolution should mean? Darwin and Watson are applying the concepts in a consistent way and setting emotion and political correctness aside, when it is deemed necessary. Darwin noted that "it is only our natural prejudice and . . . arrogance" that lead us to believe we are special in the animal world.[15]

Without an objective standard, such as that provided by the Bible, the value and dignity of human beings are left up to the opinions of people and their biased interpretations of the world around us. God tells us through His Word that each human has dignity and is a special part of the creation because each one is made in the image of God. We are all of "one blood" in a line descended from Adam, the first man, who was made distinct from all animals and was *not* made by modifying any previously existing animal (Genesis 2:7).

14. Charlotte Hunt-Grubbe, "The Elementary DNA of Dr Watson," Times Online [London], October14, 2007, www.entertainment.timesonline.co.uk/tol/arts_and_entertainment/books/article2630748.ece.
15. Darwin, *The Origin of Species and The Descent of Man*, p. 411–412.

Saltation and Punctuated Equilibrium

Contrasted with Darwin's view of a gradual process of change acting over vast ages of time, others have seen the history of life on earth as one of giant leaps of rapid evolutionary change sprinkled through the millions of years. Darwin noted that the fossil record seemed to be missing the transitions from one kind of organism to the next that would confirm his gradualistic notion of evolution. Shortly after Darwin, there were proponents of evolutionary saltation — the notion that evolution happens in great leaps. The almost complete absence of transitional forms in the fossil record seemed to support this saltation concept and this was later coupled with genetics to provide a mechanism where "hopeful monsters" would appear and almost instantaneously produce a new kind of creature (e.g., changing a reptile into a bird). These "monsters" would be the foundation for new kinds of animals.

Saltation fell out of favor, but the inconsistency between the fossil record and the gradualism promoted by Darwin and others was still a problem. The work of Ernst Mayr, Stephen J. Gould, and Niles Eldredge was the foundation for the model of "punctuated equilibrium." This model explained great periods of stasis in the fossil record punctuated with occasional periods of rapid change in small populations of a certain kind of creature. This rapid change is relative to the geologic time scale — acting over tens of thousands of years rather than millions. This idea is not inconsistent with Darwin's grand evolutionary scheme. However, it seems that Darwin did not anticipate such a mechanism, though he commented that different organisms would have evolved at different rates. Whether evolution has occurred by gradual steps or rapid leaps (or some combination) is still a topic of debate among those who hold to the neo-Darwinian synthesis of mutations and natural selection as the driving forces of evolutionary change.

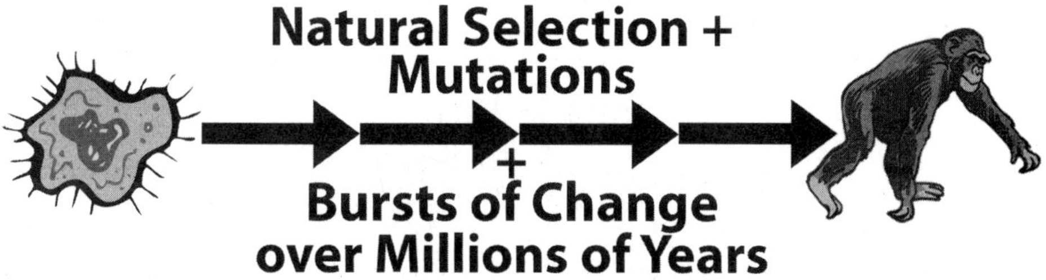

Contrary to Neo-Darwinism, punctuated equilibrium tries to account for the lack of fossil intermediates by appealing to rapid bursts of change interspersed in the millions of years. They still rely on mutations and natural selection, but at a much faster rate.

Conclusion

Sir Isaac Newton provided us with a general theory of gravity (and described laws in support of that theory) based on observational science. Even in light of modern understandings, those laws still apply today. Einstein did expand the concepts, but the functionality of Newtonian physics still applies today as much as ever.

The same cannot be said for Darwin's ideas. Darwin's hypothesized mechanism of natural selection (even with the added understanding of mutations) has failed to provide an explanation for the origin and diversity of life we see on earth today. His confident expectation that the fossil record would confirm his hypothesis has utterly failed, and the mind-boggling irreducible complexity seen in biological systems today defies the explanations of Darwin or his disciples. To say that evolutionary thinking today is Darwinian in nature can only mean that evolutionists believe that life has evolved from simpler to complex over time. Beyond that, what is called Darwinism today bears little resemblance to what Darwin actually wrote.

All of these ideas of the evolution of organisms from simple to complex are contrary to the clear teaching of Scripture that God made separate kinds of plants and animals and one kind of man, each to reproduce after its own kind. As such, these evolutionary ideas are bound to fail when attempting to describe the history of life and to predict the future changes to kinds of life in this universe where we live. When we start our thinking with the Bible, we can know we are starting on solid ground. Both the fossil record and the study of how plants, animals, and people change in the present fit perfectly with what the Bible says about creation, the Flood, and the Tower of Babel in Genesis 1–11. The Bible makes sense of the world around us.

29

What Are Some of the Best Flood Evidences?

DR. ANDREW A. SNELLING

Have you ever been "tongue-tied" when asked to provide geologic evidence that the Genesis Flood really did occur, just as the Bible describes? What follows is an overview of six geologic evidences for the Genesis Flood. Together, they will provide you with "ammunition" and a teaching tool for you and others.

Why is it that many people, including many Christians, can't see the geologic evidence for the Genesis Flood? It is usually because they have bought into the evolutionary idea that "the present is the key to the past." They are convinced that, because today's geological processes are so slow, the earth's rock layers took millions of years to form.

However, if the Genesis Flood really occurred, what evidence would we look for? We read in Genesis 7 and 8 that "the fountains of the great deep" were broken up and poured out water from inside the earth for 150 days (5 months). Plus, it rained torrentially and globally for 40 days and nights. ("The floodgates [or windows] of heaven were opened.") No wonder all the high hills and the mountains were covered, meaning the earth was covered by a global ocean. (". . . the world that then was, being overflowed with water, perished" 2 Peter 3:6; KJV.) All air-breathing life on the land was swept away and perished.

Wouldn't we expect to find billions of dead plants and animals buried and fossilized in sand, mud, and lime that were deposited rapidly by water in rock layers all over the earth? Of course! That's exactly what we find. Indeed, based on the biblical description of the Flood, there are six main geologic evidences that testify to its historicity.[1]

Evidence #1: Fossils of Sea Creatures High Above Sea Level

On every continent we find fossils of sea creatures in rock layers that today are high above sea level. For example, most of the rock layers in the walls of the Grand Canyon contain marine fossils. This includes the Kaibab Limestone at the top of the strata sequence and exposed at the rim of the canyon, which today is 7,000–8,000 feet above sea level.[2] This limestone was therefore deposited beneath lime sediment-charged ocean waters, which swept over northern Arizona (and beyond). Other rock layers of the Grand Canyon also contain large numbers of marine fossils. The best example is the Redwall Limestone, which commonly contains fossil brachiopods (a type of clam), corals, bryozoans (lace corals), crinoids (sea-lilies), bivalves (other types of clams), gastropods (marine snails), trilobites, cephalopods, and even fish teeth.[3] These marine fossils are found haphazardly preserved in this limestone bed. The crinoids, for example, are found with their columnals (disks), which in life are stacked on top of one another to make up their "stems," totally separated from one another in what can best be described as a "hash." Thus, these marine creatures were catastrophically destroyed and buried by the deposition of this lime sediment layer.

Fossil ammonites (coiled marine gastropods) are also found in limestone beds high in the Himalayas, reaching up to 30,000 feet above sea level.[4] All geologists agree that these marine fossils must have been buried in these limestone beds when the latter were deposited by ocean waters. So how did these marine limestone beds get to be high up in the Himalayas?

There is only one possible explanation — the ocean waters at some time in the past flooded over the continents. Could the continents have then sunk below today's sea level, so that the ocean waters flooded over them? No! Because the continents are made up of rocks that are less dense (lighter) than both the ocean floor

1. I want to acknowledge that these geologic evidences have also been elaborated on by my colleague Dr. Steve Austin at the Institute for Creation Research, in his book *Grand Canyon: Monument to Catastrophe* (Santee, CA: Institute for Creation Research, 1994), p. 51–52.
2. R.L. Hopkins and K.L. Thompson, "Kaibab Formation," in S.S. Beus and M. Morales, eds., *Grand Canyon Geology*, 2nd edition (New York, NY: Oxford University Press, 2003), p. 196–211.
3. S.S. Beus, "Redwall Limestone and Surprise Canyon Formation," in S.S. Beus and M. Morales, eds., *Grand Canyon Geology*, 2nd edition (New York, NY: Oxford University Press, New York, 2003), p. 115–135.
4. J.P. Davidson, W.E. Reed, and P.M. Davis, "The Rise and Fall of Mountain Ranges," in *Exploring Earth: An Introduction to Physical Geology* (Upper Saddle River, NJ: Prentice Hall, 1997), p. 242–247.

rocks and the mantle rocks beneath the continents. The continents, in fact, have an automatic tendency to rise, and thus "float" on the mantle rocks beneath, well above the level of the ocean floor rocks.[5] This is why the continents today have such high elevations compared to the deep ocean floor, and why the ocean basins can accommodate so much water. Rather, the sea level had to rise, so that the ocean waters then flooded up onto, and over, the continents. What would have caused that to happen? There had to be, in fact, two mechanisms to cause this.

First, if the volume of water in the ocean was increased, then sea level would rise. In Genesis 7:11 we read that at the initiation of the Flood all the fountains of the great deep were broken up. In other words, the earth's crust was cleaved open all around the globe and water burst forth from inside the earth. We then read in Genesis 7:24–8:2 that these fountains were open for 150 days. No wonder the ocean waters flooded up onto and over the continents.

Second, if the ocean floor itself rose, it would then have effectively "pushed" up the sea level. The catastrophic breakup of the earth's crust, referred to in Genesis 7:11, would not only have released huge volumes of water from inside the earth, but much molten rock.[6] The ocean floors would have been effectively replaced by hot lavas. Being less dense than the original ocean floors, these hot lavas would have had an expanded thickness, so the new ocean floors would have effectively risen, raising the sea level by up to more than 3,500 feet. When the ocean floors cooled and sank, the sea level would have fallen and the waters would have drained off the continents into new, deeper ocean basins.

Evidence #2: Rapid Burial of Plants and Animals

Countless billions of plant and animal fossils are found in extensive "graveyards" where they had to be buried rapidly on a massive scale. Often the fine details of the creatures are exquisitely preserved.

For example, billions of straight-shelled, chambered nautiloids (figure 1) are found fossilized with other marine creatures in a 7 feet (2 m) thick layer within the Redwall Limestone of Grand Canyon.[7] This fossil graveyard stretches for 180 miles (290 km) across northern Arizona and into southern Nevada,

5. J.P. Davidson, W.E. Reed, and P.M. Davis, "Isostasy," in *Exploring Earth: An Introduction to Physical Geology* (Upper Saddle River, NJ: Prentice Hall, 1997), p. 124–129.
6. A.A. Snelling, "A Catastrophic Breakup: A Scientific Look at Catastrophic Plate Tectonics," Answers April–June 2007, p. 44–48; A.A. Snelling, "Can Catastrophic Plate Tectonics Explain Flood Geology?" in Ken Ham, ed., *The New Answers Book 1* (Green Forest, AR: Master Books, 2006), p. 186–197.
7. S.A. Austin, "Nautiloid Mass Kill and Burial Event, Redwall Limestone (Lower Mississippian), Grand Canyon Region, Arizona and Nevada," in *Proceedings of the Fifth International Conference on Creationism*, R.L. Ivey, ed., (Pittsburgh, PA: Creation Science Fellowship, 2003), p. 55–99.

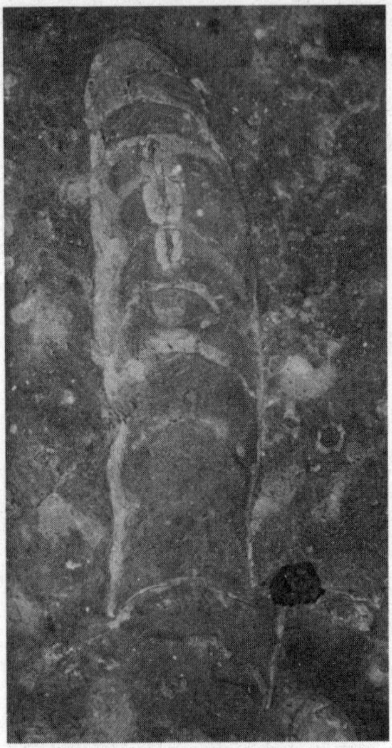

Figure 1. Fossil nautiloids, found in the Redwall Limestone were buried rapidly.

covering an area of at least 10,500 square miles (30,000 km^2). These squid-like fossils are all different sizes, from small, young nautiloids to their bigger, older relatives. To form such a vast fossil graveyard required 24 cubic miles (100 km^3) of lime sand and silt, flowing in a thick-soup-like slurry at more than 16 feet (5 m) per second (more than 11 miles or 18 km per hour) to catastrophically overwhelm and bury this huge, living population of nautiloids.

Hundreds of thousands of marine creatures were buried with amphibians, spiders, scorpions, millipedes, insects, and reptiles in a fossil graveyard at Montceau-les-Mines, France.[8] At Florissant, Colorado, a wide variety of insects, freshwater mollusks, fish, birds, and several hundred plant species (including nuts and blossoms) are buried together.[9] Bees and birds have to be buried rapidly in order to be so well preserved.

Alligator, fish (including sunfish, deep sea bass, chubs, pickerel, herring, and gar-pike 3–7 feet [1–2 m] long), birds, turtles, mammals, mollusks, crustaceans, many varieties of insects, and palm leaves (7–9 feet [2–2.5 m] long) were buried together in the vast Green River Formation of Wyoming.[10] Notice in these examples how marine and land-dwelling creatures are found buried together. How could this have happened unless the ocean waters rose and swept over the continents in a global, catastrophic Flood?

Many trillions of microscopic marine creatures had to have catastrophically buried large ammonites and other marine creatures in the chalk beds of Britain.[11] These same beds also stretch right across Europe to the Middle East, as well as into the Midwest of the United States, forming a global-scale fossil

8. B. Heyler and C.M. Poplin, "The Fossils of Montceau-les-Mines," *Scientific American*, September 1988, p. 70–76.
9. T.D.A. Cockerell, "The Fossil Flora and Fauna of the Florissant Shales," *University of Colorado Studies* 3 (1906): 157–176.
10. L. Grande, "Paleontology of the Green River Formation with a Review of the Fish Fauna," *The Geological Survey of Wyoming Bulletin* 63 (1984).
11. J.M. Hancock, "The Petrology of the Chalk," *Proceedings of the Geologists' Association* 86 (1975): 499–536; B. Smith and D.J. Batten, "Fossils of the Chalk," *Field Guides to Fossils*, no. 2, 2nd edition (London: The Palaeontological Association, 2002).

graveyard. More than seven trillion tons of vegetation is buried in the world's coal beds found across every continent, including Antarctica.

Such was the speed at which many creatures were buried and fossilized — under catastrophic flood conditions — that they were exquisitely preserved. There was no destruction of many fish, which were buried so rapidly, virtually alive, that even fine details of fins and eye sockets have been preserved. Many trilobites have been so exquisitely preserved that even the compound lens systems in their eyes are still available for detailed study.

Mawsonites spriggi, when discovered, was identified as a fossilized jellyfish (figure 2). It was found in a sandstone bed that covers more than 400 square miles (1,040 km^2) of outback South Australia.[12] Millions of such soft-bodied marine creatures are exquisitely preserved in this sandstone bed. Consider what happens to soft-bodied creatures like jellyfish when washed up on a beach today. Because they consist only of soft "jelly," they melt in the sun and are also destroyed by waves crashing onto the beach. Based on this reality, the discoverer of these exquisitely preserved soft-bodied marine creatures concluded that all of them had to be buried in less than a day!

Some sea creatures were buried alive and fossilized so quickly that they were "caught in the act" of eating their last meal, or at the moment of giving birth to a baby! One minute a huge ichthyosaur had just given birth to her baby, then seconds later, without time to escape, mother and baby were buried and "snap frozen" in a catastrophic "avalanche" of lime mud.

Figure 2. Soft-bodied marine creatures, such as this fossilized jellyfish (*Mawsonites spriggi*), are finely preserved in a sandstone bed.

These are but a few examples of the many hundreds of fossil graveyards found all over the globe that are now well-documented in the geological literature.[13] The countless billions of fossils in these graveyards, in many cases exquisitely preserved, testify to the rapid burial of plants and animals on a global scale in a watery cataclysm and its immediate aftermath.

12. R.C. Sprigg, "Early Cambrian (?) Jellyfishes from the Flinders Ranges, South Australia," *Transactions of the Royal Society of South Australia* 71 no. 2 (1947): 212–224; M.F. Glaessner and M. Wade, "The Late Precambrian Fossils from Ediacara, South Australia," *Palaeontology* 9 (1966): 599–628.
13. For example: D.J. Bottjer et al., eds., *Exceptional Fossil Preservation: A Unique View on the Evolution of Marine Life* (New York, NY: Columbia University Press, 2002).

Evidence #3: Rapidly Deposited Sediment Layers Spread Across Vast Areas

On every continent are found layers of sedimentary rocks over vast areas. Many of these can be traced all the way across continents, and even between continents. Furthermore, geologists find evidence that the sediments were deposited rapidly.

Consider the sedimentary rock layers exposed in the walls of the Grand Canyon. This sequence of layers is not unique to that region of the United States. For more than 50 years geologists have recognized that these strata belong to six megasequences (very thick, distinctive sequences of sedimentary rock layers) that can be traced right across North America.[14]

The lowest of the Grand Canyon's sedimentary layers is the Tapeats Sandstone, belonging to the Sauk Megasequence. It and its equivalents cover much of the United States. We can hardly imagine what forces were necessary to deposit such a vast, continent-wide series of deposits. Yet at the base of this sequence are huge boulders and sand beds deposited by storms. Both are evidence that massive forces deposited these layers rapidly and violently right across the entire United States. Slow-and-gradual (present-day uniformitarian) processes cannot account for this evidence, but the Genesis Flood surely can!

The Grand Canyon's Redwall Limestone belongs to the Kaskaskia Megasequence. The *same* limestones appear in many places across North America, as far as Tennessee and Pennsylvania. These limestones also appear in the exact same position in the strata sequences, and they have the exact same fossils and other features in them. What is even more remarkable is that the same Carboniferous limestone beds also appear in England, again containing the same fossils and other features.

The Cretaceous chalk beds of southern England are well known because they appear as spectacular white cliffs along the coast. The same chalk beds can be traced westward across England and appear again in Northern Ireland. In the opposite direction, these same chalk beds can be traced across France, the Netherlands, Germany, Poland, southern Scandinavia, and other parts of Europe to Turkey, then to Israel and Egypt in the Middle East, and even as far as Kazakhstan.[15]

Remarkably, the same chalk beds with the same fossils in them, and with the same distinctive strata above and below them, are also found in the Midwest United States, from Nebraska in the north to Texas in the south, and in the Perth Basin of Western Australia.

14. L.L. Sloss, "Sequences in the Cratonic Interior of North America," *Geological Society of America Bulletin* 74 (1963): 93–114.
15. D.V. Ager, *The Nature of the Stratigraphical Record* (London: Macmillan, 1973), p. 1–2.

Consider another feature — coal beds. In the northern hemisphere, the Upper Carboniferous (Pennsylvanian) coal beds of the eastern and Midwest United States are the same coal beds, with the same plant fossils, in Britain and Europe, stretching halfway around the globe, from Texas to the Donetz Basin north of the Caspian Sea in the former USSR.[16] In the southern hemisphere, the same Permian coal beds are found in Australia, Antarctica, India, South Africa, and even South America! These beds share the same kind of plant fossils across the region (but they are different from those in the Pennsylvanian coal beds).

The buff-colored Coconino Sandstone is very distinctive in the walls of the Grand Canyon. It has an average thickness of 315 feet and covers an area of at least 200,000 square miles eastward across adjoining states.[17] So the volume of sand in the Coconino Sandstone layer is at least 10,000 cubic miles!

This layer also contains physical features called cross beds. While the overall layer of sandstone is horizontal, these features are clearly visible as sloped beds. These cross beds are remnants of the sand waves produced by the water currents that deposited the sand (like sand dunes, but underwater). So it can be demonstrated that water, flowing at 3–5 miles per hour, deposited the Coconino Sandstone as massive sheets of sand, with sand waves up to 60 feet high.[18] At this rate, the whole Coconino Sandstone layer (all 10,000 cubic miles of sand) would have been deposited in just a few days!

Sediment layers that spread across vast continents are evidence that water covered the continents in the past. Even more dramatic are the fossil-bearing sediment layers that were deposited rapidly right across many or most of the continents at the same time. To catastrophically deposit such extensive sediment layers implies global flooding of the continents. And these are only a few examples.[19]

Evidence #4: Sediment Transported Long Distances

When the Flood waters swept over the continents and rapidly deposited sediment layers across vast areas, these sediments had to have been transported from distant sources.

16. Ibid., p. 6–7.
17. D.L. Baars, "Permian System of Colorado Plateau," *American Association of Petroleum Geologists Bulletin* 46 (1962): 200–201; J.M. Hills and F.E. Kottlowski, *Correlation of Stratigraphic Units of North America-Southwest/Southwest Mid-Continent Region* (Tulsa, OK: American Association of Petroleum Geologists, 1983); R.C. Blakey and R. Knepp, "Pennsylvanian and Permian Geology of Arizona," in J.P. Jenney and S.J. Reynolds, eds., "Geologic Evolution of Arizona," *Arizona Geological Society Digest* 17 (1989): 313–347.
18. A.A. Snelling and S.A. Austin, "Startling Evidence of Noah's Flood," *Creation Ex Nihilo* 15 no. 1 (1992): 46–50; S.A. Austin, ed., *Grand Canyon: Monument to Catastrophe* (Santee, CA: Institute for Creation Research, 1994), p. 28–36.
19. Ager, *The Nature of the Stratigraphical Record*, p. 1–13.

For example, as was mentioned above, the Coconino Sandstone, seen spectacularly in the walls of the Grand Canyon, has an average thickness of 315 feet, covers an area of at least 200,000 square miles, and thus contains at least 10,000 cubic miles of sand.[20] Where did this sand come from and how do we know?

The sand grains are pure quartz (a natural glass mineral), which is why the Coconino Sandstone is such a distinctive buff color. Directly underneath it is the strikingly different red-brown Hermit Formation, consisting of siltstone and shale. Sand for the Coconino Sandstone could not have come from the underlying Hermit Formation.

The sloping remnants of sand "waves" in the Coconino Sandstone point to the south, indicating the water that deposited the sand flowed from the north.[21] Another clue is that the Coconino Sandstone thins to zero to the north in Utah, but the Hermit Formation spreads further into Utah and beyond. So the Coconino's pure quartz sand had to come from a source even further north, above the red-brown Hermit.

The Grand Canyon has another layer with sands that must have come from far away — the sandstone beds within the Supai Group strata between the Hermit Formation and the Redwall Limestone. In this case, the sand "wave" remnants point to the southeast, so the sand grains had to be deposited by water flowing from a source in the north and *west*. However, to the north and west of the Grand Canyon we find only Redwall Limestone underneath the Supai Group, so there is no nearby source of quartz sand for these sandstone beds.[22] Thus, an incredibly long distance must be postulated for the source of Supai Group sand grains, probably from a source as far away as northern Utah or even Wyoming.[23]

Higher in the strata sequence is the Navajo Sandstone of southern Utah, best seen in the spectacular mesas and cliffs in and around Zion National Park. The Navajo Sandstone is well above the Kaibab Limestone, which forms the rim rock of Grand Canyon. Like Grand Canyon sandstone, this sandstone also consists of very pure quartz sand, giving it a distinctive brilliant white color, and it also contains remnants of sand "waves."

20. D.L. Baars, "Permian System of Colorado Plateau," *American Association of Petroleum Geologists Bulletin* 46 (1962): 200–201; J.M. Hills and F.E. Kottlowski, *Correlation of Stratigraphic Units of North America-Southwest/Southwest Mid-Continent Region* (Tulsa, OK: American Association of Petroleum Geologists, 1983); R.C. Blakey and R. Knepp, "Pennsylvanian and Permian Geology of Arizona," in J.P. Jenney and S.J. Reynolds, eds., "Geologic Evolution of Arizona," *Arizona Geological Society Digest* 17 (1989): 313–347.
21. Austin, *Grand Canyon: Monument to Catastrophe*, p. 36.
22. J.S. Shelton, *Geology Illustrated* (San Francisco, CA: WH Freeman, 1966), p. 280.
23. R.C. Blakey, "Stratigraphy of the Supai Group (Pennsylvanian-Permian), Mogollon Rim, Arizona," in S.S. Beus and R.R. Rawson, eds., *Carboniferous Stratigraphy in the Grand Canyon Country, Northern Arizona and Southern Nevada* (Falls Church, VA: American Geological Institute, 1979), p. 102.

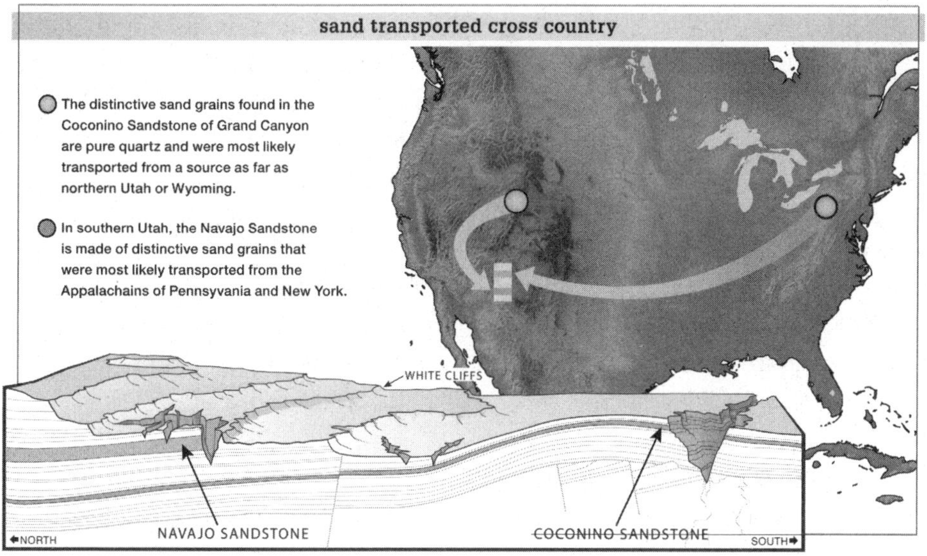

Figure 3: The deposition of these layers defies any uniformitarian explanation.

However, we have to look even farther for the original rocks that eroded to form the sand in this sandstone layer. Fortunately, within this sandstone we find grains of the mineral zircon, which is relatively easy to trace to its source because zircon usually contains radioactive uranium. By "dating" these zircon grains, using the uranium-lead (U-Pb) radioactive method, it has been postulated that the sand grains in the Navajo Sandstone came from the Appalachians of Pennsylvania and New York, and from former mountains farther north in Canada. If this is true, the sand grains were transported about 1,250 miles right across North America[24] (figure 3).

This "discovery" poses somewhat of a dilemma for conventional uniformitarian (slow-and-gradual) geologists, because no known sediment transport system, even today, is capable of carrying sand right across the entire North American continent during the required millions of years. It must have been water over an area even bigger than the continent. All they can do is postulate that some unknown transcontinental river system must have done the job. But even in their scientific belief system of earth history, it is impossible for such a river to have persisted for millions of years.

Yet the evidence is overwhelming that the water was flowing in one direction. More than half a million measurements have been collected from 15,615 localities recording water current direction indicators throughout the geologic

24. J.M. Rahl et al., "Combined Single-grain (U-Th)/He and U/Pb Dating of Detrital Zircons from the Navajo Sandstone, Utah," *Geology* 31 no. 9 (2003): 761–764.

record of North America. Based on these measurements, water moved sediments right across the continent, from the east and northeast to the west and southwest throughout the so-called Paleozoic.[25] This pattern continued on up into the Mesozoic, when the Navajo Sandstone was deposited, although some water currents shifted more southward. How could water be flowing right across the North American continent consistently for hundreds of millions of years? Absolutely impossible!

The only logical and viable explanation is the global cataclysmic Genesis Flood. Only the water currents of a global ocean, lasting a few months, could have transported such huge volumes of sediments right across North America to deposit the thick strata sequences which blanket the continent.[26]

Evidence #5: Rapid or No Erosion Between Strata

If the fossil-bearing layers took hundreds of millions of years to accumulate, then we would expect to find many examples of weathering and erosion after successive layers were deposited. The boundaries between many sedimentary strata should be broken by lots of topographic relief with weathered surfaces. After all, shouldn't periods of weathering and erosion for millions of years follow each deposition?

On the other hand, in the cataclysmic global Flood most of the fossil-bearing layers would have accumulated in just over one year. Under such catastrophic conditions, even if land surfaces were briefly exposed to erosion, such erosion (called sheet erosion) would have been rapid and widespread, leaving behind flat and smooth surfaces. The erosion would not create the localized topographic relief (hills and valleys) we see forming at today's snail's pace. So if the Genesis Flood caused the fossil-bearing geologic record, then we would only expect evidence of rapid or no erosion at the boundaries between sedimentary strata.

At the boundaries between some sedimentary layers we find evidence of only rapid erosion. In most other cases, the boundaries are flat, featureless, and knife-edge, with absolutely no evidence of any erosion, as would be expected during the Genesis Flood.

The Grand Canyon offers numerous examples of strata boundaries that are consistent with deposition during the Genesis Flood.[27] However, we will focus

25. S.R. Dickinson and G.E. Gehrels, "U-Pb Ages of Detrital Zircons from Permian and Jurassic Eolian Sandstones of the Colorado Plateau, USA: Paleogeographic Implications," *Sedimentary Geology* 163 (2003): 29–66.
26. A.V. Chadwick, "Megatrends in North American Paleocurrents," www.origins.swau.edu/papers/global/paleocurrents/default.html.
27. Austin, *Grand Canyon: Monument to Catastrophe*, p. 42–52.

here on just four, which are typical of all the others, appearing at the bases of the Tapeats Sandstone, Redwall Limestone, Hermit Formation, and Coconino Sandstone.

The strata below the Tapeats Sandstone has been rapidly eroded and then extensively scraped flat (planed off). We know that this erosion occurred on a large scale because we see its effects from one end of the Grand Canyon to the other. This massive erosion affected many different underlying rock layers —granites and metamorphic rocks — and tilted sedimentary strata.

There are two evidences that this large-scale erosion was also rapid. First, we don't see any evidence of weathering below the boundary — we don't see the expected soils.[28] Second, we find boulders and features known as "storm beds" in the Tapeats Sandstone above the boundary.[29] Storm beds are sheets of sand with unique internal features only produced by storms, such as hurricanes. Boulders and storm beds aren't deposited slowly.

Below the base of the Redwall Limestone the underlying Muav Limestone has been rapidly eroded in a few localized places to form channels. These channels were later filled with lime sand to form the Temple Butte Limestone. Apart from these rare exceptions, the boundary between the Muav and Redwall Limestones, as well as between the Temple Butte and Redwall Limestones, are flat and featureless, hallmarks of continuous deposition.

Indeed, in some locations the boundary between the Muav and Redwall Limestones is impossible to find because the Muav Limestone continued to be deposited after the Redwall Limestone began.[30] These two formations appear to intertongue (thin beds of each formation are interleaved with one another), so the boundary is gradational. This feature presents profound problems for uniformitarian geology. The Muav Limestone was supposedly deposited 500–520 million years ago,[31] the Temple Butte Limestone was deposited about 100 million years later (380–400 million years ago),[32] and then the Redwall Limestone was deposited several million years later (330–340 million years ago).[33] It is

28. N.E.A. Hinds, "Ep-Archean and Ep-Algonkian Intervals in Western North America," *Carnegie Institution of Washington Publication* 463, vol. 1, 1935.
29. A.V. Chadwick, "Megabreccias: Evidence for catastrophism," *Origins* 5 (1978): 39–46.
30. A.A. Snelling, "The Case of the 'Missing' Geologic Time," *Creation Ex Nihilo* 14, no. 3 (1992): 30–35.
31. L.T. Middleton and D.K. Elliott, "Tonto Group," in S.S. Beus and M. Morales, eds., *Grand Canyon Geology*, 2nd edition (New York, NY: Oxford University Press, 2003), p. 90–106.
32. S.S. Beus, "Temple Butte Formation," in S.S. Beus and M. Morales, eds., *Grand Canyon Geology*, 2nd edition (New York, NY: Oxford University Press, 2003), p. 107–114.
33. S.S. Beus, "Redwall Limestone and Surprise Canyon Formation," in S.S. Beus and M. Morales, eds., *Grand Canyon Geology*, 2nd edition (New York, NY: Oxford University Press, 2003), p. 115–135.

Figure 4. The flat, featureless boundary between these two layers indicates that the top layer (Coconino Sandstone) was laid down right after the bottom layer (Hermit Formation), before any erosion could occur.

much more logical to believe that these limestones were deposited continuously, without any intervening millions of years.

The boundary between the Hermit Formation and the Esplanade Sandstone is often cited as evidence of erosion that occurred over millions of years after sediments had stopped building up.[34] However, the evidence indicates that water was still depositing material, even as erosion occurred. In places, the Hermit Formation's silty shales are intermingled (intertongued) with the Esplanade Sandstone, indicating that a continuous flow of water carried both silty mud and quartz sand into place. Thus, there were no millions of years between these layers.[35]

Finally, the boundary between the Coconino Sandstone and the Hermit Formation is flat, featureless, and knife-edge from one end of the Grand Canyon to the other (figure 4). There is absolutely no evidence of any erosion on the Hermit Formation before the Coconino Sandstone was deposited. That alone is amazing.

The fossil-bearing portion of the geologic record consists of tens of thousands of feet of sedimentary layers, of which about 4,500 feet are exposed in the walls of the Grand Canyon. If this enormous thickness of sediments were deposited over 500 or more million years, then some boundaries between layers should show evidence of millions of years of slow erosion, just as erosion is

34. L.F. Noble, "A Section of Paleozoic Formations of the Grand Canyon at the Bass Trail," *U.S. Geological Survey Professional Paper* 131-B, 1923, p. 63–64.
35. E.D. McKee, "The Supai Group of Grand Canyon," *U.S. Geological Survey Professional Paper* 1173, 1982, p. 169–202; R.C. Blakey, "Stratigraphy and Geologic History of Pennsylvanian and Permian rocks, Mogollon Rim Region, Central Arizona and Vicinity," *Geological Society of America Bulletin* 102 (1990): 1189–1217; R.C. Blakey, "Supai Group and Hermit Formation," in S.S. Beus and M. Morales, eds., *Grand Canyon Geology*, 2nd edition (New York, NY: Oxford University Press, 2003), p. 136–162.

occurring on some land surfaces today. On the other hand, if this enormous thickness of sediments were all deposited in just over a year during the global cataclysmic Genesis Flood, then the boundaries between the layers should show evidence of continuous rapid deposition, with only occasional evidence of rapid erosion, or of no erosion at all. And that's exactly what we find, as illustrated by strata boundaries in the Grand Canyon.

Evidence #6: Many Strata Laid Down in Rapid Succession

The sedimentary units in the Grand Canyon are thought, by uniformitarian geologists, to have been deposited and deformed over the past 500 million years. If it really did take millions of years for these sedimentary sequences to be deposited, then individual sediment layers would not have been deposited rapidly, nor would the sequences have been laid down continuously. In contrast, if the Genesis Flood deposited all these strata in a little more than a year, then the individual layers would have been deposited in rapid succession.

Do we see evidence in the walls of the Grand Canyon that the sedimentary layers were all laid down in quick succession? Yes, absolutely! The entire sequence of sedimentary strata was still soft during subsequent folding and experienced only limited fracturing. These rock layers would have broken and shattered unless all the strata were immediately folded while the sediment was still relatively soft and pliable (figure 5).

When solid, hard rock is bent (or folded) it invariably fractures and breaks because it is brittle.[36] Rock will bend only if it is still soft and pliable — "plastic" like modeling clay or children's play-dough. If such modeling clay is allowed to dry and/or is baked in an oven, it is no longer pliable but hard and brittle, so any attempt to bend it will cause it to break and shatter.

When sediments are deposited by water in a layer, some water is trapped between the sediment grains. Clay particles may also be among the sediment grains. The pressure of other sediment layers on top of each layer squeezes the particles closer together and forces out much of the water. The internal heat of the earth may also cause additional dehydration of the sediments. Removal of the water dries the sediment layer and converts the chemicals that were in the water and between the clay particles into a natural cement. This cement transforms the originally soft and wet sediment layer into a hard, brittle rock layer.

36. E.S. Hills, "Physics of Deformation," *Elements of Structural Geology* (London: Methuen & Co., 1970), p. 77–103; G.H. Davis and S.J. Reynolds, "Kinematic Analysis," *Structural Geology of Rocks and Regions*, 2nd edition (New York, NY: John Wiley & Sons, 1996), p. 38–97.

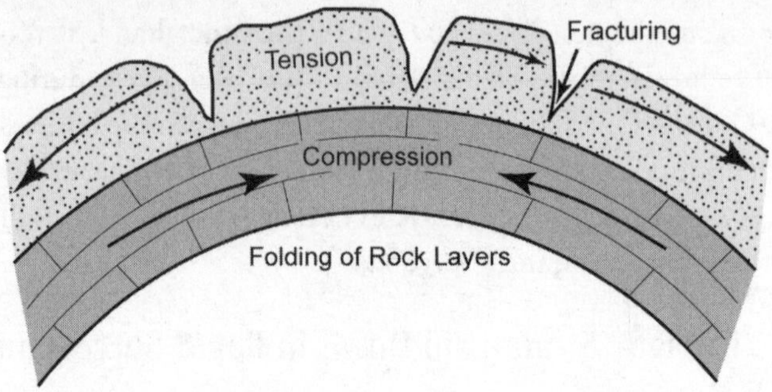

Figure 5. When solid, hard rock is bent (or folded) it invariably fractures and breaks because it is brittle. Rock will bend only if it is still soft and pliable, like modeling clay. If clay is allowed to dry out, it is no longer pliable but hard and brittle, so any attempt to bend it will cause it to break and shatter.

This process, known technically as diagenesis, can be exceedingly rapid.[37] It is known to occur within hours but generally takes days or months, depending on the prevailing conditions. It doesn't take millions of years, even under today's slow-and-gradual geologic conditions.

The 4,500-foot sequence of sedimentary layers in the walls of the Grand Canyon stands well above today's sea level. Earth movements in the past pushed up this sedimentary sequence to form the Kaibab Plateau. However, the eastern portion of the sequence (in the eastern Grand Canyon and Marble Canyon) was not pushed up as much and is about 2,500 feet lower than the height of the Kaibab Plateau. The boundary between the Kaibab Plateau and the less uplifted eastern canyons is marked by a large, step-like fold, producing what is called the East Kaibab Monocline.

It's possible to see these folded sedimentary layers in several side canyons. For example, the folded Tapeats Sandstone can be seen in Carbon Canyon (figure 6). Notice that these sandstone layers were bent 90° (a right angle), yet the rock was not fractured or broken in the fold axis or hinge line (apex) of the fold. Similarly, the folded Muav and Redwall Limestone layers can be seen along nearby Kwagunt Creek. The folding of these limestones did not cause them to fracture and break either, as would be expected for ancient, brittle rocks. The obvious conclusion is that these sandstone and limestone layers were all folded and bent while the sediments were still soft and pliable, and very soon after they were deposited.

Herein lies an insurmountable dilemma for uniformitarian (long-age) geologists. They maintain that the Tapeats Sandstone and Muav Limestone were

37. Z.L. Sujkowski, "Diagenesis," *Bulletin of the American Association of Petroleum Geologists* 42 (1958): 2694–2697; H. Blatt, *Sedimentary Petrology*, 2nd edition (New York, NY: W.H. Freeman and Company, 1992).

Figure 6. It is possible to see these folded sedimentary layers in several side canyons. All these layers had to be soft and pliable at the same time in order for these layers to be folded without fracturing. Here the folded Tapeats Sandstone can be seen in Carbon Canyon.

deposited 500–520 million years ago;[38] the Redwall Limestone, 330–340 million years ago;[39] then the Kaibab Limestone at the top of the sequence, supposedly 260 million years ago.[40] However, the Tapeats Sandstone was supposedly deposited some 440 million years before the Kaibab Plateau was uplifted, which caused the folding (supposedly only about 60 million years ago).[41] How could the Tapeats Sandstone and Muav Limestone still be soft and pliable, as though they had just been deposited, and not subjected yet to diagenesis, without fracturing and shattering when they were folded 440 million years after their deposition?

The conventional explanation is that under the pressure and heat of burial, the hardened sandstone and limestone layers were bent so slowly they behaved as though they were plastic and thus did not break.[42] However, pressure and heat would have caused detectable changes in the minerals of these rocks, telltale

38. L.T. Middleton and D.K. Elliott, "Tonto Group," in S.S. Beus and M. Morales, eds., *Grand Canyon Geology*, 2nd edition (New York, NY: Oxford University Press, 2003), p. 90–106.
39. S.S. Beus, "Redwall Limestone and Surprise Canyon Formation," in S.S. Beus and M. Morales, eds., *Grand Canyon Geology*, 2nd edition (New York, NY: Oxford University Press, 2003), p. 115–135.
40. R.L. Hopkins and K.L. Thompson, "Kaibab Formation," in S.S. Beus and M. Morales, eds., *Grand Canyon Geology*, 2nd edition (New York, NY: Oxford University Press, 2003), p. 196–211.
41. P.W. Huntoon, "Post-Precambrian Tectonism in the Grand Canyon Region," in S.S. Beus and M. Morales, eds., *Grand Canyon Geology*, 2nd edition (New York, NY: Oxford University Press, 2003), p. 222–259.
42. E.S. Hills, "Environment, Time and Material," *Elements of Structural Geology* (London: Methuen & Co., 1970), p. 104–139; G.H. Davis and S.J. Reynolds, "Dynamic Analysis," *Structural Geology of Rocks and Regions*, 2nd edition (New York, NY: John Wiley & Sons, 1996), p. 98–149.

signs of metamorphism.[43] But such metamorphic minerals or re-crystallization due to such plastic behavior[44] is not observed in these rocks. The sandstone and limestone in the folds are identical to sedimentary layers elsewhere.

The only logical conclusion is that the 440-million year delay between deposition and folding never happened! Instead, the Tapeats-Kaibab strata sequence was laid down in rapid succession early during the year of the Genesis Flood, followed by uplift of the Kaibab Plateau within the last months of the Flood. This alone explains the folding of the whole strata sequence without appreciable fracturing.

Conclusion

In this chapter we have documented that, when we accept God's eyewitness account of the Flood in Genesis 7–8 as an actual event in earth history, then we find that the geologic evidence is absolutely in harmony with the Word of God. As the ocean waters flooded over the continents, they must have buried plants and animals in rapid succession. These rapidly deposited sediment layers were spread across vast areas, preserving fossils of sea creatures in layers that are high above the current (receded) sea level. The sand and other sediments in these layers were transported long distances from their original sources. We know that many of these sedimentary strata were laid down in rapid succession because we don't find evidence of slow erosion between the strata.

Jesus Christ our Creator (John 1:1–3; Colossians 1:16–17), who is the Truth and would never tell us a lie, said that during the "days of Noah" (Matthew 24:37; Luke 17:26–27) "Noah entered the ark" and "the Flood came and took them all away" (Matthew 24:38–39). He spoke of these events as real, literal history, describing a global Flood that destroyed all land life not on the ark. Therefore, we must believe what He told us, rather than believe the ideas of fallible scientists who weren't there to see what happened in the earth's past. Thus, we shouldn't be surprised when the geologic evidence in God's world (rightly understood by asking the right questions) agrees exactly with God's Word, affirmed by Jesus Christ.

---

43. R.H. Vernon, *Metamorphic Processes: Reactions and Microstructure Development* (London: George Allen & Unwin, 1976); K. Bucher and M. Frey, *Petrogenesis of Metamorphic Rocks*, 7th edition (Berlin: Springer-Verlag, 2002).
44. Vernon, *Metamorphic Processes: Reactions and Microstructure Development*; G.H. Davis and S.J. Reynolds, "Deformation Mechanisms and Microstructures," *Structural Geology of Rocks and Regions*, 2nd edition (New York, NY: John Wiley & Sons, 1996), p. 150–202.

30

What Are Some Good Questions to Ask an Evolutionist?

MIKE RIDDLE AND DR. JASON LISLE

A football coach recruited the best defensive players he could find. His strategy was to have the best defense in the conference. All through the season the opposing teams were unable to score many points. When the season was over his team posted a record of zero wins, ten losses, and two ties. How could this happen? The answer is they had no offense.

A Christian Game Plan

This is where many Christians are in their efforts to witness to unbelievers. The Bible instructs believers to have answers when challenged by any and all who oppose the Word of God (defense — 1 Peter 3:15). The Bible also instructs believers to bring down all strongholds and anything that exalts itself against the knowledge of God (offense — 2 Corinthians 10:4–5). Sadly, while many Christians lack the knowledge to challenge unbelievers (offense), they also lack a defense.

What is meant by defense and offense in Christian witnessing? Defense means that the Christian can answer questions such as: How do you fit dinosaurs into the Bible? Where did Cain get his wife? How could Adam name all the animals in one day? What about carbon-14 dating? Does God really exist? Couldn't God have used evolution?

Offense means the Christian can ask the unbeliever questions that challenge his or her worldview. The strategy of asking good questions can be used to demonstrate to unbelievers that their belief in evolution is a sort of "blind" faith and is not something derived from empirical science. They can also illustrate to the compromised Christian (a person who professes to believe in both the Bible and ideas such as evolution or millions of years) that God's Word is a completely accurate record and is not to be modified by secular opinions of what is possible.

There are several different types of questions that are useful in apologetics; we will cover four general categories of questions in this chapter. Questions can be used to help us assess and clarify the worldview of the critic. What does he really believe, and how is he using the terms? We will call these "clarification questions." We can ask "foundation questions" about the most basic laws of science, and the beginning of first things. There are "textbook questions" — questions that can expose inconsistency in common textbook claims. These are particularly useful in public school settings. And finally, there are worldview questions — questions that can be used to show that the evolutionary worldview is utterly, intellectually defective.

Clarification Questions

These questions are used to help explain the meaning of words or terms. A definition in science needs to be clear and precise. It should include all the attributes that distinguish it from all other entities. If any of these attributes are missing, then the definition becomes ambiguous.

- What do you mean by *evolution*?
- What do you mean by *theory*?
- What is meant by a *fact* in science?

Let's examine some examples of the importance of establishing definitions.

"Evolution is change over time." This is not a legitimate definition because it includes everything in the universe.

"Evolution is genetic change in a species over time." While this may be one definition of "evolution," it is not the claim at issue in the origins debate. Such a definition includes all forms of change, including changes that both creationists and evolutionists believe in (e.g., information-decreasing mutations). Therefore, this does not adequately define the type of evolution relevant to origins; that is, Neo-Darwinian evolution that suggests that an amoeba can change into a man over millions of years.

"Evolution means both micro and macro changes." This is a common use of evolution in textbooks. Dog varieties or different beak sizes of finches thus become examples of evolution. This definition includes both variety within the kinds and Neo-Darwinian evolution (molecules to man). The definition tacitly implies that small observed changes, sometimes referred to as microevolution, will lead to large unobserved changes (macroevolution), which begs the question at issue.

From these examples we see that it is important to establish definitions of terms prior to any discussion.

Foundation Questions

These questions aim at the core, or foundation, of the unbeliever's evidence.

- What is the ultimate cause of the universe?
- How did life originate?
- Where did the dinosaurs come from?
- Where are all the millions of transitional fossils in the Precambrian and Cambrian layers?
- Since information is nonmaterial and in all observed cases always requires an intelligent sender, how did all the information contained in DNA originate?
- How do we know that is true?
- Has that ever been observed?
- Are there any assumptions in what you are describing?

In this chapter we will analyze the cause of the universe question. Analysis of the other questions can be found in the *New Answers Books 1 & 2*.

Question: What caused the universe to come into existence and where did the original energy or matter come from?

This is an important question because it aims at the very foundation or beginning of the entire evolution worldview. Without a cause (and a mass/energy source) there can be no big bang, evolution of stars, or life. Some evolutionists may scoff at such a question by stating it is not a legitimate question. Others might state that science does not deal with such questions or we can't know such things. In either case this is a "brush-off" to avoid the question. There are only three possible responses to this question:

1. The universe created itself.
2. The universe has always existed.
3. The universe had to be created.

Response 1: The universe created itself.

For something to create itself it would have to both exist (in order to have the power to act) and not exist (in order to be created) at the same time. This is a contradiction — an illogical position to take. Based on all known scientific understanding and logic we know that *from nothing, nothing comes*. Therefore, this is not a legitimate response. A person arguing this way has violated the law of non-contradiction and is ignoring good science. This now leaves two possible choices.

Response 2: The universe has always existed (no beginning).

In order to analyze this response we need to understand some basics about the second law of thermodynamics. The second law is concerned with heat — the flow of thermal energy. Everything in the universe is losing its available energy to do work. To illustrate this concept we will use the example called "No Refills."

You have just been given a new car for FREE! All expenses for the lifetime of the car are paid. Sounds like a good deal. However, there is one catch. You are only allowed to have one tank of gas and never allowed to refill the tank. Once you have driven the car and used up all the gas, the car can no longer be used for transportation. In other words, the gas (energy source) has been used up and cannot be reused to propel the car. This is what the second law of thermodynamics deals with. Usable energy is constantly becoming less usable for doing work. Unless the car obtains new fuel from an outside source, it will cease to function after it exhausts its first tank of gas.

Likewise, the universe is constantly converting useful energy into less usable forms. As one example, stars are fueled by hydrogen gas that is used up as it is converted into heavier elements. But the problem is this: for any given region of space, there is only a finite amount of available energy. There is just only so much hydrogen available per cubic meter. This means that unless the universe obtains new useable energy from an outside source, it will cease to function in a finite amount of time. Stars will no longer be possible, once the hydrogen is gone.[1]

However, there is no "outside source" available. The universe is everything, according to the secular worldview. Like the car, the universe would

1. Some stars are thought to be powered by fusion of heavier elements, eventually resulting in iron. But eventually, these heavier elements would be used up as well. Nuclear reactions of elements heavier than iron are endothermic, and cannot power a star.

cease to function after its first "tank of gas" is exhausted. But if the universe were infinitely old, it should have used up that energy a long time ago. Putting it another way, if stars have eternally been processing hydrogen into heavier elements, then there would be no hydrogen left! But there is. The fact that the universe *still* contains useable energy indicates that it is not infinitely old — it had a beginning.

Response 3: The universe had to be created.

Since the universe could not create itself and it had to have a beginning, the only logical solution is that the universe had to be created! This leaves us with the original question to the evolutionist, "Where did the matter come from to create the universe?" Any reply not recognizing that the universe was created ignores the laws of science and good logic.

When asking this question, be prepared to answer the challenge, "Where did God come from?" This question indicates a misunderstanding of the nature of God. It suggests that God is within (or "bound by") the universe and that God is part of the chain of effects within time — all of which require a cause. We should be prepared to correct the misunderstanding, and point out that God does not require a cause since He has always existed, is beyond time, and is not part of the physical universe. God is a spirit, not a sequence of energetic reactions, and so the laws of thermodynamics (which place a finite limit on the age of the universe) do not apply to Him.

> Remember the former things of old, For I am God, and there is no other; I am God, and there is none like Me (Isaiah 46:9).

Textbook Questions for the Classroom

These questions are used to help students in the classroom critically think through information in a textbook or further explore statements made by a teacher.

- While some molecules do combine to form larger structures such as amino acids, it has been shown that this always results in a mixture of left- and right-handed amino acids that is not used in life. Since this is true, is there some other explanation for how the molecules useful for life might have formed? (Be prepared for an answer involving "given enough time it could happen.")[2]
- Since oxygen is known to destroy molecular bonds, and since the lack of oxygen in the atmosphere (meaning no ozone)

2. For responses involving "given enough time it could happen," see chapter 16, "Does Evolution Have a . . . Chance?"

would cause all potential life to be destroyed by ultraviolet rays, how could life have formed? (Be prepared to follow up with a question about hydrolysis — water decomposing molecules.)

- Since water breaks down the bonds between amino acids (a process called hydrolysis), how could life have started in the oceans?
- The National Academy of Sciences defines a theory as "a comprehensive explanation of some aspect of nature that is supported by a vast body of evidence" and science as "the use of evidence to construct testable explanations and predictions of natural phenomena."[3] Does this mean scientists can reproduce how life originated or test any step of the process for how life evolved? If not, then how can evolution qualify as a theory?
- Microsoft uses intelligent programmers and complex codes to create the Windows operating system. However, information in DNA is millions of times more dense and complex. How could the process of evolution, using natural processes and chance, solve the problem of complex information sequencing without intelligence? (Be prepared for an answer involving "given enough time it could happen.")
- Bill Gates (founder and former CEO of Microsoft) recognized that the processing capabilities of DNA are "like a computer program but far, far more advanced than any software ever created."[4] Using all their intelligence and all the modern advances in science, have scientists ever created DNA or RNA in a laboratory through unguided naturalistic processes? If not, then isn't the origin of life still an unverified assumption?
- DNA, RNA, and proteins all need each other as an integrated unit. Even if only one of them existed, the many parts needed for life could not sit idle and wait for the other parts to evolve because they would dissolve or deteriorate. Is there any compelling (observable) evidence for how all these components evolved at the same time or separately over time?
- Isn't it true that whenever we see interdependent complex structures or codes we automatically assume an intelligent person had to put them together? So why do we assume that

3. National Academy of Sciences: Institute of Medicine, *Science, Evolution, and Creationism* (Washington, DC: The National Academies Press, 2008), p. 10–11.
4. Bill Gates, *The Road Ahead* (London: Penguin Books, 1996), p. 228.

DNA, or RNA, or a cell, which is more complex than any computer ever designed, happened by chance? Doesn't that seem to go against good science and logical thought?

- Is there any observed case where random chance events created complex molecules with enormous amounts of information like that found in DNA or RNA? If not, then why should we assume it happened in the past?
- A living cell is composed of millions of parts all working together and is considered more complex than any man-made machine. Then, since the process of evolution has no blueprints (cannot plan for the future) for building something, since over time things tend to deteriorate unless there is a mechanism in place to sustain them, since virtually all known mutations decrease genetic information (or are neutral), since natural selection would not be operating until the first cell formed, how could the process of evolution ever assemble something as complex as a living cell with all its information content?
- Since we started with finches and the finches stayed finches, isn't this just an example of variety within a kind?
- Since we started with bacteria, and the bacteria that became resistant to the antibiotic remained bacteria, isn't this just another example of variety within a kind?
- What naturalistic evidence could actually disprove that evolution is the explanation for life on earth (or the formation of the universe)?

It is important to remember that whenever asking questions of a teacher or instructor, asking the questions at an appropriate time and in a respectful manner is extremely important. More questions related to specific topics can also be found in the books *Evolution Exposed: Biology*[5] and *Earth Science*[6] by Roger Patterson.

Worldview Questions

These are the questions that can stop people in their tracks. A series of well-stated worldview questions can expose the inconsistency of non-biblical

5. Roger Patterson, *Evolution Exposed: Biology* (Petersburg, KY: Answers in Genesis, 2009).
6. Roger Patterson, *Evolution Exposed: Earth Science* (Petersburg, KY: Answers in Genesis, 2008).

worldviews. It is the Christian worldview alone that makes science, knowledge, and ethics possible. We can help unbelievers see this by asking the right questions.

- How do you account for the existence and nature of laws? In particular, how do you account for (1) laws of morality, (2) laws of nature, and (3) laws of logic? (Laws of morality make sense in the Christian worldview where God created human beings in His own image [according to a natural reading of Genesis] and therefore has the right to set the rules for our behavior.)
- If we are simply chemical accidents, as evolutionists contend, why should we feel compelled to behave in a particular fashion?
- If laws of morality are just what bring the most happiness to the most people, then why would it be wrong to kill just one innocent person if it happened to make everyone else a lot happier?
- If laws of morality are just the adopted social custom, then why was what Hitler did wrong? (Laws of nature make sense in the Christian worldview; God upholds the entire universe by His power. God is beyond time, and has promised to uphold the future as He has the past [Genesis 8:22].)
- In your worldview, why do the different objects in the universe obey the same laws of nature?
- Do you have confidence that laws of nature will apply in the future as they have in the past? If not, then why did you bother to answer my question? You assumed your vocal cords and my ears would work in the future as they have in the past, otherwise I could not understand your answer.
- Since you have not experienced the future, how do you know that the laws of nature will behave in the future as they have in the past? The answer "it's always been that way before" is not legitimate because it assumes that the future will be like the past, which is the very question I'm asking.
- In the Christian worldview, it makes sense to have universal, immaterial, unchanging laws of logic. These are God's standard

for correct reasoning. How do you account for the existence and properties of laws of logic?

- Do you believe laws of logic are universal (applying everywhere)? If so, why (since you do not have universal knowledge)?
- Why do we all believe laws of logic will be the same tomorrow as they are today, since we are not beyond time and have not experienced the future?
- How can you have immaterial laws if the universe is material only?
- Why does the material universe feel compelled to obey immaterial laws?
- How does the material brain have access to these immaterial laws?

If you ask these questions properly, and are prepared for the common unsound responses, you can dismantle the evolutionary worldview. There is simply no good rebuttal to the Christian position, though many will make attempts. See *The Ultimate Proof of Creation*[7] by Dr. Jason Lisle for more information on worldview apologetics, and for examples of using these kinds of questions in actual dialogues.

Conclusion

The importance of asking questions is an essential part of Christian apologetics. Jesus often used the technique of asking questions. In Mark 11:29–33 Jesus refutes the chief priests, scribes, and elders by asking them a question.

> But Jesus answered and said to them, "I also will ask you one question; then answer Me, and I will tell you by what authority I do these things: The baptism of John — was it from heaven or from men? Answer Me."
>
> And they reasoned among themselves, saying, "If we say, 'From heaven,' He will say, 'Why then did you not believe him?' But if we say, 'From men' " — they feared the people, for all counted John to have been a prophet indeed. So they answered and said to Jesus, "We do not know."
>
> And Jesus answered and said to them, "Neither will I tell you by what authority I do these things."

7. Jason Lisle, *The Ultimate Proof of Creation* (Green Forest, AR: Master Books, 2009).

Jesus used good questions to show the foolishness of those who attempt to argue with God. We can do the same, by learning to think biblically, and knowing just a few of the many inconsistencies of the evolutionary worldview.

31

What about Bacteria?

DR. JOE FRANCIS

When my children were toddlers, it seemed to my wife and me that they were always sniffling or coughing, or fighting off a cold or the flu. Many a night was spent rocking a feverish child to sleep. The two of us viewed with fear such ordinary places as the church nursery, seeing it as a breeding ground for infections.

My wife and I count our blessings, however, that our long nights were the only hardship we faced. Before the development of antibiotics and vaccines, infections were a leading cause of death among children. Most families lost at least one child to scarlet fever, diphtheria, pneumonia, measles, or smallpox.

Doctors now know that these maladies are caused by bacteria or viruses (collectively known as microbes).[1] As scientists continue to learn more about microbes, they are discovering that microbes employ intricate mechanisms to attack the human body. This raises a question: If God finished creation in six days and declared it "very good," where did these disease-causing designs come from?

Finding the answer has great potential to help mankind. A better understanding of God's original purpose for microbes could help scientists see how they have changed and to find revolutionary new ways to treat infectious diseases.

Based on the creation account in Genesis, it appears that God originally made microbes to perform only beneficial functions. If so, one would expect

1. Viruses are considered to be a separate category, but for the sake of this discussion we will include them as microbes.

many present-day bacteria to continue to perform their "very good" functions. Creation biologists predicted this and have documented examples.[2]

The Matrix

Imagine a futuristic city where vehicles are run by highly efficient acid-powered motors that produce little or no pollution. On the way home, your vehicle attaches itself to an airborne mega-transport ship, studded with hundreds of other vehicles. With the combined power of the multiple motors — complete with propellers — the mega-transport travels smoothly through rough weather and treacherous conditions to your home.

After detaching from the ship, you park in your driveway, which senses your car's dimensions and molds a raised platform to fit the car's shape, locking it securely 20 feet off the ground.

Imagine that as you're sleeping an airborne probe flies over your neighborhood and attaches to your home and car, inserting new instructions to update the operating software.

Whenever any cars in the city get the least bit outdated, tiny vehicles prowling the city track them down, attack them, and dismantle the parts. Then, using the old parts, each tiny vehicle can transform itself into a shiny new car, ready and waiting for you in the morning.

Your Gut Is a Thriving City

Futuristic city? Not really. These are just some of the things that bacteria do every day in our digestive systems. In fact, the human digestive system is the most densely populated ecosystem on earth, with hundreds of species of bacteria, yeast, and viruses interacting daily in this environment.[3]

Each species of bacteria is present in such high numbers that the total population is in the trillions. In fact, if we consider all the bacteria in the human body, there are ten for every human cell. This means that, by sheer numbers alone, you are more bacteria than human!

2. Only about 8 percent of the identified bacteria cause disease. For past predictions by creation biologists, see A. Gillen, *The Genesis of Germs* (Green Forest, AR: Master Books, 2007); J. Francis, "The Role of Virulence Factors in the Establishment of Beneficial Ecological Relationships of Vibrio cholera and Vibrio fischeri," *Occasional Papers of the BSG* 8 (2006); Francis and Wood, "The CT Toxin of Vibrio cholera, Its Structure, Function, and Origin," *Occasional Papers of the BSG* 11 (2008); and J. Francis and T. Wood, *Stadium Integrale* 16 (2009): 88, "Cholera Toxin and the Origin of Cholera Disease."
3. Steve Gill et al., "Metagenomic Analysis of the Human Distal Gut Microbiome," *Science* 312 (2006): 1355–1359.

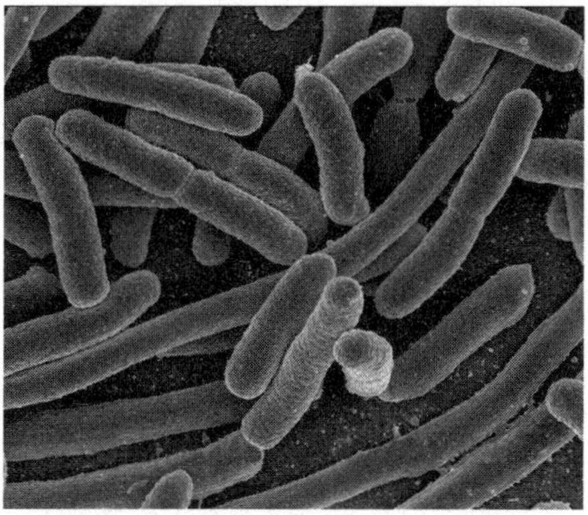

Escherichia coli (E. coli) is a bacterium commonly found in the lower intestine. Most E. coli strains are harmless and can produce vitamin K2 or prevent harmful bacteria from successfully invading the intestine. Picture copyright of Rocky Mountain Laboratories.

In the intestines, good bacteria provide nutrients, break down waste, and act as an immune system that prevents harmful bacteria from infecting our body. In fact, the human digestive system may need bacteria to be present before it can develop properly after birth.[4] Similar to the vehicles in the futuristic city above, many bacteria have elaborately designed mechanisms to move around in this dynamic, ever-changing environment. It is as though they were created to live there.

Microbes, Microbes, Everywhere

Microbes are found not only in the human body but also in every environment on earth, from high in the atmosphere to deep below the earth's surface, where they survive by eating things like oil and rocks. Microbes thrive in boiling hot springs, ice and snow, the dry heat of deserts, acids, high salt concentrations, rubber stoppers in bottles, and even hand soap.

Microbes Are Our Friends

While some microbes do cause disease, most do not. About 5,000 species of bacteria have been identified, but only about eight percent cause disease. While most species of disease-causing

Seafloor vents – On the seafloor are vents where superheated water spews toxic chemicals into the ocean. Numerous microbes, uniquely designed to withstand the extreme heat, feed on these minerals. They are the main food source for whole communities of organisms.

4. Recent research has also shown that the heart is decreased in size in animals that develop without intestinal bacteria. Peter Turnbaugh et al., "The Human Microbiome Project," *Nature* 449 (2007): 804–810.

Antarctic lake ice — Antarctica is home to numerous microbes. In fact, some organisms survive in water two miles below the continental ice sheet, where no air or light reaches. The frigid water is seven times saltier than the ocean, and the temperature falls below 14°F (-10°C).

Acidic hot springs — Hot springs, such as those in Yellowstone National Park, are home to a spectrum of microbes. They can survive temperatures well above 100°F (35°C) and acids potent enough to dissolve iron.

bacteria have been carefully identified (for obvious reasons), microbiologists estimate that 10 million other species of unidentified bacteria fill the earth. So the disease-causing species may account for only a tiny fraction of all bacterial species.

If most bacteria and other microbes don't cause disease, just what are they doing? Since the Bible states that God made everything "very good" at creation, creationists would expect to see the microbes' very good function all around us, on a grand scale.

Quite remarkably we find that microbes play a vital role in distributing and recycling nutrients all over the planet.[5] For example, every living thing needs carbon, oxygen, hydrogen, and nitrogen. Many bacteria specialize in recycling these nutrients through the air, water, and land. This crucial process, called biogeochemical cycling, takes place on an unimaginably huge scale (see "The Necessary Matrix of Bacteria").

Many, many microbes must work in concert to perform this cycling. Once thought to be a sterile wasteland, the deep earth appears to be a major chemical factory, filled with a mass of bacteria that could be greater than the combined mass of all plants and animals living on the surface.

Without the millions of different microbes, the earth's vast resources would be useless to us. We need their help to get the necessary chemicals out of the earth and into our bodies. We couldn't even eat steak or salad without bacteria

5. See J.W. Francis, "The Organosubstrate of Life: A Creationist Perspective of Microbes and Viruses," *Proceedings of the Fifth International Conference on Creationism*, 2003.

THE NECESSARY MATRIX OF BACTERIA

Bacteria are almost everywhere, busily sustaining life in ways we rarely see or appreciate. God designed bacteria, in many cases, to make inaccessible atoms available to us. A matrix of bacteria works around the clock to provide many vital ingredients of life.

OXYGEN-PRODUCING ↑

Cyanobacteria in the oceans break apart the bonds of carbon dioxide, making oxygen available to living things. Perhaps the most abundant creature on earth, these microbes may release more oxygen than all green plants combined.

NITROGEN-FIXING ↑

Our atmosphere is rich in nitrogen, but the majority is unusable because the atomic bond is too strong to break. The bacteria in the genus *Rhizobium*, which live in and around plants, fix the nitrogen, making it usable to the plants. Most animals get their nitrogen indirectly from these nitrogen-fixing bacteria.

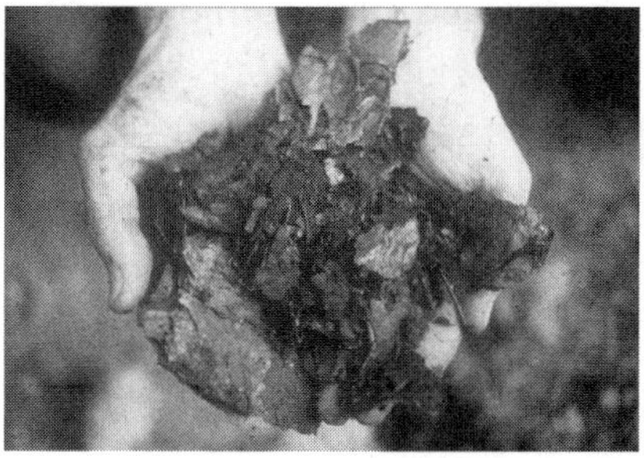

← CARBON-RECYCLING

Many different bacteria recycle carbon, an essential building block of life. One specific duty is to break down dead plant matter and sea creatures. Without these bacteria, our forests would be choked with branches and leaves, and our oceans littered with exoskeletons.

ROCK-EATING →

Many essential nutrients in the soil come from weathered rocks. A group of bacteria, known as lithotrophs, actually speed up this process by feeding on the minerals within the rocks. As these rocks break down, they enrich the soil, thus benefitting the plants that we eat.

RAIN-MAKING ↑

Some bacteria even help to make it rain! This recent discovery supports the biomatrix concept that microbes assist life cycles throughout our world.

in our stomach to help break food down. So every day, throughout the day, God displays His infinite love and wisdom, caring for every living thing even at the lowest, molecular level.

Microbes play a vital role in distributing and recycling nutrients for living things all over the planet.

Consider just one example — nitrogen recycling. Unlike the oxygen in the atmosphere, the nitrogen that we breathe is basically useless to humans and animals. The chemical bonds are just too strong. But a few bacteria and other microbes have the incredible ability to break the bonds of nitrogen and make it useful to living things.

In fact, many plants have specialized organs attached to their roots that house these nitrogen-loving bacteria. This relationship between plants and bacteria is a common phenomenon called mutualism, a form of symbiosis. It is a relationship whereby each partner benefits by living with the other partner.

Nothing Lives Alone

All creatures on earth live in symbiotic partnerships, including lowly single-celled pond-dwelling organisms. It appears that the Creator wants us to "clearly see" in these pervasive symbiotic relationships how much we depend on others — and ultimately Him — for life. From the very beginning of time, all the different creatures on earth had to be alive and working together, and we continue to depend on them (and God) for a healthy life.

So what are all these symbiotic microbes doing? Creationists have noted several major things, such as providing nutrition and influencing reproduction of insects. Let's consider just a couple of other interesting examples from the animal kingdom.

Defending plants and animals

Microbes are also involved in defending plants and animals against attack by other organisms. For example, consider that in the early 1900s a fungus almost wiped out the majestic American chestnut tree. A few trees survived the blight, however, and they were found to possess a virus that modified the blight, causing the fungus to be less potent. Now scientists are breeding resistant chestnut trees that could once again grace American forests.

It seems likely that God originally designed certain viruses as part of the immune system of plants.

Bioluminescence

Another interesting partnership is the bioluminescent (light-producing) bacteria that grow inside special light organs in creatures such as the Hawaiian bobtail squid.[6] The bioluminescence may help protect the squid against predators that swim under them at night. Perhaps the glowing squid appears as moonlight to predators lurking below, or perhaps the squid uses the light to see its way through murky water or at night. Whatever the bacteria's function, recent studies show that bioluminescent bacteria play an important role in the great depths of the ocean.

These are just a couple of the interesting symbiotic partnerships of bacteria and other microbes. Their amazing abundance and their life-supporting functions suggest that the Creator — our "living God" (Psalm 84:2) — made microbes to form a massive, life-sustaining, life-promoting biomatrix on earth.

When you look closely at the microbial world, two major themes are inescapable. One is that our living God intended to "fill the earth" with life, evidenced by the pervasive, life-sustaining biomatrix of microbes, animals, and humans. Second is the Creator's emphasis on relationships. A vast multitude of living things interact with each other as God designed it to be and as He sustains it.

> For in Him we live and move and have our being (Acts 17:28).

Good Designs Gone Bad

So what mechanism caused some of these "very good" microbes to go bad? Did God directly modify them, or did they change over time? At least three possible changes may have occurred, or a combination of all three:

1. Displacement. Microbes were originally designed to perform beneficial functions in restricted places, but after the Fall they spread to other places and began to cause disruption and disease.
2. Modification. Microbes were physically modified to become pathogenic (disease-causing).[7]
3. Uncontrolled growth. Their numbers were designed to remain within safe ranges, but now they fluctuate, causing either under- or over-population that results in disease and disruption of a once-balanced system.

6. E.G. Ruby, "Lessons from a Cooperative Bacterial-Animal Association: The Vibrio Fischeri-Euprymna Scolopes Light Organ Symbiosis," *Annual Review of Microbiology* 50 (1996): 591–624.
7. It is not yet known whether these modifications occurred directly by God or indirectly by the changed environment of the cursed post-Fall world.

Scripture hints at examples of helpful creations that have gone bad, such as thorns and thistles. Yeast is an example of a good thing that becomes invasive and harmful when it spreads too rapidly (see 1 Corinthians 5:6–8). In fact, yeast can cause severe infection, such as thrush and candidiasis in humans. Let's consider examples of other common disease-causing microbes.

Displaced Cholera

It seems that many microbes once had a good purpose but have changed as a result of the Fall and now cause disease.

Cholera is a severe intestinal illness that humans get from contaminated water or food. It leads to severe diarrhea, shock, and even death. In its most virulent form, it can kill within three hours of infection. Cholera is caused by the bacterium *Vibrio cholera*, which produces a variety of toxins. Interestingly, most species related to *Vibrio cholera* grow harmlessly on the surface of practically all shelled ocean creatures and some fish. There they perform a valuable task: breaking down chitin, the main component of the hard outside shell, or exoskeleton, of crabs, shrimp, lobsters, and many other sea creatures. Without their help, oceans and beaches would be littered with billions of shells. The breakdown of chitin also returns precious nutrients like carbon and nitrogen back to the ocean.

Even more fascinating, some of the cholera components that are toxic to the human intestines are used to break down chitin. So creationists hypothesize that *Vibrio cholera* originally broke down chitin in the ocean, but after Adam's Fall, God allowed them to spread beyond their proper place.

Disease-causing versions of *Vibrio cholera* may also have been genetically modified after the Fall. We have discovered that they have some extra DNA, apparently inserted by viruses, which allows the bacteria to produce toxin. Other types of cholera lack this DNA and are typically nontoxic.

Modified *E. Coli*

Another bacterium that appears to be modified is *Escherichia coli*. Normally, each person carries millions of harmless *E. coli* in his or her intestines, where it helps keep the digestive track running smoothly. *E. coli* is so intimately associated with the human body that health departments check for it when they want to confirm human activity in or near a waterway. Unfortunately, viruses appear to have infected some *E. coli* and introduced their own DNA into the *E. coli*'s DNA. For instance, one strain of *E. coli*[8] has an extra piece of DNA that

8. A bacterial strain is like a subspecies. *E. coli* is one species, but it has many subspecies or strains, which differ just slightly from one another. The harmless strain of *E. coli* that is found in the gut of farm animals but is toxic in humans is labeled *E. coli* O157:H7. The strain administered to newborns is *E. coli* AO 34/86.

produces lethal toxins. If you remove the offending DNA, you remove much of this bacterium's disease-causing potential.

But why would *E. coli* carry a toxin in the first place? What was this toxin created to do? No one can say for sure, but we do know that this strain of *E. coli* lives harmlessly in the gut of farm animals, where it has been shown to help protect against cancer-causing viruses.[9] So creationists hypothesize that the disease-causing abilities of this strain may have been acquired by the displacement or modification of a harmless *E. coli*.

In recent years medical researchers have also discovered that beneficial *E. coli* may protect our intestines from disease-causing bacteria. In fact, some physicians are administering a strain of *E. coli* to "at-risk" newborns to shield the babies from diarrhea-causing bacteria.[10]

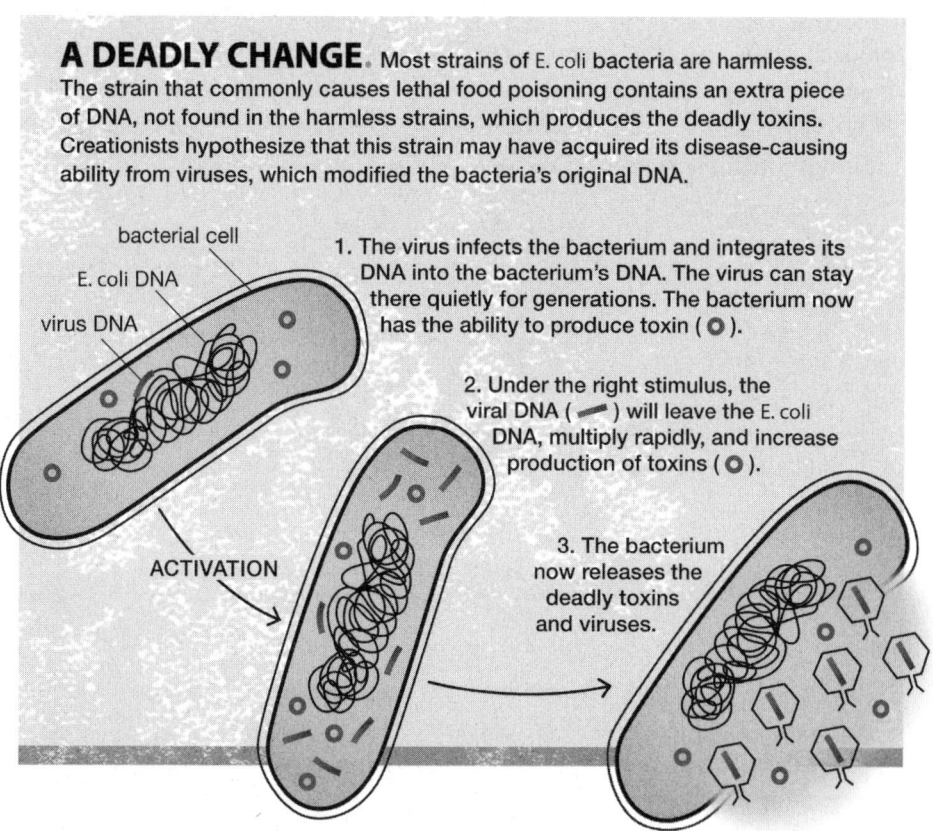

A DEADLY CHANGE. Most strains of E. coli bacteria are harmless. The strain that commonly causes lethal food poisoning contains an extra piece of DNA, not found in the harmless strains, which produces the deadly toxins. Creationists hypothesize that this strain may have acquired its disease-causing ability from viruses, which modified the bacteria's original DNA.

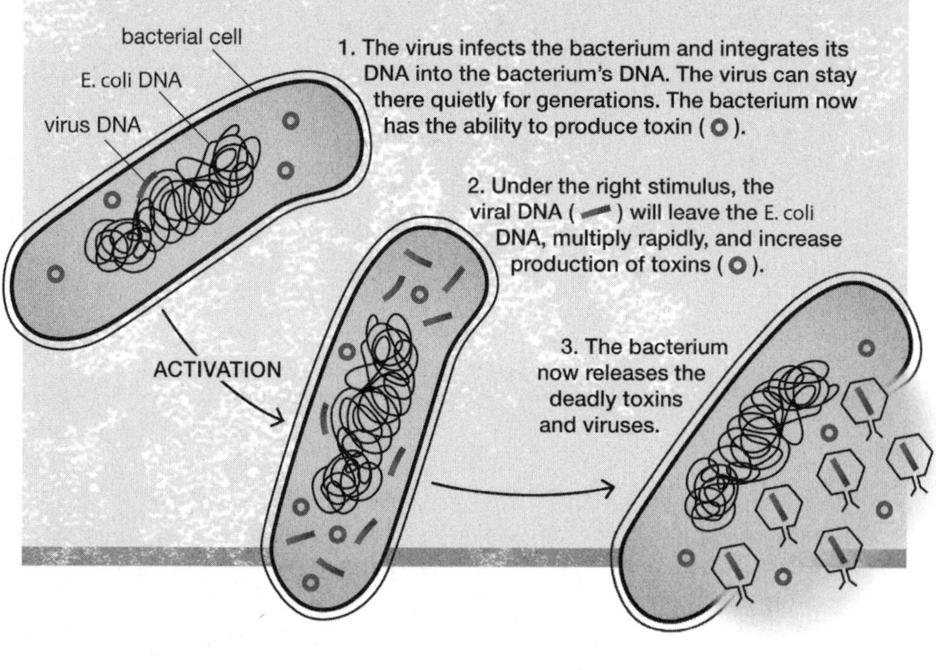
bacterial cell
E. coli DNA
virus DNA
1. The virus infects the bacterium and integrates its DNA into the bacterium's DNA. The virus can stay there quietly for generations. The bacterium now has the ability to produce toxin (O).
2. Under the right stimulus, the viral DNA (—) will leave the E. coli DNA, multiply rapidly, and increase production of toxins (O).
ACTIVATION
3. The bacterium now releases the deadly toxins and viruses.

9. C. Zimmer, *Microcosm: E. coli and the New Science of Life* (New York, NY: Pantheon, 2008).
10. See reference 8.

New Treatment Ideas

This concept, that intestinal health depends on the presence of beneficial bacteria, forms the basis of an entirely new area of medicine called probiotics. Several over-the-counter products are now available that may help boost beneficial bacteria populations in the gut.

Also, a new theory in medicine called the hygiene hypothesis is based on this idea. The proposal is that humans should be exposed to microbes early in life; and if they are not, a variety of disease conditions may result, including asthma, multiple sclerosis, and colitis.

Scripture clearly shows that plants, which now have thorns and thistles as a result of the Fall, once had only good functions. Considering this example, Christians can begin to imagine how all of God's creatures once had beneficial roles; and perhaps, in some cases, this knowledge can be used to fight disease.

At least one creation researcher is already investigating such ideas and is proposing ways in which certain bacteria can be used to fight cancer.[11] This kind of medical treatment represents a very promising and exciting new area of research. Best of all, it flows from our understanding of God's beneficent creation, which He graciously allows to persist in a fallen world.

---

11. Luke Kim, "Bacterial Attenuation and Its Link to Innate Oncolytic Potention," *Answers Research Journal* 1 (2008): 117–122.

32

Unicorns in the Bible?

DR. ELIZABETH MITCHELL

Some people claim the Bible is a book of fairy tales because it mentions unicorns. However, the biblical unicorn was a real animal, not an imaginary creature. The Bible refers to the unicorn in the context of familiar animals, such as peacocks, lambs, lions, bullocks, goats, donkeys, horses, dogs, eagles, and calves (Job 39:9–12).[1] In Job 38–41, God reminded Job of the characteristics of a variety of impressive animals He had created, showing Job that God was far above man in power and strength.[2]

Job had to be familiar with the animals on God's list for the illustration to be effective. God points out in Job 39:9–12 that the unicorn, "whose strength is great," is useless for agricultural work, refusing to serve man or "harrow (plow) the valley." This visual aid gave Job a glimpse of God's greatness. An imaginary fantasy animal would have defeated the purpose of God's illustration.

Modern readers have trouble with the Bible's unicorns because we forget that a single-horned feature is not uncommon on God's menu for animal design. (Consider the rhinoceros and narwhal.) The Bible describes unicorns skipping like calves (Psalm 29:6), traveling like bullocks, and bleeding when they die

1. In addition to Job 39:9–10, the unicorn is mentioned in Numbers 23:22, 24:8; Deuteronomy 33:17; Psalm 22:21, 29:6, 92:10; Isaiah 34:7.
2. In Job, God's list of impressive real animals goes on to discuss peacocks, ostriches, horses, hawks, and eagles. God builds up to a crescendo, commanding Job to look at the behemoth, which He had created on the same day He created man (Job 40:15). The behemoth's description matches that of a sauropod dinosaur. Following the behemoth, the list concludes with the leviathan, a powerful, fiery sea creature. See "Could Behemoth Have Been a Dinosaur?" www.answersingenesis.org/behemoth.

(Courtesy: Domenichino, Virgin and Unicorn, [working under Annibale Carracci], Fresco, 1604 – 1605, Farnese Palace, Rome)

(Isaiah 34:7). The presence of a very strong horn on this powerful, independent-minded creature is intended to make readers think of strength.

The absence of a unicorn in the modern world should not cause us to doubt its past existence. (Think of the dodo bird. It does not exist today, but we do not doubt that it existed in the past.) Eighteenth century reports from southern Africa described rock drawings and eyewitness accounts of fierce, single-horned, equine-like animals. One such report describes "a single horn, directly in front, about as long as one's arm, and at the base about as thick. . . . [It] had a sharp point; it was not attached to the bone of the forehead, but fixed only in the skin."[3]

The *elasmotherium*, an extinct giant rhinoceros, provides another possibility for the unicorn's identity. The *elasmotherium*'s 33-inch-long skull has a huge bony protuberance on the frontal bone consistent with the support structure for a massive horn.[4] In fact, archaeologist Austen Henry Layard, in his 1849 book *Nineveh and Its Remains*, sketched a single-horned creature from an obelisk in company with two-horned bovine animals; he identified the single-horned animal as an Indian rhinoceros.[5] The biblical unicorn could have been the *elasmotherium*.[6]

Assyrian archaeology provides one other possible solution to the unicorn identity crisis. The biblical unicorn could have been an aurochs (a kind of wild ox known to the Assyrians as rimu).[7] The aurochs's horns were symmetrical and

3. Edward Robinson, ed., *Calmet's Dictionary of the Holy Bible* revised edition (Boston, MA: Crocker and Brewster, 1832), p. 907–908.
4. The report in *Nature* described a 33-inch-long skull with a bony frontal protuberance more than three feet in circumference. This bony protuberance with its associated structures is thought to have supported a horn over a yard long. Norman Lockyer, "The Elasmotherium," *Nature: International Weekly Journal of Science*, August 8, 1878, p. 388.
5. Austen Henry Layard, *Nineveh and Its Remains* (London: John Murray, 1849), p. 435.
6. A margin note on Isaiah 34:7 placed in the King James Version in 1769 mentions this possible identity, and the Latin Vulgate translates the same Hebrew word as "unicorn" in some contexts and "rhinoceros" in others.
7. *Aurochs* is both singular and plural, like *sheep*.

often appeared as one in profile, as can be seen on Ashurnasirpal II's palace relief and Esarhaddon's stone prism.[8] Fighting rimu was a popular sport for Assyrian kings. On a broken obelisk, for instance, Tiglath-Pileser I boasted of slaying them in the Lebanese mountains.[9]

Extinct since about 1627, aurochs, *Bos primigenius*, were huge bovine creatures.[10] Julius Caesar described them in his *Gallic Wars* as:

> . . . a little below the elephant in size, and of the appearance, color, and shape of a bull. Their strength and speed are extraordinary; they spare neither man nor wild beast which they have espied. . . . Not even when taken very young can they be rendered familiar to men and tamed. The size, shape, and appearance of their horns differ much from the horns of our oxen. These they anxiously seek after, and bind at the tips with silver, and use as cups at their most sumptuous entertainments.[11]

The aurochs's highly prized horns would have been a symbol of great strength to the ancient Bible reader.

One scholarly urge to identify the biblical unicorn with the Assyrian aurochs springs from a similarity between the Assyrian word *rimu* and the Hebrew word *re'em*. We must be very careful when dealing with anglicized transliterated words from languages that do not share the English alphabet and phonetic structure.[12] However, similar words in Ugaritic and Akkadian (other languages of the ancient Middle East) as well as Aramaic mean "wild bull" or "buffalo," and an Arabic cognate means "white antelope."

However, the linguistics of the text cannot conclusively prove how many horns the biblical unicorn had. While modern translations typically translate *re'em* as "wild ox," the King James Version (1611), Luther's German Bible (1534), the Septuagint, and the Latin Vulgate translated this Hebrew word with words meaning "one-horned animal."[13]

---

8. Viewable at www.britishmuseum.org.
9. Algernon Heber-Percy, *A Visit to Bashan and Argob* (London: The Religious Tract Society, 1895), p. 150.
10. Brittanica Concise Encyclopedia, 2007, s.v. "Aurochs."
11. Julius Caesar, *Gallic Wars*, Book 6, chapter 28, www.classics.mit.edu/Caesar/gallic.6.6.html.
12. Elizabeth Mitchell, "Doesn't Egyptian Chronology Prove That the Bible Is Unreliable?" in *The New Answer Book 2*, Ken Ham, ed. (Green Forest, AR: Master Books, 2008), p. 245–264.
13. Some writers who hold to the two-horned identity think that the KJV translators substituted the plural *unicorns* for the singular *an unicorn* in Deuteronomy 33:17 because they were uncomfortable with the idea of a two-horned unicorn. However, the

The importance of the biblical unicorn is not so much its specific identity — much as we would like to know — but its reality. The Bible is clearly describing a real animal. The unicorn mentioned in the Bible was a powerful animal possessing one or two strong horns — not the fantasy animal that has been popularized in movies and books. Whatever it was, it is now likely extinct like many other animals. To think of the biblical unicorn as a fantasy animal is to demean God's Word, which is true in every detail.

KJV translators themselves noted the literal translation an unicorn in their own margin note. They likely chose the plural rendering to fit the context of the verse. Deuteronomy 33:17 states, "His [Joseph's] glory is like the firstling of his bullock, and his horns are like the horns of unicorns: with them he shall push the people together to the ends of the earth: and they are the ten thousands of Ephraim, and they are the thousands of Manasseh" (KJV). The verse compares the tribal descendants of Joseph's "horns," meaning descendants of his two sons Ephraim and Manasseh, with the strong horns of unicorns. "Horns" is plural because there are two sons in view, and "unicorn" is referenced because the unicorn's horn is so incredibly strong.

33

Doesn't the Bible Support Slavery?

PAUL TAYLOR AND BODIE HODGE

The issue of slavery usually conjures up thoughts of the harsh "race-based" slavery that was common by Europeans toward those of African descent in the latter few centuries. However, slavery has a much longer history and needs to be addressed biblically.[1]

Some "white"[2] Christians have used the Bible to convince themselves that owning slaves is okay and that slaves should obey their "earthly masters." Regrettably and shamefully, "white" Christians have frequently taken verses of Scripture out of context to justify the most despicable acts. In some cases, it could be argued that these people were not really Christians; they were not really born again but were adhering to a form of Christianity for traditional or national reasons. Nevertheless, we have to concede that there are genuine "white" Christians who have believed the vilest calumnies about the nature of "black" people and have sought support for their disgraceful views from the pages of the Bible.

But what does the Bible really teach?

1. It should be noted that Answers in Genesis strongly opposes both racism and slavery.
2. We are using the term "white" to refer to peoples of European origin and "black" to refer to peoples primarily of African origin. We are actually not too thrilled about these terms either since all people are really the same color just different shades, but for the sake of understanding, we will use them in this chapter.

Greek and Hebrew Words for "Slave"

The Hebrew and Greek words used for "slave" are also the same words used for "servant" and "bondservant," as shown by the following table.

	Hebrew, (Old Testament)	Greek, (New Testament)	Meaning
1	עבד ebed		Slave, servant, bondservant
2	עבד abad		Serve, work, labor
3	שפחה shiphchah		Maid, maidservant, slave-girl
4	אמה amah		Maid servant, female slave
5		δουλος doulos	Servant, slave, bondservant
6		συνδουλος sundoulos	Fellow servant, slave
7		παιδισκη paidiske	Bondwoman, maid, female slave

In essence, there are two kinds of slavery described in the Bible: a servant or bondservant who was paid a wage, and the enslavement of an individual without pay. Which types of "slavery" did the Bible condemn?

A Brief History of Slavery

It is important to note that neither slavery in New Testament times nor slavery under the Mosaic covenant have anything to do with the sort of slavery where "black" people were bought and sold as property by "white" people in the well-known slave trade of the last few centuries. No "white" Christian should think that he or she could use any slightly positive comment about slavery in this chapter to justify the historic slave trade, which is still a major stain on the histories of both the United States and the UK.

The United States and the UK were not the only countries in history to delve into harsh slavery and so be stained.

1. The Code of Hammurabi discussed slavery soon after 2242 B.C. (the date assigned by Archbishop Ussher to the Tower of Babel incident).
2. Ham's son Mizraim founded Egypt (still called Mizraim in Hebrew). Egypt was the first well-documented nation in the Bible to have harsh slavery, which was imposed on Joseph, the son of Israel, in 1728 B.C. (according to Archbishop Ussher). Later, the Egyptians were slave masters to the rest of the Israelites until Moses, by the hand of God, freed them.
3. The Israelites were again enslaved by Assyrian and Babylonian captors about 1,000 years later.
4. "Black" Moors enslaved "whites" during their conquering of Spain and Portugal on the Iberian Peninsula in the eighth century A.D. for over 400 years. The Moors even took slaves as far north as Scandinavia. The Moorish and Middle Eastern slave market was quite extensive.
5. Norse raiders of Scandinavia enslaved other European peoples and took them back as property beginning in the eighth century A.D.
6. Even in modern times, slavery is still alive, such as in the Sudan and Darfur.

We find many other examples of harsh slavery from cultures throughout the world. At any rate, these few examples indicate that harsh slavery was/is a reality, and, in all cases, is an unacceptable act by biblical standards (as we will see).

The extreme kindness to be shown to slaves/servants commanded in the Bible among the Israelites was often prefaced by a reminder that they too were slaves at the hand of the Egyptians. In other words, they were to treat slaves/servants in a way that they wanted to be treated.

Slavery in the Bible

But was slavery in the Bible the same as harsh slavery? For example, slaves and masters are addressed in Paul's epistles. The term "slave" in Ephesians 6:5 is better translated "bondservant." The Bible in no way gives full support to the practice of bondservants, who were certainly not paid the first century equivalent of the minimum wage. Nevertheless, they were paid something (Colossians 4:1) and were therefore in a state more akin to a lifetime employment contract rather than "racial" slavery. Moreover, Paul gives clear instructions that Christian "masters" are to treat such people with respect and as equals. Their employment position did not affect their standing in the Church.

Other passages in Leviticus show us the importance of treating "aliens" and foreigners well, and how, if they believe, they become part of the people of God (for example, Rahab and Ruth, to name but two). Also, the existence of slavery in Leviticus 25 underlines the importance of redemption, and enables the New Testament writers to point out that we are slaves to sin, but are redeemed by the blood of Jesus. Such slavery is a living allegory, and does not justify the race-based form of slavery practiced from about the 16th to 19th centuries.

As we already know, harsh slavery was common in the Middle East as far back as ancient Egypt. If God had simply ignored it, then there would have been no rules for the treatment of slaves/bondservants, and people could have treated them harshly with no rights. But the God-given rights and rules for their protection showed that God cared for them as well.

This is often misconstrued as an endorsement of harsh slavery, which it is not. God listed slave traders among the worst of sinners in 1 Timothy 1:10 ("kidnappers/men stealers/slave traders"). This is no new teaching, as Moses was not fond of forced slavery either:

> He who kidnaps a man and sells him, or if he is found in his hand, shall surely be put to death (Exodus 21:16).

In fact, take note of the punishment of Egypt, when the Lord freed the Israelites (Exodus chapters 3–15). God predicted this punishment well in advance:

> Then He said to Abram: "Know certainly that your descendants will be strangers in a land that is not theirs, and will serve them, and they will afflict them four hundred years. And also the nation whom they serve I will judge; afterward they shall come out with great possessions" (Genesis 15:13–14).

Had God not protected slaves/bondservants by such commands, then many people surrounding them who did have harsh slavery would have loved to move in where there were no governing principles as to the treatment of slaves. It would have given a "green light" to slave owners from neighboring areas to come and settle there. But with the rules in place, it discouraged such slavery in their realm.

In fact, the laws and regulations over slavery are a sure sign that slavery isn't good in the same way the Law came to expose and limit sin (Romans 5:13). One reverend explained it this way:

> In giving laws to regulate slavery, God is not saying it is a good thing. In fact, by giving laws about it at all, He is plainly stating it is a bad thing. We don't make laws to limit or regulate good things. After all, you won't find laws that tell us it is wrong to be too healthy or that if water is too clean we have to add pollution to it. Therefore, the fact slavery is included in the regulations of the Old Testament at all assumes that it is a bad thing which needs regulation to prevent the damage from being too great.[3]

Does the Bible Support Harsh Slavery?

There are several passages that are commonly used to suggest that the Bible condones harsh slavery. However, when we read these passages in context, we find that they clearly oppose harsh slavery.

> If you buy a Hebrew servant, he shall serve six years; and in the seventh he shall go out free and pay nothing. If he comes in by himself, he shall go out by himself; if he comes in married, then his wife shall go out with him. If his master has given him a wife, and she has borne him sons or daughters, the wife and her children shall be her master's, and he shall go out by himself. But if the servant plainly says, "I love my master, my wife, and my children; I will not go out free," then his master shall bring him to the judges. He shall also bring him to the door, or to the doorpost, and his master shall pierce his ear with an awl; and he shall serve him forever (Exodus 21:2–6).

This is the first type of bankruptcy law we've encountered. With this, a government doesn't step in, but a person who has lost himself or herself to debt can sell the only thing they have left: their ability to perform labor. This is a loan. In six years the loan is paid off, and they are set free. Bondservants who did this made a wage, had their debt covered, had a home to stay in, on-the-job training, and did it for only six years. This almost sounds better than college, which doesn't cover debt and you have to pay for it!

Regarding Exodus 21:4, if he (the bondservant) is willing to walk away from his wife and kids, then it is his own fault. And he would be the one in

3. Personal correspondence with Reverend Mathew Anderson, Ottumwa, Iowa, 2/3/2007.

defiance of the law of marriage. He has every right to stay with his family. On the other hand, his wife, since she is a servant as well, must repay her debt until she can go free. Otherwise, a woman could be deceitful by racking up debt and then selling herself into slavery to have her debts covered, only to marry someone with a short time left on his term, and then go free with him. That would be cruel to the master who was trying to help her out. So this provision is to protect those who are trying to help people out of their debt.

This is not a forced agreement either. The bondservants enter into service on their own accord. In the same respect, a foreigner can also sell himself or herself into servitude. Although the rules are slightly different, it would still be by their own accord in light of Exodus 21:16 above.

> If men contend with each other, and one strikes the other with a stone or with his fist, and he does not die but is confined to his bed, if he rises again and walks about outside with his staff, then he who struck him shall be acquitted. He shall only pay for the loss of his time, and shall provide for him to be thoroughly healed. And if a man beats his male or female servant with a rod, so that he dies under his hand, he shall surely be punished. Notwithstanding, if he remains alive a day or two, he shall not be punished; for he is his property (Exodus 21:18–21).

This passage follows closely after Moses' decree against slave traders in Exodus 21:16. We include verses 18 and 19 to show the parallel to servants among the Israelites. The rules still apply for their protection if they already have servants or if someone sells himself or herself into service.

Regarding Exodus 21:20–21, consider that many of those who sold themselves into servitude were those who had lost everything, indicating that they were often times the "lazy" ones. In order to get them up to par on a working level, they may require discipline. And the Bible does say to give discipline — even fathers were to give their children "the rod;" to withhold it is considered unloving (Proverbs 13:24, 23:13). So beating with a rod (or more appropriately "a branch") is not harsh, but required for discipline. Even the Apostle Paul reveals he was beaten with a rod three times (2 Corinthians 11:25), and he didn't die from it. In fact, the equivalent in today's culture (spanking) was commonplace in public schools until just a few years ago. Only recently has this been deemed "inappropriate."

According to verses 20–21, if an owner severely beat his servant, and the servant died, then he would be punished — that was the law. However, if the servant survived for a couple of days, it is probable that the master was punishing him and not intending to kill him, or that he may have died from another cause. In this case there is no penalty other than that the owner loses the servant who is his temporary property — he suffers the loss.[4]

Some have also complained that God is sexist in his treatment of servants (though sexism is outside the realm of this chapter, we will still address this claim).

> If a man sells his daughter as a servant, she is not to go free as menservants do. If she does not please the master who has selected her for himself, he must let her be redeemed. He has no right to sell her to foreigners, because he has broken faith with her (Exodus 21:7–8; NIV).

There is a stark delineation between male servants and the female servants in Exodus 21:7. A Hebrew male could sell himself into servitude for his labor (to cover his debts, etc.) and be released after six years. A Hebrew female could be sold into servitude, with permission of her father, not for labor purposes but for marriage. Verse 8 discusses breaking faith with her, which means that they have entered into a marriage covenant (see Malachi 2:14). If God approved of the female leaving in six years, then marriage is no longer a life-long covenant. So God is honoring the sanctity of marriage here.

Imagine what would happen if this rule wasn't in place. It would mean that men would have the free reign to marry a woman for six years and then "trade" her in for another woman. This is not approved of in the Bible. Of course, when a man buys a male servant, they are not married, and so the male servants were to be set free.

> I am the Lord your God, who brought you out of the land of Egypt, to give you the land of Canaan and to be your God. And if one of your brethren who dwells by you becomes poor, and sells himself to you, you shall not compel him to serve as a slave. As a hired servant and a sojourner he shall be with you, and shall serve you until the Year of Jubilee. And then he shall depart from you — he and his children with him — and shall return to his own family. He shall return to the possession of his fathers. For they are My servants, whom I brought out of the land of Egypt; they shall not be sold as slaves. You shall not rule over him with rigor, but you shall fear your

4. There seems to be some debate as to the proper translation of verse 21. Several versions (NIV, HCSB, NLT) translate it as ". . . if the servant recovers after a day or two," rather than "remains alive a day or two." If this is the proper translation, it obviously makes this a moot point.

> God. And as for your male and female slaves whom you may have — from the nations that are around you, from them you may buy male and female slaves. Moreover you may buy the children of the strangers who dwell among you, and their families who are with you, which they beget in your land; and they shall become your property. And you may take them as an inheritance for your children after you, to inherit them as a possession; they shall be your permanent slaves. But regarding your brethren, the children of Israel, you shall not rule over one another with rigor (Leviticus 25:38–46).

God prefaces this passage specifically with a reminder that the Lord saved them from their bondage of slavery in Egypt. Again, if one becomes poor, he can sell himself into slavery/servitude and be released as was already discussed.

Verse 44 discusses slaves that they may *already* have from nations around them. They can be bought and sold. It doesn't say to seek them out or have forced slavery. Hence, it is not giving an endorsement of seeking new slaves or encouraging the slave trade. At this point, the Israelites had just come out of slavery and were about to enter the Holy Land. They shouldn't have had many servants. Also, this doesn't restrict other people in cultures around them from selling themselves as bondservants. But as discussed already, there are passages for the proper and godly treatment of servants/slaves.

Sadly, some Israelite kings later tried to institute forced slavery, for example Solomon (1 Kings 9:15) and Rehoboam with Adoniram (1 Kings 12:18). Both fell from favor in God's sight and were found to follow after evil (1 Kings 11:6; 2 Chronicles 12:14).

> Blessed is that servant whom his master will find so doing when he comes. Truly, I say to you that he will make him ruler over all that he has. But if that servant says in his heart, "My master is delaying his coming," and begins to beat the male and female servants, and to eat and drink and be drunk, the master of that servant will come on a day when he is not looking for him, and at an hour when he is not aware, and will cut him in two and appoint him his portion with the unbelievers. And that servant who knew his master's will, and did not prepare himself or do according to his will, shall be beaten with many stripes. But he who did not know, yet committed things deserving of stripes, shall be beaten with few. For everyone to whom much is given, from him much will be required; and to whom much has been committed, of him they will ask the more (Luke 12:43–48).

As for Jesus's supposed support for beating slaves, this is in the context of a parable. Parables are stories Jesus told to help us understand spiritual truths. For example, in one parable Jesus likens God to a judge (Luke 18:1–5). The judge is unjust, but eventually gives justice to the widow when she persists. The point of that story was not to tell us that God is like an unjust judge — on the contrary, He is completely just. The point of the parable is to tell us to be persistent in prayer. Similarly, Luke 12:47–48 does not justify beating slaves. It is not a parable telling us how masters are to behave. It is a parable telling us that we must be ready for when Jesus Himself returns. One will be rewarded with eternal life through Christ, or with eternal punishment (Matthew 25:46).

> Bondservants, be obedient to those who are your masters according to the flesh, with fear and trembling, in sincerity of heart, as to Christ; not with eyeservice, as men–pleasers, but as bondservants of Christ, doing the will of God from the heart, with goodwill doing service, as to the Lord, and not to men, knowing that whatever good anyone does, he will receive the same from the Lord, whether he is a slave or free. And you, masters, do the same things to them, giving up threatening, knowing that your own Master also is in heaven, and there is no partiality with Him (Ephesians 6:5–9).

Again, Paul in Ephesians is not giving an endorsement to slavery/bondservants and masters, but gives them both the same commands, showing that God views them as equals in Christ. Again, bondservants were to be paid fair wages:

> Masters, give your bondservants what is just and fair, knowing that you also have a Master in heaven (Colossians 4:1).

Christians Led the Fight to Abolish Slavery

The slavery of "black" people by "white" people in the 16th to 19th centuries (and probably longer) was harshly unjust, like many cultures before. This harsh slavery is not discussed in Moses' writings because such slavery was forbidden in Hebrew culture. This is not surprising. Paul tells us in Romans 1:30 that people are capable of inventing new ways of doing evil. Peter even reveals that some slave owners were already being disobedient and treating slaves/bondservants harshly (1 Peter 2:18). Of course, the Bible gives no endorsement of such treatment. "White" on "black" slavery was opposed by Christians such as William Wilberforce, but not by examining passages on slavery because the slaveries were of different types.[5] "Racial" slavery was opposed because it was seen to be contrary to the value that God places on every human being, and the fact that God "has made from one blood every nation of men to dwell on all the face of the earth" (Acts 17:26). The last letter that the revival evangelist John Wesley ever wrote was to William Wilberforce, encouraging Wilberforce in his endeavors to see slavery abolished. In the letter, Wesley describes slavery as "execrable villainy."

> Reading this morning a tract wrote by a poor African, I was particularly struck by that circumstance that a man who has a black skin, being wronged or outraged by a white man, can have no redress; it being a "law" in our colonies that the oath of a black against a white goes for nothing. What villainy is this?[6]

Wesley concentrated on the value of a man, irrespective of the color of his skin. It is this principle of the value God places on human beings — a biblical principle — which was Wesley's motivation in opposing slavery.

The famous hymnwriter John Newton at one time actually captained slave ships. He did so even after his conversion to Christianity, because he was influenced by the prevailing attitudes of his society; it took time for him to realize his errors. But realize them he did — and he spent the latter part of his life campaigning against slavery. He wrote movingly and disturbingly of the suffering of slaves in the ships' galleys in his pamphlet "Thoughts upon the African Slave Trade."

> If the slaves and their rooms can be constantly aired, and they are not detained too long on board, perhaps there are not many who die; but the contrary is often their lot. They are kept down, by the weather, to breathe a hot and corrupted air, sometimes for a week: this added to the galling of their irons, and the despondency which seizes their spirits when thus confined, soon becomes fatal. . . . I believe, upon an average between the more healthy, and the more sickly voyages, and including all contingencies, one fourth of the whole purchase may be allotted to the article of mortality: that is, if the English ships purchase *sixty thousand* slaves annually, upon the

5. Paul Taylor, "William Wilberforce: A Leader for Biblical Equality," *Answers* magazine, December 2006, online at www.answersingenesis.org/articles/am/v2/n1/william-wilberforce.
6. John Wesley's letter to William Wilberforce, February 24, 1791. Wesley died six days later.

> whole extent of the coast, the annual loss of lives cannot be much less than *fifteen thousand.* [7]

Like Wesley, it was the biblical value of human life which was the deciding factor in Newton's opposition to slavery in his latter years.

The use of the term "one blood" in Acts 17:26 is very significant. If "races" were really of different "bloods," then we could not all be saved by the shedding of the blood of one Savior. It is because the entire human race can be seen to be descended from one man — Adam — that we know we can trust in one Savior, Jesus Christ (the "Last Adam").

Many other Christians could be named in the fight to abolish slavery, which seemed to culminate with Abraham Lincoln in the mid 1800s (slavery was one of the reasons for the Civil War in the United States).

Is the Bible Racist?

Some "white" Christians have assumed that the so-called "curse of Ham" (Genesis 9:25) was to cause Ham's descendents to be black and to be cursed. While it is likely that African peoples are descended from Ham (Cush, Phut, and Mizraim), it is not likely that they are descended from Canaan (the curse was actually declared on Canaan, not Ham).

However, there is no evidence from Genesis that the curse had anything to do with skin color. Others have suggested that the "mark of Cain" in Genesis 4 was that he was turned dark-skinned. Again, there is no evidence of this in Scripture, and in any case, Cain's descendants were more or less wiped out in the Flood.

Incidentally, the use of such passages to attempt to justify some sort of evil associated with dark skin is based on an assumption that the other characters in the accounts were light-skinned, like "white" Anglo-Saxons today. That assumption can also not be found in Scripture, and is very unlikely to be true. Very light skin and very dark skin are actually the *extremes* of skin color, caused by the minimum and maximum of melanin production, and are more likely, therefore, to be the genetically selected results of populations moving away from each other after the Tower of Babel incident recorded in Genesis 11.

The issue of racism is just one of many reasons why Answers in Genesis opposes evolution. Darwinian evolution can easily be used to suggest that some "races" are more evolved than others, that is, the common belief is that "blacks" are less evolved. Biblical Christianity cannot be used that way — unless it is twisted by people who have deliberately misunderstood what the Bible actually

7. J. Newton, *Thoughts upon the African Slave Trade*, 1787.

teaches. On top of this, rejecting the Bible, a book that is not racist, because one may think evolution is superior is a sad alternative. Recall Darwin's prediction of non-white "races":

> At some future period, not very distant as measured by centuries, the civilized races of man will almost certainly exterminate and replace the savage races throughout the world. At the same time the anthropomorphous apes . . . will no doubt be exterminated. The break between man and his nearest allies will then be wider, for it will intervene between man in a more civilized state, as we may hope, even than the Caucasian, and some ape as low as a baboon, instead of as now between the negro or Australian [aborigine] and the gorilla.[8]

Conclusion

Though this short chapter couldn't delve into every verse regarding slavery, the basic principles are the same. In light of what we've learned, here are a few pointers to remember:

1. Slaves under the Mosaic Law were different from the harshly treated slaves of other societies; they were more like servants or bondservants.
2. The Bible doesn't give an endorsement of slave traders but just the opposite (1 Timothy 1:10). A slave/bondservant was acquired when a person voluntarily entered into it when he needed to pay off his debts.
3. The Bible recognizes that slavery is a reality in this sin-cursed world and doesn't ignore it, but instead gives regulations for good treatment by both masters and servants and reveals they are equal under Christ.
4. Israelites could sell themselves as slaves/bondservants to have their debts covered, make a wage, have housing, and be set free after six years. Foreigners could sell themselves as slaves/bondservants as well.
5. Biblical Christians led the fight to abolish harsh slavery in modern times.

8. Charles Darwin, *The Descent of Man*, 2nd ed. (New York, NY: A.L. Burt, 1874), p. 178.

34

Why Did God Make Viruses?

DR. JEAN K. LIGHTNER

There are some fundamental differences in how creationists and evolutionists view life. Biblical creationists believe that God created various forms of life according to their kinds with the ability to reproduce and fill the earth (Genesis 1:21, 22, 24–28). This view includes the concepts that God had purpose in what He created and that it originally was very good (Isaiah 45:18; Genesis 1:31).

In contrast, evolutionists view life as all descending from a single common ancestor by chance processes. Evolutionary arguments tend to imply that life isn't really very complex or well designed. For example, 100 years ago a cell was promoted as being nothing more than a blob of protoplasm, implying that it wouldn't be difficult for it to arise by chance. This proved to be wrong; cells are incredibly complex structures.[1] At one time evolutionists argued that organs or structures with no known function actually had no function; at the time this included hundreds of organs and structures in the human body. Instead these were believed to be vestiges of evolution. This argument has become rather vestigial itself, as these organs have been found to have function.[2]

1. C. Wieland, "Chemical Soup Is Not Your Ancestor!" *Creation* 16 no. 2 (1994):46–47; see Harvard video, *Inner life of a Cell* at www.multimedia.mcb.harvard.edu/media.html.
2. D. DeWitt, "Setting the Record Straight on Vestigial Organs," www.answersingenesis.org/articles/aid/v3/n1/setting-record-straight-vestigial; see chapter 24, "Vestigial Organs — Evidence for Evolution?" in this volume.

Yet this argument reappeared in genetics. Most of the DNA in our bodies does not code for proteins, so it was labeled "junk DNA" by evolutionists who assumed it has no function. As research continues, it is becoming clear that this DNA has numerous essential functions.[3] The evolutionary worldview has a dismal track record for anticipating the astounding complexity in life uncovered by scientific research.

If God created everything good and with a purpose, why are there disease-causing bacteria and viruses in the world? It is true that we first learned about bacteria and viruses because of the problems they cause. Bacteria have been studied in considerable detail and are now recognized to be mainly helpful and absolutely essential for life on earth; bacteria that cause disease (which developed as a result of the Fall) are the exceptions, not the rule.[4] But what about viruses: what purpose could they possibly have?

What Is a Virus?

Viruses are a bit of an enigma. They contain DNA or RNA that are found in all living things. This is packaged in a protein coat. Despite this, viruses are not usually considered living because they are not made up of cells and cannot reproduce by themselves. Instead, the virus will inject the DNA or RNA into a living cell, and the cell will make copies of the virus and assemble them so they can spread.[5]

Viruses vary considerably in their ability to cause disease. Many known viruses are not associated with disease at all. Others cause mild symptoms that may often go undetected. Some, like the HIV virus that causes AIDS in people, appear to have come from another species where they do not cause disease. Given our current knowledge of viruses, it is quite reasonable to believe that disease-causing viruses are descended from viruses that were once not harmful.[6] It has been suggested that they have played an important role in maintaining life on earth — somewhat similar to the way bacteria do.[7] In fact, they may play a role in solving an intriguing puzzle that faces creationists.

3. G. Purdom, "'Junk' DNA — Past, Present, and Future, Part 1," www.answersingenesis.org/articles/aid/v2/n1/junk-dna-part-1; J. Lightner, "The Smell of Change in Our Understanding of Pseudogenes," www.answersingenesis.org/articles/aid/v3/n1/smell-of-change-pseudogenes.
4. See chapter 31, "What About Bacteria?" in this volume.
5. J. Bergman, "Did God Make Pathogenic Viruses?" *Technical Journal* 13 no. 1 (1999): 115–125.
6. J.R. Lucas and T.C. Wood, "The Origin of Viral Disease: A Foray into Creationist Virology," in *Exploring the History of Life: Proceedings of the Fifth BSG Conference* and *Occasional Papers of the BSG* 8 (2006): 13.
7. Bergman, "Did God Make Pathogenic Viruses?"; see chapter 31, "What About Bacteria?" in this volume.

A Creationist Puzzle

The biblical record tells of a global Flood when all created kinds of unclean[8] land animals were reduced to a population of two, the pair that was preserved with Noah on the ark (Genesis 7). After the Flood, these animals reproduced and filled the earth again (Genesis 8:15–19). Today many of these kinds are represented by whole families. For example, the dog family (Canidae) is believed to represent a created kind.[9] However, this is a very diverse group of animals. There are foxes that are adapted to living in the arctic, and others that live in the desert. There is incredible variety seen in modern domestic dog breeds. Where did all this variety come from? And how could it arise so quickly given that the Flood occurred around 4,300 years ago?[10]

The answer to this puzzle is probably quite complex. Some of the variety would have been carried by the pair of animals on the ark. When parents pass traits on to their offspring, these traits can appear in new combinations in the offspring (Mendelian genetics). Natural selection can weed some existing traits out of a population. However, a close examination reveals that genetic changes have also arisen in this time.[11] Many of these changes do not appear accidental and do not directly cause disease. For this reason, some creationists have proposed that God "designed animals to be able to undergo genetic mutations which would enable them to adapt to a wide range of environmental challenges while minimizing risk."[12]

Isn't That Evolution?

It is important to recognize that biologists use several distinct definitions for *evolution* that are often blurred together as if they are synonymous.[13] *Evolution* is sometimes defined as "change in the genetic makeup (or gene frequency) of a population over time." This has been observed; both creationists and evolutionists recognize this as important in building models to help us understand what

8. Unclean animals probably included all non-ruminants. See Leviticus 11; Deuteronomy 14:1–8.
9. T.C. Wood, "The Current Status of Baraminology," *Creation Research Society Quarterly* 43 no. 3 (2006): 149–158.
10. J. Ussher, *The Annals of the World*, L. and M. Pierce, trans. and ed. (Green Forest, AR: Master Books, 2003).
11. This is clear because the two animals on the ark could carry up to four alleles for any one gene. Today there are some genes where considerably more than four alleles exist in animals from the same created kind.
12. J.K. Lightner, "Karyotypic and Allelic Diversity in the Canid Baramin (Canidae)," *Journal of Creation* 23 no. 1 (2009): 94–98.
13. See "An Introduction to Evolution" on the Understanding Evolution website, www.evolution.berkeley.edu/evolibrary/search/topicbrowse2.php?topic_id=41.

likely happened in the past. A second definition of *evolution* involves the idea that all life descended from a common ancestor over millions of years through naturalistic processes. This has *not* been observed. In fact, it is in direct opposition to the testimony God (the eyewitness to creation) gives us in the Bible. The idea that all life has a common ancestor requires the *assumption* that the Bible's history is false, and the *assumption* that changes which do occur could produce the variety of life we see today from a single-celled ancestor.[14]

With regard to the first definition of evolution, creationists and evolutionists differ in the pattern of genetic changes they should expect to see. The creation model predicts that degenerative changes can occur because mankind sinned and brought death into the world (Genesis 3). It also predicts that adaptive changes could occur because God cares for His creation and intends for the earth to be inhabited (Psalm 147:8–9; Matthew 6:25–34; Isaiah 45:18). Both types of changes have been observed. The fact that some foxes are adapted to live in the arctic while others are adapted to live in the desert fits perfectly with this biblical teaching. While evolutionists accept that these types of changes occur, their model requires that most genetic changes add information to the genome. This pattern has *not* been observed. Without this pattern, they cannot account for the many organs and complex biochemical pathways that exist in animals today.[15] Scientific observations show that there is an overall pattern of decay seen in the genome, which is the opposite of what the evolutionary model would predict.[16]

Another difference is the source of the genetic change. Evolutionists assume that random mutations and natural selection can account for the genetic changes that are seen. Since the underlying mechanism is naturalistic, changes were expected to be very slow. Contrary to their expectations, rapid adaptation has been observed,[17] and evolutionists have had to adjust their thinking to accept this. Furthermore, detailed studies of the pattern in genetic differences within related animals don't make sense if mutations are assumed to always be essentially random events.[18] Something else is clearly going on here. It appears

14. See "Misconceptions about Evolution and the Mechanisms of Evolution: Evolution and Religion Are Incompatible" on the Understanding Evolution website, www.evolution.berkeley.edu/evolibrary/misconceptions_faq#d1. Note how religious beliefs are said to have nothing to do with the real (material) world; this is in stark contrast with the biblical teaching that God, as the Creator of all, is relevant to every aspect of life.
15. See L. Spetner, *Not By Chance!* (New York, NY: Judaica Press, 1998).
16. See J. Sanford, *Genetic Entropy and the Mystery of the Genome* (Lima, NY: Elim Publishing, 2005).
17. See www.answersingenesis.org/articles/aid/v3/n1/life-designed-to-adapt.
18. J.K. Lightner, "Karyotype Variability within the Cattle Monobaramin," *Answers Research Journal* 1 (2008): 77–88; J.K. Lightner, "Genetics of Coat Color I," *Answers Research Journal* 1 (2008): 109–166.

that God has placed some incredible programming into the genomes of the animals He created, and viruses may play some role in this.

Evidence of Horizontal Gene Transfer

Interestingly, there are some portions of DNA in animals that look like they came from a virus.[19] While some of these were likely originally present in the genome since they have essential functions, others may have been introduced by viruses.[20] A number of years ago, one creationist proposed that horizontal gene flow (genes picked up from somewhere in the environment rather than inherited from parents) may help to explain rapid adaptation and the interesting pattern of DNA in animals. In fact, the author lists 13 different biological phenomena that might be explained by horizontal gene flow.[21] Since viruses carry genetic material (DNA or RNA), they are the most logical agents to suspect in transferring genes. While horizontal gene transfer would not change the identity of an animal (i.e., it would still belong to the same kind), it could rapidly provide a source of genetic variability that allows for rapid adaptation. If this is the case, then viruses were created "good" (as in Genesis 1), with a support role much like bacteria are known to have.

While the evidence is largely circumstantial, further scientific investigation does seem to support these ideas.[22] In fact, a recent *PNAS* article has brought some new information to light. Previous studies had suggested horizontal transfer between closely related species. This study identified a large section of DNA (~2.9 kb) that was approximately 96 percent identical in a marsupial (opossum), several placentals (mouse, rat, bushbaby, tenrec, and little brown bat), a reptile (anole lizard), and an amphibian (African clawed frog). It was absent from the 27 other animals surveyed (which included human and Jamaican fruit bat).

19. Traditionally, this DNA has been assumed to be the result of viral infection. Recently, several creationists have presented evidence that some (RNA) viruses may actually be escapees. In other words, the genes were originally in the DNA of the animals and were able to move around within the cell (by copying on to RNA). At some point the viruses became independent and can now travel between animals. For more on this intriguing idea see Y. Liu. "The Natural History of Retroviruses: Exogenization vs Endogenization" *Answers Research Journal* 2 (2009): 97–106.
20. Y. Liu, "Were Retroviruses Created Good?" *Answers*, October–December 2006, online bonus content, www.answersingenesis.org/articles/am/v1/n2/were-retroviruses-created-good.
21. T. Wood, "The Aging Process: Rapid Post-Flood Intrabaraminic Diversification Caused by Altruistic Genetic Elements (AGES)," *Origins* 54 (2002).
22. T. Wood, "Perspectives on Aging: A Young Earth Creation Diversification Model," in *Proceedings of the Fifth International Conference on Creationism*, Robert L. Ivey, Jr., ed. (Pittsburgh, PA: Creation Science Fellowship, 2003), p. 479–489.

This sequence appears to have been incorporated into an existing functional gene in rats and mice, although its specific function is not yet known.[23] Because of the pattern observed, it appears that horizontal gene transfer was concentrated at some time in the past and perhaps occurred via a DNA virus.[24] Interestingly, several species (anole and opossum) are from Central/South America, several are restricted to Africa (bushbaby, tenrec), and the others have a wider geographical distribution.[25] This suggests that the transfer may have occurred early post-Flood or been intercontinental in scope.[26]

Since most scientists are heavily influenced by the evolutionary worldview, they often miss indicators of purpose. For example, the section of DNA discussed above is a transposon (a type of mobile genetic element or transposable element). After the putative transfer, it was copied and integrated into several different parts of the genome in the various species. This requires that the proper tools (e.g., enzymes) be in place so that the section of DNA can be incorporated into the genome initially, then modified and copied appropriately. Given that decay has occurred over time, it is not surprising to creationists that there are examples of transposons where this process doesn't work properly and disease occurs.

Diseases draw attention and research dollars, so the problems associated with transposons have been recognized before the benefits are understood (much like was true of bacteria). Many people still view these mobile genetic elements as "parasitic" or "selfish." However, they are quite widespread in the genome of plants, animals, and man. If their insertion was always purely "random," it seems they should more consistently cause problems in a complex system such

23. J.K. Pace II et al., "Repeated Horizontal Transfer of a DNA Transposon in Mammals and Other Tetrapods," *PNAS* 105 no. 44 (2008): 17,023–17,028.
24. The authors are evolutionists who carry in the assumption of common ancestry. Although creationists could argue that some kinds were created with these sequences and others were not, it appears more likely that they result from horizontal gene transfer. Also, the authors used evolutionary assumptions to estimate the time the horizontal transfer occurred (which was essentially the same for all species). When this type of estimate was done with mitochondrial DNA, the estimated mutation rate was significantly off compared to actual measured mutation rates. A. Gibbons, "Calibrating the Mitochondrial Clock," *Science* 279 no. 5347 (1998): 28–29.
25. See comment by Cedric, one of the authors of the *PNAS* article, on "Space Invader DNA Jumped Across Mammalian Genomes," www.scienceblogs.com/notrocketscience/2008/11/space_invader_dna_jumped_across_mammalian_genomes.php.
26. The creation model predicts a high concentration of horizontal gene transfer post-Flood as animals were migrating out and filling various ecological niches. There is also a chance that animals on the ark may have already carried these sequences. Further intrabaraminic comparisons may help to clarify the timing of horizontal gene flow for this particular case.

as the genome.[27] Therefore, it seems more logical to believe that transposons have purpose and were designed in a way to benefit their possessor.

The Bible Explains the Paradox

The biblical view explains an important paradox we see in the world around us. It anticipates the complexity that is constantly being uncovered by scientific research; God is an all-wise Creator and would be expected to use awesome design patterns and programming. It also explains the decay observed because mankind sinned and brought death into the world; the world is now in bondage to decay (Romans 8:20–21). This is an exciting time to be a creationist researcher, as the tremendous volume of scientific research is helping to provide answers to questions that have been asked for decades.

27. Some accidental insertions may not cause obvious problems because the genome contains a high amount of redundancy. Redundancy is a hallmark of excellent design that militates against system failure. It is also inconsistent with the notion that life arose by chance. Such accidental insertions do, however, contribute to the deterioration of the genome.

35

Wasn't the Bible Written by Mere Men?

BODIE HODGE

> All Scripture is given by inspiration of God, and is profitable for doctrine, for reproof, for correction, for instruction in righteousness, that the man of God may be complete, thoroughly equipped for every good work (2 Timothy 3:16–17).

A Bigger Problem than You Might Think

It truly is a secular age. I had the opportunity to speak to a student-led club at a government school a couple of years ago. At the end of the lecture, I began answering questions the students had. Even though there was a very negative tone coming from many of the questioners, I remained courteous in each response.

Most of the questions were common ones and fairly easy to answer. The questions began with issues related to the creation-evolution debate, such as dinosaurs and radiometric dating. After those were answered, the questions became more impassioned and were directed toward God and the Bible, such as "Who created God?" and "Isn't the Bible full of contradictions?" At the end, one statement came up that I didn't get to respond to. The bell rang and out they ran. I really wish they had brought this up sooner so I could have responded to the claim that the Bible was written by mere men. We were getting closer to the heart of the issue.

I didn't realize how important this was until I saw a statistic of young people who had walked away from the church. Out of 1,000 young adults surveyed who have left church, 44 percent of them said that they did not believe the accounts in the Bible were true and accurate. When asked what made them answer this way, the most common response (24 percent) was that *the Bible was written by men*. The rest of the results, from that 44 percent, are shown below.[1]

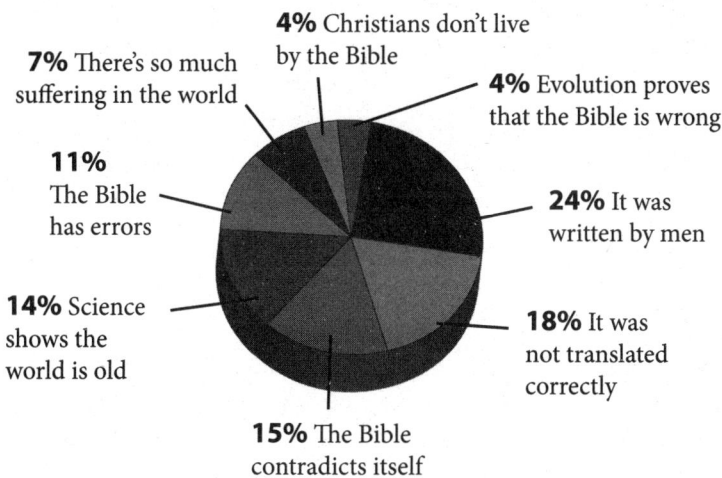

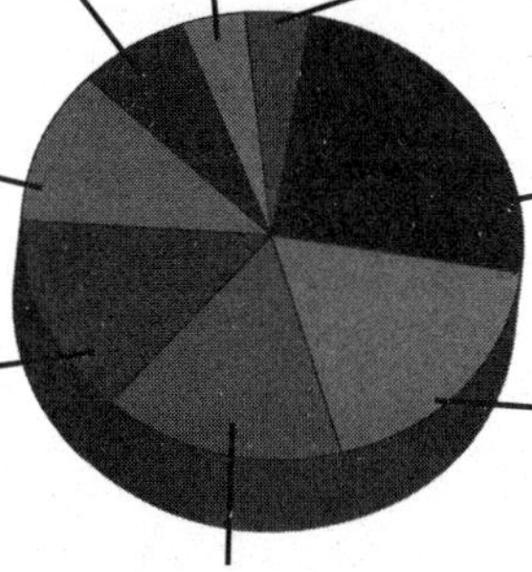

Even though 24 percent directly claimed this, take note that there are related answers such as 11 percent believing the Bible to have errors, which means God could not have been involved since God does not make errors (Psalm 12:6; Deuteronomy 32:4). Also, claiming that the Bible contradicts itself would imply that God was not involved since God cannot deny Himself (2 Timothy 2:13), and thus contradict Himself. So at least 50 percent would, in one way or another, dispute that a perfect God was responsible for the Bible!

So What Is the Answer?

When it comes to the authorship of the Bible, of course men were involved — Christians would be the first to point this out. Paul wrote letters to early churches and these became Scripture. David wrote many of the Psalms, Moses wrote the Pentateuch (the first five books of the Bible), and so on. In fact, it is estimated that over 40 different human authors were involved.[2] So this is not the real issue.

1. Ken Ham and Britt Beemer, *Already Gone* (Green Forest, AR: Master Books, 2009), p. 107.
2. Josh McDowell, *A Ready Defense* (Nashville, TN: Thomas Nelson Publishers, 1993), p. 27.

The real issue is whether God had any involvement in the authorship of the Bible. Let's think about this for a moment. When someone claims that the Bible was written by men *and not God*, this is an absolute statement that reveals something extraordinary. It reveals that the person saying this is claiming to be transcendent! For a person to validate the claim that God did not inspire the human authors of the Bible means he must be omniscient, omnipresent, and omnipotent!

1. *Omniscient:* This person is claiming to be an all-knowing authority on the subject of God's inspiration in order to refute God's claim that Scripture was inspired by Him (2 Timothy 3:16).
2. *Omnipresent:* This person is claiming that he was present, both spiritually and physically, to observe that God had no part in aiding any of the biblical authors as they penned Scripture.
3. *Omnipotent:* This person is claiming that, had God tried to inspire the biblical authors, they had the power to stop such an action.

So the person making the claim that the Bible was merely written by men alone is claiming to be God, since these three attributes belong to God alone. This is a religious issue of humanism versus Christianity. People who make such claims (perhaps unwittingly) are claiming that *they* are the ultimate authority over God and are trying to convince others that God is *subservient* to them. As we respond to claims such as these, this needs to be revealed.

What Is a Good Response?

I like to respond to this claim with a question that reveals this real issue — and there are several ways to do this. For example, referring to omnipresence, you can ask, "Do you really believe that you are omnipresent? The only way for you to make your point that God had no involvement would be if you were omnipresent." Then point out that this person is claiming to be God when he or she makes the statement that God had no involvement in the Bible.

Or, in regard to omnipotence, perhaps ask, "How is it that you are powerful enough to stop God from inspiring the authors?" Or you could direct the question to the rest of the listeners by simply asking, "Since the only way to refute the fact that God inspired the Bible is to use attributes of God such as omnipresence, omnipotence, and omniscience, do the rest of you think this person is God?" You may have to explain it further from this point so the listeners will better understand.

If you are not sure you can remember these types of questions, then remember that you can always lead the person down the path by first asking an easier question such as, "How do you know that God was not involved?" But then you will have to listen carefully to the response to know how to respond after that.

Other responses include undercutting the entire position by pointing out that any type of reasoning apart from the Bible is merely arbitrary. So the person trying to make a logical argument against the claims of the Bible (i.e., that God inspired the authors) is doing so only because he or she is assuming (though unintentionally) the Bible is true and that logic and truth exist! It is good to point out these types of presuppositions and inconsistencies.[3]

Someone may respond and say, "What if I claim that Shakespeare was inspired by God — then you would have to be omniscient, omnipresent, and omnipotent to refute it."

Actually, it is irrelevant *for me* to be omniscient, omnipresent, and omnipotent to refute such a claim. God, who is omniscient, omnipresent, and omnipotent, refutes this claim from what He has already stated in the Bible. Nowhere has God authenticated Shakespeare's writings as Scripture, unlike Christ the Creator-God's (John 1; Colossians 1; Hebrews 1) approval of the Old Testament prophetic works and the New Testament apostolic works — the cap of the canon is already sealed.[4]

Conclusion

Sadly, in today's society, children, whether churched or not, are being heavily exposed to the religion of humanism. This religion reigns in state schools. So it is logical that the younger generations are thinking in terms of humanism and applying that to their view of the Bible.

The student mentioned earlier was applying the religion of humanism (i.e., man, not God, is the authority) to the Bible when he claimed that it was written by men. He viewed himself, and not God, as the authority; and he further reasoned that there is no God at all and therefore the Bible could not have had God's involvement.

Therefore, his statement that the Bible was written by men is merely a religious claim made by a man claiming the attributes of God. It is good to point

3. Jason Lisle, "Feedback: Put the Bible Down," Answers in Genesis, www.answersingenesis.org/articles/2008/12/05/feedback-put-the-bible-down.
4. Bodie Hodge, "A Look at the Canon: How Do We Know that the 66 Books of the Bible Are from God?" Answers in Genesis, www.answersingenesis.org/articles/aid/v3/n1/look-at-the-canon.

this out as many people follow this same thought process, failing to realize the implications most of the time.

> You shall have no other gods before Me (Exodus 20:3).

If one can expose the false religion of humanism, then unbelievers may be more open to realizing that they are being deceived. After all, unbelievers are not the enemy; rather, the false principalities and dark powers that are at work to deceive are the enemy (Ephesians 6:12).

36

Isn't the God of the Old Testament Harsh, Brutal, and Downright Evil?

BODIE HODGE

Have you ever heard questions such as:

How could God kill all the innocent people, even children, in the Flood?

Why would God send Joshua and the Israelites into Canaan to exterminate the innocent Canaanites living in the land?

Do you really believe a loving God would send people to an eternal hell?

This view of God is commonly referred to in the secular media, atheistic books, and so on. There is a common claim that the God of the Old Testament (even in the New Testament) seems very harsh, brutal, and even evil.[1]

An initial response to this claim can simply be, "How can the atheist or non-Christian say God is harsh, brutal, and evil when they deny the Bible, the very book that defines harsh, brutal, and evil?" Even further, in atheistic, materialistic,

1. For example, atheist Richard Dawkins wrote that the God of the Old Testament is, "arguably the most unpleasant character in all fiction: jealous and proud of it; a petty, unjust, unforgiving control-freak; a vindictive, bloodthirsty ethnic cleanser; a misogynistic, homophobic, racist, infanticidal, genocidal, filicidal, pestilential, megalomaniacal, sadomasochistic, capriciously malevolent bully." Richard Dawkins, *The God Delusion* (Boston, MA: Houghton Mifflin Co., 2006), p. 31.

and evolutionary worldviews, such things are neither right nor wrong because there is no God in their view to establish what is right or wrong. The same people who profess to believe in a naturalistic view where animals rape, murder, and eat their own kind are those who attack the loving God of the Bible and try to call Him evil (Isaiah 5:20).

But a closer look at such claims against the God of the Bible shows that these claims have no merit. Claiming that God is evil or harsh is an attack on God's character, and every Christian should be prepared to have an answer for such attacks (1 Peter 3:15).

The intent of many of those who make such claims is to make a good God look evil in order to justify their rejection of Him, His Word, or even His existence. But if God really doesn't exist and the Bible isn't His Word, then those who attack God and His Word by calling Him harsh and evil shouldn't even care to attack Him. By attacking Him, they show that they know He exists and are simply suppressing that knowledge (see Romans 1:20–25). They are trying to justify their rebellion against God. Few that I have spoken with realize that when they attack God's character in an effort to make a case against His existence they are refuting their own position.

Some of the events in the Bible that people commonly use to justify that claim that God is harsh, include events in Genesis such as the Fall of man, the Flood, and the destruction of Sodom and Gomorrah. And then they proceed to the Canaanites, Egyptians, Benjamites, or even non-Christians in general.[2] So Genesis seems to be a good place to begin.

The Fall: Adam and Eve

Often people ask how God could sentence all of mankind to die because of Adam and Eve's sin. Adam and the Eve knew the punishment for sin (Genesis 2:17), but they sinned anyway, going against the plain commandment of God. Adam knowingly sinned (1 Timothy 2:14), so his punishment was brought upon himself. Most people fail to realize, however, that all mankind sinned in Adam as we were in the body of our ancestor when he sinned (Hebrews 7:10). Due to Adam's sin, we also receive a sin nature, and we sin ourselves (Romans 5:12). So we also die because of sin — we are no different from Adam and Eve. However, we should stop to consider the blessing that is found amidst the curse. When Adam and the woman sinned, God offered the first prophecy of Jesus Christ in Genesis 3:15. The curse of sin would be erased by the seed of a woman (the result of a virgin

2. Of course, there are other instances that can be found in Scripture where people may try to claim God is harsh, brutal, or evil, but these examples should suffice to answer this particular issue.

birth) sent to save mankind. A means of salvation was already being offered.

On top of this, the first man and woman should have died right then, but God is patient and gave them a "grace period," covering their sin by sacrificing animals (when He made coats of skins in Genesis 3:21) in their place; sin is punishable by death, so something had to die (Hebrews 9:22). Abel followed this pattern (Genesis 4:4), as did Noah (Genesis 8:20), Abraham (Genesis 22:13), and the Israelites. These animal sacrifices were not sufficient to take away sins (Hebrews 10:4) — only a perfect, sinless sacrifice, fulfilled in the death of Jesus Christ, could (Hebrews 4:15; 9:13–14). It was Christ's sacrifice alone that was sufficient to cover the sins of the whole world (1 John 2:2). The infinite Son died to pay the penalty for the infinite punishment from an infinitely Holy God.

Sacrifices made by the Lord for Adam and Eve to provide them coats of skin (Genesis 3:21)

So there are two blessings so far: a final means of salvation in Christ and a grace period of the penalty for sin being covered instead of bringing about *instant* death. But there is another blessing that few may notice without reading the rest of the Bible. By being sentenced to die, man wouldn't be forced to live in a sin-cursed world for all eternity — this is why the path to the Tree of Life was guarded (Genesis 3:22–24)! By dying in this sin-cursed world with Christ as Savior, one inherits the new heaven and new earth, which are restored to perfection, where there is no Curse, death, or suffering for eternity (Revelation 21:4, 22:3). Death will have no sting (1 Corinthians 15:53–56) for those in Christ.

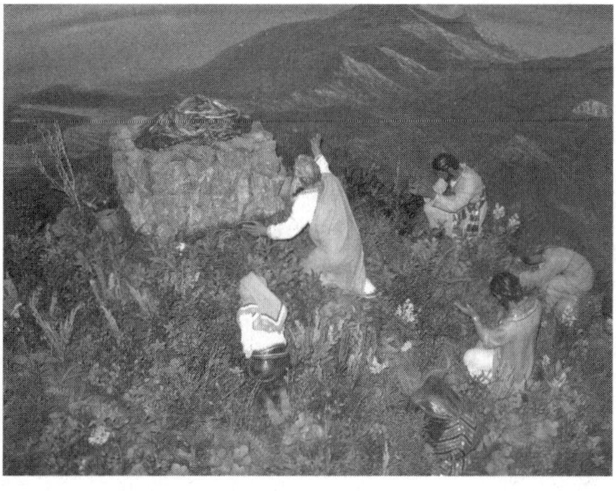

Noah offered sacrifices of clean animals after the Flood (Genesis 8:20–21).

So in this instance, man sinned and God acted justly by punishing that sin, and even went much further by offering three blessings: a grace period, a means of salvation, and a perfect place to live an eternal life without sin, death, or the Curse. Imagine if a thief went before a judge and the judge said, "You have broken the law so you deserve 50 years in jail with no parole, but I will give you 1 year in jail, and you don't have to begin serving that for six months so you may set your affairs in order. After 1 year, I'll give you a guaranteed release, and on top of that, I'll buy you a million-dollar home and prepare it for you." It seems strange that people would say that the judge would be harsh and evil for sentencing the thief to a year in jail. What would be stranger still is if the thief refused the generous offer.

The Flood

God is often attacked for killing "all the innocent people, and even children," in the Flood. In fact, some have specifically said, "But the children . . . how could God kill the little children?" The response: "If the earth was filled with violence and evil, it makes one wonder how many children were still alive anyway. After all, in today's culture, where evil has a foothold, it is children that seem to bear the brunt of much violence (e.g., hundreds of millions of abortions). Even if there were some children left, God provided the ark. Why did the parents of those children refuse to let them board? Why did they insist on putting their children in harm's way? If anyone is to blame, it is the parents and guardians who stopped them from coming to the ark."

Why blame God for something when He provided a means of salvation, which the parents refused? Imagine if a boater came to rescue a woman and her child who were on top of a roof with floodwaters rising. The boater says, "Please get in and I can save you." The woman says, "No, we will stay because I don't believe you." Then the boater patiently waits and even tries to explain what will happen, yet she continues to refuse over and over again. The boater even asks for her to send her child and she still refuses and swats the boater away . . . and then finally they drown. Is it appropriate to blame the boater for the death of the child?

People had the opportunity to come in the ark but they refused.

But consider this, judging Scripture by Scripture, it says that no one is truly innocent (Romans 3:23), and all will eventually die anyway — a repercussion of our own actions (1 Corinthians 15:22; Romans 6:23). Second, what brought such a judgment on the people before the Flood?

> Then the LORD saw that the wickedness of man was great in the earth, and that every intent of the thoughts of his heart was only evil continually (Genesis 6:5).

What a strong statement! *Every* intention and thought was evil *all* the time. Imagine the murders, rapes, thefts, child sacrifices, cannibalism, and so on. This was happening continually. Yet this was about 120 years (maximum) before the Flood (Genesis 6:3). So God was still patient, allowing time for repentance and change (1 Peter 3:20). God even called Noah to be a preacher of righteousness (2 Peter 2:5), yet people still refused to listen and continued in their evil ways.

God even went so far as to offer a way of salvation! He provided an ark through Noah and his family, and yet others didn't come. Only Noah's family was saved (2 Peter 2:5). The means of salvation, preaching of righteousness, and God's patience were there, yet everyone else refused and received their judgment.

As an aside, the claim of children dying in the Flood has always been of interest, especially when skeptics and atheists bring it up. The hypocrisy is astounding since these skeptics and atheists often support the murder of babies as we have seen in the abortion debate. If people really were evil and their thoughts evil all the time, then abortion, child murder, and child sacrifice were likely commonplace. Disobedience to God would likely mean disobeying God's command to be fruitful and multiply (Genesis 1:28). Resisting this command would result in drastically fewer children, so one could wonder if many children were even around at the time of the Flood. Noah himself had no children until he was 500 years old (lending to the view that children may have been few and far between in those days). Even so, children are sinners and can also have evil intentions and thoughts (Romans 3:23). Today, for example, we see children killing children in school, child thieves, rape among children, and so on. But if children and infants didn't make it to the ark (the means of salvation at the time), whose fault is it but their own and/or parents/guardians who refused to let them?! So why blame God when He offered them a means to be saved?

Sodom and Gomorrah

In Genesis 18:20–33, the Lord revealed to Abraham that Sodom and Gomorrah had sinned exceedingly. Their wickedness was not revealed in its entirety, but we are aware of their acts of sodomy (later in the chapter) that had overtaken them in their actions, enough to rape.

Abraham asked if God would sweep away the righteous with the wicked. He asked the Lord if there were 50 righteous, would the Lord spare it; He said yes. He asked the Lord if there were 40 righteous, would the Lord spare it; He said yes. He asked the Lord if there were 30 righteous, would the Lord spare it; He said yes. He asked the Lord if there were 20 righteous, would the Lord spare it; He said yes. He asked the Lord if there were 10 righteous, would the Lord spare it; He said yes.

This reveals how wicked and sinful the people were. They were without excuse and judgment was finally coming. This also reveals something interesting about the Flood. If God would spare Sodom and Gomorrah for only 10 righteous people, then would God have spared the earth if 10 people were righteous before the Flood? It appears that He did. Methuselah and Lamech, Noah's father and grandfather, *may* have been among those that made 10 (along with Noah, his wife, and his three sons and their wives). Of course, there may have been others who were righteous too, up until the Flood. But at the time of the Flood, we can surmise there were only eight (Methuselah and Lamech had died just before the Flood).

Lot and his family numbered less than 10 in Sodom and Gomorrah (Lot, his wife, his two daughters, his two sons-in-law, only made six). Yet, God provided a means of salvation for them — the angels helped them get to safety.

Were there children in Sodom and Gomorrah? The Bible doesn't reveal any, and homosexual behavior was rampant, so there may not have been many, if any, children. Since God made it clear that not even 10 people were righteous in the city, then even the children (if any) were being extremely sinful. But like all these situations, if the children and/or the parents/guardians refused to let them have salvation and righteous teachings, whose fault is it? It is not the fault of God, who did provide a way, but the fault of those who suppressed the truth.

God was just and gave the people of Sodom and Gomorrah, and the five cities of the plain, what they asked for (their due punishment). They wanted a life without God and His goodness . . . and God gave that to them.

The Egyptians

In this instance, God used Moses and Aaron (Exodus 5–15) to judge the Egyptians for the wickedness they were inflicting on the Israelites through harsh slavery (Exodus 1:8–14), murdering their children (Exodus 1:22), and so on. God struck the land with many plagues and disasters because Pharaoh continued to sin and the nation of Egypt followed after him in sin. It culminated with the death of the firstborn in Egypt, even though this judgment could easily have been averted had Pharaoh listened and released the Israelites from their oppression — the blood is on Pharaoh's head. Even Pharaoh and his army's final demise was on his own head, not God's. In fact, each plague could easily have been averted had Pharaoh responded to what God said through Moses and Aaron. So a means of salvation from the plagues was given, but Pharaoh and the Egyptians rejected it.

The Canaanites

As for God using people to do His bidding, this is nothing new, as we saw with Moses and Aaron and the Egyptians. God used people to build an ark, His temple, and so on. God used judges and kings to ward off attacks and to provide justice, among other functions. So the concept is nothing new. With the Canaanites, God used the Israelites to enact His judgment under Joshua's leadership.

The Canaanites were far from innocent! God was patient with them as they continued in their sin. Among the Canaanite tribes when Joshua invaded were the Amorites whose sin was prophesied to Abraham. Abraham received the prophecy that the sin of the Amorites had not reached its full measure (Genesis 15:16). During this time, Abraham met Melchizedek, a noble, kingly priest in the land of Canaan. But Melchizedek's ministry surely had an influence on the Canaanites as it took several hundred years before their sin overtook them. Had they continued to listen to what he taught, they probably wouldn't have been in this situation.

When Joshua entered the land of Canaan, the Amorites' sin *had* reached its full measure and it was time for judgment. Leviticus 18:2–30 points out the horrendous crimes that were going on in the land of Canaan. They were having sex with their mothers, sisters, and so on. Men were having sex with other men. They were giving their children to be sacrificed to Molech (vs. 21). They were having sex with animals (vs. 23). So it is impossible to make the claim that those tribes were innocent and undeserving of punishment.

But one can't neglect that children sin, too. As previously pointed out, today there are kids killing kids, kids thieving, kids raping, etc. So the innocence of children is a farce. In fact, if they were sacrificing their children, then how many children were alive when Joshua entered the Promised Land anyway?

At Jericho, both young and old were to be destroyed (Joshua 5:13–6:21), so at least Jericho had young. Yet Jericho is also the place that Christ Himself appeared as a theophany to lead Joshua into battle. Jericho must have been very bad to warrant a physical appearance of Christ to have judgment poured out on them. Perhaps all the sins listed in Leviticus 18 were going on there as well! Yet even in Jericho, there was a means of salvation as Rahab and her family were saved. She can even be found in the lineage of Christ (Matthew 1:5).

The Benjamites

The Benjamites asked for it as well (Judges 19:22–25, 20:13) and sided with the wicked. So no one can claim the Benjamites were innocent either. Sadly, the Benjamites knew the consequences of their actions prior to sinning. They were Israelites who had no excuse for not knowing what Moses wrote. They should have known better, but chose to sin deliberately (Leviticus 18, especially verses 26–30). They also brought it on themselves.

Had the Benjamites repented, the Lord would have forgiven them. The Israelites had extensive means of sacrifice to cover sin and to expel the wicked from among them. However, the Benjamites refused this means of salvation and sinned against God.

Non-Christians

When discussing eternal salvation in Christ with non-Christians, they often ask, "Do you really believe a loving God would send people to an eternal hell?" The response is: only if they sin! And the fact is, all have sinned, all fall short of the glory of God (Romans 3:23). The fascinating thing is that some will *not* spend eternity in hell. Everyone deserves that punishment, including me, but God has provided a means of salvation just as He did in the Old Testament situations described above. If one refuses to receive this salvation, can God be blamed?

Jesus Christ was born to save mankind.

There is only one God; He is God of both the Old Testament and New Testament, even though some try to suggest there are different presentations. In both the Old and New Testaments, people had the opportunity to get back to a right relationship with Him by repenting, asking forgiveness of their sin, and receiving Christ as their Lord and Savior.[3] In both Testaments, God judges sin. Mercy and patience were to be found through God's vessels: Noah, with his preaching for years, and Abraham, with his pleading for Sodom and

3. Although those alive before the time of Christ did not know His name, they still knew of the coming Messiah, as prophesied in Genesis 3:15 and many other places. Their salvation from sin was secured by their faith in the work that He would do on the Cross.

Gomorrah (even Lot urged the people not to be so wicked) — just as mercy and patience are still available today (John 7:37–38).

And He has provided a means of salvation in Jesus Christ (1 Peter 3:18), just as the ark was with the Flood and the angels were in urging Lot and his family to flee Sodom and Gomorrah. No one can blame God for not providing a merciful alternative or call Him "evil" for providing justice against sin.

Conclusion

Naturally, there are plenty of other examples in Scripture where these same principles apply. Consider the analogy of a person who steals and gets caught. When he stands before the judge, the judge finds him guilty and imposes a fine. But then the judge offers to pay the fine. Instead of accepting, the thief refuses and blames the whole mess on the judge who acted justly and even offered a way out!

This is really what is happening in today's culture. Mankind sins and gets caught. People are found guilty by a Holy God. God steps in and offers a means of salvation from the punishment of the crime (which is eternal death), even so far as to die in their place so that they can have eternal life. Yet in all this, the sinners still say no to God and then proceed to blame Him for the situation they are in! It simply doesn't make sense.

In summary:

Event/people	Were they sinning?	Did God provide justice?	Did God provide a means of salvation?
The Fall: Adam and Eve	Yes	Yes	Yes
The Flood	Yes	Yes	Yes
Sodom and Gomorrah	Yes	Yes	Yes
The Egyptians	Yes	Yes	Yes
The Canaanites	Yes	Yes	Yes
The Benjamites	Yes	Yes	Yes
Non-Christians	Yes	Yes	Yes

In light of this, God should not be blamed, but those who were punished for their sin retain the blame. God did provide a means of salvation in each of these cases even though He was not obligated to do so. God should not be blamed. Interestingly enough, individuals who say God is cruel want justice when they are wronged, for example, if someone steals from them, attacks them, or offends them in any way. They really have a double standard.

We are all sinners already under the death penalty (Romans 3:23). But again, God has provided a means of salvation in Christ. It would be nice if people realized that they should hate sin (Romans 12:9) and love God (Deuteronomy 6:5) who acts justly against sin (2 Thessalonians 1:5–10). Yet He offers abundant mercy to those who love Him (Exodus 20:6; Deuteronomy 7:9; Ephesians 2:4). Please consider this, if you haven't already.

37

Who Sinned First — Adam or Satan?

BODIE HODGE

When Christians or others speak of Adam as the first sinner, this comes from the Apostle Paul where he states:

> Therefore, just as through one man sin entered into the world, and death through sin, and thus death spread to all men, because all sinned (Romans 5:12).

It means that sin *entered the world* through Adam — that he is the one credited with sin's entrance and hence the subsequent entrance of death and suffering and the need for a Savior — a last Adam (1 Corinthians 15:45). When we look back at Genesis, it is true that Satan rebelled, and also Eve sinned, prior to Adam's disobedience.

The Sin of Eve

There were several things that Eve did wrong prior to eating the fruit. When the serpent (who was speaking the words of Satan) asked in Genesis 3:1: "Has God indeed said, 'You shall not eat of every tree of the garden'?" her response was less than perfect:

> And the woman [Eve] said to the serpent, "We may eat the fruit of the trees of the garden; but of the fruit of the tree which is in the

> midst of the garden, God has said, 'You shall not eat it, *nor shall you touch it*, lest you die' " (Genesis 3:2–3; emphasis added).

Compare this to what God had commanded in Genesis 2:16–17:

> And the LORD God commanded the man, saying, "Of every tree of the garden you may freely eat; but of the tree of the knowledge of good and evil you shall not eat, for in the day that you eat of it you shall surely die."

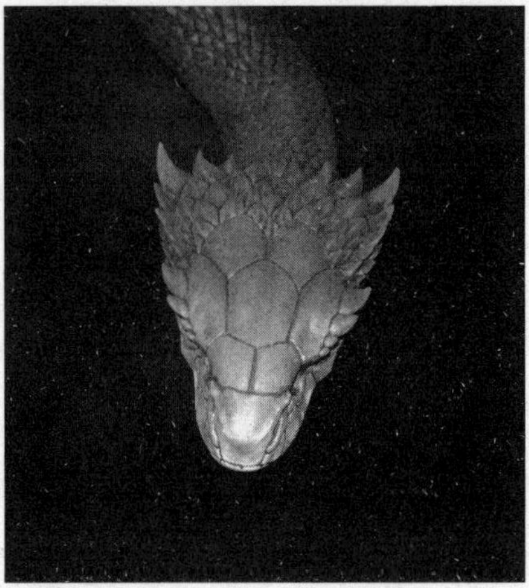

A possible depiction of the serpent as shown in the Creation Museum

Eve made three mistakes in her response:

1. She added the command not to *touch* the fruit: "nor shall you touch it." This seems to be in direct contradiction with the command of Adam to tend the Garden (Genesis 2:15), which would probably constitute touching the tree and the fruit from time to time. It also makes the command from God to be exceptionally harsh.
2. She omitted that God allowed them to *freely* eat from *every* tree. This makes God out to be less than gracious.
3. She amended the meaning of die. Let me explain. The Hebrew in Genesis 2:17 is "die die" (*muwth – muwth*), which is often translated as "surely die" or literally as "dying you shall die," which indicates the beginning of dying — an ingressive sense. In other words, if they had eaten the fruit, then Adam and Eve would have *begun to die* and would return to dust (which is what happened when they ate in Genesis 3:19). If they were meant to die right then, Genesis 2:17 should have used *muwth* only once as is used in the Hebrew meaning dead, died, or die in an absolute sense, and not *beginning to* die or *surely* die as die-die is commonly used. What Eve said was "die" (*muwth*) once instead of the way God said it in Genesis 2:17 as "die-die" (*muwth – muwth*). So she changed God's word to appear harsher again by saying they would die almost immediately.

Often we are led to believe that Satan merely deceived Eve with the statement that "You will not surely die?" in Genesis 3:4. But we neglect the cleverness and cunning that God indicates that the serpent had in Genesis 3:1. Note also that the exchange seems to suggest that Eve may have been willingly led. That is, she had already changed what God had said.

If you take a closer look, the serpent argued against Eve with an extremely clever ploy. He went back and argued against her incorrect words using the phraseology that God used in Genesis 2:17 ("die-die," muwth-muwth). This, in a deceptive way, used the proper sense of die that God stated in Genesis 2:17 against Eve's mistaken view. Imagine the conversation in simplified terms like this:

> God says: Don't eat or you will *begin* to die.
> Eve says: We can't eat or we will die *immediately*.
> Serpent says: You will not *begin* to die?

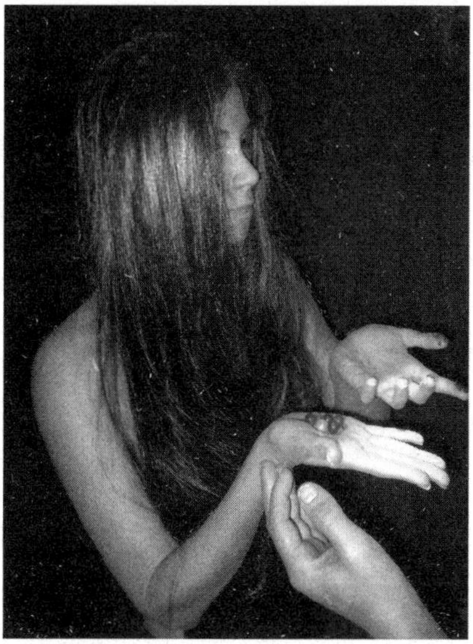

Eve offering Adam the fruit, as presented in the Creation Museum

This was very clever of Satan — using God's Words against her to deceive her. This is not an isolated incident. When Satan tempted Jesus (Matthew 4:1–11), Jesus said, "It is written" and quoted Scripture (Matthew 4:4). The second time, Satan tried quoting Scripture (i.e., God) deceptively, just as he had done to Eve (Matthew 4:5–6). Of course, Jesus was not deceived, and corrected Satan's twisted use of Scripture with a proper use of Scripture (Matthew 4:7). Because of Eve's mistaken response of God's command, it was easier for her to be deceived by Satan's misuse of what God had said.

Another point that can be brought out about Eve was her adoption of Satan's reduction of "Lord God" to simply "God" in Genesis 3:3. This mimicked the way Satan addressed God when he questioned Eve in Genesis 3:1. Satan had degraded God by not using the term God had used in Genesis 2:16–17 and Eve followed suit.

From her response, though, she started down the slope into sin by being enticed by her own thoughts about the fruit (James 1:14–15). This culminated with her eating the forbidden fruit and giving some to her husband, who also ate. Eve sinned against God by eating the fruit from the Tree of the Knowledge of Good and Evil prior to Adam eating it. However, upon a closer look at the text, their eyes were not opened until after Adam ate — likely only moments later (Genesis 3:7). Since Adam was created first (Eve coming from him, but both being created in God's image), and he had been given the command directly, and since he was the responsible party for his wife, it required his sin to bring about the Fall of mankind. When Adam ate and sinned, they knew something was wrong and felt ashamed (Genesis 3:7). Sin and death had entered into the creation.

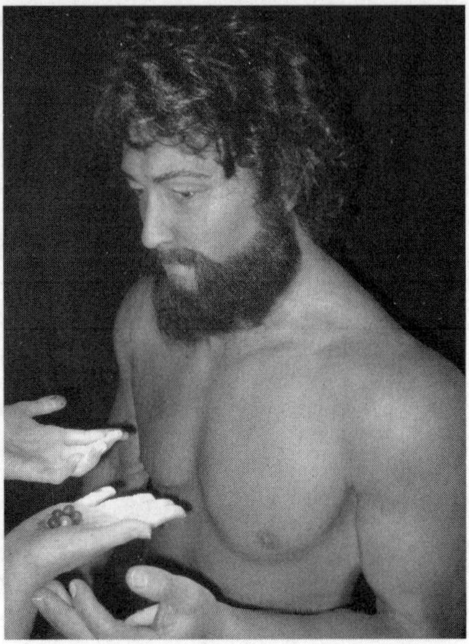
Adam taking the fruit from Eve, as depicted in the Creation Museum

The Sin of Satan

Like Eve, Satan had sinned prior to Adam's disobedience. His sin was pride in his beauty (Ezekiel 28:15–17) and in trying to ascend to be like God while in heaven (Isaiah 14:12–14). He was cast out when imperfection was found in him (Isaiah 14:12; Ezekiel 28:15) and then we find his influence in the Garden of Eden (Ezekiel 28:13; Genesis 3).

Unlike Adam, Satan was never given dominion over the world (Genesis 1:28). So his sin did not affect the creation, but merely his own person. This is likely why Satan went after those who were given dominion. Continuing in his path as an enemy of God, he apparently wanted to do the most damage, so it was likely that his deception of Eve happened soon after his own fall.

The Responsibility of Adam

Adam failed at his responsibilities in two ways. He should have stopped his wife from eating, since he was there to observe exactly what she was about to eat (Genesis 3:6). Instead of correcting the words of his wife (Genesis 3:17), he listened to her and ate while not being deceived (1 Timothy 2:14).

We could also argue that Adam failed to keep and guard the Garden as he was commanded in Genesis 2:15. God, knowing Satan would fall, gave this command to Adam, but Adam did not complete the task. God knew that Adam would fall short and had a plan specially prepared.

Many people have asked, "Why do we have to die for something Adam did?" The answer is simple — we are without excuse since we sin, too (Romans 3:23, 5:12). This has caused some to ask: "Why did we have to inherit a sin nature from Adam, causing us to sin?" We read in Hebrews 7:9–10:

> Even Levi, who receives tithes, paid tithes through Abraham, so to speak, for he was still in the loins of his father when Melchizedek met him.

If we follow this argument, then all of us were ultimately *in Adam* when he sinned. So, although we often blame Adam, the life we have was in Adam when he sinned, and the sin nature we received was because we were in Adam when he sinned. We share in the blame and the sin, as well as the punishment.

But look back further. Everyone's life (including Eve's) came through Adam and ultimately came from God (Genesis 2:17). God owns us and gives us our very being (Hebrews 1:3), and it is He whom we should follow instead of our own sinful inclinations. Since the sin of Adam, all men have had the need for a Savior, Jesus Christ, the Son of God who would step into history to become a man and take the punishment for humanity's sin. Such a loving act shows that God truly loves mankind and wants to see us return to Him. God — as the Author of life, the Sustainer of life, and Redeemer of life — is truly the One to whom we owe all things.

38

How Can Someone Start a New Life in Christ?

CECIL EGGERT

The Creator God tells us in the Book of Genesis about the origin of all things in six days — matter, light, earth, sun, moon, animals, and mankind. His desire was that all of His creation would live in a perfect world where God and man could enjoy everything He had made . . . forever. Can you imagine living in a perfect world?

There was a perfect relationship between the Creator God and man; there was no death, disease, or suffering. Fear between man and animals was non-existent, and every emotional, physical, mental, and spiritual need that Adam and Eve had was met by their Creator. The role of man was clearly defined: Adam and Eve were in charge of an orderly earth that was "very good" (Genesis 1:31)!

Just imagine! God created man in His image, to have a relationship with Him, and gave him a perfect world to care for where mankind was the pinnacle of His creation! When God created Adam and Eve, He didn't make them to be just obedient puppets; they had the freedom to choose and to make their own decisions.

The Fall

One day Adam chose to disobey God's command and go his own way.

> And the LORD God commanded the man, saying, "Of every tree of the garden you may freely eat; but of the tree of the knowledge of good and evil you shall not eat, for in the day that you eat of it you shall surely die" (Genesis 2:16–17).

> Then to Adam He said, "Because you have heeded the voice of your wife, and have eaten from the tree of which I commanded you, saying, 'You shall not eat of it': Cursed is the ground for your sake; in toil you shall eat of it all the days of your life. Both thorns and thistles it shall bring forth for you, and you shall eat the herb of the field. In the sweat of your face you shall eat bread till you return to the ground, for out of it you were taken; for dust you are, and to dust you shall return" (Genesis 3:17–19).

God called Adam's disobedience *sin.* With Adam's sin the process of death had begun. As Adam sinned and died, so do all of us. Romans 5:12 in the New Testament tells us "Therefore, just as through one man sin entered the world, and death through sin, and thus death spread to all men, because all sinned." Sin changed everything: it severed our relationship with God and introduced pain, suffering, and death into the world. This sin affected all of humanity, including you and me. The world was no longer the perfect place that God had originally created it to be.

Sin now corrupted everything (Genesis 3; Romans 8:20–22). When Adam and Eve sinned, it truly was the saddest day in the universe. But God had an eternal plan. While God, being just and holy, had to punish man's sin (or disobedience), He still desired to have a loving relationship with mankind. God made a promise to Adam and Eve. God told Satan, who had deceived Eve: "And I will put enmity between you and the woman, and between your seed and her Seed; He shall bruise your head, and you shall bruise His heel" (Genesis 3:15). This promise was that in the future, He would send a perfect sacrifice from the offspring ("seed") of Eve that would conquer Satan and restore the relationship that had been broken because of sin.

Until the perfect sacrifice was provided, animals were to be used as sacrifices for sin. The first example of this blood sacrifice was demonstrated as animals were slain and the skin was used to cover the nakedness of Adam and Eve: "Also for Adam and his wife the LORD God made tunics of skin, and clothed them" (Genesis 3:21). While this animal skin only represented a "covering" of Adam and Eve's sin, it was a picture of a coming blood sacrifice that God would provide to "cleanse" man from his sin once and for all.

Adam and Eve now had hope for a future restored relationship between them and their Creator God. From Adam to Noah to Abraham, people continued to sacrifice animals. Through Moses, God revealed His law, and the people's need for an unblemished sacrifice to be offered for sin. So, in obedience to God, the Israelites shed the blood of perfect lambs year after year for the forgiveness of sins. These temporary sacrifices only symbolized what was to come in the promised Messiah; the One who would provide the ultimate and perfect sacrifice for the sins of the world.

The Messiah

Throughout Old Testament times, the prophets declared the message of God's love, mercy, and justice, preparing the way for the coming of Messiah. Just as the prophets foretold, the Messiah came to earth, born of a virgin.

> Now in the sixth month the angel Gabriel was sent by God to a city of Galilee named Nazareth, to a virgin betrothed to a man whose name was Joseph, of the house of David. The virgin's name was Mary. . . . Then the angel said to her, "Do not be afraid, Mary, for you have found favor with God. And behold, you will conceive in your womb and bring forth a Son, and shall call His name JESUS (Luke 1:26–31).

Two thousand years ago, our loving Creator God kept His promise of Genesis 3:15 as He stepped into history in the person of Jesus Christ: "In the beginning was the Word, and the Word was with God, and the Word was God. . . . And the Word became flesh and dwelt among us, and we beheld His glory, the glory as of the only begotten of the Father, full of grace and truth" (John 1:1, 14).

He wrapped himself in the flesh of His creation to become the sinless sacrifice to die for the sins of the world.

> For God so loved the world that He gave His only begotten Son, that whoever believes in Him should not perish but have everlasting life (John 3:16).

Jesus's life was everything the prophets foretold. The sinless Son of God was born of the virgin Mary, grew in knowledge and stature, and began His public ministry when He was in His thirties.

During His ministry, Jesus healed the sick, restored the blind, raised the dead, and told them how they could receive eternal life. He did these miraculous acts to show that He truly was the Son of God.

> And truly Jesus did many other signs in the presence of His disciples, which are not written in this book; but these are written that you may believe that Jesus is the Christ, the Son of God, and that believing you may have life in His name (John 20:30–31).

The time came for Jesus to become the perfect sacrifice — to die for the sins of the world and to restore that broken relationship between God and man, once and for all. He would willingly pay the penalty that you and I would have had to pay for our sin (Romans 6:23).

While nailed to the Cross, just before He died, Jesus cried out the word *tetelestai*. This Greek word means, "the debt is paid" or "paid in full." The Cross showed God's love for us. "But God demonstrates His own love toward us, in that while we were still sinners, Christ died for us" (Romans 5:8). Christ had finished what He came to do, to become the perfect sacrifice for sin and to restore man's relationship with the Creator. This was God's eternal plan.

But it doesn't end with Jesus' death on the Cross. Jesus didn't remain in the tomb; He rose from the dead, conquering death. "He is not here; for He is risen" (Matthew 28:6). No longer was the temporary sacrifice of the unblemished animals necessary. The "Lamb of God," Jesus Christ, became the perfect and final sacrifice. You see, in Adam we all die, but in Jesus we have true life and will live forever with Him in a new heaven and new earth that God is preparing, where there will be no more sin, suffering, or death. The first man Adam brought sin and death into the world; the last Adam — Jesus Christ — brings life to the world.

> For since by man came death, by Man also came the resurrection of the dead. For as in Adam all die, even so in Christ all shall be made alive (1 Corinthians 15:21–22).

God offers us the opportunity to be forgiven, spotless, and loved. When we understand and accept what God has done through His son Jesus Christ, we have a restored relationship with our Creator. The Bible makes it clear: God's gift of salvation is offered to us, not just to hear or agree with intellectually, but to respond to in faith (John 14:6; Romans 6:23). This gift is something that we receive by faith (Eph. 2:8–9; Titus 3:5).

There Is a Decision

God calls upon men everywhere to repent of their sin and to place their faith in Christ. Those who repent and believe will have their sins forgiven, a restored relationship with their Creator, and life with Him for eternity. The Bible makes clear the eternal destiny of those who reject Him — they will be separated from Him forever in a place called hell (Revelation 20:15).

The Questions You Must Answer

Do you recognize that you are a sinner in need of salvation? Sin is disobedience to God's commands. God's commands are summarized in the Ten Commandments.

God says "do not lie." Have you ever told a lie?

God says "do not steal." Have you ever taken something that doesn't belong to you?

God says "do not covet." Have you ever been jealous of something that someone else has?

If you have disobeyed these or any other of God's commands, then you are a sinner. Your sin prevents you from having a relationship with your Creator.

Would you like to receive Jesus Christ's sacrifice on the Cross as payment for your sin, submit your life to Him, and receive the free gift of eternal life? Because this is such an important matter, let's clarify what this decision involves. You need to:

Repent. Understand that you have disobeyed the Creator's commands. Tell God you are sorry for your sins. Be willing to turn from anything that is not pleasing to Him. He will show you His plan for you as you grow in your relationship with Him and read His Word.

Receive Christ as your Savior and Lord. Believe that Jesus lived, died, and rose again in payment for your sin (John 3:16, Romans 10:9). Jesus says, "I am the door. If anyone enters by Me, he will be saved" (John 10:9).

Rely on God's strength. God does not promise that life as a Christian will be easy or that you will be healthy and wealthy. In fact, you can expect trials in life that will test your faith (James 1:2–3; 1 Peter 1:6–9). However, God promises that He will give you the strength to bear those burdens (1 Corinthians 10:13).

Jesus told His followers in Luke 14:25–33 that they should count the cost before following Him. If this is what you really want, and your desire is to make Him the center of your life, you can receive the Creator's gift (eternal life through faith in Jesus Christ) right now. The Bible tells us:

> For with the heart one believes unto righteousness, and with the mouth confession is made unto salvation. . . . For "whoever calls on the name of the Lord shall be saved" (Romans 10:10–13).

You can go to God in prayer right now, right where you are, and ask Him for this gift. Here's a suggested prayer to help you:

> Lord Jesus, I know that I am a sinner and do not deserve eternal life, but I believe You died for me and rose from the grave to pay the price for my sin. Please forgive me of my sins and save me. I repent of my sins and now place my trust in You for eternal life. I receive the free gift of eternal life. Amen.

Look at what Jesus promises to those who believe in Him: "Most assuredly, I say to you, he who believes in Me has everlasting life" (John 6:47). And, "But as many as received Him, to them He gave the right to become children of God, to those who believe in His name" (John 1:12). We want to share in your joy if you have just made this life-changing decision. Our desire is to help you grow in your understanding of Jesus Christ and God's Word. Would you please give us a call, write, or email us and tell us your story? We would love to hear from you!

We would like to know if you have made this life-changing decision or have questions on how you can receive eternal life. We also encourage you to contact a Bible-believing church in your area where the pastor accepts the accuracy and authority of the Bible from its very first verse in Genesis (including the Genesis accounts of a recent creation and a global Noah's Flood).

A Challenge as You Take This Message to Others

The C.A.R.E Factor

It has been said; "People don't care how much you know until they know how much you care." First Peter 3:15 says, "But sanctify the Lord God in your hearts, and always be ready to give a defense to everyone who asks you a reason for the hope that is in you, with meekness and fear." Colossians 4:5–6 says, "Walk in wisdom toward those who are outside, redeeming the time. Let your speech always be with grace, seasoned with salt, that you may know how you ought to answer each one."

With all the equipping and knowledge we gain through the resources available to us today, it is possible for us to have an arsenal of answers without having a heart of compassion for those who need the gospel. The following acrostic will help each of us as we share this message with heart and purpose.

The Heart

C — Compassion

Compassion is cultivated as we view each person as a soul who will spend eternity with or without God.

A — Acceptance

Accept the person as an individual who has been created in the image of God.

R — Respect

Respect each person and treat him or her with dignity. The cultivation of good listening skills is critical to the proper communication of the gospel message.

E — Encouragement

Encourage the person along the way as you help answer his or her questions.

The Purpose

C — Connecting

Connect to others in common areas of life by being yourself and being transparent. Let your heart connect to their heart.

A — Assessing

Assess the worldview of your prospect before responding so you can understand his or her questions and know how to properly answer them.

R — Responding

Responding graciously is just as important as having accurate information.

E — Evangelizing

Evangelization can only be accomplished when we share the person and work of Jesus Christ. Remember, sharing the gospel message is His mandate.

As we answer a person's questions, it may take many encounters before we are able to share the saving knowledge of Christ. A balance of grace and truth will always be in order during this process. A word of caution — grace without truth is compromise, and truth without grace is heartless. The practice of fear and meekness and grace and truth can speak as loudly as the answers we provide through His Word. As we prepare ourselves with answers to the questions of this age, let us not forget to equip ourselves with the C.A.R.E. Factor as we pray, love, and go to the lost.

Contributors

Steve A. Austin

Dr. Steven Austin earned a Ph.D. in geology from Pennsylvania State University. As a full-time scientist with the Institute for Creation Research, he participated in professional, peer-reviewed studies at Mount St. Helens and the Grand Canyon. His research has been published in the prestigious International Geology Review. Dr. Austin is now conducting additional studies in the Grand Canyon, and researching the earthquake destruction of archaeological sites in Jordan.

David A. DeWitt

Dr. David DeWitt earned a Ph.D. in neuroscience from Case Western Reserve University in Cleveland, Ohio. He is a professor of biology and director of the Center for Creation Studies at Liberty University (Lynchburg, Virginia). Dr. DeWitt's research has focused on Alzheimer's disease, and he has written a number of articles for peer-reviewed journals such as *Brain Research* and *Experimental Neurology*.

Don B. DeYoung

Dr. Don DeYoung is chairman of the department of physical science at Grace College, Indiana. He also serves on the faculty of the Institute for Creation Research, Dallas. His writings have appeared in *The Journal of Chemical Physics, The Creation Research Society Quarterly*, and elsewhere. Dr. DeYoung has also written numerous books on Bible-science topics, including object lessons for children.

Cecil Eggert

Cecil Eggert serves as advancement officer at Answers in Genesis. He has a B.S. degree from Hyles-Anderson College and is currently completing his M.S. in biblical counseling from Trinity College of the Bible and Seminary. Cecil served as an associate pastor for 29 years, culminating as pastor of outreach at Calvary Baptist Church in Covington, Kentucky, before joining the staff of Answers in Genesis.

Danny Faulkner

Dr. Danny Faulkner has a B.S. (math), M.S. (physics), M.A. and Ph.D. (astronomy, Indiana University). He is a full professor at the University of South Carolina–Lancaster, where he teaches physics and astronomy. Danny has written numerous articles in astronomical journals and is the author of *Universe by Design*.

Joe Francis

Dr. Joe Francis, professor of biological sciences at Master's College, earned his Ph.D. from Wayne State University and was a post-doctoral fellow at the University of Michigan Medical School. He serves as a board member of the Creation Biology Study Group.

Ken Ham

CEO/president of Answers in Genesis, Ken Ham is one of the most in-demand Christian speakers in North America. A native Australian now residing near Cincinnati, Ham has the unique ability to communicate deep biblical truths and historical facts through apologetics. He is the author of numerous books on evangelism, dinosaurs, and the negative fruits of evolutionary thinking, including *The Lie: Evolution* and *Already Gone*.

Bodie Hodge

Bodie attended Southern Illinois University at Carbondale and received a B.S. and M.S. in mechanical engineering. His specialty was a subset of mechanical engineering based in advanced materials processing, particularly starting powders. Currently, Bodie is a speaker, writer, and researcher for Answers in Genesis–USA.

Jean K. Lightner

Dr. Jean Lightner earned her undergraduate degree in animal science. After earning a M.S. and D.V.M., she worked for three years as a veterinary medical officer for the U.S. Department of Agriculture. From here, she resigned to stay at home to raise and teach her four children. She has contributed both technical articles and laymen articles to several creationists' magazines, journals, and websites.

Jason Lisle

Dr. Lisle received his Ph.D. in astrophysics from the University of Colorado at Boulder. He specializes in solar astrophysics and has interests in the physics of relativity and biblical models of cosmology. Dr. Lisle has published a number of books, including *Taking Back Astronomy* and *The Ultimate Proof of Creation*, plus articles in both secular and creationist literature. He is a speaker, researcher, and writer for Answers in Genesis–USA.

Tim Lovett

Tim Lovett earned his degree in mechanical engineering from Sydney University (Australia) and was an instructor for 12 years in technical college engineering courses. Tim has studied the Flood and the ark for 15 years and is widely recognized for his cutting-edge research on the design and structure of Noah's ark. He is author of the book *Noah's Ark: Thinking Outside the Box*.

David N. Menton

Now retired, Dr. David Menton served as a biomedical research technician at Mayo Clinic and then as an associate professor of anatomy at Washington University School of Medicine (St. Louis) for more than 30 years. He was a consulting editor in histology for five editions of *Stedman's Medical Dictionary*, and has received numerous awards for his teaching. Dr. Menton has a Ph.D. in cell biology from Brown University and is currently a speaker, researcher, and writer for Answers in Genesis–USA.

Elizabeth Mitchell

Dr. Mitchell earned her M.D. from Vanderbilt University School of Medicine and practiced medicine for seven years until she retired to be a stay-at-home mom. Her interests in ancient history strengthened when she began to homeschool her daughters. She desires to make history come alive and to correlate it with biblical history.

Tommy Mitchell

Dr. Tommy Mitchell is a graduate of Vanderbilt University School of Medicine. He received his M.D. in 1984 and completed his residency in internal medicine in 1987. For 20 years Tommy practiced medicine in his hometown of Gallatin, Tennessee. In 1991, he was elected a Fellow of the American College of Physicians. Dr. Mitchell has been active in creation ministry for many years. He felt the Lord's call to full time service, and in 2007 he withdrew from the active practice of medicine to join Answers in Genesis–USA as a full-time speaker and writer.

John D. Morris

Dr. John Morris is president of the Institute for Creation Research in Dallas, Texas. The author of numerous books on creation, including *The Young Earth*, Dr. Morris has led several expeditions to Mt. Ararat in search of Noah's ark. A frequent conference speaker, Dr. Morris also hosts the daily radio program *Back to Genesis*.

Terry Mortenson

Dr. Terry Mortenson earned a Ph.D. in the history of geology from Coventry University in England. His thesis focused on the "scriptural geologists," a group of men in the early 19th century who fought the rise of old-earth geological theories. A former missionary (mostly in Eastern Europe), Dr. Mortenson has researched and spoken on creation and evolution for many years. He is now a speaker, writer, and researcher with Answers in Genesis–USA.

Michael Oard

Now retired after 36 years in the U.S. National Weather Service and in research meteorology, Mike Oard holds a masters degree in atmospheric science and has published research articles in journals of the American Meteorological Association and elsewhere. An active creationist, Oard has also published articles in various creationist periodicals and in the *Proceedings of the International Conference on Creationism*. He serves on the board of the Creation Research Society.

Roger Patterson

Roger Patterson earned his B.S. Ed. degree in biology from Montana State University. Before coming to work at *Answers in Genesis*, he taught for eight years in Wyoming's public school system and assisted the Wyoming Department of Education in developing assessments and standards for children in public schools. Roger now serves on the Educational Resources team at Answers in Genesis–USA.

Georgia Purdom

Dr. Purdom received her Ph.D. in molecular genetics from Ohio State University. Her professional accomplishments include winning a variety of honors, serving as professor of biology at Mt. Vernon Nazarene University (Ohio), and the publication of papers in the *Journal of Neuroscience*, the *Journal of Bone and Mineral Research*, and the *Journal of Leukocyte Biology*. Dr. Purdom is also a member of the Creation Research Society, American Society for Microbiology, and American Society for Cell Biology.

Mike Riddle

Mike holds a degree in mathematics and a graduate degree in education. He has been involved in creation ministry for more than 25 years. Prior to getting involved in creation ministry, Mike was a captain in the U.S. Marines and a national champion in track and field. Mike also spent over 20 years in the computer field where he managed U.S. Sprint's worldwide technical training and Microsoft's worldwide engineer training. Currently, Mike is a speaker, writer, and researcher for Answers in Genesis–USA.

Andrew A. Snelling

Dr. Andrew A. Snelling received his Ph.D. (geology) from the University of Sydney (Australia). After research experience in the mineral exploration industry, he was founding editor of the *Creation Ex Nihilo Technical Journal* (Australia). He also served as a professor of geology at the Institute for Creation Research. In 2007 he joined Answers in Genesis–USA as director of research. A member of several professional geological societies, Dr. Snelling has written numerous technical papers in geological journals, and creationist publications.

Paul F. Taylor

Paul F. Taylor is the senior speaker for Answers in Genesis (UK/Europe). He holds a B.S. in chemistry from Nottingham University and a masters in science education from Cardiff University. He is the author of several books, including *Cain and Abel*, *In the Beginning*, and *The Six Days of Genesis*. Paul and his wife, Geri, have five children between them.

Tom Vail

Tom Vail has been a professional guide in the Grand Canyon since 1980. He and his wife, Paula, started Canyon Ministries in 1997 to offer Christ-centered rafting tours through the Grand Canyon. Tom is author and compiler of *Grand Canyon: A Different View*. He is also co-author of the True North Series providing biblically based guide books on our national parks, including *Your Guide to the Grand Canyon* and *Your Guide to Zion and Bryce Canyon National Parks*.

John Woodmorappe

John Woodmorappe has been a researcher in the areas of biology, geology, and paleontology for over 20 years. He has a B.A. in biology, a B.A. in geology, and an M.A. in geology. John has also been a public school science teacher. He is the author of many peer-reviewed technical articles in creationist literature.

Index

The New Answers Book 4

First printing: August 2013
Seventh printing: February 2020

Master Books®, P.O. Box 726, Green Forest, AR 72638.
Master Books® is a division of the New Leaf Publishing Group, Inc.

ISBN: 978-0-89051-788-8
ISBN: 978-1-61458-376-9 (digital)
Library of Congress Number: 2013947562

Please consider requesting that a copy of this volume be purchased by your local library system.

Printed in the United States of America

Please visit our website for other great titles:
www.masterbooks.com

For information regarding author interviews,
please contact the publicity department at (870) 438-5288.

Master Books®
A Division of New Leaf Publishing Group
www.masterbooks.com

ACKNOWLEDGMENTS AND SPECIAL THANKS

Our many thanks to the following for the work of reviewing, editing, or illustrating this book.

Dr. Jason Lisle, Dr. John Whitmore, Dr. Ron Samec, Dr. Elizabeth Mitchell, Dr. Tommy Mitchell, Dr. Andrew Snelling, Dr. Danny Faulkner, Dr. Terry Mortenson, Dr. Georgia Purdom, Dr. John Baumgardner, Gary Vaterlaus, Mike Matthews, Bob Hill, Roger Patterson, Troy Lacey, Steve Golden, Jeremy Ham, Buddy Davis, Randall Hedtke, Wayne Strasser, Mike Oard, Scot Chadwick, Erik Lutz, Dan Stelzer, Dan Lietha, Doug Rummager, Laura Strobl, Bodie Hodge, Steve Fazekas, and Diane King.

Contents

INTRODUCTION

Atheistic Devices: Spotting Them . . . but Countering Them, Too?

KEN HAM

Introduction: Atheists Using Churches to Infiltrate and Deceive

Did you know that many Christian leaders are doing exactly what the atheists are encouraging them to do? It's incredible.

You see, there's an "epidemic" that is infecting and destroying many churches around the world. It is the epidemic of Christians (including many church leaders) who are adopting man's religion of evolutionary ideas and adding them to Scripture — thus undermining the authority of the Word of God.

As we see the loss of the foundation of the authority of God's Word in our Western nations, we are also seeing a massive decline in Christian morality in society. Even the great nation of America is on a downward spiral, as we see the absolutes of Christianity being eliminated from the culture (on an almost daily basis).

We spoke to a prominent Christian leader recently. He is the pastor of a large church in a generally conservative denomination (though many of its churches allow for millions of years). He shared with us that within his denomination, he saw the next big theological debate being whether or not Adam and Eve were literal human beings!

Such re-writing of Scripture is sadly coming to this denomination. But is it all that surprising? Once the door was opened when many of its churches (and affiliated seminaries) began to compromise on the foundations of Genesis, as they added millions of years to Genesis, then the slippery slide into unbelief in other areas of Scripture began to escalate — even whether there was a real Adam.

Right into Their Hands

At AiG, we have been saying for years that as churches compromise with millions of years and evolution, eventually they will begin to compromise other parts of Scripture. They will give up on Adam and Eve and original sin — then maybe a literal hell, bodily resurrection, and virgin birth.

Sadly, we are now seeing that happening more and more in the Church. Last year, *Christianity Today* published a cover story about the battle over a literal Adam and Eve. Yes, now even *that* question is beginning to infiltrate theologically conservative churches. We also hear of Christian leaders giving up a belief in a literal hell. And there are those who are beginning to question aspects of the Resurrection and so on.

Yes, what is happening in the Church today is exactly what the atheists want to see happen. The atheists know that if they can get Christians to compromise God's Word in Genesis, eventually there will be a generational decline in the acceptance of the authority of all of God's Word.

The Trojan Horse

Last year, a professed atheist, Dr. Eugenie Scott, mailed a fundraising letter on behalf of her organization called NCSE (National Center for Science Education). This group was set up primarily to oppose biblical creation organizations like AiG.

In this letter, Dr. Scott told blatant untruths about what AiG is doing. But then again, you shouldn't be surprised when atheists don't tell the truth. After all, if they don't believe in an absolute authority, they have no basis for truth — except for how they decide to define it as such! She is obviously greatly concerned about the effect of AiG in society. Well — we can praise the Lord for that!

But in her letter, designed to cause alarm and raise funds for her anti-Christian organization and "motivate the secular troops" to oppose creationist organizations like AiG, Dr. Scott made a statement similar to the one she has made before on her website about how she seeks to recruit religious people to help her atheist group:

> Find common ground with religious communities and ally with them to promote the understanding of evolution.

And back in 2008, Dr. Scott's NCSE website made these statements in an article entitled "How You Can Support Evolution Education."[1]

One section listed these ideas:

- Suggest adult religious education projects focusing on evolution with your religious leaders.
- Encourage your religious leaders to endorse the Clergy Letter Project and to participate in Evolution Weekend.
- Encourage your religious leaders to produce educational resources about evolution and religion, and to take a formal stand in support of evolution education.

The "Evolution Weekend" referred to above was founded (and is still run) by an atheist professor. He now has thousands of clergy who have signed a statement that agrees with the concept of millions of years/evolution and have agreed to conduct an "Evolution Sunday," when they will preach the "truth" of evolution to their congregations.

An Ally of Atheists

Dr. Scott, back in September 2000, in her opening statement at the American Association for the Advancement of Science Conference entitled "The Teaching of Evolution in U.S. Schools: Where Politics, Religion, and Science Converge," said:

> You can't win this by scientific arguments . . . our best allies were members of the mainstream clergy. . . . The clergy went to school board meetings and said, evolution is okay with us . . . they didn't want the kids getting biblical literalism five days a week either, which meant they'd have to straighten them out on the weekends.

In 2005, I wrote about a supporter of AiG who attended a seminar conducted by Dr. Scott on how to teach evolution in public schools. When dealing with the issue of what to do with Christian students, she offered some sad advice. Our supporter reported:

> I attended the "Teaching Evolution" seminar yesterday led by Eugenie Scott. The teachers were advised to suggest to the Bible believers to consult their clergy who would usually assure them that belief in evolution is OK!!

1. http://ncse.com/taking-action/29-ways-to-support-science-education.

In her latest fundraising letter, this atheist continues her tactic of trying to influence churchgoers to believe in evolution/millions of years.

Atheists understand that if they can get the Church to compromise with millions of years/evolution, this will undermine the authority of the entire Bible . . . and lead to unbelief about Christianity. The atheists know that getting the Church to compromise today, then coming generations may be won over to atheism. And more of our Church leaders are doing exactly what the atheists (gleefully) want them to do.

Breaking the Yoke

One verse of Scripture I have often used to remind me of the constant battle we are in (and the stand we should be taking) is 2 Corinthians 6:14:

> Do not be unequally yoked together with unbelievers. For what fellowship has righteousness with lawlessness? And what communion has light with darkness?

When Christians compromise with the belief system of millions of years and evolution (in reality, a pagan religion), they are being unequally yoked with unbelievers.

Be Discerning

It seems almost everyone wants something free, right? Now, if you were offered a free curriculum to teach children about Genesis, would you jump at it? After all, we need to be educating young people about the authority of God's Word, correct?

Well, there is now a free curriculum for you to consider. And it's designed to teach children about Genesis. To help you in your decision-making about getting this curriculum, I'll give you some samples of what it teaches.

Now, before you read these samples (and I really urge you to look at the quotes below), consider the biblical example of the Christians at Berea who "searched the Scriptures daily to find out whether these things were so" (Acts 17:11). Okay, now read the following excerpts from this new Genesis curriculum:[2]

> During the sixth day God creates land animals, including man — Day Six began about the time the first land animals appear in the fossil record, about 250 million years ago . . . God created the land dwelling creatures on this day. . . .

2. www.oldearth.org/Day6.ppt.

> Man is clearly the ruler of earth, even though many animals are larger. God gave man the ability to think, enabling him to rule the earth. . . . Before the creation of Adam, there were other human-like animals, such as Neanderthal and *Australopithecus*. . . . Evolutionists point to them as an evolutionary path from ape to man. . . .
>
> From a Christian perspective, they were not "in the image of God" as Adam was. In other words, they did not have an eternal soul, capable of choosing eternal life with God. . . . Just how "human-like" they were is debatable, and there will always be an argument surrounding their position in God's creation. . . .
>
> Man and animals are given plants to eat. This is often misinterpreted by young-earth creationists. . . . Young-earth creationists claim there was no death before Adam's sin. They claim that only plants could be eaten based on Genesis 1:29-30. . . . First, look back at Genesis 1:28. Man was instructed to subdue the earth (and its animals). . . .
>
> It is clear from the fossil record that there was much death before Adam. . . . Day Six ends with the statement "very good." Young earth proponents say it could not be "very good" if there was death before Adam. . . . Death is a natural process of God's created world, therefore God created death. . . .

So, now would you want this free curriculum to teach your children? Absolutely not!

Twisted Scripture

I hope you will be like the Bereans. AiG supporters would realize that whoever wrote this curriculum accepts fallible man's ideas concerning evolution and millions of years and, as a result, twists and contorts the Scriptures to justify an acceptance of man's pagan religion. In other words, they are mixing the religion of the day with their Christianity just like the Israelites did with Baal in the Old Testament a number of times.

My purpose is *not* to go in-depth and critique these blatant reinterpretations of Scripture. I'm sure you can recognize the problems. But I do want to point out an increasing and related problem I see all over the Church.

Satan is very clever. However, he still uses the same tactic: to work from within the Church to lead generations of people away from the truth of God's Word and the gospel.

Truly we are in a spiritual battle — not just with the world, but also within much of the Church. The attacks on Christianity from the secular world are obvious. But the Church has many wolves in sheep's clothing (as God's Word warns us it would). And as God raises up ministries like Answers in Genesis to battle with the pagan religion of this day (evolution/millions of years) that leads people away from God's Word, Satan is actively recruiting people within the Church to try to combat what we're trying to accomplish.

Truly we are in a spiritual battle — not just with the world, but also within much of the Church. By the way, the compromising web-based curriculum I've mentioned is offered free *in the name of Christianity*. And it has an agenda for parents to teach children in a certain way about the Bible. But this website has a name that is obviously designed to mimic (and even be confused with) Answers in Genesis: it's called Answers in Creation!

Remember our book *Already Gone*? In that publication we presented the detailed research into why two-thirds of our young people are leaving the Church by college age. The major reasons came down to:

- Young people being taught to compromise Genesis with evolution and millions of years; respondents saw this as hypocrisy within the Church.
- Churches and parents not teaching children apologetics — not teaching them how to defend the Christian faith against the secular attacks of our day.

Sadly, free curricula like the new one referred to above, if used by families and churches, will lead to more young people walking away from the Church. At Answers in Genesis, we are so burdened about such sad developments that we stepped out in faith to produce a high-quality Bible curriculum for kindergarten through adult. Titled the *Answers Bible Curriculum*, it is an entirely integrated curriculum for Sunday school so the entire family (no matter the age) can discuss the material when they get home (i.e., children and parents cover the same topics — but at a different level).

Many Attacks on the Bible, Not Just Curriculum

Some days in ministry, it can be exhausting. There seem to be constant daily battles! But I remind myself that Answers in Genesis, a Bible-upholding ministry, is engaged in an ongoing spiritual war; when one battle is over, another front opens. To illustrate, here is a list of just some of the many "battles" that have involved AiG in the last year. It's not a complete list, but it still reminds us of the battles raging around us. Many are quite startling:

- An American Atheist billboard appears near the Lincoln Tunnel entrance at New York City with its message for Christmas: "You KNOW it's a Myth . . . This Season, Celebrate REASON!" The atheists are becoming more active each year, and every Christmas they ramp up their propaganda.
- Rev. Barry Lynn, president of Americans United for Separation of Church and State, and I debate on CNN's *Anderson Cooper 360* TV program over the passing of tourism incentives for our Ark Encounter project.
- The Calvin College biology department issues their "Perspective on Evolution," a statement from a *Christian* college endorsing evolution as the best scientific explanation for life on earth.
- U.S. Congressman Pete Stark of the Bay Area of California introduces a bill to proclaim February 12 as "Darwin Day."
- Bill Nye, "The Science Guy" of PBS-TV fame and well-known atheist, visits the Creation Museum for two minutes to stand in the museum driveway and take photos so he can say he has legitimacy to criticize the Creation Museum.
- Former Eastern Nazarene College physics professor Karl Giberson, and BioLogos founder Francis Collins publish *The Language of Science and Faith*, arguing for theistic evolution and against the origin of sin as taught in Genesis.
- Political activist/blogger Joe Sonka and a friend try to crash "Date Night" at our museum by pretending to be (in their words) a "flamboyantly gay" couple.
- Pastor Rob Bell publishes *Love Wins: A Book About Heaven, Hell, and the Fate of Every Person Who Ever Lived* in which the biblical view of hell is undermined.
- I'm dropped from the "Great Homeschool Conventions" programs in Cincinnati and Philadelphia for revealing the biblically compromised teachings of Peter Enns (who believes Jesus was in error), also a speaker at these conventions.
- NASA astrobiologist Richard Hoover claims life on earth may have come from other planets in the *Journal of Cosmology*.

- The office of the secretary of the Assemblies of God denomination sponsors a conference entitled Faith and Science Conference promoting theistic evolution.
- Tim Keller, well-known author and senior pastor of the Redeemer Presbyterian Church in Manhattan (New York), again endorses evolution as a possible way God created.
- Rev. Barry Lynn of Americans United for Separation of Church and State posts a YouTube video mocking our Ark Encounter project and the Bible's account of Noah.
- *Christianity Today* magazine publishes an article entitled "The Search for the Historical Adam," questioning the historicity of Adam and Eve. The cover features an "ape-man."
- In a *USA Today* article on the recent Miss USA beauty pageant winner, Alyssa Campanella shares how she believes in evolution; the article disparages Answers in Genesis, the Creation Museum, and the Ark Encounter.
- New York attacks the Bible by legalizing "gay marriage."
- A *Washington Post* blog discusses presidential candidate Michele Bachmann as an evolution-doubter and disparages the Creation Museum.
- Chinese scientist Xing Xu claims that *Archaeopteryx* is not a bird, but rather a feathered dinosaur.
- Hank Hanegraaff — the "Bible Answer Man," president of the Christian Research Institute, and host of the *Bible Answer Man* radio program — endorses William Dembski's book *The End of Christianity*, which presents an unbiblical position on the creation and evolution of humans.
- The General Presbytery of the Assemblies of God adopts a revised statement on "The Doctrine of Creation," now allowing for evolution and millions of years.
- Calvin College professor of religion John Schneider is forced to resign after casting doubt on the historical accuracy of Adam and Eve and their fall into sin.

- GOP presidential candidate Texas Governor Rick Perry is questioned about evolution by a child at a political rally and a video of it goes viral.
- Susan Brooks Thistlethwaite, professor of theology at Chicago Theological Seminary, writes an article for the *Washington Post* entitled "The Theological Case for Evolution" that criticizes the Creation Museum.
- A columnist for the United Kingdom Christian website Network Norwich calls Answers in Genesis "a cult."
- BBC TV launches a major new dinosaur series in the United Kingdom that, as expected, promotes evolution and millions of years.
- Prof. Richard Dawkins, Sir David Attenborough, and 28 other prominent UK evolutionists ask the British government to censor the teaching of creation in Britain's publicly funded schools.
- Karl Giberson, former vice president of BioLogos and former physics professor at Eastern Nazarene College, and Randall Stephens, history department chair at Eastern Nazarene College, publish *The Anointed.* Answers in Genesis is singled out at the very outset of the book as a proponent of an "anti-intellectual populism undergirding evangelical 'truth,' and that the movement takes its cues from a handful of enormously influential but only dubiously credentialed authority figures."
- *Science* magazine publishes additional articles supporting the claim that *Australopithecus sediba* was an ancestor of humans.
- Darrel Falk, president of BioLogos and biology professor at Point Loma Nazarene University, responds to my lecture on the "Anti-biblical Teachings of BioLogos" and critiques AiG's stand on Genesis by siding with the atheistic arguments against the Bible.

Did you get tired reading this list? Well, I did — and that's just the *short* list. Many of you likely have lists of attacks of your own. Christians are coming under attack from many directions in today's culture and it is good to spot these attacks so you can counter them.

Equipping the "Troops"

Amidst all this opposition, here is what Answers in Genesis is doing to counter the attacks on the Bible and equip people with effective Bible-defending "weapons":

- Provide incredible new apologetic resources on the AiG websites, with 17 million users a year accessing the sites!
- Write and publish various books [like this one in the *New Answers Book Series]*, such as *Already Gone, Already Compromised, The Fall of Satan, How Do We Know the Bible is True?, Demolishing Contradictions, The Tower of Babel, The Lie: Evolution, One Race One Blood, Coming to Grips with Genesis*, and so on.
- *Answers Bible Curriculum* for all ages (seven age levels).
- Produce new faith-defending video sets, including my 12-part *Foundations* series and Dr. David Menton's excellent *Body of Evidence* anatomy series.
- Conduct hundreds of apologetics conferences and other speaking engagements at churches and colleges in the USA and around the world.
- Announce the building of Noah's ark as part of the Ark Encounter project — a reminder that God's Word and its salvation message are true.
- Build a Creation Museum with an observatory, Special Effects Theater, Planetarium, Dinosaur Den, Bug Exhibit, and much more.
- Produce Vacation Bible School (VBS) programs, now used by thousands of churches a year!

Now this kind of list doesn't make me tired at all! It gets me excited! I often tell people that I look on the resources that AiG produces as Christian "patriot missiles," equipping believers in daily spiritual battles that seem to be heating up around the country!

Such resources are needed (and more) to help the Church be discerning to the specific attacks of our age. We need to know what the attacks on the Bible are and how to counter them. This is why this book series is so important — it gives answers in an effort to help ground Christians to have a firm foundation on the authority of the Bible.

CHAPTER 1

Does the Gospel Depend on a Young Earth?

KEN HAM

Can a person believe in an old earth and an old universe (millions or billions of years in age) and be a Christian?

First of all, let's consider three verses that sum up the gospel and salvation. 1 Corinthians 15:17 says, "If Christ is not risen, your faith is futile; you are still in your sins!" Jesus said in John 3:3, "Most assuredly, I say to you, unless one is born again, he cannot see the kingdom of God." Romans 10:9 clearly explains, "If you confess with your mouth the Lord Jesus and believe in your heart that God has raised Him from the dead, you will be saved."

Numerous other passages could be cited, but not one of them states in any way that a person has to believe in a young earth or universe to be saved.

And the list of those who cannot enter God's kingdom, as recorded in passages like Revelation 21:8, certainly does not include "old earthers."

Many great men of God who are now with the Lord have believed in an old earth. Some of these explained away the Bible's clear teaching about a young earth by adopting the classic gap theory. Others accepted a day-age theory or positions such as theistic evolution, the framework hypothesis, and progressive creation.

Scripture plainly teaches that salvation is conditioned upon faith in Christ, with no requirement for what one believes about the age of the earth or universe.

Now when I say this, people sometimes assume then that it does not matter what a Christian believes concerning the supposed millions-of-years age for the earth and universe.

Even though it is not a salvation issue, the belief that earth history spans millions of years has very severe consequences. Let me summarize some of these.

Authority Issue

The belief in millions of years does not come from Scripture, but from the fallible methods that secularists use to date the universe.

To attempt to "fit" millions of years into the Bible, you have to invent a gap of time that almost all Bible scholars agree the text does not allow — at least from a hermeneutical perspective. Or you have to reinterpret the "days" of creation as long periods of time (even though they are obviously ordinary days in the context of Genesis 1). In other words, you have to add a concept (millions of years) from outside Scripture, into God's Word. This approach puts man's fallible ideas in authority over God's Word.

As soon as you surrender the Bible's authority in one area, you "unlock a door" to do the same thing in other areas. Once the door of compromise is open, even if ajar just a little, subsequent generations push the door open wider. Ultimately, this compromise has been a major contributing factor in the loss of biblical authority in our Western world.

The Church should heed the warning of Proverbs 30:6, "Do not add to His words, lest He rebuke you, and you be found a liar."

Contradiction Issue

A Christian's belief in millions of years totally contradicts the clear teaching of Scripture. Here are just three examples:

> *Thorns.* Fossil thorns are found in rock layers that secularists believe to be hundreds of millions of years old, so supposedly they existed millions of years before man. However, the Bible makes it clear that thorns came into existence after the Curse: "Then to Adam He said, 'Because . . . you have eaten from the tree of which I commanded you, saying, "You shall not eat of it": Cursed is the ground for your sake. . . . Both thorns and thistles it shall bring forth for you' " (Genesis 3:17–18).
>
> *Disease.* The fossil remains of animals, said by evolutionists to be millions of years old, show evidence of diseases (like cancer, brain tumors, and arthritis). Thus, such diseases supposedly existed millions of years before sin. Yet Scripture teaches that after God finished creating everything and placed man at the pinnacle of creation, He described the creation as "very good" (Genesis 1:31). Certainly calling cancer

and brain tumors "very good" does not fit with Scripture and the character of God.

Diet. The Bible clearly teaches in Genesis 1:29–30 that Adam and Eve and the animals were all vegetarian before sin entered the world. However, we find fossils with lots of evidence showing that animals were eating each other — supposedly millions of years before man and thus before sin.

Death Issue

Romans 8:22 makes it clear that the whole creation is groaning as a result of the Fall — the entrance of sin. One reason for this groaning is death — the death of living creatures, both animals and man. Death is described as an "enemy" (1 Corinthians 15:26), which will trouble creation until one day it is thrown into the lake of fire.

Romans 5:12 and other passages make it obvious that physical death of man (and really, death in general) entered the once-perfect creation because of man's sin. However, if a person believes that the fossil record arose over millions of years, then death, disease, suffering, carnivorous activity, and thorns existed millions of years before sin.

The first death was in the Garden of Eden when God killed an animal as the first blood sacrifice (Genesis 3:21) — a picture of what was to come in Jesus Christ, the Lamb of God, who would take away the sin of the world. Jesus Christ stepped into history to pay the penalty of sin — to conquer our enemy, death.

By dying on a Cross and being raised from the dead, Jesus conquered death and paid the penalty for sin. Although millions of years of death before sin is not a salvation issue per se, I personally believe that it is really an attack on Jesus' work on the Cross.

Recognizing that Christ's work on the Cross defeated our enemy, death, is crucial to understanding the "good news" of the gospel: "And God will wipe away every tear from their eyes; there shall be no more death, nor sorrow, nor crying. There shall be no more pain, for the former things have passed away" (Revelation 21:4).

Rooted in Genesis

All biblical doctrines, including the gospel itself, are ultimately rooted in the first book of the Bible.

- God specially created everything in heaven and earth (Genesis 1:1).
- God uniquely created man and woman in His image (Genesis 1:26–27).
- Marriage consists of one man and one woman for life (Genesis 2:24).
- The first man and woman brought sin into the world (Genesis 3:1–24).
- From the beginning God promised a Messiah to save us (Genesis 3:15).
- Death and suffering arose because of original sin (Genesis 3:16–19).
- God sets society's standards of right and wrong (Genesis 6:5–6).
- The ultimate purpose of life is to walk with God (Genesis 6:9–10).
- All people belong to one race — the human race (Genesis 11:1–9).

False Claims

The *New York Times* on November 25, 2007, published an article on the modern biblical creation movement. The Creation Museum/Answers in Genesis received a few mentions in the article. However, I wanted to deal with one statement in the article that had the writer done just a little bit of homework, she would have found it not to be true!

The writer, Hanna Rosin, stated concerning the Creation Museum:

> The museum sends the message that belief in a young earth is the only way to salvation. The failure to understand Genesis is literally "undermining the entire word of God," Ken Ham, the founder of Answers in Genesis, says in a video. The collapse of Christianity believed to result from that failure is drawn out in a series of exhibits: school shootings, gay marriage, drugs, porn, and pregnant teens. At the same time, it presents biblical literalism as perfectly defensible science.

Note particularly the statement: "belief in a young earth is the only way to salvation." Even if a Christian believes in an old earth (and even theistic evolution), they would know that such a statement is absolutely false. The Bible makes it clear that, concerning Jesus Christ, "Nor is there salvation in any other, for there is no other name under heaven given among men by which we must be saved" (Acts 4:12). When the Philippian jailer in Acts 16:30 asked, "Sirs, what must I do to be saved?" Paul and Silas (in verse 31) replied, "Believe on the Lord Jesus Christ, and you will be saved, you and your household."

In Ephesians 2:8–9 we are clearly told: "For by grace you have been saved through faith, and that not of yourselves; it is the gift of God, not of works, lest anyone should boast." And Jesus Christ stated: "Jesus said to him, 'I am the way, the truth, and the life. No one comes to the Father except through Me' " (John 14:6).

Creation Museum/Answers in Genesis Teachings

As one walks through the Creation Museum, nowhere does it even suggest that "belief in a young earth is the only way to salvation." In fact, in the theater where the climax of the 7 C's walk-through occurs, people watch a program called *The Last Adam*. This is one of the most powerful presentations of the gospel I have ever seen. This program clearly sets out the way of salvation — and it has nothing to do with believing in a young earth.

As I often tell people in my lectures, Romans 10:9 states: "If you confess with your mouth, 'Jesus is Lord,' and believe in your heart that God raised him from the dead, you will be saved." By confessing "Jesus is Lord," one is confessing that Christ is to be Lord of one's life — which means repenting of sin and acknowledging who Christ is. The Bible DOES NOT state, "That if you confess with your mouth, 'Jesus is Lord,' and believe in your heart that God raised him from the dead — AND BELIEVE IN A YOUNG EARTH — you will be saved"!

Concluding Remarks

So it should be obvious to anyone, even our opponents, that this statement in the *New York Times* is absolutely false. Sadly, I have seen similar statements in other press articles — and it seems no matter what we write in website articles, or how often we answer this outlandish accusation, many in the press continue to disseminate this false accusation, and one has to wonder if it is a deliberate attempt to alienate AiG from the mainstream church!

I believe that one of the reasons writers such as Hanna Rosin make such statements is because AiG is very bold in presenting authoritatively what the Bible clearly states. People sometimes misconstrue such authority in the way Hanna Rosin has. It is also interesting that people who don't agree with us often get very emotional about how authoritatively we present the biblical creation view — they dogmatically insist we can't be so dogmatic in what we present!! It's okay for them to be dogmatic about what they believe, and dogmatic about what we shouldn't believe, but we can't be!

In my lectures, I explain to people that believing in an old earth won't keep someone out of heaven if they are truly "born again" as the Bible defines "born again." Then I'm asked, "Then why does AiG make an issue of the age of the earth — particularly a young age?" The answer is that our emphasis is on the authority of Scripture. The idea of millions of years does NOT come from the Bible; it comes from man's fallible, assumption-based dating methods. If one uses such fallible dating methods to reinterpret Genesis (e.g., the days of creation), then one is "unlocking a door," so to speak, to teach others that they don't have to take the Bible as written (e.g., Genesis is historical narrative) at the beginning — so why should one take it as written elsewhere (e.g., the bodily Resurrection of Christ). If one has to accept what secular scientists say about the age of the earth, evolution, etc., then why not reinterpret the Resurrection of Christ? After all, no secular scientist accepts that a human being can be raised from the dead, so maybe the Resurrection should be reinterpreted to mean just "spiritual resurrection."

The point is, believing in a young earth won't ultimately affect one's salvation, but it sure does affect the beliefs of those that person influences concerning how to approach Scripture. Such compromise in the Church with millions of years and Darwinian evolution, etc., we believe has greatly contributed to the loss of the Christian foundation in the culture.

CHAPTER 2

Do Plants and Leaves Die?

DR. MICHAEL TODHUNTER

Fall in America and throughout much of the Northern Hemisphere is a beautiful time of year. Bright reds, oranges, and yellows rustle in the trees and then blanket the ground as warm weather gives way to winter cold. Many are awed at God's handiwork as the leaves float to the ground like heaven's confetti. But fall may also make us wonder, "Did Adam and Eve ever see such brilliant colors in the Garden of Eden?" Realizing that these plants wither at the end of the growing season may also raise the question, "Did plants die before the Fall of mankind?"

Before we can answer this question, we must consider the definition of *die*. We commonly use the word *die* to describe when plants, animals, or humans no longer function biologically. However, this is not the definition of the word *die* or *death* in the Old Testament. The Hebrew word for *die* (or *death*), *mût* (or *mavet* or *muwth*), is used only in relation to the death of man or animals with the breath of life, not regarding plants.[1] This usage indicates that plants are viewed differently from animals and humans.

Plants, Animals, and Man – All Different

What is the difference between plants and animals or man? For the answer we need to look at the phrase *nephesh chayyah*.[2] *Nephesh chayyah* is used in the

1. J. Stambaugh, "Death before Sin?" *Acts & Facts*, 18 (5) (1989); http://www.icr.org/article/295/, and B. Hodge, "Biblically, Could Death Have Existed Before Sin?" *Answers in Genesis* website, March 2, 2010.
2. J.Stambaugh, " 'Life' According to the Bible, and the Scientific Evidence," *Technical Journal*, August 1, 1992; http://www.answersingenesis.org/articles/tj/v6/n2/life.

Bible to describe sea creatures (Genesis 1:20–21), land animals (Genesis 1:24), birds (Genesis 1:30), and man (Genesis 2:7).[3] *Nephesh* is never used to refer to plants. Man specifically is denoted as *nephesh chayyah*, a living soul, after God breathed into him the breath of life. This contrasts with God telling the earth on day 3 to bring forth plants (Genesis 1:11). The science of taxonomy, the study of scientific classification, makes the same distinction between plants and animals.

Since God gave only plants (including their fruits and seeds) as food for man and animals, then Adam, Eve, and all animals and birds were originally vegetarian (Genesis 1:29–30). Plants were to be a resource of the earth that God provided for the benefit of *nephesh chayyah* creatures — both animals and man. Plants did not "die," as in *mût*; they were clearly consumed as food. Scripture describes plants as *withering* (Hebrew *yabesh*), which means "to dry up." This term is more descriptive of a plant or plant part ceasing to function biologically.

A "Very Good" Biological Cycle

When plants wither or shed leaves, various organisms, including bacteria and fungi, play an active part in recycling plant matter and thus in providing food for man and animals. These decay agents do not appear to be *nephesh chayyah* and would also have a life cycle as nutrients are reclaimed through this "very good" biological cycle. As the plant withers, it may produce vibrant colors because, as a leaf ceases to function, the chlorophyll degrades, revealing the colors of previously hidden pigments.

Since decay involves the breakdown of complex sugars and carbohydrates into simpler nutrients, we see evidence for the second law of thermodynamics *before* the Fall of mankind. But in the pre-Fall world, this process would have been a perfect system, which God described as "very good."

What Determines a Leaf's Color?

When trees bud in the spring, their green leaves renew forests and delight our senses. The green color comes from the pigment chlorophyll, which resides in the leaf's cells and captures sunlight for photosynthesis. Other pigments called carotenoids are always present in the cells of leaves as well, but in the summer their yellow or orange colors are generally masked by the abundance of chlorophyll.

In the fall, a kaleidoscope of colors breaks through. With shorter days and colder weather, chlorophyll breaks down, and the yellowish colors become

3. Ibid.

visible. Various pigments produce the purple of sumacs, the golden bronze of beeches, and the browns of oaks. Other chemical changes produce the fiery red of the sugar maple. When fall days are warm and sunny, much sugar is produced in the leaves. Cool nights trap it there, and the sugars form a red pigment called anthocyanin.

Leaf colors are most vivid after a warm, dry summer followed by early autumn rains, which prevent leaves from falling early. Prolonged rain in the fall prohibits sugar synthesis in the leaves and thus produces a drabness due to a lack of anthocyanin production.

Still other changes take place. A special layer of cells slowly severs the leaf's tissues that are attached to the twig. The leaf falls, and a tiny scar is all that remains. Soon the leaf decomposes on the forest floor, releasing important nutrients back into the soil to be recycled, perhaps by other trees that will once again delight our eyes with rich and vibrant colors.

A Creation That Groans

It is conceivable that God withdrew some of His sustaining (restraining) power at the Fall to no longer uphold things in a perfect state when He said, "Cursed is the ground" (Genesis 3:17), and the augmented second law of thermodynamics resulted in a creation that groans and suffers (Romans 8:22).[4]

Although plants are not the same as man or animals, God used them to be food and a support system for recycling nutrients and providing oxygen. They also play a role in mankind's choosing life or death. In the Garden were two trees — the Tree of Life and the Tree of the Knowledge of Good and Evil. The fruit of the first was allowed for food, the other forbidden. In their rebellion, Adam and Eve sinned and ate the forbidden fruit, and death entered the world (Romans 5:12).

Furthermore, because of this sin, all of creation, including *nephesh chayyah*, suffers (Romans 8:19–23). We are born into this death as descendants of Adam, but we find our hope in Christ. "For as in Adam all die, even so in Christ shall all be made alive" (1 Corinthians 15:22, KJV). As you look at the "dead" leaves of fall and remember that the nutrients will be reclaimed into new life, recognize that we too can be reclaimed from death through Christ's death and Resurrection.

4. Of course, God still upholds all of creation. Furthermore, the second was in effect before and after the Fall, but now we are in a state where things are not upheld in perfect balance so to speak.

CHAPTER 3

Dragons . . . Were They Real?

BODIE HODGE

A Dodo of an Introduction

The dodo was a strange bird, and our understanding of its demise and extinction by 1662 is equally strange. The dodo was a flightless bird that lived on the island of Mauritius in the Indian Ocean. It was easy to catch and provided meat to sailors. There were numerous written accounts, sketches, and descriptions of the bird from the 1500s through the 1600s.

But when the dodo went extinct, no one seemed to notice. And a few years later, scientists began to promote the idea that the dodo was merely a myth. Just look at the evidence:

1. It was a very strange creature.
2. No one could find them.
3. They seemed to exist only in the old descriptions, accounts, and drawings!

Had it not been for specimens popping up in the recesses of museum collections, and finally brought to light, they could have been labeled simply as "myth" for as long as the earth endures! But in the 19th century, at last, there was vindication that the dodo was real and that it had merely gone extinct. Since then, fossils and other portions of specimens have been identified as dodo.

Drawing by Sir Thomas Herbert of a cockatoo, red hen, and dodo in 1634. Courtesy of Wikipedia Commons, http://en.wikipedia.org/wiki/File:Lophopsittacus.mauritianus.jpg.

Parallel to Dragons

So what does this have to do with dragons? Consider the following points:

1. Dragons are very strange creatures.
2. No one can find them.
3. They seem to exist only in the old descriptions, accounts, and drawings!

If we don't know our history, are we doomed to repeat it? Sadly in recent times, secular scientists have relegated dragons to myths also.

But unlike the dodo, which is just a particular type of bird, dragons are a large group of reptilian creatures. Moreover, we have descriptions, drawings, and accounts of dragons. Not just the handful like we have of the dodo, but in *massive numbers* from all over the world! And many of these descriptions and accounts are very similar to creatures known by a different name: dinosaurs. We'll consider this connection below.

Dragons in the Bible

To settle this issue of the reality of dragons, let us turn to the Word of Almighty God who knows all things.

In each case in Table 1, the verses use the word Hebrew *tannin*, or its plural form *tanninim*, which was usually translated as "dragon(s)." In some cases, you might see the translation "serpent" or "monster." There is also the word *tannim* (plural of *tan*, "jackal"), which sounds quite similar to *tannin* in Hebrew. Many previous translators viewed these creatures as dragons, too. But many scholars today suggest these are separate and that *tannim* should be translated as jackals.[1]

1. For more information see Steve Golden, Tim Chaffey, and Ken Ham, "*Tannin*: Sea Serpent, Dinosaur, Snake, Dragon, or Jackal?" Answers in Genesis, http://www.answersingenesis.org/articles/aid/v7/n1/tannin-hebrew-mean.

Table 1: Dragons in the Bible[2]

Reference	Verse
Deuteronomy 32:33	Their wine is the poison of **dragons**, and the cruel venom of asps.
Nehemiah 2:13	And I went out by night by the gate of the valley, even before the **dragon** well, and to the dung port, and viewed the walls of Jerusalem, which were broken down, and the gates thereof were consumed with fire.
Job 7:12 (YLT)	A sea-monster am I, or a **dragon**, That thou settest over me a guard?
Psalm 74:13	Thou didst divide the sea by thy strength: thou brakest the heads of the **dragons** in the waters.
Psalm 91:13	Thou shalt tread upon the lion and adder: the young lion and the **dragon** shalt thou trample under feet.
Psalm 148:7	Praise the LORD from the earth, ye **dragons**, and all deeps:
Isaiah 27:1	In that day the LORD with his sore and great and strong sword shall punish leviathan the piercing serpent, even leviathan that crooked serpent; and he shall slay the **dragon** that is in the sea.
Isaiah 51:9	Awake, awake, put on strength, O arm of the LORD; awake, as in the ancient days, in the generations of old. Art thou not it that hath cut Rahab, and wounded the **dragon**?
Jeremiah 51:34	Nebuchadnezzar the king of Babylon hath devoured me, he hath crushed me, he hath made me an empty vessel, he hath swallowed me up like a **dragon**, he hath filled his belly with my delicates, he hath cast me out.
Lamentations 4:3 (GNV)	Even the **dragons** draw out the breasts, and give suck to their young, but the daughter of my people is become cruel like the ostriches in the wilderness.[a]
Ezekiel 29:3	Speak, and say, Thus saith the Lord GOD; Behold, I am against thee, Pharaoh king of Egypt, the great **dragon** that lieth in the midst of his rivers, which hath said, My river is mine own, and I have made it for myself.
Ezekiel 32:2 (GNV)	Son of man, take up a lamentation for Pharaoh King of Egypt, and say unto him, Thou art like a lion of the nations and art as a **dragon** in the sea: thou castedst out thy rivers and troubledst the waters with thy feet, and stampedst in their rivers.
Genesis 1:21 (YLT)	And God prepareth the great monsters [***dragons***], and every living creature that is creeping, which the waters have teemed with, after their kind, and every fowl with wing, after its kind, and God seeth that it is good.[b]

2. All references are taken from the KJV except where noted.

Exodus 7:9, 10, 12	When Pharaoh shall speak unto you, saying, Shew a miracle for you: then thou shalt say unto Aaron, Take thy rod, and cast it before Pharaoh, and it shall become a serpent [***dragon***]. And Moses and Aaron went in unto Pharaoh, and they did so as the LORD commanded: and Aaron cast down his rod before Pharaoh, and before his servants, and it became a serpent [***dragon***]. . . . For they cast down every man his rod, and they became serpents [***dragons***]: but Aaron's rod swallowed up their rods.[c]

a. Some have thought this word for dragons is a copyist mistake in that tannin should be tannim and may represent another animal type (e.g., jackal). But there is no textual support for this. The argument is that reptiles today do not suckle their young. However, we know so little about extinct dragons that we can't say definitely if they suckled or not. Even some mammals were thought to only give birth to live young until we found the platypus and spiny anteaters that lay eggs, so we need to avoid making "blanket statements" about creature types based only on what we know today. We simply do not know all things about extinct creatures, and if Lamentations 4:3 does refer to dragons (or dragons of a specific type), then we would know that some did suckle.

b. Though the word here is not translated as dragon it is still the same word used of dragon elsewhere and could and likely should have been used here as well.

c. The Hebrew word translated "serpent(s)" is tannin (plural tanninim), which is typically translated "dragon." Most translate this as serpent or snake since a staff is similar in shape to a snake (i.e., serpents being a specific form of dragon). Other ancient translations render this as dragon, including the Latin Vulgate (only in v. 12), and the Greek Septuagint.

Consider also the scriptural references to "fiery serpents" or "fiery flying serpents," "leviathan," and "behemoth":

Table 2: Fiery Serpents, Leviathan, and Other Dragon-Like Creatures

Reference	Verse
Numbers 21:6, 8	And the LORD sent **fiery serpents** among the people, and they bit the people; and much people of Israel died. . . . And the LORD said unto Moses, Make thee a **fiery serpent**, and set it upon a pole: and it shall come to pass, that every one that is bitten, when he looketh upon it, shall live.
Deuteronomy 8:15	Who led thee through that great and terrible wilderness, wherein were **fiery serpents**, and scorpions, and drought, where there was no water; who brought thee forth water out of the rock of flint;

Isaiah 14:29	Rejoice not thou, whole Palestina, because the rod of him that smote thee is broken: for out of the serpent's root shall come forth a cockatrice, and his fruit shall be a **fiery flying serpent**.
Isaiah 30:6	The burden of the beasts of the south: into the land of trouble and anguish, from whence come the young and old lion, the viper and **fiery flying serpent**, they will carry their riches upon the shoulders of young asses, and their treasures upon the bunches of camels, to a people that shall not profit them.
Job 41:1	Canst thou draw out **leviathan** with an hook? or his tongue with a cord which thou lettest down?
Psalms 74:14	Thou brakest the heads of **leviathan** in pieces, and gavest him to be meat to the people inhabiting the wilderness.
Psalms 104:26	There go the ships: there is that **leviathan**, whom thou hast made to play therein.
Isaiah 27:1	In that day the LORD with his sore and great and strong sword shall punish **leviathan** the piercing serpent, even **leviathan** that crooked serpent; and he shall slay the dragon that is in the sea.
Job 40:15–24	Behold now behemoth, which I made with thee; he eateth grass as an ox. Lo now, his strength is in his loins, and his force is in the navel of his belly. He moveth his tail like a cedar: the sinews of his stones are wrapped together. His bones are as strong pieces of brass; his bones are like bars of iron. He is the chief of the ways of God: he that made him can make his sword to approach unto him. Surely the mountains bring him forth food, where all the beasts of the field play. He lieth under the shady trees, in the covert of the reed, and fens. The shady trees cover him with their shadow; the willows of the brook compass him about. Behold, he drinketh up a river, and hasteth not: he trusteth that he can draw up Jordan into his mouth. He taketh it with his eyes: his nose pierceth through snares.

These creatures could rightly be lumped among dragons. Even Leviathan is called a dragon in Isaiah 27:1.

Some have argued that the fiery flying serpents (and fiery serpents) were myth, but God clearly reveals them as real creatures, just as other creatures are real in the immediate context like scorpions, lions, vipers, donkeys, camels, and so on.

Some have argued that fiery flying serpents were real but were just venomous snakes that would leap into the air. But that would render a portion of the Scriptures redundant, as the viper, which does that very thing, is mentioned immediately before it in Isaiah 30:6. Even today there is an insect from South America called the bombardier beetle that shoots out two chemicals that essentially ignite and superheat its victim. Leviathan was also a fire breather (Job 41:1–21).

Some have suggested the behemoth as an elephant or a hippo, but neither the elephant nor the hippo eat grass like an ox, nor do they have a tail that moves like a cedar. An elephant has a tail that moves like a weeping willow, and a hippo hardly has a tail! Some have argued that behemoth and leviathan were myth, but why does God speak of real creatures (lion, raven, donkey, wild ox, ostriches, horse, locust, hawk, and eagle) in the same context as the behemoth and leviathan (Job 38–41)?

So some of what we can learn from the Bible is: (1) dragons were real creatures, and (2) the term "dragon" could include land, flying, or sea creatures.

Dragons by Ancient Historians, Literature, and Classic Commentaries

Dragons were viewed as real creatures by virtually all ancient writers who commented on them. While many references could be cited, consider these select accounts:

1. "But according to accounts from Phrygia there are Drakones in Phrygia too, and these grow to a length of sixty feet."[3]
2. "Africa produces elephants, but it is India that produces the largest, as well as the dragon."[4]
3. "Even the Egyptians, whom we laugh at, deified animals solely on the score of some utility which they derived from them; for instance, the ibis, being a tall bird with stiff legs and a long horny beak, destroys a great quantity of snakes: it protects Egypt from plague, by killing and eating the flying serpents that are brought from the Libyan desert by the south west wind, and so preventing them from harming the natives by their bite while alive and their stench when dead."[5]
4. "Among Egyptian birds, the variety of which is countless, the ibis is sacred, harmless, and beloved for the reason that by carrying the eggs of serpents to its nestlings for food it destroys and makes

3. Aelian (ca. A.D. 220), *De Natura Animalium.*
4. Pliny (ca. A.D. 70), *Natural History.*
5. Marcus Tullius Cicero (ca. 45 B.C.), *De Natura Deorum,* I, 36.

fewer of those destructive pests. These same birds meet winged armies of snakes which issue from the marches of Arabia, producing deadly poisons, before they leave their own lands."[6]

5. Gilgamesh, hero of an ancient Babylonian epic, killed a huge dragon named Khumbaba in a cedar forest.
6. The epic Anglo-Saxon poem *Beowulf* (ca. A.D. 495–583) tells how the title character of Scandinavia killed a monster named Grendel and its supposed mother, as well as a fiery flying serpent.
7. "The dragon, when it eats fruit, swallows endive-juice; it has been seen in the act."[7]

Ancient historians and writers clearly believed creatures like dragons were real. They describe seeing them first hand — often in the context of other types of animals that still live today. Some historians even describe the fiery flying serpents as real creatures in regions near where Moses and Isaiah were and point out the *winged* nature of these flying serpents. Such things are a great confirmation of the biblical text.

Interestingly, in the Beowulf account, the dragon called Grendel was known to have a heavy claw on its finger, yet had a fairly small arm. (Beowulf was famous for ripping the arm off of this dragon.) Correspondingly, we have a dinosaur with smaller arms (and its remains are found in Europe) called *baryonyx*, which literally means "heavy claw"! Its arms are actually smaller, too! The common descriptions of Grendel and baryonyx are striking.

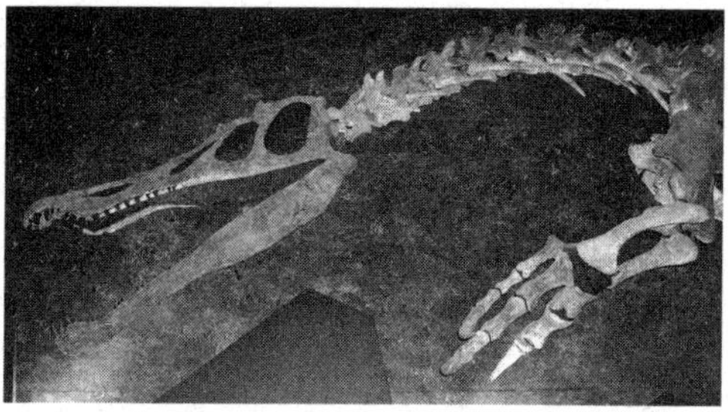

Baryonyx head and forelimb
(Ballista, Wikimedia Commons)

Classic commentators often agreed that dragons were real and spoke of them as real, and these are just a small sample of the writings these expositors of Scripture have on the subject:

1. Dr. John Gill wrote, "Of these creatures, both land and sea dragons, see Gill on 'Mic 1:8'; see Gill on 'Mal 1:3'; Pliny says the dragon has no poison in it; yet, as Dalechamp, in his notes on

6. Ammianus Marcellius (ca. A.D. 380), *Res Gestae*, 22, 15:25-26a.
7. Aristotle, *Historia Animalium*, http://etext.virginia.edu/etcbin/toccer-new2?id=AriHian.xml&images=images/modeng&data=/texts/english/modeng/parsed&tag=public&part=9&division=div2 (accessed June 14, 2013).

that writer observes, he in many places prescribes remedies against the bite of the dragon; but Heliodorus expressly speaks of some archers, whose arrows were infected with the poison of dragons; and Leo Africanus says, the Atlantic dragons are exceeding poisonous: and yet other writers besides Pliny have asserted that they are free from poison. It seems the dragons of Greece are without, but not those of Africa and Arabia; and to these Moses has respect, as being well known to him."[8]

2. John Calvin stated, "Then he says, he has swallowed me like a dragon. It is a comparison different from the former, but yet very suitable; for dragons are those who devour a whole animal; and this is what the Prophet means. Though these comparisons do not in everything agree, yet as to the main thing they are most appropriate, even to show that God suffered his people to be devoured, as though they had been exposed to the teeth of a lion or a bear, or as though they had been a prey to a dragon. "[9]

Even the artwork for John Calvin's commentary for Genesis (when translated from Latin to English in A.D. 1578) included images of dragons such as the one shown here.

3. Charles Spurgeon, when speaking of London, said, "We are not sure that Nineveh and Babylon were as great as this metropolis, but they certainly might have rivaled it, and yet there is nothing left of it, and the dragon and the owl dwell in what was the very center of commerce and civilization."[10]
4. John Trapp stated, "Anger is a short madness; it is a leprosy breaking out of a burning, and renders a man unfit for civil society; for his unruly passions cause the climate where he lives to be like the torrid zone, too hot for any to live near him. The dog days continue with him all the year long; he rageth, and eateth firebrands, so that every man that will provide for his own safety must flee

8. John Gill, Commentary notes Deuteronomy 32:33.
9. John Calvin, Commentary notes Jeremiah 51:34.
10. C. H. Spurgeon, "A Basket Of Summer Fruit" (sermon, Exeter Hall, London, England, October 28, 1860).

from him, as from a nettling, dangerous and unsociable creature, fit to live alone as dragons and wild beasts, or to be looked on only through a grate, as they; where, if they will do mischief, they may do it to themselves only."[11]

5. Church fathers, on Philip killing a dragon in Hierapolis, stated, "And as Philip was thus speaking, behold, also John entered into the city like one of their fellow-citizens; and moving about in the street, he asked: Who are these men, and why are they punished? And they say to him: It cannot be that thou art of our city, and askest about these men, who have wronged many: for they have shut up our gods, and by their magic have cut off both the serpents and the dragons."[12]

There were numerous dragon slayers in history as well. Not to belabor the point, I've simply made a table of a few:

Table 3: A Few Dragon Slayers and Capturers[13]

	Slayer/Capturer	Approximate Date	Place
1	Martha of Tarascon	A.D. 48–70	Tarasque
2	Apostles Philip and Barnabas	Before A.D. 70	Hierapolis
3	St. George	A.D. 300	North Africa
4	St. Sylvester I	A.D. 300	Italy
5	Sigurd	Before A.D. 400–500?[a]	Northern Europe
6	Beowulf	A.D. 400–500	Denmark, Sweden
7	Tristan	A.D. 700?	British Isles

a. Although the more complete account of Sigurd and the dragon is discussed in the 13th-century document called Volsunga Saga, Sigurd is mentioned in the Beowulf account, so it must have preceded it.

I could continue with hosts of other quotations from the church fathers who often spoke of dragons as real creatures, not questioning their reality. But the point is already made: people believed dragons were real.

11. John Trapp, *Complete Commentary*, s.v. Proverbs 22:24, (http://www.studylight.org/com/jtc/view.cgi?bk=19&ch=22 (accessed June 14, 2013).
12. The Acts of Philip, *Of the Journeyings of Philip the Apostle: From the Fifteenth Acts Until the End, and Among Them the Martyrdom*, http://archive.org/stream/apocryphalgospel00edin/apocryphalgospel00edin_djvu.txt (accessed June 14, 2013).
13. Bibliography for this table includes *The Golden Legend*, various texts of the church fathers, *Encyclopædia Britannica, Beowulf, Volsunga Saga*, and several others.

Dragons in Petroglyphs

It would be nearly impossible to have an exhaustive listing of dragons on walls, pottery, textiles, petroglyphs, artwork, maps, books, and so on. Here are a few, and note that some of these dragons are very similar in form to our understanding of dinosaurs.

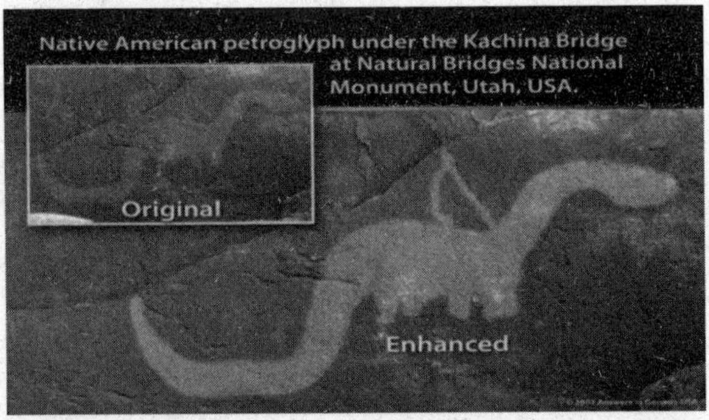

This famous petroglyph by the Anasazi natives looks strikingly like a sauropod dinosaur (i.e., dragon).[14]

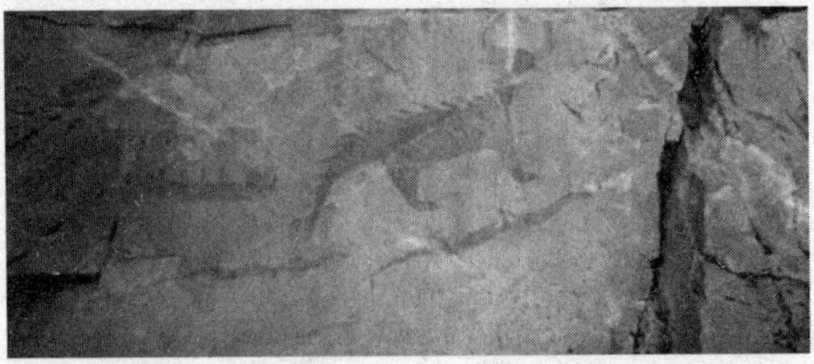

This dragon with back spines is reminiscent to a Kentrosaurus or Amarga but possibly a Lambeosaurus near Lake Superior in Canada.

This flying dragon was made by Native Americans in Utah.

14. I. Abrahams, "Feedback: Kachina Bridge Dinosaur Petroglyph," Answers in Genesis, http://www.answersingenesis.org/articles/2011/03/18/feedback-senter-and-cole.

This relief in Angkor, Cambodia, is something akin to Stegosaurus-type of dragon.[15]

Built by the order of King Nebuchadnezzar, the eighth gate of Babylon has aurochs (an extinct type of cattle) and a dragon alternating all the way up the gate. Since this dragon is a reptile (note the scales and tongue), it also has hips that raise the body off the ground; so, by definition it is also a dinosaur.

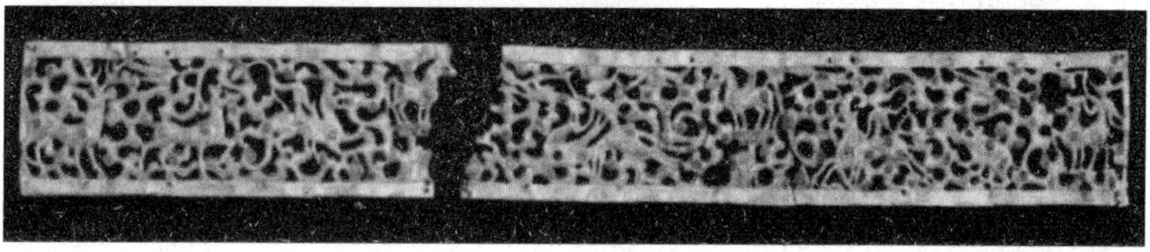

There are several animals portrayed in this ancient golden diadem from Kazakhstan. The onset of the second portion is a dragon.[16]
(http://www.kazakhembus.com/sites/default/files/documents/Nomads_and_Networks_FS_Images.pdf)

15. K.E. Cole, "Evidence of Dinosaurs at Angkor," Answers in Genesis, http://www.answersingenesis.org/articles/2007/01/15/evidence-dinosaurs-angkor.
16. Diadem (gold, turquoise, carnelian, and coral), Kargaly, Myng-Oshtaky tract, Almaty region. Photo: © The Central State Museum of the Republic of Kazakhstan, Almaty, http://www.kazakhembus.com/sites/default/files/documents/Nomads_and_Networks_FS_Images.pdf.

Dragons in Peru adorn hosts of ancient pottery, rock ark, textiles, and so on. This pottery is from the ancient Moche Culture and is on display at the Museum of the Nation in Lima, Peru.[17]

Dragons on Flags and Banners

It is fairly well known that the Welsh flag endows a dragon. But few realize that this was not the only culture to have a dragon on its flag. These cultures clearly viewed dragons as real.

The George Cross which is featured on the flag of Malta.

Even modern flags such as that of Bhutan or Malta also sport dragons referring back to previous accounts. In the case of Malta, it represents St. George killing the dragon in the upper corner.

The flag of Bhutan, though designed in 1947, heralds back to the old tradition of the *druk*, that is, dragons. They also have a national emblem that has two dragons on it.

Many other flags and banners could be added to this list, and diligent searches will turn up numerous ancient flags, banners, and emblems with such things.

17. Bodie Hodge, "The Dragons of Peru," *Answers*, September 14, 2010, http://www.answersingenesis.org/articles/am/v5/n4/dragons-peru.

Welsh flag

Royal Bavarian flag

Imperial China flag

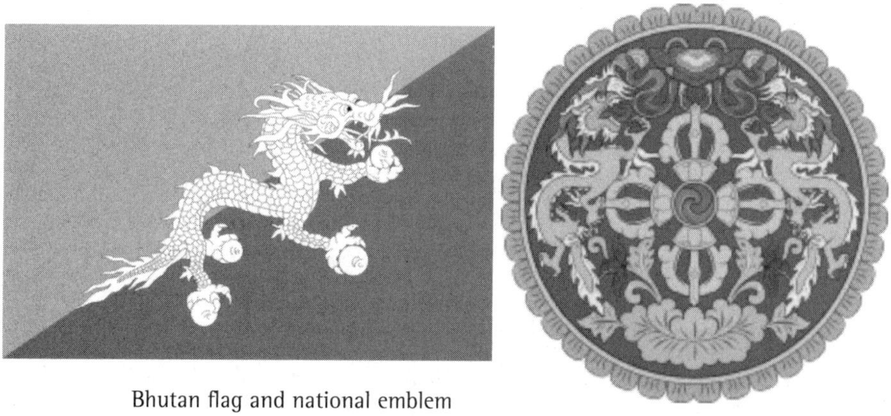

Bhutan flag and national emblem

The famous Bayeux Tapestry that depicts the Norman invasion of England has numerous animals on it. Some are dragons.

Have Dragons Been Relegated to Myths?

It was not until the 20th century that dragons were seen as myths. In 1890, a large flying dragon was killed in Arizona (in the United States), and samples were sent to universities back east. This was recorded in a newspaper under "A Strange Winged Monster Discovered and Killed on the Huachuca Desert," *The Tombstone Epitaph*, on April 26, 1890. No one seemed to entertain the idea they were myths then.

Even the 1902 edition of the *Encyclopædia Britannica,* while trying to explain away the accounts of sea dragons ("sea serpents"), concluded that they might still exist (as their numbers were few by this time):

> It would thus appear that, while, with very few exceptions, all the so-called "sea serpents" can be explained by reference to some well-known animal or other natural object, there is still a residuum sufficient to prevent modern zoologists from denying the possibility that some such creature may after all exist.[18]

Yet only eight years later, it was published that dragons were myth! In 1910, the *Encyclopædia Britannica* states the following:

> Nor were these dragons anything but very real terrors, even in the imaginations of the learned until comparatively modern times. As the waste places were cleared, indeed, they withdrew farther from the haunts of men, and in Europe their last lurking-places were the inaccessible

18. William Evans Hoyle, *Encyclopædia Britannica*, 9th ed. s.v. "Sea-Serpent" (New York, NY: The Encyclopædia Britannica Company, 1902), http://www.1902encyclopedia.com/S/SEA/sea-serpent.html.

heights of the Alps, where they lingered till Jacques Balmain set the fashion which has finally relegated them to the realm of myth.[19]

This was only about 100 years ago that the dragon first began being relegated to a mythical status. Apparently, since Jacques Balmain couldn't find one, they were deemed myth. Perhaps the idea that they went extinct was too much to consider.

Though this idea of dragons being myth still defied *Encyclopædia Britannica*'s claim even into the 1920s. They were not too eager to make such bold claims. In 1927, one dictionary consulted still viewed dragons as real but rare:

> A huge serpent or snake (now rare); a fabulous monster variously represented, generally as a huge winged reptile with crested head and terrible claws, and often as spouting fire; in the Bible, a large serpent, a crocodile, a great marine animal, or a jackal.[20]

But it makes sense as more people spread out and settled in more lands, the dragons were pushed to the brink of extinction. Many old accounts of dragons had them living underground, particularly near swamps (e.g., Beowulf). As man develops areas, those habitats are destroyed. But just like the dodo, when you can't find them any longer, they are suddenly considered "myth" instead of being seen as extinct.

Sadly, this also influenced Christians and subsequently modern translations rarely use the word dragon in the Old Testament, due, in my opinion, to these secular influences.

Dragons and Their Relation to "Dinosaurs"

Dragons include land, sea, and water reptiles. Though dragons in old forms of classification also denoted snakes, dinosaurs are more specific.

Dinosaurs are land reptiles that (by definition) have one of two kinds of hip structures that allow the creature to naturally raise itself off the ground.[21] In other words, crocodiles, komodo dragons, alligators, and so on are not seen as dinosaurs since their hip structures have their legs coming out to the side so the belly naturally rests on the ground. But neither would flying

---

19. Walter Alison Phillips, *Encyclopædia Britannica*, 11th ed. (New York, NY: The Encyclopædia Britannica Company, 1910), 8:467.
20. *The New Century Dictionary* (New York, NY: P.F. Collier & Son Corporation, reprinted in 1948), p. 456.
21. P.S. Taylor, "Dinosaur!," Films for Christ, http://www.christiananswers.net/dinosaurs/dinodef.html.

reptiles like pterodactyls or water reptiles like plesiosaurs be dinosaurs by definition either.

So all dinosaurs are dragons, but not all dragons are dinosaurs. Dinosaurs and other land dragons were made on day 6 (Genesis 1:24–31). Flying dragons and sea dragons were made on day 5 (Genesis 1:20–23).

It is important to realize that the word dinosaur did not exist until the year 1841. Sir Richard Owen invented the term "dinosaur," and it means "terrifying" or "terrible" lizard. Maybe the controversy could have been avoided if they just called dinosaur bones "dragon" bones.

But this means dinosaurs were created and lived the same time as man and went aboard the ark of Noah (Genesis 6:20). Those that did not go aboard died. Many likely rotted and decayed, and others were rapidly buried by sediment from the Flood, making them candidates for fossilization. Hence, we find many of these dragon bones (e.g., dinosaur bones) in rock layers from the Flood. Dinosaurs came off the ark and have been dying out ever since.

One dinosaur resembles a dragon so much that they named it after a dragon from a movie series.
Dracorex Hogswartsia skeleton restoration, The Children's Museum of Indianapolis (Wikimedia Commons)

Reasons for Extinction?

So why did dragons (e.g., dinosaurs) die out? The simple answer is *sin.* When Adam and Eve sinned (Genesis 3) death came into the world. Living things began to die, and many things began to die out — dragons as well as dodos were no exception.

Some specific reasons for their extinction likely include changing environments (e.g., the ice age that followed the Flood, the destruction of swamp lands by man, and so on), predation by man (cf. Genesis 10:9), diseases, genetic problems, catastrophic events, etc.[22] Keep in mind that most dragon legends end with a dragon getting killed. Like the dodo, man could have been a major factor why dragons no longer survive, as far as we know. The possibility exists that some still live in remote parts of the world or underground and only come out at certain times. This was quite common with old dragon accounts.

22. Ken Ham, gen. ed., *New Answers Book 1* (Green Forest, AR: Master Books, 2006), p. 207–219.

However, it is unlikely that we will find any living ones, in the same way that it is unlikely that we will find passenger pigeons, dodos, and many other things that have been pushed to extinction.

Conclusion: Dragons in Relation to Satan

There is much to be said about dragons, and this short chapter is just a taste. Dragons, including the specific subset of dinosaurs, were real creatures and have simply died out due to sin, just like so many other animals, including the dodo. The land-dwelling, air-breathing dragons survived on the ark of Noah, and they have been dying out ever since (Genesis 6:20, 7:21–22).

Many were surely timid creatures (especially since they are known to have inhabited old ruins), but others were known to terrorize, according to the old accounts of dragons. And when such conflicts arose, a dragon usually ended up dead by someone who could overcome it. Such conquerors were remembered in history with a powerful and strong name.

But such vicious attacks could well be the reason that Satan is metaphorically called a "dragon" in Scripture (e.g., Revelation 12:3); also consider Satan's use of a serpent in Genesis 3:1 to deceive Eve and ultimately get Adam to bring sin and death into the world (Romans 5:12).

Satan's vicious attacks leave many helpless (e.g., 2 Corinthians 2:11; 1 Peter 5:8). But Christ, the "stronger man" in Luke 11:21–22, has conquered Satan (Hebrews 2:14), and has an eternal name above all names (Philippians 2:9). For in Christ, one can have the victory over Satan, the great dragon (1 Corinthians 15:57).

With this in mind, it is good to realize the big picture. Satan *wants* people to accept the idea that dragons were myth as this is simply another attack on the authority of God's Word. Satan wants us to doubt God's Word the same way he attacked Eve using a serpent in the Garden of Eden to doubt His Word (Genesis 3:1–6; 2 Corinthians 2:11). Were dragons a myth, or did they simply die out? It's time to trust God's Word over the fallible ideas of man, who was not there and not in a position of superseding God on the subject (Isaiah 2:22).

Of course dragons were real.

CHAPTER 4

Peppered Moths . . . Evidence for Evolution?

DR. TOMMY MITCHELL

Stop me if you have heard this tale before. It's about one of the sacred cows of evolution: the peppered moth. The story of this moth has been set forth for decades as *the* prime example of evolution in action. It is a fascinating story about how, due to a combination of environmental changes and selective predation, a moth turned into, well, a moth.

The peppered moth, scientifically known as *Biston betularia*, exists in two primary forms — one light colored with spots and one almost black. As the tale goes, in the mid 1800s, the lighter variety of the moth (*typica*) predominated. During the Industrial Revolution, the lichen on tree trunks died, soot got deposited on trees, and as a result trees got darker. As this change occurred, the population of darker moths (*carbonaria*) increased, presumably due to the camouflage offered by the darker trees. Bird predators could not see the dark moths against the dark bark. As the darker moth population increased, the lighter moth population decreased.[1]

This story has been touted for years as a great example of Darwinian evolution in action. Countless textbooks are lavishly illustrated with photographs of light and dark moths resting on light and dark tree trunks to teach the wonders of evolution. "It is the slam dunk of natural selection, the paradigmatic story

1. This darkening of the wings is due to the increased amount of the pigment melanin in the wings of the *carbonaria* variety and is known as "melanism."

that converts high school and college students to Darwin, the thundering left hook to the jaw of creationism."[2]

Much of the "proof" for this evolutionary change came from the work of a man named Dr. Bernard Kettlewell, a medical doctor-turned-entomologist, at Oxford University. Dr. Kettlewell had been intrigued by changes in the relative populations of the moths. In his experiments, he set out to show that the changes were a result of natural selection in response to environmental change and selective predation.

The Work of Kettlewell

First of all, Kettlewell had to show that birds were indeed predators of these moths. Up to that time, many biologists did not consider birds the primary predators of *Biston*. Kettlewell released moths into an aviary and observed the moths being eaten as they rested. This observation settled the issue of bird predation, at least to Kettlewell's satisfaction.[3]

For the next phase of his study, Kettlewell went to a polluted woodland area near Birmingham, England. There the trees had become darkened due to pollution. In the woods, Kettlewell undertook the first of his release-recapture experiments. He released moths, 447 of the *carbonaria* variety and 137 of the *typica* variety. Traps were set to recapture the moths that night, and the numbers of each variety were assessed the next morning. A much higher percentage of darker moths than lighter moths were recovered. Kettlewell recaptured 27.5 percent of the *carbonaria*, but only 13.0 percent of the *typica*. From this data, Kettlewell concluded that "birds act as selective agents"[4] and subsequently felt that this represented evolution by natural selection.

To further examine this, Kettlewell then undertook another release-recapture experiment. This was done in a wooded area near Dorset, England. Here the trees had not been darkened by pollution. As before, both light and dark moths were released and then recaptured and counted. Here 12.5 percent of the *typica* were recaptured but only 6.3 percent of the *carbonaria*. Kettlewell anticipated this result because he hypothesized that birds would more easily prey upon the darker moths than the lighter moths due to the lighter color of the trees.

2. Judith Hooper, *Of Moths and Men: An Evolutionary Tale* (New York: W.W. Horton, 2002), p. xvii.
3. H.B.D. Kettlewell, "Selection Experiments on Industrial Melanism in the Lepidoptera," *Heredity* 9 (1955): 323–342.
4. Ibid., p. 342.

Adding credence to Kettlewell's theory, others noted that, as pollution decreased, the population of lighter moths increased in some areas. In the late 1950s, pollution control laws were enacted and air quality improved. In some places, as the lichen returned to the trees, the expected increase in the population of the *typica* variety of moth occurred.[5] Scientists believed this increase further confirmed this living example of evolution.

From this point on, there was no stopping the peppered moth bandwagon. High school and college biology textbooks heralded the peppered moth as the classic example of evolution in action. The peppered moth story has been presented to students for years as a classic case of evolution, the process by which molecules eventually turned into man.

Trouble in Paradise

Scientific claims must be confirmed through repetition, but over the years many attempts to repeat Kettlewell's studies have failed to confirm his results. These contradictory reports showed high populations of *typica* in polluted areas[6] or inordinately high numbers of *carbonaria* in lightly polluted areas.[7] Some studies failed to confirm the observation that the lighter moths increased as the lichen cover of the trees recovered. Nonetheless, the challenges failed to remove the vaunted moth from its lofty perch.

The major challenge to Kettlewell's work came in 1998 when Michael Majerus, a geneticist from Cambridge, published a book entitled *Melanism: Evolution in Action*.[8] Although many of the criticisms of Kettlewell's work had been around for years, Majerus's critique of Kettlewell's methods caused quite a stir in evolutionary circles.

In a review of this book in the journal *Nature*, Dr. Jerry Coyne said this: "My own reaction resembles the dismay attending my discovery, at the age of six, that it was my father and not Santa who brought the presents on Christmas Eve."[9] He further commented; "It is also worth pondering why there has been general and unquestioned acceptance of Kettlewell's work."[10] Things

5. Jonathan Wells, "Second Thoughts about Peppered Moths," The True.Origin Archive, http://trueorigin.org/pepmoth1.asp.
6. R.C. Stewart, "Industrial and Non-industrial Melanism in the Peppered Moth, *Biston betularia* (L.)," *Ecological Entomology* 2 (1977): 231–243.
7. D.R. Lees and E.R. Creed, "Industrial Melanism in *Biston betularia*: The Role of Selective Predation," *Journal of Animal Ecology* 44 (1975): 67–83.
8. M.E.N. Majerus, Melanism: *Evolution in Action* (Oxford: Oxford University Press, 1998).
9. J.A. Coyne, "Not Black and White," *Nature* 396 (1998): 35.
10. Ibid., p. 36.

were starting to look bad for our friend, *Biston betularia*. Then things got worse.

In 2002, a journalist named Judith Hooper published the book *Of Moths and Men: An Evolutionary Tale*. This book detailed the story of the research involving the peppered moth, including an exploration of the lives of the principal people involved. She described the lives and backgrounds of not only Kettlewell but also of E.B. Ford, Kettlewell's mentor at Oxford. The somewhat unflattering portraits of these men were disturbing and, in one sense, made for good reading — if by good reading one likes reveling in the shortcomings of other human beings.

However, it was Hooper's detailed examination of Kettlewell's experimental techniques, which fueled the most controversy. She thoroughly described the method used by Kettlewell in each of his field studies, along with an analysis of the data he collected. Her conclusions were shocking in that she suggests that Kettlewell, after obtaining disappointing data in the early phase of his study, manipulated his collection of data later in the study in order to obtain the desired result. The possibility of outright fraud was even mentioned. The scientific community was aghast. The first and foremost evidence for evolution in action, "the prize horse in our stable,"[11] was apparently in jeopardy.

What's the Problem?

Although there have been many concerns raised about Kettlewell's experimental techniques, the biggest issue seems to revolve around where moths rest during the day. In his study, Kettlewell released moths during the daytime and watched them take resting places on the trunks of trees. He then observed birds preying on the moths. During the night, he collected and counted the moths. He concluded that birds preyed more readily on the more visible moths than on the ones better hidden by their surroundings. The problem with this conclusion is that, over many years of study, it had been determined that *these moths don't rest on tree trunks during the day*! They fly only at night, and they take resting places high in the trees on the underside of branches. In these places they are much better concealed from birds than were the moths in Kettlewell's experiments. According to Howlett and Majerus, ". . . exposed areas of tree trunks are not an important resting site for any form of *B. betularia*."[12]

11. Ibid., p. 35.
12. R.J. Howlett and M.E.N. Majerus, "The Understanding of Industrial Melanism in the Peppered Moth (*Biston betularia*) (Lepidoptera: Geometridea)," *Biological Journal of the Linnean Society* 30 (1987): 40.

This is more than an insignificant criticism. Abnormal placement of the moths into a location rendering them much more visible would bring into question the validity of Kettlewell's results. First of all, the distinction between light and dark moths would be much less on the shadowy underside of a branch. Secondly, the unnaturally high concentration of moths in an unusual area might have changed the normal feeding pattern of the birds. In fact, some researchers are not convinced that birds are the primary peppered moth predators in nature — James Carey of the University of California, for example.[13] Also, some researchers (although not Kettlewell himself) have conducted experiments by using dead moths glued to tree trunks,[14] a practice that has been criticized by some observers.

Furthermore, many researchers considered the method by which Kettlewell assessed the degree of moth camouflage to be overly subjective. This bias would call into question the entire body of data. These criticisms bring into question the entire issue of selective bird predation being the driving force behind this so-called splendid example of *natural selection*. Without an observable, defined environmental factor to push the peppered moth to "evolve," the famous moth could not even be a candidate to be used as evidence to support Darwin's theory.

Was Kettlewell Wrong?

So was Kettlewell wrong? One major figure in this discussion has come to Kettlewell's defense, and that person is none other than Majerus, the man whose book fueled much of the recent controversy.

Over the last few years, Majerus has re-examined this question. He has conducted a study that apparently does not suffer from some of the supposed deficiencies of Kettlewell's experimental techniques. He was very careful to ensure that the moth's resting places mimicked those seen in nature, and the moths were released at night.[15] Also, using binoculars, he observed birds eating the moths. He claims that the results of his study validate Kettlewell's work. De Roode concludes, "The peppered moth should be reinstated as a textbook example of evolution in action."[16]

Good scientists must examine and re-examine the methods and techniques used to study our world. The experimental method itself relies on others conducting the same or similar types of investigations to see if previous conclusions

13. J. de Roode, "The Moths of War," *New Scientist* 196 no. 2633 (2007): 49.
14. Wells, "Second Thoughts about Peppered Moths," p. 7.
15. de Roode, "The Moths of War," p. 48.
16. Ibid., p. 49.

are indeed valid. As part of this quest for knowledge, flaws in the methods used by prior investigators are sometimes uncovered. After all, no one makes a perfect plan. Shortcomings in methodology can be corrected and further data collected to ensure proper conclusions are reached. To that end, all those who have questioned Kettlewell's methods should be commended. If there were problems with his methods, and apparently there were, those problems have apparently been corrected in subsequent evaluations.

Further, those who would be too critical of Kettlewell should proceed with some caution. There has been much written in both the pro-evolution and the pro-creation camps that has been very critical of Kettlewell. Some of this seems justified, but much of it does not, particularly the accusation that he falsified his data. There can be no more serious accusation made against a scientist, so it would seem that more proof is needed before that charge be made. After all, others involved in this area have collected data that validates Kettlewell's original conclusions. No one can know another's heart, so some measure of charity needs be given here. Perhaps Kettlewell's shortcomings can best be measured by this quote from a colleague who characterized him as "the best naturalist I have ever met, and almost the worst professional scientist I have ever known."[17]

So Where Are We?

So does all this debate about the validity of Kettlewell's peppered moth data really pose a problem for creationists? The evolutionist claims that the peppered moth story is such a shining example of evolution in action that to question it is to demonstrate unwillingness to accept proven science. Majerus has said, "The peppered moth story is easy to understand because it involves things that we are familiar with: vision and predation and birds and moths and pollution and camouflage and lunch and death. That is why the anti-evolution lobby attacks the peppered moth story. They are frightened that too many will be able to understand."[18]

Exactly what is it that we should be able to understand? To the creationist, it is very, very simple. Over the last 150 years, moths have changed into moths! The creationist has no difficulty with this process. The issue of Kettlewell's shortcomings notwithstanding, the creationist has no problem with the results of his (and other subsequent researchers') work. The concept that a less visible organism would survive better than a more visible one seems obvious in the extreme. What is not to understand here? According to de Roode, "The

17. J.A. Coyne, "Evolution Under Pressure," *Nature* 418 (2002): 19.
18. Ibid., p. 49.

peppered moth was and is a well understood example of evolution by natural selection."[19] The creationist would agree that this population change represents natural selection. However, this change is most certainly *not* molecules-to-man evolution. Natural selection and molecules-to-man evolution are not the same thing, and many are led astray by the misuse of these terms.

Natural selection can easily be seen in nature. Natural selection produces the variations within a kind of organism. Thanks to natural selection, we have the marvelous variety of creatures that we see in our world. However, in this process, fish change into (amazingly) fish, birds change into birds, dogs change into dogs, and moths change into moths. If, during the process of the study of peppered moths, the moths had changed into some other type of creature, a bird perhaps, then we might have something to talk about.

No amount of posturing by the evolutionist can change the fact that these moths are still moths and will continue to be moths. The variation seen is simply the result of sorting and resorting of the genetic material present in the original moths. At no time has there been any new information introduced into the genome of the moth (which is what molecules-to-man evolution would require). There is no evidence of the beginnings of an intermediate form between the present moth and the creature it is destined to evolve into. Moths stay moths, fish stay fish, and people stay people, regardless of the great variety seen within each.

Ultimately, the peppered moth story is more of the same. Although much of the clamor surrounding Kettlewell's work has made for good reading and, in some ways, has made for good science, the results are clear. There is nothing here, in even the smallest way, to provide evidence for the process of molecules-to-man evolution. That is what the creationist is "able to understand."

19. de Roode, "The Moths of War," p. 49.

CHAPTER 5

Is Evolutionary Humanism the Most Blood-stained Religion Ever?

BODIE HODGE

Introduction: Man's Authority or God's Authority . . . Two Religions

If God and His Word are not the authority . . . then by default . . . who is? *Man* is. When people reject God and His Word as the ultimate authority, then man is attempting to elevate his or her thoughts (collectively or individually) to a position of authority *over* God and His Word.

So often, people claim that "Christians are religious and the enlightened unbelievers who reject God are *not* religious." Don't be deceived by such a statement. For these nonbelievers are indeed religious . . . *very* religious, whether they realize it or not. For they have bought into the religion of humanism.

Humanism is the religion that elevates man to be greater than God. Humanism, in a broad sense, encompasses any thought or worldview that rejects God and the 66 books of His Word in part or in whole; hence *all* non-biblical religions have humanistic roots. There are also those that *mix* aspects of humanism with the Bible. Many of these religions (e.g., Mormons, Islam, Judaism, etc.) openly borrow from the Bible, but they also have mixed *human* elements into their religion where they take some of man's ideas to supersede many parts of the Bible, perhaps in subtle ways.[1]

There are many forms of humanism, but secular humanism has become one of the most popular today. Variant forms of secular humanism include atheism, agnosticism, non-theism, Darwinism, and the like. Each shares a belief in an evolutionary worldview with man as the centered authority over God.

Humanism organizations can also receive a tax-exempt status (the same as a Christian church in the United States and the United Kingdom) and they even have religious documents like the *Humanist Manifesto*. Surprisingly, this religion has free rein in state schools, museums, and media under the guise of neutrality, seeking to fool people into thinking it is not a "religion."[2]

Humanism and "Good"

Christians are often confronted with the claim that a humanistic worldview will help society become "better."[3] Even the first *Humanist Manifesto*, of which belief in evolution is a subset, declared: "The goal of humanism is a free and universal society in which people voluntarily and intelligently co-operate for the common good."

But can such a statement be true? For starters, what do the authors mean by "good"? They have no legitimate foundation for such a concept, since one person's "good" can be another's "evil." To have some objective standard (not a relative standard), they must *borrow* from the absolute and true teachings of God in the Bible.

Beyond that, does evolutionary humanism really teach a future of prosperity and a common good? Since death is the "hero" in an evolutionary framework, then it makes one wonder. What has been the result of evolutionary thinking in the past century (20th century)? Perhaps this could be a test of what is to come.

---

1. For example: in Islam, Muhammad's words in the Koran are taken as a higher authority than God's Word (the Bible); in Mormonism, they have changed nearly 4,000 verses of the Bible to conform to Mormon teachings and add the words of Joseph Smith and later prophets as superior to God's Word; in Judaism, they accept a portion of God's Word (the Old Testament) but by human standards, they reject a large portion of God's Word (the New Testament) as well as the ultimate Passover lamb, Jesus Christ.
2. Although the U.S. Supreme Court says that religion is not to be taught in the classroom, this one seems to be allowed.
3. One can always ask the question, by what standard do they mean "better"? God *is* that standard so they refute themselves when they speak of things being better or worse. In their own professed worldview it is merely arbitrary for something to be "better" or "worse."

Let's first look at the death estimates due to aggressive conflicts stemming from leaders with evolutionary worldviews, beginning in the 1900s, to see the hints of what this "next level" looks like:

Table 1: Estimated deaths as a result of an evolutionary worldview

Who/What?	Specific event and estimated dead	Total Estimates
Pre-Hitler Germany/ Hitler and the Nazis	WWI: 8,500,000[a] WWII: 70 million[b] [Holocaust: 17,000,000][c]	95,000,000
Leon Trotsky and Vladimir Lenin	Bolshevik revolution and Russian Civil War: 15,000,000[d]	15,000,000
Joseph Stalin	20,000,000[e]	20,000,000
Mao Zedong	14,000,000–20,000,000[f]	Median estimate: 17,000,000
Korean War	2,500,000?[g]	~2,500,000
Vietnam War (1959–1975)	4,000,000–5,000,000 Vietnamese, 1,500,000–2,000,000 Lao and Cambodians[h]	Medians of each and excludes French, Australia, and U.S. losses: 6,250,000
Pol Pot (Saloth Sar)	750,000–1,700,000[i]	Median estimate: 1,225,000
Abortion to children[j]	China estimates since 1971–2006: 300,000,000[k] USSR estimates from 1954–1991: 280,000,000[l] US estimates 1928–2007: 26,000,000[m] France estimates 1936–2006: 5,749,731[n] United Kingdom estimates 1958–2006: 6,090,738[o] Germany estimates 1968–2007: 3,699,624,[p] etc.	621,500,000 and this excludes many other countries
Grand estimate		~778,000,000

a. *The World Book Encyclopedia*, Volume 21, Entry: World War II (Chicago, IL: World Book, Inc.) p. 467; such statistics may have some variance depending on source as much of this is still in dispute.
b. Ranges from 60 to 80 million, so we are using 70 million.
c. Figures ranged from 7 to 26 million.
d. Russian Civil War, http://en.wikipedia.org/wiki/Russian_Civil_War, October 23, 2008.
e. Joseph Stalin, http://www.moreorless.au.com/killers/stalin.html, October 23, 2008.
f. Mao Tse-Tung, http://www.moreorless.au.com/killers/mao.html, October 23, 2008.

g. This one is tough to pin down and several sources have different estimates, so this is a middle-of-the-road estimate from the sources I found.
h. Vietnam War, http://www.vietnamwar.com/, October 23, 2008.
i. Pol Pot, http://en.wikipedia.org/wiki/Pol_Pot, October 23, 2008.
j. This table only lists estimates for abortion deaths in a few countries; so this total figure is likely very conservative, as well as brief stats of other atrocities.
k. Historical abortion statistics, PR China, compiled by Wm. Robert Johnston , last updated June 4 2008, http://www.johnstonsarchive.net/policy/abortion/ab-prchina.html.
l. Historical abortion statistics, USSR, compiled by Wm. Robert Johnston , last updated June 4 2008, http://www.johnstonsarchive.net/policy/abortion/ab-ussr.html.
m. Historical abortion statistics, United States, compiled by Wm. Robert Johnston , last updated June 4 2008, http://www.johnstonsarchive.net/policy/abortion/ab-unitedstates.html.
n. Historical abortion statistics, France, compiled by Wm. Robert Johnston , last updated June 4 2008, http://www.johnstonsarchive.net/policy/abortion/ab-france.html.
o. Historical abortion statistics, United Kingdom, compiled by Wm. Robert Johnston, last updated June 4 2008, http://www.johnstonsarchive.net/policy/abortion/ab-unitedkingdom.html.
p. Historical abortion statistics, FR Germany, compiled by Wm. Robert Johnston , last updated June 4 2008, http://www.johnstonsarchive.net/policy/abortion/ab-frgermany.html.

Charles Darwin's view of molecules-to-man evolution was catapulted into societies around the world in the mid-to-late 1800s. Evolutionary teachings influenced Karl Marx, Leon Trotsky, Adolf Hitler, Pol Pot, Mao Zedong, Joseph Stalin, Vladimir Lenin, and many others. Let's take a closer look at some of these people and events and examine the evolutionary influence and repercussions.

World War I and II, Hitler, Nazis, and the Holocaust

Most historians would point to the assassination of Archduke Francis Ferdinand on June 18, 1914, as the event that triggered World War I (WWI). But tensions were already high considering the state of Europe at the time. Darwinian sentiment was brewing in Germany. Darwin once said:

> At some future period, not very distant as measured by centuries, the civilized races of man will almost certainly exterminate and replace the savage races throughout the world. At the same time the anthropomorphous apes . . . will no doubt be exterminated. The break between man and his nearest allies will then be wider, for it will intervene between man in a more civilized state, as we may hope, even than the Caucasian, and some ape as low as a baboon, instead of as now between the negro or Australian [Aborigine] and the gorilla.[4]

4. Charles Darwin, *The Descent of Man* (New York: A.L. Burt, 1874, 2nd ed.), p. 178.

Darwin viewed the "Caucasian" (white-skinned Europeans) as the dominant "race" in his evolutionary worldview. To many evolutionists at the time, mankind had evolved from ape-like creatures that had more hair, dark skin, dark eyes, etc. Therefore, more "evolved" meant less body hair, blond hair, blue eyes, etc. Later, in Hitler's era, Nazi Germany practiced *Lebensborn*, which was a controversial program, the details of which have not been entirely brought to light. Many claim it was a breeding program that tried to evolve the "master race" further — more on this below.

But the German sentiment prior to WWI was very much bent on conquering for the purpose of expanding their territory and their "race." An encyclopedia entry from 1936 states:

> In discussions of the background of the war much has been said of Pan-Germanism, which was the spirit of national consciousness carried to the extreme limit. The Pan-Germans, who included not only militarists, but historians, scientists, educators and statesmen, conceived the German people, no matter where they located, as permanently retaining their nationality. The most ambitious of this group believed that it was their mission of Germans to extend their kultur (culture) over the world, and to accomplish this by conquest if necessary. In this connection the theory was advanced that the German was a superior being, destined to dominate other peoples, most of whom were thought of as decadent.[5]

Germany had been buying into Darwin's model of evolution and saw themselves as the superior "race," destined to dominate the world, and their actions were the consequence of their worldview. This view set the stage for Hitler and the Nazi party and paved the road to WWII.

Hitler and the Nazis

World War II dwarfed World War I in the total number of people who died. Racist evolutionary attitudes exploded in Germany against people groups such as Jews, Poles, and many others. Darwin's teaching on evolution and humanism heavily influenced Adolf Hitler and the Nazis.

Hitler even tried to force the Protestant church in Germany to change fundamental tenants because of his newfound faith.[6] In 1936, while Hitler was in power, an encyclopedia entry on Hitler stated:

5. *The American Educator Encyclopedia* (Chicago, IL: The United Educators, Inc., 1936), p. 3914 under entry "World War."
6. *The American Educator Encyclopedia*, p. 1702 under entry "Hitler."

. . . a Hitler attempt to modify the Protestant faith failed.[7]

His actions clearly show that he did not hold to the basic fundamentals taught in the 66 books of the Bible. Though some of his writings suggest he did believe in some form of God early on (due to his upbringing within Catholicism), his religious views moved toward atheistic humanism with his acceptance of evolution. Many atheists today try to disavow him, but actions speak louder than words.

The Alpha History site (dedicated to much to the history of Nazi Germany by providing documents, transcribed speeches, and so on) says:

> Contrary to popular opinion, Hitler himself was not an atheist. . . . Hitler drifted away from the church after leaving home, and his religious views in adulthood are in dispute.[8]

So this history site is not sure what his beliefs were, but they seem to be certain that he was not an atheist! If they are not sure what beliefs he held, how can they be certain he was not an atheist?[9] The fact is that many people who walk away from church become atheists (i.e., they were never believers in the first place as 1 John 2:19 indicates). And Hitler's actions were diametrically opposed to Christianity . . . but not atheism, where there is no God who sets what is right and wrong.[10]

Regardless, this refutes notions that Hitler was a Christian as some have falsely claimed. Hitler's disbelief started early. He said:

> The present system of teaching in schools permits the following absurdity: at 10 a.m. the pupils attend a lesson in the catechism, at which the creation of the world is presented to them in accordance with the teachings of the Bible; and at 11 a.m. they attend a lesson in natural science, at which they are taught the theory of evolution. Yet the two doctrines are in complete contradiction. As a child, I suffered from this contradiction, and ran my head against a wall . . . Is there a single religion that can exist without a dogma? No, for in that case it would belong to the order of science . . . But there have been human

---

7. *The American Educator Encyclopedia*, p. 1494 under entry "Germany."
8. "Religion in Nazi Germany," http://alphahistory.com/nazigermany/religion-in-nazi-germany/, April 3, 2013.
9. Romans 1 makes it clear that all people believe in God, they just suppress that knowledge, and this is also the case with any professed atheist.
10. For an extensive treatise on Hitler's (and the Nazi's) religious viewpoints, see J. Bergman, *Hitler and the Nazi Darwinian Worldview* (Kitchener, Ontario, Canada: Joshua Press Inc., 2012).

> beings, in the baboon category, for at least three hundred thousand years. There is less distance between the man-ape and the ordinary modern man than there is between the ordinary modern man and a man like Schopenhauer. . . . It is impossible to suppose nowadays that organic life exists only on our planet.[11]

Consider this quote in his unpublished second book:

> The types of creatures on the earth are countless, and on an individual level their self-preservation instinct as well as the longing for procreation is always unlimited; however, the space in which this entire life process plays itself out is limited. It is the surface area of a precisely measured sphere on which billions and billions of individual beings struggle for life and succession. In the limitation of this living space lies the compulsion for the struggle for survival, and the struggle for survival, in turn contains the precondition for evolution.[12]

Hitler continues:

> The history of the world in the ages when humans did not yet exist was initially a representation of geological occurrences. The clash of natural forces with each other, the formation of a habitable surface on this planet, the separation of water and land, the formation of the mountains, plains, and the seas. That [was] is the history of the world during this time. Later, with the emergence of organic life, human interest focuses on the appearance and disappearance of its thousandfold forms. Man himself finally becomes visible very late, and from that point on he begins to understand the term "world history" as referring to the history of his own development — in other words, the representation of his own evolution. This development is characterized by the never-ending battle of humans against animals and also against humans themselves.[13]

Hitler fully believed Darwin as well as Darwin's precursors — such as Charles Lyell's geological ages and millions of years of history. In his statements here, there is no reference to God. Instead, he unreservedly flew the banner of

11. Adolf Hitler, translated by Norman Cameron and R.H. Stevens, *Hitler's Secret Conversations*, 1941–1944 (The New American Library of World Literature, Inc., 1961).
12. *Hitler's Second Book*, Adolf Hitler, edited by Gerald L. Weinberg, translated by Krista Smith (New York: Enigma Books, 2003), p. 8.
13. *Hitler's Second Book*, p. 9.

naturalism and evolution and only mentioned God in a rare instance to win Christians to his side, just as agnostic Charles Darwin did in his book *On the Origin of Species.*[14]

One part of the Nazi party political platform's 25 points in 1920 says:

> We demand freedom of religion for all religious denominations within the state so long as they do not endanger its existence or oppose the moral senses of the Germanic race. The Party as such advocates the standpoint of a positive Christianity without binding itself confessionally to any one denomination.[15]

Clearly this "positive Christianity" was an appeal to some of Christianity's morality, but not the faith itself. Many atheists today still appeal to a "positive Christian" approach, wanting the morality of Christianity (in many respects), but not Christianity.

Christianity was under heavy attack by Hitler and the Nazis as documented from original sources prior to the end of WWII by Bruce Walker in *The Swastika against the Cross.*[16] The book clearly reveals the anti-Christian sentiment by Hitler and the Nazis and their persecution of Christianity and their attempt to make Christianity change and be subject to the Nazi state and beliefs.

In 1939–1941, the Bible was rewritten for the German people at Hitler's command, eliminating all references to Jews, and made Christ out to be pro-Aryan! The Ten Commandments were replaced with these twelve:[17]

1. Honor your Fuhrer and master.
2. Keep the blood pure and your honor holy.
3. Honor God and believe in him wholeheartedly.
4. Seek out the peace of God.
5. Avoid all hypocrisy.
6. Holy is your health and life.

---

14. In the first edition of *Origin of Species*, God is not mentioned; in the sixth edition, "God" was added several times to draw Christians into this false religion. See R. Hedtke, *Secrets of the Sixth Edition* (Green Forest, AR: Master Books, 2010).
15. Nazi Party 25 Points (1920), http://alphahistory.com/nazigermany/nazi-party-25-points-1920/.
16. B. Walker, *The Swastika against the Cross* (Denver, CO: Outskirts Press, Inc., 2008).
17. "Hitler Rewrote the Bible and Added Two Commandments," Pravda News Site, 8/10/2006; http://english.pravda.ru/world/europe/10-08-2006/83892-hitler-0/; "Jewish References Erased in Newly Found Nazi Bible," Daily Mail Online, August 7, 2006; http://www.dailymail.co.uk/news/article-399470/Jewish-references-erased-newly-Nazi-Bible.html.

7. Holy is your well-being and honor.
8. Holy is your truth and fidelity.
9. Honor your father and mother — your children are your aid and your example.
10. Maintain and multiply the heritage of your forefathers.
11. Be ready to help and forgive.
12. Joyously serve the people with work and sacrifice.

Hitler had *replaced* Christ in Nazi thought; and children were even taught to pray to Hitler instead of God![18] Hitler and the Nazis were not Christian, but instead were humanistic in their outlook, and any semblance of Christianity was cultic. The Nazis determined that their philosophy was the best way to bring about the common good of all humanity.

Interestingly, it was Christians alone in Germany who were unconquered by the Nazis, and they suffered heavily for it. Walker summarizes in his book:

> You would expect to find Christians and Nazis mortal enemies. This is, of course, exactly what happened historically. Christians, alone, proved unconquerable by the Nazis. It can be said that Christians did not succeed in stopping Hitler, but it cannot be said that they did not try, often at great loss and nearly always as true martyrs (people who could have chosen to live, but who chose to die for the sake of goodness.)[19]

Hitler and the Nazis' evolutionary views certainly helped lead Germany into WWII because they viewed the "Caucasian" as more evolved (and, more specifically, the Aryan peoples of the Caucasians), which to them justified their adoption of the idea that lesser "races" should be murdered in the struggle for survival. Among the first to be targeted were Jews, then Poles, Slavs, and then many others — including Christians, regardless of their heritage.

Trotsky, Lenin

Trotsky and Lenin were both notorious leaders of the USSR — and specifically the Russian revolution. Lenin, taking power in 1917, became a ruthless leader and selected Trotsky as his heir. Lenin and Trotsky held to Marxism, which was built, in part, on Darwinism and evolution applied to a social scheme.

18. Walker, p. 20–22.
19. Walker, p. 88.

Karl Marx regarded Darwin's book as an "epoch-making book." With regards to Darwin's research on natural origins, Marx claimed, "The latter method is the only materialistic and, therefore, the only scientific one."[20]

Few realize or admit that Marxism, the primary idea underlying communism, is built on Darwinism and materialism (i.e., no God). In 1883, Freidrich Engels, Marx's longtime friend and collaborator, stated at Marx's funeral service, that "Just as Darwin discovered the law of evolution in organic nature, so Marx discovered the law of evolution in human history."[21] Both Darwin and Marx built their ideologies on naturalism and materialism (tenants of evolutionary humanism). Trotsky once said of Darwin:

> Darwin stood for me like a mighty doorkeeper at the entrance to the temple of the universe. I was intoxicated with his minute, precise, conscientious and at the same time powerful, thought. I was the more astonished when I read . . . that he had preserved his belief in God. I absolutely declined to understand how a theory of the origin of species by way of natural selection and sexual selection and a belief in God could find room in one and the same head.[22]

Trotsky's high regard for evolution and Darwin were the foundation of his belief system. Like many, Trotsky probably did not realize that the precious few instances of the name "God" did not appear in the first edition of *Origin of Species.* These references were added later, and many suspect that this was done to influence church members to adopt Darwinism. Regardless, Trotsky may not have read much of Darwin's second book, *Descent of Man,* in which Darwin claims that man invented God:

> The same high mental faculties which first led man to believe in unseen spiritual agencies, then in fetishism, polytheism, and ultimately in monotheism, would infallibly lead him, as long as his reasoning powers remained poorly developed, to various strange superstitions and customs.[23]

20. *Great Books of the Western World,* Volume 50, *Capital,* Karl Marx (Chicago, IL: William Benton Publishers, 1952), footnotes on p. 166 and p. 181.
21. Gertrude Himmelfarb, *Darwin and the Darwinian Revolution* (London: Chatto & Windus, 1959), p. 348.
22. Max Eastman, *Trotsky: A Portrait of His Youth* (New York, 1925), p. 117–118.
23. Charles Darwin, *The Descent of Man and Selection in Relation to Sex,* Chapter III, "Comparison of the Mental Powers of Man and the Lower Animals," 1871. As printed in *Great Books of the Western World,* Volume 49, Robert Hutchins, ed. (Chicago, IL: William Benton Publishers, 1952), p. 303.

Vladimir Lenin picked up on Darwinism and Marxism and ruled very harshly as an evolutionist. His variant of Marxism has become known as Leninism. Regardless, the evolutionist roots of Marx, Trotsky, and Lenin were the foundation that communism has stood on — and continues to stand on.

Stalin, Mao, and Pol Pot, to Name a Few

Perhaps the most ruthless communist leaders were Joseph Stalin, Mao Zedong, and Pol Pot. Each of these were social Darwinists, ruling three different countries — the Soviet Union, China, and Cambodia, respectively. Their reigns of terror demonstrated the end result of reducing the value of human life to that of mere animals, a Darwinistic teaching.[24] Though I could expand on each of these, you should be getting the point by now. So let's move to another key, but deadly, point in evolutionary thought.

Abortion – the War on Babies

The war on children has been one of the quietest, and yet bloodiest, in the past hundred years. In an evolutionary mindset, the unborn have been treated as though they are going through an "animal phase" and can simply be discarded.

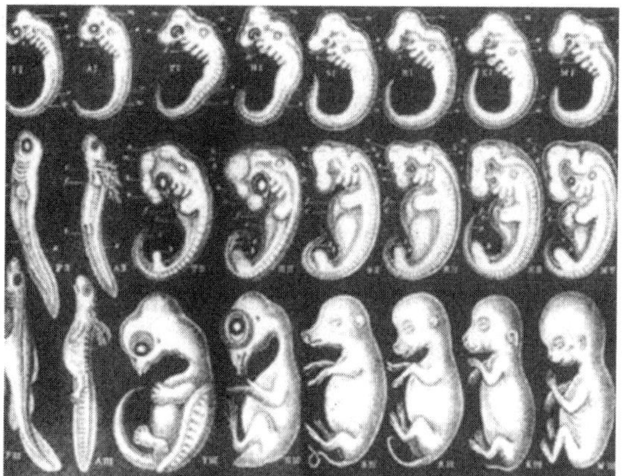

Haeckel's faked embryos: A detailed analysis of this subject will be done in chapter 26

Early evolutionist Ernst Haeckel first popularized the concept that babies in the womb are actually undergoing animal developmental stages, such as a fish stage and so on. This idea has come to be known as *ontogeny recapitulates phylogeny.* Haeckel even faked drawings of various animals' embryos and had them drawn next to human embryos looking virtually identical.

24. R. Hall, "Darwin's Impact — The Bloodstained Legacy of Evolution," *Creation* Magazine 27(2) (March 2005): 46–47, http://www.answersingenesis.org/articles/cm/v27/n2/darwin.

These drawings have been shown to be completely false.[25] Haeckel himself partially confessed as much.[26] However, this discredited idea has been used repeatedly for a hundred years! Textbooks today still use this concept (though not Haeckel's drawings), and museums around the world still teach it.

Through this deception, many women have been convinced that the babies they are carrying in their wombs are simply going through an animal phase and can be aborted. Author and general editor of this volume, Ken Ham, states:

> In fact, some abortion clinics in America have taken women aside to explain to them that what is being aborted is just an embryo in the fish stage of evolution, and that the embryo must not be thought of as human. These women are being fed outright lies.[27]

Evolutionary views have decreased the value of human life. Throughout the world, the casualties of the war on children is staggering. Though deaths of children and the unborn did exist prior to the "evolution revolution," they have increased exponentially after the promotion of Darwinian teachings.

Conclusion

Is evolution the cause of wars and deaths? Absolutely not — both existed long before Darwin was born. Sin is the ultimate cause.[28] But an evolutionary worldview has done nothing but add fuel to the fire.

In spite of the wars and atrocities caused by those who subscribed to an evolutionary worldview in recent times, there is still hope. We can end the seemingly endless atrocities against the unborn and those deemed less worthy of living, including the old and impaired.

25. Michael Richardson et al, *Anatomy and Embryology*, 196(2) (1997): 91–106.
26. Haeckel said, "A small portion of my embryo-pictures (possibly 6 or 8 in a hundred) are really (in Dr Brass's [one of his critics] sense of the word) "falsified" — all those, namely, in which the disclosed material for inspection is so incomplete or insufficient that one is compelled in a restoration of a connected development series to fill up the gaps through hypotheses, and to reconstruct the missing members through comparative syntheses. What difficulties this task encounters, and how easily the draughts — man may blunder in it, the embryologist alone can judge." "The Truth about Haeckel's Confession," *The Bible Investigator and Inquirer*, M.L. Hutchinson, Melbourne (March 11, 1911): p. 22–24.
27. Ken Ham, *The Lie: Evolution*, chapter 8, "The Evils of Evolution (Green Forest, AR: Master Books, 1987), p. 105.
28. Ken Ham, gen. ed., *The New Answers Book 1* (Green Forest, AR: Master Books, 2006), chapter 26, "Why Does God's Creation Include Death and Suffering? p.325–338; http://www.answersingenesis.org/articles/nab/why-does-creation-include-suffering.

In Egypt, Israelite boys were slaughtered by being thrown into the Nile at the command of Pharaoh (Exodus 1:20). And yet, by the providence of God, Moses survived and led the Israelites to safety, and the Lord later judged the Egyptians.

In Judea, under the Roman Empire, subordinate King Herod the Great commanded the slaughter of all the boys under the age of two in and around Bethlehem. And yet, by the providence of God, Jesus, the Son of God, survived and later laid down His life to bring salvation to mankind as the Prince of Peace. Herod's name, however, went down in history as an evil tyrant and murderer.

In this day and age, governments readily promote and fund the killing of children, both boys and girls, and sometimes command it, through abortion. By providence, however . . . you survived. While we can't change the past, we can learn from it. If we are to stop this continuing bloodshed, we must get back to the Bible and realize the bankrupt religion of evolutionary humanism has led only to death — by the millions. We need to point those who think humanity is the answer to the Savior who took the sins of humanity on Himself to offer them salvation.

CHAPTER 6

Was Charles Darwin a Christian?

DR. TOMMY MITCHELL

Much has been written about the religious views of Charles Darwin. What exactly did he believe, and when? Did he "reject" Christianity? Was he out to "destroy" Christianity, as some in the Church have come to believe?

While it is true that Darwin's ideas have caused great harm to the Church and have led many people to openly question the authority of the Bible, what did the man himself actually believe? Did he ever become a Christian?

Beginnings

Charles Darwin was born in 1809. He was part of a well-to-do family in England.

His grandfather, Erasmus, was a prominent physician, poet, and somewhat of an activist. He could best be described as a "progressive" or "free" thinker. Dr. Erasmus had a naturalistic view of origins and even promoted basic evolutionary ideas. His religious stand was as a deist, and he rejected the idea that the Bible was supernaturally inspired.

Charles never met his grandfather, who died before Charles was born. He did, however, become familiar with his grandfather's beliefs and ideas through reading his writings.

Charles's father, Robert, was also a physician. Beyond that, Robert was also a very successful investor, which provided the Darwin family with a very comfortable life.

As is often the case, the rejection of the authority of God's Word by one generation led to complete rejection of God in the next. Robert Darwin was an atheist.

In spite of Robert's lack of belief, Charles was christened in the Church of England (Anglican). This was obviously not due to any conviction that Robert had about the doctrine of the Anglican Church. It was most probably done to keep up appearances within the social order of the day.

There was, however, the influence of the mother. Susannah Darwin, Charles' mother, was a Unitarian. She took Charles to chapel for worship services. After her sudden death, Darwin's sisters took him to services at the Anglican Church.

For a year, Charles attended a Unitarian day school and later attended Shrewsbury Grammar School.

One writer has stated that Darwin "was thoroughly orthodox" during this time in his life. However, given the varied influences during his upbringing, it is difficult if not impossible, to imagine that Darwin's thinking was in any way "orthodox."

Higher Education

As was expected, Charles was to go to college to train to be a physician, like his father and grandfather. So he was soon off to Edinburgh to study medicine. That did not last long.

Darwin hated dealing with corpses, and he disliked dissection, both of which were necessary to become a doctor. To further hasten his retreat from medicine, he had developed a great interest in natural history and zoology. These pursuits began to occupy more and more of his time. His great interest in geology also took shape during these years.

It soon became clear that medicine would not be his life's work. In his autobiography Darwin wrote about this time in his life, stating, "He [Darwin's father] was very properly vehement against my turning an idle sporting man, which then seemed my probable destination."[1]

Further, Darwin wrote, "To my deep mortification my father once said to me, 'You care for nothing but shooting, dogs, and rat-catching, and you will be a disgrace to yourself and all your family.' "[2]

So at the advice of his father, it was decided that Charles would become a country clergyman. After all, this was a position with a steady income, some

1. Nora Barlow, ed., The Autobiography of Charles Darwin, 1809–1882 (New York: W.W. Norton, 1958), p. 49.
2. Ibid., p. 27.

social stature, and plenty of time to collect beetles and follow his natural history pursuits. The only thing lacking here was a genuine, heartfelt calling to the ministry.

In Darwin's own words:

> I asked for some time to consider, as from what little I had heard and thought on the subject I had scruples about declaring my belief in all the dogmas of the Church of England; though otherwise I liked the thought of being a country clergyman . . . and as I did not then in the least doubt the strict and literal truth of every word of the Bible, I soon persuaded myself that our Creed must be fully accepted. It never struck me how illogical it was to say that I believed in what I could not understand and what is in fact unintelligible.[3]

So he was off to Cambridge for his "theological" training. Unfortunately for Darwin, Cambridge was not the place to convict him of the authority of the Bible. At this time, theology training at Cambridge consisted mostly of coursework in the classics and philosophy along with a heavy emphasis on the works of William Paley — works that presented a rationalistic view of Christianity. Paley is best remembered for his arguments in favor of one of the primary theological positions of the day, natural theology. Basically, Paley held that one can know God, the Designer, by close examination of His creation, that is, nature. More simply, if it looks designed, there must be a designer. Early on, Darwin was fascinated by this argument. However, he rejected it later.

Even though Paley's theology also presented a biblical argument, this was largely ignored. During this period, the authority and historicity of the Bible had been called into question. Through the study of nature one could come to sufficiently understand God, it was believed. But the Bible was not held to be the ultimate authority. In fact, the Bible itself was being called into question, particularly regarding the actual nature of the Noahic Flood and the age of the earth. It was at this time that the idea of the earth being millions of years old was taking hold, not only in secular "scientific" circles but also in the Church itself.

Even though Darwin was at Cambridge for a degree in theology, his interest in natural science only strengthened. He attended lectures on botany, and his interest in geology grew. Most of the academics that taught Darwin in these areas were either openly critical of or outright denied the authority of the Bible.

3. Ibid., p. 49.

Again, the foundation of a system of belief was being laid that Darwin built upon in later life.

Though Darwin did get his degree in theology, he still had no heartfelt call to ministry. As reported by two of Darwin's biographers, Desmond and Moore,

> Darwin had asked Herbert whether he really felt "inwardly moved by the Holy Spirit" to enter the Church. When the Bishop put the question to him in the ordination service, what would he reply? "No," answered Herbert; he could not say he felt moved. Darwin chimed in, "Neither can I, and therefore I cannot take orders."[4]

So much for a genuine call to ministry. While some in the Church today point to Darwin's preparation for Christian ministry as evidence that he had some Christian beliefs, this is clearly not the case.

The *Beagle*

After leaving Cambridge, Darwin was presented an opportunity to participate in a South American survey expedition aboard the HMS *Beagle*. He was to join the ship's company as a naturalist and gentleman companion to the ship's captain, Robert FitzRoy.

During the voyage, Darwin was actually more interested in the geology of the lands he visited than the zoology of these new places. In fact, over half of the notes he made were geologic in nature. As he observed the geology, he became convinced that the strata were laid down over millions of years. Much of this was because he admired the works of a man named Charles Lyell. Lyell was the author of the book *Principles of Geology*.[5] As he studied Lyell's work, Darwin became convinced the uniformitarian view of geology was correct. Simply put, he came to believe that "the present is the key to the past." In other words, denying that events such as the catastrophic global Flood had a major role in shaping the earth, he believed that the ordinary geological processes we see today have always proceeded at the same rate so that the geological formations we see today required millions of years to form.

Here was a situation where a man who had already come to doubt the authority of the Bible was becoming more captivated with the secular thinking of his day. So the Bible was wrong, he decided, and the millions of years were true.

---

4. Adrian Desmond and James Moore, *Darwin* (New York: Warner Books, 1991), p. 66.
5. Incredibly, it was Capt. FitzRoy, an evangelical, who presented Darwin with Volume 1 of Lyell's book before the voyage began.

However, at that point, he had not rejected the Bible as completely untrue, rather "whilst on board the *Beagle* I was quite orthodox," he wrote, "and I remember being heartily laughed at by several of the officers (though themselves orthodox) for quoting the Bible as an unanswerable authority on some point of morality. I supposed it was the novelty of the argument that amused them."[6]

So here Darwin was using the Bible as a basis for morality although it was without any real confidence in its authority, because he then wrote, "But I had gradually come, by this time, to see that the Old Testament from its manifestly false history of the world . . . was no more to be trusted than the sacred books of the Hindoos, or the beliefs of any barbarian."[7] Later in life, he would come to understand the inconsistency in accepting biblical morality while denying its history. Unfortunately, his solution was to reject the Bible completely.

After the *Beagle*

The voyage of the *Beagle* ended October 2, 1836. Darwin soon began the process of studying the specimens he collected and pondering the observations he had made.

He was also considering spiritual things. In his autobiography, Darwin wrote, "During these two years [October 1836 to January 1839], I was led to think much about religion."[8] Unfortunately, this consideration did not lead in any way to a genuine understanding of Christianity or his need for salvation. This is obvious in his relationship to his new wife.

In January 1839, he married his first cousin, Emma. She was a very religious woman who was understandably concerned about the spiritual state of her husband. Although some have suggested that Darwin was at least a nominal Christian at that point, his own writings put that idea to rest with statements like "before I was engaged to be married, my father advised me to conceal carefully my doubts . . . some women suffered miserably by doubting about the salvation of their husbands."[9] If Darwin were saved, why would this even be an issue?

Death and Suffering

One of the most important issues in Darwin's life was his struggle with death and suffering. Perhaps it was this issue that tipped the scales for Darwin more than any other. It was a theme that he considered all his life. All around

6. Barlow, *The Autobiography of Charles Darwin, 1809–1882*, p. 71.
7. Ibid., p. 71.
8. Ibid., p. 71.
9. Ibid., p. 79.

him he saw death, disease, and struggle. And with all he saw, he doubted more and more that a caring God could exist. This is evident when Darwin said, "I cannot persuade myself that a beneficent and omnipotent God would have designedly created the *Ichneumonidae* with the express intention of their feeding within the living bodies of Caterpillars, or that a cat should play with mice."[10]

He further concluded, "This very old argument from the existence of suffering against the existence of an intelligent first cause seems to me a strong one; whereas, as just remarked, the presence of much suffering agrees well with the view that all organic beings have been developed through variation and natural selection."[11]

So if there is death and suffering, there cannot be a God that cares, he reasoned. Why would God create creatures to prey upon and kill each other? This was particularly brought home to him at the death of his daughter, Annie. She died at age ten after a brief illness. At the time, Darwin wrote, "Poor dear Annie . . . was taken with a vomiting attack, which at first thought of the smallest importance; but it rapidly assumed the form of a low and dreadful fever, which carried her off in 10 days. Thank God, she hardly suffered at all. . . . She was my favourite child. . . . Poor dear little soul."[12]

One can only imagine the grief he felt at the loss of his child. One of Darwin's major biographies states, "Annie's cruel death destroyed Charles's tatters of beliefs in a moral, just universe. Later he would say that this period chimed the final death-knell for his Christianity, St. Charles now took his stand as an unbeliever."[13]

While there is no case to be made that Darwin was in any way a Christian at that time, it is easy to understand how such an event could cause a "spiritual" person to give up his "spirituality."

Life and Faith

For much of his life, Darwin did consider issues of spirituality. Perhaps this was a result of his understanding of the logical outcome of the ideas he proposed. He did seem, at least at times, to struggle to reconcile the inconsistencies: "My theology is a simple muddle; I cannot look at the universe as the result of blind chance, yet I can see no evidence of beneficent design, or indeed of design of any kind, in the details."[14]

10. Francis Darwin, ed., *The Life and Letters of Charles Darwin*, Vol. II (New York: Appleton, 1897), p. 105.
11. Barlow, *The Autobiography of Charles Darwin, 1809–1882*, p. 75.
12. Darwin, *The Life and Letters of Charles Darwin*, Vol. I, p. 348.
13. Desmond and Moore, *Darwin*, p. 387.
14. Barlow, *The Autobiography of Charles Darwin, 1809–1882*, p. 130.

Further, he wrote, "When thus reflecting I feel compelled to look to a First Cause having an intelligent mind in some degree analogous to that of man; and I deserve to be called a Theist."

While he postulated ideas that would be a basis for an understanding of the world governed by natural processes alone, he acknowledged, for a time at least, a First Cause. This First Cause was needed to help explain the origin of life, but this "god" was detached and did not interact with man or man's affairs. This acknowledgement of even an impersonal "god" did not last.

Eventually, Darwin concluded, "The mystery of the beginning of all things is insoluble by us; and I for one must be content to remain an Agnostic."[15]

As he came to more completely realize the logical outcome of his materialist worldview, he apparently understood that his defense of the Bible while on board the *Beagle* was without basis. Those many years later he wrote, "A man who has no assured and ever present belief in the existence of a personal God or of a future existence with retribution and reward, can have for the rule of his life, as far as I can see, only to follow those impulses and instincts which are the strongest or which seem to him the best ones."[16]

As his life continued, whatever vestiges of genuine spirituality that may have existed gradually faded and died. That he never understood or accepted the basis tenets of Christianity was well described when he wrote, "I can indeed hardly see how anyone ought to wish Christianity to be true; for if so the plain language of the text seems to show that the men who do not believe, and this would include my Father, Brother and almost all my best friends, would be everlastingly punished. And this is a damnable doctrine."[17]

If Darwin ever even remotely considered that Christianity might be true, that idea was now dead. "I was very unwilling to give up my belief. . . . But I found it more and more difficult, with free scope given to my imagination, to invent evidence which would suffice to convince me. Thus disbelief crept over me at a very slow rate, but was at last complete."[18]

Did Darwin Become a Christian on His Deathbed?

One of the most popular misconceptions about Darwin is that he came to Christ on his deathbed. While it would be wonderful if it were true, unfortunately this is nothing but an urban legend.

15. Ibid., p. 78.
16. Ibid., p. 78.
17. Ibid., p. 72.
18. Ibid., p. 72.

Reports of Darwin having some sort of conversion experience began within weeks of his death. These began in England but before long at least one was reported from as far away as Canada. The most famous of them all came from a woman known as Lady Hope.

Lady Hope was born Elizabeth Reid Cotton and was the daughter of General Sir Arthur Cotton. She and her father were active evangelists in Kent, near the home of Charles Darwin. In 1877, she married Admiral Sir James Hope and thus became Lady Hope.

While attending a conference in Massachusetts in 1915, Lady Hope told of a visit that she had with Darwin some months before his death. According to her, Darwin had been bedridden for some months before he died. The report was that at the time of the visit she found Darwin sitting in bed. When she asked what he was reading, he was reported to say, "Hebrews . . . the Royal Book!" Additionally, Darwin supposedly commented, "I was a young man with unformed ideas."

Lady Hope further claimed that before her departure she was asked by Darwin to return and speak to his servants in his summerhouse. When asked about the subject on which she should speak, Darwin was said to have replied, "Christ Jesus!"

While it would be wonderful to report that this account of Darwin's conversion was true, there are just too many inconsistencies in the account. First, if Lady Hope did indeed visit Darwin, it would have been at least six months before his death. At this time Darwin was not bedridden, nor was he bedridden for an extended period of time before he died.

Second, this supposed conversion was never mentioned in any of Darwin's correspondence. Given that Darwin wrote extensively (totaling over 14,000 notes and letters), it is curious to suggest that if he did have a genuine conversion experience it was not mentioned at all in any of his writings.

Third, and most importantly, his family denied each and every report that Charles Darwin came to Christ. Certainly, a genuine conversion would be something to be celebrated and joyously shared with family and friends, especially for his wife. In 1915, Darwin's son Francis wrote, "He [Darwin] could not have become openly and enthusiastically Christian without the knowledge of his family, and no such change occurred."[19]

Also, if the story were credible, why did Lady Hope wait 33 years before relating it?

19. James Moore, *The Darwin Legend* (Grand Rapids: Baker Books, 1994), p. 144.

A close examination of this tale is fascinating because of what it does not claim. The actual report of Lady Hope's story does not say that Darwin actually renounced evolution; it merely says that Darwin speculated over the outcome of his ideas. Also, it was obviously not a deathbed meeting. It took place several months before Darwin died. Most telling is that Lady Hope never described Darwin actually professing faith in Christ. She simply reported that Darwin was reading the Bible. Even if true, this is a far cry from a saving knowledge of Jesus Christ.

So no matter how earnestly this tale is repeated in churches around the world, there is no truth to the "deathbed conversion" account.

In Conclusion

As much as we might wish it to be true, there is no evidence in the life of Charles Darwin that he was a Christian. Certainly, he struggled with spiritual issues, but that is not the same thing at all.

Many have tried to paint a picture that Darwin was a Christian, but because of circumstance or issues in his life walked away from the faith. Darwin's words themselves cause us to reject that position out of hand: "Although I did not think much about the existence of a personal God until a considerably later period of my life."[20]

There is no more personal God than Jesus Christ. If this was not a consideration for Darwin earlier in his life, then how could one even consider him to be a Christian during those years?

In a letter to F.A. McDermott dated November 24, 1880, Darwin wrote, "I am sorry to have to inform you that I do not believe in the Bible as divine revelation, & therefore not in Jesus Christ as the son of God."[21]

Charles Darwin rejected the Bible. Thus he had no basis to truly understand the world around him. He did not truly understand the geology of the world. Rejecting biblical creation, he could not answer the question of how life itself started. He never could reconcile the issue of a loving God amidst the death and suffering in the world.

Ultimately, he never acknowledged sin. He did not understand that the world is broken because of sin. Most importantly, he did not recognize that he was a sinner in need of a Savior.

Was Charles Darwin a Christian? The answer is no. More than anything else about his life, this is the tragedy. A soul lost for eternity, separated from God.

20. Barlow, *The Autobiography of Charles Darwin, 1809–1882*, p. 73.
21. Darwin Correspondence Project, letter 12851; www.darwinproject.ac.uk/entry-12851.

CHAPTER 7

Cavemen . . . Really?

DR. DAVID MENTON, DR. GEORGIA PURDOM, AND JOHN UPCHURCH

As far as stereotypes go, cavemen make easy targets — especially when transplanted into the 21st century. Their brutish way of dealing with contemporary situations earns a laugh on commercials and TV shows. They just don't understand us modern humans, and their misunderstanding strikes humor gold.

But when we cut away the laugh track and the bumbling ways, we're left with something of an enigmatic figure — a being without a settled place in our understanding of history. Perhaps, in fact, it's our discomfort with not knowing what to do with cavemen that makes us laugh. So just who were they?

Would the Real "Caveman" Please Stand Up?

Before we go spelunking, we need to limit our scope somewhat. At its most basic, the term *caveman* simply means "a person who dwells in a cave," which isn't unheard of even today. But that's rarely what we mean when we use the word. Instead, we're usually talking about a group of ancient cave hoppers who left behind animal artwork, rough-hewn weapons, and bones — at least, that's the common assumption. While the collective opinion of history and science has moved beyond considering these early humans as animal-like brutes, the term still carries with it the baggage of a being somewhat lesser than modern *Homo sapiens* (us today). And that's unfortunate — as we'll see.

Those early humans commonly classified as "cavemen" break down into several groups, scattered throughout Europe, the Middle East, Africa, and Asia. Calling these groups "cavemen" may, in fact, be somewhat misleading. Many of them simply found temporary shelter or buried their dead in caves, which tend to preserve remains and artifacts more often than houses in the open. (They probably preferred living in caves about as much as we do.)

Nevertheless, the term caveman is often used as a catch-all for peoples who lived in an earlier era in human history — the Ice Age. We'll focus on five of these groups: Neanderthals, early *Homo sapiens* (Cro-Magnon man), *Homo erectus*, Denisovans, and *Homo floresiensis*.[1] The first three have long been stalwarts of the caveman discussion, but the latter two have only recently been uncovered — the Denisovans in Siberia and *Homo floresiensis* (sometimes called hobbits) in Indonesia.[2]

Neanderthals

Neanderthals may be the most well-known of the five groups — with hundreds of individuals to study. After they served time as a separate "hominid" (human-like) species according to evolutionary scientists, DNA testing in particular has significantly trimmed their distance from *Homo sapiens*.[3] This shouldn't surprise us, considering the overwhelming evidence of their humanity.

In dozens of caves and rock shelters, for instance, we find evidence of bodies that have been carefully buried with all the care you might expect from a modern funeral.

1. The term *species* is a modern convention established by the creationist Carolus Linneaus. It traditionally refers to separate populations that are similar but can no longer produce viable offspring. But this is not the case of any humans. We need some sort of term to describe different peoples, such as Europeans or Parisians. But in this instance, some names are unfortunately scientific terms that imply "species," and there are no easily recognizable alternative names. So *Homo erectus* and early *Homo sapiens* are used in this article to describe our ancestors at certain times and places, but this is not a reference to their being different species.
2. There are some who try to take apes and lump them as humans. One needs to be discerning about such instances. Take for example *A. sediba*, which is *not* human. First let us consider just a few of the non-trivial differences between *Australopithecus sediba* and humans. *Australopithecus sediba* has a brain measuring between one-third and one-fourth the size of that of a typical human of comparable size (but well within the range of apes). A comparison of the skull of *Australopithecus sediba* with that of humans reveals that the lower face of *Australopithecus sediba* is sloped like that of apes. And, like apes, the forehead of *Australopithecus sediba* is flat, making the orbits of the eyes barely visible when viewed from the side. The mandible of *Australopithecus sediba* bears no close resemblance to that of man (or even a chimpanzee) but rather is more similar to that of a gorilla. The postcranial skeleton of *Australopithecus sediba* is also very ape-like. It has a small body with ape-like large-jointed upper and lower limbs. The arms and hands of *Australopithecus sediba* extend down to the knees, typical of long-armed knuckle walkers. In short, this is an ape, not a human, and not a caveman at all!
3. http://www.scientificamerican.com/article.cfm?id=ourneandertal-brethren.

Neanderthal remains have also been unearthed with mammoths and other big game bearing bone marks and other indicators that these animals were hunted and butchered in complex community activities. And everywhere Neanderthals are found (not always in caves), they have complex axes and other stone tools.

In fact, the title of "mere caveman" may be in jeopardy, as researchers recently unearthed a complex dwelling made from mammoth bones, which wasn't in a cave at all.[4] With all the similarities, however, Neanderthals weren't exactly like us — their physical characteristics (such as larger brows in adults and wide nasal cavities) would certainly make them stand out today.

Cro-Magnon Man

On the other hand, early *Homo sapiens* (often called Cro-Magnon man) would fit right in nowadays, though perhaps more likely on a North American football team than in an office building. The robust build, larger brain on average (1600cc vs. 1350cc), and DNA differentiate the European Cro-Magnon from modern humans.[5] However, they show a clear affinity with us.

Everything you might expect to find from the settlements of any non-industrialized people is found with Cro-Magnons. For instance, the Dzudzuana Cave in the country of Georgia contained wild flax fibers that suggest these early travelers sewed garments or wove baskets,[6] and the Lascaux caves in France long hid colorful cave paintings that may relate to phases of the moon.[7] Site after site reveals thousands of small, beautifully made javelins, arrows, and ornate artifacts, often with carvings and designs on them, such as the ivory pendant made from mammoth tusk that was found with the so-called "red lady" (actually a male) in south Wales.[8]

And the recent discovery of a buried dog's skull in Předmostí (Czech Republic) suggests that Cro-Magnon man enjoyed the company of "man's best friend."[9]

In light of these finds, the idea that these particular post-Babel humans were some mysterious "other" loses its punch.

4. http://www.physorg.com/news/2011-12-neanderthal-home-mammoth-bones-ukraine.html
5. David Caramelli et al., "A 28,000 Years Old Cro-Magnon mtDNA Sequence Differs from All Potentially Contaminating Modern Sequences," *PLoS One* 3 (2008): e2700.
6. Eliso Kvavadze et al., "30,000-Year-Old Wild Flax Fibers," *Science* 325 (2009): p. 1359.
7. http://news.bbc.co.uk/2/hi/science/nature/975360.stm.
8. http://www.britarch.ac.uk/ba/ba61/feat3.shtml.
9. Mietje Germonpré et al., "Palaeolithic Dog Skulls at the Gravettian Předmostí Site, the Czech Republic," *Journal of Archaeological Science* 39 (2012): p. 184–202.

Homo Erectus

That brings us to *Homo erectus*, a group that long held the title as most enigmatic and disputed of all early humans. As the name *erectus* implies, we're meant to be amazed at their upright, two-legged gait that allowed them to tromp across Africa, Europe, and Asia. However, the *Homo appellation* (that is, human) came later. When these ancient humans were first uncovered in Java (Indonesia), their bones were trumpeted as *Pithecanthropus erectus*, which essentially means "upright ape-man." That was certainly a misnomer.

What's truly incredible is how widespread these early humans were. They may have built fires in the Middle East (as indicated by charred bones and plant remains),[10] and they hunted across Asia and Europe, where we find many butcher sites and the stone tools they used. They must have built seafaring vessels of some sort to reach the Indonesian islands against the currents. In fact, we find their fossils before any other human remains. So we can safely say that their "primitive" ways got them pretty far. Not bad for a carless society.

Homo Floresiensis and Denisovans

Two new finds suggest that we may only be scratching the surface of the variety apparent in post-Babel humans. Recently, an unusually large tooth and a finger bone found in Denisova Cave in Altai Krai, Russia, point to a mysterious new group of wayfarers. The Denisovans, as they're being called, occupied the region around the same time as Neanderthals.

But DNA testing of the finger and two other bones indicates that this new group differed from Neanderthals.[11] Beyond that, we have only a handful of artifacts to understand these mysterious people, such as a stone bracelet that was ground and polished.

But the impact of the Denisovans has been relatively minor compared to the huge debate surrounding a group of tiny human skeletons. So far, nine members of this group have been found on the Indonesian island of Flores, giving us the tentative name *Homo floresiensis*. However, you may have heard them referred to as "hobbits," which fits their three-foot (1 m) height.

Since the discovery of the first non- fossilized skeleton in 2003, dueling scientific papers have raised, lowered, and stretched the status of these so-called hobbits — all without a single strand of DNA (which has so far eluded

---

10. http://www.answersingenesis.org/articles/am/v6/n1/camp-after-babel.
11. Pontus Skoglunda and Mattias Jakobssona, "Archaic Human Ancestry in East Asia," *PNAS* 108 (2011): p. 18301–18306.

scientists). Because access to the remains is so limited, the intrigue — and rancor — may continue for years.

Despite the debate, what's found in the dirt on Flores reveals much about the inhabitants. Numerous charred bones of the dwarf elephant Stegodon — many of them juvenile — paint the picture of a group of opportunistic hunters who roasted up the small elephant that once lived on the island — perhaps leading to its extinction.

To do so, they employed a number of advanced stone tools, quite capable of slicing and dicing tough animal skin. And while we find no evidence of their boats, these people are most similar to *Homo erectus* found on Java. Since they lived on the island of Flores, this suggests they must have built boats that could fight against strong ocean currents to get there.

The Makings of a Human

Variation among post-Babel humans has led to a great debate among evolutionists, who wonder where they fit on the roadway to being "truly human." But that way of thinking misses the fundamental truth. When God created humans, He didn't define our humanness in terms of physical characteristics. We aren't human because we have two arms or legs or skulls of a certain shape or size. Our Creator, who is spirit, made us in His spiritual image.

Genesis reveals aspects of what this implies. Our early ancestors made musical instruments and tools, farmed, built cities, and otherwise represented God as stewards of His creation (Genesis 4). With that as our standard, we can cut through the confusion and bias. All those we call "cavemen" (probably a misnomer) show the same characteristics as the first humans in the Bible.

Neanderthals buried their dead and may have worn jewelry.[12] *Homo erectus* seems to have divvied up jobs to prepare food and sailed the high seas. Even with little to go on, we can be fairly certain the Denisovans wore jewelry, and the much-maligned "hobbits" left tools useful for dicing up lunch. All uniquely human traits — traits that show creatures made in the image of God.

In other words, we can be sure that they all descended from Adam through Noah's family. These certainly aren't unique species, in the sense of being something "less than modern humans" — they're just more evidence of beautiful variations in the appearance of individuals in our one unique race. Our relatives may have looked different, but they weren't bumbling brutes. They had

12. http://www.answersingenesis.org/articles/aid/v2/n1/worthy-ancestors-2.

the very human and God-given ability to discover creative solutions in a dangerous, sin-cursed world. And they were all rebels from God, in need of His grace.

Finding a Home for Cavemen

New DNA technology has allowed scientists to peer into the past by mapping the DNA of so-called cavemen. And they have found some noticeable differences. So, what do those differences really mean — are those early people somehow less "human" than we are?

Before we can answer that question, we first need to understand two related issues. What can DNA tell us about the differences among people? And how does the biblical account of human origins shed light on these differences?

Cavemen Genetics

The ability to map DNA is an amazing feat, considering the DNA is thousands of years old! Many ancient human remains are found in equatorial regions where heat and humidity have destroyed the DNA. However, remains of the Neanderthals and another group of humans discovered in a cave in southern Siberia, the Denisovans, have been found in cold, dry, protected areas that better preserved the DNA.

When the first draft of Neanderthal DNA was published, the researchers concluded that it is 99.7 percent identical to modern human DNA. They also found that approximately 1 to 4 percent of DNA specific to Neanderthals can also be found in modern Eurasians. This led them to conclude that a very small number of Neanderthals mixed with early modern humans and produced children. Neanderthals had a wide geographic distribution in Eurasia, from Spain to southern Siberia, and from Germany to the Middle East, so it is not surprising that more of their DNA is found in modern Eurasians as opposed to other populations, such as Africans.[13]

To date, approximately 80 genes have been shown to differ between Neanderthals and modern humans.[14] These genes produce proteins that govern a wide range of functions such as metabolism (how we burn food), the growth of the skull, and skin shade. Further study of these genes may help us understand how Neanderthals were different and perhaps why they died out.

---

13. Richard E. Green et al., "A Draft Sequence of the Neandertal Genome," *Science* 328 (2010): p. 710–722.
14. Carles Lalueza-Fox and M. Thomas P. Gilbert, "Paleogenomics of Archaic Hominins," *Current Biology* 21 (2011): R1002– R1009.

For instance, one gene produces a protein involved in skin and hair color. Rare variants of this gene among modern humans lead to pale skin and red hair. The Neanderthal gene has a variation so far unknown in humans today. It is likely that this variant led to pale skin and red hair in Neanderthals.[15] If this is so, Neanderthals would have been able to absorb more sunlight than if they had darker skin. This would have been useful in producing enough vitamin D to live healthy lives in the northern regions.

Denisovan DNA is also similar to DNA in modern humans. Approximately 4 to 6 percent of DNA that is specific to Denisovans can also be found in modern Melanesians (those who live in the islands northeast of Australia).[16] As with the Neanderthals, this indicates that very few Denisovans mixed with and produced offspring with early modern humans — at least with those in Southeast Asia.[17]

Both Neanderthals and Denisovans do have small-scale differences with modern humans. Before the first draft of Neanderthal DNA, they were sometimes considered to be different human species or subspecies. But this is an arbitrary, man-made designation since two modern chimps of the same species will have more DNA variation than Neanderthals or Denisovans have to modern humans. In light of the genetic evidence, Neanderthals and Denisovans are fully human and should be classified as *Homo sapiens.*

Are the DNA Sequences Accurate?

Many difficulties must be overcome to accurately sequence ancient DNA. Sequencing DNA involves determining the correct order of the individual components (bases) that comprise the DNA. Contamination and degradation are two of the biggest obstacles.[18] Contamination comes both from bacteria found in the fossil (which can sometimes account for more than 90 percent of the DNA found!) and from bacteria transferred through handling by modern humans. Degradation occurs when the DNA is "chopped up" and certain DNA components are modified by chemical reactions. Fortunately, scientists have developed techniques that greatly limit the danger of contamination and degradation altering the actual human DNA sequence, so their impact is usually negligible.

15. Carles Lalueza-Fox, et al., "A Melanocortin 1 Receptor Allele Suggests Varying Pigmentation among Neanderthals," *Science* 318 (2007): p. 1453–1455.
16. David Reich et al., "Genetic History of an Archaic Hominin Group from Denisova Cave in Siberia," *Nature* 468 (2010): p. 1053–1060.
17. David Reich, et al., "Denisova Admixture and the First Modern Human Dispersals into Southeast Asia and Oceania," *The American Journal of Human Genetics* 89 (2011): p. 1–13.
18. Dan Criswell, "Neandertal DNA and Modern Humans," *Creation Research Society Quarterly* 45 (2009): p. 246–254.

Another issue involves the limited number of ancient individuals with viable DNA. For example, there are only two known fossil remains for Denisovans from a single cave. At the most, they represent two individuals. Compare that to the thousands of modern humans whose DNA has been sequenced. A small sampling of an ancient population may not truly reflect the full range of variety in that particular group.

The Neanderthal samples, in contrast, come from over a dozen different individuals at sites on different continents, so they are much more likely to represent the population as a whole. It is also important to acknowledge the many evolutionary assumptions that are made when comparing the DNA sequence of ancient individuals to modern humans.[19] For example, a common human-chimp ancestor was assumed. One paper stated, "To estimate the DNA sequence divergence . . . between the genomes of Neanderthals and the reference human genome sequence . . . [we used] an inferred genome sequence of the common ancestor of humans and chimpanzees as a reference to avoid potential biases."[20] Apparently the authors of the paper don't consider assumed human-chimp ancestry as a bias, but it is! Creation scientists are actively studying methods to avoid these biases so that more valid comparisons can be made.

A Biblical Perspective

Researchers studying genetics have clearly established that Neanderthals and Denisovans were fully human. Any physical differences should be viewed as nothing more than variations that can occur within the human race descended from Adam and Eve. For a time, these descendants all lived together at the Tower of Babel. Following the post-Babel migration and late into the Ice Age, differing human populations began to appear in the fossil record, such as Neanderthals and Denisovans.

The next questions for creationists are how and why these differences appeared.[21] How is much easier to answer than why! One possibility is that

19. As creation scientists have shown, bias can affect the reported similarities and differences in the DNA sequences between organisms. See Jeffrey P. Tompkins, "Genome-wide DNA alignment similarity (identity) for 40,000 chimpanzee DNA sequences queried against the human genome is 86–89%," *Answers Research Journal* 4 (2011): p. 233–241.
20. Richard E. Green et al., "A Draft Sequence of the Neandertal Genome," *Science* 328 (2010): p. 710–722.
21. Robert W. Carter, "The Neandertal Mitochondrial Genome Does Not Support Evolution," *Journal of Creation* 23 (2009): p. 40–43; Kurt P. Wise, "The Flores Skeleton and Human Baraminology," *Occasional Papers of the BSG* 6 (2005): p. 1–13; Robert W. Carter, "Neandertal Genome Like Ours," June 1, 2010, http://creation.com/neandertal-genome-like-ours.

environmental pressures, such as the Ice Age, "selected" for or against traits within the range of human genetic diversity. (In other words, those that had a particular combination of characteristics survived in that environment, and others did not.) This may have led to the specific set of features found in Neanderthal people. Many animals, following the Flood and during the Ice Age, experienced an explosion of variations that allowed them to live and function well in new environments. This could also have been true for humans.

Other possibilities include genetic effects seen mainly in small populations. Small populations would have been typical for a period of time following the breakup of the human population at Babel, as people were separated based on language. The groups that left Babel would have begun with only a few reproducing individuals and not interbred initially with other groups.

A phenomenon known as genetic drift can cause certain genetic variations to become "fixed." If the population is small, everyone with certain variations can die, without passing them down, and the survivors pass down just one variation to future generations. If no people are moving in or out of the population, characteristics like the pronounced brow ridge or the robust body form in Neanderthals can become dominant.

Another possible impact of the Babel breakup is the founder effect. The founders of each group leaving Babel might simply have differed from one another. Certain traits in one group might have been unknown among the founders of any other group. Those traits would then be unique to each group. Rather than being fixed by genetic drift, the Neanderthals' pronounced brow ridge or robust body form may have been found among the founders of only one group after they left Babel. Those people may have migrated intentionally to places where they were most comfortable (similar to human behavior today).

As time passed, the different groups would have migrated, as humans have always done. People who had the traits of modern humans possibly interbred, at times, with the other groups, such as Neanderthals and Denisovans. Yet there seems to have been a sudden loss, or a dilution, of the characteristics possessed by those other groups. The genetic makeup of modern humans became dominant.

Inbreeding can have disastrous effects on small populations by amplifying defective genes. Maybe this is why Neanderthals and Denisovans eventually became extinct. We don't know. Why this happened is still a mystery.

Conclusion

Caves have never gone out of fashion as a place to seek refuge. For instance, hermits lived in caves throughout the Middle Ages, and until recent times a clan of people were living in caves on the Mediterranean island of Malta. Even the Bible records a number of cave refugees, such as Elijah (1 Kings 19), David (1 Samuel 22:1), and Obadiah (1 Kings 18:3–4). After fleeing the destruction of Sodom and Gomorrah, Lot and his daughters found shelter in a cave (Genesis 19:30).

It seems cavemen are simply that — people who lived in caves and they have little, if anything, to do with evolution. What is not a mystery is that so-called cavemen, including Neanderthals and Denisovans, were fully human. They were among the descendants of the people scattered at the Tower of Babel — made in God's image to bring Him glory.

CHAPTER 8

Should There Really Be a Battle between Science and the Bible?

ROGER PATTERSON

There is much debate in our culture about the nature of science and religion and the interaction of the two. Some will argue that the two areas give us understanding in distinct ways that do not overlap.[1] Others suggest that science should drive our understanding of religion. Still others argue that religion should drive our scientific understanding. There are truly deep divisions in many senses as people claim different sources of authority on these issues.

But there are many contrasting ideas that are presented in the popular discussions of this topic that need to be carefully considered. Words have meanings, and we need to make sure that we are using our own words in a manner that is clear and does not hide or change the meaning of certain terms and concepts. We all recognize when a politician talks out of both sides of his mouth, but it can be a little harder to spot when we see religious leaders and scientists

1. The idea of "non-overlapping magisteria" was promoted by the late Dr. Stephen Jay Gould and proposes that science cannot answer the questions of religion, and vice versa. This forces a false dichotomy between secular and sacred, a concept that is foreign to the Bible. Christians are called to take every thought captive to the obedience of Christ (2 Corinthians 10:5), not to compartmentalize their thinking and actions into secular and sacred.

talking in the same manner.[2] While we can learn from those who have studied various ideas, we need to be careful not to accept those ideas just because some scientist, religious leader, or news analyst says something is so.

Everyone has a point he or she is trying to make! Many people will try to tell you that they do not have such biases, but it is impossible to be neutral: our thinking always begins from a specific starting point. All of the arguments that we make are based in our worldview, and our worldview is based on specific assumptions we believe to be true. The goal of this chapter is to explore some of those underlying assumptions and their implications for the arguments that are often used in the broad creation-evolution debate.

Where Did Science Come From?

What we understand today as the modern scientific method and the technologies and theories it produces has its foundations in beliefs about the nature of the universe and the God who created the universe. The scientific method is grounded in the ideas of repeatability, falsifiability, and testability. Each of these ideas assumes that there is a uniformity to the world that we live in. (This will be discussed in more detail below.) But on what grounds can we assume that the world should operate in a uniform way? Only on the grounds that God has created the universe to function according to specific laws.

Modern science blossomed in the fertile soils of Western culture where God was known as the Creator and Lawgiver of the universe. While some mathematical and technological concepts were known in the millennia prior to this time, rigorous experimentation and careful correlation of cause and effect became the focus of the discipline known then as natural philosophy. During the Middle Ages, there was much advancement in the study of nature, though it is often denigrated as a time of little advancement in the development of new ideas. These advances primarily came in the monasteries and universities that were funded and directed by the Roman Catholic Church. Surely, much of this thinking was misguided and has been corrected, but it was the notion of a Creator God who arranged an orderly universe that directed and encouraged the study of natural philosophy. It would be anachronistic to refer to these studies

2. The logical fallacy of equivocation occurs when a word is used to express an idea, but the meaning of the word changes within the argument. Similarly, the logical fallacy of special pleading is using or defining words in a way that is beneficial to the argument and not necessarily agreed upon by others. Both of these tactics are used by those arguing over the roles of religion and science. As Christians and ambassadors for Christ, we must be careful to avoid these invalid forms of argumentation because they reflect poorly on our King.

as scientific, but the foundation of scientific thinking was laid in these early centuries in the West.

And this perspective is not simply biblical creationist propaganda used to prop up a particular point of view. Dr. James Hannam, historian and physicist, writes in the conclusion to his *The Genesis of Science*:

> The starting point for all natural philosophy in the Middle Ages was that nature had been created by God. This made it a legitimate area of study because through nature man could learn about its creator. Medieval scholars thought that nature followed the rules that God had ordained for it. Because God was consistent and not capricious, these natural laws were constant and worth scrutinizing. However, these scholars rejected Aristotle's contention that the laws of nature were bound by necessity. God was not constrained by what Aristotle thought. The only way to find out which laws God had decided on was by the use of experience and observation. The motivations and justifications of medieval natural philosophers were carried over almost unchanged by the pioneers of modern science.[3]

Demonstrating that he is not interested in propping up the Bible or the existence of the Creator as truth, Hannam goes on to quote Sir Isaac Newton's insistence on God's existence to corroborate the diversity of life on earth, but states that Darwin later proved Newton wrong in this area.

"But wait, what about the Chinese in the East! They invented gunpowder!" you might protest. Developing gunpowder is one thing, but deciphering the underlying mechanics that explains how the gunpowder formed and why it explodes, even predicting how it will react with other chemicals, is an entirely different type of thinking. While a defense of this perspective is beyond the scope of this chapter, several authors have discussed this theme at length and proposed very plausible explanations for why scientific thinking and methodology did not develop in stable and flourishing cultures like China, India, and Egypt despite the talents and resources available to them.[4] Scientific thinking

3. James Hannam, *The Genesis of Science: How the Christian Middle Ages Launched the Scientific Revolution* (Washington, DC: Regnery Publishing Inc., 2011), p. 348–349.
4. For a condensed version of these theories, see Eric V. Snow, "Christianity: A Cause of Modern Science?" Institute for Creation Research, http://www.icr.org/article/427/290. For more thorough treatments of these ideas, several books have been written, though the authors are not all approaching the topic from a Christian or biblical presupposition: Nancy R. Pearcey and Charles B. Thaxton, *The Soul of Science: Christian Faith and Natural Philosophy* (Wheaton, IL: Crossway, 1994); Stanley L. Jaki, *Science and Creation*

cannot thrive in cultures where superstitions about capricious gods acting on whims influence daily events. It is only the biblical view of the nature of God and His creation that allows for the expectation of reliably discovering the underlying truths of the operation of the universe created by God. And it is only the biblical worldview that calls for a study of the creation to better understand the Creator and to properly rule the creation (Genesis 1:28) to find cures for disease, produce technology, increase food production, etc., for the good of mankind.

With that foundation, let us turn to some of the common contrasting ideas that are used to frame the discussion and examine them one at a time.

Science vs. Science

It is the very nature of language that the meanings of words change. If I had told you in 1947 that I found my missing mouse in my briefcase, you would have had a different reaction than you would today. The same is true for the word *science*. In its simplest form, science means knowledge. Examining the 1828 definition of science from Noah Webster:

> In *a general sense*, knowledge, or certain knowledge; the comprehension or understanding of truth or facts by the mind. The *science* of God must be perfect.[5]

In a general sense, science means knowledge. Interestingly, the first definition in the modern Merriam-Webster Collegiate Dictionary (2003) is not that much different:

> the state of knowing **:** knowledge as distinguished from ignorance or misunderstanding[6]

In another sense, science is the systematic study of a subject and the knowledge that is generated by that study. In the past, theology was known as the queen of the sciences (as was mathematics) and the supernatural origins of the universe and the creatures[7] on the planet were assumed to be true because they are revealed in Scripture. Today, many have hijacked science, insisting that it

(Edinburgh, Scotland: Scottish Academic Press, 1986); James Hannam, *The Genesis of Science: How the Christian Middle Ages Launched the Scientific Revolution* (Washington, DC: Regnery Publishing Inc., 2011).

5. *American Dictionary of the English Language*, 9th ed., s.v. "Science."
6. *Merriam-Webster's Collegiate Dictionary*, 11th ed., s.v. "Science."
7. Even the term *creature* naturally implies that there was a Creator who made it.

can only be done within an atheistic frame of reference (or worldview), thus completely removing God from our thinking about the physical world.

It is possible to categorize science into many different categories. Classically, the pure sciences were distinguished from the applied sciences. For an example, as we studied the pure science of how x-rays interact with matter, we were able to apply that knowledge to taking pictures of the bones inside the body. Christians understand this x-ray phenomenon as an extension of the natural laws God has programmed into the universe and employ this knowledge to exercise dominion over the earth (Genesis 1:26–28) and to reverse some of the effects of the Curse (Genesis 3) that our sin brought into the world. They do this by finding cures for disease or developing new technology. Those who hold a naturalistic worldview believe that this phenomenon is just the product of some random events culminating in some beams of radiation that can shoot through some matter and not others. All of this involves testing, observing, and repeating experiments in the present to apply that knowledge in the present.

Another important distinction to make is between operational science and historical science. Operational science employs the pure and applied methods of scientific inquiry to figure out how physical things operate or function to find cures for disease, develop new technology, or otherwise improve our standard of living. In this kind of science, researchers use observable, repeatable experiments to test hypotheses and develop our understanding of the world. Most of chemistry, physics, astronomy, biology, engineering, and medical research are in the realm of operational or experimental science. These types of things can be observed and tested by different individuals with repeatability and can be falsified if contrary evidence comes to light.

Historical science deals with questions of history and origins, such as how the Grand Canyon formed or how living creatures came into existence. Paleontology, archeology, cosmogony, much of geology, and forensics (criminal investigation) fall in the realm of historical or origin science. It looks at evidence in the present to try to figure out what happened in the unobservable, unrepeatable past to produce the evidence that we see, though there is no opportunity to repeat the initial conditions and observe their outcome. There is much conjecture involved in historical science because scientists have to make assumptions about the past. Those assumptions may or may not be correct and, in many cases, may not even be verifiable. So we must take care to understand the limits of this approach. To be clear, both creationists and evolutionists engage in historical science, but biblical creationists look to the authority of the Bible to inform their understanding of the past because it contains the eyewitness testimony of the Creator about key events in the past that explain the world we live it. But in a naturalistic (atheistic), evolutionary viewpoint, there is no eyewitness of the imagined events of millions of years ago and thus no objective standards to judge the validity of the evolutionary stories. The past cause or sequence of events that produced what we see in the present must be inferred by assuming that present processes have always operated in the same way or at the same rate as we observe today.

While operational science surely involves some levels of inference, when we move into the category of historical science, the level of inference increases greatly. Biological, geological, and cosmological evolution are all based on chains of assumptions and inferences that cannot be observed, tested, or repeated. An inference based on an inference based on an inference leaves a very weak chain.

One example of this chain of assumptions comes in the materialistic view of the age of the earth. First, the assumptions of radiometric dating must be accepted. Then, rather than dating rocks that are from earth, meteorites that are found on the earth are dated. This assumes that these meteorites formed at the same time as the earth, so they will be the same age as the earth. This then assumes that the earth formed from a cloud of dust that encircled the young, forming sun, a process known as the nebular hypothesis, and the particles collected into the earth with fragments left floating in space and later falling to earth as meteorites. The nebular hypothesis assumes that the big bang is true. This is a long chain of assumptions with no directly observed evidence. From a biblical perspective, none of this is consistent with the creation account of Genesis, the eyewitness testimony to the events of creation.

Many people try to discredit biblical creationists and say they can't be real scientists if they don't believe in evolution. However, this is a silly argument. Many will say that it is hypocritical for a biblical creationist to talk on a cell phone and take antibiotics, yet reject the "truths" of the big bang and biological evolution. But what does the big bang have to do with designing a cell phone? And what does the acceptance of a fish changing into a frog over millions of years have to do with testing bacteria in a petri dish to see what chemicals kill the bacteria? To make such claims is to confuse categories of science and appeals to the emotions by getting people to fear that technology cannot advance if people look at the world through the lens of Scripture. Knowing that many of the founders of scientific disciplines were Bible-believing scientists should give those using these scare tactics pause, but they continue to make such claims in the face of many biblical creationists carrying out scientific research and advancing our understanding of the world that God has created.

Uniformity vs. Uniformitarianism

As mentioned earlier, because God has created the universe, it follows that certain natural laws were put into place by Him. He has chosen the laws that determine how the planets orbit the sun, how water molecules form and stick to one another, how electricity travels through wires, and every other conceivable interaction of matter and energy in the universe . . . not to mention the spiritual elements of the universe. God has created a universe that operates in uniform ways, and as we study the creation we are uncovering the ways that He has ordered the universe to function or operate. Isaac Newton did not invent the laws of gravity; he simply described the way God had ordained for the universe to function. He was able to do this because God had created an orderly universe in the first place.

We see the principle of uniformity present in the early chapters of Genesis where God created the various kinds of plants and animals to reproduce after their kind. More explicitly, Genesis 8:22 communicates God's intention to uphold the earth in a way that is consistent. Connecting this to passages like Hebrews 1:3 and Colossians 1:17 provides a solid foundation for understanding why the universe is the way it is.[8]

Someone who rejects the Bible can believe that there is uniformity in the universe, but he has no reason to believe that the universe should be a place of order. He is making an arbitrary assumption about the universe with no reasoning to support that assumption. Extending that assumption, many believe in the doctrine of uniformitarianism. This doctrine is often summarized in the phrase "the present is the key to the past." As an example, the doctrine of uniformitarianism is often applied to the layers of rocks we find under our feet. We can observe the rate at which layers are forming today. If we assume that the rates we see today are the same as they were in the past, we can just look backward and see how long it took for all of the layers to form, right?

Well, the Bible makes clear that there was a global Flood that covered the entire surface of the earth about 4,350 years ago. If that is true, then that would have a major effect on the surface of the earth — the present would be dramatically different from the past. While the laws of nature were in effect during the Flood — uniformity of nature — the rates of the layers being deposited would have been dramatically different because the magnitude and duration of that

8. For a more thorough treatment of the assumptions of uniformitarianism and the illogical nature of a naturalistic, atheistic worldview, see Jason Lisle, *The Ultimate Proof of Creation* (Green Forest, Arkansas: Master Books, 2009).

catastrophic Flood far exceeded the scale of any floods, earthquakes, hurricanes, and tsunamis we observe today. The present is not the key to understanding the past. Rather, the Bible is the key to understanding both the past and the present because it gives us the key events in history to understand both!

Faith vs. Fact

Many people have bought into the myth of neutrality — the idea that people can examine ideas in a truly neutral manner. Everyone has a bias, and everyone starts their reasoning from their foundational worldview. Many people claim that those who have a naturalistic, atheistic scientific worldview, what is also called philosophical naturalism, are neutral and approach their study of the world (its operation and its history and origin) in a totally objective way. But stop and think about that . . . if you believe that there can be no supernatural influences in the world, you are biased against the supernatural.

The question becomes, which bias is the best bias to be biased by? Put another way, which worldview provides the true foundation for examining the world we live in? Every person takes these starting assumptions on faith. Faith is inescapable when we examine the world around us, regardless of whether we are Christian, Muslim, Jewish, agnostic, atheist, or whatever.

If we start from a biblical definition, faith is believing things that we have not seen or, by extension, experienced (Hebrews 11:1). "By faith we understand that the worlds were framed by the word of God, so that the things which are seen were not made of things which are visible" (Hebrews 11:3). Christians trust that God created the universe out of nothing because He has told us that He did, not because we have seen or experienced the origin of the universe. This is taken on faith in light of the truths of Scripture, which is the absolutely truthful eyewitness history from the eternal Creator. A Christian's faith does not stand isolated from evidence but is affirmed by examining evidence in light of the truths of the Bible.

On the contrary, those who believe that the big bang was the origination of the universe do so with a faith that rests on many assumptions rather than the infallible Word of God. They take on faith that which they have not seen. Despite the claim that we can "see" the beginning of the universe in the cosmic microwave background radiation and other features of the universe, that belief is based on assumptions about those observations and should rightly be called positions of faith — a faith based in naturalism rather than the testimony of our Creator God.

Likewise, the formation of the solar system by the nebular hypothesis is taken on faith. The supposed steps in the process have never been observed, but only inferred. Moving forward, the chance origin of life from non-living matter is another point that the naturalistic scientist can only hold to by faith. "It must have happened," they say, "since we are here." Within that context, the origin of the information coded in the DNA of every living organism must be taken on faith since there is no known natural mechanism that can explain its origin. Continuing on in the chain of assumptions, the evolution of one kind of organism into another different kind of organism (e.g., a reptile into a bird or mammal) must be taken on faith since it has never been observed, but is only inferred from interpreting the fossils and comparisons of biochemical molecules.

It takes a lot of faith[9] to believe in the naturalistic origins of the universe, our planet, and all of the life on it.

In many contexts, the big bang, geologic evolution, and biological evolution are referred to as scientific facts, though these are only "facts" in some redefinition of the word (special pleading). In *Science, Evolution, and Creationism*, produced in 2008 by the National Academy of Sciences Institute of Medicine, there is a page dedicated to the question of whether evolution is a theory or a fact. In the conclusion to that discussion, they state:

> In science, a "fact" typically refers to an observation, measurement, or other form of evidence that can be expected to occur the same way under similar circumstances. However, scientists also use the term "fact" to refer to a scientific explanation that has been tested so many times that there is no longer a compelling reason to keep testing it or looking for additional examples. In that respect, the past and continuing occurrence of evolution is a scientific fact. Because the evidence supporting it is so strong, scientists no longer question whether biological evolution has occurred and is continuing to occur. Instead, they investigate the mechanisms of evolution, how rapidly evolution can take place, and related questions.[10]

So, in the minds of those who believe evolution is a fact, it is a fact. Within that paragraph, we also see the subtle assertion that "scientists" no longer even question evolution. So if you question evolution, you must not be a scientist.

9. Dr. David Menton has suggested that a better term would be *credulity,* since there is no foundation for the naturalistic worldview apart from the opinions of man. The biblical position is one of faith because it is founded in the truth revealed in Scripture. However, credulity is not a word most would understand and should be reserved for the right context.
10. Francisco J. Ayala et al., *Science, Evolution, and Creationism* (Washington, DC: National Academies Press, 2008), p. 11.

This is known as the "no true Scotsman" fallacy and is simply an approach used to defame those who question or reject evolution. In fact, I have many colleagues who have earned PhDs in various scientific fields who reject evolution, so that assertion is patently false.

If you ever hear someone say, "Science says such and such," a flag should go up in your mind. Used in an argument, this is called the reification fallacy; giving personal qualities to an abstract idea. Science can't say anything, but the scientists can. Related to this idea is the use of the term "data." When you read that "the data all points to conclusion X," you should again take pause. Rather than the data (the actual observations from experimentation or measurements of a geological formation or of light from a star or galaxy) these are likely interpretations of the data. The data from the observations are facts and are the same for everyone (creationist or evolutionist), but data may not include all the relevant observations that could be made and also must always be interpreted to arrive at conclusions. In order to interpret the data, we will always apply our worldview to present an explanation that makes sense of the data. Neither science nor the data can ever truly tell us anything. Facts are always interpreted in light of faith (our unprovable worldview assumptions).

Science vs. Religion

To be very clear, there is no conflict between evolution and religion — the conflict arises between evolution and biblical Christianity. In fact, many people have made evolution a fundamental tenet of their religion. For example, Hinduism, Buddhism, animistic religions of all sorts, liberal theology, and other expressions of Christianity that do not hold to Scripture as the supremely authoritative, inerrant Word of God are perfectly compatible with evolution and millions of years. Those who call themselves humanists and look to the Humanist Manifesto III as a document with guiding principles also embrace naturalism and unguided evolutionary processes. In that document we find the following statements about how humanists understand the world we live in and how life evolved:

> Knowledge of the world is derived by observation, experimentation, and rational analysis. Humanists find that science is the best method for determining this knowledge as well as for solving problems and developing beneficial technologies. We also recognize the value of new departures in thought, the arts, and inner experience — each subject to analysis by critical intelligence.

> Humans are an integral part of nature, the result of unguided evolutionary change. Humanists recognize nature as self-existing. We accept our life as all and enough, distinguishing things as they are from things as we might wish or imagine them to be. We welcome the challenges of the future, and are drawn to and undaunted by the yet to be known.[11]

The modern Merriam-Webster's dictionary defines religion in several forms, including:

> a cause, principle, or system of beliefs held to with ardor and faith[12]

While the humanist might argue that they do not hold to these views on faith, they have no other foundation upon which to build their case. How do they know that knowledge can only come through "observation, experimentation, and rational analysis"? What experiment can be done to show that this is true? If they say they know that by rationally analyzing what they have observed, they have worked themselves into a vicious circle of thought that must be accepted by faith — the very thing they try to denounce.

In many cases, the people who are making this comparison are seeking to exclude the teaching of biblical creation from the public school classroom and other settings, believing religious views should be censored from the science classroom. However, they fail to recognize that evolution is a religious tenet of the religion of humanism and that they are forcing their own religious views into the classrooms and publicly funded museums that exclude a biblical view of the world we live in. Rather than excluding religion from the classroom, Christianity has been replaced by the religious teachings of secular humanism (which is really the religion of atheism).

The issue is not science vs. religion, but one religious view set against another. The Bible offers us an authoritative source of truth from which to begin our study of the universe. It is the only rational faith that can even explain the existence of scientific thought in the first place.[13]

Conclusion

Regardless of which of the ways the conflict is presented, Christians must always look to the Bible as the supreme source of truth and authority in every

11. American Humanist Association, "Humanist Manifesto III," http://www.americanhumanist.org/Humanism/Humanist_Manifesto_III.
12. *Merriam-Webster's Collegiate Dictionary*, 11th ed., s.v. "Religion."
13. Jason Lisle, "Evolution: The Anti-science," Answers in Genesis, http://www.answersingenesis.org/articles/aid/v3/n1/evolution-anti-science.

area. We must also call those who do not believe to look to the Creator as the truth and help them to see that we can only ultimately make sense of the world around us by starting with the truths God has revealed in the Bible.

While some people try to suggest that the facts of nature speak for themselves, a rock does not tell you how old it is — the age of the rock is an interpretation. You must make several assumptions in order to arrive at the supposed 4.5-billion-year age of the earth, including the assumptions of uniformitarianism. These assumptions are in direct conflict with the clear teaching of Scripture and deny a global Flood and the special creation of the universe only about 6,000 years ago.

Unlike rocks and fossils, the Bible does offer clear propositions and descriptions of the past. We must read the Bible much differently than we read the "book" of nature. We can only understand what nature reveals when we understand that the world we are living in has been cursed by God as a result of man's sin. We must also take into account the effects of other events like the Flood and the Tower of Babel. Ignoring these truths will naturally lead to faulty conclusions about the history of the earth and all the life on it.

Scientific thinking was born in the cradle of Christian Europe because the men who believed in the true Creator God believed they could understand the world He had created. They believed they could understand the creation because they knew God was a God of order. They believed He was a God of order because that is what the Bible clearly reveals. We must surely acknowledge that we would not have the scientific understanding that we have today apart from what God has revealed to us in the Bible. We would be fools to set aside the Bible as we continue to pursue a deeper understanding of what we see as we peer through our microscopes and telescopes or look with unaided eyes to examine God's creation.

But that is exactly what many scientists are trying to do. Having stood on the shoulders of men who trusted in God's revelation, they have denied the need for God to continue their study. It would be just as foolish for a man who has flown to the moon on a rocket to deny the rocket that took him there, claiming that he can return to earth on his own without the rocket. Sadly, he will perish there on the moon without acknowledging his need for the rocket for his safe return to earth.

Likewise, those who deny the God of the Bible as the foundation for understanding the world we live in do so at their own peril. God has created the universe, this world that we live in, and each one of us. Through the first man God created, Adam, all have become sinners. Each of us has chosen to rebel against

God and His authority as our Creator. Unless we trust what God has said about our condition in the world (that we are sinners), His just judgment against our rebellion lies on our heads and we will never know His wonderful love, mercy, and grace. Just as Scripture calls us to acknowledge God as the Creator, it also calls us to look to Jesus Christ as the only remedy for avoiding God's wrath against our sin. Each person must acknowledge those truths and look to Christ in repentant faith for the forgiveness of their sins.

As we continue to pursue scientific understanding about the universe we live in, let us do so by building on the firm foundation of what God has revealed to us in His Word. The God who has revealed Himself to us in the Bible makes science possible. Let God be true and every man a liar.

CHAPTER 9

What Did the Reformers Believe about the Age of the Earth?

DR. JOEL R. BEEKE

All Christians believe that God the Father Almighty is the Maker of heaven and earth. This belief is like a great river that runs through Christian history. It distinguishes Christianity from other forms of spirituality. Yet within this river there have been two streams of thought about how to understand Genesis: the allegorical reading and the literal reading.[1]

The Reformation of the 16th and 17th centuries marked a return to the literal reading of Scripture. The Reformers taught that God revealed in Genesis that He created all things in six ordinary days about six thousand years ago.

In this chapter, I will sketch out these two streams of thought, describe the teachings of the Reformers, and show how these teachings crystallized in their confessions of faith.

Two Views of Genesis 1 in Christian History

There have been many Christians through history who believed in a literal interpretation of Genesis 1. Basil of Caesarea (A.D. 329–379) wrote that in the context of "morning" and "evening" a "day" in Genesis 1 referred to a day of "twenty-four hours."[2] Ambrose (c. A.D. 339–397) wrote in his commentary on

1. I thank David Clayton and Paul Smalley for their research assistance on this article.
2. Basil, *Hexaemeron*, Homily 2.8, http://www.newadvent.org/fathers/32012.htm (accessed May 23, 2013).

Genesis, "The length of one day is twenty-four hours in extent."[3] The English historian and theologian Bede (c. A.D. 672–735) commented on Genesis 1:5 that the first day was "without a doubt a day of twenty-four hours."[4]

On the other hand, other Christians read Genesis 1 as an allegory or symbolic story. Origen (c. A.D. 185–254) rejected a literal interpretation of Genesis 1.[5] The great theologian Augustine (A.D. 354–430) believed that the six days were not periods of time but the way God taught the angels about creation.[6] Why did they believe this? First, they were influenced by an ancient book of Jewish wisdom that is not part of the Bible, misunderstanding it to say that God created all things in an instant.[7] Second, they wanted to reconcile Christianity with Greek philosophy much as the Jewish writer Philo of Alexandria (20 B.C.– A.D. 50) had tried to do, while not rejecting the major biblical doctrine that one God created all things.

The allegorical approach to the Bible prevailed in the Middle Ages, but some major theologians still favored a literal reading of Genesis 1. Peter Lombard (c. A.D. 1096–1164) acknowledged both ways Christians had understood the days of Genesis 1, but took the view that he believed fit Genesis better, namely, that God created everything out of nothing and shaped it into its perfected form over the period of "six days."[8] Lombard taught that the days of Genesis 1, defined by mornings and evenings, should be understood as "the space of twenty four hours."[9] Bonaventure (A.D. 1221–1274) argued that God created "in the space of six days" — a phrase that will appear later in Reformed writings.[10]

3. Ambrose, *Hexameron, Paradise, and Cain and Abel*, trans. John J. Savage, *The Fathers of the Church: A New Translation* (New York: Fathers of the Church, 1961), vol. 42 [1.37].
4. Bede, *On Genesis*, trans. Calvin B. Kendall (Liverpool: Liverpool University Press, 2008), p. 75.
5. Robert Letham, "In the Space of Six Days," *Westminster Theological Journal* 61 (1999): p. 150–51.
6. Ibid., p. 156.
7. The reference is Sirach or Ecclesiasticus 18:1, "The One who lives forever created all things together." The Latin Vulgate had *simul* or "at the same time" for "together," but the Greek reads *koine* or "in common."
8. Peter Lombard, *The Four Books of Sentences*, trans. Alexis Bugnolo, book 2, distinction 12, ch. 2, http://www.franciscan-archive.org/lombardus/opera/ls2-12.html (accessed May 29, 2013).
9. Ibid., distinction 13, ch. 4, http://www.franciscan-archive.org/lombardus/opera/ls2-13.html (accessed May 28, 2013). The word "space" translates Lombard's Latin term *spatium*, the same word later used by Calvin and the Westminster divines.
10. The Latin phrase *sex dierum spatium* appears in Bonaventure's *Commentaries on the Four Books of Sentences*, trans. Alexis Bugnolo, book 2, commentary on distinction 12, art. 1, question 2, http://www.franciscan-archive.org/bonaventura/opera/bon02295.html (accessed May 28, 2013). Bonaventure made the same argument that Calvin would that God created over a span of time "to communicate to the creature what it was able to receive."

Though they interpreted Genesis 1 in different ways, virtually all these Christians still believed that the world was only several thousand years old, in contrast to the Greek philosophical view of an eternal or nearly eternal world. They did not see creation as a process spanning long eras, but a relatively short event, whether God completed it in an instant, or in six ordinary days.[11]

The Reformation and the Interpretation of Genesis

When God brought the Reformation to the church in the 16th century, one great effect was the return to the literal sense of the Bible. For centuries the church had muddied the waters of biblical interpretation by giving each text four meanings as if the Bible consisted entirely of spiritual parables. William Tyndale (c. A.D. 1494–1536) asserted, "The Scripture hath but one sense, which is the literal sense."[12] He did not deny that the Bible uses parables and figures of speech, just as we speak and write today. But we discover the meaning of Scripture by reading it carefully in context.[13] We do not turn history into allegory.

As a result of this approach to the Bible, the Reformers embraced a literal view of Genesis. Martin Luther (A.D. 1483–1546) wrote, "We know from Moses that the world was not in existence before 6,000 years ago."[14] He relied on biblical records to compute the age of the earth, estimating that in 1540 the world was 5,500 years old.[15] He acknowledged that some people followed Aristotle's view that the world had always existed, or Augustine's view that Genesis 1 was an allegory. But Luther believed that Moses wrote Genesis in a plain sense. He said,

> Therefore, as the proverb has it, he calls "a spade a spade," i.e., he employs the terms "day" and "evening" without allegory, just as we customarily do. . . . Moses spoke in the literal sense, not allegorically or figuratively, i.e., that the world, with all its creatures, was created within six days, as the words read. If we do not comprehend the reason

11. For an overview of the views of writers through the Christian era on the origins of man, see William Vandoodewaard, *The Quest for the Historical Adam* (Grand Rapids, MI: Reformation Heritage Books, forthcoming).
12. William Tyndale, *Obedience of a Christian Man*, in *Doctrinal Treatises and Introductions to Different Portions of the Holy Scriptures*, ed. Henry Walter (Cambridge: Cambridge University Press, 1848), 304.
13. Ibid., p. 305.
14. Martin Luther, *Lectures on Genesis*, in *Luther's Works*, ed. Jaroslav Pelikan (St. Louis, MO: Concordia, 1958), 1:ix, 3.
15. Martin Brecht, *Martin Luther: The Preservation of the Church, 1532–1546* (Minneapolis, MN: Augsburg Fortress, 1993), p. 138.

for this, let us remain pupils and leave the job of teacher to the Holy Spirit.[16]

Luther's advice is sound. When the Bible speaks of God creating Adam on the sixth day, teaching Adam His command about the trees, and bringing the animals to him, these are not just spiritual parables or eternal principles but "all these facts refer to time and physical life."[17] Genesis presents itself to us not as a poem or allegory, but as an account of real history. We should accept it as such, even if we cannot answer every question one might raise about the origins of the universe. The words of the Bible are infallibly given by the Holy Spirit (2 Tim. 3:16; 2 Peter 1:21). God is the teacher, and we must be His students.

Luther understood that the world would regard Genesis as a "foolish fairy tale."[18] When he commented on the creation of Adam in Genesis 2, he said, "If Aristotle heard this, he would burst into laughter and conclude that although this is not an unlovely yarn, it is nevertheless a most absurd one."[19] But Luther said that in reality Genesis is not foolishness but wisdom, for science can only investigate what things are made of, but God's Word can reveal how they were made and for what purpose.[20]

Calvin on the Time of Creation

Though God worked through many Reformers alongside and after Luther, none is so well known as John Calvin (A.D. 1509–1564). Like Luther, he read Genesis as "the history of creation." He believed that "the duration of the world . . . has not yet attained six thousand years."[21] He also rejected Augustine's belief that creation was completed in a moment, [22] writing, "Moses relates that God's work was completed not in a moment but in six days."[23]

The Reformers were not naïve; they too faced atheistic skeptics. We should not think that only in this modern age have people tried to explain the origin of the universe and biological life without giving glory to the Creator. Calvin knew

16. Luther, *Lectures on Genesis*, in *Works*, 1:5. See also John A. Maxfield, *Luther's Lectures on Genesis and the Formation of Evangelical Identity* (Kirksville, MO: Truman State University Press, 2008), p. 41.
17. Ibid., 1:122.
18. Ibid., 1:128.
19. Ibid., 1:84.
20. Ibid., 1:124. He used the terminology of efficient and final causes.
21. John Calvin, *Institutes of the Christian Religion*, trans. Ford Lewis Battles, ed. John T. McNeill (Philadelphia, PA: Westminster Press, 1960), 1.14.1.
22. Susan E. Schreiner, "Creation and Providence," in *The Calvin Handbook*, ed. Herman J. Selderhuis (Grand Rapids, MI: Eerdmans, 2009), p. 270.
23. Calvin, *Institutes of the Christian Religion*, 1.14.2.

that the Bible's teaching of the relatively young age of the earth would provoke some to laugh and sneer, but realized that profane men will mock at almost every major teaching of Christianity.[24] He was aware that some people taught that "the world came together by chance" as "tiny objects tumbling around" formed the stars, the earth, living creatures, and human beings. Calvin believed that the excellence and artistry of the smallest parts of the human body showed such theories of random creation to be ridiculous.[25] God revealed that He created the world in six days about six thousand years ago to protect the Church from pagan fables about our origins, to glorify Himself as the only Creator and Lord, and to call us to submit our minds to God's will and Word.[26]

Calvin regarded the early chapters of Genesis as "the history of the creation of the world," and delighted in them because creation is "the splendid mirror of God's glory."[27] To be sure, the Bible does not reveal all the facts that can be discovered by astronomy — though Calvin said that astronomy is "pleasant" and "useful" for Christians.[28] Scripture records creation in words that ordinary people can understand, not technical scientific language.[29] Still, the Bible is true, and Genesis is real history. Foolish men may ridicule God's ways, but the humble know better: "Since his will is the rule of all wisdom, we ought to be contented with that alone."[30]

If someone objects that Moses was not alive at creation and so could only write fables about it, Calvin replied that Moses was not writing thoughts he invented or discovered himself, but "is the instrument of the Holy Spirit." That same Spirit enabled Moses to foretell events that would happen long after his death, such as the calling of the Gentiles to Christ. Furthermore, the Spirit helped Moses to make use of traditions handed down from Adam, Abraham, and others.[31]

24. Calvin, *Institutes of the Christian Religion*, 3.21.4.
25. John Calvin, *Sermons on Genesis: Chapters 1:1-11:4*, trans. Rob Roy McGregor (Edinburgh: Banner of Truth, 2009), p. 9, 11–12. See also his commentaries on Exod. 2:4 and Ps. 104:24. Calvin attributed such views to a form of atheism that he associated with the teachings of Epicurus (341–270 B.C.), an ancient Greek philosopher. See Nicolaas H. Gootjes, "Calvin on Epicurus and the Epicureans," in *Calvin Theological Journal* 40 (2006): p. 33–48.
26. Calvin, *Institutes of the Christian Religion*, 1.14.1–2.
27. John Calvin, *Commentaries on the First Book of Moses called Genesis*, trans. John King (Edinburgh: Calvin Translation Society, 1847), 1:xlviii; cf. 1:57.
28. Ibid., 1:79.
29. Ibid., 1:86–87.
30. Ibid., 1:61.
31. Ibid., 1:58.

Someone else might object that it makes no sense that God created light on the first day before God created the sun on the fourth day. Here too, Calvin helps us by saying that God has an important lesson for us in this: "The Lord, by the very order of creation, bears witness that he holds in his hand the light, which he is able to impart to us without the sun and moon."[32] Thus the order of the creation week reveals that God can meet all our needs even without the natural means He ordinarily uses.

Calvin was aware that some people said that the six days of Genesis 1 were a metaphor. But he believed this did not do justice to the text of Scripture. He wrote, "For it is too violent a cavil [objection] to contend that Moses distributes the work which God perfected at once into six days, for the mere purpose of conveying instruction. Let us rather conclude that God himself took the space of six days, for the purpose of accommodating his works to the capacity of men." He went on to explain that God "distributed the creation of the world into successive portions, that he might fix our attention, and compel us, as if he had laid his hand upon us, to pause and reflect."[33] Joseph Pipa writes, "Calvin's commitment to six days and the order of the days stands in bold contrast to modern theories such as the framework hypothesis and the analogical view of Genesis 1. He emphatically insists on the order of the six days as both advantageous to man and instructive about the character of God."[34]

Lutheran and Early Reformed Confessions on Creation

The Reformation was a time of tremendous rediscoveries of biblical truth. To show their faithfulness to the Scriptures and pass these truths on to future generations, evangelicals published their beliefs in confessions and catechisms.

The doctrine of creation was not a major point of disagreement between the Roman Catholic Church and the evangelical churches of the Reformation. Therefore, it did not receive much attention in the Lutheran confessions, except to affirm briefly that one God created all things.[35] The major Reformed confessions of the 16th century offered more developed statements about the creation

32. Ibid., 1:76.
33. Ibid., 1:78. See also *Sermons on Genesis*, p. 19.
34. Joseph A. Pipa Jr., "Creation and Providence," in *A Theological Guide to Calvin's Institutes*, ed. David W. Hall and Peter A. Lillback (Phillipsburg, NJ: P&R Publishing, 2008), p. 129.
35. Augsburg Confession, art. 1, and Small Catechism, part 2, art. 1, in *The Book of Concord: The Confessions of the Evangelical Lutheran Church*, trans. and ed. Theodore G. Tappert (Philadelphia, PA: Fortress Press, 1959), p. 28, 344.

of the world, angels, and mankind, but did not address the time of creation.[36] The Belgic Confession (article 14) does say that "God created man out of the dust of the earth."[37] Thus it confessed a literal understanding of Genesis 2:7, which logically contradicts the modern notion that man evolved by a natural process from other forms of life over millions of years.

Girolamo Zanchi (A.D. 1516–1590) was a professor of Old Testament and theology who taught at Strassburg and Heidelberg. A few years before he died, Zanchi published a detailed confession of faith, which said that God created the world "in the space of six days."[38] He also published a massive book titled *Concerning the Works of God in Creation during the Space of Six Days*, where he argued that Genesis 1 clearly says God created the world in six literal days.[39]

James Ussher (A.D. 1581–1656), bishop of Armagh, is now best known for his biblical history of the world, where he famously calculated the date of creation at 4004 B.C. In 1615, he led a gathering of church leaders in Dublin to adopt the Irish Articles, which say, "In the beginning of time, when no creature had any being, God by his Word alone, *in the space of six days*, created all things."[40] These words come directly from Ussher's *Principles of Christian Religion*, which he wrote around 1603.[41] Ussher was invited to participate in the Westminster Assembly, and though he declined, his writings still influenced the documents written there.

The Westminster Standards on Creation

Meeting from 1643 to 1649, British Reformed theologians wrote the Westminster Confession of Faith (WCF), Shorter Catechism (WSC), and Larger Catechism (WLC). The Westminster Standards continue to serve as

36. Belgic Confession, art. 12, Heidelberg Catechism, Q. 6, and Second Helvetic Confession, art. 7, in *Reformed Confessions Harmonized*, ed. Joel R. Beeke and Sinclair B. Ferguson (Grand Rapids, MI: Baker, 1999), p. 36–38.
37. Belgic Confession, art. 14., in *Doctrinal Standards, Liturgy, and Church Order*, ed. Joel R. Beeke (Grand Rapids, MI: Reformation Heritage Books, 2003), p. 11.
38. H. Zanchius, *Confession of Christian Religion* (London: Iohn Legat, 1599), p. 21 [5.1]. The Latin reads *intra spacium sex dierum* (H. Zanchii, *De Religione Christiana Fides* (Neostadii Palatinorvm: Matthaus Harnisch, [1588]), 17–18 [5.1]).
39. Hieron. Zanchii, *De Operibus Dei intra Spacium Sex Dierum Creatis* (1591). See Vandoodewaard, *The Quest for the Historical Adam*, for discussion.
40. Irish Articles, art. 4, sec. 18, in *Documents of the English Reformation*, ed. Gerald Bray (Minneapolis, MN: Fortress Press, 1994), p. 440, emphasis added.
41. *The Whole Works of the Most Rev. James Ussher* (Dublin: Hodges, Smith, and Col, 1864), 11:179, 183.

the confessional declarations of Presbyterian churches around the world. The Larger Catechism (Q. 17) taught a literal view of Genesis 1–2 by stating, "After God had made all other creatures, He created man male and female; formed the body of the man out of the dust of the ground, and the woman of the rib of the man."[42] The confession and both catechisms state that God created the universe in "the space of six days."[43] This same language also carried over into the confessions of the Congregationalists and Particular Baptists when they adapted the Westminster Confession for use in their own churches.[44]

What do the Westminster Standards and their daughter confessions mean by creation in "the space of six days"? Why did they not simply say, "in six days"? First, by using the word "space" they made it clear they were talking about a definite span of time, not just a metaphor with six parts. Other books from the 17th century used the words "the space of six days" to refer to the duration of six ordinary days.[45] Thus one book printed in 1693 talks about how a king conquered an entire region "in the space of six days."[46]

Second, in taking up the language of "the space of six days," the Westminster Assembly declared that it stood with previous theologians in affirming a literal six-day creation. The expression has its roots in at least four previous theologians whom the Westminster divines knew. As we have seen, the words "in the space of six days" appear in the writings of Bonaventure, Calvin, Zanchi, and Ussher.[47] Zanchi's *Confessions* may have influenced the Westminster divines, for it was a prime example of early Reformed orthodox confessions from which

42. WLC, Q. 17, in *Reformed Confessions Harmonized*, p. 39.
43. WCF 4.1, WSC Q. 9, and WLC Q. 15, in *Reformed Confessions Harmonized*, 37. The Latin phrase is *sex dierum spatium* (Philip Schaff, *Creeds of Christendom* [New York: Harper, 1877], 3:611).
44. A comparison of the WCF to the Savoy Declaration (1658) and the Second London Baptist Confession (1677/1689) may be found at http://www.proginosko.com/docs/wcf_sdfo_lbcf.html (accessed May 24, 2013).
45. *Journals of the House of Lords* (1642), 5:535; Nathan Bailey, "Founday," in *An Universal Etymological English Dictionary* (London: for R. Ware et al, 1675); *The Laws and Acts Made in the First Parliament of Our Most High and Dread Soveraign James VII*, ed. George, Viscount of Tarbet (Edinburgh: Andrew Anderson, 1685), p. 141; Pierre Danet, "Judaei," in *A Complete Dictionary of the Greek and Roman Antiquities* (London: for John Nicholson et al., 1700).
46. *The History of Polybius the Megapolitan*, 2nd ed. (London: Samuel Briscoe, 1693), 2:128.
47. Bonaventure, *Commentaries on the Four Books of Sentences*, book 2, distinction 12, art. 1, question 2; Calvin, *Commentaries on Genesis*, 1:78; Zanchius, *Confession of Christian Religion*, 21 [5.1]; *De Operibus Dei intra Spacium Sex Dierum Creatis*; Ussher, *Works*, 11:183.

to draw.[48] Certainly the Irish Articles of Ussher influenced the Westminster Confession.[49]

Research into the writings of several members of the Westminster Assembly has confirmed that they believed in a relatively young earth and a literal six-day creation.[50] In 1674, Thomas Vincent wrote the following in his explanation of the Westminster Shorter Catechism: "In what time did God create all things? God created all things in the space of six days. He could have created all things together in a moment, but he took six days' time to work in."[51] Thus, we have good reason to conclude that the Westminster Confession, Larger Catechism, and Shorter Catechism teach us to regard Genesis 1 as a real week of time in history.

Some godly men who love the Westminster Confession disagree with me, arguing that "the space of six days" is ambiguous and it was only meant to exclude the idea of creation in an instant.[52] But the Westminster Standards do more than reject instantaneous creation. They also affirm creation over a specified period of time: "the space of six days."

Conclusion

Though all Christians believe that God created the world, through the history of the Church a literal reading of Genesis has competed with an allegorical reading. In the Reformation, Luther and Calvin embraced the literal reading of Genesis, with the result that they believed in a six-day creation some six thousand years ago. We also find evidence of the literal view in the Belgic Confession, the *Confession of Faith* by Zanchi, the Irish Articles, and the Westminster Confession of Faith.

48. Richard Muller, *Post-Reformation Reformed Dogmatics, Volume Two, Holy Scripture: The Cognitive Foundation of Theology*, 2nd ed. (Grand Rapids, MI: Baker Academic, 2003), p. 85.
49. Benjamin B. Warfield, *The Westminster Assembly and Its Work* (New York: Oxford University Press, 1931), p. 127, 148, 169–74.
50. David W. Hall, "What Was the View of the Westminster Assembly Divines on Creation Days?" in *Did God Create in Six Days?* ed. Joseph A. Pipa, Jr., and David W. Hall (Taylors, SC: Southern Presbyterian Press, 1999), p. 41–52.
51. Thomas Vincent, *An Explicatory Catechism: Or, An Explanation of the Assembly's Shorter Catechism* (New Haven, CT: Walter, Austin, and Co., 1810), p. 42, on WSC Q. 9.
52. "Westminster Theological Seminary and the Days of Creation," Westminster Theological Seminary, http://www.wts.edu/about/beliefs/statements/creation.html (accessed May 28, 2013); R. Scott Clark, *Recovering the Reformed Confession* (Phillipsburg, NJ: P&R Publishing, 2008), p. 49. A critique of some of Hall's conclusions may be found in William S. Barker, *Word to the World* (Ross-shire, UK: Christian Focus Publications, 2005), p. 259–270. This article also appeared in *Westminster Theological Journal* 62 (2000): p. 113–120. I note, however, that Barker does not offer examples of Westminster divines who rejected creation in six literal days.

But in this modern era, an increasing number of evangelical and Reformed Christians are turning back to the old error of embracing a symbolic view of Genesis, albeit often in new forms. I believe that we face a double danger here. First, we are in danger of losing our confidence that words can clearly communicate truth. There seems to be a hermeneutical issue at stake here, namely, the perspicuity of Scripture. It is fascinating that, generally speaking, the same Reformed scholars who argue for some kind of allegorical interpretation of the plain and literal words of Genesis 1 tend to reinterpret the plain and literal words of the Westminster Confession when it states that creation took place "in the space of six days." If plain words can take on allegorical or alternative meanings so easily so that they do not mean what they plainly state, how do we know what anything means? The resulting uncertainty that such interpretations convey leads into the second danger, that of doctrinal minimalism. If we cut back the meaning of our confessions by saying their statements merely stand against some specific error, then we lose the richness of what the confessions positively affirm. Similarly, if we reduce Genesis 1 to the bare truth that "God created everything," then we lose the richness of what God reveals in the whole chapter.

An uncertain and minimalist approach to the doctrine of creation opens the door for serious errors to enter the church, such as the evolution of man from animals or the denial that Adam and Eve were real, historical people. Happily, a robust doctrine of creation provides a strong foundation for our faith.

CHAPTER 10

What Are Some of the Best Evidences in Science of a Young Creation?

DR. ANDREW A. SNELLING, DR. DAVID MENTON,
DR. DANNY R. FAULKNER, AND DR. GEORGIA PURDOM

The earth is only a few thousand years old. That's a fact, plainly revealed in God's Word. So we should expect to find plenty of evidence for its youth. And that's what we find — in the earth's geology, biology, paleontology, and even astronomy.

Literally hundreds of dating methods could be used to attempt an estimate of the earth's age, and the vast majority of them point to a much younger earth than the 4.5 billion years claimed by secularists. The following series of articles presents what Answers in Genesis researchers picked as the ten best scientific evidences that contradict billions of years and confirm a relatively young earth and universe.

Despite this wealth of evidence, it is important to understand that, from the perspective of observational science, no one can prove absolutely how young (or old) the universe is. Only one dating method is absolutely reliable — a witness who doesn't lie, who has all evidence, and who can reveal to us when the universe began!

And we do have such a witness — the God of the Bible! He has given us a specific history, beginning with the six days of creation and followed by detailed genealogies that allow us to determine when the universe began. Based on this history, the beginning was only about six thousand years ago (about four thousand years from creation to Christ).

In the rush to examine all these amazing scientific "evidences," it's easy to lose sight of the big picture. Such a mountain of scientific evidence, accumulated by researchers, seems to obviously contradict the supposed billions of years, so why don't more people rush to accept the truth of a young earth based on the Bible?

The problem is, as we consider the topic of origins, all so-called "evidences" must be interpreted. Facts don't speak for themselves. Interpreting the facts of the present becomes especially difficult when reconstructing the historical events that produced those present-day facts, because no humans have always been present to observe all the evidence and to record how all the evidence was produced.

Forensic scientists must make multiple assumptions about things they cannot observe. How was the original setting different? Were different processes in play? Was the scene later contaminated? Just one wrong assumption or one tiny piece of missing evidence could totally change how they reconstruct the past events that led to the present-day evidence.

When discussing the age of the earth, Christians must be ready to explain the importance of starting points. The Bible is the right starting point.

That's why, when discussing the age of the earth, Christians must be ready to explain the importance of starting points and assumptions. Reaching the correct conclusions requires the right starting point.

The Bible is that starting point. This is the revealed Word of the almighty, faithful, and true Creator, who was present to observe all events of earth history and who gave mankind an infallible record of key events in the past.

The Bible, God's revelation to us, gives us the foundation that enables us to begin to build the right worldview to correctly understand how the present and past are connected. All other documents written by man are fallible, unlike the "God-breathed" infallible Word (2 Timothy 3:16). The Bible clearly and unmistakably describes the creation of the universe, the solar system, and the earth around six thousand years ago. We know that it's true based on the authority of God's own character. "Because He could swear by no one greater, He swore by Himself" (Hebrews 6:13).

In one sense, God's testimony is all we need; but God Himself tells us to give reasons for what we believe (1 Peter 3:15). So it is also important to conduct scientific research (that is part of taking dominion of the earth, as Adam was told to do in Genesis 1:28). With this research we can challenge those who reject God's clear Word and defend the biblical worldview.

Indeed, God's testimony must have such a central role in our thinking that it seems demeaning even to call it the "best" evidence of a young earth. It is, in truth, the only foundation upon which all other evidences can be correctly understood!

Following are the ten best evidences from science that confirm a young earth.

#1 Very Little Sediment on the Seafloor

If sediments have been accumulating on the seafloor for three billion years, the seafloor should be choked with sediments many miles deep.

Every year, water and wind erode about 20 billion tons of dirt and rock debris from the continents and deposit them on the seafloor[1] (figure 1). Most of this material accumulates as loose sediments near the continents. Yet the average thickness of all these sediments globally over the whole seafloor is not even 1,300 feet (400 m).[2]

Some sediments appear to be removed as tectonic plates slide slowly (an inch or two per year) beneath continents. An estimated 1 billion tons of sediments are removed this way each year.[3] The net gain is thus 19 billion tons per year. At this rate, 1,300 feet of sediment would accumulate in less than 12 million years, not billions of years.

This evidence makes sense within the context of the Genesis Flood cataclysm, not the idea of slow and gradual geologic evolution. In the latter stages of the year-long global Flood, water swiftly drained off the emerging land, dumping its sediment-chocked loads offshore. Thus most seafloor sediments accumulated rapidly about 4,350 years ago.[4]

Rescuing Devices

Those who advocate an old earth insist that the seafloor sediments must have accumulated at a much slower rate in the past. But this rescuing device doesn't "stack up"! Like the sediment layers on the continents, the sediments on the continental shelves and margins (the majority of the seafloor sediments)

1. John D. Milliman and James P. N. Syvitski, "Geomorphic/Tectonic Control of Sediment Discharge to the Ocean: The Importance of Small Mountainous Rivers," *The Journal of Geology* 100 (1992): p. 525–544.
2. William W. Hay, James L. Sloan II, and Christopher N. Wold, "Mass/Age Distribution and Composition of Sediments on the Ocean Floor and the Global Rate of Sediment Subduction," *Journal of Geophysical Research* 93, no. B12 (1998): p. 14,933–14,940.
3. Ibid.
4. For a fuller treatment and further information see John D. Morris, *The Young Earth* (Green Forest, AR: Master Books, 2000), p. 88–90; Andrew A. Snelling, *Earth's Catastrophic Past: Geology, Creation and the Flood* (Dallas, TX: Institute for Creation Research, 2009), p. 881–884.

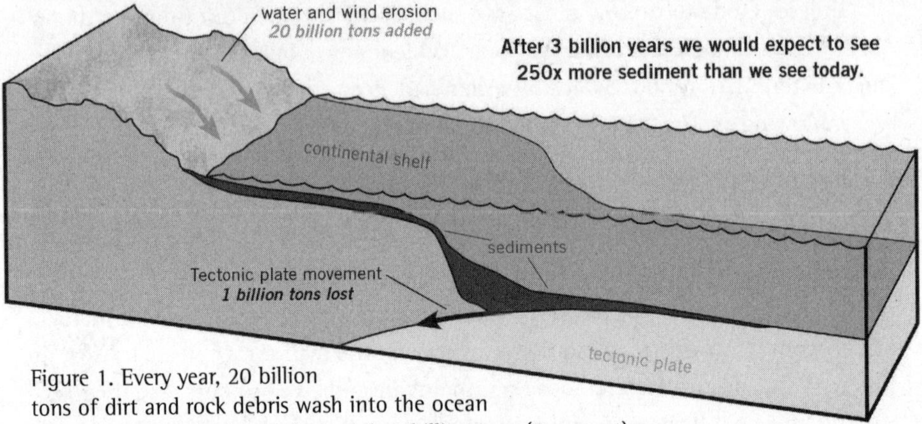

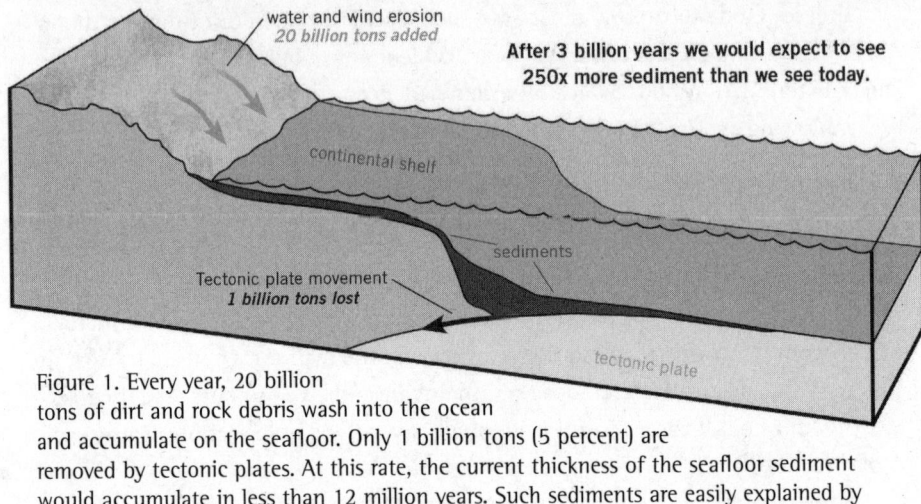

Figure 1. Every year, 20 billion tons of dirt and rock debris wash into the ocean and accumulate on the seafloor. Only 1 billion tons (5 percent) are removed by tectonic plates. At this rate, the current thickness of the seafloor sediment would accumulate in less than 12 million years. Such sediments are easily explained by water draining off the continents toward the end of the Flood.

have features that unequivocally indicate they were deposited much faster than today's rates. For example, the layering and patterns of various grain sizes in these sediments are the same as those produced by undersea landslides, when dense debris-laden currents (called turbidity currents) flow rapidly across the continental shelves and the sediments then settle in thick layers over vast areas. An additional problem for the old-earth view is that no evidence exists of much sediment being subducted and mixed into the mantle.

#2 Bent Rock Layers

In many mountainous areas, rock layers thousands of feet thick have been bent and folded without fracturing. How can that happen if they were laid down separately over hundreds of millions of years and already hardened?

Hardened rock layers are brittle. Try bending a slab of concrete sometime to see what happens! But if concrete is still wet, it can easily be shaped and molded before the cement sets. The same principle applies to sedimentary rock layers. They can be bent and folded soon after the sediment is deposited, before the natural cements have a chance to bind the particles together into hard, brittle rocks.[5]

The region around Grand Canyon is a great example showing how most of the earth's fossil-bearing layers were laid down quickly and many were folded

5. R.E. Goodman, *Introduction to Rock Mechanics* (New York: John Wiley and Sons, 1980); Sam Boggs Jr., *Principles of Sedimentology and Stratigraphy* (Upper Saddle River, NJ: Prentice-Hall, 1995), p. 127–131.

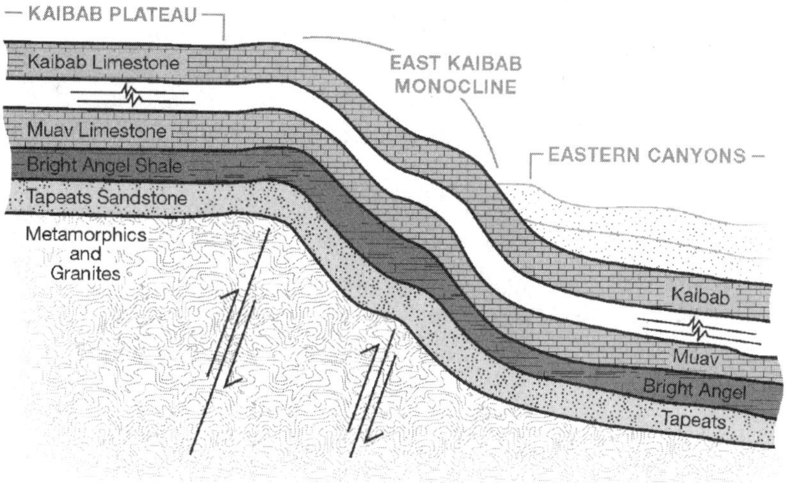

Figure 2. The Grand Canyon now cuts through many rock layers. Previously, all these layers were raised to their current elevation (a raised, flat region known as the Kaibab Plateau). Somehow this whole sequence was bent and folded without fracturing. That's impossible if the first layer, the Tapeats Sandstone, was deposited over North America 460 million years before being folded. But all the layers would still be relatively soft and pliable if it all happened during the recent, global Flood.

while still wet. Exposed in the canyon's walls are about 4,500 feet (1,370 m) of fossil-bearing layers, conventionally labeled Cambrian to Permian.[6] They were supposedly deposited over a period lasting from 520 to 250 million years ago. Then, amazingly, this whole sequence of layers rose over a mile, around 60 million years ago. The plateau through which Grand Canyon runs is now 7,000–8,000 feet (2,150–3,450 m) above sea level.

Think about it. The time between the first deposits at Grand Canyon (520 million years ago) and their bending (60 million years ago) was 460 million years!

Look at the photos on the following page of some of these layers at the edge of the plateau, just east of the Grand Canyon. The whole sequence of these hardened sedimentary rock layers has been bent and folded, but without fracturing (figure 2).[7] At the bottom of this sequence is the Tapeats Sandstone, which is 100–325 feet (30–100 meters) thick. It is bent and folded 90° (photo 1). The Muav Limestone above it has similarly been bent (photo 2).

6. Stanley S. Beus and Michael Morales, eds., *Grand Canyon Geology*, 2nd edition (New York: Oxford University Press, 2003).
7. Andrew A. Snelling, "Rock Layers Folded, Not Fractured," *Answers* 4, no. 2 (April–June 2009): p. 80–83.

Photo 1. The whole sequence of sedimentary layers through which Grand Canyon cuts has been bent and folded without fracturing. This includes the Tapeats Sandstone, located at the bottom of the sequence. (A 90° fold in the eastern Grand Canyon is pictured here.)
(Photo courtesy of Andrew Snelling)

Photo 2. All the layers through which Grand Canyon cuts – including the Muav Limestone shown here – have been bent without fracturing.
(Photo courtesy of Andrew Snelling)

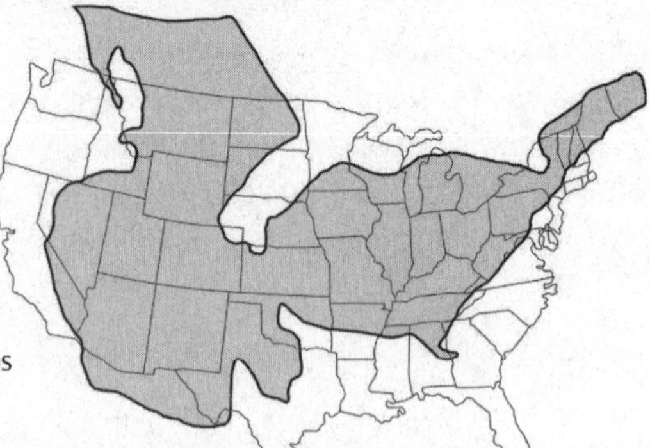

Figure 3. This phenomenon was not regional. The Tapeats Sandstone spans the continent, and other layers span much of the globe.

However, it supposedly took 270 million years to deposit these particular layers. Surely in that time the Tapeats Sandstone at the bottom would have dried out and the sand grains cemented together, especially with 4,000 feet (1,220 m) of rock layers piled on top of it and pressing down on it. The only viable scientific explanation is that the whole sequence was deposited very quickly — the creation model indicates that it took less than a year, during the global Flood cataclysm. So the 520 million years never happened, and the earth is young.

Rescuing Devices

What solution do old-earth advocates suggest? Heat and pressure can make hard rock layers pliable, so they claim this must be what happened in the eastern Grand Canyon, as the sequence of many layers above pressed down and heated up these rocks. Just one problem. The heat and pressure would have transformed these layers into quartzite, marble, and other metamorphic rocks. Yet Tapeats Sandstone is still sandstone, a sedimentary rock!

But this quandary is even worse for those who deny God's recent creation and the Flood. The Tapeats Sandstone and its equivalents can be traced right across North America (figure 3),[8] and beyond to right across northern Africa to southern Israel.[9] Indeed, the whole Grand Canyon sedimentary sequence is an integral part of six megasequences that cover North America.[10] Only a global Flood cataclysm could carry the sediments to deposit thick layers across several continents one after the other in rapid succession in one event.[11]

#3 Soft Tissue in Fossils

Ask the average layperson how he or she knows that the earth is millions or billions of years old, and that person will probably mention the dinosaurs, which nearly everybody "knows" died off 65 million years ago. A recent discovery by Dr. Mary Schweitzer, however, has given reason for all but committed evolutionists to question this assumption.

Bone slices from the fossilized thigh bone (femur) of a *Tyrannosaurus rex* found in the Hell Creek Formation of Montana were studied under the

8. F. Alan Lindberg, *Correlation of Stratigraphic Units of North America (COSUNA)*, Correlation Charts Series (Tulsa, OK: American Association of Petroleum Geologists, 1986).
9. Andrew A. Snelling, "The Geology of Israel within the Biblical Creation-Flood Framework of History: 2. The Flood Rocks," *Answers Research Journal* 3 (2010): p. 267–309.
10. L.L. Sloss, "Sequences in the Cratonic Interior of North America," *Geological Society of America Bulletin* 74 (1963): p. 93–114.
11. For a fuller treatment and further information see Morris, *The Young Earth*, p. 106–109; Snelling, *Earth's Catastrophic Past: Geology, Creation and the Flood*, p. 528–530, 597–605.

microscope by Schweitzer. To her amazement, the bone showed what appeared to be blood vessels of the type seen in bone and marrow, and these contained what appeared to be red blood cells with nuclei, typical of reptiles and birds (but not mammals). The vessels even appeared to be lined with specialized endothelial cells found in all blood vessels.

Amazingly, the bone marrow contained what appeared to be flexible tissue. Initially, some skeptical scientists suggested that bacterial biofilms (dead bacteria aggregated in a slime) formed what only appear to be blood vessels and bone cells. Recently, Schweitzer and co-workers found biochemical evidence for intact fragments of the protein collagen, which is the building block of connective tissue. This is important because collagen is a highly distinctive protein not made by bacteria.[12]

Some evolutionists have strongly criticized Schweitzer's conclusions because they are understandably reluctant to concede the existence of blood vessels, cells with nuclei, tissue elasticity, and intact protein fragments in a dinosaur bone dated at 68 million years old. Other evolutionists, who find Schweitzer's evidence too compelling to ignore, simply conclude that there is some previously unrecognized form of fossilization that preserves cells and protein fragments over tens of millions of years.[13] Needless to say, no evolutionist has publically considered the possibility that dinosaur fossils are not millions of years old.

An obvious question arises from Schweitzer's work: is it even remotely plausible that blood vessels, cells, and protein fragments can exist largely intact over 68 million years? While many consider such long-term preservation of tissue and cells to be very unlikely, the problem is that no human or animal remains are known with certainty to be 68 million years old (figure 4). But if creationists are right, most dinosaurs were buried in the Flood 3,000 to 4,000 years ago. So would we expect the preservation of vessels, cells, and complex molecules of the type that Schweitzer reports for biological tissues historically known to be 3,000 to 4,000 years old?

The answer is yes. Many studies of Egyptian mummies and other humans of this old age (confirmed by historical evidence) show all the sorts of detail Schweitzer reported in her *T. rex*. In addition to Egyptian mummies, the Tyrolean iceman, found in the Alps in 1991 and believed to be about 5,000 years old according to long-age dating, shows such incredible preservation of DNA and other microscopic detail.

12. See Schweitzer's review article, "Blood from Stone," *Scientific American* (December 2010): p. 62–69.
13. Marcus Ross, "Those Not-So-Dry Bones," *Answers* (Jan–Mar 2010): p. 43–45.

We conclude that the preservation of vessels, cells, and complex molecules in dinosaurs is entirely consistent with a young-earth creationist perspective but is highly implausible with the evolutionist's perspective about dinosaurs that died off millions of years ago.

Figure 4. A little skin: a largely intact dinosaur mummy, named Dakota, was found in the Hell Creek Formation of the western United States in 2007. Some soft tissue from the long-necked hadrosaur was quickly preserved as fossil, such as the scales from its forearm shown here.

#4 Faint Sun Paradox

Evidence now supports astronomers' belief that the sun's power comes from the fusion of hydrogen into helium deep in the sun's core, but there is a huge problem. As the hydrogen fuses, it should change the composition of the sun's core, gradually increasing the sun's temperature. If true, this means that the earth was colder in the past. In fact, the earth would have been below freezing 3.5 billion years ago, when life supposedly evolved.

The rate of nuclear fusion depends upon the temperature. As the sun's core temperatures increase, the sun's energy output should also increase, causing the sun to brighten over time. Calculations show that the sun would brighten by 25 percent after 3.5 billion years. This means that an early sun would have been fainter, warming the earth 31°F (17°C) less than it does today. That's below freezing!

But evolutionists acknowledge that there is no evidence of this in the geologic record. They even call this problem the faint young sun paradox. While this isn't a problem over many thousands of years, it is a problem if the world is billions of years old.

Rescuing Devices

Over the years, scientists have proposed several mechanisms to explain away this problem. These suggestions require changes in the earth's atmosphere. For instance, more greenhouse gases early in earth's history would retain more heat, but this means that the greenhouse gases had to decrease gradually to compensate for the brightening sun.

None of these proposals can be proved, for there is no evidence. Furthermore, it is difficult to believe that a mechanism totally unrelated to the sun's brightness could compensate for the sun's changing emission so precisely for billions of years.

#5 Rapidly Decaying Magnetic Field

The earth is surrounded by a magnetic field that protects living things from solar radiation. Without it, life could not exist. That's why scientists were surprised to discover that the field is quickly wearing down. At the current rate, the field and thus the earth could be no older than 20,000 years old.

Several measurements confirm this decay. Since measuring began in 1845, the total energy stored in the earth's magnetic field has been decaying at a rate of 5 percent per century.[14] Archaeological measurements show that the field was 40 percent stronger in A.D. 1000.[15] Recent records of the International Geomagnetic Reference Field, the most accurate ever taken, show a net energy loss of 1.4 percent in just three decades (1970–2000).[16] This means that the field's energy has halved every 1,465 years or so.

Creationists have proposed that the earth's magnetic field is caused by a freely decaying electric current in the earth's core. This means that the electric current naturally loses energy, or "decays," as it flows through the metallic core. Though it differs from the most commonly accepted conventional model, it is consistent with our knowledge of what makes up the earth's core.[17] Furthermore, based on what we know about the conductive properties of liquid iron, this freely decaying current would have started when the earth's outer core was formed. However, if the core were more than 20,000 years old, then the starting energy would have made the earth too hot to be covered by water, as Genesis 1:2 reveals.

Reliable, accurate, published geological field data have emphatically confirmed the young-earth model: a freely decaying electric current in the outer core is generating the magnetic field.[18] Although this field reversed direction several

14. A.L. McDonald and R.H. Gunst, "An Analysis of the Earth's Magnetic Field from 1835 to 1965," *ESSA Technical Report*, IER 46-IES 1 (Washington, DC: U.S. Government Printing Office, 1967).
15. R.T. Merrill and M.W. McElhinney, *The Earth's Magnetic Field* (London: Academic Press, 1983), p. 101–106.
16. These measurements were gathered by the International Geomagnetic Reference Field. See D. Russell Humphreys, "The Earth's Magnetic Field Is Still Losing Energy," *Creation Research Society Quarterly* 39, no. 1 (2002): p. 1–11.
17. Thomas G. Barnes, "Decay of the Earth's Magnetic Field and the Geochronological Implications," *Creation Research Society Quarterly* 8, no. 1 (1971): p. 24–29; Thomas G. Barnes, *Origin and Destiny of the Earth's Magnetic Field, Technical Monograph no. 4*, 2nd edition (Santee, CA: Institute for Creation Research, 1983).
18. D. Russell Humphreys, "Reversals of the Earth's Magnetic Field During the Genesis Flood," in *Proceedings of the First International Conference on Creationism*, vol. 2, R.E. Walsh, C.L. Brooks, and R.S. Crowell, eds. (Pittsburgh, PA: Creation Science Fellowship, 1986), p. 113–126.

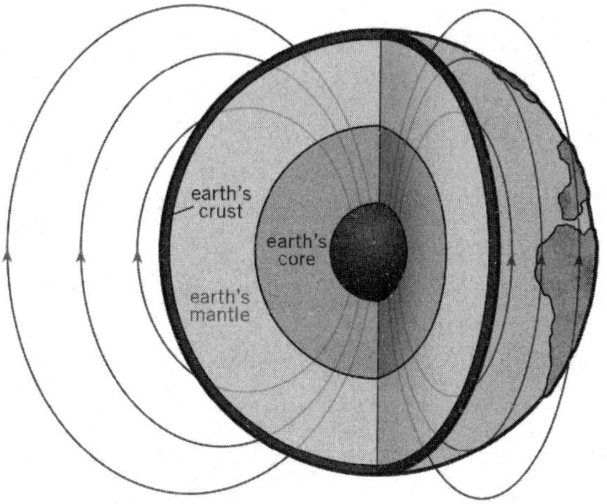

Figure 5. Creationists have proposed that the earth's magnetic field is caused by a freely decaying electric current in the earth's core. (Old-earth scientists are forced to adopt a theoretical, self-sustaining process known as the dynamo model, which contradicts some basic laws of physics.) Reliable, accurate, published geological field data have emphatically confirmed this young-earth model.

times during the Flood cataclysm when the outer core was stirred (figure 5), the field has rapidly and continuously lost total energy ever since creation (figure 6). It all points to an earth and magnetic field only about 6,000 years old.[19]

Rescuing Devices

Old-earth advocates maintain the earth is over 4.5 billion years old, so they believe the magnetic field must be self-sustaining. They propose a complex, theoretical process known as the dynamo model, but such a model contradicts some basic laws of physics. Furthermore, their model fails to explain the modern, measured electric current in the seafloor.[20] Nor can it explain the past field reversals, computer simulations notwithstanding.[21]

To salvage their old earth and dynamo, some have suggested the magnetic field decay is linear rather than exponential, in spite of the historic measurements and decades of experiments confirming the exponential decay. Others have suggested that the strength of some components increases to make up for other components that are decaying. That claim results from confusion about the difference between magnetic field intensity and its energy, and has been refuted categorically by creation physicists.[22]

19. For a fuller treatment and further information see Morris, *The Young Earth*, p. 74–85; Snelling, *Earth's Catastrophic Past: Geology, Creation and the Flood*, p. 873–877.
20. L.J. Lanzerotti et al., "Measurements of the Large-Scale Direct-Current Earth Potential and Possible Implications for the Geomagnetic Dynamo," *Science* 229, no. 4708 (1985): p. 47–49.
21. D. Russell Humphreys, "Can Evolutionists Now Explain the Earth's Magnetic Field?" *Creation Research Society Quarterly* 33, no. 3 (1996): p. 184–185.
22. D. Russell Humphreys, "Physical Mechanism for Reversal of the Earth's Magnetic Field During the Flood," in *Proceedings of the Second International Conference on Creationism*, vol. 2, p. 129–142.

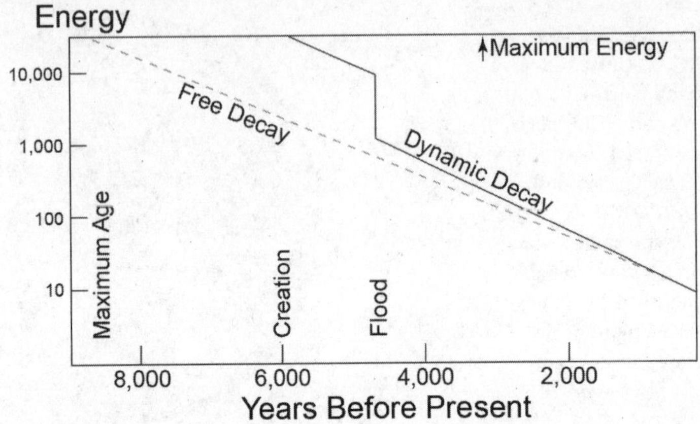

Figure 6: The earth's magnetic field has rapidly and continuously lost total energy since its origin, no matter which model has been adopted to explain its magnetism. According to creationists' dynamic decay model, the earth's magnetic field lost more energy during the Flood, when the outer core was stirred and the field reversed direction several times.

#6 Helium in Radioactive Rocks

During the radioactive decay of uranium and thorium contained in rocks, lots of helium is produced. Because helium is the second lightest element and a noble gas — meaning it does not combine with other atoms — it readily diffuses (leaks) out and eventually escapes into the atmosphere. Helium diffuses so rapidly that all the helium should have leaked out in less than 100,000 years. So why are these rocks still full of helium atoms?

While drilling deep Precambrian (pre-Flood) granitic rocks in New Mexico, geologists extracted samples of zircon (zirconium silicate) crystals from different depths. The crystals contained not only uranium but also large amounts of helium.[23] The hotter the rocks, the faster the helium should escape, so researchers were surprised to find that the deepest, and therefore hottest, zircons (at 387°F or 197°C) contained far more helium than expected. Up to 58 percent of the helium that the uranium could have ever generated was still present in the crystals.

23. R.V. Gentry, G.L. Glish, and E.H. McBay, "Differential Helium Retention in Zircons: Implications for Nuclear Waste Containment," *Geophysical Research Letters* 9, no. 10 (1982): p. 1129–1130.

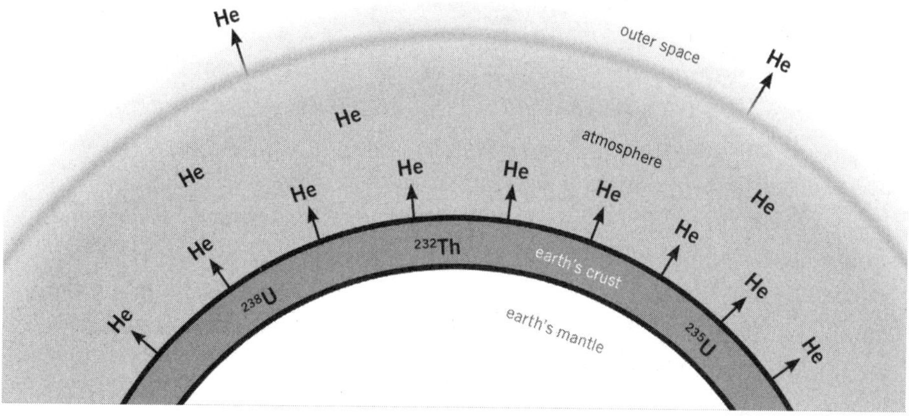

Figure 7. Radioactive elements in rocks produce a lot of helium as they decay; and this gas quickly slips away into the atmosphere, especially when the rocks are hot. Yet radioactive rocks in the earth's crust contain a lot of helium. The only possible explanation: the helium hasn't had time to escape!

The helium leakage rate has been determined in several experiments.[24] All measurements are in agreement. Helium diffuses so rapidly that all the helium in these zircon crystals should have leaked out in less than 100,000 years. The fact that so much helium is still there means they cannot be 1.5 billion years old, as uranium-lead dating suggests. Indeed, using the measured rate of helium diffusion, these pre-Flood rocks have an average "diffusion age" of only 6,000 (± 2,000) years.[25]

These experimentally determined and repeatable results, based on the well-understood physical process of diffusion, thus emphatically demonstrate that these zircons are only a few thousand years old. The supposed 1.5-billion-year

24. S.W. Reiners, K.A. Farley, and H.J. Hicks, "He Diffusion and (U-Th)/He Thermochronometry of Zircon: Initial Results from Fish Canyon Tuff and Gold Butte, Nevada," *Tectonophysics* 349, no. 1–4 (2002): p. 297–308; D. Russell Humphreys et al., "Helium Diffusion Rates Support Accelerated Nuclear Decay," in *Proceedings of the Fifth International Conference on Creationism*, R.L. Ivey Jr., ed. (Pittsburgh, PA: Creation Science Fellowship, 2003), p. 175–196; D. Russell Humphreys, "Young Helium Diffusion Age of Zircons Supports Accelerated Nuclear Decay," in *Radioisotopes and the Age of the Earth: Results of a Young-Earth Creationist Research Initiative*, L. Vardiman, A.A. Snelling, and E.F. Chaffin, eds. (El Cajon, CA: Institute for Creation Research, and Chino Valley, AZ: Creation Research Society, 2005), p. 25–100.
25. Humphreys et al., "Helium Diffusion Rates Support Accelerated Nuclear Decay"; Humphreys, "Young Helium Diffusion Age of Zircons Supports Accelerated Nuclear Decay."

age is based on the unverifiable assumptions of radioisotope dating that are radically wrong.[26]

Another evidence of a young earth is the low amount of helium in the atmosphere. The leakage rate of helium gas into the atmosphere has been measured.[27] Even though some helium escapes into outer space, the amount still present is not nearly enough if the earth is over 4.5 billion years old (figure 7).[28] In fact, if we assume no helium was in the original atmosphere, all the helium would have accumulated in only 1.8 million years even from an evolutionary standpoint.[29] But when the catastrophic Flood upheaval is factored in, which rapidly released huge amounts of helium into the atmosphere, it could have accumulated in only 6,000 years.[30]

Rescuing Devices

So glaring and devastating is the surprisingly large amount of helium that old-earth advocates have attempted to discredit this evidence.

One critic suggested the helium didn't all come from uranium decay in the zircon crystals but a lot diffused into them from the surrounding minerals. But this proposal ignores measurements showing that less helium gas is in the surrounding minerals. Due to the well-established diffusion law of physics, gases always diffuse from areas of higher concentration to surrounding areas of lower concentration.[31]

Another critic suggested the edges of the zircon crystals must have stopped the helium from leaking out, effectively "bottling" the helium within the zircons. However, this postulation has also been easily refuted because the zircon crystals are wedged between flat mica sheets, not wrapped

26. Andrew A. Snelling, "Radiometric Dating: Back to Basics," *Answers* 4, no. 3 (July–Sept. 2009): p. 72–75; Andrew A. Snelling, "Radiometric Dating: Problems With the Assumptions," *Answers* 4, no. 4 (Oct.–Dec. 2009): p. 70–73.
27. G.E. Hutchinson, "Marginalia," *American Scientist* 35 (1947): p. 118; Melvin A. Cook, "Where Is the Earth's Radiogenic Helium?" *Nature* 179, no. 4557 (1957): p. 213.
28. J.C.G. Walker, Evolution of the Atmosphere (London: Macmillan, 1977); J.W. Chamberlain and D.M. Hunten, *Theory of Planetary Atmospheres*, 2nd edition (London: Academic Press, 1987).
29. Larry Vardiman, *The Age of the Earth's Atmosphere: A Study of the Helium Flux Through the Atmosphere* (El Cajon, CA: Institute for Creation Research, 1990).
30. For a fuller treatment and further information see Morris, *The Young Earth*, p. 83–85; DeYoung, *Thousands . . . Not Billions*, p. 65–78; Snelling, *Earth's Catastrophic Past: Geology, Creation and the Flood*, p. 887–890.
31. D. Russell Humphreys et al., "Helium Diffusion Age of 6,000 Years Supports Accelerated Nuclear Decay," *Creation Research Society Quarterly* 41, no. 1 (2004): p. 1–16.

in them, so that helium could easily flow between the sheets unrestricted.[32] All other critics have been answered.[33] Thus all available evidence confirms that the true age of these zircons and their host granitic rock is only 6,000 (± 2,000) years.

#7 Carbon-14 in Fossils, Coal, and Diamonds

Carbon-14 (or radiocarbon) is a radioactive form of carbon that scientists use to date fossils. But it decays so quickly — with a half-life of only 5,730 years — that none is expected to remain in fossils after only a few hundred thousand years. Yet carbon-14 has been detected in "ancient" fossils — supposedly up to hundreds of millions of years old — ever since the earliest days of radiocarbon dating.[34]

Figure 8. A sea creature, called an ammonite, was discovered near Redding, California, accompanied by fossilized wood. Both fossils are claimed by strata dating to be 112–120 million years old but yielded radiocarbon ages of only thousands of years.

Even if every atom in the whole earth were carbon-14, they would decay so quickly that no carbon-14 would be left on earth after only 1 million years. Contrary to expectations, between 1984 and 1998 alone, the scientific literature reported carbon-14 in 70 samples that came from fossils, coal, oil, natural gas, and marble representing the fossil-bearing portion of the geologic record, supposedly spanning more than 500 million years. All contained radiocarbon.[35] Further, analyses of fossilized wood and coal samples, supposedly spanning 32–350 million years in age, yielded ages between 20,000 and 50,000

32. Humphreys, "Young Helium Diffusion Age of Zircons Supports Accelerated Nuclear Decay."
33. D. Russell Humphreys, "Critics of Helium Evidence for a Young World Now Seem Silent," *Journal of Creation* 24, no. 1 (2010): p. 14–16; D. Russell Humphreys, "Critics of Helium Evidence for a Young World Now Seem Silent?" *Journal of Creation* 24, no. 3 (2010): p. 35–39.
34. Robert L. Whitelaw, "Time, Life, and History in the Light of 15,000 Radiocarbon Dates," *Creation Research Society Quarterly* 7, no. 1 (1970): p. 56–71.
35. Paul Giem, "Carbon-14 Content of Fossil Carbon," *Origins* 51 (2001): p. 6–30.

years using carbon-14 dating.[36] The fossilized sea creature and wood in figure 8 both yield radiocarbon ages of only thousands of years. Diamonds supposedly 1 to 3 billion years old similarly yielded carbon-14 ages of only 55,000 years.[37]

Even that is too old when you realize that these ages assume that the earth's magnetic field has always been constant. But it was stronger in the past, protecting the atmosphere from solar radiation and reducing the radiocarbon production. As a result, past creatures had much less radiocarbon in their bodies, and their deaths occurred much more recently than reported!

So the radiocarbon ages of all fossils and coal should be reduced to less than 5,000 years, matching the timing of their burial during the Flood. The age of diamonds should be reduced to the approximate time of biblical creation — about 6,000 years ago.[38]

Rescuing Devices

Old-earth advocates repeat the same two hackneyed defenses, even though they were resoundingly demolished years ago. The first cry is, "It's all contamination." Yet for 30 years, AMS radiocarbon laboratories have subjected all samples, before they carbon-14 date them, to repeated brutal treatments with strong acids and bleaches to rid them of all contamination.[39] And when the instruments are tested with blank samples, they yield zero radiocarbon, so there can't be any contamination or instrument problems.

The second cry is, "New radiocarbon was formed directly in the fossils when nearby decaying uranium bombarded traces of nitrogen in the buried fossils." Carbon-14 does form from such transformation of nitrogen, but actual calculations demonstrate conclusively this process does not produce the

36. John R. Baumgardner et al., "Measurable ^{14}C in Fossilized Organic Materials: Confirming the Young Earth Creation-Flood Model," in *Proceedings of the Fifth International Conference on Creationism*, R.L. Ivey, Jr., ed. (Pittsburgh, PA: Creation Science Fellowship, 2003), p. 127–142.
37. John R. Baumgardner, "^{14}C Evidence for a Recent Global Flood and a Young Earth," in *Radioisotopes and the Age of the Earth: Results of a Young-Earth Creationist Research Initiative*, p. 587–630.
38. For a fuller treatment and further information see Don B. DeYoung, *Thousands . . . Not Billions*, p. 45–62; Snelling, *Earth's Catastrophic Past: Geology, Creation and the Flood*, p. 855–864; Andrew A. Snelling, "Carbon-14 Dating — Understanding the Basics," *Answers* 5, no. 4 (Oct.–Dec. 2010): p. 72–75; Andrew A. Snelling, "Carbon-14 in Fossils and Diamonds — an Evolution Dilemma" *Answers* 6, no. 1 (Jan.–Mar. 2011): p. 72–75; Andrew A. Snelling, "50,000-Year-Old Fossils — A Creationist Puzzle," *Answers* 6, no. 2 (April–June 2011): p. 70–73.
39. Andrew A. Snelling, "Radiocarbon Ages for Fossil Ammonites and Wood in Cretaceous Strata near Redding, California," *Answers Research Journal* 1 (2008): p. 123–144.

levels of radiocarbon that world-class laboratories have found in fossils, coal, and diamonds.[40]

#8 Short-Lived Comets

A comet spends most of its time far from the sun in the deep freeze of space. But, once each orbit, a comet comes very close to the sun, allowing the sun's heat to evaporate much of the comet's ice and dislodge dust to form a beautiful tail. Comets have little mass, so each close pass to the sun greatly reduces a comet's size, and eventually comets fade away. They can't survive billions of years.

Two other mechanisms can destroy comets — ejections from the solar system and collisions with planets. Ejections happen as comets pass too close to the large planets, particularly Jupiter, and the planets' gravity kicks them out of the solar system. While ejections have been observed many times, the first observed collision was in 1994, when Comet Shoemaker-Levi IX slammed into Jupiter.

Given the loss rates, it's easy to compute a maximum age of comets. That maximum age is only a few million years. Obviously, their prevalence makes sense if the entire solar system was created just a few thousand years ago, but not if it arose billions of years ago.

Rescuing Devices

Evolutionary astronomers have answered this problem by claiming that comets must come from two sources. They propose that a Kuiper belt beyond the orbit of Neptune hosts short-period comets (comets with orbits under 200 years), and a much larger, distant Oort cloud hosts long-period comets (comets with orbits over 200 years).

Yet there is no evidence for the supposed Oort cloud, and there likely never will be. In the past 20 years, astronomers have found thousands of asteroids orbiting beyond Neptune, and they are assumed to be the Kuiper belt. However, the large size of these asteroids (Pluto is one of the larger ones) and the difference in composition between these asteroids and comets argue against this conclusion.

#9 Very Little Salt in the Sea

If the world's oceans have been around for three billion years as evolutionists believe, they should be filled with vastly more salt than the oceans contain today.

40. Baumgardner, "^{14}C Evidence for a Recent Global Flood and a Young Earth," p. 614–616.

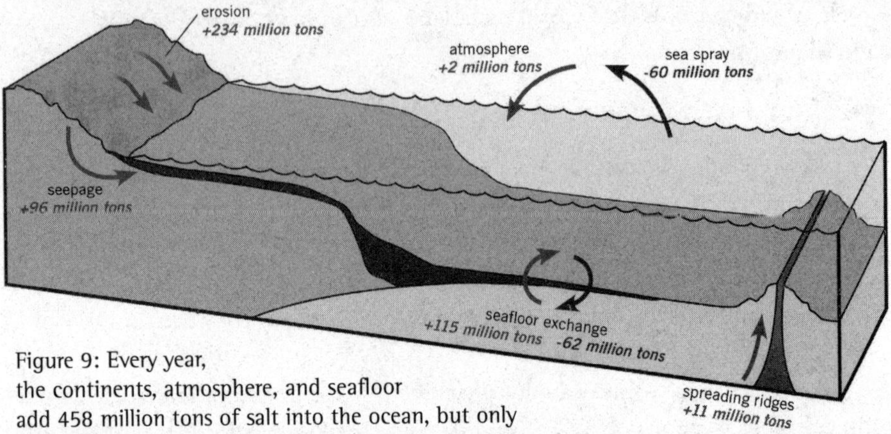

Figure 9: Every year, the continents, atmosphere, and seafloor add 458 million tons of salt into the ocean, but only 122 million tons (27 percent) is removed. At this rate, today's saltiness would be reached in 42 million years. But God originally created a salty ocean for sea creatures, and the Flood quickly added more salt.

Every year rivers, glaciers, underground seepage, and atmospheric and volcanic dust dump large amounts of salts into the oceans (figure 9). Consider the influx of the predominant salt, sodium chloride (common table salt). Some 458 million tons of sodium mixes into ocean water each year,[41] but only 122 million tons (27 percent) is removed by other natural processes.[42]

If seawater originally contained no sodium (salt) and the sodium accumulated at today's rates, then today's ocean saltiness would be reached in only 42 million years[43] — only about 1/70 the three billion years evolutionists propose. But those assumptions fail to take into account the likelihood that God created a saltwater ocean for all the sea creatures He made on day 5. Also, the year-long global Flood cataclysm must have dumped an unprecedented amount of salt into the ocean through erosion, sedimentation, and volcanism. So today's ocean saltiness makes much better sense within the biblical time scale of about six thousand years.[44]

41. M. Meybeck, "Concentrations des eaux fluvials en majeurs et apports en solution aux oceans," *Revue de Géologie Dynamique et de Géographie Physique* 21, no. 3 (1979): p. 215.
42. F.L. Sayles and P.C. Mangelsdorf, "Cation-Exchange Characteristics of Amazon with Suspended Sediment and Its Reaction with Seawater," *Geochimica et Cosmochimica Acta* 43 (1979): p. 767–779.
43. Steven A. Austin and D. Russell Humphreys, "The Sea's Missing Salt: A Dilemma for Evolutionists," in *Proceedings of the Second International Conference on Creationism*, p. 17–33.
44. For a fuller treatment and further information see Morris, *The Young Earth*, p. 85–87; Snelling, *Earth's Catastrophic Past: Geology, Creation and the Flood*, p. 879–881.

Rescuing Devices

Those who believe in a three-billion-year-old ocean say that past sodium inputs had to be less and outputs greater. However, even the most generous estimates can only stretch the accumulation time frame to 62 million years.[45] Long-agers also argue that huge amounts of sodium are removed during the formation of basalts at mid-ocean ridges,[46] but this ignores the fact that the sodium returns to the ocean as seafloor basalts move away from the ridges.[47]

#10 DNA in "Ancient" Bacteria

In 2000, scientists claimed to have "resurrected" bacteria, named Lazarus bacteria, discovered in a salt crystal conventionally dated at 250 million years old. They were shocked that the bacteria's DNA was very similar to modern bacterial DNA. If the modern bacteria were the result of 250 million years of evolution, its DNA should be very different from the Lazarus bacteria (based on known mutation rates). In addition, the scientists were surprised to find that the DNA was still intact after the supposed 250 million years. DNA normally breaks down quickly, even in ideal conditions. Even evolutionists agree that DNA in bacterial spores (a dormant state) should not last more than a million years. Their quandary is quite substantial.

However, the discovery of Lazarus bacteria is not shocking or surprising when we base our expectations on the Bible accounts. For instance, Noah's Flood likely deposited the salt beds that were home to the bacteria. If the Lazarus bacteria are only about 4,350 years old (the approximate number of years that have passed since the worldwide flood), their DNA is more likely to be intact and similar to modern bacteria.

Rescuing Devices

Some scientists have dismissed the finding and believe the Lazarus bacteria are contamination from modern bacteria. But the scientists who discovered the bacteria defend the rigorous procedures used to avoid contamination. They claim the old age is valid if the bacteria had longer generation times, different mutation rates, and/or similar selection pressures compared to modern bacteria. Of course these "rescuing devices" are only conjectures to make the data fit their worldview.

45. Austin and Humphries, "The Sea's Missing Salt: A Dilemma for Evolutionists."
46. Glenn R. Morton, pers. comm., Salt in the sea, http://www.asa3.org/archive/evolution/199606/0051.html.
47. Calculations based on many other seawater elements give much younger ages for the ocean, too. See Stuart A. Nevins (Steven A. Austin), "Evolution: The Oceans Say No!" *Impact* no. 8. (Santee, CA: Institute for Creation Research, 1973).

CHAPTER 11

Have People Always Been Brilliant or Were They Originally Dumb Brutes?

DON LANDIS[1]

Many Christians today have unanswered questions about the authority of the Bible due to their acceptance of an evolutionary time-line of history and, in particular, their view of mankind within that time-line. If the claims that mankind emerged from the slow process of evolution are true, then the Bible must be wrong, because the biblical record tells us that men were intelligent since the day of their creation (e.g., able to converse with God, able to work, and so on).

Yet our modern society believes we are just reaching the height of human intelligence and capabilities. If we accept this evolutionary view, what do we do with the biblical account? Is it completely unfounded and simply a myth? Or is the Bible true and verifiably so, thus making the evolutionary time-line errant?[2]

Most secular historians have not completely ignored the record of the Bible. However, they cite it as simply a source of information (e.g., a document of men, without God). In doing so, they undermine the authority of Scripture by

1. With the ancient man research team from Jackson Hole Bible College
2. For a discussion of philosophical issues of the Bible's truthfulness, see Ken Ham and Bodie Hodge, gen. eds., *How Do We Know the Bible Is True?* Vol. 1 (Green Forest, AR: Master Books, 2011).

not placing the Bible in its rightful place. Many Christians unwittingly accept this abuse of God's Word and furthermore even promote it! When it is assumed that the Bible is only one of many records of early man and it is placed in a time-line alongside the other legends predating it, two key points are missed:

1. God existed before creation and is the infinite, omniscient, and omnipresent Creator and, as such, He is the ultimate authority above all things (Genesis 1:1; Isaiah 40:28; Isaiah 40:14; Proverbs 15:3; Psalm 24:1).
2. The Bible is the inspired, inerrant, infallible, and authoritative Word of God, given to us as God spoke through human writers (2 Peter 1:21; 2 Timothy 3:16).

Therefore, God's account of what happened at the beginning of time, and since then, is accurate and true, and it is fallible man's accounts of history that is subject to error. No matter *when* in human, historic time the Bible was actually penned or by whom, it has priority over any other account. In our book, *The Genius of Ancient Man,*[3] we refer to this idea as the "priority of God in sequence and time." God predates the universe and all human history; He was actually there, and His account (which He revealed to us in His Holy Word) is the accurate one.

What about Legends, Myths, and Pagan Histories?

All non-biblical records and legends of the beginning come from oral (or written) traditions passed down through the descendants of Adam and Eve.[4] They are often mutilated and distorted while still containing some elements of the original truth concerning human history. Unfortunately, some of these accounts are given more historical "credit" because they predate the writing of the Bible. Historians tend to give priority to older documents. Television, movies, books, and modern education continue to undermine the validity and authority of Scripture by quoting the Bible as a late source. Even the Christian world is being fed a secular time-line of historical events. The Bible may be accurately quoted, but it is not given proper authority over all other pieces of historical data.

3. Don Landis and a team from Jackson Hole Bible College compiled an in-depth study of ancient man according to the biblical historical record. Their research is presented in the book *The Genius of Ancient Man* (Green Forest, AR: New Leaf Press, 2012).
4. In some cases, there may be completely "homemade" stories to try to counter others of ancient man's day and age, but all false ideas of origins originate in the mind of fallen beings, such as mankind.

For example, secular historians give the Code of Hammurabi superiority over God's Law found in the Pentateuch (the first five books of the Bible, written by Moses). Hammurabi, an ancient Babylonian king, wrote this set of laws about 340 years before Moses.[5] The Code appears to have a moral basis similar to that of the Pentateuch. The assumption is that since Hammurabi wrote before Moses, then Moses's writings are a copy or revision of this previously written moral code. Some historians theorize that Moses even stole or edited many such codes that predated him. So the authority of the Bible is undermined because God is no longer the original author of moral law! Moses is depicted as a compiler of good thoughts and morals that are essentially without truth or integrity. However, if God revealed the truth to Moses, then it was the authoritative, original truth. All the previous allusions to morality or history are distortions of the original and diluted with man's fallible ideas.

If the secular world only presented this occasionally it might not have a wide effect, but we are literally flooded with this idea from different avenues of the media. Without thinking it through, the average Christian subconsciously assumes it is true and thus their confidence in the text of God's Word is devalued. Because Christians have become accustomed to merely accepting these things without challenge, they are in danger of rejecting scriptural authority as a whole.

The study and correct evaluation of ancient man according to the biblical time-line becomes an apologetic vehicle of argument concerning the truthfulness, credibility, and authority of the Scriptures.

Two Views of History

Evolutionary History (Man Is the Authority): If evolution is accurate, then the sequence of life forms transitioning from single-celled organisms to humankind demands simplicity leading to complexity. This means the early animals would be weaker in mental ability and awareness. As early pre-man developed, he would be a simple thinker with limited ability to contemplate life. In modern terms, pre-man would have been stupid and illiterate. Ancient men would have lived as evolution depicts them, eating raw meat and dragging their women around by the hair and living in caves. Then, as man continued to evolve, he became more intellectual and aware and led us to where we are today: sophisticated 21st-century man.

5. Hammurabi's law code was written in 1786 B.C. For more information, see Don Landis, "Hammurabi or Moses — Who's the Authority?" *Answers*, January–March 2012, p. 80–81. See also Landis, *The Genius of Ancient Man*, p. 16–17.

Biblical History (God Is the Authority): The first humans were made in God's image and therefore created very intelligent (Genesis 1:27). The early chapters of Genesis tell us that Adam and Eve were moral beings who could communicate with God and each other, rationalize, name things, and work. Their descendants were gardeners (Genesis 4:2), musicians (Genesis 4:21), builders (Genesis 4:17), and metal workers (Genesis 4:22). Man was made in the image of an intelligent, moral, and creative God.

These two opposing views (based on presuppositions and biases) are easy to understand and follow to their logical conclusions. Yet the implications of each are profound. For if evolution is true, then the further back one studies into human history, the simpler and less sophisticated man should be (not to mention that nothing ultimately matters in this worldview). But if we hold that creation is true, then the evidence we find should portray great intelligence and advanced ancient cultures. So what is it that we find through scientific and historical discoveries?

Using Evidence, Which Model of History Is Correct?

The truth is, there is a vast amount of evidence, much of it ignored by scholars who are working from the paradigm of evolution, which clearly shows that early man could build, think, and design very complex cities and empires. They could create with technology that is still unexplainable. They had structured cultures and societies that show an appreciation for beauty and order. They were adept astronomers, fundamentally religious, and dedicated builders.

Ancient man's intelligence is proven by data that is now becoming available to anyone. It directly contradicts the stereotypical view of early barbaric man, dressed in animal skins and searching for the formula for fire. Unfortunately, much of the archeological evidence is basically ignored because it does not fit the evolutionary time-line. There are some who acknowledge the evidence of ancient genius, but they hold the rather mystical view that aliens from outer space endowed ancient peoples with their inexplicable knowledge and ability.[6] But the ever-growing list of new discoveries reveals that the data best fits the biblical paradigm.

Just as the fossil record attests to the authority and accuracy of the biblical text (e.g., a global Flood), so does the study of early man.[7] When the picture of ancient

6. Proponents of this theory include Richard Dawkins, David Childress, and Robert Bauval.
7. For further information on how the fossil record supports biblical authority, see Dr. Andrew Snelling, "Order in the Fossil Record," Answers in Genesis, http://www.answersingenesis.org/articles/am/v5/n1/order-fossil-record, and John D. Morris and Frank J. Sherwin, *The Fossil Record: Unearthing Nature's History of Life* (Dallas, TX: Institute for Creation Research, 2010).

man is clearly seen and the evidence evaluated from a proper perspective, there is no alternative: man was intelligent from the beginning just as the Bible indicates.

Examples of Ancient Man's Genius

As more researchers stop ignoring the data, more evidence is being reported and catalogued. The amount of information continues to grow, and we can use it as a good confirmation of biblical history. In light of this new research and evidence, we can confidently state that ancient peoples had exceptional capabilities in construction, astronomy, and transportation. Their architectural skill is still an unsolved mystery. Without the use of modern power tools or machines, early man constructed large buildings with incredible precision. Many of these ancient structures were built in line with astronomical events such as solstices and equinoxes (this is known as "archaeoastronomy"). As far as we know, ancient civilizations did not possess computer technology and yet demonstrated an advanced understanding of the heavens.

The following information gives strong evidences for ancient man's genius around the world.

Puma Punku

Part of a large ancient city known as Tiwanaku in Bolivia, this archeological site displays one of the greatest examples of advanced stone-cutting techniques. The blocks are cut and shaped so well they fit together perfectly. In fact, they are so well cut that even robots today would have trouble making the stones so precise.[8]

Puma Punku
(photo: Wikimedia Commons)

Palace of Knossos

Part of the Minoan civilization on the island of Crete, existing between 2100 B.C. and 1450 B.C., the palace is a highly advanced structure that is perhaps the most impressive ancient structure in Europe; it possesses a water and drainage

Palace of Knossos
(photo copyrights: Ken Zuk)

8. See Landis, *The Genius of Ancient Man*, p. 52.

system and was built to withstand earthquakes and to use sunlight to brighten rooms deep within the palace.[9]

La Bastida
(http://www.murciatoday.com/images/articles/13378_la-bastida-totana_1_large.jpg)

Great Pyramid of Gaza
(photo: Wikimedia Commons)

Baalbek
(photo: Wikimedia Commons)

La Bastida

A fortress located in the Agaric region of Spain dating to around 2200 B.C., it displays that the people who built it possessed advanced military techniques, as well as the oldest arch in the world.[10]

Great Pyramid of Giza

Perhaps the most famous ancient structure in the world, the Great Pyramid is the pinnacle of ancient man's ability to construct advanced buildings. It is not only massive in size; it is precisely aligned with true north within 3/60 of a degree. Its base is only 7/8 of an inch out of level and it covers an area of 13 acres. It, along with the two neighboring pyramids, may be aligned with Orion's belt. Another factor exhibiting its advancement is that the mortar used was stronger than rock, most with less than 1/50 of an inch between them. Over a million stones were used in its construction, averaging 2.5 to 15 tons each. The heaviest weighs around 80 tons![11]

Baalbek

A temple in Lebanon, it was designed to withstand earthquakes.

9. Matthew Zuk, "The Genius of Ancient Man: The Minoan Civilization: Proof of Advanced Nature"; http://geniusofancientman.blogspot.com/.
10. Matthew Zuk, "The Genius of Ancient Man: La Bastida: Europe's Most Formidable City"; http://geniusofancientman.blogspot.com/. http://geniusofancientman.blogspot.com/2013/03/la-bastida.html.
11. See Landis, *The Genius of Ancient Man*, p. 63, as well as the book's blog at www.geniusofancientman.blogspot.com.

The foundation of the temple has no known origin. This site possesses the largest stones ever cut. The three large stones (made of limestone) are labeled the Trilithon stones, and each weighs 800 tons. The lower layers are made up of smaller stones (though still very large), which allowed them to move with the earth during earthquakes, thereby making them stable.[12]

Stonehenge
(photo: Wikimedia Commons)

Stonehenge

Located in England, this is one of the greatest examples of archaeoastronomy in the ancient world. It was likely used to predict when the solstices, equinoxes, and cross quarter days would occur each year. It is yet a mystery as to how these stones were moved, but they are a perfect example of ancient man's knowledge of the sky.[13]

Cuzco
(photo: Wikimedia Commons)

Cuzco

Located in Peru, Cuzco was the first Incan capital. The structures at the site are so well put together not even a knife blade can fit between the stones, yet no mortar is used as a seal. The stones used are also very large and cut at odd angles, but that did not detract from the seamless construction.[14]

City of Alexandria
(photo: Wikimedia Commons)

City of Alexandria

An example of archaeoastronomy, the entire city of Alexandria was originally laid out so that the sun was aligned with the main street on Alexander the Great's birthday.[15]

12. See Landis, *The Genius of Ancient Man*, p. 69-71.
13. Ibid.
14. Ibid.
15. Stephanie Pappas, "Ancient Egypt City Aligned With Sun on King's Birthday," Live Science, http://www.livescience.com/23994-ancient-city-alexandria-sun.html (accessed April 18, 2013).

Harappa

Harappa
(photo: Wikimedia Commons)

Located in the Indus valley of Pakistan, it dates back to around 2300–1900 B.C., and it is yet unknown as to why it fell, but it began a rapid decline around 1900 B.C. There are several large domiciles, including a citadel and baths, and large, strong walls. It was primarily a city culture, and used bricks in construction. It also had advanced systems of agriculture, irrigation, and sanitation. There is little evidence for warfare, monarchies, temples or religious deities, slavery, or class distinctions. However, they did have precise measuring systems as well as a form of writing.[16]

Antikythera Mechanism

Antikythera Mechanism
(photo: Wikimedia Commons)

Found off the island of Antikythera, Greece, it dates to around the second century B.C.; it is one of the most advanced artifacts ever found. Around the size of a shoebox, it is believed to have predicted movements of the sun, moon, 12 signs of the zodiac, and possibly five planets. It also tracked the four-year cycle of the Olympic games. Thirty of its gears are still intact, but it may have once had 37.[17]

Ancient Man and the Historic Ice Age

A relatively new and very recent series of discoveries is bringing shock waves to the archeological world. Underwater archeological sites showing evidence of Ice Age civilizations are being discovered at an ever-increasing number.

16. T.A. Kohler, "Week 16: Indus Valley (Harappan) Civilization," Washington State University, http://public.wsu.edu/~tako/Week16.html (accessed April 18, 2013).
17. See Landis, *The Genius of Ancient Man*, p.49.

Since we believe the Bible is true and therefore the Flood of Noah actually took place, we see how perfectly these new discoveries fit into the time-line of the text.

There are studies that indicate that the Flood was followed by a global Ice Age.[18] There are also legends around the world that describe an Ice Age in earth's history.[19] It is theorized that there was a massive buildup of ice in the polar regions of the world during this time, and this would have lowered the water levels of earth's oceans. In Genesis 11, God scattered man because of the rebellion against Him in the building of the Tower of Babel. The city was built to keep man together (Genesis 11:4), directly defying God's command to "fill the earth" in Genesis 9:1. Due to the Ice Age, lower water levels would have allowed man to travel greater distances as the people scattered, settling in new lands, often along subtropical coastlines using land bridges to cross to the Americas, England, Japan, and so on.[20] But these water levels would have risen again as the earth began to warm and the ice caps melted, and slowly covered the coastal cities. This is most likely why many ancient structures have been found largely intact under the earth's oceans.

Examples of underwater sites include Yonaguni near Japan, Dwarka near India, and Yarmuta near Lebanon; there is also evidence of extensive urban civilization off the coasts of both Cuba and Greece.[21] This is a very exciting new field of study that will continue to confirm the truth of Scripture.

The Implications of the Tower of Babel

Along with the truth of intelligent ancient man, the biblical account demands the truth of the city of Babel. The Bible records that mankind gathered together:

> And they said, "Come, let us build ourselves a city, and a tower whose top is in the heavens; let us make a name for ourselves, lest we be scattered abroad over the face of the whole earth" (Genesis 11:4).

18. The Ice Age is not explicitly discussed in the Bible, but there are a few passages that imply there was a cooler climate after the Flood. There is also ample geologic evidence to suggest that it did indeed occur. Also, the possibility of an Ice Age does not conflict with the chronology presented in the Bible. See Michael Oard, "Where Does the Ice Age Fit?" Answers in Genesis, http://www.answersingenesis.org/articles/nab/where-does-ice-age-fit; Dr. Andrew Snelling and Mike Matthews, "When Was the Ice Age in Biblical History?" *Answers*, April–June 2013, p. 46–52; and Landis, *The Genius of Ancient Man*, p. 96–97.
19. See Landis, *The Genius of Ancient Man*, p. 77.
20. Though let's not forget that ancient man was also adept at building boats and likely went to many places by ship. There were the coastline (island or maritime) peoples in Genesis 10:5, and Noah and his sons were excellent ship builders and lived extensively after the Flood, passing along this technology.
21. See Landis, *The Genius of Ancient Man*, p. 71.

In rebellion against God, they formed their own unified government, a counterfeit religion, and a man-centered philosophical system of thought. Some of these concepts were carried throughout the world by those dispersed from Babel.

Thus, the characteristics of Babel are reflected in the ancient empires and cultures all around the globe. The astonishingly advanced civilizations show incredible similarities. Their religious practices such as pagan sacrifice, sun and star worship, and devotion to a false trinity all have their roots in Babel. The pyramids and ziggurats and mounds built around the world are likely examples of man-made mountains, built in rebellion against God, just like what was introduced at the Tower of Babel.[22] It is fascinating to study the ancient cultures and recognize the elements of biblical truth that were present, as well as the perversions that were introduced throughout history.

The Decline of Early Man

Some skeptics might question: if ancient man was intelligent and built such amazing, highly developed civilizations, where are they today? Why is there such a large segment of human history showing men with little ability or technological progress?[23]

It is clearly written in Genesis that Adam and Eve were not created infinite, but "very good." Their sin against God brought an abrupt end to that innocent state, ultimately resulting in death.

When Adam sinned, God placed a curse on the ground that affected the whole universe, a punishment of death instigating a downward spiral to all of creation, including man's being (Genesis 3; Romans 8). Adam and Eve began to die physically (their bodies would deteriorate with age until they died), they died spiritually (they were separated from God), they died mentally (the superior intelligence and capabilities of their minds were weakened), and they died socially (they hid their nakedness from each other). This picture of history is not one of early man moving up a gradual ladder of development via evolution but of the first man, in his created state, rebelling against God and degenerating downhill, "devolving," if you please.

22. See presupposition 3 in chapter 2 of Landis, *The Genius of Ancient Man,* for more detail on the evidence of Babel around the earth. A map of the distribution of man-made mountains worldwide is found on p. 65 of *The Genius of Ancient Man.*
23. From history, we know that man did at times live in caves and in a somewhat barbaric fashion. There are records of "stone age" type living conditions and uncivilized cultures. But this does not invalidate the text of the Bible. It is important to remember that where a person lives does not necessarily reflect the intelligence of that person. Was Jesus an illiterate, dumb brute because he had "nowhere to lay His head"(Matthew 8:20)?

The Bible also records that ancient men lived extraordinarily long lives (Genesis 5 and 11). Until the effects of the Curse became more severe and their life expectancy dropped dramatically, they were able to pass on their knowledge to the next generations.

It is true that if one only goes back to the mid-history of early man, there is evidence of a lack of knowledge and skill (often when pagan religions started bearing their fruit and suppressed such things), but if one jumps over this period to even earlier times, the intelligence is remarkable. In our research for *The Genius of Ancient Man,* it became clear that some of the knowledge from these highly advanced, early generations was passed on, although much was eventually lost as time progressed.[24]

Aren't We More Intelligent Now Than We Ever Were?

In present times, we are again *amassing* vast amounts of knowledge and data. In this, modern man takes pride and in fact assumes it is an evidence for evolution. But this is not true. It has taken hundreds of years for our knowledge of technology and science to reach where it is today.

Man's inherent inquisitive nature, evident in an ongoing pursuit of education, testifies that something was lost in Eden. Mankind longs to know things and to discover. He longs for intelligence. He longs to improve himself. This is because he is trying to get back to the way things were (and also hints back to the Fall and man's desire to be like God recorded in Genesis 3).

Though the massive amount of technical data we have accumulated seems impressive, an honest evaluation of our society today still reveals a barbaric inhumanness. For example, in recent times millions were killed by Hitler, Stalin, Lenin, Mao Tse-tung, and other despots.[25] Man is not evolving upward into something better. Rather, these recent events confirm the depraved heart of man and not the ascendency of the human spirit. We think we have reached great heights of technology, but the wisdom of man has led only to an intellectual and moral insanity.

Conclusion

Christians have nothing to be ashamed of when it comes to the time-line of history and ancient man. We need not "hide our heads in the sand" on any truth supposedly "proven" by secular archeology or science or by any discovery

24. Reasons for this loss are highly speculative. For more detail about ancient technology and the mystery surrounding it, see chapter 6 in Landis, *The Genius of Ancient Man.*
25. See chapter 5 in this volume on the results of an evolutionary worldview.

— past, present, or future. Observational science and history continue to confirm, support, and validate our faith.

The truth is, man *was* brilliant — brilliant in all the splendor of unspoiled creation, brilliant in intellect and imagination, brilliant in creativity and invention. But with the entrance of sin and the Curse, man began and continues a downward spiral in his rebellion. Without the hope of salvation through the Lord Jesus Christ, man is ultimately doomed to the wrath of God. But we who are believers in Christ have this hope that one day, because of Christ's atoning work, we will again be brilliantly glorified with Him (Romans 8:16–30).

The Bible is true, in far more ways and detail than even imagined by today's believers. Do not undermine its authority. Do not doubt its inerrancy. Stand firm, "Test all things; hold fast what is good" (1 Thessalonians 5:21).

CHAPTER 12

What about Living Fossils?

DR. JOHN WHITMORE[1]

When Charles Darwin published the first edition of the *Origin of Species* in 1859, he imagined a large evolutionary "tree" of organisms that were continuously connected by various transitional forms. Furthermore, he envisioned life constantly changing through time as various environmental and climatic conditions changed — with only the fittest and best adapted offspring surviving. At the time, paleontology was still a relatively young science and Darwin realized that the fossil record did not yet support his theory. Subsequently, he predicted that numerous fossil "intermediate links" would be found, gradually leading to the animals that we have today. Darwin did not predict that organisms at the lowest taxonomic levels would remain unchanged for long periods of time.[2] He thought that their morphology (or body shape) would change (or evolve) over time.

What Are Living Fossils?

Initially, the term "living fossil" doesn't make much sense. How could something be alive and a fossil at the same time? "Living fossils" are organisms that

1. Professor of Geology, Cedarville University, Cedarville, OH 45314 johnwhitmore@cedarville.edu.

2. Carolus Linnaeus developed the system that biologists still use to classify animals: kingdom, phylum, class, order, family, genus, and species. These are known as "taxonomic levels" and a group(s) within a taxonomic level is referred to as a "taxon" (singular) or "taxa" (plural). The species is the most basic taxonomic level and contains only a single type of organism or taxon. Genera are similar groups of species. Families are similar groups of genera and so on. Most creationists think the Genesis "kind" approximates the family level of the Linnaean system of classification.

The classification system of life developed by Carolus Linnaeus.

Kingdom
Phylum
Class
Order
Family
Genus
Species

Many creation biologists believe the "Genesis kind" is near the family level of classification and that the various genera and species that we have today developed from the original Genesis kinds.

can be found both living in the world today and also found preserved in the rock record as fossils, with the living animals showing little if any difference from their fossil counterparts. Studying and comparing fossils to modern organisms is important because we can see how (or if) they have changed over time. The study of these organisms has implications for both evolutionary and biblical models of earth history. An organism is considered a "living fossil" if it has fossil representatives that are from the same taxonomic level — usually in the same genus or species group. Living fossils are impressive from an evolutionary perspective with some animal genera existing for nearly the entire range of the Phanerozoic[3] record — that's more than half of a billion years! From a biblical perspective, no fossils are much older than the time of the Flood, about 4,300 years ago, so a creationist might predict living fossils would be more common. Many famous examples of living fossils are found in the Cenozoic rocks, or ones that were made after the Flood was over.

What Are Some Examples of Living Fossils, and How Many Are There?

Notable examples of living fossil genera (plural for the classification level of a genus that can be further divided into distinct species), that have conserved the characteristics of their genus for millions of years (from an evolutionary perspective, these organisms appeared millions of years ago: MYA), include the ginkgo tree (*Ginkgo*, 252 MYA–present), the coast redwood (*Sequoia*, 151MYA–present), horsetails (*Equisetum*, 361MYA–present), a brachiopod (*Lingula*, 513 MYA–present), an annelid marine worm (*Spirorbis*, 488 MYA–present), the cockroach (*Periplaneta*, 49 MYA–present), the chambered nautilus (*Nautilus*, 340 MYA–present), and a sea mussel (*Mytilus*, 419 MYA–present).[4] Some living genera have very close sister taxa in the fossil record (animals in related groups

3. This is a conventional time period lasting from 542 million years ago to the present. It contains the Paleozoic, Mesozoic, and Cenozoic Eras of geological time.
4. When a word is in italics and capitalized it refers to the taxonomic level of the genus. The evolutionary ages are represented as millions of years ago (MYA). The abbreviation "Ma" means millions of years before the present or "mega-annum" and is used in more technical literature. The ranges are conventional ages for these taxa obtained from the Paleobiology Database (pbdb.org).

Coelacanth from the London Museum of Natural History (Photo by John Whitmore)

whose body shapes are very similar) including the coelacanth fish (*Latimeria)* with *Coelacanthus* (318–247 MYA), the horseshoe crab (*Limulus*) with *Limuloides* (419–416 MYA), and the Tuatara lizard (*Sphenodon)* with *Cynosphenodon* (190–183 MYA). There are many more examples of living fossils, many of which can be found published in various issues of creationist periodicals.

The standard geological time column is divided up into three main fossil-bearing portions: the Paleozoic, the Mesozoic, and, the most recent period of time, the Cenozoic (which contains the Neogene and Quaternary Periods). Most creation geologists believe the Paleozoic and Mesozoic portions represent rocks that were formed during Noah's Flood and that the Cenozoic represents post-Flood rocks. A recent query of the online Paleobiology Database (pbdb.org) was completed to find how many living fossils have been reported from each of these three periods of time. In this database, the genus is the lowest taxonomic group for which large amounts of data are available. From the Paleozoic, 99 living fossil genera were found; from the Mesozoic, 548 living fossil genera were found; and from the Cenozoic 2,594 living fossil genera were found. This is a total of 3,241genera that can be found both living today and fossilized in the rock record![5] The database is updated daily by paleontologists as they find new fossils and catalog old ones, so this figure is surely an underestimate. The data were plotted using conventional 10-million-year (Ma) time bins (figure 1). The graph shows a "flat" distribution of living fossil genera during the Paleozoic and Mesozoic[6] and then a "spike" in the number at about the Mesozoic/Cenozoic[7]

5. Organisms (genera) were counted if they had a fossil record greater than 2.6 Ma (or older than the Quaternary Period) and their range extended into the present time (or into the Quaternary Period, 2.6 Ma or less).
6. Several statistical techniques were employed to see if there were any significant trends in the Paleozoic and Mesozoic data or in all the combined data. No significant trends could be found (where R2 values > 0.90), other than the observation that the number of living fossils increases dramatically toward the present time. Finding R2 values is a mathematical technique that can be used to test whether a predictable trend is present or not.
7. A logarithmic or exponential trend can be demonstrated in the Cenozoic data, having R2 values of 0.90 or greater. A logarithmic or exponential curve is one that increases rapidly, going from flat to almost vertical.

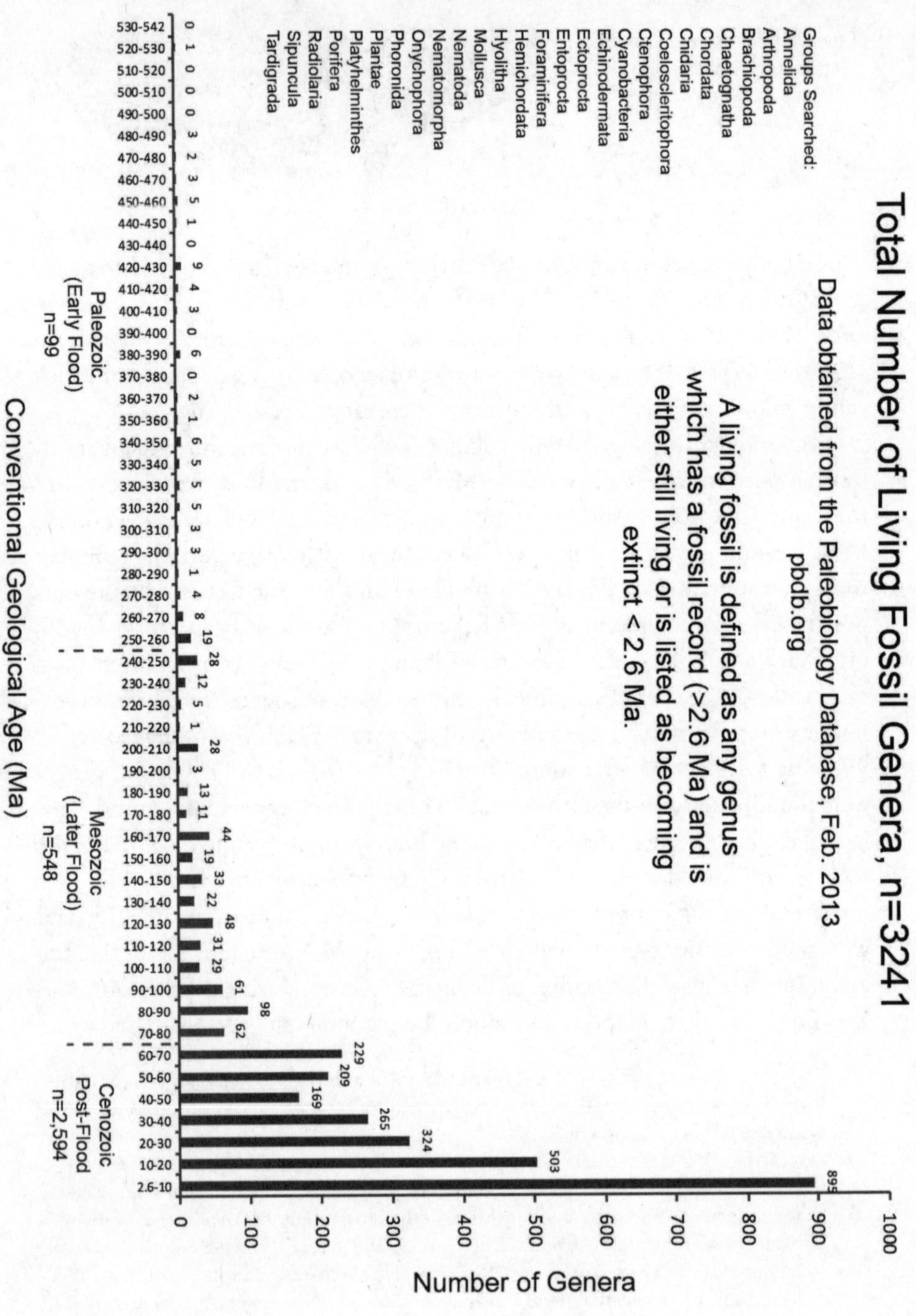
Total Number of Living Fossil Genera, n=3241
Data obtained from the Paleobiology Database, Feb. 2013
pbdb.org
A living fossil is defined as any genus which has a fossil record (>2.6 Ma) and is either still living or is listed as becoming extinct < 2.6 Ma.
Groups Searched:
Annelida
Arthropoda
Brachiopoda
Chaetognatha
Chordata
Cnidaria
Coeloscleritophora
Ctenophora
Cyanobacteria
Echinodermata
Ectoprocta
Entoprocta
Foraminifera
Hemichordata
Hyolitha
Mollusca
Nematoda
Nematomorpha
Onychophora
Phoronida
Plantae
Platyhelminthes
Porifera
Radiolaria
Sipuncula
Tardigrada
Paleozoic
(Early Flood)
n=99
Mesozoic
(Later Flood)
n=548
Cenozoic
Post-Flood
n=2,594
Conventional Geological Age (Ma)
Number of Genera
Figure 1

boundary, a time during which most creation geologists think approximately marks the end of the Flood in the rock record.[8]

Do Living Fossils Support the Theory of Evolution?

As mentioned above, Darwin predicted that organisms would change over time. However, the number of living fossil genera (albeit small compared to the large number of extinct fossil genera) is troubling from a naturalistic perspective. Perhaps organisms could resist evolutionary change over long periods of time if their environment or climate did not change; however, this is very unrealistic. From an evolutionary perspective, continents have come together and broken apart several times, there have been several "ice ages," multiple mass extinctions, and many changes in sea level during the time intervals examined. All of these factors have been claimed as impetuses for evolutionary change. In other words, these events have been cited as causes for extinction and evolutionary change every time they occur.[9] Clearly, living fossils do not support the theory of gradual evolution (sometimes called "gradualism") as proposed by Darwin.

To their credit, some paleontologists have recognized that gradualism is not the main pattern in the fossil record.[10] "Stasis" is when organisms remain unchanged and have no recognizable evolutionary change for long periods of time. Gould argued[11] that living fossils might be explained as organisms that have persisted through time and do not have very many different kinds of species within their respective genera and families. Since species diversity was low, the groups therefore lacked the genetic diversity to evolve and the group remained pretty much unchanged through time. This hypothesis might be successful in explaining some small groups like the coelacanths and lungfishes that still look similar after more than 300 million years of geological time. However, there are

8. See J.H. Whitmore and P. Garner, "Using Suites of Criteria to Recognize Pre-Flood, Flood, and Post-Flood Strata in the Rock Record with Application to Wyoming (USA)," in A.A. Snelling (ed.), *Proceedings of the Sixth International Conference on Creationism* (Pittsburgh, PA: Creation Science Fellowship and Dallas, TX: Institute for Creation Research, 2008), p. 425–448.
9. For example see M. Foote and A.I. Miller, *Principles of Paleontology*, 3rd ed. (New York: W.H. Freeman and Company, 2007), or S.M. Stanley, *Earth System History*, 3rd ed. (New York: W.H. Freeman and Company, 2009), or C. Patterson, *Evolution,* 2nd ed. (Ithaca, NY: Comstock Publishing Associates, 1999).
10. Steven Jay Gould (1941–2002) was probably the most prominent paleontologist espousing this view, arguing that stasis in the fossil record was *data* that needed to be explained. See his discussion, for example, on p. 759 of his book *The Structure of Evolutionary Theory* (Cambridge, MA: The Belknap Press of Harvard University Press, 2002). Technically, his evolutionary arguments were for stasis at the taxonomic level of the species.
11. Gould, *The Structure of Evolutionary Theory*, p. 816–817.

two problems with this explanation: 1) smaller groups (like the coelacanths) should consistently be favored for extinction (since they have low diversity[12]), yet somehow they continue to persist for millions of years through many climate changes and extinction events, and 2) the explanation fails to explain why groups that are quite diverse, like the cockroaches (which number over 3,700 described species in multiple genera and families and whose group has been around for 300 million years[13]) fail to evolve. Cockroaches are a group with great genetic diversity; yet living fossils persist within it.

Do Living Fossils Support the Biblical Account of Biology and Geology?

If Paleozoic and Mesozoic fossils primarily represent organisms that were buried during the Flood and if the Cenozoic fossils primarily represent the post-Flood era, several hypotheses can be suggested to explain the patterns in the fossil record of these times. We might predict that rocks deposited during the Flood would lack clear sequences of transitional fossils because most fossils in those rocks would have been from organisms that were alive on the day the Flood began. This could explain the apparent phenomenon of "stasis" that is so common in the fossil record, especially in pre-Cenozoic rocks.[14] We might expect a large number of pre-Flood taxa to become extinct, especially those that lived in marine environments (like trilobites or plesiosaurs) because they were not protected on the ark. Other organisms became extinct because they lived in ecosystems that were permanently destroyed during the Flood (like the floating forests proposed by Wise[15]). Living fossil taxa from the pre-Flood world would then be organisms that found comparable ecosystems in which to live after the Flood and had at least several representatives that survived the Flood. Apparently, not many genera were able to survive the Flood unchanged; there are only 647 Paleozoic and Mesozoic living fossil genera.

12. Low genetic diversity within a group is often touted as a cause for extinction.
13. Richard C. Brusca and Gary J. Brusca, *Invertebrates* (Sunderland, MA: Sinauer Associates, 1990).
14. Refer to K. Wise's 1989 paper on reasons for stasis and abrupt appearance in the fossil record: Punc Eq Creation Style, *Origins* v. 16(1): 11-24. Note that I am using the term "stasis" at the genus level (in referring to living fossils), where Wise uses "stasis" at the species level (referring to punctuated equilibrium). It is not yet possible to evaluate stasis at the species level in the paleobiology database (pbdb).
15. See K.P. Wise, K.P., "The Pre-Flood Floating Forest: A Study in Paleontological Pattern Recognition,: in R.L Ivey, Jr. (ed.), *Proceedings of the Fifth International Conference on Creationism* (Pittsburgh, PA: Creation Science Fellowship, 2003), p. 371–381.

In the post-Flood times (the Cenozoic), four times the number of living fossils can be found compared to that of the pre-Cenozoic; and in the immediate post-Flood interval (60–70 MYA) there are more than twice the living fossils of any previous interval. In a Flood model, the Cenozoic would have been the longest period in earth history (lasting more than 4,300 years).[16] The spike in the numbers of living fossils in the Cenozoic may be due to the rapid diversification of organisms immediately following the Flood,[17] and the ability of those organisms to establish themselves in the new niches they were filling. In other words, organisms changed quickly after the Flood (producing many new genera within Genesis kinds[18]) and once they became well-adapted to one of the many new niches after the Flood, change stopped. It is interesting that the graph shows a huge spike in the last interval of time, which may indicate additional diversification due to the climate changes that occurred at the beginning of the "Ice Age" or the Pleistocene Epoch.

Conclusion

From an old earth/evolutionary perspective, "living fossils" are an unexpected problem. Evolutionary change is predicted over time, but some genera remain unchanged for tens or hundreds of millions of years. Furthermore, why are the numbers of living fossil taxa fairly "flat" in the Paleozoic and Mesozoic times with a sudden spike during Cenozoic times? A creation-Flood model might answer this as rapid diversification of organisms following the Flood. Only a few select genera that were alive before the Flood were able to survive with their body shape unchanged. After the Flood, rapid diversification occurred probably because of climate changes and/or the opportunity for organisms to fill new niches. We think that all of these changes happened within the context of the "kinds" God created in Genesis 1. In other words, there was a lot of change, but within Genesis kinds.

---

16. The later part of the Cenozoic, the Quaternary Period, is probably the longest period of earth history from a biblical perspective. The Quaternary Period includes the Pleistocene and Holocene Epochs and is fully contained within the Cenozoic Era. We believe the Pleistocene probably begins after the Tower of Babel since this is the time where we begin to see widespread human fossils in the rock record.
17. For examples see J.H. Whitmore and K.P. Wise, "Rapid and Early Post-Flood Mammalian Diversification Evidenced in the Green River Formation," in A.A. Snelling (ed.), *Proceedings of the Sixth International Conference on Creationism* (Pittsburgh, PA: Creation Science Fellowship and Dallas, TX: Institute for Creation Research, 2008), p. 449–457.
18. It is estimated that there were less than 300 mammal kinds on the ark that diversified into all the mammal species that we have today. See K. Wise, 2009, "Mammal Kinds: How Many Were on the Ark?" in T. Wood and P. Garner (eds.), *Genesis Kinds: Creationism and the Origin of Species*, Center for Origins Research Issues in *Creation*, no. 5, p. 129–161.

Fossils of Organisms also Found Living in the World Today

Sand dollar

Wasp

Brittlestar

Dragonfly

Crab

Fossil photos by
Bodie Hodge

CHAPTER 13

What Is the State of the Canopy Model?

BODIE HODGE

If there is one thing you need to know about biblical creationists . . . they can be divided on a subject. This isn't necessarily a bad thing. Though we all have the same heart to follow Christ and do the best we can for the sake of biblical authority and the cause of Christ, we can have differences when it comes to details of models used to explain various aspects of God's creation.

When divisions occur over scientific models, this helps us dive into an issue in more detail and discover if that model is good, bad, needs revision, and so on. But note over *what* we are divided; it is not the Word of God nor is it even theology — it is a division over a *scientific model*.

This is where Christians can rightly be divided on a subject and still do so with Christian love, which I hope is how each Christian would conduct themselves — in "iron-sharpening-iron" dealings on a model while still promoting a heart for the gospel (Proverbs 27:17).

The debate over a canopy model is no different — we are all brothers and sisters in Christ trying to understand *what the Bible says and what it doesn't say* on this subject (2 Timothy 2:15). It is the Bible that reigns supreme on the issue, and our scientific analysis on the subject will always be subservient to the Bible's text.

What Is the Canopy Model(s)?

There are several canopy models, but they all have one thing in common.[1] They all interpret the "waters above" the expanse (firmament) in Genesis 1:7 as some form of water-based canopy surrounding the earth that endured from creation until the Flood.

> Then God said, "Let there be a firmament [expanse] in the midst of the waters, and let it divide the waters from the waters." Thus God made the firmament [expanse], and divided the waters which were under the firmament [expanse] from the waters which were above the firmament [expanse]; and it was so (Genesis 1:6–7).

Essentially, the waters above are believed to have formed either a vapor, water (liquid), or ice canopy around the earth. It is the vapor canopy that seemed to dominate all of the proposed models.[2] It is suggested that this canopy was responsible for several things such as keeping harmful radiation from penetrating the earth, increasing the surface atmospheric pressure of oxygen, keeping the globe at a consistent temperature for a more uniform climate around the globe, and providing one of the sources of water for the Flood.

Some of these factors, like keeping radiation out and increasing the surface atmospheric pressures of oxygen, were thought to allow for human longevity to be increased from its present state (upwards of 900 years or so as described in Genesis 5). So this scientific model was an effort to explain several things, including the long human life span prior to the Flood. Other potential issues solved by the models were to destroy the possibility of large-scale storms with reduced airflow patterns for less extreme weather possibilities, have a climate without rain (such as Dillow's model, see below) but instead merely dew every night, and reduce any forms of barrenness like deserts and ice caps. It would have higher atmospheric pressure to possibly help certain creatures fly that may not otherwise.

---

1. This is not to be confused with canopy ideas that have the edge of water at or near the end of the universe (e.g., white hole cosmology), but instead the models that have a water canopy in the atmosphere, e.g., like those mentioned in J.C. Whitcomb and H.M. Morris, *The Genesis Flood* (Phillipsburg, NJ: Presbyterian and Reformed Publishing, 1961); J.C. Dillow, *The Waters Above: Earth's Pre-Flood Vapor Canopy*, Revised Edition (Chicago, IL: Moody Press, 1981); or John C. Whitcomb, *The World that Perished* (Winona Lake, IN: BMH Books, 2009).
2. This is in large part due to the influence of Joseph Dillow, whose scientific treatise left only the vapor models with any potential. He writes on page 422 of his treatise: "We showed that only a vapor canopy model can satisfactorily meet the requirements of a the necessary support mechanism." Dillow, *The Waters Above: Earth's Pre-Flood Vapor Canopy*, .

A Brief History of Canopy Models

Modern canopy models can be traced back to Dr. Henry Morris and Dr. John Whitcomb in their groundbreaking book *The Genesis Flood* in 1961.[3] This book triggered a return to biblical authority in our age, which is highly commendable and much is owed to their efforts. In this volume, Whitcomb and Morris introduce the possibility of a vapor canopy as the waters above.

The canopy models gained popularity thanks to the work of Dr. Joseph Dillow,[4] and many creationists have since researched various aspects of these scientific models, such as Dr. Larry Vardiman with the Institute for Creation Research.

Researchers have studied the possibility of solid canopies, water canopies, vapor canopies, thick canopies, thin canopies, and so on. Each model has the canopy collapsing into history at the time of the Flood. Researchers thought it could have provided at least some of the water for the Flood and was associated with the 40 days of rain coming from the "windows of heaven" mentioned along with the fountains of the great deep at the onset of the Flood (Genesis 7:11).

However, the current state of the canopy models have faded to such an extent that most researchers and apologists have abandoned the various models. Let's take a look at the biblical and scientific reasons behind the abandonment.

Biblical Issues

Though both will be discussed, any biblical difficulties that bear on the discussion of the canopy must *trump* scientific considerations, as it is the authority of the Bible that is supreme in all that it teaches.

Interpretations of Scripture Are Not Scripture

The necessity for a water-based canopy about the earth is not directly stated in the text. It is an *interpretation* of the text. Keep in mind that it is the *text* that is inspired, not our interpretations of it.

Others have interpreted the waters above as something entirely different from a water-based canopy about the earth. Most commentators appeal to the waters above as simply being the clouds, which are water droplets (not vapor) in the atmosphere. For they are simply "waters" that are above.

But most do not limit this interpretation as simply being the clouds, but perhaps something that reaches deep into space and extends as far as the *Third Heaven* or *Heaven of Heavens*. For example, expositor Dr. John Gill in the 1700s said:

3. Whitcomb and Morris, *The Genesis Flood.*
4. Dillow, *The Waters Above: Earth's Pre-Flood Vapor Canopy.*

> The lower part of it, the atmosphere above, which are the clouds full of water, from whence rain descends upon the earth; and which divided between them and those that were left on the earth, and so under it, not yet gathered into one place; as it now does between the clouds of heaven and the waters of the sea. Though Mr. Gregory is of the opinion, that an abyss of waters above the most supreme orb is here meant; or a great deep between the heavens and the heaven of heavens. . . .[5]

Gill agrees that clouds were inclusive of these waters above but that the waters also extend to the heaven of heavens, at the outer edge of the universe. Matthew Poole denotes this possibility as well in his commentary in the 1600s:

> . . . the expansion, or extension, because it is extended far and wide, even from the earth to the third heaven; called also the firmament, because it is fixed in its proper place, from whence it cannot be moved, unless by force.[6]

Matthew Henry also concurs that this expanse extends to the heaven of heavens (third heaven):

> The command of God concerning it: Let there be a firmament, an expansion, so the Hebrew word signifies, like a sheet spread, or a curtain drawn out. This includes all that is visible above the earth, between it and the third heavens: the air, its higher, middle, and lower, regions — the celestial globe, and all the spheres and orbs of light above: it reaches as high as the place where the stars are fixed, for that is called here the firmament of heaven Ge 1:14,15, and as low as the place where the birds fly, for that also is called the firmament of heaven, Ge 1:20.[7]

The point is that a canopy model about the earth is simply that . . . an interpretation. It should be evaluated as such, not taken as Scripture itself. Many respected Bible interpreters do not share in the interpretation of the "waters above" being a water canopy in the upper atmosphere of earth.

Stars for Seasons and Light and other Implications

Another biblical issue crops up when we read in Genesis 1:14–15:

5. John Gill, *Exposition of the Bible*, Genesis 1:7.
6. Matthew Poole, *A Commentary on the Holy Bible*, Genesis 1:7.
7. Matthew Henry, *A Commentary on the Whole Bible*, Genesis 1:7.

> Then God said, "Let there be lights in the firmament [expanse] of the heavens to divide the day from the night; and let them be for signs and seasons, and for days and years; and let them be for lights in the firmament [expanse] of the heavens to give light on the earth"; and it was so.[8]

The stars are intended by God to be used to map seasons. And they were also to "give light on the earth." Though this is not much light, it does help significantly during new moon conditions — that is, if you live in an area not affected by light pollution.

Water

If the canopy were liquid water, then in its various forms like mist or haze, it would inhibit seeing these stars. How could one see the stars to map the seasons? It would be like a perpetually cloudy day. The light would be absorbed or reflected back to space much the way fog does the headlights of a car. What little light is transmitted through would not be sufficiently discernable to make out stars and star patterns to map seasons. Unlike a vapor canopy, clouds are moving and in motion, one can still see the stars to map seasons when they moved through. Furthermore, if it was water, why didn't it fall?[9]

Ice

If it were ice, then it *is* possible to see the stars but they would not appear in the positions one normally sees them, but still they would be sufficient to map seasons. But ice, when kept cool (to remain ice), tends to coat at the surface where other water molecules freezes to it (think of the coating you see on an ice cube left in the freezer). This could inhibit visibility, as evaporated water from the ocean surface would surely make contact — especially in a sin-cursed and broken world.

Vapor

But if an invisible vapor canopy existed in our upper atmosphere, then it makes the most sense. But there could still turn out to be a problem. As cooler vapor nears space, water condenses and begins to haze, though as long as the vapor in the upper atmosphere is kept warm and above the dew point, it could remain invisible. But there are a lot of "ifs." In short, the stars may not serve their purpose to give light on the earth with some possibilities within these models.

---

8. See also Genesis 1:17.
9. Would one appeal to the supernatural? If so, it defeats the purpose of this scientific model that seeks to explain things in a naturalistic fashion.

But consider, if there were a water *vapor* canopy, what would stop it from interacting with the rest of the atmosphere *that is vapor*? Gases mix to equilibrium, and that is the way God upholds the universe.[10] If it was a vapor, then why it is distinguished from the atmosphere, which is vapor?

The Bible uses the terms *waters* above, which implies that the temperature is between 32°F and 212°F (0°C and 100°C). If it was meant to be vapor, then why say "waters" above? Why not say vapor (*hebel*), which was used in the Old Testament?

Where Were the Stars Made?

If the canopy really was part of earth's atmosphere, then all the stars, sun, and moon would have been created within the earth's atmosphere. Why is this? A closer look at Genesis 1:14 reveals that the "waters above" may very well be much farther out — if they still exist today.

The entirety of the stars, including our own sun (the greater light) and moon (lesser light) were made "***in*** the expanse." Further, they are obviously not in our atmosphere. Recall that the waters of verse 7 are above the expanse. If the canopy were just outside the atmosphere of the young earth, then the sun, moon, and stars would have to be in the atmosphere according to verse 14.

Further, the winged creatures were flying *in the face of* the expanse (Genesis 1:20; the NKJV accurately translates the Hebrew), and this helps reveal the extent of the expanse. It would likely include aspects of the atmosphere as well as space. The Bible calls the firmament "heaven" in Genesis 1:8, which would include both. Perhaps our understanding of "sky" is similar or perhaps the best translation of this as well.

Regardless, this understanding of the text allows for the stars to be in the expanse, and this means that any waters above, which is beyond the stars, is not limited to being in the atmosphere. Also, 2 Corinthians 12:2 discusses three heavens, which are likely the atmosphere (airy heavens), space (starry heavens), and the heaven of heavens (Nehemiah 9:6).

Some have argued that the prepositions in, under, above, etc., are not in the Hebrew text but are determined from the context, so the meaning in verses 14 and 17 is vague. It is true that the prepositions are determined by the context, so we must rely on a proper translation of Genesis 1:14. Virtually all translations have the sun, moon, and stars being created *in* the expanse, not *above* as any canopy model would require.

10. Again, would one appeal to the supernatural? If so, it defeats the purpose of this scientific model that seeks to explain things in a naturalistic fashion.

In Genesis 1, some have attempted to make a distinction between the expanse in which the birds fly (Genesis 1:20) and the expanse in which the sun, moon, and stars were placed (Genesis 1:7); this was in an effort to have the sun, moon, and stars made in the second expanse. This is not a distinction that is necessary from the text and is only necessary if a canopy is assumed.

From the Hebrew, the birds are said to fly "across the face of the firmament of the heavens." Looking up at a bird flying across the sky, it would be seen against the face of both the atmosphere and the space beyond the atmosphere — the "heavens." The proponents of the canopy model must make a distinction between these two expanses to support the position, but this is an arbitrary assertion that is only necessary to support the view and is not described elsewhere in Scripture.

Expanse (Firmament) Still Existed Post-Flood

Another issue that is raised from the Bible is that the waters above the heavens were mentioned *after* the Flood, when it was supposedly gone.

> Praise Him, you heavens of heavens, and you waters above the heavens! (Psalm 148:4).

> So an officer on whose hand the king leaned answered the man of God and said, "Look, if the LORD would make windows in heaven, could this thing be?" And he said, "In fact, you shall see it with your eyes, but you shall not eat of it" (2 Kings 7:2; see also 2 Kings 7:19).

> "Bring all the tithes into the storehouse, that there may be food in My house, and try Me now in this," says the LORD of hosts, "If I will not open for you the windows of heaven and pour out for you such blessing that there will not be room enough to receive it" (Malachi 3:10).

The biblical authors wrote these in a post-Flood world in the context of other post-Flood aspects. So, it appears that the "waters above" and "windows of heaven" are in reference to something that still existed after the Flood. So "the waters above" can't be referring to a long-gone canopy that dissipated at the Flood and still be present after the Flood. This is complemented by:

> The fountains of the deep and the windows of heaven were also stopped, and the rain from heaven was restrained (Genesis 8:2).

Genesis 8:2 merely points out that the two sources were stopped and restrained, not necessarily *done away* with. The verses above suggest that the windows of heaven remained after the Flood. Even the "springs of the great deep" were stopped but did not entirely disappear, but there may have been residual waters trapped that have slowly oozed out since that time; clearly not in any gushing, spring-like fashion.[11]

Is a Canopy Necessary Biblically?

Finally, is a canopy necessary from the text? At this stage, perhaps not. It was promoted as a scientific model based on a possible interpretation of Genesis 1 to deal with several aspects of the overall biblical creation model developed in the mid-1900s. I don't say this lightly for my brothers and sisters in the Lord who may still find it appealing. Last century, I was introduced to the canopy model and found it fascinating. For years, I had espoused it, but after further study, I began leaning against it, as did many other creationists.

Old biblical commentators were not distraught at the windows of heaven or the waters not being a canopy encircling the earth. Such an interpretation was not deemed necessary in their sight. In fact, this idea is a recent addition to scriptural interpretation that is less than 100 years old. The canopy model was a scientific interpretation developed in an effort to help explain certain aspects of the text to those who were skeptical of the Bible's accounts of earth history, but when it comes down to it, it is not necessary and even has some serious biblical issues associated with it.

Scientific Issues

Clearly, there are some biblical issues that are difficult to overcome. Researchers have often pointed out the scientific issues of the canopy model, as well. A couple will be denoted below.

This is no discredit to the *researchers* by any means. The research was valuable and necessary to see how the model may or may not work with variations and types. The development and testing of models is an important part of scientific inquiry, and we should continue to do so with many models to help us understand the world God has given us. So I appreciate and applaud all the work that has been done, and I further wish to encourage researchers to study other aspects to see if anything was missed.

Temperatures

To answer the question about how the earth regulates its temperature without a canopy, consider that it may not have been that much different than the way

11. I would leave open the option that this affected the ocean sea level to a small degree but the main reasons for changing sea level was via the Ice Age.

it regulates it today — by the atmosphere and oceans. Although there may have been much water underground prior to the Flood, there was obviously enough at or near the surface to sustain immense amounts of sea life. We know this because of the well-known figure that nearly 95 percent of the fossil record consists of shallow-water marine organisms. Was the earth's surface around 70 percent water before the Flood? That is a question creationist researchers still debate.

An infinitely knowledgeable God would have no problem designing the earth in a perfect world to have an ideal climate (even with variations like the cool of the day Genesis 3:8) where people could have filled the earth without wearing clothes (Genesis 2:25, 1:28). But with a different continental scheme that are remnants of a perfect world (merely cursed, not rearranged by the Flood yet), it would surely have been better equipped to deal with regulated temperatures and climate.

A vapor canopy, on the other hand, would cause major problems for the regulation of earth's temperature. A vapor canopy would absorb both solar and infrared radiation and become hot, which would heat the surface by conduction downward. The various canopy models have therefore been plagued with heat problems from the greenhouse effect. For example, solar radiation would have to decrease by around 25 percent to make the most plausible model work.[12] The heat problem actually makes this model very problematic and adds a problem rather than helping to explain the environment before the Flood.[13]

The Source of Water

The primary source of water for the Flood was the springs of the great deep bursting forth (Genesis 7:11). This water in turn likely provided some of the water in the "windows of heaven" in an indirect fashion. There is no need for an ocean of vapor above the atmosphere to provide for extreme amounts of water for the rain that fell during the Flood.

For example, if Dillow's vapor canopy existed (40 feet of precipitable water) and collapsed at the time of the Flood to supply, in large part, the rainfall, the latent heat of condensation would have boiled the atmosphere!

12. For more on this see "Temperature Profiles for an Optimized Water Vapor Canopy" by Dr. Larry Vardiman, a researcher on this subject for over 25 years at the time of writing that paper; http://static.icr.org/i/pdf/technical/Temperature-Profiles-for-an-Optimized-Water-Vapor-Canopy.pdf.
13. Another issue is the amount of water vapor in the canopy. Dillow's 40 feet of precipitable water, the amount collected after all the water condenses, has major heat problems. But Vardiman's view has modeled canopies with 2 to 6 feet of precipitable water with better temperature results and we look forward to seeing future research.

And a viable canopy would not have had enough water vapor in it to sustain 40 days and nights of torrential global rain as in Vardiman's model (2–6 feet of precipitable water). Thus, the vapor canopy doesn't adequately explain the rain at the Flood.

Longevity

Some have appealed to a canopy to increase surface atmospheric pressures prior to the Flood. The reasoning is to allow for better healing as well as living longer and bigger as a result. However, increased oxygen (and likewise oxidation that produces dangerous free radicals), though beneficial in a few respects, is mostly a detriment to biological systems. Hence, antioxidants (including things like catalase and vitamins E, A, and C) are very important to reduce these free radicals within organisms.

Longevity (and the large size of many creatures) before and after the Flood is better explained by genetics through the bottlenecks of the Flood and the Tower of Babel as opposed to pre-Flood oxygen levels due to a canopy. Not to belabor these points, this idea has already been discussed elsewhere.[14]

Pre-Flood Climate

Regardless of canopy models, creationists generally agree that climate before the Fall was perfect. This doesn't mean the air was stagnant and 70°F every day, but instead had variations within the days and nights (Genesis 3:8). These variations were not extreme but very reasonable.

Consider that Adam and Eve were told to be fruitful and multiply and fill the earth (Genesis 1:27). In a perfect world where there was no need for clothes to cover sin (this came after the Fall), we can deduce that man should have been able to fill the earth without wearing clothes, hence the extremes were not as they are today or the couple would have been miserable as the temperatures fluctuated.

Even after the Fall, it makes sense that these weather variations were minimally different. But with the global Flood that destroyed the earth and rearranged continents and so on, the extremes become pronounced — we now have ice caps and extremely high mountains that were pushed up from the Flood (Psalm 104:8). We now have deserts that have extreme heat and cold and little water.

14. Ken Ham, ed., *New Answers Book 2* (Green Forest, AR: Master Books, 2008), p. 159–168; Bodie Hodge, *Tower of Babel* (Green Forest, AR: Master Books, 2013), p. 205–212.

Biblical Models and Encouragement

Answers in Genesis continues to encourage research and the development of scientific and theological models. However, a good grasp of all biblical passages that are relevant to the topic must precede the scientific research and models, and the Bible must be the ultimate judge over all of our conclusions.

The canopy model may have a glimmer of hope still remaining, and that will be left to the proponents to more carefully explain, but both the biblical and scientific difficulties need to be addressed thoroughly and convincingly for the model to be embraced. So we do look forward to future research.

In all of this, we must remember that scientific models are not Scripture, and it is the Scripture that we should defend as the authority. While we must surely affirm that the waters above were divided from the waters below, something the Bible clearly states, whether or not there was a canopy must be held loosely lest we do damage to the text of Scripture or the limits of scientific understanding.

CHAPTER 14

Are There Transitional Forms in the Fossil Record?

DR. DAVID MENTON

The central idea of evolution is that all of the kinds of living organisms on earth share a common ancestor and that over time they have evolved one from another by an unplanned and unguided natural process. This unobserved sort of "amoeba-to-man" evolution extending over hundreds of millions of years is called macroevolution to distinguish it from the relatively small-scale variations we observe among the individuals of a species. Evolutionists like to refer to these small variations as "microevolution" with the tacit assumption that over eons of time they add up incrementally to produce macroevolution. Thus, evolutionists look for evidence of these incremental steps, often referring to them as "transitional forms," suggesting that they represent stages of transformation of one organism into a different kind of organism.

Since macroevolution is not observable in the time frame of human observers, evolutionists often invoke microevolution as both evidence for macroevolution as well as its presumed mechanism. But as any animal or plant breeder knows, the limited variation that is observed among the individuals of a species has not been observed to lead to the essentially limitless process of macroevolution. In 1980, a group of evolutionists met in Chicago to discuss the relationship of micro- and macroevolution. Roger Lewin summed up this meeting in the journal *Science* as follows:

> The central question of the Chicago conference was whether the mechanisms underlying microevolution can be extrapolated to explain the phenomena of macroevolution. At the risk of doing violence to the positions of some of the people at the meeting, the answer can be given as a clear No.[1]

The lack of a clear relationship between microevolution and macroevolution has continued to be a problem for evolutionists.[2]

No matter what mechanism one might postulate for macroevolution, in the course of presumed evolutionary history there would have been an unimaginably vast number of transitional forms revealing at least some of the incremental stages of macroevolution. Thus evolutionists typically turn to the fossil record in an effort to identify transitional stages in the macro evolutionary process. When this fails, they turn to currently living biological organisms in the hope of "reconstructing" evolutionary transitional stages from living examples. When an appearance of progress is lacking among living organisms and their organs, evolutionists turn to artists who obligingly illustrate what they believe must surely have been the missing transitional stages of evolutionary progress. And, finally, when even artistic imagination fails to produce plausible intermediates of evolutionary progress, some evolutionists simply deny that there even is a vector of progress in evolution! However, evolutionists never question that there is a naturalistic evolutionary process of some kind that explains the origin of all living things.

"Transitional" Fossils – The Missing Links

Evolutionists begin with the unquestioned assumption that evolution has occurred, starting with some primordial life form and progressing over time in a purely naturalistic way to produce all the kinds of living organisms on earth, past or present. Thus for "evidence" of evolution they need only to examine available fossils and attempt to arrange them in a sequence that appears to show progress over time. But a plausible sequential progression of intermediate stages is rarely, if ever, observed in the fossil record, which explains why we hear so

1. Roger Lewin, "Evolutionary Theory Under Fire," *Science* 210, no. 4472 (1980): p. 883–887.
2. D.L. Stern, "Perspective: Evolutionary Developmental Biology and the Problem of Variation," *Evolution* 54, no. 4 (2000): p. 1079–1091; R.L. Carroll, "Towards a New Evolutionary Synthesis," *Trends in Ecology & Evolution* 15, no. 1 (2000): p. 27–32; A.M. Simons, "The Continuity of Microevolution and Macroevolution," *Journal of Evolutionary Biology* 15, no. 5 (2002): 688–701.

much about "missing links." Even Darwin himself was aware of this problem and said in his *Origin of Species*:

> The number of intermediate varieties, which have formerly existed on the earth, [must] be truly enormous. Why then is not every geological formation and every stratum full of such intermediate links? Geology assuredly does not reveal any such finely graduated organic chain; and this, perhaps, is the most obvious and gravest objection which can be urged against my theory.[3]

Why, indeed! For example, no one has observed progressive stages of "pre-bats" in the fossil record showing a mouse-like mammal gradually evolving into a bat with its long fingered wings. Evolutionists concede that what they consider to be the oldest bat fossils are 100 percent bats with some even showing evidence of sonar navigation.[4] G.K. Chesterton put it simply: "All we know of the Missing Link is that he is missing — and he won't be missed either."

Many evolutionists now concede the dearth of transitional forms in the fossil record and feel obliged to come up with some sort of explanation for it. The late evolutionist Steven J. Gould bluntly admitted, "the extreme rarity of transitional forms in the fossil record persists as the trade secret of paleontology."[5]

Again, Eldridge and Gould noted, "Most species during their geological history, either do not change in any appreciable way, or else they fluctuate mildly in morphology, with no apparent direction."[6]

Gould even goes so far as to concede that not only are transitional stages not found in the fossil record, but in many cases we are not even able to imagine such intermediates:

> The absence of fossil evidence for intermediate stages between major transitions in organic design, indeed our inability, even in our imagination, to construct functional intermediates in many cases, has been a persistent and nagging problem for gradualistic accounts of evolution.[7]

3. Charles Darwin, *The Origin of Species* (1859; repr., New York: Avenel Books, Crown Publishers, n.d.).
4. G. Jepsen, "Bat Origins and Evolution," in *Biology of Bats,* W. Wimsatt, ed. (New York: Academic Press, 1970), p. 1–64; G.L. Jepsen, "Early Eocene Bat from wyoming," *Science* 154, no. 3754 (1966): p. 1333–1339.
5. Stephen J. Gould, "Evolution's Erratic Pace," *Natural History* 86, no. 5 (1977): p. 12–16.
6. N. Eldredge and Stephen J. Gould, "Punctuated Equilibria: The Tempo and Mode of Evolution Reconsidered," *Paleobiology* 3, no. 2 (1977): p. 145–146.
7. Gould, "Is a New and General Theory of Evolution Emerging?" *Paleobiology* 6, no 1 (1980): p. 127.

This conspicuous lack of fossil evidence for intermediate or transitional stages of evolution led Gould to a highly speculative rescuing hypothesis for evolution called "punctuated equilibrium," or as it is sometimes called, the "hopeful monster theory." In this scenario, the lack of fossil transitional forms is explained away by claiming that the transitional stages (hopeful monsters) being both unlikely and unstable occurred rarely and relatively quickly (on a geological time scale) leaving no fossil evidence. So what we actually see is stasis, i.e., no change over long periods of geological time![8] No wonder some evolutionists have argued that ancestor descendent relationships simply cannot be determined from fossils. For example, with regard to human evolution, Richard Lewontin said, "Despite the excited and optimistic claims that have been made by some paleontologists, no fossil hominid species can be established as our direct ancestor."[9]

"Transitional" Living Organisms and Organs – Looking for the Dead Among the Living

When the fossil evidence fails to provide expected transitional stages, evolutionists often turn to living organisms in an attempt to arrange them in a way that appears to show a sequential process of evolution. An advantage of living organisms is that they allow the evolutionist to create an evolutionary scenario for the soft organs of the body. While we are becoming increasingly aware of evidence of soft tissue in fossils, most fossils show only hard tissue such as shells, teeth, and bones. Hard tissues represent a relatively small part of a living organism compared to their soft tissues. So with a bit of imagination, living organisms can sometimes be selectively arranged in a way to give the impression of an evolutionary sequence for soft tissue organs such as eyes, hearts, and kidneys.

In an effort to show evolutionary progress among living organisms, evolutionists look for structures or functions that appear to be intermediate in some way to those of other living organisms. These intermediate structures are then extrapolated to represent "transitional" stages in a sequential evolutionary progress. But while an organ or organism may be considered intermediate in appearance between two other organs or organisms, it does not necessarily mean that it represents an evolutionary transition between the two. Declaring something to be intermediate with regard to some arbitrary structure or character is merely

8. Stephen J. Gould, "The Return of the Hopeful Monsters," *Natural History* 86 (1977): p. 22–30.
9. Richard Lewontin, *Human Diversity* (San Francisco, CA: W.H. Freeman & Company, 1995), p. 179.

an organizational decision, whereas declaring it to be transitional presumes an evolutionary or transformational process.

Living organisms are often used in an effort to explain the evolution of the eye. Darwin conceded in *The Origin of Species* that to suppose the eye could have evolved by natural selection "seems absurd in the highest degree," and that to support his theory it would be necessary to demonstrate the existence of "numerous gradations" from the most primitive eyes to the most advanced ones. Since the fossil record provides no evidence for this, evolutionists attempt to arrange the eyes of present-day living invertebrates and vertebrates into what appears to be a progressive evolutionary sequence. For example, in a journal devoted to giving evolutionary support for teachers, Lamb claims to have evidence from living hagfish that the vertebrate eye evolved through numerous subtle changes:

> The great majority of the gradual transitions that did occur have not been preserved to the present time, either in the fossil record or in extant species; yet clear evidence of their occurrence remains. We discuss the remarkable "eye" of the hagfish, which has features intermediate between a simple light detector and an image-forming camera-like eye and which may represent a step in the evolution of our eye that can now be studied by modern methods.[10]

But a recent study of microRNA expression patterns in the hagfish and lamprey showed that the cyclostomes are closely related.[11] This leaves evolutionists arguing whether the relatively simple hagfish eye is really a precursor of the more complex lamprey type eye or a degenerate form of that type eye. From what then did the vertebrate eye evolve? There is a bewildering array of eyes found among the invertebrates. One of the world's most distinguished experts on the eye, Sir Duke-Elder, said in volume one (*The Eye in Evolution*) of his monumental 15-volume work, *System of Ophthalmology*, that the eyes of invertebrates do not show a series of transitional stages:

> The curious thing, however, is that in their distribution the eyes of invertebrates form no series of continuity and succession. Without obvious phylogenetic sequences, their occurrence seems haphazard;

10. T. Lamb, E. Pugh, and S. Collin, "The Origin of the Vertebrate Eye," *Evolution: Education and Outreach* 1, no. 4 (2008): p. 415–426.
11. Alysha M. Heimberg et al., "MicroRNAs Reveal the Interrelationships of Hagfish, Lampreys, and Gnathostomes and the Nature of the Ancestral Vertebrate," *PNAS* 107, no. 45 (2010): p. 19379–19383.

> analogous photoreceptors appear in unrelated species, an elaborate organ in a primitive species (such as the complex eye of the jelly-fish Charybdea) or an elementary structure high in the evolutionary scale (such as the simple eyes of insects), and the same animal may be provided with two different mechanisms with different spectral sensitivities subserving different types of behavior.[12]

Duke-Elder was not even convinced that we ever will find a solution for the evolution of the eye:

> Indeed, appearing as it does fully formed in the most primitive species extant today, and in the absence of transition forms with which it can be associated unless by speculative hypothesis with little factual foundation, there seems little likelihood of finding a satisfying and pragmatic solution to the puzzle presented by its (the eye's) evolutionary development.[13]

With about 1.5 million named and categorized living species (and possibly several times more species unnamed or categorized) we might reasonably expect to see at least some evidence of a series of transitional stages among living organisms, but such is not the case. In his book *Patterns and Processes of Vertebrate Evolution*, evolutionist Robert Carroll concedes that very few examples of intermediate organisms or organs have been proposed:

> Although an almost incomprehensible number of species inhabit Earth today, they do not form a continuous spectrum of barely distinguishable intermediates. Instead, nearly all species can be recognized as belonging to a relatively limited number of clearly distinct major groups, with very few illustrating intermediate structures or ways of life.[14]

"Transitional" Drawings and Illustrations – Making Your Own Data

When all else fails, there are always artists who will make a picture or model of any missing link the evolutionist might desire. Sadly, laymen are often

12. S.S. Duke-Elder, *The Eye in Evolution*, vol. 1, *System of Ophthalmology* (London: Henry Kimpton, 1958), p. 178.
13. Ibid., p. 247.
14. Robert L. Carroll, *Patterns and Processes of Vertebrate Evolution* (New York: Cambridge University Press, 1997), p. 9.

strongly influenced by such fanciful illustrations. Consider the famous "March of Progress" monkey-to-man drawing commissioned by Time Life Books,[15] one of the most famous and recognizable science illustrations ever produced. This drawing presumed to compress 25 million years of imagined human evolution into a row of progressively taller and more erect primates until finally a human walks away with a marine drill sergeant posture and gait.

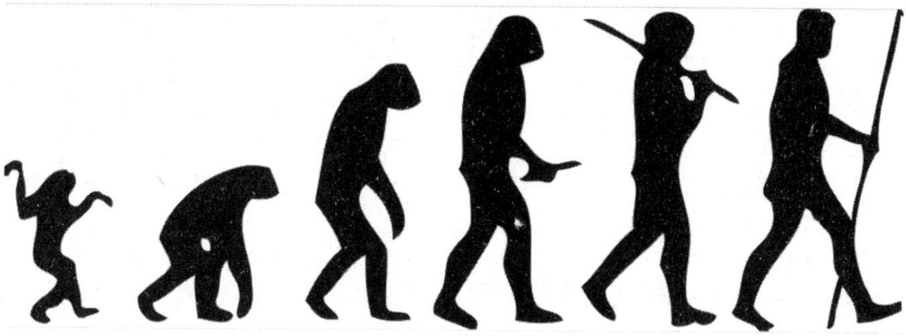

Many evolutionists have expressed their disapproval over this illustration showing a triumphalist linear progression of evolution that simply does not exist. Nonetheless, this "March of Progress" illustration has probably done more to convince uncritical laymen of the bestial origin of man than any other evidence.

Several years ago, the popular evolutionist Carl Sagan was on a television program where he showed a video clip of a rapid series of cartoon illustrations purporting to show amoeba-to-man evolution while a harpsichord solemnly played in the background. At the conclusion, the audience applauded enthusiastically, seemingly convinced that they had actually seen the whole sweep of amoeba-to-man evolution in a few minutes. We are living in an age where many are careless in distinguishing artistic license from scientific evidence.

But not all pictorial evidence for the imagined transitional stages of evolution is found in the popular literature meant for laymen. Imaginative drawings and illustrations are frequently found in the scientific literature intended for the specialist. An example of artistic license passing for "evidence" of transitional stages of evolution may be seen in efforts to explain the evolution of feathers.

Now that evolutionists are dead certain that dinosaurs evolved into birds (with many insisting that birds are in fact dinosaurs) they are left with the unenviable task of showing how reptile scales evolved into feathers. For years,

15. F. Clark Howell, *Early Man* (New York, NY: Time-Life Books, 1965).

evolutionists have insisted that feathers and scales are very similar, but nothing could be further from the truth.[16] Scales are essentially continuous folds in the epidermis while feathers grow from individual follicles. This is why the reptile must shed its entire skin to replace its scales while a bird sheds its feathers individually from feather follicles (in matched left-right pairs in the case of primary feathers). It is hard to imagine two cutaneous appendages more profoundly different than scales and feathers; they share almost nothing in common. In fact, feathers and their follicles show far more similarity to hairs and hair follicles than they do to reptilian scales, but there is no evolutionary scenario that relates the phylogeny of birds to mammals, so this is ignored by evolutionists. So evolutionists are stuck with making feathers out of scales and to do so they must employ artists to illustrate transitional stages not seen in fossils or living creatures.

An attempt was made by Xu et al to show the hypothetical stages of evolution from scale to feather.[17] Their artist illustrates an elongated hollow scale first becoming a frayed or branching structure. This then somehow becomes a compound branching structure (see step II to IIIA below). To accomplish this, a structure with a simple branching pattern (all branching from one node) must implausibly become a compound branching structure (branching from several different nodes). The compound branching structure then undergoes another order of branching to give a superficial resemblance to a feather.

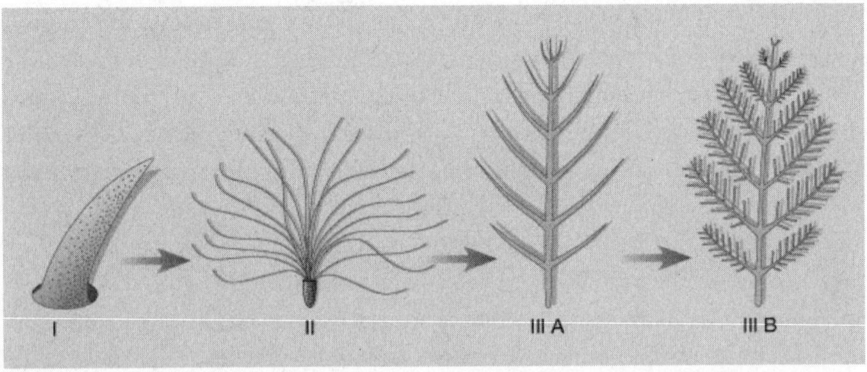

Unmentioned is that in real life, all feather development must occur inside a follicle, where the feather is folded up in a sheath like a ship in a bottle. But then this presents no restrictions for an artist's imagination and drawing.

16. David N. Menton, *Formed to Fly*, DVD (Hebron, KY: Answers in Genesis, 2007).
17. X. Xu, Z. Zhou, and R.O. Prum, "Branched Integumental Structures in Sinornithosaurus and the Origin of Feathers" *Nature* 410, no. 6825 (2001): p. 200–204.

What Do "Transitional" Stages Mean If There Is No Progress?

Can we have transitions or intermediates without progress? Many evolutionists are coming to the conclusion that there is neither purpose nor progress in evolution. In a recent survey of over 150 of the nation's most influential and prestigious evolutionists (all members of the National Academy of Science), it was revealed that nearly 42 percent believe that evolution shows neither purpose nor progress.[18] But if there is no purpose or progress in evolution, how can one identify incremental transitional changes in the process? Another 48 percent of these distinguished evolutionists believe evolution shows progress but no purpose. But how can there be progress without purpose? The English Wordnet dictionary defines progress as "an anticipated outcome that is intended or that guides your planned actions" and the Merriam-Webster dictionary defines progress as "a forward or onward movement as to an objective or to a goal." Since nearly 80 percent of the evolutionists in the survey describe themselves as atheists, it is not surprising that they shun the notion of purpose in evolution. Purpose suggests the Creator (and accountability to the Creator), and that is unthinkable to these professional atheist/evolutionists.

Isn't It Great to Be a Christian and Recognize God's Purpose in Creation?

As Bible-believing Christians, we can gladly recognize the obvious that there is overwhelming evidence of intelligent design and purpose in God's creation. Some evolutionists concede that they are aware of this evidence for design, but as the Bible says, they "suppress the truth in unrighteousness" (Romans 1:18). No better example of this suppression of the truth can be seen than the ardent atheist/evolutionist Richard Dawkins who wrote in the first page of his book titled *The Blind Watchmaker: Why the Evidence of Evolution Reveals a Universe Without Design*:

> Biology is the study of complicated things that give the appearance of having been designed for a purpose.[19]

Dawkins concedes that this obvious appearance of design in biological systems cries out for some kind of explanation:

18. G. Graffin, "Evolution and Religion," The Cornell Evolution Project, http://www.polypterus.com/.
19. Richard Dawkins, *The Blind Watchmaker: Why the Evidence of Evolution Reveals a Universe Without Design* (New York: Norton & Company, 1986), p. ix.

> The complexity of living organisms is matched by the elegant efficiency of their apparent design. If anyone doesn't agree that this amount of complex design cries out for an explanation, I give up.[20]

But the only explanation the atheist evolutionist can offer is that somehow nature "counterfeits" intelligent design. How sad.

20. Ibid.

CHAPTER 15

Could the Flood Cataclysm Deposit Uniform Sedimentary Rock Layers?

DR. ANDREW A. SNELLING

This is definitely a legitimate question to ask concerning the nature of the evidence one would expect to be left behind by the Flood cataclysm. Because the waters of local floods today are often full of sediments and are fast-moving, it is commonly thought that neat, uniform sediment layers are not deposited under such conditions. So this question needs close examination, starting by looking at what the evidence is that we see in the rock record.

Do We Find Neat Uniform Sedimentary Rock Layers in the Geologic Record?

Whether looking into the Grand Canyon from one of the rim overlooks or traversing through the Grand Canyon on foot or by raft, the answer to this question is obviously yes. The fossil-bearing sedimentary layers deposited by the Flood can be seen exposed in the walls, stacked on top of one another like a huge pile of pancakes. And the view is much the same no matter where one views the Grand Canyon. So at the regional scale in the Grand Canyon area it is clearly evident that the sedimentary rock layers deposited there during the Flood cataclysm are neat and uniform.

Similar observations can be made in many other places across the earth's surface. This pattern is often seen in road cuts and in mountainous areas where erosion has exposed the constituent rock layer sequences. So it is hardly necessary to defend the assertion that the fossil-bearing sedimentary layers that were deposited during the Flood cataclysm are generally neat and uniform and stacked in a sequence that is exposed to view in many places across the earth's continents.

The assertion that these fossil-bearing sedimentary layers were deposited during the Flood cataclysm is easy to defend.[1] The obvious observation to make is that many of these fossil-bearing sedimentary layers contain fossils of creatures that today live on the shallow ocean floors fringing the continents, and not on the continents where countless billions of them are buried in these sedimentary layers. Indeed, sedimentary rock layers containing the same fossils are not found on the ocean floors today, nor are they found in comparable dimensions on the continental shelves fringing the continents. But the vast marine-fossil-bearing sedimentary layers we find spread right across the continents today are thus consistent with the ocean waters having flooded over the continents on a global scale, tearing marine creatures from their shallow ocean floor habitats and picking up sediments, then burying those creatures in those sediments up and across the continents in vast sedimentary layers. This is consistent with the biblical description of the Flood.

Many geologists are already aware that there are six thick sequences of fossil-bearing sedimentary strata, known as megasequences, which can be traced right across the North American continent. This was documented five decades ago in 1963[2] and subsequently verified by numerous observations so that it is now well recognized. In the early 1980s, the American Association of Petroleum Geologists (AAPG) conducted a project in which all the local geologic strata "columns" derived from the mapping of outcrops in local areas, supplemented by drill-hole data, were put on charts to show the sequences of fossil-bearing sedimentary rock layers right across the North American continent (figure 1).[3]

The rationale used to identify these megasequences was based on mapping the preserved rock record across the North American continent. These thick sequences or packages of fossil-bearing sedimentary rock layers were easily identified because they were bounded by erosion surfaces (called unconformities)

1. Andrew Snelling, "What Are Some of the Best Flood Evidences?" in Ken Ham, ed., *The New Answers Book 3* (Green Forest, AR: Master Books, 2010), p. 283–298.
2. Laurence L. Sloss, "Sequences in the Cratonic Interior of North America," *Geological Society of America Bulletin* 74, no. 2 (1963): p. 93–114.
3. F.A. Lindberg, *Correlation of Stratigraphic Units of North America (COSUNA): Correlation Charts Series* (Tulsa, OK: American Association of Petroleum Geologists, 1986).

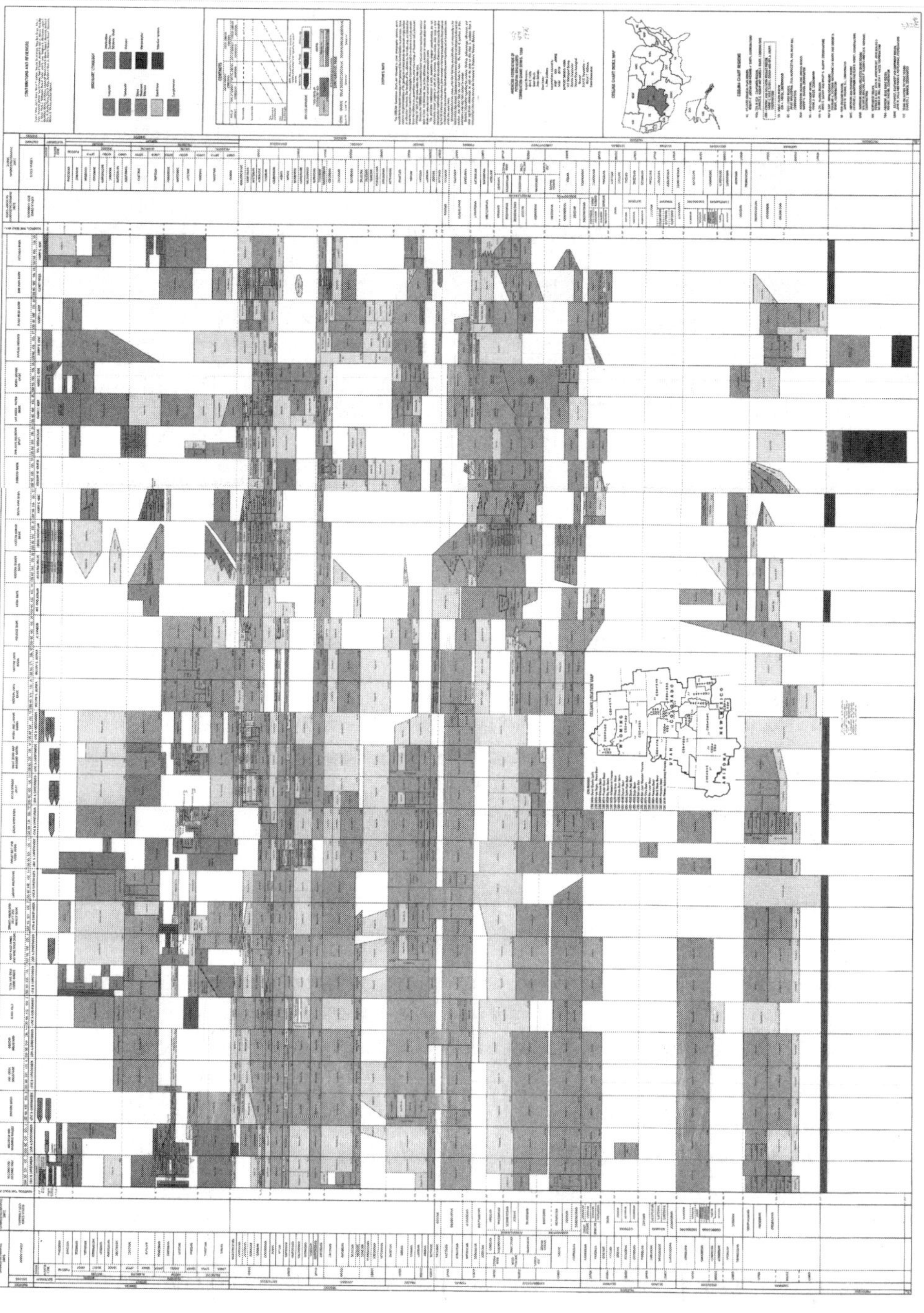

Figure 1. An example of one of the charts produced during the AAPG project showing the local strata columns in the central and southern Rockies region of the USA.

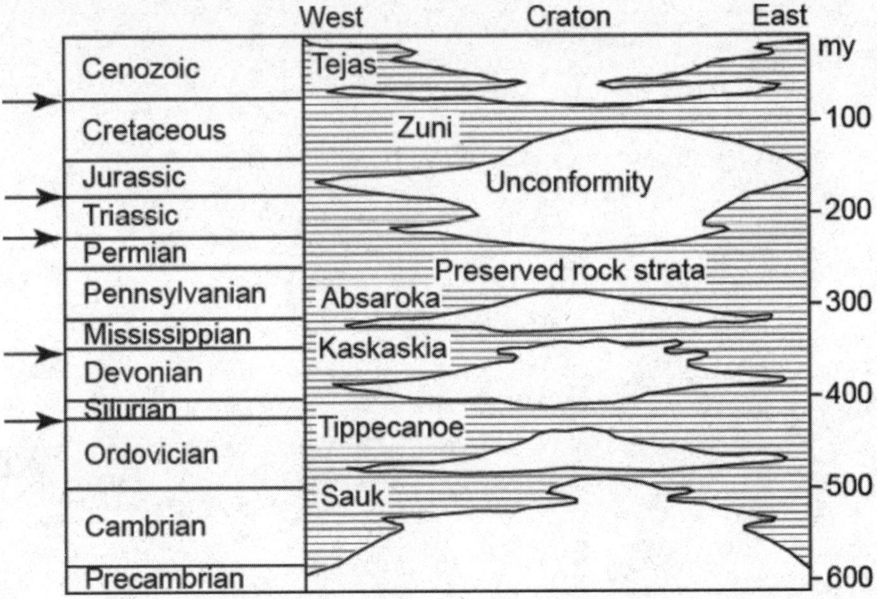

Figure 2. The preserved rock record, consisting of named megasequences, between major unconformities and mass extinctions (arrowed) across the North American continent.

due to the actions of the ocean waters as they advanced over the continent depositing the sedimentary rock layers before retreating again (figure 2).[4] These unconformities therefore coincide with rising and falling water levels as ocean waters oscillated across the continent and back again after depositing their sediment loads, often also coinciding with the mass burial of creatures in what evolutionary geologists have called mass extinctions. Significantly, some of the fossil-bearing sedimentary layers in these megasequences can also be traced beyond North America to other continents.[5]

Within each megasequence are various named strata units. For example, the first (lowermost) of these megasequences, called the Sauk Megasequence, in the Grand Canyon area consists of the Tapeats Sandstone, the Bright Angel Shale, and the Muav Limestone. Thorough geologic mapping was initially only done locally, so the rock units identified and mapped were given names locally. Therefore, even if a rock unit stretched into adjoining local areas and beyond, it often had different names in adjoining local areas. Thus, in the 1980s, when the

4. Leonard R. Brand, *Faith, Reason, and Earth History* (Berrien Springs, MI: Andrews University Press, 1997).
5. Andrew Snelling, "Transcontinental Rock Layers," *Answers* (July–September 2008): p. 80–83.

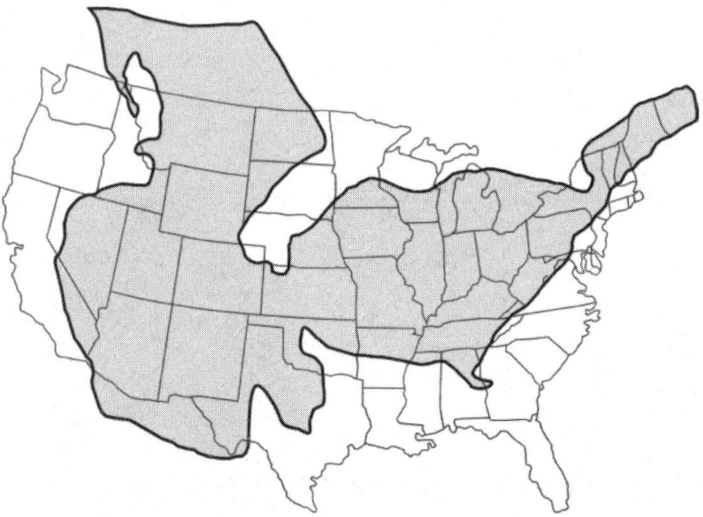

Figure 3. The distribution of the Tapeats Sandstone and its equivalents across North America, constructed from the local geologic columns compiled in the COSUNA charts produced by the AAPG.

American Association for Petroleum Geologists (AAPG) tabulated all the local strata columns across the continent, it became possible to see how some specific rock units, which had been given different names in different local areas, actually were the same unit, which could be traced vast distances across the continent. The Tapeats Sandstone in the Grand Canyon area is one of those units, and it can be traced all across Arizona northward to the Canadian border and beyond, northeastward right across the USA as far as Maine (figure 3).[6] The same sandstone unit in exactly the same geologic strata position is also found in southern Israel, from where it can be traced across to Jordan and into Egypt, and then right across north Africa.[7] Thus the Tapeats Sandstone represents one unit within one megasequence that is easily identified over vast continental scale areas due to its uniform makeup.

However, while some units within megasequences traverse continents, many others are only recognizable and able to be traced over regions, though still vast in extent compared to one's local area. In the Grand Canyon area, for example, the Coconino Sandstone, within the fourth of the megasequences, known as the Absaroka Megasequence, can be traced from northern and central Arizona

6. Andrew Snelling, *Earth's Catastrophic Past: Geology Creation and the Flood* (Dallas, TX: Institute for Creation Research, 2009), p. 1082, figure 45.
7. Andrew Snelling, "The Geology of Israel within the Biblical Creation-Flood Framework of Earth History: 2. The Flood Rocks," *Answers Research Journal* 3 (2010): p. 267–309; available online at http://www.answersingenesis.org/articles/arj/v3/n1/geology-of-israel-2.

across New Mexico into Colorado, Kansas, Oklahoma, and Texas over an area approaching 200,000 square miles, though an isolated remanent in southwestern Arizona indicates the unit previously had a wider distribution that has been reduced by erosion (figure 4).

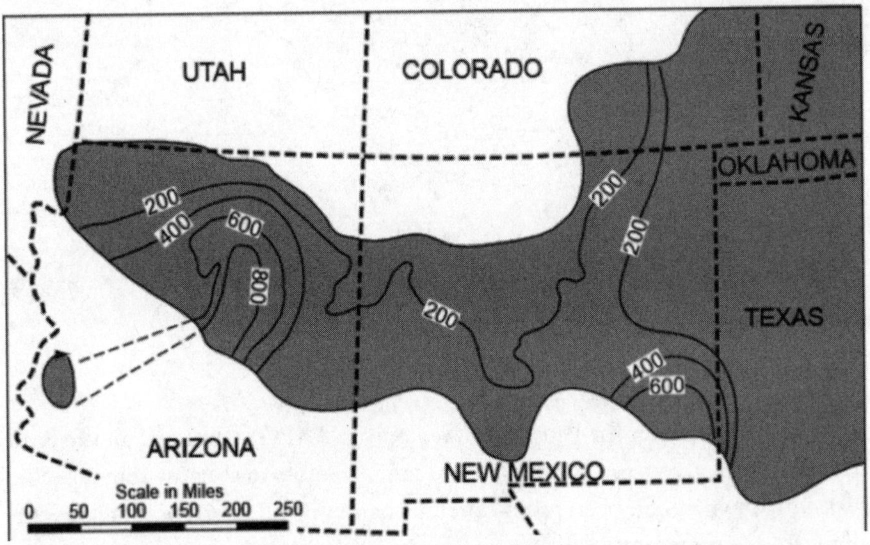

Figure 4. The distribution of the Coconino Sandstone and its equivalents from northern Arizona into adjoining states, showing the variations in its thickness (contour lines in feet) (after Austin[8]).

Nevertheless, not all the strata units are uniform, the character of the rock units changing due to later variations. For example, the Toroweap Formation is a limestone in the Grand Canyon area, but laterally to the southwest it changes into sandstone, along with local variations that include beds of gypsum to the west.[9] Indeed, many strata units change their rock character laterally, which reflects both the type and composition of the sediments within the mixture carried by the ocean waters over the continent to deposit them. Furthermore, not only is the sediment composition related to the source of the sediments, but changes in the sediment composition can occur. As the ocean waters carried sediments up and across the continent, they sometimes eroded underlying

8. Steven A. Austin, *Grand Canyon: Monument to Catastrophe* (Santee, CA: Institute for Creation Research, 1994), p. 36, figure 3.13.
9. Christine E. Turner, "Toroweap Formation," in Stanley S. Beus and Michael Morales, eds., *Grand Canyon Geology*, 2nd ed. (New York, NY: Oxford University Press, 2003), p. 180–195.

sediment layers of different compositions, adding them to their sediment loads before eventually depositing them.

Another aspect of this question is the thickness of the fossil-bearing sedimentary rock layers deposited across the continents. Even on local scales, variations in the thicknesses of strata units can be seen, as well as sometimes even compositional changes. So, for example, even though the Coconino Sandstone averages a thickness of 315 feet in the Grand Canyon area, it changes its thickness through the length of the Grand Canyon, thinning to the west and thickening even up to 1,000 feet toward the southeast (see figure 4). Furthermore, some rock units are made up of beds of alternating compositions, such as within some of the strata units in the Cincinnati area which consist of alternating beds of limestone and shale (figure 5).[10] Sometimes these thinner beds thicken and thin even within the outcrop scale of a road cut. So whereas we do find neat, uniform fossil-bearing sedimentary rock layers across the continents as a record of the Flood, the depositional processes produced and left behind local variations, both in thicknesses of the layers and beds within the named strata units, but also variations in compositions, from local to regional scales.

Were the Fast-Moving Flood Waters Also Churning?

During the Flood cataclysm, there were four main causes for generating the surging flows of water currents that picked up and carried sediments onto and across the continents to deposit the fossil-bearing sedimentary rock layers there.

First, there was the normal ebb and flow of the rising and falling tidal oscillations. The effect of these approximately twice-daily tidal surges would have increased as the Flood waters became global. It has been shown that on a global ocean there would have been a resonating effect by which the tidal surges would have progressively built in height and, therefore, in the strength and impact of each surge, due to the close overlapping of the tidal peaks and troughs in the approximate 12–13 hour spacing between successive highs and lows.[11]

Superimposed on those tidal flows and surges, there would have been repeated tsunamis generated by earthquakes caused by repeated catastrophic earth movements. The "fountains of the great deep" were broken up (Genesis

10. Andrew Snelling, "Cincinnati — Built on a Fossil Graveyard," *Answers* (July–September 2011): p. 50–53.
11. M.E. Clark, and H.D. Voss, "Resonance and Sedimentary Layering in the Context of a Global Flood," in *Proceedings of the Second International Conference on Creationism*, vol. 2, Robert E. Walsh and Christopher L. Brooks, eds. (Pittsburgh, PA: Creation Science Fellowship, 1990), p. 53–63.

Figure 5. Alternating beds of limestone (hard) and shale (soft) in the Fairview Formation in a road cut in the Cincinnati area of northern Kentucky. (Photograph: Andrew A. Snelling)

7:11), initiating the catastrophic plate tectonics that drove the Flood event.[12] The earth's crust was broken up around the globe, producing massive earthquakes, followed by the accelerated motion of the crustal fragments (called plates) across the earth's surface at many-feet-per-second speed. As the Flood event progressed, plates collided with one another, or some plates were pushed under the edges of other plates. All these earth movements would have generated many catastrophic earthquakes that in turn would have repeatedly produced massive tsunamis. As these tsunamis moved, they would have surged toward and onto the continents.

Furthermore, superimposed on the tides and tsunamis would have been the progressive raising of the ocean floor. As the ocean floor plates were pushed apart, molten rock rose from inside the earth to generate new ocean floor rocks. The new warm ocean floor, being less dense, would steadily rise, thus pushing

12. Steven A. Austin, John R. Baumgardner, D. Russell Humphreys, Andrew Snelling, Larry Vardiman, and Kurt P. Wise, "Catastrophic Plate Tectonics: A Global Flood Model of Earth History," in *Proceedings of the Third International Conference on Creationism*, Robert E. Walsh, ed. (Pittsburgh, PA: Creation Science Fellowship, 1994), p. 609–621; Andrew Snelling, "A Catastrophic Breakup: A Scientific Look at Catastrophic Plate Tectonics," *Answers* (April–June 2007): p. 44–48.

up the sea level. This raising of the sea level would have in turn caused a surge of ocean waters toward the continents to flood them.

The net result would have been huge fluctuations in the water levels combined with catastrophic surges of walls of water moving from open ocean areas toward and onto the continents and across them. Yet another force at work driving these surging water currents would have been super-storms. These would have been generated in the atmosphere as a result of the supersonic steam jets at the crustal fracture zones, catapulting ocean waters aloft before they fell back to the earth's surface as global torrential rainfall (the "windows or floodgates of heaven" were opened, Genesis 7:11). It is estimated that such super-storms and their winds moving across the surface of the Flood waters would have driven water currents at speeds of 100 miles an hour or more.[13]

So there is no doubt that there were adequate mechanisms for driving fast-moving, catastrophically powerful water currents and surges from the oceans toward and onto the continents. These were thus capable of carrying the sediments and creatures to be buried in the fossil-bearing sedimentary rock layers deposited across the continents, stacked up in sequence as a result of the fluctuating water levels and the ebb and flow of the water.

Just as is observed today, in the open ocean there are no major effects on the ocean surface from the passage of tsunamis, tidal surges, and fast-moving water currents apart from waves. It is at the base of the water column deep below the surface where the moving and surging water picks up loose sediments from the ocean floor, or scours and erodes sediments from the ocean floor, and then transports them in a slurry of sediment-laden water.

What was happening at the base of the water column of these surging, fast-moving water currents during the Flood would have depended on a number of factors, which in turn would have produced differing results. Though somewhat oversimplified, if the water was flowing over uneven ocean floor topography, then turbulent flow (churning water) could be generated. But if the water was flowing over a flat surface, then the flow would be laminar and sheet-like, and any erosion would result from cavitation, a process in which the fast water flow generates vacuum bubbles that hammer rock surfaces, pulverizing the rock rapidly. If there were loose sediments on the surface being traversed, once the

13. John R. Baumgardner, and Daniel W. Barnette, "Patterns of Ocean Circulation Over the Continents during Noah's Flood," in *Third International Conference on Creationism*, p. 77–86; Larry Vardiman, "Numerical Simulation of Precipitation Induced by Hot Mid-Ocean Ridges," in *Proceedings of the Fourth International Conference on Creationism*, Robert E. Walsh, ed. (Pittsburgh, PA: Creation Science Fellowship, 1998), p. 595–606.

water reached a critical speed it would pick up those loose sediments and carry them. Often, once the process is started, if there is even the slightest of downward slopes on the surfaces being traversed, then gravity takes over to produce debris flows. Many strata units in the rock record bear testimony to having been deposited by gravity-driven underwater debris flows.

The quantity and type of sediments transported would depend on the composition and particle sizes in those loose sediments, so that generally the faster the water flow, the greater the sizes of the particles that could be picked up and transported. Below a critical speed, no sediments would be picked up and carried by the water flows. And that critical speed would likely be lower for turbulent flow and higher for laminar flow, except where gravity is driving the water's ability to pick up sediments to produce debris flows. At higher speeds and carrying more sediment, the water at the base of the water column would become more erosive. The more sediments the water carried, the more they would add to the water's abrasive and erosive power. At the highest water speeds though, when the amount of sediment in the water is greater than the amount of water in the slurry mixture, the density of the slurry is so great that even boulders are transported, suspended in the slurry.

Fast-moving waters are certainly capable of depositing sediments, and many strata layers in the rock record of the Flood would have been deposited in that way, as witnessed by the strata layers that were deposited right across continents. Additionally, once the water started to slow down in its passage over the continents, the water would start to drop the rest of its sediment load and deposit it in more sediment layers, also burying the creatures that had been carried by the water. An example is the postulated progressive simultaneous deposition of the Tapeats Sandstone, Bright Angel Shale, and Muav Limestone across Nevada, Arizona, and New Mexico as the Flood waters advanced, the bottom current speed decreasing in the returning underflow so sediments of decreasing grain sizes were progressively deposited.[14] As the water slowed it would also be less likely to erode previously deposited sediment layers, especially where the surface of those previously deposited sediment layers had started to cohere, and cementation had begun to bind the sediment particles (the first stage of the hardening process).

The net result would be that the Flood waters at the base of the flow would tend to erode in source areas as the current flow increased, and then started switching to depositional mode as the water currents flowed over the continents and started to deposit their loads. Thus, when the water currents subsequently

14. Steven A. Austin, *Grand Canyon: Monument to Catastrophe* (Santee, CA: Institute for Creation Research, 1994), p. 69, figure 4.12.

slowed as they continued further sediment deposition, they would not be eroding at the same time. The outcome would be to deposit uniform sediment layers during their passage across the continents as they progressively spread out and deposited their sediment loads. Of course, there could be lateral variations in sediment types. Sometimes as the waters slowed, the heavier particles would settle out first. Then at slower speeds finer particles would be deposited, so that the sediment particle sizes could change laterally as the one rock unit was deposited across the continent. In some strata layers the grading of the sediment particle sizes is the inverse. But many layers do not exhibit any graded bedding. Instead, the changes between water flow surges meant changes in sediment loads, with sediments of different compositions and types, each consisting of uniform similar particle sizes being deposited, such as lime mud versus quartz sand, as in the example of the Toroweap Formation in the Grand Canyon area being deposited on top of the Coconino Sandstone, as has already been mentioned.

Natural and Experimental Examples

In 1960, Hurricane Donna created surging ocean waves that flooded inland up to 5 miles along the coast of southern Florida for 6 hours.[15] As a result, the hurricane deposited a neat, uniform 6-inch-thick mud layer, with numerous thin laminae within it, across the area traversed by the flood waters. In June 1965, a storm in Colorado produced flooding of Bijou Creek, which resulted in the deposition from the fast-moving waters of new sediment layers containing fine laminations.[16] Then on June 12, 1980, an eruption of Mount St. Helens produced a hurricane-velocity, surging-flow of volcanic ash, which accumulated in less than five hours as a neat, uniform 25-foot-thick layer of laminated volcanic ash, including uniform neat, alternating laminae of coarse and fine sediment grains (figure 6).[17]

In a detailed study of a seven-foot-thick bed within the Redwall Limestone in the Grand Canyon area, Austin[18] has convincingly argued that the evidence

15. M.N. Ball, B.A. Shinn, and K.W. Stockman, "The Geological Effects of Hurricane Donna in South Florida," *Journal of Geology* 75 (1967): p. 583–597.
16. E.D. McKee, E.J. Crosby, and H.L. Berryhill Jr., "Flood Deposits, Bijou Creek, Colorado, June 1965," *Journal of Sedimentary Petrology* 37, no. 3 (1967): p. 829–851.
17. Steven A. Austin, "Mount St. Helens and Catastrophism," in *Proceedings of the First International Conference on Creationism*, vol. 1 (Pittsburgh, PA: Creation Science Fellowship, 1986), p. 3–9.
18. Steven A. Austin, "Nautiloid Mass Kill and Burial Event, Redwall Limestone (Lower Mississippian), Grand Canyon Region, Arizona and Nevada," in *Proceedings of the Fifth International Conference on Creationism*, Robert L. Ivey, Jr., ed. (Pittsburgh, PA: Creation Science Fellowship, 2003), p. 55–99.

Figure 6. The 25-foot-thick deposit is exposed in the middle of the cliff. The fine layering within this deposit was produced within hours at Mount St. Helens on June 12, 1980, by hurricane-velocity surging flows from the crater of the volcano. (Photograph: Steven A. Austin)

is consistent with the bed's deposition by a gravity-driven debris flow. In the middle section of this bed, which has been traced over more than 11,600 square miles, are billions of straight-shelled nautiloid fossils of various lengths. Though mostly buried and fossilized horizontally, some are found at various angles, and some are even vertical. These and the ubiquitous vertical fluid evulsion structures are consistent with rapid burial in a debris flow that turbulently tossed some of the nautiloids around during this mass kill event. Yet the bed overall is neat and uniform over this large area.

The three observed examples described above demonstrate that local-regional natural catastrophes do deposit neat, uniform sedimentary rock layers, even though in most instances the flow of water and air respectively was rapid and sometimes turbulent (churning). It should also be noticed that in two of the three examples the surging, fast-moving sediment-laden flows did not erode into the surfaces they flowed over, even though those surfaces consisted of loose materials (soils and sands, and previously deposited volcanic ash, respectively). Instead, the flows left smooth, neat, uniform boundaries at the bases of the neat, uniform sediment layers they deposited. These sediment layers resulting from these local-regional natural catastrophes closely mirror at a smaller scale the neat, uniform sedimentary layers left behind by the Flood waters, stacked up neatly on top of one another with smooth, uniform boundaries between them.

Not only do we have numerous modern examples where local-regional natural catastrophic events have resulted in the rapid accumulation of neat uniform sedimentary layers, but we have numerous laboratory experiments that have allowed researchers to document the same processes. For example, using a circular flume, it was demonstrated that high-velocity water currents sort and deposit sediment grains by weight, density, and shape, and that as the fast-moving current loses its velocity, the segregation of grains produces a succession of thin, parallel laminae in the resultant neat uniform sediment layer.[19] Other linear flume experiments with water swiftly carrying sand grains have demonstrated how a neat uniform sand layer is progressively deposited as the sand-carrying water current advances.[20] These examples demonstrate that water moving at upper (high) flow regime speeds produces planar beds rapidly. Indeed, the results of such flume experiments correlate closely with the observed natural sedimentation processes from swift-flowing water in tidal channels, floods, and other catastrophic events, and also accurately replicate at a smaller scale the features seen in the neat uniform sedimentary rock layers preserved in the continental geologic record.

The difference between the flume experiments and the observed local-regional natural catastrophes on the one hand, and between the observed local-regional natural catastrophes and the global Flood cataclysm on the other, is in both instances the scale of the sedimentation. However, it is a progressive increase in scale from the flume experiments to the observed local-regional natural catastrophes, and then to the scale of the global cataclysmic Flood. Nevertheless, it has been demonstrated that both the flume experiments and the local-regional natural catastrophes produce neat, uniform sediment layers by deposition from the laminar (sheet) flow of fast-moving waters, rather than from turbulent (churning) flow. Thus, because the continental-scale sedimentary rock layers deposited during the Flood cataclysm are neatly uniform across the continents, it is evident that even under global cataclysmic Flood conditions it was the laminar flow of fast-moving waters, and not turbulent or churning

19. P.V.H. Kuenen, "Experimental Turbidite Lamination in a Circular Flume," *Journal of Geology* 74, no. 5 (1966): p. 523–545.
20. Pierre Y. Julien, Yongqiang Lan, and Guy Berthault, "Experiments on Stratification of Heterogeneous Sand Mixtures," *Creation Ex Nihilo Technical Journal* 8, no. 1 (1994): p. 37–50.

waters, that were responsible for the deposition of these neat, uniform sedimentary rock layers.

Conclusion

In answer to the question that was posed, namely, how could neat uniform sedimentary rock layers be deposited during the Flood cataclysm with all the fast-moving waters, we have seen that the observed sedimentation processes in both flume experiments and larger scale (local-regional) natural catastrophes result in neat, uniform sediment layers being deposited from fast laminar (sheet)-flowing waters. Thus it has been argued that the observed neat, uniform sedimentary rock layers found deposited across the continents as a result of the global Flood cataclysm can be envisaged to have also been the result of the same sedimentation processes from similarly fast laminar-flowing waters. In other words, we can confidently extrapolate the orders of magnitude to the enormous scale of the global Flood cataclysm. Though the flume experiments have been conducted at various small scales, the orders of magnitude extrapolation to the observed natural catastrophes over large regions still results in the same observed pattern of uniform sediment layers deposited neatly in succession by fast-moving waters. This makes us confident that at the global scale of the Flood cataclysm the same sedimentation processes would have also been responsible for the neat, uniform sedimentary rock layers we observe to have been stacked on top of one another and preserved in the continental geologic record, even though the Flood waters were often fast-moving.

CHAPTER 16

Should We Be Concerned about Climate Change?

DR. ALAN WHITE

There is good evidence that global temperatures have been slowly climbing for the past four centuries and were slowly declining for many centuries prior to that. But are these temperature changes a serious threat to our way of life or are they just a part of normal variation to which we can readily adjust? Sadly, our lives are going to be affected whether global warming is a real threat or not. Global warming has been blamed for almost every ill in our society.[1] In his State of the Union speech in 2013, President Obama said this:

> It's true that no single event makes a trend. But the fact is, the 12 hottest years on record have all come in the last 15. Heat waves, droughts, wildfires, floods — all are now more frequent and more intense. We can choose to believe that Superstorm Sandy, and the most severe drought in decades, and the worst wildfires some states have ever seen were all just a freak coincidence. Or we can choose to believe in the overwhelming judgment of science — and act before it's too late.[2]

Within this short quote, many of the common issues related to climate change are raised — recent events that are not necessarily indicative of a long-term trend,

1. For example, see Michael E. Mann and Lee R. Kump, *Dire Predictions Understanding Global Warming* (New York: DK Publishing, 2008), p. 108–139.
2. "Transcript of Obama's State of the Union Address," ABC News, http://abcnews.go.com/Politics/OTUS/transcript-president-barack-obamas-2013-state-union-address/story?id=18480069#.

a claim that the "science" is settled, and a warning that we must act right now. The president followed these words by vowing that, if legislation were not forthcoming, he would do all he could by executive order.

These new policies will almost certainly raise the cost of energy. Higher energy costs will lower the standard of living for all, particularly the poorest among us. Is a disastrous change in the climate looming? Is man responsible? Let's begin our journey to answer those two questions by defining our terms.

What Is Climate Change?

The *Oxford English Dictionary* defines climate change as a change in global or regional climate patterns, in particular a change apparent from the mid to late 20th century onward and attributed largely to the increased levels of atmospheric carbon dioxide produced by the use of fossil fuels. [3] Other dictionary definitions are much more succinct and do not specify cause, direction, or time frame. It is not surprising that there is some disparity in the definitions. With controversial subjects, people often disagree on exactly what the words mean. For the purpose of this chapter, the phrase "climate change" will be used to mean long-term changes in climate (mainly temperature) without implying any cause for, or direction in, the change.

Do *Climate Change* and *Global Warming* Mean the Same Thing?

Some use these phrases interchangeably, and others do not. Those who see the global temperature as going only in one direction often use them interchangeably. However, the phrase "global warming" was much more popular before 2006 and 2007 when the average global temperature declined significantly. "Climate change" is much more commonly used today and seems much less prejudicial. Therefore, "climate change" will be used herein.

How Could There Be So Much Disagreement over a *Scientific* Issue?

When there is a lack of good data and when people view the data from two very different perspectives, it is easy to have disagreement.

A Lack of Good Data

Measuring the average temperature of the earth is very difficult. At any point in time, different parts of the earth are experiencing different conditions; for example, day and night, summer and winter, cloudy and clear, arid and humid, and windy and calm. This level of variability requires frequent measurements

3. *Oxford English Dictionary*, s.v. "climate change."

to be made in many places over many years in order to calculate an average global temperature. Temperature measurements have been made at land-based weather stations since 1880. Two main factors have made those measurements less accurate than they need to be — drastic changes in the immediate area around some of these weather stations and poor distribution of weather stations around the earth. These facts led scientists to push for temperature measurements from satellites.

Satellites are able to provide much-improved data over land-based systems. But even the satellite measurements, which began in 1979, are not without their issues. In 2002, the satellite orbits were adjusted so the measurements could be made at a consistent place and time of day.[4] Clearly, only a few years of useful measurements are not enough to give us a good understanding of climate change. That's not even enough time for us to be sure that these new satellite measurements are sufficiently accurate. Lord Kelvin said, "to measure is to know." We will never have a clear understanding of climate change until we are able to accurately measure the earth's temperature for decades, if not centuries.

The lack of accurate measurements has not stopped scientists from interpreting the data they do have. No problem. That is how science works. Scientists do their best to gather accurate data and propose theories based on those measurements. They test those theories by doing further experiments to see if the new measurements are consistent with the latest theory. In the process of using this scientific method, scientists learn how to do better experiments, make more accurate measurements, and propose better theories. The problem here is that we are in a very early stage in the process of understanding climate change. In early stages, researchers have a strong tendency to develop theories based on their own worldview and to run experiments designed to prove their theory rather than test it. The current bias toward global warming will likely lengthen the time required to construct more accurate climate models.

Two Different Views of the World

To those who believe that the universe is the result of the supposed big bang, where invisible particles somehow came into being and randomly organized themselves into atoms, molecules, stars, and planets, there would be no reason to expect that the earth's temperature would be controlled within a specific range. That life exists at all should be considered exceedingly unlikely from this perspective. Stephen J. Gould, an evolutionist, put it this way, "We are here

4. Roy W. Spencer, *The Great Global Warming Blunder* (New York: Encounter Books, 2010), p. 13.

because one odd group of fishes had a peculiar fin anatomy that could transform into legs for terrestrial creatures; because the earth never froze entirely during an ice age; because a small and tenuous species, arising in Africa a quarter of a million years ago, has managed, so far, to survive by hook and by crook. We may yearn for a 'higher' answer — but none exists."[5]

To those who believe that the heavens and the earth were designed and created by a "higher" power, there is ample reason to expect that earth's temperature will remain in a range to support life. In fact, God gives us that promise in Genesis 8:22:

> While the earth remains,
> Seedtime and harvest,
> Cold and heat,
> Winter and summer,
> And day and night
> Shall not cease.

Within this worldview it makes perfect sense that the earth would have a temperature control system just like our bodies do, since God designed them both.

Has the Media Accurately Reported on Climate Change?

"When a dog bites a man that is not news, but when a man bites a dog that is news."[6] Likewise, a stable climate is not news, but a dramatically changing one is.

In the late 1970s, numerous popular media outlets were reporting dire warnings about impending climate change. An April 28, 1975 article in *Newsweek* began with this phrase, "There are *ominous* signs that the earth's weather patterns have begun to change *dramatically* and that these changes may portend a *drastic* decline in food production," and ended, "The longer the planners delay, the more difficult will they find it to cope with the climatic change once the results become a *grim reality*" (emphases mine).[7] Sounds familiar, doesn't it? We hear similar pronouncements today. For example, then-Senator Barack Obama said in 2006, "Not only is it [global climate change] real, it's here, and

5. Stephen Jay Gould, quoted in James A. Haught, *2000 Years of Disbelief, Famous People with the Courage to Doubt* (New York: Prometheus Books, 1996), p. 290; or the original reference is S.J. Gould in "The Meaning of Life," *Life Magazine* (Dec. 1988), p. 84.
6. *Bartlett's Familiar Quotations*, 16th ed., ed. Justin Kaplan (Boston, London, and Toronto: Little, Brown, 1992), p. 554.
7. Peter Gwynne, "The Cooling World," *Newsweek*, April 28, 1975; available online at http://denisdutton.com/newsweek_coolingworld.pdf.

its effects are giving rise to a *frighteningly* new global phenomenon: the man-made natural *disaster*" (emphases mine).[8]

The surprising thing is that the *Newsweek* article in the 1970s was referring to global **cooling,** and then-Senator Obama was referring to global **warming**. Yes, that's right. The panic in the 70s was that the earth's temperature was **declining** and would continue to **decline**. Today, the concern is the earth's temperature is **rising** and that it will continue to **rise**.

How Could Predictions about the Direction of Climate Change Be So Different after Only 30 Years?

If, in the 1970s, you considered the data from only the previous 30 years, it would have been possible to conclude that the short-term trend is cooling, particularly if you extrapolate well into the future expecting that trend to continue (figure 1). Interpolation of data, trying to estimate a value within a range you have studied, is challenging enough. But extrapolation of scientific data into a region that you know nothing about is not wise.

If today you again take the perspective of the last 30 years and extrapolate far into the future, it is possible to conclude that the short-term trend is warming (figure 1).[9] Actually, over the last century, it appears that the temperature rose from 1900 to 1940, declined slightly from 1940 to 1970, and increased from 1980 to around 2000. It is easy to make headlines by drawing sweeping conclusions from small ranges of data; however, it is still unclear whether these short-term trends add up to an unprecedented rise in global temperature. Some climatologists claim that the science was not settled in the 1970s and that they were not in agreement with the popular press at that time.[10] Today, those climatologists are convinced that the latest data, now that it has been corrected, is reliable, and the earth is warming.[11]

Very recently, a few people have begun to conclude that we may actually be in the early stages of another cooling trend.[12] Those who suspect this generally

8. Barack Obama, "The Coming Storm: Energy Independence and the Safety of Our Planet" (campaign speech, Chicago, IL, April 3, 2006).
9. Data from the Goddard Institute for Space Studies, National Aeronautics and Space Administration. http://data.giss.nasa.gov/gistemp/graphs_v3/Fig.A2.txt. These data are updated from the data in J. Hansen, Mki. Sato, R. Ruedy, K. Lo, D.W. Lea, and M. Medina-Elizade, "Global temperature change," *Proc. Natl. Acad. Sci.*, 103 (2006) 14288-14293, doi:10.1073/pnas.0606291103.
10. Mann and Kump, *Dire Predictions Understanding Global Warming*, p. 45.
11. Ibid., p. 38–39.
12. For example, see http://www.icr.org/article/new-evidence-for-global-cooling/ and http://www.icr.org/article/will-solar-inactivity-lead-global-cooling.

Figure 1

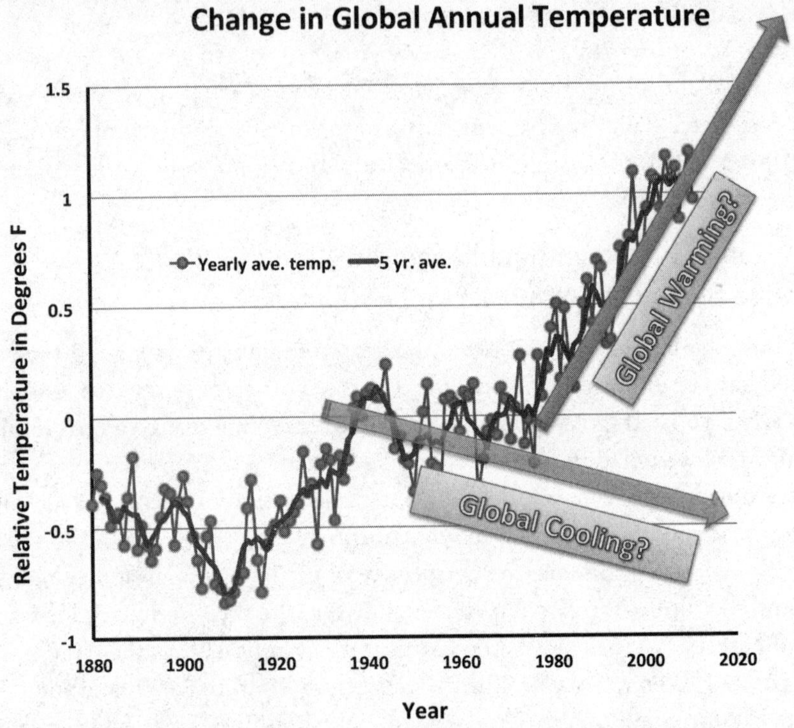

fall in one of two camps. Some are looking at a specific, narrow range of time (1998 to 2012) where there has certainly been no increase in global temperature. Others are focused on solar activity. They are convinced that the sun is the major factor in determining global temperature. This, of course, is a very reasonable conclusion since almost all our energy comes from the sun. In fact, the number of observed sunspots in this latest sunspot cycle is expected to be the lowest in many decades, and the earth did experience the Little Ice Age at a point in time when sunspot activity was very low.[13] Has the global temperature started to decline after having increased for about 400 years? Only time will tell. Frankly, with our limited understanding of the major factors that affect global temperature, no one should be confident in predicting the future global temperature.

What Are the Politics of Climate Change?

At present a number of expert climatologists and the IPCC (Intergovernmental Panel on Climate Change) appear to be in agreement that the earth's

13. http://solarscience.msfc.nasa.gov/SunspotCycle.shtml

temperature is rising and will continue to rise. However, it is hard to know what the scientific judgment of these individuals would be in the absence of overwhelming political pressure. Their funding and their livelihoods are clearly affected by their stance on this issue.

We scientists want to believe that we are unbiased — that we are strictly interpreting the data and are not swayed by other factors. Are scientists different from all other human beings in this regard? Obviously not. We are swayed by our emotions and our beliefs, just like everyone else. So beware when scientists become emotionally attached to their theories, ignore the uncertainties in their data, or claim that "all reputable scientists agree" or that "the science is settled."[14] When one or more of these is true, you can be sure that the issue being discussed is not purely scientific. When "the science" really is settled, the evidence will be overwhelming, and there will be no need to *claim* that the science is settled.

While investigating any subject, it is interesting to follow the money. There is big money in climate change issues. The person that is the most closely associated with "global warming" is Al Gore. "Critics, mostly on the political right and among global warming skeptics, say Mr. Gore is poised to become the world's first 'carbon billionaire,' profiteering from government policies he supports that would direct billions of dollars to the business ventures he has invested in."[15] "Mr. Gore says that he is simply putting his money where his mouth is."[16] Gore's many multi-million dollar investments in green energy projects and his purchase of a $9M ocean-view home in California are clear evidence of his financial success in this arena. He will certainly have a good vantage point from which to watch a possible rise in sea level!

Is the Truth about Climate Change Really Inconvenient?

It is tempting for each of us to focus only on what has happened in our lifetime. However, for questions related to climate, we need a much longer-term perspective. Have the global temperatures in the last few decades been significantly higher than in the distant past? Unfortunately, there is no way to know for sure. No temperature measurements are available before 1880. Scientists have tried to correlate other scientific data with global temperature, but estimating temperatures in this way is fraught with difficulties. Correlation of ice

14. For a similar discussion, see Roy W. Spencer, *Climate Confusion* (New York: Encounter Books, 2008), ch. 2.
15. John W. Broder, "Gore's Dual Role: Advocate and Investor," *The New York Times*, http://www.nytimes.com/2009/11/03/business/energy-environment/03gore.html?_r=0.
16. Ibid.

Figure 2

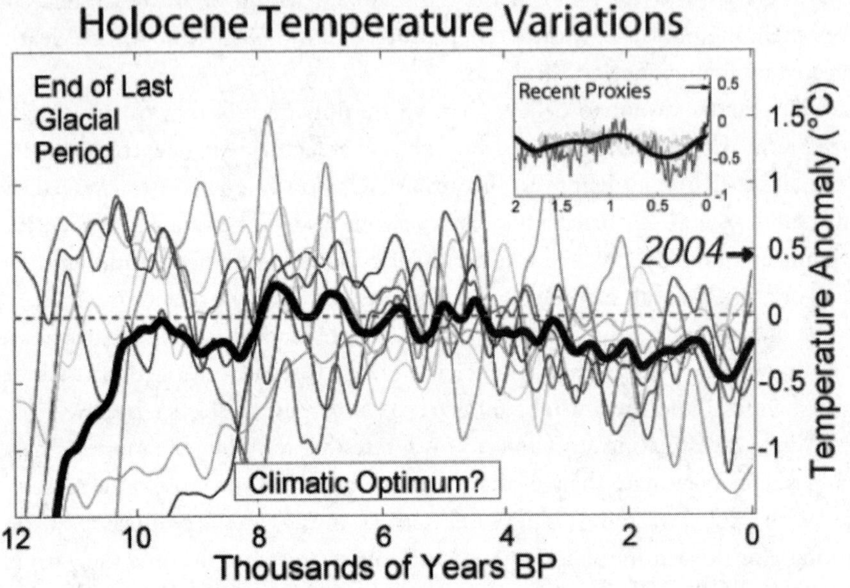

core or tree ring data to global temperatures is full of assumptions that cannot be verified. Figure 2 shows eight different attempts that were made to predict global temperature.[17] The dark line is the average of these data for what they presume to be the last 12,000 years of earth history. Confused as to why anyone would be convinced by these data? You should be. The most recent reconstructions are shown in the insert of figure 2 for the last 2,000 years. These data have led many climatologists to conclude that the climate is much warmer now than in the last 2,000 years.

Historical evidence provides a different perspective on global temperatures during the last two millennia. There is good evidence that the climate in the Northern Hemisphere was warmer about a thousand years ago — the Vikings were able to farm in Greenland. After a few hundred years, they stopped farming due to a cooler climate. The temperature continued to decline for a few hundred more years, and the Thames in London began to regularly freeze.[18] The decline in temperature reversed course in about A.D. 1700. If this warming trend continues, it may again be possible to farm in Greenland, and the sea ice in the north

17. The original literature references for all these data can be found at Wikimedia Commons, "File: Holocene Temperature Variations.png," http://commons.wikimedia.org/wiki/File:Holocene_Temperature_Variations.png (GNU free documentation license).
18. Spencer, *The Great Global Warming Blunder*, p. 2 and references.

Figure 3

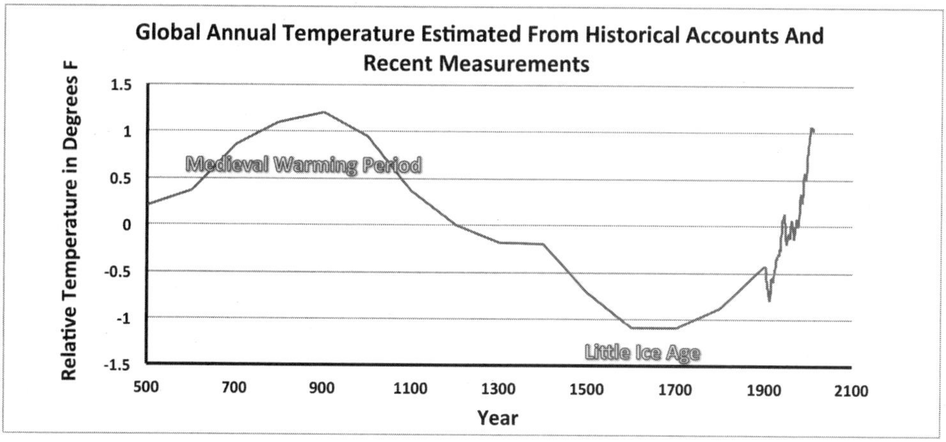

Atlantic may again be scarce. Figure 3 is an estimation of the relative global temperature from historical observations before 1900 and from weather station data after 1990. While we cannot be certain about what was true in ancient times from either historical or scientific data, the historical observations seem more reliable in this instance. From these limited data, it appears that the global temperature cycles around a mean temperature and has been slightly warmer in recorded history than it is today. There is no reason to panic.

Are We the Cause of the Rise in Temperature Since the Little Ice Age?

Many believe that this recent rise in temperature is caused by an increase in carbon dioxide due to our burning of more fossil fuels. Let's look at some facts about carbon dioxide and examine the evidence of its effect on global temperature.

The presence of carbon dioxide in the air is essential to life on earth. Without carbon dioxide, there would be no plant life, and without plant life there would be no animal life. Despite this, Lisa Jackson of the Environmental Protection Agency declared that carbon dioxide was a *pollutant* under the Clean Air Act and deemed that it was a *hazard* to human health.[19] So is CO_2 essential to life or a pollutant? The government apparently thinks that it is both — essential at low levels and harmful at high levels. But is there a level at which CO_2 is too high? As with most government regulations, this regulation preceded our

19. John Broder, "EPA Clears Way for Greenhouse Gas Rules," *The New York Times*, http://www.nytimes.com/2009/04/18/science/earth/18endanger.html.

understanding of the science. While CO_2 does influence the global temperature, the exact relationship has not been established nor has the maximum CO_2 concentration in air.

We do know that carbon dioxide is a greenhouse gas. Greenhouse gases act as a blanket over the earth. When sunlight heats the earth's surface, the warm earth radiates some of that heat into the atmosphere. Greenhouse gases slow the escape of that radiated heat. You have been led to believe that the most important greenhouse gas is carbon dioxide. It is not. Water vapor and clouds are actually responsible for about 80 to 90 percent of the total greenhouse effect. That's right, *at least 80 percent.* That is why clear mornings are usually much colder than cloudy mornings. On clear mornings, we do not have that blanket of clouds to hold in the heat. The percentage of the greenhouse effect attributable to CO_2 is believed to be as high as 20 percent by some and as low as 4 percent by others.[20] Almost everyone agrees that the percent of CO_2 that is man-made is only about 4 percent of total CO_2. Therefore, the greenhouse effect caused by man-made CO_2 is *less than* 1 percent of the total and may be a small fraction of 1 percent.

Despite this, many scientists today claim that the rise in man-made CO_2 is the major cause of the rise in global temperatures over the past century. Just because global temperature and CO_2 concentrations have risen over the past several decades does not mean that one caused the other. Figure 4 shows that the correlation between the CO_2 concentration and global temperature is not strong, particularly between 1900 and 1950. The temperature profile in figure 3 also does not match well with man-made CO_2 levels because man-made CO_2 was not high during the Medieval Warming Period. These data are not convincing.

Is the Global Temperature Nearly Out of Control?

Climatologists' greatest concern is that a temperature increase during the last few decades might be amplified by positive feedback causing the global temperature to spiral out of control. They are worried, for example, that a higher temperature on the earth could melt more of the permafrost, release more CO_2, and cause a greater greenhouse effect. On the other hand, a higher temperature on earth could cause more evaporation, more cloud formation, and more sunlight to be reflected away from the earth. This negative feedback could moderate the global temperature. Which type of feedback is more influential? Scientists

20. Spencer, *The Great Global Warming Blunder*, p. 44; G.A. Schmidt, R.A. Ruedy, Ron L. Miller, and A.A. Lacis, "Attribution of the Present-day Total Greenhouse Effect," *Journal of Geophysical Research* 115 (2010): D20106.

Figure 4

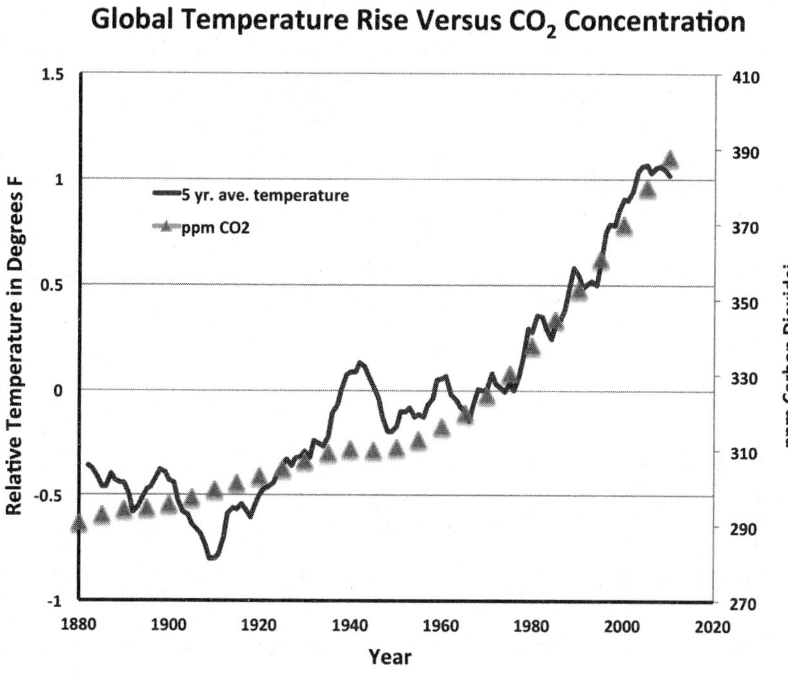

are currently not able to quantify them well enough to know whether the negative feedback outweighs the positive.

Engineers familiar with control systems are well aware that control systems dependent on positive feedback easily go out of control whereas those based on negative feedback generally do not. Since the earth's temperature has been relatively stable for many centuries, it seems more likely that the earth's climate is moderated by more powerful negative feedback systems.

It appears that a brilliant designer has designed a molecule that is both essential to human life and essential for controlling the climate of the earth. Water is a polar molecule that is able to dissolve salts, proteins, and DNA that are essential for our cells to function and for life to exist. Water's other physical properties are just as critical to controlling the earth's climate. It takes more heat to change water from a solid to a liquid or from a liquid to a gas than any other common molecule. The 310,000,000 cubic *miles* of water on the earth's surface are able to hold a tremendous amount of heat and provide great temperature stability to the earth. Water can readily transfer heat from the earth's surface to the air by evaporation and condensation, which is the basis of the hydrological cycle and much of our weather. Cloud formation may also be the key to a negative feedback system that helps moderate temperature changes in the earth's atmosphere. Without water, the range of temperature from day to night and from the earth's surface to the upper atmosphere would be much greater. Clearly water is critical to human life in many, many ways.

How Should We Then Live?

In the first chapter of the first book of the Bible, God commands us to subdue the earth (see Genesis 1:28). Most interpret this to mean that we should take care of the earth and be good stewards of its natural resources. If it were true that the burning of coal, oil, and natural gas did have a significant negative effect on our environment, it would make sense for us to modify our behavior. But it appears that we are just in the upper range of a natural temperature cycle. It is not at all clear that the small amount of additional CO_2 produced by the burning of fossil fuels is detrimental to the environment. It is humbling to remember that when God was judging the earth with a global flood that He was creating inexpensive fuel sources for future generations. Let's obey God's command and use our scientific knowledge to be good stewards of our natural resources and preserve our environment for the next generation until He comes again. [21]

21. For further information on this issue, see Michael Oard, "Is Man the Cause of Global Warming?" in *The New Answers Book 3*, Ken Ham, gen. ed. (Green Forest, AR: Master Books), p. 69–79.

CHAPTER 17

What about Creation, Flood, and Language Division Legends?

TROY LACEY (WITH BODIE HODGE)

Introduction

Nearly every culture around the world has a creation legend and just as many have worldwide flood legends, and, believe it or not, there are even many language division legends around the world in different and diverse cultures.

In today's highly secularized culture, there are attacks on the Bible using these legends. Those who do not trust what the Bible plainly says often speculate that the Bible's discussion about creation, the Flood, and the Tower of Babel are just more legends and determine that the Bible cannot be trusted. What these attackers fail to realize is that these legends are a great confirmation of the Bible and that the Bible retains the true account recorded by God in His Word. From an historical perspective, this makes perfect sense and is consistent with a biblical worldview, but it is hard to explain all these legends from a secular evolutionary worldview — people supposedly evolved and slowly filled the earth with a gradual changing of languages and no global Flood. Why then should we find so many common threads in so many accounts from all over the world? The evolutionary explanation fails to provide a reason for commonality, whereas the biblical one does.

The Nature of "Legends"

Of course, many of these legends have been distorted and have become highly mythologized and embellished over time; and this is to be expected as people dispersed from Babel and the knowledge of God and mankind's early history was forgotten or turned into folklore. Many have common themes, involving mankind being created from clay; a remnant understanding of God (i.e., a "god") as angry with mankind for some reason; large boats (or rafts) being constructed to survive a coming flood, often foretold to the hero by this "god"; animals being collected by the hero in order to survive the coming deluge, and so on. Many of these legends sometimes still bear striking resemblances in many particulars to biblical accounts.

Many, though, are drastically different and show corruption from an original account, as one would expect from an orally passed-down story. Others show details that seem to be in direct contrast to the biblical accounts of creation and the Flood. We see numerous examples of "gods" being killed to create the physical earth and/or heavens; mankind given power by the "gods" to create the animals; mankind re-creating the earth after the Flood; animals that rescue people from the Flood, and so on. Some of these may be the result of distortion over time, while others may be a deliberate attempt by post-Babel peoples to reshape the world and the "gods" in their own image. Romans chapter 1 clearly shows that human hearts and minds willfully suppress the true God and make up one in their own image, or in the image of animals (see also Genesis 8:21; Exodus 32:4–8; 1 Kings 12:28–33; and Jeremiah 17:5–9).

Legends from Genesis 1–11 That Confirm the Bible

There is no way to exhaustively cover this topic in such a short article, as there are literally hundreds of books detailing these creation, Flood, and language division legends. Rather, ten of each will be discussed in the following tables.

Creation Legends

	Who	Where	Quote
1	Mosotene	Bolivia, South America	"Dobitt created the world. He made it in the shape of a great raft which floats in space supported by innumerable spirits. Then Dobitt created mankind to live in the world. He made images out of clay and gave them life, and then went off to live in the sky. Later Dobitt returned and made animals and birds. He

			carried a big basket full of water and spilled it out here and there over the earth to make the rivers." Authority: Leach[a]
2	Lenape or Delaware Native Americans	United States, North America	"In the beginning, forever, lost in space was the Great Manito. He made the earth and the sky. He made the sun, moon, and stars. He made everything move in harmony. Then the wind blew violently, it became lighter and water flowed strongly and from afar, and groups of islands emerged and remained. Once again the Great Manito spoke, one Manito to other Manitos, to mortal creatures, spirits and all. All creatures were friendly with one another at that time." Authority: Maclagan[b]
3	Zuni Native Americans	United States, North America	"The creator Awonawilona conceived in himself the thought and the thought took shape and got out into space, and through this it stepped out into the void, into outer space and from this came nebulae of growth and mist, full of power and growth." Awonawilona then made "mother-earth" and "father-sky." Authority: Maclagan
4	Ona	Argentina, South America	"Temaukl always existed. He created the earth and sky, and there was no time when Temaukl was not. Kenos was the first man, sent into the world by Temaukl to put things in order. So Kenos created the plants and animals and gave the Ona their land." Authority: Leach
5	Ekoi	Nigeria, Africa	"One day in the beginning of the world, Obassi Osaw made a man and a woman and brought them down to live upon the earth. He placed them here in the green world and then went back into the sky. He returned to see how they were getting along. 'What have you eaten? What have you had to drink?' Obassi asked them. 'Nothing' they replied. Then Obassi dug a ditch, drew forth a jar full of water and poured the water into the ditch. This was the first river. The next thing he did was to plant a palm kernel which he carried in his hand. 'Drink the water. Take care of the Palm tree.' So the man and woman watched the palm tree grow and tended it with care and love. After a while great clusters of yellow fruit ripened. When Obassi saw this, 'This is your food' he said to the man and woman." Authority: Leach

6	Lozi	Zambia, Africa	"In the beginning, Nyambi lived on earth with his wife Nasi-lele. As god, he made the birds and all of the animals and fishes. One thing Nyambi made was different, and it was a man. The first man's name was Kamonu. One day Kamonu fixed a spear for himself. He killed an antelope and did not stop there. He killed again and again. 'Man!' Nyambi shouted 'What you do is wrong. You are killing your brothers' So Nyambi gave him a place to plant and grow things to calm man." Authority: Hamilton[c]
7	Norse	Scandinavia and Northern Europe	"In Norse mythology, a foggy void between the lands of fire (Muspell) and ice (Niflheim) produced a primeval cow Audumbla, and the Frost Giant Ymir. The cow licked at ice and eventually uncovered the 'god' Buri. Ymir produced frost giants as he slept, and Buri married one of Ymirs daughters. Ymir was later killed by Odin, a grandson of Buri. Ymir's flesh became the earth, his bones the mountains, his teeth became rocks and his blood became rivers, lakes, and seas. Mankind was created later by three gods; Odin gave them life, Vili gave them intelligence, and Ve gave them the five senses." Authority: Cotterel and Storm[d]
8	Babylon	Middle East	"Apsu (primeval water) and Tiamat (chaos and salt-water) created the great gods, who begat other gods. These gods danced and made noise, so that Apsu wanted to destroy them so that he and Tiamat could rest. One of the gods Ea cast a spell on Apsu which caused him to sleep, and then Ea killed him. Tiamat produced monsters so that they could avenge Apsu on the gods. Ea's son Marduk promised to kill Tiamat if he was given supreme power by the other gods. They agreed and Marduk trapped Tiamat in a net and killed her with an arrow and a whirlwind. He cut her body in half and with the two halves, made the sky and the earth. Later he made man out of the blood of Tiamat's second husband Kingu." Authority: Hamilton

9	Tahitians	Polynesia	"Ta-aroa lived alone in a shell shaped like an egg. The egg revolved in dark empty space for ages. Then came a new time when Ta-aroa broke out of the egg. Being so by himself, he made the god Tu. Tu became Ta-aroa's great helper in the wonderful work of creation. Ta-aroa and Tu made gods to fill every place. They made the universe and they brought forth land and creatures. Last, they created man to live on the earth." Authority: Hamilton
10	Greece	Southern Europe	"In the days of old there was only chaos. Out of chaos arose the earth Gaia, and out of earth rose Uranus the sky god with whom she mated to produce the Titans who ultimately deposed Uranus." The Titans were later led by Cronos who was overthrown by Zeus and the rest of the classical Greek pantheon of gods. Much later, a Titan named Prometheus created man out of clay and water; while even later, Zeus ordered Hephaistos to create woman, more or less as a punishment for mankind for having been given fire by Prometheus. Authority: Leach; Cotterell and Storm

a. Maria Leach, *The Beginning: Creation Myths Around the World* (New York: Funk and Wagnalls, 1956), p. 127–139, 234–235.
b. David Maclagan, Creation Myths: Man's Introduction to the World (London: Thames and Hudson, 1977), p. 15, 78–79.
c. Virginia Hamilton, *In the Beginning: Creation Stories from Around the World* (New York: Harcourt, Inc., 1988), p. 65–67, 83–85, 101–103.
d. Arthur Cotterell and Rachel Storm, *The Ultimate Encyclopedia of Mythology* (Lanham, MD: Anness Publishing, 1999).

Flood Legends

1	Aztec	Central America	"Humanity was wiped out by a flood, but one man Coxcoxtli and one woman Xochiquetzal escaped in a boat, and reached a mountain called Colhuacan." Authority: Sheppard[a]
2	Hindu (Sanskrit)	India	"The God Brahma, in the form of a fish told Manu who had cared for him many years: 'the dissolution of all moving and unmoving things of earth is near. This deluge of the worlds is approaching . . . all ends in violent water. A boat is to be built by you, furnished with a sturdy cord. There with the seven Rsis, sit Great Manu and take with you all

			the seeds, preserving them in portions.' " Manu did as instructed and waited in the boat after it was built, and he and the Rsis were preserved in the boat by Brahma while the entire earth was flooded. The boat landed on a peak of the Himalayas. Authority: Martin[b]
3	Karina	Venezuela South America	"The sky-god Kaputano came down to the kingdom of the Karina. 'Children, hear me well. Soon a great rain will fall on the earth and will cover all with water.' Only four couples listened, as the rest scoffed, declaring that there wouldn't be any flood. Kaputano and the eight people began building a very large canoe, and when they were done they went around gathering two of every animal to put on board. They also brought seeds from every plant on earth. Once they were done they got on board and it began to rain, and it rained for many, many days. Soon the entire earth was flooded. Eventually the rain stopped and the water began to recede and the land began to dry. The four couples exited their canoe but the world was destroyed. Kaputano asked how the Karina would like the world remade, and they asked for rivers and hills and trees, which Kaputano made for them." Authority: Martin
4	Babylon	Middle East	Utnapishtim related to Gilgamesh how the god Ea told him to build a boat to escape a worldwide flood the other gods were sending to wipe out mankind. It was to be a 30x30 cubit boat in the shape of a cube. He was instructed by Ea to also bring two of every animal, and water and provisions. He obeyed and after loading the boat, with his cargo, his wife and a captain to pilot the boat, the rains came and lasted for 7 days. All, the earth was flooded and destroyed, but 12 days later dry land began to appear. Utnapishtim waited 7 more days then sent out a dove, then a swallow which both returned, then a raven which did not. After this, Utnapishtim unloaded all the animals from the Ark. He offered a sacrifice to the gods and he and his wife were granted immortality. Authority: Martin

5	Bahnars	China	"Once a crab and a kite had an argument. The kite pecked the crab so hard that he pierced the crab's shell. To avenge this great insult, the crab caused the waters of the sea to swell. They swelled so much that everything on earth was destroyed, except for a brother and a sister, who survived by locking themselves in a huge chest. Because they were afraid that everything would perish forever, they brought on board two of every animal. After 7 days they heard a rooster crowing outside the chest (which the ancestors had sent) and knew it was safe to come out." Authority: Martin
6	Greece	Southern Europe	"One Day Prometheus came to Deucalion and told him 'Zeus was going to destroy all the men of this bronze age. Build yourself a chest of wood so that you and your wife may survive.' Deucalion did just that and after he had provisioned it, took his wife aboard with him. Zeus caused a great flood which destroyed everything." Authority: Martin
7	Hareskin Tribe	NWT, Western Canada	"Kunyan (Wise Man) resolved to build a great raft. When his wife asked him why he would build it, he said 'If there comes a flood, as I foresee, we shall take refuge on the raft.' He told his plan to other men on the earth and they laughed at him saying, 'If there is a flood we shall take refuge in the trees.' But Kunyan made a great raft, joining the logs together by ropes made of roots. Suddenly there came a flood such that the like of it had never been seen before. Men climbed up into trees, but the water rose after them and all were drowned. But Kunyan, his wife and his son floated safely on his strong raft. As he floated he thought of the future, and he gathered two of all the animals he met with on his passage. 'Come up on my raft, for soon there will be no more earth' he said. Indeed the earth disappeared under the water. The Wise Man told a beaver to dive down into the waters and see what he could find. The beaver returned with a piece of mud. Kunyan took the mud into his hand and breathed on it and it began to grow. So he laid it on the water, kept it from sinking and watched as it continued to grow.

			Later, after a long time, it grew to the size of the land as it was before the flood. Then Kunyan, his wife and son, and all the animals came off the raft and repopulated the world." Authority: Frazier[c]
8	Rotti	West Timor, Indonesia	"Once, the sea-god became angry with mankind and decided to flood the whole earth. In fact the entire earth was destroyed except for the peak of one mountain. A man and his sister along with several animals had escaped to the high mountain and there survived. However, there was nowhere to go, so they asked the sea-god to bring the waters back down. He refused unless they could find a creature whose hairs he could not number. After throwing into the waters a pig, goat, dog, and hen, all of whose hairs the sea-god could number, they finally threw in a cat and the sea-god gave up and agreed that the waters could recede. He then caused an osprey to fly over the mountain, sprinkling dirt on the water. The dirt became dry land and the man, his sister, and the animals were able to descend the mountain." Authority: Martin
9	Montagnais Tribe	Quebec and Labrador, East Canada	"A race of giants was destroying the earth, and God, angry with them for it, commanded a man to build a very large canoe. The man did as he was told and as soon as he entered in the water rose on all sides until no land could be seen in any direction. Bored with the scenery, the man told an otter to dive down into the waters and see what he could find. The otter returned with a piece of earth. The man took the earth into his hand and breathed on it and it began to grow. So he laid it on the water, kept it from sinking and watched as it continued to grow. As it grew the man saw that it was becoming an island. The man decided that the earth was not yet large enough so he continued to blow on it. In time all of the lakes, mountains, and rivers were formed, and the man knew it was time to leave the canoe." Authority: Martin

10	Lake Tyers Aborigines	Victoria, Australia	"Once upon a time, a huge frog swallowed all the water of the world and everyone was thirsty. Because of this all the animals took a poll and decided that the best way to make the frog give the water back was to make him laugh. So they all took turns playing pranks and cutting up in front of him. They were so hilarious that everyone else would have died laughing, but the frog did not even smile. Finally, as a last resort the eel wriggled about dancing and swaying as it stood on its tail. Not even the glum frog could watch this without laughing. He laughed and laughed until tears ran down his cheeks. The water poured from his mouth and soon became a flood. The waters rose killing many people. In fact all of mankind would have drowned, if the pelican had not paddled about in a canoe, rescuing survivors as he went." Authority: Martin

a. Pam Sheppard, "Tongue Twisting Tales," Answers in Genesis, http://www.answersingenesis.org/articles/am/v3/n2/tongue-twisting-tales.

b. Charles Martin, *Flood Legends: Global Clues of a Common Event* (Green Forest, AR: Master Books, 2009), p. 126–129, 131–143.

c. James George Frazier, *Folklore in the Old Testament: Studies in Comparative Religion* (Whitefish, MT: Kessinger Publishing, 2010), p. 310–312.

Language Splitting/Tower of Babel Legends

1	Maidu Natives	Western North America	"suddenly in the night everybody began to speak in a different tongue except that each husband and wife talked the same language. . . . Then he called each tribe by name, and sent them off in different directions, telling them where they were to dwell." Authority: Sheppard[a]
2	Quiches	Central America	"when the tribes multiplied and left their old home to a place called Tulan. Here the language changed, and the people sought new homes in various parts of the world as a result of not being able to understand each other." Authority: Sheppard
3	Wa-Sania	East Africa	"that of old all the tribes of the earth knew only one language, but that during a severe famine the people went mad and wandered in all directions, jabbering strange words, and so the different languages arose." Authority: Sheppard

4	Mikir	Northeastern India	"Higher and higher rose the building, till at last the gods and demons feared lest these giants should become the masters of heaven, as they already were of earth. So they confounded their speech, and scattered them to the four corners of the world. Hence arose all the various tongues of mankind." Authority: Sheppard
5	Greece	Southern Europe	"for many ages men lived at peace, without cities and without laws, speaking one language, and ruled by Zeus alone. . . . At last Hermes introduced diversities of speech and divided mankind into separate nations." Authority: Sheppard
6	Polynesia	Pacific Island of Hao	"they made an attempt to erect a building by which they could reach the sky, and see the creator god Vatea [Atea]; but the god in anger chased the builders away, broke down the building, and changed their language, so that they spoke diverse tongues." Authority: Sheppard
7	Sumerians	Middle East	"In those days . . . the whole universe, the people in unison. . . . Enki, the Lord of abundance. . . . Changed the speech in their mouths, and [brought?] contention into it, Into the speech of man that [until then] had been one." Authority: Sheppard
8	Gaikho	Southeast Asia	"In the days of Pan-dan-man, the people determined to build a pagoda that should reach up to heaven. . . . When the pagoda was half way up to heaven, God came down and confounded the language of the people, so that they could not understand each other. Then the people scattered, and Than-mau-rai, the father of the Gaikho tribe, came west, with eight chiefs, and settled in the valley of the Sitang." Authority: Sheppard
9	Greece	Southern Europe	"In the days of old the gods had the whole earth distributed among them by allotment. There was no quarrelling; for you cannot rightly suppose that the gods did not know what was proper for each of them to have, or, knowing this, that they would seek to procure for themselves by contention that which more properly belonged to others. They all of them

			by just apportionment obtained what they wanted and people to their own districts. . . . Now different gods had their allotments in different places which they set in order."[b]
10	Inca	Western South America	"In the story of the creator god Virachocha, he created the second race of human beings from clay — the earth. Having painted his creations with distinctive clothes and given them different languages and customs that would distinguish them, he breathed life into them and caused them to descend into the earth and disperse."[c]

a. Pam Sheppard, "Tongue Twisting Tales," Answers in Genesis, http://www.answersingenesis.org/articles/am/v3/n2/tongue-twisting-tales.
b. Plato, *Critias, in Great Books of the Western World*, vol. 7, ed. Robert Maynard Hutchins (Chicago, IL: University of Chicago, 1952), p. 479.
c. David M. Jones, *The Lost History of the Incas* (Leicester: Hermes House, 2007), p. 198.

Conclusion

The Bible records the true account of creation, the Flood, and the Tower of Babel. The more we find legends from cultures around the world that contain elements of these actual events, the more excited the Christian should be to connect these confirmations to the Bible's truth. As people left Babel, they took their history with them. Therefore, we would expect to find cultures with this history of Creation, Catastrophe, and Confusion, and we would expect it to be corrupted, unlike the Bible, whose word will never pass away (Luke 21:33).

Even many atheists have a massive flood legend. The problem with their flood legend is that it is said to have happened on Mars, while insisting that there is not enough water on earth for a global Flood! The primary reason they reject a global flood on earth is because it gives credence to truthfulness of God's Word, which they do not want due to their religious convictions.

CHAPTER 18

How Big Is the Universe?

DR. DANNY R. FAULKNER

Introduction

The universe appears to be very large — billions of light years across. Since this is far larger than a few thousand light years, people frequently ask how we can see objects this far away if the universe is only thousands of years old, as the Bible seems to imply. This is a very good question — so good that we have given this question a name: the light travel time problem. There are a number of proposed solutions to this problem, but I will not discuss those here.[1] Instead, I will address the question of whether the universe really is as big as is often claimed. The short answer is, yes, the universe most certainly is that large. To explain this conclusion, I will describe some of the methods astronomers use to measure distances of astronomical bodies.

Distances

I emphasize that there are three realms of astronomical distances: those within the solar system, those within the galaxy, and those of objects outside of our galaxy. The techniques used in these realms are different, and there is little overlap between the techniques used in those realms. The first distance measurements in astronomy were within the solar system, and they were done by geometric means as planets orbited the sun. This largely was replaced by more accurate radar measurements in the latter half of the 20th century. I will not

1. See Chapter 21 in this volume for more on the distant starlight models.

discuss solar system distances, for the light travel times involved here amount to mere hours at most, and thus are not a problem for recent creation. The sun and all the stars that we can see are members of the Milky Way galaxy, a vast collection of more than 100 billion stars spanning nearly 100,000 light years. The term "stellar distance" normally refers to measuring distances of stars within the galaxy. The first stellar distance measurement was in 1838. There are billions of many other galaxies, each of them being millions or even billions of light years away. We say that the distances of other galaxies are extra-galactic. The first extra-galactic distance measurement was in 1924.

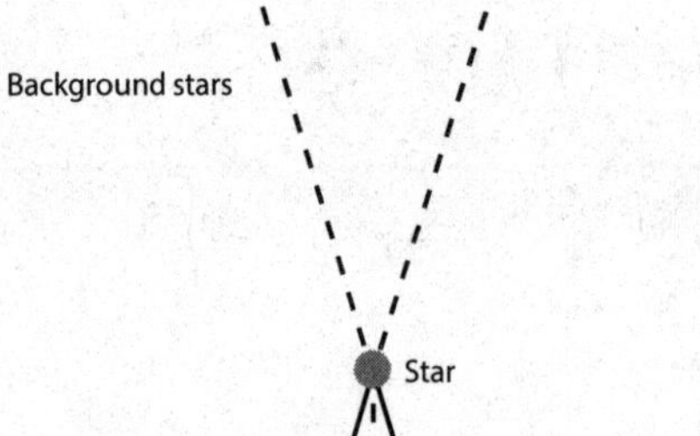

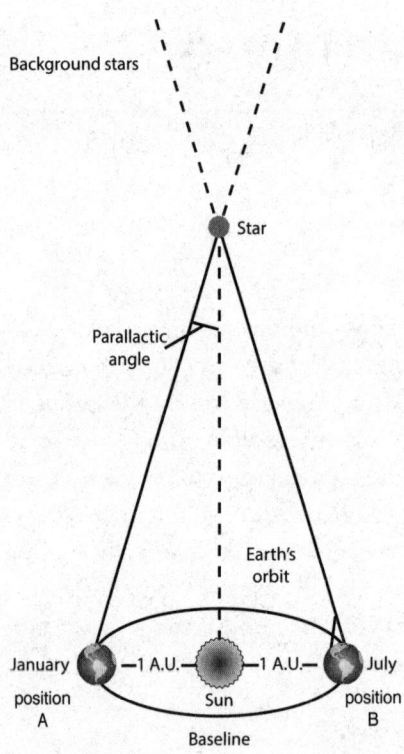

Figure 1. An illustration of parallax

The first stellar distance measure used trigonometric parallax. Trigonometric parallax employs the principle of the apparent shifting position of an object due to our changing location from where we view the object. You can illustrate this by looking at your upheld thumb at arm's length. Close one eye and note the position of your thumb with respect to background objects. Now open that eye and close the other eye. You'll notice that your thumb appears to shift position. We call this apparent shift in position parallax. If you hold your thumb closer to your eyes or if you try this with a more distant object, you will discover that the amount of parallax depends upon the distance of the object — the greater the distance, the less the parallax. The amount of parallax also depends upon the length of the baseline, in this case the distance between your eyes. For a given distance, a greater baseline produces a larger parallax. Surveyors have long used this technique to measure the distances of faraway objects and altitudes of mountains. They set up a transit (in ancient times a dioptra) to view a distant object and measure the angles that the object made at either end of the baseline. Using the baseline and angles, one can use simple trigonometry to measure the distance to the object.

In similar manner, astronomers use the baseline of the earth's orbit around the sun to measure the apparent shift in the positions of nearby stars with respect to more distant stars. An astronomer at location A on one side of the earth's orbit measures the position of a star. Six months later, when the astronomer has arrived at position B on the other side of the earth's orbit, he re-measures the position of the star. The total shift in apparent position is a very small angle, so we normally express it in seconds of arc.[2] Notice that the baseline is the diameter of the earth's orbit, which is twice the distance of the earth from the sun. The average distance of the earth from the sun is a standard unit of distance that astronomers call the astronomical unit (AU for short). Astronomers express the baseline as one AU, so the parallax angle is defined to be one-half the total measured shift. The closest star, Proxima Centauri, has the largest parallax, but its parallax is less than one arc second. And its distance is about 26 trillion miles, so use of normal trigonometric relationships would be quite cumbersome. To avoid this and use a very simple formula, astronomers have defined their own units. If d is the distance of the star and π is the parallax, then the formula is:

$$\pi = 1/d$$

Note that π here is a variable and does not refer to the ratio of the circumference of a circle to its diameter. For this equation to work, astronomers have defined a new unit of distance, the parsec (abbreviated pc). The parsec is the distance that a star would have if its parallax were one second of arc.[3] Since we normally measure the parallax and then compute the distance, we can re-write the equation:

$$d = 1/\pi$$

Friedrich Bessel was the first to measure a star's parallax in 1838 (the star was 61 Cygni). In the early 20th century, astronomers began to use photography for parallax work. The techniques of the time allowed reasonably accurate measurements of stellar distances (within 20 percent) out to about 20 pc (65 light years) and thus included a few hundred stars. The primary limitation of this work was the blurring of the earth's atmosphere. To avoid this problem, the European Space Agency (ESA) launched the Hipparcos satellite in 1989. The Hipparcos mission accurately measured the parallaxes of more than 100,000 stars, providing good distances to about 600 light years. ESA has scheduled the launch of

2. A degree is divided into 60 minutes of arc, and each minute is divided into 60 seconds, so there are 3,600 seconds in one degree.
3. The name comes from <u>par</u>allax of one <u>sec</u>ond of arc. A parsec is 3.26 light years.

Gaia, a much-improved mission, in late 2013. If successful, Gaia will accurately measure distances of millions of stars to tens of thousands of light years. Obviously, parallax data obtained so far are not a problem for recent creation, but the Gaia data could be a problem for a creation that is only 6,000 years old.

Trigonometric parallax is the only direct method of measuring stellar distances, but astronomers have developed other indirect means. Many of these indirect methods involve the use of "standard candles." A standard candle is a star or other object for which we have a good idea of how bright it actually is. Astronomers use magnitudes to express stellar brightness. A star's apparent magnitude, m, is how bright a star appears to us. Its absolute magnitude, M, is a measure of how bright a star actually is, defined to be the apparent magnitude a star would have if it were ten pc away. A star's apparent magnitude depends upon the star's absolute magnitude and its distance. We can directly measure m, and if we think that we know M, we can form the distance modulus m – M. We can find the distance, d, in pc, by inserting the distance modulus in the following formula:

$$d = 10^{(m-M+5)/5}$$

The best example of a standard candle is the use of Cepheid variables. Cepheid variables are pulsating giant and super giant stars with temperatures similar to the sun. As these stars pulsate, their diameters alternately increase and decrease while their temperatures cyclically change. The changes in size and temperature cause a Cepheid regularly to vary in brightness over a definite period. The periods of Cepheid variables range from two days to two months. About a century ago, the astronomer Henrietta Leavitt discovered that Cepheid variables follow a period-luminosity relation. That is, the longest-period Cepheid variables have the greatest average brightness. In observing a Cepheid variable, an astronomer obtains the star's average m and period. Knowing the period, the period-luminosity relation reveals the Cepheid variable's absolute magnitude,

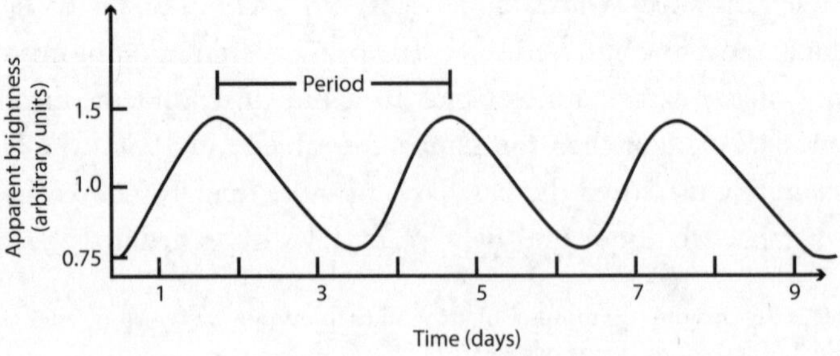

Figure 2. A Cepheid variable light curve showing the period

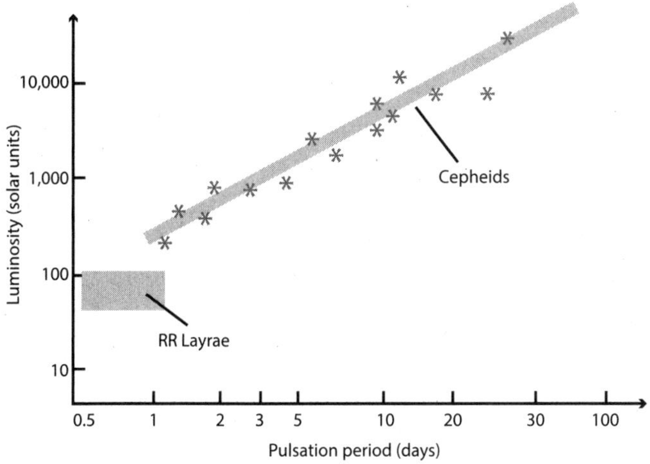

Figure 3. The period-luminosity diagram

which yields the distance modulus and hence the distance. Astronomers had used some other indirect methods to calibrate the period-luminosity relation, but now the Hipparcos mission has allowed direct calibration, in good agreement with the indirect methods. How do we know that a particular variable star indeed is a Cepheid? Cepheid variables have unique characteristics, such as their temperature and the shape of their light curves. The physics of the pulsation is well understood, and from the theory we would expect them to follow the period-luminosity relation.

In addition to standard candles, astronomers can compute distances of stars by estimating their intrinsic luminosities. The spectra of stars contain dark absorption lines that form in the atmospheres of stars. Absorption lines normally are very narrow, but various mechanisms can broaden spectral lines. One of the most important broadening mechanisms in stellar spectra is pressure broadening. The physics of pressure broadening is well understood, with the result being there is an inverse relationship between the amount of pressure broadening and the size of a star. That is, the largest stars have the narrowest lines. Astronomers can estimate the size of a star (expressed by radius, R) by how broad its spectral lines are. We can also determine a star's temperature, T, expressed in Kelvin, a number of different ways. The total luminosity, L, of a star may be expressed as:

$$L = 4\pi R^2 \sigma T^4$$

In this equation, σ is the Stefan-Boltzmann constant. When combined with a model atmosphere, we can convert the luminosity to an absolute magnitude. If we measure the star's apparent magnitude, we know the distance modulus, and we can use the previous equation to find the distance of the star.

Figure 4. Crab Nebula
(Shutterstock.com)

There are some specialized distance determination methods. A good example of this is use of the expansion rate of the Crab Nebula to find its distance. The Crab Nebula is the remnant of a supernova that the Chinese recorded seeing on July 4, 1054. Modern photographs taken a few decades apart reveal that knots of material near the periphery of the remnant are moving outward. Measurement of the motion of those knots allows astronomers to estimate the age of the remnant, an age consistent with the known age. In addition, emission lines in the spectrum of the remnant have both positive and negative Doppler motions along our line of sight. The negative Doppler motion comes from gas moving toward us on the near side of the remnant, while the positive Doppler motion comes from gas on the far side moving away from us. We combine this Doppler motion with the aforementioned expansion of the knots to measure the size and distance of the Crab Nebula. This last step assumes that the expansion is uniform in all directions. The nebula is elongated on photographs, showing that the expansion is not exactly uniform, but this simple assumption probably introduced less than 25 percent error in the final results. We find that the Crab Nebula is about 6,000 light years away.

It is most fortunate that the Crab Nebula also contains a pulsar. Pulsars are radio sources that periodically flash radiation with very regular periods. Astronomers think that pulsars are rapidly spinning neutron stars. There are now more than 2,000 known pulsars, with periods ranging from

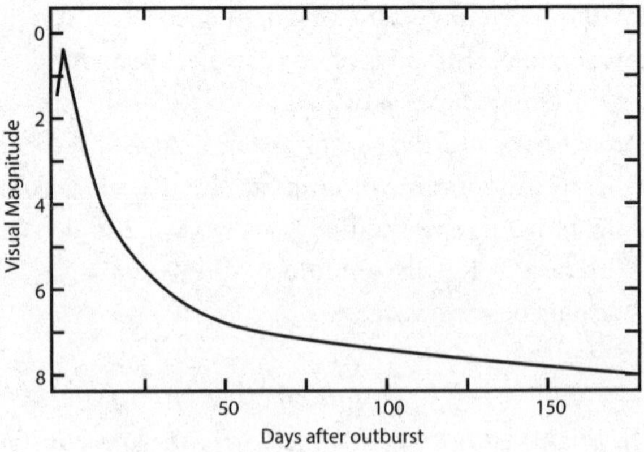

Figure 5. A nova light curve

about a thousandth of a second to a few seconds. Pulsar timings are done in the radio spectrum, and astronomers have found that the pulses are slightly delayed, or dispersed, depending upon the frequency of observation. Dispersion is a well-understood phenomenon, and in addition to dispersion depending upon the frequency, dispersion also depends upon the number density of electrons, n, in the interstellar medium (ISM). The dispersion of the Crab Pulsar and its known distance allow astronomers to measure the average value of n in the ISM along our line of sight to the Crab Nebula. Given the great distance of the Crab Nebula, this appears to be a good average of n in the ISM, which in turn allows radio astronomers to measure the distance of any pulsar with dispersion measurements. This has been done with nearly all pulsars. One of the closest is PSR J0108-1431, about 400 light years away. Pulsars are found throughout the galaxy, with distances up to tens of thousands of light years. Furthermore, astronomers have found pulsars in the Large and Small Magellanic Clouds, two small satellite galaxies of the Milky Way, about 160,000 and 200,000 light years away.

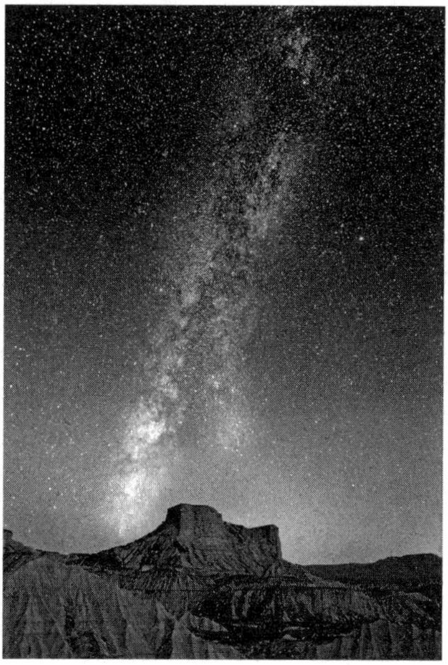

Figure 6. Milky Way over the desert of Bardenas, Spain (Shutterstock.com)

The various methods of finding distances within the Milky Way galaxy help establish the size of the Milky Way, about 100,000 light years across. Since these distances are greater than 6,000 light years, there is some tension between recent creation and these distances. I now turn my attention to extra-galactic distances. Since Cepheid variables are such bright stars, we see them in other galaxies, so this is the one overlap between these realms. The only difference between galactic Cepheids and extra-galactic Cepheid variables is that the ones in other galaxies are much fainter than the ones that we see in our own galaxy. It follows from their faint apparent magnitudes that these Cepheid variables, and hence their host galaxies, are millions of light years away.

In addition to Cepheid variables, astronomers have developed other standard candles for extra-galactic use. They include:

1. Novae
2. Bright super giant stars
3. Bright Globular star clusters
4. Bright HII regions
5. Type Ia supernovae

Novae are eruptions on white dwarf stars in close binary systems, and they have a large range in brightness. But the brightest, classical novae appear to have a narrow range in maximum absolute magnitude. If we happen to observe a nova in another galaxy near its peak brightness, we can measure the apparent magnitude, find the distance modulus, and use the distance formula to find the distance. In similar manner, it appears that the brightest super giant stars in galaxies of the same type have about the same absolute magnitude, allowing an estimate of distances. Large spiral galaxies like the Milky Way have about 200 globular star clusters. The largest and brightest have about the same absolute magnitude. In addition, globular clusters have some appreciable diameter, so that they show up as extended objects on photographs. The largest globular clusters have about the same diameter, so their apparent size can be used as a method of finding distances to them and hence their host galaxies. HII regions are glowing clouds of ionized hydrogen gas surrounding hot, bright stars. While HII regions vary in size and brightness, like globular clusters, there appears to be uniformity among the biggest and brightest ones. These methods now work out to a distance of nearly 50 million light years. Within this range, astronomers like to use several methods and average the results. The variance gives an idea of the errors involved.

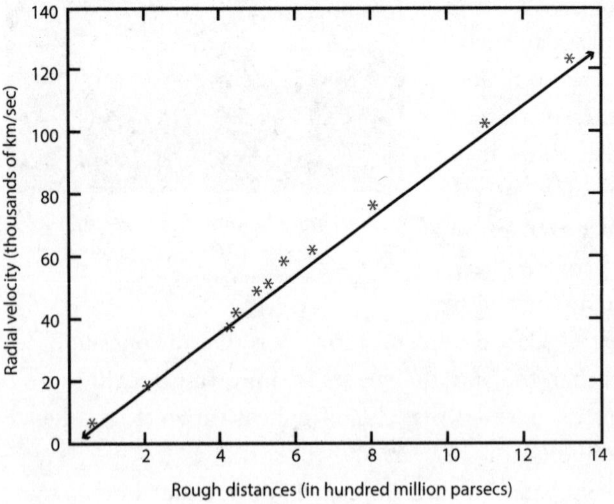

Figure 7. Hubble relation

Type Ia supernovae characteristics are distinctive from other types of supernovae, so they are easy to identify. In recent years, they have stood out as one of the most powerful methods of

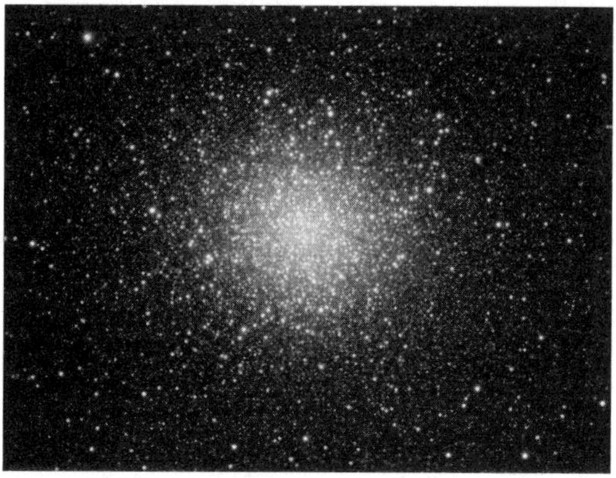

Figure 8. Globular cluster (Shutterstock.com)

finding extra-galactic distances. Astronomers think that a type Ia supernova results from the disintegration of a white dwarf star in a close binary system. The white dwarfs involved in this scenario are so similar that the resulting explosions are nearly identical. This means that at peak brightness all type Ia supernova have the same absolute magnitude, so measurement of the apparent magnitude easily yields the distance. Supernovae are rare events in any particular galaxy, but in recent years astronomers have automated robotic telescopes to look for type Ia supernovae in other galaxies. This has resulted in the discovery of a vast number of these supernovae and hence distances to the host galaxies. In 2013, the HST detected a type Ia supernova about ten billion light years away. In 1999, data from type Ia supernovae played a key role showing that the rate of expansion in the universe may be speeding up, an effect attributed to dark energy. The only restriction on this method is that it works only for galaxies that have type Ia supernova that we happen to observe.

Finally, since Edwin Hubble's 1928 discovery of the expansion of the universe, astronomers have used the Hubble relation to find the distances of galaxies. Redshift is a measure of how much the lines in the spectrum of a galaxy are shifted toward longer (more red) wavelengths.[4] Hubble showed that there is a relationship between redshift and distance, something that one would expect if the universe is expanding. There is some scatter in the data, but the trend generally holds. Mathematically, we can represent the Hubble relation as:

$$v = Hd$$

In this equation, v is redshift expressed as velocity in km/s, d is the distance in Mpc (megaparsec = one million pc), and H is the Hubble constant. This is the

4. People often liken the redshift due to expansion to a Doppler effect of an object moving away from us. Though they observationally are not distinguishable, they are not the same thing.

equation of a line with H being the slope. H is difficult to determine, but once we get that, we can find distances by re-writing the Hubble relation:

$$d = v/H$$

As long as there is enough light to obtain a spectrum of a galaxy, we can measure the galaxy's redshift, and we may use the Hubble relation to find its distance.

Conclusion

I have described here some of the simpler and more often used methods of finding distances to galaxies. In each case, they produce distances that are millions and even billions of light years. While all of these methods suffer from error, those errors would not reduce the distances down to just thousands of light years. The universe is very large, much larger than people can really comprehend. Douglas Adams probably said it best in *The Hitchhiker's Guide to the Galaxy*:

> You just won't believe how vastly hugely mind-bogglingly big it is. I mean, you may think it's a long way down the road to the chemist, but that's just peanuts to space.[5]

Many recent creationists worry about the light travel time problem and entertain possibilities of the universe being far smaller than generally thought as a way out of this dilemma. But this stumbles over something that ought to be obvious. Only a truly powerful Creator could conceive and make such a large universe. It is as if He created the world so large that we finite creatures upon seeing His handiwork ought to fall down prostrate in worship of Him. It may not be possible for a mere human to truly grasp the immensity of the universe, and understanding the power required to create such a universe is infinitely beyond that, but we creationists accept that fact. Yet we so often stumble over how God could have brought the light here so that we could see the universe. Compared to creation, the light travel time problem is trivial. Chapter 21 briefly discusses some of the proposed solutions to the light travel time problem.

5. Douglas Adams, *The Hitchhiker's Guide to the Galaxy* (New York: Ballentine Books, 2005), p. 76.

CHAPTER 19

Could Noah's Ark Have Been Made of Wood?

TIM LOVETT

There's a biblical ark that rode out the Flood, and it was no bathtub. Noah built it somehow, with or without some mysterious ancient technology or extreme gopherwood. Does this mean God had to suspend the laws of physics to keep Noah afloat?

Let's say He didn't. In that case, could Noah get through the whole ark operation?

Building the Ark

Constructing an ark of biblical proportions would take time, resources, and know-how.

Time: Noah had plenty of time — 120 years in fact. In Genesis 6:3, the Lord said, "My Spirit shall not strive with man forever, for he is indeed flesh; yet his days shall be one hundred and twenty years." Some take this as God setting the human lifespan to 120 years. There's a problem with that: every patriarch from Noah to Amram broke God's new "ruling." Noah made it to 950 years of age, his son Shem was 600, and even Abraham died "full of years" at 175.

It's not about lifespans, but about God giving Noah 120 years warning of the Deluge. That's a long time to build a boat; too long, in fact. At that pace, Noah would still be chipping away at the stern while the bow had been exposed to the weather for a century. It makes more sense that Noah spent a lot of this time in preparation until, with everything prepared, he organized a serious barn-raising.

This is where the pitch comes in. The pitch for Noah's ark was probably not bitumen but the gum-based resins extracted from certain trees (such as pitch pine). Wooden ships were routinely waterproofed in this way. The difference here is that God directed Noah to apply the icky goo *inside* as well as out. That's a lot of pitch, so no doubt God had a good reason. Here are two: pitch stabilizes the moisture content of the wood and acts as a preservative. This is ideal for a larger-than-average wooden ship that takes a decade or more to assemble, not just the typical year or so.

Resources: Did Noah need help? A pit-sawing team (of two) would take many decades to cut the wood for one ark. That's cutting it close. Noah and sons had other things to focus on, so it makes sense that labor was hired, or that processed materials like sawn lumber were purchased. Noah should have been extremely wealthy having lived 480 years before the project even began, probably with the help of his grandfather Methuselah,[1] who lived to see the ark constructed.

His world had abundant resources (particularly timber and food), and bronze and steel technology had been around for generations ever since Tubal-Cain first got into working bronze and iron (Genesis 4:22). With such long life spans, technology could rapidly increase in the 1,656 years from Adam to the Flood.

But let's not get too carried away. There are limits to the technology of the pre-Flood world. The ark was made of wood, not metal, which is better for ship hulls. There were also no other survivors in ships (or space-stations for that matter!). The civilizations immediately after the Babel dispersion give us some clues. They excelled at building big things in difficult materials but were not industrialized in the modern sense. An appropriate estimate for the level of technology in Noah's day might be something on par with ancient Egyptians, Greeks, Romans, Chinese, etc. The Egyptians could drill and cut granite, the Greeks could build huge ships with furniture-like precision. These were very ingenious, accomplished builders, experts in crafting metals, ceramics, and other materials — but without the industrialized manufacturing made possible with electricity and heat engines (i.e., steam or combustion engines) implying high-precision machine tools.

We will treat such industrialization as missing from the pre-Flood world as we describe the following construction materials and techniques.

1. Methuselah, the oldest man recorded in the Bible, died at 969 in the same year as the Flood (Genesis 5:21–27; 1 Chronicles 1:3). Dr. Henry Morris said his name may mean, "When he dies, judgment."

Permitted materials and hardware: (Technology of ancient civilizations) Wood: Accurately sawn to fixed sectional dimensions. An up/down saw driven by flowing water or animal draft power, for instance. Sawing is a key technology. Metals: bronze and iron (cast and/or hand forged). Ceramics: fired and glazed pots, oil lamps, stoneware, small glass panes. Other: leather, bone, animal, and resin glues. Fasteners: wooden pegs, metal rods, spikes, and straps. Basic processing/cooking/distilling of pitch/glues. Hand tools in bronze and iron: Drilling auger or spade bit, hand saws, axes, chisels. Measurement: basic surveying, water levels. Lifting and carrying devices: cranes, winches, wheels, rollers, rope, and pulleys. Special long lead-time methods: Planting and harvesting old-growth trees, training trees into shapes (arborsculpture), breeding and training of animals.

Excluded materials and hardware: (Technology after the Industrial Revolution) Electrical power machines, heat engines (steam or internal combustion), threaded bolts and screws, rolled steel plate, metallic films and sheet, processed polymers, highly oxidizing metals like aluminum and titanium, stainless steel, electronics, advanced chemical processing, engineered wood products such as finger jointed and glulam beams, bulk dressed lumber (planed), plate glass (laminated or tempered), steel rope and drawn steel wire, advanced adhesives like epoxy.

Know-how: There are many examples in Scripture where God called people to tackle things outside their expertise, so Noah may not have had much experience in shipbuilding. This is rather unlikely at age 480, but on a 120-year project he could afford to do decades of research.

Having lived for around five centuries, Noah may have been perfectly capable of designing the ark all on his own. The ark is briefly specified in only three verses (Genesis 6:14–16), even lacking crucial data such as the number of animals or amount of food. Perhaps Noah was given more detail, just like Moses received the tabernacle instructions that included exact dimensions and even the number of curtain rings. There's a hint given in Genesis 6:22: "Noah did *everything* just as God commanded him" (NIV), which strongly parallels Exodus 39:32: "The Israelites did *everything* just as the Lord commanded Moses" (NIV). Perhaps this "everything" was more than three verses we have recorded for our benefit, or maybe this is all he had to go on.

Either way, Noah had to get it right the first time — there were no second chances. As far as miracles are concerned, there is one "miracle" recorded; God gave instructions, however brief.

Launching the Ark

The launch of the ark was not meant to be an extreme sport. Noah needed a safe way to launch during earthquakes and strong currents.

The Flood started suddenly when "on that day, all the fountains of the great deep were broken up, and the windows of heaven were opened" (Genesis 7:11). Any flood that rapidly inundates the world (in 40 days or less) will involve massive high-speed currents that would dwarf any modern tsunami. In fact, no modern flood lays down sediments anything like the huge, fossil-filled rock layers deposited all over the world. Such a sudden inundation would pulverize everything in its path, including all shipping and coastal settlements.

How could the ark survive? One solution is to launch from the highest point. This keeps Noah out of the violence of the initial inflows of ocean water as predicted by the Catastrophic Plate Tectonics model.[2] The Flood went on to drown every mountain in the pre-Flood world (Genesis 7:19). Since modern oceans contain enough water to drown the planet to a depth of about 1.8 miles (2.85 km),[3] the pre-Flood terrain was probably limited to within this elevation.[4] By the time the water reached the ark, the currents would have slowed to manageable levels before the launch.

Noah, whether by acumen or divine guidance, may have selected an elevated site where temperate conditions could support a pine forest. Pine, a possible candidate for the mysterious "gopherwood," is especially suited to both shipbuilding and pitch production. This original location is unknown to us today because "the ark moved about" (Genesis 7:18) before finally coming to rest in the Middle East. Gopherwood doesn't have to be a desert acacia, or even a cedar of Lebanon. The very fact that gopherwood is never mentioned again suggests the wood had vanished too. It may have been alive and well on the other side of the world, be it Douglas fir, yellow pine, or even teak.

Here is a quick rundown of a possible construction plan.

2. John Baumgardner. Email correspondence. May 3, 2004. "In regard to the pre-Flood topography, my strong suspicion was that there were highland areas that were not destroyed until late into the cataclysm. . . . It is my feeling that the wave action where the water was shallow was extremely violent throughout the 40 days. It would seem to me that however the ark was launched, it had to get into deeper water very quickly to avoid being destroyed by such violent wave activity."
3. S L Polevoy, *Water Science and Engineering* (UK: Chapman Hall, 1996), p. 192.
4. A. Snelling, ". . . it is likely the pre-Flood hills and mountains were nowhere near as high as today's mountains, a sea level rise of over 3,500 feet (1,067 m) would have been sufficient to inundate the pre-Flood continental land surfaces."

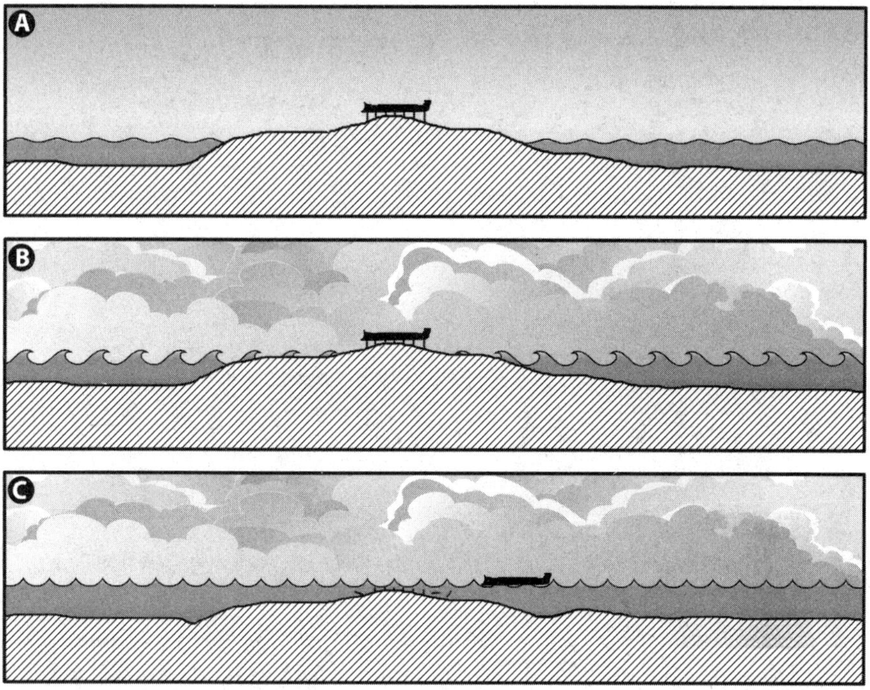

Figure 1. (A). The ark constructed at a high elevation (by pre-Flood standards). (B). Violent inundation devastates lowlands. (C). Water surface less severe once oceans meet.

Noah clear-cuts the hilltop expanse. A foundation is prepared with massive stone walls running transversely to support the hull. Large stones give resistance to strong currents, and tapered ends avoid snagging. Besides all that, don't ancient people always seem to baffle us with their stonework — oversized and outrageously precise? Those ancient civilizations were not a great many generations after Noah himself (Genesis 10).

The three keels laid on the foundation walls help to:

- form a base to erect frames while the bottom can still be planked from underneath;
- hold the hull upright without shoring (handy when planking multiple layers);
- absorb earth tremors and turbulent water (sliding at wall/keel interface);
- reduce rocking in waves (increased roll damping);

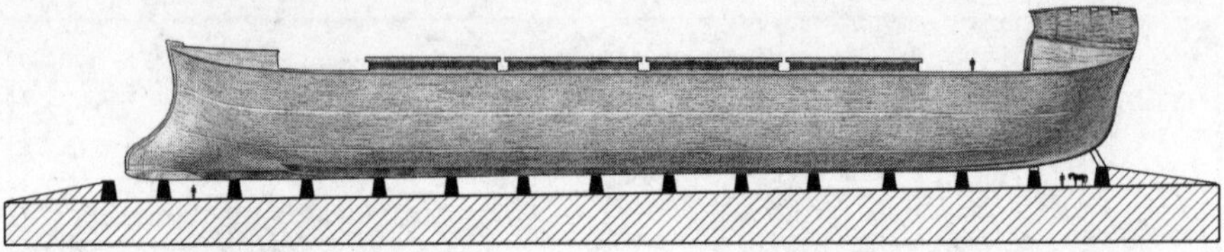

Figure 2. The ark built on pedestal walls to provide underside access and a safe launch.

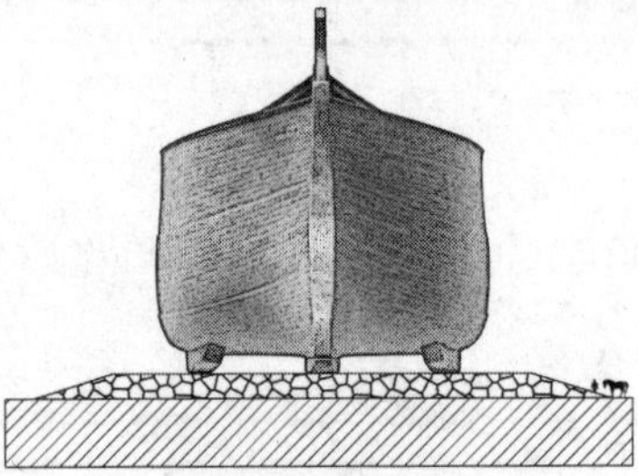

Fig. 3. Ark built on pedestal walls: very heavy stonework to resist erosion during launch.

- improve direction-keeping in winds (keel gets deeper toward the stern);
- resist abrasion (multiple sacrificial layers — false keel/keel shoes);
- keep the ark level when beached (sloping floors would be annoying for seven months!).

These massive keels are built up by laying beams and pinning them together (edge bolted). The lower members within the keels are not scarfed in order to manage stresses.[5]

Ships are normally launched on a slipway, but in Noah's case "the waters increased and lifted up the ark" (Genesis 7:17). Extra safeguards would be prudent, such as releasable mooring ropes to keep the ark from moving away until properly buoyed. There should be no solid obstacles higher than the skid platform — including tree stumps.

The Ark on the Floodwaters

Once afloat, the depth of the water would average almost two miles (three km),[6] shielding the ark from tectonic activity. Deep water is safe in a

5. As it rides over waves, the hull acts as a longitudinal beam under alternating bending (hogging and sagging). The bending stress increases with distance from the center (neutral plane), so the lowermost keels and uppermost skylight should be deliberately flexible or discontinuous to avoid overstressing them.
6. "Ocean Facts: Did you know?" http://www.ocean-expeditions.com/ocean-facts/. "If all the land in the world was flattened out, the earth would be a smooth sphere completely covered by a continuous layer of seawater 2,686 metres [over a mile] deep."

tsunami.[7] The ark had to survive the ocean surface, not the massive sediment flows at the seabed.

But the surface was no picnic either. Later in the voyage, God sent a wind (Genesis 8:1), and wind creates waves, so rough seas are at least part of the five-month voyage. Since the proportions of the ark (Genesis 6:15) are ideal[8] for an ocean-going vessel, it was obviously meant to behave like a ship. With such proportions, the necessary stability and sea kindliness can be achieved even for extreme seas,[9] by a suitable coordination of hull shape and load distribution.

But is it even possible for a wooden vessel as large as Noah's ark to survive the stresses at sea?

The Trouble with Carvel Hulls

The largest wooden ships in recent history (1800s and early 1900s) tended to flex in rough seas, making them prone to leakage. These ships were carvel-built, a plank-on-frame construction method that lacks inherent resistance to racking.[10] The stiffness of the hull depended almost entirely on the tightness of caulking between planks.

Figure 4. Racking: Without bracing, a plank-on-frame structure distorts to a parallelogram under shear loading.

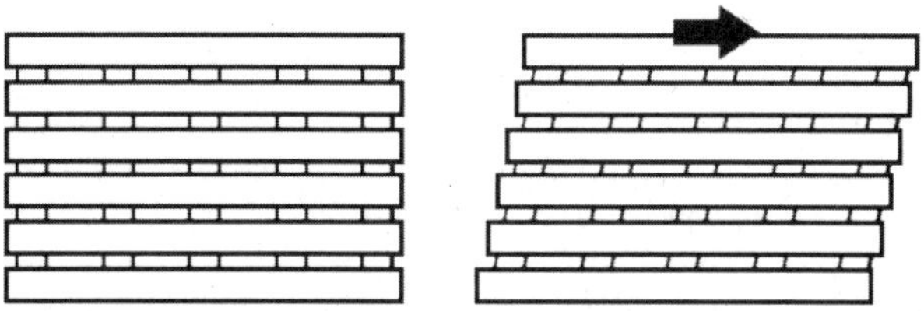

Carvel[11] planking dominated wooden shipbuilding in the last few centuries. The method was simple and quick, but a new ship did not stay a "tight ship" for

7. "What You Should Do," National Weather Service, http://www.nws.noaa.gov/om/brochures/tsunami6.htm. "If there is time to move your boat or ship from port to deep water. . . ."
8. S.W. Hong et al. "Safety Investigation of Noah's Ark in a Seaway," *CEN Technical Journal* 8(1) (1994):26–36. Comparison of 12 arks of various proportions showed the biblical specification (300x50x30) gave the optimal combination of stability, seakeeping, and strength.
9. Ibid., figure 5: Vertical accelerations (43 m significant wave height), figure 6: Roll stability (47.5 m)
10. D.L. Dennis, *The Deficiencies of Wooden Shipbuilding* (London, UK: Mariner's Mirror, 1964), p. 50, 62–63.
11. A carvel hull is formed by parallel horizontal planking fixed to parallel vertical frames (usually by spikes, trunnels, or bolts) to form a smooth outer surface. Lengthwise joints between the planks are typically caulked with fiber and sealed.

very long. Even fitting two pins in each plank gave little improvement.[12] Larger ships were subject to higher forces, accelerating the loosening of the caulked planks. This led to reinforcement with iron straps.

These diagonal straps certainly helped improve a bad design and gave the single layer of carvel planking some much-needed shear resistance. But the steel straps were pinned (bolted) to softer wooden frames, a considerable stress concentration, especially at the ends of the straps.

This led to the next patch-up — steel plates at the top and bottom to secure the diagonal bracing. That kept the hull sides intact, but now the problem was transmitted to extremities, like the top deck.[13]

Later, during World War I, steel was scarce and wooden supply ships[14] were being built in a hurry. Naval architects revisiting the carvel hull-bending problem made big increases to keelson depth[15] and upper deck reinforcement (using clamp and shelf strakes).[16] One design aimed to "produce a boat which will have strength equivalent to that of a steel hull without using excessive amounts of timber."[17] It had a double layer of diagonal planking under the standard planks. That is not a carvel hull, it is cold-molded just like the wooden minesweepers built in the 1990s.[18]

---

12. H.R. Milner and J. Peczkis, "Wooden Ship Hulls as Box Girders with Multiple Interlayer Slip," *Journal of Structural Engineering* 133, no. 6 (June 2007): 855–861. "In frame-built construction, there is usually no direct lateral plank-to-plank connection: There is only the friction provided by the oakum rammed between the planks . . . even carvel construction that employs two rows of densely spaced fasteners (instead of the usual single row) fails to achieve complete composite action."
13. Milner and Peczkis, "Wooden Ship Hulls as Box Girders with Multiple Interlayer Slip," p. 859. ". . . the asymmetric cross section of traditionally built wooden hulls, in which too much timber is already situated in the sides and bottom, and not enough in the deck "flange."
14. Harvey Cole Estep, *How Wooden Ships Are Built: A practical treatise on modern American wooden ship construction, with a supplement on laying off wooden vessels* (Cleveland, OH: The Penton Publishing Co., 1918).
15. Ibid., figure 36, 37.
16. Ibid., p. 6.
17. Ibid., figure 40.
18. The last Avenger-class wooden minesweeper was commissioned in 1994. The 224-foot (68 m) hull was framed in wood and planked with diagonal layers of fir, then covered with fiberglass. Wood was used to minimize the magnetic signature of the vessel. USS *Guardian* (MCM-5), launched in 1987, and ran aground near the Philippines on Tubbataha Reef on January 17, 2013. The hull was holed but remained intact for months before being cut into sections and lifted off the reef by crane ships by March 30, 2013.

So the shortcomings of a carvel hull are not easily corrected.[19] A better way is to use a planking system with inherent shear strength, akin to a house frame braced with plywood instead of clapboards (lap siding or weatherboards).

The claim that Noah's ark is an impossible size for a wooden ship is based on the apparent limiting size of documented wooden ships of the 1800–1900s; around 330 feet (100 m) even with iron bracing. In comparison, using one of the most reliable ancient cubits, the Royal Egyptian Cubit at 20.6 inches (0.523 m), Noah's ark would be 515 feet long, 86 feet wide, and 51.5 feet high (157 m x 26 m x 15.7 m).

That makes it about 50 percent longer than the longest wooden ships in modern records.

Working with Wood

Is this proof positive that the laws of physics must be suspended to keep Noah afloat? This assumes that Noah's ark is built like a carvel hull, or worse. Wood may be an ancient building material, but it still has a competitive strength-to-weight ratio, even compared to metals. For large structures like buildings, bridges, and ships, the problem is not the strength properties of the wood itself, but the manner of *joining.*[20]

Using the strength properties of wood, calculations can determine the required thickness for a vessel the size of Noah's ark operating in extreme seas. Naval architects at the world-class ship research center KRISO (renamed MOERI in 2005) in Korea, studied Noah's ark in 1992 and declared the biblical specifications sound. They used a planking layer 12 inches (0.3 m) thick, taken as a shear resistant "plate structure." Internal structural framework comprised of beams 20 inches (0.5 m) square.

This structure was assessed to determine the stresses on the hull under increasingly severe ocean conditions, with irregular (random) waves up to 30 meters (98 feet).

19. A. Shimell (SP-High Modulus) and H. Ten Have (Dykstra & Partners Naval Architects), Symposium paper: "Structural Design of S/Y Dream Symphony: The Largest Wooden Ship Ever Built," 22nd International HISWA Symposium on Yacht Design and Yacht Construction, Nov. 13, 2012. Referring to steel reinforcement of large carvel hulls, "But even with these reinforcements, the lack of rigidity was never fully solved."
20. Ibid., p. 4. "The main reason for these problems lies in the traditional carvel-planking building method of these ships. As their size increased, the thickness of their members also grew. However, being limited to pin or nail connections, the stresses around the joints and connections became very high. This caused the wood to crack and give way around the connections of members and seams of the shell, resulting in large deflections and ultimately structural failure."

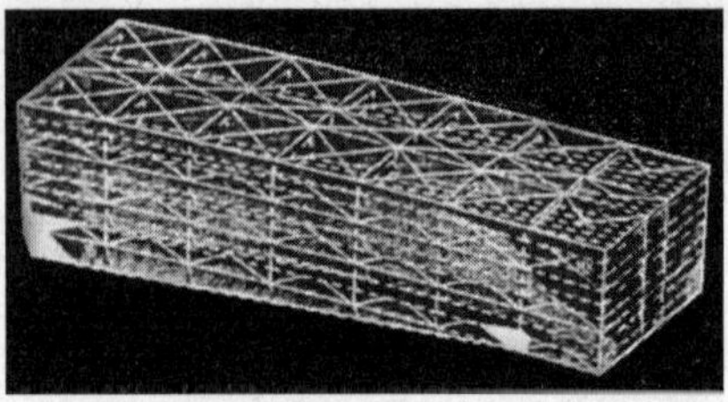

Figure 5. Lattice of 2 feet (0.5m) square beams. The ark may well have been constructed by joint structures of frames and plates. The frame structure of thick beams (50 cm x 50 cm) could have been installed in longitudinal, transverse, and diagonal directions, and connected to each other at each end. The plate structure may have been attached to the frame structure to make the shell, deck, and compartments using thick boards (30cm).

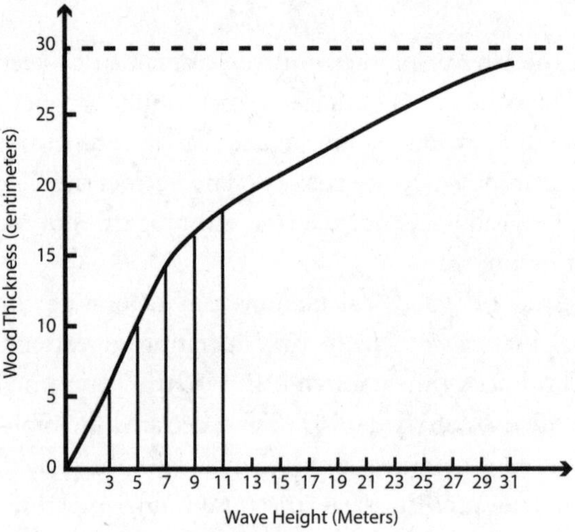

Figure 6. Plate structure (planking) thickness vs. wave height. To calculate the voyage limit from the structure viewpoint, the required thickness of the wood was plotted for varying wave heights. This showed that the ark's voyage limit was more than 30 meters if the thickness of the wood was 30 cm, which was quite a reasonable assumption.

Planking

There are several ways to create this integrated "plate structure." Carvel is not one of them:

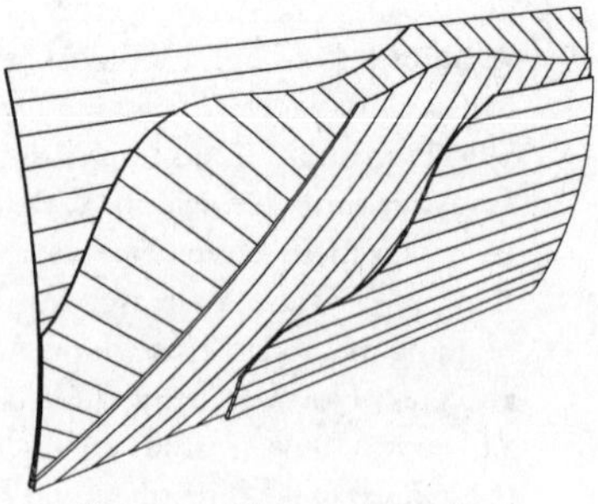

Figure 7. Diagonal planking

- **Diagonal planking.** The definitive way to build a strong wooden hull is to use multiple diagonal layers. Used for U.S. Navy minesweepers (1990s) and PT boats (1940s), diagonal planking also appeared in the design of World War I wooden steamers.[21] However, the British beat them to it with multi-layered diagonal planking in Aberdeen ships such as *Vision*[22] (1854), *Schomberg*[23] (1855), and *Chaa-Sze*[24] (1860), and even "Queen Victoria's new yacht." In 1998, another old ship, the USS *Constellation*, was switched from carvel to diagonal planking to avoid clumsy steel beams to fix hogging strains.[25] In 2012, naval architects proposed a wooden hull laminated in diagonal layers for the 463 feet (141 meters) yacht *Dream Symphony.*[26]

- **Mortise and tenon planking.** A spectacular (almost unbelievable) solution to shearing between planks includes mortise and tenon attachments. Characteristic of Greek and Roman ships, this method was in use well before the 14th century before Christ,[27] then faded away around A.D. 500 to be forgotten until recently rediscovered through underwater archaeology. This lends credence to the records of ark-sized wooden ships of antiquity. For example, *Athenaeus* discussed a large warship that was 427 feet (130 m) long. It

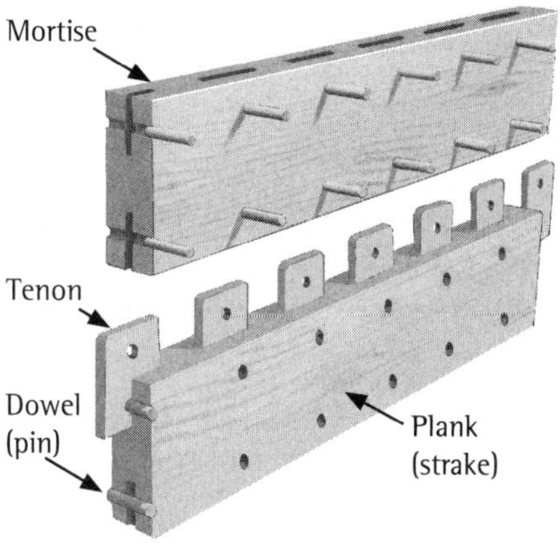

Figure 8. Mortise and tenon planking

21. Estep, *How Wooden Ships Are Built*, p. 3., fig 4, with description on p. 5.
22. *Vision* (diagonal planked), built in 1854 by A. Hall & Co., Aberdeen, *Aberdeen Journal*, 11/10/1854: "The New Principle of Building Wooden Vessels Diagonally."
23. http://www.dpcd.vic.gov.au/heritage/maritime/shipwrecks/victorian-shipwreck-dive-sites/schomberg-shipwreck.
24. *Chaa-Sze* built in 1860 by A. Hall & Co., Aberdeen. Diagonal build. Laid down as a steam whaler. The teak frames were from 4–6 ft. apart with a triple thickness of planking binding the whole together in 9 inches of solid teak. The planking was fastened with screw-treenails (patented in June 1853).
25. Andrew Davis and Keith Gallion, "A Cold Molded Shell for the USS *Constellation*," http://www.maritime.org/conf/conf-davis.htm, October 21, 2007.
26. A. Shimell and H. Ten Have, *Structural Design of S/Y Dream Symphony:* Length overall 141m, beam 18m.
27. Lionel Casson, *The Ancient Mariners*, second edition (Princeton, NJ: Princeton University Press, 1991) p. 108. The practice of joining planks with mortise and tenon joints "certainly goes back to the 14th century B.C. and very much likely before that."

was built by Ptolemy Philopater around 250–200 B.C.[28] It proved itself worthy, even in war. Then there was the *Leontifera* — based on the specification of 8 tiers of oarsmen, it is estimated at about 393 feet (120 m) long.[29]

- **Multiple layers of planking.** Simple but effective. This method was clearly used by Chinese shipbuilders,[30] which would include the treasure ships of Zheng He (A.D. 1400s), with a reported length of 444 chi (137m or 450 feet). It was also seen in Greek and Roman ships (80s B.C.).[31] More recently (A.D. 1800s) multiple layers were employed for impact with floating ice.[32] Each successive layer of overlapping planking dramatically increases the shear resistance of the planking system. Even a double layer is "vastly superior to single carvel."[33]
- **Edge bolted.** The easy way to do mortise and tenon is to use vertical pins (drift bolts) to connect horizontal members (strakes) together. By the sixth century A.D., iron spikes had replaced the painstaking mortise and tenon for edge joining of planks.[34] This technique was used by American shipbuilders[35] to fasten ceiling strakes and keelsons together.[36]

---

28. Athenaeus, *The Deipnosophists*, trans. Charles Burton Gulick, Loeb Classical Library 208 (Cambridge, MA: Harvard University Press, 1987), Book 5, Section 203f–204b, (2:421–425).
29. Memnon, *Excerpts*, c. 14, 15, as cited by James Ussher, *The Annals of the World*, second printing, trans. Larry and Marion Pierce (Green Forest, AR: Master Books, 2003), p. 354.
30. G. Deng, *Maritime Sector, Institutions, and Sea Power of Premodern China* (Westport, CT: Greenwood Publishing Group, 1999). "Chinese ships were maintained once a year by adding a layer of planks to the hull. As a rule, when six layers were added, the ships were half-retired from the ocean-going fleet to coastal services due to the ship's loss of speed." Claims of an ark-like scale of Zheng He's treasure ships have drawn skepticism (mostly by non-Chinese commentators), but it is agreed they were built with two or three layers of planks.
31. Lionel Casson, *Ships and Seamanship in the Ancient World* (Baltimore, MD: The Johns Hopkins University Press, 1995). "The Mahdia wreck had double planking with bands of impregnated cloth between the layers."
32. Loney, *Wrecks Along The Gippsland Coast.*
33. David H. Pascoe, "Surveying Wood Hulls, Part 3: Appendix," http://marinesurvey.com/surveyguide/wood3.htm. "Double Planked: Same as carvel only uses light inner layer with heavier outer layer, parallel longitudinal. Vastly superior to single carvel. Much less prone to leaking, working, and fastener failure."
34. Casson, *Ships and Seamanship in the Ancient World,* p. 208.
35. *Rules Relative to the Construction of Lake Sail and Steam Vessels, 1866.* "Ceiling on sides of vessels of 300 tons and upward, must be edge-bolted between each frame"; http://www.maritimehistoryofthegreatlakes.ca/Documents/Rules1866/default.asp.
36. Estep, *How Wooden Ships Are Built,* p. 13. "In modern wooden vessels built on the coasts of the United States, considerable use is made of edge-bolting to fasten the various keel

Figure 9. Multi-layered planking

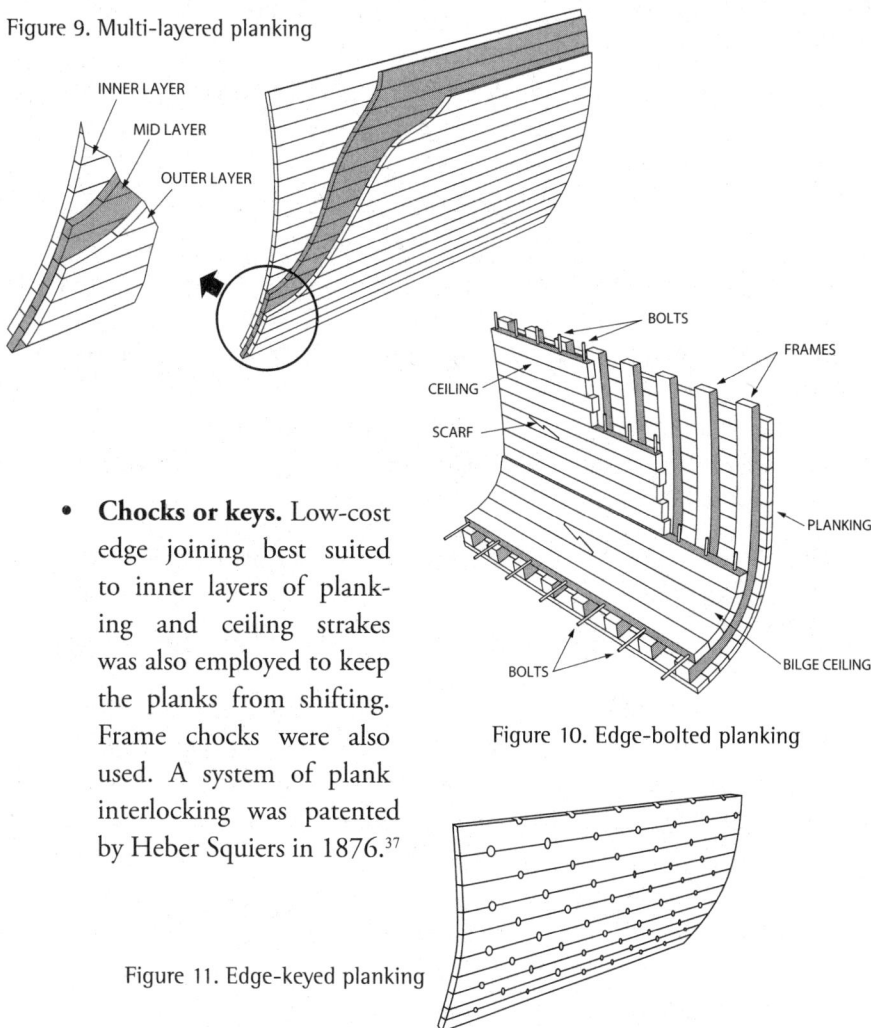

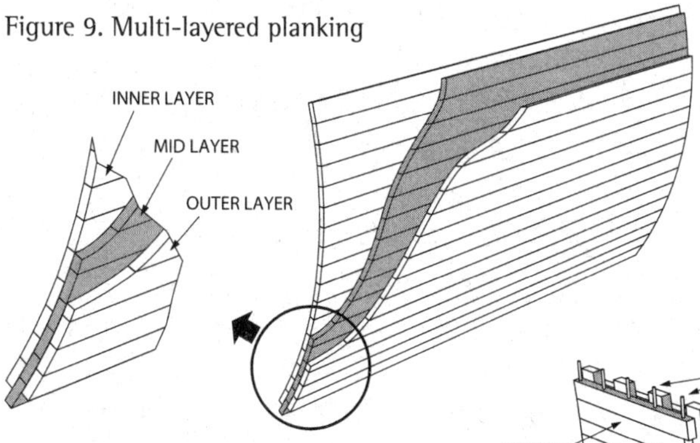

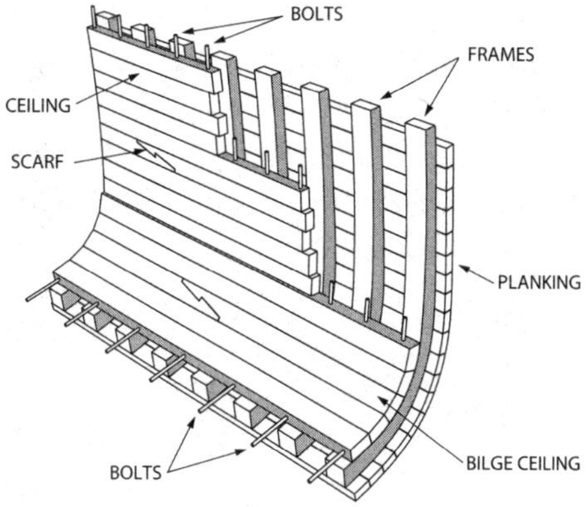

Figure 10. Edge-bolted planking

- **Chocks or keys.** Low-cost edge joining best suited to inner layers of planking and ceiling strakes was also employed to keep the planks from shifting. Frame chocks were also used. A system of plank interlocking was patented by Heber Squiers in 1876.[37]

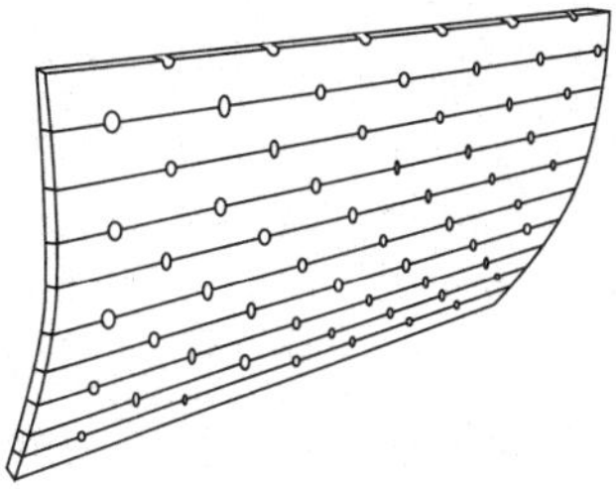

Figure 11. Edge-keyed planking

and keelson elements and the strakes of ceiling together. Edge bolting means fastening the pieces together longitudinally. In other words, the ceiling strakes are bolted through and through to each other, as well as being bolted to the frame timbers. There is no doubt that this form of fastening adds greatly to the strength of the hull structure, particularly in a longitudinal direction, offering resistance to hogging strains. In fact, some experts go as far as to say that the edge-bolting is all that prevents the largest of wooden ships from breaking-up in a seaway. This is probably an exaggeration, although it has been demonstrated that timbers well edge-bolted at least approximate the strength of single pieces of the size of the members so combined."

37. Heber Squier, "Heber Squier's New Method of Fastening and Strengthening Wooden Ships : patented, Aug. 29th, 1876. Gives the Most Strength for the Least Money, and Applies Equally well to Vessels Old and New," Grand Haven, MI, August 1877 (Clarke Historical Library Central Michigan University).

Internal Framing

The Hong study (see footnote 8) also included frames and beam members (50 cm x 50 cm) in their structural analysis. These beams need to be joined together somehow, a critical detail especially in joints that could undergo completely reversed loading. Due to the wave loadings and accelerations at sea, joints that normally sustain compression forces can also go into tension. These joints are the most difficult to achieve in wood, but the full tensile strength of a 0.5 m square beam is an unlikely requirement. Joints must be designed to handle various combinations of compression, tension, twisting (torsion), and possibly bending.

There are a number of structural options for joining large beams. All are held together by metal rods (called bolts) driven into pre-drilled holes, or spikes (large nails). Metal fasteners are also found in large ancient ships.

- **Knees.** A knee is a reinforcing elbow made from a natural bend, typically in large oak branches (crooks). A hanging knee, based on American clippers, used iron "bolts" driven through and clenched.
- **Clamps and Shelves.** A shelf was reinforced with thick longitudinal beams (shelf/clamp) bolted through both frames and beams. Detail based on WWI wooden motor ship.
- **Straps.** A cast or hand-forged bronze strap was held by spikes to opposing members to take tensile forces. Iron straps (or stirrups) were typically used to reinforce connections where axial forces dominate (such as stanchion to deck beam). Straps can also accommodate complex members like diagonal braces.
- **Lamination.** A shear wall performs the dual role of bracing the frame and tying framing members together. This is also the most effective form of bracing.

Each of these framing joints has its own merits and is suitable for different tasks, so several of these methods can be found on any one ship.[38]

The familiar mortise and tenon framing joint is conspicuously absent in primary ship structures. It is too weak, especially in tension.

38. Estep, *How Wooden Ships Are Built*, p. 6, fig 7, "Midship Section: Standard Wooden Steamer for Government."

Figure 12. Knee type framing joint

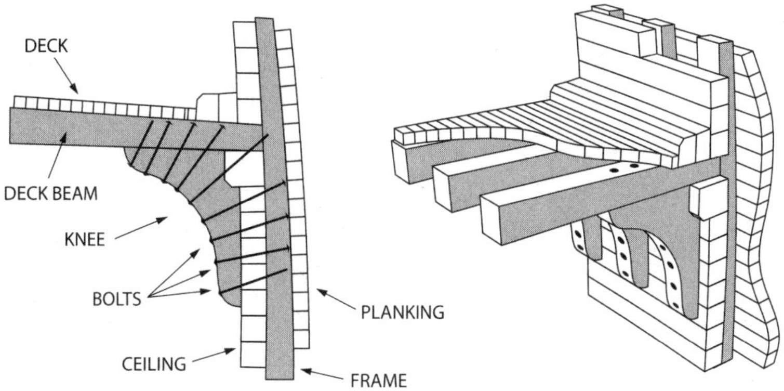

Figure 13. Clamp type framing joint

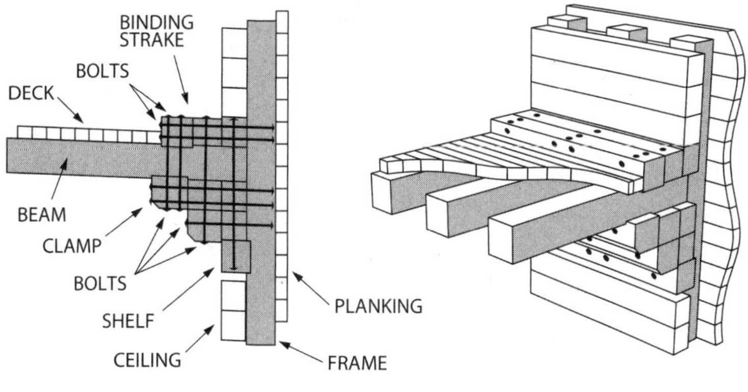

Figure 14. Strap type framing joint

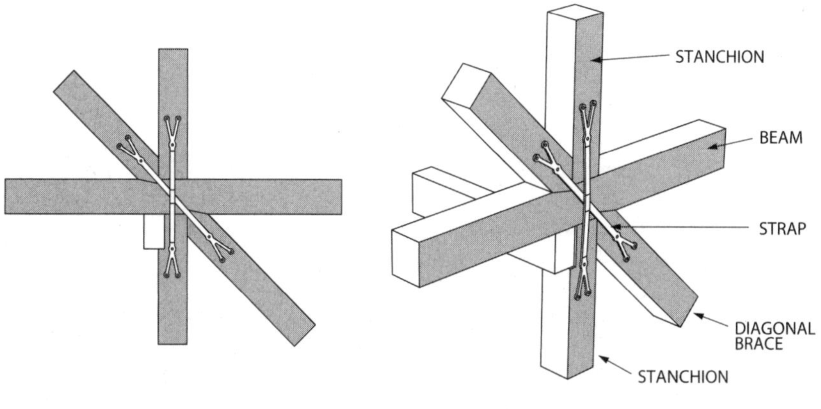

Figure 15. Bulkhead type framing joint

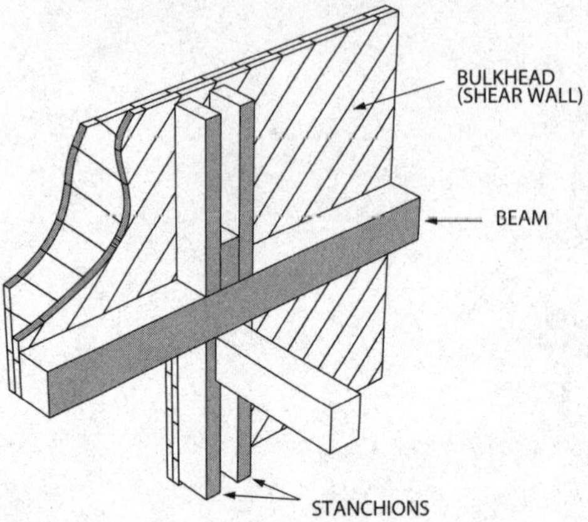

Figure 16. Mortise and tenon framing joint

Bulkheads

Another problem for these "oversized" carvel ships was weak frames[39] or "ribs." The curved frame profiles were built up of many short segments bolted together, but this made them flex and go out of shape. Modern wooden frames are laminated, but the best fix is to use bulkheads — lateral shear walls at regular intervals along the hull. The Chinese were doing that at least 12 centuries[40] before Benjamin Franklin suggested it in 1787.

Extensive use of internal walls actually suits the ark. It was never meant to have a cavernous interior; in fact, quite the opposite. Noah was directed to build "nests" for the animals, not cattle yards. Private enclosures are appropriate for the transport and care of live animals as it helps to keep them calm. From a structural viewpoint, this could mean plenty of bulkhead structures (walls) in both transverse and longitudinal directions. This all adds to the structural integrity of the hull.

39. Milner and Peczkis, "Wooden Ship Hulls as Box Girders with Multiple Interlayer Slip," p. 856. "Frames consisting of many small timbers bolted together are readily deformed, and have frequently been identified as a major weakness of traditional wooden ships."
40. L. Xi, X. Yang, and X. Tang, eds., *The History of Science and Technology in China: Transportation.* (Beijing, China: Science Press, 2004), p. 58. "It can now be deduced that the first watertight bulkheads appeared around 410." The Chinese design was a deliberate shear wall, complete with dowel pins and ledges for shear resistance and even limber holes for maintenance. They form watertight compartments to keep the boat afloat if damaged.

Conclusion

While 330 feet (100 m) may well be the practical limit for a carvel-built hull with a single layer of planking, more appropriate construction methods would extend that boundary by at least 50 percent.

As for the compulsory miracles: God gave instructions to Noah, He brought the animals, He closed the door, and He even sent a wind. But was supernatural intervention the only thing holding Noah's ark together?

Not necessarily. Maybe Noah used ancient bulkheads and ancient planking to build a ship that was more than able to withstand the stresses it faced during the Flood.

CHAPTER 20

What about Environmentalism?

DR. E. CALVIN BEISNER[1]

Introduction

Environment and economy: One of the great difficulties of addressing these two challenges together is that many people think economic development puts the created environment at risk, on the one hand, and environmental protection puts economic development at risk, on the other hand. And indeed, sometimes economic development does cause environmental damage, and sometimes environmental protection does impede economic development. The great challenge is learning how to pursue both at once, for the benefit of men, women, and children, and for the good of animals and plants, of earth, water, and air, all to the glory of God our loving, wise, all-powerful Creator.

While some, like Dr. Michael Nortcott, think — as he expresses it repeatedly in his recent book *A Moral Climate: The Ethics of Global Warming* — that we must choose between people's rising out of poverty and protecting the environment, as if either prevented the other (a bifurcation fallacy), we believe the two are not exclusive alternatives but mutually interdependent. A clean,

1. Dr. Beisner is the national spokesman for the Cornwall Alliance for the Stewardship of Creation. This chapter is based on a lecture presented originally at Creation Care Colloquium: Perspectives in Dialogue, Southeastern Baptist Theological Seminary, Wake Forest, North Carolina, August 28, 2009.

healthful, beautiful environment being a costly good, and wealthier people being able to afford more of a costly good than poor people, it follows that growing wealth — accompanied by ethics and values informed by Scripture, and in the context of a just civil social order — can protect and improve our surroundings (the real meaning, by the way, of the word *environment*) rather than degrade them.

While Dr. Northcott and others prescribe abandonment of industrial civilization, or what Dr. Northcott calls "the machine world," and a return to a hunter-gatherer, or at most a "primitive," subsistence agricultural social order, as the solution to environmental problems,[2] we believe a technologically advanced society and ecological well-being can co-exist, and indeed that they must co-exist if humanity is to fulfill the stipulation of Genesis 1:28 to multiply and to fill, subdue, and rule the earth — a stipulation not repealed after the Fall but repeated in God's covenant with Noah (Genesis 9:1–17).[3]

Let us look at some foundational principles in Scripture, beginning at the beginning, with the biblical record of creation and early history in Genesis 1–9. It will be impossible to touch on, let alone to expound in detail, all the relevant truths in these chapters, but we can notice some of the most prominent.

The Doctrines of Creator and Creation

"In the beginning God created the heavens and the earth. . . . [and] God saw all that He had made, and behold, it was very good" (Genesis 1:1, 31).[4] The

2. Michael S. Northcott, *A Moral Climate: The Ethics of Global Warming* (Maryknoll, NY: Orbis Books, 2007), p. 113, 124–126, 129–130, 175, 232–241, and elsewhere.
3. I argued for this complementarity rather than opposition of economic development and environmental stewardship in my books *Prospects for Growth: A Biblical View of Population, Resources, and the Future* (Wheaton, IL: Crossway Books, 1990) and *Where Garden Meets Wilderness: Evangelical Entry into the Environmental Debate* (Grand Rapids, MI: Eerdmans/Acton Institute, 1997), and in a monograph published by the Institute on Religion and Democracy and available online titled *What Is the Most Important Environmental Task Facing American Christians Today?* (Washington, DC: Institute on Religion and Democracy, 2008), and similar arguments have been made by such scholars as Julian L. Simon and Indur Goklany: Julian L. Simon, *Population Matters: People, Resources, Environment, & Immigration* (New Brunswick, NJ: Transaction Publishers, 1990), Simon ed., *The State of Humanity* (New York and London: Blackwell, 1995), and Simon, *The Ultimate Resource 2* (Princeton, NJ: Princeton University Press, 1996); Ronald Bailey, ed., *The True State of the Planet* (New York: Free Press, 1995); Indur M. Goklany, *The Improving State of the Planet: Why We're Living Longer, Healthier, More Comfortable Lives on a Cleaner Planet* (Washington, DC: Cato Institute, 2007).
4. Scripture quotations in this chapter, unless otherwise identified, are from the *New American Standard Version*.

first and last verses of Genesis 1 immediately set forth the eternity, omnipotence, and sovereign righteousness of God the Creator and the temporality, finitude, and dependence of all created things. They affirm all of creation, material and spiritual alike, as God's work and therefore neither evil — contrary to Gnosticism and much Eastern philosophy, such as that underlying yoga, which sees nature, or *pakruti*, as evil because it traps the soul, *parusa* — nor value neutral, as presumed by the materialist worldview.[5] Between those verses we have a record of God:

1. creating light and, separating it from darkness, establishing the cycles of day and night (verses 2–5);
2. making sky and sea, with their liquid and gaseous waters, and separating them from each other (verses 6–8);
3. gathering the waters of the sea into one place, separating them from the dry land, and causing vegetation to sprout from the land (verses 9–12);
4. establishing the heavenly bodies, especially sun and moon, to rule and separate day and night (verses 13–19);
5. making living creatures and separating their domains into water and sky (verses 20–23); and finally
6. making living creatures to inhabit the dry land, and, on that same day, making mankind and separating it from all other living creatures by endowing it with His own image.

On that sixth day, having made man, male and female, in His image, crowned with glory and honor (as we learn from Psalm 8), God "blessed them; and God said to them, 'Be fruitful and multiply, and fill the earth, and subdue it; and rule over the fish of the sea and over the birds of the sky and over every living thing that moves on the earth'" (verse 28). The verse is pregnant with implications.

The first implication is that human beings are different from all other creatures on earth. *Like* all other creatures, they're not God, they're creatures. But *unlike* all other creatures, they are God's image. *Like* all other living things, they are to reproduce after their kind. But *unlike* all others, they are to fill not just "the waters in the seas" (fish, verse 22), not just the air (birds, verse 20), but the whole earth (verse 28). And *like* all other living things, they are to obey their

5. Vishal Mangalwadi, " 'Cap-and-Trade' Legislation: Secularizing Sin?" draft paper, October 27, 2009.

Creator (implicit in His commanding them), but *unlike* all others, people are to have rule over other living creatures, and over the earth itself.

And what is it for them to bear the image of God? It is partly what we have just noticed: to rule over other creatures. And from the New Testament and elsewhere we learn that it is for them to have rational and moral capacity (Ephesians 4:24; Colossians 3:10). But we must not neglect what the immediate context reveals about the image of God in man. It is what it reveals about God Himself in verses 1–25: that He is a Maker — indeed, a prolific, even extravagant Maker. People, too, are to be makers — not makers of things *ex nihilo*, "out of nothing," which is the province of God alone, but *ex quispiam*, "out of something." That is, people, made in God's image, are to make new things out of what God puts before them — and, as God made all things of nothing, so people more fully express this creative aspect of His image as they make more and more out of less and less.

The second implication is that the earth and the various living creatures in it — in its seas, in its air, on its ground (*ha-adamah*, related to the name for man; *adam*, who was taken from it, fashioned by God, who then breathed into him the *neshamah hayyim*, the breath of life) — the earth and all in it, while "very good" (verse 31), were not yet as God intended them to be. They needed filling, subduing, and ruling.

Was this because there was something evil about them? No. We have already seen that the biblical doctrine of creation rules out notions of the inherent evil of the material world, whether derived from the Hindu and Buddhist view of matter and spirit as antithetical (in opposition), or from the Platonic and neo-Platonic doctrine of a hierarchical "great chain of Being" from God (who has most being) to nothing (which has none). It was not that there was something evil about the earth and its non-human living creatures. It was that they were designed as the setting, the circumstance — the *environment*, if you will (that word coming from the French *envirroner*, "to surround") — they were designed as the *surroundings* in which Adam and Eve and their descendants are to live out their mandate as God's image bearers.

As God created it, the earth and all its constituents were very good. They were perfect — not terminally perfect, but circumstantially perfect, perfectly suited as the arena of man's exercise of the *imago Dei* (image of God) in multiplying, filling, subduing, and ruling according to the knowledge and righteousness that most essentially constitute the *imago*.

Already we can recognize some important distinctions between a biblical ethic of creation stewardship, on the one hand, and secular and pagan religious environmentalisms, on the other.[6] The common environmentalist vision of human beings as chiefly consumers and polluters, using up earth's resources and degrading it through their waste (a view expressed by Paul Ehrlich and others in the famous formula I=PAT, that is, environmental impact [which is always harmful] is a function of population, affluence, and technology). They claim that an increase in any of those factors inevitably brings more harm to the earth. This vision of man as essentially consumer and polluter confronts the biblical view that people are designed to be producers and stewards, capable of transforming raw materials into resources through ingenuity and hard work, making more resources than they consume, so that each generation can pass on to the next more of the material blessings than it received, and through godly subduing and ruling of the earth actually improving the environment.

In Genesis 2, a parallel account of creation that focuses more specifically on mankind on day 6, we learn that God placed Adam in the Garden of Eden, stipulating that he was to "cultivate it and keep it" (2:15). Almost as an aside, both this and the mandate of 1:28 to multiply and to fill, subdue, and rule the earth are not *solely* commands but *also* stipulations — God's speaking to them ensuring their fulfillment just as surely as His saying "Let there be light" ensured that light would be.

We should note that this means God's intention that mankind multiply and fill, subdue, and rule the earth, and that he cultivate and keep the Garden, is not conditioned on mankind's remaining morally perfect. We *shall* multiply, we *shall* fill, we *shall* subdue, we *shall* rule, we *shall* cultivate, and we *shall* guard — none of that is uncertain. *How* we shall do these things — *that* is what is in question: whether we shall do them wisely and righteously, or foolishly and wickedly. Our Fall into sin unquestionably influences *how* we do these things, but it neither does nor can prevent our doing them or relieve us of the duty imposed by these mandates.

Although some Christian environmental writers attempt to use Genesis 2:15's stipulation of cultivating and keeping the Garden to define Genesis 1:28's stipulation of subduing and ruling the earth, that is surely mistaken, for

6. For example, Pantheism, and Gnostic, illusionist, and Manichaean dualisms, and Platonic idealism and Aristotelian materialism, and modern Marxist and secular humanist naturalisms — all of these fall before this biblical worldview. Epistemological and moral relativism, antinomian utilitarianism, existentialism's claim that we define morality by our choice, postmodernism's rejection of hierarchy and transcendence and enduring meaning — these, too, fall.

two reasons. First, the Garden is not the whole earth; it is a specific, limited geographical location, "toward the east, in Eden" (2:8). Just as we saw separation of light from darkness, heavens from earth, waters from land, life from non-life, animal life from vegetable, and human life from non-human life, so also there is a separation of Garden from the rest of the earth — a distinction that will later be developed between wilderness and Promised Land.

Second, the language in the stipulations differs radically. In 1:28, God told Adam and Eve to "subdue and rule" (*kabash* and *radah*), the words meaning, respectively, to subdue or bring into bondage, and to have dominion or rule. In 2:15, God told Adam to "cultivate and keep" (*abad*, and *shamar*), the words meaning, respectively, to work or till, and to keep, watch, or preserve. What God assigned Adam to do in the *Garden* (to cultivate and keep) was not the same thing He assigned him to do in the *earth* (to subdue and rule). Some environmental writers have also suggested that the command to cultivate, or till, the Garden should be translated "to serve," and then, by equating Garden with earth, have inferred that humankind is to serve the earth. But this is not only to equate Garden and earth, which Scripture expressly distinguishes, but also to misuse the Hebrew *abad*, which, although it *may* bear the sense of serve when followed by an accusative of *person*, does *not* bear that sense when followed by an accusative of *thing*.[7]

From these two stipulations — to subdue and rule the earth, and to cultivate and keep the Garden — it follows:

1. that humans are not aliens, much less a cancer or a plague on the earth, but its rightful, God-ordained rulers;
2. that it is not wrong in principle but right that they should subdue and rule the earth;
3. that their *cultivating* the Garden (to increase its fruitfulness) and *keeping* it (to protect it against degradation) are not mutually exclusive but complementary; and
4. that their cultivating and keeping are not antithetical to but additional and complementary to their subduing and ruling the earth and everything in it.

It follows also that the beliefs, common among many environmentalists, that "nature knows best," that nature is best untouched by human hands, that nature's unaided fruitfulness is all that is right and sufficient for mankind, and

7. Francis Brown, S.R. Driver, and Charles A. Briggs, eds., *A Hebrew and English Lexicon of the Old Testament* (Oxford: Clarendon Press, 1907, 1953, 1978), s.vv.

that, as Dr. Northcott puts it in *A Moral Climate,* "the move from the hunter-gathering lifestyle of Eden to the agrarian life on the plains [was] a fall from grace,"[8] are all contrary to the biblical worldview and to the binding stipulations/commands given to mankind at creation.

Adam and Eve did not abandon their post in the Garden and strike out into the wilderness of their own accord and so come under God's judgment. Rather, they disobeyed the probationary command not to eat of the fruit of one particular tree in the Garden, and in response God banished them from the Garden into the wilderness — where, consistent with the stipulatory character of the commands to multiply and to fill, subdue, and rule the earth, they would indeed do so.

Indeed, Dr. Northcott's assertion that Edenic society would have been hunter-gatherer rather than agrarian is explicitly contradicted by the command to *cultivate* the Garden. His claim, again, that "Just as the story of Genesis is that of a Fall from the Garden to an imperious and idolatrous urban culture, so the story of redemption in Exodus is of an urban prince who leads his people in a revolt against the slavery imposed by the city, back out to the levelling nomadic lifestyle of the wilderness,"[9] is also mistaken, for Israel's destination in the exodus was not the *wilderness*, where God forced it to spend 40 years as chastisement for its rebellion, but the *Promised Land*, where the Israelites would possess and settle in cities and houses that they did not build (Deuteronomy 19:1).

And contrary to the common environmentalist notion that cities are essentially bad, God names some of them as places of refuge (Deuteronomy 19:1–10); chooses one city, Jerusalem, as the special abode of His Temple; and ultimately describes the completed and perfected Church, the Bride of Christ, as the holy city, the New Jerusalem, descending out of heaven (Revelation 21:2, 10). Thus, the biblical history of creation, Fall, Curse, redemption, and consummation begins in a Garden, makes its way through a wilderness, and ends in a Garden City, and it becomes clear that the command/stipulation of Genesis 1:28 to multiply and to fill, subdue, and rule the earth was a command/stipulation to go forth from the Garden of Eden into the rest of the earth to transform wilderness into Garden City.[10]

8. Northcott, *A Moral Climate: The Ethics of Global Warming*, p. 233.
9. Ibid., p. 235.
10. Indeed, as I have argued in *Where Garden Meets Wilderness*, it was precisely by doing this that mankind were to guard, or keep, the Garden, for what threatened it — even before Adam and Eve's fall into sin — was encroachment by the wilderness.

Thus far we have taken only little notice of a very significant statement at the end of God's creative activity: Genesis 1:31, "God saw all that He had made, and behold, it was very good." We have seen that this does not entail that it was *terminally* perfect but that it was the perfect setting for man's probation and for his exercising the *imago Dei*. Let's draw one other implication from this brief and simple sentence. A crucial element of the environmentalist worldview is that the earth and its habitats and inhabitants are extremely fragile and likely to suffer severe and perhaps even irreversible damage from human action. Let us for now ignore the implicit assumption here that humans are aliens, that they alone among all living things are prohibited from transforming their surroundings. Rather, what are we to think of the explicit thrust: that the earth and its various ecological subsystems are fragile? That element of environmentalism contradicts this verse. It is difficult to imagine how God could have called "very good" the habitat of humanity's vocation in a millennia-long drama if the whole thing were prone to collapse like a house of cards with the least disturbance.

Now, I have encountered an objection to this reasoning, pointing out that, after all, some things in this world *are* fragile — a fly's wing, for instance. But there are two mistakes in this rejoinder. First, it confuses the part with the whole. That some inhabitants of the earth are fragile doesn't entail that the whole earth is, and that the wings of individual flies are fragile doesn't entail that therefore the genus *Drosophila*, or even the species *Drosophila melanogaster*, is fragile. Though many individual flies lose their wings and all flies die, the genus and even the species endure.

Second, it neglects that, seen in proportion, what deprives a fly of its wing is not, in proportion to the fly and its wing, a tiny disturbance. The fly's wings serve quite well for their normal purposes and in the absence of *proportionally* overwhelming impingement. To speak of the whole biosphere, or even of extensive ecosystems, as extremely fragile is both to neglect the force of Genesis 1:31 and to ignore the testimony of geologic history, which includes the recovery of vast stretches of the Northern Hemisphere from long coverage by ice sheets several miles thick — which certainly wiped out more ecosystems more thoroughly than human action has come close to doing — not to mention the recovery, according to Genesis, of the whole earth from a Flood that destroyed all air-breathing, land-dwelling life but the few representatives rescued in Noah's ark and the curse in Genesis 3.

Let me apply this insight to the most controversial environmental issue of our day — indeed, of the whole history of environmentalism to date – anthropogenic global warming. Briefly put, the fear is that human emissions of carbon dioxide and other "greenhouse gases" (a sadly misleading metaphor since greenhouses work not by absorbing infrared radiation, as do these gases, but by preventing the movement upward of warm air) — that our emissions of these gases have caused, by increasing the rate of absorption of infrared radiation bouncing back from earth's surface toward space, or will soon cause, sufficient warming of the earth's surface to set off a series of positive feedback mechanisms (for example, more evaporation and hence more water vapor, which then absorbs yet more infrared radiation). The feedbacks will warm the surface still more, thus instituting a positive feedback loop that leads to a runaway greenhouse effect that eventually makes the earth uninhabitable, at least to human beings, and particularly to human beings living in modern civilization. (As an aside, one wonders why those environmentalists who despise industrial society mourn the prospect of its collapse due to global warming. One would expect them to celebrate it as judgment instead.)

Clearly, this scenario rests upon precisely the assumption of the fragility not of individual elements but of the whole of the bio-/geosystem. That an increase in carbon dioxide from one molecule in every 3,704 in the atmosphere to one molecule in every 2,597 — from 270 to 385 parts per million — from 0.027 percent to 0.0385 percent — should cause catastrophic damage to the biosphere, or even set off a positive feedback loop ("runaway global warming") that will cause such damage — particularly when carbon dioxide's infrared absorption is logarithmic (each new unit absorbing less than the previous one) — is fundamentally inconsistent with the biblical worldview of the earth as the "very good" product of the infinitely wise Creator. That biblical worldview instead suggests that the wise Designer of the earth's climate system, like any skillful engineer, would have equipped it with balancing positive and negative feedback mechanisms that would make the whole robust, self-regulating, and self-correcting.[11]

Perhaps more importantly, they should prompt Christians to praise God for the way in which the earth, like the human body, is "fearfully and wonderfully made" (Psalm 139:14). In some senses this planet, like the eye, may be fragile, but overall it is, by God's wise design, more resilient than many fearful environmentalists can imagine even in a sin-cursed world.

11. For more on global warming and climate change, please see chapter 16 in this volume and chapter 7 in the *New Answers Book 3*. See also *A Renewed Call to Truth, Prudence, and Protection of the Poor: An Evangelical Examination of the Theology, Science, and Economics of Global Warming* (Burke, VA: Cornwall Alliance, 2009; online at http://www.cornwallalliance.org/docs/a-renewed-call-to-truth-prudence-and-protection-of-the-poor.pdf).

The Doctrines of Fall, Curse, and Redemption

As we move along in these early chapters of Genesis, we come to the account of mankind's fall into sin. It is not, as we have already noted, a sin of moving from the idyllic hunter-gatherer life of the Garden on the mountain to the urban life of the plain (against which God had given no command, and "sin is lawlessness" [1 John 3:4]), but disobedience to a specific command: not to eat of the fruit of the tree that was in the midst of the Garden. The aetiology (study of causation) of this sin is significant for our discussion of environmental ethics: it came about when Eve, who as bearer of the *imago Dei* was supposed to rule over every living thing that moved on the earth, abdicated her rule and instead bowed to the serpent, "more crafty than any beast of the field which the LORD God had made" and then Adam, to whom Eve was to be a helper rather than a ruler, bowed to Eve (Genesis 3:1–6). The rejection of human rule over the animal world, common to many environmentalists, reflects Eve's abdication, and it is not right. This ultimately led to Adam's sin as well.

In response to their sin, God pronounced judgment on Adam and Eve: pain for her in childbirth, and a frustrated desire to rule over her husband; pain for him in cultivating the ground; and death for both of them (Genesis 3:16–19). Yet at the very same time, "God said to the serpent, 'Because you have done this, cursed are you . . . and I will put enmity between you and the woman, and between your seed and her seed; he shall bruise you on the head, and you shall bruise him on the heel" (verses 14–15), and "God made garments of skin for Adam and his wife, and clothed them" (verse 21), spilling the blood of animals to cover over the now-embarrassing nakedness of these sinners, typifying the sacrificial system of Judaism and the ultimate sacrifice of His incarnate Son on the cross.

Judgment and the promise of redemption met in that moment. And then "God sent him [that is, the man generically — Adam and Eve together, the human race] out from the garden of Eden, to cultivate the ground from which he was taken" (verse 23). Despite the Fall, the God-ordained vocation of cultivation remained — only now it would be cultivation in a more difficult, less cooperative environment — instead of the Garden, the wilderness, a term consistently associated in Scripture with curse. Yet the stipulation that Adam and Eve should multiply and fill, subdue, and rule the earth, transforming wilderness into Garden, remained, and indeed the next chapter recounts the beginning of the fulfillment of that stipulation in Adam and Eve's bearing of children; the eruption of enmity between the seed of the woman and the seed of the serpent in the farmer Cain's murder of his sheep-herding brother Abel; a new pronouncement of curse on Cain, frustrating his cultivation of the earth and making him "a vagrant and a wanderer on the earth" (a description that well fits the hunter-gatherer life admired by some environmentalists), and yet again God's gracious extension of life despite sin (Genesis 4:1–17).

For space's sake let's skip over the detailed accounts of the descendants of Cain and Seth and come to Noah, in whose day the wickedness of mankind reached such a height that God "was sorry that He had made man on the earth, and He was grieved in His heart," and He said, "I will blot out man whom I have created from the face of the land, from man to animals to creeping things and to birds of the sky; for I am sorry that I have made them," for "the earth was corrupt in the sight of God, and the earth was filled with violence" (Genesis 6:6–7, 11). "But Noah found favor in the eyes of the Lord" (verse 8), i.e., God looked on him with grace, and God instructed him to construct an ark to rescue remnants of all flesh from the Flood He decreed.

God then rained His judgment on the earth and wiped out all air-breathing land-dwelling life, excepting only those few on the ark. It must be admitted that the event brought ecological devastation on a scale unmatched by anything man has done. And yet that devastation was done by God due to disobedience to God's Word. This, it seems to me, is difficult to reconcile with environmentalist notions of inherent as opposed to imputed value in nature and the condemnation of any action that harms any of it.

Following the Flood, we read, Noah built an altar to God and sacrificed birds and animals on it, and God "smelled the soothing aroma; and the Lord said to Himself, 'I will never again curse the ground on account of man, for the intent of man's heart is evil from his youth; and I will never again destroy every living thing, as I have done. While the earth remains, seedtime and harvest, and cold and heat, and summer and winter, and day and night shall not cease" (Genesis 8:18–22). The Hebrew poetic *merism* in verse 22 uses pairs of opposites to express the inclusion of all things of the sort mentioned. The implication is that God has promised to Himself that He will sustain the cycles on which human and other life on earth depend as long as the earth itself remains. This promise of God to Himself is, it seems to me, difficult to reconcile with fears that some human action will send the climate into irreversible, catastrophic disruption, threatening mass species extinctions and the destruction of human civilization or perhaps even human extinction.

And then God makes a promise to Noah and his sons, repeating the command/stipulation first given to Adam and Eve in Genesis 1:28: "Be fruitful and multiply, and fill the earth." But this time He continues, "The fear of you and the terror of you will be on every beast of the earth and on every bird of the sky; with everything that creeps on the ground, and all the fish of the sea, into your hand they are given. Every moving thing that is alive shall be food for you; I give all to you, as I gave the green plant" (Genesis 9:1–3). This passage forever invalidates the claim that vegetarianism is ethically superior to meat eating. God has permitted people to kill and eat "every moving thing that is alive." The Apostle Peter would later write of "unreasoning animals, born as creatures of instinct to be captured and killed" (2 Peter 2:12).

And finally, God re-establishes His covenant with Noah and, through him, the whole human race, and even with "every living creature": ". . . all flesh shall never again be cut off by the water of the flood, neither shall there again be a flood to destroy the earth." He ordains the rainbow as the sign of the covenant, and says, "when I bring a cloud over the earth . . . the bow will be seen in the cloud, and I will remember My covenant . . . and never again shall the water become a flood to destroy all flesh" (Genesis 9:9–15). We find this language reflected later in Psalm 104:5–9, which says that after the Flood God "set a boundary, that [the waters] may not pass over, so that they will not return to cover the earth."

Conclusion

In a stunning passage, the prophet Jeremiah compares the stubborn and rebellious people of Judah with the waves of the sea (Jeremiah 5:21–25) due to their lack of fear of the Lord. Just as the sea could not overcome the boundaries God set for it following the Flood, so the people of Judah could not overcome the boundaries God had set for them. Rage against His laws as they might, they would still face His judgments. I will conclude with two observations on this passage.

First, like Psalm 104:5–9, what it says about the boundaries God has set for the sea is difficult to reconcile with fears of catastrophic sea level rise. While there is evidence that sea level was once much higher than what it is now, the sea has never again prevailed against the land. This is best interpreted in the light of the Flood of Noah's day — a never-to-be-repeated, cataclysmic judgment of God that would have been followed by an ice age (accompanied by much reduced sea level as water was stored in vast ice sheets on land) as the atmosphere lost its high water vapor content and so cooled rapidly, and then a gradual recovery as water vapor (which accounts for over 95 percent of the greenhouse effect) rose to approximately its present concentration (accompanied by a gradual sea level rise to near-present levels as the continental glaciers melted).

This does not mean that sea level cannot rise (and likewise fall) gradually over long periods as earth warms and cools through natural cycles. But it is inconsistent with the fear of catastrophic sea level rise driven by anthropogenic global warming, which also finds no support in sound science. The IPCC reduced its estimate of likely 21st-century sea level rise from about 35 inches in its 2001 report to just 17 inches in its 2007 report, in which it also projected that there would be no significant melting of the Greenland ice sheet for several millennia — and then only if the world remained at least 2°C warmer than today throughout those millennia (an unlikely scenario granted historical temperature cycles driven by cycles in solar radiance). While the IPCC included no sea level experts among its authors, one of the world's leading experts on sea level, Nils-Axel Mörner, head of the sea level commission of the International Union for Quaternary Research, concluded in the study "Estimating Future Sea Level Changes from Past Records" that 21st-century sea level rise would be much lower than even the revised IPCC estimates:

> In the last 5000 years, global mean sea level has been dominated by the redistribution of water masses over the globe. In the last 300 years, sea level has been [in] oscillation close to the present with peak rates in the period 1890–1930. Between 1930 and 1950, sea [level] fell. The late 20th century lack[ed] any sign of acceleration. Satellite altimetry indicates virtually no changes in the last decade. Therefore, observationally based predictions of future sea level in the year 2100 will give a value of + 10 ± 10 cm (or +5 ± 15 cm) [0 to + 7.88 inches, or −3.94 to + 7.88 inches], by thus discarding model outputs by IPCC as well as global loading models. This implies that there is no fear of any massive future flooding as claimed in most global warming scenarios.[12]

Recent data from sea level monitoring stations around the southwest Pacific confirm that sea level rise during the last 30 years, despite widespread claims to the contrary and fears of the impending submersion of island nations like Tuvalu and Kiribati, has been slight to nonexistent and certainly not significantly greater than its long-term rate.[13]

12. Nils-Axel Mörner, "Estimating Future Sea Level Changes from Past Records," *Global and Planetary Change* 40 (2004): p. 49–54.
13. Cliff Ollier, "Sea Level in the Southwest Pacific is Stable," *New Concepts in Global Tectonics Newsletter*, no. 50 (June 2009), accessed online September 7, 2009, at http://nzclimatescience.net/images/PDFs/paperncgtsealevl.pdf.

Second, God's words through Jeremiah make it clear what is the real root of fears of natural catastrophes like droughts: the absence of the fear of the Lord, manifested in persistent sins like those named so frequently throughout Jeremiah: idolatry (1:16; 2:5; 3:6; 7:9, 18; 8:19; 10:2; 11:10; 16:18; 17:2), forsaking God (Jahweh) and worshiping pagan gods, which God called spiritual adultery (1:16; 2:11, 17, 20; 3:1, 2-3, 9, 20; 5:7, 18; 7:30; 9:2, 13; 11:10, 17; 13:10, 25, 27; 14:10; 15:6; 16:11), prophets speaking in the name of false gods (2:7), absence of the fear of God (2:19), rejecting and killing God's prophets (2:30), forgetting God (2:32), murder (2:34; 4:31; 7:9), injustice (5:1; 7:5), falsehood and lies (5:1, 12; 6:13; 7:9; 8:8, 10; 9:3), deception (9:8), oppression (5:25–29, 6:6; 7:6; 9:8; 17:11), fraud (5:27), false priests and prophets "and My people love to have it so" (5:30; 14:15), rejection of God's Word (6:10, 19; 8:9; 9:13; 11:10; 13:10), covetousness (6:13; 8:10), religious formalism and presumption (7:3-4), stealing (7:8–9), sexual adultery (7:9; 9:2), general disobedience to God's law (7:28), child sacrifice (7:31), worship of nature (8:2), covenant breaking (11:3), general wickedness (12:4), complaint against God (12:8), pride (13:8), trusting in man instead of in God (17:5), and Sabbath breaking (17:21).

It is significant that, in contrast to some Christian environmentalists' claims that God sent Israel and Judah into exile because they defiled the land, never once do the prophets describe the sins for which God punishes them as unsustainable farming practices, pollution, or similar things. Oh, the people defile the land, true. But how? "[T]hey have polluted My land: they have filled My inheritance with the carcasses of their detestable idols and with their abominations" (16:18). It is precisely because the people of Judah do not fear God (and so practice all kinds of sin) that they come to fear that the spring and autumn rains will fail.

Fear of environmental catastrophe grows out of the lack of the fear of God. That, I would argue, is the real root of the environmental scares that have plagued the modern world.[14] And such fears will continue — with or without scientific basis[15] — until people repent and fear the Lord. "Cursed is the man

14. For catalogues and exposés of such, see Julian L. Simon, ed., *The State of Humanity* (New York and London: Blackwell, 1995); Aaron Wildavsky, *But Is It True? A Citizen's Guide to Environmental Health and Safety Issues* (Cambridge, MA: Harvard University Press, 1995); Ronald Bailey, *Eco-Scam: The False Prophets of Ecological Collapse* (New York: St. Martin's Press, 1993); Christopher Booker and Richard North, *Scared to Death: From BSE to Global Warming: Why Scares Are Costing Us the Earth* (London: Continuum, 2007).
15. Charles Mackay, *Extraordinary Popular Delusions and the Madness of Crowds* (1841; reprint ed., Hampshire, UK: Harrman House Ltd., 2007); Booker and North, *Scared to Death: From BSE to Global Warming: Why Scares Are Costing Us the Earth.*

who trusts in mankind, and makes flesh his strength, and whose heart turns away from the Lord. . . . Blessed is the man who trusts in the Lord and whose trust is the Lord. For he will be like a tree planted by the water, that extends its roots by a stream, and will not fear when the heat comes; but its leaves will be green, and it will not be anxious in a year of drought nor cease to yield fruit" (Jeremiah 17:5, 7–8).

A Christian should be aware of the unchristian roots and philosophies underlying the environmental religious movement today. It is important to get back to God's Word as the ultimate authority and rely on God and His Word as the solution to such issues.

CHAPTER 21

What about Distant Starlight Models?

DR. DANNY R. FAULKNER AND BODIE HODGE

Distant starlight is seen as one of the biggest difficulties to trusting God's Word about a young universe and earth. When adding up genealogies back to creation week, there are about 4,000 years from Christ to Adam.[1] With six normal-length days in creation week, there is no room for the idea of billions of years (Exodus 20:11)!

In *The New Answers Book 1*, astrophysicist Dr. Jason Lisle tackled the subject of distant starlight by looking at the various assumptions behind the issue.[2] This complementary chapter discusses the various models that have been proposed for distant starlight by creationists in an effort to show how this alleged problem can be overcome.

But we would like to give some background to make sure that readers understand the issues at stake.

Why Is Distant Starlight a Problem in the First Place?

Usually, the way this issue is couched to Bible-believing Christian is this: "So how do you get starlight billions of light years away to earth in only about 6,000 years?"

1. Bodie Hodge, "How Old Is the Earth?" in *The New Answers Book 2*, Ken Ham, gen. ed. (Green Forest, AR: Master Books, 2008), p. 41–52.
2. Jason Lisle, "Does Distant Starlight Prove the Universe Is Old?" in *The New Answers Book 1*, Ken Ham, gen ed. (Green Forest, AR: Master Books, 2006), p. 245–254.

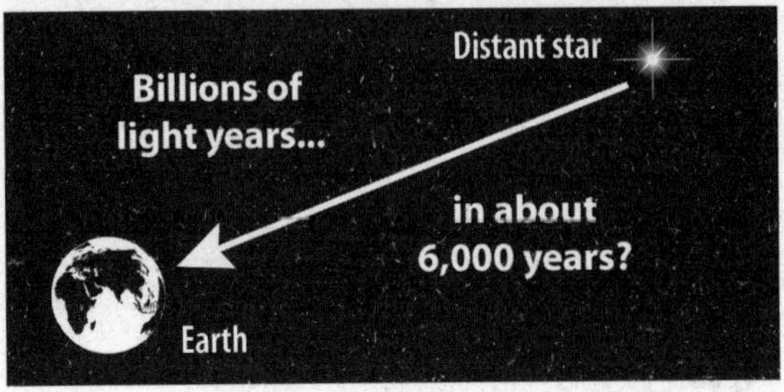

Most Christians are at a loss as to how to answer this question. Some try to say that the distances are not that accurate. But we would disagree. The distances really are that far.[3] That should give you an inkling of the mind of God!

There are ways to measure the distances such as parallax and the Hubble relation. We will not belabor these points, as they are already discussed in chapter 18 in this volume.

But the issue is even more difficult than many may think. We are not just trying to get light billions of light years away to earth in only 6,000 years, but we are trying to get light to earth in only two days. Why? The stars were created on day 4, and Adam was created on day 6. Starlight needs to arrive for Adam to be able to use the stars to mark the passage of time, which is one of the purposes of stars listed in Genesis 1:14.

The Secularists Have the Same Sort of Problem

The opposition rarely realizes that they have a starlight problem, too. In the big-bang model, there is the "Horizon Problem," a variant of the light-travel time problem.[4] This is based on the exchange of starlight/electromagnetic radiation to make the universe a constant temperature.

In the supposed big bang, the light could not have been exchanged and the universe was expected to have many variations of temperature, but this was not the case when measured. Such problems cause many to struggle with the big-bang model, and rightly so.

(1) Early in the alleged big bang, points A and B start out with different temperatures.
(2) Today, points A and B have the same temperature, yet there has not been enough time for them to exchange light.

3. See Danny Faulkner, "Astronomical Distance Determination Methods and the Light Travel Time Problem," *Answers Research Journal* 6 (2013): p. 211–229, http://www.answersingenesis.org/articles/arj/v6/n1/astronomical-distance-light-travel-problem.
4. Robert Newton, "Light-Travel Time: A Problem for the Big Bang," *Creation*, September–November 2003, p. 48–49, http://www.answersingenesis.org/articles/cm/v25/n4/light-travel-time.

Inflation

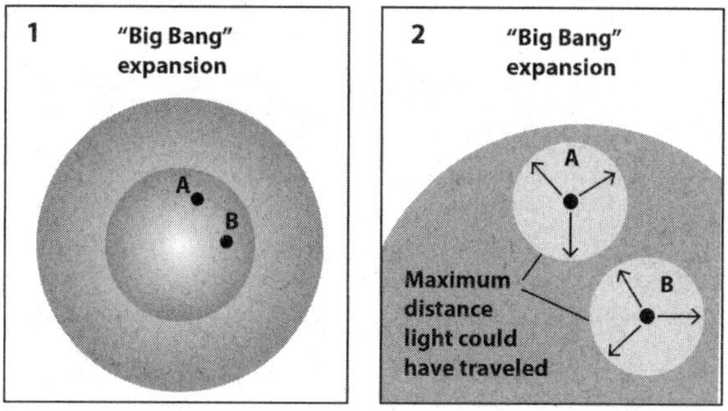

How did secularists try to solve it? In laymen's terms, they appealed to "inflation of the universe" in big-bang models as an *ad hoc* explanation. In other words, very quickly after the big bang, the fabric of space in the universe supposedly expanded very quickly (faster than the speed of light), then instantly slowed to the rate we see today. But what caused all that?

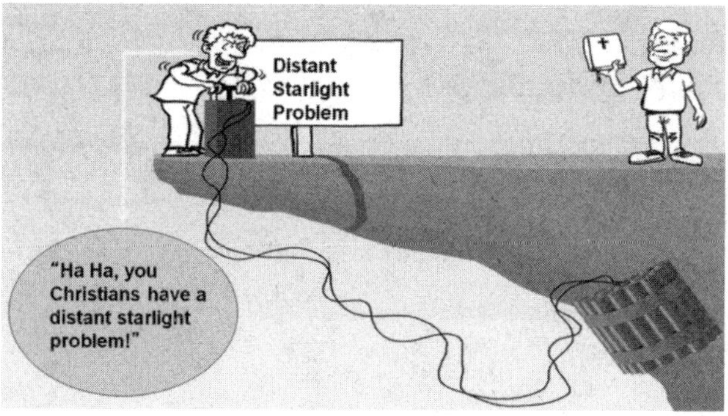

They suggest that some field existed that caused inflation. There is no direct evidence of inflation; that is, there is no independent evidence. Inflation was invented to solve the horizon problem and another problem (the flatness problem, but that will not be addressed in this chapter).

Researchers recognize there are problems with inflation and the big bang. Some physicists and astronomers have been "jumping ship" from the big-bang model in recent times, and this movement has continually gained steam since an open letter with respected signatories was published in the magazine *New Scientist* in 2004.[5] However, the majority of old universe believers still adhere to the big bang.

The hope of many who opposed the big bang was to revise the big bang and inflation to avoid the many problems. More recently, *New Scientist* ran an article called "Bang Goes the Theory."[6] The article quotes two leading cosmologists, Drs. Paul Steinhardt and Max Tegmark:

5. Eric Lerner, "Bucking the Big Bang," *New Scientist*, May 22, 2004, p. 20. To view the signers of this statement, visit http://www.cosmologystatement.org.
6. Amanda Gefter, "Bang Goes the Theory," *New Scientist*, June 30, 2012, p. 32–37.

> We thought that inflation predicted a smooth, flat universe. . . . Instead, it predicts every possibility an infinite number of times. We're back to square one.[7]
>
> Inflation has destroyed itself. It logically self-destructed.[8]

To boil it down, some researchers recognize there are problems with inflation and the big bang, and they are questioning aspects of these ideas, such as:

1. the big bang and its type
2. nothing to something
3. what started and stopped inflation
4. the starlight problem and recognizing how bad it is

Inflation and the big bang have their problems, and honest scientists fully admit this.

Potential Models to Solve the Problem

Interestingly, biblical creationists have known about the distant starlight problem for a while and have been working on solutions. The popular ideas include:

1. Light in transit (or mature creation)
2. Speed of light decay (cdk[9])
3. Relativistic models
4. Alternate Synchrony Conventions
5. *Dasha* Solution

Let's take a look at each of these in brief.

Light in Transit

Light in transit: This is the idea that God created the universe mature, or fully functioning. The functions of the stars (Genesis 1: 14–17; Psalm 19:1–2) required that Adam see them right away, so God created starlight in transit when He created the stars. Many reject this particular model today.

The reason many do not accept the light in transit idea is because starlight contains a tremendous amount of detailed information about stars. For instance, stars have been known to blow up into supernovas like *SN 1987a*.

7. Ibid., p. 35.
8. Ibid., p. 35.
9. cdk = c decay, where c is the symbol that physicists use for the speed of light.

Had this merely been starlight in transit, then what we saw would *not* have represented a star or a supernova, but instead merely light arriving at our eye to *appear* as a star and then a supernova. In other words, the star that was observed before the supernova could not have come from the actual star. If the light in transit idea is correct, then the light was encoded on the way to earth to make it look like an actual star. In that case, the supernova itself did not really happen but amounted to an illusion, sort of like a movie.

Many have suggested that if this were the case, then most stars are not stars. The implication is that God would be deceptively leading us to believe they were stars, when in fact they are illusions of stars. The idea of light in transit was widely popular among creationists for some time, but now many reject this idea because it seems far too deceptive.

Speed of Light Decay (cdk)

Speed of light decay (spearheaded by Barry Setterfield): This is the idea that the speed of light was much faster in the past and has been slowing down primarily in a uniform fashion (but possibly in steps) to what we observe today.

Most creationists reject this idea now, but we encourage researchers to keep working on it. In the end though, it appears to have problems with other constants in the universe that are tied to it. If the speed of light were to change, then these constants would change, too. Those constants govern the structure of matter so that matter would drastically change as the speed of light changed.

Evidence for a reduced speed of light decay is also lacking and in centuries past, the accuracy of such measuring devices has been limited. Furthermore, as people really researched the speed of light over the past three centuries, it really was not changing as previously thought, but has remained largely the same.[10]

In recent times, secularists such as John Moffat, Andreas Albrecht, and Joao Magueijo have appealed to the speed of light decay (VSL or Variable Speed of Light) as a possible solution to the secular starlight problem.[11] Perhaps as secular scientists do further research, they will see that there are some problems with this model. Either way, creation scientists are "light years" ahead of them in the research (pun intended).

10. Gerald A. Aardsma, "Has the Speed of Light Decayed?" Institute for Creation Research, http://www.icr.org/article/has-speed-light-decayed/ (accessed June 17, 2013).
11. Andrew Sibley, "Variable Speed of Light Research Gets a Boost," *Journal of Creation* 20, no. 1 (2006): p. 16–18, http://creation.com/images/pdfs/tj/j20_1/j20_1_16-18.pdf.

Relativistic Models

White Hole Cosmology[12]

Dr. Russell Humphreys has a model dubbed the "White Hole" cosmology. A white hole is like a black hole, except that matter flies outward from a white hole whereas matter falls into a black hole. Near the boundary of a black hole or a white hole, space and time are distorted. According to Einstein's theory of general relativity, this distortion can be described as stretching the fabric of space, and time progresses at different rates depending upon where you are.

So this theory plays off general relativity to solve the distant starlight problem with gravitational time dilation. From an overview perspective, Dr. Humphreys challenges the commonly held assumption that the universe has no boundary. Running a bounded cosmos through general relativity results in a model that is not at all like the big bang and consistent with biblical creation.

Essentially, in the White Hole cosmology, all the matter in the universe flew out of this "white hole." This would have occurred during creation week, and the white hole would have vanished some time during that week. As matter left the white hole, gravitational time dilation occurred. The earth was near the center of the white hole, so time on earth passed much more slowly than time near the boundary of the white hole.

Though there are still problems with this issue, such as blue shifts and red shifts not matching what they should be,[13] this model also holds some promise, and so we encourage further work on this model.

Hartnett Model (Carmelian Physics)[14]

Physicist Dr. John Hartnett has two different models. One of those models utilizes relativity, similar but different from Dr. Humphreys. Like the Humphreys model, the Hartnett model also relies on time dilation — a massive amount on earth. He postulates that most of this occurred on day 4 of creation week resulting from space expansion as God was creating galaxies. So time was running at different rates with six days passing on earth but more time passing elsewhere. Much of this dilation of time would have occurred during creation

12. D. Russell Humphreys, *Starlight and Time* (Green Forest, AR: Master Books, 1994).
13. John G. Hartnett, "Look-back Time in Our Galactic Neighbourhood Leads to a New Cosmogony," *Technical Journal* 17, no. 1 (2003): p. 73–79.
14. Hartnett, "A New Cosmology: Solution to the Starlight Travel Time Problem," *Technical Journal* 17, no. 2 (August 2003): 98–102; Hartnett, "Starlight, Time, and the New Physics," in Proceedings of The Sixth International Conference on Creationism, Andrew A. Snelling, ed. (Pittsburg, PA: Creation Science Fellowship, Inc., and Institute for Creation Research, 2008), p. 193–204.

week, as opposed to Humphrey's model where it occurred all along at a more steady rate. Hartnett has produced some interesting results. Both the Humphreys and the Hartnett models are still being developed.

The other method is based on a solution utilizing Carmelian physics (named for Moshe Carmeli). In a different approach than four dimensions, this has assumed five dimensions (utilizing Carmeli's approach).

Alternate Synchrony Conventions

Lisle-Einstein Convention[15]

This model derives from passages like Genesis 1:17 that states that the stars were to "give light on the earth." For a God who created all things, having distant stars give light on earth is no problem. Astrophysicist Dr. Jason Lisle (also writing under the pen name of Robert Newton) led the research on this model.

From the concept of light being given from stars to light the earth, Dr. Lisle derived the Lisle-Einstein Synchrony Convention, otherwise known as the Anisotropic Synchrony Convention (ASC), which is based on an alternative convention that is *position-based* physics as opposed to *velocity-based* physics. Einstein left open both options but did most of his work on velocity based, and so have most physicists since him.

Einstein pointed out that time is not constant in the universe, so our simple equation [Distance = Speed x Time] is not so simple anymore. But this starlight model is based on something quite "simple." Dr. Jason Lisle built on this position-based physics and the *one* direction speed of light (which cannot be known), and it solves distant starlight.

In laymen's terms, think of it like this: You leave on a jet from New York at 1 P.M. and you land in L.A. at 1 P.M. But you might say, "The flight took about five hours on the jet." Here is the difference: according to Einstein, when you approach the speed of light, time goes to zero. So if you rode on top of a light beam from a star that was billions of light years away from earth, it took *no time* for you to get here. So that five-hour flight was a "no hour" flight for light. It was an instantaneous trip.

15. For more, see Robert Newton, "Distant Starlight and Genesis: Conventions of Time Measurement," *Technical Journal* 15, no. 1 (April 2001): p. 80–85, http://www.answersingenesis.org/articles/tj/v15/n1/starlight; Jason Lisle, "Anisotropic Synchrony Convention — A Solution to the Distant Starlight Problem," *Answers Research Journal* 3 (2010): p. 191–207, http://www.answersingenesis.org/articles/arj/v3/n1/anisotropic-synchrony-convention; Lisle, "Distant Starlight — Anisotropic Synchrony Convention," Answers, January–March 2011, p. 68–71, http://www.answersingenesis.org/articles/am/v6/n1/distant-starlight.

Based on this convention-based model, light left distant stars and arrived on earth in no time. This fulfills God's statement that these lights were to give light on the earth in Genesis 1:14. Of course, the physics is more complicated than this, but this analogy should give you an idea of how this model might work. However, it does not appear that we could perform an experiment to see if the ASC solution is true.

Dasha Solution

We would leave open miraculous options (as this was creation week). One particular form is by co-author Dr. Danny Faulkner (astronomer) dubbed the *Dasha* Solution.[16] *Dasha* is the Hebrew word for "sprout" as found in Genesis 1:11. Many processes during creation week were done at rates uncommon today.

While some things were created *ex nihilo* (out of nothing) during creation week (Genesis 1:1), many things during that week probably were made of material created earlier in the week. For instance, the day 3 account tells us something about how God made plants (Genesis 1:11–12). The words used there suggest that the plants shot up out of the ground very quickly, sort of like a time-lapse movie. That is, there may have been normal growth accomplished abnormally quickly. The result was that plants bore fruit that the animals required for food two to three days later. The plants had to mature rapidly to fulfill their function.

God made stars on day 4, but to fulfill their functions the stars had to be visible by day 6 when Adam was on the scene. As the normal process of plant development may have been sped up on day 3, the normal travel of starlight may have been sped up on day 4. If so, this rapid thrusting of light toward earth could be likened to the stretching of the heavens already mentioned.

Some people may want to equate this stretching of starlight with some physical mechanism such as cdk or relativistic time effects, but this would not explain the abnormally fast development of plants on day 3. This also overlooks the fact that much about the creation week was miraculous, hence untestable today. If one were to attempt to explain the light travel time problem in terms of a physical mechanism, one might as well look for a physical mechanism for the virgin birth or Resurrection.

---

16. Danny Faulkner, "Astronomical Distance Determination Methods and the Light Travel Time Problem," *Answers Research Journal* 6 (2013): p. 211–229, http://www.answersingenesis.org/articles/arj/v6/n1/astronomical-distance-light-travel-problem.

Conclusions

When all is said and done, this alleged problem of distant starlight does not seem as problematic for the biblical creationist. Researchers have several options that can solve this problem, so it is not a problem for a young universe. Furthermore, we want to encourage researchers currently working on these projects.

But from a big picture standpoint, no one outside of God completely understands all the aspects of *light* (or *time* for that matter). It acts as a particle and in other instances acts as a wave, but we simply cannot test both at the same time. This dual behavior is still an underlying mystery in science that is simply accepted in practice. The more light is studied, the more questions we have, rather than finding answers.

Such things are similar in the theological world with the deity of Christ (fully man and fully God). Even the Trinity is a unique yet accepted mystery (Father, Son, and Holy Spirit; one God but three persons). And in science, there is the "triple point" of water, where at one temperature and pressure, water is solid, liquid, and gas at the same time.

Light is truly unique in its makeup and properties, and with further study perhaps we can be "enlightened" to understand this issue in more detail. Regarding the distant starlight issue, there are plenty of models that have some promising elements to solve this alleged problem, and we would leave open future models that have not been developed yet (and we would also leave open the miraculous).

But as we consider the light travel time problem, we frequently overlook the immensity of the creation itself. The sudden appearance of space, time, matter, and energy is a remarkable and truly miraculous event. This is something that we humans cannot comprehend at all. Compared to creation, the light travel time problem is not very big at all.

CHAPTER 22

What Are the Tactics of the New Atheists?

DR. ELIZABETH MITCHELL[1]

Following the April 29 opening of their documentary *The Unbelievers* at Toronto's Hot Docs Film Festival, outspoken atheists Richard Dawkins and Lawrence Krauss discussed the merits of their approaches to "ridding the world of religion." In a recent interview with Steve Paikin,[2] they made it clear that, despite their sometimes different personas, they have the same agenda — getting people to get rid of their belief in God. Yet they both say that Christians should not feel "threatened" by their efforts to expunge religion from human history.

The Goal of *The Unbelievers* Documentary

Evolutionary biologist Dawkins and theoretical physicist Krauss recounted that when they first met they had a heated debate about, as Dawkins said, "Whether we should have a kind of full-on attack on religion or whether we should, as Lawrence preferred, seduce them."[3] Krauss explained that this is really

1. Footnotes are by Bodie Hodge.
2. http://ww3.tvo.org/video/190768/rise-new-atheists.
3. Of course, Dawkins means all religions but his own. He is very religious, being a secular humanist. He is a signer of the Humanist Manifesto III. Humanism comes in various flavors like "agnosticism," "traditional atheism," "new atheism," etc. When someone says he is "not religious" in this context, it is a fancy way of saying he adheres to the religion of humanism in one form or another. Dawkins' religious viewpoint is "new atheism," distinguished from traditional atheism in that it actively proselytizes for the atheistic viewpoint, whereas adherents of traditional atheism believe that nothing matters and so see no reason to proselytize.

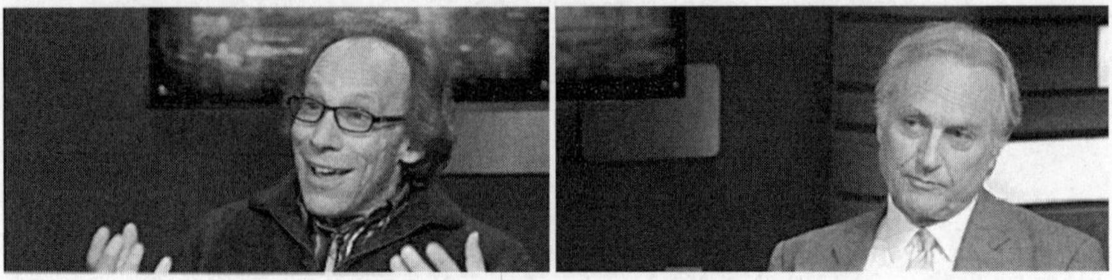

Outspoken atheists Lawrence Krauss and Richard Dawkins, costars of the documentary *The Unbelievers,* discuss their strategy for ridding the world of religion in general and Christianity in particular. They consider Christianity "demeaning" and wish to re-design society "the way we want it." (Image:screen shots from interview with Steve Paikin on http://ww3.tvo.org/video/190768/rise-new-atheists.)

"a strategic question."[4] They agree that both approaches have merit depending on the nature of the people being targeted. However, expressing general agreement with the more confrontational approach of the often-irascible Dawkins, Krauss said, "You've got to confront silly beliefs by telling them they are silly," adding, "If you're trying to convince people, pointing out that what they believe is nonsense is a better way to bring them around."[5]

4. We have known about their strategic attacks for some time. They have tried to force the religion of humanism in the classroom and now elsewhere. In 1983, humanist John Dunphy also spoke of this strategy — to put their atheistic religion into schools — when he said: "I am convinced that the battle for humankind's future must be waged and won in the public school classroom by teachers who correctly perceive their role as the proselytizers of a new faith: a religion of humanity that recognizes and respects the spark of what theologians call divinity in every human being. These teachers must embody the same selfless dedication as the most rabid fundamentalist preachers, for they will be ministers of another sort, utilizing a classroom instead of a pulpit to convey humanist values in whatever subject they teach, regardless of the educational level — preschool, daycare, or large state university. The classroom must and will become an arena of conflict between the old and the new — the rotting corpse of Christianity, together with all its adjacent evils and misery, and the new faith of humanism." John Dunphy, "A Religion for a New Age," quoted in John Dunphy, "The Book that Started It All," Council for Secular Humanism, http://www.secularhumanism.org/index.php?section=library&page=dunphy_21_4.
5. Yet these atheists do not realize the silliness of their own views. Dawkins himself admits that it is possible that aliens designed and seeded life on earth — yes, really! Krauss and Dawkins both believe that all people ultimately came from a rock — clearly this is in violation of the law of biogenesis. Both believe that everything is material; therefore, from their view, logic, truth, and knowledge, which are non-material, cannot exist. By thus laying claim to logic, truth, and knowledge, they inadvertently borrow from a Christian worldview — how silly for their religion to borrow from its enemy! Dawkins argues there is no morality and then tries to say Christians are immoral. Both believe that nothing ultimately matters; yet they both seem to think it matters a great deal to force this belief on others. Neither Krauss nor Dawkins seem to realize that in an atheistic worldview, the atheist is actually claiming to be "God" (because to know there is no God, one must be omnipresent and omniscient, which are attributes of God alone), which refutes their own atheism. This short list should suffice. Such silliness should be embarrassing to an atheist.

Despite their great hostility toward religious beliefs (other than their own) and avowal that they hope this film will help in their efforts to eradicate all religion worldwide, the atheist pair indicates that belief or non-belief in a deity is not what really matters to them. Krauss declares that what is actually important to them is that "everything should be open to question and that the universe is a remarkable place."[6] By contrast, he says, "This is more important to us than not believing in God — that's not important at all." Dawkins and Krauss both expressed grudging tolerance for evolutionists who want to keep their religious beliefs in order to keep the good things religion offers them — "spirituality," "consolation," and "community" — so long as they do not then reject evolution.[7] They said that people are "hard-wired" to seek something spiritual, but by "spiritual" they refer to a sort of emotional high. And they declare that science offers a better kind of spirituality, "a sense of oneness with the universe."[8] Therefore science,[9] they maintain, can meet the inmost needs of people better than religion of any sort.

"Spirituality is a sense of awe and wonder at something bigger than oneself,"[10] Krauss explained, adding that being "insignificant is uplifting."[11] And while some people cling to their religion to satisfy some spiritual need,[12] he says, "The spirituality of science is better than the spirituality of religion because

6. Interestingly, Christians believe in asking questions and seeking answers to all sorts of tough questions — including the scientific and the theological. And Christians certainly recognize that the universe is a remarkable place, but we know it was created by God. So the opposition to Christianity on this ground is completely without warrant by their own criteria.
7. Evolution (and millions of years, or geological evolution) is the real key. These are tenets of the Humanist Manifestos, so humanists do not want to give up this key aspect. They must fight for this in their religion. But underlying all of this is the idea that man is the ultimate authority, not God.
8. "Oneness with the universe" is a tenant of Buddhism, which is strange, considering they are arguing to oppose Buddhism along with all other religions.
9. What they mean by "science" here is not the observable and repeatable science that makes discoveries about how things work and applies that knowledge, but instead a "science" that embraces naturalism and evolution as absolutely axiomatic. Therefore, what Dawkins and Krauss mean when they say science is not just how things work but their own naturalistic, unverifiable, dogmatically held ideas about where everything came from. By science, they really mean their religion of humanism.
10. If one believes there is something greater than oneself in atheism, then it means that he is not atheistic. Hence, this is self-refuting.
11. If being insignificant is so great, then why waste time seeking popularity by speaking out against Christianity by making documentaries?
12. This is oddly similar to what the religious atheist is doing, per the very context.

it is real."[13] Both of course vigorously deny that their own atheistic position is one of "belief," saying "we don't define ourselves by what we don't believe in."

Dawkins and Krauss Want to Rid the World of All Religion Except Their Own

Like most atheists, Dawkins and Krauss fail to recognize the worldview-based nature of the interpretations they define as "real." They repeatedly refer in the interview to accepting the "evidence of reality" concerning origins when they are actually equating their worldview-based interpretations with reality. Furthermore, the atheistic belief that there is no God is actually a "religion."

There really is no such thing as a person without a religion — you either believe that there is or is not a god. You are either for Christ or against Him (Luke 11:23), and you base your interpretation of origins, morality, and the meaning of life on that belief. The belief that there is or is not a god is essential to how one explains existence, the nature of authority, and our place in the universe. Krauss's belief that the atoms in his body originated billions of years ago in stardust, for instance, is the "religious" way he explains his existence without God and the way he experiences what passes for spirituality by knowing the "fantastic" truth that he is "intimately connected to the cosmos."

Atheists do claim to be non-religious, but they use their set of beliefs as a way to explain life without God — they worship and serve the creation (e.g., the universe) rather than the Creator (Romans 1:25). Krauss extols the profound sense of wonder he gets studying the cosmos and Dawkins enjoys the "poetry of science," but they tie their love for science to their belief in atheistic evolution and their sheer joy in shaking their fists at the possibility of a Creator's existence.

The Reason Behind the Hostility toward Religion

And frankly, the point here is not whether a person defines his worldview as a religion or not, or whether he believes in a "god." Christianity is unique — it is the truth — and, perhaps for that reason as much as any other, is the especial target for Dawkins and most others. Those who love "darkness" (e.g., sin, rebellion against God, and rejection of Jesus Christ) will naturally attack the light (John 3:19–21). Based on Scripture, we know that God looks at the heart to see how each person stands in relation to Jesus Christ (Romans 10:9–10; cf.

13. This is a "No True Scotsman" fallacy, meaning that the arguer has defined the terms in a biased way to protect his argument from rebuttals.

1 Samuel 16:7). Again, Jesus made clear that a person is either for or against Him (Matthew 12:30, 25:46).

Dawkins and Krauss reserve their greatest hostility for young-earth creationists. They indicated that all debate about origins has been completely and unequivocally settled by "Darwin and his successors"[14] or else by big-bang cosmology,[15] which Krauss describes as "the last bastion of God — I mean there are some fundamentalists of course who say the earth is 6,000 years old and don't believe in evolution — but rational 'theologians' have moved away from that debate."[16]

Design in Nature

Furthermore, even Dawkins admits that nature — in particular, biology — appears to be specially designed. We see, for instance, precise irreducible complexity everywhere we look, from major anatomical features to biological processes at the molecular level. Dawkins agrees that "special creation" is "intuitive" — a look at nature in essence screams that there must have been a Creator. But Dawkins says that he is thankful to Darwin for coming up with a very "*non-intuitive*" way to explain nature without God. Darwinian belief basically builds a theoretical guess about biological origins by appealing to a series of billions of tiny, unobservable changes over billions of unobserved years.[17] Yet neither Darwin nor his successors have through scientific observation shown how either abiogenesis or the evolution of biological complexity is possible.

Dawkins explains that both biology and physics (cosmology) are complementary fields that supplant belief in God.[18] But he indicated that biology,

14. It is sad that they appeal to Darwin, a racist, who went so far as to say that the more evolved Caucasians would eventually exterminate everyone else (Charles Darwin, *The Descent of Man* [New York: A.L. Burt, 1874, 2nd ed.], p. 178). Even James Watson, a co-discoverer of the structure of DNA, also has underlying racist attitudes. But note that they appeal to man as the ultimate authority.
15. Which big-bang model (open models, closed model) do they think is true, and why are the others wrong?
16. The atheists simply do not like the fact that Christians actually believe God when He speaks. They really want us to compromise God's Word with theirs like Eve did in the Garden and to deny God's Word in Genesis in favor of their fallible sinful words. The issue is not mere distaste for creationists, but rather their distaste for God's Word. Note this: the conflict is not between atheists and creationists; it is between atheists and God.
17. Note what replaced God in their religion. It was time, chance, and death. Without these, evolution is meaningless. These are the "god" for an evolutionary worldview.
18. Yet science comes out of a Christian worldview, where God upholds the universe in a particular fashion, and this all-knowing God has told us so (e.g., Genesis 8:22 and others). In the humanistic view, how can man know that the laws in the universe will be the same

because design is so apparent, was the first battleground in the war against a Creator:

> Historically biology, I suppose, has been the most fertile ground for thosc who wish to make a supernatural account because living things are so fantastically complicated and beautiful and elegant, and they carry such an enormous weight of apparent design. They really look as though they're designed.
>
> So historically biology has been the most fertile ground for theological arguments. That's all solved now. Darwin and his successors solved that.
>
> I think the spotlight in a way has shifted to physics and to cosmology where we're less confident I think about how the universe began — in one way more confident because there's a lot of detailed mathematical modeling going on — but there are some profound questions remaining to be answered in that field and that's where cosmologists like Lawrence come in. We are complementary.

In typical fashion, Krauss and Dawkins believe that anyone who disagrees with their own interpretations about origins is irrational and out of touch with reality. And as happens with most lay people, anything that can be "mathematically modeled" is accepted as truth because numbers surely do not lie. Yet mathematical models concerning cosmology (like the big bang) and the long-age interpretations ascribed to radiometric dating are based on unverifiable, worldview-based assumptions.[19] Dawkins and Krauss say that they hope that viewers of their film will be inspired by the wonders of science to critically evaluate their beliefs and to acknowledge that they are "silly." As discussed below, however, from a biblical worldview, a careful study of the wonders of science only affirms what God reveals in the Bible and actually glorifies the Creator (Psalm 19:1; Colossians 1:16–17).

Biblical creationists understand that God created all the various kinds of living organisms about *6,000 years ago* (based on the genealogies listed in the Bible). According to Genesis 1, God equipped each to reproduce "after their kinds." There is no indication in Scripture that God used evolutionary processes

---

in the future? According to man, from the big bang to today, the laws have changed. How does one know they will not change tomorrow? If one says, "Because they always have," he is arbitrarily begging the question.

19. Such methods are classic cases of begging the question; they are using long-age assumptions to prove long ages. We could just as easily do the same thing by using young-age assumptions to prove a young earth, but this simply shows the arbitrariness of their uniformitarian claims.

or that He made organisms able to evolve through random processes into new and increasingly complex kinds of creatures. We also do not see this happen in biology. As many articles on the Answers in Genesis website explain, organisms vary within their kinds (e.g., variations in dogs or in cats) but do not evolve into new, more complex kinds of organisms (e.g., amoebas into dogs or cats). Bacteria remain bacteria, canines remain canines, apes remain apes, and humans remain humans — though there is much biodiversity among each *created kind.* This diversification within kinds is observable. But evolution of new kinds is not, and biological observation can offer no actual *mechanisms* by which *this can happen.*[20]

Further, biological observation confirms that living things do not spring into existence through the random interaction of non-living components, despite evolutionary claims about abiogenesis. This is consistent with the biblical account of our origins. Thus, biblical history — God's eyewitness account of what He did when He created us and what sort of biology He put in motion — does *not* differ from biological observations. There is nothing "irrational" about recognizing that observable science is consistent with biblical history.[21]

Can Dawkins and Krauss Really "Rid This World of Religion"?

The interviewer concluded by asking the pair, *"Is it your hope or expectation that you can, in your words, rid this world of religion?"*

"I'm not sure how soon," Dawkins answered. "I think that religion is declining, that Christianity is declining throughout Christendom."[22] Looking to the future, he adds, "And I think that that's going to continue. If we look at

20. The two proposed mechanisms of evolution are called: (1) natural selection, a creationist concept by the way, and (2) mutations. In both cases, they are losing information (i.e., it is going in the wrong direction for evolution). For example, natural selection filters out already existing information; mutations lose information quickly, or in many cases it remains nearly neutral. See http://www.answersingenesis.org/articles/nab/is-natural-selection-evolution and http://www.answersingenesis.org/articles/nab2/mutations-engine-of-evolution.
21. Isn't it fascinating that humanists who are materialistic by their very admission appeal to logic and claim we are irrational, when rational thought is only possible if nonmaterial things exist like concepts, truth, logic, and so on? Yet these atheists (materialists, humanists) must reject it because if they leave open an immaterial realm (i.e., a spiritual realm), then God could exist and they cannot be atheistic or humanistic (i.e., humans are the ultimate authority).
22. Yet Christianity is still the fastest growing religion. Please see http://fastestgrowingreligion.com/numbers.html; it is merely declining or stagnant in certain places, like Western Europe and the United States.

the broad sweep of history, it's clear that the trend is going in the right direction. I'm not so optimistic that it will be in my lifetime, but it will happen."[23]

And what do Dawkins and Krauss hope to accomplish by getting rid of Christianity? Why do they care what others believe? Why are they so eager to expedite God's exit from human history? Dawkins summed up the proud position of humanism when he said that he wants to see us "intelligently design our society, our ethics, our morality — so that we live in the kind of society we want to live in rather than in the kind of society that was laid down in a book written in 800 B.C."[24] Krauss added that accepting the ideas of "Iron Age peasants" is "demeaning."[25]

Though Dawkins and Krauss disparage the ideas of biblical peasants, their notions of social planning really sound very much like the post-Flood population who built the Tower of Babel in rebellion against God's command to replenish the earth. In their pride (Psalm 10:4; Proverbs 16:8), those people said, "Let us make a name for ourselves" (Genesis 11:4). Indeed, how arrogant does a person have to be to assume that everyone who disagrees with him is either ill-informed or irrational? Is it any wonder that God hates pride, for through humanistic pride people not only reject God's ways but "suppress the truth" (Romans 1:18) of His very existence?

Dawkins and Krauss seem to want to redesign the world and society for the rest of us according to their own vision, making certain that God is written out of the picture. Yet those of us who know and trust God and accept the Bible as His revealed Word believe wholeheartedly that Jesus Christ, our Creator and Savior, possesses all true wisdom and knowledge (Colossians 1:16–17, 3:2). And we not only accept the history in God's Word but also God's declaration that we

23. Did you catch that Dawkins just made a prophecy? He predicted that religion would cease. God disagrees with him (Matthew 16:18; Daniel 2:44).
24. Satan, in the Bible, sinned with his pride of wanting to ascend to God's position (Isaiah 14:14). It appears clear that Dawkins wants to replace God, too, as the "intelligent designer" no less, albeit of society rather than the universe. (We suppose even Dawkins knows he has some limitations!) Interestingly, Dawkins does seem to believe in a form of intelligent design because he has said he considers it a possibility that aliens designed life here (per his comments in the documentary *Expelled* with Ben Stein, not in this interview). Furthermore, it is unclear what book Dawkins is talking about, though he is surely alluding to the Bible with a prejudicial conjecture about the timing. The Bible was written over the course of about 1450 B.C. to about A.D. 68–95. (Christians do debate this.) Take note of the irony here though; Dawkins wants people to follow what he says in his books, but not follow God's book! Again, he is trying to replace God (2 Corinthians 2:11), and in his own mind, he already has.
25. Note the straw man fallacies these atheists are committing. They are trying to make Christianity look silly, but because they cannot even get basic facts correct, they look silly by default.

are all sinners in need of the grace of Jesus Christ. By contrast, those who, like Dawkins and Krauss, refuse to even acknowledge the testimony of the "design" they themselves see in nature (Romans 1:18–22) and their own consciences (Romans 2:12–16), much less God's Word, are — according to God — "fools" (Psalm 14:1, 53:1). *"Professing to be wise, they became fools"* (Romans 1:22).

In answer to the interviewer's final question about the prospects for the imminent demise of religion, Krauss said, "I would have thought that by now religion would be gone. I thought religion was on the way out [in the 1960s], so I was kind of surprised and disappointed in some ways by the resurgence of fundamentalism in my country [the United States]."[26] Speaking of the future he expects, he adds, "But I do think that it's obvious that access to information and knowledge is decreasing" the number of people who say they are religious worldwide and that "inevitably knowledge and wonder of the real universe will supplant" religion.[27] Answers in Genesis exists to make knowledge available to help people make informed decisions about the claims of atheistic evolutionists so that they will see that they can trust God's Word from the very first verse.

Both Krauss and Dawkins think it unreasonable that people feel "threatened" by their efforts to rid the world of religion.[28] Dawkins said, "Where religion is concerned if you speak clearly it sounds threatening" and "if you say something clearly and distinctly and truthfully there are people who will take that as threatening." He said that religion is so entrenched that it "gets a free ride" and that "very mild criticism" and "questioning" shouldn't be regarded as threatening.[29]

26. This is reminiscent of atheist Friedrich Nietzsche who declared "God is dead" several times in the 1800s. It is sad that atheists like Krauss know so little about God's Word that they fail to realize a dominating principle: the power of God in the Resurrection. When the Jews had Christ crucified, even Christ's disciples thought the Son of God was dead. But God is known for His Resurrection. Though Nietzsche is dead, God continues to live and gives to all life and breath. And Christianity continues to grow by the power of the Holy Spirit.
27. Note here that Krauss has now prophesied the same sort of thing as Dawkins. He is predicting that universe worship, like his atheistic view, will come to destroy religion. But this would naturally fail, as atheism and universe worship are a form of religion, making Krauss's prediction inherently contradictory.
28. Actually, Christians should find it a blessing. Matthew 5:11 says, "Blessed are you when they revile and persecute you, and say all kinds of evil against you falsely for My sake."
29. Again, Christians do not fear questioning, nor do we get a free ride or mild criticism. Christians in various parts of the world are murdered for their beliefs, attacked and beaten for their beliefs, abused for their beliefs, and lied about because of their beliefs. If one is not a Christian, like Dawkins, why assume such people actually adhere to the Ten Commandments, which say not to lie? Dawkins claimed that there is no morality in his debate with Lanier. So why trust him to tell the truth? With this in mind, notice Dawkins's deception here. He wants the freedom to question, but he does not want us to respond. Nor does he want Christians to question things like evolution or the big bang —

Conclusion: Man's Word vs. God's Word

Krauss and Dawkins repeatedly refer to the "evidence of reality" in this interview. Yet they, like other evolutionary scientists, fail to distinguish between testable scientific reality — experimental science — and the untestable, unobservable, and unverifiable assumptions on which the scientific claims of evolutionary origins science are based. What they claim as "reality" is interpreted through their own worldview, a worldview that is clearly hostile toward God.[30] And while they oppose "all" religion, it is clear they particularly oppose Christianity and the Bible. They firmly believe that anyone who fails to accept their worldview is irrational. They admit that religion meets the needs of some people for "spirituality," but their concept of spirituality is a purely emotional response.[31]

And lest this "response" be deemed defensive (a point made not only in this interview but also by a number of atheists who have recently written in to this ministry), let me hasten to point out that if "just asking a question" should not be seen as "threatening," then neither should just answering one. If saying "something clearly and distinctly and truthfully" should not be seen as threatening when Dawkins speaks, then neither should the truth from God's Word be taken that way. It should not be threatening when we question evolution, big bang, millions of years, humanism, or even Dawkins and Krauss themselves. In fact, they would welcome it in every forum, if they were consistent.

Krauss and Dawkins do have one thing in common with most biblical creationists — a sense of awe and wonder at what we can learn from experimental science about the world around us. Krauss and Dawkins appreciate the "poetry of science," but superimpose their own rhapsodic notions about the

---

especially in classrooms! If he did welcome responses, he would be happy for Christians to question evolution, the big bang, naturalism, and so on, and to respond to his false claims about Christianity in a proper forum, like the classroom, which is a place for learning. But Dawkins is adamant that Christians should have no say, no response, and no questioning of the evolutionary view in the state schools. Dawkins wants only his religion taught in schools and only his religion is permitted to question others. This is a double standard.

30. Remember, they assume long ages to prove long ages — an arbitrary begging-the-question fallacy.
31. They are trying to demote all religions to being materialistic (underlings to their religion). This is why they say spiritual is not immaterial, but merely emotion (e.g., chemical reaction in the brain). They are trying to change the definition of spirit and spiritual. They want to make God (who is spirit, John 4:24) into part of the universe or place Him in a position that is lower than the universe. Hence, the universe can be the unofficial "god" to the atheist, next to *man*, of course.

atoms in our bodies being derived from stardust billions of years old.[32] Biblical creationists, however, examine the actual facts of science — the observable and repeatable ones, not evolutionary story telling and conjectures — in light of God's revealed truth and see that there actually is no contradiction between the history revealed in the Bible and science (Romans 1:18–22).

Krauss and Dawkins hope their film will prompt Christians to ask questions and to critically examine their beliefs in light of science. At Answers in Genesis we encourage people — both believers and unbelievers — to ask questions and to critically examine scriptural revelation and scientific facts. We provide help in finding answers to those questions. Sadly, one example Dawkins provided was a young-earth creationist who came to his lectures on evolution and was very impressed, having never heard the evolutionary point of view. We do not encourage ignorance about evolutionary positions[33] but instead want to equip people with the information they need to discern the difference between *observable experimental science and historical science*, between that which can be tested and that which can only be imagined, between what can actually be seen in the world through science and the claims of evolutionists.

We want to equip children, teens, and adults with the tools they need to help them trust God's Word and see through false religions like atheism, so that they will then be able to trust Jesus Christ as their Savior and the Lord of their lives. The very name of our ministry, Answers in Genesis, makes it clear we are not encouraging people to have blind faith. On the contrary, we are providing reasonable, scientific, and biblical answers for questions on origins. And we do so with confidence that the Bible has the answers to explain the world we live in — scientifically, morally, and theologically.

---

32. When Krauss attacks the Bible with his famous mantra, "Forget Jesus, the stars died so you can be here today," he is promoting a mere fairy tale and stories to satisfy a meaningless atheistic worldview.
33. This is why we teach people about each evolutionary view and its problems. In brief, there are five main views: (1) The Epicurean evolutionary view, which has its roots in Greek mythology. This is where evolution came from. The newer forms we have today are just rehashes of this mythology that Paul refuted in Acts 17. (2) Lamarckian evolution, which taught that animals can acquire new traits through interactions with their environments, and then pass them on to the next generation. (3) Traditional Darwinism, where natural selection and time are the primary factors for change. (4) Neo-Darwinism, where natural selection and time are combined with mutations as the primary factors for evolution. (5) Punctuated Equilibrium, which tries to explain the lack of fossil evidence for transitional forms. This view assumes that evolution occurred in bursts and is not recorded in the fossil layers; it still relies on natural selection, mutations, and time. For more, see Roger Patterson and Dr. Terry Mortenson, "Do Evolutionists Believe Darwin's Ideas about Evolution?" *New Answers Book 3*, Ken Ham, gen. ed. (Green Forest, AR: 2010), p. 271–282.

The Bible attests not only to the true history of our origins but also the truth about humanity's rebellious and sinful nature.[34] Dawkins and Krauss consider biblical truth restrictive and demeaning. The Bible does make it clear that all people are sinners who have rebelled against the omniscient, omnipotent, and holy God. Dawkins and Krauss personify this rebellious spirit in declaring their desire to redesign the world the way "we" — in other words, "they" — want it to be. But evil men and seducers will, according to Scripture, get worse and worse (2 Timothy 3:13), so much so that Jesus said *"Nevertheless, when the Son of Man comes, will He really find faith on the earth?" (Luke 18:8).* As Christians, meanwhile, we are commanded to respond to the "nonthreatening threats" volleyed at us by skeptics and by sincere questioners by providing answers (1 Peter 3:15, KJV; 2 Timothy 2:22–26), including the answer to people's sin problem (Romans 3:23, 6:23) — salvation through the shed blood of Jesus Christ. But the final end of humanity's destiny is not the end prophesied by Dawkins and Krauss, for the same Jesus Christ that rose from the dead will indeed come again (Revelation 22:20). Dawkins and Krauss may be leading the charge to eradicate Christianity, but it is the Lord Jesus Christ who will surely have the last word.

For more information:

http://www.answersingenesis.org/articles/aid/v4/n1/morality-and-irrationality-evolutionary-worldview.

http://blogs.answersingenesis.org/blogs/georgia-purdom/2011/10/04/the-magic-of-reality-or-unreality/.

http://www.answersingenesis.org/articles/am/v5/n2/variety-within-kinds.

http://www.answersingenesis.org/articles/arj/v5/n1/evolution-myth-biology.

http://www.answersingenesis.org/articles/2011/06/03/feedback-search-for-historical-adam.

34. It is important to note that in the beginning, God called His creation "very good" (Genesis 1:31; Deuteronomy 32:4). It is because of man's sin that death, suffering, and disease came into the creation. God did not make the world like it is today (full of suffering) but subjected it to this due to man's sin. We have essentially been given a taste of what life is like without God. But Christ did not leave us to perish; instead, He took the punishment that we deserve on the Cross, once for all. Christ, the God-man, took the infinite punishment that is demanded by the very nature of God, who is infinite. God then offers the free gift of salvation, and promises a new heavens and new earth that will not be subjected to death, suffering, and decay. See http://www.answersingenesis.org/articles/2009/04/21/what-does-it-mean-to-be-saved.

http://www.answersingenesis.org/articles/2012/06/01/feedback-evolutionary-call-to-arms.

http://www.answersingenesis.org/articles/2011/06/24/feedback-huffing-and-puffing.

http://www.answersingenesis.org/articles/2012/04/12/teacher-protection-academic-freedom-act.

http://www.answersingenesis.org/articles/2012/08/30/bill-nye-crusade-for-your-kids.

http://blogs.answersingenesis.org/blogs/ken-ham/2013/01/08/teaching-on-hell-worse-than-child-sexual-abuse/.

http://blogs.answersingenesis.org/blogs/ken-ham/2013/02/14/biblical-creation-and-child-abuse/.

http://blogs.answersingenesis.org/blogs/ken-ham/2012/09/18/origins-and-child-abuse/.

http://www.answersingenesis.org/articles/nab2/does-big-bang-fit-with-bible.

http://www.answersingenesis.org/articles/nab/is-natural-selection-evolution.

http://www.answersingenesis.org/articles/nab2/do-rock-record-fossils-favor-long-ages.

http://www.answersingenesis.org/articles/am/v5/n1/order-fossil-record.

http://www.answersingenesis.org/articles/am/v4/n3/radiometric-dating.

http://www.answersingenesis.org/articles/am/v4/n4/assumptions.

http://www.answersingenesis.org/articles/am/v5/n1/patterns.

http://www.answersingenesis.org/articles/am/v1/n1/radioactive-dating.

http://www.answersingenesis.org/store/product/does-biology-make-sense-without-darwin/.

**Thanks to Bodie Hodge, AiG–U.S., for his helpful and insightful additions in the footnotes.

CHAPTER 23

Were There Any Volcanoes, High Mountains, and Earthquakes before the Flood?

DR. ANDREW A. SNELLING

The Scriptures are silent on the issue of whether there were any volcanoes or earthquakes in the world before the Flood, but we do know there were mountains. The opening chapters of Genesis only have an abbreviated description of the earth's early history (only six chapters describing more than 1,650 years). However, it is still possible to glean hints from the scriptural record, and to a subordinate extent infer details from the geologic record, to demonstrate that the pre-Flood earth was likely very stable with no major catastrophes.

Springs and Rivers

We are told in Genesis 7:11 that the Flood began with the breaking up of "the fountains of the great deep," a vivid description of catastrophic geologic activity. This implies that whatever caused this "breaking up" was restrained in the pre-Flood world. While the Hebrew phrase translated "the great deep" is used in Scripture to refer to and describe sub-oceanic waters, some uses also refer to subterranean waters (Isaiah 51:10 and Psalm 78:15, respectively).[1]

1. David M. Fouts and Kurt P. Wise, "Blotting and Breaking Up: Miscellaneous Hebrew Studies in Geocatastrophism," in *Proceedings of the Fourth International Conference on Creationism*, Robert E. Walsh, ed. (Pittsburgh, PA: Creation Science Fellowship, 1998), p. 217–228.

So "the fountains of the great deep" in Genesis 7:11 would have likely been primarily oceanic springs, although the possibility of these also including terrestrial springs that tapped waters residing within the earth's crust cannot be ruled out. Thus, the geologic activity referred to by the term "breaking up" must imply deep fracturing of the earth's crust accompanied by dramatic earth movements, volcanic eruptions, and devastating earthquakes. Such catastrophic geologic activity on a global scale must therefore have been restrained and thus absent in the pre-Flood world.

Genesis 2:6 describes a mist that went up from the earth and watered the whole face of the ground. The Hebrew word usually translated as "mist" is *ed*, but old translations such as the Septuagint, Syriac text, and the Vulgate all translate the word as "spring."[2] Such a translation would seem relevant in the light of other biblical evidence for the existence of terrestrial and oceanic springs. In Revelation 14:7, an angel declares, "Worship Him who made heaven and earth, the sea, and the springs of waters," which suggests that fountains or springs were an integral part of the created earth. It would have been the same fountains that were then "broken up" at the beginning of the Flood (Genesis 7:11, "all the fountains of the great deep were broken up"). The connotation in both the Greek and Hebrew words used in these verses, respectively, is of gushing springs where water burst forth from inside the earth. It is also the connotation of a different Hebrew word used in Job 36, usually translated as "springs."

Some have contended that Genesis 2:5 implies that there was definitely rain in the pre-Flood era, just no rain before Adam was created, as stated in the verse. They have thus suggested that the river that flowed through the Garden of Eden to water it, and then split into four rivers (Genesis 2:10–14), was fed by these fountains or springs.[3] Of course, the biblical record does not specifically say that there was a connection between these fountains or springs and the rivers on the pre-Flood earth. However, since the existence of these springs and fountains on both the land surface and the ocean floor are clearly mentioned in the Scriptures, then it is not unreasonable to expect that at least some of the

2. Gordon J. Wenham, Genesis 1–15, vol. 1, *Word Biblical Commentary* (Waco, TX: Word Books, 1987), p. 58; Victor P. Hamilton, *The Book of Genesis: Chapters 1–17, The New International Commentary on the Old Testament* (Grand Rapids, MI: William B. Eerdmans, 1990), p. 154.
3. Joachim Scheven, "The Geological Record of Biblical Earth History," *Origins* (Biblical Creation Society UK) 3, no. 8 (1990): 8–13; Joachim Scheven, "Stasis in the Fossil Record as Confirmation of the Belief in Biblical Creation," in *Proceedings of the Second International Conference on Creationism*, vol. 1, Robert E. Walsh and Christopher L. Brooks, eds. (Pittsburgh, PA: Creation Science Fellowship, 1990), p. 197–215.

rivers on the pre-Flood earth were fed by springs. Furthermore, even though the Hebrew word *ed* in Genesis 2:6 is probably correctly translated as "mist," the existence of springs and fountains on the pre-Flood earth is clearly mentioned in other passages.

Nevertheless, we cannot be dogmatic that there was no rain for the entire pre-Flood era, even though Genesis 2:6 indicates that the mist "watered the whole face of the ground." In this way, the pre-Flood land surface must have been well watered and have produced lush vegetation. The latter is, of course, attested to by the huge volume of fossilized vegetation in the coal beds in the geologic record, which was destroyed and buried by the Flood.[4] Thus, climatic conditions in the pre-Flood era would seem to have been ideal for animal and human habitation across the face of the earth and must have been generally warm and humid. Though the Scriptures are silent on the subject, it could perhaps be inferred that there may not have been the same extremes of weather conditions that we experience on today's post-Flood earth. If this were the case, then it might also be inferred that there were not the same extremes in topography across the pre-Flood earth as there are today, because high mountains do affect weather patterns and conditions, for example, causing "rain shadows" and inducing snowfalls.

Topography and Mountains

While we are given no specific statements about the topography of the pre-Flood earth and how much it varied, we are given some hints. For example, the Garden of Eden must have been at a relatively high elevation, because we are told that the river flowing from it divided into four other rivers as it flowed downhill (Genesis 2:10–14). Furthermore, that there were mountains on the pre-Flood earth's land surface is clearly specified in Genesis 7:19–20, where we are told that the Flood waters prevailed exceedingly on the earth so that all the high hills under the whole of the heaven were covered, and then the mountains were covered. The difference between these topographic terms "hills" and "mountains" are somewhat subjective and arbitrary, but they do indicate a difference in sizes and elevations. So while we cannot be specific about the elevation differences on the pre-Flood land surface, we could potentially infer from all these descriptions that the topographic relief then was, for instance, not as enormously different and varied as it is today, and therefore was much more subdued. After all, today's high mountain ranges were produced during and

4. Andrew A. Snelling, "How Did We Get All This Coal?" *Answers* (April–June 2013): p. 70–73.

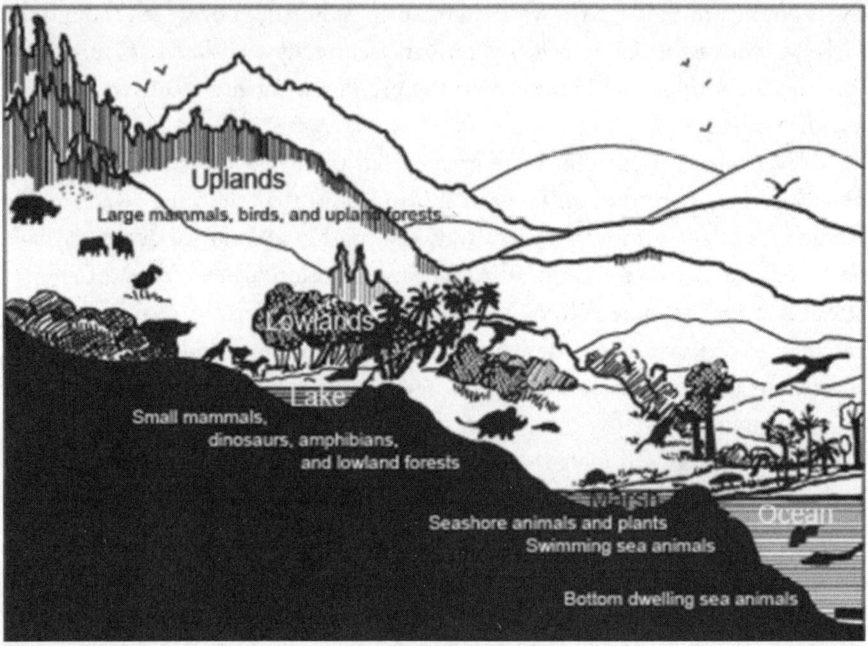

Figure 1. Diagrammatic presentation of likely ecological zonation in the pre-Flood world, illustrating how animals and plants could then be buried in a roughly predictable order by the rising Flood waters.

soon after the Flood catastrophe, because they often consist in part of Flood-deposited, fossil-bearing sedimentary rock layers that have been buckled due to catastrophic crustal plate collisions during the Flood, followed immediately by major rapid uplift due to post-Flood isostatic (vertical crustal weight balance) adjustments.[5]

Other clues not only come from the text of Scripture, but also from the geologic record of the Flood. It has been amply argued that the fossil-bearing sedimentary rock layers in the geologic record resulted from the Flood waters rising up over the continents and progressively burying different pre-Flood eco-systems and biological communities (figure 1).[6] Just as today there are different

5. Steven A. Austin, John R. Baumgardner, D. Russell Humphreys, Andrew A. Snelling, Larry Vardiman, and Kurt P. Wise, "Catastrophic Plate Tectonics: A Global Flood Model of Earth History," in *Proceedings of the Third International Conference on Creationism*, Robert E. Walsh, ed. (Pittsburgh, PA: Creation Science Fellowship, 1994), p. 609–621.
6. Harold W. Clark, *The New Diluvialism* (Angwin, CA: Science Publications, 1946); Harold G. Coffin, *Origin by Design* (Hagerstown, MD: Review and Herald Publishing

biological communities at different elevations because they are suited to those different micro-climates, it was likely the same in the pre-Flood world. Today, the rims of the Grand Canyon are covered in pine forests with squirrels, deer, and other animals, but as one hikes down into the canyon, with the loss in elevation and increasing temperatures, the biological communities gradually change, until at the bottom of the canyon the predominant vegetation is cacti typical of a desert, with different animals, such as ringtails, and seasonally, bighorn sheep.

In the fossil record, for example, dinosaur fossils are primarily found only in association with "naked seed" plants (gymnosperms) that do not have flowers, such as cycads and gingkoes. Flowering plants (angiosperms) are only rarely found fossilized with dinosaurs, and instead are found higher in the fossil record buried with mammal fossils. This potentially suggests that in the pre-Flood world there was a mammal-angiosperm biological community at higher elevations, geographically separated from a dinosaur-gymnosperm biological community at lower elevations.[7] We can conclude this difference in elevations between these two different biological communities (biomes) in the pre-Flood world because the dinosaur-gymnosperm biome would have been buried first as the Flood waters rose higher over the continents. Also, this difference in elevations between these two biomes thus likely not only reflects different elevations, but different climatic conditions for each biological community.

Furthermore, it is clear from the description of Adam's life in the Garden of Eden that the garden contained fruit trees (angiosperms) and beasts of the field that he named (mainly mammals). As well, the inference has already been noted that the garden was at a higher elevation, because the river running through it flowed downhill out of it and divided into four other rivers. Thus the mammal-angiosperm biological community must have been at the generally cooler higher elevations in the pre-Flood world. That would have meant the geographically separated dinosaur-gymnosperm biological community was

Association, 1983); Andrew A. Snelling, "Doesn't the Order of Fossils in the Rock Record Favor Long Ages?" in *The New Answers Book 2*, Ken Ham, gen. ed. (Green Forest, AR: Master Books, 2008), p. 341–354; Kurt P. Wise, "Exotic Communities Buried by the Flood," *Answers* (October–December 2008): p. 44–45; Andrew A. Snelling, "Order in the Fossil Record," *Answers* (January–March 2010): p. 64–68; Andrew A. Snelling, "Paleontological Issues: Deciphering the Fossil Record of the Flood and Its Aftermath," in *Grappling with the Chronology of the Genesis Flood*, Steven W. Boyd and Andrew A. Snelling, eds. (Green Forest, AR: Master Books, 2013).

7. Kurt P. Wise, *Faith, Form, and Time* (Nashville, TN: Broadman and Holman Publishers, 2002), p. 173–174.

likely found in generally warmer lowland areas. Of course, the Scriptures are clear that dinosaurs and humans lived at the same time, because dinosaurs as land animals were created on day 6 of the creation week, just before man was created.

So based on all of this discussion of what we can glean from Scripture and from the geologic record about the pre-Flood world, we can answer part of the posed question. Clearly, there were mountains in the pre-Flood world, because it was those mountains that Genesis 7:20 describes as being eventually covered by the Flood waters. However, while we cannot dogmatically say that those mountains were not high, the scriptural evidence would suggest that the pre-Flood mountains were not as high as today's mountains. The latter were formed and thrust up to their current elevations by the catastrophic mountain-building processes during the Flood, when some fossil-bearing Flood-deposited sedimentary rock layers were buckled and then elevated. The hints in Scripture suggest that there were conducive climatic conditions around the globe to support the lush vegetation worldwide that was subsequently buried *en masse* and fossilized to form the coal beds during the Flood. This would likely have precluded high mountains and major elevation and climate differences in the pre-Flood world, as would also the lack of mention of any ice or snow in the Scriptures describing the pre-Flood world.

Volcanoes and Earthquakes

The issue of whether there were any volcanoes and earthquakes in the pre-Flood world is a lot harder to discern because there is no mention of them in the scriptural account, unlike the mountains. Today volcanic eruptions and earthquakes often result in destruction and loss of life, including *nephesh*-bearing animal life. However, prior to the Fall and the resultant Curse, we would have to assert that there were no physical events that would have resulted in the death of any *nephesh*-bearing creatures. In Genesis 1:31, at the end of day 6 of the creation week, God declared that all He has made was "very good," with animals and man eating only plants. And Paul reminds us in Romans 8:20–22 that today's world is subject to corruption and death because of man's sin. So these scriptural details would seem to preclude the possibility of volcanoes and earthquakes in the pre-Flood world.

It is certainly true that there are no fossils of *nephesh*-bearing animals in the geologic record below the fossil-bearing sedimentary rock layers that represent such powerful evidence of the Flood cataclysm, when the ocean waters flooded over the continents and all land-dwelling, air-breathing, *nephesh*-bearing animals outside the ark perished (Genesis 7:17–24). However, the geologic record may still give us some clues.

The rocks found below where the evidence of the Flood begins are very thick and extensive. They are foundational to the structure of today's continents. Yet they also represent the astounding results of God's creative activity during the creation week to build the land on which would be man's home, followed by the minor, non-destructive geologic activity of the pre-Flood world. Obviously, there could not have been any catastrophic geologic activity across the earth after God created the dry land on day 3 of the creation week, because any such catastrophic geologic activity would have impacted the sea creatures, the land creatures, and man that God created on days 5 and 6. This matches the lack of any such fossils in the geologic record of the creation week and pre-Flood eras.

What we do see in some pre-Flood era sedimentary rocks that is relevant to understanding the topography and environmental conditions in the pre-Flood world are occasional fossilized stromatolites, layered structures probably built by algal mats.[8] Today's rare living stromatolites are usually found in intertidal zones and on the shallow sea floor where the algal mats trap and bind sediment particles to build these structures. The fossilized stromatolites found in pre-Flood era sedimentary rocks usually occur in thick sequences of limestones and related rocks, including cherts, unusual rocks likely produced from hot water springs. Thus, it has been proposed that in the pre-Flood world there could have been a unique ecosystem consisting of stromatolite reefs built in association with hydrothermal springs on the shallow ocean floor some distance from, and fringing, the coastline of the pre-Flood supercontinent and enclosing a wide, shallow lagoon inhabited by now-extinct unusual marine invertebrates.[9] Confirming evidence of just such a stromatolite reef has been documented in what have been interpreted as pre-Flood shallow ocean floor sedimentary rock layers now exposed in the eastern Grand Canyon.[10]

---

8. Georgia Purdom and Andrew A. Snelling, "Survey of Microbial Composition and Mechanisms of Living Stromatolites of the Bahamas and Australia: Developing Criteria to Determine the Biogenicity of Fossil Stromatolites," in *Proceedings of the Seventh International Conference on Creationism*, Mark F. Horstemeyer, ed. (Pittsburgh, PA: Creation Science Fellowship, 2013).
9. Kurt P. Wise, "The Hydrothermal Biome: A Pre-Flood Environment," in *Proceedings of the Fifth International Conference on Creationism*, Robert L. Ivey, ed. (Pittsburgh, PA: Creation Science Fellowship, 2003), p. 359–370.
10. Kurt P. Wise and Andrew A. Snelling, "A Note on the Pre-Flood/Flood Boundary in the Grand Canyon," *Origins* (Geoscience Research Institute) 58 (2005): p. 7–29.

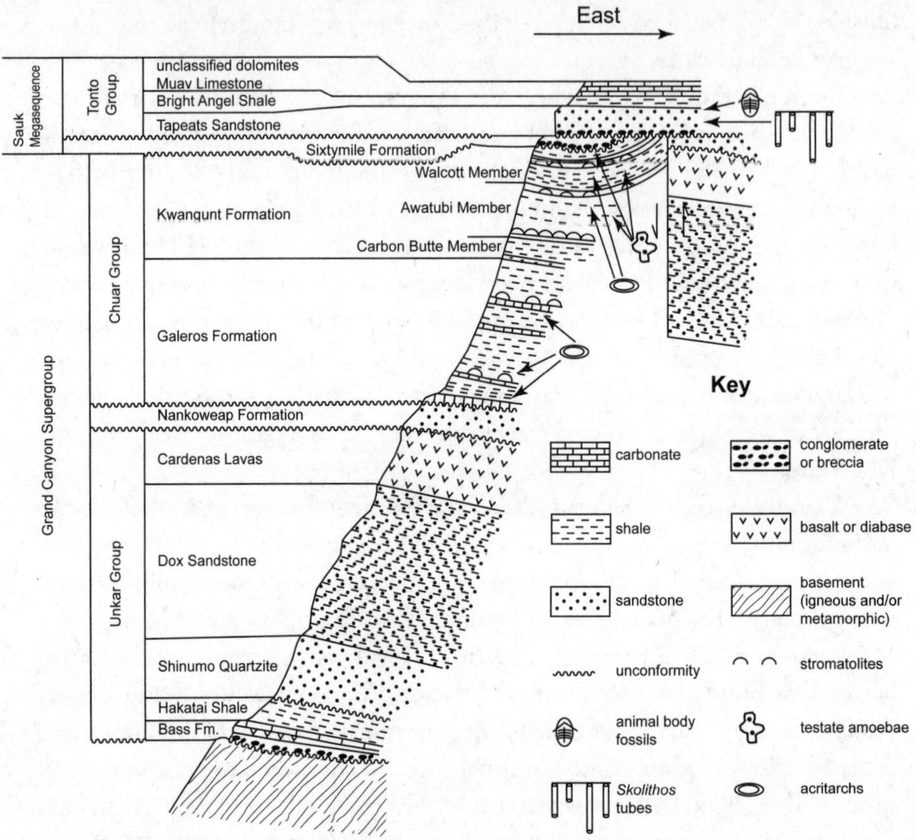

Figure 2. The strata sequence of Cambrian (earliest Flood) and Precambrian (pre-Flood) sedimentary rock layers of eastern Grand Canyon, schematically showing the relative position of the Cardenas Basalt lavas and the strata thicknesses to scale.

However, none of these details from both Scripture and the geologic record precludes the possibility of minor volcanic eruptions of a non-explosive nature on the deeper ocean floor of the pre-Flood world, well away from the creatures that inhabited the shallow ocean floor surrounding the pre-Flood supercontinent. For example, in the eastern Grand Canyon exposed among the pre-Flood rocks are lava flows of the Cardenas Basalt (figure 2).[11] They outcrop not

11. Steven A. Austin and Andrew A. Snelling, "Discordant Potassium-Argon Model and Isochron 'Ages' for Cardenas Basalt (Middle Proterozoic) and Associated Diabase of Eastern Grand Canyon, Arizona," in *Proceedings of the Fourth International Conference on Creationism*, p. 35–51; J.D. Hendricks and G.M. Stevenson, "Grand Canyon Supergroup: Unkar Group," in Stanley S. Beus and Michael Morales, eds., *Grand Canyon Geology*, 2nd ed. (New York, NY: Oxford University Press, 2003), p. 39–52.

far below the pre-Flood/Flood boundary in the Grand Canyon strata record. Above the Cardenas Basalt lava flows are the sedimentary rock layers containing evidence that they accumulated on the pre-Flood ocean floor, including the fossilized stromatolite reef originally built by algal mats on the shallow sea-floor adjacent to hydrothermal springs. Because there are no shallow marine creatures in the rocks above and below the Cardenas Basalt lava flows, the latter would appear to have erupted on the deeper ocean floor. As basalt eruptions are not explosive and these erupted on the deep ocean floor, then no destruction of animal life would have resulted. So there would have been no impact from these volcanic eruptions on the pre-Flood land surface to affect land animals and man.

Nevertheless, earthquakes usually accompany the lead-up to volcanic eruptions, due to the molten rock moving up inside the earth into the throat of the volcano. It is because of such earthquakes that volcanologists are able to predict and warn of impeding volcanic eruptions. Thus, since there were likely such volcanic eruptions during the pre-Flood era, we cannot rule out the possibility of accompanying earthquakes. Whether they were felt by the people living at the time, we have no indication whatsoever. But if such volcanic eruptions were only on the deep ocean floor, far away from the pre-Flood supercontinent, then it is not likely the people noticed any of the accompanying earthquakes. From what we can glean from the scriptural comments about life in the pre-Flood era, people were so engrossed in the pursuit of pleasure and sin (Genesis 6:5, 11–12), as well as the normal routines of living (as Jesus said in Matthew 24:37–39; Luke 17:26–27), ignoring Noah's preaching (2 Peter 2:5), that they had no premonition of the Flood coming from any earthquakes or volcanic eruptions until it was too late — "the Flood came, and took them all away" (Matthew 24:39).

Conclusions

While the Scriptures are silent on the issue of whether there were any volcanoes or earthquakes in the world before the Flood, we do know there were mountains. Since all the high hills and mountains are specifically mentioned as being inundated by the waters of the Flood as they prevailed during the first 40 days of that global cataclysm, then mountains must have existed in the pre-Flood world. But those mountains were likely not nearly as high as today's mountains (formed out of buckled Flood-deposited rock layers that were then uplifted), because the pre-Flood mountains were evidently upland areas like the Garden of Eden inhabited by flowering plants, mammals, and people.

On the other hand, we have to infer rather sketchily from the geologic record that there were likely some volcanic eruptions accompanied by earthquakes in the pre-Flood world, but these occurred far away from human habitations out on the deep ocean floor, where they had no impact on people or animals. In all probability, there were no mountainous volcanoes across the pre-Flood land surface like we have scattered across today's world, and thus no devastating earthquakes and volcanic eruptions. Since there are no fossils of *nephesh*-bearing creatures in pre-Flood rocks, the pre-Flood earth was likely very stable with no major catastrophes.

CHAPTER 24

What about Beneficial Mutations?

DR. GEORGIA PURDOM

Many claim that beneficial mutations provide examples of "evolution in action." These mutations supposedly result in the formation of "major innovations" and "rare and complex traits"[1] that over time have resulted in the evolution of all living things from a common ancestor. However, analyses of these mutations show they only result in variations in *pre-existing* traits, traits that organisms already possess, and cannot result in the *origin of* novel traits necessary for molecules-to-man evolution.

All You Need Is Novelty!

For a simple, single-celled ancestor to evolve into a human over billions of years, novel traits must be gained. New anatomical structures — like brains, arms, and legs — and new functions — like cardiovascular and muscle activities — must be acquired. Regardless of whether this is proposed to occur through beneficial mutations that result in the addition of new DNA, changes in existing DNA, or through other mechanisms, there must be a way to add novel traits. However, all observed mechanisms, including beneficial mutations, do just the opposite — they cause the loss of or slight variation in *pre-existing*

1. Bob Holmes, "Bacteria Make Major Evolutionary Shift in the Lab," *New Scientist* (June 2008), http://www.newscientist.com/article/dn14094-bacteria-make-major-evolutionary-shift-in-the-lab.html.

traits.[2] Beneficial mutations and other mechanisms cannot account for the *origin of* novel traits of the type necessary for molecules-to-man evolution. In a paper entitled "A Golden Age for Evolutionary Genetics? Genomic Studies of Adaptation in Natural Populations," the authors (who are evolutionists) agree that the lack of mechanisms to add novel traits is a problem: "Most studies of recent evolution involve the loss of traits, and we still understand little of the genetic changes needed in the origin of novel traits."[3]

In this paper, the scientists give many examples of variations in organisms such as pattern changes in butterfly wings, loss of bony structures in stickleback fish, loss of eyes in cavefish, and adaptations to temperature and altitude. But none of these examples involve the *origin of* novel traits necessary to evolve into a different kind of organism. Again, they realize this problem and state, ". . . over the broad sweep of evolutionary time what we would really like to explain is the gain of complexity and the origins of novel adaptations."[4]

Their frustration with the lack of evidence for "novelty-gaining" mechanisms like beneficial mutations sinks to apparent desperation when they state, "Of course, to some extent the difference between loss and gain could be a question of semantics, so for example the loss of trichomes [hair-like appendages on flies] could be called gain of naked cuticle."[5] The authors have decided that the whole loss/gain issue is merely one of semantics! In order to get the gain required by molecules-to-man evolution they will just change the wording and say it is a "gain of loss."

That's equivalent to a person who has suddenly lost all their money saying, "I've not lost money; I've just gained poverty!" While it makes the person sound optimistic, it doesn't change the fact that they have lost all their money. In the same way, an organism doesn't gain novel traits needed to evolve into something else — instead, organisms lose traits or develop variations in *pre-existing* traits. It doesn't matter how evolutionists choose to say it; there is still no mechanism that results in the *origin of* novel traits required for molecules-to-man evolution.

2. Kevin L. Anderson and Georgia Purdom, "A Creationist Perspective of Beneficial Mutations in Bacteria," in the *Proceedings of the Sixth International Conference on Creationism* (Pittsburgh, PA: Creation Science Fellowship, 2008): p. 73–86.
3. Nicola J. Nadeau and Chris D. Jiggins, "A Golden Age for Evolutionary Genetics? Genomic Studies of Adaptation in Natural Populations," *Trends in Genetics* 26 (2010): p. 484–492.
4. Ibid.
5. Ibid.

Do Beneficial Mutations Exist?

While beneficial mutations may not result in the *origin of* novel traits necessary to go from molecules to man, they do exist . . . sort of. Let me explain. It is more appropriate to say that some mutations have beneficial *outcomes* in certain environments. Mutations are context dependent, meaning their environment determines whether the outcome of the mutation is beneficial. One well-known example of a proposed beneficial mutation is antibiotic resistance in bacteria.[6] In an environment where antibiotics are present, mutations in the bacterial DNA allow the bacteria to survive. However, these same mutations come at the cost of damaging the normal functions of the bacteria (such as the ability to break down nutrients). If the antibiotics are removed, the antibiotic resistant bacteria typically do not fare as well as the normal (or wild-type) bacteria that have not been affected by mutations. Thus, the benefit of any given mutation is not an independent quality, but rather a dependent quality based on the environment.

Another common example of a supposed beneficial mutation, this time in humans, is individuals that are resistant to infection with HIV. These people have a mutation that prevents HIV from entering the white blood cells and replicating, making them unlikely to develop AIDS. However, studies have shown these individuals may be at a higher risk of developing illness associated with West Nile virus[7] and hepatitis C[8] (also caused by a virus). Again, we see that the mutations are only beneficial in a given environment, such as if the person were exposed to HIV. It is possible that the mutations would not be beneficial in other environments such as if the person were exposed to West Nile virus. The benefit of any given mutation is a dependent quality based on the environment.

There is no question that mutations can be beneficial in certain environments, but do they lead to the *origin of* novel traits of the type necessary for molecules-to-man evolution? Let's look at several examples commonly used to support this idea and the problems with them.

---

6. Georgia Purdom, "Is Natural Selection the Same Thing as Evolution?" in *The New Answers Book 1*, ed. Ken Ham (Green Forest, AR: Master Books, 2006), p. 271–282.
7. William G. Glass et al., "CCR5 Deficiency Increases Risk of Symptomatic West Nile Virus Infection," *The Journal of Experimental Medicine* 203 (2006): p. 35–40.
8. Golo Ahlenstiel, et al., "CC-chemokine Receptor 5 (CCR5) in Hepatitis C- at the Crossroads of the Antiviral Immune Response?" *Journal of Antimicrobial Chemotherapy* 53 (2004): 895–898.

Proposed Beneficial Mutations in Bacteria

Richard Lenski and the Citrate Mutation in *E. coli*

In 1988, Dr. Richard Lenski, an evolutionary biologist at Michigan State University, began culturing 12 identical lines of *Escherichia coli* (a common gut bacteria). Over 50,000 generations and 25 years later, the experiment continues. Lenski has observed many changes in the *E. coli* as they adapt to the culture conditions in his lab. For example, some lines have lost the ability to break down ribose (a sugar),[9] some have lost the ability to repair DNA,[10] and some have reduced ability to form flagella (needed for movement).[11] In other words, they've gotten lazy as they've adapted to life in the lab! If they were grown in a natural setting with their wild-type (normal) counterparts, they would not stand a chance in competing for resources.

In 2008, Lenski's lab discovered another change in one of their lines of *E. coli*. A *New Scientist* writer proclaimed, "A major innovation has unfurled right in front of researchers' eyes. It's the first time evolution has been caught in the act of making such a rare and complex new trait."[12] But was this change really the formation of "a rare and complex new trait"?

Normal *E. coli* has the ability to utilize citrate as a carbon and energy source when *oxygen levels are low*. They transport citrate into the cell and break it down. Lenski's lab discovered that one of their *E. coli* lines could now utilize citrate under *normal oxygen levels*.[13] It's easy to see that this was not "a major innovation" or the "making of a rare and complex new trait" because the normal *E. coli* already has the ability to transport citrate into the cell and use it! This was simply a beneficial outcome of mutations that changed under what conditions citrate was used by *E. coli*.[14] The mutations caused the alteration of a *pre-existing* system, not the *origin of* a novel one. There is a lot of citrate in the medium that the bacteria are grown in, and since other

9. Vaughn S. Cooper et al., "Mechanisms Causing Rapid and Parallel Losses of Ribose Catabolism in Evolving Populations of *E. coli* B," *Journal of Bacteriology* 183 (2001): 2834–2841.
10. Paul Sniegowski et al., "Evolution of High Mutation Rates in Experimental Populations of *E. coli*," *Nature* 387 (1997): 703–705.
11. Tim F. Cooper et al., "Parallel Changes in Gene Expression after 20,000 Generations of Evolution in *E. coli*," *PNAS* 100 (2003): 1072–1077.
12. Holmes, "Bacteria Make Major Evolutionary Shift in the Lab."
13. Zachary Blount et al., "Historical Contingency and the Evolution of a Key Innovation in an Experimental Population of *Escherichia coli*," *PNAS* 105 (2008): 7899–7906.
14. Zachary Blount et al., "Genomic Analysis of a Key Innovation in an Experimental *Escherichia coli* Population," *Nature* 489 (2012): 513–518.

carbon sources are not plentiful, the bacteria have merely adapted to the lab conditions.

Lenski stated, "It is clearly very difficult for *E. coli* to evolve this function. In fact, the mutation rate of the ancestral strain . . . is immeasurably low. . . ."[15] If developing the ability to utilize citrate under different conditions by altering the *pre-existing* citrate system is so rare, then how much more improbable is it to believe that similar beneficial mutations can lead to the *origin of* novel traits necessary for dinosaurs to evolve into birds!

Nylon-Digesting Mutation in Bacteria

In the mid-1970s, bacteria (*Arthrobacter* sp. K172) were discovered in ponds with wastewater from a nylon factory that could digest the byproducts of nylon manufacture. Nylon is a synthetic polymer that was first produced in the 1940s, thus, the ability of bacteria to break down nylon must have been gained in the last few decades. Many evolutionists touted that the bacteria's ability to break down nylon occurred through the gain of new genes and proteins. In a 1985 article entitled "New Proteins Without God's Help," the author explained testing that supposedly showed the bacteria's ability to break down nylon was due to the formation of new proteins, not the modification of pre-existing ones.[16] In conclusion he stated, "All of this demonstrates that . . . the creationists . . . and others who should know better are dead wrong about the near-zero probability of new enzyme formation. Biologically useful macromolecules are not so information-rich that they could not form spontaneously without God's help."[17]

Does this mean that biblical creationists should run screaming and stick our heads into the sand? No. In 2007, genetic analyses of *Arthrobacter* sp. K172 showed that no new genes or proteins had been added that resulted in the ability of the bacteria to break down nylon.[18] Instead it was discovered that mutations in a *pre-existing* gene resulted in a protein that is capable of breaking down nylon. The protein, known as EII, normally breaks down a substance very similar to nylon. Slight alterations in what is called the "active site" of the protein (where the activity of breaking down the substance occurs) changed its

15. Blount et al., "Historical Contingency and the Evolution of a Key Innovation in an Experimental Population of *Escherichia coli*," 7899–7906.
16. William M. Thwaites, "New Proteins without God's Help," *Creation Evolution Journal* 5 (1985): 1–3.
17. Ibid.
18. Seiji Negoro et al., "Nylon-oligomer Degrading Enzyme/Substrate Complex: Catalytic Mechanism of 6-aminohexanoate-dimer hydrolase," *Journal of Molecular Biology* 370 (2007): 142–156.

specificity such that it could now also break down nylon. No changes occurred of the type necessary to go from molecules to man, just a "tweak" in a gene and protein whose normal function is to break down something very similar to nylon. Again, we see the alteration of a *pre-existing* gene and protein, not the *origin of* new ones. Information-rich molecules like DNA and protein cannot spontaneously form — they do need "God's help."

Barry Hall and the *ebg* Mutation in *E. coli*

Beginning in the 1970s and continuing into the 1990s, Dr. Barry Hall, professor emeritus of the University of Rochester, New York, did extensive work in the field of what has been termed adaptive or directed mutations. According to evolutionary ideas, mutations are random changes in the DNA that may or may not be beneficial to an organism in its environment. However, research from scientists like Hall has indicated that adverse environmental conditions, like starvation, may initiate mechanisms in bacteria that result in mutations that *specifically* allow the bacteria to survive and grow in a given environment. These changes do not appear to be random in respect to the environment, thus the term directed or adaptive mutations.

There are two reasons why adaptive mutations are problematic for evolution. First, the mechanisms in bacteria for generating adaptive mutations are specifically responding to the environment. The changes are goal-oriented, allowing the organism to adapt and survive by alteration of *pre-existing* traits. A second reason is that the mechanisms resulting in adaptive mutations (which appear to be a very common type of mutation in bacteria) set limits on the genetic change possible and cannot account for the *origin of* novel traits.

E. coli can break down the sugar lactose to use as a food source. Hall was able to mutate a strain of *E. coli* such that it lost the ability to break down lactose.[19] He then put the mutant *E. coli* in a starvation situation where lactose was the only food source. In order to survive, the *E. coli* either had to develop the ability to break down lactose or die. After a period of time, *E. coli* developed the ability to break down lactose. How did *E. coli* do this? Were new genes and proteins added to allow this to happen?

No. Genetic analyses showed that mutations had occurred in a group of *pre-existing* genes named *ebg*. These genes are in normal *E. coli* and produce proteins that very weakly break down lactose. The genes were also present in

19. Georgia Purdom and Kevin L. Anderson, "Analysis of Barry Hall's Research of the *E. coli ebg* Operon," in the *Proceedings of the Sixth International Conference on Creationism* (Pittsburgh, PA: Creation Science Fellowship, 2008): p. 149–163.

Hall's mutant *E. coli* (he mutated only the primary set of genes used for lactose breakdown *not* the *ebg* genes). In response to the starvation conditions, mechanisms were initiated in the bacteria that resulted in mutations in the *ebg* genes that produced proteins with enhanced ability to break down lactose well enough that the mutant bacteria could survive. No new or novel traits were gained, there was merely the alteration of a *pre-existing* trait that allowed the bacteria to adapt and survive.

Interestingly, Hall theorized that if both the primary set of genes needed for lactose breakdown and the *ebg* genes were made non-functional (through mutations) that adaptive mutations would occur in *other* genes resulting in *E. coli* once again developing the ability to break down lactose.[20] However, "despite extensive efforts," Hall was unable to get *E. coli* that could survive on lactose. They did not survive because adaptive mutations only make *limited* changes. *Ebg* genes in *E. coli* already possess the ability to break down lactose, adaptive mutations enhanced this ability. Adaptive mutations cannot make possible the *origin of* lactose breakdown from genes whose functions are not as similar.

Despite the evidence, Hall concluded this aspect of his research by saying, "Obviously, given a sufficient number of substitutions, additions, and deletions, the sequence of any gene can evolve into the sequence of any other gene."[21] But Hall's own experiments showed otherwise — a gene cannot just become a completely different gene; adaptive mutations are limited. Mutations can cause changes in *pre-existing* traits, but observable mechanisms, such as adaptive mutation, cannot account for the *origin of* novel traits necessary for molecules-to-man evolution.

Proposed Beneficial Mutations in Animals

TRIM5-CypA **Mutation in Monkeys**

The *TRIM5* gene is found in humans, monkeys, and other mammals. The protein produced from this gene binds to the outer covering (capsid) of retroviruses (like HIV) and prevents them from replicating inside cells, thus essentially preventing the spread of infection. A portion of the *TRIM5* gene (C-terminal domain) seems especially variable and may confer resistance to different types of viruses.[22] In 2004, it was discovered that owl monkeys (*Aotus* sp.) have a unique

20. Barry G. Hall, "Evolutionary Potential of the *ebgA* Gene," *Molecular Biology and Evolution* 12 (1995): 514–517.
21. Ibid.
22. Welkin E. Johnson and Sara L. Sawyer, "Molecular Evolution of the Antiretroviral *TRIM5* Gene," *Immunogenetics* 61 (2009): 163–176.

version of the *TRIM5* gene that appears to be a fusion of this gene to the nearby *CypA* gene.[23] The *CypA* gene can produce a protein that also binds to the outer covering of viruses, including HIV. Thus, the *TRIM5-CypA* fusion protein has the antiviral activity of *TRIM5* coupled to the HIV recognition of *CypA* and the fused protein was able to prevent infection from HIV. (A similar fusion gene/protein has also been discovered in certain species of macaques.)[24]

New Scientist writer Michael Le Page in an article entitled "Evolution Myths: Mutations Can only Destroy Information," stated in regard to this mutation, "Here, a single mutation has resulted in a new protein with a new and potentially vital function. New protein, new function, new information."[25] But is this really a new protein with a new function?

No. *TRIM5-CypA* is the fusion of two *pre-existing* genes producing a fused protein. The fusion doesn't change the function of *TRIM5* or *CypA*, so there is no new function. The addition of *CypA* merely allows *TRIM5* to recognize a different group of viruses and exert its antiviral activity against those viruses. This fusion does not result in the *origin of* a novel trait of the type necessary for molecules-to-man evolution.

Gene Duplication, Mutation, and "New" Genes and Functions

Evolutionists often cite gene duplication, followed by subsequent mutation of the duplicated gene, as a mechanism for adding new genes with new functions to organisms. The idea is that the duplicated gene is free to mutate and gain new functions because the original copy of the gene can still perform the original function. Evolutionary biologist Dr. Sean Carroll, referring to his work on gene duplication in yeast, stated, "This is how new capabilities arise and new functions evolve. This is what goes on in butterflies and elephants and humans. It is evolution in action."[26] However, a deeper look at a couple of examples of gene duplication and mutation show exactly the opposite — the complete impotence of these mechanisms to explain the *origin of* novel traits necessary for molecules-to-man evolution.

23. Sébastien Nisole, "A Trim5-cyclophilin A Fusion Protein Found in Owl Monkey Kidney Cells Can Restrict HIV-1," *PNAS* 101 (2004): 13324–13328.
24. Cheng-Hong Liao et al., "A Novel Fusion Gene, *TRIM5-cyclophilin A* in the Pig-tailed Macaque Determines Its Susceptibility to HIV-1 Infection, *AIDS* 21 (2007): S19–S26.
25. Michael Le Page, "Evolution Myths: Mutations Can only Destroy Information," *New Scientist*, April 2008, http://www.newscientist.com/article/dn13673-evolution-myths-mutations-can-only-destroy-information.html.
26. Terry Devitt, "A Gene Divided Reveals Details of Natural Selection," *University of Wisconsin-Madison News*, October 10, 2007, http://www.news.wisc.edu/14276.

RNASE1 and *1B* in Monkeys

The diet of most monkeys consists of fruit and insects; however, the colobine monkeys predominantly eat leaves. These monkeys have a special foregut that harbors symbiotic bacteria that help in the digestion of the leaves. *RNASE1* is a digestive enzyme in colobines that breaks down RNA from the bacteria in the foregut. This results in the efficient recycling of phosphorus and nitrogen that are used in the production of the monkey's own proteins and nucleic acids like DNA and RNA.

It has been shown that some colobines have two RNASE genes—*RNASE1* and *RNASE1B*.[27] *RNASE1B* is proposed to be a duplication of the gene *RNASE1*. There are several differences in the genes and the proteins produced, however, the function remains the *same*. Both enzymes break down RNA, but the changes in RNASE1B allow it to break down RNA in more acidic conditions such as those found in the foregut of the monkeys. The authors of one study of the *RNASE1* genes commented, "Gene duplication has long been thought by evolutionary biologists to be the source of novel gene function. . . . We believe our data to be another example that do not support this hypothesis."[28] Other authors of similar research indicate: "Taken together, our results provide evidence of the important contribution of gene duplication to adaptation of organisms to their environments."[29] The differences (caused by mutations) in the *RNASE1B* gene appear to enhance the *pre-existing* function of the original *RNASE1* gene, resulting in adaptation, and do not represent the type of mutation necessary for the *origin of* novel traits needed for molecules-to-man evolution.

Antifreeze Proteins in Fish

Antifreeze proteins (AFPs) prevent the growth of ice crystals in organisms that live in very cold environments such as the Arctic and Antarctica. There are five classes of these proteins found in fish. AFP type III is found in the Antarctic zoarcid fish. The *AFPIII* gene is proposed to be a duplication of a portion of the *SAS* (sialic acid synthase) gene.[30] The *SAS* gene is responsible for the synthesis

27. John E. Schienman, et al., "Duplication and Divergence of 2 Distinct Pancreatic Ribonuclease Genes in Leaf-eating African and Asian Colobine Monkeys," *Molecular Biology and Evolution*, 23 (2006): 1465–1479.
28. Ibid.
29. Jianzhi Zhang, et al., "Adaptive Evolution of a Duplicated Pancreatic Ribonuclease Gene in a Leaf-eating Monkey," *Nature Genetics* 30 (2002): p. 411–415.
30. Cheng Deng, "Evolution of an Antifreeze Protein by Neofunctionalization under Escape from Adaptive Conflict," *PNAS* 107 (2010): 21593–21598.

of sialic acids (found on cell surfaces) but *also* has an antifreeze function. Mutations in the *AFPIII* gene (a duplicate copy of a portion of the *SAS* gene) appear to have further enhanced the antifreeze function.

One of the authors of the study on the formation of the *AFPIII* gene commented, "This is the first clear demonstration . . . [of] the underlying process of gene duplication and the creation of a completely new function in one of the daughter [duplicate] copies."[31] But the *AFPIII* protein does not have a "completely new function"! Instead the *AFPIII* gene is likely the result of a duplication of a portion of the *pre-existing SAS* gene with mutations that enhanced the *SAS* gene's *pre-existing* antifreeze function. Once again, we see that the differences (caused by mutations) in the *AFPIII* gene appear to enhance the *pre-existing* antifreeze function of the original *SAS* gene resulting in adaptation to the environment and do not represent the type of mutation necessary for the *origin of* novel traits needed for molecules-to-man evolution.

Beneficial Mutations from a Biblical Creation Perspective

The previous examples show that there can be beneficial outcomes to mutations. However, these mutations can only alter *pre-existing* traits; they cannot result in the *origin of* novel traits necessary for molecules-to-man evolution. In every example, it appears that the mutations help organisms in adapting to a specific environment. This is easily seen in bacteria when they are faced with limited food choices and must gain the ability to break down a different nutrient or die. It is also seen in animals like monkeys and fish that have essentially become more specialized for eating a particular diet or living in a particular environment.

But are these mutations random in respect to the environment? On the Evolution 101 website, sponsored by the University of California Museum of Paleontology, it states:

> The mechanisms of evolution — like natural selection and genetic drift — work with the *random* variation generated by mutation. (emphasis in original)
>
> For example, exposure to harmful chemicals may increase the mutation rate, but will not cause more mutations that make the organism resistant to those chemicals. In this respect, mutations are random —

31. Diana Yates, "Researchers Show how One Gene Becomes Two (with Different Functions)," University of Illinois at Urbana-Champaign News Bureau, January 12, 2011, http://www.news.illinois.edu/news/11/0112genes_cheng.html.

> whether a particular mutation happens or not is generally unrelated to how useful that mutation would be.[32]

The basis of molecules-to-man evolution is random mutation in conjunction with other mechanisms like natural selection. However, mutations with beneficial outcomes do not appear to be random or at least the mechanisms generating the mutations are not random. From a biblical creation perspective, this could be a type of adaptive variation that God has designed in organisms to allow them to survive in a world dramatically changed by the Fall and Flood. Rather than the changes being random, organisms have been "pre-programmed" to change in response to their environment.

These types of adaptive traits may be the result of what creationists have termed *mediated design*. Several creation scientists describe it this way:

> God specifically designed the created kinds with genes [in the DNA] that could be turned on to help them adapt to new environments. In other words, the Creator continues to accomplish His purpose for organisms after creation, not by creating something new, but by working through existing parts that were designed during Creation Week. An analogy is the manufacturer of a fully equipped Swiss army knife, who stores within the knife every tool a camper might need as he faces the unknown challenges of wilderness living.[33]

God designed adaptive traits to be expressed only under certain conditions to allow microbes, animals, plants, and humans to fill the earth as environments changed over time (Genesis 1 and 8:16–19). Thus, God programmed organisms with mechanisms that would be triggered under certain conditions that would then modify *pre-existing* traits to allow organisms to survive and thrive in new environments. Possible mechanisms to accomplish this are seen in the previous examples with directed mutations (*ebg* and *E. coli*) and duplication followed by mutation (*RNASE1* and *1B* in monkeys). Another exciting area of modern genetics research is the role of epigenetics in modifying how genes and, thus, the physical traits, are expressed. Epigenetic markers, chemical tags on DNA, have been shown to be heritable and may be a way to pass on modified traits to future generations (see postscript). Understanding the God-given ability of organisms to change and adapt is an active area of creation research.

32. "Mutations Are Random," Evolution 101, http://evolution.berkeley.edu/evosite/evo101/IIIC1aRandom.shtml.
33. Tom Hennigan, Georgia Purdom, and Todd Charles Wood, "Creation's Hidden Potential," *Answers*, January–March 2009, p. 70–75.

But what adaptive variations can't do is change one kind of organism into a completely different kind of organism because they do not result in the *origin of* novel traits needed for this type of change. This is consistent with Scripture because God created animals and plants according to their kind (usually at the family level in modern classification schemes).[34] The inference from Scripture is that animals were to reproduce according to their kind (Genesis 1, 6, and 8). We observe mechanisms that allow animals and plants to adapt but not evolve into different kinds of organisms.

So why, in spite of all the evidence to the contrary, do many scientists, who are unbelievers, argue that beneficial mutations are a valid mechanism (as evidenced by their quotes) to account for the *origin of* novel traits resulting in molecules-to-man evolution? Paul says that God can be known through what He has created (Romans 1:20), but just before that Paul states why people don't acknowledge God as the Creator: "For the wrath of God is revealed from heaven against all ungodliness and unrighteousness of men, who *suppress the truth in unrighteousness*" (Romans 1:18, emphasis added). Just as Pharaoh hardened his heart repeatedly (1 Samuel 6:6), today, people's hearts have been hardened in their willful rebellion against God. They want to continue in their sin and will go to extremes to "deny the obvious" and reject God as the Creator.

God, in His mercy, compassion, and grace, designed living organisms with the ability to adapt and fill, survive and thrive in a fallen world. We look forward to the day when all life will be restored to perfection and the wolf will live with the lamb, the lion will eat straw like an ox, and a baby will play by the cobra's hole (Isaiah 11:6–8).

Postscript: Epigenetics – Inheriting More Than Genes

All our lives, we've heard that our physical makeup is determined by our genes, not environment. But the science of epigenetics is forcing scientists to rethink their assumptions.

You're probably familiar with the phrase, "You are what you eat." But did you know that you are also what your mother and grandmother ate? The budding science of epigenetics shows that our physical makeup is about much more than inheriting our mother's eyes or our father's smile.

We are accustomed to thinking that the only thing we inherit from our parents are genes — packets of information in DNA that give instructions for

34. Jean K. Lightner et al., "Determining the Ark Kinds," *Answers Research Journal* 4 (2011): 195–201.

proteins. These genes determine our physical traits such as hair and eye color, height, and even susceptibility to disease.

But we also inherit specific "modifications" of our DNA in the form of chemical tags. These influence how the genes express our physical traits. The chemical tags are referred to as "epigenetic" markers because they exist outside of (*epi-*) the actual sequence of DNA (*-genetics*).

Let me use an analogy to explain. The following sentence can have two very different meanings, depending on the punctuation used. "A woman, without her man, is nothing" or "A woman: Without her, man is nothing." Perhaps it's a silly illustration, but it gets the point across.

The words of both sentences are the same, but the meaning is different because of the punctuation. The same is true for DNA and its chemical tags. The sequence of DNA can be identical but produce different results based on the presence or absence of epigenetic markers. For example, identical twins have the same DNA sequence but can have different chemical tags leading one to be susceptible to certain diseases but not the other.

Parents can pass down epigenetic markers for many generations or their effect can be short-lived, lasting only to the next generation. Either way, the changes are temporary because they do not alter the sequence of DNA, just the way DNA is expressed.

What does this mean in practice? Your behavior, including the food you eat, could change how your body expresses its DNA. Then those changes — for good or bad — could be passed to your children! If you do something to increase your susceptibility to obesity or cancer or diabetes, your children could inherit that from you.

In one experiment, mice from the same family, which were obese because of their genetic makeup, were fed two different diets. One diet consisted of regular food. The other diet consisted of the same food but contained supplements that were known to alter the chemical tags on DNA.

Normally when these mice eat regular food they produce fat offspring. However, the mice that ate the same food with the supplements produced offspring that were normal weight. The parents' diet affected their offspring's weight!

Scientists are still trying to understand the details. The epigenetic markers that were modified by the food supplements appear to have "silenced" genes that encourage appetite. The parents' environment — in this case, the food they ate before becoming parents — affected the weight of their offspring.

Certain types of medicine have also been suspected of causing changes in epigenetic markers, leading to cancer in the offspring of women who took the

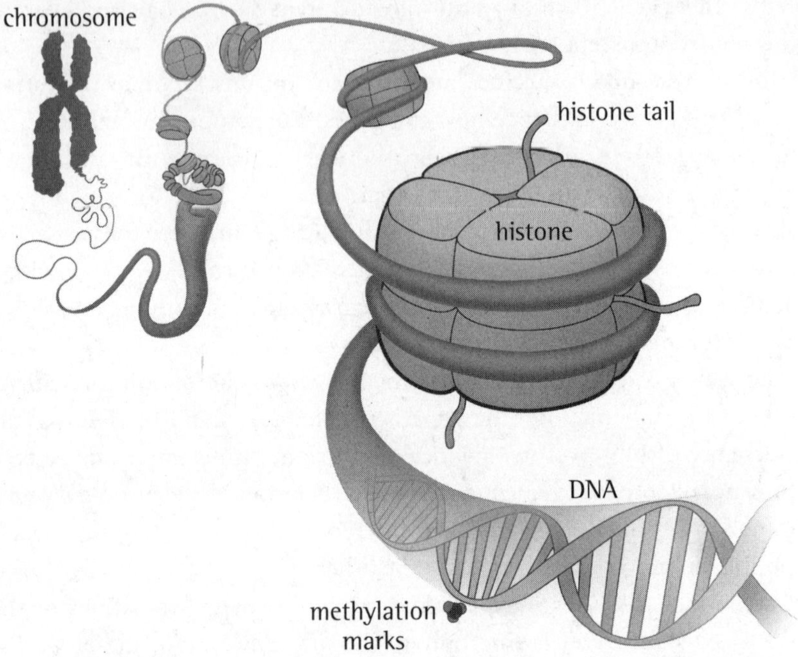

medicine. For example, a type of synthetic estrogen prescribed to prevent miscarriages has been linked to an increased number of cancers in their daughters' and granddaughters' reproductive organs.

Studies point to changes in the epigenetic markers related to the development of reproductive organs, which the mothers passed down to their daughters. This finding affirms the adage that "you are what your mother — or grandmother — ate."

Tagalongs to Our Genetic Code

Our DNA includes additional components, which may sometimes be passed from parent to child at the same time as the genetic code. First are molecules attached to the DNA, called methylation marks, that turn genes on and off. Second are balls of proteins composed of histones, which the DNA wraps around. Histones and a portion of these proteins, called histone tails, regulate how the DNA is folded (and thus what is turned on or off).

The food you eat and other aspects of your environment can change these tagalongs. Then they can be passed down to your children and even your grandchildren, affecting the genes that are turned on.

Epigenetics: A Problem for Evolution?

Until these findings, many evolutionists dismissed the ideas of Charles Darwin's contemporary, Jean-Baptiste Lamarck, who believed that animals could acquire new traits through interactions with their environment and then pass them to the next generation. For instance, he believed giraffes stretching their necks to reach leaves on trees in one generation would cause giraffes in the next generation to have longer necks. Many science textbooks today reject Lamarck's ideas, but epigenetics is a form of Lamarckianism.

Of course this is contrary to classic Darwinian evolution. The theory of evolution is based on random changes or mutations occurring in DNA. If a change happens to be beneficial, then the organism will survive via natural selection and pass this trait to its descendants.

Although evolutionists do not deny the reality of epigenetics, its existence is hard to explain! Epigenetic changes are not random; they occur in response to the environment via complex mechanisms already in place to foster these changes.

These non-random epigenetic changes imply that evolution has a "mind." Creatures appear to have complex mechanisms to make epigenetic changes that allow them to adapt to *future* environmental challenges. But where did this forward-thinking design come from? Evolution is mindless; it cannot see the future. So how could it evolve mechanisms to prepare for the future?

But God does! God is omniscient (all-knowing), and He foreknew Adam and Eve would sin. He would judge that sin (Gen. 3) and the world would be cursed (Rom. 8:22). God knew that organisms would need the ability to adapt in a world that was no longer "very good." God likely designed organisms with epigenetic mechanisms to allow them to change easily and quickly in relation to their environment. These types of changes are much more valuable than random mutation and natural selection because they can produce immediate benefits for offspring without harming the basic information in the actual sequence of DNA.

Although we often hear that "nothing in biology makes sense except in the light of evolution," it should be said that "nothing in biology makes sense without the Creator God." Epigenetics is an exciting field of science that displays the intelligence and providence of God to help organisms adapt and survive in a fallen world.

CHAPTER 25

What about the Hebrew Language and Genesis?

DR. BENJAMIN SHAW

Introduction

A number of years ago, I heard a noted New Testament scholar relate a story about teaching a Sunday school class. As would be expected, he was using an English translation. At one point, one of the students in the class asked, "What does it say in the Greek?" The teacher's response was, "The same thing it says in the English." His point was not that there is no difference between Greek and English; only that in that passage the English gave an accurate and adequate presentation of the Greek.

It is the same in the Old Testament with Hebrew. Often, the Hebrew text says just what it does in English. That is not to say that there are not differences between Hebrew and English. There are, and frequently those differences pose difficulties for the translator. But in many places that is not the case. That is the reason that if you take a number of the more literal English translations (such as the KJV, NASB, NKJV, and ESV) and compare them verse-by-verse you will often see very little difference among them.

Why Are the Original Languages Important in Studying Genesis?

Vocabulary

To qualify my opening statement, there are many differences between Hebrew and English, and those differences can make it difficult to convey some of the subtleties of Hebrew in an English version. These differences are of various kinds. Some of them have to do with vocabulary. Two examples here might suffice.

One is the Hebrew word *hesed.* It can be translated "steadfast love," "lovingkindness," "mercy," "faithfulness," and some other words as well. According to *Strong's Concordance,* the KJV translates it into about 12 different words or phrases. The point is that the range of meaning for *hesed* is wider than that of any of the English words used to translate it.

A second example is the word *shalom.* It is usually translated "peace" in English versions, but again, the range of meaning of the Hebrew word is much wider. It can mean health, well-being, and satisfaction, as well as simply absence of conflict (at least seven different English words are used to translate it in the KJV).

Grammar and Syntax

Other differences have to do with grammar and syntax. Grammar, as I use it here, has to do with the form and function of words, whereas syntax has to do with the structure of sentences. As an example for the differences in grammar, the English verb system is time-based. That is, English has past, present, and future tenses (and variations on each of those), and the primary consideration is *when* the action took place. Hebrew verbs, on the other hand, have an aspect-based system. That is, the verb form can vary depending on whether the action is viewed as a whole, or viewed as incomplete or repeated. Thus, a particular verb in Hebrew may be translated past tense, present tense, or even future, depending on the context. The one consistency among the three would be that in each case the aspect from which the action is viewed is of primary importance. Hebrew verbs do have tense, but it is simply indicated by the context rather than by the form of the word. English tense is indicated (usually) by the form of the word. We know that "see" is present tense, while "saw" is past tense.

Another example would be in the use of the definite article (the). Hebrew will sometimes use the article in places where English would not, and vice versa. So, for example, in Genesis 28:10 the English says, "So he came to a certain place." In Hebrew, it says, "and he came to the place." In English, the use of "the" in such a context implies that the place had already been introduced, whereas that is not the case in Hebrew. In Hebrew, the definite article is regularly used to refer to something that has not been previously introduced but is definite in the mind of the narrator. This explains the English rendering "a certain place" in Genesis 28:10.

As for syntax, the normal word order in English is subject-verb-object: John (subject) saw (verb) the ball (object). In Hebrew, the normal word order, at least in narrative, is verb-subject-object. If that word order is changed, it is a clue to the reader that something other than straightforward narrative is taking place, or that some explanatory comment is being inserted into the narrative.

These differences between Hebrew and English vocabulary, grammar, and syntax mean that there are always some subtleties that are lost in translation. We find this in the Greek New Testament as well. As an example here, in John 2:4, Jesus says to Mary, "Woman, what does your concern have to do with Me?" For most English readers, that may sound as if Jesus is being rude to His mother. But in fact, He is simply being formal. Understanding this is largely a matter of vocabulary, knowing the various nuances that the noun "woman" may have in a particular context. For these reasons, in any detailed study of the Bible, it is important to have recourse to the original languages.

Problems That Arise in Today's Debates Due to Lack of Hebrew Knowledge

Today, there are many study helps and lexicons that can aid a layman and professional scholar. I suppose in some sense that the real problems here are not so much due to a lack of knowledge of Hebrew, though that may often be the case with laymen, nor with scientists who are knowledgeable in their own field but ignorant in the biblical languages.

Rather, the most serious problems are with those who know Hebrew, many of them fluent in it, yet because Genesis 1–2 is special (especially in today's debate over millions of years and evolution), all the ordinary rules of Hebrew vocabulary, grammar, and syntax seem to be thrown out the window! Essentially, it seems that outside ideas are influencing people to reinterpret Genesis 1–2 instead of reading it in a straightforward fashion in the normal sense of grammar, syntax, and vocabulary. Let's define some non-traditional, modern views of Genesis:

1. Day-Age: Days of Genesis are long periods of time to accommodate the secular concepts of long ages.
2. Framework Hypothesis: Days 1–3 parallel days 4–6 in many aspects, so this sets up a literary style so Genesis 1 is denoting importance, not history, and long ages can therefore be incorporated into Genesis 1.
3. Gap Theory: Separate Genesis 1:1 and 1:2 and put a large gap of time between these verses to accommodate long ages.

4. Theistic Evolution: Essentially reinterpret Genesis 1–11 as myth with some truth value and replace it with an evolutionary worldview, picking up the biblical narrative with Abraham.

Matters Having to Do with Vocabulary

Yom/Day

We might as well begin here with the common "problem" of the definition of "day" (*yom* in Hebrew). According to the Brown-Driver-Briggs Hebrew lexicon (dictionary), *yom* has six basic uses in the Old Testament. The first is day as opposed to night as in Genesis 1:4, where the light period is called "day," and the dark period is called "night." The second is day as a division of time, so for example, "three days journey" as in Genesis 30:36 or Exodus 3:18. Under this sense, day is defined by evening and morning, where the dictionary cites Genesis 1–2.

Third is the particular phrase "the day of the Lord." Fourth is the use of the plural "days" to refer to the life of someone (Genesis 6:3; Deuteronomy 22:19). Fifth is the use of the plural to indicate an indefinite period (Genesis 27:44, 29:10). Finally, there is the use of "day" (again, primarily in the plural) to indicate "time." So, for example, in Proverbs 25:13, "the day of harvest" refers to harvest time, not to a single day. See also Genesis 30:14 or Joshua 3:15. Other Hebrew dictionaries, including the most recent, set out essentially the same range of meanings for the word *yom*.

It is clear from the discussion in the dictionary that *yom* in reference to the days of creation discussed in Genesis 1–2 refers to ordinary days. However, many scholars are unwilling to take it in that sense because of the "special" character of these chapters as viewed by modern scholars and their response to things like "millions of years."

In part, this contributed to the development of the day-age view of Genesis 1 (as well as other long-age views). It gave the developers of the view a way of reading Genesis 1 that allowed them to hold to the old age of the earth that was being put forth by secular geologists at that time.

It is important to note, however, that the definition of day in Genesis 1 as an ordinary day is not limited to the standard dictionaries. It is also the case with many of the classic commentaries on Genesis such as John Gill, John Calvin, Jamieson-Fausset-Brown, H.C. Leopold, and others. This is also true of some modern commentators. For instance, Gordon Wenham, commenting on Genesis 1:5, says,

> There can be little doubt that here "day" has its basic sense of a 24-hour period. The mention of morning and evening, the enumeration of the days, and the divine rest on the seventh show that a week of divine activity is being described here.[1]

Claus Westermann doesn't even discuss the possible range of meaning of *yom*. He says,

> What is essential for P [sadly, *Westermann presumes that this part of Genesis has come from the so-called "Priestly source" from the outdated and refuted Documentary Hypothesis]* is only the chronological disposition of the works of creation. The alternation between night and day is not conceived as a period of 24 hours, as a unity with a precise beginning; the 24 hours comprise two parts. The constantly recurring sentence which concludes the work of each day plots the regular rhythm of the passage of time, and gives P's account of creation the character of an event in linear time which links it with history.[2]

In short, the interpretation of *yom* in Genesis 1 as anything other than an ordinary day appears to be special pleading on the part of interpreters in an attempt to avoid the clear implication of the passage that what we have here is an ordinary week at the very beginning of time.

Firmament/Expanse

Another term that comes in for frequent discussion is the word "firmament." In Hebrew, the word is *raqiya'*. It is derived from a verb that means "to hammer out" or "to flatten." It is usually used in reference to metal that has been flattened out by hammering or beating. As a result, most scholars take the view that the *raqia'* is a solid expanse. Westermann says, "In earlier times the heavens were almost always regarded as solid."[3] However, it may also be the case that what is in view is the idea of something being stretched out. Psalm 104:2 refers to God as the one "who stretch[es] out the heavens like a curtain." A different verb is used here than in Genesis 1:6, but the idea is the same. In verse 8,

1. Gordon Wenham, *Word Biblical Commentary* (Dallas, TX: Word Incorporated, 1987), Genesis 1–15, p. 19.
2. Claus Westermann, *Genesis1-11: A Commentary*, trans. J.J. Scullion (Minneapolis, MN: Augsburg Publishing House, 1984), p. 115. For an explanation of the Documentary Hypothesis, see *How Do We Know the Bible Is True?* Volume 1, Ken Ham and Bodie Hodge, gen. eds. (Green Forest, AR: Master Books, 2011), chapter 8: "Did Moses Write Genesis?" p. 85–102.
3. Ibid., p. 117.

the firmament is called "heavens." Thus, while it may be the case that ancient societies saw the heavens as something solid, it does not appear that that view is necessarily being taught in Genesis 1:6. Many translations today use the word "expanse" to denote this.

One other element having to do with vocabulary should also be discussed here. That is the use of a figure of speech called a "hendiadys." The word comes from Greek and literally means one-through-two. It is the use of two related terms to identify one idea. Some examples in English are: law and order, assault and battery, and kith and kin.

In the Bible, there are numerous examples. In Leviticus 24:47, the phrase "stranger and sojourner" means "resident alien." In Lamentations 2:9, the phrase "destroyed and broken" means "totally ruined." In Genesis 1, there is one important example of hendiadys. In verse two, the phrase "without form and void" does not indicate two separate things, but one thing. Wenham translates it as "total chaos" and makes the following comment: " 'Total chaos' is an example of hendiadys."[4] Similarly, Westermann says, "E.A. Speiser describes the phrase as 'an excellent example of hendiadys'; it means the desert waste and is used as the opposite of creation."[5]

If this phrase is indeed a hendiadys, it seriously undercuts one aspect of the framework hypothesis.[6] The framework hypothesis generally takes the phrase as two separate words, the first meaning "unformed" and the second meaning "unfilled." Days 1–3 then deal with the forming of the various elements of creation, while days 4–6 deal with their filling. Such hair-splitting of the terms is unlikely.

Matters Having to Do with Syntax

Here the primary syntactical observation is the use of what is called the vav-consecutive in Hebrew (sometimes denoted as a "waw-consecutive"). As was mentioned above, Hebrew verbs function somewhat differently than do English verbs. The vav-consecutive is a verb construction that is the ordinary verb form used for relating a narrative. The verb form also appears in poetry, but it is a matter of dispute among Hebrew grammarians whether the form has the same function in poetry as it does in narrative. It is conceded by all that Genesis 1 is

4. Wenham, *Word Biblical Commentary*, p. 15.
5. Westermann, *Genesis1-11: A Commentary*, p. 103.
6. For a refutation of the framework hypothesis, see *How Do We Know the Bible Is True?* Volume 1, Ken Ham and Bodie Hodge, Gen. Eds., "Chapter 17: Framework Hypothesis?" (Green Forest, Arkansas: Master Books, 2011), pp. 189–200.

narrative. Some want to qualify that by calling it "poetic narrative" or "elevated narrative." However, it is still narrative.

Not only does the repeated use of the vav-consecutive indicate that a passage is narrative, but it also indicates sequence. That is, the action of the second verb follows the action of the first verb in sequence; the third follows the second, and so forth. That is the standard character of the vav-consecutive in other biblical narratives, such as the stories in the books of Samuel and Kings. The vav-consecutive appears approximately 50 times in Genesis 1:1–2:4. This emphatically characterizes the passage as narrative, and it traces an extended sequence of actions throughout the section. This consideration is particularly damaging to the framework hypothesis, which sees days 1–3 as paralleled in days 4–6. Thus, days 4–6 do not follow days 1–3 in sequence, but take place at the same time. If that were the case, there would be no good reason for the repeated use of the vav-consecutive, since there would be no sequence of events to report.

A second consideration having to do with syntax deals with the transition from Genesis 1:1 to Genesis 1:2. Though the gap theory[7] probably originated in some form well before the 19th century, it became popular in that century as a way to provide concordance between the reading of Genesis 1 and the idea of an old earth (much older than five or six thousand years) that was being put forward by the secular geologists of the day. It later gained great popularity, particularly in fundamentalist circles, through its inclusion in the *Scofield Reference Bible*.

An essential element of this theory is the idea that there is a gap between Genesis 1:1 and Genesis 1:2. Genesis 1:1 is taken as a statement regarding the original creation of the totality of the universe. Verse two is then translated "and the earth *became* formless and void." The idea is that there was an original creation, perhaps many millennia ago, perhaps even millions of years ago. Then, in more recent time, the earth became formless and void.

Part of the defense of this view is the use of the identical phrase in Jeremiah 4:23, where the formless and void state is a result of judgment. This consideration is strengthened by the fact that in Jeremiah 4:23 there is the additional statement that the earth had no light. The reasoning then is that the earth being dark, formless, and void in Genesis 1:2 is the result of some catastrophic judgment. From that point, gap theorists develop an explanation of what took place in that "gap" period to bring about such a catastrophic judgment that the earth had to be entirely recreated.

7. For a discussion on the problems with the gap theory, see *The New Answers Book 1*, Ken Ham ed. (Green Forest, Arkansas; Master Books, 2006), chapter 5: "What about the Gap and Ruin-Reconstruction Theories?" p. 47–63.

There are two fundamental problems with this view. The first is that it makes Genesis 1:2 dependent on Jeremiah 4:23, while the opposite is the case. Genesis was written well before the time of Jeremiah, and Jeremiah is borrowing the imagery from Genesis to express the severity of the judgment that is about to befall the nation of Judah. The people have persisted in their idolatry and their rebellion against God, and He is about to bring judgment on the land. The judgment will be so severe that it is as if the earth will be returned to its primordial state, before God began to order the creation.

The second problem is with the translation of the verb as "became." The verb used here can indeed mean become, or come into being, as in Genesis 2:7, "and man became a living being." More commonly, however, it simply means to happen. The definition of the verb itself does not answer the question. The issue here is the syntax. How does this verb relate to the verb in the preceding verse? In English, we do not often think of how one verb may be related to preceding or following verbs. English is full of adverbs and prepositions that indicate how one statement relates to preceding or following statements.

This is similar in the case of Greek, too. So, for example, the reader may well have heard a preacher say that when we see a "therefore" in one of Paul's letters, we need to ask what it's there for. Hebrew does not have the same structure as English, and it does not have the large number of conjunctions, adverbs, and prepositions that English has.

Instead, the relation of one verb to preceding and following verbs is regularly indicated by two things. Hebrew indicates the relationship between clauses and sentences first by the form of the verb; and second, by the placement of the verb in the sentence. The verb "created" in Genesis 1:1 is in the perfect state (not to be confused with the perfect in English), as is ordinarily the case with the beginning of a narrative. We would then expect the next verb to be at the beginning of the next sentence, and to be the vav-consecutive form. This would indicate the continuation of the narrative sequence. However, neither of those two things is true of the verb "was" in Genesis 1:2.

First, the verb is not in first place in verse two. In verse 2, the subject comes first (and the earth). Second, the verb is in the perfect state. The combination of these two factors indicates that verse 2 is a descriptive clause about the noun (usually referred to as a nominal clause). It is making some further statement about the last element in verse 1 before the narrative sequence is continued. Thus, verse 2 is very closely related to verse 1, and this close relationship does not allow for the gap needed by the gap theory.

An expanded translation of the two verses, indicating this relationship, would be something like this: "*In the beginning God created the heavens and the earth. As for the earth, it was without form and void. . . .*" The narrative begins with a general statement about the heavens and the earth. It then moves to focus on the earth, giving the reader information about the state of the earth at the very beginning of time. In order for the gap theory to work at this point, the reader would simply have to ignore this standard element of Hebrew syntax. As Wenham says, "And + noun (=earth) indicates that v 2 is a disjunctive clause. It could be circumstantial to v 1 or v 3, but for reasons already discussed the latter is more probable."[8]

Limitations to the Use of Hebrew Grammar and the Work of Hebrew Experts

In the material already discussed, there has been a fair amount of unity in the views of Hebrew experts. However, Hebrew experts are not agreed on all matters Hebrew. For example, while most view "without form and void" as a hendiadys, not all do.

It is at this point, for example, that I would take issue with the NKJV. It translates the beginning of verse 2 this way: "The earth was without form, and void." By putting the comma between the two words, the translators indicate that they do not see the two words as a hendiadys. In this, it follows the KJV, but it is the only modern translation to do so.

In Genesis 1, however, the deepest disagreement among Hebrew experts has to do with the way the first three verses are translated. Aside from the issue of "formless and void," the NKJV is representative of most modern English versions. It translates verses 1–3 as follows:

> In the beginning God created the heavens and the earth. [2]The earth was without form, and void; and darkness *was* on the face of the deep. And the Spirit of God was hovering over the face of the waters. [3]Then God said, "Let there be light"; and there was light.

Some other translations will give the reader a sense of the different ways some translators understand the verses.

> When God began to create the heavens and the earth — [2]the earth was without shape or form, it was dark over the deep sea, and God's wind swept over the waters — [3]God said, "Let there be light." And so light appeared. (Common English Bible)

8. Wenham, *Word Biblical Commentary*, p. 15.

> In the beginning when God created the heavens and the earth, [2]the earth was a formless void and darkness covered the face of the deep, while a wind from God swept over the face of the waters. [3]Then God said, "Let there be light"; and there was light. (New Revised Standard Version)

> When God began to create heaven and earth — [2]the earth being unformed and void, with darkness over the surface of the deep and a wind from God sweeping over the water — [3]God said, "Let there be light"; and there was light. (Tanak: The New Jewish Publication Society translation)

A careful reading of these versions shows that the Hebrew is being read differently by Hebrew experts. All of them are grammatically and syntactically possible, though each of the three after the NKJV require some playing around with the text. It demonstrates that the translation and interpretation of a Bible passage does not depend on a knowledge of vocabulary, syntax, and grammar alone. As I sometimes tell my Hebrew students, "A detailed knowledge of Hebrew grammar will not answer all your questions."

It's important for the reader to know what is going on with above variant translations. This explanation is summarized from that of Wenham, who gives a clear and fair presentation of the evidence.[9]

There are four ways of understanding the syntax of Genesis 1:1–3 that have been defended by various Hebrew experts. The first is that verse 1 is a temporal clause that is subordinate to verse 2, which is the main clause. That is, "When God created . . . the earth was without form." The second view is that verse 1 is a temporal clause subordinate to the main clause in verse 3, while verse 2 is a parenthetical comment. That is, "When God created . . . (the earth being formless and void) . . . God said." The third view is that verse 1 is a separate main clause, serving as a title to the remainder of the section. The actual creation then begins with verse 2. The last view is that verse 1 is the main clause. It indicates the first act of creation, which is then continued in the following verses.

The first view was first set forth by one of the medieval Jewish rabbis by the name of Ibn Ezra, but not many have adopted his view. The second view was adopted by the medieval rabbi Rashi, though it may have been set out earlier. It is represented by all three of the alternate translations given above. The third and fourth views are represented by the standard translations such as the NKJV,

9. Ibid., p. 11–13.

the NASB, and the ESV. View three and four are distinguished only by interpretation, not by translation.

The third and fourth views clearly do not understand verse 1 as a temporal clause, while the other two do. The main point of contention is the very first word in the verse, which is usually translated as "in the beginning." Some grammarians have observed that the first word in verse 1 does not have the definite article (the). As a result, in their view it should be translated as the start of a temporal clause ("when God began to create," or, more literally, "in beginning of God's creating"). However, there are other examples where this same word is used without a definite article, yet it is clearly definite in sense (see Isaiah 46:10, where even the NRSV translates: "declaring the end from the beginning").

The idea that Genesis 1:1 should begin with this kind of temporal clause (when God began to create) has also been defended by the fact that one of the Babylonian creation myths, the *Enuma Elish*, begins "when the heavens had not been named." The idea here is that the author of Genesis (not Moses, in the view of those who hold to this theory) was influenced by the way in which the Babylonian myths began. However, more recent scholarship has seen little influence of Babylonian mythology on the organization of Genesis 1. Further, the ancient translations, such as the Septuagint (the Greek translation of the Old Testament that was done before the time of Christ), translate Genesis 1:1–3 in just the same way as our modern, literal translations do.

The grammar and syntax of the Hebrew in Genesis 1:1–3 allow for the differing translations provided above. However, the first two options at least leave room for, and probably demand, the idea of matter existing before creation. That is, God and matter are both eternal. However, that view is inconsistent with the theology taught in the remainder of the Scriptures — that God is the sole source of all that is, and that nothing existed but God before creation (e.g., Exodus 20:11; Nehemiah 9:6; Colossians 1:16). That leaves us, then, with the traditional translation of Genesis 1:1–3 as best representing the vocabulary, grammar, and syntax, as well as the theology, of the Hebrew text.

Conclusion

A knowledge of Hebrew vocabulary, grammar, and syntax is important for providing the basis for an accurate understanding of what the opening chapters of the Bible teach. The standard, traditional Christian understanding of the teaching of these chapters is not based on English mistranslations and misinterpretations. Instead, it has a solid foundation in the Hebrew language itself. But it is important for the reader who knows only English to realize that faulty theology can be as damaging to understanding Genesis as a faulty understanding of Hebrew. It is only when we are faithful to the teaching of the whole Bible that we can be confident that we have not misrepresented the teaching of any one part.

CHAPTER 26

The Recapitulation of Recapitulation: Does Embryology Prove Evolution?

DR. ELIZABETH MITCHELL

Introduction

Do human embryos replay the evolutionary history of their species as they develop? This idea has led many people to believe that what is in a woman's womb is merely an animal that can be simply disposed of by abortion at its fish stage. The false portrayal of embryonic development has tragically convinced countless people that the evolutionary worldview must be true, that humans are just highly evolved animals, and that abortion is acceptable.

Summed up in many textbooks by the popular and pithy declaration "Ontogeny recapitulates phylogeny," *recapitulation theory* (also known as the *biogenic law*) was popularized by evolutionist Ernst Haeckel's famous (or infamous) 19th-century illustrations intended to demonstrate how embryos pass through stages reminiscent of their evolutionary ancestors.

While the inaccuracy of Haeckel's drawings became apparent almost immediately, they have continued to be presented in textbooks, museums, and the secular media as "proof" of evolution even into this century. Evolutionary biologists who freely acknowledge the inaccuracy of the drawings continue to debate the validity of the "theory"[1] and its variants. Applications of recapitulation

1. A theory in science usually has little if anything against it, and a scientific "law" should have no exceptions. In light of this, recapitulation is more like a failed hypothesis since

theory are widely accepted in other disciplines such as linguistics and developmental psychology.

To many people, the evolutionary principles underlying recapitulation theory are fundamental truths, so the theory retains its authority in their thinking even when it requires substantial modification to exist alongside observable facts. Moreover, in recent years even Haeckel's evolutionary critics have shifted gears and begun to rehabilitate his reputation and his work. Forgiving the "liberties" he took, some now consider him positively brilliant for manufacturing pictures to prove what he "knew" must be true.

Many creationists are under the impression that evolutionists have abandoned recapitulation theory. Its persistence in the educational system, however, testifies to its usefulness even in the hands of those who believe that it has some problems. It remains a tool to explain evolutionary principles to students and to convince them that evolution is true.

Furthermore, many still believe that recapitulation theory (in some form or other) is sufficiently true to count as convincing evidence for evolution. And in the world of professional evolutionists, while some debate which variations of it they accept, others consider it a valid predictor of evolutionary stages and use it to unravel the secrets and subtleties of an evolutionary past shrouded by deep time and an incomplete fossil record. Thus, recapitulation theory continues to fuel the evolutionary thinking of students from the cradle to college, the lay public, and many academic professionals.

Big Words

"Ontogeny recapitulates phylogeny." The way that phrase rolls off of the tongue, combined with the compelling visual images that usually accompany it, appeal to the ear, the eye, and the mind. After all, how could big words that rhyme so well convey an untruth? But what do all those big words mean?

Ontogeny

Ontogeny means development from the earliest stages to maturity. In biology, *ontogeny* is roughly synonymous with embryologic development. Certainly, a fertilized egg must pass through a number of stages as it develops into a mature organism ready for life outside its mother's womb or its egg. A developing embryo changes its shape dramatically as it grows and morphs into its mature form.

---

it has so much against it. But since this is the recognized terminology, we will continue to refer to "recapitulation theory" and the "biogenic law" in this chapter for the sake of understanding.

Some anatomical structures first appear in the embryo in an apparently simple form and develop complexity. (That morphological simplicity is generally only a superficial impression, but the illusion of simplicity fits the evolutionary story that embryology supposedly tells.) Some embryonic anatomical structures disappear completely or remain only as *vestiges* (literally, "footprints") in the final mature product. *Vestigial organs* are commonly (and erroneously) viewed as "useless" anatomical structures leftover from our evolutionary past.

Phylogeny

Phylogeny refers to evolutionary ancestry. It is based on the presumption that all living organisms evolved from simpler forms through natural processes. The phylogenetic tree of life is a metaphor for the branching of the earliest life forms into stem branches, which, through the ongoing development of complexity and continued divergence into more and more branches, eventually produced the life forms we see today. Moreover, Haeckel, like many evolutionists then and now, maintained that this phylogeny is *monophyletic* — that all animal life can be traced back to a single common ancestor.

Recapitulation

Recapitulation refers to summarizing, repeating, or restating something. Thus, "ontogeny recapitulates phylogeny" is the claim that the developing embryo goes through stages that resemble, at least structurally, the various animals on that organism's ancestral trip up the tree of life.

Simply stated, Haeckel claimed that the embryonic forms of an animal resembled the adult organisms in its evolutionary ancestry. Because observation shows that developing embryos do not resemble the adults on the evolutionary tree of life, a modified form of the theory holds that an embryo only resembles the embryos of its evolutionary ancestors. A more recent reinterpretation of Haeckel's claims credits him with only claiming recapitulation applies to individual traits, rather than to entire embryonic stages.[2]

Seen and Unseen

Ontogeny is observable. Embryonic development of an organism can be studied through the lens of actual scientific methodology. Even the development of the human embryo has been studied in great detail.[3] The anatomy of

2. M. Richardson and G. Keuck, "Haeckel's ABC of Evolution and Development," *Biological Reviews of the Cambridge Philosophical Society* 77 no. 04 (2002): p. 495–528.
3. "Feedback: Embryo Protection" Answers in Genesis (July 22, 2011) http://www.answersingenesis.org/articles/2011/07/22/feedback-embryo-protection.

each stage of human embryonic development and that of many animals has been examined, sketched, and photographed.

When Haeckel's embryo drawings were published, they purportedly showed a comparison of the embryos of a number of vertebrates. Some see Haeckel's illustrations as blatant frauds, and others say he took artistic liberties to emphasize a point. Regardless, the images were almost immediately shown to be inaccurate by comparison with observable reality.

Phylogeny is *not* observable. No amount of scientific achievement makes it possible to see back through time to observe the purported upward evolution of life. Neither does biological research reveal any mechanism by which a simpler kind of organism can acquire the genetic information to become a more complex kind of organism.

Furthermore, no such transformation has ever been observed. Fossils labeled "transitional forms" are actually just animals with a variety of characteristics interpreted through an evolutionary imagination that connects the dots through time.

Thus, phylogeny is a figment intended to explain life without God. The claim that "ontogeny recapitulates phylogeny" is a claim that the observable steps in embryonic development are similar to and therefore reveal the unobservable evolutionary past of that organism.

Because the unobservable evolutionary past is not amenable to scientific examination, it is impossible to "test" the recapitulation claim. But because "evolution" is presented to students and to the public and held by the majority of mainstream scientists to be indisputable fact, recapitulation theory becomes a tool for education, a visually appealing bit of evidence, and a paleontological predictor to order fossils into the "right" lineages.

History

While Haeckel's drawings are the expression of recapitulation theory most familiar to modern schoolchildren, college students, and adults, the idea did not originate with Haeckel or even with Darwin. The germs of recapitulation theory can be found in the ancient world, but it gradually acquired its more modern form in the 19th century, with contributions by J.F. Meckel (1811), Karl Ernst von Baer (1828), Charles Darwin (1859), and finally Ernst Haeckel (1866).

Haeckel was a professor of zoology in Germany. He was particularly moved by Darwin's *Origin of Species,* and actively promoted Darwinian evolution to the public and to academia. As he taught how humans gradually developed

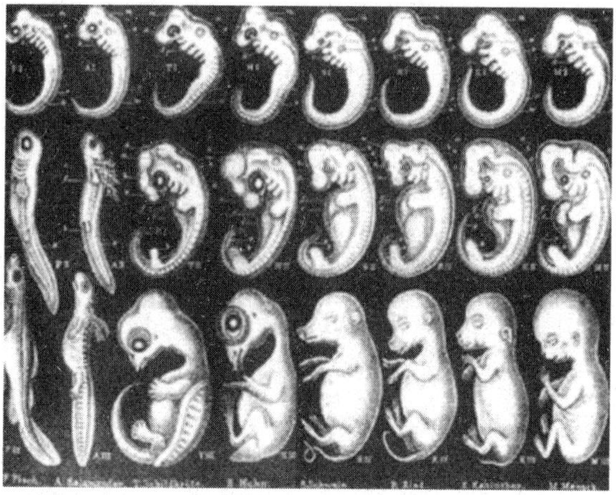

Haeckel's famous (infamous) set of 24 drawings purporting to show eight different embryos in three stages of development, as published by him in *Anthropogenie*, in Germany, 1874. This is the version of his drawings most often reproduced in textbooks. Left to right are shown embryos of a fish, salamander, turtle, chicken, pig, cow, rabbit, and human. Top to bottom depicts three stages of development. The drawings contain errors intended to emphasize embryonic similarity and support recapitulation theory. IMAGE: from M. Richardson and G. Keuck, "Haeckel's ABC of Evolution and Development," *Biological Reviews of the Cambridge Philosophical Society*, 77 no. 04 (2002): p. 495–528.

through upward evolution along a tree of life, he presented hypothetical simple organisms as if they were real, an ape-man for which he had no evidence, and his infamous doctored embryo sketches.

Haeckel's version of the "biogenetic law" held that embryos looked like the adult forms of their evolutionary forebears. He wrote that embryonic development paralleled phylogenetic (evolutionary) history — that "embryonic development is a short and rapid re-run, or recapitulation, of evolution."[4] To support his claim, in his book *Natürliche Schöpfungs-geschichte* (Germany, 1868; published in English in 1876 as *The History of Creation*), Haeckel included sketches of embryos substantially altered to make his point. "His drawings are also highly inaccurate, exaggerating the similarities among embryos, while failing to show the differences," explains embryologist Michael Richardson, lead author of a famous 1997 article refuting Haeckel's claims.[5]

4. Richardson and Keuck, *Biological Reviews of the Cambridge Philosophical Society*, p. 495–528.
5. M. Richardson et al., "There is no highly conserved embryonic stage in the vertebrates: implications for current theories of evolution and development," *Anatomy and Embryology* 196 no. 2 (1997): p. 91–106.

Soon after publication, Haeckel's 19th-century contemporaries spotted the fraud and publicized it. For instance, in 1874, William His, after critiquing Haeckel's ideas and demonstrating that many of the embryo figures were "invented," concluded, "The procedure of Professor Haeckel remains an irresponsible playing with the facts even more dangerous than the playing with words criticized earlier."[6]

For over a century, criticism from the evolutionary scientific community has continued. "Scientific objections to Haeckel's drawings . . . include charges of:

> (i) doctoring (the alteration of images during copying);
> (ii) fabrication (the invention of features not observed in nature); and
> (iii) selectivity (the use of a misleading phylogenetic sample)."[7]

The most generous and gracious modern assessments have been unable to allay charges of falsification, and Haeckel even admitted to some of the accusations. For instance, to the charge that he printed a woodcut of a single turtle embryo three times, altered to represent three different species, he confessed to "an imprudent folly" necessitated by a shortage of time.[8]

Despite the almost immediate rejection of Haeckel's evidence by much of the scientific community, his rather impressive fabrications did their job: they found their way into textbooks as evidence illustrating evolutionary claims for over a century. Countless children and adults — and young women coaxed to proceed with abortion — have been told that the human embryo goes through a fish stage, an amphibian stage, and a reptilian stage. Attesting to the sometimes-disputed fact that these fraudulent "teaching tools" persisted in the educational system despite their known errors and general rejection in the scientific community, leading evolutionist Stephen Gould in the year 2000 wrote:

> Haeckel had exaggerated the similarities by idealizations and omissions. He also, in some cases — in a procedure that can only be called fraudulent — simply copied the same figure over and over again. At certain stages in early development, vertebrate embryos do look more alike, at least in gross anatomical features easily observed

6. Richardson and Keuck, *Biological Reviews of the Cambridge Philosophical Society*, p. 495–528.
7. Ibid.
8. Ibid.

> with the human eye, than do the adult tortoises, chickens, cows, and humans that will develop from them. But these early embryos also differ far more substantially, one from the other, than Haeckel's figures show. Moreover, Haeckel's drawings never fooled expert embryologists, who recognized his fudgings right from the start.
>
> At this point, a relatively straightforward factual story, blessed with a simple moral story as well, becomes considerably more complex, given the foils and practices of the oddest primate of all. Haeckel's drawings, despite their noted inaccuracies, entered into the most impenetrable and permanent of all quasi-scientific literatures: standard student textbooks of biology. . . . We should therefore not be surprised that Haeckel's drawings entered nineteenth-century textbooks. But we do, I think, have the right to be both astonished and ashamed by the century of mindless recycling that has led to the persistence of these drawings in a large number, if not a majority, of modern textbooks![9]

In a succinct summation of Haeckel's work, Gould concluded that Haeckel, who used his doctored diagrams as data to support his scientific hypotheses, committed the "academic equivalent of murder."[10]

A 1997 study of comparative embryology, published in the journal *Anatomy and Embryology* by embryologist Michael Richardson, then of London's St. George's Hospital Medical School, also called attention to the persistent acceptance of Haeckel's fraudulent diagrams. He found that Haeckel had resized embryos and eliminated limb buds and heart bulges to enhance similarity. He wrote, "These drawings are still widely reproduced in textbooks and review articles, and continue to exert a significant influence on the development of ideas in this field."[11] Gould quotes Richardson saying, "I know of at least fifty recent biology textbooks which use the drawings uncritically."[12]

While some excuse Haeckel's diagrams as mere schematics, these "schematics" were clearly meant to systematically and deceptively improve on nature. For instance, he selectively removed limbs on one of his embryos while rendering

9. Stephen Jay Gould, "*Abscheulich!* (Atrocious!)," *Natural History*, 109 no. 2 (2000): p. 44–45. Quoted in Revisiting Those Pesky Embryo Drawings — Evolution News & Views http://www.evolutionnews.org/2010/06/revisiting_those_pesky_embryo035741.html.
10. Richardson and Keuck, *Biological Reviews of the Cambridge Philosophical Society*, p. 495–528.
11. Richardson et al., *Anatomy and Embryology*, p. 91–106.
12. Stephen Jay Gould, *Natural History*, p. 44—45.

others perfectly, commenting that they were similar with "no trace of limbs or 'extremities' in this stage."[13] According to Richardson, the "intent [of these systematic alterations] is to make the young embryos look more alike than they do in real life."[14]

Despite overwhelming evidence that has been used to refute Haeckel's claims and the manufactured data he used to support them, Richardson and colleagues write, "The idea of a phylogenetically conserved stage has regained popularity in recent years."[15] To assess the merits of recapitulation theory and Haeckel's work, they conducted a systematic examination of embryos from all sorts of vertebrates, noting that modern textbooks typically confine their attention to the frog, the chick, and the "typical" mammal.

They compared the most *phylotypic* stage of each — the stage at which vertebrate embryos possess comparable characteristics such as a notochord, pharyngeal arches ("gill slits"), a neural tube, somites (segments of undifferentiated blocks of embryonic mesoderm), and a postanal tail (a posterior extension of the embryo's developing musculoskeletal structures beyond the anus).

Richardson et al. in 1997 confirmed that even the earliest stages of embryonic development vary greatly between vertebrate species. They attributed these differences to evolution, as they hold an evolutionary worldview. But their paper demonstrated, on the basis of rigorous comparative embryology, that the "biogenetic law" as commonly understood is false.[16]

A quick Internet search today will produce many references to recapitulation theory as "inadmissibly simplified,"[17] "outdated" and "buried,"[18] "refuted," "defunct" and "largely discredited." Haeckel's drawings are recognized by many as "fraudulently modified"[19] "misinformation."[20] Embryologist Michael Richardson was quoted in a 1997 issue of *Science* magazine saying Haeckel's work was "turning out to be one of the most famous fakes in biology."[21] So has

13. Ernst Haeckel, *Anthropogenie oder Entwickelungsgeschichte des Menschen. Keimes- und Stammesgeschichte* (Engelmann, Leipzig, 1903); quoted in Michael Richardson and Gerhard Keuck, "A Question of Intent: When Is a 'Schematic' Illustration a Fraud?" *Nature* 410 no. 144 (2001).
14. Ibid.
15. Richardson et al., *Anatomy and Embryology*, p. 91–106.
16. Ibid.
17. http://www.frozenevolution.com/haeckel-s-recapitulation-theory.
18. http://education.stateuniversity.com/pages/2026/Hall-G-Stanley-1844-1924.html.
19. http://www.thematrix.co.uk/texttopic.asp?ID=31.
20. https://en.wikipedia.org/wiki/Recapitulation_theory.
21. E. Pennisi, "Haeckel's Embryos: Fraud Rediscovered," *Science* 277 (1997):1435. Quoted in http://home.uchicago.edu/~rjr6/articles/Haeckel--fraud%20not%20proven.pdf.

Haeckel's work — so heavily criticized even in the evolutionary community — dropped off the scene? No. Why is that?

Despite over a century of widespread acknowledgement that Haeckel faked his pictures, Haeckel's claims and even colorized adaptations of his diagrams still show up in the popular press and even textbooks. For instance, the cover story of *Time* magazine (November 11, 2002) reported that the human embryo at 40 days "looks no different from that of a pig, chick or elephant. All have a tail, a yolk sac and rudimentary gills."[22] Even 21st-century textbooks perpetuate this 19th-century fraud. Sylvia Mader's 2010 edition of *Biology*, for instance, features colorized Haeckel-ish embryos and teaches, "At these comparable developmental stages, vertebrate embryos have many features in common which suggests they evolved from a common ancestor."[23]

In a world where evolutionary educators decry any effort to "teach the controversy" in public schools — allowing students to be exposed to facts that reveal problems with evolutionary dogma — the convenient foot-dragging on the removal of this compelling lie from curricula is telling.

Those Fishy Gill Slits

Our embryonic "gill slits" are possibly the most oft-cited anatomical "proof" of our fishy ancestry. *Inside the Human Body*, a popular 2011 BBC1 program hosted by Dr. Michael Mosley, provides a typical example. The program features a state-of-the-art high-quality video of human embryonic development called "Anatomical clues to human evolution from fish."[24] The video was produced by digitally splicing scans taken in early pregnancy. Mosley interprets the developing features as anatomical proof of fish in our evolutionary past. Among these are "gill-like structures," a reference to the "gill slits."[25]

The poorly named "gill slits" in human embryos are not anything at all like gills and are not even slits, just folds of tissue destined to develop into various anatomical parts of the head and neck. They never have a function or a structure remotely resembling gills. They don't even turn into anything having to do with the lungs. Never in the course of development does a human embryo absorb oxygen from water as fish do with gills.

22. *TIME*, November 11, 2002, http://www.time.com/time/magazine/article/0,9171,1003653,00.html.
23. "Current Textbooks Misuse Embryology to Argue for Evolution," *Evolution News Views*, http://www.evolutionnews.org/2010/06/current_textbooks_misuse_embry035751.html.
24. Available online at http://www.bbc.co.uk/news/health-13278255.
25. See "Vestigial Hiccups, Folding Fish-eyes, and Other Fables: Our Fishy Forebears . . . Again!" at http://www.answersingenesis.org/articles/aid/v6/n1/fishy-fables.

Evolutionist Steven Jay Gould writes, "In Haeckel's evolutionary reading, the human gills slits *are* (literally) the adult features of an ancestor" (emphasis in original).[26] In later writings, Haeckel did not ascribe a respiratory function to these structures in the non-fish embryo. He still maintained that there were actual gill slits and gill arches in the non-fish embryos but that they had evolved into other structures. He wrote in 1892 that "we never meet with a Reptile, Bird or Mammal which at any period of actual life breathes through gills, and the gill-arches and openings which do exist in the embryos are, during the course of their ontogeny, changed into entirely different structures, viz. into parts of the jaw-apparatus and the organ of hearing."[27] And by 1903 he wrote of the "total loss of respiratory gills," saying that "in the embryos of amniotes there is never even a trace of gill lamellae, of real respiratory organs, on the gill arches."[28]

Evolutionists consider homologies in fish gills, fish jaws, reptilian jaws, and mammalian ear bones to be sequential evolutionary developments that demonstrate the common evolutionary ancestry of fish, reptiles, and mammals. *Homologous* structures are the different anatomical structures that form from a similar embryonic structure. Meckel's cartilage, for instance, has different destinies in different creatures. Meckel's cartilage supports the gills in cartilaginous fish. It ossifies to form the jaws of bony fish and reptiles. And in mammalian embryos, Meckel's cartilage helps shape the middle ear bones and the mandible; then it virtually disappears. But each creature has its own kind of DNA directing the process, and at no time in science do we see DNA of one creature mutating to produce new information that can change the organism into a new kind. And at no point do these so-called mammalian "gill slits" have anything to do with gills or respiratory structures.

Mammalian "gill slits" are folds in the region of the tiny embryo's throat. By the 28th day of life, the embryo's brain and spinal cord seem to be racing ahead of the rest of the body in growth. Therefore, for a time, the spinal cord is actually longer than the body, forcing the body to curl and flexing the neck area forward. (This curled embryo with the long spinal cord is mistakenly accused by

26. Stephen Jay Gould, *Ontogeny and Phylogeny* (Cambridge, MA: Belknap Press, 1977), p. 7.
27. Ernst Haeckel, *The History of Creation* [translation of the 8th German Edition of *Natürliche Schöpfungsgeschichte*]. (ed. E.R. Lankester) (Kegan Paul, London: 1892). Quoted in Richardson and Keuck, "Haeckel's ABC of Evolution and Development," *Biological Reviews of the Cambridge Philosophical Society*: p. 495–528.
28. Ernst Haeckel, *Anthropogenie oder Entwickelungsgeschichte des Menschen*; Quoted in Richardson and Keuck, "Haeckel's ABC of Evolution and Development," *Biological Reviews of the Cambridge Philosophical Society*: p. 495–528.

some people of having an animal's tail.) Just as many people develop a double chin when bending the neck forward, so the embryo has folds in its neck area due to this flexing.

Gill slits, thus, is a misleading name, since these folds are neither gills nor slits. Another popular name, *branchial arches*, is just as deceptive because *branchial* comes from the Greek word for "gills." Somehow the name *neck folds* just isn't fancy enough for our scientific minds, so these folds are called *pharyngeal arches*, since they are arch-shaped folds near the throat. (*Pharyngeal* is the scientific word for things having to do with the throat. When you say you have a sore throat, your doctor says you have pharyngitis.) The creases between the folds are called *pharyngeal clefts*, and the undersides of the folds are called *pharyngeal pouches*. The pouches and clefts are not connected by an opening. Each fold shapes itself into specific structures, none of which are ever used for breathing. The outer and middle ear as well as the bones, muscles, nerves, and glands of the jaw and neck and even the immune system's thymus gland develop from these folds as tissues differentiate in compliance with the blueprint in human DNA.

Nevertheless, the meaning-packed terms *gill slits* and *gill-like structures* persist. But mammalian pharyngeal arches are no more related to gills — ancestrally or otherwise — than stars are to streetlights.

Even texts that refer to these folds by correct names sometimes perpetuate the powerful gill slit myth. For instance, Mader's *Biology* (2007 edition) correctly describes the ultimate anatomic destiny of each pharyngeal arch component and then asks:

> Why should terrestrial vertebrates develop and then modify such structures like pharyngeal pouches that have lost their original function? The most likely explanation is that fishes are ancestral to other vertebrate groups.[29]

What "lost original function"? No one has ever documented that pharyngeal pouches in the embryos of terrestrial vertebrates function as gills or that adult terrestrial vertebrates ever had gills. Preserved in textbooks and the media, the fishy ancestral myth persists. Our unseen and unverified fishy past still surfaces regularly in the assumptions that the pouches/folds/slits, or whatever-they-get-called, are leftovers from a fish ancestor.

In a chilling application of this misinformation, many abortionists have used Haeckel's embryologic falsehoods to assuage the guilt of women seeking

29. Sylvia Mader, *Biology*, 9th ed. (New York: McGraw-Hill, 2007), p. 97.

abortion, telling them they're only removing something like a fish, not a baby. The late Dr. Henry Morris observed, "We can justifiably charge this evolutionary nonsense of recapitulation with responsibility for the slaughter of millions of helpless, pre-natal children — or at least for giving it a pseudo-scientific rationale."[30]

The Current Debate

Given all the data researchers have used to refute recapitulation theory, do real scientists still cling to its discredited notions? After all, it's one thing to foist a fabricated over-simplified bit of evolutionary evidence on the gullible public and generations of children and college students, but do professionals hang on to these notions, too?

While some professional evolutionary scientists have given up on recapitulation theory altogether, many continue to cling to various permutations of it.

Some distance the beloved recapitulation dogma from Haeckel and look back a bit further to Karl Van Baer's 1828 version that claimed embryonic stages only recapitulate the embryonic stages of their evolutionary ancestors. Neither version has ever truly explained embryologists' observations, however. And as Richardson's work has clearly demonstrated, vertebrate embryos have discernible differences even at the earliest stages, an observation that finally strips the underpinnings of both versions. Thus to make the theory work, some evolutionary biologists who wish to keep it have modified it, choosing which parts they can make the best case for.

Ernst Mayr's modification, laid out in "Recapitulation Reinterpreted: The Somatic Program," appeared in 1994 in the *Quarterly Review of Biology.* He wrote that despite "the disrepute into which Haeckel's claims had fallen . . . every embryologist knew that there was a valid aspect to the claim of recapitulation."[31] A 2012 paper co-authored by Richard Lenski, "Ontogeny Tends to Recapitulate Phylogeny in Digital Organisms," notes that Mayr's "sentiment is still widely held today, and the idea that ontogeny recapitulates phylogeny in some form has its modern proponents."[32]

30. Henry Morris, T*he Long War Against God* (Ada, Michigan:Baker Book House, 1989), p. 139.
31. Ernst Mayr, "Recapitulation Reinterpreted: The Somatic Program," *Quarterly Review of Biology* (2002).
32. J. Clune et al., "Ontogeny Tends to Recapitulate Phylogeny in Digital Organisms," *The American Naturalist*, 180 no. 3 (2012): p. E54–E63.

Making It Work

Recapitulation theory is just too appealing to abandon for many evolutionists. Lenski's group wrote, "At a minimum, the fact that the debate has continued for so long lends credence to Mayr's view that there is at least some validity to recapitulation."[33]

Perhaps the most dramatic rehabilitation of Haeckel has come at the hands of one of his best-known modern critics, Michael Richardson. In the 2002 paper "Haeckel's ABC of Evolution and Development," published in *Biological Reviews of the Cambridge Philosophical Society,* Richardson and Gerhard Keuck re-examined Haeckel's work. They wrote:

> Haeckel recognized the evolutionary diversity in early embryonic stages, in line with modern thinking. He did not necessarily advocate the strict form of recapitulation and terminal addition commonly attributed to him. Haeckel's much-criticized embryo drawings are important as phylogenetic hypotheses, teaching aids, and evidence for evolution. While some criticisms of the drawings are legitimate, others are more tendentious. . . . Despite his obvious flaws, Haeckel can be seen as the father of a sequence-based phylogenetic embryology.[34]

Richardson and Keuck conclude that the biogenetic law is valid after all, if applied to the evolution of "single characters only" and not entire embryonic and evolutionary stages.[35] In other words, so long as only single traits are followed through evolutionary time and embryonic development, Richardson is now aboard the recapitulation bandwagon.

Richardson and Keuck's analysis of Haeckel's work was not able to expunge the charge of falsification, but they clearly have granted him absolution. They and others support "Haeckel's practice of filling in gaps in the embryonic series by speculation,"[36] even though "Haeckel presented the embryo drawing as data in support of his hypotheses"[37] and not just helpful teaching aids.

Haeckel's artistic liberties are clearly not the result of any lack of observation skills or artistic ability. One of his latter-day apologists has even praised Haeckel's diagrams of single-celled radiolarians, noting their resemblance to

33. Ibid.
34. Richardson and Keuck, *Biological Reviews of the Cambridge Philosophical Society*, p. 495–528.
35. Ibid.
36. Ibid.
37. Ibid.

modern light microscope images and electron micrographs.[38] Haeckel was a skilled illustrator able to render what he observed with accuracy and detail when he wanted to. But when real observation failed to confirm what he needed to be true in order to support his worldview-based beliefs about the evolutionary past and its parallels in the present, he opted to draw his own version of "reality."

The ultimate excuse for Haeckel's graphic concoctions has come from those who wish to honor what they see as his cognitively pure prescience coupled with a somewhat liberal view of the purpose of scientific illustration. "Haeckel's own views on art stressed the primacy of interpretation over pure observation,"[39] write Richardson and Keuck. They note that Haeckel's own writings reveal that he knew early embryos of various species have a lot of differences. They assert that Haeckel therefore never intended for his pictures to depict his actual observations but rather to show what he deemed to be "a true reproduction of the really existing natural produce."[40] And fabrications though some of these drawings clearly were, Haeckel intended them as support for his recapitulation theory. Yet because Richardson and Keuck maintain that recapitulation theory is true so long as it is viewed in a certain way — one trait at a time, with allowances for traits that have disappeared over time — they believe "Haeckel's embryo drawings are important as phylogenetic hypotheses, teaching aids — *even scientific evidence*" (emphasis ours).[41]

But Why?

What recapitulation believers still struggle with, however, is some reason recapitulation should be true. What evolutionary advantage would it have? If embryos really recapitulate their evolutionary past, what is the evolutionary advantage of anatomic structures that develop and ultimately don't get used? Why would unused "gill slits," for instance, stick around across the evolutionary time scales through organisms that did not need gills until they could evolve a non-respiratory purpose?

Some embryologic structures only serve temporary purposes in the embryo and then disappear or regress. If these represent footprints of an evolutionary

38. Ibid.
39. Ibid.
40. Ernst Haeckel, (1904). *Kunstformen der Natur. Bibliographisches Institut, Leipzig und Wien*; quoted in M. Richardson and G. Keuck, "Haeckel's ABC of evolution and development," *Biological Reviews of the Cambridge Philosophical Society*, p. 495–528.
41. Richardson and Keuck, *Biological Reviews of the Cambridge Philosophical Society*, p. 495–528.

past, why would structures that don't get used in the mature organism persist purposelessly through millions of year of evolutionary history?

In an attempt to answer this question, some expand on Gould's idea of "terminal addition," proposing that successful earlier evolutionary innovations are not lost but allowed to keep functioning while new developments are added. To undo earlier developments before they have served their place-holding purpose in the newly evolving organism would disrupt subsequent add-ons. While this describes exactly what happens in a developing *embryo* whose development is directed by its DNA blueprint, however, how can mindless random evolution "know" it needs to keep a useless structure in place for millions of years?

Phylogeny and the Return of Haeckel

Haeckel's diagrams do not represent observable embryologic reality, and Haeckel knew that they didn't when he made them. Nevertheless, he intended them — doctored though they were — to be data in support of his evolutionary ideas. He intentionally falsified scientific observations in order to use "embryonic resemblance as proof of evolution"[42] and "recapitulation as proof of the Biogenetic Law."[43] Yet he receives praise for his insight into the evolutionary past and his ability to reconstruct the observable present to prove what evolutionists believe.

Rigorous comparative embryology confirms "there is no evidence from vertebrates that entire stages are recapitulated."[44] Thus, Haeckel's claims about embryonic development are not supported by actual observation. Even if embryonic development did proceed as he claimed, of course, it would not prove anything about a hypothetical evolutionary past.

But that aside, why are evolutionary scientists and educators so keen to use inaccurate diagrams for "phylogenetic hypotheses, teaching aids, and evidence for evolution"? Why do Haeckel's modern apologists strain at his work, repackaging it to show how it could be true so long as it is viewed a certain way, such as one trait at a time?

Embryology, because it outlines successful steps that produce fully functional, mature organisms, tells the evolutionist what to look for. And because whole organisms don't often fill the needs of the evolutionary story, evolutionists can now justify tracing single traits through deep time and seeking parallels in embryology. A fossil that seems to possess a trait in any of the ways it appears

42. Ibid.
43. Ibid.
44. Ibid.

in an embryological developmental sequence can be claimed as a representative of its evolutionary sequence and assigned its spot in history.

If fossils seeming to fit the step-wise nature of different embryological stages can be found, they are lined up as evidence for evolution. But fossils do not demonstrate evolutionary transitions. Neither do embryologic stages. Yet by claiming that both actually do represent evolutionary sequences, evolutionists concoct visually compelling evidence and tie it together through a comforting knot of circular reasoning.

The controversy about the evolutionary origin of the turtle shell illustrates both of these points. Evolutionists have long debated the origin of the turtle shell. Until recently all the turtle fossils found had been fully equipped with modern-appearing shells. Therefore, evolutionists have debated whether the shell evolved over millions of years by following the sequence seen inside the turtle egg or whether it evolved as a modification of external scales.

Now that two varieties of turtle with seemingly less developed parts of the shell have been identified, evolutionary researchers have noted that these shell variations more or less mirror shell developmental stages in the embryo. They therefore are asserting that turtle embryology predicted those forms successfully, proving on the one hand that those turtles are genuine transitional forms and on the other hand that ontogeny of turtle shells really does recapitulate phylogeny.[45]

In reality, no evolution from non-turtles is seen here, only two varieties of turtles. What these turtle fossils reveal is not a series of non-turtles evolving into turtles but just varieties of turtles. Mutations alter genetic information, and it is likely that these two extinct turtles are merely variations that developed from the original turtle kind God created about 6,000 years ago. Finally, as teaching aids, teachers and textbook manufacturers can now once again return in good conscience to teaching the mantra, "ontogeny recapitulates phylogeny," that is — those that ever actually stopped in the first place. For many who accept evolution as unquestioned fact, any evidence that can be used to indoctrinate the young or the gullible is acceptable, even fraudulent concoctions from a man who was in the habit of manufacturing whatever counterfeits and forgeries he needed in order to promote evolution with the evangelistic zeal of a missionary.

Thus, despite their inaccuracies, Haeckel's sometime critic-turned-defender concludes, "Haeckel's embryo drawings are important as phylogenetic

45. "Turtle in the Gap," Answers in Genesis (June 29, 2013), http://www.answersingenesis.org/articles/2013/06/29/turtle-gap.

hypotheses, teaching aids — even scientific evidence. . . . The drawings illustrate embryonic similarity, recapitulation, and phenotypic divergence."[46]

Recapitulation's Future

Just because something is proven blatantly false, like recapitulation theory, doesn't mean people are *persuaded.* These controversies can be expected to continue, not because there is proof that all life evolved from simpler ancestral forms, but because there is a popular widespread worldview-based belief in molecules-to-man evolution.

Believing that life must be explained as the product of natural evolutionary processes, evolutionary scientists must seek natural explanations wherever they can. Yet embryonic development is observable, and evolutionary phylogeny is not. Their supposed parallelism and the notion that such parallelism would constitute evolutionary proof are popular and powerful lies.

The observable wonders of embryology — surely a showcase of God's design — were hijacked by Haeckel and continue to be much too valuable components of the evolutionary toolkit to relinquish. Recapitulation has therefore been resurrected and repackaged to teach and to convince. Haeckel's "liberties" are excused with a nod that would never be extended to any modern scientist who faked his findings.

Recapitulation theory will doubtless continue to serve a prominent place in classrooms and on television documentaries aimed at convincing the public of the "obvious" truth of evolution. Moreover, as illustrated by the case of the turtle shell, highly trained evolutionary scientists, seeking to answer not "whether" things evolved but "how," will find recapitulation theory to be a convenient tool to provide the circular reasoning to justify the theory of the moment.

46. Richardson and Keuck, *Biological Reviews of the Cambridge Philosophical Society*, p. 495–528.

CHAPTER 27

Is Speciation Evidence for Creation or Evolution?

DR. GARY PARKER

In a debate at a major Texas university, the creationist was challenged with this claim: Hawaiian fruit flies that could once all interbreed had changed into numerous reproductively isolated species, and that, said the challenger to considerable applause, "proved evolution." The creationist responded (also to considerable applause) that such a change would be the opposite of evolution. Losing the ability to interbreed, each "new species" would have less genetic variability, less ability to meet changes in its existing environment, and less ability to explore new environments — all suggesting decline and demise rather than the expansion of genetic potential required for what Darwin called "the production of higher animals."

Which of these views is more consistent with our present understanding of genetic science and with the biblical record of earth history?

Basic Genetics

The Bible records several key events in early earth history that suggest concepts geneticists can test scientifically. Genesis 1 states that God created many distinct "kinds." We infer from a plain reading of Scripture that animals and plants were created to reproduce within the boundaries of their kinds (Genesis 1, 6, and 8). A created kind is typically equivalent to the level of family in modern classification schemes as many members of a family can interbreed and produce offspring. The kinds were also "to fill" (scatter, move into) earth's varied environments (Genesis 1:22, 8:17). Multiple biological mechanisms accounted for this filling and resulted in variation within kinds, or speciation. Do these fundamental concepts in God's Word — discrete created kinds, or baramins (Hebrew: *bara* = create and *min* = kind), having broad but limited variability — help scientists understand the genetic changes in organisms and speciation found in God's world? Indeed, they do!

The complete set of DNA specifying a kind is called its genome. The human genome includes approximately 20,000 to 25,000 protein-coding chromosomal segments commonly called genes. The genes and the information they encode are largely responsible for the set of biological traits that distinguish human beings from other kinds of life. All humans have essentially the same genes, and they are over 99 percent similar in all seven billion of us; hence, geneticists refer to the human genome and have concluded that we are all members of one race, the human race (as the Apostle Paul preached in ancient Greece, Acts 17:26).

The similarity among all human beings is obvious, but so is the tremendous variation! The genes we share in the human genome make us all the same (100 percent human); but different versions of these shared genes, called alleles, produce the spectacular variation that makes each individual unique. For any given gene, God could have created it in four different allelic varieties (two in both Adam and Eve). Genetic alterations occurring since sin corrupted creation have introduced many new alleles, but no new genes.

The human genome, for example, has genes for producing hair and controlling its shape; allelic versions of these genes result in individuals with straight, wavy, curly, and tightly-curled hair; all variations within the human kind. Although, genetically speaking, skin color is more complex, the variation in human skin tone can be described as the action of two pairs of genes with different alleles (A/a and B/b) that influence the production of the skin pigment melanin. As shown in figure 1,[1] two people with medium-brown skin tone and genes AaBb could have children with the full range of skin tones — from very dark (AABB), to dark (AABb or AaBB), to medium (like AaBb), to light (like Aabb), to very light (aabb). That would certainly be "change through time" but a lot of change in a little time (one generation!) With no genes added, this is just variation within a kind.

Mutations, changes in DNA that occurred after man's sin corrupted God's creation, do not produce new genes. Rather, mutations only produce alleles, variations in pre-existing genes. Alleles are not different genes in the sense that

1. Gary Parker, *Building Blocks in Life Science* (Green Forest, AR: Master Books, 2011), p. 9.

A couple with **melanin** control genes
AaBb (Adam and Eve?)
would have "**medium**" skin tone, and each
would make four kinds of reproductive cells,
as shown along top and side of this
"genetic square":

genes in mother's egg cells

genes in father's sperm cells	AB	Ab	aB	ab
AB	**AA BB**	AA Bb	Aa BB	Aa Bb
Ab	AA Bb	AA bb	Aa Bb	Aa bb
aB	Aa BB	Aa Bb	aa BB	aa Bb
ab	Aa Bb	Aa bb	aa Bb	aa bb

Each box in the larger "Punnett square" shows the gene
combination possible in a child.

As shown in the pictures below,
children of "medium" parents could have
most to least melanin and
color darkest to lightest with 4, 3, 2, 1, 0 "capital letter"
genes indicated with each picture.

Figure 1. Inheritance of melanin skin color

genes for skin color and genes for making sickle-cell hemoglobin (resulting in sickle cell anemia) are. Similarly, the sickle-cell gene is a different allele (version) of the hemoglobin gene in the sense that it was not present at creation, but it is only a different harmful version of a *pre-existing* gene. In fact, the allele for sickle-cell hemoglobin differs in sequence in only one position out of several hundred from the normal gene for making hemoglobin. Again, we see mutations leading to different versions of *pre-existing* genes resulting in a variety of alleles but not the creation of brand new genes encoding novel proteins with novel functions of the type necessary for molecules-to-man evolution.

Variation within a Kind

All the genes in one generation available to be passed on to the next are called the gene pool. Members of the same kind may also be defined as organisms that share the same gene pool. The number of genes for different kinds of traits, the number in a complete genome, can be called the depth of the gene pool. The human gene pool is around 20,000–25,000 genes deep. The width of the gene pool refers to the amount of its allelic variation. Among dogs, for example, the width of a greyhound's gene pool is very narrow; crossing purebred greyhounds just gives you more greyhounds, all very similar in speed, color, intelligence, hair length, nose length, etc. Crossing two mongrels, however, can give you big dogs and small dogs, dark and light and splotchy-colored dogs, dogs with long and short hair, yappy and quiet dogs, mean and affectionate dogs, and the list goes on! The width of the mongrel's gene pool (its allelic variability) is quite large compared to the greyhound's, but the depth of the gene pool (the number of genes per genome) is the same for both dogs.

A kind is defined in terms of depth of the gene pool, which is the total number of different genes in a genome and a list of traits they encode for. Variation within a kind is defined in terms of the width of the gene pool, the number of possible alleles at each gene site (locus).

Geneticists call the shuffling of *pre-existing* genes recombination. Perhaps you have played a game with a common deck of 52 cards that includes four groups (hearts, diamonds, clubs, and spades), each with 13 different numbers or "faces" (2–10 plus J, Q, K, A). In a game called bridge, each of four players gets a "hand" of 13 cards. You can play bridge for 50 years (and some people do!) without ever getting the same group of 13 cards. The hands you are dealt are constantly changing, and each is unique — but the deck of cards remains always the same.

Although the comparison is not perfect, a deck of cards illustrates the concept of variation within a created kind. The bridge hands dealt are unique, different, and constantly changing, like the individual members of a population. But the deck of 52 cards remains constant, never changing, always the same, like the kind. Individual variation plus group constancy equals variation within a created kind.

Faith in Man Versus Faith in God's Word

Based on faith in Darwin's words, evolutionists assume that all life started from one or a few chemically evolved life forms with an extremely small gene pool. For evolutionists, enlargement of the gene pool by Darwinian selection (struggle and death) among random mutations is a slow, tedious, grim process that burdens each type with a staggering "death load" and "genetic load" of harmful mutations and evolutionary leftovers. Based on faith in God's Word, creationists assume each created kind began with a large gene pool, designed to multiply and fill the earth with its tremendous ecological and geographic variety.

Neither creationists nor evolutionists were there at the beginning to see how it was done, of course, but the creationist can build on the Word of the One who was there "in the beginning" (Genesis 1:1; John 1:1–3). Furthermore, the creationist mechanism is consistent with scientific observation. The evolutionary mechanism doesn't work, and is not consistent with present scientific knowledge of genetics and reproduction. As a scientist, I prefer ideas that do work and do help to explain what we can observe, and that's biblical creation!

Since animals were commanded to multiply and fill the earth, we can infer that the created kinds were "endowed by their Creator" with tremendous allelic variability and allelic potential in very wide gene pools. Geneticists now know, for example, that alleles for the full range of normal human variation — darkest to lightest skin tone, Pygmy to Watusi heights, wide to thin lips, hair from straight to wavy to curly to tightly-curled, eyelids producing round to oval shapes, etc. — are possible, beginning with just two people. Genetics problems solved by high school students (figure 1) show how such parents could produce children with traits from darkest to lightest, shortest to tallest, with hair of any style, and eyes and lips of any shape in just one generation — all with NONE of the deep time, chance mutations, and ceaseless struggle to the death that evolutionists use to explain variation in beak sizes in finches or amounts of black pigment in moth wings.

What Does This Awesome Variability within Kinds Mean?

For one thing, such awesome variation reflects God's creativity. God created the first man from the dust of the ground and the first woman from a rib from his side (Genesis 2:7, 21–22). Then, God rested from His creative acts at the end of the creation week (Genesis 2:1–2). But we still see God's creativity unfolding before our very eyes in a different way in the birth of each child. As they relate to the genetic potential God created in our first parents, we may not yet have seen the fastest runner or the greatest mathematical or musical genius. Genes were not produced one at a time by evolutionary processes — time, chance, mutations, struggle, and death over millions of years. This unfolding of genetic variability in *pre-existing* genes is all stunning variation within a kind, but it is NOT the formation of new genetic information of the type required for molecules-to-man evolution.

As the descendants of each created kind multiplied to fill the earth, we see their genetic potential unfolding. God created the bear kind, for example. But as bears moved into different environments around the world after the Flood, their built-in variability and ability to genetically change came to visible expression in black bears, brown bears, grizzly bears, polar bears, etc. The created dog kind diversified into specialized subtypes: wolves, coyotes, domestic dogs, etc. Think also about the tremendous genetic variability brought to visible expression in the cat kind, rose kind, tomato kind, etc.

There is a strong tendency, both in nature and in experimental breeding, for generalized, adaptable organisms to produce a variety of specialized, adaptable subgroups. Figure 1, discussed earlier, showed that if Adam and Eve, for example, had a variety of alleles for skin tone (AaBb) they could have children with skin tones from darkest to lightest. However, some of that initial genetic variability would be lost when subgroups of the human population moved apart and remained reproductively isolated, as they did at the Tower of Babel (Genesis 11). Some language groups may have included only A and B alleles, losing a and b; in such AABB subgroups, parents could only have children with very dark skin. Subgroups without the A and B alleles (only a and b) would produce only very light-skinned children, and either AAbb or aaBB subgroups would always be medium brown. AaBb subgroups would continue to produce the entire color range, like some groups in India still do today.

Darwin thought otherwise, but scientists now recognize that people groups who express only part of the full range of melanin color variation (such as very dark skin) are 100 percent human. But among animals and plants, both in nature and from selective breeding, subgroups of some kinds may become so different (e.g., size, courtship ritual, mating season, chromosomal rearrangements, aggressiveness, etc.) that they can no longer interbreed (even though their identity as members of the same created kind can still be confirmed by genetic testing). Such reproductive isolation was once used as the key criterion for defining species.

What? Two or more specialized species descended from one generalized ancestral kind? Doesn't that prove evolution after all! Exactly the opposite.

Speciation, yes; evolution, no. Molecules-to-man evolution requires a net *increase* in novel genetic information, the addition of genes for new trait categories to a genome. Reproductive isolation and subsequent speciation results in a *loss* of genetic variability (alleles), converting a large gene pool into subgroups with smaller gene pools (i.e., "new species" with less ability to meet changes in their environment, restricted ability to explore new environments, and reduced prospects for long-term survival). Indeed, evolutionists now regularly use the term "over specialization" in speciation as an explanation for extinction versus evolutionary progress.

The Florida panther, for example, is considered an endangered species. What endangers it? The small, inbred population was so riddled with mutations that no cubs could survive to reproductive age. The cure? Since it is only a species within a kind, it was bred with western panthers (members of the same kind) having different post-Fall mutations. The former Florida panther is now recovering from its flirt with extinction and being restored to health.

Distinctive genetic diseases and abnormalities characterize many purebred dogs, which have often reached the end of the line, genetically speaking. Each has all the genetic information in its genome to be 100 percent dog (so each has the same gene pool depth), but the allelic variability (gene pool width) could be reduced ultimately to 0 percent (only one allele per locus in a population). Therefore, crossing purebred poodles with poodles, for example, would produce only poodles and would not be a promising path for recapturing the ancestral wolf or generalized dog kind. If a "poodle plague" wiped out the poodle, however, poodles could be brought back again over several generations through breeding wolves or mongrel dogs. Even the quagga, an extinct subspecies of zebra, is being brought back through cross breeding varied members of the horse kind.

The Wrong Kind of Change

Speciation is moving in the wrong direction to support the evolutionary belief in upward changes between kinds, or molecules-to-man evolution. Speciation produces only variation within kinds as a result of the subdivision and/or alteration of *pre-existing* genetic variability. Speciation also brings to visible expression the magnificent variability and potential for variation that God programmed into the members of each of the original created kinds.

After man's sin, mutations introduced many "negative variations," helping scientists to explain the origin of birth defects and disease. Evolutionists had hoped mutations would provide the new genetic information required to move organisms up the so-called evolutionary tree. But mutations only produce variation in *pre-existing* genes, which are alleles that only make a gene pool wider rather than deeper. So mutations result in variation within a kind and not the formation of new and different kinds, which Darwin called the "production of higher animals."

Uncritical acceptance of evolution has so stunted scientific thinking that people give mutations god-like qualities. They act as if a cosmic ray striking a cell can cause a mutation that somehow assembles over 1,500 DNA nucleotides into a brand new gene, regulators and all, that suddenly begins producing a brand-new protein responsible for a brand-new trait, raising the lucky mutated organism to the next higher limb on the evolutionary tree! NOTHING remotely like that has ever been observed, nor will it be!

Mutations are NOT genetic "script writers"; they are merely typographic alterations in a genetic script that has already been written. Typically, a mutation changes only one letter in a genetic sentence averaging 1,500 letters long. To make evolution happen — or even to make evolution a theory fit for scientific discussion — evolutionists desperately need some kind of genetic script writer to create novel genetic information, increasing the size of a genome and the depth of a gene pool. Mutations have no ability to compose genetic sentences, no ability to produce novel genetic information, and, hence, no ability to make evolution happen, at all.

Yet molecules-to-man evolution requires phenomenal expansion of genetic information. It would take thousands of mutations adding novel information to change simple cells into invertebrates, vertebrates, and mankind. The evolutionist's problem is with the fundamental nature of information itself. The information in a book, for example, cannot be reduced to nor derived from the properties of the ink and paper used to write it. Similarly, the information in the genetic code cannot be reduced to nor derived from the properties of matter or the allelic variations caused by mutations. Its message and meaning originated instead in the mind of its Maker, Jesus Christ, the Author of life (John 1:1–3). What we see in God's world agrees with what we read in God's Word.

CHAPTER 28

Are Genetically Modified Organisms (GMOs) Wrong?

DR. ANDREW FABICH

I don't like food, I love it!" — Anton Ego in *Ratatouille*

We all like food. Some of us like food more than others. Food is more popular today than it was 20 years ago. There are even several TV channels devoted to food and a full-length animated film about food. Unfortunately, our love of food goes to many unhealthy extremes. So we have organizations like the Food and Drug Administration (FDA) to help oversee our food supply. The FDA is supposed to make sure our food is safe to eat, even providing guidelines on what to eat or what not to eat. Even with FDA approval, we have an abundance of "safe" food products. Occasionally, the FDA has to move things from the safe list to the unsafe list.

About ten years ago, the food battle waged against artificial sweeteners like those found in Sweet'N'Low (i.e., the chemical aspartame). In addition to tasting bad, some claim that Sweet'N'Low causes cancer. More recently, the FDA has appropriately recalled foods like beef tainted with deadly *E. coli.* Warnings have been placed on cigarettes, which cause lung cancer. In those instances, the FDA has acted responsibly by removing food products and labeling foods that are dangerous to eat. But there has been a shift in food battles lately. Today's food battle typically wages against seemingly wholesome foods containing "corn, soybean, cotton, wheat, canola, sorghum, and sugar cane seeds."[1] What

1. http://www.monsanto.com/products/Pages/default.aspx, accessed 04-12-13.

is common to all these seemingly wholesome foods is that they typically are genetically modified in the US — their DNA has been changed. Currently, the FDA has no requirement to label foods made with these ingredients and there have been no recalls. But have they acted in a safe and responsible fashion? Or is there anything really wrong with these common "all natural" products?

Let me give you some background. In the old days, farmers used to breed plants together and make "hybrids" — think of a corn hybridized from crossing two different varieties of corn. This was done to enhance the corn to make it bigger or healthier and so on. They would do this with other farm commodities like breeding various cattle together as well. But corn is a great example. Corn is found in the American food supply in the form of high fructose corn syrup. We find this high fructose corn syrup in many household products as a general additive. To understand how much high fructose corn syrup you are consuming, just check the ingredients label in your pantry. (Really, if you're reading this and haven't ever looked, quickly carry your book to the pantry and look for yourself.) The ingredients are listed in the order of abundance, so the first ingredient is most abundant in the food you eat. You may be surprised to find all the products that have high fructose corn syrup in them (let alone how much of it) — especially soft drinks. Even the ethanol additive in our gasoline at the gas pump was produced from corn products! You may begin wondering: what doesn't have corn in it?

The biggest surprise for most people is that most Americans have consumed a vegetable product, including corn, that has been genetically altered . . . without even knowing it. This brings us to *genetically modified organisms* (GMOs).[2] They are any organism (like plants — specifically here, corn) that has been modified with DNA from another organism. Instead of cross-pollinating corn to make it better, like the old days, they are now taking genes from one organism and forcing them into the DNA (or genome) of a different organism to make it better. Essentially, scientists have added some genes from something else to improve the crop (e.g., to make food grow bigger, taste better, etc.). For the sake of this chapter, I will focus on the GMOs in the American food supply.

There are large lobbies interested in whether GMOs should be in the food supply or not.[3]

2. For our discussion in this chapter, we will primarily be looking at GMOs that involve the artificial transfer of genetic information from one kind of organism to another. This is the area that raises the most ethical concerns and is the primary focus of the GMO food debate.
3. I receive no benefits from any GMO producers or from any non-GMO organizations. My primary concern is for the future ecosystem and the health of my children.

1. The first lobby interested in GMOs is for the use of GMOs and includes major corporations like Monsanto. Monsanto is one of the largest agricultural companies that sells "seeds, traits developed through biotechnology, and crop protection chemicals." They have been at the center of some recent U.S. Supreme Court decisions (e.g., Bowman v. Monsanto Company).[4]
2. The second lobby interested in GMOs is against the use of GMOs and includes the Non-GMO Project. "The Non-GMO Project is a non-profit organization committed to preserving and building sources of non-GMO products, educating consumers, and providing verified non-GMO choices."[5]
3. The third lobby that should be interested in GMOs is the unaware majority of Americans having already consumed a GMO without knowing it.

But is ignorance bliss? As a trained scientist who has done the research and also as a dad, let me first scrutinize these GMOs using the Scriptures then scientifically evaluate GMOs to determine if there is anything wrong with using them.

Do Scriptures Teach against GMOs?

Since the structure of the DNA double helix was discovered only recently (1953), the human authors of the Bible could not use the term "genetically engineered" like we use it today. The lack of GMOs in Scripture does not invalidate Scripture nor does it mean that these genetic engineering concepts are not addressed in Scripture, leaving us without a guide through the 21st century. (Keep in mind that the word *dinosaur* was not invented until the 1800s and so it, too, is not found in Scripture even though God created dinosaurs.)

To the contrary, some important words that also define biblical Christianity and yet do not appear in Scripture include (but are not limited to) the Trinity and the hypostatic union. Significant words always discussed in the GMO debate like "drought-resistant crops" and the active herbicide found in RoundUp™ (the chemical glyphosate) are hardly found in normal people's vocabulary and were not in our vocabulary until recently. But even though

4. For reference, see http://www.monsanto.com last accessed 06-18-13 and http://www.nytimes.com/2013/05/14/business/monsanto-victorious-in-genetic-seed-case.html?_r=0 last accessed 06-18-13.
5. http://www.nongmoproject.org.

drought-resistant crops and the herbicide glyphosate are certainly not biblical, they are directly related to the biblical subject of man's dominion over the earth.

Both the image of God and man's dominion are first mentioned in Scripture simultaneously. When God creates the first humans on day 6, Scripture tells us:

> Then God said, "Let Us make man in Our image, according to Our likeness; let them have dominion over the fish of the sea, over the birds of the air, and over the cattle, over all the earth and over every creeping thing that creeps on the earth." So God created man in His own image; in the image of God He created him; male and female He created them. Then God blessed them, and God said to them, "Be fruitful and multiply; fill the earth and subdue it; have dominion over the fish of the sea, over the birds of the air, and over every living thing that moves on the earth" (Genesis 1:26–28).

It is abundantly clear that these verses teach what is traditionally referred to as the dominion mandate. God gave the dominion responsibility to those who bear His image and to nothing else. Since we bear His image, we must understand the responsibility of dominion over organisms, their seeds, and their DNA so that we act according to God's desires. Furthermore, we must guard against the abuse and misuse of God's creation.

> The works of the Lord are great, studied by all who have pleasure in them (Psalm 111:2).

> You have made him to have dominion over the works of Your hands; You have put all things under his feet (Psalm 8:6).

When using any part of God's creation, we must be found good stewards. Our dominion should be taken seriously, but also not neglected (cf. Luke 19:11–27). Since we are entrusted with creation, we have the God-given responsibility to care for it. Some people have taken Leviticus 19:19, "You shall not sow your field with mixed seed," out of context to interpret seed to mean the genetic material of one organism should not be mixed with that of another organism. The text says mixing *seeds* (kil'ayim, which also appears in Deuteronomy 22:9 in the same context) is wrong, not the mixing of *kinds* (miyn) (where the biblical term kind is usually synonymous with the family level in modern classification schemes). While the word "seeds" falls in the semantic range encompassed by the word "kinds," the converse is not true (i.e., "kinds" are not "seeds").

In today's modern technological world, we often find ourselves enjoying God's creation because of different technologies. But as any technology changes new challenges arise. When Noah built the ark, the technology included tools made of stone, bronze, and/or iron. When Moses was writing the Law, the Egyptians were repairing devastation. Nebuchadnezzar finished his hanging gardens during the lifetime of Daniel. All roads were headed to Rome while Jesus walked this planet. Everyone should realize that using technology is not wrong in and of itself, but can be problematic when someone uses the technology in a *wrong* way (e.g., Nazis' inventions for the destructions of Jews, Poles, Slavs, and others). Building pyramids, hanging gardens, and road construction are technologies in their own right, but can this be true for scientists today genetically modifying our food?

Since technological innovations are developed by real-world, problem-solving scientists, then Christians should not be afraid of properly using technology (e.g., cell phones, spaceships, or the computer I used to write this chapter). GMOs are intended, like any technology, to potentially improve humanity when used properly, but they may also bring harm.

So picking on GMOs because they are new technology is a bad argument because there have been new technologies since the beginning of time. In fact, is it any wonder that it has taken us this long since Adam to invent GMOs? Of all people, today's Christians live with more information available, have the complete Word of God, and so should "have an answer" (1 Peter 3:15) for GMOs because they directly relate to the dominion mandate. Essentially, GMOs are like any technology that should be used consistent with what the Scriptures teach. While there is no specific verse teaching against GMOs, is there a scriptural principle that teaches GMOs violate the dominion mandate?

Do Scriptural Principles Teach against GMOs?

The Bible contains several very interesting examples of biotechnology without using the words DNA or GMOs. Genesis 30 records an exchange between Jacob and his father-in-law Laban. The exchange includes Jacob negotiating Laban's daughter to be his wife for an unusual price. The unusual price was for taking care of Laban's livestock; in exchange, Jacob would marry one of Laban's daughters. At the same time, Jacob was cunning enough to secure some livestock to provide for his future wife. All newlyweds start off with very little wealth and so Jacob asked for Laban's undesirable livestock to provide for his future wife. In exchange for those undesirable livestock, Jacob also promised to take care of Laban's desirable livestock. Specifically, the undesirable livestock that Jacob requested were "speckled and spotted among the goats, and brown among the lambs" (Genesis 30:33). Even though Jacob was deceived, he made the best of the situation by performing an odd technique that we still do not understand today: "Jacob took for himself rods of green poplar and of the almond and chestnut trees, peeled white strips in them, and exposed the white which was in the rods" (Genesis 30:37). This passage about using "rods of green poplar" (among others) implies that Jacob was artificially selecting (i.e., breeding) desirable traits from his newly acquired undesired animals. While Jacob worked with animals, the techniques he used are based on the same principles used to make GMOs.[6] So Jacob used the biotechnology of his day to artificially select certain desirable traits among his livestock (similar to dog breeding today). Not exactly a GMO by today's definition, but Jacob never compromised the dominion mandate in what he did.

Later in the New Testament, Paul writes to the Romans to describe important heavenly truths using an earthly example from the science of plant cultivation. Paul uses the term "graft" six times in Romans 11 to describe the spiritual truth that the Gentiles were to spiritually flourish essentially because God did so with the nation Israel. When Paul was writing in the first century, the term "graft" was often used to describe taking a slice of an olive branch and placing the cut branch into a fresh olive tree. GMOs and grafting are similar because they combine two separate sources of DNA. Grafting was a common practice in the ancient world and still used today to cultivate particular foods like seedless grapes. Paul used common language about grafting biotechnology (GMOs) to convey a spiritual truth.[7] Since olive trees do not bear the image of God and cutting a tree branch does not cause them to go extinct, then Paul's point did not suggest an abuse of the dominion mandate.

These two biblical examples of common practices when the Scriptures were written demonstrate that the concepts of genetic engineering and biotechnology do not necessarily violate any biblical principles. Modern genetic engineering principles and biotechnology practices are modified forms of

6. The way in which GMOs relate to animal breeding is that we look within a population of traits and select the ones we're interested in for breeding purposes. While this example of Jacob's goats explicitly refers to the same species, it is relatively easy to discuss movement of traits within a biblical kind. The trait does not necessarily have to be identified by its DNA in one species before moving it to another species — all within a created kind. The term GMO is usually set at the species level. Further discussion of moving genes between created kinds is discussed in subsequent sections with regard to grafting and previously discussed in terms of seeds/kinds.
7. See footnote 4 for the logic, but applied to grafting.

ancient animal breeding and plant grafting (as described in Scripture), which are simply a form of artificial selection. Scripture never says artificial selection is wrong, but actually uses examples of artificial selection to convey spiritual truths. No one can point to any verse or idea to suggest that artificial selection is wrong, let alone GMOs. Therefore, nothing is wrong with the process of genetically modifying any organism, even in a "very good" creation, so long as it glorifies God (all the more so now that we live in a fallen world). Whether Noah or Adam "artificially selected" anything is purely conjecture because Scripture is silent, but it is interesting to speculate nonetheless. In one sense, the animals were brought on Noah's ark due to a form of supernatural selection that gave us variation in the original gene pool necessary for all species existing today (cf. Genesis 7:16). So there is no specific verse teaching against GMOs, nor is there a biblical principle being violated. But is producing GMOs a valid scientific endeavor?

Is the Science Supporting GMOs Flawed?

Making a GMO is a long process that begins by identifying a feature of an organism to improve. Knowing which feature to improve then simplifies finding another organism with the desirable feature. Before we go further, let's hypothetically consider faster-growing crops as the feature we desire in our slower-growing crops. Let's continue, hypothetically, saying that we know certain weeds grow fast because of a faster-growing gene, and farmers could potentially benefit from placing the faster-growing weed gene into corn seeds to produce faster-growing corn (see figure 1 for a general overview of the process to make a GMO). To make this hypothetical situation happen, we first need to make copies of the faster-growing weed gene before introducing it into the slower-growing corn. Once the faster-growing weed gene is introduced into the slower-growing corn, we officially have our genetically modified corn and the corn is then tested in a controlled situation. Simply because the hypothetically faster-growing corn has a weed gene does not make it a weed and vice versa (see the previous comment about Leviticus 19:19). No one selling a GMO is going to under-deliver on the benefits claimed for their new product (in this case, faster growth of the corn). So the hypothetical company tests their product in controlled conditions until they feel it is safe. But when the faster-growing corn is sold, will it overtake all the traditional corn (not genetically modified) in the world?

To understand whether faster-growing corn is bad science depends on our understanding of natural selection and artificial selection. Natural selection

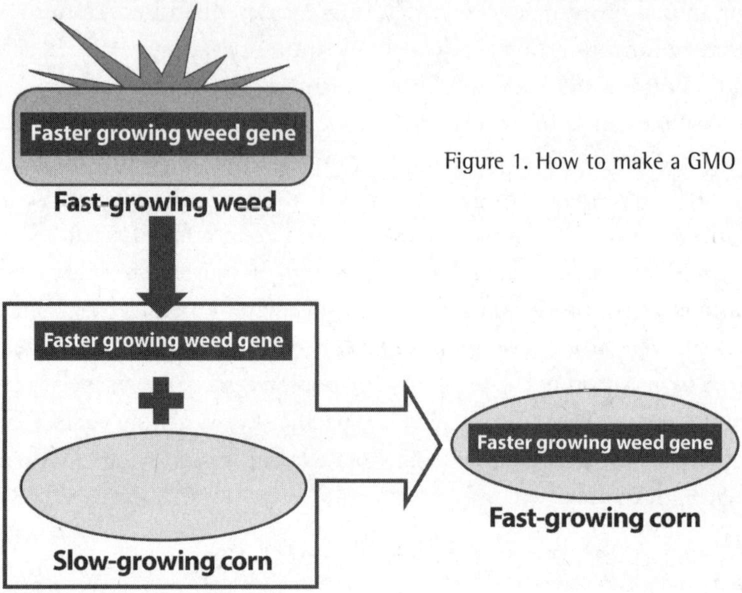

Figure 1. How to make a GMO

is the process designed by God that preserves the genetic makeup of a created kind. (Regrettably, many people incorrectly think that natural selection is equivalent to molecules-to-man evolution. Natural selection and evolution are not the same thing; they are very different.[8]) Artificial selection is the process humans use to choose certain desirable features within created kinds. Natural selection helps explain the diversity of Darwin's finches in the Galápagos, while artificial selection explains diversity among dog breeds. We have Great Danes, Doberman pinschers, dachshunds, and (yes) poodles as a result of artificial selection by humans from the original dog kind on Noah's ark. Whether talking about the artificial selection of dogs or plants, it is best to understand artificial selection as simply selective breeding. Ultimately, GMOs are a really sophisticated form of selective breeding. GMOs are slightly different from traditional selective breeding because we artificially introduce the desirable features from another organism in a single generation using technology. Even though certain features have moved between organisms, we are still involved in the selection process (i.e., this is still artificial selection). So the scientific methods of making GMOs does not violate biblical principles, but are GMOs safe for the environment and for human consumption?

8. See also *The New Answers Book 1*, question #22 "Is Natural Selection the Same Thing as Evolution?" (Green Forest, AR: Master Books, 2006).

If Nothing Is Wrong with GMOs Scripturally or Scientifically, Then What Is Holding Us Back?

The immediate benefits of GMOs include "increased pest and disease resistance, drought tolerance, and increased food supply."[9] Even with all those potential benefits, many countries have already banned the production and sale of GMOs. The Non-GMO Project is staunchly against GMOs and quite politically active against them. According to the Non-GMO Project:

> Most developed nations do not consider GMOs to be safe. In nearly 50 countries around the world, including Australia, Japan, and all of the countries in the European Union, there are significant restrictions or outright bans on the production and sale of GMOs. In the U.S., the government has approved GMOs based on studies conducted by the same corporations that created them and profit from their sale. Increasingly, Americans are taking matters into their own hands and choosing to opt out of the GMO experiment.[10]

Many people within the Non-GMO Project and its supporters want to educate the public and raise awareness about GMOs, and I couldn't agree more that education is important. So what does the actual research show about GMOs? All indications suggest that GMOs released in the United States are approved by the FDA, meeting significant scrutiny by multiple rounds of testing. Contrary to the claims that GMOs are unhealthy, the number of actual scientific reports in the scientific literature is very small that say GMOs cause cancer or other disease. The study titled "Long Term Toxicity of a Roundup Herbicide and a Roundup-tolerant Genetically Modified Maize"[11] has significant flaws and should not be considered authoritative. The flaws of the research include facts like the rodents fed increasing amounts of GMOs had better survival rates than those fed a smaller amount of GMOs. Additionally, their research mice that were fed non-GMO foods died at an alarming rate. According to the non-GMO lobby, the rodents fed non-GMO food should not have died under

9. http://www.webmd.com/food-recipes/features/are-biotech-foods-safe-to-eat, accessed 3-27-13.
10. http://www.nongmoproject.org/learn-more/ accessed 03-27-13.
11. Gilles-Eric Séralinia, Emilie Claira, Robin Mesnagea, Steeve Gressa, Nicolas Defargea, Manuela Malatestab, Didier Hennequinc, Joël Spiroux de Vendômoisa, "Long term Toxicity of a Roundup Herbicide and a Roundup-tolerant Genetically Modified Maize," *Food and Chemical Toxicology*, Volume 50, Issue 11, November 2012, Pages 4221–4231, http://www.sciencedirect.com/science/article/pii/S0278691512005637.

the same conditions as the rodents fed the GMOs; however, the non-GMO lobby's hypothesis was not supported by their own data and the mice fed non-GMO food also died. All this goes without mention that their sample size was extremely small and unrealistic to represent the 7 billion people of the world.

"Over three trillion servings of foods with [GMO] ingredients have been consumed, and in almost 20 years of experience with [GMO] crops, there has not been a single confirmed instance of harm to human health or disruption of an ecosystem."[12] There are no obvious warning signs that we should neither mass produce nor completely ban GMOs, contrary to the extreme positions of Monsanto supporters or the Non-GMO Project, respectively. More experimentation must happen to determine long-term consequences of GMOs in nature before we prematurely conclude that all GMOs are either greatly beneficial or extremely harmful in our food supply. We must remember that the science developing GMOs is the same science behind modern medical marvels such as antibiotics, vaccines, chemotherapy, pain relievers, antiseptics, blood transfusions, and many more. Those arguing wholeheartedly against GMOs must consider their logic and take care that they are not arguing against all forms of modern medicine at the same time.

Along those same lines, many accuse GMOs of being unhealthy foods that should not be sold without warning labels. Often, these accusations are unfounded. In reality, the real problem is not usually the GMO itself, but the actual food product. For instance, the high fructose corn syrup previously mentioned is unhealthy for you regardless of whether it comes from a natural/organic source or a GMO.[13] For every other food that includes a GMO, there are no legitimate reports of the GMOs damaging human health. Americans consume too much of everything and need to cut back on everything in general. We were never made to worship the material creation (i.e., our food) like an idol and overindulge.

As different world powers discuss GMOs, well-respected individuals are on both sides of this debate for a variety of legitimate reasons. All the biblical

12. http://www.genengnews.com/gen-articles/anti-ge-activism-will-it-ever-end/4825, accessed 04-22-13.
13. http://www.mayoclinic.com/health/high-fructose-corn-syrup/AN01588 is a site that demonstrates how having too many empty calories, like those found in high fructose corn syrup, increases the risk for obesity. http://www.webmd.com/food-recipes/features/are-biotech-foods-safe-to-eat emphasizes again that the current GMOs are 100% safe (even when entertaining all the supposed risks). When looking at the traditional soda pop with 39 g of sugar, that is equivalent to approximately 10 sugar cubes added to 12 ounces of liquid. I don't know anyone that adds 10 sugar cubes to a cup of coffee (let alone water) and maintains a healthy body mass index.

creationists are not on one side or the other; neither are the evolutionists. Creationists and evolutionists are on both sides of the argument, which is expected when some recently developed GMOs (like corn, soy, and rice) have not clearly violated either Scripture or secular principles. Ironically, the famed atheist Richard Dawkins offers advice based on biblical principles about GMOs. Dawkins says,

> I am undecided about the politics of GM foods, torn between the potential benefits to agriculture on the one hand and precautionary instincts on the other. But one argument I haven't heard before is worth a brief mention. The American grey squirrel was introduced to Britain by a former Duke of Bedford: a frivolous whim that we now see as disastrously irresponsible. It is interesting to wonder whether taxonomists of the future may regret the way our generation messed around with genomes. . . . The whole point of the precautionary principle, after all, is to avoid future repercussions of choices and actions that may not be obviously dangerous now.[14]

While Dawkins is a vehement atheist, his point about GMOs ultimately makes sense because he is unknowingly using biblical principles. The paraphrase of Proverbs 25:8 in *The Message* captures what to do with situations where there is no clear biblical direction: "Don't jump to conclusions — there may be a perfectly good explanation for what you just saw." Dawkins' argument is essentially what Solomon wrote thousands of years ago. In this instance, Dawkins acknowledges that we do not fully understand potential problems with GMOs in nature. He knows of no problem with GMOs in the lab. So he suggests some precautionary actions taken to not jump to a *hasty* decision. Public perception of GMOs is much worse than they deserve. It would be prudent to occasionally experiment with GMOs, collect the data, and then decide what to legislate before losing what we have on a global scale. GMOs are not problematic scientifically; the potential problem with GMOs is whether they harm God's creation in a way that cannot be fixed. If anyone should conclusively demonstrate a problem with a GMO, then that GMO should not be given to the public. Until potential harmful effects of GMOs are clearly documented scientifically, they should be used within reason and tested accordingly.

14. Richard Dawkins, *The Greatest Show on Earth* (New York: Simon and Schuster, 2009), p. 304.

Conclusions

Modified Organisms

	Instance	Result
1	Jacob and the flocks (e.g., Genesis 30)	Separating out the DNA
2	Grafting branches (e.g., Romans 11)	Mixing DNA
3	Hybridizing crops	Bringing DNA together
4	Artificial selection and breeds (e.g., Deuteronomy 32:14 with ram breeds)	Separating out DNA[a]
5	Natural variation	Separating out DNA
6	GMOs	Separating, mixing, bringing together DNA at a genomic level instead of an organismal level

a. In some cases, there could be a bringing together to form certain breeds as well. This would be the same for natural variations.

The question for this chapter remains: are GMOs wrong? I cannot give a biblical or scientific reason to wholeheartedly support or completely reject GMOs. Imaginary problems with GMOs arise when people take extreme positions on GMOs without using a biblical worldview. Too many Christians get too involved with picking sides on this debate when there is no clear violation of Scripture. Please stop the name-calling, develop a biblical worldview, and let's do good science to figure out the long-term effects of GMOs before picking an extreme (unbiblical) position.

In the meantime, if big business monopolizes the common farmer, then let the political process rectify the plight of the common farmer. If people are hungry because countries ban the sale of GMOs, then let the political process rectify the plight of the hungry people. Christians should obey the law of the land, work hard within their local church to help people, and be involved in the political process by making an informed vote. Ultimately, the Lord will rectify all injustice (Revelation 14:7) and redeem His creation (Revelation 21:1). In the meantime, the world will watch how America handles GMOs . . . and so should Christians.

We should do more research on GMOs to fully see their strengths or weaknesses. The intent of this chapter is to honestly examine our current knowledge of GMOs. At the end of the day, some people are opposed to eating GMOs and others are fine with GMOs. Regardless of whether we eat GMOs, we must keep a Christian attitude among the brethren and recall what Paul wrote while waiting for the research to finish: "So let no one judge you in food or in drink, or regarding a festival or a new moon or sabbaths, which are a shadow of things to come, but the substance is of Christ" (Colossians 2:16–17).

CHAPTER 29

What about Design Arguments Like "Irreducible Complexity"?

DR. STUART BURGESS

What Is the Design Argument?

The design argument says that design reveals a designer and the attributes of the designer. In the same way that the intricate design of an aircraft shows the skill and care of a human designer, so the intricate design of creation shows the skill and care of the divine Designer.

There are many verses in the Bible that contain the design argument. The most famous verse is Romans 1:20 which says, "For since the creation of the world His invisible attributes are clearly seen, being understood by the things that are made, even His eternal power and Godhead, so that they are without excuse." This verse teaches that God's handiwork in creation is clear for everyone to see and no one has an excuse not to believe in a Creator.

Another example of the design argument can be found in Hebrews 3:4 where we read, "For every house is built by someone, but He who built all things is God." In the same way that a house requires intricate design to make it suitable for humans to live in, so the earth requires intricate design to make it fit for human habitation. In fact, Isaiah 45:18 says that God deliberately designed the earth to be inhabited.

The Book of Job contains many verses on the wonder of creation, including the design of fish, birds, animals, dinosaurs, rain, snow, clouds, and the stars. The Book of Job speaks of how creation is so wonderfully designed that it is beyond human comprehension (Job 9:10 and 37:5). The Psalms also give glory to God for His creation. Psalm 139:14 speaks of the wonder of the design of the human body and how God deserves our praise for His workmanship.

Christians have used the design argument in preaching and writing down through the ages. The Apostle Paul used the design argument when he preached to the Athenians in Acts 17. In 1692, the Puritan preacher Thomas Watson used the following design argument in his writing:

> If one should go into a far country and see stately edifices he would never imagine that they could build themselves, but that there had been an artificer to raise such goodly structures; so this great fabric of the world could not create itself, it must have some builder or maker, and that is God.[1]

In 1802, William Paley wrote a famous book called *Natural Theology* in which he argued that in the same way that a mechanical watch must have a human designer, so the natural world must have a divine Designer. In recent times, creationists have written many books and articles on how creation is wonderfully designed. Creationists have explained how there are specific hallmarks of design such as irreducible complexity, common design, over-design and added beauty, which defy evolution. The following sections give a brief introduction to these arguments for design.

Irreducible Complexity

Irreducible complexity is an evidence for design that represents a key scientific test for evolution. Irreducible complexity is the term applied to a structure or mechanism that requires several precise parts to be assembled simultaneously for there to be a useful function for that structure or mechanism. Irreducible complexity cannot be produced by evolution because evolution is restricted to step-by-step change where every change must give a survival advantage. Evolution has no ability to bring about the many precise design changes that are necessary to make the leap from one design concept to another. If there are examples of irreducible complexity in nature, then the theory of evolution absolutely breaks down.

1. Thomas Watson, *The Creation* (London: Banner of Truth, 1965), ch. 13, "A Body of Divinity," p. 114.

Charles Darwin himself knew full well that irreducible complexity was a key test for evolution. Even though Darwin did not use the term "irreducible complexity," he said:

> If it could be demonstrated that any complex organ existed, which could not possibly have been formed by numerous, successive, slight modifications, my theory would absolutely break down. But I can find out no such case.[2]

Creation scientists have shown that creation actually does contain many cases of irreducible complexity. In microbiology there are many irreducible structures like the living cell and bacterial flagellum and there are irreducible processes like blood clotting.[3] Other examples of irreducible complexity are the eye,[4] human knee joint,[5] and the upright stature of humans.[6] Creationists have also shown how design requires information to be specified and that information must come from an intelligent source.[7] It would be fascinating to know if Charles Darwin would still believe his theory of evolution if he were here today and able to see the many case studies of irreducible complexity!

The Irreducible Human Arched Foot

Human feet represent a clear example of irreducible complexity.[8] Human feet have a unique arch structure that is completely different from the flat feet of apes. Arched feet are very important for the upright stature of humans because they allow fine control of the position of the body over the feet. When standing upright, a person can maintain balance by adjusting the relative pressures on the heels and balls of the feet.

Human feet have an arch between the heel and the ball of the foot, as shown in figure 1. The equivalent engineering arch is also shown in figure 1.

---

2. Charles Darwin, *The Origin of Species: A Facsimile of the First Edition* (Cambridge, MA: Harvard University Press, 1964), p. 189.
3. M.J. Behe, *Darwin's Black Box: The Biochemical Challenge to Evolution* (New York, NY: The Free Press, 1996), p. 46.
4. J. Sarfati, "Stumbling Over the Impossible: Refutation of Climbing Mt. Improbable," *Journal of Creation* 12(1) (1998): p. 29–34.
5. S.C. Burgess, "Critical Characteristics and the Irreducible Knee Joint," vol. 13, no. 2 of the *Creation Ex Nihilo Technical Journal*, 1999.
6. S.C. Burgess, "Irreducible Design and Overdesign: Man's Upright Stature and Mobility," *Origins, Journal of the Biblical Creation Society*, vol. 57 (2013): p 10–13.
7. A.C. McIntosh, *Genesis for Today: Showing the Relevance of the Creation/Evolution Debate for Today's Society* (Leominster, United Kingdom: Day One Publications, 1997).
8. Burgess, "Irreducible Design and Overdesign," p. 10–13.

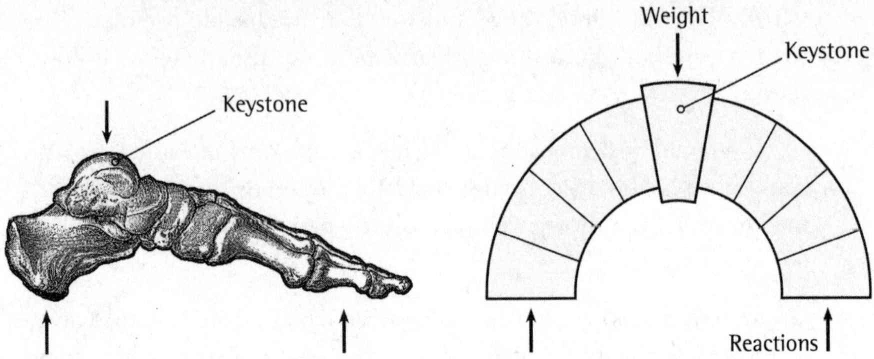

Figure 1. The irreducible human arched foot and equivalent man-made arch

The human foot has 26 precisely shaped bones, together with many ligaments, tendons, and muscles. Several of the bones are wedge-shaped so that a strong arch is formed. There are several parts in the foot that must be in place and correctly designed before the foot can function properly. In other words, the human foot cannot evolve step by step from a non-arched structure like a hand.

It is well known in engineering that an arched structure is an irreducible structure. An arch needs the right components, like a keystone and wedge-shaped blocks, to be in place to work, as shown in figure 1. Since the human foot has parts equivalent to a keystone and wedge-shaped blocks, the human foot must be an irreducible structure. Only an intelligent designer has the ability to think ahead and plan all the features needed to make an arch like the foot.

The arched structure of the human foot is a perfect design for giving humans upright mobility. In contrast to humans, apes have very flexible feet that are effectively a second pair of hands for gripping branches. In consequence, apes have very limited abilities for two-legged standing, walking, and running.

The Fossil Record Confirms Irreducible Complexity

The fossil record confirms the biblical truth that organisms have not gradually evolved step by step. One of the reasons we know that humans have not evolved from a type of ape-like creature is that there has never been a fossil of a foot that is a transitional form between the flat ape foot and the human arched foot. All fossils of so-called ape-men have either fully ape feet or fully human feet, showing that they are either fully ape or fully human, respectively.

The prominent evolutionist Stephen J. Gould has admitted that fossil evidence supports the creation worldview:

> The absence of fossil evidence for intermediary stages between major transitions in organic design, indeed our inability, even in our imagination, to construct functional intermediates in many cases, has been a persistent and nagging problem for gradualistic accounts of evolution.[9]

The human foot, is a clear example of a structure where evolutionists cannot imagine what intermediate forms would look like. The reason for this is that there are no physically plausible intermediate structures due to the need for the foot to be assembled simultaneously.

Common Design

Common design is another important evidence for design that is a challenge for evolution. Common design is where the same design solution is used in different situations by a common designer. Human designers often carry out common design because it represents good design practice. For example, a designer will select nuts and bolts as a method for joining parts together in different products such as bicycles, cars, and spacecraft because this is the best design solution in each case. In the case of the common design of nuts and bolts, this is not an evidence of evolution but evidence of the careful work of a designer.[10]

The eye is a good example of common design by the common Designer in creation. The eye is seen in very different types of creatures like mammals, birds, fish, amphibians, and reptiles. In each case, there are specialized light-sensitive cells, nerve pathways for conveying the signals to the brain, and a part of the brain for processing the signals. In addition, there is usually some form of lens for directing the light onto the light-sensing cells. When you consider the great differences between different classes of creatures, it is remarkable how the eye for each creature is so similar in design. The similarity in design is just what would be expected from the common Designer, because He would know it is the best solution in each case. Interestingly, the Bible tells us in Proverbs 20:12 that the Lord "made the seeing eye."

The similarity of the eye in different classes of creature is not what would be expected from evolution, because evolution has no ability to coordinate designs in different applications. The evolutionist has to believe that the eye

9. Stephen Jay Gould, "Is a New and General Theory of Evolution Emerging?" *Paleobiology*, vol. 6(1) (January 1980): p. 127.
10. S.C. Burgess, *Hallmarks of Design*, 2nd Ed (Leomimster, UK: Day One Publications, 2008).

Figure 2. The eye is an example of common design by the common Designer.

evolved independently around 30 times.[11] It takes a lot of faith to believe that the same basic layout of eye evolved independently so many times. Some evolutionists argue that a common ancestor would help explain why structures like the eye appear in different creatures. However, the eye is found in such diverse creatures and has such similar design that common ancestry is not a credible explanation for the common design of the eye, even within the evolutionary worldview.

There is also a remarkable pattern in the design of the face across the whole animal kingdom with the easily recognizable features of two eyes, a nose, and a mouth. Such a common pattern is just what would be expected from the Creator who wanted to create an ordered and beautiful creation. A recognizable face also helps people to enjoy the company of animals like dogs, cats, and horses.

The principle of common design shows that it is wrong for secular biology books to use commonality of features in organisms as an evidence for evolution (sometimes referred to as homology). At the very least, biology books should mention that common design can be seen as evidence for the Creator *or* evidence for evolution. But the most accurate statement is that common design is more an evidence for creation than evolution.

Over-design

Over-design is another hallmark of design and a big challenge to evolution. Over-design involves design features that are above and beyond what is needed for survival. Human designers often carry out over-design, especially in luxury products like expensive cars where the aim is to greatly exceed the basic requirements.[12] Over-design should not be produced by evolution because, with evolution, every aspect of design must be capable of being explained in terms of what is needed to survive.

11. Sarfati, "Stumbling Over the Impossible: Refutation of Climbing Mt. Improbable," p. 29–34.
12. Ibid.

One area where we clearly see over-design in creation is in the design of the human being.[13] Humans are over-designed with skills and creativity that are far beyond what is needed to survive. The survival abilities of apes include the ability to find food and water, climb trees, build a den, defend territory, find a mate, and reproduce. The fact that humans have abilities that are vastly beyond these basic survival tasks provides great evidence that humans have not evolved from an ape-like creature but have been specially created to be beings of great skill and intelligence.

Over-design of Man

One aspect of over-design in humans is the ability to make facial expressions. Humans have around 25 unique facial muscles, as shown in figure 3. These muscles are dedicated to making expressions like smiling, grinning, and frowning, as shown in figure 4. Such expressions convey emotions such as happiness, pleasure, concern, anger, worry, and surprise. Researchers have found that humans have the amazing ability to make up to 10,000 different facial expressions![14] Facial expressions are very important in human communication even though we are often unaware that we are making expressions and responding to expressions. Smiling is one of the first things a baby does in its first few weeks of life, and one of the first things a baby can recognize.

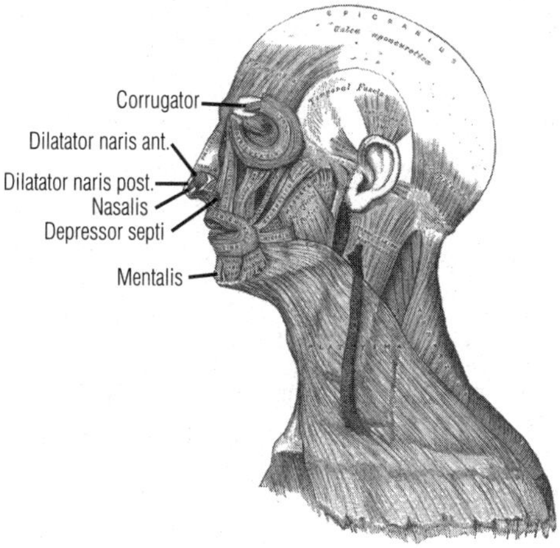

Figure 3. Muscles used for facial expressions

According to evolution, facial muscles and facial expressions came about because there was a survival advantage. But evolution cannot adequately explain what survival advantage comes from smiling or frowning. However, such

13. S.C. Burgess, The *Design and Origin of Man: Evidence for Special Creation and Over-design*, 2nd ed. (Leominster, UK: Day One Publications, 2013).
14. Paul Ekman and Wallace Friesen, Facial Action Coding System (Human Interaction Laboratory, Department of Psychiatry, University of California Medical Centre, San Francisco, Psychologist Consulting Press Inc., 577 College Avenue, Palo Alto, CA 94306, 1978.)

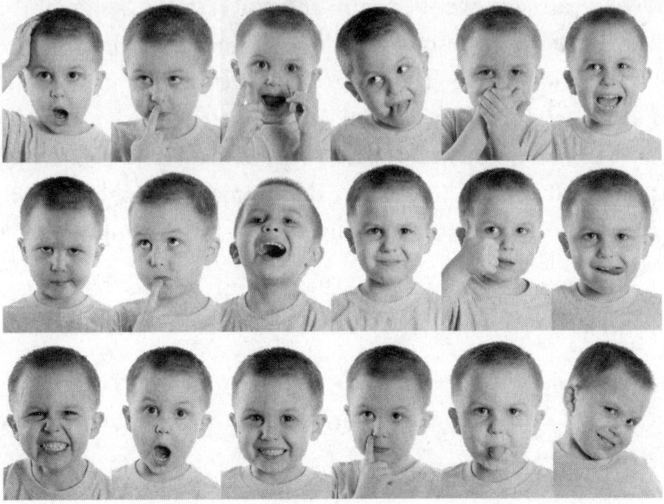

Figure 4. Examples of facial expressions in a young boy.

expressions are just what would be expected since God has created humans to be emotional beings made in the image of God.

Skillful hands are another example of over-design in humans. According to evolution, human hands have evolved to perform survival tasks such as throwing spears, building dens, and making simple clothes. However, human hands are capable of so much more than these basic tasks. Humans have potential for immense skill in areas like playing music, carpentry, medicine, engineering, and craftwork. Evolution has no credible explanation for why humans are able to hold a pen and other instruments in a perfect tripod grip with thumb, index finger, and middle finger. In contrast, the dexterity of human hands is just what would be expected since man is made in the image of God as a creative being.

There are several other areas where over-design can be seen in human beings. For example, humans have the ability to think and communicate complex thoughts through intricate languages due to the specialized design of the throat, tongue, and brain. However, there is no credible reason why such ability was ever essential for survival. Also, humans have a uniquely fine skin that helps them enjoy the sense of touch. However, the ability to enjoy the sense of touch does not help survival. Perhaps the greatest example of over-design is in the human brain that is so much more powerful than is needed for a person to simply survive.

The over-design of man is just what would be expected since God had created humans to be creative beings, able to appreciate beauty, develop technology, create works of art, play sports, and be stewards of creation. As spiritual and creative beings, God had to equip humans with special skills and intelligence that are far beyond what is needed to *survive*. One of the reasons why humans are fearfully and wonderfully made (Psalm 139:14) is that they are over-designed.

Added Beauty

Added beauty is another powerful evidence for design. Human designers often add beauty solely for beauty's sake in architecture and engineering in order to create pleasing aesthetics. An example of added beauty is the embellishments in classical architecture. The intricate patterns on pillars and walls in classical architecture represent compelling evidence for design because there is no physical purpose for the intricate design. Of course, beauty is subjective and cannot be quantified. However, there are real and clearly recognizable features that produce beauty such as patterns, borders, embellishments, surface textures, colors, and variety. To produce intricate beauty requires not just creativity but also design information, and that design information has to come from somewhere.

Evolution cannot produce added beauty because, as with over-design, evolution can only produce what is needed for survival. Many evolutionists realize that beauty is a big problem for evolution. One leading evolutionist, Dr. John Maynard Smith, said:

> No topic in evolutionary biology has presented greater difficulties for theorists [than beauty]. [15]

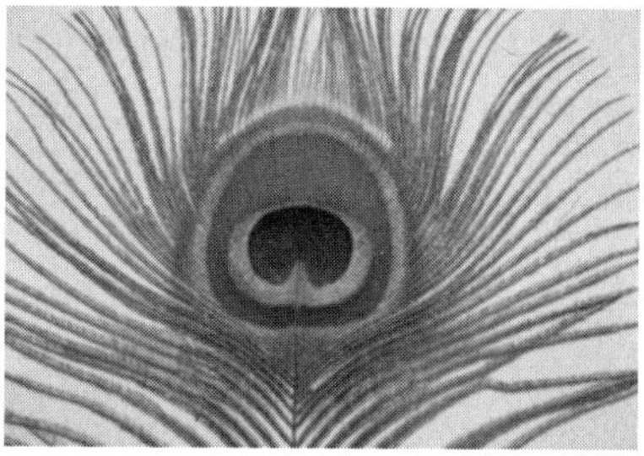

Figure 5. The peacock tail feather

Some of the clearest examples of added beauty in creation are the brightly colored feathers of birds like peacocks, as shown in figure 5. There are several intricate design features in the peacock tail feather, such as the multi-layered segments that reflect light to produce bright and iridescent colors. These segments are so precisely designed and co-ordinated that amazing digital patterns are produced. There are also subtle features like multiple borders and a lack of stem in the eye pattern.

The peacock tail feather is a big problem for evolution because the only function of the feather is to create a beautiful display. The feather does not help the bird in any physical way. In fact, the feather makes flying harder and it

15. John Maynard Smith, "Theories of Sexual Selection," *Trends Ecol. Evol.*, 6 (1991): p. 146–151.

even makes the bird easier for predators to see. Evolutionists say that birds like peacocks need display feathers to attract a mate, but that does not explain the need for beauty. Most animals make very basic calls to attract a mate, showing that intricate beauty is not required. The fact that peacocks display their tails to attract mates is just what would be expected form the Creator who wanted the beauty of peacocks to be visible to humans.

Darwin was well aware that there was beauty for beauty's sake in creation. He said:

> A great number of male animals have been rendered beautiful for beauty's sake.[16]

Since the beauty of bird feathers contradicted Darwin's theory of evolution, he created another theory called the theory of sexual selection. However, that theory has been shown to be totally inadequate for giving a naturalistic explanation of the origin of beauty.[17]

There are many other areas of creation where we see added beauty such as birdsong, flowers, tropical fish, and the human being. Even though we live in a fallen world with death and decay, we still see glimpses of outstanding beauty that point to the Creator. We can also look forward to heaven, which is the perfection of beauty (Psalm 50).

The Effect of the Fall

Genesis 3 teaches that God cursed creation as a result of Adam's sin and rebellion. As a consequence, creation was changed very significantly, including the design of plants and animals. Thorns and hard-to-control plants appeared, and these made farming and gardening much more difficult. Evolutionists argue that thorns evolved as a way of protecting plants. However, the fact that many plants come with and without thorns, like blackberries, raspberries, and palm trees, shows that thorns are not necessary for survival.

Some creatures became carnivores, and this introduced violence and suffering into creation. Predators like cats and dogs were vegetarian before the Fall but became meat-eaters after the Fall. Predators may have had new design features for killing introduced at the time of the Fall, or they may have developed features through natural selection (or a combination of both). In

16. H. Cronin, *The Ant and the Peacock* (Cambridge, UK; New York: Cambridge University Press, 1991), p. 183.
17. S.C. Burgess, "The Beauty of the Peacock Tail and Problems with the Theory of Sexual Selection," *Journal of Creation* 15 (2) (August 2001): p. 94–102.

other cases, the designs may have been used for a different purpose such as vegetarianism.

There is no doubt that everything was beautiful in the Garden of Eden, because the Bible tells us that God made everything beautiful in its time (Ecclesiastes 3:11). The Curse that followed the Fall had the effect of tarnishing the beauty of creation. Some plants and creatures became marred with sin and reflected an "ugliness" where the predator-prey relationship meant many animals had to be camouflaged, thus reducing the number of brightly colored creatures in creation. Violence and suffering has also reduced the beauty of creation.

The negative effects of the Fall will not last forever. The Book of Isaiah teaches that in heaven, predators will be changed back to being harmless and pleasant creatures. Isaiah 11:16 says that predators like wolves, leopards, and lions will live peacefully with gentle animals like lambs and goats. In heaven, the full beauty of creation will be restored, because heaven is the perfection of beauty (Psalm 50:2).

What Is the Intelligent Design Movement?

The Intelligent Design (ID) movement argues the case for intelligent design without any reference to the identity of the Creator and without any reference to the Bible. The ID movement is helpful in some ways because it publicizes examples of design arguments like irreducible complexity and shows the weaknesses of evolution. However, there are limitations to the ID movement.[18]

One limitation is that it does not give an explanation for the origin of death and suffering in nature. This can be a problem because people always want to know why a Designer would design some creatures to kill. When people do not know the biblical origin of suffering, a result of man's sin, they may find it hard to believe there is a creator, or they may have an incorrect view of the Creator. Only with the right biblical understanding of the Fall can people understand that God is a loving Creator who cares deeply for His creation, including mankind.

A second limitation of the ID movement is that it does not promote a biblical worldview. Instead it attempts to be neutral, with no doctrinal agenda. However, it is impossible to be completely neutral, and everyone has a worldview that is ultimately biblical or non-biblical.

18. An important limitation is that it takes the glory due Jesus Christ as the Creator (Hebrews 1:1–4; Colossians 1:13–18) and gives that glory to some vague idea of an intelligent creator that could fit in with Islam, Hinduism, deism, and many forms of theism.

Conclusion

According to evolution, creation should contain designs that are inferior to the designs of humans because of the limitations of step-by-step evolution compared to intelligent design. However, the reality is clearly different; creation contains vastly superior designs to human designs showing that God must exist.

Creation reveals the Designer who is powerful (Romans 1:20), caring (Matthew 6:30), and perfect in knowledge (Job 37:16). I have personally worked with some of the best engineering designers in the world in America, Japan, and Europe, but it is clear that all of them are limited in their knowledge. This is why so many engineers today are keen to copy solutions from creation to make better airplanes, materials, and other products in order to utilize the brilliant designs that God has placed before us.

There has been a sad change of worldview in the majority of the scientific community. In past ages, most scientists acknowledged God and gave glory to God for His creation. That is no longer the case. However, there are still many scientists who are prepared to face criticism and even demotion by giving glory to the Creator. In addition, there are many believers today who have the joy of knowing, personally, the one true Creator God.

It is not possible to scientifically prove the truth about origins, as science is vastly limited in this area. Only God was there at the foundation of the world (Job 38:4), and so we rely on the testimony of His written Word to find out how the world was made. We also need faith to believe God's Word. This is why the Bible says, "By faith we understand that the worlds were framed by the Word of God" (Hebrews 11:3).

Keep in mind that it is important to realize that the faith of the Christian is not blind faith. God has left His fingerprints and hallmarks on His creation so that His existence and attributes are clear for all to see. However, faith is important, because without faith it is impossible to please God (Hebrews 11:5). The origins debate is ultimately about faith versus faith. The atheist has great faith in chance, and the Christian has faith in a great God who has given us eternal life through His Son, Jesus Christ.

CHAPTER 30

What about the Origin of the Solar System and the Planets?

DR. DANNY R. FAULKNER

Genesis 1 tells us that God created the earth "in the beginning." It is not until three days later on day 4 that God made the sun, moon, and stars. The Hebrew word for stars includes the planets,[1] their satellites, comets, and asteroids, so we can infer that the rest of the solar system was made after the earth was. This is very different from the evolutionary view of the origin of the solar system. Most scientists today think that the earth formed about the same time as the sun and everything else in the solar system — about 4.6 billion years ago. The solar system supposedly formed gradually from the collapse of a cloud of gas and dust. Obviously, this idea is at odds with the biblical creation narrative.

We can trace the origin of the modern theory of solar system formation to Emmanuel Swedenborg in 1734, but it was Emmanuel Kant who developed the idea in 1755. Pierre-Simon Laplace proposed a similar model in 1796. This nebular hypothesis was that the solar system began as a contracting and cooling proto-solar nebula. As the nebula contracted, it flattened into a disk, and most of the material fell to the center. The material in the center formed the sun, and the material in the disk eventually coalesced to form the planets. Any remaining material formed the satellites of the planets, asteroids, and comets. The nebular hypothesis enjoyed wide support throughout the 19th century, but eventually

1. Our English word *planet* comes from *asters planetai*, ancient Greek for "wandering stars." Ancient languages defined a star as any luminous object in the sky other than the sun and the moon.

astronomers realized there was an angular momentum problem. While the sun has more than 99 percent of the mass in the solar system, the planets possess more than 99 percent of the angular momentum. If the solar system formed via the nebular hypothesis, the distribution of angular momentum ought to be proportional to the distribution of mass. Because of this problem, astronomers abandoned the nebular hypothesis in the early 20th century.

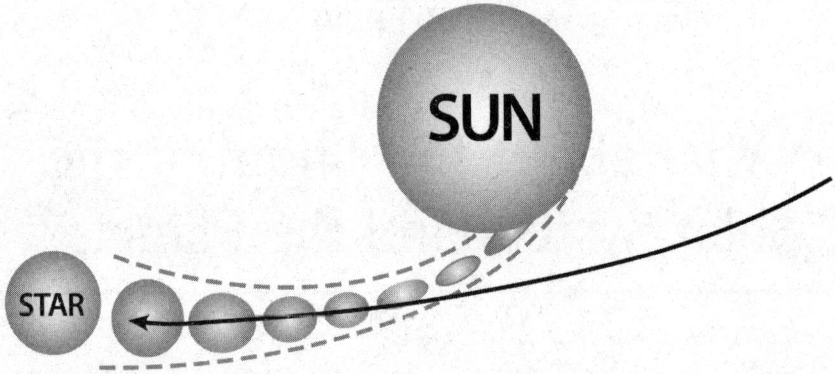

The first replacement theory was the tidal hypothesis of Thomas Chamberlin and Forest Ray Moulton in 1905. They suggested that shortly after the sun formed, another star passed very close to the sun, raising tidal bulges on the solar surface. The tidal bulges combined with solar prominences to eject material from the sun that produced two spiral-like arms. Much of the material in the spiral arms fell back onto the sun, but some coalesced into planets. As before, leftover matter formed the satellites, asteroids, and planets.

With both the nebular and tidal hypotheses, astronomers looked for confirmation elsewhere, and they thought that they found it in photographs of "spiral nebulae." The word *nebula* comes from the Latin word for cloud. A nebula is a cloudy, indistinct, luminous object in the night sky. A few were known to the ancients, but many more were discovered after the invention of the telescope. The telescope also revealed that many nebulae actually were star clusters in which the individual stars are too faint to be seen with the eye alone. Many other nebulae remained indistinct, from which astronomers concluded that they truly were clouds of gas in space. Today, we reserve the use of the word *nebula* to refer to one of these, and the true nebulae probably inspired Kant and Laplace in their ideas. Many of the "nebulae" appeared flattened with bulges in their centers, and many sported spiral arms. This appearance certainly inspired Kant and Laplace, but also Chamberlin and Moulton. In fact, a century ago first drawings and later photographs of the "spiral nebulae" often were used as proof of these naturalistic theories of the solar system's origin. I keep putting "spiral nebulae" in quotes because in 1924 Edwin Hubble showed that these were not nebulae at all, but instead were galaxies, vast collections of many billions of stars that are millions of light years away from us. Being so far away, the stars in other galaxies appear very faint to us. Astronomers up to that time had failed to recognize that the "spiral nebulae" were distant galaxies similar to our Milky Way galaxy, because their telescopes were not large enough to reveal any individual stars in them. However, in 1924, Hubble, using what was then the largest telescope in the world, was able to photograph a few of the brightest individual stars in a couple of these "nebulae." Since 1924, it has not been proper to refer to these objects as "spiral nebulae," though that term continued being used for decades afterward. It is important to note that for years astronomers used these objects as proof of the evolutionary view of the formation of the solar system, though astronomers eventually were forced to abandon this proof.

There were variations on the tidal interaction theme suggested by Chamberlin and Moulton. For instance, in 1918 Sir James Jeans and Sir Harold Jeffreys suggested that solar prominences were not involved and that a near miss by a passing star raised a single filament of material from the sun from which the planets and other bodies in the solar system formed. The tidal theory enjoyed broad support for much of the first half of the 20th century, but by 1940 problems had developed. One problem was that any column drawn out of the sun would dissipate rather than condense. Another problem was that material drawn out with sufficient speed to account for the angular momentum of the planets (especially Jupiter) would have left the solar system entirely, so the angular momentum problem remained. Consequently, during the middle of the 20th century there was no agreed-upon theory for the formation of the solar system.[2]

In the 1960s, astronomers began to revive a form of the old nebular hypothesis; though, I suppose in an attempt to dissociate it from the original, that name is not used to describe the modern version. As before, the solar system supposedly formed from the collapse of a gas cloud that flattened and concentrated in its center, with the sun forming from the central condensation and the planets forming from material in the disk. The modern theory borrows a term

2. Though it was far out of date, this theory was in science texts used in my elementary school in the mid 1960s. Those books were not very old, but this theory was probably included because there was no other alternative.

coined by Chamberlin, *planetesimal* (from the words "planet" and "infinitesimal"). A planetesimal is a small body amalgamated from microscopic particles. Planetesimals supposedly grew within the proto-planetary disk to form bodies large enough to begin gravitationally attracting other planetesimals to form the planets. As before, leftover planetesimals formed planetary satellites, asteroids, and comets. And, as before, the angular momentum problem remained. The most common explanation for that problem now is that magnetic effects removed angular momentum from the inner part of the nebula and transferred it to the outer portions of the nebula in the form of spiral arms or through jets extending fore and aft out of the disks.

The modern nebular hypothesis has other problems as well. What causes the microscopic bits of matter to coalesce into planetesimals? Gravity will work only when planetesimals have grown to kilometer size. Various mechanisms have been proposed to get the planetesimals up to that size. One mechanism is that static electricity attracted particles together. Another is that sticky, organic goo coated microscopic dust particles so that they stuck together when they happened to touch. Another idea is that gaseous molecules in space froze onto solid particles. Of course, none of this is actually observed, but astronomers generally assume that it must have happened somehow, or else how did those planets get here? Another problem is what caused the gas cloud to contract to begin with. This is the long-standing problem of star formation in general. One might answer that gravity drove the process. Gas clouds do have gravity, but they also possess gas pressure, and that pressure very effectively counteracts gravity. Early in the 20th century, Jeans showed that if a gas cloud is contracted down to a certain size, gravity can take over to complete the process. The problem is that all gas clouds that we see are far larger than Jean's length. Compression or cooling is needed to further contract the gas cloud so that gravity could complete the process. Some astronomers have suggested cooling from dust particles, but astronomers do not think that dust is primordial. Where did dust come from? This theory requires that several generations of stars must first create dust before stars could form by this mechanism. One might suppose that a gas cloud could get a sort of jump-start by an outside agent that compresses the cloud. For instance, a shock front from the explosion of a nearby supernova or associations of hot stars with strong UV radiation and stellar winds might do this, but this does not tell us where stars ultimately came from, because it requires that at least one star first exist. All theories of pre-stellar collapse of gas clouds suffer from this chicken-and-egg problem — stars must first exist to produce stars.

The modern theory of solar system formation has been refined with the addition of magnetic fields. If a gas cloud contained any magnetic field initially, the magnetic field would intensify as the cloud contracted. And as the cloud contracted it would have heated and ionized some of the gas. This produces plasma. In the swirling environment of the contracting cloud, models suggest that electromagnetic effects propel material outward in the two directions along the axis perpendicular to the plane of the disk. Astronomers call this bi-polar flow, a phenomenon found in some stars and in many galaxies and quasars. In more recent years, astronomers have created computer simulations supposedly to show how the solar system might have formed. One might question if the success of the simulation merely proves that the programmer was especially good at writing a program to produce the intended outcome.

In addition to improved models, since the early 1970s astronomers have made much progress in the development of technology, such as in the infrared (IR) part of the spectrum and the superb clarity of telescopes in space. These have resulted in observations of objects that astronomers generally think are stars and solar systems in the process of forming. For instance, in 1995 the Hubble Space Telescope took the stunning "Pillars of Creation" photograph of a dark dust and gas region in the Eagle Nebula that astronomers think is the site of active star formation. Orion is a region where astronomers think new stars are forming or recently formed. In this region, astronomers have used IR telescopes to detect star-like sources embedded in clouds of dust and gas, which are regions where they think stars likely form. Astronomers have observed bi-polar flows from stars or star-like objects in environments supposedly conducive for star formation, suggesting that these are stars that have nearly formed or very recently formed. Some stars have IR excess that suggest that they are surrounded by dust. The star β Pictoris was the first star discovered to have what appears to be a disk of dust surrounding it. This was interpreted as a proto-planetary disk that may yet condense into planets. More recently, astronomers have found disks of material around other stars that astronomers think are very young. All of these sorts of things have been put forth as proof of the prevailing theory of solar system formation.

But is this proof? The process of solar system formation is supposedly a very slow one, progressing far too slowly for us to witness any real change even in many human lifetimes. So these data all amount to sorts of snapshots of various stars and other astronomical bodies supposedly in various stages of the process of stellar and planetary formation but with no real evidence that these objects are actually undergoing the alleged processes. Rather, these snapshots are interpreted in terms of the ruling paradigm of solar system formation, and then they are offered up as proof of that paradigm. This amounts to circular reasoning, which is no proof at all. Remember that a century ago astronomers used the photographs of "spiral nebulae" as proof of the then-prevailing ideas of solar system formation. At the time, nearly everyone was convinced of the correctness of this view, but later observations proved otherwise. The supposedly iron-clad proof of solar system formation today could be interpreted very differently in just a few years. In fact, the history of science strongly suggests that the current paradigm of solar system formation eventually will be discarded.

Do stars form today? Biblically, we do not have a clear answer. Some recent creationists think that since Genesis 1 records that God made the stars on day 4, no more stars are being made. But Genesis 1 also tells us that God made horses on day 6, but new horses are born every day. While on one has never observed the formation of new stars, there is no reason why stars could not form today (e.g., it would not be inconsistent with a biblical worldview). The question is whether the star formation rate today is nearly great enough as required by the evolutionary paradigm.

In the 1990s, astronomers first discovered planets orbiting other stars. Since then the number of extra-solar planets has grown tremendously. This has shown that planets must be common in the universe, and hence planetary system formation must be common in the universe today. However, this conclusion stems entirely from an evolutionary worldview. That is, the assumption is made that planetary systems can arise only through natural means apart from a Creator. Therefore, if planetary systems are common, then all of them must have come about through evolutionary processes. Since planetary systems are common, planetary formation must be simple and straightforward, which proves that our solar system must have formed through such a process. Therefore, the solar system formed pretty much the way astronomers think that it did. Of course, this is circular reasoning, and no such inference of naturalism legitimately can be drawn. A creationist could just as easily state that since all things were made by God, then anything that exists was made by Him. Since so many other planetary systems exist, then God must have made them all, just as He made our solar system. Therefore, this proves creation. Of course, evolutionists would violently disagree with this conclusion, for it disagrees with their starting premise of naturalism. This illustrates that the data alone do not allow for a definite conclusion about the ultimate origin of planetary systems, including our own.

The purpose of looking for extra-solar planets is to show how common planets are and how typical our solar system is. But is our solar system common?

The evidence thus far suggests otherwise. In our solar system, the large gas giant planets are far from the sun and the small rocky planets are close to the sun. Planetary scientists have developed models of how this might have happened, and those theories indicate that the large gas giant planets ought to be far from the sun, as is the case in the solar system. But extra-solar planets tend to be very large and very close to their parent stars,[3] the opposite of the situation in the solar system, and contrary to the prevailing theories of planetary formation. Scientists have concocted multiple encounters of planets (again using computer simulations) to show how extra-solar planets might have formed far from their stars but then migrated inward. Evolutionists must devise these explanations because the observations defy their theories.

Evolutionary ideas of planetary formation are fraught with problems. Man's ideas about the origin of the solar system have changed, and they will continue to change. However, the Word of God does not change. While the Bible does not tell us much about how the solar system came into being, it does give us some information about when the earth and the rest of the solar system came into existence. The Christian has confidence that what God has revealed to us is true, so we ought to compare man's ideas to the revealed truth. The current thinking of solar system formation disagrees with the Genesis creation account, so we know that it is not correct.

3. With time, this situation might change. Observational bias is in favor of finding massive planets close to their host stars. With improvements in technology, we may eventually find smaller and more distant planets more easily.

CHAPTER 31

Did Noah Need Oxygen Tanks on the Ark?

BODIE HODGE

Why would someone ask this question? Let's back up and look at this from a big picture. Consider what the Bible says about the voyage of the ark:

> The water prevailed more and more upon the earth, so that all the high mountains everywhere under the heavens were covered. The water prevailed fifteen cubits higher, and the mountains were covered (Genesis 7:19–20).[1]

People look at the earth *today* and note that the highest mountain is Mt. Everest, which stands just over 29,000 feet above sea level. Then they put two and two together and say that Noah's ark floated at least 15 cubits above Mt. Everest — and at such high altitude, people need oxygen![2]

It sounds like a straightforward argument, doesn't it? But did you notice that I emphasized the word *today*? In light of this, the solution is quite simple: the Flood did not happen on today's earth, but rather on the earth of nearly 4,300 years ago (according to Ussher).

1. Scripture is taken from the *New American Standard Bible* for this chapter.
2. For cubit studies and lengths see (for laymen) Bodie Hodge, "How Long Was the Original Cubit? *Answers magazine*, March 19, 2007, http://www.answersingenesis.org/articles/am/v2/n2/original-cubit , and (semi-technical); T. Lovett, "A More Likely Cubit for Noah's Ark?" WorldwideFlood.com website, June 2005, http://www.worldwideflood.com/ark/noahs_cubit/cubit_paper.htm.

The world today is not the same as it was before the Flood, or even during the Flood. For instance, if the mountains, continents, and oceans basins of today's earth were more leveled out (as would be expected in a global Flood), the planet's surface water alone would cover the earth an estimated 1.66 miles deep — about 8,000 feet. Yet when I visited Cusco, Peru, which is around 11,000 feet above sea level, I didn't need an oxygen tank.

Furthermore, atmospheric air pressure is relative to sea level. So as rising sea levels pushed the air column higher, the air pressure at sea level would stay the same.

Psalm 104:6–9: Creation or the Flood?

Beginning on day 150 of the Flood, mountains began overtaking the water again, as the mountain-building phase had begun (Genesis 8:2–4). Poetic Psalm 104 gives further hints of this mountain building as the valley basins sank down:

> You covered it with the deep as with a garment; the waters were standing above the mountains. At Your rebuke they fled, at the sound of Your thunder they hurried away. The mountains rose; the valleys sank down to the place which You established for them. You set a boundary that they may not pass over, so that they will not return to cover the earth (Psalm 104:6–9).

This section of the Psalm is obviously speaking of the Flood, as water would no longer return to cover the earth — if this passage were speaking of creation week (as some commentators have stated), then God would have erred when the waters covered the whole earth during the Flood.

Consider this overview, as the entire Psalm continues down through history:

Psalm 104:1–5	Creation Week
Psalm 104:6–9	Flood
Psalm 104:10–35	Post-Flood

It makes sense that, because the Psalm is referring to the earth and what is in it, it begins with earth history (creation week). But mentions of donkeys (verse 11) and goats (verse 18) show variation within the created kind, which shows this would have taken place after the Flood. Also, a post-Flood geographic location is named (Lebanon, verse 16) as well as ships (verse 26) that indicate this Psalm was not entirely a look at creation week.

Lost in Translation?

While everyone agrees that Psalm 104:1–5 is referring to creation week, what of the argument — made by many commentators from the 1600s onward — that attributes Psalm 104:6–9 to creation week? One could suggest that much of this is due to the translation being viewed. Two basic variants of the translation of the Hebrew in Psalm 104:8 read:

1. "They went up over the mountains and went down into the valleys."
2. "Mountains rose and the valleys sank down."

In fact, a variety of translations yield some variant of one of these two possibilities.

Table 1. Translations of Psalm 104:8a[3]

Translation	Agrees with: "They went up over the mountains and went down into the valleys"	Agrees with: "Mountains rose and the valleys sank down"
New American Standard		X
New International Version	X	
King James Version	X	
New King James Version	X	
English Standard Version		X
Holman Christian Standard		X
English translation of the Septuagint	X	
Revised Version (UK)	X	
Amplified Bible		X
Good News Bible	X	
New English Bible	X	
Revised Berkley		X
J.N. Darby's		X

3. Data was taken from two sources: (1) Charles Taylor, "Did Mountains Really Rise According to Psalm 104:8?" *TJ* 12(3) (1998): p. 312–313; and (2) looked up individually on Online Bible, Larry Pierce, February 2009, or looked up separately.

Living Bible		X
New Living Translation		X
Jerusalem Bible	X	
R.G. Moulton	X	
Knox Version		X
The Holy Scriptures according to the Masoretic Text (a new translation by the Jewish Publication Society)		X
Revised Standard Version		X
Young's Literal Translation	X	
King James 21st Century Version	X	
Geneva Bible		X
New Revised Standard Version	X	
Webster's Bible	X	
New International Children's Version		X
Interlinear Bible		X

Obviously, there is no consensus on translation among these English versions. Looking at other languages, we see how the Hebrew was translated.

Table 2. Some Foreign Translations of Psalm 104:8[4]

Foreign translation	Agrees with: "They went up over the mountains and went down into the valleys"	Agrees with: "Mountains rose and the valleys sank down"
Luther's German		X
Menge's German		X
French Protestant Bible (Version Synondale)		X
Italian Edizione Paoline		X
Swedish Protestant		X

4. Ibid.

Spanish Reina Valera		X
Latin Vulgate (by Jerome)		X
La Bible Louis Segond 1910 (French)		X
Septuagint (Koine Greek)		X

Notice that there doesn't seem to be a discrepancy. Of course, there are many translations, so one cannot be dogmatic, but the point is that many foreign translations agree with "mountains rising and valleys sinking down."

Hebrew

In Hebrew, which reads right to left, the phrase in 104:8a is literally four words. Translated into English, the phrase in question is:

biq'ah	*yarad*	*har*	*alah*
valleys	down go/sink	mountains	up go/rise/Ascend

Take note that there are no prepositions like "over" or "into." It is literally "up go mountains, down go valleys." It makes sense why many translations, including non-English translations, use the phrase "mountains rose and the valleys sank down" — this is what it should be.

Why Would Commentators Miss This?

Commentaries could easily misinterpret this passage if they were based on translations that agree with "they went up over the mountains and went down into the valleys." For example, the most popular English translation for several hundred years, the King James Version, reads this way.

Furthermore, from a logical perspective, water doesn't flow uphill over mountains, but rather the opposite. Given language like this, commentators likely attributed this to a miraculous event during creation week, when many miracles were taking place anyway; also, creation week was referenced earlier in the chapter. Of course, the problems came when reading the rest of the context. One excellent commentator, John Gill, regarding verse 9 and the waters not returning to cover the earth, stated:

> That they turn not again to cover the earth; as they did when it was first made, #Ps 104:6 that is, not without the divine leave and power;

> for they did turn again and cover the earth, at the time of the flood; but never shall more.[5]

Gill was forced to conclude that the waters *did* return to cover the earth, and he justified their return on "divine leave and power"! Yet this would mean that God breaks promises. Because we know that God does not break promises, this must be referring to the end of the Flood.

That said, we should understand the difficulty in commenting on the passage: it is a psalm of praise to God, and thus it is not as straightforward as literal history. It is difficult to determine where the shift from creation to the Flood occurs and where the shift from Flood to post-Flood occurs. However, there are a few more hints in the text.

A Few More Comments

We should use clear passages in Scripture to help interpret unclear passages. Consider that God's "rebuke" would not exist in a perfect world, where nothing would need rebuking or correcting. (Remember, a perfect God created a perfect world — Genesis 1:31, Deuteronomy 32:4.) One should expect nothing less of such a God.[6]

Therefore, during creation week when everything was good, there would be no need for any rebuking. If Psalm 104:6–9 were referring to creation week (specifically day 3), then why the rebuke in Psalm 104:7? This implies an imperfect, *not* very good creation. But if Psalm 104:6–9 is referring to the Flood, then of course a rebuke would exist in a fallen world where the judgment of water had overtaken the earth.

Additionally, note that Psalm 104:9 is clearly referencing Genesis 9:8–16 in saying that the waters would not return to cover the earth. (Some have asked how mountains and valleys could move up and down when the foundations are identified as immovable in Psalm 104:5. Keep in mind that mountains and valleys are not the foundation, but like the seas, they all sit above the foundation.)

Lastly, note that when the land appeared in Genesis 1 on day 3, the land that was being separated from the water was *dry*, not wet. The text in Genesis says that the waters were gathered into one place *and then* the dry land appeared. It says nothing of water flowing over the land to make it wet; otherwise, wet

5. J. Gill, Commentary notes, Psalm 104:9.
6. It was due to man's sin that the world is now imperfect and fallen.

land would have appeared and then *become* dry.[7] But during the Flood, the land was indeed overtaken by water that eventually stood above the land.

Conclusion

The Hebrew phrase in Psalm 104:8a is the basis for the correct translation of mountains rising and valleys sinking. This shows that mountains and valleys during the Flood were not the same height as they are today. Even today, mountains and valleys are changing their height; volcanic mountains, for instance, can grow very quickly, such as Surtsey or Paricutin (a volcanic mountain in Mexico that formed in 1943).

Therefore, with mountains and continents leveled out and ocean basins nowhere near the depth they are today, it makes perfect sense that Noah was not at the height of modern-day Mt. Everest. Instead, the ark would have been at sea level, where oxygen would have been nearly the same as today at sea level. Noah and those aboard the ark would not have required oxygen.

7. I understand some scientific models are built on this principle that the land and water separated and then the land *became* dry. But the text of Scripture, I suggest, leans in the direction of dry land appearing as a more supernatural occurrence, as opposed to naturalistic; especially considering the context of a supernatural creation week.

CHAPTER 32

The Image of God

DR. COREY ABNEY

You are special. Perhaps you've heard this from a parent, teacher, or member of your family. You received numerous compliments as the result of a special talent or accomplishment. Someone encouraged you because of your education and expertise. You were honored for a significant contribution. Or you grew up in a family where your grandmother reminded you of your "special" status every time you spent the night at her home (I can relate to this one)! Based on your background, personality, and life experience, you have a concept of what it means to be significant, and if you're like many people, you base how special you are on talent, education, or accomplishment. In other words, you look to yourself and to others. You play the comparison game. You try to measure up.

Unfortunately, many people aren't measuring up. The self-help industry is a multi-billion dollar industry with thousands of books published each year. Suicide is one of the leading causes of death among teenagers and young adults. Euthanasia is legal in the Netherlands, Belgium, and Luxembourg, with assisted suicide now legalized in Switzerland and in the U.S. states of Washington, Oregon, Vermont, and Montana. Humanism is weaving its way into the fabric of culture and institutions of higher learning. Many scholars believe and teach that human beings are no different than animals or plants. On this basis, human life is viewed in some circles as disposable, insignificant, or meaningless. Consider the teaching of Julian Huxley, a famous humanist, who writes,

> I use the word "humanist" to mean someone who believes that man is just as much a natural phenomenon as an animal or a plant;

> that his body, mind, and soul were not supernaturally created, but are products of evolution, and that he is not under the control or guidance of any supernatural being or beings, but has to rely on himself and his own power.[1]

Joseph Krutch, an American author, critic, and naturalist, says, "There is no reason to suppose that man's own life has any more meaning than the life of the humblest insect that crawls from one annihilation to another."[2]

Even the former Chief Justice of the United States Supreme Court, Oliver Wendell Holmes, states, "I see no reason for attributing to man a significance different in kind from that which belongs to a baboon or a grain of sand."[3]

No wonder so many people struggle with identity and significance. If humans are no different than animals, plants, or grains of sand, one could argue that we aren't so special after all.

The Image of God Established

Thankfully, the Bible presents a very different picture of humanity. You are special, but not primarily as the result of your talents, accomplishments, education, or upbringing. Your significance is not tied to how you measure yourself, how you compare with others, or how others view you; rather, your significance is tied to how your *Creator* views you. And here's the good news: your Creator views you as special . . . significant . . . unique. Human beings are special in the eyes of God because we are unique in the order of creation. You see, when God created the heavens and the earth, He also created every creature after its own kind (Genesis 1:20–25). He created sea creatures, crawling creatures, birds, livestock, and wildlife, pronouncing that such animals were "good" (Genesis 1:25). But when God created mankind, He said,

> "Let Us make man in Our image, according to Our likeness; let them have dominion over the fish of the sea, over the birds of the air, and over the cattle, over all the earth and over every creeping thing that creeps on the earth." So God created man in His own image; in the image of God He created him; male and female He created them (Genesis 1:26–27).

1 Julian Huxley, *The Humanist Frame* (New York: Harper & Brothers, 1961).
2 Joseph Wood Krutch, *The Modern Temper* (New York: Harcourt Brace, 1929).
3 Richard Posner, *The Essential Holmes: Selections from the Letters, Speeches, Judicial Opinions, and other Writings of Oliver Wendell Holmes* (Chicago, IL: Chicago University Press, 1992).

God created human beings in His *image* and *likeness*. These words are used interchangeably in the Book of Genesis, but only when referring to mankind (Genesis 1:26, 5:1, 9:26). No animal or plant is made in God's image or likeness. For this reason, human beings should be viewed as the crowning jewel of God's creative activity. After God made man, He looked over His creation and declared everything "very good" (Genesis 1:31).

A Unique Dignity

According to Genesis 1:26–30, mankind has a unique dignity. Moreover, Genesis 5:1 states, "In the day that God created man, He made him in the likeness of God." Human beings have a special dignity because men and women are God's image-bearers. This does not mean we reflect the physical appearance of God, because God is spirit and not represented in a human form (John 4:24).

Rather, bearing God's likeness points to the *spiritual*, not the physical. To be created in the divine image includes having an interpersonal relationship with God. Anthony Hoekema says,

> In this way human beings reflect God, who exists not as a solitary being but as a being in fellowship — a fellowship that is described at a later stage of divine revelation as that between the Father, the Son, and the Holy Spirit.[4]

People can know God, love God, and worship God. We can also think, reason, and choose between right and wrong. We have the capacity to look at the world and deduce that everything has a Creator (Romans 1:19–20). The image of God is the defining mark of humanity that sets us apart from animals, plants, and grains of sand. You can teach an animal tricks, but only man can learn truth. You can make an animal work, but it is man who can worship. Animals can see the sun, but man can glorify God for the beauty of a sunset. Mankind has a unique dignity that is seen primarily in the spiritual ability to fellowship with God and others. Both animals and man were created material and immaterial,[5] but only man was created with a spiritual component as well.[6]

4 Anthony Hoekema, *Created in God's Image* (Grand Rapids, MI: Eerdmans, 1994).

5 Many animals were created with *nephesh* in Hebrew. This is often translated as living *creature* or living *soul*. Man is also described as *nephesh*, but unlike animals, our spiritual component is made in God's image.

6. Editorial note: There are three views of the nature of the human being but this paper is not the place to discuss this theological topic. For the astute reader, the three positions are (1) *dichotomous* [body and soul/spirit; where soul and spirit are merely interchangeable words of the same substance], (2) *trichotomous* [body, soul, and spirit; where each are

Human beings are special because we have a unique dignity that enables us to have a relationship with God.

A Unique Dominion

Not only does mankind have a special dignity; we also have a unique dominion. God created human beings to rule over the fish, birds, cattle, and everything that creeps on the earth (Genesis 1:26). Moreover, God commanded the first man and woman to exercise dominion over every living creature on the planet:

> Then God blessed them, and God said to them, "Be fruitful and multiply; fill the earth and subdue it; have dominion over the fish of the sea, over the birds of the air, and over every living thing that moves on the earth." And God said, "See, I have given you every herb that yields seed which is on the face of all the earth, and every tree whose fruit yields seed; to you it shall be for food. Also, to every beast of the earth, to every bird of the air, and to everything that creeps on the earth, in which there is life, I have given every green herb for food"; and it was so (Genesis 1:28–30).

Mankind is the lord of creation who represents the ultimate Lord in a formal sense. He is God's caretaker on the earth and is expected to maintain order and unity. God provides fruit and vegetation for both man and animals to eat, but man alone is charged with the responsibility to "subdue" and "have dominion over" the created order.[7] Human beings are commanded to rule the earth for God and to develop a culture that glorifies the Creator. Many years after our first parents were created and commanded to exercise this unique dominion, King David reflected upon mankind's role in the world. He wrote,

> When I consider Your heavens, the work of Your fingers, the moon and the stars, which You have ordained, what is man that You are mindful of him, and the son of man that You visit him? For You have made him a little lower than the angels, and You have crowned

truly separate and unique (1 Thessalonians 5:23)], or (3) *modified* [the spirit would be a modified *aspect* of the soul, like a flip side of the same coin. There is one coin, but two unique sides to it. In other words, our soul is specially fashioned with a spiritual aspect, like duality. So soul and spirit could almost be used interchangeably (being two parts to the same "coin"), which we find in Scripture (Luke 1:36–47). Yet soul and spirit could be seen as unique (two sides of the "coin"), which we also find in Scripture (1 Thessalonians 5:23; Hebrews 4:12)].

7. Plants are not seen as living creatures in the Bible, unlike animals and humans, so they could not die in a biblical sense.

> him with glory and honor. You have made him to have dominion over the works of Your hands; You have put all things under his feet, all sheep and oxen — even the beasts of the field, the birds of the air, and the fish of the sea that pass through the paths of the seas (Psalm 8:3–8).

David is overwhelmed by God's grace and kindness toward humanity. As he surveys the mysteries of the heavens, the moon and the stars, and the existence of angels, he is amazed that God created man with glory and charged him with the responsibility of caring for the created order. David understood that human beings possess a unique dignity and dominion that set us apart from all other created beings.

The Image of God Tarnished

The image of God in mankind enables us to fellowship with our Creator and to exercise dominion over the earth. Sounds like a solid game plan, doesn't it? When you read the second chapter of Genesis, everything is certainly going according to plan. Initially, our first parents experienced unbroken communion with God and a peaceful relationship with each other (Genesis 2:21–25). Death was not a part of the world. The first man and woman did not experience distrust or disappointment. The question is, what happened?

Sin happened. The first man and woman (Adam and Eve) disobeyed God and rebelled against His will for their lives. God told them to eat from any tree on the earth with the exception of the tree of the knowledge of good and evil. Genesis 2:16–17 says, "And the LORD God commanded the man, saying, 'Of every tree of the garden you may freely eat; but of the tree of the knowledge of good and evil you shall not eat, for in the day that you eat of it you shall surely die.' "

Satan, a fallen angel who rejected God and His sovereign reign over the universe (Isaiah 14:12–14; Ezekiel 28:12–18; Luke 10:18), tempted the woman through the use of a serpent.[8] Adam and Eve yielded to the temptation and sinned by eating, and forever changed the course of human history:

> Then the serpent said to the woman, "You will not surely die. For God knows that in the day you eat of it your eyes will be opened, and you will be like God, knowing good and evil." So when the woman saw that the tree was good for food, that it was pleasant to the eyes, and a tree desirable to make one wise, she took of its fruit and ate. She also gave to her husband with her, and he ate (Genesis 3:4–6).

8. Bodie Hodge, *The Fall of Satan* (Green Forest, AR: Master Books, 2011), p. 43–45.

The rebellion of Adam and Eve plunged humanity into a sinful state where death, pain, and suffering entered the world. Moreover, the image of God in man was tarnished and broken from that point forward. Human beings now search for significance in themselves and their accomplishments instead of finding significance in the Creator whose image we bear. We remain rational, spiritual beings, but our rationality and spirituality no longer impart a true knowledge of God.

We are still relational people who possess the capacity to fellowship with God and others, but the outworking of our relationships no longer reflects the relationship between Father, Son, and Holy Spirit. In other words, mankind continues to reflect a unique dignity as God's image-bearers that no other creature enjoys, but the dignity is damaged significantly by the consequences of sin. Similarly, human beings continue to exercise dominion over the earth, but in many ways are selfish dictators who rule over nature for selfish gain, working against the will of God in the world.

The Image of God Restored

The image of God in man was tarnished, but not beyond repair. God the Father, in His infinite mercy and grace, reached out to Adam and Eve in the midst of their rejection and rebellion. Adam and Eve experienced consequences for their sin, but God also issued a promise of hope and restoration:

> So the Lord God said to the serpent: "Because you have done this, you are cursed more than all cattle, and more than every beast of the field; on your belly you shall go, and you shall eat dust all the days of your life. And I will put enmity between you and the woman, and between your seed and her Seed; He shall bruise your head, and you shall bruise His heel" (Genesis 3:14–15).

God promised to send a man who will conquer Satan and put an end to the reign of sin and death. According to this first of many prophetic statements in Genesis 3, God will send a deliverer who will save His people from their sins; He will send a healer who is able to restore the image of God in mankind.

Christ the Image of God

The New Testament makes it clear that Jesus Christ is the Son of God and the promised seed of the woman who dealt a fatal blow to death through His crucifixion and Resurrection. Satan bruised Jesus on the Cross, but Jesus crushed Satan's head when He rose from the dead (Genesis 3:15; Colossians

2:13–15). As a result, salvation from sin and death is found in Christ alone through faith alone (Ephesians 2:8–9). Moreover, the image of God is redefined in terms of Christ Himself as the true image. For example, Christ is called the *image* of God in three New Testament passages:

> But even if our gospel is veiled, it is veiled to those who are perishing, whose minds the god of this age has blinded, who do not believe, lest the light of the gospel of the glory of Christ, *who is the image of God*, should shine on them (2 Corinthians 4:3–4, emphasis added).

> He is *the image of the invisible God*, the firstborn over all creation. For by Him all things were created that are in heaven and that are on earth, visible and invisible, whether thrones or dominions or principalities or powers. All things were created through Him and for Him (Colossians 1:15–16, emphasis added).

> God, who at various times and in various ways spoke in time past to the fathers by the prophets, has in these last days spoken to us by His Son, whom He has appointed heir of all things, through whom also He made the worlds; who being the brightness of His glory and *the express image of His person*, and upholding all things by the word of His power, when He had by Himself purged our sins, sat down at the right hand of the Majesty on high (Hebrews 1:1–3, emphasis added).

Jesus is the true image of God. He is equally God and not a mere copy of the original. He *is* the original image. The Apostle Paul says, "For it pleased the Father that in Him all the fullness should dwell" (Colossians 1:19). Jesus shows us the glory of God (John 1:14), and when He comes again in His glorified humanity, He will be manifested directly as the true image of God. The Apostle John states, "Beloved, now we are children of God; and it has not yet been revealed what we shall be, but we know that when He is revealed, we shall be like Him, for we shall see Him as He is" (1 John 3:2).

The Image of Christ Restored in Us

The image of God in man needs restoration and renewal. In order for the image of God to be restored, however, we must look to Christ for salvation and sanctification. The Apostle Paul says, "Do not lie to one another, since you have put off the old man with his deeds, and have put on the new man who is renewed in knowledge according to the image of Him who created him" (Colossians 3:9-10). He also writes, "For whom He foreknew, He also predestined to be conformed to the image of His Son, that He might be the firstborn among many brethren" (Romans 8:29). We must pursue Christ and His image, knowing that we will be like Him in the new creation:

> And as we have borne the image of the man of dust, we shall also bear the image of the heavenly Man (1 Corinthians 15:49).

> But we all, with unveiled face, beholding as in a mirror the glory of the Lord, are being transformed into the same image from glory to glory, just as by the Spirit of the Lord (2 Corinthians 3:18).

Jesus Christ is our only hope for restoration. Without saving faith in the life, death, Resurrection, and return of Jesus, the image of God in us will remain tarnished by our sin and rebellion. We will continue to search for significance in ourselves, but we will never find it there, because our significance is ultimately tied to the image of God within us.

Conclusion

So, you really are special — not because of what you do, but because of who God created you to be. You are more than a plant, an animal, or a grain of sand. God created you in His image! Furthermore, despite your sin and rebellion that leads to the tarnishing of His image, God sent His Son to die for your sin, in your place, as a righteous substitute who satisfied the demands of the holy Judge. Three days later, God raised His Son from the dead, ensuring salvation and eternal life for all who believe in Him. Jesus Christ is the true image of God; therefore, when you submit your life to Jesus, God works through the power of His Holy Spirit to restore His broken image in you. And I can't think of anything more special than that.

CHAPTER 33

Dear Atheists . . . Are You Tired of It All?

BODIE HODGE

Are you tired of all the evil associated with the philosophy of atheism — Stalin, Hitler, Pol Pot, and so on?[1] After all, most murderers, tyrants, and rapists are not biblical Christians, and most have rejected the God of the Bible. Even if they claim to believe in the God of the Bible, they are not really living like a true Christ follower (who strives to follow God's Word), are they?

Do you feel conflicted about the fact that atheism has no basis in morality (i.e., no absolute right and wrong; no good, no bad)? If someone stabs you in the back, treats you like nothing, steals from you, or lies to you, it doesn't ultimately matter in an atheistic worldview, where everything and everyone are just chemical reactions doing what chemicals do. And further, knowing that you are essentially no different from a cockroach in an atheistic worldview (since people are just animals) must be disheartening.

Are you tired of the fact that atheism (which is based in materialism,[2] a popular worldview today) has no basis for logic and reasoning? Is it tough trying to get up every day thinking that truth, which is immaterial, really doesn't exist? Are you bothered by the fact that atheism cannot account for uniformity in nature[3]

1. B. Hodge, "The Results of Evolution," Answers in Genesis, July 13, 2009, http://www.answersingenesis.org/articles/2009/07/13/results-evolution-bloodiest-religion-ever.
2. J. Lisle, "Atheism: An Irrational Worldview," Answers in Genesis, October 10, 2007, http://www.answersingenesis.org/articles/aid/v2/n1/atheism-irrational.
3. J. Lisle, "Evolution: The Anti-science," Answers in Genesis, February 13, 2008, http://www.answersingenesis.org/articles/aid/v3/n1/evolution-anti-science.

(the basis by which we can do real science)? Why would everything explode from nothing and, by pure chance, form beautiful laws like $E=MC^2$ or $F=MA$?[4]

Do you feel like you need a weekend to recoup, even though a weekend is really meaningless in an atheistic worldview — since animals, like bees, don't take a day of rest or have a weekend? So why should atheists? Why borrow a workweek and weekend that comes from the pages of Scriptures, which are despised by atheists? Weeks and weekends come from God creating in six literal days and resting for a literal day; and then the Lord Jesus resurrected on the first day of the week (Sunday). And why look forward to time off for a holiday (i.e., holy day), when nothing is holy in an atheistic worldview?

For professing atheists, these questions can be overwhelming to make sense of within their worldview. And further, within an atheistic worldview, atheists must view themselves as God. Essentially, atheists are claiming to be God. Instead of saying there *may not* be a God, they say there is *no* God. To make such a statement, they must claim to be omniscient (which is an essential attribute of the God of the Bible) among other attributes of God as well.[5] So by saying there is no God, the atheist refutes his own position by addressing the question as though he or she were God!

Do you feel conflicted about proselytizing the faith of atheism, since if atheism were true then who cares about proselytizing? Let's face it, life seems tough enough as an atheist without having to deal with other major concerns like not having a basis to wear clothes, or no basis for marriage, no consistent reason to be clean (snails don't wake up in the morning and clean themselves or follow other cleanliness guidelines based on Levitical laws), and no objective reason to believe in love.

Are you weary of looking for evidence that contradicts the Bible's account of creation and finding none?[6] Do the assumptions and inconsistencies of dating methods weigh on your conscience when they are misrepresented as fact?[7] Where

---

4. K. Ham, Gen. Ed., *New Answers Book 1*, J. Lisle, J., "Don't Creationists Deny the Laws of Nature? (Green Forest, AR: Master Books, 2006), p. 39–46; http://www.answersingenesis.org/articles/nab/creationists-deny-laws-of-nature.
5. If one claims that God may exist or that there may be a spiritual realm, then that person is not an atheist, but an agnostic, at best. The agnostic says that one cannot know whether God exists, but how can they know that for certain apart from being omniscient themselves? Additionally, the Bible says in 1 John 5:13 that we can *know* for certain that we have eternal life. So an agnostic — who claims we cannot know — does not hold a neutral position regarding the biblical God.
6. K. Ham, "Missing? or Misinterpreted?" Answers in Genesis, March 1, 2004, http://www.answersingenesis.org/articles/cm/v26/n2/missing.
7. K. Ham, *New Answers Book 1*, M. Riddle, "Does Radiometric Dating Prove the Earth Is

do you suppose those missing links have gone into hiding? Surely the atheist sees the folly and hopelessness of believing that everything came from nothing.

In fact, why would an atheist care to live one moment longer in a broken universe where one is merely rearranged pond scum and all you have to look forward to is . . . death, which can be around any corner? And in 467 trillion years, no one will care one iota about what you did or who you were or how and when you died — because death is the ultimate "hero" in an atheistic, evolutionary worldview. Of course, as a Christian I disagree, and I have a basis to see you as having value.

Invitation

I invite you to reconsider that the false religion of atheism is simply that. I'm here to tell you that atheism is a lie (Romans 1:25).[8] As a Christian, I understand that truth exists because God exists, who is the Truth (John 14:6),[9] and we are made in His image.[10] Unlike an atheist, whose worldview doesn't allow him to believe in truth or lies, the Bible-believer has a foundation that enables him to speak about truth and lies. This is because believers in God and His Word have an authority, the ultimate authority on the subject, to base statements upon.

There is a God, and you are also made in His image (Genesis 1:26; 9:6).[11] This means you have value. Whereas consistent atheists teach that you have no value, I see you differently. I see you as a relative (Acts 17:26)[12] and one who — unlike animals, plants, and fallen angels — has the possibility of salvation from death, which is the result of sin (i.e., disobedience to God; see Romans 6:23).[13]

Old? (Green Forest, AR: Master Books, 2006), p. 113–134; http://www.answersingenesis.org/articles/nab/does-radiometric-dating-prove.

8. "Who exchanged the truth of God for the lie, and worshiped and served the creature rather than the Creator, who is blessed forever. Amen" (Romans 1:25)
9. "Jesus said to him, 'I am the way, the truth, and the life. No one comes to the Father except through Me' " (John 14:6).
10. Keep in mind that Christians, including me, do fall short due to sin and the Curse, but God never fails.
11. "Then God said, 'Let Us make man in Our image, according to Our likeness; let them have dominion over the fish of the sea, over the birds of the air, and over the cattle, over all the earth and over every creeping thing that creeps on the earth' " (Genesis 1:26); "Whoever sheds man's blood, by man his blood shall be shed; for in the image of God He made man" (Genesis 9:6).
12. "And He has made from one blood every nation of men to dwell on all the face of the earth, and has determined their preappointed times and the boundaries of their dwellings" (Acts 17:26).
13. "For the wages of sin is death, but the gift of God is eternal life in Christ Jesus our Lord" (Romans 6:23).

We have all fallen short of God's holy standard of perfect obedience thanks to our mutual grandfather, Adam (Romans 5:12).[14] And God sees you differently, too (John 3:16).[15] While you were *still* a sinner, God stepped into history to become a man to die in your place (Romans 5:8)[16] and offer the free gift of salvation (Ephesians 2:8–9).[17]

Atheists have no consistent reason to proselytize their faith, but Christians like me do have a reason — Jesus Christ, who is the Truth, commands us to (Matthew 28:19).[18] We want to see people repent of their evil deeds and be saved from death (Acts 8:22, 17:30).[19] What a wonderful joy (Luke 15:10).[20]

Where atheists have no basis for logic and reason (or even for truth, since truth is immaterial), Bible believers can understand that mankind is made in the image of a logical and reasoning God who is the truth. Hence, Christians can make sense of things because in Christ are "hidden all the treasures of wisdom and knowledge" (Colossians 2:3).[21] Christians also have a basis to explain why people sometimes don't think logically due to the Fall of mankind in Genesis 3. The most logical response is to give up atheism and receive Jesus Christ as Lord and Savior to rescue you from sin and death (Romans 10:13).[22] Instead of death, God promises believers eternal life (1 John 2:25; John 10:28)[23] and in 467 trillion years, you will still have value in contrast to the secular view of nothingness.

Christians do have a basis to wear clothes (to cover shame due to sin; see Genesis 2:25, 3:7),[24] a reason to uphold marriage (God made a man and a

14. "Therefore, just as through one man sin entered the world, and death through sin, and thus death spread to all men, because all sinned" (Romans 5:12).
15. "For God so loved the world that He gave His only begotten Son, that whoever believes in Him should not perish but have everlasting life" (John 3:16).
16. "But God demonstrates His own love toward us, in that while we were still sinners, Christ died for us" (Romans 5:8).
17. "For by grace you have been saved through faith, and that not of yourselves; it is the gift of God, not of works, lest anyone should boast" (Ephesians 2:8–9).
18. "Go therefore and make disciples of all the nations, baptizing them in the name of the Father and of the Son and of the Holy Spirit" (Matthew 28:19).
19. "Repent therefore of this your wickedness, and pray God if perhaps the thought of your heart may be forgiven you" (Acts 8:22); "Truly, these times of ignorance God overlooked, but now commands all men everywhere to repent" (Acts 17:30).
20. "Likewise, I say to you, there is joy in the presence of the angels of God over one sinner who repents" (Luke 15:10).
21. "In whom are hidden all the treasures of wisdom and knowledge" (Colossians 2:3).
22. "For 'whoever calls on the name of the Lord shall be saved' " (Romans 10:13).
23. "And this is the promise that He has promised us — eternal life" (1 John 2:25); "And I give them eternal life, and they shall never perish; neither shall anyone snatch them out of My hand" (John 10:28).
24. "And they were both naked, the man and his wife, and were not ashamed" (Genesis 2:25);

woman; see Genesis 1:27; Matthew 19:4–6),[25] a reason to be clean (Leviticus contains many provisions to counter diseases in a sin-cursed world), and a source of real love (since God made us in His loving image; see 1 John 4:8).[26] As Christians, we have a solid foundation for saying things like back-stabbing, theft, and lies are wrong (see the Ten Commandments in Exodus 20).

I invite you to leave the false religion of atheism and its various forms and return to the one true God who came to rescue you (John 17:3).[27] Jesus Christ, who is God the Son, loved you enough to come down and die in our place so we can experience God's goodness for all eternity instead of the wrath of God for all eternity in hell (Matthew 25:46).[28] And we all have sentenced ourselves to judgment because of our disobedience to God and rejection of Him (John 3:17–18).[29]

The day is coming when we all will give an account before God for our actions and thoughts (Romans 14:12).[30] Will you repent and receive Christ as your Lord and Savior today so that you will join Christ in the resurrection from the dead (John 11:25; Romans 6:5)?[31] I invite you personally to become an ex-atheist, join the ranks of the saved through Jesus Christ, and become a new creation (2 Corinthians 5:17)[32] as we continue to advance with the gospel in peace that only God can provide (Romans 5:1).[33]

"Then the eyes of both of them were opened, and they knew that they were naked; and they sewed fig leaves together and made themselves coverings" (Genesis 3:7).

25. "So God created man in His own image; in the image of God He created him; male and female He created them" (Genesis 1:27); "And He answered and said to them, 'Have you not read that He who made them at the beginning "made them male and female," and said, "for this reason a man shall leave his father and mother and be joined to his wife, and the two shall become one flesh"? So then, they are no longer two but one flesh. Therefore what God has joined together, let not man separate' " (Matthew 19:4–6).
26. "He who does not love does not know God, for God is love" (1 John 4:8).
27. "And this is eternal life, that they may know You, the only true God, and Jesus Christ whom You have sent" (John 17:3).
28. "And these will go away into everlasting punishment, but the righteous into eternal life" (Matthew 25:46).
29. "For God did not send His Son into the world to condemn the world, but that the world through Him might be saved. He who believes in Him is not condemned; but he who does not believe is condemned already, because he has not believed in the name of the only begotten Son of God" (John 3:17–18).
30. "So then each of us shall give account of himself to God" (Romans 14:12).
31. "Jesus said to her, 'I am the resurrection and the life. He who believes in Me, though he may die, he shall live' " (John 11:25); "For if we have been united together in the likeness of His death, certainly we also shall be in the likeness of His resurrection" (Romans 6:5).
32. "Therefore, if anyone is in Christ, he is a new creation; old things have passed away; behold, all things have become new" (2 Corinthians 5:17).
33. "Therefore, having been justified by faith, we have peace with God through our Lord Jesus Christ" (Romans 5:1).

Contributors

Ken Ham is the president and CEO of Answers in Genesis (USA). He has authored several books, including the best-seller *The Lie: Evolution*. He is one of the most in-demand speakers in the U.S. and has a daily radio program called *Answers...with Ken Ham,* which is heard on over 850 stations in the US and over 1,000 worldwide. Ken has a BS in applied science (with an emphasis in environmental biology) from Queensland Institute of Technology in Australia. He also holds a diploma of education from the University of Queensland (a graduate qualification for science teachers in the public schools in Australia). Ken has been awarded two honorary doctorates: a Doctor of Divinity (1997) from Temple Baptist College in Cincinnati, Ohio, and a Doctor of Literature (2004) from Liberty University in Lynchburg, Virginia.

Dr. Corey Abney earned a BA from Cedarville University, and a MDiv and PhD from Southern Baptist Theological Seminary in Louisville, KY. Dr. Abney has served the Kentucky Baptist Convention as both a member of the executive board and as a past president of the Pastors' Conference. He is also involved in the Southern Baptist Convention as a trustee of Golden Gate Baptist Theological Seminary located in Mill Valley, CA. Currently, he is the senior pastor at Florence Baptist Church in Florence, KY.

Dr. E. Calvin Beisner, PhD, is founder and national spokesman of the Cornwall Alliance for the Stewardship of Creation, author of several books on biblical economics and environmental ethics, and a former Christian college and seminary professor. He has testified as an expert witness on climate policy before committees of the U.S. Senate and House of Representatives and lectured on the subject at the Vatican, for International Conferences on Climate Change, and for colleges, seminaries, and churches.

Dr. Joel R. Beeke is president and professor of systematic theology and homiletics at Puritan Reformed Theological Seminary, a pastor of the Heritage Netherlands Reformed Congregation in Grand Rapids, Michigan, editor of *Banner of Sovereign Grace Truth,* editorial director of Reformation Heritage Books, president of Inheritance Publishers, and vice-president of the Dutch Reformed Translation Society. He has written, co-authored, or edited 70 books.

Dr. Stuart Burgess is professor of design and nature in the department of mechanical engineering at Bristol University-UK. He is author of three books:

Hallmarks of Design, He Made the Stars Also, and *The Origin of Man*, all published by Day One Publications (www.dayone.co.uk).

Dr. Andrew Fabich earned a BS from Ohio State University and PhD from the University of Oklahoma. He began his teaching career at Tennessee Temple University. Dr. Fabich is currently an assistant professor of microbiology at Liberty University. Dr. Fabich did his dissertation on understanding how good and bad *E. coli* colonize the mammalian intestine to cause disease. He continues working on gastrointestinal research using molecular techniques in animal models. His professional memberships include American Society of Microbiology and Virginia Academy of Sciences.

Dr. Danny R. Faulkner has a BS (math), MS (physics), MA and Ph.D. (astronomy, Indiana University). He is full professor at the University of South Carolina–Lancaster, where he teaches physics and astronomy. He has published about two dozen papers in various astronomy and astrophysics journals.

Bodie Hodge earned a BS and MS in mechanical engineering at Southern Illinois University at Carbondale in 1996 and 1998, respectively. His specialty was in materials science working with advanced ceramic powder processing. He developed a new method of production of submicron titanium diboride. Bodie accepted a teaching position as visiting instructor at Southern Illinois in 1998 and taught for two years. After this, he took a job working as a test engineer at Caterpillar's Peoria Proving Ground. Bodie currently works at Answers in Genesis (USA) as a speaker, writer, and researcher after working for three years in the Answers Correspondence Department.

Troy Lacey earned his bachelor of natural sciences (biology/geology) degree from the University of Cincinnati. Troy is the correspondence representative and chaplain services coordinator for Answers in Genesis–USA.

Don Landis is pastor of Community Bible Church in Jackson, Wyoming. He is founder and president of Jackson Hole Bible College (www.jhbc.edu), a one-year program with special emphasis on creation for young adults. Don is also the founding chairman of the board for Answers in Genesis-USA.

Tim Lovett earned his degree in mechanical engineering from Sydney University (Australia) and was an instructor for 12 years in technical college engineering courses. Tim has studied the Flood and the ark for 15 years and is widely recognized for his cutting-edge research on the design and structure of Noah's ark. He is author of the book *Noah's Ark: Thinking Outside the Box*.

Dr. David Menton was an associate professor of anatomy at Washington University School of Medicine from 1966 to 2000 and has since become Associate Professor Emeritus. He was a consulting editor in histology for *Stedman's Medical Dictionary*, a standard medical reference work. David earned his PhD from Brown University in cell biology. He is a popular speaker and lecturer with Answers in Genesis (USA), showing complex design in anatomy with popular DVDs such as *The Hearing Ear and Seeing Eye* and *Fearfully and Wonderfully Made*. He also has an interest in the famous Scopes Trial, which was a big turning point in the creation/evolution controversy in the USA in 1925.

Dr. Elizabeth Mitchell earned her MD from Vanderbilt University School of Medicine and practiced medicine for seven years until she retired to be a stay-at-home mom. Her interest in ancient history strengthened when she began to homeschool her daughters. She desires to make history come alive and to correlate it with biblical history.

Dr. Tommy Mitchell graduated with a BA with highest honors from the University of Tennessee–Knoxville in 1980 with a major in cell biology and a minor in biochemistry. He subsequently attended Vanderbilt University School of Medicine in Nashville, where he was granted an MD degree in 1984. Dr. Mitchell's residency was completed at Vanderbilt University Affiliated Hospitals in 1987. He was board certified in internal medicine, with a medical practice in Gallatin, Tennessee (the city of his birth). In 1991, he was elected to the Fellowship in the American College of Physicians (F.A.C.P.). Tommy became a full-time speaker, researcher, and writer with Answers in Genesis (USA) in 2006.

Dr. Gary Parker is a biologist who was with Answers in Genesis as a senior lecturer since AiG's first year (1994, and remained full-time until 1999). Dr. Parker was also the head of the science department at Clearwater Christian College (CCC) in Florida. For 12 years, he served on the science faculty of the Institute for Creation Research (ICR) in the San Diego area. En route to his BA in biology/chemistry, MS in biology/physiology, and EdD in biology/geology from Ball State, Dr. Parker earned several academic awards, including admission to Phi Beta Kappa (the national scholastic honorary), election to the American Society of Zoologists (for his research on tadpoles), and a 15-month fellowship award from the National Science Foundation. He has published five programmed textbooks in biology and six books in creation science (the latter translated into a total of eight languages), has appeared in numerous films and television programs, and has lectured worldwide on creation.

Roger Patterson earned his BS Ed degree in biology from Montana State University. Before coming to work at Answers in Genesis, he taught for eight years in Wyoming's public school system and assisted the Wyoming Department of Education in developing assessments and standards for children in public schools. Roger now serves on the Educational Resources team at Answers in Genesis–USA.

Dr. Georgia Purdom received her PhD in molecular genetics from Ohio State University in 2000. As an associate professor of biology, she completed five years of teaching and research at Mt. Vernon Nazarene University in Ohio before joining the staff at Answers in Genesis (USA). Dr. Purdom has published papers in the *Journal of Neuroscience,* the *Journal of Bone and Mineral Research*, and the *Journal of Leukocyte Biology*. She is also a member of the Creation Research Society, American Society for Microbiology, and American Society for Cell Biology. She is a peer-reviewer for *Creation Research Society Quarterly*. Georgia has a keen interest and keeps a close eye on the Intelligent Design movement.

Dr. Benjamin Shaw has earned a MDiv from Pittsburgh Theological Seminary and a ThM from Princeton Theological Seminary. He went on to earn a Ph.D. from Bob Jones University. Currently, Dr. Shaw is an associate professor in Hebrew and Old Testament and academic dean at Greenville Presbyterian Theological Seminary. He is a regular author and blogger.

Dr. Andrew A. Snelling is currently Director of Research with Answers in Genesis (US). He received a BS in applied geology with first-class honors at the University of New South Wales in Sydney, and earned his PhD in geology at the University of Sydney for his thesis entitled "A Geochemical Study of the Koongarra Uranium Deposit, Northern Territory, Australia." Between studies and since, Andrew worked for six years in the exploration and mining industries in Tasmania, New South Wales, Victoria, Western Australia, and the Northern Territory as a field, mine, and research geologist. Andrew was also a principal investigator in the RATE (Radioisotopes and the Age of The Earth) project hosted by the Institute for Creation Research and the Creation Research Society.

Dr. Michael Todhunter, who earned his doctorate in forest genetics from Purdue University, has spent 14 years in forest genetic research, industrial tree-breeding research, and project management. Dr. Todhunter has also published numerous works in his field of expertise.

John UpChurch serves as the editor for Jesus.org and is a contributor to the Answers in Genesis website. He graduated summa cum laude from the University of Tennessee with a BA in English.

Dr. Alan White earned his BS in chemistry from the University of Tennessee and his PhD in organic chemistry from Harvard University in 1981. He worked for 30 years at Eastman Chemical Company and reached the rank of Research Fellow. Alan spent his career at Eastman working in research and development in the fields of organic and polymer chemistry. Among his achievements are the discovery of a commercial biodegradable polymer and the improvement of many commercial polymer processes, including the new Integrex™ technology for producing PET, the soft drink bottle polymer. Alan has been granted 41 U.S. patents and is an author on 18 scientific publications. He has recently retired to spend more time in research, writing, and speaking in the area of creation science.

Dr. John Whitmore received a BS in geology from Kent State University, an MS in geology from the Institute for Creation Research, and a PhD in biology with paleontology emphasis from Loma Linda University. Currently an associate professor of geology, he is active in teaching and research at Cedarville University. Dr. Whitmore serves on the board of Creation Research Science Education Foundation located in Columbus, Ohio, and he is also a member of the Creation Research Society and the Geological Society of America.

Index

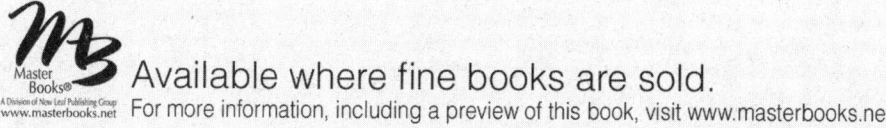